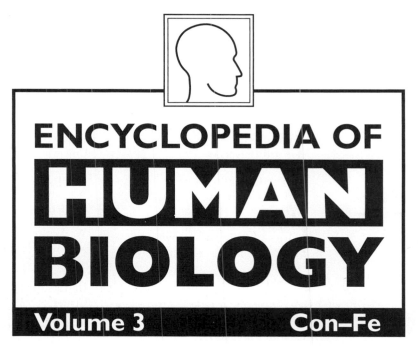

ENCYCLOPEDIA OF
HUMAN
BIOLOGY

Volume 3   **Con–Fe**

Second Edition

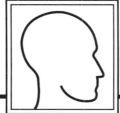

ENCYCLOPEDIA OF HUMAN BIOLOGY

Volume 3   Con–Fe

Second Edition

Editor-in-Chief
RENATO DULBECCO
The Salk Institute
La Jolla, California

ACADEMIC PRESS

San Diego   London   Boston   New York   Sydney   Tokyo   Toronto

This book is printed on acid-free paper. ∞

Academic Press
a division of Harcourt Brace & Company
525 B Street, Suite 1900, San Diego, California 92101-4495, USA
http://www.apnet.com

Academic Press Limited
24-28 Oval Road, London NW1 7DX, UK
http://www.hbuk.co.uk/ap/

Library of Congress Cataloging-in-Publication Data

Encyclopedia of human biology / edited by Renato Dulbecco. -- 2nd ed.
            p.   cm.
        Includes bibliographical references and index.
        ISBN 0-12-226970-5 (alk. paper: set). -- ISBN 0-12-226971-3 (alk.
    paper: v. 1). -- ISBN 0-12-226972-1 (alk. paper: v. 2). -- ISBN
    0-12-226973-X (alk. paper: v. 3). -- ISBN 0-12-226974-8 (alk. paper:
    v. 4). -- ISBN 0-12-226975-6 (alk. paper: v. 5). -- ISBN
    0-12-226976-4 (alk. paper: v. 6). -- ISBN 0-12-226977-2 (alk. paper
    : v. 7). -- ISBN 0-12-226978-0 (alk. paper: v. 8). -- ISBN
    0-12-226979-9 (alk. paper: v. 9)
        1. Human biology--Encyclopedias.   I. Dulbecco, Renato, date.
        [DNLM: 1. Biology--encyclopedias.   2. Physiology--encyclopedias.
    QH 302.5 E56 1997]
    QP11.E53   1997
    612'.003-dc21
    DNLM/DLC
    for Library of Congress   97-8627
                                                                CIP

PRINTED IN THE UNITED STATES OF AMERICA
        98 99 00 01 02 EB 9 8 7 6 5 4 3 2

F

A GUIDE TO USING THE ENCYCLOPEDIA

The *Encyclopedia of Human Biology, Second Edition* is a complete source of information on the human organism, contained within the covers of a single unified work. It consists of nine volumes and includes 670 separate articles ranging from genetics and cell biology to public health, pediatrics, and gerontology. Each article provides a comprehensive overview of the selected topic to inform a broad spectrum of readers from research professionals to students to the interested general public.

In order that you, the reader, derive maximum benefit from your use of the Encyclopedia, we have provided this Guide. It explains how the Encyclopedia is organized and how the information within it can be located.

ORGANIZATION

The *Encyclopedia of Human Biology, Second Edition* is organized to provide the maximum ease of use for its readers. All of the articles are arranged in a single alphabetical sequence by title. Articles whose titles begin with the letters A to Bi are in volume 1, articles with titles from Bl to Com are in Volume 2, and so on through Volume 8, which contains the articles from Si to Z.

Volume 9 is a separate reference volume providing a Subject Index for the entire work. It also includes a complete Table of Contents for all nine volumes, an alphabetical list of contributors to the Encyclopedia, and an Index of Related Titles. Thus Volume 9 is the best starting point for a search for information on a given topic, via either the Subject Index or Table of Contents.

So that they can be easily located, article titles generally begin with the key word or phrase indicating the topic, with any descriptive terms following. For example, "Calcium, Biochemistry" is the article title rather than "Biochemistry of Calcium" because the specific term *calcium* is the key word rather than the more general term *biochemistry*. Similarly "Protein Targeting, Basic Concepts" is the article title rather than "Basic Concepts of Protein Targeting."

TABLE OF CONTENTS

A complete Table of Contents for the *Encyclopedia of Human Biology, Second Edition* appears in Volume 9. This list of article titles represents topics that have been carefully selected by the Editor-in-Chief, Dr. Renato Dulbecco, and the members of the Editorial Advisory Board (see p. ii for a list of the Board members). The Encyclopedia provides coverage of 35 specific subject areas within the overall field of human biology, ranging alphabetically from Behavior to Virology.

In addition to the complete Table of Contents found in Volume 9, the Encyclopedia also provides an individual table of contents at the front of each volume. This lists the articles included within that particular volume.

INDEX

The Subject Index in Volume 9 contains more than 4200 entries. The subjects are listed alphabetically and indicate the volume and page number where information on this topic can be found.

ARTICLE FORMAT

Articles in the *Encyclopedia of Human Biology, Second Edition* are arranged in a single alphabetical list by title. Each new article begins at the top of a right-hand page, so that it may be quickly located. The author's name and affiliation are displayed at the beginning of the article. The article is organized according to a standard format, as follows:

- Title and author
- Outline
- Glossary
- Defining statement
- Body of the article
- Bibliography

OUTLINE

Each article in the Encyclopedia begins with an outline that indicates the general content of the article. This outline serves two functions. First, it provides a brief preview of the article, so that the reader can get a sense of what is contained there without having to leaf through the pages. Second, it serves to highlight important subtopics that will be discussed within the article. For example, the article "Gene Mapping" includes the subtopic "DNA Sequence and the Human Genome Project."

The outline is intended as an overview and thus it lists only the major headings of the article. In addition, extensive second-level and third-level headings will be found within the article.

GLOSSARY

The Glossary contains terms that are important to an understanding of the article and that may be unfamiliar to the reader. Each term is defined in the context of the particular article in which it is used. Thus the same term may appear as a Glossary entry in two or more articles, with the details of the definition varying slightly from one article to another. The Encyclopedia includes approximately 5000 glossary entries.

DEFINING STATEMENT

The text of each article in the Encyclopedia begins with a single introductory paragraph that defines the topic under discussion and summarizes the content of the article. For example, the article "Free Radicals and Disease" begins with the following statement:

A FREE RADICAL is any species that has one or more unpaired electrons. The most important free radicals in a biological system are oxygen- and nitrogen-derived radicals. Free radicals are generally produced in cells by electron transfer reactions. The major sources of free radical production are inflammation, ischemia/reperfusion, and mitochondrial injury. These three sources constitute the basic components of a wide variety of diseases. . . .

CROSS-REFERENCES

Many of the articles in the Encyclopedia have cross-references to other articles. These cross-references appear within the text of the article, at the end of a paragraph containing relevant material. The cross-references indicate related articles that can be consulted for further information on the same topic, or for other information on a related topic. For example, the article "Brain Evolution" contains a cross reference to the article "Cerebral Specialization."

BIBLIOGRAPHY

The Bibliography appears as the last element in an article. It lists recent secondary sources to aid the reader in locating more detailed or technical information. Review articles and research papers that are important to an understanding of the topic are also listed.

The bibliographies in this Encyclopedia are for the benefit of the reader, to provide references for further reading or research on the given topic. Thus they typically consist of no more than ten to twelve entries. They are not intended to represent a complete listing of all materials consulted by the author or authors in preparing the article.

COMPANION WORKS

The *Encyclopedia of Human Biology, Second Edition* is one of an extensive series of multivolume reference works in the life sciences published by Academic Press. Other such works include the *Encyclopedia of Cancer, Encyclopedia of Virology, Encyclopedia of Immunology,* and *Encyclopedia of Microbiology,* as well as the forthcoming *Encyclopedia of Reproduction.*

Conditioning

W. MILES COX
University of Wales

GLOSSARY

Classical conditioning Simple form of learning in which a stimulus, initially incapable of evoking a response, acquires the ability to do so through repeated pairing with a stimulus that already elicits the response

Extinction In classical conditioning, the weakening or elimination of a conditioned response when the conditioned stimulus is no longer paired with the unconditioned stimulus. In instrumental or operant conditioning, the weakening or elimination of a conditioned response when the response is no longer followed by reinforcement.

Instrumental (or operant) conditioning Form of conditioning in which the consequences of a response (viz., reward or punishment) determine the likelihood of that response occurring again in the future

Learning Relatively permanent change in behavior that often results from reinforced practice. Conditioning is the simplest form of learning

Motivation Internal state of an organism that corresponds to observable behavior that is directed toward achieving a goal—either to get something that the organism wants or to get rid of something that it does not want

Schedule of reinforcement In operant conditioning, the plan according to which reinforcements are delivered, in terms of either the number of responses that have been emitted or an interval of time that has elapsed

Social learning Process by which a person learns new behavior simply by observing another person perform the behavior

Spontaneous recovery Following a period of rest, the reappearance of a conditioned response that has been extinguished

Stimulus generalization Phenomenon whereby stimuli that are similar to a conditioned stimulus also elicit the conditioned response

CONDITIONING IS THE SIMPLEST, MOST BASIC FORM of learning. It is the type of learning that psychologists study in the psychological laboratory, for doing so allows them to identify principles underlying more complex forms of behavior. There are two major types of conditioning. Classical conditioning occurs when a stimulus (the conditioned stimulus) that is initially neutral acquires the ability to elicit a response because it is presented with another stimulus (the unconditioned stimulus) that already elicits a response. Instrumental (or operant) conditioning occurs when organisms emit responses that are followed closely by reinforcement; such responses become more likely to occur again in the future. Social learning (another type of learning that is studied experimentally) occurs when people acquire a new behavior merely by observing other people perform the behavior and imitating them in doing so. The principles of learning and conditioning have been useful in explaining why people acquire new behaviors (both desirable and undesirable ones), and therapists have developed conditioning techniques to help people change their behavior.

I. INTRODUCTION

Behavior is malleable. It can undergo distinct changes with time as a result of people's experiences and their interactions with the environment. Sometimes these changes are for the better. People can acquire new desirable behaviors, or they can rid themselves of undesirable behaviors. At other times, changes in behavior are undesirable. People may acquire new, undesirable habits, or they may stop performing desirable behaviors that they once performed regularly. Regardless of which type of change in behavior is being considered, conditioning is the mechanism that is often responsible for the change.

Our purposes in this article are to examine the various forms of conditioning and the principles by which each of them operates, and to demonstrate how the principles of conditioning can be used to understand, predict, and change people's behavior. There are two major types of conditioning that we will consider: classical conditioning and instrumental (or operant) conditioning. In addition, we will discuss social learning, another type of learning that is often studied in the laboratory.

A. Learning versus Conditioning

To psychologists, the term learning does not mean exactly the same as in everyday language. According to its common usage, learning refers to the acquisition of knowledge and is considered to be desirable. By contrast, learning in a psychological sense refers to a change in behavior that may be either desirable or undesirable. More specifically, learning is a relatively permanent change in behavior that often results from reinforced practice. Learning, therefore, is distinguished from (a) temporary changes in behavior, such as those that occur when a person is under the influence of a psychoactive drug, and (b) changes that are permanent but do not result from reinforced practice, such as changes resulting from injury to the body. Conditioning is the simplest form of learning, and psychologists often study conditioning in simple laboratory situations and in organisms simpler than humans, because doing so allows them to understand the principles underlying more complex forms of human behavior. [*See* Learning and Memory.]

II. CLASSICAL CONDITIONING

For centuries, people have known that when two stimuli (objects or events) are closely associated with each other, they become linked in people's minds. In fact, associationism, a philosophical doctrine that was popular in the eighteenth and nineteenth centuries, held that people form concepts through principles of association. Two of these are the principle of similarity (i.e., ideas are formed with objects or events that are similar to each other) and the principle of contiguity (i.e., ideas are formed with objects or events that occur closely together in time or space). Classical conditioning extends associationism to include connections between stimuli and responses. This form of conditioning occurs when a stimulus that originally elicits no response at all comes to do so regularly because it has been associated with another stimulus that already elicits a response. A connection, or bond, is said to be formed between the old stimulus and the new response.

The original experiments on classical conditioning were performed by Ivan P. Pavlov, a famous Russian physiologist who was awarded a Nobel Prize in 1904 for his work on digestion. In fact, Pavlov's work in classical conditioning grew out of his research on digestive functions. While examining the neural mechanisms responsible for food-elicited salivation in dogs, Pavlov became curious about why the dogs in his experiments began to salivate as soon as they saw the food (or even when the person feeding them entered the room), rather than wait until food was actually placed into their mouths. Pavlov believed that the anticipatory salivation that he observed could not be explained physiologically, and he referred to it as a "psychic secretion." He began, therefore, to redirect his research to investigate the mechanisms controlling this psychological phenomenon.

Using dogs as subjects, Pavlov performed his experiments by pairing neutral auditory or visual stimuli with meat. He referred to the originally neutral stimulus (e.g., a tone or a light) as the conditioned stimulus (CS). He found that when a CS was repeatedly presented just before a dog was given food, the CS itself—when later presented alone—would gradually come to elicit salivation. He called the salivation elicited by the CS the conditioned response (CR), because the ability of the CS to elicit the response was dependent on its being repeatedly paired with food. Pavlov referred to the food itself as the unconditioned stimulus (UCS), because its ability to elicit salivation occurred automatically and was not dependent on the dog's training. The salivation that occurred in response to the food was called the unconditional response (UCR). Although the terms "conditioned" and "unconditioned" have come to be used in English to refer to the two types of stimuli and responses observed in classical conditioning, these terms actually are erroneous translations of the Russian words meaning "conditional" and "unconditional." The more exact translation would more accurately convey the intended meaning of these terms.

It was no revelation that dogs salivate in anticipation of receiving food; people had probably known that for centuries. What was ingenious about Pavlov's work, however, is that he identified the basic principles, or laws, by which organisms learn to make new responses to old stimuli. Because Pavlov was one of the original scientists to study conditioning, the type of conditioning that he studied has come to be called classical conditioning, or Pavlovian conditioning. Classical conditioning experiments have three major features: acquisition, experimental extinction, and spontaneous recovery.

The acquisition phase of a classical conditioning experiment is the phase during which the CR is being acquired. Each time that the CS is paired with the UCS is referred to as a trial, and across successive trials of the experiment, there is a gradual increase in the ability of the CS to elicit the CR. During the course of the experiment, the experimenter measures the strength of the CR by occasionally administering a test trial, that is, a trial on which the CS is presented alone without the UCS. When the results of a classical conditioning experiment are displayed on a graph, successive conditioning trials are plotted on the abscissa, or horizontal axis, and the strength of the CR (e.g., the number of drops of saliva) is plotted on the ordinate, or vertical axis. On such a graph, we usually see a negatively accelerated learning curve. That is, across successive trials, the strength of the CR increases, but the magnitude of the increase becomes less and less, until eventually there is no further increase at all. At that point, the learning curve is said to have reached asymptote.

Extinction, the next phase of a classical conditioning experiment, is the one during which the experimenter repeatedly presents the CS in the absence of the UCS. As this occurs, the strength of the CR gradually extinguishes. For example, if an animal has been conditioned to salivate each time that a tone is presented, and the tone is then repeatedly presented in the absence of food, salivation as a CR to the tone will gradually decline. Even after a CR has become completely extinguished, however, it may regain some of its original strength without any reconditioning by the experimenter. This phenomenon is called spontaneous recovery, and is observed following a period of rest after the extinction phase of a classical conditioning experiment. In fact, the amount of recovery of the CR, as a general rule, increases as the length of the rest increases.

Four additional classical conditioning phenomena that Pavlov identified are of interest: stimulus generalization, conditioned discrimination, experimental neurosis, and higher-order conditioning. Stimulus generalization is observed when stimuli that are similar to, but not identical with, the original CR also elicit that response. A gradient of stimulus generalization is observed, such that the greater the similarity between the original CS and the new test stimulus, the greater is the response to the test stimulus. For example, if the original CS were a tone of a certain frequency (say, 1000 cycles per second), tones of either a lower or a higher frequency (e.g., 800 or 1200 cps) would still elicit a CR, although the response would be weaker than one elicited by the CS itself. Tones still further removed from the original CS (e.g., 600 or 1400 cps) would elicit still weaker CRs. Stimulus generalization can be overcome through conditioned discrimination. That is, if the experimenter continues to consistently pair the original CS (called an S+) with the UCS, while repeatedly presenting a generalized stimulus (called an S−) alone, the organism will learn to discriminate between the S+ and S−, consistently responding to the S+ while failing to respond to the S−.

While investigating stimulus generalization and conditioned discrimination, Pavlov discovered another interesting phenomenon called experimental neurosis. He first conditioned dogs to consistently salivate in the presence of a circle (a S+), and not to salivate in the presence of an ellipse. On successive trials thereafter, he made the circle more and more like an ellipse, and the ellipse more and more like a circle, until eventually the dog could no longer distinguish between the two stimuli. At this point, the dogs became emotional and acted "neurotic." They were unable to stand quietly in the restraining harness of the conditioning apparatus, barking and struggling to get away. In fact, the dogs eventually became so disturbed that they resisted being taken into the conditioning laboratory.

Higher-order conditioning is another interesting feature of classical conditioning experiments. Once a CS comes consistently to elicit a CR, that CS can function as a UCS for the organism to acquire still another CR. For instance, if a tone has been repeatedly paired with food, so that the tone consistently elicits salivation, the tone could now be paired with a new neutral stimulus (say, a light) until the light also comes to elicit salivation. The original pairing of the CS (tone) and UCS (food) is called "first-order" conditioning. The pairing of the new CS (light) with the

UCS is called "second-order" conditioning. If the tone were now paired with yet another CS, "third-order" conditioning would have occurred. Even second-order conditioning, however, can be difficult to establish because at the same time the original CS is functioning as a UCS, it is no longer being paired with the original UCS (e.g., food). Hence, the CR to the original CS begins to undergo extinction.

A. Using Classical Conditioning to Understand and Change Human Behavior

The principles of classical conditioning discovered in the laboratory have direct applications to human behavior. These principles, for instance, have been useful in helping psychologists understand how people acquire irrational fears and how they can be helped to rid themselves of such fears. People often become fearful of an object or event that was originally neutral because it occurred in the presence of a noxious stimulus. For example, after being involved in an automobile accident, a person may develop a strong fear of driving an automobile. Similarly, a person may develop a phobia for snakes, bugs, heights, or almost any other object or event that occurs in close proximity with an aversive stimulus.

In the same way that people acquire irrational fears through classical conditioning, they can be helped to rid themselves of such fears with the aid of conditioning techniques. One such technique is called "flooding." It requires that the phobic individual be exposed to (i.e., "flooded with") the stimulus that elicits his or her fear for an extended period of time in the absence of any actual danger. As exposure continues, with no actual danger occurring, the fear response undergoes extinction.

Another technique for eliminating fears is called systematic desensitization, the goal of which is to "systematically desensitize" a person to stimuli that he or she fears by conditioning a response to the fear-eliciting stimuli that is incompatible with fear. Systematic desensitization involves several steps. First, the therapist and client together devise a hierarchy of stimuli that elicit the client's fear, ranging from the least to the most fear-arousing stimulus. Second, the therapist teaches the client how to relax, usually by using progressive muscle relaxation techniques. Next, while the client is completely relaxed, the therapist introduces a fear-arousing stimulus that is very low on the patient's hierarchy of fear-arousing stimuli. Once the client is able to completely relax in the presence of the stimulus that once elicited a minimal amount of fear, the thera-

pist then introduces the next stimulus in the hierarchy and the process is repeated. Eventually, the client will be able to relax in the presence of the most anxiety-arousing stimulus in the hierarchy. In other words, at that point, the client's fear will have become extinguished. It should be noted that both systematic desensitization and flooding are typically conducted using imagined rather than the actual feared stimuli. Nevertheless, both techniques have been useful in helping clients eliminate unwanted fears.

Yet another classical conditioning procedure for helping people to overcome undesirable behaviors is called aversive conditioning, which has been used with some success with alcoholics. The procedure involves administering a drug that induces nausea (an emetic) to alcoholics at the same time that they are given their favorite alcoholic beverage to drink. After repeatedly pairing the alcoholic beverage with nausea, the beverage itself becomes aversive, and the alcoholic avoids it. One difficulty, however, in using aversive conditioning with alcoholics is that the aversive properties that a particular alcoholic beverage has acquired may not generalize to other alcoholic beverages. Moreover, aversive conditioning does not correct the underlying problems that caused an alcoholic to overindulge in the first place. Thus, aversive conditioning in the treatment of alcoholism should be viewed as an adjunct to other treatment techniques.

III. INSTRUMENTAL CONDITIONING

Instrumental conditioning generally involves a different category of responses than those that are classically conditioned. Whereas a CR or UCR occurs automatically when an organism encounters a CS or UCS, responses that are instrumentally conditioned are voluntary, because the organism can choose whether or not to emit them. (It should be noted, however, that there are some exceptions to this generalization. Under certain circumstances, involuntary responses can also be instrumentally conditioned.) A response becomes instrumentally conditioned (i.e., stronger or more likely to occur again) when it leads to reinforcement. If the response does not lead to reinforcement, it becomes weaker or less likely to occur. In other words, the response itself is instrumental in producing the consequences that will determine whether or not it is learned.

The study of instrumental conditioning began with the work of Edward L. Thorndike, an American psychologist who was a contemporary of Pavlov. The original purpose of Thorndike's work was to devise experiments to study animal intelligence. A well-known series of experiments that Thorndike designed for this purpose involved placing a hungry animal (e.g., a cat) inside a cage that had been specially designed so that, if the animal made a correct response (e.g., pulled a latch), the cage door would open and the animal would gain access to food outside. When the animal was initially placed inside this "puzzle box," it made various random movements, but eventually made the response that allowed the door to open. Thorndike believed that the animal's successful escape from the box caused an associative bond to be formed between the stimuli inside the box and the response that led to escape. Furthermore, each time that the correct response was made in the presence of these stimuli, the association between the two was strengthened. What this meant in terms of the hungry animal's behavior was that with each successive trial of being placed inside the puzzle box, it emitted fewer incorrect responses, and the correct response was made within a shorter period of time.

On the basis of these findings, Thorndike formulated his famous law of effect. It states that if a response is made in the presence of a stimulus and is followed by a "satisfying state of affairs," the association between the stimulus and the response will be strengthened. If the response is followed by an "unsatisfying state of affairs," the association will be weakened. The law of effect is considered the cornerstone of instrumental conditioning; if a response is instrumental in producing satisfying consequences, that response becomes conditioned.

A. Subsequent Research on Instrumental Conditioning

Thorndike's work paved the way for experimental psychologists' dominant interest in instrumental conditioning during the first half of the twentieth century. However, the experimental procedure that they favored for studying instrumental conditioning was to train white rats to traverse a maze in order to obtain a food reward. In these experiments, various parameters for the food reward were systematically manipulated by the experimenter to study the resulting changes in the rats' behavior. Such variations included (a) the size of the reward that was placed in the goal box at the end of the maze, (b) the length of the delay between the rat's entry into the goal box and the presentation of the food reward, and (c) the consistency with which the reward was provided (i.e.,

whether it was given at the end of every trial or only on some portion of them).

The foremost learning theorist during this era was Clark L. Hull. He devised an elaborate, mathematical theory of learning consisting of formal theorems, postulates, and corollaries from which he deduced various hypotheses to test empirically by conditioning white rats to traverse a maze for a food reward. On the basis of his experiments, Hull accepted certain aspects of his theory as experimentally confirmed, and he modified other aspects of the theory to take contradictory evidence into account. In Hull's view, his theory would not only help to identify the laws that govern the behavior of white rats, but also would be useful for understanding the behavior of humans.

One important way in which Hull changed his theory was to assign an increasingly important role to motivational factors as determinants of behavior. Hull came to realize that whether or not an organism performs a given behavior is determined not simply by whether or not it has been instrumentally conditioned to do so. The occurrence of the behavior also vitally depends on the organism's motivation to perform it. One key determinant of motivation, in turn, is the value of the incentive that the organism receives for performing the behavior. As a general rule, the more valuable, or attractive, an incentive is to an organism, the more highly motivated it will be to perform the behavior that is necessary to obtain the incentive.

IV. OPERANT CONDITIONING

The experimental paradigm for studying operant conditioning is the same as the one for instrumental conditioning. Operant conditioning occurs when an operant response (i.e., a voluntary response that operates on the environment) is followed closely by a reinforcing stimulus. When this occurs, the response becomes more likely to occur again. If the response is not reinforced, or is punished, it becomes less likely to recur. These are exactly the same circumstances under which instrumental conditioning occurs. In fact, the similarity between instrumental and operant conditioning is so close that some authors make no distinction between them; they use the two terms interchangeably.

According to our use of the terms, however, there are two key differences between instrumental and operant conditioning. These include (a) the procedure for performing a conditioning experiment and (b) the measure of conditioning that is taken. An instrumental conditioning experiment involves discrete trials. Whether a cat, for example, is being conditioned to open a puzzle box, or a rat is being trained to run through a maze for a food reward, the experimenter administers individual trials to the subject in succession. The strength of the conditioning can be measured by whether or not the animal makes a correct response and how rapidly it makes the response. By contrast, in an operant conditioning experiment, the experimenter places the animal in a situation where it is free to make an unlimited number of responses (such as press a lever for a food reward) in a given period of time. The rate of responding (or number of responses per unit of time) is the measure of conditioning in the free-responding, operant-conditioning experiment.

The study of operant conditioning was begun by the influential American psychologist B. F. Skinner, who published his first book "The Behavior of Organisms," in 1938 and who has continued to be prolific until his recent death. In large part, Skinner devoted his work to the functional analysis of behavior—a simple, laboratory analog of the way rewards and punishments control people's behavior in everyday life. He was especially careful to describe the variables controlling behavior as objectively as possible, assiduously avoiding theoretical and mentalistic concepts to account for the behavior that he observed. Hence, Skinner's atheoretical approach was in marked contrast to that of Hull.

To functionally analyze behavior, Skinner devised an apparatus called the operant conditioning chamber (or a "Skinner box"). This apparatus is simply a soundproof box that contains (a) a manipulandum (e.g., a lever for a rat to press or a lighted key for a pigeon to peck) and (b) a mechanism for delivering reinforcement (usually a dispenser for food pellets). Depending on the particular organism that is being studied, the experimenter decides on a particular response (e.g., lever pressing for rats and key pecking for pigeons) to reinforce or not to reinforce. The equipment attached to the operant conditioning chamber automatically records each time the designated response is made and each time a reinforcement is delivered. From the record that is produced, the experimenter can identify relationships between the number of responses that the animal made (and the pattern in which they were made) and the plan, or schedule, according to which the reinforcements were delivered.

Skinner's definition of a reinforcer differs markedly from that of Thorndike and Hull. In Skinner's view,

a reinforcer is simply any stimulus that follows a response and increases its probability of occurring again in the future. He does not specify that a reinforcer be "satisfying" or "pleasurable" to the organism that receives it. Nevertheless, Skinner divides reinforcers into two major classes. A positive reinforcer is a stimulus (e.g., food or water) whose presentation increases the probability of the response that preceded it. A negative reinforcer is a stimulus (e.g., electric shock, a bright light, or a loud noise) whose cessation increases the likelihood of the response that preceded its removal. Punishment, on the other hand, is the presentation of an aversive stimulus (or removal of an attractive one) that reduces the likelihood of the response that preceded it. (It should be noted that the term negative reinforcer is frequently used incorrectly to mean punishment.) From his research, Skinner concluded that using positive and negative reinforcers to control behavior is more effective than using punishment. Reinforcers can be used to bring about enduring changes in behavior. Punishment, on the other hand, may temporarily suppress a response, but will not permanently extinguish it.

One of the best-substantiated findings from research on operant conditioning is that the rate of responding depends on the schedule according to which the reinforcement is delivered. If an organism is reinforced after every response, it is said to be on a continuous schedule of reinforcement. If an organism is never reinforced, it is said to be on an extinction schedule of reinforcement. When the organism is reinforced, but not after every response, it is on an intermittent schedule of reinforcement. There are four basic schedules of intermittent reinforcement, each of which is associated with a characteristic manner of responding. These schedules vary according to whether delivery of the reinforcement is based on (a) the number of responses that have been emitted since the last reinforcement was delivered or (b) the interval of time that has elapsed since the last reinforcement was delivered. Moreover, within each of these, the schedule can be either fixed or variable.

First, let us consider the fixed-ratio (FR) schedule of reinforcement. On this schedule, the organism is reinforced after having made a fixed number of responses. For example, a rat being conditioned to press the lever in an operant-conditioning chamber might be on a FR15 schedule, meaning that whenever it presses the lever 15 times it will receive a reinforcement. This schedule produces a steady, high rate of responding because it is to the animal's advantage to make the required number of responses as rapidly as possible in order to receive its reinforcement. In everyday life, a factory worker who is paid according to the number of units of work that he or she completes would be on a fixed-ratio schedule.

Next, consider the variable-ratio (VR) schedule. On this schedule, the organism again is reinforced for having made a certain number of responses, but in this case, the number of responses that is required varies around a constant value. For instance, if an animal is on a VR10 schedule, only part of the time will it be reinforced for having made exactly 10 responses. At other times, it might be reinforced for having made, say, 5 responses of 15 responses. By the end of the conditioning session, however, the average number of responses for which it will have been reinforced will equal 10. In other words, on this schedule the exact number of responses that is required to receive a reinforcement is unpredictable. As might be expected, therefore, variable-ratio schedules result in very high rates of responding. In fact, animals have sometimes been observed to respond so rapidly on variable-ratio schedules that the amount of food that they receive as reinforcement is not sufficient to compensate for the great energy that they expend in responding. In the real world, gambling devices (e.g., slot machines in Atlantic City and Las Vegas) are programmed to "pay off" (i.e., to deliver monetary reinforcements) on variable-ratio schedules. Thus, always expecting that their very next response might lead to the desired reinforcer (i.e., money), gamblers respond at very high rates, and their behavior often is very difficult to extinguish.

A third schedule of reinforcement is the fixed-interval (FI) schedule. On this schedule, the organism is reinforced for the first response that it emits after a fixed interval of time (e.g., one minute) has elapsed. Regardless of how rapidly the animal might respond, it is impossible for it to receive a reinforcer earlier. Hence, when on a fixed-interval schedule, the animal responds at a very low rate—or stops responding altogether—just after a reinforcer has been delivered. This phenomenon is known as the postreinforcement pause. As the waiting interval draws to a close, however, the rate of responding becomes more rapid and continues at a rapid rate until the next reinforcer is delivered. Because of the erratic rate of responding, FI schedules are associated with the lowest overall rate of responding of any of the schedules discussed here. In real-life situations, studying for examinations in order to get a good grade is an example of behavior that operates according to a fixed-interval schedule. As the day of an examination draws near, many stu-

dents' studying behavior (i.e., their rate of responding) increases markedly. Then, immediately after the examination, many students study considerably less, or not at all. The same pattern repeats itself throughout the school term.

The final schedule that we will consider is the variable-interval (VI) schedule. On this schedule, a reinforcer again becomes available after an interval of time has elapsed, but as with the variable-ratio schedule, the interval varies around a constant value. Thus, exactly when a reinforcement will be available is unpredictable. As might be expected, therefore, variable-interval schedules result in more steady rates of responding than do fixed-interval schedules. In real life, mail delivery is an example of reinforcement (or perhaps punishment, if the mail contains unwanted bills) that is delivered according to a variable-interval schedule. Mail is delivered at an approximate (or average) time each day (say, 2:00 P.M.). On some days the carrier arrives early, whereas on other days he or she arrives late. As 2:00 o'clock approaches, a person might begin to check his or her mailbox, although it is very unlikely that he or she would do so as early as 10:00 A.M. Thus, although variable-interval schedules eliminate the exact predictability associated with fixed-interval schedules, they still allow for a certain degree of predictability.

The principles of operant conditioning are used to help people eliminate undesirable behaviors (e.g., a bad habit such as smoking) or to acquire desirable behaviors (e.g., social skills) in which they are deficient. Operant techniques are often used in institutions both to help maintain order and to help patients acquire desirable behaviors. One example of such an operant technique is the token economy. This is a system that allows patients to earn tokens for emitting desirable behaviors (e.g., making one's bed, arriving at group therapy sessions on time, making appropriate social responses) and to lose tokens for emitting undesirable behaviors. Tokens earned can later be exchanged for desired commodities or special privileges. The token economy, which at first might appear somewhat artificial, in many respects resembles the everyday world in which most people live and work. For example, by regularly going to work and performing adequately on the job, people earn money (i.e., tokens) with which they can later purchase items they want. On the other hand, if one violates a law (e.g., exceeds the speed limit while driving) a portion of his or her money (i.e., tokens) might have to be relinquished. It should be noted, however, that in society the value of "tokens" is always relative. For instance, having to pay a $50 speeding ticket would most likely represent a greater punishment for a person who earns $20,000 per year than for one who earns $50,000.

V. SOCIAL LEARNING

Although psychologists recognize the value of classical and instrumental conditioning in explaining and controlling behavior, those who have studied social learning believe that the laws of conditioning are incapable of fully accounting for human behavior. Social psychologists, as this group of scientists is sometimes called, emphasize that human behavior is often learned from social interactions among people, and is not merely the result of trial and error. According to social learning theorists, people's cognitions (i.e., the thought processes that occur in their minds) are a strong determinant of which stimuli they attend to, the meaning that they attribute to those stimuli, and how they respond to them.

Social psychologists have devised experiments to demonstrate that people can acquire a new behavior by observing other people perform the behavior. A series of these experiments conducted by Albert Bandura and his colleagues at Stanford University during the 1960s stimulated the initial interest in social learning. In one experiment, one group of young children observed an adult model display aggressive behavior toward an inflated doll (a "Bobo" doll), whereas a control group observed the adult displaying innocuous behavior. Later, when the children themselves were allowed to play in the room with the Bobo doll, they imitated the behavior of the adult model whom they had observed; that is, only those children who had observed aggressive behavior were aggressive themselves. In another experiment, each of two groups of children observed a model being aggressive, but one group saw the model being rewarded for being aggressive, and the second group saw the model being punished for being aggressive. Later, the children who had observed the model being rewarded for being aggressive were themselves more aggressive than those who saw the model being punished. However, when the children were themselves then offered a reward for imitating the behavior of the model, both groups displayed aggressive responses toward the Bobo doll. These experiments illustrate two important principles of social learning. First, reinforcement is not necessary for a person to acquire new behavior. It can be acquired merely by observing somebody else perform the behavior. Second, reinforcement and punishment often determine if a particular behavior is performed and whether or not it is learned.

In everyday life, a person is often influenced by the actual or perceived behavior of other people. For example, when people enter an ambiguous social situation (e.g., a gathering at their new employer's house) they tend to proceed cautiously until they have had a chance to observe others and gain an understanding of the dynamics of the situation. For example, in the situation just described, the newcomer may look to others to determine if it is appropriate to drink alcohol or call the boss by his or her first name. In fact, a new employee might well have asked a senior employee about the appropriate dress code for the party in advance of the event itself. Thus, by modeling his or her behavior after that of others, the new employee would be able to avoid the embarrassment of arriving at the party dressed differently from the other partygoers.

Advertisers often use the principles of social learning in their attempts to influence consumers' behavior. For example, to change consumers' attitudes about a particular product (and thus their decision to buy the product), the advertiser may depict somebody (i.e., a model) using the product who epitomizes the everyday-life role (e.g., career woman or mother) of persons at whom the advertisement is directed. The goal of the advertiser is to get consumers to change their attitude about the product by imitating the behavior of the model. In the event that a consumer already uses the product, the advertisement is designed to reaffirm his or her decision to do so.

Advertisements aimed at preventing drug abuse are another example of attempts to change people's behavior by having them imitate the behavior of a model. The goal of these advertisements is to convey vicariously to young people information about the negative consequences of drug use, without their having to acquire the information through their own experiences. For these approaches to succeed, however, the person hearing such an advertisement must be willing to accept the information communicated by the model in the advertisement. For this reason, models whom young people feel positively about and whom they trust are recruited to discuss the negative aspects of using drugs. The models include individuals of high status, such as athletes or rock stars, and people who have recovered from drug abuse through treatment.

tioned to respond to or the responses that they have been instrumentally or operantly conditioned to emit.

VI. CONCLUSIONS

The study of conditioning in the laboratory has enabled psychologists to identify the principles according to which animals and humans acquire new behavior or change existing behavior. Through classical conditioning, organisms learn to make new responses to neutral stimuli, or to extinguish responses that they have previously learned. Through instrumental or operant conditioning, the probability of organisms' emitting particular responses either increases and decreases, depending on whether or not a particular response is followed by reinforcement. Therapists have used the principles of conditioning identified in the laboratory to help people either rid themselves of unwanted behaviors or acquire new, desirable behaviors. In short, classical conditioning, instrumental conditioning, and operant conditioning have all been useful in helping us to understand, predict, and control behavior. Nevertheless, as the research on social learning indicates, conditioning principles alone do not account for the distinction between organisms' learning a particular behavior and their motivation to perform that behavior. For this reason, psychologists have grown increasingly aware that people's behavior is strongly influenced by motivational factors and is not determined simply by the stimuli to which people have been classically condi-

BIBLIOGRAPHY

Bootzin, R. R., Loftus, E. F., and Zajonc, R. B. (1983). Learning. *In* "Psychology Today: An Introduction" (R. R. Bootzin, E. F. Loftus, and R. B. Zajonc, eds.), 5th Ed., pp. 177–197. Random House, New York.

Bower, G. H., and Hilgard, E. R. (1981). "Theories of Learning," 5th Ed. Prentice–Hall, Englewood Cliffs, NJ.

Domjan, M., and Burkhard-Ebin, B. (1993). "The Principles of Learning and Behavior," 3rd Ed. Brooks/Cole, Moneterey, CA.

Hill, W. F. (1990). "Learning: A Survey of Psychological Interpretations," 5th Ed. Harper Collins, New York.

Hull, C. L. (1943). "Principles of Behavior." Appleton–Century–Crofts, New York.

Hull, C. L. (1952). "A Behavior System." Yale Univ. Press, New Haven, CT.

Kantawitz, B. H., and Roediger, H. L. (1984). Conditioning and learning. *In* "Experimental Psychology: Understanding Psychological Research" (B. H. Kantawitz and H. L. Roediger, eds.), 2nd Ed., pp. 201–226. West Publishing, St. Paul, MN.

Klein, S. B. (1991). "Learning: Principles and Applications." McGraw–Hill, New York.

Klinger, E. (1977). "Meaning and Void." Univ. of Minnesota Press, Minneapolis.

Pavlov, I. (1927). "Conditioned Reflexes." Oxford Univ. Press, Oxford, England.

Petri, H. (1991). "Motivation: Theory, Research, and Applications," 3rd Ed. Wadsworth, Belmont, CA.

Skinner, B. F. (1938). "The Behavior of Organisms. An Experimental Analysis." Appleton–Century–Crofts, New York.

Thorndike, E. L. (1898). Animal intelligence: An experimental study of the associative processes in animals. *Psychol. Rev. Monogr. Suppl.* **2,** 1–109.

Connective Tissue

KENNETH P. H. PRITZKER
Mount Sinai Hospital, University of Toronto

GLOSSARY

Aggrecan Large chondroitin sulfate proteoglycan present in cartilage

Basement membrane Connective tissue layer between histologically dissimilar tissues

Collagen Major group of structural proteins that form connective tissue fibers

Elastin Insoluble structural protein. Elastin and fibrillin molecules become organized to form elastic fibers

Extracellular matrix All material in connective tissue outside of cells

Fibronectin Structural glycoprotein that can bind specifically to collagen, proteoglycans, and cell membranes and that mediates many cell–extracellular matrix interactions, including cell migration

Glycosaminoglycan Structural sugar polymer containing repeating disaccharide units

Laminin Family of large glycoproteins present in basement membranes. Laminins are involved in cell attachment and migration

Parenchymal cell Cells intrinsic to the functions of an organ, excluding connective tissue cells

Perlecan Large heparan sulfate proteoglycan present in basement membranes

Proteoglycan Complex structural macromolecule composed of glycosaminoglycan chains attached to a protein core

Versican Large chondroitin sulfate proteoglycan involved with cell–matrix interaction

I. DEFINITION AND CLASSIFICATION

Connective tissue can be defined as the tissue that occupies the space between parenchymal organs and within organs, as well as between parenchymal cells. The name "connective tissue" and the preceding topological definition imply functions of structural support and physiological integration of cells to tissues, of tissues to organs, and of organs to the body. Tissues are collections of cells and intracellular substances forming definite visible structures. Accordingly, connective tissues are classified by the distinctive naked eye appearance (color, texture, and consistency) of their predominant components (Table I). The density of capillary blood vessels in the reddish-brown color of muscle, dissolved lipid pigments in yellow fat, chromophore molecules in orange/yellow elastic fibers, and the dense arrays of collagen fibers in white fascia or ligament demonstrate the variety of factors that contribute to connective tissue color. The relative concentration of amorphous macromolecules to fibrillar macromolecules and the orientation of fibrillar macromolecules provide textural appearances as varied as the smooth clear structure of the cornea, the glasslike hyaline cartilage, and the fibrous cable of tendon. The consistency of connective tissues may range from soft gelatinous or myxoid tissues and fatty tissues through firmer tissue such as fibrous, elastic, and cartilaginous tissue to the mineralized tissue of bone. Each tissue has a characteristic structure adapted to specific functions.

Connective tissues are both varied and complex in nature. For example, except for cartilage and cornea, vascular tissue is normally intermixed in most connective tissues. Vascular tissues contain elastic fibers in specific architectural arrangements. Elastic fibers may also be present in fibrous tissue or cartilage. Adipocytes (fat cells) may be present in fibrous tissue, in muscle tissue, or within bone. The soft connective

ENCYCLOPEDIA OF HUMAN BIOLOGY, Second Edition, VOLUME 3.

TABLE I
Connective Tissue Classification

Type	Examples
Myxoid	Nucleus pulposus (intervertebral disc), vitreous humor (eye), synovium, areolar connective tissue
Fibrous	Tendon, ligament, fascia
Elastic	Ligamentum nuchae, artery
Adipose tissue	White fat, brown fat
Muscle	Striated muscle, smooth muscle
Vascular tissue	Artery, capillary, vein
Cartilage	Hyaline cartilage, fibrocartilage, elastic cartilage
Bone	Compact bone, trabecular bone

tissue between organs that contains a mixture of fibrous and gelatinous tissues is often termed areolar connective tissue. Moreover, connective tissues themselves may be arranged into distinct organs, that is, discrete tissues with specific architecture composed of multiple connective tissues arranged for specific functions. For example, a bone is an organ composed of bone, cartilage, adipose, and often hematopoietic (blood) tissues; a muscle contains muscle, vascular, and fibrous connective tissues. This article will provide an overview of the biological principles common to the structure and function of all connective tissues.

II. STRUCTURE AND COMPOSITION

Connective tissue is derived embryonically from mesenchyme and ultimately from mesoderm. Connective tissues are the predominant bulk component of body mass and accommodate many diverse functions. This tissue is extremely heterologous both in its cellular population and in the composition of the intercellular material substance commonly called extracellular matrix. [See Extracellular Matrix.]

Connective tissue cells are of two general types. First, there are cells that produce, resorb, and maintain extracellular matrix. Cells of fibrous tissue (fibroblasts), joint lining (synovial) cells, cartilage cells (chondrocytes), and bone cells (osteoblasts, osteocytes, osteoclasts) have some or all of these activities. Second, there are cells that synthesize and maintain specialized internal structures or functions. These cells include myocytes with their specialized contractile apparatus and adipocytes, which are adapted for vacuolar cytoplasmic storage of fat. Vascular cells (myofibroblasts and endothelial cells) share characteristics of both cell types. These cells have contractile functions as well as actively produce extracellular matrix.

Each connective tissue contains extracellular matrix with distinct compositional properties. This matrix is a complex composite material that varies not only from one matrix to another but also from domain to domain within each matrix. In connective tissues where intracellular functions predominate, for example, muscle and adipose tissue, extracellular matrix is a minor component, occupying less than 5% of the tissue. In connective tissues where most cells are specialized for extracellular matrix synthesis, for example, tendon, cartilage, and bone tissues, the extracellular matrix may occupy up to 95% of tissue volume. Most connective tissue matrices are intimate composites of three phases. First, there is a freely diffusible fluid phase consisting of water and low-molecular-weight solutes. Less diffusible, medium-weight solutes and suspended noncollagenous proteins and lipids comprise the second phase. Both of these phases exist within and are constrained by a structural macromolecular phase composed of fibers such as collagen and elastin, as well as nonfibrillar soluble macromolecules such as nonfibrillar collagens, glycosaminoglycans, and proteoglycans. Although some macromolecules such as elastin of elastic fibers are products of a single gene, most macromolecules, including the collagens and proteoglycans, are products of multiple genes. Moreover, there is extensive heterogeneity among these macromolecules. Currently, over 16 different collagen types are recognized, with component molecules encoded by at least 30 different genes.

In bone, teeth, calcified tissue, and pathological calcifications, a fourth solid phase is present. This phase is composed of the mineral, calcium apatite $[(Ca)_{10}(PO_4)_6(OH_2)]$.

The components of connective tissue matrices are derived variably from synthesis by cells within the matrix and from diffusion from the plasma of the circulating blood (Table II).

The structure of each connective tissue matrix is determined predominantly by the composition and arrangement of the fibrillar and amorphous macromolecules. Each tissue structure is highly related to mechanical function. For example, in tendon, parallel arrays of type I collagen fibers facilitate its tensile function. In cartilage, a tissue that is adapted to resist compressive forces, the tissue is a dense hydrogel com-

TABLE II
Connective Tissue Composition

Component	Examples
I. Cells	Fibroblast, adipocyte, chondrocyte, osteocyte
II. Extracellular matrix	
1. Water	
2. Freely diffusible solutes	Gases (O_2, CO_2, N), ions, hormones, vitamins, cytokines, cell metabolites
3. Less diffusible substances	Immunoglobulins, structural glycoproteins, fibronectin, laminin, enzymes, enzyme inhibitors, lipids
4. Structural macromolecules	Fibrillar molecules: collagens I, II, III, V, VI, VII, XI, elastin
	Amorphous soluble polymers: proteoglycans, glycosaminoglycans, nonfibrillar collagens
5. Mineral	Calcium apatite (bone, teeth)

TABLE III
Connective Tissue Matrix Macromolecules: Arrangements and Interactions

Arrangement	Examples
Molecular orientation	Collagen in tendon or bone
Molecular polymerization and aggregation	Collagen fibril aggregation, collagen intermolecular cross-linkages, proteoglycan aggregation
Heteromolecular interaction	Collagen/proteoglycan binding, elastin/fibrillin binding, collagen/fibronectin/cell binding, proteoglycan/laminin/cell membrane binding, collagen and calcium apatite binding
Macromolecule/ion interaction	Ca^{2+}, Mg^{2+}, and glycosaminoglycans

posed of hydrated type II collagen and amorphous proteoglycan macromolecules. In arterial blood vessels, where the wall must modulate pulsatile pressures, elastic fibers are characteristic. In bone, where rigidity and strength are crucial, the matrix is a complex composite material composed mainly of type I collagen and the mineral, calcium apatite. [See Cartilage.]

The arrangements and interactions among extracellular matrix molecules are highly controlled and specify the functional capacity of the tissue. Some major intermolecular arrangements, discussed briefly in the following, are listed in Table III. Fibrillar macromolecules have specific orientations that are adapted to the ambient mechanical forces. For example, collagen fibers in tendon are aligned in parallel groups or fascicles to provide an arrangement that maximizes tensile strength. Collagen fibers in dermis (skin) are arranged in a mat-like or reticular network. This pattern becomes oriented to applied forces yet retains proteoglycans and other amorphous compoents within the tissue. In articular cartilage, the collagen fibers, though oriented in a nonrandom fashion, remain aligned in multiple directions with application of forces. This arrangement provides enhanced resistance to compression. [See Articular Cartilage and the Intervertebral Disc; Skin.]

Collagen is secreted from the connective tissue cells as a large molecule, procollagen. Within the matrix, procollagen molecules are cleaved by an enzyme to form collagen monomers. These molecules align themselves in a quarter-staggered fashion to form fibrils that are then stabilized by cross-links that form between fibrils. To a large extent, fibrillogenesis and the radial growth of collagen fibers and controlled by electrostatic binding of proteoglycans to specific sites on the collagen molecules. Similarly, glycosaminoglycans, long-chain sugars with repeating disaccharide units, associate with core polypeptides to form proteoglycans. These molecules in turn associate with hyaluronic acid glycosaminoglycan molecules to form proteoglycan aggregates. The functional state of proteoglycan polymers and aggregates determines many characteristics of connective tissue, including resistance to compression forces and diffusion of molecules through the matrix. Elastic fibers are a composite material consisting of the intimate interaction of an amorphous protein, elastin, and a fibrillar glycoprotein component, fibrillin. Elastic fibers, as the name implies, are deformable tissue elements that permit reversible absorption of energy by connective tissues without loss of tissue shape.

An important class of interactions is molecular binding of cells to structural macromolecules. These molecules contribute to the organization of connective tissue macromolecules around the cell. Fibronectin, a molecule that has separate binding sites for collagen, proteoglycan, and cell membranes, is an important molecule of this type. Laminin, a molecule that binds proteoglycan and cell membranes, serves this function in basement membrane structures that are present between parenchymal cells and connective tissues. The relationship of calcium apatite to bone collagen fibers provides an excellent illustration of the exquisite regulation of the connective tissue matrix. First, calcium apatite crystals that form within bone are very similar to each other in size and crystallinity. However, the density in which the apatite crystals are packed within the matrix varies considerably within bone domains and between bones. Moreover, the calcium apatite crystals are deposited anisotropically, that is, with the axis of each crystal specifically oriented along the collagen fibril. [*See* Collagen; Extracellular Matrix.]

Some critical interactions in connective tissue involve the relations among elemental ions, small charged molecules, and proteoglycans. Proteoglycan molecules can be thought of as negatively charged ion-exchange resins with charges usually counterbalanced by positive sodium ions and low-molecular-weight proteins. Ions diffusing through connective tissue are hydrated, that is, associated with water molecules. Highly charged ions such as calcium ions will preferentially bind to glycosaminoglycans. This binding may change the conformation of the glycosaminoglycans as well as release water. Both of these activities will alter the local properties of the matrix, including molecular diffusion. [*See* Proteoglycans.]

Molecular diffusion is dependent quantitatively on the binding characteristics of the matrix molecules and the volume of water within the matrix. Matrix water content varies from tissue to tissue. More than 99% of the vitreous within the eye is water compared to less than 10% in dentin. Matrix water itself can be considered as a two-phase system. There is a free phase in equilibrium with interstitial fluid elsewhere and with blood plasma, as well as a bound or restricted phase associated with macromolecules and ions.

III. CONNECTIVE TISSUE FUNCTIONS

The functions of connective tissue are dependent on the matrix as a composite material. Connective tissue function depends not only on the quantity of the various components but also on the qualities of their orientation and molecular interaction. Perhaps the most important ideas concerning the varied functions of connective tissue are that each tissue has multiple functions and that the functions are nested within each tissue. Matrix domains of specific composition have specific characteristics for each function. Variation in matrix composition results in variation of each functional capacity of the matrix. Control of matrix structure and hence control of connective tissue function reside with the regulation of connective tissue cell metabolism. Common to all connective tissues are functions that can be grouped as follows: structural support and mechanical adaptation, transport and barrier, storage, cell modulation, tissue morphogenesis, and repair (Table IV).

TABLE IV

Connective Tissue Functions

Function	Examples
I. Structural and mechanical	Support and protection of cells and organ arrangements, adaptation to external forces (tension, compression, shear, etc.), lubrication, cell shape determination
II. Transport barrier functions	Nutrients and wastes, hormones and paracrine molecules, immunoglobulins, microorganisms
III. Storage	Intracellular: carbohydrates, amino acids, lipids, lipid-soluble substances Extracellular: water, ions, Ca^{2+}, PO_4^{3-}, enzymes and inhibitors, immunoglobulins, growth-stimulating molecules and inhibitors, matrix degradation products, cell wastes
IV. Cell modulation	Growth, proliferation and differentiation, adhesion and migration, matrix synthesis and degradation
V. Tissue morphogenesis and repair	Parenchymal tissue organization, parenchymal tissue repair, connective tissue repair and regeneration

A. Structural and Mechanical Functions

Like other solid materials, connective tissue resists three types of stress forces: compression, tension, and shear. Connective tissue mechanical properties can be described in a similar manner to those of other materials, namely, by their quantitative change with the application of mechanical forces. Terms used include strain: the change in dimension of tissue compared to initial dimension; extensibility, the maximal strain at which tissue fails (tears); stress: the force/unit area; stiffness: the change in stress/change in strain; and toughness: the strain energy absorption/area beneath the stress–strain curve. A characteristic of connective tissues is that these relationships change over the time that the force is applied. For tensile forces this is termed viscoelastic behavior. The change in properties of connective tissues over time and with increasing load results from realignment of collagen and elastic fibers under load. The low resistance of the matrix to such rearrangements depends on the fluid properties of the proteoglycan hydrogel.

Connective tissue can be grouped by mechanical characteristics as tensil or rope-like materials, pliant or deformable materials, and rigid materials that resist stress with little deformation. As applied to connective tissues, tendon is adapted to high tensile strength and stiffness. The dermis of skin has moderate tensile strength but can be repeatedly stretched and relaxed without loss of shape. Arterial vessel wall stretches easily at low loads and becomes increasingly stiff at higher loads. Cartilage is moderately strong and stiff and has high extensibility, whereas bone is very strong, stiff, and inextensible.

Lubrication is a subset of connective tissue mechanical properties. It is the ability of tissues to accommodate to shear forces. Lubrication occurs not only at joints but at all surfaces where one tissue moves against another, whether this be at a macroscopic or microscopic level. The mechanisms and molecules involved in lubrication vary depending on the forces and the deformability of the tissues. For soft tissues, the glycosaminoglycan hyaluronic acid appears as the critically important molecule, whereas in cartilage, specific glycoproteins or lipids may be implicated in this function.

Connective tissues maintain parenchymal cells in functionally effective conformations. These arrangements are adapted to support and protect the parenchymal cells to resist ambient forces, including those induced by gravity and locomotion. Less obvious, but equally important, is the interaction of cells with specific cell matrix molecules, including proteoglycans and glycoproteins, which determines the shape and frequently the functional state of cells.

As materials, connective tissues are extremely complex and can degrade rapidly when studied out of the body. Therefore, although mechanical behavior at macroscopic and microscopic levels can be described in material science and mathematical terms, the functional understanding of connective tissue mechanical properties remains incomplete.

B. Transportation and Barrier Functions

The transport and barrier functions of connective tissue are associated mainly with the hydrogel macromolecular structure. This structure consists of water constrained by negatively charged (anionic) macromolecules counterbalanced by low-molecular-weight, positively charged cations. In this gel, gases and low-molecular-weight molecules diffuse more rapidly than larger molecules; cationic molecules diffuse faster than uncharged or anionic substances. Conversely, molecules bind with different affinities and capacities to connective tissue matrix components. An example of the complexity of the diffusion process is that attachment to negatively charged solutes may enhance the permeability and decrease binding of some molecules to connective tissue macromolecules. The diffusion of particular molecules varies from tissue domain to tissue domain related primarily to the charge density and hydration of tissues. These factors are to some extent regulated by the connective tissue cells. Functionally, this means that connective tissues are permeable to cell nutrients and wastes but are relatively impermeable to high-molecular-weight proteins and microorganisms. In basement membrane matrices, perlecan, a heparan sulfate proteoglycan, is important for the charge-dependent ultrafiltration characteristics of the matrix. In tissues with a very high density of anionic aggrecan proteoglycans, such as cartilage, proteins secreted by the cells such as enzyme inhibitors diffuse slowly and bind avidly. This leads to their accumulation within the matrix, which makes cartilage resistant to enzymatic degradation from external cells. One consequence of this property is the resistance of cartilage to penetration by blood vessels. Many processes in connective tissues may be related by the relative diffusion of molecules through the matrix. Cell activity is modulated by the relative diffusion of hormones and cytokines compared to diffusion of their degradative enzymes. Collagen and elastic fiber formation is dependent on the constrained diffusion and orientation of protein monomers. As another example, mineralization of connective tissues depends on relative diffusion and binding to achieve the critical concentrations of calcium and phosphate necessary to initiate calcium apatite crystal deposition.

C. Storage Functions

Connective tissues have important storage functions for cell nutrients, major and trace elements, water, and molecules with special functions such as immunoglobulins. For many substances, connective tissues are the major storage sites because of the mass and relative stability of these tissues. The storage function may be general, for the organism as a whole, or local, adapted to the requirements of cells and matrixes within connective tissue domains. Most cell nutrients are stored intracellularly following anabolic reactions that simultaneously store energy whereas other substances are stored extracellularly. Adipocytes, for example, store large molecules of neutral lipid in cytoplasmic vacuoles. However, within these vacuoles, these cells also store lipid-soluble substances, both physiological substances such as fat-soluble vitamins and nitrogen and pathological compounds such as fat-soluble pesticides. Within muscle cells, amino acids are stored as contractile proteins and carbohydrate as glycogen. All of these substances may be released during periods of exercise or undernutrition, when catabolic activity exceeds anabolism. [See Adipose Cell.]

Extracellularly within the hydrogel of soft tissues, water, ions, and diverse specific molecules such as immunoglobulins, growth factors, and enzyme inhibitors accumulate. At these sites, stored molecules may have roles in local modulation of cells and modification of matrix properties. The extracellular compartment of bone stores many substances extensively. Not only is 99% of the calcium phosphate stored, but also large amounts of magnesium, carbonate, and trace elements such as fluoride. This is another example of nested functions. Bone tissue serves as a reservoir for mineral; bone mineral contributes to bone strength and rigidity. Many organic substances adhere to the bone mineral calcium apatite. Further deposition of apatite results in long-term retention of these substances. A maker of this storage capacity is the fluorescent antibiotic tetracycline, which can be seen years later in bone that formed at the time of drug administration.

D. Cell Modulation

Connective tissues provide much more than a passive material support for parenchymal cells. Extracellular matrix molecules are active in regulating essential processes such as cell adhesion/migration and cell proliferation/differentiation. Two classes of molecules appear to be involved: glycoproteins such as fibronectin and laminin and specialized proteoglycans such as perlecan and versican. In general, regulation is mediated by charge-dependent binding of glycosaminoglycans to cell adhesion molecules and growth factors or by binding of specific glycosaminoglycan sugar sequences to glycoproteins. These intermolecular linkages, which include linkages to the cell membrane, enhance adhesion by promoting organization of actin intermediate filaments within the cell. Similarly, by blocking the binding sites on either the glycosaminoglycans or the cell membrane, soluble proteoglycans can promote cell mobility. Glycosaminoglycans such as heparan sulfate inhibit cell proliferation, possibly by binding growth factors such as fibroblast growth factor and transforming growth factor β. This binding may in fact be a storage mechanism for growth factors. Matrix degradation, occurring in processes such as remodeling or inflammation, may release intact growth factors that in turn stimulate cell proliferation.

E. Tissue Morphogenesis and Repair

From the formation of mesoderm, an embryonic tissue consisting of primitive fibroblast cells that produce predominantly proteoglycans and glycosaminoglycans, the connective tissue extracellular matrix modulates the form of organogenesis. Three-dimensional form and the organization of other tissue elements appear to be controlled by the secretion of specific proteoglycans and glycoproteins at specific intervals of development. These molecules regulate the rate of cell proliferation, cell differentiation, and cell migration, as well as the development of collagen and elastic fibers within the matrix. The exact recapitulation of this process following injury is termed regeneration, a process that occurs under very limited circumstances within human connective tissues. It is possible to obtain complete regeneration in the adult human, the foremost example being bone regeneration following fracture. More commonly following injury, connective tissue reacts in a process termed repair, which leads to a less functional result than regeneration. The repair process is a program in which, following resorption of tissue by the inflammatory response, there is a timed sequence of restorative cellular events. The repair process includes the orderly elaboration of proteoglycans, vascular invasion, collagen fibrillogenesis, collagen fiber polymerization and cross-link-

PREFACE TO THE SECOND EDITION

The first edition of the *Encyclopedia of Human Biology* has been very successful. It was well received and highly appreciated by those who used it. So one may ask: Why publish a second edition? In fact, the word "encyclopedia" conveys the meaning of an opus that contains immutable information, forever valid. But this depends on the subject. Information about historical subjects and about certain branches of science is essentially immutable. However, in a field such as human biology, great changes occur all the time. This is a field that progresses rapidly; what seemed to be true yesterday may not be true today. The new discoveries constantly being made open new horizons and have practical consequences that were not even considered previously. This change applies to all fields of human biology, from genetics to structural biology and from the intricate mechanisms that control the activation of genes to the biochemical and medical consequences of these processes.

These are the reasons for publishing a second edition. Although much of the first edition is still valid, it lacks the information gained in the six years since its preparation. This new edition updates the information to what we know today, so the reader can be confident of its full validity. All articles have been reread by their authors, who modified them when necessary to bring them up-to-date. Many new articles have also been added to include new information.

The principles followed in preparing the first edition also apply to the second edition. All new articles were contributed by specialists well known in their respective fields. Expositional clarity has been maintained without affecting the completeness of the information. I am convinced that anyone who needs the information presented in this encyclopedia will find it easily, will find it accessible, and, at the same time, will find it complete.

Renato Dulbecco

E

CONTENTS OF VOLUME 3

Contents for each volume of the Encyclopedia appears in Volume 9.

PREFACE TO THE FIRST EDITION

We are in the midst of a period of tremendous progress in the field of human biology. New information appears daily at such an astounding rate that it is clearly impossible for any one person to absorb all this material. The *Encyclopedia of Human Biology* was conceived as a solution: an informative yet easy-to-use reference. The Encyclopedia strives to present a complete overview of the current state of knowledge of contemporary human biology, organized to serve as a solid base on which subsequent information can be readily integrated. The Encyclopedia is intended for a wide audience, from the general reader with a background in science to undergraduates, graduate students, practicing researchers, and scientists.

Why human biology? The study of biology began as a correlate of medicine with the human, therefore, as the object. During the Renaissance, the usefulness of studying the properties of simpler organisms began to be recognized and, in time, developed into the biology we know today, which is fundamentally experimental and mainly involves nonhuman subjects. In recent years, however, the identification of the human as an autonomous biological entity has emerged again—stronger than ever. Even in areas where humans and other animals share a certain number of characteristics, a large component is recognized only in humans. Such components include, for example, the complexity of the brain and its role in behavior or its pathology. Of course, even in these studies, humans and other animals share a certain number of characteristics. The biological properties shared with other species are reflected in the Encyclopedia in sections of articles where results obtained in nonhuman species are evaluated. Such experimentation with non-human organisms affords evidence that is much more difficult or impossible to obtain in humans but is clearly applicable to us.

Guidance in fields with which the reader has limited familiarity is supplied by the detailed index volume. The articles are written so as to make the material accessible to the uninitiated; special terminology either is avoided or, when used, is clearly explained in a glossary at the beginning of each article. Only a general knowledge of biology is expected of the reader; if specific information is needed, it is reviewed in the same section in simple terms. The amount of detail is kept within limits sufficient to convey background information. In many cases, the more sophisticated reader will want additional information; this will be found in the bibliography at the end of each article. To enhance the long-term validity of the material, untested issues have been avoided or are indicated as controversial.

The material presented in the Encyclopedia was produced by well-recognized specialists of experience and competence and chosen by a roster of outstanding scientists including ten Nobel laureates. The material was then carefully reviewed by outside experts. I have reviewed all the articles and evaluated their contents in my areas of competence, but my major effort has been to ensure uniformity in matters of presentation, organization of material, amount of detail, and degree of documentation, with the goal of presenting in each subject the most advanced information available in easily accessible form.

Renato Dulbecco

ing, and elastic fiber formation, followed by dehydration and contraction of the tissue to form a dense fibrous scar. This repair tissue, though very effective in uniting tissues separated by injury, may be less functional than the original tissue for several reasons.

To illustrate, reparative connective tissues, being noncontractile, lessen muscle strength. In skin, the less hydrated reparative tissue inhibits diffusion of nutrients and waste and contains poorly oriented macromolecules, thereby being less adapted to mechanical forces. Further, the rearrangement of connective tissue fibers in repair may prevent regeneration of structures, including the arrangement of parenchymal cells within internal organs, muscle fibers, and nerve axons. The processes of morphogenesis, regeneration, and repair appear to share a similar program of cellular activity and matrix biomolecular synthesis. However, repair is distinguished from morphogenesis by the lesser degree of regulation. This is because the repair process of tissue is proceeding in domains with much greater separation between connective tissue cells and with abundant preexisting matrix molecules, some partially degraded. These molecules are capable of storing and releasing various stimulatory and inhibition factors accumulated during normal growth and maturation of the tissue under conditions that are less fully controlled than in the pristine environment of embryonic development.

IV. CONNECTIVE TISSUE REGULATION

Although connective tissues are highly complex composite materials that can vary considerably from domain to domain, each domain retains its overall structure and exquisite adaptation to multiple functions over long periods. Furthermore, with growth, repair, and aging, connective tissues undergo orderly changes in structure and function. Except for muscle and endothelial cells, tissue regulation occurs among cells with very limited intercellular contact. Mature connective tissue cells can produce differing amounts and types of extracellular matrix components depending on external stimuli. The most versatile cell is the fibroblast, which, depending on circumstances, can synthesize type I or type III collagens, elastic fibers, or proteoglycans. All of these products share with other cell products the highly regulated processes of protein formation through gene activation, transcription, and translation. This is followed by intracellular protein modification (e.g., hydroxylation or glycosylation)

and translocation through the Golgi apparatus and the cell membrane to the extracellular space. For most other kinds of cells, metabolism is modulated by endocrine and neural stimuli mediated by receptors and by signal transduction mechanisms within the cell membrane.

The regulation of connective tissue cells appears to be more complex. This complexity is related to the further modification of connective tissue macromolecules in the extracellular space and their integration by orientation and bonding into a composite material. Moreover, connective tissue cells are shielded from short-term fluctuations in plasma hormones by the differential diffusion properties of the extracellular matrix. In addition, connective tissue cells are capable of transducing mechanical and electrical signals present at the matrix–cell interface to modulate cell metabolism. Transduction of mechanical and electrical signals by individual connective tissue cells surrounded by matrix is still poorly understood. However, in fibroblasts, chondrocytes, and osteocytes, the solitary cilium with its prominent microtubular organization is thought to be implicated as the cell receptor organelle.

Endocrine hormones, as well as paracrine and cytokine regulators, are differentially permeable and are selectively bound and released by components of extracellular matrix. Consequently, to affect cells within connective tissue matrices, their substances must diffuse through the matrix to the cell. Similarly, cell metabolites in a feedback loop must pass through the matrix to other cells or to the bloodstream. In articular cartilage and intervertebral disk, the total diffusion distance may be up to several centimeters in length. The difference in connective tissue composition induced by hormones can be striking, the best examples being the most visible gender differences between women and men.

Adaptation of connective tissues to changing environments leads to changes in matrix composition, a process regulated by the connective tissue cells. In these circumstances, the cells secrete matrix-degrading enzymes such as glycosaminoglycanases, collagenases, and other proteases. The activity of these enzymes is controlled by ambient ions, cofactors, and inhibitors. The controlled degradation of matrix by pericellular proteolysis releases growth factors that stimulate cell proliferation and cell differentiation and manufacture of new matrix macromolecules. These macromolecules polymerize in the matrix in orientations constrained by preexisting molecules and ambient mechanical forces.

V. AGING

Aging is reflected in connective tissues by structural changes in bulk, color, texture, and consistency, as well as functional changes. Tissues decrease in size; translucent tissues become opaque; white tissues develop yellow pigmentation; smooth tissues become rougher; and gelatinous tissues become firm. Between the resiliency of the bouncing baby and the homeostenotic and structural brittleness of the aged are connective tissue changes that trend from a gel state toward a crystal phase. These changes reflect the accumulation of extraneous substances within the matrix, loss of hydration, and increased intermolecular binding, in particular collagen cross-linking. All of these changes are interrelated and must be considered as failure of the connective tissue cells to regulate effectively their matrix domains.

To place the sequence of aging events in perspective, we must recall that solutes bind differentially within the matrix. Among the solutes are degradation products of matrix components and cell membranes produced during growth and remodeling, as well as from reactions to injury and repair. Substances diffusing through the matrix may be local in origin or derived from distant sites and transported to the matrix via the blood plasma. Ideally, the proteolytic enzymes elaborated by connective tissue cells, together with the ability of these cells to synthesize and reconstitute matrix components, would restore matrices to an optimal functional state. In aging, this ability decreases over time. Various molecules accumulate within the matrix, the most visible of which is a yellow oxidized lipid pigment, lipofuscin. The changes in molecular composition decrease the capacity of connective tissue to retain water. Dehydration in turn brings connective tissue macromolecules closer together. This facilitates molecular bonding, including collagen cross-linking and heterotopic mineralization, such as calcium phosphate crystal deposits in vessel walls or calcium pyrophosphate dihydrate crystal deposition in cartilage.

Among the substances that can accumulate in matrix are hormones and growth factors. The interactions of these substances with connective tissue cells led to stimulation of proteolytic enzyme production, such as elastase in vessel walls and lung alveoli. Elastase degrades elastic fibers, fibronectin, and possibly cell-associated proteoglycans. These cell growth factors stimulate matrix macromolecule production

by cells, modulated toward the production of fibrous molecules such as collagen rather than water-retaining molecules such as proteoglycans. The net effect of these changes is loss of tissue bulk and loss of functional capacity. This is reflected in manifestations of aging such as thinning and decreased elasticity of dermis, decreased vessel elasticity, loss of muscle bulk and strength, serous atrophy of fat, and osteoporosis.

Driving the rate of these changes in connective tissues is the biological clock of circadian hormonal and nutritional rhythms and their effects on connective tissue cells. Less well understood are the consequences of changes in the connective tissues, including the permeability of hormones and feedback regulating molecules on the oscillations of hormone production by endocrine cells. It is likely that a functional understanding of connective tissue aging will involve the intimate interaction between the endocrine and paracrine effects on connective tissue function and the connective tissue matrix influence on these functions. In this paradigm, processes that allow or promote matrix degradation and resorption with matrix molecule accumulation will accelerate aging; processes that promote matrix hydration and clearance of solutes may retard the aging process. [See Circadian Rhythms and Periodic Processes; Endocrine System.]

VI. DISEASE

Disease in connective tissue results from biological processes that reversibly or permanently alter the structure and functional capacity of these tissues to the functional disadvantage of the body. Given the ubiquity and mass of connective tissue, disease affecting these tissues can be classified as primary diseases affecting connective tissue molecules and diseases secondary to other disease processes that occur within connective tissue. Connective tissue disease can be also divided topographically into generalized disease and localized disease. Generalized disease affects all connective tissue in a particular organ or particular connective tissues in the body as a whole. Localized disease involves selected regions in particular tissues (Table V).

Primary generalized connective tissue diseases affect the biosynthesis, structure, or degradation of a single or multiple classes of connective tissue molecules. These alterations adversely affect the macromolecular organization of the tissue, leading to defective

TABLE V
Connective Tissue Diseases

Type/topology	Generalized (affects sites throughout the body)	Localized (affects specific sites within the body)
Primary (affects all of one or more classes of connective tissue molecules)	Primary/generalized	Primary/localized
Secondary (connective tissue structure affected by another disease process)	Secondary/generalized	Secondary/localized

function, particularly biomechanical functions that present as clinical symptoms. A relatively well-studied example is the group of diseases known as osteogenesis imperfecta. These diseases are characterized by inherited defects in the structure of type I collagen, resulting in brittle bones, thin sclera and dermis, and weak ligaments. Within the four clinical variants of this disorder, several single defects of amino acid deletions, insertions, or substitutions have been found in both primary polypeptide chains that comprise type I collagen fibers. What must be emphasized is that although there are similar patterns of molecular defects, the potential heterogeneity of defects in these large molecules is immense and only partially correlates with the severity of the clinical syndrome. This reflects the observations that the defects may be variably expressed in the molecules and that the connective tissue cells may modify the structure of the collagen macromolecular organization to compensate for the weakness of fibrils that are defective at a single site. Another example of generalized connective tissue disease is scurvy, which is related to vitamin C (ascorbic acid) deficiency. Ascorbic acid is a factor required in the promotion of extracellular maturation and cross-linking of collagen fibers and the formation of elastic fibers. Ascorbic acid deficiency results in defective fibril formation and maintenance, resulting in increased fragility, particularly of structures that involve both components such as small blood vessels. Yet another example is myxedema, in which deficiency of thyroid hormone results in the presence of excess proteoglycan in fibrous and areolar connective tissues. [See Ascorbic Acid.]

The secondary generalized involvement of connective tissues in diseases is observed dramatically in disorders of interstitial fluid volume regulation in which dehydration or edema may ensue. Another example is diabetes mellitus, in which the increase of glucose within the extracellular matrix of connective tissues adversely alters the permeability and binding characteristics of other solutes.

Primary localized connective tissue diseases affect one or more class of connective tissue molecules at local specific sites. An illustration of these diseases is the degenerative change in dermal collagen and elastic fibers following prolonged excessive exposure to sunlight. A second common example is the degenerative change in articular cartilage matrix following repeated impact injuries, such as occurs with excessive athletic activity.

Secondary localized diseases of connective tissue embrace an immense spectrum of chronic disease involving both connective tissue structures and connective tissue within parenchymal organs. First, the disease process may be localized to a single site typified by a scar following local injury. As noted previously, the scar may be dysfunctional because it is not as permeable or as contractile as normal tissue. For this group of disorders, the implication is that the disease process involves cells within or adjacent to connective tissues. The reaction in connective tissue leads to local changes in structural organization that disrupt the functional capacity of the tissue. These changes are also observed in the invasive process of malignancy, where tumor cells may elaborate proteolytic enzymes that facilitate tumor cell invasion. Alternatively, tumor cells may induce dense collagen fiber formation or excessive proteoglycan production, features that inhibit the access of antibodies and other antitumor substances to the tumor cells.

Secondary diseases may involve an entire class of sites. Certain autoimmune diseases provide illustration of this phenomenon. Antigen–antibody complexes localizing in vessel walls induce inflammation that increases the permeability of vessels, leading to mucinous edema, a noted feature of "collagen diseases." Antibodies directed against connective tissue basement membrane antigens may give rise to inflammation that depending on circumstances, can provoke blistering of skin or inflammation of kidney glomeruli. Third, connective tissue repair at multiple sites within a parenchymal organ may contribute to parenchymal organ failure. Two examples are offered. The lack of contractility in a scar within the heart with expansion of the scar from an aneurysm can contribute to heart failure. Scarring of kidney glomeruli or tubules following scattered inflammatory events may lead to decreased kidney function. For most secondary processes involving parenchymal organs, the connective tissue reaction is appropriate to the stimuli inducing change in the connective tissue organization. Disease ensues because the connective tissue organization or location is inappropriate for the normal functioning of the parenchymal tissue. [See Antibody–Antigen Complexes: Biological Consequences.]

BIBLIOGRAPHY

Ayad, S., Boot-Handford, R., Humphries, M. J., Kadler, K. E., and Shuttleworth, A. (1994). "The Extracellular Matrix Facts Book." Academic Press, London/Toronto.

Comper, W. D. (ed.) (1996). "Extracellular Matrix" Volumes 1 and 2. Harwood Academic Publishers, Australia.

Evered, D., and Whelan, J. (eds.) (1986). "Functions of Proteoglycans" Ciba Foundation Symposium No. 124. John Wiley & Sons, New York/Toronto.

Hay, E. D. (ed.) (1991). "Cell Biology of Extracellular Matrix." Plenum, New York/London.

Nimni, M. (1988–1992). "Collagen," Vols. I–V. CRC Press, Boca Raton, FL.

Pritzker, K. P. H. (1997). Articular skeletal system, muscle, fat and other connective tissues. In "Functional Endocrine Pathology," 2nd Ed., (S. L. Asa and K. Kovacs, eds.). Blackwell, Boston.

Silver, F. H. (1987). "Biological Materials: Structure, Mechanical Properties and Modeling of Soft Tissue." New York Univ. Press, New York.

Vitto, J., and Pereuda, A. J. (1987). "Connective Tissue Disease." Dekker, New York.

Vogel, S. (1988). "Life's Devices." Princeton Univ. Press, Princeton, NJ.

Yurchenco, P. D., Birk, D. E., and Mecham, R. P. (eds.) (1994). "Extracellular Matrix Assembly and Structure." Academic Press, San Diego/Toronto.

Copper, Iron, and Zinc in Human Metabolism

ANANDA S. PRASAD
Wayne State University

GLOSSARY

Ataxia Unsteady gait

Dysarthria Difficulty in speech

Erythropoiesis Production of red blood cells

Geophagia Clay- or dirt-eating

Hypogeusia Decreased taste acuity

Interleukin-2 Cytokine essential for proliferation of T helper cells

Neutropenia Decreased number of granulocytes in peripheral blood

Thymulin Zinc-dependent thymic hormone

Total parenteral nutrition All essential nutrients are supplied by intravenous feeding

THE IMPORTANCE OF IRON AND IODINE FOR human health has been known for over a century. The importance of other trace elements such as zinc, copper, manganese, selenium, and chromium, however, has been realized only in the past two to three decades. Prior to 1961, it was considered unlikely that zinc deficiency in humans could pose a significant clinical problem, inasmuch as zinc is ubiquitous in the environment. Reports from the Middle East established that zinc deficiency could occur under practical dietary conditions, and zinc deficiency is now recognized to occur in both developing and developed countries under various clinical conditions. Some investigators believe that a deficiency of zinc may be one of the most common nutritional problems in the world.

During the past two decades, considerable progress has been made in the field of trace elements. Impressive advances have been observed in the clinical, biochemical, immunological, and molecular biological areas as related to trace elements.

Three events appear to have contributed to these advances. The first, and perhaps the most important, event was the documentation that deficiency of trace elements in human subjects can occur under practical dietary situations and may be present in many disease states as well. The second event was the development and availability of the atomic absorption spectrophotometer in the 1960s, which simplified the assessment of trace element status in humans. The third event was the recognition of the many biochemical and immunological functions of trace elements that have been observed since 1960. For instance, zinc is now known to regulate the activities of at least 200 enzymes. It is known to be required for many functions of the lymphocytes and, as zinc finger proteins, to be involved in gene expression. These are indeed exciting developments.

Of all the trace elements known to be essential to humans, knowledge of the functions for human health is most complete for iodine, copper, iron, and zinc. The clinical importance and essentiality of copper, iron, and zinc for human subjects are well established. Distinct deficiency disorders of chromium and selenium have yet to be characterized, and clinical deficiency of manganese remains to be documented convincingly.

In this article, clinical, biochemical, and metabolic effects of copper, iron, and zinc are addressed. A general overview of trace elements is provided in another article in this encyclopedia. [*See* Minerals in Human Life.]

I. COPPER

The presence of copper in plant and animal tissues was first recognized almost 170 years ago. In 1847, copper was shown to be present in the blood proteins of snails, and a copper-containing pigment (hemocyanin) was recognized to function as a respiratory compound in 1878. The first evidence of the dietary importance of copper in rats was reported in 1925 and, in 1928, it was observed that copper, in addition to iron, was necessary for blood formation in rats. Later, naturally occurring copper deficiency in livestock was recognized. Subsequent studies showed that copper-deficient pigs are unable to absorb iron, mobilize it from the tissues, and utilize it for hemoglobin synthesis.

A. Copper Deficiency in Humans

Copper deficiency has been implicated in the etiology of three clinical syndromes in the human infant. In the first of these, hypochromic microcytic anemia, hypoproteinemia, hypoferremia, and hypocupremia are present, and combined therapy with both iron and copper is required for complete recovery. The second syndrome is known to affect malnourished infants being rehabilitated on high caloric and low-copper-containing diets who exhibit hypochromic microcytic anemia, neutropenia, diarrhea, osteopenia, and hypocupremia. The third clinical condition in infants is Menkes kinky hair syndrome, which is a genetic disorder. In addition to these syndromes, copper deficiency resulting from prolonged administration of total parenteral nutrition (TPN) without copper supplementation has been reported. In adults, copper deficiency due to TPN and secondary to prolonged zinc administration in high dosage has been observed. Copper deficiency in humans has been induced experimentally by dietary manipulation (0.83 mg/day dietary intake). Increase in plasma cholesterol level and abnormal glucose metabolism and a decrease in the activity of superoxide dismutase (a copper- and zinc-containing enzyme) in the red cells were observed. These studies demonstrate the importance of copper in clinical medicine.

I. Manifestations of Copper Deficiency

In animals, besides hypochromic microcytic anemia and leukopenia, hypopigmentation, defective wool keratinization, and abnormal hair texture, the following have been reported as a result of copper deficiency: ataxia from abnormal defective myelination in newborn lambs, abnormal bone formation, reproductive failure, heart failure, spontaneous fractures, and arterial rupture due to aneurysm caused by abnormal elastin formation.

In humans, hypochromic anemia and neutropenia are consistent features of copper deficiency. The anemia appears to be related to defects in iron utilization due to a deficiency of copper. The pathogenesis of neutropenia remains poorly understood. In infants with copper deficiency and in patients with Menkes disease, osteoporosis and pathological bone fractures as well as "scurvy-like" bone abnormalities have been reported. Arterial tortuosity and aneurysms have reportedly occurred in patients with Menkes disease. The bony and vascular abnormalities in copper deficiency appear to be related to defective cross-linking in collagen associated with reduced activity of the copper-dependent enzyme lysyl oxidase. Hypercholesterolemia and abnormal glucose metabolism were observed in humans in whom a copper deficiency was induced by experimental means.

In patients with Menkes disease, hypopigmentation of hair has been reported. This may be due to decreased activity of the copper-dependent enzyme tyrosinase, which is needed for normal melanin synthesis. The other abnormality observed in Menkes disease, kinky hair, may be related to defective formation of disulfide bonds.

Neurological manifestations have not been described in cases of nutritional copper deficiency in humans, although a number of neurological symptoms have been observed in patients with Menkes disease. These include hypotonia, hypothermia, episodic apnea, seizures, and mental retardation. These manifestations may be due to either decreased cytochrome oxidase activity or impaired catecholamine synthesis in the central nervous system. Whether or not alterations of lipid composition or some other mechanisms are involved in this syndrome remains to be elucidated.

Other abnormalities in copper deficiency include hypoproteinemia and hypercholesterolemia in experimental animals. The mechanisms of these alterations are not well understood.

2. Menkes Disease

Menkes disease was first described in 1962 as a syndrome characterized by growth retardation, focal cerebral and cerebellar degeneration associated with mental retardation, and white hair with peculiar twisting, brittleness, and breakage. Later studies showed that intestinal copper absorption was impaired and hepatic copper content was reduced, but that red cell copper was normal and intravenous copper was handled normally. Anemia and neutropenia, which are consistently present in cases of nutritional deficiency of copper, are not seen in Menkes disease. Also, the use of a wide range of copper preparations by various routes was found to be ineffective therapy, although serum copper and ceruloplasmin levels showed an increase. Subcutaneous administration of copper–histidine was reported to be an effective treatment in a few cases.

Copper content of duodenal mucosal cells has been shown to be increased, and copper appeared to be localized in the mucosal brush border in Menkes disease. In cultured fibroblasts, copper accumulation at the plasma membrane of cells has been observed. Other studies with ^{64}Cu have shown that efflux of copper from cultured fibroblasts is also abnormal. Whether or not copper is bound abnormally to intracellular protein such as metallothionein or another ligand remains to be settled.

Most of the clinical features of Menkes disease can be explained by a deficiency of copper-dependent enzymes, most likely due to a defect of an intracellular copper transport protein that binds and transports copper to different enzymes. The Menkes disease gene is located on the X chromosome (X, q, 13.3) and is recessive. The gene product is a member of a cation-transporting P-type ATPase subfamily with an N terminus containing six 23-residue repeats, each with a GMTCXXC motif probably involved in copper binding.

B. Copper Excess

The signs and symptoms of acute copper poisoning are the result of direct irritation to the gastrointestinal tract by copper. Pancreatitis resulting from acute copper toxicity has also been observed. Intravascular hemolysis with resultant jaundice, hemoglobinuria, and acute renal failure has been observed in cases of acute copper toxicity. The precise mechanism of acute hemolysis is not known, but it is believed to be associated with decreased glycolysis and decreased activity of glucose-6-phosphate dehydrogenase. Increased osmotic fragility of red cells incubated with copper has been observed; however, it is not known if this mechanism plays a role *in vivo* in clinical hemolysis.

The clinical manifestations of chronic copper toxicity are less well defined. In those patients in whom copper is retained in the liver, sequestration of copper

in lysosomes apparently results in protection from toxicity. Later, when this mechanism is overwhelmed, accumulation of cytosol copper results in hepatitis, which may subsequently lead to cirrhosis of the liver. Copper may also accumulate in renal tubules, cornea, brain, and other organs, and later damage to these structures takes place.

1. Wilson Disease (Hepatolenticular Degeneration)

Sak Wilson described a series of patients with the autosomal recessive genetic disorder in the 1912 volume of "Brain"; however, only in 1948 was its association with increased copper recognized. It is believed that in Wilson disease the accumulation of copper in the liver begins at birth (stage I). The subject remains asymptomatic in the early stages and later the only physical finding may be the presence of a Kayser-Fleischer ring (copper deposition in the cornea). In most patients at about the age of adolescence, the capacity of the liver to store copper may be exceeded, and there is sudden hepatic copper release and redistribution (stage II). At this time, the patient may exhibit a hemolytic episode or experience hepatic failure, with findings suggestive of chronic aggressive hepatitis or even hepatic necrosis. Subsequently, macronodular cirrhosis develops. If the patient survives, he/she enters stage III, in which cerebral copper accumulation begins, until in stage IV the patient becomes symptomatic with neurological disease, including dysarthria, gait disturbance, tremor, and loss of fine motor coordination. These cerebral changes due to excessive copper deposition are associated with degeneration and cavitation of the putamen as well as other cerebral structures.

Laboratory features of this disease include hypoceruloplasminemia, increased non-ceruloplasmin-bound serum copper, impaired incorporation of radio copper into ceruloplasmin, and increased urinary copper excretion.

It has been suggested that ceruloplasmin probably plays a central role in the pathophysiology of this disease; however, this is obscured by the observation of patients with familial hypoceruloplasminemia who have no other evidence of abnormal copper metabolism, whereas other patients with Wilson disease have been noted to have normal ceruloplasmin levels. Extensive biochemical studies from many laboratories have failed to demonstrate any structural, immunochemical, or enzymatic defect of ceruloplasmin in patients with Wilson disease. Cultured fibroblasts from patients with Wilson disease have elevated copper concentration, but decreased cytoplasmic copper protein ratio in comparison to the normal controls, the significance of which remains unknown.

Recent studies indicate that the Wilson disease gene is on chromosome 13, band q 14.3, whereas the ceruloplasmin gene is on chromosome 3. The sequence of the Wilson disease gene forms part of the P-type ATPase gene, which is very similar to the Menkes disease gene, with six putative metal-binding regions similar to those found in prokaryotic heavy metal transporters. The expression patterns of Wilson and Menkes disease genes are very different. Menkes gene is expressed in lung, skeletal muscle, and heart, but is hardly detectable in the liver or kidney. Wilson disease gene, on the other hand, is expressed mainly in the liver and kidney.

Menkes disease and Wilson disease (hepatolenticular degeneration) are both caused by a disruption in copper transport. However, these two diseases affect different tissues. In X-linked Menkes disease, copper export is defective in many tissues but is normal in the liver. Copper is taken up by the intestinal cells, but is not transported further. In Wilson disease, there is a failure to incorporate copper into ceruloplasmin in the liver, and failure to excrete copper from the liver into bile. This defect in Wilson disease results in toxic accumulation of copper in the liver, kidney, brain, and cornea.

The Wilson disease gene protein appears to have the potential to play a direct role in copper incorporation into ceruloplasmin, and in copper excretion, by maintaining the metal ion in the correct redox state (CUI).

The usual treatment of Wilson disease is penicillamine, a copper-chelating agent, which is not well tolerated by all subjects and results in several side effects, some of which may be serious. Research over the past decade has shown that zinc therapy produces a negative copper balance in patients with Wilson disease. Zinc absorbed by intestinal mucosal cells induces synthesis of the metal-binding cysteine-rich protein metallothionein in the intestinal cells and then exchanges the zinc for copper, which it binds firmly, preventing its movement through the cell and into the body. Zinc also effectively blocks the intestinal uptake of oral ^{64}Cu. Oral zinc therapy thus appears to be a good alternative treatment for Wilson disease on a long-term basis, inasmuch as zinc is considerably less toxic than penicillamine. Zinc therapy may be an excellent approach for prevention of copper damage to liver, brain, and other organs in subjects who may be considered to be potentially affected by this disorder.

C. Biochemistry and Metabolism

1. Biochemistry

Neonatal ataxia and other neurological signs are commonly observed among several species that have been subjected to copper deficiency. Lack of myelination has been observed in both nutritional deficiency and Menkes disease. Some observers have suggested that the reduction in myelinated axons is consistent with the observed degree of nerve cell death. A marked decrease in the activity of the myelin markers 2′,3′-cyclic nucleotide-3′-phosphohydrolase in weaning rats whose mothers were deprived of copper was observed. Whether the reduction in myelin is the result of a specific metabolic defect or simply failure of nerve cell production and survival remains unknown.

A more specific metabolic defect in the brain relates to the metabolism of catecholamines. Nutritional copper deficiency results in depressed levels of norepinephrine in lambs and rats, but the serotonin levels are not affected. The low norepinephrine level can be explained by impaired activity of the copper-dependent enzyme dopamine hydroxylase. On the other hand, the depressed dopamine levels are not reversed by copper repletion.

The copper- and zinc-dependent cytosolic superoxide dismutase enzyme (SOD) catalyzes the dismutation of the superoxide anions and forms hydrogen peroxide, which is subsequently transformed to water by the enzyme glutathione peroxidase. In copper deficiency, the activity of SOD is decreased.

The activity of cytochrome oxidase, a copper-dependent enzyme, is lowered in most tissues by copper deficiency in animals. The depressed activity has been postulated to play a role in the genesis of central nervous system pathology, but evidence for a specific effect is lacking.

Cardiovascular disorders are evident in almost all species subject to copper deficiency of either genetic or nutritional origin. The metabolic defect responsible for most, if not all, of the pathology, such as dissecting aneurysms and angiorrhexis, is failure of cross-link formation in the connective tissue proteins collagen and elastin. The cross-links in the proteins are formed from lysine or hydroxylysine after they are incorporated into the soluble precursor proteins. Lysyl oxidase, a copper-dependent enzyme, catalyzes the oxidation with oxygen, serving as the electron acceptor. In most species, when copper is limiting, lysyl oxidase activity is decreased and cross-linking fails, which leads to loss of rubber-like elasticity of elastin and decreased tensile strength of collagen.

The emphysema-like lung that occurs during development under condition of copper deficiency appears to result from the failure of cross-link formation. Other factors, such as low SOD activity in copper deficiency, may also contribute to peroxidative damage of lung tissue.

The metabolic defect in genetic copper deficiency (Menkes disease) remains poorly understood. The defect may involve abnormal intracellular binding of copper or failure to transport it out of the cell.

2. Metabolism

It has been estimated that a normal 70-kg human contains 80 mg of total copper; however, many factors may affect this estimate. Differences in dietary and environmental copper exposure have been noted, although their effect on total body copper content is not well documented. Liver and brain are especially rich in copper and represent about one-third of total body copper. Skeletal muscle also represents one-third of total body copper, although the concentration of copper in muscle is relatively low. In adults, homeostasis is apparently maintained by the absorption of approximately one-third or more of the estimated daily intake of <2.0 mg/day dietary copper. Analysis of copper in 15 American diets has revealed a daily intake ranging from 0.2 to 3.48 mg/day, with 8 diets containing <1.0 mg/day. Of total daily copper losses, from 0.5 to 1.3 mg is excreted in the bile, 0.1–0.3 mg passes directly into the bowel, 0.01–0.06 mg appears in the urine, and small amounts are lost in sweat, skin, and hair. In copper balance studies, a minimum of 1.3 mg of copper intake was required to maintain balance. These estimates were minimal because surface loss of copper (i.e., for hair, desquamated skin, and sweat) was not included in the balance equation. The recommended dietary allowance (RDA) (United States National Academy of Sciences, Food and Nutrition Board) for copper has been set at 2.0 mg/day. Clearly many individuals in the United States do not reach this level of intake. Whether or not this poses any health problems remains to be determined, although, as mentioned earlier, under experimental conditions, human copper deficiency showed an increase in plasma cholesterol level and abnormal glucose tolerance.

II. IRON

Iron is essential for many biochemical processes. It exists in both ferric and ferrous states, and its impor-

tance in the oxygen and electron transport systems concerned with cellular energy production is well established. Iron was considered to be of celestial origin in ancient civilizations of the eastern Mediterranean region. The "metal of heaven" was used in Egypt and Mesopotamia for therapeutic purposes, and its use was described in the Ebers Papyrus, an Egyptian pharmacopeia dating around 1500 B.C. Therapeutic use of iron was also mentioned in ancient Indian (Hindu) history dating around 500 B.C.

Iron became an accepted method of treatment of a disease known as chlorosis following Thomas Sydenham's studies published in 1681, although the mechanism of its action was not understood and its use remained controversial. In 1893, it was demonstrated convincingly that iron administration to women with chlorosis resulted in an increase in the hemoglobin level. In 1938, it was unequivocally demonstrated that inorganic iron is incorporated quantitatively into hemoglobin. Biochemical studies later showed that iron is intimately involved in oxygen utilization by the tissues, as well as in oxygen transport as part of the hemoglobin molecule.

A. Deficiency of Iron

Iron is essential for many biochemical processes. Although the earth's crust consists of almost 4% iron and the human diet may contain plenty of iron, ferric iron is insoluble; thus, its availability for human metabolism is low. The body is limited in the adjustments it can make to loss of iron due to hemorrhage. Indeed, iron deficiency is the most common cause of anemia throughout the world.

When the body is in a state of negative iron balance, the iron is mobilized from the body storage pool for the synthesis of hemoglobin. Iron absorption is increased when stores are reduced, and serum ferritin is decreased before anemia develops. Serum ferritin is an iron storage compound that correlates with the total iron stored in the body. When the stores are entirely depleted, the stainable iron in the bone marrow is no longer present. The next parameter to be affected is the serum transferrin saturation, which falls to <15%. Transferrin, a glycoprotein, is an iron-binding protein responsible for transport of iron to precursors of erythroid cells for hemoglobin synthesis. Recent studies indicate that as a result in iron deficiency, even though the subjects may not be anemic, there may be impaired exercise tolerance and physical work capacity and reduced cognitive function in children.

If the negative iron balance continues, anemia develops and the patient may show breathlessness, pallor, and tachycardia. The red cells become hypochromic and microcytic. The serum total iron binding capacity increases, and the serum iron decreases. The reticulocyte count is low. Partial villous atrophy, with minor degrees of malabsorption of xylose and fat, reversible by iron therapy, has been reported in infants suffering from iron deficiency; this has not been observed in iron-deficient adults. Iron-containing enzymes such as the cytochromes, catalase, and tryptophan pyrrolase are usually better preserved in the tissues than other iron-containing compounds. In severe iron deficiency, however, the activities of various iron enzymes are decreased. Poor activation of lymphocyte transformation, diminished cell-mediated immunity, and impaired intracellular killing of bacteria (white blood cells) by neutrophils have been reported by some investigators, but detailed immunological studies are not available.

Iron deficiency is seen in subjects with chronic blood loss, malabsorption syndrome, and diets from which iron is poorly bioavailable and in subjects with increased requirement of iron. Iron deficiency anemia is easily correctable by oral or parenteral supplementation of iron.

Iron and zinc deficiencies are often associated in certain populations. Phytate inhibits retention of both iron and zinc. Subjects whose diets consisted of predominantly cereal proteins high in phytate showed evidence of both iron and zinc deficiency in many parts of the world. Blood loss in premenopausal women also contributes to negative balance for both iron and zinc, inasmuch as red cells contain high amounts of both these elements. A recent study in premenopausal women in the United States indicates that a low serum ferritin concentration is suggestive of both iron and zinc deficiency.

I. Sideroblastic Anemias

Sideroblastic anemias comprise a group of refractory anemias in which, although the peripheral blood may show the presence of cells with reduced hemoglobin content, their precursor, the normoblasts in the bone marrow, contain an excess of iron. In a significant number of normoblasts, the iron accumulates around the nucleus and these are called ringed sideroblasts.

The congenital sideroblastic anemia is a rare X-linked disorder affecting mainly males in childhood or adolescence. The females are heterozygotes with only one altered gene in their X chromosomes. In advanced cases, spleen and liver are enlarged; the se-

rum iron is very high, and iron-binding capacity is decreased. The anemia is severe.

Primary acquired sideroblastic anemia usually affects middle-aged and elderly subjects and is considered to be a variety of altered development of white blood cells. Many of these subjects represent examples of prelymphoma, premyeloma, or preleukemia. The clinical symptomatology of primary sideroblastic anemias is related to severe anemia and iron overload. The treatment is not very effective.

Secondary sideroblastic anemias may be seen in conditions with abnormalities of vitamin B_6 metabolism, leading to disorder in heme synthesis. These include antituberculous chemotherapy, celiac disease, hemolytic anemia, and alcoholism. Secondary sideroblastic anemia seen in lead poisoning, alcoholism, and use of chloramphenicol is also due to disturbance of heme synthesis or mitochondrial function. Secondary sideroblastic anemia in patients with collagen diseases, malignancy, and megaloblastic anemias is due to ineffective iron utilization by the normoblasts.

2. Anemia of Chronic Disorders

One of the most common types of anemia is the anemia of chronic disorders. This is due to chronic infections such as tuberculosis, malignant diseases, chronic inflammatory diseases, and renal failure. The pathogenesis of this disorder is not well understood. The marrow fails to mount appropriate erythropoiesis in spite of anemia. Whether the production of erythropoietin is decreased or erythroblast response to erythropoietin is reduced in this disease is not known. The changes in iron metabolism are consistent with a block in the release of macrophage iron. The decreased serum iron level is believed to be caused by increased lactoferrin production from granulocytes and increased apoferritin synthesis by macrophages. A mild shortening of red cell life span has also been observed in these cases. There is an imbalance between plasma iron supply and the erythroid marrow requirements in these conditions.

B. Iron Excess

In idiopathic hemochromatosis, there is an excessive parenchymal iron storage, particularly in the liver, leading to tissue damage. Clinically, the disease is manifested by cirrhosis of the liver, endocrine disorders, skin pigmentation, and cardiac failure. The excess iron store is believed to be due to an unknown genetic defect of iron metabolism, leading to an increased absorption of iron. Human lymphocyte antigen (HLA)-linked hemochromatosis is inherited as an autosomal recessive trait, affecting approximately 1 in 300 of the Caucasian population. The responsible gene has not yet been identified, but it is known to be located on the short arm of chromosome 6 (6P) very close to the HLA class I complex.

The basic defect leading to an inappropriately high iron absorption by intestinal mucosal cells in hemochromatosis is not known. Theoretically the defect could reside in the iron proteins transferrin or ferritin or their receptors. Whether or not hemochromatosis gene product abnormality involves a member of a cation-transporting P-type ATPase family remains to be established.

Other causes of iron overload include ineffective red blood cell development, such as thalassemia, sideroblastic anemias, and congenital dyserythropoietic anemias. Excessive intake of iron, such as that seen in certain African groups from drinking local beers prepared in iron drums and from excessive blood transfusion, results in iron overload. Other disorders, such as alcoholic cirrhosis, idiopathic pulmonary hemosiderosis, rheumatoid arthritis, and paroxysmal nocturnal hemoglobinuria, may be associated with iron overload in tissues.

Excess iron in tissues leads to peroxidation of cell membrane lipids due to iron-catalyzed hydroxyl radical formation and oxidation of intracellular proteins, which ultimately result in tissue damage. Damage to lysosomal membranes by hemosiderin has also been shown to occur, with release of lysosomal enzymes into other cell compartments. Liver damage in transfused patients may also be due to hepatitis B virus or non-A, non-B hepatitis.

The main treatment is iron chelation to solubilize iron by use of iron chelators such as desferrioxamine or repeated blood drawing, which ultimately leads to removal of iron from the body.

C. Biochemistry and Metabolism

I. Biochemistry

Hemoglobin is one of the most important functional iron-containing proteins. Hemoglobin (molecular weight 64,500) contains four iron-containing heme groups linked to four globin chains and can bind four molecules of oxygen. The total amount of iron in the hemoglobin pool is 2500 mg. Myoglobin present in muscles (molecular weight 17,000) contains approximately 10% of body iron as a single heme group attached to its one polypeptide chain. The total

amount of iron in this pool is 300 mg. Myoglobin has higher affinity for oxygen than does hemoglobin. The mitochondria contain a series of heme and nonheme iron proteins. These include the cytochromes a, b, and c, succinate dehydrogenase, and cytochrome oxidase, which form an electron transport pathway responsible for the oxidation of intracellular substrates and the simultaneous production of adenosine triphosphate, the energy currency of cells. [*See* Hemoglobin.]

Cytochrome P-450 is present in the endoplasmic reticulum and is involved in detoxification of various chemicals by the liver. The iron-containing enzymes catalase and lactoperoxidase are involved in peroxide breakdown. Tryptophan pyrrolase, an iron enzyme, is needed for the oxidation of tryptophan to formylkynurenine. Xanthine oxidase, aconitase, and nicotinamide adenine dinucleotide dehydrogenase are iron-reduced and sulfur-containing proteins. Iron is also necessary for the enzyme ribonucleotide reductase, an important enzyme for DNA synthesis. [*See* Cytochrome P-450; DNA Synthesis.]

Heme consists of a protoporphyrin ring with an iron atom at its center. The porphyrin ring consists of four pyrrole groups united by methene bridges. The hydrogen atoms in the pyrrole groups are replaced by four methyl, two vinyl, and two propionic acid groups. Heme is synthesized from the precursors succinic acid and glycine. The enzyme ferrochelatase and glutathione are required for the incorporation of iron into the protoporphyrin molecule. Once the heme is synthesized, it combines with globin chains to make hemoglobin.

Transferrin (molecular weight 79,500) is the vital iron transport protein and accounts for only 4 mg of body iron. It is a β-globulin glycoprotein, is present in plasma and extravascular spaces, and has a plasma half-life of 8–11 days. The transferrin gene is on chromosome 3, and the protein is synthesized in the liver, with synthesis inversely related to iron stores. Plasma level is usually 1.8–2.6 g/liter. Two atoms of ferric iron may be attached to each molecule. The binding sites most likely contain three tyrosine, two histidine, and an arginine group. Binding of iron to transferrin also involves attachment of an anion, usually bicarbonate.

Lactoferrin (molecular weight 77,000) is a related glycoprotein and also binds two atoms of iron per molecule. It is present in milk and other secretions. Its presence in neutrophils is believed to have a bacteriostatic effect by depriving the offending organisms of the iron needed for their growth.

Ferritin (molecular weight 480,000) is the primary storage compound for iron. It is made up of a spherical shell enclosing a core of ferric hydroxyphosphate (up to 4000 iron atoms). Human ferritin has 24 subunits of two immunologically distinct chain types H and L. The small amount of ferritin present in human serum contains little iron and consists almost exclusively of L subunits. It has been suggested that serum ferritin may be secreted by macrophages and/or hepatocytes, which have been stimulated by iron to synthesize ferritin.

Hemosiderin is a noncrystalline and water-insoluble iron storage compound. In normal subjects, storage iron is about two-thirds ferritin and one-third hemosiderin, but in iron overload the proportion of hemosiderin increases considerably.

Iron enters dividing cells of higher eukaryotes via the transferrin receptor (TfR). Once inside the cell, the iron may enter one of the three pools. Iron may be utilized in a variety of metabolic processes or may be sequestered in ferritin. Intracellular iron may also play a regulatory role such that when iron is abundant (such as in hemin-treated cells) the common trans-acting protein (IRE-BP) is in its 4Fe-4s state, which has high aconitase enzymatic activity but low affinity for IREs (cis-acting RNA element, IRE-BP$_{off}$). When iron is scarce (such as in desferrioxamine-treated cells), the IRE-BP is in its high-affinity state for RNA binding (IRE-BP$_{on}$) with negligible aconitase activity. The IRE-BP$_{on}$ state results in decreased translation of ferritin mRNA. Thus iron functions uniquely such that the translation of the mRNA encoding ferritin is attenuated by iron deprivation as is the degradation of the mRNA encoding the transferrin, demonstrating an overlap in posttranscriptional regulation of the expression of two genes.

2. Metabolism

Iron absorption depends on the dietary content of iron, its bioavailability, and the body's need for iron. A normal Western diet should provide approximately 15 mg/day iron. Only 5–10% of dietary iron is absorbed, and heme iron is better absorbed than nonheme iron. The absorption of heme iron is 20–30%, whereas only 1–5% of nonheme iron is absorbed in normal subjects. Ferric iron salts are less absorbed than ferrous compounds. Phytates, phosphates, tannates, oxalates, phosphopeptides, and products of Maillard browning decrease the availability of iron for absorption. Iron absorption is increased by the following conditions: iron deficiency, increased erythropoiesis, ineffective erythropoiesis, pregnancy, and anorexia. Ascorbic acid increases iron absorption by reducing ferric iron to the ferrous form. Iron absorption is decreased when the body is overloaded with iron and in acute and chronic infections.

Iron absorption may be regulated both at the stage of mucosal uptake (possibly by varying the number of brush border iron receptors) and at the stage of transfer to the blood. Iron is maximally absorbed from the duodenum. Iron uptake by mucosal cells appears to involve binding to specific receptors on the brush borders followed by an energy-dependent transfer across the cell membrane. At higher doses, there may be a passive diffusion as well. Heme enters the mucosal cells intact, whereupon heme oxygenase breaks up heme and releases iron intracellularly. Iron then may be transferred to the portal circulation or may enter ferritin to be eventually lost with the exfoliation of the mucosal cell. The nature of the intracellular iron pool and transport across the cell membrane remains unknown.

Iron from the mucosal cell is transferred to the plasma transferrin, which transports iron to bone marrow for synthesis of heme by the erythroblasts. The binding of iron to transferrin is facilitated by ceruloplasmin, a ferrooxidase. Transferrin picks up iron not only from the intestinal cells but also from macrophages and liver.

The recently described pathway of iron absorption across the duodenal mucosal cell is as follows: (1) mucin binds iron at acid pH to solubilize it for absorption in duodenum; (2) integrin (90/150 kDa), a transmembrane protein, facilitates iron transfer through the microvilli; and (3) mobilferrin (56 kDa), a cytosolic protein, serves to monitor iron in an appropriate redox state and ultimately transports it to plasma transferrin. This absorptive process appears to be driven by a cascade, so that iron moves from luminal mucin to integrin and mobilferrin and finally to plasma transferrin, probably mediated by differences in the dissociation constant for iron of each of these proteins.

Senescent red cells are phagocytized by macrophages, whereupon iron is released by the action of heme oxygenase. As ferrous iron, it can then either enter ferritin, where it is oxidized to ferric iron by the ferritin protein, or be released into plasma, where it binds to transferrin and then is transported to the bone marrow. The mechanism of iron donation to transferrin is poorly understood.

Uptake of iron from transferrin by developing red cell or other cells requires the presence of specific receptor sites. It has been reported that some of the receptors are shed into the plasma and their level correlates with the overall activity of the erythron. These receptors can be assayed by immunoassay techniques. The transferrin receptor is a transmembrane protein consisting of two monomers linked by a disulfide bridge; each subunit is able to bind one transferrin molecule. The human transferrin is identified by the monoclonal antibody OK T9. It has a much higher affinity for fully saturated, diferric transferrin than for monoferric transferrin. Some investigators believe that the transferrin receptor complex remains on the cell surface during iron release. Some evidence suggests that in both reticulocytes and hepatocytes, a receptor-mediated endocytosis may occur, that is, the transferrin receptor complex is engulfed into the cells in small vesicles. The iron is released inside the cell perhaps by a fall in pH within the vesicle. Whether or not a specific intracellular carrier, either a protein or low-molecular-weight chelate, is then involved in transport of iron to the mitochondria or to ferritin remains uncertain.

About 90% of the iron used by the bone marrow cells each day is derived from red cells that are lysed by the reticuloendothelial cells, and only 5–10% is derived from the slow turnover of the storage pool and from absorption from the gut.

III. ZINC

The importance of zinc for human health has been recognized only the past 25 years. During this time, remarkable progress has been made in the clinical, biochemical, and immunological aspects of the role of zinc in humans. Although the essentiality of zinc for plants and animals was known, its ubiquity made it seem improbable that human deficiency occurred or that alteration in zinc metabolism could lead to significant problems in clinical medicine. Research has shown that prediction to be in error. The first documentation of human zinc deficiency was reported by A. Prasad and his associates from the Middle East in 1963.

Deficiency of zinc was suspected to occur for the first time in 1958, in an Iranian 21-year-old male who looked like a 10-year-old boy. In addition to severe growth retardation and anemia, he had hypogonadism, hepatosplenomegaly, rough and dry skin, mental lethargy, and geophagia. He ate only bread (wheat flour) without any intake of animal protein and consumed approximately 500 g of clay every day. The habit of geophagia is fairly common among the villages in that part of the world. There was no evidence of blood loss; the anemia was due to iron deficiency. Inasmuch as this syndrome was fairly prevalent in Iranian villages, hypopituitarism as an explanation for growth retardation and hypogonadism was ruled out.

It was difficult to explain all of the clinical features solely on the basis of tissue iron deficiency. Because growth retardation and testicular atrophy are not seen in iron-deficient experimental animals, the possibility that zinc deficiency may have been present was considered. Zinc deficiency was known to produce retardation of growth and testicular atrophy in animals; thus, it was speculated that some factors responsible for decreased availability of iron in these patients with geophagia may also have decreased the availability of zinc. Many other cases followed. Zinc deficiency was recognized based on the following: the zinc concentration in plasma, red cells, and hair was decreased and ^{65}Zn studies revealed that the plasma zinc turnover rate was greater, the 24-hr exchangeable pool was smaller, and the excretion of ^{65}Zn in stool and urine was less in patients than in control subjects.

Further studies showed that the rate of growth was greater in patients who received zinc as compared with those who received iron instead and those receiving only an adequate animal protein diet. Pubic hair appeared in all cases within 7–12 weeks after zinc supplementation was initiated. Genitalia size became normal and secondary sexual characteristics developed within 12–24 weeks in all patients receiving zinc. No such changes occurred in a comparable length of time in iron-supplemented groups or in the group on an animal protein diet. Thus, the growth retardation and gonadal hypofunction in these subjects was related to a deficiency of zinc.

Nutritional deficiency of zinc affecting growth in children and adolescents is fairly widespread throughout the world. Zinc deficiency is expected to occur in countries where cereal proteins are primary in local diet.

A. Clinical Spectrum of Human Zinc Deficiency

During the past two decades, a spectrum of clinical zinc deficiency in human subjects has been recognized. On the one hand, the manifestations of zinc deficiency may be severe; on the other, zinc deficiency may be mild or marginal.

A severe deficiency of zinc has been reported to occur in patients with acrodermatitis enteropathica

(AE), following TPN without zinc, excessive use of alcohol, and penicillamine therapy.

The manifestations of severe zinc deficiency in humans include bullous pustular dermatitis, alopecia, diarrhea, emotional disorder, weight loss, intercurrent infections due to cell-mediated immune dysfunctions, hypogonadism in males, neurosensory disorders, and problems with healing of ulcers. If this condition is unrecognized and untreated, it becomes fatal.

A moderate level of zinc deficiency has been reported in a variety of conditions. These include nutritional deficiency due to dietary factors, malabsorption syndrome, alcoholic liver disease, chronic renal disease, sickle cell disease, and chronically debilitated conditions. The manifestations of a moderate deficiency of zinc include growth retardation and male hypogonadism in adolescents, rough skin, poor appetite, mental lethargy, delayed wound healing, cell-mediated immune dysfunctions, and abnormal neurosensory changes.

Although the clinical, biochemical, and diagnostic aspects of severe and moderate levels of zinc deficiency in humans are fairly well defined, the recognition of mild levels of zinc deficiency has been difficult. Zinc assays in plasma, urine, and hair have been proposed as potential indicators of body zinc status. Currently, plasma zinc appears to be the most widely used parameter for assessment of human zinc status, and it is known to be decreased in cases of severe and moderate zinc deficiency. However, several physiological and pathological conditions may also affect zinc levels in the plasma and urine; thus, a reduced plasma or urine zinc level cannot be taken necessarily as an indicator of low body zinc status. Zinc in hair and erythrocytes does not reflect active or recent status of body zinc, inasmuch as these tissues are slowly turning over. Furthermore, in the cases of mild deficiency of zinc in humans, the plasma levels of zinc may remain normal and, clinically, overt evidence of zinc deficiency may not exist, thus creating a difficult diagnostic problem.

Therefore, assay of zinc in more rapidly turning over blood cells such as lymphocytes, granulocytes, and platelets as indicators of zinc status in human subjects has been utilized. With the use of these data, mild deficiency of zinc in humans was recognized.

A mild deficiency of zinc was experimentally induced in human volunteers by dietary means. Measurable effects occurred in zinc concentration of cells such as lymphocytes, neutrophils, and platelets. Some subjects showed abnormal dark adaptation, decreased lean body mass, and hypogeusia, which were corrected with zinc supplementation. Natural killer cells as well as interleukin-2 activity decreased. Sperm count declined slightly during zinc restriction. Clearly, these observations show that a mild zinc deficiency in humans adversely affects clinical, biochemical, and immunological functions.

B. Biochemistry and Metabolism of Zinc

I. Biochemistry

Several studies have shown that zinc is required for protein synthesis. Zinc supplementation has been found to be beneficial for normal healing in deficient subjects. Collagen is the main fibrous protein of the connective tissue and is largely responsible for the development of tensile strength in tissue as well as in the healing wound. In zinc-deficient rats, a significant reduction in total collagen in sponge connective tissue (bone) was observed, apparently caused by a generalized effect on protein synthesis and nucleic acid metabolism, rather than specifically on collagen synthesis, possibly through decreased proliferation of fibroblasts. Also, glucose tolerance of zinc-deficient animals appears to be impaired, possibly through a reduction of the rate of insulin secretion in response to glucose stimulation. [See Collagen, Structure and Function.]

Total insulin-like activity and immunoreactive insulin have been reported to be decreased in zinc-deficient animals. One possible explanation is that insulin destruction is increased in zinc deficiency. [See Insulin and Glucagon.]

The role of zinc in gonadal function was investigated in rats. Body weight gain, zinc content, and testes weights were significantly lower in the zinc-deficient rats than in the controls. The serum luteinizing hormone (LH) responses of the pituitary gonadotropins and follicle-stimulating hormone (FSH) to gonadotropin-releasing hormone (GNRH) administration were higher in the zinc-deficient rats, but serum testosterone response was lower than in the restricted fed controls, suggesting an impairment of gonadal function through alteration of testicular steroidogenesis. Similar results have now been reported in experimentally induced zinc-deficient human subjects. Supplementation with zinc resulted in reversal of testicular failure in such cases. [See Testicular Function.]

Zinc has been shown to improve filterability through a 3.0-μm nucleopore filter of sickle cells, red blood cells containing an altered form of hemoglobin, probably through an effect of zinc on the red cell membrane. Recent studies show that the process of formation of irreversibly sickled cells involves the cell membrane. Calcium and/or hemoglobin binding may promote the formation of irreversibly sickled cells, thus hindering their filterability. Zinc may block the calcium and/or hemoglobin binding to the membrane.

2. Zinc and Immunity

Recent studies clearly indicate that zinc is required for lymphocyte transformation. The effect of zinc appears to be that of a mitogen, and the kinetics of these influences most closely approximate the effects of antigen stimulation on lymphocyte culture. Currently available data suggest a direct stimulatory influence of zinc on DNA replication. Direct cell-surface effects of zinc cannot be ruled out, however. It is conceivable that zinc could be operating at several different levels. [*See* Lymphocytes.]

Assessment of the role of zinc in the development and functions of different lymphoid cell populations strongly indicates that this element has an effect predominantly on T lymphocytes. Recent studies have shown that thymulin, a thymic hormone, is zinc dependent. In zinc deficiency the serum level of active thymulin is decreased, which may adversely affect the functions of T lymphocytes. In AE, zinc deficiency exerts a profound and apparently specific effect on the thymus, thymocytes, and cellular immune functions, which are reversible with zinc repletion.

Granulocytes from chronic uremics who are zinc deficient show significantly impaired mobility, both chemotactic and chemokinetic, in comparison with subjects who are supplemented with zinc. Furthermore, a significant correlation between granulocyte chemotaxis and both plasma and granulocyte zinc concentrations among all patients supports a pathophysiological relationship between the severity of impaired granulocytic chemotactic response and zinc deficiency in these patients. Abnormal granulocyte chemotaxis, corrected by zinc supplementation, has also been observed by others in nonuremic patients with AE.

3. Zinc and Metallothionein

Highly purified metallothionein isolated from equine and human kidney contains 26 SH groups per mole, and the protein binds cadmium and zinc as well as copper. In the liver, kidneys, and possibly other organs, metallothionein apparently mediates the absorption of copper and other trace elements. The antagonism among copper, cadmium, and zinc may result from competition for binding sites on metallothionein.

It is believed that zinc induces the synthesis of metallothionein *de novo*, although one cannot exclude the possibility that zinc stabilizes the apoprotein that is being continually synthesized but normally has a sufficiently short turnover time to prevent its accumulation in the liver. An overall homeostatic mechanism is proposed wherein metallothionein synthesis is controlled at the transcriptional level by body zinc status.

4. Zinc in Gene Expression

Recently, the importance of zinc-binding finger loop domains in DNA-binding proteins as regulators of gene expression has been recognized. The first zinc finger protein to be recognized was the transcription factor-III A of *Xenopus laevis*, which contained tandem repeats of segments with 30 amino acid residues, including pairs of cysteines and histidines. The presence of zinc in these proteins is essential for site-specific binding to DNA and gene expression. The zinc ion apparently serves as a strut that stabilizes folding of the domain into a finger loop, which is then capable of site-specific binding to double-stranded DNA. The zinc finger loop proteins provide one of the fundamental mechanisms for regulating the gene expression of many proteins. In humans, the steroid hormones (and related compounds, such as thyroid hormones, vitamin D_3, and retinoic acid) enter cells by facilitated diffusion and combine with respective receptors (which contain the DNA-binding domain of two zinc finger loops), either before or after entering the nucleus. Complexation of a hormone by its specific receptor evidently initiates a conformational change that exposes the zinc finger loops, so that they bind to high-affinity sites on DNA and regulate gene expression.

5. Metabolism of Zinc

The zinc content of a normal 70-kg male is estimated as approximately 1.5–2.0 g. Liver, kidney, bone, retina, prostate, and muscle appear to be rich in zinc. In humans, zinc content of testes and skin has not been determined accurately, although clinically it appears that these tissues are sensitive to zinc depletion.

Zinc in the plasma is mostly present as bound to albumin, but other proteins such as α_2-macroglobulin, transferrin, ceruloplasmin, haptoglobin, and gammaglobulins also bind significant amounts of zinc. Besides the protein-bound fraction, a small proportion of zinc (2–3% of overall zinc) in the plasma also exists as ultrafilterable fractions, mostly bound to amino acids, but a smaller fraction is present in the ionic form. Histidine, glutamine, threonine, cysteine, and lysine appear to have significant zinc binding affinity. Whereas amino acids competed effectively with albumin, haptoglobin, transferrin, and IgG for binding of zinc, a similar phenomenon was not observed with respect to ceruloplasmin and α_2-macroglobulin, suggesting that the latter two proteins exhibited a stronger binding affinity for zinc.

Approximately 10–30% of ingested dietary zinc is absorbed. Data on both the site(s) of absorption in humans and the mechanism(s) of absorption, whether it be active, passive, or facultative transport, are meager. Zinc absorption is variable in extent and is highly dependent on a variety of factors. Zinc is more available for absorption from meat and meat products; it is poorly available from cow's milk, but the availability from human milk is very good. Among other factors that might affect zinc absorption are body size, level of zinc in the diet, and presence in the diet of other potentially interfering substances, such as phosphate, phytate, hemicellulose, products of Maillard browning, lignin, phosphopeptides, casein, and other binding substrates.

According to a proposed model, a portion of the dietary zinc entering the lumen of the small intestine is transported across the mucosal brush border membrane by a process probably requiring adenosine triphosphate. Within the intestinal cells, newly acquired cytoplasmic zinc equilibrates with the "zinc pool" and is transferred either to high-molecular-weight proteins and/or metallothionein or to the plasma.

In animals, if zinc status is adequate, a significant amount of zinc is transferred to the plasma. If dietary zinc is high, the plasma zinc concentration and the *de novo* synthesis of metallothionein are concomitantly increased. A reduction in zinc absorption is directly correlated with the uptake of orally administered zinc into newly formed thionein polypeptides. In view of the interactions that dietary zinc undergoes during transit through the intestinal cells, it is conceivable that information regarding zinc status programs the rate and extent of zinc absorption, in part via changes in the concentration of inducible metallothionein.

Normal zinc intake in a well-balanced American diet with animal protein is approximately 10–12 mg/day, although the RDA for zinc is 15 mg/day for men and 12 mg/day for women. Urinary zinc loss is approximately 0.5 mg/day. Loss of zinc by sweat may be considerable under certain climatic conditions. Under normal conditions, approximately 0.5 mg of zinc may be lost daily by sweating. Endogenous zinc loss in the gastrointestinal tract may amount to 3–5 mg/day. [*See* Zinc Metabolism.]

IV. CONCLUDING REMARKS

Copper, iron, and zinc are essential elements for human health. Copper is required for many biochemical functions, including its vital role in iron utilization, collagen cross-linking, free radical reactions, and neurotransmitters. Deficiency of copper in humans is thought to be uncommon.

Iron is essential for hemoglobin synthesis, and iron deficiency anemia is very prevalent throughout the world. Iron is also required for many enzymatic functions. Defective utilization of iron results in sideroblastic anemia and anemia of chronic disorders. Hemochromatosis is a genetic disorder, characterized by excessive iron absorption and accumulation of excess iron, ultimately leading to organ damage.

Zinc regulates the activities of over 200 enzymes and is also essential for the gene expression of various proteins. Zinc is a growth factor, and several hormones appear to be zinc dependent. Growth retardation is seen in zinc-deficient infants, children, and adolescents. Zinc is essential for cell-mediated immune functions. Deficiency of zinc is prevalent throughout the world.

BIBLIOGRAPHY

Brewer, G. J., Yuzbasiyan, V. A., Iyengar, V., Hill, G. M., Dick, R. D., and Prasad, A. S. (1988). Regulation of copper balance and its failure in humans. *In* "Essential and Toxic Trace Elements in Human Health and Disease" (A. S. Prasad, ed.), pp. 95–104. Liss, New York.

Brock, J. H., Halliday, J. W., Pippard, M. J., and Powell, L. W. (eds.) (1994). "Iron Metabolism in Health and Disease." Sanders, London.

Bull, P. C., Thomas, G. R., Rommens, J. M., Forbes, J. R., and Cox, D. W. (1993). The Wilson disease gene is a putative copper transporting P-type ATPase similar to the Menkes gene. *Nature Genetics* **5**, 327–337.

Conrad, M. E. (1993). Regulation of iron absorption. *In* "Essential and Toxic Elements in Human Health and Disease: An Update" (A. S. Prasad, ed.), pp. 203–220. Wiley–Liss, New York.

O'Dell, B. L. (1982). Biochemical basis of the clinical effects of copper deficiency. *In* "Clinical, Biochemical, and Nutritional Aspects of Trace Elements" (A. S. Prasad, ed.), pp. 301–314. Liss, New York.

Pippard, M. J., and Hoffbrand, A. V. (1989). "Iron in Postgraduate Hematology" (A. V. Hoffbrand and S. M. Lewis, eds.), 3rd Ed. pp. 26–54. Heinemann Professional Publishing, Oxford, England.

Prasad, A. S. (1978). "Trace Elements and Iron in Human Metabolism." Plenum, New York.

Prasad, A. S. (1982). Clinical and biochemical spectrum of zinc deficiency in human subjects. *In* "Clinical, Biochemical, and Nutritional Aspects of Trace Elements" (A. S. Prasad, ed.), pp. 3–62. Liss, New York.

Prasad, A. S. (1988). Clinical spectrum and diagnostic aspects of human zinc deficiency. *In* "Essential and Toxic Trace Elements in Human Health and Disease" (A. S. Prasad, ed.), pp. 301–314. Liss, New York.

Prasad, A. S. (1993). "Biochemistry of Zinc." Plenum, New York.

Vulpe, C., Levinson, B., Whitney, S., Packman, S., and Gitschier, J. (1993). Isolation of a candidate gene for Menkes disease and evidence that it encodes a copper transporting ATPase. *Nature Genetics* 3, 7–13.

Williams, D. M. (1982). Clinical significance of copper deficiency and toxicity in the world population. *In* "Clinical, Biochemical, and Nutritional Aspects of Trace Elements" (A. S. Prasad, ed.), pp. 277–399. Liss, New York.

Cortex

KARL ZILLES
University of Düsseldorf

GLOSSARY

Brain stem Lowest part of the brain immediately rostral to the spinal cord

Cytoarchitecture Differences in cell densities and cell types among different areas, nuclei, or layers of the brain

Neurotransmitter Chemical compound that can be released at synapses and that effects neurotransmission

Ontogeny Growth and differentiation of organs during embryonic, fetal, and early postnatal periods

Synapse Special contact between two nerve cells, where information is transmitted from one cell to another

Telencephalon Most anterior part of the brain, which comprises the cortex, underlying white matter, and part of the basal ganglia

Thalamus Part of the diencephalon (i.e., the most caudal part of the forebrain), which consists of modality-specific (e.g., vision, hearing, pain, tactile sensations, and taste) and nonspecific nuclei

THE CEREBRAL CORTEX OF ALL MAMMALIAN SPEcies consists of nerve cells and support (glial) cells arranged in a stratified manner and represents the superficially exposed part of the telencephalon. The human cortex is folded, causing its typical macroscopic appearance with gyri (ridges) separated by sulci (furrows). The microscopically visible strata (cortical layers) are defined by variations in the packing densities of nerve cell bodies and by differences in predominating cell types. Although the largest part of the human cerebral cortex consists of six layers (the isocortex), a minor part has a different lamination (the allocortex). Moreover, there are further subdivisions of the isocortex and the allocortex into numerous areas. These architectonic studies lead to a cortical map with areas that differ in connection and function from other brain regions. The most pronounced increase in volume during evolution of the human brain is found within the isocortex. Since this part of the cortex is of relatively recent evolution, it is also called the neocortex.

The neocortex comprises the areas where the sensory input from the receptors of the skin, taste buds, inner ear, and eye terminates, as well as motor control areas and association areas (the largest part of the human neocortex). Association areas receive major input from numerous other cortical areas. Higher functions of the central nervous system (e.g., understanding of verbal or written information and memory) are localized in these association areas.

The allocortex comprises target areas of olfactory input, which are collectively called the rhinencephalon, or paleocortex, and areas involved in diverse functions such as emotional activities, learning, and memory. The latter part of the allocortex is called the archicortex and is often collectively called the cortical part of the limbic system.

A cortical area receives direct input from the ascending sensory systems via the thalamus, from other cortical areas of the same or the opposite hemisphere, and is also connected with the basal ganglia and regions in the lower brain stem and the spinal cord.

The cortex is not only stratified by layers parallel to the surface, but is also subdivided by small vertically oriented cell clusters, which are restricted

ENCYCLOPEDIA OF HUMAN BIOLOGY, Second Edition, VOLUME 3.

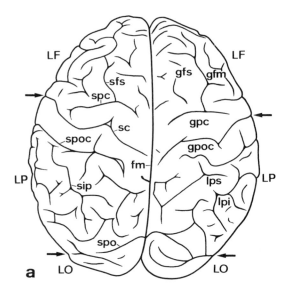

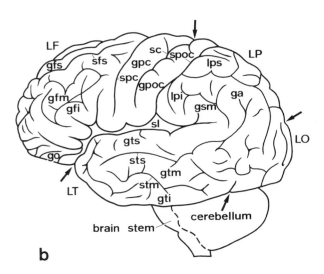

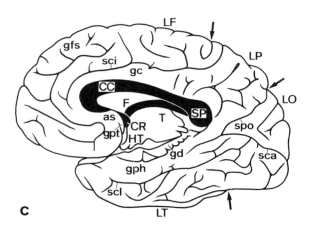

to one layer or cross all layers. These periodically arranged units are called modules, columns, stripes, or blobs.

This article ends with a short review of the distribution and localization of some neurotransmitters and their receptors.

I. MACROSCOPIC ANATOMY

A. Outer Appearance

The cerebral hemispheres consist of the cortex, the underlying white matter, parts of the deeply located basal ganglia, and the lateral ventricles, which contain the cerebrospinal fluid. Cortex and white matter constitute the pallium (i.e., the mantle of the brain). The hemispheres are partially separated from each other by the median fissure (Fig. 1). A large connection between both hemispheres, called the corpus collosum, is found in the central part. Two smaller interhemispheric connections, the rostral (anterior) and the hippocampal commissures, are located beneath the most anterior and posterior ends of the corpus callosum, respectively. [*See* Hemispheric Interactions.]

Some of the sulci of the human cortex can be used as landmarks for the division of the hemispheres into six lobes. The frontal lobe is demarcated from the parietal lobe by the central sulcus. The parietooccipital sulcus separates the parietal from the occipital lobe. The temporal lobe is bordered by the lateral sulcus from both the frontal and parietal lobes. Only the border between the occipital and temporal lobes is not defined by a sulcus and remains arbitrary. The insular lobe is completely covered by parts of the

---

FIGURE I Human forebrain in dorsal (a), lateral (b), and medial (c) views. CC, corpus callosum; CR, rostral commissure; F, fornix; HT, hypothalamus; LF, frontal lobe; LO, occipital lobe; LP, parietal lobe; LT, temporal lobe; SP, splenium; T, thalamus; as, subgenual area; fm, median fissure; ga, angular gyrus; gc, cingulate gyrus; gd, dentate gyrus; gfi, inferior frontal gyrus; gfm, medial frontal gyrus; gfs, superior frontal gyrus; go, orbital gyri; gpc, precentral gyrus; gph, parahippocampal gyrus; gpoc, postcentral gyrus; gpt, paraterminal gyrus; gsm, supramarginal gyrus; gti, inferior temporal gyrus; gtm, medial temporal gyrus; gts, superior temporal gyrus; lpi, inferior parietal lobule; lps, superior parietal lobule; sc, central sulcus; sca, calcarine sulcus; sci, cingulate sulcus; scl, collateral sulcus; sfs, superior frontal sulcus; sip, intraparietal sulcus; sl, lateral sulcus; spc, precentral sulcus; spo, parietooccipital sulcus; spoc, postcentral sulcus; stm, medial temporal sulcus; sts, superior temporal sulcus. Arrows indicate the borders between the different lobes of the hemisphere.

frontal, parietal, and temporal lobes and can be found only after an artificial dilatation of the lateral sulcus. The limbic lobe is not well defined. It encircles the rostral part of the brain stem and consists of small portions of the frontal, parietal, temporal, and occipital lobes. This subdivision of the hemispheres also reflects a functional aspect. Besides other functions (see Section V), the frontal lobe comprises the motor centers, the parietal lobe comprises the somatosensory centers, the temporal lobe comprises the auditory centers, and the occipital lobe comprises the visual centers. Some other sulci are also consistent features in all human brains (Fig. 1). Most of the sulci, however, vary greatly among the brains of different individuals.

B. Quantitative Data

The volume of the cortex of one hemisphere shows a considerable variability among individuals, ranging from 190 to 360 cm³ depending on sex, body size, age, and various other reasons. The mean volume of the cortex in males is larger by about 30 cm³ than in females. The difference can be explained at least partially by the larger average body size of the males. In both sexes the cortex of the two hemispheres represents 46% of the total brain volume. The volume of the white matter is about 10% smaller than that of the cortex. The mean volume of the cortex is 6% smaller in people over 80 years of age compared with individuals younger than 50 years. The most extensive reduction is found in the frontal lobe (−12%).

The total surface of the cortex amounts to 1600–1700 cm² (i.e., both hemispheres), but nearly two-thirds of the surface is buried in the depth of the sulci. Comparative quantitative data corroborate that the human cortex has the highest degree of folding compared with other primates and most of the other mammals. The degree of folding reaches the highest values over the association regions of the brain. The estimated number of nerve cells in the cortex of one hemisphere varies from 7 to 9 × 10⁹ cells. The number of glial cells is about 10 times larger. Although exact data on the number of synaptic contacts are lacking for the human cortex, measurements derived from cortices of other mammals give an estimate of 2×10^{15} synapses in the human cortex of one hemisphere.

It is generally believed that the number of nerve cells decreases with aging. However, recent observations show that the total number of cortical nerve cells is fairly constant between the 20th and 100th years. A reason for the believed decrease could be the effect of the slow increase in brain size during the last 100 years. This means that in a sample of brains from individuals of a wide age range, older persons have smaller brains and, therefore, a lower number of cortical cells.

II. MICROSCOPIC ANATOMY

A. Laminar Pattern

The stratification of the cortex into cell layers parallel to the surface is the most important architectonic feature. The largest part (approximately 95% of the total cortical volume) of the cortex has a six-layered structure (the isocortex). The smaller part of the human cortex (the allocortex) shows a greatly varying laminar structure (mostly less than six layers).

The six layers of the isocortex can be delineated in histological sections in which the nerve cell bodies are visualized with Nissl stain (Fig. 2). The most superficial layer is lamina I (the molecular layer), which contains only a low number of nerve cells, or neurons. The second layer (the outer granular layer) is densely packed with small nerve cell bodies (granular cells). Layer III (the outer pyramidal layer) has a lower pack-

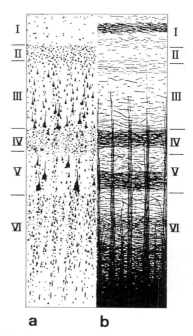

FIGURE 2 The laminar pattern of the human isocortex with the delineation of six layers (I–VI). (a) Nissl stain for the demonstration of cell bodies. (b) Myelin stain for the demonstration of myelinated nerve fibers.

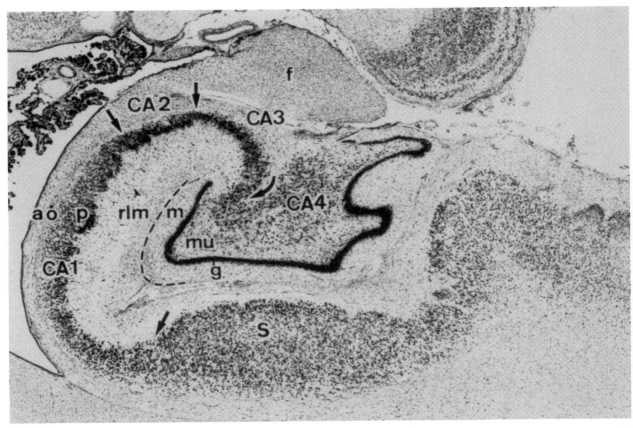

FIGURE 3 Nissl stain showing the cytoarchitecture of the human hippocampus, magnification ×10. a, alveus; CA1–CA4, regions of Ammon's horn; f, fimbria hippocampi; g, granular layer; m, molecular layer; mu, multiform layer; o, oriens layer; p, pyramidal layer; rlm, radiatum–lacunosum–moleculare layer; S, subiculum. The arrows indicate the borders of the subdivisions (CA1–CA4) of Ammon's horn.

ing density. The cell bodies are medium sized and many of them display a triangular (i.e., pyramidal) shape. The fourth layer (the inner granular layer) has the highest packing density of all layers. Small and mostly round cell bodies are found here. The fifth layer (the inner pyramidal layer) contains the largest cell bodies, and most of the cells have a pyramidal shape. The packing density is low in this layer. The innermost, sixth, layer displays an increasing cell packing density with bodies of greatly varying shapes (the polymorphic layer).

The myelin stain demonstrates cellular sheaths wrapped around the axons of nerve cells. The sheaths are formed by glial cells and contain a high concentration of lipids (i.e., myelin). Myelin-stained sections of the isocortex demonstrate its stratified structure, in addition to many vertically oriented bundles of myelinated axons (Fig. 2).

A silver impregnation procedure called the Golgi method reveals neuronal cell bodies together with their cell processes. Since this method is capricious and allows the impregnation of only a minor portion of the total cell population, it visualizes selected aspects of the cortical structure.

A description of the widely differing laminar patterns of all allocortical areas is not within the scope of this article. The cytoarchitecture of the hippocampus, which represents the major part of the archicortex, is illustrated as one example (Fig. 3). This simpler three-layered structure contrasts clearly with the isocortical lamination pattern. [See Hippocampal Formation.]

B. Cell Types

The cortex is composed of many different cell types. A fundamental difference is found when the dimensions of the efferent cell processes (i.e., axons) are analyzed. The axon of a cortical neuron can be short and terminates in the vicinity of its cell body, or is long and travels to target areas far from the place of origin. The short axon cells are, therefore, the elements of the local cortical circuitry, whereas the neurons with the long axons are the cellular basis for far-reaching projection systems. The former cells are called interneurons; the latter are projection neurons.

All pyramidal cells (Fig. 4) belong to the class of projection neurons. They have a roughly triangular cell body with a thick dendrite leaving at the apical part of the cell body and numerous dendrites originating at the basis of the cell body. The dendrites are arborizing in different layers of the cortex. They bear some thousands of so-called spines, which are the postsynaptic parts of synapses. The dendrites offer the major portion of postsynaptic contacts on which the presynaptic terminals of numerous other neurons end. Most of the synapses on dendrites are excitatory, but inhibitory synapses are also found. Pyramidal cells have exclusively inhibitory synapses on the cell body and the initial axon segment.

The class of interneurons comprises many different cell types, all of which are characterized by dendrites with few or no spines at all. The cell bodies are of greatly varying size and shape. They can be larger than small- or medium-sized pyramidal cells and can be round, oval, or irregularly shaped. The dendrites and axons form different, but for each cell type characteristic, territories.

III. ONTOGENY

The ontogeny of the human cortex starts with bilateral evaginations of the dorsolateral walls of the forebrain during the fifth embryonic week (Fig. 5). A cerebral vesicle comprises the primordia for the pal-

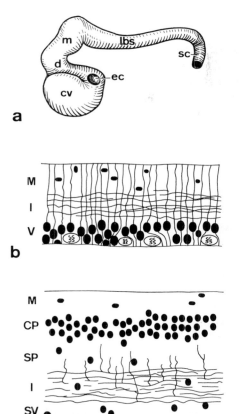

FIGURE 5 Ontogeny of the human cortex. (a) Cerebral vesicle (cv) of the brain during the fifth embryonic week. d, diencephalon; ec, eye cup; m, mesencephalon; lbs, lower brain stem; sc, spinal cord. (b) Three-layered structure of the primordial pallium with the marginal (M), intermediate (I), and ventricular (V) zones. (c) Cortical plate (CP) and cortical stratification during fetal stages with the marginal (M), intermediate (I), subventricular (SV), and ventricular (V) zones and the subplate layer (SP).

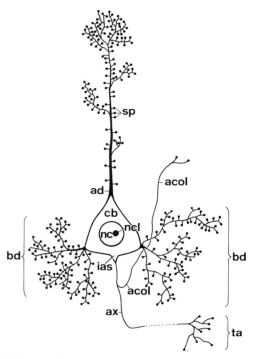

FIGURE 4 A Golgi-impregnated pyramidal neuron in the isocortex. acol, axonal collateral; ad, apical dendrite; ax, main stem of the axon; bd, basal dendrites; cb, cell body; ias, initial axonal segment; nc, cell nucleus; ncl, nucleolus; sp, spines; ta, terminal arborization of the axon.

lium (neo-, archi-, and paleopallium) and parts of the basal ganglia.

The primordial pallium is a thin neuroepithelial layer. Each epithelial cell spans the whole thickness of the wall of the vesicle. The first afferent nerve fibers from lower brain regions arrive in the primitive paleopallium and extend during the following week into the neopallial primordium. These generate a three-layered structure with an abundance of nerve fibers and some immature neurons in a superficial layer, the marginal plexiform lamina. This layer is separated from the densely packed neuroepithelium (the matrix or ventricular zone) by the primordial white matter (the intermediate zone) (Fig. 5).

The number of mitoses in the ventricular zone increases during the seventh and eighth weeks, and the first descending corticofugal fibers are detectable in the primordial white matter. The newly generated immature neurons of the ventricular zone move in an opposite direction into the marginal zone. These cells migrate in a highly ordered manner along processes of radial glial fibers, which span the whole width of the primordial pallium. The immature neurons stop their migration in the center of the marginal zone, forming an initially thin, but rapidly thickening, new layer, the cortical plate (Fig. 5). The marginal zone is divided by this developmental process into a superficial layer (the later layer I of the cortex) and a subplate zone (the later layer VIB). Other observers have defined the subplate zone as a superficial part of the intermediate zone.

The cortical plate gives rise to adult cortical layers II–VIA. The first cells arriving in the cortical plate form the lamina VIA of the adult cortex. The following waves of migrating neurons traverse this layer and are located more superficially; the result is an inside-out layering, with the latest wave of migrating neurons forming the adult layer II. The first synapses are found in layers I and VIB at the time of the formation of the cortical plate. The first afferent fibers to the cortical plate arrive around the 15th week. During the following week the arrival of thalamocortical, callosal, and other corticocortical fibers and the differentiation of nerve cells proceed from the lower to the more superficial parts of the cortical plate.

The generation of neurons in the ventricular zone stops around the time of birth. Since these neurons migrate into the cortical plate, the ventricular zone disappears. Only a single layer of epithelial cells, the ependyma, separates the adult white matter from the ventricular cavity.

The adult volume of the primary visual cortex (area

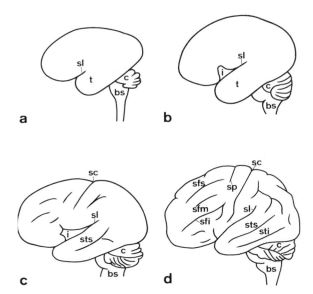

FIGURE 6 Ontogeny of the sulcal pattern of the lateral cortical surface. (a) Smooth cortical surface during the 16th embryonic week. The temporal lobe and the lateral sulcus are visible. (b) The temporal lobe and the lateral sulcus are enlarged, together with the frontal and pareital lobes, during the 25th embryonic week. The surface of the prospective insular lobe is not growing as fast as the other lobes, leading to a retrusion of the insular region. (c) A group of typical sulci is recognizable during the 33rd embryonic week, and (d) the adult pattern is visible around the time of birth. bs, brain stem; c, cerebellum; i, insular lobe; sc, central sulcus; sfi, inferior frontal sulcus; sfm, medial frontal sulcus; sfs, superior frontal sulcus; sl, lateral sulcus; sp, precentral sulcus; sti, inferior temporal sulcus; sts, superior temporal sulcus; t, temporal lobe.

17) is reached at the beginnning of the fifth postnatal month. At about the same time the volume proportion of the neuropil—the space between the cell bodies of nerve cells and glial cells containing cell processes of both types and most of the synaptic connections—reaches adult values.

The cortical surface is smooth during the first 23 weeks of ontogeny. At this time the enlargement of the surface leads to the appearance of the first cortical sulci. The adult degree of folding is acquired at the time of birth (Fig. 6).

IV. COMPARATIVE ANATOMY AND EVOLUTION

A comparison of the surface areas of the cortices between different mammals reveals the progressive development of the human cortex. Its surface is not only absolutely larger than that of most nonprimates, but surpasses the cortical surface of all living primates. Since total brain and cortex volumes are associated with body size, an analysis of the cortical volume independent of body size can prove these findings. The volume of the paleocortex of a hypothetical insectivore scaled to human body size is more than three times that of the human paleocortex. The volume of the human archicortex, however, surpasses that of the hypothetical insectivore four times and that of a rhesus monkey of equal body size almost twice. Even more pronounced is the evolution of the neocortex. The human neocortical volume is 156 times larger than in the insectivore and 45 times larger than that of the rhesus monkey in relation to body size. The progressive evolution of the human cortex is, therefore, characterized mainly by the growth of the neocortex (i.e., neocorticalization). [*See* Comparative Anatomy.]

The size of the association regions of the neocortex dominates the primary sensory and motor regions, indicating that human cortical evolution favors higher levels of information processing. The association regions of the frontal, parietal, and temporal lobes comprise more than two-thirds of the neocortical surface. Within the occipital lobe only 25% of the cortical surface is occupied by the primary visual area, whereas 75% is represented by association areas.

V. AREAL PATTERN, FUNCTIONAL PARCELLATION, AND CONNECTIVITY

A. Paleocortex

The human paleocortex comprises the olfactory bulb, retrobulbar region, olfactory tubercle, piriform cortex, septum, amygdala, and parts of the insular cortex (Fig. 7).

The olfactory bulb is less differentiated than in any other primate. The most important input comes from the olfactory epithelium through the fila olfactoria of the first cranial nerve. The olfactory bulb projects to the other paleocortical areas via an olfactory fiber tract at the basal surface of the frontal lobe. In contrast to other mammals, an accessory olfactory bulb, which is connected with the vomeronasal organ, cannot be found in the adult human brain. Destruction of the olfactory bulb leads to loss of the sense of smell. [*See* Olfactory Information Processing.]

The retrobulbar region has often been termed an anterior olfactory nucleus. Despite the low degree of

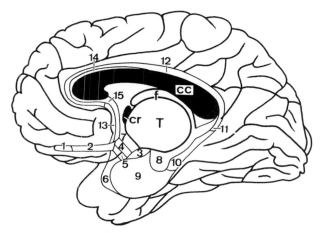

FIGURE 7 Medial view of the human telencephalon with paleocortical and archicortical areas. cc, corpus callosum; cr, rostral commissure; f, fornix; T, thalamus; 1, olfactory bulb; 2, retrobulbar region and olfactory tract; 3, amygdala; 4, olfactory tubercle; 5, piriform cortex; 6, peripaleocortical region; 7, septum; 8, retrocommissural hippocampus; 9, entorhinal area; 10, pre- and parasubiculum; 11, retrosplenial cortex; 12, periarchicortical cingulate areas; 13, subgenual area; 14, supracommissural hippocampus; 15, precommissural hippocampus.

architectonic differentiation of this area in the human brain, comparative anatomical studies have corroborated the cortical character of this rhinencephalic region. The retrobulbar region is found at the transition of the olfactory tract into the basal frontal cortex. Its main afferent fibers originate in the olfactory bulb. The retrobulbar region is an important relay station for olfactory information.

The olfactory tubercle is located in the anterior perforate substance, which borders the retrobulbar region. Although heavily reduced in the human brain, the cortical structure of this area can be recognized. The olfactory tubercle has reciprocal connections with the other paleocortical areas.

The piriform cortex is also located in the anterior perforate substance of the basal forebrain. A three-layered cortical structure is visible. The most superficial layer contains the lateral part of the olfactory tract. The piriform cortex is reciprocally connected with the other paleocortical regions and projects to the hippocampus and the hypothalamus. Besides its olfactory function, the piriform cortex plays a role in sexual behavior.

The septum is found on the medial side of the olfactory tubercle and extends onto the medial surface of the hemisphere. It is classified as part of the paleocortex, because comparative anatomical observations have shown that at least parts of the human septum are equivalent to paleocortical areas in other primates. The septum receives input from paleocortical areas, the hippocampus, and isocortical and subcortical regions. A strong projection leaves the medial part of the septum via a fiber bundle beneath the corpus callosum, called the fornix, and terminates in the hippocampus and surrounding archicortical areas, where the axonal terminals release acetylcholine as a neurotransmitter. This renders the septum part of the magnocellular basal forebrain region, the source of cholinergic innervation in cerebral cortex. Alzheimer's disease is associated with a remarkable degeneration of acetylcholine-containing neurons in the basal forebrain, including the septum.

The amygdala is a huge agglomeration of neurons in the temporal lobe just in front of the lower horn of the lateral ventricle. The hippocampus caudally borders the amygdala. One part of the amygdala shows a clear cortical structure, whereas the major part represents a subcortical nuclear formation. Reciprocal connections exist to paleo-, archi-, and isocortical regions and the hypothalamus. The cortical parts of the amygdala influence nutritional and sexual behavior, whereas the subcortical parts are involved in anxiety and aggression.

B. Archicortex

The archicortex comprises the hippocampus and the periarchicortical entorhinal, pre- and parasubicular, retrosplenial, and cingulate areas.

The major (i.e., retrocommissural) part of the hippocampus extends from the medial wall of the lower horn of the lateral ventricle to the caudal end of the corpus callosum (called the splenium corporis callosi). The hippocampus continues around the splenium to the dorsal surface of the corpus callosum (the supracommissural part) and follows this structure to its rostral end, where the precommissural hippocampus is located. The supracommissural and precommissural parts are small structures that do not show the typical lamination pattern nor regional subdivisions of the retrocommissural hippocampus. Further descriptions are relevant for the retrocommissural hippocampus.

This archicortical region can be subdivided into the Ammon's horn, dentate gyrus, and subiculum (Fig. 3). Ammon's horn and the dentate gyrus show a basic three-layered cortical pattern with the strata oriens, pyramidale, and radiatum–lacunosum–moleculare in the former and the strata moleculare, granulosum, and multiform in the latter subregion. The hippocampus is covered with a thin layer of myelinated nerve fibers on the ventricular surface, called the alveus. The alveus contains afferent and efferent fibers. The alvear fibers converge in the fimbria hippocampi, which continues as the fornix at the level of the splenium. The width and cell packing density of the pyramidal layer are the basis for further parcellation of Ammon's horn into CA1–4 regions. The CA4 region is often considered together with the multiform layer of the dentate gyrus as the hilus. The main cell type of the pyramidal layer of CA1–4 is the pyramidal cell. The main cell type of the dentate gyrus is the granule cell. All of the other neurons of the hippocampus are interneurons. The subiculum is a transitory area between Ammon's horn and the adjoining periarchicortical regions.

Three major systems of afferent fibers originating in cortical areas reach the hippocampus: one system comes from the septum (see earlier) and terminates mainly in the stratum radiatum of Ammon's horn; the other two systems (i.e., perforant path and the alvear tract) originate in the entorhinal area, the most important entrance to the hippocampus, and terminate mainly in the molecular layer of the dentate gyrus. The granule cells give rise to a bundle of axons (mossy fibers), which form synapses with the apical dendrites of the pyramidal cells in CA3. These cells send their axons via the fornix to the septum, the hippocampus of the contralateral side, or other target areas. Collaterals of these axons (i.e., Schaffer collaterals) travel to the apical dendrites of the pyramidal cells in CA1. These cells again form a major efferent pathway to the subiculum or via the fornix to the septum.

The activity in the intrahippocampal pathways is highest during slow-wave sleep and is strongly inhibited during rapid eye movement sleep or waking states. Although an association of the hippocampus with memory seems to be an important functional aspect, we are presently far from understanding sufficiently the functional impact of the human hippocampus.

The presubicular, parasubicular, and entorhinal areas are characterized by a cytoarchitectonic peculiarity, a nearly cell-free layer in the central part of the cortical width. Therefore, the term "schizocortex" has been coined for these periarchicortical areas. They are located on the parahippocampal gyrus between the subiculum and the laterally adjoining neocortex. The entorhinal area receives input from the total neocortex and gives a summary of all of this information into the hippocampus.

The retrosplenial areas are found in the caudal parts of the cingulate gyrus. The cingulate areas are located in the rostral parts of the cingulate gyrus. Both areas have lamination patterns in the regions near the archicortex, which allow classification as the periarchicortex. However, the laminar patterns of those parts of the retrosplenial and cingulate areas, which are nearer the neocortex, show an increasing differentiation. Therefore, classification as the proisocortex seems legitimate. The rest of the cingulate cortex near the cingulate sulcus is characterized by the typical six layers of the isocortex. The retrosplenial and anterior cingulate areas are part of the Papez circuit, which connects the anterior thalamus via the cingulate gyrus and the schizocortex with the hippocampus. Since the Papez circuit has been claimed to be important for memory, the retrosplenial and anterior cingulate areas could serve this complex function. Additionally, changes in affectivity and lowering of spontaneous activity are observed after lesions of the cingulate gyrus.

C. Neocortex

The most widely accepted map of the human neocortex into areas of specific architecture and function has been proposed by neuroanatomist Korbinian Brodmann in his classic monography from 1909. Although numerous corrections of details seem necessary, the more general aspects of his cortical map are of great value even today. His map (Fig. 8) is therefore the basis for the following descriptions of the areal pattern of the human neocortex. [*See* Neocortex.]

The areas in the neocortex of the frontal lobe can be subdivided architectonically into a group with a clearly visible inner granular layer (layer IV) and a group lacking this layer. The agranular part shows the six-layered structure of the typical isocortex only during the fetal period, but loses layer IV during the early postnatal period. This subdivision coincides roughly with the definition of a prefrontal (granular) and a motor (agranular) cortex in the frontal lobe. The prefrontal cortex comprises Brodmann's areas 8–11 and 44–47. The motor cortex comprises primary motor area 4 and premotor area 6.

The prefrontal areas are reciprocally connected with nearly all brain regions. The connection with the mediodorsal thalamic nucleus in the diencephalon is of special interest, because this system defines and delineates all prefrontal areas. A strong efferent projection of the prefrontal areas terminates in the basal ganglia.

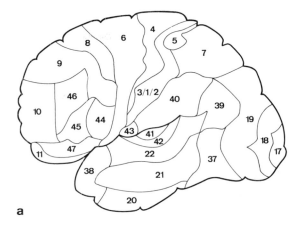

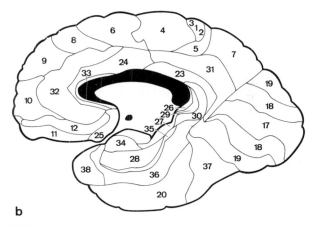

FIGURE 8 Areal map of the human cortex, according to Brodmann (1909). (a) Lateral and (b) medial views of the hemisphere.

Lesions of the prefrontal cortex in humans lead to attention disorders connected with cognitive deficits and preceptual distortions. Humans with destructions in the dorsolateral part of the prefrontal cortex show apathy and a lack of spontaneous movement. Lesions in the basal (orbital) part lead to hyperkinesia, euphoria, and disinhibition. Patients with frontal lesions cannot organize new and deliberate sequential behavior, owing to deficits in short-term memory, planning, and interference control. These deficits show that the prefrontal cortex is clearly an association region of the highest order. It might be a crucial structure for the high adaptive capacity of humans (i.e., their ability to develop new behavioral strategies).

Area 8 is at least partially equivalent to the frontal eye field found in other primates. This area contains cells responsive to visual stimuli, which are active before eye movement. Dominance of the right frontal eye field is found when nonverbal signals are analyzed.

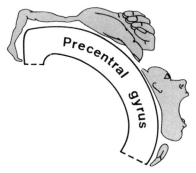

Area 44 and, eventually, area 45 represent Broca's motorical speech region. Lesions in this area lead to motorical aphasia, which is characterized by a slow and effortful speech delivery. The speech lacks normal fluidity and continuity, and the articulation of words is disturbed. It is interesting that this area is functionally active only in the left hemisphere of most humans.

The primary motor cortex (area 4) is found in the rostral wall of the central sulcus. Brodmann has defined area 4 by the presence of huge pyramidal cells (Betz cells) in layer V. Area 4 receives input from thalamic nuclei and many cortical areas. The main efferent pathway of area 4 is part of the pyramidal tract. These axons terminate in the brain stem and the spinal cord, where they control motor neurons, which give rise to cranial and spinal nerves. Area 4 is therefore the primary motor center for voluntary movements. This area shows a conspicuous somatotopy; that is, the muscles of the body are represented at distinct sites in the cortex. These sites have the same topological relationships as the muscles of the body. However, the extents of their cortical representation areas are correlated not with the dimensions of the muscles, but with their functional importance. In fact, there is an overproportionate representation of the hand muscles, especially the muscles of the thumb, and a small cortical representation of the buttock muscles (Fig. 9). Lesions of area 4 cause disruptions of motor functions on the contralateral side of the body. In an initial stage the tonus and the reflexes of muscles are lost, but after some time spasticity and hyperreflexia appear.

Area 6 is located between the primary motor and the prefrontal cortices and can be subdivided into a supplementary motor area and a premotor area. The small supplementary motor area is found on the medial side of the hemisphere. The premotor cortex covers the rest of area 6. Both subdivisions have a similar cytoarchitecture (i.e., agranular isocortex), but are connected to different parts of the ventrolateral thalamic nucleus in the diencephalon. They receive afferent fibers from nearly all cortical areas and many subcortical areas. The efferent projections of the supplementary motor area reach the spinal cord directly, whereas the projections of the premotor cortex arrive via a synaptic connection in the brain stem. Both subfields project to the basal ganglia, cerebellum, and some other brain stem nuclei. A bilateral activation of these subfields is found, when a person is imagining, but not executing, complex voluntary movements. A destruction of the premotor area impairs mainly the coordination of proximal muscles of the limbs. A lesion of the supplementary motor area leads to forced grasping, a reduction in spontaneous movements of the contralateral hand, and impairment of bilaterally coordinated activities. A bilateral destruction of this area results in a permanent akinesia.

The isocortical areas of the parietal lobe can be subdivided into primary and association areas. The anterior part of the parietal lobe roughly coincides with the postcentral gyrus, where the primary somatosensory cortex (i.e., areas 3, 1, and 2) is found. The posterior part consists of superior and inferior parietal lobules separated by the intraparietal sulcus (Fig. 1). The superior lobule is composed of areas 5 and 7, whereas the inferior parietal lobule includes areas 39 and 40. The whole parietal cortex shows the basic six-layered structure of the isocortex.

The primary somatosensory cortex has a conspicuous layer IV. Area 3 is bordered by area 4 rostrally and area 1 caudally. Area 3a is found at the transition from area 3 to area 4 and has a broad layer V and a small layer IV. This area receives its main input from the muscle spindles informing about the actual degree of muscle tonus. Information from slowly adapting mechanoreceptors of the skin arrive in area 3. Area 1 is informed by rapidly adapting mechanoreceptors of the skin, whereas the respective peripheral organs of area 2 are the mechanoreceptors of joints. As in area 4, the body is represented in a somatotopic manner in the primary somatosensory cortex.

Recent observations have shown that there are multiple cortical representations of one part of the body. The major input of areas 3, 1, and 2 comes from specific thalamic nuclei. Additionally, afferent fiber systems from other cortical areas of the same and the opposite hemisphere (through the corpus callosum) terminate in the primary somatosensory cortex. Efferent fibers project to thalamic nuclei, brain stem regions, and the primary motor and prefrontal cortices. A lesion of the primary somatosensory cortex leads to paresthesia, tactile agnosia, and impairment of tactile discrimination.

Areas 5 and 7 of the superior parietal lobule are association areas. They are reciprocally connected with the prefrontal cortex and the inferior parietal lobe. Afferent fibers arrive from most of the isocortical areas. The efferent projections of this part of the parietal cortex terminate in many cortical areas, as well as in the basal ganglia, the thalamus, and regions of the lower brain stem.

Areas 39 and 40 of the inferior parietal lobule are reciprocally connected with various areas of the frontal, temporal, and occipital cortices. Additional input arrives from the hippocampus, retrosplenial cortex, and brain stem regions. The efferent projections terminate in various cortical areas, basal ganglia, thalamus, and further brain stem regions.

Lesions in the posterior parietal lobe lead to impairment in visual perception of space, size, and distance. Fixation of gaze and an inability to move the eye voluntarily toward a target are also found, together with losses of the capacity to write, right–left orientation, and calculation. The maintenance of a spatial reference system for goal-directed movements seems to be the main function of this association region.

The isocortical part of the temporal lobe comprises Brodmann's areas 41, 42, 20–22, 37, and 38. The primary auditory cortex and its belt region are represented by areas 41 and 42, whereas the other areas form the auditory association (area 22) and nonauditory association cortices. All of these areas reveal the basic isocortical lamination pattern, with the highest cell density in layer IV of the primary cortex.

Areas 41 and 42 (auditory cortex) are located on the dorsal plane of the temporal lobe (Fig. 10). Their position is recognizable by the appearance of a transverse gyrus (i.e., Heschl's gyrus). A massive fiber bundle ascending from the medial geniculate body of the thalamus is the most important input to the primary auditory cortex. Efferent projections terminate in the auditory association area, other isocortical regions, the medial geniculate body, and lower brain stem nuclei. Observations with positron emission tomography, which detects functionally active areas in a living subject, have revealed a tonotopic (i.e., different sites for different tones) organization in the primary auditory area.

The auditory association cortex (area 22) is found on the lateral surface of the superior temporal gyrus and on the dorsal surface of the temporal lobe poste-

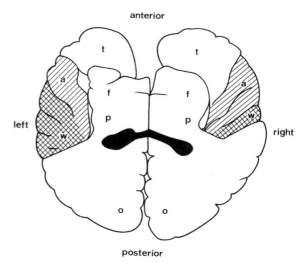

FIGURE 10 Dorsal plane of the temporal lobe with the primary auditory cortex (a) and part of Wernicke's area (w). The frontal (f), parietal (p), and occipital (o) lobes have been partially removed to give a view of the surface of the temporal (t) lobe from above. Wernicke's area of the left side is much larger than that of the right side (representing the dominance of the left hemisphere for the speech function).

rior to Heschl's gyrus. The latter region is called the planum temporale. The most posterior part of laterally exposed area 22, together with the planum temporale, resembles a cortical field in which the sensory speech region of Wernicke is located. It receives input from the auditory and visual association cortices and gives rise to a strong fiber bundle, which terminates in Broca's area. Some authors include the posterior parts of area 40 and the whole of area 39 in their definition of Wernicke's area. A destruction of this area leads to a lost understanding of verbal information and an inability to write. This defect was described in the last century as sensoric aphasia and agraphia. It has been shown that the planum temporale of the left hemisphere is much larger in most human brains than that of the right side (Fig. 10). This lateralization of an anatomical structure is associated with the same lateral preference of function.

The nonauditory association cortex comprises all other isocortical areas of the temporal lobe. Efferent connections of this region lead to the prefrontal and premotor areas and to the hippocampal region. The afferent fibers originate in sensory association cortices. Lesions of this area correlate with memory deficits for more complex and abstract visual patterns and deficits in verbal learning.

The occipital lobe includes areas 17–19. The hu-

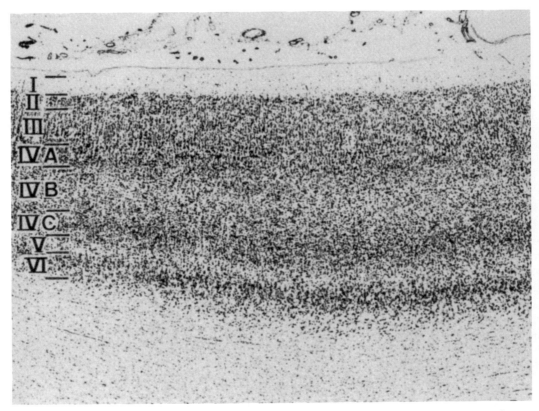

FIGURE 11   Nissl stain showing the lamination pattern of the primary visual cortex. Original magnification ×13. Roman numerals indicate the different isocortical layers in this highly differentiated area.

man primary visual cortex (area 17) shows the most differentiated laminar structure of all isocortical areas (Fig. 11). A conspicuous structure of this area is a stripe of myelinated fibers in layer IVB, which runs parallel to the cortical surface (i.e., Gennari's stripe). This stripe can be seen with the naked eye on unstained sections. Therefore, area 17 is also called the striate area. It extends from the occipital pole over the total length of the calcarine sulcus. The major part of this area is buried in the sulcus. The largest portion of area 17 receives input from both eyes (binocular part). The monocular part, which receives information only from the contralateral eye, is small and located in the most rostral extension of area 17. The whole cortical area is the target of information from the entire contralateral visual hemifield and a small part of the ipsilateral hemifield.

The strongest afferent fiber bundle to area 17 is the visual radiation, which originates in the lateral geniculate body of the thalamus of the same side. Since input from both eyes is already present in one lateral geniculate body, the information arriving in the area 17 of one side is also from both eyes. The optic radiation retains a clear separation of the input from both eyes. This is also conserved in the visual cortex, because the visual inputs from both eyes terminate separately in alternating stripes in layer IVC. The stripes consist of densely packed small neurons, synapses between these neurons, and the geniculocortical axon terminals. They have been demonstrated in many mammals, including primates, by axonal tracing techniques (Fig. 12). These stripes appear as columns in cross sections through the visual cortex and have been termed ocular dominance columns. Additional types of vertically oriented periodical structures (i.e., modules) can be found in area 17 (modules of orientation-selective or color-specific cells). Efferent projections of the primary visual cortex terminate in the secondary visual area (area 18), tertiary visual area (area 19), and other isocortical areas, the lateral geniculate body, the pulvinar complex of the thalamus, and regions in the brain stem. Removal of the striate area leads to a loss of visual perception.

Area 18 is bordering the striate area. The strongest

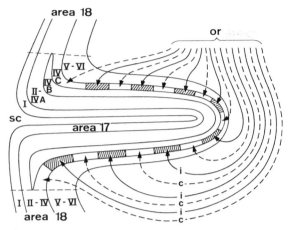

FIGURE 12 A coronal section through areas 17 and 18 with the ocular dominance columns (hatched and open areas in layer IVC) in the primary visual cortex. The ocular dominance columns are areas of alternating input from both eyes. The geniculocortical afferent fibers [optic radiation (or)] contain fibers from the ipsilateral (i) or contralateral (c) eye, which terminate in separate areas of layer IVC. Layer IVB is the stripe of Gennari, which disappears at the border between areas 17 and 18. Roman numerals indicate isocortical layers. sc, calcarine sulcus.

input to this area originates in the pulvinar and the primary visual cortex. The efferent projections terminate in area 19 and other isocortical areas; the pulvinar and regions, in the lower brain stem.

Area 19 is a single cortical area in Brodmann's map, but more recent observations have demonstrated that it contains numerous separate areas, each representing a complete visual half-field. Therefore, the term "third visual tier" has been introduced. All subdivisions of this region are strongly visually influenced. Areas 18 and 19 can be summarized as the extrastriate visual cortex. Electrical stimulation of the extrastriate areas leads to complex optical hallucinations, in contrast to simple visual sensations after stimulation of area 17.

Despite areal differences in laminar structure and input–output specificities, some general layer-specific aspects of connectivity in the isocortex can be stated. Table I summarizes these findings for the most important connections.

VI. NEUROCHEMISTRY

A. Neurotransmitters and Neuropeptides

All classical neurotransmitters and neuropeptides as well as their receptors can be demonstrated in the cortex. Their regional and laminar distribution varies greatly between different cortical areas and layers. This indicates a highly differentiated neurochemical organization, which is summarized in the next section, considering only the major afferent systems of the neocortex and the neurotransmitter receptor distribution in the striate area.

The human neocortex has no acetylcholine-containing neurons, but receives a cholinergic innervation from the basal nucleus of Meynert, which is an area in the basal forebrain region at the transition between the telencephalon and the diencephalon. Over 90% of the neurons in the basal nucleus of Meynert contain acetylcholine. The number of these neurons and the acetylcholine content of the cortex are markedly reduced in patients with Alzheimer's disease. Noradrenaline-containing fibers originate in the locus coeruleus, which is a nuclear configuration in the brain stem near the floor of the fourth ventricle. They are most densely packed in the primary somatosensory area and are found in all cortical layers, with the

TABLE I

Isocortical Layers as Origins or Targets of Afferent and Efferent Fiber Systems, Respectively[a]

Origin of afferent fibers	Target of efferent fiber	Layer
Thalamus and cortex	—	I
Cortex	Cortex	II
Thalamus and **cortex**	Cortex	III
Thalamus and cortex	Cortex and basal ganglia	IV
Thalamus and cortex	**Spinal cord, brain stem**, thalamus, **basal ganglia, cortex**	V
Thalamus and cortex	Thalamus, **claustrum**, cortex	VI

[a]The most important origins or targets are shown in boldface type.

lowest density in layer I. Dopamine-containing fibers arrive from the ventral tegmental area of the midbrain. They reach the highest densities in the motor and prefrontal cortices. Layers I and V–VI of the neocortex show maximal innervation densities. The axons of neurons in the raphe nuclei of the midbrain provide the serotonin innervation of the neocortex. A relatively uniform distribution is described over all areas, with peak values in layer IV.

γ-Aminobutyric acid (GABA) is the major inhibitory neurotransmitter of the cortex. GABA-containing neurons are found in great numbers within the cortex. Nearly all of them are local circuit neurons. Glutamate is the major excitatory transmitter of the cortex and is produced by a vast number of cortical neurons, especially pyramidal cells. These cells project to the basal ganglia, the thalamus, and brain stem nuclei. The vasoactive intestinal polypeptide, cholecystokinin, and corticotropin-releasing factor show the highest concentrations of neuropeptides in the cortex. Many of the peptides are colocalized in one neuron. Peptides might also be colocalized with classical neurotransmitters in the human cortex. A colocalization of GABA, for example, with vasoactive intestinal polypeptide has been described for the rat neocortex, but comparable studies in the human cortex are still lacking. [See Peptides.]

B. Receptors

All of these transmitters act on pre- or postsynaptically localized receptors during neurotransmission. The receptors are protein molecules traversing the cell membrane. Some of these directly regulate the opening or closing of ion channels; others are connected with secondary messenger systems. Drugs have been developed that are active at these receptors and can, therefore, influence neurotransmission. Table II gives a short summary of the laminar distribution of some neurotransmitter receptors in the human striate area. This table and results from many other cortical areas reveal that the single receptor shows a considerable laminar specificity (i.e., distribution pattern). [See Neurotransmitter and Neuropeptide Receptors in the Brain.]

A comparison of the laminar distribution patterns of different receptors in the human primary visual cortex demonstrates that some receptors have similar laminar distributions (i.e., M1 receptors with glutamate binding sites; glutamate binding site with GABA$_A$, 5-HT$_1$, and 5-HT$_2$ receptors; GABA$_A$ with 5-HT$_1$ and 5-HT$_2$ receptors; 5-HT$_1$ with 5-HT$_2$ recep-

TABLE II
Laminar Distribution of Some Neurotransmitter Receptors in the Human Primary Visual Cortex[a]

Layer	Receptor types
I	D1, α_1
II	M1, Glut, GABA$_A$, 5-HT$_1$, **D1**, **α_1**
III	**M1**, **Glut**, GABA$_A$, 5-HT$_1$, 5-HT$_2$
IV	M1, M2, Glut, GABA$_A$, 5-HT$_1$, 5-HT$_2$, D1
V	**M2**, D1
VI	—

[a]Only the layers with high receptor densities are listed. Boldface type indicates the cortical layers with the highest densities of a specific receptor. M1, M2, muscarinic cholinergic receptors; Glut, glutamate receptors; GABA$_A$, GABA receptor; 5-HT$_1$, 5-HT$_2$, serotonin receptors; D1, dopamine receptor; α_1, noradrenaline receptor.

tors), but other receptors (e.g., M2 and α_1) do not show this laminar codistribution. The codistribution is a structural prerequisite for functional interaction of different transmitter systems within a cortical area.

Presynaptically localized receptors (e.g., M2 receptor) reduce the transmitter release, whereas postsynaptic receptors influence the neuronal excitability. It has been shown that M2 receptors are found on the axonal terminals of basal forebrain neurons projecting into the cortex. Consequently, the density of M2 receptors is lowered in the cortices of brains from Alzheimer's disease patients, which frequently show a massive neuronal degeneration in the basal forebrain complex. Since certain drugs can mimic the effects of natural transmitters at specific receptor sites, a more detailed analysis of the receptor distributions in normal and pathologically changed human cortices is necessary for the development of new therapeutic strategies in neurological disorders.

BIBLIOGRAPHY

Armstrong, E., Zilles, K., and Schleicher, A. (1993). Cortical folding and the evolution of the human brain. *J. Hum. Evol.* 25, 387–392.

Brodmann, K. (1909). "Vergleichende Lokalisationslehre der Grosshirnrinde." Barth Verlag, Leipzig, Germany.

Fuster, J. M. (1989). "The Prefrontal Cortex," 2nd Ed. Raven, New York.

Nieuwenhuys, R. (1985). "Chemoarchitecture of the Brain." Springer-Verlag, Berlin.

Peters, A., and Jones, E. G. (eds.) (1984–1995). "Cerebral Cortex," Vols. 1–11. Plenum, New York.

crease binding of, and activation by, low concentrations of thrombin (<0.4 nM). GPIb exists in a complex comprising the disulfide-link d subunits GPIbα (M_r 140,000) and GPIbβ (M_r 27,000) and the noncovalently bound GPIX (M_r 22,000); this GPIb–IX complex is associated with GPV (M_r 82,000) in a ratio of 2 to 1. However, of the approximately 30,000 copies of the GPIb complex present on the platelet surface, only about 50 that are present in a supercomplex with a functional molecular weight of ~900,000 as determined by radiation inactivation and target analysis are able to bind thrombin with high affinity. The thrombin binding domain of GPIbα has been localized to the sequence Gly^{271}–Glu^{285} in the carbohydrate-poor amino-terminal domain and sulfation of three tyrosine residues in this region appears to modulate thrombin binding.

All of the components of the GPIb–IX complex have been cloned and shown to belong to a gene family characterized by repeated leucine-rich sequences, but the mechanism by which the binding of α-thrombin to the supercomplex initiates platelet activation through the high-affinity pathway is not known. Patients suffering from Bernard-Soulier syndrome lack the components of the GPIb–IX complex and do not respond to low concentrations of α-thrombin.

The moderate-affinity thrombin receptor (M_r 66,000) has recently been cloned and shown to be a member of the seven-transmembrane family of G-protein-coupled receptors. Thrombin activates this receptor by an unusual mechanism involving cleavage of the Arg^{41}/Ser^{42} peptide bond in the amino-terminal extracellular sequence of the receptor, giving rise to a new amino terminus for the receptor that has the initial sequence S^{42}FLLRN and that then reacts intramolecularly with the receptor itself to effect platelet activation: a synthetic peptide with this sequence can itself induce platelet activation. A second protease-activated seven-transmembrane domain G-protein-coupled receptor has recently been characterized and termed protein-activated receptor 2 (PAR-2), suggesting that the moderate-affinity thrombin receptor may be similarly termed PAR-1. Platelets from "knockout" mice lacking the gene for the moderate affinity receptor have recently been shown to have a normal hemostatic response indicating that this receptor is not obligatory for platelet activation by α-thrombin.

J. von Willebrand Factor

In the regions of the microvasculature where the blood moves with high velocity, the adhesion of platelets to subendothelium appears to be mediated by vWF bound to collagen and, possibly, to other subendothelial components. The initial interaction with platelets is dependent on the presence of GPIb on the platelet surface, and the vWF-binding domain of GPIb has been localized as being in the sequence Ser^{251}–Tyr^{279}, which overlaps the binding site for thrombin. Thus patients with Bernard-Soulier syndrome, whose platelets lack the GPIb–IX complex, have deficient platelet adhesion as well as a deficient response to α-thrombin. Similar deficient platelet adhesion is seen in Glanzmann's thrombasthenia owing to the absence of vWF from these patients' plasma. Agglutination of formalin-fixed platelets by vWF in the presence of the antibiotic ristocetin has been a useful laboratory test for this GPIb-dependent interaction.

This initial attachment of platelets to the vessel wall is mediated by the interaction of GPIb with the sequence V^{449}–Q^{460} in the amino-terminal fourth of the vWF polypeptide chain. This results in the activation of a phospholipase A$_2$, which leads to the expression of GPIIb/IIIa in a form that binds to the Arg^{1794}-Gly-Asp^{1796} sequence present in the carboxy-terminal third of the vWF molecule. These steps provide the clearest evidence in platelets of different receptors mediating adhesion and anchorage to the same adhesive substrate.

BIBLIOGRAPHY

Greco, N. J., and Jamieson, G. A. (1991). High and moderate affinity pathways for α-thrombin-induced platelet activation. *Proc. Soc. Exp. Biol. Med.* **198**, 792–799.

Greenwalt, D. E., Lipsky, R. H., Ockenhouse, C. F., Ikeda, H., Tandon, N. N., and Jamieson, G. A. (1992). Membrane glycoprotein CD36: A review of its roles in aherence, signal transduction and transfusion medicine. *Blood* **80**, 1105–1115.

Kaushansky, K. (1995). Thrombopoietin: The primary regulator of platelet production. *Blood* **86**, 419–456.

Lopez, J. A. (1994). The platelet glycoprotein Ib–IX complex. *Blood Coag. Fibrinolys.* **5**, 97–119.

Mills, D. C. B. (1996). ADP receptors on platelets. *Thromb. Haemostas.* **76**, 835–856.

State of the Art Lectures (1995). *Thromb. Haemostas.* **74**, 1–579.

Cranial Nerves

J. E. BRUNI
University of Manitoba

GLOSSARY

Cranial ganglia Groups of nerve cell bodies located outside the brain and associated with certain of the cranial nerves. They represent either cell bodies of sensory neurons or the location of postganglionic parasympathetic neurons

Cranial nerve lesion Injury from any of a number of causes to a cranial nerve, its nuclei of origin or termination, and/or its supranuclear projections that results in some pathological alteration in its function

Cranial nerve nuclei Groups or columns of nerve cells within the brain stem representing the sites of origin and termination of the various cranial nerves

Cranial nerves Twelve pairs of nerves that originate from the brain and provide sensory and motor innervation to regions of the head, neck, thorax, and abdomen

TWELVE PAIRS OF CRANIAL NERVES EMERGE FROM the brain and innervate regions of the head, neck, thorax, and abdomen. Like the 31 pairs of spinal nerves, they consist of afferent and/or efferent fiber components. Efferent fibers (Latin, *ex*, "out"; *ferre*, "to carry") convey impulses in a centrifugal direction away from the central nervous system (CNS), whereas afferents (Latin, *ad*, "to"; *ferre*, "to carry") carry nervous impulses from various sensory receptors in a centripetal direction toward the CNS.

The afferent and efferent fiber components of the cranial nerves are further categorized on the basis of the structures they supply. Fibers that innervate skin, skeletal muscle, tendons, joints, and ligaments are referred to as somatic. Those that innervate internal organs, smooth muscle, vessels, and glands are referred to as visceral and are components of the autonomic nervous system (ANS).

The cranial nerves contain (1) general somatic afferent (GSA) fibers, which convey information from the skin and mucous membranes located in the head; (2) general visceral afferent (GVA) fibers, which transmit information arising from the internal organs (pharynx, larynx, trachea, esophagus, thoracic and abdominal viscera) of the body; (3) general somatic efferent (GSE) fibers, which arise from multipolar motor neurons within the brain stem and which innervate striated skeletal musculature in the head (extraocular and lingual muscles) derived from myotomes; and (4) general visceral efferent (GVE) fibers, which are all parasympathetic and innervate smooth muscle, cardiac muscle, and glands. All of the aforementioned fiber types are also found in spinal nerves. [*See* Visceral Afferent Systems Signaling and Integration; Autonomic Nervous System.]

Because of the peculiar way the head and its associated special sense organs develop, some of the cranial nerves contain fiber components that are not found in spinal nerves. Nerve fibers contained in some cranial nerves innervate striated muscles of the head derived from the branchial or gill arches and are referred to as special visceral efferent (SVE) fibers. These include the muscles of mastication, facial muscles, and mus-

ENCYCLOPEDIA OF HUMAN BIOLOGY, Second Edition, VOLUME 3.

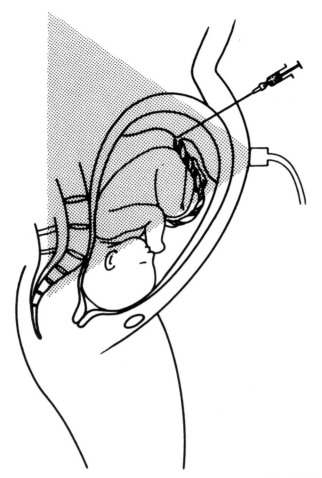

FIGURE I Umbilical cord blood sampling using a needle guided with ultrasound.

B. Results

Over 10,000 FBSs have been carried out to this day in the world. At the 37th Annual Meeting of the American College of Obstetricians and Gynecologists, held in May 1989, American and European teams obtained the following results. In Europe, 6336 samplings were carried out. The total fetal loss was 2.59% (ranging from 0.7 to 6.46%) and the procedure-related risk was evaluated at 1.05% (ranging from 0.4 to 2.98%). In the United States, 1610 samplings were performed with a procedure-related risk of 1.56% (ranging from 0 to 6%). The complications most frequently encountered were infection, membrane rupture, hemorrhage at the point of the puncture on the umbilical vein, severe slowdown of the fetal heart, and blood clotting.

C. Specific Technical Problems

When practiced by a well-trained team, FBS usually takes little time (<10 min in 92% of cases), but there are exceptions to the rule. When the mother is obese, it is necessary to use longer needles, which are consequently more difficult to guide, and the quality of the ultrasound image may be poor. High-resolution ultrasonography is indispensable, quality being an essential factor to simplify the technique and make it successful. In cases of fetal mobility, excessive fetal movements may occasionally block access to the cord. The problem can usually be overcome with a little patience. In rare instances, an adequtae view of the cord insertion may not be possible, and the needle must be introduced near the abdominal insertion (the other fixation point) or, in exceptional circumstances, on a loop of the cord.

D. Purity of the Blood Samples

Controlling the quality of the sample is an absolute prerequisite before the fetal biological results can be interpreted. It has to be done immediately after sampling through different tests. The sample may be contaminated with maternal blood or diluted with amniotic fluid. For instance, a 1/1000 contamination with maternal blood may lead mistakenly to conclude the presence of IgM in the fetal blood and incorrectly to diagnose congenital infection, or substantial dilution by amniotic fluid may suggest the presence of fetal anemia; on the other hand, a slight dilution of amniotic fluid could induce the activation of coagulation and bring about false information about the blood-clotting apparatus of the fetus.

II. PRENATAL DIAGNOSIS

Prenatal diagnosis of fetal pathology (i.e., affections) was first reduced to simple diagnoses such as the study of the presence or absence of a given protein or enzyme in fetal blood when a risk of hereditary diseases such as hemophilia or hemoglobinopathies exists. The knowledge of normal fetal biology now makes it possible to recognize more complex illnesses contracted in the course of a pregnancy: infections, anoxemia, and feto-maternal immunological conflict (Table I). [*See* Embryofetopathology.]

A. Fetal Infection

Affection of a fetus by an infection contracted by the mother during pregnancy may have very serious

TABLE I

Indications of Fetal Blood Samplings

Prenatal diagnosis	• Rapid chromosome analysis	Late pregnancy
		Amniotic fluid mosaicism
		Ultrasound detected pregnancy abnormality
	• Genetic disorders	Fragile X chromosome
		Hemoglobinopathies
		Coagulation disorders
		Miscellaneous
	• Congenital infection	Toxoplasmosis
		Rubella
		CMV—varicella
		Other viral infection
Assessment of fetal welfare	• Alloimmunizations	Rhesus system
		Platelet systems
	• Idiopathic thrombocytopenic purpura	
	• Growth retardation	Anoxemia
		Nutritional assessment
Fetal therapy	• Intravascular transfusions	Red blood cells
		Platelets
	• Drugs	Curare
		Digitoxin
		Thyroxin
	• Assessment of maternal therapy on fetal state	Steroids
		IgG
		Antibiotics
		Oxygen
Prenatal pharmacology	• Placental transfer of drugs	
	• Fetal effects	

consequences for the fetus. Roughly, the earlier in the course of pregnancy the infection occurs, the more serious it will be. However, from Weeks 18–20 of pregnancy, the fetus is able to produce its own specific antibodies in sufficient quantities, and the sequelae will therefore be less serious after that period.

Many congenital infections can be diagnosed *in utero* (e.g., toxoplasmosis, rubella, cytomegalo-virus (CMV), varicella, parvovirus, chlamydiae). As for adults, a diagnosis of fetal infection relies on two types of biological indices.

I. The Specific Indices of an Infection

a. Identifying the Pathogenic Organism in Fetal Blood or in the Amniotic Fluid

Parasites and viruses, and any live pathogenic organism, can be revealed by means of cell culture or by inoculating an animal with the buffy coat of the fetal blood or with the residue from centrifugation of the amniotic fluid. According to the agent considered, the identification is usually easily obtained and more or less constant.

A culture of a rubella virus is almost always positive in fetal blood when the fetus is infected, irrespective of the period when the fetal blood was sampled, whereas the cytomegalovirus is not always found in fetal blood in a case of congenital infection. For *Toxoplasma gondii,* the parasite is not always found in fetal blood or in amniotic fluid (Table II). The parasite is more frequently found in fetal blood when the congenital infection took place shortly before the sampling of fetal blood. Apparently, fetal parasitemia ceases rapidly, and the parasite is later eliminated through fetal urine and can be found in the amniotic fluid.

b. The Presence of Specific IgM Antibodies

The possibility for the fetus to synthesize IgM antibodies appears progressively in the course of fetal life. Total IgM can exceptionally be found in fetal blood before 17 weeks of pregnancy, its rate increasing rapidly between 17 and 22 weeks, and fetuses are capable of antigenic response of good quality from Week 22 of amenorrhea onward.

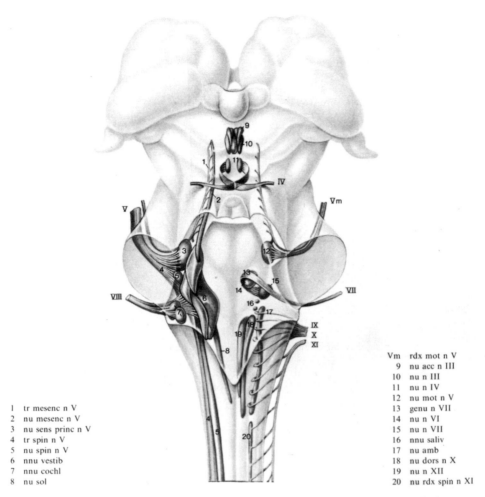

1 tr mesenc n V
2 nu mesenc n V
3 nu sens princ n V
4 tr spin n V
5 nu spin n V
6 nnu vestib
7 nnu cochl
8 nu sol

Vm rdx mot n V
9   nu acc n III
10 nu n III
11 nu n IV
12 nu mot n V
13 genu n VII
14 nu n VI
15 nu n VII
16 nnu saliv
17 nu amb
18 nu dors n X
19 nu n XII
20 nu rdx spin n XI

FIGURE 3 Diagrammatic representation of the dorsal surface of the brain stem showing the location of the various cranial nerve nuclei. Sensory nuclei are on the left and motor nuclei are on the right. [Reproduced with permission from R. Nieuwenhuys, J. Voogd, and C. van Huijzen (1988). "The Human Central Nervous System: A Synopsis and Atlas," 3rd Ed. Springer-Verlag, Berlin.]

oculomotor nerves arise from either side of the ventral midline of the midbrain along its attachment to the pons (Fig. 2).

The pons (Latin, "bridge") is a short segment of the brain stem that lies between the midbrain and the medulla oblongata. The ventral part of the pons, called the basilar pons, is especially prominent in the human brain and forms a stout bridge of gray and white matter that links the cerebral cortex on one side to the cerebellar cortex on the contralateral side. Within a region of the pons called the pontine tegmentum, the nuclei of origin and termination of four of the cranial nerves are found in whole or in part: the trigeminal (cranial nerve V), abducens (cranial nerve VI), facial (cranial nerve VII), and vestibulocochlear (cranial nerve VIII). Cranial nerves VI, VII, and VIII

emerge ventrally along the transverse plane between the pons and medulla. The trigeminal nerve arises from the pons at the point of origin of the middle cerebellar peduncle from the basilar pons (Figs. 1 and 2).

The medulla oblongata is the rostral continuation of the spinal cord at the level of the foramen magnum of the skull. It extends forward from the cervical cord to the caudal border of the pons. From its ventral surface along a shallow longitudinal depression just lateral to the region known as the medullary pyramids, many small rootlets of the hypoglossal nerve (cranial nerve XII) emerge (Fig. 2). From its ventrolateral surface, just dorsal to a prominent oval swelling called the olive, a linear series of rootlets belonging to the glossopharyngeal (cranial nerve IX), the vagus (cranial nerve X), and the spinal accessory (cranial nerve XI) nerves emerge from the medulla (Figs. 1 and 2).

The oculomotor, trochlear, and abducens nerves lack sensory components. Together they innervate the six extrinsic or extraocular muscles that are responsible for eye movements. [*See* Eye Movements.]

II. OCULOMOTOR NERVE (CRANIAL NERVE III)

The oculomotor is the largest of the three cranial nerves supplying the extrinsic muscles of the eye. Efferent fibers of this nerve arise from the oculomotor nuclear complex situated on either side of the ventral midline within the periaqueductal gray matter of the rostral midbrain (Fig. 3). This nucleus is a complex of cell groups that provide motor fibers to four of the extraocular muscles (superior rectus, inferior rectus, medial rectus, and inferior oblique) as well as to the levator palpebrae superioris muscle. The medial rectus muscle rotates the eye medially and the superior and inferior recti muscles direct gaze upward and downward, respectively, and medially rotate the eye. The inferior oblique muscle rotates the eye upward and laterally and the levator palpebrae superioris muscle raises the upper eyelid.

The oculomotor nuclear complex also includes an additional small component, called the accessory oculomotor (or Edinger-Westphal) nucleus, in which cell bodies of fibers belonging to the parasympathetic division of the ANS reside. Preganglionic parasympathetic fibers originate from the Edinger-Westphal nucleus, are distributed with the oculomotor nerve, and synapse in the ciliary ganglion located lateral to the optic nerve within the orbit. The short postganglionic parasympathetic fibers arising from the ciliary ganglion then supply the intrinsic smooth muscle of the iris (sphincter pupillae muscle) and the ciliary muscle of the eye. Upon contraction, the former constricts the pupil, reducing the amount of light passing to the retina. The latter alters the curvature of the lens of the eye and thus its refractive power as required to focus objects on the retina for near vision—part of a reflex known as accommodation. [*See* Autonomic Nervous System.]

Fibers of the oculomotor nerve course ventrally through the midbrain from their origin and emerge from the interpeduncular fossa medial to the cerebral peduncles (Fig. 2). They then pass laterally and forward through the lateral wall of the cavernous sinus

to enter the orbit through the superior orbital fissure. [*See* Meninges.]

A. Oculomotor Nerve Lesions

Lesions can involve the nucleus of the nerve itself, its central connections, and/or the nerve in its course through the midbrain or externally en route to the eye; each lesion with distinct clinical manifestations. Lesions that affect the oculomotor nerve can be of vascular, inflammatory, traumatic, or compressive origin and they may be complete or incomplete. A complete lesion of the nerve involving the nucleus itself or interruption of its nerve fibers results in (1) the inability to move the eye on the affected side up, down, or inward; (2) a lateral deviation (or external strabismus) of the affected eye because of the unopposed action of the superior oblique and lateral rectus muscles that are innervated by cranial nerves IV and VI, respectively; (3) ptosis (drooping) of the upper eye lid due to paralysis of the levator palpebrae superioris muscle; (4) dilation of the pupil (mydriasis) on the affected side due to denervation of the sphincter pupillae muscle and the consequent unopposed action of the sympathetically innervated dilator muscle of the pupil; (5) loss of the accommodation and pupillary light reflexes; and (6) double vision (diplopia). Paralysis of an extraocular muscle disturbs the alignment of the eyes and results in double vision. This is a characteristic feature with interruption of any of cranial nerves III, IV, or VI.

III. TROCHLEAR NERVE (CRANIAL NERVE IV)

The trochlear (Latin, "pulley-like") nerve arises from the trochlear nucleus, which is located in the ventral midline of the periaqueductal gray matter at the level of the inferior colliculi (Fig. 3). As the axons arise from the motor neurons of this nucleus they pass toward the dorsal surface of the midbrain, where they cross to the opposite side before emerging from the dorsal midbrain immediately behind the inferior colliculi. The trochlear nerve is the slenderest of the cranial nerves and the only one to emerge from the dorsal surface of the brain stem.

After emerging from the dorsal surface, the trochlear nerve then passes around the midbrain to a ventrolateral position and then forward within the lateral wall of the cavernous sinus to enter the orbit via the superior orbital fissure. [*See* Meninges.]

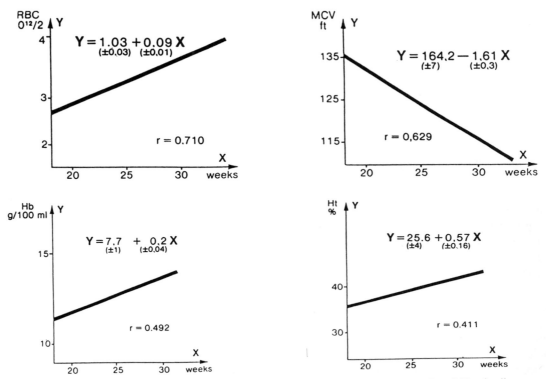

FIGURE 2 Evolution of some hematological parameters throughout the pregnancy. RBC, red blood cell count; MCV, mean corpuscular volume; Hb, hemoglobin; Ht, hematocrit.

cental passage is made beforehand. Figure 4 shows that fetal creatininemia is of no diagnostic or prognostic interest, because fetal and maternal creatininemias are always evenly balanced, whereas no correlation exists between the lactic dehydrogenase (LDH) activities of the mother and her fetus; therefore, this parameter can be considered as a reliable reflection of the fetus's condition.

IV. FETAL THERAPY AND PRENATAL PHARMACOLOGY

The fetus can be treated either through its mother or directly by intravenous infusion. In either case, FBS is essential. When the fetus is treated through its mother, repeated fetal sampling makes it possible to check the effectiveness of the treatment on the fetus.

Injecting drugs directly into the fetal vascular compartment is also possible. At present, the most frequent intravenous therapies are transfusion of red blood cell concentrates in cases of Rhesus immuniza-

tion and transfusions of platelet concentrates in cases of PLA_1 alloimmunizations. By evaluating the amount of red blood cells in the fetal blood before and at the end of transfusion, we can almost perfectly monitor the exact count and frequency of transfusions to avoid severe fetal anemia, which might entail irreversible damage.

Umbilical Vein

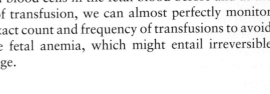

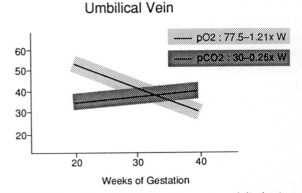

FIGURE 3 Evolution of fetal blood gases in umbilical vein according to the stage of the pregnancy.

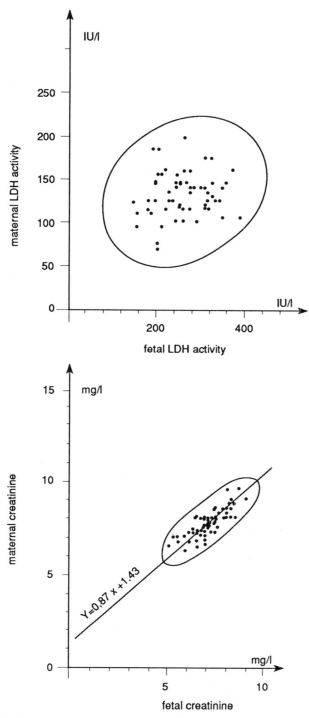

FIGURE 4 Spearman coefficient between maternal and fetal values. Lactic dehydrogenase does not cross the placenta, whereas creatinine does.

The possibility of having simultaneous access to either side of the placental barrier, that is, to fetal and maternal blood, has also opened exceptional prospects in prenatal pharmacology, provided these

TABLE VI

Transplacental Passage of Vitamin K_1 during 2nd and 3rd Trimesters of Pregnancy after Maternal Supplementation with Vitamin K^a

	Not supplemented (pg/ml)	Supplemented (pg/ml)
2nd trimester		
Mother	565	81,130
Fetus	30	765
Birth		
Mother	395	45,190
Fetus	21	783

^{a}Supplemented daily with 20 mg 3–7 days before sampling.

studies are carried out in accordance with appropriate ethical conditions. The study of the transplacental passages of vitamin K_1 is a suitable example.

Two questions arise in matters of prenatal pharmacology, and the answers to such questions are not necessarily univocal:

- Does the drug permeate through the placental barrier?
- Does it have some effect on the physiology of the fetus?

Tables VI and VII show that if vitamin K_1 easily crosses through the placental barrier, it has no effect on the synthesis of the coagulation factors that depend on vitamin K_1; they remain very low in the fetus, not because of a lack of vitamin K_1, as was formally believed, but because of the immaturity of the fetal synthesis in the liver.

TABLE VII

Prothrombin Activity^a in Fetal Plasma during 2nd and 3rd Trimesters of Pregnancy According to Maternal Supplementation

	Not supplemented	Supplemented
2nd trimester	16.5	16.9
Birth	49.5	47.8

^{a}Expressed as percentage of that in normal adult plasma.

oid muscles, and the posterior belly of the digastric muscle.

The facial nerve also has a parasympathetic component that arises from neuronal cell bodies in the superior salivatory and lacrimal nuclei, located medial to the facial motor nucleus in the caudal pons (Fig. 3). Preganglionic parasympathetic fibers from these nuclei travel within the nervus intermedius and the chorda tympani nerve to the pterygopalatine and submandibular ganglia suspended from the maxillary (V$_2$ division) and lingual (V$_3$ division) nerves, respectively. Postganglionic fibers originating from the pterygopalatine ganglion innervate the lacrimal glands and mucous glands located in the nasal and oral cavities. Those from the submandibular ganglion are distributed to the submandibular and sublingual salivary glands via the chorda tympani nerve, where they stimulate secretion.

A. Facial Nerve Lesions

Lesions of the central or peripheral divisions of the facial nerve result in various sensory and motor deficits depending on the location and extent of the lesion. The most obvious deficits encountered are paresis or paralysis of the muscles of facial expression and loss of taste sensation from the anterior two-thirds of the tongue. The former is the principal feature of Bell's palsy, a condition of unknown cause that affects the nerve within its course through the temporal bone.

If a lesion of the nerve occurs at the point where the motor root emerges from the bony stylomastoid foramen, as sometimes occurs with lacerations of the neck, there would be paralysis and atrophy of all the ipsilateral muscles of facial expression. Such an individual would show facial distortion: drooping of the eyebrow and corner of the mouth, widening of the palpebral fissure, and flattening of the nasolabial and forehead folds. On the affected side, the eye would not close because of denervation of the orbicularis oculi muscle, and thus the corneal reflex would also be absent even though sensation from the cornea indicative of an intact ophthalmic nerve would still be felt.

If the lesion were to occur within the facial canal of the temporal bone, distal to the geniculate ganglion but proximal to the stylomastoid foramen, not only would all motor function be lost as just described but the sensation of taste from the anterior two-thirds of the ipsilateral tongue would also be lost. In addition, there would be impairment of submandibular and

sublingual salivary secretion and an excessive sensitivity to sound (hyperacusis) on the affected side because of a loss of the dampening effect of the stapedius muscle on the middle ear ossicles. With lesions of the facial nerve proximal to the geniculate ganglion, a loss of the ability to produce tears ipsilaterally would also be added to the constellation of motor and sensory deficits already enumerated here.

It should be noted that the motor nucleus of the facial nerve supplying the muscles of the upper face receives input from both cerebral hemispheres, whereas the part that innervates the lower facial muscles receives innervation from the contralateral cortex only. Therefore, a unilateral lesion involving motor neurons (upper) that project to the facial motor nucleus does not affect the muscles of the upper face, but results in weakness/paralysis of the muscles, on the contralateral lower half of the face only. Although voluntary movement is lost, the lower facial musculature usually remains responsive to emotional stimulation. Spontaneous laughing or crying can still occur, indicating that emotional and voluntary inputs to the nucleus follow different pathways.

VII. VESTIBULOCOCHLEAR (STATOACOUSTIC) NERVE (CRANIAL NERVE VIII)

The vestibulocochlear nerve, also known as the statoacoustic nerve, emerges from the brain stem along the transverse sulcus between the pons and medulla just lateral to the facial nerve (Figs. 1 and 2). It is an entirely sensory nerve that consists of two distinct components, the vestibular and the cochlear nerves, which both convey information from the inner ear.

The cell bodies of vestibular nerve neurons are situated in the vestibular ganglion (of Scarpa) within the internal acoustic meatus. The peripheral processes of these neurons end around the hair cells of the sensory receptors called the macula of the saccule and utricle and the cristae of the three semicircular canals. This vestibular labyrinth is stimulated by linear and angular acceleration and gravitational forces all of which are concerned with the position of the head and body in space. The central processes of these afferents enter the brain stem and terminate in the vestibular nuclei located beneath the lateral floor of the fourth ventricle at the junction of the pons and the medulla oblongata (Fig. 3). A small number of vestibular afferent fibers concerned with maintaining balance and equilibrium pass uninterrupted directly to the cortex of the flocculonodular lobe of the cerebellum.

The vestibular nuclear complex consists of four nuclei from which a number of secondary pathways originate and project to the cerebellum, to motor nuclei of cranial nerves, and to the spinal cord. These connections provide for the adjustment of muscle tone as required to support the head in various positions, to maintain posture and balance, for conjugate eye movements, and to coordinate head and eye movements to maintain gaze. Vestibular information also reaches areas of the cerebral cortex, where spatial orientation is consciously perceived.

The cell bodies of cochlear nerve neurons are located in the cochlear (or spiral) ganglion within the bony modiolus of the cochlea (inner ear). Distal processes of these ganglion cells terminate around the base of the sensory hair cells of the organ of Corti. Central processes of the ganglion cells traverse the internal auditory meatus, enter the brain stem, and terminate in the dorsal and ventral cochlear nuclei. These nuclei are located on the nerve at the root of the inferior cerebellar peduncle (Fig. 3). Second-order ascending fibers leave the cochlear nuclei to form the auditory pathway called the lateral lemniscus, which ascends in the brain stem. This is a bilateral multisynaptic pathway that relays in the inferior colliculus and medial geniculate nucleus before ultimately reaching the auditory cortex (Heschl's convolutions) in the temporal lobe of the brain, where sound is consciously appreciated. In addition to this central pathway, fibers from the cochlear nuclei also project to various brain stem centers, including motor nuclei of cranial nerves for the purpose of mediating reflex responses to auditory stimuli (startle, rapid head/eye movement in response to loud sudden sound, or dampening of the inner ear ossicle/tympanic membrane movement in response to loud sounds, etc.).

A. Vestibulocochlear Nerve Lesions

The signs of dysfunction of the vestibular system include vertigo, imbalance, and spontaneous abnormal eye movements known as nystagmus. Nystagmus is a rhythmic conjugate movement of the eyes characterized by a slow movement in one direction and a fast movement in the opposite direction.

Lesions of the cochlear ganglia and/or nerve before it enters the brain stem will result in complete deafness in the corresponding ear. The significant bilaterality of the auditory system within the brain stem, however, ensures that damage to the central pathway on one side produces only a partial loss of hearing in both ears rather than a total loss in either. The hearing loss, however, is greatest in the ear opposite the lesion.

Because of their proximity, disease may affect both the vestibular and auditory components of the nerve. Meniere's disease is a condition of unknown cause that is characterized by bouts of vertigo, ringing in the ears, and progressive deafness. Nausea and vomiting are also frequently accompanying symptoms.

VIII. GLOSSOPHARYNGEAL NERVE (CRANIAL NERVE IX)

The glossopharyngeal, vagus, and spinal accessory nerves can be conveniently grouped together because they share some brain stem nuclei. They (except the spinal divisions of cranial nerve XI) emerge from the medulla along a horizontal sulcus (retro-olivary) dorsal to the olive, and they all exit the skull together through the jugular foramen.

The glossopharyngeal (Greek, pertaining to the tongue and pharynx) is primarily a sensory nerve conveying information from the tongue and pharynx as the name suggests, although it also has a small motor component. It emerges as a series of four to five rootlets from the rostral part of the retro-olivary sulcus of the medulla (Fig. 2). The cell bodies of the afferent fiber components of the nerve are located in two ganglia along the nerve: a small superior ganglion containing cell bodies of the GSA fibers and an inferior (petrosal) ganglion containing the cell bodies of GVA and special visceral afferent fibers.

The glossopharyngeal provides the sensory innervation to the taste buds on the posterior third of the tongue. The cell bodies of these SVA neurons are located in the inferior ganglion of nerve IX; their central processes enter the medulla and terminate in the rostral part of the nucleus of the tractus solitarius (Fig. 3) of the medulla in company with taste fibers from the facial and vagus nerves.

General sensory information (pain, temperature, and touch) from the external ear, a small area of skin behind the ear, the acoustic meatus, and tympanic membrane is also conveyed via a branch of the glossopharyngeal nerve. The central processes of these neurons located in the superior ganglion of the nerve are believed to terminate within the spinal nucleus of the trigeminal nerve along with fibers from the same location conveyed via the facial and vagus nerves.

Once infectious with HIV, it usually takes at least 22 days before enough antibody is formed to be detected. The HIV antigen test (see below) detects evidence of HIV-1 infection at 16 days following infectivity with the virus. It is during this 16-day "window" period that HIV-infected blood donations can slip through the safety net and be transfused to a patient. In addition, errors in specimen handling, error in testing, or results reporting may allow an unacceptable unit of blood to be used for transfusion. At the time of this writing, the risk per unit of contracting HIV from blood transfusion was about 1 in 750,000 compared to a risk estimate in 1989 of 1 in 38,000 to 1 in 150,000.

B. HIV Antigen Testing

The testing of donor blood for HIV-related proteins, such as an antigen designated HIV p24, shortens the "window" period between infectivity and a positive test for infection to about 16 days.

In March of 1996, the FDA licensed the use of donor screening tests for the HIV p24 antigen.

C. HTLV-I Antibody Test

This antibody test is used by blood banks to detect exposure to another retrovirus known as HTLV-1. Rarely, infection with HTLV-1 may result in leukemia or neurologic disease. It is estimated that approximately 1% of HTLV-1-infected individuals develop leukemia following a 20- to 30-year latency period. With the advent of the HTLV-1 antibody test, the per unit risk of contracting HTLV-1 from blood transfusion is about 1 in 50,000 to 1 in 70,000.

D. Hepatitis Tests

Viral hepatitis is the most frequent transfusion-transmitted infectious disease. Because the hepatitis A virus is rarely transmitted by blood transfusion (fewer than five cases have been reported since 1980), blood banks do not test for this virus. Hepatitis B (HBV) and hepatitis C (HCV) viruses both cause transfusion-transmitted disease, with most cases being caused by HCV (formerly called non-A, non-B hepatitis agent).

Hepatitis B virus infection can cause a serious disease that is occasionally fatal. The per unit risk is about 1 in 100,000. Although 90% of adults infected with this virus recover completely, almost 10% remain chronically infected. Individuals with chronic HBV infection lasting 20–30 years are at increased risk of developing cirrhosis and liver cancer (hepato-

cellular carcinoma). A child infected with hepatitis B has a 90% chance of becoming a chronic carrier of HBV and is likely to suffer the sequelae of this disease, including chronic active hepatitis, cirrhosis, and hepatocellular carcinoma. The testing of donated blood by the HBsAg test and anti-HBc test helps prevent posttransfusion HBV infection.

Hepatitis C virus infection accounts for the vast majority of posttransfusion hepatitis cases. The per unit risk is about 1 in 100,000. This virus is associated with a chronic carrier state in approximately 50% of infected patients. Furthermore, 10% of these patients develop cirrhosis, and deaths secondary to this virus have been reported. The testing of donated blood with the anti-HCV test helps prevent posttransfusion HCV infection.

E. Serologic Test for Syphilis

Transfusion-transmitted syphilis is extremely rare, partly because a serologic test for syphilis is routinely used by blood bank laboratories to screen donated blood and partly because the spirochete which causes syphilis (*Treponema pallidum*) dies when cooled to 4°C (the temperature at which blood is stored).

F. Cytomegalovirus Antibody Testing

Symptomatic cytomegalovirus virus infection is an unusual transfusion risk when cellular blood products are administered to individuals who have a normal immune system. However, some patients (such as very low birth weight neonates and immunocompromised adults) may suffer severe complications if infected by CMV following a blood transfusion. Consequently, blood banks do not routinely test all blood donors for exposure to cytomegalovirus (CMV). Rather, when a patient at high risk of severe CMV infection needs a blood transfusion, donated blood is tested to select CMV antibody negative units or, if the antibody status of the blood product cannot be determined, the blood products may be depleted of white blood cells, which removes CMV virus. CMV resides within white blood cells, and the removal of white blood cells from blood products renders those products essentially free of CMV infection risk. Patients who benefit from CMV risk-free transfusions include neonates weighing less than 1200 g at birth, pregnant women who are CMV negative, HIV-infected individuals who are CMV antibody negative, and CMV-negative transplant patients who receive organs or tissues from CMV-negative donors. CMV-negative blood products are of no proven benefit once a patient is known to be CMV antibody positive.

G. Alanine Aminotransferase Testing

ALT testing is a surrogate or indirect test for infection with a hepatitis-causing virus. This test used to be performed on all donated blood products, but was no longer required when data showed that specific tests for hepatitis virus infection were adequate in preventing posttransfusion hepatitis.

II. DONOR HISTORY AS A SCREENING TEST

Since it is not possible to eliminate the risk of infectious disease transmission from blood transfusion by testing alone, attempts are made to increase the safety of transfusion by using a careful predonation history to weed out unacceptable donors, by efforts to recruit volunteer rather than paid blood donors, and by avoiding improper donor incentives and coercion, which could alter the truthfulness of some donors.

Each individual blood donor is required to read information about blood safety and is encouraged to leave, without explanation, if he or she recognizes that giving blood would be inappropriate. Potential donors are also asked a series of questions about their health and life-style (including direct questions on sexual behavior designed to identify high-risk activities) and undergo a miniphysical before being allowed to donate. The questions and examinations are designed to prevent individuals who are at high risk for HIV, hepatitis, and other infectious diseases from donating blood. This process is continually refined in order to ensure that blood is drawn from the most appropriate individuals.

Blood donors may be offered a confidential opportunity to exclude their blood from use in transfusion by attaching stickers to the paperwork identifying the collected unit for use or withdrawal. If a donor knows of any reason why his or her blood should not be used for transfusion, he or she places the sticker indicating that the unit should not be transfused on the label. This is done to ensure that no pressure is exerted on the donor to give blood.

Every donation is checked against existing records. If a donor was indefinitely deferred, the collected unit is withdrawn from circulation and potential use. This process acts as a barrier to prevent the release of any

blood from a donor who was previously judged to be indefinitely unacceptable. When a specific test for a disease does not exist, a surrogate test for a similar disease entity is sometimes used as an alternative screening method.

III. TESTING IN 1997 AND BEYOND

A. Babesiosis

Babesiosis is a malaria-like illness caused by *Babesia microti*. The organism is an intraerythrocytic parasite transmitted by the tick *Ixodes dammini*. A few cases of babesiosis transmitted by blood have been reported. At this time the blood supply is not tested for babesiosis.

B. Chagas' Disease

Chagas' disease is caused by a parasite (*Trypanosoma cruzi*) endemic in parts of Mexico, and South and Central American countries. Some authorities believe that transfusion-transmitted Chagas' disease could become a problem in certain areas within the United States in which there are large number of immigrants from endemic regions. At this time, reliable tests for this blood parasite have been developed, but none of these have been licensed by the FDA for blood donor screening.

C. Hepatitis G Viruses

A new hepatitis virus has recently been described. Research on this virus is ongoing, including what risk, if any, the virus presents to the U.S. blood supply.

BIBLIOGRAPHY

Busch, M. P., and Alter, H. J. (1995). Will human immunodeficiency virus p24 antigen screening increase the safety of the blood supply and, if so, at what cost? *Transfusion* 35(7), 536–539.

Busch, M. P., Lee, L. L. L., Satten, G. A., *et al.* (1995). Time course of detection of viral and serologic markers preceding human immunodeficiency virus type 1 seroconversion: Implications for screening of blood and tissues donors. *Transfusion* 35(2), 91–97.

Cumming, P. D., Wallace, E. L., Schorr, J. B., and Dodd, R. Y. (1989). Exposure of patients to human immunodeficiency virus through the transfusion of blood components that test antibody-negative. *N. Engl. J. Med.* 321, 941–946.

Dodd, R. Y. (1992). The risk of transfusion transmitted infection. *N. Engl. J. Med.* 327, 419–420.

FDA memorandum (1995). Recommendations on donor screening with a licensed test for HIV antigen.

innervates both the intrinsic and extrinsic (styloglossus, hyoglossus, and genioglossus) muscles of the tongue.

A. Lesions of the Hypoglossal Nerve

Lesions of the hypoglossal nucleus or interruption of the nerve fibers, as may occur with traumatic injuries, tumors, inflammation, vascular disorders, or degenerative diseases, results in paralysis of the muscles on the ipsilateral half of the tongue. The muscles on the affected side undergo fasciculation and become flaccid and atrophic. With the loss of all voluntary and reflex movement, some disturbances in speech, chewing, and swallowing may be seen. Attempts to protrude the tongue result in its deviation toward the paralyzed side because of the unopposed force of the intact genioglossus muscle that pulls the tongue forward on the opposite side.

In contrast, lesions of the nervous system above the level of the hypoglossal nucleus (supranuclear) are distinguished by paralysis or weakness of the contralateral tongue and its deviation to the side opposite the lesion. In addition, atrophy and fasciculation are absent and reflex tongue movements are preserved.

BIBLIOGRAPHY

Brodal, A. (1965). "The Cranial Nerves. Anatomy and Anatomico-Clinical Correlations," 2nd Ed. Blackwell Scientific, Oxford, England.

Carpenter, M. B. (1991). "Core Text of Neuroanatomy," 4th Ed. William & Wilkins, Baltimore.

Montemurro, D. G., and Bruni, J. E. (1988). "The Human Brain in Dissection," 2nd Ed. Oxford Univ. Press, New York.

Ross, R. T. (1985). "How to Examine the Nervous System," 2nd Ed. Medical Examination Publishing Co., New York.

Wilson-Pauwels, L., Akesson, E. J., and Stewart, P. A. (1988). "Cranial Nerves. Anatomy and Clinical Comments." Decker, Toronto.

Craniofacial Growth, Genetics

LUCI ANN P. KOHN

Washington University School of Medicine

GLOSSARY

Craniofacial complex Pertaining to structures of the head, including the face and braincase

Finite element methods Method from continuum mechanics, recently applied to morphology, that allows the estimation of form change local to landmarks on the skull, as the deformation necessary to produce a resultant form from an initial form

Form change Sum of the size and shape change necessary to explain the difference between two forms

Genotype Collection of genes possessed by an individual

Growth Result of form change between two ages; it can be summarized by the sum of the size and shape changes necessary to develop a resultant form from an original form

Landmarks Homologous anatomical structure that can be reliably identified in all individuals of a sample

Phenotype Observable properties of a trait, produced by a combination of the genetic make-up of the individual, as well as nongenetic factors (environment)

Roentgenographic cephalometry Method of estimating form change from X rays in which an individual's head is in a standardized position and a constant distance from the X-ray film and source; size change is estimated by the differences between linear measurements between landmarks on two or more forms, whereas shape change is estimated as the difference (or change in angle) between intersecting lines through landmarks

THE HEAD IS AN IMMENSELY COMPLEX STRUCTURE that performs numerous diverse functions. The brain and its closely associated special sense organs (i.e., the eyes, ears, and olfactory region) receive and interpret information from the outside world. Mastication, respiration, and vocalization are performed by the oral, nasal, pharyngeal, and laryngeal portions of the head. The factors that determine normal craniofacial growth are of interest to individuals in a number of scientific disciplines. The orthodontist and plastic surgeon are concerned with recognition and treatment of abnormal dental and facial growth patterns. The anatomist and the biological anthropologist are interested in normal human variation, inferences on human evolution, and further understanding of the underlying "mechanisms" of growth events. A geneticist is interested in the inheritance of craniofacial form for purposes of developmental and evolutionary studies and for genetic counseling.

I. PRENATAL FORMATION OF THE CRANIOFACIAL COMPLEX

The bony head and face begin to form during the embryonic period, at approximately 22 days after conception, after the formation of other neural tissues such as the brain, eyes, and peripheral nerves. The tissues that will develop into the bones and cartilages of the craniofacial complex are from two tissue sources: mesenchyme derived from neural crest cells, and mesenchyme from the cranial somites. Neural crest mesenchyme covers the head region and invades the branchial arches, with the first two branchial arches contributing to the face. Many of the neural crest cells have their origins distant to the face and must migrate into their proper location and merge with the comparable bilateral structures. Imperfection

ENCYCLOPEDIA OF HUMAN BIOLOGY, Second Edition, VOLUME 3.   Copyright © 1997 by Academic Press.   All rights of reproduction in any form reserved.

Puberty General term covering the period of time of the rapid growth of the adolescent growth spurt and development of the secondary sex characteristics before menarche in girls

Testosterone Androgen produced by the testis. Testosterone regulates the development of male genitalia and male characteristics of the skeleton and muscular system

TOO LITTLE BODY FAT AND TOO MUCH BODY FAT are both associated with the disruption or delay of the reproductive ability of women. Evidence gathered from women with moderate or extreme weight loss caused by injudicious dieting, intensive exercise, or both indicates that this association is causal and that the large amount of body fat (26–28% of body weight) stored by the human female at maturity influences reproduction directly.

Many young girls who diet excessively or who are well-trained athletes or dancers have a delayed menarche (the first menstrual cycle). Menarche may be delayed until as late as 19 or 20 years. If their athletic training begins after menarche, girls may have anovulatory menstrual cycles, irregular cycles, or a complete absence of cycles (secondary amenorrhea).

In addition to these extreme effects of weight loss and athletic training on the menstrual cycle, women who train moderately or who are regaining weight into the normal range may have a menstrual cycle that appears to be normal, but that actually has a shortened luteal phase or is anovulatory.

All these disruptions of reproductive ability are usually reversible after varying periods of time after weight gain, decreased athletic training, or both.

Secondary amenorrhea occurs in dieting girls and women or in athletes and dancers when weight loss is 10–15% of normal weight for height, which is equivalent to a loss of about one-third of body fat. Primary amenorrhea (absence of menarche at age 16 or older) also occurs in association with excessive thinness. These data suggest that a minimum level of body fat (i.e., stored, easily mobilized energy), in relation to the lean body mass, is necessary for the onset and maintenance of regular ovulatory menstrual cycles. Both the absolute and the relative amounts of fat are important because lean mass and fat must be in a particular absolute range, as well as a relative range (i.e., the individual must be big enough to reproduce successfully).

Data on obese women show that excessive fatness is also associated with infertility, as was described for cattle and mares almost a century ago. Loss of weight restores fertility of the women and of the animals. Too little or too much fat are thus both associated with infertility.

I. WHY FAT?

Reproduction costs calories; a pregnancy requires about 50,000 calories above normal metabolic requirements, and lactation requires about 1000 calories a day. In premodern times, lactation was an essential part of reproduction; storage of fat when food supplies are uncertain, as they were in our prehistory, would therefore be of selective advantage to the female.

Whatever the reason, during the adolescent growth spurt that precedes menarche, girls increase their body fat by 120% compared with a 44% increase in lean body mass. Changes in body composition can be monitored by direct measurements of body water. Because fat contains only about 10% water compared with about 80% water in muscle and viscera, an increase in fatness results in a decrease in body water as a percentage of total body weight. Direct measurements of body water of girls from birth to completion of growth show a continuous decline in the proportion of body water because of the large relative increase in body fat (Fig. 1). At menarche, girls average about 24% of their body weight as fat (11 kg, 24 lb). At the completion of growth, between ages 16 and 18

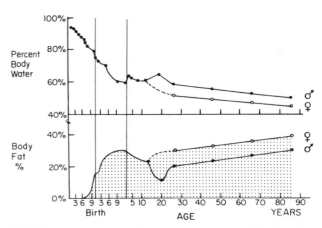

FIGURE 1 Changes in body water as percentage of body weight throughout the life span and corresponding changes in the percentage of body fat. [Adapted from Friis-Hansen, B. (1965). Hydrometry of growth and aging. *In* "Human Body Composition" (J. Brožek, ed.), Vol. VII, pp. 191–209. Pergamon Press, Oxford. Reprinted with permission.]

TABLE I

Total Body Water as Percentage of Body Weight, an Index of Fatness: Comparison of an 18-Year-Old Girl and a 15-Year-Old Boy of the Same Height and Weight

Variable	Girl	Boy
Height (cm)	165.0	165.0
Weight (kg)	57.0	57.0
Total body water (liter)	29.5	36.0
Lean body weight (kg)[a]	41.0	50.0
Fat (kg)	16.0	7.0
Fat/body weight (%)[b]	28.0	12.0
Total body water/body weight (%)	51.8	63.0

[a]Lean body weight = total body water/0.72.

[b]Fat/body weight % = 100 − [(total body water/body wt %)/0.72].

years, the body of a well-nourished young girl in the United States contains on average about 26–28% fat (16 kg, 35 lb) and about 52% water. In contrast, the body of a boy at the same height and weight contains about 12% fat and 63% water (Table I). At the completion of growth, men are about 15% fat and about 61% water. The main function of the 16 kg of stored female fat, which is equivalent to 144,000 calories, may be to provide energy for a pregnancy and for about 3 months lactation.

A. Is the Amenorrhea of Athletes and Underweight Girls Adaptive?

Infant survival is correlated with birth weight, and birth weight is correlated with the prepregnancy weight of the mother and, independently, with her weight gain during pregnancy. An underweight woman is therefore at high risk of an unsuccessful pregnancy. As Dr. J. M. Duncan observed more than a century ago, if a seriously undernourished woman could get pregnant, the chance of her giving birth to a viable infant, or herself surviving the pregnancy, is small. Therefore, the amenorrhea of underweight girls and women can be considered adaptive.

II. HOW ADIPOSE TISSUE MAY REGULATE FEMALE REPRODUCTION

There are at least four mechanisms by which body fat (adipose tissue) may directly affect ovulation and the menstrual cycle, hence fertility: (1) Conversion of androgen to estrogen takes place in adipose tissue of the breast and abdomen, the omentum, and the fatty marrow of the long bones. Adipose tissue therefore is a significant extragonadal source of estrogen. (2) Body weight, hence fatness, influences the direction of estrogen metabolism to the most potent or least potent forms. (3) In fatter girls and women, the capacity of serum sex hormone-binding globulin to bind estradiol is diminished, resulting in an elevated percentage of free serum estradiol. (4) Adipose tissue stores steroid hormones.

Changes in relative fatness might also affect reproductive ability indirectly through disturbance of the regulation of body temperature and energy balance by the hypothalamus. Lean amenorrheic women, both anorectic and nonanorectic, display abnormalities of temperature regulation at the same time that they have delayed response, or lack of response, to exogenous gonadotropin-releasing hormone (GnRH).

III. HYPOTHALAMIC DYSFUNCTION, GONADOTROPIN SECRETION, AND WEIGHT LOSS

It is now known that the amenorrhea of underweight and excessively lean women, including athletes, is due to hypothalamic dysfunction. Consistent with the view that this type of amenorrhea (hence infertility) is adaptive, the pituitary–ovarian axis is apparently intact and functions when exogenous GnRH is given. [See Hypothalamus.]

Girls and women with this type of hypothalamic amenorrhea have both quantitative and qualitative changes in the secretion of gonadotropins, luteinizing hormone (LH), and follicle-stimulating hormone (FSH): (1) LH and FSH are low as are estradiol levels. (2) The secretion of LH and the response to GnRH are reduced in direct correlation with the amount of weight loss. (3) Underweight patients respond to exogenous GnRH with a pattern of secretion similar to that of prepubertal children; the FSH response is greater than the LH response. The return of LH responsiveness is correlated with weight gain. (4) The maturity of the 24-hr LH secretory pattern and body weight are related; weight loss results in an age-inappropriate secretory pattern resembling that of prepubertal or early pubertal children. Weight gain restores the postmenarcheal secretory pattern.

IV. METHODS OF ANALYSES OF NORMAL HUMAN CRANIOFACIAL GROWTH AND MORPHOLOGY

Numerous analyses of normal human craniofacial morphology and growth have been conducted since the late 1930s and the introduction of standardized cranial X rays, or roentgenographic cephalograms (RCM). The X rays are recorded with the head in a standard position a fixed distance from the X-ray anode and the X-ray film. Bony landmarks are then located on the cephalograms, and their locations are recorded. Figure 1 presents a cephalogram and five commonly identified landmarks whose definitions are found in Table II. With our recent respect for decreased exposure to radiation, additional collection by these methods is unlikely.

Growth studies may be performed on either cross-sectional or longitudinal data. In cross-sectional growth studies, individuals are observed only once. Cross-sectional growth studies can provide information on the average size of individuals at a given age or, in a sample of related individuals, the influence of genetic parameters on craniofacial morphology.

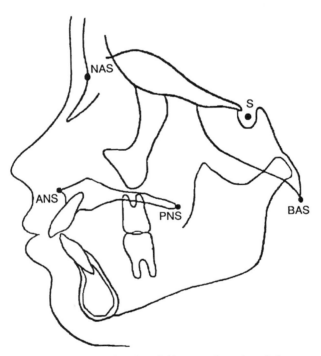

FIGURE 1 Landmarks identifiable on a lateral cephalogram. ANS, anterior nasal spine; PNS, posterior nasal spine; S, sella; NAS, nasion; BAS, basion. Landmark locations are defined in Table I. [From D. H. Enlow (1990). "Handbook of Facial Growth." Saunders, Philadelphia.]

TABLE II
Landmark Names, Abbreviations, and Definitions

Landmark name	Abbreviation	Definition
Anterior nasal spine	ANS	Tip of anterior nasal spine on maxilla
Posterior nasal spine	PNS	Tip of posterior nasal spine on palatine
Sella	S	Midpoint of sella turcica
Nasion	N	Most anterior point on nasofrontal suture
Basion	B	Median point of anterior margin of foramen magnum

Although large samples of cross-sectional data are relatively easy to collect, individual patterns of variation are obscured. Such studies concentrate on the influence of genetic parameters on craniofacial morphology and the results of craniofacial growth. However, these studies ignore the developmental mechanisms that produce the observed patterns. The heterogeneous age structure of cross-sectional family studies is usually handled in one of two ways. These analyses are generally limited to adults, or individuals of various ages are included but are statistically adjusted to a single age before analysis. Both of these methods control for the effect of age on craniofacial morphology.

Alternatively, participants are observed repeatedly during an extended period of time in longitudinal growth studies. This provides the advantage of information on individual variation and the change of variation with age. However, particularly with human studies, such analyses take a great deal of time and money and are greatly dependent on participant loyalty to the project. Therefore longitudinal analyses of human growth, especially those that include family members, are rare.

In traditional roentgenographic cephalometry, craniofacial morphology has been characterized by linear distances between landmarks, with size change estimated by the change in the linear distances with age. The angle formed by the intersection of these lines is identified as a measure of shape, which may also be estimated at different ages. It is assumed that little or no growth occurs at one landmark, usually sella, and that only linear growth occurs between two given landmarks, usually along the sella–nasion line. Measures are then expressed with reference to this landmark and line, a practice also known as registration.

Objections to RCM concern the representation of craniofacial growth and form by linear segments and angles, and the registration of cephalograms during analyses. The assumptions surrounding the registration of cephalograms have been questioned. There are no landmarks of the skull where growth does not occur. Therefore, no landmark, including sella, can be considered to be an origin or location of no change, because its growth forces a change in its relation to other landmarks in the craniofacial complex. In addition, growth along the sella–nasion line is not simply a linear increase, but rather bends during growth.

An additional problem occurs because of the estimation of growth as the change in linear measurement between two landmarks, which best estimates the response of the two landmarks to growth. However, many traditional measures involve landmarks on separate regions of the craniofacial complex, forcing the line segment to cross more than one region of growth and remodeling. RCM analyses do not allow us to distinguish where growth is occurring or the magnitude of form change local to each landmark.

By representing craniofacial growth and form as change in linear segments and angles between linear segments and by analyzing cephalograms within registered systems, RCM may not adequately model the biological processes of craniofacial growth, obscuring the growth processes or underlying genetic control of growth. RCM is not inappropriate for all models of craniofacial growth, as it may appropriately be used for the description of generalized differences between two forms at a single point in time.

Few alternative methods have been proposed for use in analyzing morphological change, and many of these have their own associated errors. One of the earliest proposals of morphometric techniques came from D'Arcy Thompson, who suggested that form change between two morphs could be modeled by the transformation of cartesian coordinates of one form, the initial form, into the target form. Forms were registered on all landmarks, and no landmark was left unaltered in the transformation. Unfortunately, the methods were incompletely described and difficult to quantify, and they remained inaccessible until the methods of biorthogonal grids and finite element scaling were introduced.

Before finite element analyses, homologous landmarks are identified on all individuals. Lines connect landmarks at vertices to form finite elements, which may be triangular, quadrilateral, hexahedral, or other forms. A complex biological form may be divided into a number of elements, with each element representing

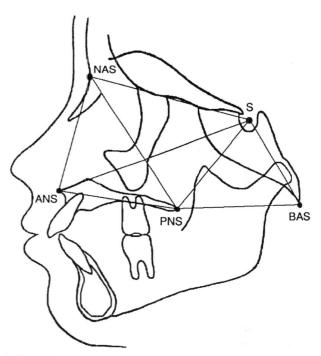

FIGURE 2 Triangular finite elements on a lateral cephalogram. ANS, anterior nasal spine; PNS, posterior nasal spine; S, sella; NAS, nasion; BAS, basion. Landmark locations are defined in Table I. [From D. H. Enlow (1990). "Handbook of Facial Growth." Saunders, Philadelphia.]

that part of the organism included within the bounded landmarks (Fig. 2). Registration occurs simultaneously on all landmarks, making the analyses free of registration bias. Growth is then estimated local to each landmark, as the size and shape change necessary to deform an individual at one age into a subsequent age. Size and shape are estimated independently by the finite element methods. Thus the estimates of form change are inherent to the biological form and can be localized.

V. GENETICS OF CRANIOFACIAL GROWTH

Although numerous studies have found that adult craniofacial morphology has a large genetic component to observed variation, only three analyses have focused on genetic influences on craniofacial growth. The earliest longitudinal genetic analysis was carried out on a sample of 12 monozygotic and 10 dizygotic twin pairs observed during a 10-year period. This analysis showed that only 4 of 15 linear and angular

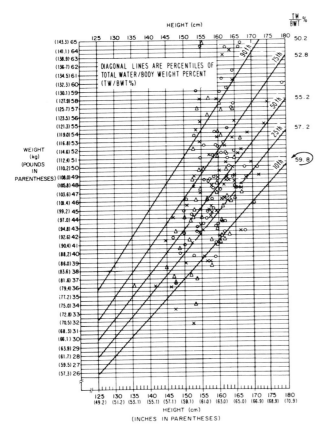

FIGURE 4 Minimal weight necessary for a particular height for *onset* of menstrual cycles is indicated on the weight scale by the 10th percentile diagonal line of total water/body weight percentage, 59.8%, as it crosses the vertical height lines. Height growth of girls must be completed or approaching completion. For example, a 15-year-old girl whose completed height is 160 cm (63 in.) should weigh at least 41.4 kg (91 lb) before menstrual cycles can be expected to start. [From Frisch, R. E., and McArthur, J. W. (1974). *Science* 185, 949–951.]

menarche. For example, a 15-year-old girl whose completed height is 160 cm (63 in.) should weigh at least 41.4 kg (91 lb) before menstrual cycles can be expected to start. The minimum weights for menarche would also be used for girls who become amenorrheic as a result of weight loss shortly after menarche, as is often found in cases of anorexia nervosa.

The weights at which menstrual cycles ceased or resumed in postmenarcheal women were 10% heavier than the minimal weights for the same height observed at menarche. In accord with this finding, the body composition data showed that both early and late maturing girls gained an additional 4.5 kg (10 lb) of fat from menarche to age 18 years. Almost all this gain was achieved by age 16 years, when mean fat is 15.7 kg (35 lb), 27% of body weight. At age 18 years,

mean fat is 16.0 kg, 28% of the mean body weight of 57 kg (125 lb). Reflecting this increased fatness, the total water/body weight percentage decreases from 55.1% at menarche to 52.1% at age 18 years.

The prediction of the minimum weights for height is from total water as percent of body weight, not fat/percent of body weight, indicating that the size of the lean mass is important in relation to the amount of fat (i.e., the prediction is based on a lean : fat ratio).

Other factors, such as emotional stress, affect the maintenance or onset of menstrual cycles. Therefore, menstrual cycles may cease without weight loss and may not resume in some subjects even though the minimum weight for height has been achieved. Also, these minimal weight standards apply thus far only to Caucasian women in the United States and Europe. Different races have different critical weights at menarche, and it is not yet known whether the different critical weights represent the same critical body composition of fatness.

Some amenorrheic athletes, such as shot-putters, oarswomen, and some swimmers, are not lightweight because they are very muscular; muscles are heavy (80% water) compared with fat (about 10% water). The cause of their amenorrhea is, nevertheless, most probably an increased lean mass and a reduced fat content of their bodies; gaining body fat or ceasing exercise usually restores menstrual cycles. In relation to athletic amenorrhea, it is important to note that body composition may change without any change in body weight. A woman may increase muscle mass by increasing training and at the same time lose fat, without a perceptible change in body weight. Direct measurements of body composition with magnetic resonance imaging showed that well-trained athletes had 30–40% less fat than comparable sedentary women, although the body weights of the two groups of women did not differ.

VI. PHYSICAL EXERCISE, DELAYED MENARCHE, AND AMENORRHEA

Does intense exercise cause delayed menarche of dancers and athletes or do late maturers choose to be dancers and athletes? In accord with many reports, the mean age of menarche of college swimmers and runners was significantly later (13.9 ± 0.3 years) than that of the general population (12.8 ± 0.05 years). However, the mean menarcheal age of the athletes whose training began *before* menarche was 15.1 ± 0.5 years, significantly later than the mean menarcheal age (12.8 ± 0.2 years) of the athletes whose training began after their menarche. The latter mean age was similar to that of the college controls (12.7 ± 0.4 years) and the general population. Therefore, training, not preselection, is the delaying factor. Each year of premenarcheal training delayed menarche by 5 months (0.4 year). The training also disrupted the regularity of the menstrual cycles in both pre- and postmenarcheal trained athletes. Athletes with irregular cycles or amenorrhea had hormonal levels confirming lack of ovulation, hence infertility.

VII. LONG-TERM REGULAR EXERCISE DECREASES THE RISK OF SEX HORMONE–SENSITIVE CANCERS

The amenorrhea and delayed menarche of the college athletes raised the question: Are there differences in the long-term reproductive health of moderately trained athletes compared with nonathletes?

A study of 5398 college alumnae aged 20–80 years, 2622 of whom were former athletes and 2776 nonathletes, showed that the former athletes had a significantly lower lifetime occurrence (prevalence rate) of breast cancer and cancers of the reproductive system (Fig. 5) compared with nonathletes; 82.4% of

the former college athletes began their training in high school or earlier compared with 24.9% of the nonathletes. The analysis controlled for potential confounding factors including age, age of menarche, age of first birth, smoking, and family history of cancer. The former college athletes were leaner in every age group compared with the nonathletes. [See Breast Cancer Biology.]

Although we can only speculate at present as to the reasons for the lower risk of sex hormone–sensitive cancers among the former athletes, the most likely explanation is that the former athletes had lower levels of estrogen because they were leaner and more of the estrogen was metabolized to a nonpotent form of estrogen (catechol estrogens). The former college athletes also had a lower prevalence of benign tumors of the breast and reproductive system and a lower prevalence of diabetes, particularly after age 40, compared with the nonathletes.

VIII. DOUBLE-MUSCLED CATTLE AND OTHER PERTINENT ANIMAL DATA

Carcass analyses of meat animals and of experimental animals provide important data unattainable from

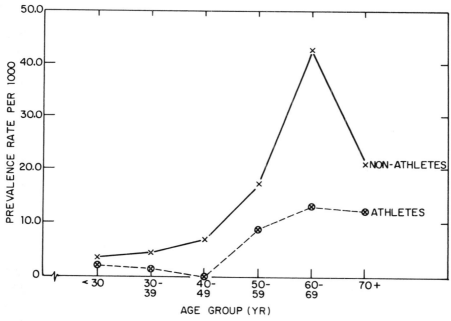

FIGURE 5 Prevalence rate (lifetime occurrence rate) of cancers of the reproductive system for athletes and nonathletes by age group. [From Frisch *et al.* (1985). *Br. J. Cancer* 52, 885.]

Byard, P. J., Lewis, A. B., Ohtsuki, F., Siervogel, R. M., and Roche, A. F. (1984). Sibling correlations for craniofacial measurements from serial radiographs. *J. Craniofac. Genet. Dev. Biol.* **4**, 265–269.

Cheverud, J. M., Lewis, J. L., Bachrach, W., and Lew, W. D. (1983). The measurement of form and variation in form: An application of three-dimensional quantitative morphology by finite-element methods. *Am. J. Phys. Anthropol.* **62**, 151–165.

Dudas, M., and Sassouni, V. (1973). The hereditary components of mandibular growth, a longitudinal twin study. *Angle Orthod.* **43**, 314–323.

Enlow, D. H. (1990). "Handbook of Facial Growth." Saunders, Philadelphia.

Falconer, D. S. (1981). "Introduction to Quantitative Genetics." Longman, New York.

Kohn, L. A. P. (1989). A genetic analysis of craniofacial growth using finite element methods. Ph.D. thesis, Department of Anthropology, University of Wisconsin, Madison.

Kohn, L. A. P. (1991). The role of genetics in craniofacial morphology and growth. *Annu. Rev. Anthropol.* **20**, 261–278.

Moore, W. J. (1981). "The Mammalian Skull." Cambridge Univ. Press, Cambridge, England.

Moss, M. L., and Salentijn, L. (1969). The primary role of functional matrices in facial growth. *Am. J. Orthod.* **55**, 566–577.

Moyers, R. E., and Bookstein, F. L. (1979). The inappropriateness of conventional cephalometrics. *Am. J. Orthod.* **75**, 599–617.

Nakata, M. (1985). Twins in craniofacial genetics: A review. *Acta Genet. Med. Gemello.* **34**, 1–14.

Richtsmeier, J. T., and Cheverud, J. M. (1986). Finite element scaling analysis of normal growth of the craniofacial complex. *J. Craniofac. Genet. Dev. Biol.* **6**, 289–323.

Richtsmeier, J. T., Cheverud, J. M., and Lele, S. (1992). Advances in anthropological morphometrics. *Annu. Rev. Anthropol.* **21**, 283–305.

Thompson, D. (1961). "On Growth and Form." Cambridge Univ. Press, Cambridge, England.

Van Limborgh, J. (1982). Factors controlling skeletal morphogenesis. *In* "Factors and Mechanisms Influencing Bone Growth" (A. D. Dixon and B. G. Sarnat, eds.), pp. 1–17. Liss, New York.

Craniofacial Growth, Postnatal

BERNARD G. SARNAT

University of California and Cedars–Sinai Medical Center, Los Angeles

GLOSSARY

Alveolar process Bone surrounding and supporting a tooth

Ectodermal dysplasia Clinical condition wherein a number of structures (e.g., teeth, sweat and sebaceous glands, and hair) of ectodermal origin are deficient

Endosteum Fibrous connective tissue lining the marrow cavities of bone

Periosteum Fibrous connective tissue covering the outer surfaces of bone

Ramus Posterior vertical part of the mandible. The superior portion, the condyle, articulates with the temporal bone (see Fig. 5)

Vital marker Substance that, when administered to a living animal, marks its tissues

THE BLUEPRINT OF A BONE IS INHERENT. POSTNAtal craniofacial growth is but a continuation of prenatal craniofacial growth interrupted by the event of birth. *In utero* the fetus is subjected to the vicissitudes of the maternal environment. After birth the individual is subjected to the effects of the general environment, thereby altering the external form and internal architecture of a bone or a complex of bones. Changes in craniofacial form are related to the synchronous, harmonious coordination of three-dimensional, multiple, differential, skeletal growth site activities, and associated structures.

I. BONE GROWTH

Although significant reports in regard to bone growth appeared in the literature 200 years ago, many ques-

tions are still unanswered. What are some of the problems in need of study? What are the inherent difficulties? Any determination of bone growth must concern itself with one or more of the following questions: What are the sites? Are they primary or secondary? What are the amounts? What are the rates? Do they vary? When? What are the directions? What are the changes in proportion? What factors are influential?

Growth of bone(s) occurs in three ways: cartilaginous, sutural, and appositional and resorptive (remodeling) (Table I). Growth is change with time. A basic physiological concept is that, throughout life, bone is in a continuous state of apposition and resorption. Consequently, skeletal size and shape are always subject to change. When skeletal mass increases, as in children, apposition is more active than resorption. Cartilaginous and sutural growth are active (i.e., positive growth). When skeletal mass is constant, as in the adult, apposition and resorption, although active, are in equilibrium (i.e., neutral growth). Cartilaginous and sutural growth have ceased. When the skeletal mass decreases, as in old age, resorption is more active than apposition (i.e., negative growth). [*See* Bone Density and Fragility, Age-Related Changes.]

The physiological stability of the bony components is the result of many interrelated factors, normal functional use being a prominent one. Well recognized are the effects of either excessive use, with hypertrophy (i.e., an increase in the mass of bone), or disuse, with atrophy (i.e., a decrease in the mass of bone). Thus, modifications in the functions of a part are reflected in alterations in the form of the part.

Bony tissue, despite its hard, semirigid, supporting, mineralized nature, and by virtue of the highly sensitive periosteal and endosteal membranes, is dynamic and ever-changing, adaptable to every nuance of tension and pressure. The basic and dual response of resorption and apposition is evident in the reaction

ENCYCLOPEDIA OF HUMAN BIOLOGY, Second Edition, VOLUME 3.

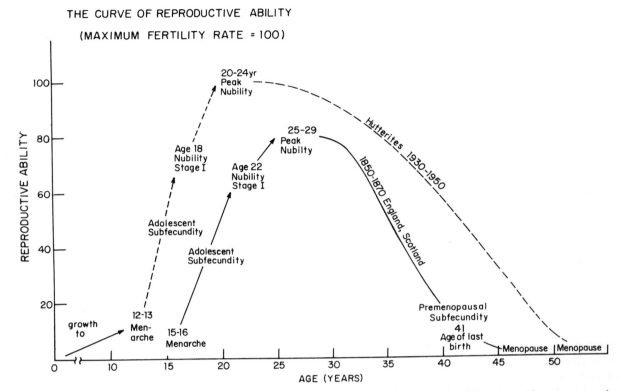

FIGURE 7   Mid-19th century curve of female reproductive ability (variation of the rate of child bearing with age) compared with that of the well-nourished, noncontracepting modern Hutterites. Whereas the Hutterite fertility curve results in an average of 10–12 children, the 1850–1870 fertility curve results in about 6 to 8 children. [From Frisch, R. E. (1978). *Science* **199**, 22–30.]

Taiwan, which accompanied recent improvements in health and nutrition. Therefore, the need for family planning programs, when there is the hoped-for progress in standards of living, may be much greater than realized heretofore during the transitional period to a desired lower family size.

BIBLIOGRAPHY

Friis-Hansen, B. (1965). Hydrometry of growth and aging. *In* "Human Body Composition" (J. Brožek, ed.), Vol. VII, pp. 191–209. Pergamon, Oxford.

Frisch, R. E. (1975). Demographic implication of the biological determinants of female fecundity. *Soc. Biol.* **22**, 17–22.

Frisch, R. E. (1978). Population, food intake and fertility. *Science* **199**, 22–30.

Frisch, R. E. (1985). Fatness, menarche and female fertility. *Perspect. Biol. Med.* **28**, 611–633.

Frisch, R. E. (1988). Fatness and fertility. *Sci. Am.* **258**, 88–95.

Frisch, R. E., and McArthur, J. W. (1974). Menstrual cycles: Fatness as a determinant of minimum weight for height necessary for their maintenance or onset. *Science* **185**, 949–951.

Frisch, R. E., Snow, R. C., Johnson, L. A., *et al.* (1993). Magnetic resonance imaging of overall and regional body fat, estrogen metabolism, and ovulation of athletes compared to controls. *J. Clin. Endo. Metab.* **77**, 471–477.

Frisch, R. E., Wyshak, G., and Vincent, L. (1980). Delayed menarche and amenorrhea of ballet dancers. *N. Engl. J. Med.* **303**, 17–19.

Siiteri, P. K. (1981). Extraglandular oestrogen formation and serum binding of estradiol: Relationship to cancer. *J. Endocrinol.* **89**, 119p–129p.

Vigersky, R. A., Anderson, A. E., Thompson, R. H., and Loriaux, D. L. (1977). Hypothalamic dysfunction in secondary amenorrhea associated with simple weight loss. *N. Engl. J. Med.* **297**, 1141–1145.

Wyshak, G., and Frisch, R. E. (1982). Evidence for a secular trend in age of menarche. *N. Engl. J. Med.* **306**, 1033–1035.

TABLE II
Approximate Information Obtained from Various Methods Used to Study Postnatal Bone Growth[a]

Method[b]	Growth[c]				Type of study	Limitations
	Site	Amount	Rate	Direction		
Direct measurements						
Osteometry						
Skeletal remains	0	X	X	X	Cross-sectional	Material of unknown history, posthumous distortion
Living	0	X	X	X	Longitudinal	Soft tissues restrict accurate measurement
Vital staining[2]	XXX	XXX	XX	X	Longitudinal	Toxicity, method requires refinement
Implant markers[1,2]	XX	XXX	X	X	Longitudinal	Local reaction to implants, requires re-operation
Histological and histochemical methods	XXXX	X	0	X	Cross-sectional	Sections show conditions at time of death of tissue only
Indirect measurements						
Impressions and casts	0	XX	XX	XX	Longitudinal	Soft tissues restrict accuracy of impression
Photographs	0	X	X	X	Longitudinal	Two-dimensional study of three-dimensional process
Serial radiography	0	XXX	XX	XX	Longitudinal	Must obtain stable landmarks, three-dimensional information not entirely accurate, radiation exposure
Measurements in combination						
Serial radiography and implantation[1,2]	XXX	XXX	XXX	XXX	Longitudinal	Three-dimensional information not entirely accurate, radiation exposure
Serial radiography and metaphyseal bands[3]	XXX	XXX	XXX	XXX	Longitudinal	Record of a toxic process, rate of growth not normal, radiation exposure
Radioautographs	XXX	0	0	0	Cross-sectional	Primarily of qualitative value

[a]Modified from B. G. Sarnat and B. J. Gans (1952). *Plast. Reconstr. Surg.* 9, 140–160, with permission.
[b]Can be used to measure: 1, sutural growth; 2, apposition and resorption; 3, endochondral growth.
[c]0, gives no information; X, shows trends; XX, relatively accurate; XXX, grossly accurate; XXXX, microscopically accurate.

calcifying substances might be obtained by a single intraperitoneal or intravenous injection of a 2% solution of alizarin red S (color index 1034, Coleman & Bell Company, Norwood, OH).

For microscopic studies, ground sections are prepared, since the staining effects are lost by the action of the acid used in preparing decalcified sections. Ground sections (25–50 μm thick) show sharp red lines that may be studied with a dissecting microscope under reflected light. Under higher magnification and strong transmitted illumination, the red lines (5–20 μm in width) are readily counted, and the distance between them can be accurately measured with a micrometer.

Injections of alizarin have been used to determine the rate of calcification of bone under both control and experimental conditions. They have also been used in studies of calcification in other conditions such as healing of fractures, kidney casts, calcified plaques of an atheromatous aorta, and calcified scars. The eggshell, the shell of the turtle, and the cenmentum and dentin of teeth are also stained by alizarin red S.

Other agents, such as the procion dyes, tetracycline, fluorochrome, lead acetate, trypan blue, and sodium fluoride, have been shown to be of value, all with advantages and disadvantages. One or more different vital markers have been used in the same animal.

2. Implants

Implants as reference markers have been utilized in the study of growth of bones (Table III). John Hunter inserted two pellets along the length of the shaft of the tarsus (a foot bone) of a young pig and measured

TABLE III

A Brief Historical Review of Implant Markers Used in the Longitudinal Study of Postnatal Bone Growth[a]

Investigator	Year	Material used	Bone studied	Animal
Gross (direct) studies				
Hales	1727	Holes	Tibia	Chicken
Duhamel	1742	Silver stylets	Long bone	Pigeon, dog
Hunter	1770	Lead shot	Tibia, tarsometatarsal	Pig, chicken
Humphry	1863	Wires	Mandible	Pig
Gudden	1874	Holes	Parietal, frontal	Rabbit
Wolff	1885	Metal	Frontal, nasal	Rabbit
Giblin and Alley	1942	Was	Parietal, frontal, etc.	Dog
Roy and Sarnat	1956	Stainless-steel wire, black silk suture	Rib	Rabbit
Gross (direct) and/or serial radiographic (indirect) studies				
Dubreuil	1913	Metal	Tibia	Rabbit
Gatewood and Mullen	1927	Shot	Femur	Rabbit
Troitzky	1932	Silver wires	Skull	Dog
Levine	1948	Dental silver amalgam	Frontal, nasal	Rabbit
Gans and Sarnat	1951	Dental silver amalgam	Various facial	Monkey
Sissons	1953	Metal	Femur	Rabbit
Selman and Sarnat	1955	Dental silver amalgam	Frontal, nasal	Rabbit
Robinson and Sarnat	1955	Dental silver amalgam	Mandible	Pig
Björk	1955	Tantalum	Various facial	Human
Elgoyhen et al.	1972	Tantalum	Various facial	Monkey
Sarnat and Selman	1978	Dental silver amalgam	Nasal	Rabbit
Sarnat and McNabb	1981	Tantalum	Plastron	Turtle

[a]From B. G. Sarnat (1986). *Am. J. Orthodont. Dent. Orthop.* **90**, 221–233.

the distance between the pellets. When the tarsus was fully grown, he found that the distance between the pellets had remained the same and that the bone had increased in length at the ends. Thus, this experiment demonstrated that there was no interstitial growth of bone. Implantation of gold, silver, dental silver amalgam, stainless steel, vitallium, or tantalum in the form of screws, pegs, pins, clips, or wires within a single bone can be used for studying the total amount of bone growth by measuring the increase in distance between the implants and the outer borders of the bone. This direct study, however, does not yield serial data without reoperation or killing of the animal. In addition, this method has been used to determine sutural growth by placing implants on either side of the suture (see Section II,B,2).

B. Indirect Methods

1. Radiographs

Cephalometric radiography is an indirect method of studying growth of the skull (see Fig. 4). A refinement in the cross-sectional method was made by the use of radiography and the superimposing of tracings of the serial radiographs over various supposedly stable bone landmarks, to obtain the pattern of growth. The accuracy of this method depends on standardization of the technique. Selection of a stable anatomical base, however, for superimposing the radiographic tracings is the key to reliable findings, since any shift of the baseline distorts the true direction of growth.

Cephalometric radiography eliminated the most serious deficiencies of the anthropological technique. It permits a dynamic study of the growing child, that is, the increase in size and the change in proportion of the same growing bone (e.g., the mandible) or a group of bones forming a complex (e.g., the middle third of the face and neurocranium). It reveals the rate, the amount, and relative direction of bone growth. This method does not, however, reveal either the sites or the mode of bone growth. Of course, this is a two-dimensional study of a three-dimensional subject.

2. Serial Cephalometric Radiography and Implanatation

Use of a combination of serial radiography and radiopaque implants is a more accurate and reliable approach for a dynamic longitudinal study of bone growth (Tables II and III). This method has been used in the growth study of the face in both lower animals (see Fig. 5) and humans. The serial radiographs demonstrate the increase in size and the change in proportion. In addition, a stable base for superimposing the serial radiographic tracings is obtained by inserting two or more radiopaque implants. Thus, the ensuing growth can be accurately determined and measured by superimposing tracings of radiographs over the images of the metallic implants. To avoid foreshortening, implants must lie in a plane parallel to the X-ray film. This approach is particularly useful in studying the growth pattern of the mandible or other single bones if they are clearly outlined on the radiograph.

Another advantage of this combined method is the ability to measure the amount of new bone formation and resorption that occurred from one period to another without reoperation, or euthanizing the subject. There is also no interference with the normal diet, such as occurs in madder-fed animals. A disadvantage is that the radiograph demonstrates the sum total of apposition and resorption at that particular time without the detailed intervening changes as shown with vital markers and histological sections. In addition to the study of the growth pattern of a single bone, this combined method is valuable in the study of sutural growth by placement of implants on either side of a suture.

In addition to the foregoing armamentarium, digital subtraction radiography, computerized tomography, ultrasonics, and magnetic resonance imaging can offer more detailed and accurate information. Further, newer methods for describing craniofacial growth that are relatively independent of registration and superimposition (e.g., biorthogonal grids, elliptical Fourier functions, and finite elements) might be of future value.

III. NORMAL CRANIOFACIAL GROWTH

The craniofacial skeleton changes in size and shape in all three planes: height, width, and depth. It grows, however, in these three dimensions of space differentially in both time and rate (Figs. 1–4). Many sites contribute to the multidirectional growth. The dynamics and details of normal postnatal growth, coordination and simultaneity, and change and nonchange of the craniofacial skeletal system in both the young and adult are fascinating, complex, and incompletely understood problems in the field of biology. [See Craniofacial Growth, Genetics.]

Craniofacial bones grow in three principal ways (Table I). One is cartilaginous at the nasal septum and endochondral growth (i.e., the replacement of cartilage by bone) at the base of the skull at the spheno-occipital and sphenoethmoidal junctions. In addition, endochondral growth of bone occurs at the septopresphenoid joint and at the mandibular condyle. These bones are joined by cartilage (synchondroses). A second way is by sutural growth, where bones are united by connective tissue (synarthroses). This is found only in the skull. Sutures grow differentially and by apposition without resorption. The amount of growth may vary on either side of a suture, the rate varies for different sutures at a particular time, and the same suture grows differentially at different times. These sites, as well as the endochondral sites, are of limited growth and usually cease activity upon reaching adulthood. A third type is appositional and resorptive growth (i.e., remodeling), which occurs on the outer surfaces (periosteal) or inner surfaces (endosteal) of bone throughout life. The differential responses and interrelationships of these processes are important.

The size and shape of the skull is determined not only by the growth of bone(s) but also by its cavities. Increases in the contents of the neurocranial and orbital cavities of the skull influence the growth of adjoining bones and sutures. This occurs by a combination of resorption and deposition of bone on the surfaces and adjustments at the sutures. The neurocranium and the masticatory facial skeleton are integrated into an anatomical and biological unit. The masticatory skeleton, however, is partially dependent on muscular influences, growth of the tongue, and the dentition. These two parts of the skull follow different paths of development and the timing of their growth rates is entirely divergent. Nevertheless, growth of any one part of the skull is coordinated to the growth of the whole. The air-containing maxillary, frontal, ethmoid, and sphenoid sinuses also increase in size (Fig. 2).

Thus, the skull, a complex of bones, has proved to be a rich source of study, particularly since the combination of different types of bone growth, increase in size of various cavities, and growth, calcification, and eruption of teeth are not found elsewhere in the body. Neurocranial and orbital growth occurs predominantly early in life, whereas facial growth occurs predominantly later in life, mostly during the periods of the growth and eruption of the primary

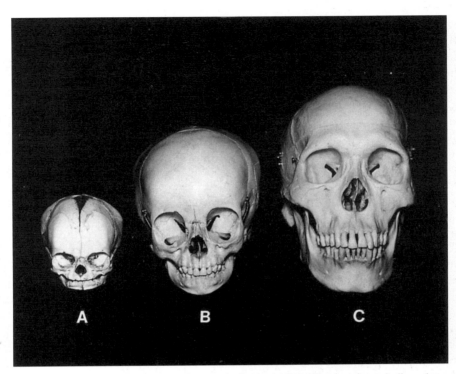

FIGURE 1 Normal growth of the human skull. (A) A clinically edentulous skull at about birth; (B) skull of a child with completely erupted deciduous primary dentition; (C) the skull of an adult with completely erupted permanent secondary dentition. Note that in the infant the neurocranium is prominent, whereas the face is much less so, representing a lesser amount of the total skull size. Also note that the orbit makes up a large part of the face. In the adult the face is prominent and represents a large part of the total skull size. The orbit makes up considerably less of the total face in the adult than in the infant. Differential growth takes place at different times and rates in various parts of the skull. [From B. G. Sarnat (1983). *Angle Orthodont.* 53, 263–289.]

and secondary dentitions. Serial cephalometric radiography has been a great aid in these studies (Fig. 4). A number of excellent references are available that describe the anatomical structures and the details of craniofacial growth and movement.

A. Lower Face

Mandibular growth occurs primarily in two ways. One is appositional and resorptive with differential bone remodeling at various periosteal and endosteal surfaces. In the ramus, the posterior border is a partic-ularly active site of bone apposition, whereas the anterior border is a particularly active site of bone resorption (Fig. 5). The second type of mandibular bone growth is endochondral at the condyle. *Growth* of the condyle and ramus, the most prolific sites, is in a superior and posterior direction. Because the condyle articulates with the mandibular fossa of the temporal bone, the final result is a downward and forward *movement* of the mandible. Another consideration is that the mandible is distracted by the inframandibular muscles with secondary adaptive growth at the condyle.

FIGURE 2 Lateral cephalometric radiographs of the skulls shown in Fig. 1. (A) In the infant skull, note the presence of unerupted teeth in the jaws. (B) In the child skull, the primary dentition is fully erupted and the permanent teeth are forming within the jaws. (C) In the adult skull, the permanent teeth are fully erupted. The maxillary (m) and frontal (f) sinuses are not evident in the infant skull, are in early development in the child skull, and are fully developed in the adult skull. Note the open actively growing suture(s) in the infant neurocranium in contrast to the closed inactive suture(s) in the adult neurocranium. st, sella turcica; o, orbit; h, head holder apparatus. (With assistance of Stuart C. White.)

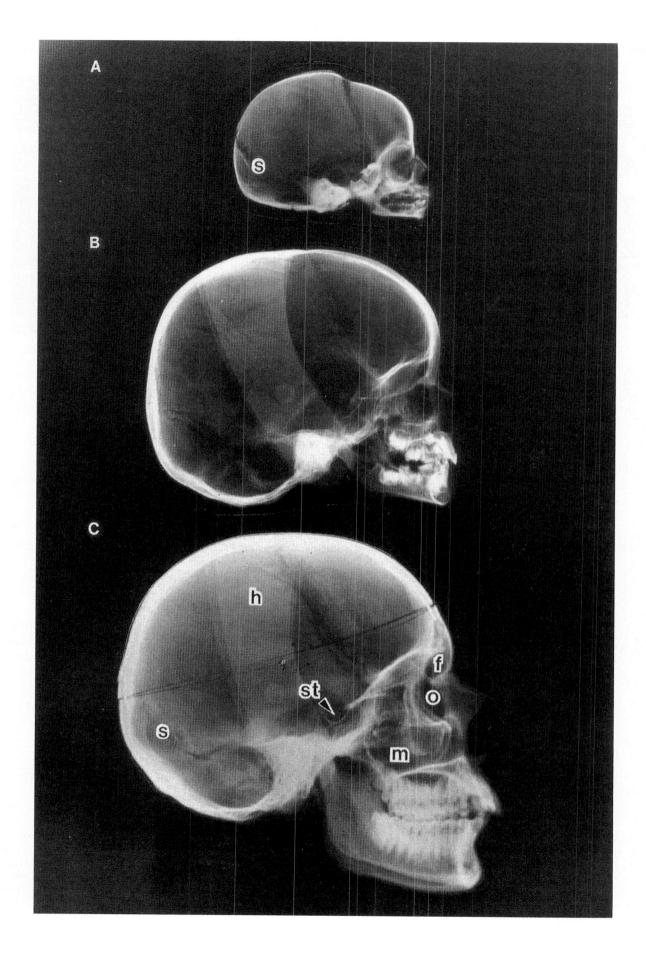

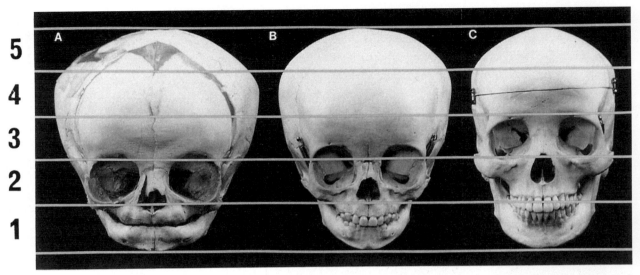

FIGURE 3 Frontal view photographs of the skulls shown in Figs. 1 and 2 enlarged here to about the same skull height and oriented in the Frankfurt horizontal plane. Note the differences in form and proportions of the total skull and its components. The distance between the lower border of the mandible to the superior border of the orbit represents about 40% of the skull height in the infant and 60% in the adult. Orbital height is nearly the same in all three skulls. Neurocranial height represents about 60% of the skull height in the infant and 40% in the adult. Skull height is divided into fifths. [From B. G. Sarnat (1983). *Angle Orthodont.* **53**, 263–289.]

The condyloid process of the mandible grows by the replacement of cartilage by bone (i.e., endochondral bone growth). Microscopic examination of a growing condyle reveals the presence of three zones: (1) chondrogenic, (2) cartilaginous, and (3) osteogenic (Fig. 6). The condyle is capped not by hyaline cartilage, as are nearly all articular surfaces, but by a narrow layer of avascular fibrous tissue that contains a few cartilage cells. The inner layer of this covering is chondrogenic, giving rise to the hyaline cartilage cells that constitute the second zone. In the third zone, destruction of the cartilage and formation of bone around the cartilage scaffolding occur. [*See* Cartilage.]

This cartilage is not a derivative of a cartilaginous model, as are epiphyseal and articular cartilages in long bones. Wheras the latter are derived from the primary cartilaginous skeleton, the condylar cartilage of the mandible can be classified as a secondary growth cartilage. Appositional and interstitial growth

FIGURE 4 Superimposed (on sella turcica and nasion) tracings of cephalometric radiographs of a normal person at 5 years, 5 months (·····); 10 years, 5 months (----); and 15 years, 7 months (———). Note in particular the downward and forward movement of the face and jaws. [From B. G. Sarnat, A. G. Brodie, and W. H. Kubacki (1953). *Am. J. Dis. Child.* **86**, 162–169.]

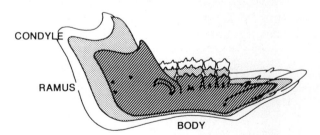

FIGURE 5 Serial tracings of lateral cephalometric radiographs of pig superimposed on outlines of four implants. Note the direction and differential growth that have occurred between various periods along all borders except the anterior border of the ramus, which has been resorbed. ▨, 8 weeks; ▢, 16 weeks; ☐, 20 weeks. [From I. B. Robinson and B. G. Sarnat (1955). *Am. J. Anat.* **96**, 50.]

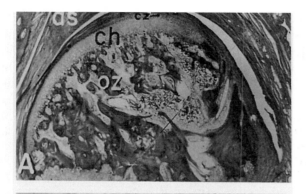

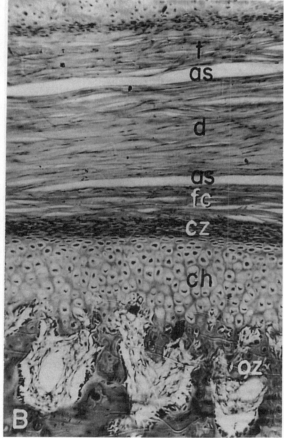

FIGURE 6 Sagittal hematoxylin-eosin-stained section through the left temporomandibular joint of a growing monkey. t, temporal bone; as, articular space; d, disk; fc, fibrous connective tissue covering of the condyle; cz, chondrogenic zone; ch, cartilaginous zone; oz, osteogenic zone. (A) Original magnification ×20; (B) original magnification ×120. [From B. G. Sarnat (1957). *Am. J. Surg.* **94,** 19–30.]

of the cartilaginous portion of the condyle contributes directly to the increase of mandibular height and length. This cartilage, however, is not homologous to an epiphyseal cartilage because it is not interposed between two bony parts, and it is not homologous to articular cartilage because the free surface bounding the articular space is covered by fibrous tissue, rather than hyaline cartilage. Thus, defects in the condylar articular surface are repaired more rapidly and completely than similar injuries to other articular surfaces such as those in long bones.

The growth pattern of the pig mandible has been studied with serial cephalometric radiographs in combination with metallic implants (Fig. 5). The condylar region contributed about 80% to total ramus height of the mandible, whereas the posterior border contributed as much as 80% to total length of the mandible. Since the amount of apposition at the posterior border was about twice the amount of resorption at the anterior border, ramus width increased. Resorption of the anterior border of the ramus played an important role in lengthening the body of the mandible in the molar region. Thus, it could be stated that in the growing mandible the ramus of today will be the body of tomorrow.

As the anterior border of the ramus continued to be resorbed, it not only created more mandibular body but also simultaneously permitted exposure of the crowns of erupting permanent molars. There is usually a harmonious relationship between dental development and growth of the mandible. This growth, however, was not dependent on the developing teeth. Yet normal eruption of the permanent molars was dependent in part on the normal growth of the mandibular ramus.

The roles played by the alveolar processes and mandibular bone proper were quite different. In growth of the mandible, the only part subservient to erupting teeth was alveolar bone. The contribution of the growth of the alveolar bone to the increase in the height of the mandibular body was about 60%. The important feature of alveolar bone was its ability to serve the needs of the ever-changing requirements of the teeth. In contrast, however, the formed mandibular bone remained relatively constant except for changes in proportion. Serial studies have been made of prenatal growth of the mandible in the living fetus.

B. Upper Face

Many sites contribute to growth of the upper facial skeleton, a three-dimensional mosaic of bones and cavities. It is influenced by endochondral growth at the cranial base by the spheno-occipital and the sphenoethmoidal synchondroses. Sites of growth for the maxillary complex are the maxillary tuberosities and three sutures on each side: the frontomaxillary, zygomaticomaxillary, and pterygopalatine. This maxillary complex moves downward and forward. Transverse growth at the median palatine suture, which is affected by the downward and divergent growth of the pterygoid processes, is both simultaneous and correlated with the widening of the downward-shifting maxillary complex. Anteroposterior growth occurs along the transverse palatine suture between the horizontal plate of the maxilla and the palatine bone and along the posterior margin of the palatine bone.

The downward shift of the hard palate, which increases the size of the nasal cavity, is by resorption of bone on its nasal surface and apposition on its oral surface. This shift could also be related to growth of the cartilaginous nasal septum. The downward, forward, and lateral growth of the subnasal part of the maxillary body is accompanied by the eruption of the teeth and apposition of bone at the free borders of the alveolar process. This contributes to the increase in height and width of the upper facial skeleton.

The upper facial skeleton also enlarges with increase in size of the orbital cavities, the nasal cavity, and the air-containing maxillary, frontal, and ethmoid sinuses (Fig. 2). Concurrent with all of these, growth and movement processes of the upper face are continuous, differential apposition and resorption of bone. This was noted in particular with the dental amalgam implant markers in the study of facial sutural (and mandibular) growth. Implants placed on the lateral borders of several craniofacial bones were found on the medial borders at the end of the study. These findings illustrate lateral apposition and medial resorption of bone. According to the functional matrix hypothesis, these are secondary sites of growth dependent on the relationships to the adjoining soft tissue.

C. Orbit

Facial growth is related to orbital growth. Shape and size of the orbit result from the balance of a number of genetic and epigenetic factors that may function on a systemic, regional, and local basis. Is there a key factor(s) that influences orbital growth? Is there a correlation between orbital size and intraorbital mass? If so, what role do the muscles, other extraocular structures, the globe, the vitreous and the aqueous humors, and the lens play?

The relative capacity of the orbit and the size of the eye diminish with increase in body weight and size. In the fetus and the newborn, the eyeball is so large in relation to the socket that it projects consider-

ably beyond the orbital rim so that a normal fetal exopthalmos exists. In infants, the eyes are not only larger in proportion to body weight than in the adult, but also in proportion to the size of the orbit. In humans, orbital height at birth is about 55% of the adult and 79% of total growth at 3 years of age (Fig. 4). At 7 years of age, orbital height is about 94% of the adult, whereas facial height is still only 80% (Figs. 1–4).

D. Nasal Bones

The growth pattern of the nasal bone region in the rabbit was studied using the same method for studying the growth pattern of the mandible, also a single bone. Two radiopaque implants were inserted into each left and right nasal bone. Ventrodorsal cephalometric radiographs were obtained in a specially designed head holder. From these radiographs, separate tracings were made on matte acetate paper of the left and right nasal bone regions, including the radiopaque implants for two different periods. Since the implants maintained the same relationship to each other in each nasal bone, they served as stable reference sites. The differences in the two established outlines represented the changes in size and shape in two dimensions. Growth at the proximal and distal borders was about equal, about twice that of the lateral border, and about five times that of the medial border. What factors are active to produce this differential in growth?

E. Relationship of Teeth to the Jaws and the Face

The importance of the deciduous and permanent teeth in the growth of the jaws and face has been a much debated question. For example, the anteater, born without teeth, has long jaws. A patient with complete anodontia and ectodermal dysplasia presented an unusual opportunity to study this problem. Serial cephalometric radiographs were taken from 21 months to 16 years of age, a period of growth and eruption of the deciduous and permanent dentitions. Study of the superimposed tracings of the radiographs indicated that growth was within small normal limits and that complete absence of teeth did not significantly impair development of the face and jaws, with the exception of the alveolar bone. Five sets of full upper and lower artificial dentures were designed, constructed, and delivered during this time. Each successive denture was larger and contained more and larger teeth to accom-

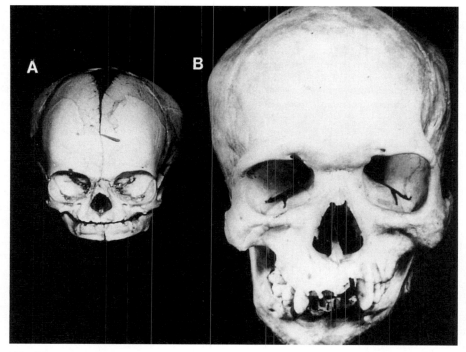

FIGURE 7 Comparison of (A) a normal symmetrical infant human skull with (B) an abnormal asymmetrical adult human skull. Note undergrowth primarily of the right condylar region, mandible, orbit and face, as well as an ankylosis of the right temporomandibular joint and a malocclusion. Trauma to, or infection of, the temporomandibular region during infancy or early childhood could have led to this deformity. [From B. G. Sarnat (1979). *Surg. Gynecol. Obstet.* 148, 659–669, by permission of *Surgery, Gynecology & Obstetrics.*]

modate the increase in the size of the jaws. Thus, in instances of surgical procedures upon the jaws, disturbance of the teeth or tooth buds could lead to loss, malformation, or malposition of teeth with changes in alveolar bone. The general size and shape of the jaws and face, however, will not be affected because of damage to the teeth. [*See* Dental and Oral Biology, Anatomy.]

IV. ENVIRONMENT AND GROWTH

Throughout our lives we are constantly reacting to our environment. Variations in temperature, light, humidity, atmospheric pressure, terrestial and extraterrestial radiation, and gravity affect us. In addition, the vast number of noxious chemical, physical, or biological agents, intentionally or unintentionally ingested in our food and water and inhaled in our air, determine our destinies. Deficiencies in essential nutrients might also play a part. It is important to consider the effect of our environment on the skeletal growth

sites and the resulting changes in size and shape of the skull (Fig. 7) and the body.

Young rats exposed to cold stresses had a less large skull, a longer face in relation to the cranial vault, a narrower nose, a rounder neurocranium, and a shorter femur. Populations living at high altitudes and exposed to the environmental stresses of hypoxia and cold have a slower postnatal growth and less of an adolescent growth spurt than do other groups. On earth, gravity is considered normal or $1.0g$. What skeletal and other changes might occur in environments of hypogravity (the moon, $0.18g$) or hypergravity (Jupiter, $2.65g$)? These and other factors are of great interest to the field of cosmic biology and should be of both interest and concern to everyone.

BIBLIOGRAPHY

DuBrul, E. L. (1988). "Sicher's Oral Anatomy," 8th Ed. Ishiyaku Euro America, St. Louis.
Enlow, D. H. (1990). "Handbook of Facial Growth," 3rd Ed. Saunders, Philadelphia.

Johnston, L. E. (1976). The functional matrix hypothesis: Relfections in a jaundiced eye. *In* "Factors Affecting Growth of the Midface" (J. A. McNamara, Jr., ed.), Craniofacial Growth Series, Monograph 6, pp. 131–168. Univ. of Michigan, Ann Arbor.

Krogman, W. M. (1974). Craniofacial growth and development: An appraisal. *Yearb. Phys. Anthropol.* **18,** 31–64.

Sarnat, B. G. (1983). Normal and abnormal craniofacial growth: Some experimental and clinical considerations. *Angle Orthodont.* **53,** 263–289.

Sarnat, B. G. (1984). Differential growth and healing of bones and teeth. *Clin. Orthop. Relat. Res.* **183,** 219–237.

Sarnat, B. G. (1986). Growth pattern of the mandible: Some reflections. *Am. J. Orthodont. Dent. Orthop.* **90,** 221–233.

Sarnat, B. G. (1997). Some methods of assessing postnatal craniofaciodental growth: A retrospective of personal research. *Cleft Palate–Craniofac. J.* **34,** 160–173.

Sarnat, B. G., and Laskin, D. M. (eds.) (1991). "The Temporomandibular Joint: A Biological Basis for Clinical Practice," 4th Ed. Saunders, Philadelphia.

Thompson, D. (1958). "On Growth and Form." *Cambridge Univ. Press,* Cambridge, England.

Crime, Delinquency, and Psychopathy

RONALD BLACKBURN
University of Liverpool

I. Biology and Criminology
II. Genetics and Criminality
III. Psychophysiological and Biochemical Correlates
IV. Higher Nervous Functions
V. Conclusions

GLOSSARY

Electroencephalogram Electrical record of brain rhythms that are conventionally divided into four frequency bands of delta (0.5–3 Hz), theta (4–7 Hz), alpha (8–13 Hz), and beta (14–30 Hz) activity

Socialization Process by which social values and moral prohibitions become internalized in the form of conscience, or superego

Temperament Emotional and expressive characteristics distinguishing an individual

Wechsler tests Commonly used measures of intellectual ability, or intelligence quotient (IQ), divided into verbal and nonverbal or performance components

CRIMINAL BEHAVIOR IS ACTION PROHIBITED BY law, but since laws vary over time and among societies, no act is inherently criminal. Delinquency refers to acts of juveniles that would be criminal if committed by an adult, but also misbehavior that derives from the status of being a minor. A criminal or a delinquent is strictly someone who has been subject to the official sanctions of the criminal justice system, but much criminal behavior remains undetected. Modern criminological research therefore also includes as criminals those who admit to criminal acts in self-report measures. Criminality hence can be regarded as a tendency or a disposition to engage in criminal behavior, which varies in degree throughout the population at large. Psychopathy, in contrast, is a psychological concept denoting abnormality of personality. The term lacks a universally agreed meaning, but relates to a disregard for the rights and feelings of others.

Psychopathic personalities are identified by deficiencies in emotional experiences that normally restrain harmful behavior and, hence, by characteristics such as callousness, lack of empathy and guilt, and impulsive, often criminal, actions. Common synonyms for psychopathic personality are sociopathic or antisocial personality disorder, the latter being the currently preferred term in American psychiatry. Those described as psychopaths by clinicians are not homogeneous, and a distinction is therefore frequently made between primary and secondary psychopaths, the former characterized by assertiveness and an absence of guilt and anxiety, the latter by social anxiety and poor self-esteem. Psychopathy, like criminality, can be construed as a continuum, but not all criminals display psychopathic personality traits and not all of those showing such traits violate the law. However, more extreme criminality overlaps with psychopathy, and theories of psychopathy are relevant to the explanation of serious and persistent criminal behavior.

I. BIOLOGY AND CRIMINOLOGY

A. Study of Crime

Criminology is of interest to many disciplines, but has been dominated by sociology, psychiatry, and psychology. Sociologists are concerned with identifying which sections of society are more likely to be criminal. Crime rates are higher, for example, among males, in urban centers, in inner city slums, and in the United States among black minorities. Also, the peak rates of crime occur among those aged 15 to 18, although only a minority of delinquents become adult criminals. Sociological theories account for these variations in terms of unequal economic opportunities, the effects of different subcultures, and the influence of the controlling social forces within the family, school, and community that discourage violation of social rules. Sociologists view crime as a creation of society and seek the formal causes of crime in terms of the activities of powerful groups that determine what behavior is to be defined as criminal.

Psychiatry and psychology, on the other hand, aim to identify which individuals are more likely to violate laws, and hence emphasize the efficient, or antecedent, causes of criminal behavior. These disciplines assume that social forces are insufficient to explain criminal acts, since, among sections of society with high crime rates, only a minority become criminal. Variations between individuals are therefore invoked as causal factors. Psychiatry focuses on abnormal mental states that might be correlated with the commission of crimes, although only a small minority of criminals show serious mental disorders. Although psychological theories of crime overlap with psychiatric perspectives, they are more typically concerned with criminal acts as variations of normal behavior and seek explanations in terms of personal attributes and behavioral tendencies and the processes of development and learning through which such attributes arise, rather than in terms of discrete abnormalities.

The notion that biological endowment could contribute to crime has always been represented in psychiatric and psychological theories. A biological approach inquires about variations in physiological and neurochemical functioning that might differentiate those who violate laws. The most extreme view was that of Italian physician and anthropologist Cesar Lombroso, who in 1876 proposed the concept of the "born criminal," a remnant of early human ancestry, characterized by primitive physical and psychological development. This was discredited by early anthropometric studies of prisoners, and biological perspectives on crime have subsequently been viewed with disfavor. Such perspectives are criticized for disregarding the apparently overwhelming role of social institutions in defining what is criminal and in creating conditions conducive to crime.

However, current biological approaches do not postulate a gene for crime, and they focus instead on the contribution of normal biological variation to adaptive learning, including the learning of society's rules. They view criminal behavior in terms of a tendency to break rules rather than the commission of specific criminal acts. Since learning depends on a social environment, biological approaches are also concerned with biosocial interaction and do not assert a unidirectional biological determinism. This view therefore recognizes the influence of the family, school, or social group in shaping rule-breaking behavior, but challenges an environmental determinism that assumes that individuals are merely passive recipients of these influences.

B. Theories of Criminality

Psychological theories of criminality traditionally focus on the way in which individuals are socialized through interactions with significant others, and child-rearing practices clearly play a significant role in this process. Parents of antisocial children commonly fail to punish deviant behavior with consistency and moderation, do not reward socially acceptable behavior, and do not themselves behave as models of conformity and achievement that the child can imitate. However, the child's own contribution to this process remains debated. Since there is evidence from behavior genetics studies that ability and temperament variables have a genetic origin, the child's personal attributes might well influence the causal chain leading to undersocialization and the subsequent drift to delinquency. For example, infant temperament might significantly affect the response of the parents.

Not all theories take into account the potential influence of biological variation. Psychoanalysis sees crime in terms of a failure to tame biologically determined erotic and aggressive instincts during the early years of life through the formation of the superego. The psychopath fails to develop a superego and, hence, engages in antisocial and often violent acts without moral qualms. The "neurotic" delinquent, in contrast, has too strict a superego and acts out repressed instinctual wishes in symbolic form as antisocial behaviors that invite punishment. Criminal acts are thus the expression of biological forces, but the psychoanalytic theory emphasizes biological similarities among people in the form of universal human instincts. [See Psychoanalytic Theory.]

Learning theorists reject psychodynamic instinct theory. Social learning theorists propose that much criminal behavior is learned in the same way as other behavior, through imitation and the rewarding influences of peers. Others, however, notably Hans Eysenck, retain the notion of a generalized conscience that restrains antisocial acts, but attribute failure of socialization to biological differences among people.

Eysenck's theory does not propose that crime is biologically determined, but rather that biological variation influences the capacity to learn control of self-gratification. Variations in human temperament are attributed to three independent personality dimensions of neuroticism (N), extroversion (E), and psychoticism (P), each of which has a biological basis. N is held to reflect greater reactivity in the limbic and autonomic nervous systems and determines emotional responses to stress. Underlying E is the level of cortical arousal or arousability, governed by activity in circuits linking the reticular system of the brain stem and the cerebral cortex. Extroverts have low arousal, relative to introverts, and are predicted to form conditioned responses less readily, and to require more intense stimulation from the environment. More tentatively, P is related to circulating androgens.

Conscience formation depends on the classical conditioning of anxiety reactions. Punishment of antisocial behavior by parents, teachers, or peers generates anxiety in the child, and cues associated with punishment, including internal stimuli, become conditioned stimuli that themselves arouse anxiety. The child avoids anxiety by resisting the temptation to indulge in the punished behavior, and through stimulus generalization a generalized conscience is acquired. Since extroverts form conditioned responses slowly, they will be less well socialized than introverts and, hence, more likely to engage in antisocial behavior. Their need for stimulation also leads to impulsive "thrill-seeking," which might take the form of prohibited acts. Additionally, however, the traits of the P dimension (i.e., hostility, insensitivity, and cruelty) make antisocial behavior more likely. Eysenck, therefore, proposed that criminals in general, and psychopaths in particular, show the more extreme characteristics associated with E and P.

Research has established a significant association between personality measures of P and criminality, and the suggested greater need of criminals for stimulation also receives support in studies of sensation-seeking. The link with E is less consistently supported, and though self-reported delinquency does seem to be correlated with E, officially defined criminals have not been found to be significantly more extroverted than nonoffenders. The postulated links between E and arousal and conditionability, and between P and androgen level, remain to be substantiated. [See Sensation-Seeking Trait.]

This theory has been criticized for assuming that different physiological response systems condition in a uniform fashion. One alternative view is that much criminal behavior arises primarily from deficiencies in child-rearing, but that some individuals (e.g., psychopaths) are relatively resistant to such training as a result of deficits in passive avoidance learning. Passive avoidance involves learning to inhibit punished behavior, but it is dependent on the conditioning of anxiety responses specifically, rather than on conditionability in general. [See Conditioning.]

It has been proposed that individual differences are based on specific forebrain systems, notably the behavioral inhibition system (BIS), a septohippocampal system responsive to conditioned stimuli for punishment or the absence of an anticipated reward, which mediates passive avoidance. The BIS interacts with the behavioral activation system (BAS), which mediates responsiveness to conditioned stimuli for reward or nonpunishment, although the biological substrate for this is less clear. Psychopaths are believed to have a weakly reactive BIS, leading them to be insensitive to threat stimuli. No dysfunction of the BAS is suggested, but reward-seeking might be disinhibited in psychopaths through the failure of punishment stimuli to inhibit rewarding behavior.

II. GENETICS AND CRIMINALITY

Much research has focused on the general proposition that there are hereditary influences on criminality, rather than on the specific characteristics of the nervous system that might mediate antisocial behavior. This does not assume that complex human actions are "preprogrammed," since genetic effects on behavior are polygenic and probabilistic. Genotypes give an initial direction to development by providing basic elements of behavior, which are incorporated into larger adaptive units through learning. They influence phenotypes through the combination of genes and environments supplied by parents, through differential reactions from others to biologically different individuals, and through differential selection of environments by these individuals. Genetic studies of criminality attempt to isolate innate influences on these complex pathways through several research designs.

A. Family Studies

Family studies compare the distribution of antisocial behavior in the biological relatives of offenders and nonoffenders. Criminal parents tend to be more likely to have criminal children. Family studies of delinquent females also suggest that they have more socially deviant relatives than do other females and have more familial pathology than do antisocial males, although familial factors seem to be of equal importance for the development of criminality in both sexes. However, family studies do not permit any clear separation of genetic and environmental influences.

B. Twin Studies

Monozygotic (MZ) twins have the same genotypes, whereas dizygotic (DZ) twins are no more alike, genetically, than other siblings. Studies of twins therefore use the logic that phenotypic differences between same-sex MZ and DZ twin pairs are likely to reflect genetic influences, assuming similar rearing conditions. The relevant differences are most commonly expressed as the percentage of criminals with twins whose twin is criminal (i.e., pairwise concordance).

About 12 twin studies of varying sample sizes have been reported. Earlier studies found that, on average, MZ twins showed 60% concordance for criminal history, whereas DZ showed 30%, but differences are most apparent for adults, and twin studies of officially defined juvenile delinquents suggest minimal genetic influence. However, recent research reports greater similarity of self-reported delinquency for MZ than for DZ twins, while also indicating significant gene–environment interaction.

Earlier studies used selected samples of twins and unreliable methods of determining zygosity. Lower concordance rates emerge from recent Scandinavian studies, which relied on unselected samples of twins from national registers and determined zygosity from blood samples. For example, among Danish males who became criminal, pairwise concordance rates were 35% (MZ) and 13% (DZ), whereas for females they were 21% (MZ) and 8% (DZ). The high discordance rates indicate substantial nongenetic effects, but the MZ concordance rate is nevertheless substantially higher than the DZ rate. However, similar research in Norway found male concordances of only 26% (MZ) and 15% (DZ). The difference is not significant and raises doubts about hereditary effects on crime. Nevertheless, all studies find differences between MZ and DZ concordance rates in the same direction, and the difference between the Norwegian and Danish studies remains unexplained.

One common argument is that phenotypic similarities of MZ twins represent a greater similarity of treatment by parents. Although it is equally plausible that any experience of environmental similarity by MZ twins might be an effect of their genetic similarity, this remains a possible confound in all studies of twins reared together. Without data on the criminality of twins reared apart, twin studies remain suggestive, rather than conclusive.

C. Adoption Studies

If children adopted shortly after birth resemble biological more than adoptive parents in some attribute, this is strong evidence for genetic influence. Research on criminality in adoptees uses two designs. The first identifies criminal parents who have given up their offspring for adoption, and compares their children with the adopted offspring of noncriminal biological parents. Studies using this design provide support for a genetic contribution to crime. For example, more of the adopted offspring of female offenders had acquired a criminal record in comparison with controls in one American study, while in another, adoptees whose biological parents were diagnosed as antisocial personalities showed a greater probability of receiving this diagnosis as adults.

The second design begins with a heterogeneous sample of adoptees and compares criminal and noncriminal adoptees in terms of criminality in biological and adoptive parents. Several studies have been reported during the past two decades from Scandinavia, and it has been found that the biological fathers of adoptees who are psychopathic show a higher incidence of psychopathy than do the adoptive fathers. It has also been shown that the likelihood of an adoptee's becoming criminal increases significantly if the biological father was also criminal, but that it increases further if both biological and adoptive fathers have criminal records. These data point to a genetic influence on criminality, but also to a genotype–environment interaction. However, recent work emphasizes the heterogeneity of criminals. Research in Stockholm suggests a genetic involvement for petty criminality, but not for violent crime. For some criminals, violent behavior is symptomatic of alcohol abuse, which could itself have a genetic basis.

In none of these studies has more than a minority of the adopted children of criminal parents become criminal or antisocial, and what is implied by the results is a genetic contribution to crime, given certain environments, rather than any overwhelming genetic determination. However, preadoption influences are not solely genetic, but include perinatal complications determined by the living conditions of the mother. Whether these might account for the relationship between criminality in biological parents and their adopted children remains unclear.

D. Chromosome Anomalies

Variations from the normal complement of 23 chromosome pairs usually arise from errors in cell division, and represent genetic factors that are innate, but not inherited. These rare anomalies have been linked with behavior disorder, and particular interest has been shown in sex chromosome patterns that depart from the usual configurations of 46,XY in males, and 46,XX in females. Most research has focused on males showing a 47,XYY complement, or the 47,XXY configuration (Klinefelter's syndrome). [See Chromosome Anomalies.]

The XXY pattern has long been known to be more frequent among the mentally retarded, but research in the 1960s also indicated elevated frequencies of XYY males among prisoners and mentally disordered offenders. Most were of above average height, had a low intelligence quotient (IQ), and had personality disorders. Subsequent research has confirmed the higher prevalence of the XYY complement in institutionalized antisocial populations, particularly mentally disordered offenders, but has also shown an incidence in the general population of about 0.1%, which is higher than was previously thought.

The initial studies appeared to illustrate a genetic determination of criminal behavior and were also interpreted in terms of what might be contributed to crime by the normal Y chromosome. It was suggested that the sex difference in crime reflected the masculine traits contributed by the Y chromosome, and that the extra Y exaggerated masculinity and violence. However, possession of the extra X chromosome also correlates with criminality, and characteristics of the XYY male have typically been inferred from small samples in institutions.

A study reported from Copenhagen in 1976 avoided the problem of sampling bias and identified 12 XYY and 16 XXY individuals in a large birth cohort. Although 42% of the XYYs, 19% of the XXYs, and 9% of controls had criminal records, confirming an association of the extra Y with criminality, the offenses of the former groups were mainly petty and nonviolent, and criminal behavior was confined to a minority. Since both the XYYs and the XXYs were also of lower intelligence, and the former showed electroencephalogram (EEG) abnormalities, it would appear that chromosome abnormality makes a nonspecific contribution to criminality through the medium of genetic disorganization and developmental failure. Interest in this rare phenomenon has subsequently declined, although there have been some speculations about the length of the Y chromosome in criminals.

E. Physique and Appearance

Although body constitution is influenced by the environment, it is fixed by the genotype. Lombroso claimed that criminals showed primitive bodily features, but this rested on statistically unsound observations. More modern research has nevertheless indicated a correlation between unattractive physical appearance and antisocial behavior. The association could, however, reflect a self-fulfilling prophecy, in that those who are facially unattractive are judged negatively, and react accordingly.

There is a long European tradition of attempting to link physique to temperament and psychiatric disorder, which has focused on body build, or somatotype. Modern studies of body build in delinquents derive from American research by W. H. Sheldon in the 1940s. Sheldon assessed somatotype in terms of three embryologically derived concepts: endomorphy (i.e., fat and circular), mesomorphy (i.e., muscular and triangular), and ectomorphy (i.e., thin and linear). These were proposed to relate to specific temperament components. Rating somatotype components from photographs, Sheldon found that delinquents were significantly more mesomorphic and less ectomorphic than were students. A 30-year follow-up of Sheldon's delinquents has extended these findings, and has also shown that more serious criminals had higher ratings on andromorphy, a measure of masculinity of secondary sex characteristics.

Sheldon's research was criticized for the subjectivity of his ratings and his imprecise criteria of delinquency, and some argue that somatotype might be more appropriately assessed by height and width of body build. Nevertheless, subsequent studies confirm that those who violate legal rules are more likely to be muscular and less fragile in physique. This relationship might be mediated by associated temperament factors, since mesomorphs have been found to describe themselves as significantly more active and aggressive, and ectomorphs as more socially avoidant. The combination of mesomorphy and andromorphy also suggests higher testosterone levels. Alternatively, delinquent peer groups might differentially reward toughness, whereas criminal justice agents might react more negatively to tough appearance.

III. PSYCHOPHYSIOLOGICAL AND BIOCHEMICAL CORRELATES

A. Theoretical Issues

Genotypes set limits on phenotypic variation through the intermediaries of enzymes, hormones, and neurons, and probably contribute to rule-breaking tendencies through the medium of neurochemical functions associated with learning and temperament. Processes associated with stimulation-seeking, passive avoidance, conditionability, and emotional responsiveness have provided the main potential links between genetically determined characteristics of the nervous system and antisocial behavior, and correlates of criminality have therefore been sought in peripheral recordings of cortical and autonomic activity or biochemical assays. The concept of arousal has been prominent in this research. Although this term refers broadly to the level of physiological activation of the nervous system, there is no single psychophysiological measure of general arousal. The EEG provides the closest approximation, but the arousal level is often inferred from activity in particular autonomic systems, which might not reflect what is going on in other parts of the nervous system.

In Eysenck's theory, extroverts have low levels of arousal and, hence, a higher optimal level of stimulation. The notion that psychopaths and delinquents generally have a suboptimal level of arousal has been popular, since much antisocial behavior appears to represent risk-taking consequent on boredom. Nevertheless, arousal remains an ambiguous concept, and recent research on sensation-seeking suggests that the optimal level of stimulation might have more to do with activity in brain catecholamine systems involved in information processing than with the nonspecific arousal level. However, more specific psychophysiological processes have also been implicated in undersocialized behavior. For example, recent proposals suggest that deficient activity in Gray's BIS is manifest in hyporesponsiveness of the electrodermal system. Since such variations should be most clearly shown by those who are least socialized, research has focused particularly on psychopaths.

B. Electrocortical Correlates

Clinical interest in the EEG centers on abnormalities in the wave form, but the notion of "abnormality" is somewhat arbitrary. It includes unusual discharges, but more predominantly refers to diffuse slow wave activity or excesses of theta activity in temporal areas. However, these features are apparent in about 15% of normal adults, and more than one-quarter of young children. Although they are interpreted variously as indicating developmental delay, limbic system abnormalities, or cortical arousal level, their functional significance remains unclear.

There is an extensive literature on EEG correlates of behavior disorder, and high frequencies of abnormalities have been reported in aggressive and psychopathic samples. In one early report of psychiatrically disordered combat troops, for example, EEG abnormalities were displayed in the records of 65% of aggressive psychopaths, 32% of inadequate psychopaths, 26% of neurotics, and 15% of nonpatient controls. There are, nevertheless, many inconsistencies in the literature, which is generally marred by unreliability in EEG analysis and subject classification.

Recent research using replicable measurement of psychopathy and EEG quantification has raised doubts about the extent of abnormalities among antisocial personalities, some studies finding no differences between psychopaths and nonpsychopaths. However, when psychopaths have been divided into primary and secondary subgroups, it is the latter group that is distinguished by more theta activity and lower alpha frequency. EEG abnormalities and, by implication, low cortical arousal are not, then, uniformly characteristic of psychopaths. Though there is slightly more consistent evidence of abnormalities in violent adults and hyperactive children, it is not unequivocal.

Although EEG abnormalities have also been reported in samples of incarcerated delinquents, the evidence is again inconsistent. However, prospective studies of Scandinavian boys have found that those who became delinquent were more likely than nondelinquents to show an excess of slow alpha activity in records taken in early adolescence. These delinquents were predominantly property offenders, and the results therefore indicate that EEG abnormalities are not specific to violent offenders. However, they are also not specific to offenders, and the presence of high-amplitude slow waves in the EEG might be associated with an increased risk for social maladjustment generally, rather than antisocial behavior specifically.

Recent attention has focused on event-related potentials not readily analyzable from the raw EEG trace. Of particular interest is the question of whether psychopaths show idiosyncrasies in information processing. A few studies have examined the contingent negative variation (CNV), or "expectancy wave," a slow negative potential elicited in forewarned reaction time experiments. Early suggestions of smaller-amplitude CNVs in psychopaths have not been confirmed, and there is probably no direct relationship between psychopathy and CNV amplitude. Although there have been several investigations of sensory-evoked potentials in psychopaths, the diversity of methods, subjects, and components examined precludes any firm conclusions. However, there is some indication in this research that psychopaths allocate more attentional resources to events of immediate interest.

Individual differences in evoked potentials are also observed in the modulation of stimulus intensity, or augmenting–reducing. With increasing stimulus intensity some individuals show an increase in the amplitude of the early wave component (augmenters), whereas others show a decrease (reducers). Augmenting is thought to reflect a less "sensitive" nervous system, and there is some evidence that sensation-seekers and male delinquents show an augmenting response. However, this area has not been extensively investigated.

C. Autonomic Nervous System Activity in Psychopaths

In experimental tasks involving avoidance learning, psychopaths do not show any deficit in active avoidance of punishment stimuli, but their postulated deficit in passive avoidance learning has been found when electric shock was used as a punishment paradigm. In contrast, under punishment conditions such as social disapproval or loss of money, psychopaths perform similarly to nonpsychopaths and do not, therefore, appear to have any generalized deficit in passive avoidance. However, it has recently been suggested that the passive avoidance deficit might be confined to situations in which there are competing cues of both reward and punishment, and experiments have confirmed that psychopaths overfocus on reward, which interferes with attention to punishment cues. Since this failure to alter a dominant response set parallels the effects of septal lesions in rats, these findings are consistent with deficient functioning of the BIS.

Research on electrodermal (ED) and cardiac responses of psychopaths prior to noxious stimulation confirms a deficiency in anticipating punishment. For example, when forewarned that a particular stimulus in a series will be followed by a shock stimulus, psychopaths produce less anticipatory ED arousal. They also show weaker acquisition of conditioned ED responses when shock is the unconditioned stimulus, and poorer vicarious ED conditioning has been found for both primary and secondary psychopaths observing aversive stimulation delivered to others. However, a study that involved both noxious and pleasant conditioned stimuli, and that recorded both ED and heart rate responses, indicated that deficient conditionability in psychopaths is confined to anticipatory responses to noxious stimuli and to the ED system. These results contradict Eysenck's hypothesis of a generalized deficit in conditionability in psychopaths, but are consistent with a deficit in the BIS.

However, threatening aversive stimuli that elicit smaller ED reactions in psychopaths simultaneously produce larger heart rate responses. It has been suggested that this dissociation of cardiac and ED responding reflects a "cortical tuning" mechanism involving a cardiovascular-induced reduction of cortical arousal via baroreceptors in the carotid sinus. This model of a protective mechanism that reduces the emotional impact of aversive stimulation has similarities to the concept of augmenting–reducing, but remains to be investigated further.

There is no consistent evidence that psychopaths in general are hyporesponsive in their orienting reactions (ORs) to simple nonaversive stimuli, but this has been observed in secondary psychopaths and schizoid offenders and could relate to attentional inefficiency. The rate of habituation of the ORs in normal ED records is correlated with the appearance of nonspecific fluctuations (NSF), and higher rates of these fluctuations and slower habituation of the ED ORs are associated with superior vigilance and more effective allocation of attentional resources. Some research suggests that psychopaths habituate more slowly in cardiac ORs to auditory stimuli, but more rapidly in ED ORs, and also show lower rates of NSF. This dissociation of the ED and cardiac systems is related to drowsiness, and these findings point to lower cortical arousal. However, there is some evidence that this pattern is more prominent in secondary than in primary psychopaths.

One other ED variable of interest is response recovery time. Slow autonomic recovery could lead to a failure of fear reduction to reinforce passive avoidance responses, and slower recovery has been demonstrated in criminals and psychopathic delinquents. However, the functional significance of recovery time remains debated.

Psychopaths do not exhibit generally lower autonomic arousal levels, but some prospective research suggests lower cardiac rates in children who subsequently become delinquent. Children of criminal fathers have also been found to have lower heart rates than do children of noncriminals, although this does not characterize children of psychopaths. If, as some suggest, low tonic heart rate is associated with low cortical arousal, these findings imply that lower arousal in general might be a correlate of petty delinquency, but not psychopathy.

The foregoing research supports the view that psychopaths are underresponsive in the ED system and, hence, in the BIS. There is less clear evidence of low arousal in psychopaths, and recent EEG research contradicts this proposal. However, any conclusions must be tempered by inconsistencies in the research to date, which could reflect the heterogeneity of "psychopaths." There is, for example, some indication that lower arousal might be characteristic of secondary psychopaths.

D. Biochemical Correlates

Hormones secreted by the endocrine glands influence behavior through their role in normal development as well as their effects on the temporary state. Correlates of criminal behavior and related temperament variables have been sought in variations in the production and level of hormones secreted by the gonads (i.e., androgens and estrogens), the adrenal glands (i.e., adrenaline and noradrenaline), and the pancreas (i.e., insulin). [See Hormonal Influences on Behavior.]

Androgens are crucial in sexual differentiation and the appearance of secondary sexual characteristics at puberty, and testosterone level has therefore been considered a possible factor for the universal correlation of criminal behavior with gender and age and for the greater aggressiveness of males. Although initial studies found that both the rate of production and the level of testosterone correlated positively with hostility in younger nonoffender subjects, subsequent studies have failed to confirm this. Results from offenders have been equally inconclusive. Higher testosterone levels have been reported in aggressive prisoners and extremely violent rapists, but there are several negative findings. The evidence to date is therefore not sufficient to indicate a direct influence of testosterone.

Testosterone might, however, have an indirect effect on behavior through the activity of brain neurotransmitters. It inhibits the activity of the enzyme monoamine oxidase (MAO), which metabolizes several neurotransmitters, and this might allow monoamines (e.g., noradrenaline) to accumulate to higher levels in the brain. Blood platelet MAO activity is lower in human males than in females, and lower MAO levels have been found to be associated with "disinhibitory" temperament variables (e.g., impulsivity, sensation-seeking, and undersocialization) in nonoffenders. Low MAO activity is also correlated with low concentrations of 5-hydroxyindoleacetic acid (5-HIAA), a derivative of the neurotransmitter serotonin, in the cerebrospinal fluid, and low 5-HIAA has also been reported to be correlated with impulsivity and sensation-seeking in nonoffenders and with habitually aggressive offending.

Female crime has long been thought to relate to hormonal changes occurring in the menstrual cycle, particularly premenstrual tension (PMT). PMT symptoms include depressed mood, lethargy, and headaches, and PMT might make females more vulnerable to deviant behavior as a consequence of irritability. Studies have found that female prisoners are more likely to have committed their crimes during the 8 days preceding or during menstruation (i.e., paramenstruum), this being most marked for theft, and among those experiencing PMT. Disciplinary problems among female prisoners and mentally disordered females have also been found to occur more frequently in the premenstrual week. However, the peaking of antisocial behavior during the paramenstruum is not consistently associated with PMT, and it is not specific to violent behavior. Also, since severe PMT symptoms occur in up to 40% of women, they are obviously neither necessary nor sufficient to account for female offending.

Criminal acts have been reported to occur in states of hypoglycemia, and recent research indicates an association of violence with dysfunction in glucose metabolism. Hypoglycemia is related to increased insulin secretion that could follow from starvation or alcohol ingestion, and impairs cerebral function. Aggressive behavior in psychiatric patients has been found to be associated with glucose dysfunction and EEG abnormalities, and slower recovery from experimentally induced hypoglycemia has similarly been observed in habitually violent adult offenders diagnosed as antisocial personalities, particularly those with a history of violence under the influence of alcohol. Hypoglycemia proneness also correlates with self-reported aggressiveness among nonoffenders.

Secretion of the catecholamines adrenaline and noradrenaline has been of interest because of earlier hypotheses linking adrenaline increase with fear and noradrenaline increase with aggession. These hypotheses are now seen as too simple, but a few studies have examined urinary catecholamine in relation to stress responses in antisocial individuals. Among offenders awaiting trial, for example, psychopaths show less increase in adrenaline immediately prior to trial, suggesting less stress responsiveness. Lower adrenaline secretion under stress has also been reported in studies of persistent bullies, and adolescents who later obtained criminal convictions. However, adrenaline increase might be more related to cortical alertness than specifically to stress. [See Catecholamines and Behavior.]

These findings might be integrated in terms of autonomic balance. Insulin secretion is normally opposed by adrenomedullary hormones, whereas testosterone increases catecholamine output by lowering MAO activity. Low cortical arousal, lower heart rate, hypoglycemia proneness, and low adrenaline output would all be consistent with a dominance of the vagoinsulin system, or parasympathetic balance. However, the data are not wholly consistent, since hypoglycemia proneness is associated with violence, but other features are more characteristic of nonviolent delinquents.

IV. HIGHER NERVOUS FUNCTIONS

A. Brain Pathology and Crime

Damage to the brain can result from perinatal complications, head injuries, tumors, infections, or exposure to toxic substances (e.g., atmospheric lead), but such conditions do not invariably lead to structural damage. Brain damage must be severe before significant psychological disturbance ensues, and unsocialized behavior is not a necessary consequence. For example, a survey in Finland found that less than 6% of 507 veterans who had sustained head wounds received a criminal conviction leading to imprisonment in the 30 years following injury.

In the absence of known cerebral insult, brain pathology remains difficult to detect. Most research relies on indirect signs, such as medical history, "soft" neurological signs, EEG records, or neuropsychological tests, which detect brain dysfunction, and not necessarily tissue damage. Moreover, causal implications are not always clear. Although antisocial acts might be positive symptoms of brain disorganization, representing the release of subcortical activity from inhibitory control, they might alternatively be a negative symptom of brain pathology, in that cerebral impairment results in a deficiency of functions necessary for cognitive development and socialization.

Some 0.5% of the population suffers from epilepsy, a symptom of brain disorganization. Although confined largely to incarcerated samples, the evidence indicates significantly higher rates among offenders. However, this does not necessarily indicate a direct biological effect on behavior, since social problems of sufferers from epilepsy often reflect the experience of the stigma resulting from societal reactions. [See Epilepsy.]

Although there is legal interest in the question of whether crimes can be committed automatically and without awareness (i.e., automatism), automatism rarely accounts for the offenses of criminals with epilepsy, and the evidence indicates that criminal behavior that goes beyond the fragmentary character of automatisms is unlikely to reflect a seizure. It is widely believed that temporal lobe epilepsy increases the likelihood of violence, but the evidence remains controversial. Rage reactions or aggressive outbursts have been reported in more than one-third of adult and child patients with this form of epilepsy, but these estimates might reflect referral biases. In some studies, prisoners with epilepsy were no more likely to have committed violent crimes than were those without epilepsy, although other research has found a correlation between violent history and epileptic symptoms among delinquents. Disparate findings could reflect differing criteria of both violence and epilepsy.

Also controversial is the concept of the dyscontrol syndrome, in which brain lesions are assumed to be causes of violence, even though observable seizures are absent. Dyscontrol is characterized by explosive outbursts of violent behavior with minimal provocation, with evidence of adequate adjustment between episodes, and such behavior is believed to reflect epileptic-like discharges from sites of focal abnormality in the limbic system. Evidence for these abnormalities, however, rests largely on indirect indicators such as EEG abnormalities or soft neurological signs, which are common in patients referred for neurological or psychiatric evaluation following recurring episodes of unprovoked rage. However, dyscontrol simply describes a correlation between a vaguely defined behavioral syndrome and signs of brain pathology whose validity is not clearly established, and histories of childhood deprivation and family violence might equally account for the violence of these patients. Specific neural triggers for aggression have not, in fact, been clearly established in laboratory research, and recent studies of neurological patients suggest that the involvement of brain dysfunction in violence is indirect and nonspecific.

B. Minimal Brain Dysfunction, Hyperactivity, and Learning Disabilities

Many child behavior disorders are assumed to be symptomatic of brain dysfunction resulting from cerebral trauma sustained in the perinatal period or early infancy, and histories of head and facial injuries are found with some frequency in serious delinquents. Such histories have encouraged use of the concept of minimal brain dysfunction (MBD), which refers to a symptom complex of motor restlessness, impulsivity, deficits in attention and learning, and soft neurological signs. "Hyperactivity" has been used interchangeably with "MBD" to denote a similar symptom complex, currently referred to as attention deficit hyperactivity disorder. However, the validity of these concepts remains controversial, and firm evidence that hyperactivity or MBD is symptomatic of brain damage has not been forthcoming.

Although hyperactivity has long been associated with child conduct problems (e.g., aggression or stealing), recent studies suggest that children displaying hyperactive behavior are not uniformly antisocial. However, follow-up studies indicate that childhood hyperactivity is associated with antisocial behavior in adolescence and adulthood and that hyperactive children are more likely to meet criteria for antisocial personality disorder later in life. Apart from the problems of social training presented by a child who is restless and inattentive, the nature of the contribution of hyperactivity to criminality remains unclear. Delinquent outcomes in hyperactive children seem to be related to both early biological aspects and later social factors, including the reactions of parents and others.

Diagnoses of MBD and hyperactivity often overlap with learning disabilities (LDs), which refer to a discrepancy between what is expected of a child on the basis of established ability and actual educational achievements, and which are assumed to have a constitutional basis. The role of LDs in delinquency is of interest in view of the established correlation between poor school performance and delinquency. Retrospective prevalence estimates of LDs in delinquents have ranged from 26% to 73%, but the link remains tenuous. Although learning problems are probably common among delinquents, "LD" subsumes heterogeneous disorders that are unlikely to have a single etiology. Recent attempts to distinguish among LDs indicate varying forms that are differentially related to both academic achievement and behavior disorder, and one recent study found that only 14% of a delinquent sample showed a pattern overlapping with that predominating in a nondelinquent LD group.

C. Neuropsychological Dysfunctions

Performance of offenders on neuropsychological tests has been cited as evidence for brain dysfunction in a number of studies. Consistent evidence of deficits comes from findings with intelligence tests, notably the Wechsler scales, on which delinquents not only score lower than nondelinquents, but also tend to produce larger discrepancies between performance (P) and verbal (V) IQ in favor of the former. The PIQ > VIQ sign is generally interpreted in terms of deficient verbal skills, rather than superior nonverbal abilities, and might be a correlate of LD. VIQ and PIQ are also thought to have some association with the differential functions of the left (i.e., linguistic processing and sequential analysis) and right (i.e., spatial and qualitative analyses) cerebral hemispheres, respectively. The PIQ dominance has therefore been interpreted in terms of reduced left hemispheric lateralization in delinquents. [See Cerebral Specialization.]

Recent studies of both officially defined and self-reported delinquency have found that delinquents produce not only a larger PIQ > VIQ sign, but also poorer scores on a variety of neuropsychological tests. The pattern of deficits shown by delinquents suggests only minimal dysfunction in motor skills or gross sensory functioning, but significant deficiencies in problem-solving abilities requiring verbal, perceptual, and nonverbal conceptual skills, although female delinquents tend to differ from males.

Other studies, however, have suggested that lateralized deficits in the left (i.e., dominant) hemisphere, particularly in frontal and temporal areas, are characteristic of violent and psychopathic offenders. Frontal lobe damage has long been associated with impulsive poorly planned behavior, and the frontal lobes are thought to be involved in the integration and direction of voluntary behavior. Frontolimbic connections are also part of Gray's BIS. Recent findings point to less specialized lateralization of language functions in psychopaths, suggesting that their cerebral organization of language might be marked by poorer integration of affective and other components linking cognition and behavior. This might account for deficiencies in emotional experiences.

Although neuropsychological studies of offenders have so far been limited by an overemphasis on institutionalized populations, impairment of language-related skills and regulative functions controlled by the frontal lobes has been found with some consistency. Current views associate these with development failure, rather than neurological damage, and suggest that delinquents, particularly those identified as violent and impulsive, might have a relative inability to use inner speech to modulate attention, affect, thought, and behavior under conditions of stress.

V. CONCLUSIONS

Children and adults who are more prone to violate social rules differ from those who are more socialized in a number of bodily functions and processes that are likely to reflect innate or congenital influences. Inconsistencies of results in several areas might be attributed to the relatively modest and indirect role played by biological processes in socialization and to the heterogeneity of antisocial groups, and it is unlikely that biological factors are equally relevant for all classes of deviant individuals. Nevertheless, the evidence to date indicates that criminology cannot ignore the relevance of a biological level of analysis. Although no one would now claim that there are "born criminals" or deny that crime is critically dependent on social environments, biological factors appear to increase the likelihood that some individuals will become criminals in some environments.

BIBLIOGRAPHY

Blackburn, R. (1993). "The Psychology of Criminal Conduct: Theory, Research, and Practice." John Wiley & Sons, Chichester, England.

Eysenck, H. J., and Gudjonsson, G. H. (1989). "The Causes and Cures of Criminality." Plenum, New York.

Fishbein, D. J. (1990). Biological perspectives in criminology. *Criminology* **28**, 27–72.

Miller, L. (1988). Neuropsychological perspectives on delinquency. *Behav. Sci. Law* **6**, 409.

Moffitt, T. E., and Mednick, S. A. (eds.) (1988). "Biological Contributions to Crime Causation." Nijhoff, Dordrecht, The Netherlands.

Rowe, D. C., and Osgood, D. W. (1984). Heredity and sociological theories of delinquency: A reconsideration. *Am. Sociol. Rev.* **49**, 526.

Zuckerman, M. (1991). "Psychobiology of Personality." Cambridge Univ. Press, New York.

Cryotechniques in Biological Electron Microscopy

MARTIN MÜLLER

MARCUS YAFFEE*

*Laboratory for EMI and *Biochemistry I, ETH-Zürich*

GLOSSARY

Cryofixation Stabilization of cryoimmobilized, biological structures at low temperatures by cross-linking agents (aldehydes, O_sO_4, embedding resins). The term "cryofixation" is frequently misused in place of "cryoimmobilization"

Cryoimmobilization Near immediate arrest of cellular processes and solidification of a biological specimen by very rapid cooling. Cryoimmobilized samples are stable only at very low temperatures, e.g., 170–100 K

Cryotechniques Biological specimen preparation procedures for electron microscopy utilizing the solid state of water at low temperature

Segregation pattern Compartments formed during ice crystal growth by exclusion of solutes from the crystal lattice. Segregation patterns are the most obvious artifact of microscopical cryotechniques, reflecting insufficient cooling rates

Vitrification Solidification of specimen water in a glass-like amorphous state

CRYOTECHNIQUES HAVE BECOME VERY USEFUL tools in modern biology; macromolecules, cells, and even entire organs can be stored under conditions that preserve viability. Cryotechniques also contribute to microscopical ultrastructural studies. Procedures required to maintain the vital functions of biological matter differ in many ways from those needed to preserve the structural integrity necessary for ultrastructural studies. This article examines the role of cryotechniques applied to biological electron microscopy.

I. INTRODUCTION

One of the important tasks of biological electron microscopy is to provide structural information with which one may correlate structure and function. It is currently the only methodology with the inherent potential to observe structures down to molecular dimensions within the context of complex biological systems. Specimen preparation techniques should be directed toward preserving the smallest significant details in order to fully exploit this unique, integrating feature of biological electron microscopy, thereby complementing the progress of the techniques used in cell biology, biochemistry, and molecular biology.

Problems encountered during preparation of biological specimens for electron microscopy arise from the necessity to transform the hydrated biological sample into a solid state in which it can resist the physical impact of the electron microscope (e.g., high vacuum, electron beam irradiation). Ideally one would prefer specimen preparation procedures that simultaneously guarantee absolute preservation of the dimensions and spatial distribution of diffusible elements. Antigens, receptors, lectin-binding sites, and so on should become demonstrable through immunochemical and other cytochemical techniques. A universal specimen preparation procedure will, perhaps, remain a dream. One must, nevertheless, attempt to realize it so that the integrating power of electron microscopy can further develop into a complementary tool in modern biological research. [See Electron Microscopy.]

The immobilization of biological material kept under optimally controlled physiological conditions is the first and most critical step in attempting to preserve the complex interactions of organelles, macromolecules, ions, and water in a close relationship to the living state. Immobilization must be sufficiently rapid, trapping dynamic events at membranes (e.g., membrane fusion and exocytosis, which occur on a millisecond time scale) and preventing the lateral displacement of lipids and proteins within the membranes. A lateral diffusion coefficient of fluorescent lipid analogs of 3×10^{-11} m^2 sec^{-1} would allow a lateral displacement of approximately 300 nm within 1 msec as well as the displacement of ions, which, depending on their interaction with macromolecules and water molecules, have a lateral diffusion coefficient of up to 10^{-9} m^2 sec^{-1}.

In this respect, techniques based on chemical immobilization (aldehyde fixation) seem to approach their limits, although their use will remain indispensable in solving many relevant biological problems. Chemical fixatives react relatively slowly and cannot preserve all cellular components. Most of the diffusible ions are lost or redistributed during sample preparation. Fixation influences the diffusion properties of the membranes and results, therefore, in alterations of shape, volume, and content of the cell and its components. It becomes evident that the initial potential of electron microscopy, to serve as an integrating source of primary information on the structural complexity at cellular to macromolecular dimensions, can hardly be approached by techniques based on chemical fixation. Immunocytochemical methods represent an important exception because, for many questions, it is irrelevant whether the antigen is detected, for example, in a structurally preserved or a distorted organelle. Electrophysiological experiments demonstrated the effect of 0.2% glutaraldehyde on electrically coupled homokaryons of $BICR/MIR_k$ cells (a permanently growing tumor cell line in monolayer culture derived from a spontaneous mammary tumor of the Marshall rat). A collapse of membrane potentials was found after approximately 3 min. It was concluded from the measurements that glutaraldehyde cannot preserve gap junction channels in their open state.

Cryoimmobilization represents a further reaching alternative. High cooling rates (10^4–10^6K sec^{-1}) are required to prevent the formation and growth of ice crystals, which would affect the structural integrity. The high cooling rates, at the same time, rapidly arrest physiological events (i.e., a high time resolution for dynamic processes in the cell and consequently structural immobilization closely related to the living state).

II. CRYOIMMOBILIZATION

The structural integrity of biological material is guaranteed only if cryoimmobilization brings about solidification of water or solutes in a vitreous or microcrystalline state in the absence of any chemical pretreatment. Heat can be extracted only through the surface of the sample. Heat transfer from deeper within the specimen is limited by the low thermal conductivity of water and the developing solid layer. Extremely high cooling rates can be achieved at the surface of the sample, thus leading to the immobilization of a thin surface layer in the vitreous state. Insufficient cooling rates allow ice crystals to form deeper in the sample. More heat is produced by ice crystal formation with increasing depth than is transferred through the ice to the cooled surface. This progressively reduces the cooling rate and results in increasing ice crystal dimensions.

A. Vitrified Samples

Freezing cellular water in the amorphous state (vitrification) would be ideal because the basic nature of the liquid phase would be preserved. Vitrification of cellular water, however, requires very high cooling rates, present only in thin layers at the specimen surface. These layers are usually too thin (100 nm) to encompass the bulk of the sample, and further processing for electron microscopy entails some difficulties. True vitrification of biological solutions has been demonstrated by low-temperature electron diffraction using the "bare grid" technique. Thin aqueous layers (<100 nm) of suspensions of viruses, phages, liposomes, and macromolecules are formed by this technique within the meshes of an electron microscope grid, either bare or coverted with a perforated carbon foil, vitrified by immersion into liquid ethane, and observed in the microscope at low temperatures using a cold stage.

B. Rapid Freezing Techniques

Aqueous layers of suspensions of cells, microorganisms, organelles, and tissue cultures up to a thickness of approximately 10–20 μm are adequately cryoimmobilized by several rapid freezing techniques. These techniques include plunge freezing, wherein the thin sample layer is sandwiched between two metal platelets and plunged into liquid coolants (propane and ethane are preferentially used); propane-jet freezing, wherein two jets of liquid propane are directed onto the specimen sandwich from opposite directions simultaneously; and spray freezing, wherein small droplets of the sample are sprayed into the liquid coolant. Impact freezing allows adequate cryoimmobilization of a thin layer at the surface of suspension droplets or tissue pieces by forced contact with a polished metal surface kept at liquid–helium or liquid–nitrogen temperature. The freezing rate achievable in aqueous layers of 10–20 μm may be too low for vitrification but high enough to prevent growth of damaging ice crystals. Solutes, however, are excluded from the crystal lattice when ice crystals grow and are concentrated between neighboring ice crystals to form a eutectic. The eutectic appears in electron micrographs as a network of segregation compartments in freeze-substituted or freeze-fractured biological material that has been cryofixed with insufficient cooling rates. In practice, the absence of segregation patterns is a generally accepted indication of adequate cryoimmobilization, reflecting high cooling rates but not necessarily vitrification. The concentration of solutes by crystallization leads to drastic local changes in osmolality and pH, which may induce conformational changes (e.g., of proteins). Structural details may, therefore, be altered even if no segregation patterns are visible.

The appropriate rapid freezing procedure can reproducibly yield adequate freezing of very thin samples, immobilizing the specimens in a state most closely related to the living situation, and thus allowing one to investigate significant structural details as a function of the physiological state. Rapid freezing procedures are, however, generally not suitable for cryoimmobilization of more complex samples (e.g., animal and plant tissues). Useful structural information in a thin superficial zone at the natural or cut surface of tissue samples can be obtained by impact freezing. The thickness of the zone, in which no segregation patterns are visible, depends on the concentration of cellular components that exhibit cryoprotective properties and may often reach 20 μm. This is generally, however, still insufficient to analyze intact tissue cells.

C. High-Pressure Freezing

Thicker, more complex systems can be studied by cryofixation-based electron microscopy only if the physical properties of the cellular water are manipulated in a way that adequate cryoimmobilization is achieved with much slower cooling rates. The impregnation of larger samples with cryoprotectants, usually in combination with aldehyde fixation, is frequently employed. Numerous artifacts, however, can be introduced by these procedures.

The sole aim of cryofixation is to physically immobilize the specimen. The development of an alternative method based on the application of high hydrostatic pressure was started in the mid-1960s by Moor and co-workers and adequate instrumentation subsequently became commercially available. Freezing increases the volume of water. This expansion is hindered by high pressure, thereby increasing the viscosity. This effect is demonstrated by a lowering of the freezing point and reduced rates of nucleation and ice crystal growth. Consequently, less heat is produced by crystallization, and less heat has to be extracted per unit time by cooling. This means that adequate immobilization can be achieved with reducing cooling rates. At a pressure of 2045 bars, the melting point of water is lowered to 251K, and the temperature of homogeneous nucleation is reduced to 181K, as deduced from the phase diagram of water.

High-pressure freezing is the only practical way of cryofixing larger unpretreated samples. It permits structural analysis of more complex systems, i.e., plant and animal tissue, tissue culture cells grown on established substrates (e.g., glass coverslips), and larger amounts of cellular suspensions. The sample thickness that can be adequately cryoimmobilized greatly depends on the specimen itself, e.g., on the concentration of components that exhibit cryoprotective activities. It may reach several hundreds of micrometers (e.g., 500 μm in special cases such as young plant leaves and root tips). Practical experience and refined calculations of the heat flux in aqueous samples, however, indicate that adequate cryoimmobilization is generally only achieved in samples not exceeding a thickness of 200 μm. These advantages are somewhat reduced by the relatively slow cooling rates (approximately 5000K sec^{-1}) achieved in the center of a 200-μm-thick sample. These rates may be too slow to catch dynamic events at membranes or to prevent structural alterations due to lipid-phase transition and segregation phenomena. However, the transition temperature of membrane lipids is raised by approximately 20K kbar^{-1}. This means that by applying a pressure of more than 2 kbar (attained in approximately 25 msec), the membrane lipids may be immobilized very quickly purely by the action of high pressure. Experimental data supporting this assumption, as well as other short-lived high-pressure effects on biological material, are not yet available. Possible reactions of biological specimens exposed to high pressures for periods in excess of 1 min have been observed.

Unsatisfactorily low yields of adequately frozen biological samples frequently hamper attempts to correlate structure and function. This has been especially true for high-pressure freezing until Studer and colleagues improved the transfer of pressure and cold to the biological specimen sandwiched between two metal supports that contain an appropriate cavity. This method gives high yields by quickly immersing the samples in 1-hexadecene prior to freezing. We believe that 1-hexadecene removes the free water surrounding the specimens, thereby reducing the danger of extraspecimen ice nucleation. It also completely fills the space unoccupied by the specimen in the cavity of the metal sandwich, assuring a good thermal contact with the metal platelets.

Excision of tissue specimens, as well as preparation of tissue samples suitably sized to fit into the metal supports for high-pressure freezing or to be mounted on special holders for plunge freezing or impact freezing, requires some time (e.g., 2–3 min), during which physiological changes and structural alterations undoubtedly occur. These changes may reach an extent that prohibits any attempt to correlate structure and function at high resolution. This problem, while inherent to most ultrastructural studies, may only be overcome by *in situ* freezing approaches, which are as yet unavailable. The problem of tissue excision and sizing is particularly severe for high-pressure freezing because the sample must not exceed 200 μm. A step in the right direction has been described by Hohenberg and co-workers. It has become possible to produce undistorted tissue slabs of suitable size and to freeze them under high pressure in significantly less than 1 min by means of a commercially available "fine needle microbiopsy" system.

Suspensions of cells (microorganisms, blood cells, tissue culture cells) and smaller organisms (nematodes, euglena) are frozen successfully at high pressure in porous cellulose capillary tubes that have an inner diameter of 200 μm. The organisms are filled into the tubes by capillary attraction, i.e., by dipping the tubes into the suspension.

III. FOLLOW-UP PROCEDURES

Successfully cryoimmobilized samples have to be processed further for electron microscopic analysis by various follow-up procedures, each of which yields different information and poses different technical problems. Subsequent processing has to be performed at sufficiently low temperatures at which devitrification and secondary ice crystal growth are avoided. The devitrification range for vitreous, amorphous water was determined to be approximately 140K through means of low-temperature electron diffraction. Vitreous water recrystallizes into cubic ice at higher temperatures. In this state, no effects of phase separation due to ice crystal formation are visible when employing the most frequently used follow-up procedures (i.e., freeze-fracturing, freeze-substitution). At still higher temperatures, however (e.g., above 190K), cubic ice may be transformed into hexagonal ice in which modification ice crystals may rapidly grow and alter the specimen.

A. Physical Procedures

Low-temperature electron microscopy of vitrified, thin aqueous layers and of cryosections, as well as freeze-fracturing and freeze-drying, are considered to be direct, purely physical procedures. They provide reliable structural information most closely related to the living state. Cryosectioning of untreated, cryofixed biological material is a very demanding technique, and the sections are often too thick to provide structural information at high resolution. Cryosections observed in the microscope at low temperatures, however, either frozen hydrated or freeze-dried, represent the best, if not the only, means of obtaining qualitative and quantitative information on the spatial distribution of diffusible ions by X-ray microanalysis.

Freeze-fracturing is the easiest and best-established physical technique to obtain a "safe" representation of structural details (down to a resolution of approximately 5 nm). Many reviews treat this technique in detail. Freeze-fracturing exposes specific structural aspects depending on the fracturing behavior of the sample and its components. It is especially suited to the characterization of membranes because, for energetic reasons, the fracture plane proceeds through the hydrophobic interior of the membranes, thus providing information about size and distribution of intramembraneous particles (IMP). The pattern formed by the IMPs is characteristic for each specific membrane fracture face. Alterations of these specific patterns may reflect dynamic processes at membranes or, if the sample is cryoimmobilized too slowly, the occurrence of phase transitions and segregation phenomena of the lipid phase. The nature of IMPs is still under discussion. They may indicate the positions of transmembrane or intramembranous proteins but may also be lipid in nature. Freeze-fracturing is performed in a high vacuum at very low temperatures where the fracture planes can be replicated by shadowing with platinum/carbon (2 nm) followed by a carbon backing layer (2 nm). Outside the vacuum, the biological material is removed so that the replica can be examined in a transmission electron microscope (TEM). Identically freeze-fractured and metal-shadowed samples can also be imaged in scanning electron microscopes (SEM), thereby eliminating the need to remove the biological material from the replica. The samples have to be transferred to the cold stage of the SEM at low temperature either immersed in liquid nitrogen or by a vacuum cryotransfer system. The signal of the back-scattered electrons is used for imaging. An "in-lens" field emission SEM equipped with a highly sensitive detector for the back-scattered electrons is required in order to achieve a resolution of structural details that is comparable to the examination of a replica in the TEM. None of the aforementioned direct physical procedures is suitable for immunocytochemical work unless mild chemical fixation and cryoprotection precede cryofixation as immunocytochemical reactions must be performed at room temperature. [See Low-Temperature Scanning Electron Microscopy.]

B. Hybrid Techniques

Freeze fracturing is a straightforward, easily handled technique that provides reliable structural information. Its major disadvantages are that it can be used for hardly anything other than structural description and that the fracture plane proceeds at will. These problems can be partially overcome by hybrid techniques, which combine the advantages of cryofixation with those of conventional plastic embedding and thin-sectioning procedures. Freeze-substitution and freeze-drying are frequently employed. Both procedures are essentially dehydration processes: freeze-substitution dissolves the ice in a cryoimmobilized specimen with an organic solvent whereas freeze-drying eliminates the frozen water by sublimation in a vacuum chamber. Freeze-drying and freeze-substitution must be executed well above the devitrification range of amorphous water (~140K). Because of low vapor pressure, freeze-drying at temperatures below approximately 150K lead to impractically long drying times. The temperature limits during freeze substitution are set by the melting point of the solvent used and the amount of water the solvent can take up at low temperatures. Generally, temperatures of 180–190K are considered to be "safe" for cryoimmobilized biological samples because of the rather high natural cryoprotective activity of many cellular components. After completion of the dehydration process, the samples are warmed to room temperature, infiltrated with the embedding resin, and heat polymerized. With respect to the preservation of structural integrity, these hybrid techniques are much more obscure than the purely physical follow-up procedures such as cryosectioning or freeze-fracturing (discussed earlier), as effects of the organic solvents (e.g., lipid extraction) and the embedding chemistry cannot be excluded. The structural description provided by the physical procedures, in which the water remains in the specimen, represents, therefore, the standard by which all the other procedures have to be measured. Freeze-substitution and freeze-drying may allow an accurate control of the dehydration process, but because of our incomplete knowledge of cellular water, they are still insufficiently understood. Our present knowledge about the role of water in the cell and the effects of its removal are summarized by the following statements:

1. Water in the cell exhibits different physicochemical properties and is classified into two major groups: "bulk" or "free water," in contrast to anomalous water referred to as "bound water" or "nonfreezable water." This anomalous water is supposed to be closely associated with surfaces (e.g., of macromolecules, membranes, and ions) and is sometimes termed "hydration shell," "surface-modified water," or "vicinal water."

2. This surface-modified water is extremely important for all metabolism as well as for the maintenance of the structural integrity of proteins and other cell constituents.

3. One may conclude that bulk water is more easily removed and affects the preservation of structural

integrity less than the water of hydration shells during the dehydration process in biological electron microscopy.

These assumptions are supported by the nonlinear shrinkage behavior of cells and tissues during conventional dehydration of chemically fixed material by organic solvents at room temperature. Cells start to shrink when approximately 70% of cellular water is replaced by an organic solvent. Fully dehydrated, they shrink up to 30–70% of their initial volume. This is a primary indication that a specific fraction of the cellular water can be removed that does not introduce gross dimensional changes. There is, however, water closely associated with the cellular structures. Removing this residual water may lead to conformational changes of cellular components (collapse) and aggregation. The temperatures above which different types of macromolecules collapse when exposed to dehydrating agents, such as organic solvents and vacuum, have been determined. These temperatures range from 215 to 263K and appear to depend solely on the temperature and the polarity of the dehydrating agent. Experimenters have used an ultrahigh vacuum apparatus to study the freeze-drying of test specimens containing deuterium oxide (D_2O) instead of H_2O and followed its escape with a mass spectrometer. They observed that a first peak of D_2O evaporating in the temperature range of 180 to 190K approached zero after 2 hr at 190K. A second peak of D_2O was observed only after further heating the specimen and reached a maximum between 220 and 230K. Thus, the water is held in the tissue by different forces, and it may be concluded that some of the specimen water is bound, therefore, needing a higher energy for evaporation than the free water. The temperature at which the second peak of D_2O was observed is within the range of the collapse temperature, and it may be speculated that it corresponds to the water of the hydration shells.

Ideally, freeze-drying and freeze-substitution could be used to control the residual water content (i.e., how much water must remain in order that cells maintain their structural and functional integrity, and how much water must be removed to permit successful plastic embedding). Experiments have shown, however, that the hydration shells can prevent an efficient copolymerization between biological material and resin. Strongly hydrated organelles may therefore not become embedded at all, and resin and biological material may separate very easily along membranes. Freeze-drying, in contrast to freeze-substitution, has not yet found wide application for dehydrating cryofixed samples for subsequent plastic embedding. It is, however, used very successfully in combination with metal shadowing in transmission as well as in scanning electron microscopy. The organic solvents used to freeze-substitute cryofixed biological samples frequently contain fixatives (e.g., O_sO_4, uranyl ions, aldehydes). One assumes that these stabilize the biological structures at the ice solvent interphase or during the gradual or stepwise increase of the temperature, but little is known about their reactivity at these high-subzero temperatures. Experimental evidence shows, however, that uranyl ions react and prevent the extraction of phospholipid by solvent at even the lowest temperatures (180K). A reaction of O_sO_4 with the double bonds of unsaturated fatty acids has been reported to occur at 203K. Glutaraldehyde starts to effectively cross-link proteins at 223K. Fixatives may be necessary to reduce solvent effects (e.g., the loss of lipid and other low molecular constituents). They might also help reduce the effects of conformational changes and aggregations of macromolecules and supramolecular structures, which inevitably occur as the hydration shells are removed at higher temperatures (cf. collapse temperatures). The much more homogeneous finer-grained appearance of the cytoplasm after freeze-substitution, when compared to conventionally dehydrated samples, supports this assumption. The presence of fixatives in the substitution medium is essential if freeze-substitution is followed by conventional embedding at room temperature and heat polymerization at 335K. Samples prepared in this way exhibit excellent structural detail with conservation of the dimensions comparable to freeze-fracturing, yielding valuable new ultrastructural information and identifying many artifacts associated with the conventional procedures based on chemical fixation and dehydration. All membranes possess smooth contours, which, after chemical fixation and dehydration, frequently appear undulated. Plant vacuoles appear fully turgescent. Mitochondria exhibit a close apposition of the two membranes, which is reflected by a frequent deflection of the fracture plane between the inner and the outer membrane after freeze-fracturing and a multilaminar appearance after freeze-substitution. Figure 1 illustrates the different appearance of the ultrastructure of rat liver after chemical fixation with glutaraldehyde and osmium tetroxide (O_sO_4) (Fig. 1A) and after high-pressure freezing followed by freeze-substitution in acetone/osmium tetroxide (Fig. 1B). Both samples were embedded in Araldite/Epson at room temperature and heat polymerized at

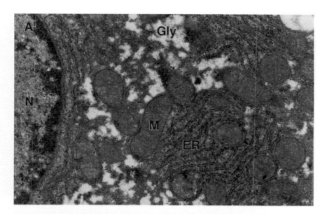

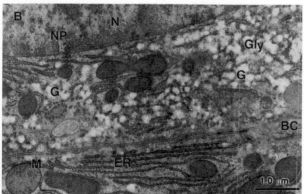

FIGURE 1   The appearance of rat liver ultrastructure after chemical fixation and dehydration (A) and that of high-pressure cryoimmobilization, followed by freeze-substitution (B). Note the various aspects (e.g., membranes). N, nucleus; NP, nuclear pore; M, mitochondrion; ER, endoplasmic reticulum; G, Golgi complex; Gly, glycogen; BC, bile caniculus.

333K. Excellent structural preservation is obtained by freeze-substitution in acetone- or methanol-containing fixatives like osmium tetroxide, uranyl ions, or aldehydes either alone or combined, followed by conventional embedding and heat polymerization. Samples prepared in this way, however, are hardly useful for cytochemical studies. Specimens that exhibit both an optimal structural description and the labeling of intracellular antigens are obtained by combining freeze-substitution with low-temperature embedding in Lowicryl. Freeze-substitution followed by low-temperature embedding and polymerization is currently under study in many laboratories and will undoubtedly find more applications due to the possibility of high-pressure cryoimmobilization of larger and more complex samples. The label efficiency of freeze-substituted and low-temperature-embedded samples is mainly affected by osmium tetroxide as a stabilizing additive. It should therefore be avoided.

Uranyl ions and aldehydes at low concentration often show no untoward effects with respect to label efficiency; they again help to minimize effects of the organic solvent and the Lowicryls. Furthermore, they may improve the stainability of the biological structures. Figure 2 presents an example. A primary culture of adult rat heart muscle cells, grown for 7 days on a carbon-coated, 50-μm-thick glass substrate, was high-pressure frozen and freeze-substituted in ethanol containing 0.5% uranylacetate. Freeze-substitution was performed at 183K for 8 hr followed by 3 hr at 223K. Infiltration with Lowicryl HM 23 and polymerization by UV light were executed at the same temperature. Ultrathin sections were cut parallel to the glass substrate. Figure 2A shows an overview and illustrates adequate structural preservation and identification.

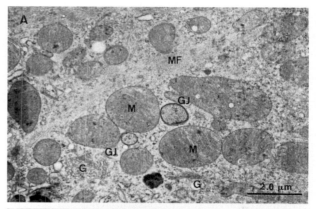

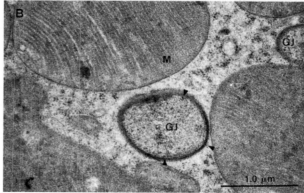

FIGURE 2   A portion of a high-pressure cryoimmobilized rat heart muscle cell in culture, freeze-substituted in the absence of fixatives, low-temperature embedded, and polymerized at 223K. Thin sections of material prepared in this way frequently permit successful immunolabeling and display adequate structural preservation (A). An internalized gap-junction vesicle (GJ), labeled with an antibody against a gap-junction protein, is shown in (B) at higher magnification. The arrowheads point to 8-nm colloidal gold particles used to localize the primary antibody. G, Golgi complex; M, mitochondria; MF, myofilament.

The thin section is treated with an antibody against a gap-junction protein. The primary antibody is visualized by 8-nm gold colloids, coupled to protein A from *Staphylococcus aureus*. The internalized gap-junction vesicle, shown in Fig. 2B at higher magnification, is clearly identified by the gold colloids. Freeze-substitution in a pure solvent can be combined with low-temperature embedding at low temperatures under carefully controlled conditions.

IV. CRYOTECHNIQUES IN SCANNING ELECTRON MICROSCOPY

The factors affecting the preservation of the structural integrity are identical in both scanning and transmission electron microscopy. Shrinkage due to complete dehydration and drying may be even more pronounced because the removed water is usually not replaced by an embedding resin. Techniques based on cryofixation help overcome the major problems.

A. Low-Temperature Scanning Electron Microscopy

Low-temperature scanning electron microscopy (LTSEM) embodies the direct, physical approach to structural studies. Modern LTSEM equipment consists of a high-vacuum preparation chamber attached directly to the scanning microscope. The specimen can be kept in the preparation chamber at controlled temperatures. It may be retained intact, fractured, or dissected and kept fully frozen hydrated, partially freeze-dried ("etched"), or fully freeze-dried. The samples may be coated for subsequent observation in the SEM. The gate valve between the preparation chamber and the microscope is then opened, and the sample is transferred onto a temperature-controlled stage in the SEM where it is examined at low temperatures (e.g., 100K). Uncoated samples may be repeatedly dissected or "etched" if necessary. They may also be partially freeze-dried under visual control in the microscope. LTSEM experiments illustrate various dimensional and structural changes that occur in partially or completely freeze-dried specimens, suggesting that the traditional classification of "free" or "bulk" water as opposed to "bound water" or water of hydration shells might provide a much simplified view of cellular water. Models that more comprehensively describe the complex interactions of macromolecules,

membrane surfaces, ions, and water are supported. LTSEM is currently bound to lower magnifications (e.g., up to 10,000×), mainly due to technical limitations of the equipment (e.g., stability of cold stages, moderate resolving power of conventional SEM instruments). A safe representation of structural facts is, however, always more valuable than the high-resolution detection of insignificant structural details. LTSEM will undoubtedly develop toward improved resolution. Reasonably priced, easily handled, high-resolution SEM instruments, equipped with reliable field emission guns as well as more sophisticated cryo-preparation attachments, are now commercially available. Furthermore, high-pressure freezing offers adequate cryofixation of samples of significant size. [*See* Scanning Electron Microscopy.]

B. Freeze-Drying and Freeze-Substitution for SEM Observation at Room Temperature

Freeze-drying of cryofixed samples can be performed under well-controlled conditions only if the temperature is homogeneous throughout the sample. This is achieved only in very thin samples that remain in perfect thermal contact with the stage of the freeze-drier (e.g., membranes, frozen hydrated cryosections, macromolecular solutions). Evidence shows that such specimens can be kept partially freeze-dried at 193K without apparent shrinkage. A certain amount of water is removed from the specimen at this temperature. More water is released only when the specimen is warmed to the temperature range between 223 and 243K and is referred to as the water of hydration shells. Shrinking occurs in this temperature range, which is well in accordance with the collapse temperatures (263–223K) for freeze-dried model solutions. Removal of the hydration shells most certainly leads to conformational changes of proteins and macromolecules. Whether or to what extent it is the main factor responsible for dimensional alterations (e.g., shrinkage) is not yet clear. In a cell, water may play a much more complex role than in the aforementioned model experiments and its controlled removal may be very difficult.

Only fully freeze-dried samples can be examined in a SEM at room temperature. Specimens, therefore, have always suffered from shrinkage and collapse. Nevertheless, fully freeze-dried samples can provide useful information because, in contrast to the conventional critical point drying procedure, any interaction with organic solvents is avoided. Such a solvent effect is illustrated by the different appearance of the yeast cell surface after conventional and cryofixation-based preparations for electron microscopy. The cells reveal a smooth surface after chemical fixation and dehydration in graded ethanol series followed by critical point drying. Hair-like structures called "fimbriae," some of which are identified as acid phosphatase, a highly glycosylated glycoprotein, are detected in SEM after cryoimmobilization followed by complete freeze-drying and in TEM after partial freeze-drying ("deep etching") and replication or freeze-substitution. Freeze-drying should be performed with a clean solvent (e.g., distilled water) in order to avoid the deposition of solutes onto the specimen surface during drying. Aldehyde prefixation, therefore, is frequently required to render the specimen resistant against treatment with distilled water. Freeze-substitution followed by critical point drying helps partially overcome this problem in the medium resolution range. Various freeze-substitution media have been systematically tested for stabilizing the metachronal wave of the cilia of paramecium in the SEM.

C. A Step Beyond: Structural Analysis of Cryoimmobilized Phage, Protein, and DNA

Ultrahigh resolution scanning electron microscopes of the "in-lens" type are now commercially available. Equipped with a field emission gun, they allow one to examine the samples with a diameter of the electron probe of <1 nm. Adequate structural preservation is required to make full use of this resolving potential. Experiments using the T_4-polyhead mutant as a model specimen showed that only rapid freezing followed by freeze-drying revealed the ring-like structure of the capsomers of 8 nm diameter composed of six subunits (Fig. 3). Conventional critical point drying after chemical dehydration of the aldehyde-fixed samples by ethanol, as well as after freeze-substitution, failed to preserve structural details at this level. Figure 3 shows a secondary electron image of a rapidly frozen, freeze-dried T_4-polyhead preparation. It was freeze-dried at 188K for 30 min, coated with a thin (1–2 nm), continuous film of chromium, warmed to room temperature, and examined in a ultrahigh resolution scanning electron microscope (Hitachi S-900) at a primary magnification of 200,000×.

These studies helped establish methodology applied to the following cryopreparation and low-

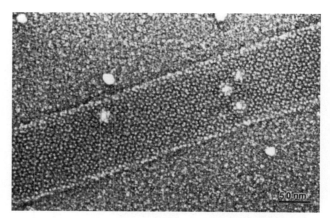

FIGURE 3 The resolution that can be obtained with modern "in-lens" scanning electron microscopes, provided that adequate specimen preparation procedures are available. T_4-polyheads were adsorbed to a carbon foil and rapidly frozen by plunging into liquid ethane. Freeze-dried at 188K, the samples were coated with a thin (1–2 nm), continuous film of chromium. The capsomers consisting of six subunits are readily identified.

temperature observations of protein and DNA structures.

The resolution of a tenascin molecule, down to the level of constituent domains, was achieved by analysis of cryo-fixed samples. The fibronectin-like repeat region of tenascin is revealed as a staggered array (Fig. 4a), which confirms predictions from crystallographic studies (Fig. 4a, insert). The elucidation of iron complexes bound to double-stranded mitochondrial DNA *in situ* was achieved through the analysis of liver mitochondrial DNA exposed to various doses of Fe(III):gluconate. Figure 4b exhibits a whole mitochondrial DNA molecule in a supercoiled, type I form. This molecule was extracted from Wistar rat liver mitochondria. DNA isolated from mitochondria incubated with Fe(III):gluconate complexes exhibited iron complexes directly bound to double-stranded regions: the metal character was elucidated through BSE detection and was localized in concert with ultrahigh, resolution topological SE imaging of the mitochondrial DNA.

Figures 3 and 4 illustrate some of the analytical possibilities that scanning electron microscopy, optimized with respect to both structural preservation and instrumentation, can presently achieve.

V. CONCLUSIONS

Cryofixation alone can immobilize biological structures close to the living state. The application of cryo-

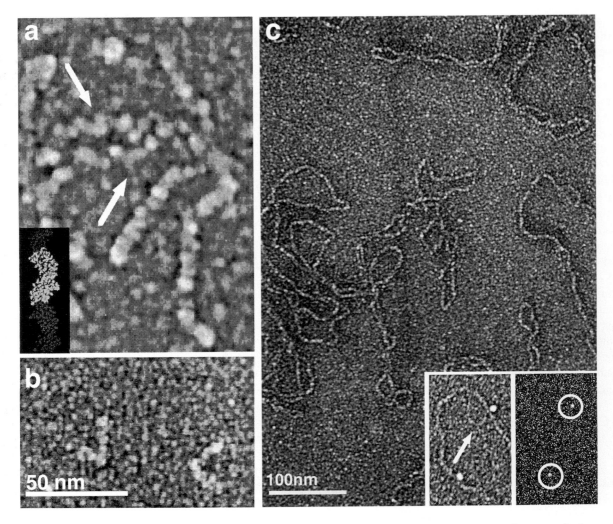

FIGURE 4 High-resolution, field emission, in-lens scanning electron microscopy (HR-SEM) imaging of tenascin displays the fn-repeat domains clearly in this hexamer (a, arrows). The neuronal ligand of tenascin, contactin/F11, displays a similar offset in the IgG-like region (b). This reoccurring motif of offset elbows has been suggested to facilitate side-by-side binding of receptors and ligands. HR-SEM imaging permits direct observation of tight twisting in the three-dimensional surface topology of supercoiled DNA obtained from control pVC21 plasmide (c). Surface topography and BSE metal imaging permitted structural analysis of mtDNA from mitochondria incubated in Fe(III):gluconate. Complexes are clearly visible in the SE (c, left insert) and BSE (c, right insert) images. The arrow cites a putative displacement or D-loop (c, left insert), and BSE iron complexes are circled (c, right insert). Molecules were sprayed onto an oxygen plasma-etched carbon foil, immediately frozen by plunging in liquid nitrogen, and kept at 170K, 10^{-7} Mbar (10^{-5} Pa) for 4 hr prior to DARS coating with 10 Å chromium. The background deposition of chromium has a fine and random texture. Primary images were recorded at 30 kV on Kodak T-Max film in a Hitatchi S-900 SEM equipped with a Gatan cryostage.

fixation-based preparative procedures is mandatory for the correlation of structure and function at the ultrastructural level. Purely physical techniques provide safe structural description and preserve the spatial distribution of diffusible elements. Hybrid techniques (e.g., freeze-substitution) facilitate thin sectioning and, in combination with low temperature embedding, immunocytochemical studies of intracellular antigens. The quality of structural preservation, however, has to be gauged by purely physical procedures such as cryosectioning and freeze-fracturing. Adequate procedures to cryofix suspensions and thin superficial layers of tissues are available. High-pressure freezing offers the opportunity to cryofix samples up to a thickness of approximately 0.2 mm. Samples that can be prepared in thin layers are successfully cryoimmobilized by the rapid freezing techniques under controlled physiological conditions. The relatively long time needed to excise tissue samples and to size them suitable for cryoimmobilization frequently prohibits the correlation of structure and function at high resolution.

ACKNOWLEDGMENTS

The authors thank C. Richter, P. B. Walter, A. Wasserfallen, D. Studer, M. Michel, R. Hermann, H. Hohenberg, P. Walther, and P. Schwarb for their contributions.

BIBLIOGRAPHY

Beckett, A., and Read, N. D. (1986). Low temperature scanning electron microscopy. *In* "Ultrastructure Techniques for Microorganisms" (H. C. Alrich and W. J. Todd, eds.), pp. 45–86. Plenum Press, New York/London.

Clegg, J. S. (1979). Metabolism and the intracellular environment: The vicinal-water network model. *In* "Cell-Associated Water" (W. Drost-Hansen and J. S. Clegg, eds.), pp. 363–413. Academic Press, London/New York.

Dubochet, J., Adrian, M., Chang, J. J., Homo, J. C., Lepault, J., McDowall, A. W., and Schultz, P. (1988). Cryo-electron microscopy of vitrified specimens. *Quart. Rev. Biophys.* **21**(2), 129–228.

Gross, H. (1987). High resolution metal replication of freeze-dried specimens. *In* "Cryotechniques in Biological Electron Microscopy" (R. A. Steinbrecht and K. Zierold, eds.), pp. 205–215. Springer, Berlin.

Hohenberg, H., Mannweiler, K., and Müller, M. (1994). High-pressure freezing of cell suspensions in capillary tubes. *J. Microsc.* **175**, 34–43.

Hohenberg, H., Tobler, M., and Müller, M. (1996). High-pressure freezing of tissue obtained by fine needle biopsy. *J. Microsc.* **183**, 133–139.

Huelser, D. F., Paschke, D., and Greule, J. (1989). Gap junctions: Correlated electrophysiological recordings and ultrastructural analysis by fast freezing and freeze-fracturing. *In* "Electron Microscopy of Subcellular Dynamics" (H. Plattner, ed.), pp. 33–49. CRC Press, Boca Raton, FL.

Kellenberger, E. (1987). The response of biological macromolecules and supramolecular structures to the physics of specimen cryopreparation. *In* "Cryotechniques in Biological Electron Microscopy" (R. A. Steinbrecht and K. Zierold, eds.), pp. 35–63. Springer, Berlin.

Knoll, G., Verkleij, A. J., and Plattner, H. (1987). Cryofixation of dynamic processes in cells and organelles. *In* "Cryotechniques in Biological Electron Microscopy" (R. A. Steinbrecht and K. Zierold, eds.), pp. 258–271. Springer, Berlin.

Moor, H. (1987). Theory and practice of high pressure freezing. *In* "Cryotechniques in Biological Electron Microscopy" (R. A. Steinbrecht and K. Zierold, eds.), pp. 175–191. Springer, Berlin.

Negendank, W. (1986). The state of water in the cell. *In* "The Science of Biological Specimen Preparation" (M. Müller, R. P. Becker, A. Boyde, and J. J. Wolosewick, eds.), pp. 21–32. SEM, AMF O'Hare, Illinois.

Pinto da Silva, P., and Kan, F. W. K. (1984). Label fracture: A method for high resolution labelling of cell surfaces. *J. Cell Biol.* **99**, 1156–1161.

Plattner, H. (1989). "Electron Microscopy of Subcellular Dynamics." CRC Press, Boca Raton, FL.

Plattner, H., and Bachmann, L. (1982). Cryofixation: A tool in biological ultrastructural research. *Int. Rev. Cytol.* **79**, 237–304.

Robards, A. W., and Sleytr, U. B. (1985). "Low Temperature Methods in Biological Electron Microscopy." Elsevier, Amsterdam.

Sitte, H., Edelmann, L., and Neumann, K. (1987). Cryofixation without pretreatment at ambient pressure. *In* "Cryotechniques in Biological Electron Microscopy" (R. A. Steinbrecht and K. Zierold, eds.), pp. 87–113. Springer, Berlin.

Steinbrecht, R. A., and Müller, M. (1987). Freeze-substitution and freeze-drying. *In* "Cryotechniques in Biological Electron Microscopy" (R. A. Steinbrecht and K. Zierold, eds.), pp. 149–172. Springer, Berlin.

Steinbrecht, R. A., and Zierold, K. (1987). "Cryotechniques in Biological Electron Microscopy." Springer, Berlin.

Studer, D., Michel, M., and Müller, M. (1989). High pressure freezing comes of age. *Scan. Microsc. Suppl.* **3**, 253–269.

Tanford, C. (1980). "The Hydrophobic Effect: Formation of Micelles and Biological Membranes," 2nd Ed. Wiley, New York.

Tokuyasu, K. T. (1984). Immunocryoultramicrotomy in immunolabelling for electron microscopy. *In* "Immunolabelling for Electron Microscopy" (J. M. Polak and I. M. Varndell, eds.), pp. 71–82. Elsevier, Amsterdam.

Walther, P., Wehrli, E., Hermann, R., and Müller, M. (1995). Double-layer coating for high resolution low-temperature scanning electron microscopy. *J. Microsc.* **179**, 229–237.

Wildhaber, I., Gross, H., and Moor, H. (1982). The control of freeze-drying with deuterium oxide (D_2O). *J. Ultrastruct. Res.* **80**, 367–373.

Yaffee, M. I., Walter, P., Richter, C., and Müller, M. (1996). Direct observation of iron-induced conformational changes of mitochondrial DNA by high-resolution field-emission in-lens scanning electron microscopy. *Proc. Nat. Acad. Sci. U.S.A.* **93**, 5341–5346.

Cystic Fibrosis, Molecular Genetics

ERIC J. SORSCHER

University of Alabama at Birmingham

GLOSSARY

Cystic fibrosis transmembrane conductance regulator Gene that is abnormal in cystic fibrosis

Gene product Protein encoded by a particular gene

Vector Vehicle (viral or nonviral) for delivering a gene to a targeted cell

CYSTIC FIBROSIS (CF) LEADS TO ABNORMALITIES IN the salt metabolism of exocrine secretions and is predominantly a disease of secretory tissues. The gene responsible for the disease, the cystic fibrosis transmembrane conductance regulator (CFTR), functions as an epithelial chloride channel and may govern many other important aspects of epithelial cell biology. The CFTR contains 1480 amino acids; over 500 mutations have been described that are associated with clinical disease. The most common mutation, accounting for approximately 70% of defective CF alleles, is omission of a single phenylalanine residue at CFTR position 508 (ΔF508). Progress in understanding the mechanisms by which CFTR mutations cause disease has led to new approaches to CF therapy, including the possibility of gene replacement.

I. CLINICAL MANIFESTATIONS OF CYSTIC FIBROSIS

An autosomal recessive disease, cystic fibrosis is most frequent among Caucasians. One in 2000 to 3000 Caucasian babies in the United States develop the disease. From the standpoint of the pediatrician, cystic fibrosis often manifests itself as a failure to thrive, that is, a failure to meet the normal development milestones for height and weight. A positive sweat test confirms the diagnosis. In cystic fibrosis, the levels of sweat sodium and chloride are severalfold higher than those in non-cystic fibrosis individuals. Substantial elevations of sweat electrolytes (sodium and chloride) are virtually pathognomonic for the disease; there are very few other conditions that cause sweat to have such high salt content. The failure to thrive is usually attributed to functional defects in the exocrine pancreas, leading to defective secretion of the pancreatic enzymes necessary for normal digestion of proteins, fats, and carbohydrates. Histopathologically, the affected pancreas is characterized by cytolysis, fatty, dense fibrotic tissue, and loss of the normal microscopic architecture of the gland.

The failure to secrete pancreatic digestive enzymes is treated by administering these same enzymes (e.g., derived from the pig pancreas) at meal time. Malabsorption and many of the nutritional aspects of CF can be managed in this way. Later in life, however, other exocrine tissues become damaged or destroyed by the disease process. At some point during the lives of most patients, thick, hyperviscous secretions in the airways lead to chronic damage, fibrosis, and pulmo-

ENCYCLOPEDIA OF HUMAN BIOLOGY, Second Edition, VOLUME 3.

nary infections. These airway manifestations are the major cause of morbidity and mortality in CF. Most patients die of pulmonary complications by age 30, although exceptions to this rule and a much longer life expectancy have been observed in many individuals. Cystic fibrosis has been termed a generalized exocrinopathy, since many or most of the exocrine glands may be involved. The liver, salivary glands, reproductive glands in both males and females, and the gastrointestinal (GI) tract may all be compromised. The first stool may be thick and hyperviscous, and intestinal blockage termed meconium ileus may complicate the first days of life in newborns with CF.

II. STUDIES OF THE CYSTIC FIBROSIS DEFECT PRIOR TO THE IDENTIFICATION OF THE CF GENE

In autosomal recessive diseases, one gene is classically responsible for all of the manifestations of the clinical syndrome. Among the estimated 50,000–100,000 gene products encoded within the genome, therefore, one protein abnormality must underlie the clinical findings in cystic fibrosis. Unifying features of the organ system abnormalities in CF include a disease of secretory glands and a defect in the movement of electrolytes. Initial attention to understand the disease focused on the physiology of epithelial glandular tissue, and in particular the transport of sodium and chloride in these tissues.

A generalized exocrine gland is shown in Fig. 1. Epithelial glands produce a primary secretion in a structure termed the acinus or coil. The primary secretion travels down the duct, where it is modified. The secretion then emerges from the duct into the GI tract, airways, or surface of skin, for example, depending on the particular gland. In all cases, the salt concentration is regulated by individual proteins that reside in the plasma membranes of glandular cells.

Proteins that regulate ion transport in epithelial cells constitute an important focus of the experiments designed to understand the molecular defects in CF. Even before the gene responsible for the disease was identified, these proteins were characterized and scrutinized to determine whether they functioned normally in the disease. Whenever charged particles such as sodium or chloride ions are transported across a barrier (such as a layer of cells or a cell monolayer), this movement creates a current (and, by Ohm's law, a potential difference). Bioelectric measurements of

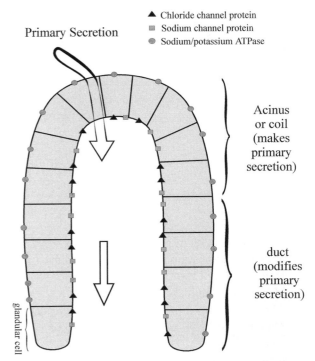

Primary Secretion

▲ Chloride channel protein
■ Sodium channel protein
● Sodium/potassium ATPase

Acinus or coil (makes primary secretion)

duct (modifies primary secretion)

glandular cell

FIGURE 1 An idealized exocrine gland. Examples of generalized ion channel proteins and ion pumps are shown distributed in the plasma membranes of individual glandular cells. Each specific type of exocrine gland has a unique complement of ion transport proteins and a unique distribution of these proteins within the epithelial cells.

current or voltage can therefore serve as a proxy for the movements of ions such as sodium or chloride. When sweat ducts from CF patients were perfused *in vitro* with solutions containing different salt compositions, a defect concerning transport of chloride out of the duct was identified. This finding led to the hypothesis that the normal sweat duct removed chloride from the primary secretion and that a failure of the CF sweat duct to perform this function resulted in elevated chloride in the final sweat on the surface of the skin. In this model, sodium was also elevated in the final sweat, since sodium served as a counterion for chloride and would remain in the sweat duct to maintain electrical neutrality if chloride could not be absorbed.

To further characterize the chloride abnormality in cystic fibrosis, several investigators studied individual chloride channel proteins present in the plasma membranes of CF cells. If a single gene product was responsible for the overall CF defect, work in the sweat duct suggested that a chloride channel was a leading candidate for the abnormal protein. Individual ion channels can be studied using the "patch-clamp" technique (Fig. 2). In this method, a patch of plasma membrane containing an individual ion channel protein of interest is excised from the cell. This patch is situated across the tip of a pipette in such a way as to contain the individual ion channel. The solution and bioelectric conditions across the membrane patch can be modified to define the bioelectric properties of individual ion channels, including the ions that can pass through a particular channel, the freedom with which they are conducted, and the drugs that open or close the channel. Airway cells taken from CF patients and studied in this way demonstrated a clear defect in single chloride channels in the disease. In summary, an effort toward understanding the molecular physiology of cystic fibrosis evolved from clinical observations of abnormal glandular function to sodium and chloride transport abnormalities in mounted tissues (e.g., sweat glands), and ultimately to the identifica-

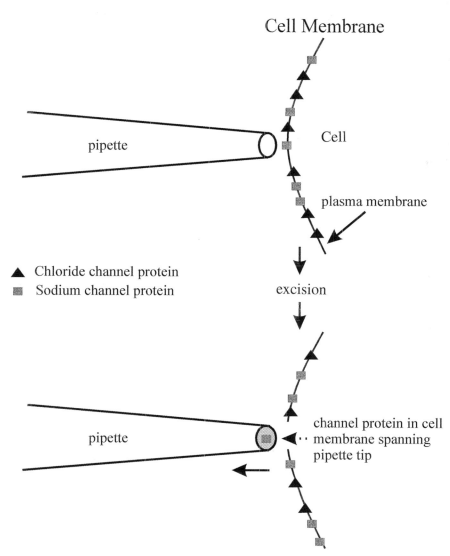

FIGURE 2   Patch-clamp analysis of individual ion channels in epithelial cells. A patch of plasma membrane is situated across the tip of a pipette in such a way as to contain an ion channel protein of interest. The solutions bathing the patch and other conditions can be modified in order to study the bioelectric characteristics of single ion channels.

tion of a single chloride channel abnormality in the disease.

III. IDENTIFICATION AND CHARACTERIZATION OF THE CYSTIC FIBROSIS TRANSMEMBRANE CONDUCTANCE REGULATOR

At about the same time that a chloride channel defect in cystic fibrosis had become well established, the gene responsible for CF was identified. The identification of the gene was accomplished by a combination of molecular genetic techniques, including DNA restriction fragment polymorphism and chromosomal walking and jumping. [*See* DNA Methodologies in Disease Diagnosis.] The gene encompasses approximately 230,000 base pairs on the long arm of chromosome 7, and the cDNA is approximately 4500 nucleotides in length. The predicted protein has 1480 amino acids in its full-length form. CFTR, the name given to the gene, reflected some initial questions concerning overall function. Although it was clear that the CFTR could act as a regulator of chloride conductance across epithelial cell membranes and tissues, would the CFTR gene act as a chloride channel itself, or would it instead be a regulator of endogenous Cl⁻ channels in epithelial cells?

Based on its primary amino acid structure and homology to other proteins, a predicted CFTR topology has been proposed (Fig. 3). In this topology,

CFTR includes two membrane spanning domains (TM-1 and TM-2, each composed of six predicted membrane spanning α helices), two domains that bind or hydrolyze nucleotides (nucleotide binding domains 1 and 2, or NBD-1 and NBD-2), and a large polyvalent domain (regulatory or R-domain) with multiple consensus phosphorylation sites for protein kinase A and protein kinase C. The overall sequence homology places CFTR firmly within the ATP Binding Cassette (ABC) gene family, in which most members do not act as ion channels. Classically, the members of this gene family act as ATP-dependent pumps that import or export small solutes across the plasma membrane. Other members of the ABC gene family include the multidrug resistance gene that transports chemotherapeutic drugs out of tumor cells (and mediates chemotherapy refractoriness in human tumors), the chloroquine resistance gene of *Plasmodium*, and the gene responsible for adrenoleukodystrophy.

Several types of experiments have now confirmed that CFTR itself functions as an epithelial chloride channel. First, expression studies of CFTR cDNA in many different cell types led to the appearance of a new chloride permeability pathway. Second, small mutations in the transmembrane or other domains of CFTR lead to subtle differences in the bioelectric properties of single chloride channels associated with CFTR overexpression. Finally, when purified CFTR protein is reconstituted in an artificial cell membrane, a chloride channel is observed even in the absence

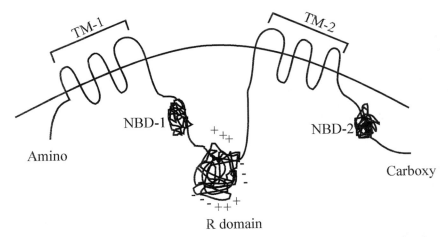

FIGURE 3 Predicted topology of CFTR. CFTR includes two membrane spanning domains (TM-1 and TM-2, each composed of six predicted membrane spanning α helices), two domains believed to bind or hydrolyze nucleotides (nucleotide binding domains 1 and 2, or NBD-1 and NBD-2), and a large polyvalent domain (regulatory or R-domain) with multiple consensus phosphorylation sites for protein kinase A and protein kinase C.

of other cellular proteins. These experiments provide conclusive proof that CFTR functions as a chloride channel, rather than solely as a regulator of other ion channels within epithelia.

IV. MUTATIONS IN CFTR AND THE RELATIONSHIP TO CLINICAL DISEASE

More than 500 mutations have been described in CFTR that may be associated with clinical disease. The majority of these mutations occur in the first CFTR nucleotide binding domain. For example, more than 70% of defective CFTR alleles contain a single phenylalanine deletion at CFTR position 508 (termed ΔF508). Interestingly, the chloride channel function of ΔF508 CFTR appears to be largely retained when the protein is studied in cell-free systems, and considerable evidence suggests that the ΔF508 mutation causes disease by initiating premature degradation of CFTR and failure to insert the ΔF508 protein in the plasma membrane. The machinery in the cell responsible for protein biogenesis recognizes the ΔF508 protein as abnormal and detains the mutant protein in the endoplasmic reticulum. Although the ΔF508 protein appears functional, it is marked for rapid degradation in the cell.

Other CFTR mutations cause disease through distinct mechanisms. A glycine → aspartic acid replacement at CFTR position 551 (termed G551D), for example, results in a full-length CFTR that is normally inserted in the plasma membrane but lacks normal chloride channel function. G551D accounts for approximately 5 to 10% of defective CFTR alleles in the United States. Nucleotide binding domains containing G551D do not bind ATP analogs normally, whereas the ΔF508 mutation does not appear to interfere with binding in these assays. Other mutations lead to shortened, poorly functioning CFTR. These include R553X and G542X (mutations that place protein termination signals in place of arginine or glycine at CFTR positions 553 and 542, respectively). These truncated CFTR proteins possess only TM-1 and part of NBD-1, and therefore lack many of the domains likely to be important for preventing disease.

In addition to information revealed by naturally occurring mutations, our understanding of CFTR has been enhanced by well-designed studies of the domains of CFTR and their importance in chloride channel function. For example, omission of the regulatory domain results in a constitutively active CFTR chloride channel that cannot be further activated by interventions that stimulate the wild-type protein. Deletion of the first four α-helical segments of CFTR TM-1 does not disrupt Cl$^-$ channel function. Omission of the entire second half of CFTR results in a functional chloride channel, perhaps because the truncated molecule dimerizes to regain function. Similarly, truncated CFTR containing R553X and G542X retain some chloride channel activity, although this function *in vivo* appears to be at a very low level, perhaps because messenger RNA associated with these mutations is unstable.

Several of these studies are summarized in Fig. 4. Overall, the available information suggests that CFTR chloride channel function can remain intact even when large segments of the protein are omitted. Moreover, experiments suggest that the construction of the pore necessary for chloride transport may be contained within the fifth and sixth α helices of TM-1, or the NBD-1. An improved understanding of the amino acid residues that constitute the Cl$^-$ pore in CFTR is an important aspect of characterizing overall CFTR function. [*See* DNA Markers as Diagnostic Tools.]

V. HOW NEW MOLECULAR INFORMATION ABOUT CYSTIC FIBROSIS MIGHT BE USED TO DEVELOP THERAPIES FOR THE DISEASE

Despite a growing comprehension of mechanisms by which CFTR mutations disrupt normal function, the pathogenic chain of events that leads from defective chloride channel activity to clinical disease is not completely understood. A leading hypothesis concerning the pathophysiology of CF lung disease, for example, is that airway chloride secretion drives water secretion in the airways, and therefore that a defect in chloride secretion in CF airway epithelial cells leads to a failure to secrete water. This causes poor hydration of airway secretions, hyperviscous mucous, and the pulmonary sequelae of the disease. An alternative hypothesis argues that levels of sodium and chloride in CF airways (as in sweat glands) may actually be increased. An increase in airway electrolytes could predispose CF patients to pulmonary infection since endogenous antimicrobial agents secreted by airway epithelium are inactivated in the presence of elevated salt. Additional

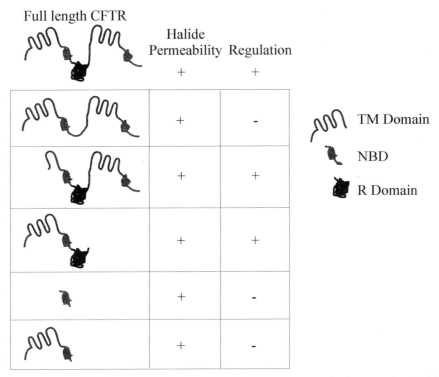

FIGURE 4 Summary of selected experiments designed to understand the function of particular CFTR domains in overall function.

information will be necessary in order to distinguish between these models.

Studies in cystic fibrosis mice suggest that non-CFTR chloride channels (i.e., Cl⁻ channels that serve an overlapping function with CFTR) may ameliorate pulmonary or GI disease. This observation supports the notion that chloride channel abnormalities are in fact the basic problem in CF. On the other hand, CFTR clearly maintains other important functions that may be independent of chloride channel activity. The CFTR regulates epithelial sodium channels and may contribute to cellular processes such as exocytosis, endocytosis, and Golgi acidification. It has been suggested that CFTR exports ATP out of the cell, and that extracellular ATP is responsible for regulatory effects on other cell membrane proteins. For example, the abnormal chloride channel first identified by patch-clamp techniques in CF has a single channel signature that is quite different from CFTR itself. It has recently become clear that the presence of CFTR in the apical cell membrane regulates the activity of this and possibly other alternate chloride permeability pathways.

Strategies for correcting cystic fibrosis defects have been developed based on the growing perception of how CF mutations cause disease (Table I). For example, it has been shown that CFTR truncations can be suppressed *in vitro* by treating epithelial cells with aminoglycoside antibiotics. This has led to clinical trials designed to translate full-length CFTR in patients with R553X and G542X. Efforts are also under way to develop improved aminoglycoside analogs that will be optimal for suppression of premature termination codons in CFTR. Drugs such as butyrate, glycerol, and dimethyl sulfoxide have been shown to augment processing of ΔF508 CFTR to the cell membrane *in vitro*. Although the mechanisms by which these agents act are not completely understood, clinical trials designed to augment ΔF508 CFTR maturation have begun. It is reasonable to imagine that efforts to identify small molecules capable of augmenting ΔF508 processing will become an important part of the development of new therapies for cystic fibrosis. Finally, new ways in which chloride permeability pathways in the plasma membranes of CF cells can be activated have been established. Small amounts of ΔF508 CFTR may exist in the plasma membrane *in vivo*. Drugs designed to open these residual CFTR channels have shown encouraging results *in vitro* and in early studies of ΔF508 cystic fibrosis mice. More

TABLE I
Genetic Causes of Cystic Fibrosis

Localization of CFTR defect	Alleles that disrupt activity (examples)	Prevalence in the United States (% defective alleles)	Functional defects	Examples of interventions in clinical trials
Apical cell membrane	G551D, all others	~5–10% (G551D)	Cl⁻ and Na⁺ transport abnormalities	Nucleotide activators of Cl⁻ secretion
Endocytic and exocytic vesicles			? Failure to internalize *Pseudomonas aeruginosa*	
ER/Golgi	ΔF508	~70%	Retention and degradation of CFTR	Butyrate analogues
Transcription/translation	R553X, G542X	5–10%	Defective truncated protein activity	Aminoglycoside antibiotics

extensive human experience has accumulated with the use of nucleotides such as ATP or UTP to activate alternative (non-CFTR) chloride permeability pathways present in CF airways. Such experiments are intended to activate functions that might mimic CFTR *in vivo*.

VI. GENE REPLACEMENT THERAPY FOR CYSTIC FIBROSIS

The CFTR serves many functions in the epithelial cell. As described earlier, it is not certain how these functions converge to cause all of the manifestations of the disease. For example, it is not known with certainty whether chloride transport, sodium transport, vesicle recycling, or more direct effects on *Pseudomonas* infection are the most important functions of CFTR as they relate to human disease. Even in the absence of a complete understanding of cystic fibrosis pathophysiology, however, it is reasonable to imagine that if sufficient amounts of CFTR could be safely transferred to the appropriate target cells (airway surface epithelial cells and glandular cells) using emerging gene replacement techniques, important manifestations of cystic fibrosis would be ameliorated. Major hurdles to the success of this overall goal include (1) avoiding inflammatory responses in the lung after gene administration, (2) achieving persistence of CFTR gene expression, and (3) clarifying the target cell and the levels of expression (on a per cell or per tissue basis) that are required in order to correct various cystic fibrosis functional defects. [*See* Gene Targeting Techniques.]

Two main strategies have evolved for the delivery of the CFTR to airway epithelium. In the first strategy, recombinant viruses containing CFTR are used to transfer the gene to airway epithelial cells. A growing number of studies with recombinant adenoviral vectors in the nose and lower airways have now been completed. In general, first-generation adenoviral vectors are limited by an inflammatory response to the virus itself and (at least in the nasal airway) low efficiency of gene transfer. A considerable effort has therefore been dedicated to developing adenoviral constructs that are less immunogenic and more efficient at transferring genes to CF target cells. Another viral vector approach uses the potentially less inflammatory and longer-acting adeno-associated virus and is presently being tested in clinical trials. The adeno-associated virus expression may lead to longer persistence of the CFTR in airway epithelial cells, although this remains to be established.

The second major category of vectors for CFTR gene transfer are nonviral in nature. The best studied among this group are cationic liposomes, which are highly efficient in transferring genes to cystic fibrosis cells in culture. Cationic liposomes mediate reasonable levels of gene transfer in the airways of a variety of mammalian models, and in the nasal airways of cystic fibrosis patients. Although these vectors appear to be much less proinflammatory than adenoviruses, the levels of expression that can be obtained with liposome-based CFTR gene transfer may be below that which is required for full correction of cystic fibrosis defects. Newer-generation lipids developed specifically for pulmonary epithelial gene transfer are now available, and these represent promising reagents for cystic fibrosis research. However, even if such reagents are shown to transiently correct cystic fibrosis defects in human airways, the maneuver results in effects of short duration (on the order of several days), and a mechanism for achieving persistence of the delivered gene is likely to be critical if the strategy is to become a useful part of therapy.

VII. CONCLUSIONS

Considerable progress has been made concerning our understanding of the molecular genetics of cystic fibrosis. This understanding has led to an appreciation of the different mechanisms by which CFTR mutations may cause disease. An improved understanding of the electrophysiology, structure, and function of CFTR has led to novel approaches toward treatment. Gene replacement therapy offers a means by which the defect caused by any CFTR mutation might ultimately be overcome.

BIBLIOGRAPHY

Collins, F. S. (1992). Cystic fibrosis: Molecular biology and therapeutic implications. *Science* **256,** 774–779.

Fuller, C., and Benos, D. J. (1992). CFTR! *Am. J. Physiol.* **263,** C267–C286.

Jilling, T., and Kirk, K. (1996). The biogenesis, traffic and function of CFTR. *Int. Rev. Cytol.* **172,** 193–241.

Quinton, P. M. (1983). Chloride impermeability in cystic fibrosis. *Nature* **301,** 421–422.

Rommens, J. M., Iannuzzi, M. C., Kerem, B. T., Drumm, M. L., Melmer, G., Dean, M., Rozmahel, R., Cole, J. L., Kennedy, D., Hidaka, N., Zsiga, M., Buchwald, M., Riordan, J. R., Tsui, L. C., and Collins, F. S. (1989). Identification of the cystic fibrosis gene: Chromosome walking and jumping. *Science* **245,** 1059–1065.

Sferra, T. J., and Collins, F. S. (1993). The molecular biology of cystic fibrosis. *Annu. Rev. Med.* **44,** 133–144.

Welsh, M. J., and Smith, A. E. (1993). Molecular mechanisms of CFTR chloride channel dysfunction in cystic fibrosis. *Cell* **73,** 1251–1254.

Welsh, M. J., Tsui, L.-C., Boat, T. F., and Beaudet, A. L. (1995). Cystic fibrosis. *In* "The Metabolic and Molecular Bases of Inherited Disease" (C. R. Scriver, A. L. Beaudet, W. S. Sly, and D. Valle, eds.). McGraw–Hill, New York.

Cytochrome *P*-450

FRANK J. GONZALEZ
National Cancer Institute, National Institutes of Health

GLOSSARY

Allele One of two or more alternating forms of a gene occupying corresponding sites on homologous chromosomes; multiple alleles for a single gene can exist in a population, and these can be recognized by DNA changes within or flanking a gene or by differing properties of a gene product; DNA changes of various types can give rise to mutations that inactivate a gene—these are known as mutant alleles

Complementary DNA In this article, refers to a double-stranded DNA that is copied from a messenger RNA (mRNA); the primary amino acid sequence of the mRNA-encoded protein can be deduced from the sequence of the complementary DNA (cDNA)

Concerted evolution Following a gene duplication, two or more related genes diverge, as during divergent evolution, and then become more similar to each other by exchanging DNA or genetic information. This phenomenon is recognized when a subfamily is examined in two species (A and B) and the two genes (1 and 2) in a single species are known to have existed prior to the divergence of the two species; as a result of concerted evolution, gene 1 will be more similar in sequence to gene 2 in species A than to gene 1 in species B

Divergent evolution Following a gene duplication, the two daughter genes begin to diverge by accumulating random base substitutions and, hence, become less similar to each other

Gene conversion Generally thought to be the physical mechanism of convergent evolution, gene conversion is a nonreciprocal recombination event in which a segment of one gene replaces the corresponding segment of a related gene

Genetic polymorphism Occurrence together in a population of two or more genetically determined phenotypes in such proportions that the rarest of them cannot be maintained merely by recurrent mutation

Ligand (1) Organic molecule that donates electrons to form coordinate covalent bonds with metallic ions such as iron; (2) compound that specifically binds to a protein receptor

Orthologous A gene or protein in one species is said to be orthologous to a gene or protein in another species if both are believed to have evolved from a single gene present at the time of divergence of the species

Xenobiotic Compound that is foreign to the body, such as a drug or an environmental pollutant (*xeno-*, foreign; *-biotic*, of biological origin)

THE TERM CYTOCHROME *P*-450 REFERS LITERALLY to a colored substance in the cell that absorbs light at around 450 nm, within the visible spectrum (*cyto-*, cell; *-chrome*, color; *P*, pigment; *450*, wavelength). In general, a cytochrome is a cellular heme-containing protein (hemoprotein) whose principal function is electron transport. The heme moiety is a molecule called protoporphyrin IX complexed with an iron ion, the latter of which can undergo reversible changes in valency. The protein environment surrounding the heme governs the selective properties of hemoproteins in the organism or cell. For example, hemoglobin serves to transport O_2 within the body, whereas *P*-450 is a monooxygenase using one atom from O_2 and two electrons to oxidize chemical substrates. The *P*-450s are ubiquitous enzymes that have probably been present since very early in biological evolution and are found in bacteria, fungi, and throughout the plant and animal kingdoms. *P*-450s exist in multiple forms and are classified into a superfamily of enzymes, all of which include a conserved cysteine-containing region near their carboxy termini, which serves to bind the heme molecule. These enzymes use electrons from nicotinamide adenine dinucleotide (NADH) and/or nicotinamide adenine dinucleotide phosphate (NADPH) and O_2 to oxidize their substrates. In humans, *P*-450s perform three critical functions: (1) five are involved as key enzymes in pathways leading to steroid synthesis; (2) some *P*-450s convert arachidonic acid, in the epoxygenase pathway, to signal metabolites that mediate a wide range of biological processes; and (3) many other *P*-450s catalyze the oxidation of foreign compounds or xenobiotics ingested in the form of plant metabolites, drugs, and environmental contaminants. These enzymes are the primary interface between humans and the chemical environment. Human deficiencies of both the steroidogenic and xenobiotic-metabolizing *P*-450s are responsible for a severely debilitating and sometimes fatal disease called congenital adrenal hyperplasia. Deficiencies in xenobiotic-metabolizing *P*-450s are associated with clinical drug oxidation polymorphisms that result in toxic reactions to some prescription drugs.

I. BIOCHEMISTRY OF *P*-450s

A. Physical Properties

Cytochrome *P*-450s are a large group of enzymes that have several common features. They all contain a single iron ion that is bound to a protoporphyrin IX molecule in a complex called heme. This green pigment is noncovalently bound to the enzyme and found in many other proteins, including hemoglobin, myoglobin, and cytochrome *c*, and in enzymes such as thyroid peroxidase, xanthine oxidase, and myeloperoxidase. Collectively, these enzymes can all bind and/or metabolize oxygen. Thus, *P*-450s also metabolize oxygen in the form of atmospheric O_2. Generally, the concentration of O_2 that is in equilibrium with water in living cells is more than sufficient to support the catalytic activity of *P*-450s. Every *P*-450 found in bacteria, yeast, plants, and animals possesses heme and metabolizes oxygen in the same manner.

The heme-binding region can be readily identified in any *P*-450 sequence. This region is found within the carboxy-terminal third of the *P*-450 polypeptide chain and centers around the amino acid cysteine (Fig.

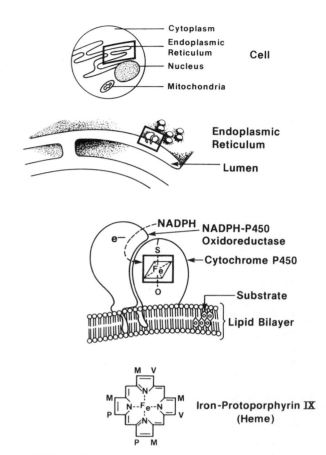

FIGURE I Diagrammatic representation of cytochrome *P*-450 and its relationship to other components of the cell. The elements enclosed by rectangles are expanded in more detail in the lower panels.

1). The sulfur or thiolate portion of this amino acid serves as the fifth ligand to the heme iron and is absolutely required for *P*-450 enzymatic activity. The precise positioning of this thiol group determines the spectral properties of the *P*-450 in the presence of electrons [which reduce the iron from a valency of Fe^{3+} (ferric form) to Fe^{2+} (ferrous form)]. In the presence of the strong ligand carbon monoxide (CO) (which binds to the heme in the sixth coordinate position to the iron, in the same way as O_2, but is not metabolized), the enzyme displays a unique absorption of light at around 450 nm of the visible spectrum. Hemoglobin, cytochrome b_5, cytochrome *c*, and other heme proteins absorb at a lower wavelength of about 420 nm. This difference in absorption properties reflects the polypeptide sequence that surrounds the heme molecule and, more importantly, the positioning of the cysteine residue. Other hemoproteins appear to have basic amino acids in the fifth position, which in *P-450* is occupied by cysteine. In fact, with denaturation, *P-450* no longer absorbs light at 450 nm but, instead, at 420 nm, suggesting that the heme is displaced from the cysteine and that another, perhaps basic amino acid, is the fifth ligand to the iron. [*See* Hemoglobin.]

All *P-450*s are similar in size, ranging from 48,000 to 60,000 Da, and all contain approximately 500 amino acids, except the smaller bacterial *P-450*$_{cam}$, found in *Pseudomonas putida,* and one unusually large enzyme found in the bacterium *Bacillus megaterium* (see Section III). The bacterial *P-450*s isolated to date are soluble in water, whereas the yeast, plant, and animal enzymes are hydrophobic molecules that are bound to cell membranes. In animals, most *P-450*s are part of an intracellular membrane network called the endoplasmic reticulum (Fig. 1). This membrane network extends throughout the cell and consists of large sheets or tubes containing an inner space called the lumen. The outer region of the membrane faces the cytoplasm, and this is where *P-450*s are found. In fact, *P-450*s are frequently referred to as microsomal enzymes. Microsomes are actually fragments of endoplasmic reticulum that form vesicles when a cell is disrupted. These vesicles can be separated from soluble cellular components and other macromolecular organelles by differential high-speed centrifugation.

A few of the *P-450*s involved in steroid biosynthetic and metabolic pathways are found in the inner membranes of mitochondria. These enzymes are encoded by genes located in the cell nucleus. The endoplasmic reticulum-bound and mitochondrial *P-450*s are inserted into their respective cellular compartments by distinctly different mechanisms. The mitochondrial enzyme must be transported through the outer membrane of this organelle to be embedded ultimately in the inner membrane. This transport process is mediated by the system that is responsible for the incorporation of many inner mitochondrial enzymes. The endoplasmic reticulum-bound *P-450*s are inserted directly into the endoplasmic reticulum lipid bilayer by a complex pathway that involves recognition of a hydrophobic peptide (the signal sequence) located at the amino terminus of the *P-450*. [*See* Cell Membrane Transport.]

The predicted structure of an endoplasmic reticulum-bound *P-450* is shown in Fig. 1. The enzyme is anchored to the phospholipid membrane bilayer of this membrane system by the amino-terminal portion of its polypeptide chain. The bulk of the enzyme faces the cytoplasm of the cell or the outside of the endo-

plasmic reticulum network. This cellular location is ideally suited for the *P-450* to bind to and metabolize hydrophobic substrates such as the polycyclic aromatic hydrocarbon that dissolves in lipid (Fig. 1). The endoplasmic reticulum also contains other enzymes, such as the conjugating enzymes or transferases, which are crucial in the further metabolism and elimination of *P-450* metabolites resulting from *P-450*-catalyzed oxidations of these types of substrates.

B. Enzymology

The *P-450*s have been called mixed-function monooxygenases because only one atom from O_2 is inserted into the substrate, in contrast to dioxygenases such as cyclooxygenase and lipoxygenase, which insert both atoms from O_2 into their substrates. *Mixed function* refers to the fact that these enzymes are capable of incorporating one oxygen atom into water and the other into the substrate, the latter of which includes numerous structurally diverse chemicals. *P-450*s use O_2 and electrons, supplied by the cofactor NADPH, to add a single oxygen atom to the substrate, usually in the form of a hydroxyl group or an epoxide. This process can be seen in the diagram of the *P-450* catalytic cycle (Fig. 2). The substrate "RH" binds to the enzyme near the heme residue. The heme iron is then reduced from Fe^{3+} to Fe^{2+} by the addition of an electron from NADPH, which, in microsomal *P-450*s, is donated by the flavin-containing enzyme NADPH-*P-450* oxidoreductase (note that the electron is trans-

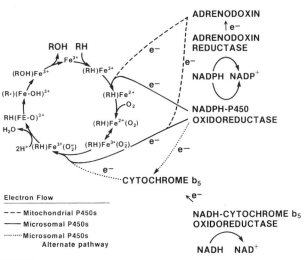

FIGURE 2 Catalytic cycle of cytochrome *P-450*s. Two electrons are individually introduced into the cycle via participation of other enzymes.

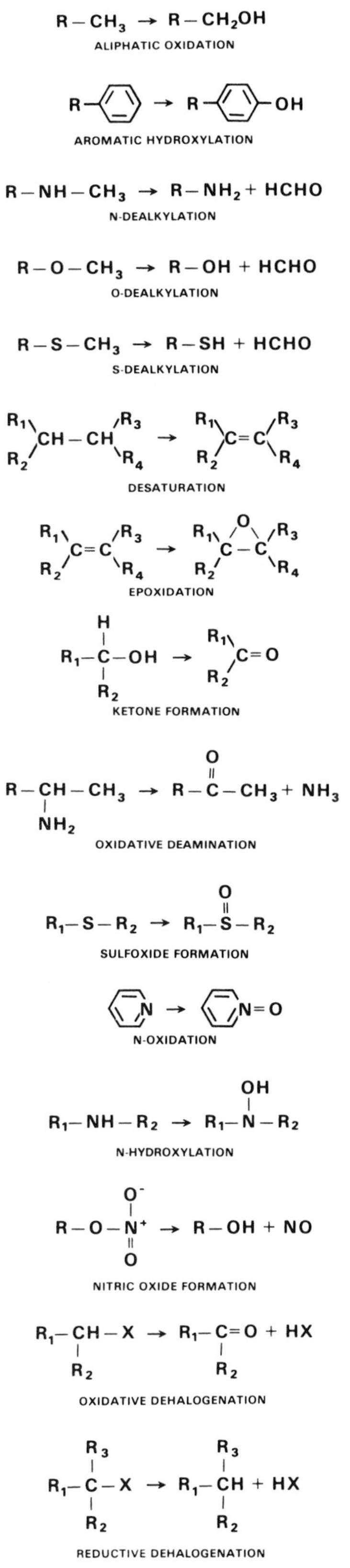

$R-CH_3 \rightarrow R-CH_2OH$

ALIPHATIC OXIDATION

$R-\langle\bigcirc\rangle \rightarrow R-\langle\bigcirc\rangle-OH$

AROMATIC HYDROXYLATION

$R-NH-CH_3 \rightarrow R-NH_2 + HCHO$

N-DEALKYLATION

$R-O-CH_3 \rightarrow R-OH + HCHO$

O-DEALKYLATION

$R-S-CH_3 \rightarrow R-SH + HCHO$

S-DEALKYLATION

$\begin{array}{c} R_1 \\ R_2 \end{array}CH-CH\begin{array}{c} R_3 \\ R_4 \end{array} \rightarrow \begin{array}{c} R_1 \\ R_2 \end{array}C=C\begin{array}{c} R_3 \\ R_4 \end{array}$

DESATURATION

$\begin{array}{c} R_1 \\ R_2 \end{array}C=C\begin{array}{c} R_3 \\ R_4 \end{array} \rightarrow \begin{array}{c} R_1 \\ R_2 \end{array}C-C\begin{array}{c} R_3 \\ R_4 \end{array}$ (with O bridge)

EPOXIDATION

$R_1-\overset{H}{\underset{R_2}{C}}-OH \rightarrow \begin{array}{c} R_1 \\ R_2 \end{array}C=O$

KETONE FORMATION

$R-\underset{NH_2}{CH}-CH_3 \rightarrow R-\overset{O}{\underset{}{C}}-CH_3 + NH_3$

OXIDATIVE DEAMINATION

$R_1-S-R_2 \rightarrow R_1-\overset{O}{\underset{}{S}}-R_2$

SULFOXIDE FORMATION

$\langle\bigcirc\rangle N \rightarrow \langle\bigcirc\rangle N=O$

N-OXIDATION

$R_1-NH-R_2 \rightarrow R_1-\overset{OH}{\underset{}{N}}-R_2$

N-HYDROXYLATION

$R-O-\overset{O^-}{\underset{O}{N^+}} \rightarrow R-OH + NO$

NITRIC OXIDE FORMATION

$R_1-\underset{R_2}{CH}-X \rightarrow R_1-\underset{R_2}{C}=O + HX$

OXIDATIVE DEHALOGENATION

$R_1-\overset{R_3}{\underset{R_2}{C}}-X \rightarrow R_1-\overset{R_3}{\underset{R_2}{CH}} + HX$

REDUCTIVE DEHALOGENATION

ferred as a hydride ion, H⁻). After reduction, an O_2 molecule binds to the heme, and a second electron, usually donated by another molecule of NADPH (again through NADPH-P-450 oxidoreductase), is added. However, with certain substrates, the second electron may come from NADH via cytochrome b_5 and NADH-cytochrome b_5 oxidoreductase. The mitochondrial enzymes receive both electrons from NADPH using the iron–sulfur protein adrenodoxin and adrenodoxin reductase. Like the NADPH-P-450 oxidoreductase, adrenodoxin reductase is also a flavin-containing enzyme.

The addition of a second electron results in the splitting of oxygen and the release of one oxygen atom as water. The second oxygen atom is then in such a state that it can be inserted into the substrate to form an epoxide or alcohol "ROH." Prior to the release of substrate, an active intermediate "R·" is sometimes formed; the rearrangement of this intermediate may result in the loss of stereospecificity and the formation of a number of other products. In many cases, rearrangement results in the removal of a hydrogen rather than the addition of an oxygen.

Many activities are associated with P-450-catalyzed reactions, depending on the substrate and the nature of the radical intermediate. Some of these are shown in Fig. 3, where R_1 and R_2 represent alkyl and aryl side chains. The high-energy intermediates of some P-450-catalyzed reactions are the principal basis for cell toxicity and cell transformation mediated by chemical carcinogens. In general, these chemicals yield high-energy intermediates that cannot undergo intramolecular rearrangement. In these cases, the intermediates, usually electrophiles, can exit the P-450 active site and react with other cellular compounds, particularly endogenous chemicals with nucleophilic properties. Among the most nucleophilic cellular constituents are ribonucleic acid (RNA) and deoxyribonucleic acid (DNA), the genetic information source of the cell. When DNA or specific genes such as oncogenes or tumor-suppressor genes become damaged or mutate, cell transformation and cancer can result. Thus, P-450-mediated carcinogen activation is the first step in chemically induced cancer. Cigarette smoking-associated lung cancer is one example of a cancer that results from exposure to carcinogens.

FIGURE 3 Reactions catalyzed by cytochrome P-450s. Note that no attempt has been made to balance these reactions.

II. P-450 NOMENCLATURE

The nomenclature system for *P-450s* is based solely on primary amino acid sequence relatedness among *P-450* forms. Computer programs capable of paired comparisons of multiple primary amino acid sequences are used to generate percentage sequence similarities between *P-450s*. By examining sequence similarities for multiple *P-450s*, arbitrary percentage cutoff values have been assigned to distinguish among gene families and gene subfamilies. By definition, *P-450* from one gene family displays ≤40% sequence similarity to a *P-450* in any other gene family. With only a couple of minor exceptions, *P-450s* within individual families are >40% identical in sequence. Within these families, *P-450s* that display >59% sequence similarity have been assigned to subfamilies.

The individual human *P-450* genes are named using the root symbol *CYP* (cytochrome *P-450*), an Arabic number to denote the gene family, a capital letter to designate the subfamily, and another Arabic number to represent the individual gene. For example, the gene encoding a *P-450* that oxidizes polycyclic aromatic hydrocarbons is *CYP1A1*. The *CYP1A1* gene product or enzyme is also designated CYP1A1, except without italics. *P-450s* are sometimes identified by their trivial names (Fig. 4). In some cases, the numbering of *P-450* gene families was based on certain reactions carried out by their encoded enzymes. For example, the *P-450* involved in the synthesis of 11-deoxycorticosterone from progesterone oxidizes the latter steroid at the 11-carbon position, and hence its family was assigned the name *CYP11*. In fact, all of the five families of steroidogenic *P-450s* were named in this fashion. The four families containing the *P-450s* involved in drug and fatty acid oxidations were named 1–4. To allow for newly discovered families in higher eukaryotes, the numbering scheme began at 51 for yeast *P-450* families, at 71 for plant *P-450s*, and at 101 for bacterial *P-450s*.

The *P-450s* that are known to be present in humans are listed in Table I. Surely others are yet to be discovered. In some cases, as with the steroidogenic *P-450s*, the human enzyme, complementary DNA (cDNA), or gene has not been sequenced but is known to exist. These cases are indicated by asterisks. The cDNA for all human *P-450s* listed in the table have been sequenced to deduce their protein sequences. Substrates are also listed; however, as will be discussed later, the *P-450s* within families 1–3 can metabolize many different compounds. Therefore, only partial lists of the best-studied substrates are included.

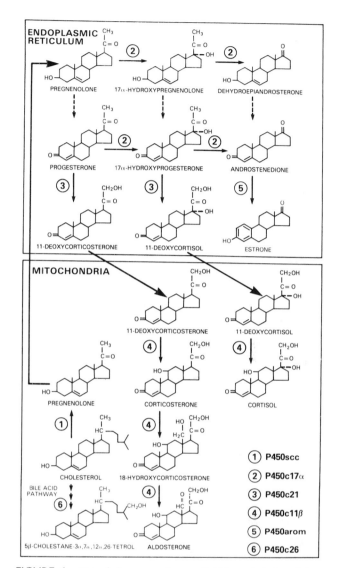

FIGURE 4 List of those cytochrome *P-450s* and their cellular compartments that carry out reactions in steroidogenic pathways. These enzymes are identified by their trivial names.

The most confusing aspect of the nomenclature system is the naming of individual genes. The complexity of this system reflects difficulties in assigning orthologous *P-450s* between species and is particularly evident in families 2 and 3. As will be discussed, *P-450s* are evolving quite rapidly and in a species-specific manner. That is, *P-450s* within one subfamily in rat are evolving quite differently than those within the same subfamily in humans. This evolutionary diversity is probably a consequence of species-specific adaptation as a result of environmental and dietary exposures.

TABLE I

List of Human P-450s, Tissues in Which They Are Expressed, and Some of Their Substrates

Gene[a]	Tissue[b]	Substrates[c]
CYP1A1	Liver Lung Placenta Other	Benzo(a)pyrene (C) 7,12-Dimethylbenz(a)- anthracene (C)
CYP1A2	Liver	Acetaminophen (D) 2-Acetylaminofluorene (C) Aflatoxin B$_1$ (C) Heterocyclic arylamines (C) Phenacetin (D)
CYP1B1	Adrenal Ovaries	Similar to CYP1A1
CYP2A6	Liver (P) Nasal epithelium Other	Coumarin N-Nitrosodiethylamine (C)
CYP2B6	Liver (P)	7-Ethoxy-4-trifluoromethyl- coumarin Aflatoxin B$_1$ (C)
CYP2B7	Lung	
CYP2C8	Liver	Tolbutamide (D) R-Mephenytoin (D)
CYP2C9	Liver	Tolbutamide (D) R-Mephenytoin (D) Warfarin (D)
CYP2C19	Liver	S-Mephenytoin
CYP2D6	Liver (P) Kidney	Bufuralol (D) Debrisoquine (D) Dextromethorphan (D) Nortriptyline (D) Propranolol (D)
CYP2E1	Liver Other	Acetoacetate (E) Acetol (E) Acetaminophen (D) Ethanol (D) Halothane (D) N-Nitrosodiethylamine (C)
CYP2F1	Lung Liver Other	7-Ethoxycoumarin Testosterone (E)
CYP3A4	Liver GI tract Kidney Other	Aflatoxin B$_1$ (C) Cortisol (E) Cyclosporine (D) Erythromycin (D) Midazolam (D) Nifedipine (D) Warfarin (D) Testosterone (E)
CYP3A5	Liver (P) Other	Similar to CYP3A4
CYP3A7	Liver (fetal)	Dehydroepiandrosterone-3- sulfate (E)
CYP4A9	Liver Other	Lauric acid Arachidonic acid (E) Other fatty acids

TABLE I   (Continued)

Gene[a]	Tissue[b]	Substrates[c]
CYP4A11	Liver Other	Similar to CYP4A9
CYP4B1	Lung Other	
CYP7A1	Liver	Cholesterol (E)
CYP11A1	Adrenal gland Ovary Testis	Cholesterol (E)
CYP11B1	Adrenal gland Ovary Testis	17α-Hydroxyprogesterone (E) Progesterone (E)
CYP17A1	Adrenal Ovary Testis	Dehydroepiandrosterone (E) 17α-Hydroxyprogesterone (E) 17α-Hydroxypregnenolone (E) Progesterone (E)
CYP19A1	Ovary Placenta	Androstanedione (E)
CYP21A2	Adrenal gland Ovary Testis	Progesterone (E) 17α-Hydroxyprogesterone (E)
CYP26A1[d]	Liver	Cholesterol (E)

[a]The amino acid sequences of all P-450s listed have been determined using human cDNA libraries, except for CYP26A1, which encodes the cholesterol 26-hydroxylase. Because this enzyme is a member of a cascade of enzymes involved in bile acid formation, it is presumed to exist in humans.
[b]The tissues listed are known to express P-450s, however, not every human tissue has been carefully examined. On the basis of studies in rodents, some P-450s are believed to be expressed in other tissues. Some genes are polymorphically expressed (P).
[c]The substrates listed fall into the classes of carcinogens (C), drugs (D), and endogenous compounds (E). The unmarked substrates are chemicals that happen to be substrates but are not drugs or carcinogens. The carcinogenicity of these compounds in rodents varies considerably from the potent aflatoxin B$_1$ to the weak heterocyclic arylamines. In most cases, the carcinogenic potency in humans is unknown. This list is not inclusive. Several of these P-450s are known or presumed to metabolize many other compounds. It should also be noted that a single substrate can be metabolized by multiple P-450 forms (e.g., aflatoxin B$_1$, 7-ethoxycoumarin).
[d]Present in rodents and believed to be present in humans.

The CYP2D subfamily can be used to illustrate species differences in P-450 genes. There are five genes in the rat CYP2D subfamily, and these have been designated CYP2D1 through CYP2D5. Two genes and an inactive pseudogene, designated CYP2D6, CYP2D7, and CYP2D8P (P stands for pseudogene), exist in humans. None of the human CYP2D P-450s has a high degree of amino acid sequence similarity to any of the five rat CYP2D P-450s. Therefore, it is practically impossible to determine which rat P-450 gene is the evolutionary equivalent of the human

CYP2D6 gene. In these cases, *P*-450s have been named as separate entities, even though it is believed that all *CYP2D* genes evolved from the same gene or genes in the predecessor of rats and humans more than 80 million years ago. In some instances, however, the assignment of orthologues is more straightforward. For example, the human *CYP2A6* gene product displays a high level of sequence similarity to one of the three rat *CYP2A P*-450s, *CYP2A3*. Hence, both rats and humans have three *CYP2A P*-450s and two can clearly be assigned as orthologues.

It is important to note that individual *P*-450s are *forms* of *P*-450, not isozymes. These enzymes cannot formally be considered isozymes because, even though they have similar catalytic cycles and all metabolize O_2, they do not necessarily carry out the same reactions on the same substrates.

III. EVOLUTION OF *P*-450s

Primary amino acid sequences have been determined from individual *P*-450s found in bacteria, yeast, plants, and animals. More than 150 different *P*-450 sequences representing 28 families were known at the time this article was written, and it is presumed that many more sequences will be determined, particularly those of plant *P*-450s. These sequence data can be analyzed by computer programs that align multiple sequences simultaneously, yielding percentage differences between a single *P*-450 and all others. The resultant percentages are then converted into accepted point mutations (PAMs), using an evolution table developed from M. O. Dayhoff's protein sequence atlas. Phylogenetic trees are constructed using these data and methods such as the unweighted pair group method of analysis (UPGMA), in which an arbitrary unit, termed the evolutionary distance, is used to plot divergence times between *P*-450s. The evolutionary distance can be converted to a real time scale using data derived from fossil evidence of divergence times between, for example, birds and mammals or, for more recent divergence times, the mammalian radiation that occurred approximately 75 million years ago. An example of a *P*-450 phylogenetic tree and further details concerning the derivation of such trees can be found within readings in the bibliography.

By examining the *P*-450 phylogenetic tree it becomes apparent which *P*-450s existed before the divergence of eukaryotes and before the mammalian radiation. By using these observations in conjunction with the known catalytic activities of *P*-450s, it is possible to speculate about the evolution of these enzymes.

The early *P*-450s appear to have been more clearly related to the modern *P*-450 enzymes that are known to metabolize cholesterol and fatty acids. These may have been required to maintain membrane integrity in primordial organisms. Two types of eukaryotic *P*-450s emerged, and these differed in their intracellular location and electron source. The mitochondrial *P*-450s, involved in steroid and bile acid biosynthetic pathways, receive electrons from the iron–sulfur protein adrenodoxin via the flavoprotein adrenodoxin reductase. These seem to be the most closely related to the well-studied bacterial *P*-450$_{cam}$, which metabolizes camphor and also receives electrons from an iron–sulfur protein. The eukaryotic and prokaryotic iron–sulfur proteins are both reduced by NADPH using an evolution-related, flavin-containing enzyme. Therefore, the mitochondrial *P*-450s appear to have evolved from bacterial *P*-450$_{cam}$. A second bacterial *P*-450, expressed in *Bacillus megaterium*, CYP102, is a *P*-450 hemoprotein that is fused with an electron transport flavoprotein related to the NADPH-*P*-450 oxidoreductase. This bacterial enzyme may be the precursor of the microsomal *P*-450s and the flavoprotein NADPH-*P*-450 oxidoreductase, which transfers electrons directly to the *P*-450 (without the aid of an iron–sulfur protein). Therefore, the mitochondrial and microsomal *P*-450s and their electron carriers appear to have evolved from separate origins.

Although all mitochondrial *P*-450s catalyze reaction steps in critical biosynthetic pathways, only four microsomal *P*-450s carry out reactions in steroid biosynthesis. The majority of the microsomal *P*-450s are involved in xenobiotic metabolism. It is in these enzymes that significant species differences and intraspecies polymorphisms are evident (see the following).

All of the known mammalian *P*-450 families were in existence 400 million years ago. An explosion in the number of *P*-450 genes, especially those involved in xenobiotic metabolism, has occurred within the past 75–100 million years. What caused this large diversification in *P*-450s? Current thinking suggests that these enzymes were needed as a defense against plants, which typically produce chemicals that are highly toxic to animals. The *P*-450s may have evolved to enable land animals to feed on plants. (Aquatic animals also have *P*-450s but, in contrast to terrestrial animals, many lipid-soluble toxic compounds can be eliminated through their gills.) Plants would then have developed new toxins and animals would have evolved or made use of new *P*-450 genes. This animal–plant *warfare* is an excellent example of coevolution and is probably the major reason for the large number of P-450s and the diverse substrate specificities of these enzymes.

IV. STEROIDOGENIC P-450s

Among the most critical P-450s in humans are those involved in steroid biosynthetic pathways. Three endoplasmic reticulum-bound P-450s and two mitochondrial P-450s participate as key enzymes in the synthesis of steroid hormones, including aldosterone, androstanedione, cortisol, estrogen, and progesterone (Fig. 4). These enzymes are encoded by genes within five separate families and are expressed in various tissues of the body that produce steroids, such as the adrenal gland, which manufactures aldosterone and cortisol, and the gonads, which make sex steroids. Two P-450s, CYP7A1 and CYP26A1, are expressed in liver and are involved in the pathway by which cholesterol is converted to bile acids, the major route of cholesterol elimination from the body. [See Bile Acids; Cholesterol; Steroid Hormone Synthesis.]

The steroidogenic enzymes are of crucial medical importance. Approximately one in 7,000–15,000 livebirths results in a condition called congenital adrenal hyperplasia, in which the child is deficient in the production of aldosterone, cortisol, and other steroids. A lack of these steroids can cause severe debilitation, including growth and developmental abnormalities and a loss of salt from the body, which can result in death within the first 2 weeks of life. In cases involving other, milder conditions or partial deficiencies, symptoms do not occur until later in life. [See Steroids.]

Congenital adrenal hyperplasia results, in most cases, from mutations in genes encoding the steroid-synthesizing P-450s shown in Fig. 4. By far the most common deficiency is a lack of functional P-450c21 (CYP21A2), commonly called steroid 21-hydroxylase. However, mutations have also been found in genes encoding P-450scc (CYP11A1), P-450c17α (CYP17A1), and P-450c11β (CYP11B1), albeit at much lower frequencies than the mutation affecting P-450c21 (CYP21A2). The symptoms of this deficiency result not only from a lack of steroid hormone production but also from an accumulation of intermediates in the steroid-synthesizing pathways. These intermediates are shunted into other biochemical pathways (e.g., androgen synthesis, resulting in the masculinization of females). The accumulation of intermediates in the steroid pathway results from loss of negative feedback regulation of the pituitary, resulting in excess production of the peptide hormone adrenocorticotropic hormone (ACTH). Excess ACTH

production in the pituitary is accompanied by an increase in other peptide hormones, such as melanocyte-stimulating hormone.

The genetic defects responsible for congenital adrenal hyperplasia can sometimes be detected using cloned gene probes for CYP21A2. More importantly, it is possible to detect prenatally the presence of two copies of a defective gene. The precise diagnosis of all mutant genes is difficult because the genetic lesions responsible for the disruption of the CYP21A2 gene are quite diverse. For example, mutations in one gene may be caused by a deletion, whereas another gene becomes inactivated as a result of an amino acid substitution.

P-450$_{arom}$, the key enzyme in estrogen synthesis (Fig. 4), may also be of importance in the treatment of breast cancer. Because many breast cancer cells are estrogen dependent, their growth can be arrested by inhibiting the synthesis of estrogen. This can be accomplished by treatment with agents such as 4-hydroxyandrostein-3,17-dione and 10-propargy-lestr-4-ene-3,17-one and related drugs that inhibit the activity of P-450$_{arom}$, resulting in a cessation of estrogen synthesis and a halt in the growth of estrogen-dependent tumors. These agents are also effective in treatment of androgen-dependent prostatic cancers. [See Breast Cancer Biology.]

Other P-450 enzymes (e.g., CYP26, Fig. 4) are of critical importance because of their role in the degradation of cholesterol, which has been shown to contribute to arteriosclerosis and heart disease. The cholesterol 7α-hydroxylase P-450 (CYP7) may also be important in lowering cholesterol levels.

A mitochondrial P-450 that catalyzes the 1'-hydroxylation of vitamin D_3 is found in the kidney. This enzyme, which is known to exist but has not yet been purified, is critical in the production of active hormone.

Finally, a yeast P-450 called CYP51, or lanosterol 14α-demethylase, is involved in the biosynthetic pathway leading to the production of the steroid ergosterol. When production of this enzyme is inhibited, lanosterol and other 14-methylsterol metabolites accumulate in the cells, resulting in growth inhibition. Some potent antifungal agents, including diniconazole and ketoconazole, inhibit CYP51.

V. FATTY ACID-METABOLIZING P-450s

A group of P-450s in the CYP4A subfamily metabolize fatty acids. These enzymes catalyze the oxidation of fatty acids, including palmitic acid and arachidonic

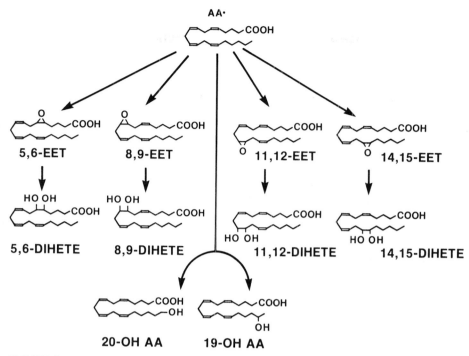

FIGURE 5 Metabolism of arachidonic acid by cytochrome *P*-450s, commonly known as the epoxygenase pathway of arachidonic acid metabolism. AA, arachidonic acid; DIHETE, dihydroxyeicosatrienoic acid; EET, epoxyecosatrienoic acid. [Reproduced with permission from F. A. Fitzpatrick and R. C. Murphy (1988). Cytochrome *P*-450 metabolism of arachidonic acid: Formation and biological actions of epoxygenase-derived eisosanoids. *Pharmacol. Rev.* 40, 229–241.]

acid. The *P*-450s within the *CYP4A* subfamily are expressed in many tissues, a finding that is consistent with their possible role in fatty acid metabolism. Other, yet uncharacterized, *P*-450s are involved in the synthesis of arachidonic acid epoxides or epoxyeicosatrienoic acid (EET), many of which are biologically active (Fig. 5). Among the many known biological activities of these compounds are (1) to promote the release of peptide hormones from brain tissue and insulin from pancreas, (2) to regulate ion transport in a variety of tissues, (3) to stimulate vasodilation, and (4) to stimulate platelet aggregation. An CYP4A or related *P*-450 is also expressed in polymorphonuclear leukocytes, where it converts the potent chemotactant leukotriene B$_4$ to a hydroxylated derivative that is biologically inactive.

VI. XENOBIOTIC-METABOLIZING *P*-450s

A. General Characteristics

The majority of *P*-450s are involved in the metabolism of foreign compound's (xenobiotics) such as drugs,

environmental pollutants, and plant metabolites. These enzymes appear to exist solely for this purpose. If an individual did not ingest drugs, plants containing toxic chemicals, or other foreign chemicals, he or she could probably survive without xenobiotic-metabolizing *P*-450s.

The xenobiotic-metabolizing *P*-450s are found in families 1, 2, and 3. In humans, more than 30 *P*-450s may be expressed in family 2 and perhaps six in family 3. In general, the oxidation of drugs by these enzymes results in the abolition of the compound's therapeutic activity, although in some instances a drug may be activated to a therapeutically active metabolite. Oxidation of lipid-soluble drugs by *P*-450s can render them more water-soluble, aiding in their elimination through the kidney by reducing reabsorption. Without *P*-450s, some drugs would be retained by the body for long periods.

The xenobiotic-metabolizing *P*-450s are noteworthy for their capacity to metabolize numerous structurally diverse compounds. A single *P*-450 can oxidize a large number of drugs. CYP2D6, for example, can metabolize compounds with a wide range of structures and clinical uses (Fig. 6), including β-blocking agents, tricyclic antidepressants, cardiovascular

β-ADRENERGIC BLOCKING AGENTS

Bufuralol

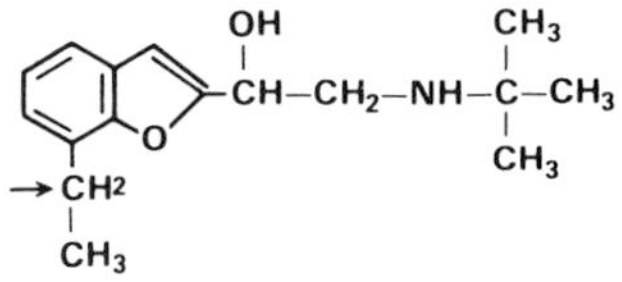

Metoprolol

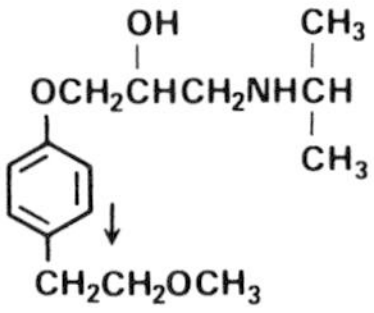

Propanolol

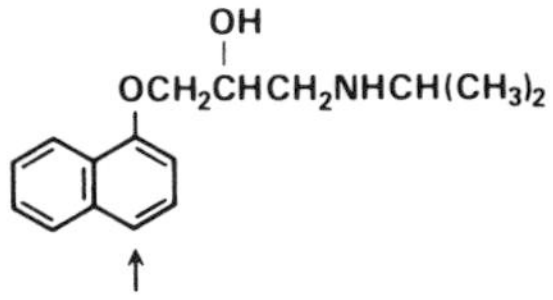

Timolol

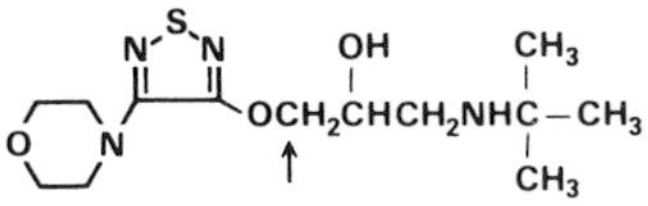

TRICYCLIC ANTIDEPRESSANTS

Amitriptyline

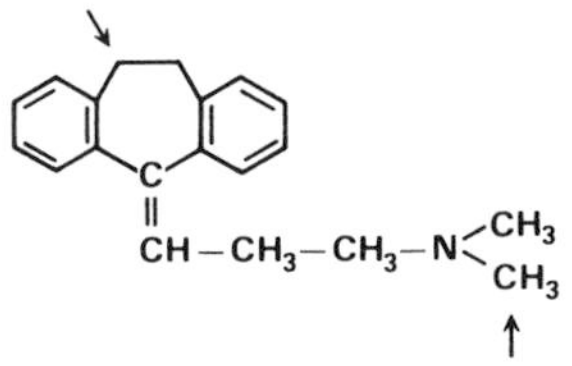

Nortriptyline

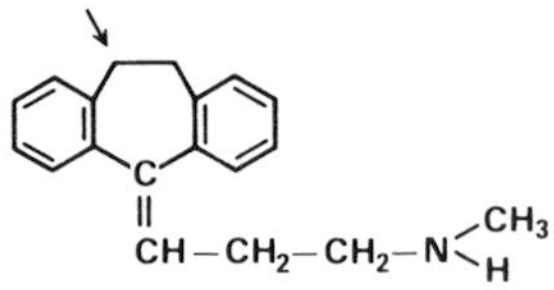

Desmethylimipramine

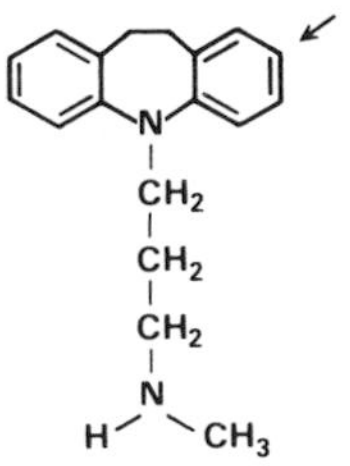

CARDIOVASCULAR DRUGS

Debrisoquine

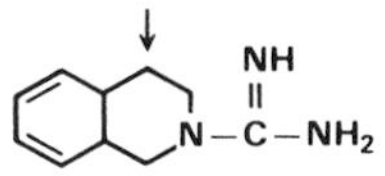

Sparteine

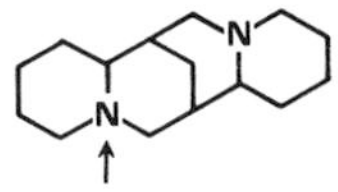

Guanoxan

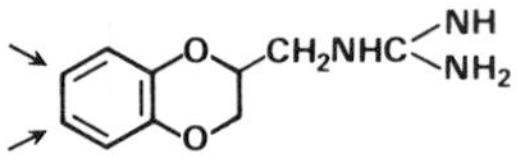

Propafenone

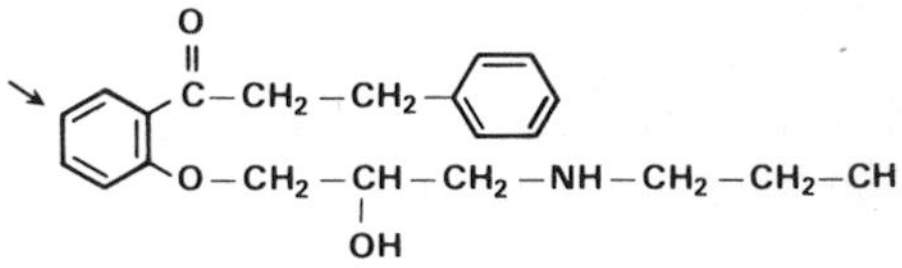

Encainide

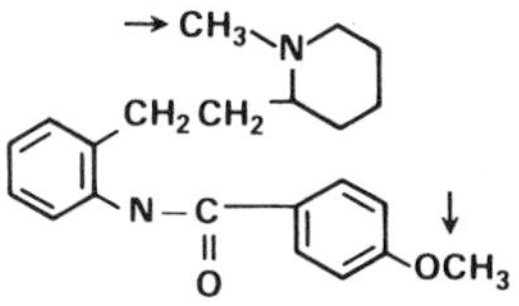

Perhexiline

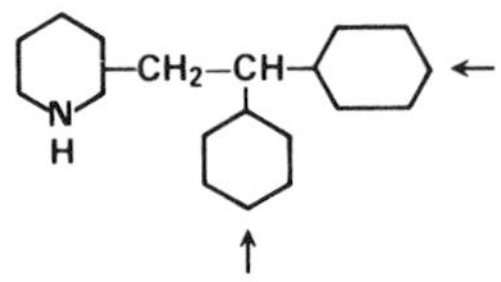

N-Propylajmaline

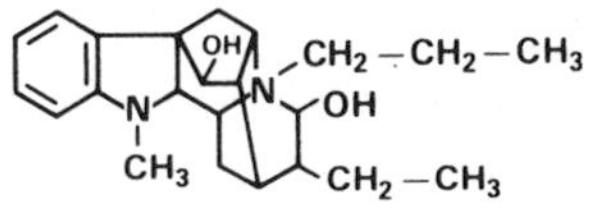

MISCELLANEOUS

Dextrometorphan

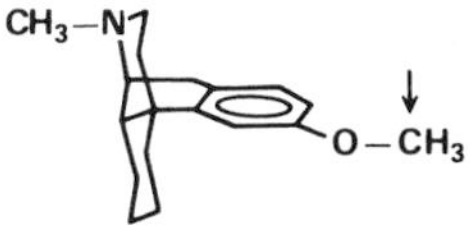

Methoxyamphetamine

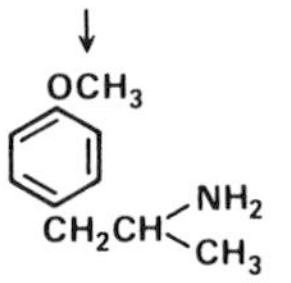

Phenformin

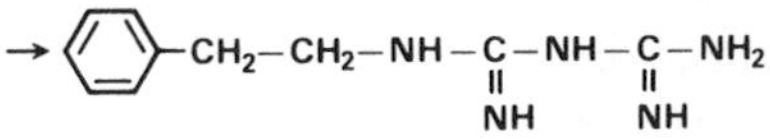

Codeine

Morphine

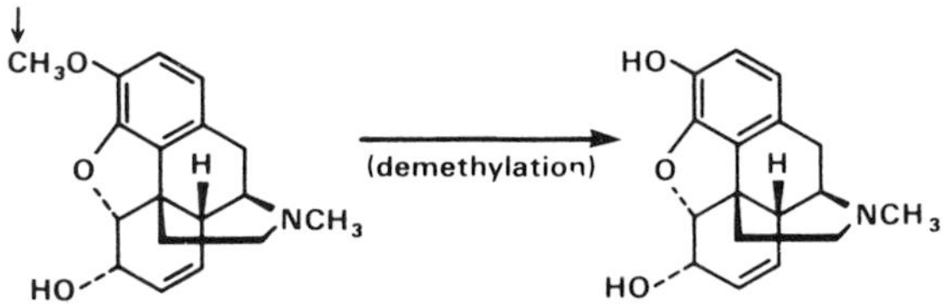

(demethylation)

drugs, and other miscellaneous drugs, including dextromethorphan, the active component of over-the-counter cough medicine. This enzyme is also capable of converting the inactive analgesic codeine to its active derivative, morphine (Fig. 6).

P-450s also have overlapping substrate specificities. A single compound can be metabolized at the same position by multiple P-450s. The affinity of different P-450 forms can vary significantly, such that one form may be active against a compound at low concentrations, whereas another P-450 may require high substrate concentrations. When a drug is first ingested, many P-450s may contribute to its metabolism, but later, when the drug is present at lower concentrations, only a single P-450 may be able to inactivate the substance.

A single compound or drug can also be metabolized at different positions by one or more P-450s, and this is particularly true of large compounds with complex structures. For example, note the positions of hydroxylation of the cardiovascular drugs guanoxan and encainide, shown in Fig. 6. In these cases, the P-450 can bind and hydroxylate these compounds at different positions. Frequently, oxidation at one site inhibits substrate binding and oxidation at another site on the same molecule. It is also possible that a single compound can be metabolized at different sites by different P-450 forms.

Finally, it is fairly common to find that a single substrate or drug is metabolized primarily by a single form of P-450. Therefore, when this P-450 is not expressed, the drug cannot be metabolized. This condition can lead to a drug oxidation polymorphism.

B. P-450 Polymorphisms

Genetic polymorphisms in P-450-mediated reactions occur in humans. These have been identified clinically as drug oxidation polymorphisms. For example, the debrisoquine polymorphism affects about 10% of European and North American Caucasians, who cannot metabolize the cardiovascular drug debrisoquine and numerous other compounds that are specific substrates for a single P-450 form, CYP2D6. However, such clinical polymorphisms do not affect the metabolism of all drugs that act as substrates for CYP2D6 (Fig. 6), possibly because other P-450s can also metabolize these compounds when CYP2D6 is not expressed.

The debrisoquine polymorphism results from mutations in the structural gene encoding CYP2D6, the CYP2D6 gene. Several mutant alleles of the CYP2D6 gene exist that produce CYP2D6 primary RNA transcripts that are incorrectly spliced, and consequently P-450 protein synthesis does not occur. Other mutations have also been detected, including these insertions that disrupt the coding region of the mRNA and amino acid changes that result in an enzymatically inactive CYP2D6 protein.

The clinical consequences of P-450 polymorphisms can be quite severe. When debrisoquine was used in Europe in the early 1970s, many patients suffered exaggerated side effects, including dramatic decreases in blood pressure, because they were unable to metabolize and inactivate the drug. This variability in clinical response led to the discontinuation of debrisoquine use. Fortunately, other drugs could be substituted for debrisoquine. The metabolism of some other drugs (e.g., bufuralol) has also been found to be affected by the polymorphism in CYP2D6 and, as a consequence, these agents have not been widely used.

Several other P-450s are also known to be polymorphically expressed in humans but do not complicate the clinical use of drugs, in part because many drugs can also be metabolized by other P-450s or the drugs metabolized have broad therapeutic windows such that marked differences in their metabolism do not cause increased toxicity or decreased efficacy. CYP2C19 could be considered such a P-450. It should also be recognized that some potential drugs that are subjected to polymorphic metabolism in humans are never taken beyond the development or preclinical testing stage. It is possible, therefore, that many potentially useful compounds are eliminated because of P-450 polymorphisms. Uncertainties regarding drug metabolism will decrease considerably with the development of tests to detect mutant P-450 genes and, hence, to identify individuals who are incapable of metabolizing a certain drug. P-450 gene testing could be undertaken using DNAs derived from small blood samples. These tests will allow physicians to tailor drug prescriptions to an individual's P-450 phenotype.

Certain rare toxicities are sometimes associated with the use of drugs. For example, acetaminophen hepatotoxicity occurs very rarely and is usually fatal.

FIGURE 6 Substrates for CYP2D6. The hydroxylation positions of each substrate are indicated by arrows. The conversion of the analgesic codeine to its active derivative morphine is shown at the bottom of the figure. (The author thanks Dr. Urs A. Meyer for use of this figure.)

This common analgesic compound is known to be metabolized by a P-450, and it is therefore possible that a P-450 defect is responsible for its toxicity. Such a rare defect would not be considered a polymorphism but might be a spontaneous germ line mutation in a gene encoding a specific P-450 form. Alternatively, a regulatory defect in a gene may result in increased levels of a certain P-450 that produces toxic metabolites.

The fact that P-450s metabolize and activate carcinogens suggests that they may be involved in the susceptibility or resistance to human cancer. For example, it is well established that cigarette smoking produces a 10-fold increase in the relative risk for lung cancer. On the average, about one in five heavy smokers develops lung cancer. Because the primary cancer-causing agents in tobacco smoke are chemical carcinogens that must be activated by P-450s, it is possible that susceptibility to lung cancer may be associated with the presence in some individuals of a P-450 that activates a chemical carcinogen to a DNA-damaging and mutagenic metabolite. Conversely, a P-450 that is polymorphically expressed may be able to convert a chemical carcinogen to an inactive metabolite. The role of these enzymes as risk factors in susceptibility to cancers of the lung and other organs is still unknown and is under investigation. [See Tobacco Smoking, Impact on Health.]

Each year the drug and chemical industries produce tens of thousands of new chemicals during the development of drugs, insecticides, plastics, polymers, and other products. Many of these are tested for toxicity and carcinogenicity in animal-based systems. For example, rats and mice will be given a specific chemical and observed for toxic reactions. If a chemical harms a rodent, it is usually regarded as harmful to humans. However, as discussed earlier, rodents can have P-450s and enzyme activities that are distinctly different from those of humans. In these cases, the development of potentially useful chemicals will be discontinued even before it has been determined that they are actually harmful to humans. Today, cell culture systems are being developed in which P-450s derived from cloned human cDNAs can be used to test chemicals. It is also possible to produce transgenic animals that have human P-450 genes. [See Carcinogenic Chemicals.]

C. P-450 Regulation

P-450s are regulated by a number of xenobiotics. In many cases, the cellular content of a P-450 is elevated by the same chemical that it metabolizes. The induction of P-450s by certain chemicals is known to proceed via a receptor-mediated mechanism. This type of regulation is common in biological systems involving hormones. The chemical or ligand enters the cell and binds to a receptor protein. This binding confers on the receptor an ability to specifically interact with regulatory regions of the P-450 gene, resulting in the activation of gene transcription and a subsequent increase in mRNA and protein levels. P-450 then rapidly degrades the chemical.

The induction of other P-450s is governed by a substrate-induced stabilization of P-450 levels. In the presence of substrate, P-450 levels are stabilized, or the rate of P-450 degradation is decreased. The cellular level of enzyme is then increased until the chemical or substrate is metabolized and its concentration lowered. In some cases, the mRNA encoding a P-450 becomes stabilized against degradation, resulting in an increase in mRNA level and a subsequent increase in P-450 synthesis.

Induction of P-450s has important implications for clinically significant drug interactions. For example, oral contraceptives have been found to fail in patients taking the antibiotic rifampicin. This antibiotic is known to induce P-450s that can metabolize 17α-ethynylestradiol, the principal component of birth control pills. The synthetic steroid dexamethasone and the anticonvulsant agent phenobarbital can also induce P-450s and impair usage of other drugs.

BIBLIOGRAPHY

Fitzpatrick, F. A., and Murphy, R. C. (1988). Cytochrome P-450 metabolism of arachidonic acid: Formation and biological actions of epoxygenase-derived eicosanoids. *Pharmacol. Rev.* 40, 229.

Gonzalez, F. J. (1988). The molecular biology of cytochrome P-450s. *Pharmacol Rev.* 40, 243.

Gonzalez, F. J. (1992). Human cytochromes P-450: Problems and prospects. *Trends Pharmacol. Sci.* 13, 346.

Gonzalez, F. J., and Nebert, D. W. (1990). Evolution of the P-450 gene superfamily: Animal–plant "warfare," molecular drive and human genetic differences in drug oxidation. *Trends Genet.* 6, 182.

Gonzalez, F. J., Skoda, R. C., Kimura, S., Umeno, M., Zanger, U. M., Nebert, D. W., Gelboin, H. V., Hardwick, J. P., and Meyer, U. A. (1988). Characterization of the common genetic defect in humans deficient in debrisoquine metabolism. *Nature* 331, 442.

Guengerich, F. P. (ed.) (1987). "Mammalian Cytochromes P-450," Vols. I and II. CRC Press, Boca Raton, FL.

Lechner, M. C. (ed.) (1994). "Cytochrome *P-450*: Biochemistry, Biophysics and Molecular Biology." John Libbey Eurotext, Montrouge, France.

Miller, W. L. (1988). Molecular biology of steroid hormone synthesis. *Endocr. Rev.* **9**, 295.

Nelson, D. R., Koymans, L., Kamataki, T., Stegeman, J. J., Feyereisen, R., Waxman, D. J., Waterman, M. R., Gotoh, O., Coon, M. J., Estabrook, R. W., Gunsalus, I. C., and Nebert, D. W. (1996). The *P-450* superfamily: Update on new sequences, gene mapping, accession numbers and nomenclature. *Pharmacogenetics* **6**, 1.

Ortiz de Montellano, P. R. (1989). Cytochrome *P-450* catalysis: Radical intermediates and dehydrogenation reactions. *Trends Pharmacol. Sci.* **10**, 354.

Cytokines and the Immune Response

PHILIP J. MORRISSEY

Immunex Corporation

GLOSSARY

Antigen A foreign substance that stimulates an immune response

Cytokines Proteins produced and secreted by cells that influence (both in a positive and negative manner) the function of the immune system

Granulocytes White blood cells that are produced from immature cells in the bone marrow. Three distinct types exist based on the staining of cytoplasmic granules: neutrophils, basophils, and eosinophils. These cells play a role in innate or immediate resistance

Lymphocytes Another type of bone marrow-derived white blood cell that functions in the specific arm of the immune response

Transgenic animals Mice in which the natural genome has been manipulated either by inserting a functional gene that would be naturally expressed (transgenic overexpressor mice) or by deleting a specific gene through the process of homologous recombination (gene knockout mice)

CYTOKINES ARE POLYPEPTIDES (PROTEINS) PROduced largely by cells of the immune system. These proteins are for the most part secreted (a few can be effective while bound to the surface of the cell that produces them). As soluble proteins they diffuse through the immediate environment and bind to specific receptor proteins on the membranes of other cells. As a consequence of this binding, a signaling pathway is triggered within that cell and a certain biological response, such as cell activation, differentiation, or division, is induced. Many, but not all, cytokines have been given an "interleukin" designation (abbreviated IL) in the order of their discovery. This simplified the nomenclature that existed at the time in which cytokines were known by functionally descriptive terms. Some factors, e.g., tumor necrosis factor (TNF), still retain their descriptive terminology. The receptors for these cytokines are designated with "R". Thus, the receptor for IL-1 would be written IL-1R. A number of cytokines are being used clinically to treat disorders ranging from anemia to autoimmune disease. Many more are in clinical trials. Augmenting host resistance to infectious disease or suppressing abnormal immune responses as in arthritis through the use of cytokines is a significant advance in clinical medicine and holds great promise for the future.

I. HOST RESISTANCE

The immune system exists to protect us from invasion and colonization by other organisms such as viruses, bacteria, or fungi. Our interior environment, which has a neutral pH, significant amounts of nutrients, and a temperature conducive to growth, is a very hospitable and inviting medium for microorganisms. Normally, the immune system functions very efficiently in its task and we only need to observe the continual problems people with compromised immune systems, such as in acquired or inherited immune deficiency diseases, have to appreciate this.

The immune system has at its disposal a large array of weaponry to limit the spread and to eventually eradicate invading microbes. These can be thought of

ENCYCLOPEDIA OF HUMAN BIOLOGY, Second Edition, VOLUME 3.

as two distinct systems that are interrelated. The first system is the rapid response arm that consists predominantly of two cell types, the granulocyte and macrophage. The function of these cells is to migrate quickly (within hours) from the bloodstream to the area where the infection has been first detected and attempt to eradicate it or at least limit its spread. Two types of signals alert these cells to the presence of microbes. One signal involves soluble proteins such as cytokines that would be produced by infected or damaged cells. The other signal is the expression of adhesion molecules by the vascular endothelial cells. Granulocytes and macrophages traveling through the bloodstream attach to the vascular adhesion molecules and migrate through the vessel wall to the damaged or infected tissue. Once at the site of infection, these cells become activated and secrete an array of substances in an attempt to eradicate the infectious agent. The rapid large-scale accumulation of granulocytes and macrophages at a site of infection leads to the phenomenon of inflammation which is characterized by swelling, redness, and pain. This arm of the immune system is nonspecific in nature. The array of biochemical weaponry unleashed to control the microbe damages the cells of the host as well as the invader. It appears that specificity is sacrificed in exchange for the rapidity of the response. [See Immune System; Macrophages.]

The other arm of the immune system involves the participation of lymphocytes. These cells become activated in response to signals from the cells participating in the early inflammatory response. Lymphocytes uniquely respond in a specific manner to the invading microorganism. Lymphocytes achieve this through the use of proteins (known as antigen receptors) that are made from genes that undergo rearrangement or shuffling. This results in a clonal distribution of unique amino acid sequences in the business end (i.e., antigen binding) of the antigen receptors. This process of gene rearrangement occurs during lymphocyte development and results in greater than 10^{10} unique antigen-binding domains. There are generally two distinct classes of antigen receptors: the immunoglobulin molecule on the surface of B cells and the T-cell antigen receptor. The surface immunoglobulin molecule on B cells can be secreted after the B cell has become activated and this is the basis of the antibody response. The T-cell antigen receptor binds to small peptides (about 10–20 amino acids) that are bound to cell surface proteins encoded by the major histocompatibility locus. T cells are divided into two classes depending on whether they express cell surface molecules known as CD4 or CD8. Importantly, one function of CD4$^+$ T cells is to regulate the immune response. The specific population that does this is known as "T helper" or Th cells. CD8$^+$ T cells are effector cells specializing in cytotoxic activity. [See Lymphocytes; T-Cell Receptors.]

For many years it was observed that the immune system was able to activate weaponry designed preferentially for extracellular pathogens [e.g., antibodies) or intracellular parasites (e.g., cytotoxic T cells or delayed-type hypersensitivity responses (DTH)]. This mystery was explained by investigators who were studying the array of cytokines produced by clonal populations of T helper cells. They found that these cells produced either IL-2, TNF-β, and interferon (IFN)-γ or IL-4, IL-5, and IL-10. Interestingly, the first pattern of cytokine production leads to the preferential development of DTH whereas the second pattern results in the preferential development of an antibody response. As a consequence, T helper cells were designated either Th1 (IL-2, IFN-γ producing) or Th2 (IL-4, IL-5 producing) (Fig. 1). Interestingly, Th1-type cytokines inhibit the proliferation of Th2 cells and vice versa. It is still not entirely clear how the immune system determines what type of response to develop. It is thought that T cells that have not encountered antigen (known as Th0) have the capacity to become either Th1- or Th2-type cells. Early signals that include cytokines from activated macrophages as well as natural killer (NK), natural T cells (NT), mast cells, or basophils are probably responsible for developing the cytokine array that influences the development of a Th1 or Th2 response. For instance, the production of IL-4 by mast cells and IL-10 by macrophages will lead to the activation of a Th2 response. Conversely, IFN-γ production by NK cells and IL-12 production by macrophages will lead to a Th1 response.

There are a few important principles regarding the influence of cytokines on immune function.

1. Normally, the immune system exists in a quiescent state.
2. Activation is partly achieved by antigen binding to antigen receptors on lymphocytes.
3. Other signals are required for the full activation of the immune system and for the development of effector function (e.g., antibody production or cytotoxic T-cell development).
4. These signals can be soluble cytokines as well as membrane-bound molecules known as costimulatory molecules which exert their effects through cell–cell contact.

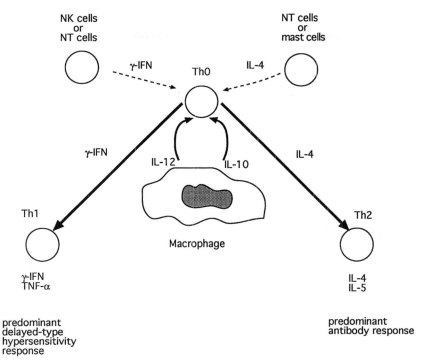

FIGURE 1   Schematic representation of helper T-cell function.

II. CYTOKINES

Cytokines are secreted proteins produced to activate and influence the function of the immune system. More than 24 cytokines have been described. They regulate every phase of host resistance, including inflammation, hematopoiesis, and immune function. Through recombinant DNA technology, the genes encoding these proteins have been identified and cloned. The availability of recombinant DNA-derived cytokines produced in bacteria or yeast and purified to homogeneity has led to significant advances in the understanding of the precise role of the cytokines in immune function. Also, experimental approaches to manipulating the genome of animals (largely mice) to purposefully overexpress a cytokine (transgenic animals) or have a cytokine or cytokine receptor deleted through homologous recombination (gene knock-out animals) have also been valuable sources of information.

A number of important principles have emerged regarding cytokine function:

1. Cytokines are plieotropic (they have many effects).
2. Cytokines are redundant (different cytokines have similar functions).
3. Cytokines are produced as a result of cellular activation.
4. The biological activity of cytokines depends on which cells express receptors for these cytokines.

In addition, recent progress has revealed that cytokines can be grouped together in families depending on structural similarity not only of the cytokine but in the receptors to which they bind (Table I). Some cytokines actually share receptor subunits. In these cases the unique subunit would impart receptor specificity for the binding of a particular cytokine. An example of this is the IL-2R system (Fig. 2) in which the IL-2Rγ chain is shared with IL-4, IL-7, IL-9, and IL-15. IL-15 also uses the IL-2Rβ chain. Cytokines and cytokine receptors can also be grouped with respect to structural similarities. The TNF/TNFR family is the best example of this. For instance, molecules in the TNFR family are characterized by the presence of multiple cysteine-rich repeats of about 40 amino acids in the extracellular domain. This family also includes molecules that function as cell bound rather than as secreted entities. Thus, cell–cell contact is necessary for signal transduction between a cell-bound receptor and its ligand. Cell-bound molecules such as CD40 and its cognate ligand CD40L are co-stimulatory molecules for T- and B-cell activation.

TABLE I

Commonalities among Cytokines

Cytokines that share receptor subunits	
IL-2, IL-4, IL-7, IL-9, IL-15	IL-2Rγ chain
IL-3, IL-5, GM-CSF	β$_c$ chain
IL-6, IL-11, LIF, OSM, CNTF	gp130
Structurally related families of cytokines and cytokine receptors	
TNF ligand family	TNF-α, LTα, LTβ, FasL, CD40L, CD30L, CD27L, 41BBL
TNF receptor family	TNFRp55, TNFRp75, TNFR-RP, Fas, CD30, CD40, CD27, OX40, 41BB
Hemopoietin receptor family	EPOR, G-CSFR, IL-2R, IL-4R, IL-3R, IL-5R, GM-CSFR

Members of the hemopoietin superfamily have similar binding domains consisting of approximately 200 amino acid molecules that show a distinctive conservation of four cysteine residues in the N-terminal half and a 5 amino acid motif near the C-terminal end. Thus, many of these proteins may have been derived from a common ancestral precursor.

III. CYTOKINES, HEMATOPOIESIS, AND LYMPHOPOIESIS

The production of mature, functional lymphocytes, macrophages, and granulocytes as well as red blood cells (RBC) and platelets proceeds by an ordered process of differentiation from immature precursor cells

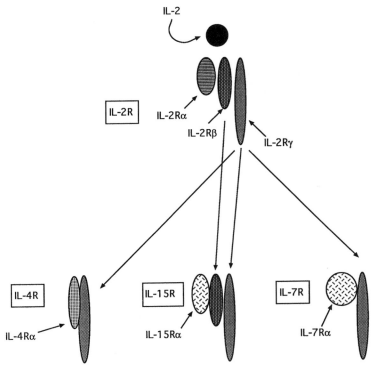

FIGURE 2 Sharing of IL-2R components among receptors for related cytokines.

(Fig. 3). Significant numbers of RBC, granulocytes, B lymphocytes, and platelets are produced by the cells of the bone marrow on a daily basis. T lymphocytes also originate from a precursor in the bone marrow but their maturation depends on a period of residence in the thymus. The vast majority of T cells are produced early in life. Removal of the thymus of adult mice has a minimal impact on the number of T cells in the body. A number of soluble factors have been identified that stimulate either the proliferation or the differentiation of precursors to the just-mentioned cells. These are listed in Table II. These factors are largely secreted by the stromal cells of the bone marrow which support the process of hematopoiesis, but others such as IL-3 and GM-CSF are also produced by T cells upon activation. A number of factors are known as colony-stimulating factors (CSF) because of their ability to induce the growth of immature hematopoietic cells as small colonies in semisolid media. For instance, G-CSF induces the growth of granu-locyte colonies, CSF-1 induces the growth of macro-phage colonies, and GM-CSF induces the growth of mixed colonies containing granulocytes and macro-phages. Two of these factors are used clinically (G-CSF and GM-CSF) to treat neutropenia as a consequence of chemotherapy or bone marrow transplantation. Other factors such as erythropoietin and thrombopoietin are lineage-specific growth factors that generally stimulate the near final process of maturation. Erythropoietin is used clinically to treat patients with anemia. A number of cytokines such as IL-1, IL-3, IL-4, IL-5, IL-6, and IL-11 can augment hematopoietic production; however, their mechanism of action is not clear. Lymphocyte development seems to depend on IL-7. IL-7 and IL-7R knockout animals have markedly diminished levels of mature B and T cells and the defect can be traced to a block in lymphocyte development. IL-7-overexpressing mice have lymphoid hyperplasia. Thus, IL-7 seems to play an integral role in the expansion of both B and T cells.

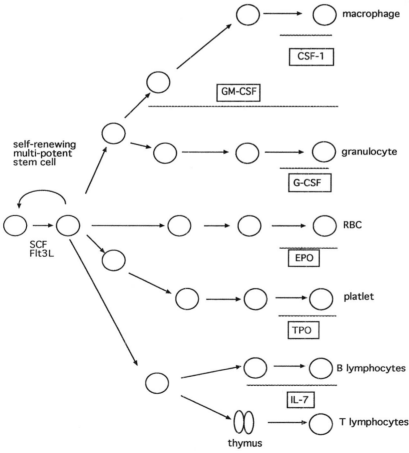

FIGURE 3 The process of hematopoeisis.

TABLE II

Cytokines Involved in Hematopoiesis

Cytokine	Biological activity
Early factors	
Stem cell factor (SCF)	Augments activity of early stem cells and promotes mast cell production
FIt3L	Involved in the maintenance of early stem cell populations
Colony-stimulating factors	
GM-CSF	Induces proliferation of cells with potential to become macrophages or granulocytes
G-CSF	Induces proliferation/differentiation of cells committed to becoming granulocytes
M-CSF	Induces proliferation/differentiation of cells committed to becoming macrophages
Lineage-specific factors	
EPO	Induces proliferation/differentiation of cells committed to becoming red blood cells
TPO	Induces proliferation/differentiation of cells committed to becoming platlets
Lymphoid-specific growth factor	
IL-7	Acts in the proliferative/expansion phase of cells committed to becoming either T cells or B cells.
Other cytokines that modulate hematopoiesis	
IL-1, IL-3, IL-4, IL-5, IL-6, IL-9, IL-11	Act to augment the response of immature cells to CSFs or lineage-specific growth factors; may promote the production of a certain type of cell or lineage

Of all the cytokines that have been shown to stimulate the proliferation of thymocytes *in vitro,* only IL-7 knockout animals have diminished thymic cellularity. Normal thymocyte numbers and phenotype are seen in IL-1R, TNF-αR, IL-2, and IL-4 gene knockout animals. Thus, it appears that these cytokines are not necessary for thymocyte development or that they affect thymocyte development in a manner not yet appreciated by our current state of knowledge. [*See* Hemopoietic Systems.]

IV. PROINFLAMMATORY CYTOKINES

A. Chemokines

An early response of cells to infection or injury is the elaboration of factors that alert host resistance mechanisms. A number of these factors have been described (Table III) and classified according to their mode of action. One group, termed the chemokines, consists of two structurally related families of small proteins between 8 and 10 kDa in size. Proteins of both families have four conserved cysteine residues that form two disulfide bridges, and the subfamilies can be distinguished according to the position of the first two cysteines that are either separated by one amino acid (C-X-C) or are adjacent (C-C). The C-X-C family has also been known as intercrine α factors and the C-C family as intercrine β family. The genes for the C-X-C family are clustered on human chromosome 4 and the genes for the C-C family are similarly clustered on human chromosome 17. A number of these factors (PF4, PBP) are stored in granules in platelets and are released on aggregation. These factors are also produced by many different cell types such as macrophages, fibroblasts, and endothelial and epithelial cells. It is quite possible that most cell types in the body can produce at least one of the chemokines when appropriately stimulated. The known stimuli include bacterial by-products, immune complexes, and other cytokines such as IL-1, TNF, and GM-CSF.

TABLE III
Cytokines Involved in the Inflammatory Process

		Activity
Chemokines		
CXC family	IL-8	Chemotaxis/activation of granulocytes and macrophages
	GROα	
	GROβ	
	GROγ	
	NAP-2	
	ENA-78	
	PF4	
CC family	MCP-1	
	HC14	
	MCP-3	
	MIP-1α	
	MIP-1β	
Interleukin 1		Acute-phase response/fever/cell activation
TNF-α		
IL-6		

The main activity of the secreted chemokines is to stimulate neutrophil and macrophage migration toward the site of chemokine production (chemotaxis). Some of the chemokines may also act to attract lymphocytes, but this seems to be a lesser function. They also have the properties of activating some cellular functions that are important in host defense. Thus, production of these proteins results in the migration and activation of substantial numbers of granulocytes and macrophages from the bloodstream into the surrounding tissue where damage or infection has occurred. Also, in conditions of chronic inflammation, such as psoriasis and arthritis, some of the chemokines such as IL-8 are expressed excessively.

B. Interleukin 1

The cytokine IL-1 is involved in nonspecific inflammation as well as in specific immune function. There are two related IL-1 proteins, IL-1α and IL-1β. These proteins, which are approximately of the same size (15 kDa), are the products of two distinct but linked genes on human chromosome 2. At the protein level they are homologous at approximately 26% of the amino acids. Both IL-1α and IL-1β are produced by cleavage from a larger protein. Interestingly, the larger form of IL-1α has biological activity whereas the larger form of IL-1β does not. Both IL-1α and IL-1β bind to the same cell surface receptor and there is

no known difference in the spectrum of biological activities of IL-1α and IL-1β. IL-1 is produced by many different cell types, such as keratinocytes, endothelial cells, lymphocytes, and macrophages. The production of IL-1 by macrophages is thought to play a central role in the initiation of immune function. IL-1 also has systemic activities in many different physiological systems. IL-1 induces fever, induces or increases the expression of adhesion molecules on vascular endothelial cells, induces the production of acute-phase proteins by the liver, induces the production of CSFs to stimulate hematopoiesis, and alters cardiovascular homeostasis as well as intermediary metabolism. The overproduction of IL-1 during systemic persistent bacterial infection (sepsis) is believed to contribute to severe hypotension, organ failure, and, if not resolved, death.

C. Tumor Necrosis Factor

The cytokine TNF was discovered as an activity in the serum of mice that were injected with bacterial lipopolysaccharide (LPS). This activity caused the hemorrhagic necrosis of tumor cells *in vivo* and thus was known as tumor necrosis factor or TNF-α. Lymphotoxin (LT) is a related molecule secreted by T cells and is also known as TNF-β (see Section V,R). TNF-α is a protein of approximately 17 kDa and the gene encoding it is on human chromosome 6. Secreted TNF-α exists as a homotrimer. TNF-α also can be cell membrane bound and it has biological activity in this state. Two receptors for TNF-α are known as p55 and p75 (based on their molecular weight). TNF-α shares many biological properties with IL-1, although there is little homology between the two proteins. Thus TNF-α induces fever, induces and increases vascular adhesion molecule expression, and induces hepatic acute-phase protein production and the production of CSFs. TNF-α can also induce death in susceptible cells through a process of apoptosis. TNF-α binds to specific receptors that are distinct from the receptors for IL-1. TNF-α as well as IL-1 can be considered inflammatory cytokines as they induce many of the changes seen in the early physiological response to infection or tissue injury. Excessive production of TNF-α as well as IL-1 is also thought to contribute to the pathology of sepsis. TNF-α is vital to host resistance to infectious agents, however. This is best demonstrated by the studies on TNFRp55 knockout mice. These mice are very resistant to the injection of bacterial-derived LPS which mimics sepsis; however, they are unable to resist an infection by *Listeria monocytogenes*.

V. INTERLEUKINS AND OTHER CYTOKINES THAT INFLUENCE LYMPHOCYTE GROWTH AND FUNCTION

A. Interleukin 1

IL-1 was discussed in detail in the previous section as an inflammatory cytokine. IL-1 also plays a role in lymphocyte activation. IL-1 stimulates proliferation of B cells that have been activated by the cross-linking of their surface immunoglobulin. IL-1R are also expressed on T-cell clones of the Th2 phenotype and thus IL-1 production by macrophages is thought to drive the immune response toward antibody production.

B. Interleukin 2

Of all the interleukins, IL-2 is perhaps the most potent stimulator of T-cell growth. In fact, IL-2 was initially called the T-cell growth factor. This polypeptide is secreted by almost all T cells on activation; resting T cells do not produce IL-2. It binds to a specific cell surface receptor that consists of three distinct protein chains. Expression of this receptor is also induced by some prior activation event. The main effect of IL-2 binding to IL-2R is to stimulate the proliferation, activation (e.g., cytotoxic function), and production of cytokines. Although IL-2 mainly effects T cells, NK cells also become activated in response to IL-2, and IL-2 can also influence the function of activated B cells and macrophages. Given the potent T-cell growth and activation properties of IL-2, there was initially great interest in using IL-2 in various clinical settings, such as in cancer and infectious disease. However, significant toxicities were seen with repeated IL-2 administration. Despite this, IL-2 treatment is effective in a small proportion of patients (generally <25%) with metastatic melanoma and renal carcinoma. Also, there is continual interest in the IL-2-driven *ex vivo* expansion of T cells from patients with cancer and then reinfusing the expanded, stimulated cells that would eradicate the cancer. Again, while some remarkable therapeutic responses were seen early on, this treatment does not appear to work in a large proportion of patients. However, at this time the clinical uses of IL-2 are still an area of active research.

Interestingly, in mice that lack a functional IL-2 gene, lymphopoiesis appears normal. These mice develop severe autoimmune disease (hemolytic anemia and colitis) at a relatively young age. Treatment of these mice with IL-2 significantly lessens the severity of the autoimmune disease. Thus, it is postulated that the function of IL-2 *in vivo* is to somehow regulate immune function that would prevent the development of autoimmune disease. [*See* Interleukin-2 and the IL-2 Receptor.]

C. Interleukin 3

IL-3 is a cytokine that is produced by almost all T cells after activation. Its activities appear to be mostly on hematopoietic lineage cells, although it does not directly stimulate hematopoietic cell growth or colony formation. Of interest is the observation that one component of the IL-3 receptor is shared with the receptors for IL-5 and GM-CSF. Also, the genes encoding these three cytokines, IL-3, IL-5, and GM-CSF, are clustered on human chromosome 5 or mouse chromosome 11. These associations somehow suggest a common functionality. Evidence exists that these cytokines regulate the production of eosinophils. This class of granulocytes (WBC) is thought to be important in mediating resistance to parasites such as *Schistosoma mansoni*. Mice genetically engineered to overexpress IL-3 develop pathological levels of myeloid cells that result in the destruction of normal tissue. IL-3 is being tested in clinical trials for its efficacy in stimulating platelet production following the suppressive effects of chemotherapy and ablative treatment prior to bone marrow transplantation.

D. Interleukin 4

IL-4 was initially known for its ability to act as a costimulatory molecule in B-cell activation. After the gene encoding IL-4 was cloned and purified recombinant IL-4 was produced, it was found quite surprisingly to be a potent stimulator of T cells as well as B cells. Indeed, IL-4 was used to augment *ex vivo* IL-2-driven proliferation and activation of LAK cells in cancer patients, and the efficacy of IL-4 as an antitumor agent is being investigated in the clinic. IL-4 is produced by some T cells after activation as well as by basophils, mast cells, and natural T cells (a subset of NK cells). IL-4 is important in determining the development of immune system effector function. It is important in the development of Th2-like responses that result in predominant antibody production. As mentioned earlier, IL-4 is a potent inducer of the proliferation of B cells that have been activated through the cross-linking of their surface IgM. As well as driving proliferation, IL-4 can also act to induce immunoglobulin secretion and isotype class switching by activated B cells. In this regard, IL-4 is believed to be important in the production of IgG. Excessive or inappropriate IgE production by the immune system can lead to allergic reactions to otherwise innocuous substances (e.g., bee venom, grass, pollen). IL-4 also has anti-inflammatory properties that seem to act through the down-regulation of macrophage function.

E. Interleukin 5

IL-5 was initially studied as a T-cell-derived B-cell growth factor. It had activities similar to IL-4, but there were significant differences in various experimental systems. IL-5 appears to be important for the growth of a small population of B cells (B-1) that may be important in immunoregulation and in the development of autoimmunity. Also, it appears that IL-5 may act in synergy with other cytokines to promote proliferation and differentiation of other B-cell populations. Again it is important to note that the activity of IL-5 on these B cells requires prior activation of those cells. As mentioned previously, IL-5 shares a receptor subunit with IL-3 and GM-CSF and thus has hematopoietic activities. Evidence shows that IL-5 can drive the production of eosinophiles, again in synergy with other cytokines. Mice that constitutively overexpress IL-5 have elevated levels of eosinophiles and treatment of mice infected with protozoan parasites with antibodies that neutralize IL-5 activity prevent the development of eosinophilia.

F. Interleukin 6

IL-6 is a cytokine with an extraordinary range of activities. For instance, early independent studies on an unidentified factor with antiviral activity, hybridoma/plasmacytoma growth activity, hepatocyte-stimulating activity, and B-cell differentiation activity were all attributable to IL-6. IL-6 is produced by many cell types in the body such as macrophages, fibroblasts, and T cells in response to inflammatory signals such as IL-1 and TNF-α. IL-6 can act as a costimulator of T-cell activation and enhances the development of cytotoxicity by T cells. IL-6 is important in B-cell growth and immunoglobulin production and this is thought to be a late stage event. For instance, terminally differentiated antibody-produc-

ing B cells (plasma cells) respond to IL-6. Also, mice engineered to overexpress IL-6 in their B cells develop plasmacytomas, and there is interest in finding an inhibitor of IL-6 activity for patients with multiple myeloma. IL-6 also has systemic effects such as the induction of acute-phase proteins by hepatocytes, stimulation of bone resorption by osteoclasts, and fever induction as well as hematopoietic-potentiating activity. IL-6 levels in serum are also elevated in sepsis and this has been associated with a poor prognosis. IL-6 is being tested in separate clinical trials as an antitumor agent and as a stimulator of platelet production.

G. Interleukin 7

IL-7 was initially discovered as an activity that stimulated the growth of a population of B-cell precursors. After cloning of the gene, studies with purified recombinant IL-7 revealed that it was a potent costimulator in T-cell activation assays, a direct stimulator of some long-term T-cell lines, and a potent inducer of early thymocyte proliferation. Mice lacking a functional IL-7 gene are severely lymphopenic in well-defined stages of lymphopoiesis, and mice that overexpress IL-7 develop lymphoid hyperplasias. The involvement of IL-7 in immune responses is unclear at this time. Evidence suggests that IL-7 may act on peripheral lymphocytes to prevent death via apoptosis. Thus, in addition to being a growth factor for immature lymphocytes, IL-7 may act as a maintenance or survival factor for mature lymphocytes.

H. Interleukin 8

IL-8 is a member of the proinflammatory chemokine family and was discussed earlier. Certain members of the chemokine family have been reported to have chemoattractant activity for lymphocytes.

I. Interleukin 9

IL-9 is a cytokine that was identified as an activity in the conditioned supernatant of a murine helper T-cell line that stimulated the growth of other T-cell lines in the absence of antigen or antigen-presenting cells. Evidence indicates that IL-9 is produced by Th2 cells and acts to amplify the Th2 immune response. In addition, other activities have been attributed to this cytokine, including mast cell proliferation, Ig production, and erythroid differentiation. Interestingly, IL-9 stimulates the growth of murine thymic lymphoma cell lines and prevents the induction of dexamethasone-induced apoptosis in these lines. Mice genetically engineered to overproduce IL-9 develop thymic lymphomas.

J. Interleukin 10

IL-10 was originally described as an activity that inhibited the production of IFN-γ by T-cell clones as well as stimulating the proliferation of activated thymocytes. The gene for IL-10 was isolated from a T-cell clone that produced the inhibitory activity. Quite interestingly, both the murine and the human IL-10 gene show considerable homology to an open reading frame from the Epstein–Barr virus (EBV) genome, and evidence shows that this gene is translated and that the viral gene product is secreted. IL-10 is produced by many cell types such as T cells, B cells, activated macrophages, and mast cells. Purified recombinant IL-10 significantly influences the type of immune response that develops in response to antigen. IL-10 elaborated by Th2 cells inhibits the proliferation of Th1 cells by inhibiting the expression of costimulatory molecules, as well as the production and response to IFN-γ and IL-2. IL-10 is thought to act in concert with IL-4 in promoting the development of a Th2 response. IL-10 does increase proliferation and Ig production by B cells, especially in concert with IL-4. It is in this context that the EBV IL-10 makes some sense in that viral survival and propagation would be best served by a Th2 and not a Th1 immune response. Cytotoxic T cells that would be produced in a Th1 environment would be effective in eliminating the virus. IL-10 also has significant effects on down-regulating macrophage function. Mice with a disrupted IL-10 gene develop severe colitis. It is thought that this occurs due to the absence of the dampening effect of IL-10 on macrophage function. Because of the large numbers of lumenal bacteria residing in the large intestine, inappropriate macrophage activation could lead to a chronic inflammation and tissue destruction. IL-10 is being evaluated in clinical trials on patients with steroid refractory colitis.

K. Interleukin 11

IL-11 is a recently discovered cytokine that has hematopoietic properties. Its mode of activity is to synergize with other cytokines or CSFs. Injection of mice with IL-11 results in greater numbers of neutrophils and platelets. Also, IL-11 has been reported to influence B-cell growth and activation in the presence of other cytokines. IL-11 is also considered to play a role in inflammation since it induces acute-phase protein production by the liver.

L. Interleukin 12

IL-12 was originally described as an activity that stimulated NK cells and cytotoxic T-cell function. IL-12 is a heterodimer consisting of p35 and p40 subunits. It is produced mainly by macrophages with B cells being a secondary source. The receptor for IL-12 is expressed on T cells and NK cells. IL-12 has been shown to play a pivotal role in the early stages of the immune response. It is produced by macrophages that have encountered, ingested, or otherwise been stimulated by antigen. Its prime function appears to be the potent stimulation of NK cells that results in IFN-γ production. The consequences of this are two-fold. First, elaboration of IFN-γ results in the up-regulation of macrophage antigen processing and priming for intracellular cytotoxic function. Second, IFN-γ affects the subsequent development of cytokine profiles by the activated T helper cells. Thus, IL-2 production is increased and IL-4 and IL-10 production is decreased and, as a consequence, results in the predominant develompent of a Th1 response characterized by strong DTH activity. IL-12 administration to experimental animals has increased antitumor immunity and resistance to infectious disease. Thus, IL-12 shows promising preclinical potential as an immunomodulator. Significant hepatotoxicity has been seen in mice treated long term with IL-12, and its toxicity in the clinic could limit its potential as a therapeutic.

M. Interleukin 13

IL-13 was first cloned from a library of induction-specific mRNA derived from a murine Th2 cell line. Interestingly, in both human and mouse, IL-13 maps to a region that encodes the IL-3, IL-4, GM-CSF, and IL-5 genes. The IL-13 gene shares a common intron–exon structure with GM-CSF, IL-4, and IL-5. IL-13 is produced by activated CD8+ and CD4+ T-cell clones. In humans, IL-13 is produced by Th1, Th2, and Th0 T-cell clones. Unlike IL-4, IL-13 does not activate T cells. IL-13 down-regulates the cytotoxic and inflammatory function of macrophages in both human and mouse. IL-13 enhances the proliferative response of human B cells that have been stimulated with anti-IgM or anti-CD40 antibodies. IL-13, like IL-4, can induce Ig switching to IgE in human B cells. IL-13 does not appear to act on mouse B cells. Thus, IL-13 is a cytokine with considerable, but not total, functional overlap with IL-4.

N. Interleukin 14

IL-14 was recently described as a high molecular weight B-cell growth factor. The protein was purified from culture supernatant of PHA-stimulated Namalava cells. A cDNA library was screened with an antibody raised against the purified protein. The gene identified encodes a protein of 483 amino acids (53 kDa). It is distinct from IL-4 and IL-5, two other cytokines that promote B-cell growth. Recombinant-derived IL-14 induced a five-fold enhancement of B-cell proliferation to the bacterial stimulant *Staphylococcus aureus* Cowan. It also inhibited Ig production. These activities are consistent with a cytokine that augments B-cell proliferation.

O. Interleukin 15

IL-15 was originally identified as a background activity in CTLL cells (an IL-2-dependent cell line) grown in the presence of culture supernatants of the simian kidney epithelial cell line CV-1/EBNA. The gene encoding IL-15 has been cloned from these cells and encodes a 162 amino acid protein of 14–15 kDa. There is minimal sequence homology between IL-15 and IL-2, although computer modeling supports a four helix bundle-like structure that is typical of the helical cytokine family that includes IL-2. Interestingly, IL-15 seems to perform many of the functions of IL-2. IL-15 can induce the proliferation of T-cell lines and secondary T-cell blasts. It increases cytotoxic T-cell activity as well. Uniquely, IL-15 is not produced by T cells as is IL-2. Initial reports indicate that it is produced by many other cell types that are not considered members of the immune system (epithelial cells, myocytes). IL-15 and IL-2 share receptor components (IL-2Rβ and γ chain) and there is a unique IL-15 receptor subunit (IL-15Rα chain). It is of interest to speculate why these molecules with similar profiles of bioactivity were conserved. The evidence thus far suggests that it allows a wide variety of cell types other than lymphocytes to influence and modulate immune reactivity.

P. Transforming Growth Factor β (TGF-β_1)

Historically, transforming growth factors were identified as products of virally transformed cells that induced phenotypic transformation of nonneoplastic cells. One of these factors, TGF-β, has been shown to have pronounced immunomodulatory activity. Three TGF-β genes have been identified and they are structurally similar. The effects of TGF-β_1 on immune function have been studied in great detail. Much less is known about the immunoregulatory activities of TGF-β_2 and TGF-β_3. TGF-β_1 has a variety of different effects on immune function. It is generally considered to be highly suppressive of activated T cell, B-cell growth, and macrophage function. However, it can act to augment the proliferation and cytokine secretion of naive lymphocytes and macrophages. TGF-β_1 also induces B cells to switch the isotype of the secreted immunoglobulin to IgA. Interestingly, the induction of oral tolerance, i.e., the immunological unresponsiveness induced to an antigen by feeding, is mediated by TGF-β_1 secreted by T cells. Clinical trials that involve feeding the suspected antigen are currently underway in patients with autoimmune disease. For instance, patients with multiple sclerosis are fed myelin basic protein daily in the hopes of inducing nonresponsiveness. The multiple effects of TGF-β_1 are clearly demonstrated by the phenotype of genetically engineered mice. TGF-β_1 gene knockout animals die at a very early age due to severe colitis and inflammatory infiltrates in many organ systems. Thus, in the absence of the suppressive effect of TGF-β_1, chronic inflammation results. In mice engineered to overexpress TGF-β_1 in the liver, excessive collagen synthesis and liver cirrhosis leading to death are observed. It is apparent that the proper regulation of TGF-β_1 expression is necessary for the proper function of the immune system as well as other organ systems. [*See* Transforming Growth Factor, Beta.]

Q. Interferons

IFNs represent a class of polypeptides that are noted for potent inhibition of viral replication. There are three classes of IFNs: α, β, and γ. IFN-α and -β are also known as the type I IFNs and are structurally related. IFN-γ is known as type II interferon. The genes for IFN-α and -β, but not IFN-γ, are linked on human chromosome 9 and they bind to same cell surface receptor. There are a multiplicity of genes (over 20) encoding IFN-α and there is one gene encoding IFN-β. Production of the type I IFNs is induced in many cell types, such as WBC, macrophages, fibroblasts, and epithelial cells after viral infection. IFN-α has been approved for the treatment of hepatitis B and C, genital warts caused by papilloma virus, hairy cell leukemia, and kaposi's sarcoma. IFN-β is approved for the treatment of relapsing-remitting multiple sclerosis. IFN-γ is unrelated to IFN-α and IFN-β. The single gene encoding IFN-γ is on human chromosome 12. IFN-γ binds to a different cell surface receptor than the type I IFNs. IFN-γ is produced by lymphocytes and NK cells after activation. Its main activity is to activate macrophages which includes cytokine production, costimulatory molecule expression, and anti-microbial effector function. IFN-γ is also an integral cytokine in the development of a Th1-type response of the immune systesm. IFN-γ is approved for the treatment of chronic granulomatous disease. All of the interferons are being tested for efficacy in many different disease states. [*See* Interferons.]

R. Lymphotoxin

Lymphotoxin has also been known as TNF-β despite poor sequence homology to TNF-α. Two forms of LT have been identified; LTα and LTβ. The genes for these proteins are linked to the gene for TNF-α on human chromosome 6. LTα is secreted by activated T cells as a homotrimer and binds to the receptors for TNF-α. The restricted expression of LTα in comparison to TNF-α suggests that its main role is in immune modulation and not inflammation, and evidence exists that LTα can be a costimulatory factor for B cells and T cells. LTβ is also expressed by activated T cells and it exists as a cell surface bound protein that complexes with LTα. LTα_1/LTβ_2 complexes seem to be the predominant form, but LTα_2/LTβ_1 forms have been described. These complexes are thought to remain cell surface bound. A receptor specific for the LTα_1/LTβ_2 heterotrimer has been described (LTβR). This receptor does not bind TNF-α or LTα homotrimers. Of interest is the observation that mice with a disrupted LTα gene do not form lymph nodes and have a disrupted splenic architecture. This is despite normal numbers of lymphoid cells and monocytes in the peripheral blood and suggests that signaling through the LTβR is required for the organogenesis of peripheral lymphoid organs.

VI. CONCLUSION

The use of gene cloning technology to produce interleukins purified to homogeneity has led to remarkable advances in our knowledge of their function in inflammation, lympho/hematopoiesis, and immune responses. As a consequence, the use of cytokines and interleukins has now moved into the therapeutic arena where their efficacy in treating human disorders is being assessed. The plieotropic and redundant nature of most cytokines, however, makes predicting their actual effect difficult. It should be pointed out that most cytokines are produced in the microenvironment of the lymphoid organs or at the site of infection. Current clinical trials where these potent biologics are administered systemically do not mimic their physiological role. Also, the production of cytokines during an immune response changes with time. It might be predicted that the efficacy of administering a cytokine to influence the immune response would depend on the temporal stage of the response. In addition, the best therapeutic response by treatment with cytokines may be achieved by using cytokines in combination and not as single agents. Thus, there are significant challenges that need to be overcome before clinical medicine benefits from the use of these molecules. However, the future holds much promise.

BIBLIOGRAPHY

Interleukin 1

Dinarello, C. A. (1994). The interleukin 1 family: 10 years of discovery. *FASEB J.* 8, 1314–1324.

Interleukin 2

Bruton, J. K., and Koeller, J. M. (1994). Recombinant interleukin 2. *Pharmacotherapy* 14, 635–656.

Interleukin 3

Gianelli-Borradori, A. (1994). Present and future clinical relevance of interleukin 3. *Stem Cells* 12 (Suppl. 1), 241–248.

Interleukin 4

Ricci, M. (1994). IL-4: A key cytokine in atopy. *Clin Exp. Allergy* 24, 801–812.

Sonoda, Y. (1994). Interleukin 4-a dual regulatory factor in hematopoiesis. *Leuk. Lymphoma* 14, 231–240.

Interleukin 5

Takatsu, K., Takaki, S., and Hitoshi, Y. (1994). Interleukin 5 and its receptor system: Implications in the immune system and inflammation. *Adv. Immunol.* **57**, 145–190.

Interleukin 6

Akira, S., Taga, T., and Kishimoto, T. (1993). Interleukin 6 in biology and medicine. *Adv. Immunol.* **54**, 1–78.

Interleukin 7

Appasamy, P. M. (1993). Interleukin 7: Biology and potential clinical applications. *Cancer Invest.* **11**, 487–499.

Interleukin 8

Baggiolini, M., Dewald, B., and Moser, B. (1994). Interleukin 8 and related chemotactic cytokines-CXC and CC chemokines. *Adv. Immunol.* **55**, 97–179.

Interleukin 9

Renauld, J. C., Houssiau, F., Louahed, J., Vink, A., Van Snick, J., and Uytenhove, C. (1993). Interleukin 9. *Adv. Immunol.* **54**, 79–97.

Interleukin 10

Mosman, T. R. (1994). Properties and functions of interleukin 10. *Adv. Immunol.* **56**, 1–28.

Interleukin 11

Du, X. X., and Williams, D. A. (1994). Interleukin 11: A multifunctional growth factor derived from the hematopoietic microenvironment. *Blood* **83**, 2023–2030.

Interleukin 12

Trinchieri, G. (1995). Interleukin 12: A proinflammatory cytokine with immunoregulatory functions that bridge innate resistance and antigen-specific immunity. *Annu. Rev. Immunol.* **13**, 251–276.

Interleukin 13

Zurawski, G., and de Vries, J. E. (1994). Interleukin 13, an interleukin 4-like cytokine that acts on monocytes and B cells, but not on T cells. *Immunol. Today* **15**, 19–26.

Interleukin 14

Ambrus, J. L., Jr., Pippin, J., Joseph, A., Xu, C., Blumenthal, D., Tamayo, A., Claypool, K., McCourt, D., Srikiatchatochorn, A., and Ford, R. J. (1993). Identification of a cDNA for a human high-molecular-weight B-cell growth factor. *Proc. Natl. Acad. Sci. USA* **90**, 6330–6334.

Interleukin 15

Giri, J. G., Anderson, D. M., Kumaki, S., Park, L. S., Grabstein, K. H., and Cosman, D. (1995). IL-15, a novel T cell growth factor that shares activities and receptor components with IL-2. *J. Leukocyte Biol.* **57**, 763–766.

Tumor Necrosis Factor

Tracey, K. J., and Cerami, A. (1994). Tumor necrosis factor; a pleiotropic cytokine and therapeutic target. *Annu. Rev. Med.* **45**, 491–503.

Transforming Growth Factor β

McCartney-Francis, N. L., and Wahl, S. M. (1994). Transforming growth factor β: A matter of life and death. *J. Leukocyte Biol.* **55**, 401–409.

Interferons

Johnson, H. M., Bazer, F. W., Szente, B. E., and Jarpe, M. A. (1994). How interferons fight disease. *Sci. Am.* **270**, 68–75.

Lymphotoxin

Warzocha, K., Bienvenu, J., Coiffier, B., and Salles, G. (1994). Mechanisms of action of tumor necrosis factor and lymphotoxin ligand-receptor system. *Eur. Cytokine Netw.* **5**, 83–96.

Paul, N. L., and Ruddle, N. H. (1988). Lymphotoxin. *Annu. Rev. Immunol.* **6**, 407–438.

Dementia in the Elderly

ALICIA OSIMANI
Ben Gurion University of the Negev and Kaplan Hospital

MORRIS FREEDMAN
Baycrest Centre for Geriatric Care, Mount Sinai Hospital, and University of Toronto

Procedural memory Learning of tasks or skills that is not dependent on conscious awareness (e.g., learning how to ride a bicycle). It has been defined as "learning how" as opposed to "learning what"

Progressive supranuclear palsy Degenerative disease of unknown cause affecting several structures of the upper brain stem. The clinical features consist mainly of impaired ocular movements, impaired gait, frequent falls, dysarthria, and dementia

GLOSSARY

Aphasia Disorder of language and/or comprehension caused by damage to the language zone of the dominant cerebral hemisphere (usually the left)

Apraxia Ideomotor apraxia refers to an impairment in the performance of pantomimed movements to command and possibly to imitation, but not when using the real objects. Ideational apraxia is the inability to perform sequential movements correctly with the actual object

Dysarthria Disorder of speech caused by neurogenic impairment of the structures involved in respiration, phonation, and/or articulation

Fronto-temporal dementia and Pick's disease Degenerative diseases of the brain affecting primarily the frontal and temporal lobes. The relation between fronto-temporal dementia and Pick's disease is unclear

Huntington's disease Familial disease affecting the basal ganglia, characterized clinically by involuntary movements, changes in muscular tone, and dementia

Hypophonia Decreased volume in speech

Paraphasia Substitution of one word for another with a similar meaning or a similar sound (verbal paraphasia), or omission, addition, and/or substitution of letters within one word (literal paraphasia)

DEMENTIA HAS BEEN DEFINED AS AN ACQUIRED impairment of intellectual function with compromise in at least three of the following spheres of mental activity: language, memory, visuospatial skills, emotion, personality, and manipulation of acquired knowledge (abstraction, calculation, judgment).

In the United States it has been estimated that approximately 10 to 15% of the population over age 65 are demented. Ten percent are mildly to moderately impaired, whereas 5% are severely demented. The prevalence of dementia is quite homogeneous throughout the world in areas where studies have been done. Any variation in prevalence rates is largely due to differences in the age of distributions of the populations. Because dementia affects predominantly older individuals, prevalence rates are higher in populations with a greater proportion of elderly.

Women are more susceptible than men to some dementing illnesses [e.g., Alzheimer's disease (AD) and Pick's disease], whereas multi-infarct dementia (MID) and Parkinson's disease (PD) with dementia affect men more than women.

I. ETIOLOGY

Dementia may affect the brain as a primary disorder, or it may be a secondary manifestation of general medical disease. In the elderly, primary dementia represents 80% of the total cases, while the remainder are due to a miscellaneous group of metabolic, endocrine, and systemic diseases.

The most common cause of primary dementia is AD. Other causes of dementia include MID, the Lewy body variant of Alzheimer's disease, fronto-temporal lobar degeneration, progressive supranuclear pulsy, PD, Huntington's disease, chronic alcoholism, depression, drug intoxication, and normal pressure hydrocephalus. Most of these conditions are of unknown cause. Some are familial (e.g., Huntington's disease). About 50% of cases with fronto-temporal lobar degeneration are familial and some cases of AD are also familial. [See Alzheimer's Disease; Huntington's Disease; Parkinson's Disease.]

Parkinson's disease is a degenerative disease that produces primarily a motor disorder, but it may produce dementia. Although some authors report that the incidence of dementia is as much as 93% in Parkinson's patients, others claim that cognitive impairment is seen in only 4% of these patients.

Among the causes of secondary dementia, depression and drug intoxication must always be taken into consideration in the elderly, as they are reversible. The elderly tend to be sensitive to many drugs. Doses that may be appropriate for younger people may be toxic for the elderly. Two mechanisms account for this exaggerated sensitivity: altered pharmacokinetics in the elderly, with prolonged times of metabolism or excretion, and an enhanced response to certain drugs, caused by involutional changes in the central nervous system. Combinations of drugs may have deleterious effects on mental function in the elderly, causing confusion and behavioral changes that are often interpreted as dementia, but which gradually disappear once the drug is discontinued. Alcoholic brain damage may also be arrested or partially reversed if diagnosed in the early stages. Chronic alcohol abuse may also produce Korsakoff's syndrome, which is an irreversible amnestic disorder with loss of past memories and impaired ability to learn new information. [See Alcohol Toxicology; Pharmacokinetics.]

Normal pressure hydrocephalus is a condition that may be secondary to a previous disease of the brain (e.g., subarachnoid bleeding, meningitis), but it may appear without any previous history of neurologic disease. The diagnosis of this condition is important, because it can be neurosurgically treated with a shunt and is thus one of the treatable causes of dementia.

II. PATHOLOGY

A wide variety of changes have been described in aging and in the different dementing processes seen in the elderly. Several changes may be found in the brain of normal elderly individuals without any clinical manifestation of neurologic disease during life (normal aging). The brain loses weight, there is a loss of neurons and myelin, and there is accumulation of amyloid in the vascular walls of the brain. All of these phenomena are age-related, but they may be more prominent in degenerative dementing processes.

Granulovacuolar degeneration, neurofibrillary tangles, senile plaques, and Hirano bodies may occur in small numbers and in relatively restricted areas of the normal aged brain; their density and distribution are much increased in Alzheimer's disease.

Although all degenerative disease processes affect the brain in a selective fashion, there are differences in the distribution of the lesions. In fronto-temporal lobar degeneration and Pick's disease, for example, the degenerative process affects the anterior temporal lobes and frontal lobes. In AD, however, the degeneration affects primarily the cortical association areas in the temporal and parietal lobes, as well as the amygdala, hippocampus, and basal forebrain. The nucleus basalis of Meynert is part of the basal forebrain and is the source of 90% of the cortical cholinergic input to the cerebral cortex. Because the cholinergic system has been related to cognition, lesions in the nucleus basalis may be extremely important in the pathogenesis of dementia. In AD, the primary motor and sensory cortices as well as the cerebellum are usually intact. The nucleus basalis of Meynert may be affected in other degenerative diseases (e.g., PD). This predilection for specific areas of the brain explains the characteristic symptomatology that is seen in the different forms of dementia. [See Brain.]

Each dementing disorder is characterized by specific neuropathologic features, although there may be some overlap between certain disorders. In AD, there are senile plaques, neurofibrillary tangles, and Hirano bodies; in Pick's disease, there are Pick's bodies, which are intracytoplasmatic inclusions within the nerve cells; in Creutzfeldt–Jacob disease, there is spongiform degeneration; in PD, there are Lewy's bodies, which are intracytoplasmatic inclusions within nerve cells in the substantia nigra and locus ceruleus of the brain stem.

III. CLINICAL FEATURES

Dementia is a clinical neurobehavioral syndrome. The diagnosis of dementia can, therefore, only be made on the basis of a thorough clinical history and mental status examination.

A sudden onset and a stepwise deterioration are characteristic of MID, although in some cases infarctions (i.e., ischemic strokes) have been small so that neither the family nor the patient can pinpoint the onset or specific episodes of neurologic damage as manifested by sudden episodes of deterioration.

Degenerative diseases (e.g., AD) usually present insidiously and progress gradually over months or years. In some cases, an external cause that seems totally unrelated may precipitate the symptoms. An elderly patient with some mild memory problems may show a rapid deterioration after moving to a new house or after suffering a mild febrile disease or after an episode of confusion caused by drug overdose or dehydration. In other cases, the initial symptom is depression. The patients become apathetic, show no initiative, and display no interest in anything that surrounds them.

As defined earlier, dementia is a deterioration of cognitive function involving several areas. The following is a brief summary of the findings that may be seen in these areas.

A. Attention and Concentration

One of the early signs in most dementing disorders is a decrease in attention and concentration. Patients cannot focus their attention for more than a few seconds at a time. They are often distracted by external stimuli, losing track of their conversation or the tasks they are performing.

B. Memory

Memory is one of the most complex areas of cognition. It includes several processes and is also dependent on other functions, particularly attention and concentration. It is related to the ability to develop strategies and to associate ideas. These processes are normally somewhat reduced in the elderly, but they are particularly altered in dementia.

Episodic memory refers to personal memory for events and their location in time. Remembering a face and placing it in the context of when it has been seen or remembering an event read in the newspaper some time before are part of the process of episodic memory. Semantic memory, however, refers to the fund of knowledge the individual has accumulated throughout life, such as knowing that an elephant is an animal or that water freezes at 32 degrees.

Patients with amnesia (i.e., patients with memory loss but with relative preservation of other cognitive areas) show a loss of episodic memory with intact semantic memory. Demented patients show impairment of both processes. Another kind of memory that has lately been the focus of attention is procedural memory. Some amnestic patients have shown an intact ability to learn motor tasks and other tasks (e.g., mirror reading) without being consciously aware of this learning.

Although memory loss is considered an early symptom in dementia, this is true in most but not in all cases. Some patients with Pick's disease, for instance, may show severe language problems for a long time, with little or no changes in memory. [See Learning and Memory.]

C. Language

The language examination is perhaps one of the most valuable diagnostic tools in dementia. In so-called subcortical dementia (e.g., in PD, Huntington's disease, and progressive supranuclear palsy), speech impairment is an early sign. There is frequently dysarthria and hypophonia, but there are seldom signs of linguistic impairment in the use of language itself. In other dementing diseases, however, language disorders are common. In AD, for instance, speech is neither dysarthric nor hypophonic, but there is usually an early impairment of language production manifested by word finding difficulty and later by frequent paraphasias. At this stage, comprehension and repetition are usually unimpaired. In the more advanced stages of this disease, comprehension is affected. In final stages, language production is poor, with many paraphasias, unfinished sentences, and poor comprehension.

In Pick's disease, language impairment may be the only salient sign of dementia during the early stages, or it may be accompanied by severe behavioral changes. It later evolves to other areas, affecting memory, attention, and concentration.

Reading and writing are also affected in dementia. In PD, reading is preserved until advanced stages. Writing is altered not only by the motor problems that produce a tremulous handwriting but also by micrographia. In AD and Pick's disease, reading and writing are altered at some stage in the disease. Misspellings are frequent, with omissions and substitutions of letters and frequent perseverations. [*See* Speech and Language Pathology.]

D. Praxis

Ideational apraxia may be an early sign in some dementing disorders (e.g., AD) although it is frequently absent in others. In daily life, patients have difficulty in activities requiring sequential use of objects (e.g., preparing a cup of coffee) or in the manipulation of certain objects (e.g., peeling an orange with a knife or unwrapping a package). Ideomotor apraxia is also frequent in some dementias, but it is more a diagnostic examination tool than an obstacle in daily activities. [*See* Motor Control.]

E. Visuospatial Skills

Visuospatial skills have a tendency to decrease with age. These skills are especially altered in many dementing disorders (e.g., AD, PD, progressive supranuclear palsy, and Huntington's disease). There may be some perceptual difficulties, but there are also problems of performance. Altered perception may sometimes underlie some of the hallucinations of the demented patients.

F. Manipulation of Acquired Knowledge

This area includes the different aspects of cognition and thought, requiring logical thinking, abstraction, and calculation. Deterioration in this area is frequently an early sign of dementia, and it seriously interferes with normal activities. Patient's thoughts become concrete, with increasing difficulty in manipulating old knowledge. Judgment is also impaired. Because the dementing process is usually gradual, it often goes unnoticed in the course of daily routine by the family or the persons surrounding the patients. It may become evident, however, when abnormal circumstances arise (e.g., a fire, an accident, or a traffic problem).

G. Behavioral and Mood Changes

Depression may be associated with early dementia, especially in AD and PD. In other disorders, depression may not be prominent. In fact, patients may seem euphoric, without any concern for their problem. Mood may also be normal.

Behavioral changes may occur early in dementia. Some patients with fronto-temporal dementia and Pick's disease may show only personality changes for some time until the full picture of dementia develops. Social graces are often lost early, and the changes are of such magnitude that they frequently require psychiatric hospitalization. In AD, on the contrary, social graces are usually preserved until late in the disease.

Delusions, hallucinations, and paranoid ideation are frequent in dementia. Patients may believe their family is trying to harm them or rob them, thus creating secondary problems in behavior and in their relations. [*See* Depression; Mental Disorders.]

H. Frontal Signs

A special section will be dedicated to frontal signs because they are so frequent and so prominent in many dementing disorders. The frontal lobes constitute almost two-thirds of the brain.

Personality changes are frequently related to frontal lobe dysfunction. Disinhibition, impulsivity, and irritability are common manifestations. Patients may show perseveration (i.e., a tendency to persist on the same task, the same thought, or the same movement, with an inability to shift set). If they are drawing a square, for instance, and are asked to draw a flower, they may continue drawing squares. Fragments of previous sentences may sometimes contaminate the new sentences, or movements may be repeated again and again. A common sign is what has been called being "stimulus bound." The patients' attention is pulled by the stimulus perceived at that moment, so that they act in consequence and forget to focus on their previous plan of action. Patients may, for instance, leave their room to go to the dining room because they feel hungry, but seeing a door on the way they may enter another room and forget their initial plan. In some cases, apathy is the central feature. Difficulty in initiation of speech may also be part of the syndrome.

The patterns of impairment vary according to the etiology of dementia. In AD, there is usually an early impairment of memory, language, and manipulation of acquired knowledge. Progressive deterioration leads to impairment of most of the other areas without physical signs in the general neurological examination until late in the disease.

In fronto-temporal lobar degeneration and Pick's disease, changes in behavior are prominent in the early stages of the disease. Other areas of cognition (e.g., memory) are often affected later in the process.

In PD, speech is altered, but the linguistic processes are usually intact; there is a loss of attention and concentration, some memory problems, and often visuospatial problems. Frontal symptoms are frequent. The most salient sign is the slowness of thought. Other areas of cognition are affected only late in the disease. This dementia occurs in a context of motor signs of PD.

In alcoholic dementia, frontal signs are prominent; poor attention and concentration, memory loss, and decreased manipulation of acquired knowledge are early features, as well as impulsivity and irritability. In all of these cases, the course may be variable. Some patients develop dementia with a slow course, so that they are still functioning at home 8 or 10 years after onset, whereas others are bedridden and totally dependent within a year or 2. In AD, the average survival after the diagnosis of dementia is about 8 years. Death is usually caused by intercurrent diseases (e.g., infections, aspiration pneumonia).

In MID, the symptomatology varies in relation with the localization of the strokes within the brain. There is usually dysarthria, attention and memory problems, and frontal signs, but the rest of the features vary. The process may be arrested if no new strokes occur.

IV. DIAGNOSIS

The identification of dementia is based on clnical history and mental status examination. In most cases, the etiology is usually evident on the basis of the clinical examination.

Mental status examination can be further complemented by formal neuropsychological testing. Modern neuropsychological batteries include sensitive tests that explore different areas of cognition and compare results to norms standardized for age, sex, and education. They may detect subtle impairments and contribute to the diagnosis, and they may also be used as objective measures in the follow-up of a dementing process.

Laboratory tests and neuroimaging techniques are helpful in determining the cause of dementia. In the case of some of the degenerative diseases, the cause of dementia can be suspected but cannot be definitely proven, except by pathology studies.

Neuroimaging techniques, such as X rays, are helpful tools in the differential diagnosis of etiology of dementia. Brain tumors, hemorrhages, strokes, and normal pressure hydrocephalus may be diagnosed by computerized tomography of the brain (CT) or magnetic resonance imaging (MRI). Modern techniques of nuclear medicine [e.g., PET and SPECT (single photon emission computerized tomography)] have become promising tools for the understanding of dementia. PET is an expensive test, which has so far been used for experimental and research purposes but has not been used as a routine diagnostic tool. SPECT provides a simpler, less expensive means of studying brain function through the measurement of cerebral blood flow. It is based on the concept that cerebral blood flow is usually directly related to the brain metabolism, which, in turn, is related to brain function. Areas of lower perfusion are therefore indicative of lower function. In AD, SPECT classically shows a pattern of low perfusion affecting in particular the posterior parietotemporal areas and frontal lobes. In frontotemporal dementia and Pick's disease, the typical pattern is low perfusion of the anterior temporal lobes and frontal lobes. Areas of infarction (i.e., ischemic stroke) are seen on SPECT as areas without perfusion. The technique is not always sensitive, but it is helpful in confirming the clinical diagnosis or sometimes in revealing strokes that were not seen on CT.

As to other laboratory tests, they are performed to rule out metabolic, endocrine, or systemic disease that may be the underlying cause of dementia.

In very early stages of degenerative dementia, the diagnosis may not be clear. Two circumstances pose special diagnostic difficulties: depression and normal aging. Depression in the elderly is often accompanied by cognitive impairment, and some dementing disorders may present with depression. Neuroimaging techniques may show no pathology in early stages of dementia, and both mental status examination and neuropsychological testing may show cognitive impairment. It may sometimes be difficult to distinguish between those two conditions on these grounds. In this case, a therapeutic trial with antidepressants may be helpful. Cognitive improvement after treatment suggests depression, whereas the opposite is indicative of early dementia.

The changes of normal aging in mental status examination and CT scan are variable and in studies of large series the findings overlap those of early stages of dementia. This is especially so in the ninth decade and beyond, for which standardized tests are lacking. The diagnosis of degenerative dementia must be made cautiously and after several follow-up examinations during periods ranging from 6 months to 1 year or more.

V. MANAGEMENT

Some dementing processes are reversible if treated in the early stages. Hence the diagnosis of dementia and the underlying process should be identified as early as possible. Nutritional and endocrine causes should be treated. Normal pressure hydrocephalus and chronic subdural hematoma can be surgically treated.

Multi-infarct dementia may be arrested by preventing new infarcts. This can often be achieved by managing high-risk factors (e.g., hypertension, diabetes) or treating any underlying cardiovascular disease.

Degenerative diseases have no currently curative medical treatment, and they represent a difficult problem for caregivers. A neurobehavioral assessment is helpful in detecting intact cognitive areas that may be used in the management of the patient. It should also provide a guide and support to the caregivers. Psychotropic drugs are often needed for treatment of depression, agitation, or paranoid ideation, but these drugs should be used very cautiously because they often have marked side effects in the elderly and, in particular, in the brain-damaged patient.

BIBLIOGRAPHY

Albert, M. L. (1994). "Clinical Neurology of Aging," 2nd Ed. Oxford Univ. Press, New York.

Brodal, A. (1969). "Neurological Anatomy." Oxford Univ. Press, New York.

Cummings, J. L., and Benson, D. F. (1992). "Dementia: A Clinical Approach." Butterworth, Boston.

Heilman, K. M., and Valenstein, E. (1993). "Clinical Neuropsychology." Oxford Univ. Press, New York.

Luria, A. R. (1980). "Higher Cortical Functions in Man." Basic Books, New York.

Mesulam, M. M. (1985). "Principles of Behavioral Neurology," F. A. Davis, Philadelphia.

Plum, F., and Geiger, S. R. (eds.) (1987). "Handbook of Physiology, Vol. V, Higher Functions of the Brain, Part 2" (V. B. Mountcastle, section ed.). American Physiological Society, Bethesda, Maryland.

Reisberg, B. (1983). "Alzheimer's Disease." Free Press Collier Macmillan, New York.

Snowden, J. S., Neary, D., and Mann, D. M. A. (1996). "Frontotemporal Lobar Degeneration: Fronto-temporal Dementia, Progressive Aphasia and Semantic Aphasia." Churchill Livingstone, New York.

Weiner, W. J., and Lang, A. E. (1995). "Behavioural Neurology of Movement Disorders," Advances in Neurology, Vol. 65. Raven, New York.

Dental and Oral Biology, Anatomy

A. DAVID BEYNON
University of Newcastle upon Tyne

I. Oral Cavity
II. Teeth
III. Dental Tissues
IV. Mandible, Maxilla, and Temporomandibular Joint
V. Oral Mucosa
VI. Salivary Glands

GLOSSARY

-Blast Denoting a formative cell (e.g., ameloblast, an enamel-secreting cell; odontoblast, a dentine-secreting cell)

Cusp Rounded or pointed elevation on or near to the masticatory surface of a tooth

Deciduous dentition Primary dentition consisting of 20 teeth, subsequently replaced by the permanent dentition

Distal Away from the midline along the dental arch

Fossa Depression or concavity on the masticatory surface of a tooth, into which cusps on opposing teeth usually fit

Mesial Toward the midline along the dental arch

Occlusion Functional contact relations between teeth in opposing jaws; anatomical alignment of teeth within jaws

Permanent dentition Secondary dentition consisting of 32 teeth, which replaces the deciduous dentition

DENTAL AND ORAL ANATOMY INCLUDES THE structure, both gross and microscopic, development, and function of oral and dental tissues. Teeth are essential for initial food processing, and their early loss may have life-threatening consequences in primitive societies. They consist of durable mineralized tissues, including enamel and dentine, which are supported in tooth sockets by a fibrous sling (periodontal ligament) anchored on the tooth side by cementum and to the jaw bone by specialized bone of attachment. The interface between the tooth and gingiva is highly specialized, forming an attachment site that is vulnerable to damage from microorganisms in the oral cavity. Oral health depends on a healthy mucosal lining, which is maintained by the secretions of salivary glands. Salivary glands are differentiated on the basis of their types of secretions, which perform specific functions within the oral cavity.

I. ORAL CAVITY

The oral cavity is a space devoted to food preparation and quality control in the form of taste reception and to the modification (articulation) of sounds formed initially in the larynx (phonation). It is a space bounded for the most part by highly mobile structures. The lips form the anterior boundary, forming an essential seal to permit chewing and swallowing. The cheeks lie laterally, and it is roofed by the palate. The floor of the mouth is occupied by the highly mobile tongue, and its posterior limit is the anterior tonsillar pillar containing the palatoglossus muscle, which acts as a posterior seal. The palate comprises a rigid anterior segment (i.e., the hard palate, supported by the palatal processes of the maxilla and the palatine bone) and a moveable posterior segment (i.e., the soft palate, which forms a seal between mouth and nose and between mouth and pharynx during swallowing and speech articulation). The teeth lie toward the periphery of the oral cavity, dividing it into a smaller peripheral vestibule bounded laterally by the lips and cheeks, with the large oral cavity proper internally.

The teeth are supported by the alveolar processes of the mandible and maxilla. The lips and cheek muscles are attached to the lateral surfaces of those bones

ENCYCLOPEDIA OF HUMAN BIOLOGY, Second Edition, VOLUME 3.   Copyright © 1991 by Academic Press.   All rights of reproduction in any form reserved.

and serve to bring contents in the vestibule between the teeth to allow efficient chewing. Inadequate neuromuscular control leads to cheek biting. The oral cavity is lined with mucosa, which is an epithelium moistened by salivary secretions. Mucosa is organized in three categories based on their microscopic anatomy and functions. Masticatory mucosa is found where the mucosa is abraded by food processing, typically around the necks of the teeth and on the hard palate, where mucosal ridges (or rugae) are found. Rugae may play a minor role in food breakdown before swallowing. Lining mucosa covers the mobile lips and cheeks in the vestibule, reflecting onto the alveolar process, and is present internally on the floor and on the underside of the tongue. Specialized or gustatory mucosa is present on the tongue and is characterized by the presence of papillae, which comprise four forms.

The nerve supply to the oral cavity is complicated, arising from five cranial nerves. The motor nerve supply to the cheeks and lips is through the facial (VII) nerve, the soft palate from the pharyngeal branch of the vagus (X), and tongue by the hypoglossal (XII) nerve. Common sensation from the oral cavity returns to the central nervous system via the trigeminal (V) nerve. Special visceral (taste) sensation from the anterior two-thirds of the tongue and the palate return with the facial nerve (nervus intermedius), while the posterior one-third of the tongue is innervated by the glossopharyngeal (IX) nerve. The blood supply to the oral cavity arises from branches of the external carotid artery, including the lingual, facial, and several branches of the maxillary artery, including the inferior and superior alveolar branches, palatine arteries, and infraorbital branches. Lymphatic drainage is to the submental, submandibular, and deep cervical lymph nodes.

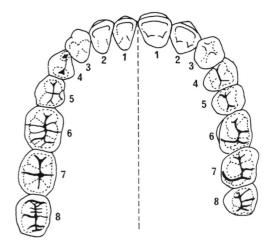

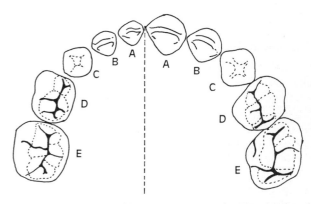

FIGURE 1 Diagram of human permanent dentition (top) and deciduous dentition (bottom). The mandibular crown morphology is indicated on the left side, and maxillary morphology on the right side. Permanent teeth are numbered (1, central incisor; 2, lateral incisor; 3, canine; 4, first premolar; 5, second premolar; 6, first molar; 7, second molar; 8, third molar). Deciduous teeth are indicated by capital letters (A, central incisor; B, lateral incisor; C, canine; D, first molar; E, second molar).

II. TEETH

Humans possess two dentitions, with the primary or deciduous dentition being replaced during childhood by the secondary or permanent dentition (Fig. 1). The deciduous dentition consists of two incisors, a canine, and two molars in each quadrant of the two dental arches (upper and lower), each arch being split into two quadrants (left and right) (Table I). The permanent dentition contains two incisors, one canine, two premolars or bicuspids, and three molars in each quadrant (Table II). The successional secondary dentition replaces the primary dentition with exact corre-

spondence of tooth number in the incisor, canine, and deciduous molar/premolar sectors. The three permanent molar teeth erupt behind the deciduous molars during childhood, with the first molar at $\simeq 6$ years, the second molar at $\simeq 12$ years, and the third molar (wisdom tooth) at $\simeq 18$ years, although this tooth is highly variable in its eruption times. The deciduous teeth are smaller, whiter, and have more bulbous crowns, and the incisors and canines have much smaller roots than their permanent successors. The deciduous molars have short but widely divergent roots to accommodate the developing successional premolar teeth. The enamel on deciduous teeth is rela-

TABLE I

Deciduous Dentition

	Central incisor (upper/lower)	Lateral incisor (upper/lower)	Canine (upper/lower)	First molar (upper/lower)	Second molar (upper/lower)
First calcification (months I.U.)	4/4.5	4.4	5.2/5	5	6
Crown complete (months)	4	4/4.2	9	6	10–12
Eruption (mean) (months)	7.5/6.5	6.5/7	16–20	12–16	20–30
Root complete (years)	1.5–2	1.5–2	2.5–3	2.5–3	3

tively thinner and is less highly mineralized, allowing relatively rapid tooth wear to take place.

Teeth may be grouped into three functional sets. Incisors have a narrow elongated incisal edge, with uppers overlapping lower incisor teeth, giving rise to a cutting or shearing action suitable for nipping off fragments of food. Canine teeth are pointed and, although now smaller than in our earlier ancestors, still play a minor role in puncturing and tearing food held in the anterior part of the mouth. Premolars and molars have complicated crown morphologies, with elevations (cusps) on one tooth entering into depressions (fossa) on opposing tooth surfaces, similar to a mortar and pestle relation. These teeth are specialized for crushing and grinding of food in the posterior part of the mouth, where the biomechanical relations permit the generation of the greatest biting forces on the tooth surfaces.

In both permanent and deciduous dentitions the largest incisor is the upper central, the smallest being the lower central, with the upper and lower lateral incisors being more similar in size. The incisors are wedge-shaped and have a single conical root. Canine teeth have a single pointed cusp with sloping arms, of which the mesial is generally shorter relative to the distal arm. The permanent premolar teeth have two main cusps, a buccal and lingual cusp separated by the mesiodistal occlusal fissure. The upper premolars have a flattened oval morphology, and the two cusps are more equal in height, although the buccal cusp is always the highest. The lower premolars in contrast have a more rounded occlusal morphology, with relatively much reduced lingual cusps particularly in the first premolar. The second premolar may possess two lingual cusps. Premolars tend to have a single mesiodistally flattened root, although it is common for the upper first premolar to have two roots, buccal and lingual. The permanent molar teeth show the greatest differences between the upper and lower dentitions. The upper molars characteristically have a rhomboid-

shaped crown with two cusps (mesiolingual and distobuccal) in broad contact, separating mesiobuccal and distolingual cusps widely, and also giving rise to an H-shaped fissure pattern. Upper molars have three roots: two buccal roots, which are usually flattened mesiodistally, and one larger rounded palatal root. There is a progressive tendency when passing distally along the tooth row from first to third molar for the distolingual cusp to be reduced, with the crown eventually becoming three cusped and with progressive merging and fusion of the three roots. Lower molars, in contrast, have a rectangular crown with four or five cusps separated by an approximately cross-shaped fissure pattern. They have two large roots, placed mesially and distally, and these are flattened in a mesiodistal direction. The first molar has five cusps, with a distal cusp usually being present. The second molar generally has only four cusps, with the loss of the distal cusp. The third molar shows a tendency to have five cusps, which may be supplemented by extra cusplets and fissures. The deciduous molar teeth show specific differences in crown morphology, most particularly in the first deciduous molar teeth. In both upper and lower first deciduous molars, the mesiobuccal corner is elongated (the molar tubercle) giving the crown an irregular quadrilateral shape. The buccal and lingual cusps tend to converge toward the midline of the tooth. The upper first deciduous molar has the simplest morphology, with a mesiodistal fissure dividing the occlusal surface into two cusps, buccal and lingual. The buccal cusp is long and narrow, forming a cutting edge, whereas the lingual cusp is more pointed. There is a prominent buccal cingulum extended mesially to form the molar tubercle. This tooth, together with the second deciduous molar, shares the upper molar characteristic of having three roots. The lower first deciduous molar is elongated mesiodistally and is divided into buccal and lingual halves by a mesiodistal fissure. The buccal half consists of two cusps, mesiobuccal and distobuccal, sepa-

TABLE II

Permanent Dentition

	Central incisor (upper/lower)	Lateral incisor (upper/lower)	Canine (upper/lower)	First premolar (upper/lower)	Second premolar (upper/lower)	First molar (upper/lower)	Second molar (upper/lower)	Third molar (upper/lower)
First calcification (months)	3–4	10/3–4	4–5	17/21	30	birth	30	≃9 yr
Crown complete (years)	4–5	4–5	8–9/6–7	5–6	6–7	2.5	7–9/7–8	≃ 12
Eruption (mean) (years)	7.2/6.5	8.5/7.5	11.2/10.5	10.5/10.5	11/11.2	6.7/6.5	12/12	≃18
Root complete (year)	10/9	11/10	13–15/12–14	12–13	12–14/13–14	9–10/9–10	15	≃21

rated by a shallow buccal groove, of which the mesio-buccal cusp is larger. The lingual half is narrower and comprises two cone-shaped cusps, of which the mesiolingual cusp is again larger. The mesiolingual and mesiobuccal cusps are commonly connected by a transverse ridge dividing the fissure system into a mesial pit and distal fissure. This tooth has a mesial and distal root, like all lower molars. The upper and lower second deciduous molars closely resemble in morphology the first permanent molar tooth, which erupts behind them.

III. DENTAL TISSUES

A. Enamel

Enamel is the hardest structure in the body, with a hardness value similar to that of mild steel. Its color ranges from white to gray-yellow, depending on its thickness, which allows the yellow color of underlying dentine to show through under thinner enamel. It consists of 96% by weight of inorganic material, largely composed of hydroxyapatite (HA), with the remainder consisting of less than 1% of organic material and water. The organic material consists principally of a specific enamel protein called *enamelin*. The mature tissue in the mouth is without cells, and therefore it is incapable of repair, unlike all other calcified tissues. Enamel structure is wholly dependent on the orientation of HA crystallites within it. These crystalites are packed in a highly ordered fashion, producing rod (or prism) structural units, each of which is about 5 μm across (Fig. 2). Prisms viewed from their ends have a keyhole shape, with a relatively large body, tapering down to a constricted waist with a final minor enlargement in the tail, which is directed toward the neck of the tooth. Crystals running in the body of the prism are broadly parallel to its length, but passing into the tail region, the crystalites diverge progressively more from this direction, terminating at an acute angle in the base of the tail adjacent to the adjacent inferior prism. The prismatic structure of enamel is a consequence of changes in crystal orientation at each prism boundary. Individual crystalites are hexagonal in cross-striation and are relatively long, indeed their actual length is unknown in enamel.

Crystals in enamel are approximately 10 times larger than those in calcified connective tissues including dentine, cementum, and bone; this may be a consequence of differences in the organic matrix. Enamel

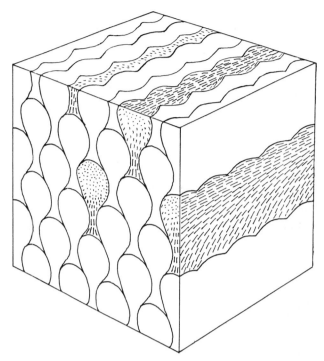

FIGURE 2 Diagram of human enamel prisms. Right-hand vertical face shows prisms cut in vertical longitudinal section along their lengths. Left-hand vertical face shows prisms cut in vertical cross section showing the characteristic keyhole shape. The top face shows a horizontal section passing through heads and tails of keyhole-shaped prisms. Crystal direction in four prisms is indicated. Crystals run in the long axis of the body of the prism, with crystals diverging from this longitudinal axis passing into the tail to an increasing extent, producing a marked change in crystal orientation at the boundary between prisms.

proteins are mobile and labile and are partially removed during development, in contrast to the more rigid collagen scaffolding in other mineralized tissues. This rigid cross-linked structure of collagen possibly restrains crystalite growth during development in calcified tissues. The ultimate hardness of enamel is a function of both the relatively large size of enamel crystals and the relatively small amount of proteins remaining in the tissue. Enamel formation is a continuous process punctuated by repetitive fluctuations over short and longer periods. The short-period fluctuations are caused by the circadian rhythm, which affects the rate of secretion of enamel, giving rise to alternating varicosities and constrictions along the length of enamel prisms. Longer-period fluctuations in enamel formation give rise to striae of Retzius, which are orientated broadly perpendicular to the prism direction, running obliquely toward the enamel–dentine junction (EDJ). These accentuated markings represent the position of the whole secretory ameloblast sheet at the time of the disturbance. Striae manifest themselves in two different ways, illustrating differing aspects of enamel growth. Over the cusp tip, they run obliquely from one side of the tooth to the other, without reaching the enamel surface, and represent successive layers of enamel deposited during the phase of secretion of full-thickness enamel over the cusp tip. Once this cuspal thickness is achieved, ameloblasts continue to secrete enamel on the sides of the crown, often for relatively large distances over an extended period, to complete the formation of the full crown height. In this lateral enamel, striae run obliquely from the EDJ toward the surface, producing a tile-like arrangement of layers of enamel (imbricational enamel). These produce alternating transverse ridges (perikymata) and depressions. There is accumulating evidence that striae have a systemic origin and repeats over a circaseptan period (6–9 days, with a modal value of around 8).

Prisms do not follow a straight course from within outward, which would be prone to fracture. They follow a sinuous or undulating course, which is most marked over the cusp tips in the form of gnarled enamel. Here prisms undergo frequent and sharp changes in direction, giving rise to a complex interlocking structure. In the bulk of the lateral enamel, the prisms undulate two or three times from side to side in a horizontal plane, straightening in the outer one-third to one-quarter of the enamel. These sinuosities move progressively out of phase at differing horizontal levels, reaching the opposite phase approximately after 20 prism levels, approximately 50–100 μm apart. This decussating structure causes the dispersion of cracks as they propagate into the deeper enamel and acts as an energy absorbing device, limiting further crack extension.

The internal surface at the EDJ is scalloped, with long shallow undulations extending over a distance of approximately 50 μm. These depressions in the dentine surface are mirrored by shallow elevations on the enamel surface. This arrangement may represent a device to increase the area of enamel–dentine surface contact, which may help to prevent less elastic enamel shearing away from the elastic underlying dentine. Extensions of the odontoblastic processes commonly pass through the EDJ into the enamel, particularly over the dentine horn under the cusps. These extensions are called *enamel spindles.*

The external surface of lateral enamel is characterized by the presence of surface ridges, or perikymata, which reflect the imbricational enamel structure. Recent interest has focused on the possibility of counting the numbers of perikymata on lateral enamel in teeth to gain an estimate of how long crown formation takes to be completed, using nondestructive methods. Crystalites in subsurface enamel are commonly orientated parallel to one another, without prism boundaries. This aprismatic enamel layer is typically 20–30 μm thick.

B. Dentine

Dentine is the second most highly mineralized tissue in the mature tooth, 70% of which (by weight) is HA, and 20% organic material, principally collagen. Dentine, which is yellow in color, differs from bone and cementum in that cells are never found within the body of the tissue, although processes from the odontoblasts extend throughout the tissue. The odontoblast cell bodies are located internally at the periphery of the dental pulp. This specialized cell–tissue relation enables dentine to respond to injury and irritation by either laying down mineralized tissue within the tubule containing the processes (peritubular dentine) or to mount a pulpal repair response. Collagen fibers in peripheral dentine, particularly in the crown, are orientated perpendicular to the EDJ; this arrangement distinguishes this most peripheral (mantle) dentine from the remaining circumpulpal dentine, in which the collagen fibers are arranged broadly parallel to the tooth surface. In circumpulpal dentine, the collagen fibers form a meshwork at right angles to the dentinal tubules, which traverse the tissue. In the early development of dentine, the dentinal tubules are relatively large and filled by the odontoblastic process, which extends to the periphery of the dentine. With the passage of time there is deposition of hypermineralized peritubular dentine within the original tubule, associated with a retraction of the process. This peritubular dentine is distinguished from the intervening intertubular dentine by the absence of a collagen matrix and by its relatively high mineral level. The organic constituents of peritubular dentine appear to be glycosaminoglycans, although these have not been properly characterized. There are also differences in the mineral phase, peritubular dentine consisting of octocalcium phosphate. Earlier studies suggested that the odontoblastic process did not extend to the periphery in mature teeth, but according to recent evidence they extend through the breadth and length of the tubules where they remain patent. Some dentinal tubules are penetrated by nerve fibers for relatively short distances, although these nerves do not extend throughout dentine. [*See* Collagen, Structure and Function.]

Incremental lines are present in dentine and are broadly similar to those in enamel in their periodicity. Short period lines, equivalent to cross-striations and related to the circadian cycle, are lines of von Ebner. Long period lines are represented by contour lines of Owen, which are variable in their spacing and expression and generally inconstant. Hypomineralized defects occur in dentine, the larger ones occurring in the coronal dentine in the form of interglobular dentine. These are caused by failure of fusion of globular calcification masses called calcospherites. Interglobular dentine is rare in the root, possibly because the mineralization of this part of the tooth occurs linearly. In root dentine, however, there is a characteristic defect called *Tomes granular layer,* just within the peripheral dentine. The individual defects in this layer resemble interglobular dentine but are much smaller. This granular layer is usually present throughout all root dentine and is bounded externally by a structureless (hyaline) layer, about 20 μm thick.

On the pulpal side of dentine is a uniform uncalcified layer called *predentine.* Its thickness varies between 20 and 40 μm and is equivalent to unmineralized bone (osteoid). The odontoblasts secrete collagen and ground substance into this layer, which then undergoes mineralization to become dentine.

Dentine responds to irritation in two ways. Mild stimuli, including gradual dentine exposure during tooth wear, may result in the obliteration of tubules by peritubular dentine. This has the beneficial effect of sealing off the pulp from the oral cavity using a hypermineralized tissue, which will resist further tooth wear. Severe irritation, for example, arising from carious attack on the tissue, may cause the odontoblasts to die and loss of processes from the tubules. A secondary response beneath dead tracts causes the deposition of irregular or reparative secondary dentine, which seals off the pulp at the pulpal side. Another form of secondary dentine occurs with tooth aging and involves a continuous deposition of dentine throughout life. The pulp chamber becomes progressively smaller, until the original odontoblasts can no longer be accommodated at its periphery. A proportion of the odontoblasts die, and the remaining ones regroup and secrete irregular secondary dentine, which is characterized by a reduction in the number of dentinal tubules and by an abrupt change in their direction.

C. Pulp

The pulp is a specialized connective tissue whose function is to support and supply nutrients to dentine.

The odontoblasts are located at the periphery of the pulp with processes extending into dentine. They are tall columnar cells, being approximately 30 μm in length, and are polarized: the nucleus is located away from the predentine, whereas a well-developed rough endoplasmic reticulum is situated near the predentine in the distal part of the cell. Recent evidence suggests that the odontoblasts are joined to one another by a continuous zone of tight junctions at the secretory end of the cell. This junction may serve to restrain the free passage of mineral ions into forming dentine. Between the odontoblasts and the pulp there is first a cell-free zone and then a cell-rich zone. The majority of the pulp cells are fibroblasts, which secrete collagens and abundant ground substance. The rate of tissue turnover in pulp is high, particularly for the ground substance. With increasing age the fibroblasts tend to become flatter and to deposit more collagen fibers. Apart from fibroblasts, two other principal cell types are associated with neural and vascular elements.

Nerves are present in relatively large numbers, comprising both unmyelinated and myelinated nerve fibers. Nerves in pulp are either sensory, transmitting only the sensation of pain, or efferent sympathetic fibers, responsible for altering blood flow in vessels. Sensory nerve fibers form a network below the odontoblasts, called the *plexus of Raschkow,* and terminate on and around the odontoblast cell bodies. In certain parts of the pulp, particularly toward the pulpal horns, sensory nerve fibers accompany the odontoblastic cell processes in the dentinal tubules for a distance of 300–400 μm. Although nerves are restricted to the inner part, the whole dentine, even the outer layer, is extremely sensitive to pain. Various theories have been proposed to account for this. The most plausible explanation (hydrodynamic theory) is that disturbance to outer dentine causes fluid flow through the numerous dentinal tubules present in dentine (approximately 40,000/mm^2 in coronal dentine). The resulting movement and displacement of odontoblasts and their processes may then trigger a response in the nerve plexus surrounding these cells.

Other cells of the dental pulp include macrophages and lymphocytes. Macrophages reside in perivascular locations and act as pulpal scavenger cells. Lymphocytes are defense cells, which are responsible for producing antibodies, thus playing a role in cellular immunity. [*See* Lymphocytes.]

The vascular supply to the dental pulp is extensive. One or two larger vessels (arterioles) enter through the apical foramen and proceed upward centrally through the pulp, giving off a network of branches, which extends toward the subodontoblastic area, forming an extensive capillary network. Some capillaries penetrate a short distance into the odontoblast cell layer. Discontinuities (fenestrations) are present in the walls of these capillaries, allowing rapid transfer of nutrients to the extracellular space. Venules collect the returning blood in relatively large vessels. Arteriovenous shunts (i.e., anastomoses) within the pulp direct blood from the arterial inflow into the venous outflow, diverting it away from the capillary circulation. Lymphatic vessels arise as blind thin-walled vessels in the coronal part of the pulp and drain out through the apical foramen.

The pulp chamber is effectively a rigid closed box, with one or several narrow entry points from the apical foramen(ina). The pulp is therefore susceptible to damage when there is an inflammatory response secondary to infection, which commonly arises from dental caries. Inflammation causes an increase in tissue fluid pressure; when this exceeds the low pressure in the venous outflow, the blood supply to the pulp is cut off, ultimately leading to pulpal necrosis and death. [See Dental Caries; Inflammation.]

D. Periodontium

The periodontium comprises cementum, periodontal ligament, the alveolar bone of the tooth socket, and the gingiva applied to the tooth (dentogingival junction). The function of the periodontium is primarily to achieve a mobile but strong tooth supporting system, to enable teeth to move slightly during biting and mastication, and also to form a seal between the tooth-supporting tissues and the oral cavity. This sealing function is the primary responsibility of the dentogingival junction.

1. Cementum

Cementum is a mineralized connective tissue that closely resembles bone. It is mineralized to about 50% by weight, the remainder consisting of organic material, principally collagen, and water. Cementum is avascular and depends for its nutrition on blood vessels in the periodontal ligament. It does not normally undergo remodeling. Successive layers are added onto preexisting layers to provide new functional attachment for periodontal ligament fibers during the life of the tooth. A recent classification of cementum divides it into afibrillar and fibrillar forms. Afibrillar cementum is found only, in small amounts, in the cervical part of the tooth. The bulk of cementum is fibrillar, consisting of collagen fibrils. Fibrillar cemen-

tum is further subdivided: one part, which is located predominantly in the cervical third of the root, consists primarily of extrinsic (or Sharpey's) fibers (i.e., is engaged in tooth attachment); the other part, located in the more apical parts of the root, also has intrinsic fibers. This classification is a functional one: a high proportion of extrinsic fiber content implies an exclusively attachment role; a lower proportion indicates a filler function.

An older classification involves two categories: acellular cementum (broadly equivalent to extrinsic fiber) and cellular cementum (mixed fiber). Acellular cementum is formed relatively slowly and is restricted to the neck region of the tooth. Cellular cementum is principally formed toward the root apex, and also in the bifurcation region of multirooted teeth. In both locations it serves to compensate for loss of crown material caused by attrition and/or jaw growth. With increasing age, teeth wear both occlusally (i.e., at the surface of contact with opposite teeth) and approximally between teeth. To avoid progressive spacing developing between adjacent teeth and between jaws, teeth migrate mesially and occlusally. This occlusal movement, into the mouth cavity, is compensated by elongation of the tooth root through deposition of rapidly formed cellular cement.

2. Alveolar Bone

The function of alveolar bone is tooth support, achieved by the anchoring of periodontal ligament fibers into its most superficial layers. The embedded parts of the ligament fibers are the Sharpey's fibers (see earlier). Bone containing such extrinsic fibers, called *bundle bone*, is relatively deficient in intrinsic fibers secreted by osteoblasts. The deeper layers of alveolar bone are lamellar and provide the greatest strength per unit mass of any bone tissue type. Alveolar bone forms the wall of the socket and is perforated by numerous canals (for further details see Section VI).

3. Periodontal Ligament

The periodontal ligament forms a fibrous sling, connecting the cementum to the alveolar bone. Its width is approximately 0.2 mm in juveniles and may become thinner with increasing age. Its primary function is to convert forces, placed on the teeth during biting of hard objects, into tensile forces on the alveolar bone. It also prevents excessive displacement of the tooth apically, which would damage the nerves and blood vessels entering the pulp. It serves an important proprioceptive function in that nerves present in the periodontal ligament monitor tooth contact positions and assist in the control of masticatory movements.

The ligament is unusual in several respects. It differs from other typical ligaments in that the collagen fibers do not form straight, closely approximated bundles, but follow an undulating course and tend to be dispersed between the fibroblasts. It is also unique as a connective tissue in its high rate of metabolic activity and collagen turnover, with a half-life on the order of only 1–2 days, much shorter than in other tissues. This high rate of turnover is achieved through the activities of the periodontal fibroblasts, which constantly secrete collagen, at the same time digesting and degrading previously secreted fibers.

Collagen fibers are arranged in a series of groupings called *principal fibers*. These include gingival fibers, which are principally attached to cervical cementum and support the gingiva. Alveolar crest and horizontal fibers are attached to the alveolar bone crests and insert into cementum. The oblique group is the most abundant, and acts to prevent extreme intrusion of the tooth. Other minor groups include the apical group at the root apex and interradicular fibers found at the bifurcation in multirooted teeth. A system of unusual connective fibers, called *oxytalan fibers*, resembling preelastin fibers, follows a cervico-apical course. They originate either in the gingiva or in the cervical cementum and run downward and outward across the oblique collagen fibers to terminate at the periphery of the ligament, usually in association with blood vessels. Their function remains disputed, but it is likely that they play some role in the regulation of blood flow within the ligament. The principal blood supply comes from branches of the dental arteries. Vessels entering the pulp give off branches supplying the apical part of the ligament, and the remaining ligament is supplied by arteries that penetrate the alveolar cribriform (perforated) plate to supply the bulk of the periodontal ligament. Blood vessels running within the gingiva supplement this supply around the tooth neck and are responsible for supplying nutrients to the dentogingival tissues.

4. Dentogingival Junction

The dentogingival junction lies between the gingiva and the tooth, which is essential for the maintenance of the integrity of the tooth-supporting tissues. It consists of the mucosa facing toward the tooth and is divided into the coronally directed (or crownward) part, called *sulcular* or *crevicular epithelium*, lining the gingival sulcus, and inferiorly junctional epithelium, which is attached to both the underlying connective tissues and the enamel. The gingival sulcus is a shallow groove between the gingiva and the tooth, with a depth of approximately 1–2 mm in health. The sulcus provides a sheltered environment that is difficult to cleanse, in which oral microorganisms thrive, causing tissue irritation and damage with their products. These damaging effects are to some extent counteracted by the outward flow of crevicular fluid, which helps to wash out noxious substances. [See Gum (Gingiva).]

The junctional epithelium forms an adherent cuff around the tooth, wider toward the sulcus and narrow toward the cervix. Junctional epithelium originally derives from enamel epithelium, which formed enamel, but is replaced by downgrowing oral epithelium approximately 3 years after tooth emergence. The epithelium does not contain fibrous keratin and the structural features of the cells reflect its functions of attachment and cell replacement. Epithelial cells are attached to each other by desmosomes, although these are much fewer in junctional epithelium than in typical oral epithelium, and in consequence the intercellular spaces are larger. The basal cells adjacent to the basal lamina at the border with the connective tissue are attached to it by hemidesmosomes. On the tooth side the epithelial cells are attached by hemidesmosomes, in a comparable manner, to a thin organic layer, or primary cuticle, on the surface of the enamel. Specialized junctions preventing fluid passage between cells are absent, and membrane-coating granules, which are thought to play a role in "waterproofing" epithelia, are few. These features probably account for the leakiness of this tissue, which allows the outflow of crevicular fluid. This fluid crosses the basal lamina, which appears to act as a semipermeable filter to the passage of proteins outward through the junctional epithelium. Cell turnover rate in this tissue is high, on the order of 6 days. Daughter cells produced at all levels migrate toward the oral cavity and are shed into the gingival sulcus. Polymorphonuclear leukocytes and monocytes are also commonly observed migrating through the junctional epithelium, to be shed into the mouth. They may play some role in the defense against bacteria and their products in this local environment.

IV. MANDIBLE, MAXILLA, AND TEMPOROMANDIBULAR JOINT

The mandible is a U-shaped bone consisting of two vertically orientated rami located posteriorly, linked by a stout elliptically shaped body. The ramus possesses an anterior coronoid process, which is an attachment for the temporalis muscle, separated by a mandibular notch from the posterior condylar process, which forms the inferior articulation of the temporomandibular joint. The posterioinferior angle of the mandible is the insertion site of the masseter muscle exteriorly and of the medial pterygoid muscle internally. The temporalis, masseter, and medial pterygoid are the principal jaw elevator muscles and are responsible for jaw closure. The lateral pterygoid muscle, which inserts into the neck of the condylar process, is responsible for pulling the condyl forward during jaw opening and is the principal muscle capable of jaw protrusion.

Muscles attached anteriorly to the body of the mandible include the digastric and mylohyoid muscles, as well as the inferior origins of the muscles of facial expression. The digastric is the principal jaw depressor muscle but may be assisted by the mylohyoid when the hyoid bone is fixed by contraction of infrahyoid muscles. Near the midline of the mandible, which is called the *symphysis,* there is the head of the genioglossus muscle, which runs backward into the body of the tongue and is responsible for protrusion of that organ. Between this muscle and the mylohyoid is the geniohyoid, which also helps to protrude the tongue. Other muscles attached posteriorly to the body of the mandible include the buccinator muscle, which is responsible for maintaining the cheek in close approximation to the posterior teeth, and part of the origin of the superior constrictor muscle, which encloses the pharynx and assists in swallowing. The body of the mandible is a relatively dense bone, because considerable stresses are generated in it during biting. Unilateral biting on posterior teeth produces powerful shearing forces adjacent to the loaded teeth, as well as considerable torsional forces across the midline symphyseal region.

The mandible develops by intramembranous ossification at about 40 days of intrauterine development. The site of initial ossification is near the bifurcation of the inferior dental nerve into the incisive and mental branches, on both sides. Ossification spreads from both points forward toward the symphysis, where a midline joint persists until about the first year of postnatal life. Ossification backward from the initial centers causes the development of the ramus and formation of the coronoid and condylar processes. In these two latter sites, as well as in the symphysis and transiently in the angle, secondary cartilages appear, possibly to stabilize these structures initially. The cor-

onoid cartilage disappears before birth, and the symphyseal one within the first years of life, leaving the coronoid process as the principal growth site in the mandible until adulthood. [*See* Craniofacial Growth, Postnatal; Dental Embryology and Histology.]

The maxilla is attached directly to the skull, with processes extending up toward the frontal and zygomatico-temporal regions. The maxilla abuts posteriorly against the pterygoid plates, and the palatal processes of the maxilla meet in the midline, across the roof of the palate, where they are supported above by the vomer and ethmoid bones. These wide maxillary attachments to the skull disperse forces generated during mastication, and accordingly the construction of the maxilla is much less heavy than that of the mandible. It also lacks attachment of powerful muscles of mastication, although most of the small muscles of facial expression gain their superior origin from this bone. The maxilla develops, like all the facial bones, intramembranously and has minor contributions from secondary cartilages, which, if present, are obliterated before birth.

Alveolar processes develop on both the maxilla and the mandible initially as outgrowths at about 60 days of uterine life, forming primitive crypts enclosing tooth germs. True alveolar processes do not develop until the eruption of the teeth, which begins shortly after birth with the deciduous incisors and terminates in late adolescence with the eruption of the third permanent molar tooth. The alveolar process consists of an external continuation of the cortical plate of the two bones, which is reflected back into the socket. This internal alveolar plate is perforated by numerous canals called the *cribriform plate* and allows the passage of blood vessels and nerves into the periodontium. The bone lining the tooth socket consists of bundle bone, which is bone of attachment containing embedded Sharpey's fibers. In the mature jaw the underlying alveolar bone, including that of the outer cortex, is lamellar, with circumferential lamellae on the external surfaces.

Alveolar bone is dependent on tooth function for its proper development and maintenance. If teeth are congenitally absent it fails to develop, and if a tooth is lost, it undergoes resorption. During life, alveolar bone undergoes active remodeling, at a higher rate than any other bone in the adult. This active reworking is probably a consequence of the fact that the teeth undergo slow but continuous movement throughout life, to compensate for tooth wear on both occlusal and contact zones between teeth. The latter process is accentuated along the length of the dental arch in populations who chew particularly vigorously; for example, in Eskimos it may amount to as much as 10 mm during a lifetime. Tooth wear of this fashion, if uncompensated for, would leave spaces between the teeth, and damage to the gums from food impaction. Spacing is prevented by the process of mesial drift, where all the posterior teeth move toward the midline, maintaining close contact between individual teeth. This process is believed to be mediated through the activity of fibroblasts in the transeptal fiber group, which join the teeth into one continuous arch. [*See* Bone Remodeling, Human.]

The temporomandibular joint is a complex articulation between the cranium and mandible. In a functional sense the joints of the two sides should be considered together, because movement in one causes movement in the other. The joints are unusual in that there is a fibrous disc between the cranial articulation on the temporal bone and the condylar process of the mandible. The disc separates the joint space into a superior compartment in which movements are largely of a translatory forward type, and in an inferior compartment they are predominantly rotary. The joint is usually described as being fibrocartilaginous in structure, although this is an inaccurate description. The tissue on all its articulating surfaces is fibrous, and there is normally no cartilage in the temporal bone or disc elements, except possibly in old age. On the condylar surface the fibrous articulating layer is separated from an underlying hyaline cartilage layer by a functionally distinctive cell layer, called the *intermediate cell zone*. This layer is the source of proliferating cells, of which the majority pass downward and differentiate into chondroblasts. They are responsible for most of the postnatal growth in face height. A minority of cells migrate upward to replace effete fibroblasts in the surface articulating layer. This pattern of appositional growth of the condylar cartilage is unique postnatally: in other cartilaginous joints it is usually interstitial. Growth of the condylar cartilages ceases at about 20 years of age, but some believe that it can be restarted in acromegaly, by excessive production of growth hormone; then proliferation of the condylar cartilage gives rise to the protrusive bulldog-like jaw characteristic of untreated cases. [*See* Articulations, Joints between Bones; Cartilage.]

The joint is enclosed in a capsule that is strengthened laterally by the temporomandibular ligament. This ligament consists of an oblique superficial component and horizontally orientated deep part. Its function is to prevent the posterior displacement of the mandible, and it is sufficiently strong to cause fracture of the condylar neck rather than backward displacement of the jaw as a result of force exerted on the chin. Also associated with the joint are the sphenomandibular and stylomandibular ligaments. The sphenomandibular ligament is the embryological remnant of the perichondrium surrounding the primary cartilage of the lower jaw, Meckel's cartilage. The stylomandibular ligament extends from the tip of the styloid process toward the angle of the mandible and is part of the fascia enclosing the posterior margins of the parotid gland. Neither ligament exerts a significant role in controlling and stabilizing jaw movements. The capsule is lined on its inner surface by a synovial membrane, which is absent over the articulating surfaces.

Jaw opening involves the depressor muscles in the floor of the mouth and also the lateral pterygoids, which pull the condyl forward onto the articular eminence. This jaw movement involves both a rotatory component and a forward translatory component, which take place in the inferior and superior joint spaces, respectively. Jaw closure is produced by the contraction of elevator muscles and by the controlled relaxation of the major inferior head of the lateral pterygoid muscle. The minor superior head of the pterygoid inserted into the articular disk contracts during this phase, showing that its movements are separate from those of the jaw. Protrusion is achieved by contraction of both lateral pterygoid muscles, and the jaw is pulled back by the horizontal posterior fibers of temporalis as well as digastric muscles. Unilateral jaw shifts, as occur in chewing, take place around an axis located posteriorly to the condyle on the displacement side, with the opposite condyle being shifted anteromedially by its lateral pterygoid muscle.

V. ORAL MUCOSA

Oral mucosa, which lines the oral cavity, has properties intermediate between skin and the lining of the rest of the digestive tract. It consists of a superficial epithelial layer and an underlying connective layer, the lamina propria. This mucosa forms an envelope, protecting the underlying connective tissues, and has sensory functions, including taste.

Oral epithelium is divided into keratinized (or parakeratinized) and nonkeratinized epithelium. In keratinized epithelium, the cells of a surface layer are completely filled with keratin, and intracellular organelles, including nuclei, are absent. In para-keratinized epithelium, the surface cells are filled with keratin, but some nuclei are still present. Nonkeratinized epithelium contains nuclei in superficial levels, and keratin is scarce. Keratinized cells are impermeable to the passage of materials. Diffusion of substance between cells is also prevented by glycoproteins secreted by immediate level cells. Epithelium in the floor of the mouth, however, is relatively permeable, allowing certain drugs to be administered parenterally via this route. Other nonepithelial cells include pigment cells (melanocytes) that contain the pigment melanin, the nonpigmented Langerhans cells, which are thought to be members of the macrophage series important in presentation of antigenic material to lymphocytes, and Merkel cells, round clear cells that are usually in close approximation to intraepithelial nerves and could have a sensory function.

Oral epithelium is divided into three main categories: masticatory, lining, and specialized or gustatory.

A. Masticatory

Masticatory mucosa, which is usually keratinized or para-keratinized, is found in sites subjected to abrasion, including the gingival tissues around the teeth and the hard palate. The interface between the epithelium and the lamina propria is deeply indented, producing epithelium ridges and troughs filled with prolongations (papillae) from the lamina propria. Masticatory mucosa is firmly bound to the underlying periosteum covering the alveolar processes around the teeth and on the hard palate. In the palate there is a submucosal layer containing fat cells anteriorly and mucous cells posteriorly.

B. Lining

In contrast to masticatory mucosa, lining mucosa is usually nonkeratinized and has a thicker epithelium and a less indented interface with the underlying connective tissues. It covers the lips and cheeks in the oral vestibule, reflecting onto the base of the alveolar process, and meets masticatory mucosa at the muco–gingival junction. Within the oral cavity proper, lining mucosa is restricted to the floor of the mouth.

C. Gustatory

Specialized or gustatory mucosa, which is found on the superior and the lateral aspects of the tongue, is characterized by the presence of papillae, of which four types are recognized. The two smaller types project from the surface of the tongue, one mushroom-shaped (fungiform papillae) with associated taste buds and the other conical keratinized (filiform papillae). Foliate papillae are occasionally found on the sides of the tongue and consist of alternating vertical clefts, lined with epithelium containing taste buds. Circumvallate papillae, which are the most distinctive, have taste buds on their inner walls and respond principally to bitter taste sensation. They are found at the boundary between the anterior two-thirds and posterior one-third of the tongue, anterior to the sulcus terminalis. They are typically 2–3 mm across and are surrounded by a circular ditch or vallum, into which serous glands (of von Ebner) drain.

VI. SALIVARY GLANDS

The oral cavity is kept moist by secretions from salivary glands. There are three pairs of major glands located external to the oral cavity, with long ductal systems. Small minor glands are present on the inner surface of the lips and cheek and also in the tongue. Saliva consists principally of water (>99%), and it contains large numbers of epithelial cells and polymorphonuclear leukocytes.

The total volume of saliva secreted in 24 hr is approximately 0.6–1.2 liters, of which the great majority is produced on stimulation from masticatory and gustatory processes associated with eating. The functions of saliva are numerous, including lubrication, protection, and as a solvent for food substances to be transmitted to taste buds. The most characteristic components are mucins, which are predominantly carbohydrate in composition and are strongly hydrophilic. Saliva contains various proteins with antibacterial properties, including lysozyme and other substances such as lactoferrin, and the antibody immunoglobulin A, which aggregates oral bacteria. Saliva has important buffering properties (due to bicarbonate, phosphate ions, and amino groups on salivary proteins), which reduce the effects of acid production by oral microorganisms. It is saturated with calcium and phosphate ions, which are prevented from precipitating by the presence of specific proteins.

Salivary glands are formed by initial secretory units and a ductal system. Serous or seromucous secretory units, secreting carbohydrate components, are organized into spherical acini, consisting of cells with a central nucleus and abundant rough endoplasmic reticulum (RER) arranged principally in the apical pole, where the small Golgi apparatus is also located. Salivary proteins, including amylase, are synthesized on ribosomes, packed into secretory granules in the Golgi apparatus, and secreted apically. Mitochondria are dispersed through the cell to provide energy for this process. Mucous secretory units have cells arranged in elongated tubular structures called *alveoli*. They contain a flattened nucleus, pushed toward the cell base, around which are small amounts of RER and mitochondria. There is a prominent Golgi complex from which large secretory granules containing mucin are secreted apically. Both types of primary secretory unit drain into intercalated ducts, with a lining of low cuboidal epithelium, and contain the antibacterial proteins lysozyme and lactoferrin. These ducts lead into striated ducts, which have a lining of columnar cells with a centrally placed nucleus and highly infolded external membrane associated with numerous mitochondria. This greatly enlarged cell surface, with its associated ion pumps, plays an important role in ion transfer into and from primary salivary secretions. The primary secretion is isotonic but is made hypotonic by resorption of sodium ions, with some replacement by potassium ions. These cells are the probable source of the important buffering ion bicarbonate, which is secreted in large amounts at high flow rates. Striated ducts terminate in excretory ducts, which combine to unite the secretions of the different parts of the gland. [*See* Salivary Glands and Saliva.]

A. Major Glands

The parotid gland contains almost exclusively seromucous secretory units, relatively long intercalated ducts, draining into prominent striated and excretory ducts, which converge into a single duct entering the mouth opposite the second maxillary molar. The parotid secretion is the most watery (or serous) and contains relatively little mucin but abundant salivary amylase, which may help to digest starch dietary products from around the teeth. The submandibular gland is mixed with a predominantly seromucous product. It contains serous acini, mucous alveoli, and mucous alveoli capped with crescents of serous cells. Intercalated ducts are shorter and striated ducts longer than in the parotid. The main secretory duct terminates in the floor of the mouth, below the tip of the tongue. The sublingual gland consists predominantly of mucous alveoli, with relatively few serous cells usually present as crescentic caps. Both intercalated and striated ducts are short, and the gland drains by multiple ducts into the anterolateral floor of the mouth.

B. Minor Glands

Minor salivary glands in the cheek, lips, and anterior tongue are small mucous glands with short ductal systems. The glands of von Ebner are exceptional, being serous and draining into the ditch around the circumvallate pipillae.

BIBLIOGRAPHY

Moss-Salentijnan, L., and Hendricks-Klyvert, M. (eds.) (1990). "Dental and Oral Tissue: An Introduction," 3rd Ed. Williams & Wilkins, Baltimore.

Woelfel, J. B. (ed.) (1990). "Dental Anatomy: Its Relevance to Dentistry," 4th Ed. Williams & Wilkins, Baltimore.

Dental and Oral Biology, Pathology

JAMES H. P. MAIN
University of Toronto

GLOSSARY

Ameloblast Cell of epithelial origin primarily involved in the formation of dental enamel

Bullae Blister or bleb on a body surface, larger than a vesicle

Cavernous sinus Paired blood sinus lying on the superior surface of the sphenoid bone in the base of the cranial cavity

Dentine Calcified tissue that constitutes the central and larger part of a tooth

Hypoplasia Defective formation or incomplete development of a part

Periodontium Tissues surrounding a tooth including the periodontal membrane and alveolar bone

ORAL PATHOLOGY IS THE STUDY OF DISEASES OF the mouth. Several diseases and abnormalities are unique to the mouth (e.g., dental caries and its sequelae, periodontal disease, and congenital disorders such as cleft lip and palate). In a number of systemic diseases there are specific oral manifestations, and in some cases the symptoms may occur only in the mouth (e.g., in lichen planus). Many other diseases may occur in the mouth as they may do elsewhere in the body (e.g., tumors). The study of all these conditions comprises oral pathology.

I. DEVELOPMENTAL ABNORMALITIES

A. Cleft Lip and Palate

These congenital anomalies occur in about 1 in 1000 births with racial variations. Types include (1) cleft lip, (2) cleft palate, and (3) cleft lip and palate (unilateral or bilateral). Milder related deformities also occur (e.g., notching of the lip and bifid uvula).

The pathogenesis is believed to be due to the lack of growth of mesodermal tissue at around the eighth week of embryonic development with consequent failure of fusion of the processes involved in facial formation. This results in persistence of epithelium, cleft formation, and a deficiency of tissue.

B. Hypoplasia of Enamel

During formation of the enamel, the ameloblasts may be damaged by a number of local or systemic factors of which the following are common:

- Apical infections on a deciduous predecessor
- Trauma
- Exanthematous diseases
- Hypovitaminosis D
- Fluoride in drinking water at concentrations of more than 1 ppm
- Congenital syphilis

These can cause enamel defects in one or many teeth.

ENCYCLOPEDIA OF HUMAN BIOLOGY, Second Edition, VOLUME 3.

C. Amelogenesis Imperfecta

This condition includes a group of hereditary disorders of enamel formation that cause defects in the quantity or quality of the enamel in both deciduous and permanent dentitions. The dentine is normal.

D. Dentinogenesis Imperfecta

In this hereditary disorder, the dentine is defective and the teeth are discolored brown. The enamel is of normal structure but flakes off the dentine readily. The condition may occur on its own or as part of osteogenesis imperfecta.

II. PATHOLOGY OF DENTAL CARIES

In this common disease, the hard tooth tissues are destroyed, and eventually the soft tissues in the middle of the affected tooth (the dental pulp) become infected by oral bacteria. It is initiated by the action of bacterial products formed in dental plaque. The microbiology and epidemiology are described elsewhere. [*See Dental Caries.*]

A. Caries of the Enamel

The first visible sign of the caries is a white spot on the enamel caused by a loss of translucency. Some parts of the tooth are more susceptible than others, notably pits and fissures and the interproximal surfaces. Histologically the lesion is seen in cross section as a triangular area with the base parallel to the enamel surface, which is itself intact. In the middle of this altered area, "pores" form because of loss of minerals, and it is surrounded by a hypercalcified margin. At this stage remineralization from saliva or medications is possible under certain conditions.

The lesion increases in size until the apex of the triangle reaches the amelo–dentinal junction. At this stage surface cavitation develops and bacterial ingress into the defect.

B. Dentinal Caries

Rapid lateral spread of the caries occurs along the amelo–dentinal junction, undermining the enamel and giving rise to retrograde enamel caries. Bacteria invade the dentinal tubules, and dissolution of the hard tissues occurs with loss of mineral salts. The dental pulp reacts to caries by laying down reparative

dentine at the pulpal ends of affected tubules and by producing calcification of the odontoblast processes within the tubules, so sealing them. This reaction is unavailing in the untreated tooth, and in time the rate of carious destruction of the dentine outstrips the rate of production of reparative dentine, resulting in bacterial invasion of the pulp.

III. SEQUELAE TO DENTAL CARIES

Untreated dental caries eventually results in inflammation of the tooth pulp (pulpitis), the commonest cause of toothache. Irritant fillings or tooth cutting operations are other causes. Untreated pulpitis leads to pulpal necrosis and gangrene.

Apical infection results from pulpal gangrene. This is a pyogenic inflammation with mixed oral flora and with pus and granulation tissue formation. Apical infection may remain quiescent for months or years or if the balance of host resistance/organismal virulence is disturbed, it may lead to spreading infection. Commonly the infection spreads buccally, penetrating the alveolar bone, to form a submucosal vestibular abscess. Less often it may spread to a number of potential tissue spaces around the jaws, causing one or more space infections. Infection may also spread into the bone marrow, causing osteomyelitis.

The draining lymph nodes (submandibular, submental, and deep cervical) are usually involved. If a hemolytic streptococcus is present in apical infection, it can cause a serious cellulitis of the neck (Ludwig's angina). Rarely apical infections in the maxilla spread to involve veins communicating with the cavernous sinus and cause cavernous sinus thrombophlebitis.

IV. PERIODONTAL DISEASE

The various forms of periodontal disease are all believed to be caused by a disturbance of the balance between host resistance and noxious products of bacterial dental plaque. The progression of periodontal disease is episodic and site specific with extended periods of equilibrium between the host reaction and the bacteria. There is ongoing discussion as to whether the bacterial agents are specific or not.

Chronic gingivitis consists of inflammation of the marginal gingivae with no deeper tissue destruction.

Chronic periodontitis is a later development in which there is irregular destruction of the periodontium with the formation of pockets between the teeth and adjacent tissues that are lined by epithelium. If left untreated, this leads to loosening and ultimately loss of teeth and/or formation of abscesses in the pockets.

Juvenile periodontitis is an adolescent disease in which rapid destruction of the periodontium occurs around the permanent incisors and/or first molars.

Acute necrotizing ulcerative gingivitis (ANUG) is characterized by necrosis of the interdental papillae. The ulceration may spread widely. Large numbers of *Bacillus fusiformis* and *Borrelia vincentii* are present in the slough. Uncommon in healthy people, it is more common and more serious in the malnourished or under stressful conditions [e.g., trenches in warfare (trench mouth)].

V. IMPACTED TEETH

A tooth is impacted when its eruption is prevented by another tooth or by bone. Impaction is of common occurrence; the third molars being most frequently affected followed by the maxillary canines.

When part of an impacted tooth emerges into the mouth, a mixed infection by oral commensals gains access to the potential space around the crown causing pericoronitis. Infection from this condition may spread as described earlier.

VI. SPECIFIC INFECTIONS

A. Herpes Simplex Virus (HSV) Infection

Spread is usually via direct contact with a latent period of 3–5 days. Primary herpetic stomatitis, usually HSV-1, occurs commonly in children and in susceptible adults. The clinical features are of acute painful vesiculo-bullous disease of the oral mucosa with regional lymphadenopahty and mild pyrexia.

The disease resolves spontaneously in 7–10 days in healthy individuals, but the virus remains latent in the cells of the trigeminal ganglion. Recurrent lesions, often induced by minor trauma or sunshine, occur on the skin–vermillion border of the lips (herpes labialis).

Occasional cases of HSV-2 oral infection occur by oro–genital contact with the same clinical features. [*See* Herpesviruses.]

B. Herpes Zoster (Shingles)

This disease is caused by the varicella-zoster virus (VZV), a member of the herpes group. Zoster is be-

lieved to be a reactivation of latent VZV caused by immunosuppression of various causation and occurs chiefly in older adults.

The clinical features are those of unilateral acute painful vesiculo-bullous disease in the distribution of a sensory nerve lasting 2–4 weeks. Involvement of the second or third division of the trigeminal results in disease affecting the face and mouth. Postherpetic neuralgia occasionally follows resolution.

C. Herpangina

This infection is caused by a coxsackie virus and presents as a vesicular eruption of the soft palate and oropharynx. Constitutional disturbances, lymphadenitis, fever, etc., are usually mild.

D. Human Immunodeficiency Virus

The oral manifestations of human immunodeficiency virus (HIV) infection include:

- Hairy leukoplakia
- Candidiasis
- HIV-associated periodontal disease
- Kaposi's sarcoma
- Lymphoma

Hairy leukoplakia occurs on the sides of the posterior part of the tongue, usually bilaterally, and is asymptomatic. It appears as white wrinkled lines that may fuse to form plaques. Epstein–Barr virus (EBV) and *Candida* are usually present in the epithelium but not HIV. The presence of this lesion is associated with subsequent development of the acquired immunodeficiency syndrome fairly quickly in a high percentage of patients.

Oral candidiasis is a common finding in HIV-infected persons.

HIV-associated periodontal disease may present as severe ANUG, with heavy infection by the *B. fusiformis* and *B. vincentii*, or as a rapidly progressive chronic periodontitis with alveolar bone destruction.

The tumors listed are those most commonly associated with HIV infection and may present intraorally as elsewhere in the body, with Kaposi's sarcoma being common intraoraly. [*See* Acquired Immunodeficiency Syndrome (Virology).]

E. Candidiasis

Candida albicans is an oral commensal that becomes pathogenic in immunodepression, general debility, or as a result of chronic local tissue damage. Four types occur:

- Acute pseudomembraneous candidiasis (thrush)
- Chronic hypertrophic candidiasis
- Atrophic candidiasis
- Angular cheilitis (perleche)

The infection will clear with appropriate antibiotic therapy but will recur unless the predisposing factor(s) is eliminated.

F. Actinomycosis

This infection is caused by *Actinomyces israelii,* often in combination with other bacteria, and occurs in the abdomen, thorax, and cervicofacial region. Infections occur as complications of surgery (e.g., tooth extractions) or secondary to trauma.

The clinical features are those of a persistent swelling around the mandibular angle with draining skin sinuses. The pus contains yellow "sulfur granules" that are colonies of the organism.

G. Syphilis

All stages of syphilis may affect the mouth. In congenital syphilis, screwdriver-shaped incisor teeth and mulberry molars are seen. The primary chancre may occur intraorally or on the lip as a painless indurated ulcer teeming with *Treponema pallidum* that lasts for about a month. Secondary syphilis causes a skin rash and "snail track" ulcers on the oral mucosa. Among the many manifestations of tertiary syphilis are glossitis and palatal perforations caused by gumma formation. [*See* Sexually Transmitted Diseases.]

H. Tuberculosis

Oral tuberculosis is rare and usually occurs secondarily to open pulmonary infection in the form of chronic ulceration, the result presumably of bacterial implantation from infected sputum.

I. Other Fungal Infections

Histoplasmosis, coccidioidomycosis, cryptococcosis, and blastomycosis may also involve the mouth, usu-

ally as chronic ulcers and generally secondarily to lung infection.

VII. CYSTS

A cyst is a pathological cavity in the tissues lined by epithelium and containing fluid or semifluid material. Cysts are common in the jaws, the epithelium being derived usually from remnants of odontogenic epithelium. Cysts are slow-growing lesions, causing swelling and displacement of teeth. They may become infected.

The commoner types of cysts are

1. Radicular cyst. This arises in a chronic periapical inflammation in which epithelial rests are stimulated to proliferate by the infection. The cyst may remain *in situ* after the tooth has been extracted (residual cyst).
2. Dentigerous cyst. Cysts can arise from the odontogenic epithelium around the crown of an impacted or unerupted tooth, with the tooth crown projecting into the lumen.
3. Odontogenic keratocyst. The pathogenesis of this lesion is unknown. They can grow to large size, and "daughter" cysts may bud off the epithelial lining, causing a high recurrence rate. They sometimes occur as part of the basal cell naevus syndrome.
4. Inclusion cysts. Epithelium may be enclaved during fusion of the embryonic processes that form the face and jaws. In later life this epithelium may form cysts. The most common type is the incisive canal cyst that lies at the point of fusion of the primitive palate and the two palatal shelves.

VIII. MUCOSAL DISEASES

Lichen planus is a chronic inflammatory condition occurring in adults that may affect skin or mucous membranes or both and that arises because of an abnormality of the immune system. Intraorally it results in hyperkeratosis usually bilaterally and in the form of an interlacing network of fine lines or dots, occasionally as plaques. The adjacent mucosa may be inflamed or ulcerated.

Discoid lupus erythematosus may cause oral lesions that are erythematous or ulcerated with adjacent hyperkeratosis.

Pemphigus vulgaris usually affects the oral mucosa and may start on it. The oral lesions are multiple vesicles that rupture within a few hours, leaving ulcers that may become secondarily infected.

Mucous membrane pemphigoid is a chronic vesiculo-bullous disease usually of the elderly in which the bullae rupture and become secondarily infected. If left untreated, the ulcers may persist for years.

Erythema multiforme may occur on skin or mucous membranes or on both and is most common in young male adults. It has an acute onset and intraorally results in large serpiginous ulcers; the edges are crusted with blood clots and almost always involves the lips. It resolves spontaneously in about 3 weeks and may be an allergic response to herpes simplex or to some other antigen.

Epidermolysis bullosa is an inherited condition that occurs with varying levels of severity and presents in childhood. It affects skin and mucous membranes; the lesions are bullae that form after minor friction so that they frequently form on areas of oral mucosa subject to masticatory stress.

Scleroderma is a generalized disease in which collagen is formed in excessive quantities in the soft tissues and elsewhere. In the mouth it chiefly affects the cheeks, which become stiff and immobile so that oral hygiene measures and dental treatment become difficult, resulting in caries and periodontal disease.

IX. FIBRO-OSSEOUS CONDITIONS

Fibrous dysplasia is a developmental abnormality present in childhood, and often affecting the jaws, in which areas of bone are affected by a defect in metabolism that causes resorption of mature bone and its replacement by cellular connective tissue and excessive amounts of immature bone. This results in swelling, deformity, and displacement of teeth. In many cases the condition resolves spontaneously around puberty.

Cherubism is a subtype of fibrous dysplasia in which only the molar areas of the mandible and, less often, the maxilla are affected. In early childhood, mild involvement produces a cherubic facies but later is disfiguring.

Cleidocranial dysplasia is a congenital abnormality of bones formed in membrane affecting the skull and clavicles, which are often absent. In this condition the teeth do not erupt.

X. INFLAMMATORY POLYPS

Polyps of irritational origin are common in the mouth and consist of varying proportions of granulation and fibrous tissue containing inflammatory cell infiltrates and are covered by stratified squamous epithelium that may be hyperkeratotic, hyperplastic, or ulcerated. Many names have been used for these lesions, but all can be classified as fibro-epithelial polyps or, when composed largely of granulation tissue, as pyogenic granulomata. They have minimal neoplastic potential. Generalized fibrous hyperplasia of the gingivae occurs idiopathically and as a result of chronic dilantin therapy.

XI. TUMORS

A. Epithelial

1. Papilloma

Small villous exophytic squamous papillomas occur on all parts of the oral mucosa and lips. Some of them are verruca vulgaris and others may be caused by human papilloma virus (HPV) infection. Condyloma acuminatum and focal epithelial hyperplasia also occur because of HPV infection. [See Papillomaviruses and Neoplastic Transformation.]

2. Squamous Cell Carcinoma

This is the most common oral malignancy. Its incidence varys greatly in different populations and is highest in India where it is associated with the habit of chewing pan (a mixture of tobacco, slaked lime, betel nut and leaf, and other additives). In Western countries, oral cancer accounts for about 4% of all malignancies. It can occur anywhere in the mouth but is most frequent on the sides of the tongue and on the lower lip, where it is associated with actinic damage to the tissues. Excessive consumption of alcohol and tobacco are the only common predisposing factors.

It is predominantly a disease of the elderly. It presents clinically as an area of hyperkeratosis, as a lump, as an ulcer with raised rolled margins, as a velvety red patch, or as a combination of these. If left untreated it can cause great local tissue damage and spreads by local invasion and by metastases to the regional lymph nodes in the neck but seldom farther.

3. Mesenchymal and Neurogenic Tumors

Virtually all of the benign and malignant tumors of mesenchymal and neurogenic origin may occur in the mouth but with no locally distinctive features.

4. Salivary Tumors

A range of epithelial tumors are found in the major (parotid, submandibular, and sublingual) salivary glands and in the numerous minor glands found all around the mouth. The parotid is by far the most common site and in it the great majority of tumors are benign. The proportion of malignant tumors is much higher in the submandibular, sublingual, and minor glands.

All benign tumors present clinically as slow-growing masses. Malignant tumors grow more rapidly and may ulcerate. In the parotid they may invade the facial nerve, causing unilateral facial paralysis. Metastases usually develop late except for the adenocarcinoma. The adenoid cystic carcinoma is notable for its infiltrative tendencies, often along nerve sheaths.

A simplified list of salivary tumors consists of

- Pleomorphic adenoma
- Monomorphic adenoma (various types) — Benign
- Mucoepidermoid tumor
- Acinic cell tumor — Low-grade malignancy
- Adenoid cystic carcinoma
- Carcinoma expleomorphic adenoma
- Adenocarcinoma — Malignant

5. Odontogenic Tumors

These tumors are site specific as they arise from epithelial, mesenchymal, or both tissues involved in the process of tooth formation. Histologically, they consist of cells or cell products, including enamel, dentine, and cementum, that are seen in normal tooth formation. They demonstrate a range of behavior from completely benign lesions that are essentially hamartomas to rare metastasizing malignancies. Most are benign. The ameloblastoma, most common of the group, and the odontogenic myxoma are slow-growing infiltrative neoplasms that cause great local destruction and, if arising in the maxilla, may invade the base of the skull and brain. The calcifying epithelial odontogenic tumor contains deposits of amyloid.

Benign tumors
 Epithelial
  Ameloblastoma
  Calcifying epithelial odontogenic tumor
  Adenomatoid odontogenic tumor
  Squamous odontogenic tumor
 Mixed (epithelial and mesenchymal)
  Odontoma
  Ameloblastic fibroma
  Ameloblastic fibro-odontome
 Mesenchymal
  Odontogenic myxoma
  Odontogenic fibroma
  Cementifying fibroma
  Cementoblastoma
Malignant tumors
 Epithelial
  Malignant ameloblastoma
  Ameloblastic carcinoma
 Mesenchymal
  Ameloblastic fibrosarcoma

BIBLIOGRAPHY

Cawson, R. A., Binnie, W. H., and Eveson, J. W. (1993). "Color Atlas of Oral Disease, Clinical and Pathological Correlations," 2nd Ed. Wolfe Publishing, Hong Kong.

Kramer, I. R. H., Pindborg, J. J., and Shear, M. (1992). "Histological Typing of Odontogenic Tumours," 2nd Ed. Springer-Verlag, Berlin.

Lucas, R. B. (1984). "Pathology of Tumours of the Oral Tissues," 4th Ed. Churchill Livingstone, Edinburgh.

Nikiforuk, G. (1985). "Understanding Dental Caries," Vol. 1. Karger, Basel.

Regezi, J. A., and Sciubba, J. J. (1993). "Oral Pathology:" Clinical-Pathologic Correlations," 2nd Ed. Saunders, Philadelphia.

Shear, M. (1992). "Cysts of the Oral Regions," 3rd Ed. Wright, Oxford.

Wright, B. A., Wright, J. M., and Binnie, W. H. (1988). "Oral Cancer: Clinical and Pathological Considerations." CRC Press, Boca Raton, FL.

Dental and Oral Biology, Pharmacology

University of Texas Dental Branch

I. Local Anesthetics
II. Antibiotics
III. Pharmacologic Control of Pain and Anxiety
IV. Fluorides and Antiplaque Agents
V. Other Drugs Used in Dental Practice

GLOSSARY

Amide local anesthetic Local anesthetic consisting of an aromatic ring linked to an alkyl chain with an amino terminus by an amide bond

Benzodiazepine One of a group of sedative–hypnotic drugs composed of a benzene ring coupled to a seven-member ring referred to as a diazepine ring

Broad spectrum An antibiotic that is equally effective against both gram-positive and gram-negative microorganisms

Ester local anesthetic Local anesthetic consisting of an aromatic ring linked to an alkyl chain with an amino terminus by an ester bond

Narrow spectrum An antibiotic that is limited to a relatively small range of microorganisms (e.g., only gram-positive ones)

Prostaglandin Twenty carbon fatty acid containing three to five double bonds and a five member, substituted ring; the several classes of prostaglandins serve a variety of roles as autacoids

Vasoconstrictor Drug used to elevate systemic blood pressure or to reduce bleeding by inducing contraction of vascular smooth muscle with a consequent reduction in blood vessel diameter

THE ADMINISTRATION AND THE PRESCRIPTION OF drugs in dentistry have become considerably more complex over the years, due in large part to the devel-opment of a vast array of medically prescribed drugs that may interact with drugs prescribed by the dentist as well as to the introduction of many new drugs for dentistry. The mid-1980s brought about a virtual revolution in the concepts of the treatment of pain and anxiety in dental patients with the use of new analgesic and sedative drugs, as based on evolving research findings in neuropharmacology.

This same evolution has occurred in the therapy of periodontal disease, which is now increasingly being treated with nonsurgical methods, including antibiotics and other, topically applied antimicrobial agents.

This section addresses the state of the art in the most important drug groups used in dental practice (local anesthetics, antibiotics, analgesics, sedatives, and preventive agents).

I. LOCAL ANESTHETICS

Modern dental local anesthesia utilizes several agents of the amide class of local anesthetics (Table I), almost entirely replacing the ester-type agents (e.g., procaine or Novocain). Procaine is the prototypical drug in the ester class of local anesthetics. The other major class, amides, is typified by the drug lidocaine (Xylocaine and other trade names). Local anesthetics are drugs that, when placed on a nerve, penetrate the nerve membrane and prevent the transient increase in so-dium permeability, which ordinarily accompanies chemical or mechanical stimulation of the nerve. Be-cause depolarization (the change of the electrical po-tential across the nerve membrane) is prevented in the affected portion of the nerve, impulse generation and conduction are locally blocked and pain impulses can-not be transmitted to the central nervous system. This mechanism of action differs from that of opioid anal-gesics, for example, which affect the perception of impulses within the brain. [*See* Pain.]

169

ENCYCLOPEDIA OF HUMAN BIOLOGY, Second Edition, VOLUME 3.   Copyright © 1997 by Academic Press.   All rights of reproduction in any form reserved.

TABLE I

Dental Local Anesthetic Agents Used for Injection

Drug	Trade name	Concentration (%)	Vasoconstrictor
Lidocaine	Xylocaine	2	Epinephrine
Mepivacaine	Carbocaine	2	Levonordefrin
		3	—
Prilocaine	Citanest	4	—
	Citanest Forte	4	Epinephrine
Bupivacaine	Marcaine	0.5	Epinephrine
Etidocaine	Duranest	1.5	Epinephrine

At present, local anesthetics are thought to affect nerve membrane sodium conductance through the internal aspect of the sodium channel within the nerve membrane. Combination of the positively charged local anesthetic molecule with the channel theoretically induces the inactivation of a gate controlling the internal opening of the sodium channel, stabilizing the inactivated, or nonconducting, channel configuration. This model of local anesthetic action also accounts for the phenomenon of use dependence, in which more frequent nerve depolarizations enhance the rate of nerve block by the local anesthetics. This is presumably due to greater accessibility of the sodium channel to the local anesthetic during repeated opening of the channels.

Because of their ability to block inward sodium currents in electrically excitable tissues, local anesthetics can, at toxic blood levels, adversely affect the heart, blood vessels, and central nervous system and, most seriously, can cause central stimulation followed by depression and cardiorespiratory arrest. Toxic reactions to local anesthetics may be associated with overdosage, accidental intravascular injection, or an inability of the patient to effectively metabolize or excrete the local anesthetics. Correct dosages of local anesthetics are determined by the patient's body weight. For example, the maximum recommended dose of lidocaine is 4.4 mg/kg and a total dose of no more than 300 mg in individuals weighing >70 kg.

In clinical usage, local anesthetics are generally employed in combination with a vasoconstrictor such as epinephrine. The rationale for adding a small amount of such a drug to a local anesthetic is to counteract the vasodilation that is ordinarily produced by the local anesthetic, as described earlier. This prolongs the duration of anesthesia as well as reduces the rate of absorption of the anesthetic, thereby reducing the overall potential for a toxic reaction.

Vasoconstrictors themselves can produce toxic reactions and may be especially dangerous in patients with cardiovascular disease. Because they constrict small arteries and arterioles, vasoconstrictors can elevate blood pressure, resulting in possible cerebrovascular hemorrhage and an increased workload on the heart, with the potential consequence of myocardial infarction in an already diseased myocardium. The doses of vasoconstrictors in cardiovascular patients must be reduced and occasionally they must be eliminated in these patients.

The side effects of vasoconstrictors include elevations in blood pressure, cardiac arrhythmias, and increases in cardiac work. In general, vasoconstrictors do not cause adverse effects in dental patients and greatly prolong the duration and increase the depth of local anesthesia. The types of vasoconstrictors used in local anesthetic solutions are found in Table I.

II. ANTIBIOTICS

A. Treatment of Dental Infections

Ever since the early 1980s, the nature of dental infections has been further elucidated to establish the prominent role of obligate anaerobic bacteria (which can only exist in the absence of oxygen) as major infective agents, usually mixed with facultative organisms (which can live with or without oxygen). While up to 264 bacterial groups or species may exist within the oral cavity, several prominent genera (*Staphylococci, Streptococci, Bacteroides, Peptostreptococci,* and *Fusobacterium*) play a major role in dental infections and apparently produce pathological changes through synergistic interactions. This synergism is apparently facilitated by some temporary breakdown in the immune defenses in the host.

B. Antibiotics Used in Dentistry

Penicillin remains the drug of choice for most odontogenic infections. This drug offers the advantages of being bactericidal and inexpensive, and its narrow spectrum (gram-positive) is well suited for application to dental infections. Penicillin kills bacteria by producing a defect in their cell walls, which ultimately allows osmotic forces to rupture the bacterium.

In most situations, antibiotic therapy is instituted on an empirical basis with first-choice agents. Treatment of dental infections should be based on culture and sensitivity testing, although this is frequently not feasible and, in many cases, is unnecessary because mild infections of dental origin frequently respond to first- or second-choice agents when those antibiotics are administered in conjunction with drainage of the infected area and appropriate dental treatment (e.g., root canal therapy, tooth extraction). Culture and sensitivity testing serve as a basis for changing antibiotic therapy if agents selected initially prove to be ineffective.

Unfortunately, penicillin cannot be used in many patients because they are allergic to it, although in nonallergic patients it produces few, if any, side effects. Penicillin is available in oral and injectable forms. The potassium salt of phenoxymethyl penicillin (penicillin V) is usually selected as the oral form, with doses of 250–500 mg administered every 6 hr.

In cases of penicillin allergy, an alternative antibiotic must be selected for initial therapy. Erythromycin has been generally accepted as the best alternate antibiotic for use in routine outpatient infections. Its spectrum is narrow but very similar to that of the penicillins and is, therefore, useful in oral infections. Its major disadvantages include a bacteriostatic, rather than a bactericidal, mechanism of action, fairly rapid development of microbial resistance to the drug, and a relatively high incidence of gastrointestinal side effects, including nausea and vomiting. Several oral forms of erythromycin are available under numerous trade names, including erythromycin base, erythromycin ethylsuccinate, erythromycin estolate, and erythromycin stearate. Clindamycin is being used more frequently as an alternative to penicillin. Other alternate antibiotics are summarized in Table II.

An important concept in the use of antibiotics in dentistry is that of infection prevention, or prophylaxis. Prophylactic use of antibiotics is generally directed at the prevention of bacterial endocarditis, which can result when microorganisms (especially *Streptococci*) infect diseased heart tissue, such as heart valves damaged by an episode of rheumatic fever, or prosthetic implant materials, such as artificial heart valves. The American Heart Association has developed prophylactic antibiotic regimens for the prevention of such problems. Dental patients with the aforementioned medical complications should receive antibiotic prophylaxis before any procedure that will introduce bacteria into the blood (bacteremia), including routine teeth cleaning (prophylaxis). [*See* Antibiotics.]

III. PHARMACOLOGIC CONTROL OF PAIN AND ANXIETY

A. Nonnarcotic Analgesics

Recognition of the role of prostaglandins in the genesis of pain and inflammation has resulted in the now widespread use of nonsteroidal, anti-inflammatory drugs (NSAIDs) for the treatment of dental pain. Aspirin is the prototypical drug in this class, but its use is limited to mild levels of pain in dentistry. When combined with 8, 15, 30, or 60 mg of codeine, it is regarded as a reliable analgesic for relief of moderate dental pain. The newer agents (Table III) are capable of relieving mild to moderate levels of pain, with side effects that, while qualitatively similar to those of aspirin, appear to have a lower incidence and/or reduced severity than those produced by aspirin. The side effects of these drugs include gastrointestinal irritation, inhibition of platelet aggregation (increased bleeding time), allergic reactions of varying severity, fluid retention and edema, and, occasionally, visual

TABLE II

Alternate Oral Antibiotics Used in Dentistry[a]

Drug	Usual adult dose (mg)[b]
Erythromycin[c]	250–500
Clindamycin	150–300
Ampicillin	500
Tetracycline	250–500

[a]Only the most common alternates are listed. Products are available under numerous trade names.
[b]Dosage every 6 hr.
[c]Base form. Like erythromycin, various salts of antibiotics are available but are not listed.

TABLE III

Oral Nonsteroidal Anti-inflammatory Drugs Used in Dentistry

Drug	Usual adult dosage
Ibuprofen	400 mg, every 6 hr p.r.n.[a]
Diflunisal	1000 mg initially, then 500 mg every 12 hr p.r.n.
Naproxen[b]	550 mg initially, then 275 mg, every 6–8 hr p.r.n.
Ketoprofen	25–50 mg, every 6–8 hr p.r.n.

[a]p.r.n. indicates "as needed" (for pain).
[b]Dosage listed is for sodium salt.

and hearing disturbances. The NSAIDs are cross-aller-genic with aspirin and should not be used in aspirin-allergic individuals. All NSAIDs, including aspirin, can interact with other drugs that prolong bleeding time, irritate the gastrointestinal tract, or lower blood pressure. Protein-bound drugs, such as oral antidiabetic drugs, may be potentiated by drugs in this group.

Acetaminophen is a commonly used analgesic and, in dentistry, is regarded as an alternative to aspirin for the treatment of mild pain or, in combination with codeine, for the treatment of moderate pain. It is usually selected for aspirin-allergic patients or patients with gastrointestinal or bleeding disorders.

B. Opioid Analgesics

The opioid analgesics are generally reserved for use in treating moderate to severe levels of pain in dentistry. Their use for moderate pain has declined with the increasing use of NSAIDs, but they are still widely used for the treatment of severe levels of pain. Codeine is widely used in combination with aspirin or acetaminophen, as described earlier for moderate pain, as are pentazocine and hydrocodone. Oxycodone, meperidine, methadone, and dihydromorphone are prescribed for oral treatment of severe pain, whereas morphine is usually reserved for the parenteral treatment of the most severe types of pain.

The use of opioid analgesics is limited by their side effects, the most notable of which in dental applications include nausea, vomiting, and depression of the central nervous system. All the opioid analgesics mentioned thus far are classified as "controlled substances" by federal and state laws and, as such, have the potential to produce tolerance, physical dependence, and psychological dependence. As central depressants, these drugs can dangerously interact with other drugs, including alcohol, antihistamines, sedative–hypnotics, and others.

In addition to their use in controlling dental pain, opioid analgesics are widely used in combination with sedative–hypnotics to produce enhanced analgesia and patient cooperation during intraoperative, intravenous sedative procedures. A newer group of drugs, including butorphanol and nalbuphine, are partial narcotic antagonists and are being used in intavenous sedation to limit the degree of respiratory depression encountered with classical opioids, such as morphine or meperidine.

C. Sedative–Hypnotic Drugs

This group of drugs constitutes the major group of agents used in dentistry to control pre- and intraoperative anxiety. In the past, barbiturates such as pentobarbital saw widespread use for dental sedation, but because of their attendant problems (respiratory depression, laryngospasm and abuse potential), they have been replaced by the benzodiazepine-type agents (Table IV). These drugs represent a significant advantage in producing less respiratory depression than that of the barbiturates. They are also useful as orally administered, preoperative sedative–hypnotic drugs. Some drugs in this group (flurazepam, lorazepam, temazepam, and triazolam) are better for hypnosis (induction of sleep on the night before a dental appointment), whereas others (alprazolam, chlordiazepoxide, chlorazepate, diazepam, and oxazepam) are preferred for their antianxiety effect during waking hours. Until recently, diazepam was the preferred benzodiazepine for intravenous sedation in dentistry; however, because it is not water soluble and its vehicle (propylene glycol) produced thrombophlebitis in some patients, midazolam (a water-soluble benzodiazepine) is now widely used for this application.

The benzodiazepines are classified as controlled substances because long-term use can produce tolerance and dependence. Their principal side effects include drowsiness and other types of central depression, paradoxical central excitation, allergy, weight gain, nausea, vomiting, and xerostomia. They can, of course, interact adversely with other central depressant drugs. They are contraindicated in allergic individuals, pregnancy, some types of glaucoma, liver or kidney disease, and in chronic obstructive pulmonary diseases.

TABLE IV
Benzodiazepine Drugs

Official name	Trade name
Alprazolam	Xanax
Chlordiazepoxide	Librium
Clonazepam	Klonopin
Clorazepate	Tranxene
Diazepam	Valium
Flurazepam	Dalmane
Lorazepam	Ativan
Midazolam	Versed
Oxazepam	Serax
Przepam	Centrax
Temazepam	Restoril
Triazolam	Halcion

In addition to barbiturates and benzodiazepines, a number of other drugs may be used for sedation and anxiety control in dentistry. In children, hydroxyzine, an antihistamine, and chloral hydrate are widely used and can be administered by the oral route. Other nonbarbiturate agents include carbamates (ethinamate and meprobamate), ethchlorvynol, paraldehyde, glutethimide, and methyprylon. The oral administration of a drug to produce sedation is limited because it relies on patient compliance, the absorption of the drug can be very erratic, the drug effect cannot be titrated, and the duration of action tends to be prolonged.

Nitrous oxide (laughing gas) is used frequently to produce relief of anxiety and control of gagging in dental patients. Nitrous oxide is administered by inhalation of 20–50% concentrations in combination with 80–50% pure oxygen, respectively. Nitrous oxide has a rapid onset of action and can be rapidly titrated to the level of sedation needed; it is very safe, nonallergenic, and not metabolized by the liver. Nitrous oxide is rapidly excreted by the lungs so that patient recovery is extremely rapid, unlike other forms of sedation. Because nitrous oxide lacks significant effects on the cardiovascular system and other organs, it is very useful in patients with systemic disease. It is contraindicated in patients with some types of respiratory diseases, pregnancy, and in uncooperative patients. Long-term exposure to nitrous oxide can produce a polyneuropathy and, in females, has been implicated in increasing the risk of spontaneous abortion and fetal malformations.

IV. FLUORIDES AND ANTIPLAQUE AGENTS

The beneficial effects of fluoridation of drinking water on dental caries have been known since the 1950s. The systemic administration of optimal fluoride levels in infants, children, and young adolescents results in the incorporation of fluoride ion into the enamel crystals of developing teeth, rendering them significantly less soluble in the acids produced by dental plaque bacteria. Where drinking water concentrations are suboptimal (<0.7–1 parts per million), systemic supplementation may be accomplished by the prescription of sodium fluoride (in liquid or tablet form). The recommended total dietary fluoride ion intake is 0.25 mg for infants (neonates to age 2 years), 0.5 mg for ages 2–3 years, and 1 mg/day for ages 3–13 years.

Fluoride also confers reduced acid solubility when applied topically to teeth. Prescription-strength preparations for in-office or home application include 0.2% sodium fluoride; 0.5–1.23% acidulated phosphate fluoride gels, sodium fluoride, and phosphoric acid topical solutions and gels; and 0.4% stannous fluoride gels. Fluoride, as sodium fluoride, is available over the counter in a variety of dentifrices as well as in 0.05% concentrations in mouth rinses.

More recently, agents specifically targeted at dental plaque have become available. Of these, chlorhexidine is now available in prescription form as a 0.2% mouth rinse and acts as a broad-spectrum antibacterial. Chlorhexidine reduces the formation of plaque, and its side effects are limited to a bitter taste, reversible staining of teeth and restorations, and occasional irritation of the oral mucosa.

V. OTHER DRUGS USED IN DENTAL PRACTICE

A. Corticosteroids and Antihistamines

Corticosteroids, such as prednisolone, are used to a limited degree in dentistry to control inflammation and as secondary agents in the treatment of immediate-type (anaphylactic) allergic reactions. Antihistamines, including diphenhydramine, are prescribed for the symptomatic treatment of mild, delayed-type allergic reactions.

B. Emergency Drugs

Epinephrine is recommended for the emergency treatment of acute, life-threatening allergic reactions and is administered subcutaneously or intravenously in this application. Other emergency drugs typically found in a dental office include nitroglycerin (antianginal), diazepam (anticonvulsant), spirits of ammonia (smelling salts), naloxone (opioid antidote), glucose (for hypoglycemia), and oxygen with a positive-pressure ventilation assist.

BIBLIOGRAPHY

Cooper, S. A. (1983). New peripherally-acting oral analgesic agents. *Annu. Rev. Pharmacol. Toxicol.* **23**, 617–647.

Council on Dental Therapeutics (1984). "Accepted Dental Therapeutics," 40th Ed. American Dental Association, Chicago.

Gangarosa, L. P., Ciarlone, A. E., and Jeske, A. H. (1993). "Pharmacotherapeutics in Dentistry," Medical College of Georgia, Augusta, GA.

Dental Anthropology

University of Alaska, Fairbanks

GLOSSARY

Antimere Corresponding teeth in the left and right sides of a jaw (e.g., left and right upper first molars)

Carabelli's trait Morphological character derived from cingulum of mesiolingual cusp of upper molars

Cingulum Bulge or shelf passing around the base of the tooth crown

Cusp Pointed or rounded elevation on the occlusal (chewing) surface of a tooth crown

Isomere Corresponding teeth in the upper and lower jaws (e.g., left upper and lower first molars)

Labret Ornament worn in and projecting from a hole(s) pierced through the upper and lower lips and cheeks

Phenetics Classificatory method for adducing relationships among populations on the basis of phenotypic similarities

Quadrant One half of the upper or lower dentition

Shoveling Mesial and distal marginal ridges enclosing a central fossa on the lingual surface of the incisors

DENTAL ANTHROPOLOGY IS A FIELD OF INQUIRY that utilizes information obtained from the teeth of either skeletal or modern human populations to resolve anthropological problems. Given their nature and function, teeth are used to address several kinds of questions. First, teeth exhibit variables with a strong hereditary component that are useful in assessing population relationships and evolutionary dynamics.

Given their role in chewing food, dental pathologies and patterns of tooth wear can indicate kinds of food eaten and other aspects of dietary behavior, including food preparation techniques. Teeth can also exhibit incidental or intentional modifications, which reflect patterns of cultural behavior. Finally, as the process of tooth formation is highly canalized (i.e., buffered from environmental perturbations), developmental defects provide a general measure of environmental stress on a population. Researchers in several disciplines, including physical anthropology, archeology, paleontology, dentistry, genetics, embryology, and forensic science, conduct research that falls directly or indirectly within the province of dental anthropology.

I. THE HUMAN DENTITION

A. Terms and Concepts

A tooth has two externally visible components, crown and root, and is made up of three distinct hard tissues, enamel, dentine, and cementum, and one soft tissue, the pulp, which provides the blood and nerve supply to the crown and root. Teeth are anchored in the bony alveoli of the upper and lower jaws by one or more roots and the periodontal membrane. Terms of orientation for teeth are mesial (toward the anatomical midline, or the point between the two central incisors), distal (away from the midline); buccal (toward the cheek), labial (toward the lip), lingual (toward the tongue), and occlusal (the chewing surface of a tooth). [*See* Dental and Oral Biology, Anatomy.]

The reptilian dentition is homodont (generally uniform, single-cusped, conical teeth for grasping food objects) and polyphyodont (multiple generations of teeth). By contrast, the mammalian dentition is heterodont (four types of teeth, each performing different functions) and biphyodont (only two generations of teeth, primary and permanent). Thus, humans share with other mammals the presence of four distinctive tooth types (i.e., incisors, canines, premolars, and molars) and two successive dentitions. In the human primary dentition, there are two incisors, one canine, and two molars per quadrant. In the permanent dentition, there are two incisors, one canine, two premolars, and three molars per quadrant.

In the human dentition, the four tooth types are found on both the left and right sides of the upper jaw (maxilla) and lower jaw (mandible). Antimeres exhibit mirror imaging, but isomeres differ in both size and morphology. Incisors and canines, referred to collectively as anterior teeth, have one cusp and a single root. Premolars exhibit one buccal cusp, one lingual cusp, and a single root. Upper molars are characterized by four major cusps and three roots, whereas lower molars exhibit five cusps and two roots. The multicusped premolars and molars are referred to collectively as posterior or cheek teeth. Variations on these normative characterizations of cusp and root number fall within the area of dental morphological variation.

An important concept that relates to the different types of teeth in mammals is Butler's field theory. When this concept was adapted to the human dentition by A. A. Dahlberg, he used the phrase tooth districts to describe eight morphological classes corresponding to the four types of teeth in the two jaws. Within each tooth district, there is a "key" tooth, which shows the most developmental and evolutionary stability in terms of size, morphology, and number. For humans, the key tooth in a given tooth district is usually the most mesial element (e.g., upper central incisor, lower first molar); the only exception is in the lower incisor district where the lateral incisor is the key tooth. In a given tooth district, variation increases with distance from the key tooth. In the two molar districts, for example, the first molar is the most stable (least variable), while the third molar is highly variable (least stable); second molar variation falls between these extremes. The implication is that the key teeth best reflect the genetic-developmental programs controlling tooth development, whereas the distal elements of a field are more susceptible to environmental effects. This may be related to the relatively protracted period of tooth development in humans; early developing teeth (e.g., M1) are the most stable, whereas late developing teeth (e.g., M3) exhibit more environmentally induced variability.

B. Why Study Teeth?

For the resolution of anthropological problems, a number of advantages are associated with the study of human dental variation.

1. Preservability

Teeth preserve exceptionally well in the archeological record (due in part to the chemical properties of enamel) and are frequently the best represented part of a skeletal sample. This is evident in both Holocene archeological series and Pleistocene hominid fossil remains.

2. Observability

For the most part, human biologists interested in biochemical polymorphisms, dermatoglyphics, and other anatomical and physiological variables are limited to the study of living populations. Likewise, most variables of interest to human osteologists can be observed only in prehistoric and protohistoric skeletal remains. Teeth, on the other hand, can be directly observed and studied in both skeletal and living populations (e.g., through intraoral examinations, permanent plaster casts, extracted teeth). Because teeth are observable in both extinct and extant human groups, they provide a valuable research tool for the analysis of short-term and long-term temporal trends.

3. Variability

Because teeth are critical in food-getting and food-processing behavior, their development is controlled by a relatively strict set of genetic-developmental programs. On the other hand, as the dentition interfaces directly with the environment, teeth are also modified postnatally by physical factors associated with mastication and disease factors related to the interplay of dietary elements and a complex oral microbiota. Thus, a wide variety of dental variables are available for analysis; some provide information on the genetic background of a population, whereas others reflect environmental and behavioral factors that impinge on individuals in a given population.

C. Teeth as Indicators of Age

An accurate determination of age and sex is fundamental to any inquiry relating to human skeletal remains in both archeological and forensic contexts. One characteristic of the dentition, which makes teeth useful in aging individual skeletons, is a predictable sequence of developmental events, including crown and root formation, calcification, and eruption. As this genetically controlled sequence of events varies to only a limited extent among recent human populations, the principles of aging children by stages of dental development can be applied to all human groups. Before the age of 12 years, teeth are the best and most readily available indicator of age. During adolescence, teeth provide a useful adjunct to patterns of epiphyseal fusion in age determination. After the permanent dentition is completed with the eruption of the third molars (ca. 18 years of age), degree of crown wear and gradients of wear between the first, second, and third molars allow researchers to estimate adult age by decade or within broader age categories (e.g., young, middle, and old adult). Because tooth wear in adulthood has a strong cultural component, it is necessary to apply different standards to spatially and temporally circumscribed populations. For example, medieval Europeans exhibited much greater degrees and rates of crown wear than modern Europeans, so tooth wear standards for modern Europeans would not be applicable to their medieval forebearers.

II. DENTAL PHENETICS AND PHYLOGENY: INFERRING HISTORY FROM TEETH

Traditionally, physical anthropologists have been interested in human population origins and relationships. Historical questions can be posed on a broad geographic scale (e.g., where did Native Americans come from?; who are they most closely related to?; when did they arrive in the New World?) or have more regional focus (e.g., what is the relationship between peoples of the Jomon culture and modern Japanese and Ainu populations?). Many biological traits have been employed to address such questions, including general morphology, body size, pigmentation, dermatoglyphic patterns, genetic markers of the blood, mitochondrial DNA, and metric and nonmetric skeletal traits. Although variation in human tooth size and morphology has long been recognized, historical inferences based on teeth have been limited until recently.

The derivation of historical relationships from dental data requires variables with a significant genetic component. Variables that meet this requirement fall under the broad headings of tooth size, crown and root morphology, hypodontia (missing teeth), hyperodontia (supernumerary teeth), and eruption sequence polymorphisms. As most historical analyses focus on tooth size and morphology, this discussion is limited to metric and morphologic variables.

A. Tooth Size

In studies of human tooth size variation (odontometrics), the measurements reported most commonly are maximum crown length [mesiodistal (MD) diameter] and maximum crown breadth [buccolingual (BL) diameter]. In some instances, measurements are reported for crown height and intercuspal distances, but crown wear must be minimal or the landmarks used for measurement are obliterated. Recently developed techniques to measure the volume of individual teeth are promising, but little comparative volumetric data are currently available.

In human populations, there is a modest sex dimorphism in tooth dimensions. A comparison of individual crown diameters within a single population usually shows that male teeth are 2–6% larger than those of females. This dimorphism is most pronounced in canine dimensions. Because there is a slight but consistent dimorphism in tooth size, workers generally present data separately for male and female tooth diameters.

In the human dentition, a high degree of dimensional intercorrelation exists, i.e., the size of one tooth is not independent of the size of all other teeth. Correlation matrices among the 32 MD and BL diameters (not including antimeres) show that interclass correlation coefficients vary from about 0.30 to 0.60 between specific tooth dimensions. Principal component analyses usually reveal that three to seven underlying components account for 50–75% of the covariance among the 32 crown dimensions. Although different populations may not exhibit comparable component loadings, some of the primary latent structures involve (1) general size; (2) anterior vs posterior teeth; (3) MD vs BL diameters; and (4) premolars vs molars. Typically, component loadings for maxillary and mandibular isomeres are similar. Although the exact significance of the latent structures underlying the major tooth size components remains unknown, they probably reflect some aspect of the genetic programs that moderate tooth crown development.

In addition to interdimensional correlations, crown diameters are also associated with other dental variables, including hypodontia, hyperodontia, and, to some extent, crown morphology. Within European populations, large-toothed (megadont) individuals are more likely to have supernumerary teeth, whereas small-toothed (microdont) individuals are more likely to have missing teeth. There is also a detectable but weak relationship between crown size and the expression of certain morphologic traits (e.g., the hypocone and Carabelli's trait of the upper molars).

B. Crown and Root Morphology

Teeth exhibit two types of morphological variation. First, there is variation in the form of recurring structures (e.g., labial curvature of the upper central incisors). However, most morphological crown and root traits that have been operationally defined take the form of presence–absence variables. That is, within a population, some individuals exhibit a particular structure while others do not. For tooth crowns, such structures may be exhibited as accessory marginal or occlusal ridges (cf. Fig. 1, incisor shoveling; Fig. 2, deflecting wrinkle), cingular derivatives (cf. Fig. 1, *tuberculum dentale*), and supernumerary cusps (cf. Fig. 2, cusps 6 and 7). Morphological root traits are most often defined in terms of variation in root number; lower molars, for example, can exhibit one, two, or three roots. For most crown and root traits manifested as presence–absence variables, presence expressions vary in degree from slight to pronounced.

Although some morphological variables exhibit significant sex differences (e.g., the canine distal accessory ridge), the majority of these traits show similar frequencies and class frequency distributions for males and females. For this reason, population frequencies are generally reported for combined data on males and females.

Crown morphology is characterized by high within-field correlations in trait expression (cf. Fig. 1; incisor shoveling on U11 and U12; *tuberculum dentale* on UI1, UI2, UC). Shoveling, which can be expressed on all four upper and four lower central incisors, should

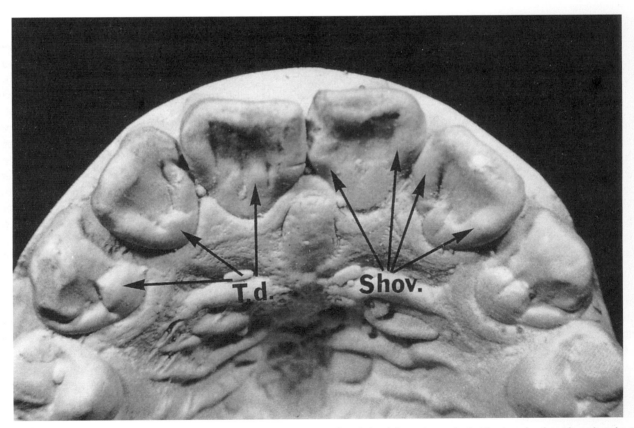

FIGURE 1   Upper anterior teeth exhibiting shoveling (Shov.; mesial and distal lingual marginal ridges) and *tuberculum dentale* (T.d.; cingular ridges and tubercles).

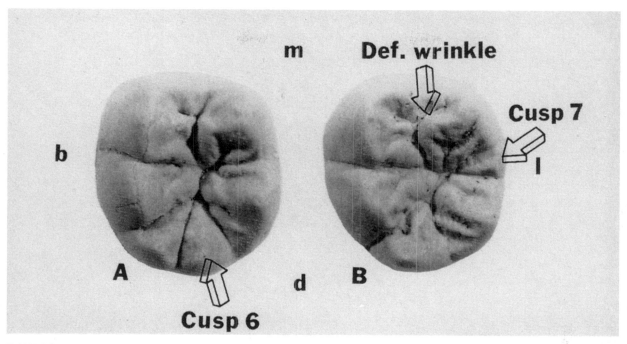

FIGURE 2 Two left lower first molars: tooth A exhibits five major cusps plus a supernumerary cusp (cusp 6); tooth B exhibits an occlusal ridge variant on the mesiolingual cusp (deflecting wrinkle) and a supernumerary cusp between the mesiolingual and distolingual cusps (cusp 7). Orientation: m, mesial; d, distal; b, buccal; l, lingual.

not be viewed as four or eight distinct traits. Rather, it is a single trait that may be expressed on all teeth of the upper and lower incisor districts. Although there are a few exceptions (e.g., Carabelli's trait and the protostylid), different morphological traits usually show little or no correlation among themselves.

C. Dental Genetics

Monozygotic (MZ) twins share identical genotypes, a fact widely exploited by geneticists interested in determining the relative genetic and environmental components of variance underlying the expression of biological traits. Although there are complicating factors, such as common prenatal and postnatal environments, traits with a strong genetic component should show similar expressions between MZ twin pairs. The dentitions of identical twins, in fact, often show remarkable parallels in crown size, the timing and sequence of tooth eruption, general crown form, and small morphological details.

Two matched MZ twin pairs, shown in Fig. 3, illustrate close similarity in tooth size, form, and morphology. However, the expression of one morphologic feature in these twin pairs can be used to illus-

trate a general point. Carabelli's trait, when present, is manifested on the mesiolingual cusp of the deciduous second molars and permanent molars. In Fig. 3, this cingular trait is expressed to a moderate degree on the permanent first molars of all four twins. However, in one set of twins (B1-B2), there is almost perfect concordance in the form and expression of this trait, while degree of expression differs between A1 and A2. As MZ twins have the same genotype, such differences in expression are environmental in origin. In general, discordance between MZ twins for any dental variable is attributable to environmental effects. The differences in tooth structure and form between MZ twins may be caused by environmental factors similar to those that produce asymmetry between the left and right sides of the dentition.

As a series of metric variables, the hereditary basis of human tooth size has been assessed through the methodology of quantitative genetics. Workers who have analyzed tooth size in twins and families estimate that the heritability of crown dimensions is relatively high (ca. 0.60–0.80). These estimates indicate a strong genetic component in the development of tooth size, but environmental factors, including maternal effects, also have some influence. Studies showing significant differences in tooth size between generations also

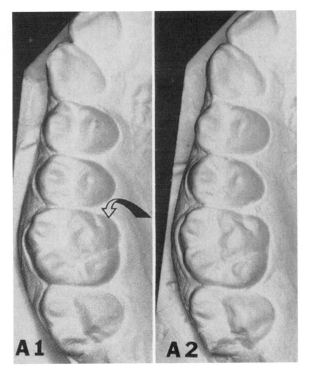
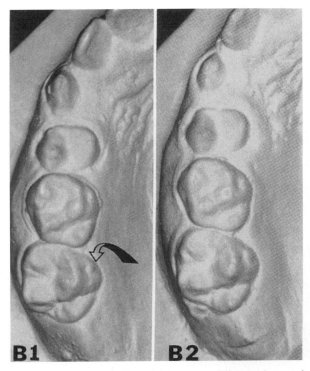

FIGURE 3   Right upper dentitions of two pairs of monozygotic twins; Carabelli's trait (indicated by arrows) differs in degree of expression between A twins (A1, A2) but is identical between B twins (B1, B2) on both the deciduous second molar and the permanent first molar.

point to an environmental component in dental development.

Considering the presence–absence nature of crown and root traits, early workers thought these variables might follow simple dominant–recessive patterns of inheritance. It now seems likely that morphologic traits, like other threshold traits, have polygenic modes of inheritance similar to continuous variables. That is, the genotypic distribution underlying trait expression is continuous (with multiple loci and/or alleles), but there is an underlying scale (absence) and a visible scale (presence) associated with this distribution, with the two scales separated by the presence of a physiological threshold. Genotypically, individuals who fall below the threshold fail to express the trait, whereas those just beyond the threshold exhibit slight expressions. With increasing distance from the threshold along the visible scale, there is an associated increase in degree of expression.

Although complex segregation analysis suggests that major gene effects may be associated with the expression of some dental morphological variables, it is still not possible to reduce this variation to gene frequencies. Currently, dental morphological data are presented in the form of class frequency distributions (i.e., the frequency of absence and each degree of trait presence) or total trait frequencies.

D. Tooth Size and Population History

Teeth from many human populations, skeletal and living, have been measured for mesiodistal and buccolingual crown diameters. These basic tooth crown dimensions are often broken down into two components for between-group odontometric comparisons. First, there is a size component that centers on the absolute dimensions of the tooth crowns. The second component, shape, is a measure of among-tooth proportionality. Between-group analyses of odontometric variation often utilize multivariate distance statistics, which are designed to take into account both absolute dimensions and among-tooth proportions. In particular, the newly developed tooth crown apportionment method may prove to be a powerful method for utilizing crown size to assess affinities among human populations. However, for a general characterization of tooth size variation, two synthetic variables can serve to illustrate dimensional and proportional variation among the major subdivisions of humankind.

To summarize human tooth size variation on a global scale, summed cross-sectional crown areas of upper and lower premolars and molars (excluding M3) were calculated for males from 75 recent skeletal and living samples representing 12 geographic regions or population groupings. The derived means and ranges for posterior crown areas (see Fig. 4) show four general divisions of humankind: (1) <700 mm², this small-toothed grouping includes two broad geographic populations, India and the Middle East, and two relatively small groups from Europe (Saami) and South Africa (San); (2) 700–750 mm², this grouping includes Asian (East Asia, Southeast Asia) and Asian-derived (Eskimo-Aleut, American Indian) populations and Europeans (including European-derived groups); (3) 750–800 mm², this range includes sub-Saharan Africans (excluding San) and Melanesians; and (4) >800 mm², with a mean posterior tooth crown area of 864.3 mm² and a range of 835.9–912.8 mm², native Australians fall well beyond any other regional population in tooth size. It should be noted that the limited range for three of the regional populations (San, Saami, Melanesia) is due primarily to the small number of samples used in the calculations. Moreover, the distantly related yet small-toothed San and Saami populations also share small body size. The question of tooth size–body size scaling is an important avenue of inquiry that should be further explored in different geographic contexts.

Although absolute tooth dimensions provide useful information on relative population relationships, odontometric comparisons are even more discriminating when tooth shape is also taken into account. To illustrate how information on shape differs from that of size, the population variation of the incisor length index (UI1 MD diameter/UI2 MD diameter × 100) has been summarized on a global scale. A low index indicates that the upper lateral incisor is broad relative to the upper central incisor, whereas a high index signifies a relatively narrow lateral incisor.

The means and ranges of the UI1/UI2 index, derived from 105 samples, are shown in Fig. 5 for the same 12 geographic areas (or populations) compared for overall crown dimensions. The worldwide sample range for this index is 113.5–139.4, with group means between 119.3 and 130.0. Of the first five groups with the lowest mean indices, four represent Asian and Asian-derived groups (five if one includes Melanesians, but their historical status is less certain). At the other extreme, Asiatic Indians, Middle Easterners, and Europeans have the highest indices. Sub-Saharan Africans, the San, and native Australians fall in the middle of the global range.

When the major geographic subdivisions of humankind are analyzed on the basis of simple genetic markers, population geneticists find that (1) Africans are the most highly differentiated from all other regional populations; (2) Asiatic Indians, Middle Easterners, and Europeans form a coherent genographic grouping; and (3) mainland Asian and Asian-derived groups in the Americas and the Pacific cluster together at low to intermediate levels of differentiation. Australians remain the most enigmatic population from a genetic standpoint, with hints of distant historical ties to both Southeast Asia and Africa.

Geographic variation in absolute tooth dimensions and the UI1/UI2 index broadly parallel the relationships indicated by measures of genetic distance. East Asians, Southeast Asians, American Indians, and the Eskimo-Aleut clustered for both tooth size and the incisor length index. Europeans clustered with Asiatic Indians and Middle Easterners for the incisor length index, although tooth size is somewhat larger in Europe. The Saami, with exceptionally small teeth, fall between European and Asian groups for the incisor

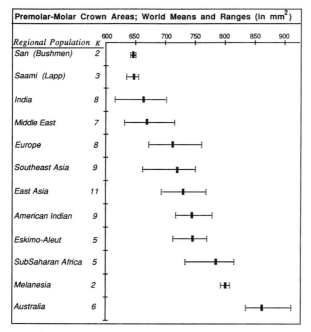

FIGURE 4 Global variation in cross-sectional crown areas of male upper and lower premolars and molars (excluding third molars). Vertical bars denote means, and horizontal lines show ranges within geographic regions. K, number of samples used in calculations.

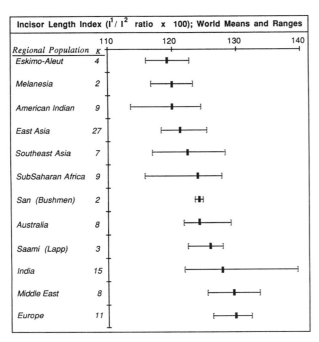

Incisor Length Index (I^1 / I^2 ratio x 100); World Means and Ranges

Regional Population	K
Eskimo-Aleut	4
Melanesia	2
American Indian	9
East Asia	27
Southeast Asia	7
SubSaharan Africa	9
San (Bushmen)	2
Australia	8
Saami (Lapp)	3
India	15
Middle East	8
Europe	11

FIGURE 5 Global variation in incisor length index. Vertical bars represent means, and horizontal lines show ranges within geographic regions. K, number of samples used in calculations.

length index as they do for many other genetic and biological variables. Australians, who reflect a world extreme for tooth size, are in the middle of the range for the incisor length index, more closely aligned with African than with European or Asian groups. Africans are, however, less distinctive odontometrically than they are genetically. Based on this summary, they show more dental similarity to Melanesia and Australia than to either Europe or Asia.

In general, when both tooth size and shape are taken into account, odontometric data provide a useful tool for assessing population relationships. Of course, this discussion focused on the distributions of sample means for some of the major subdivisions of humankind. There is still some debate among researchers on the discriminatory power of odontometric data when comparisons are made among closely related populations. Improved analytical methods (e.g., tooth crown apportionment) and refinements in measuring tooth mass (e.g., volumetrically) should ultimately increase resolution in studies of tooth size microdifferentiation.

In addition to addressing questions of population affinities, tooth size variation has also been used to assess temporal trends in recent human evolution. In many different parts of the world, temporally divided

Holocene skeletal collections show a significant decrease in tooth size over the past 12,000 years. Frequently, this decrease in tooth size parallels the origins and development of agriculture. Some workers have explained this trend in terms of natural selection (either reduced selection and concomitant reduction in tooth mass or positive selection for smaller and morphologically simpler teeth), but environmental factors may also play a role. Some domestic animals of the early Neolithic also tend to have smaller teeth than their wild ancestors. The correlates of sedentism (increased population density, endemic and epidemic diseases, between group warfare and other forms of general stress), along with the development of new food preparation techniques and the greater reliance on plant domesticates, may have produced a similar effect in the "domestication" of human populations. [*See* Evolution, Human.]

E. Dental Morphology and Population History

Although human populations exhibit a great deal of within- and between-group variation in the frequencies of various crown and root traits, this fact was not fully exploited in studies of population relationships and origins until recently. This can be attributed to (1) improved and expanded standards of observation increasing intra- and interobserver reliability in scoring morphologic trait expression; (2) the ready availability of computers and statistical packages for calculating biological distance measures and performing cluster and components analysis; and (3) the demonstration that tooth morphology is useful in assessing population relationships at levels of differentiation below the major geographic race.

The utility of dental morphology in resolving questions of population history is well illustrated by a problem that has concerned anthropologists for decades: the question of Native American origins. Many years ago, A. Hrdlička argued that American Indians were most closely related to Asian populations; one biological feature used to support this position was incisor shoveling (Fig. 6). Asians and Native Americans expressed this trait in high frequencies and often pronounced degrees, whereas Europeans and Africans were characterized by lower frequencies and slight degrees of trait expression. Genetic markers and other biological traits eventually vindicated this position; the similarity between Asians and Native Americans in shoveling was an indication of homology, not anal-

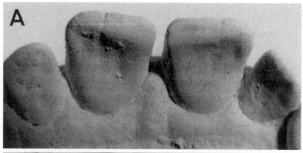

FIGURE 6 Variation in expression of upper incisor shoveling; slight shoveling shown in A and B characteristic of Europeans, while pronounced expressions in C and D limited largely to north Asians and Native Americans.

ogy, and could reasonably be attributed to common ancestry.

The analysis of population variation in another trait, three-rooted lower first molars (3RM1), led to a further refinement of knowledge on Native American origins. After making observations on samples of Aleut, Eskimo, and American Indian skeletal remains, C. G. Turner II noted that the frequencies of 3RM1 fell into two distinct clusters; Eskimos and Aleuts were

characterized by high frequencies (20–40%), whereas American Indians showed uniformly low frequencies (ca. 5%). The only exceptional group among American Indians was a small Navajo sample, representing the Na-Dene language family, with an intermediate frequency of 27%. On the basis of this single trait, Turner hypothesized that recent populations of the New World (i.e., American Indians, or Macro-Indians, south of the Canadian border; Na-Dene speakers of Alaska and Canada; Eskimo-Aleuts) represent the descendant groups of three major migrations out of Asia. Subsequent observations on 23 crown and root traits among dozens of samples representing North and South American native populations supported the initial model based on 3RM1 frequencies. One slight modification was to expand the Na-Dene grouping beyond this language family to include other Indian groups residing in the Pacific Northwest.

The peopling of the Pacific was also addressed by Turner through the use of dental morphologic data. However, to unravel the biological diversity of Oceanic populations, a data base was first established for mainland Asian populations. The examination of numerous Asian skeletal series showed a major dichotomy among populations combined traditionally under the general term of Mongoloid (or Asian). In contrast to the broad Mongoloid Dental Complex, defined primarily on the basis of Japanese dental morphology, Turner defined two complexes: Sinodont (North Asian) and Sundadont (Southeast Asian).

Several historical patterns emerged following the definitions of the Sinodont and Sundadont complexes. First, the suite of variables that characterized the Sinodont complex of north Asians also characterized all Native American populations. Although there is dental variation among New World populations, it appears that all were derived from ancestral populations in North Asia. Second, Polynesians and Micronesians exhibited crown and root trait frequencies in accord with the Sunadont pattern, so the historical inference is that these groups were ultimately derived from Southeast Asian populations. Moreover, the Ainu of Japan exhibit a Sundadont rather than Sinodont pattern, so it is possible that the range of Sundadont populations may have extended further to the north at one time, eventually to be displaced and/or replaced by north Asian groups. In this matter, comparisons of both tooth size and morphology support the position that the Ainu are the descendants of the aboriginal Jomon population of Japan, while the modern Japanese population, allied dentally with other Sinodont groups of the Asian mainland, are descended from relatively recent migrants to the Japanese archipelago.

In addition to assessing broad patterns of historical relationships, dental morphology has also been used to measure microdifferentiation among local populations within circumscribed geographic regions (e.g., Middle East, Solomon Islands, Yanomama, American Southwest). At these levels of divergence, trait frequency differences and the associated measures of biological distance are relatively small. Still, population relationships indicated by dental morphology show close correspondance to those suggested by simple genetic markers, other biological systems, and language.

As dentally based biological distance estimates among closely related populations are small when compared with distances between groups with a more remote common ancestry, it may be possible to estimate times of divergence from these values. Turner demonstrated the potential of this approach in a comparison of dental distances among 85 samples with a primary focus on Asian and Asian-derived groups. For the most part, there is close agreement between dentally derived times of divergence and independent estimates from archeology, geology, and linguistics. Further development of this method, labeled dentochronology appears warranted in light of current interest surrounding the origins and dispersal of modern *Homo sapiens*.

III. THE ENVIRONMENTAL INTERFACE: TEETH AND BEHAVIOR

Once a tooth crown is fully developed, enamel and dentine are subject only to physicochemical changes. Interest here is with alterations of the tooth crown, which indirectly reflect four classes of human behavior: (1) dietary, (2) implemental, (3) incidental cultural, and (4) intentional cultural.

A. Dietary Behavior

Since the early 1980s, isotope and trace element analyses of bone collagen and apatite have been widely used to infer general characteristics of the diet of earlier human populations. Inferring dietary constituents and their relative proportions is also a goal of those who study the natural processes that result in the cumulative loss of enamel and dentine from the occlusal surfaces of the teeth, referred to as crown wear.

Crown wear, which occurs on both the chewing surfaces of the teeth (occlusal wear) and at the contact points between adjacent teeth (approximal wear), is generated by the combined action of attrition and abrasion. The process of attrition results from direct tooth-on-tooth contact. Within- and between-group variation in attrition may reflect the nature of foodstuffs being consumed (how much chewing is required for particular foods), the amount of energy brought to bear between the upper and lower teeth by the muscles of mastication, the thickness and quality of crown enamel, and the nonmasticatory grinding of one's teeth (e.g., bruxism). Abrasion, the second component of wear, is caused by the introduction of foreign material (e.g., grit) into the foods being consumed. Abrasive elements may be inherent in foods (e.g., silicate phytoliths in plants, grit in shellfish), incidentally generated by certain food preparation techniques (e.g., a fine grit is added to the flour when seeds are processed by grinding stones), or introduced accidentally into foods by external sources (e.g., windborne silt, sand).

In all human groups, crown wear is produced by both attrition and abrasion. Regarding general subsistence levels, it appears that crown wear in earlier agricultural groups had a significant abrasive component because of the reliance on stone-ground grains and the greater amounts of windborne grit due to lands cleared for farming. Because meat is less abrasive than plant foods, crown wear in hunter-gatherer groups may have a relatively higher attrition component than in agricultural groups, due in part to a more powerful chewing musculature. However, the potential for introduced abrasive elements in the diet is present in all human populations, regardless of subsistence level. Both early hunter-gatherer and agricultural populations are characterized by rapid rates and pronounced degrees of crown wear, although the relative contributions of attrition and abrasion to this wear was probably highly variable (Fig. 7). It has been suggested that angle of crown wear, rather than absolute degree of wear, may distinguish groups practicing different subsistence economies, i.e., agriculturalists seem to exhibit a steeper angle of wear on the posterior teeth in contrast to the flatter wear plane of hunter-gatherers.

As many factors are involved in crown wear, it is difficult to generalize from comparisons among distantly related populations residing in contrasting environmental settings. The most fruitful lines of study are those that focus on closely related populations from similar environments or involve temporal com-

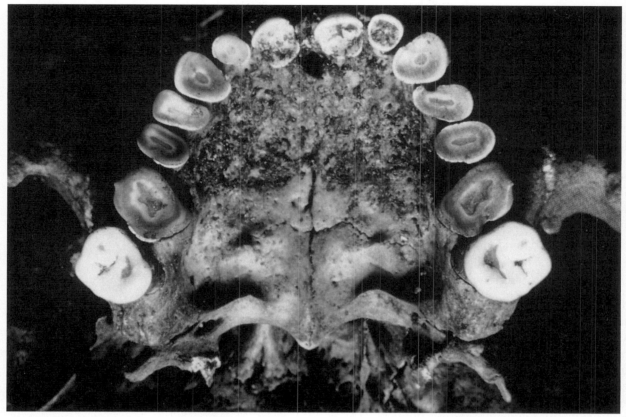

FIGURE 7 Pronounced crown wear in the upper dentition of a medieval Norwegian from St. Gregory's Church, Trondheim, Norway. The relative roles of attrition and abrasion in producing wear in this individual cannot be discerned.

parisons among groups living in a circumscribed locale. As this approach controls for some of the variables contributing to crown wear (e.g., craniofacial morphology, tooth size, plant and animal resources), measurements of degree, rate, and angle of wear can reveal significant differences and/or changes in resource utilization and diet.

In addition to crown wear, certain dental pathologies can be utilized to make inferences about dietary and other cultural behavior. Dental caries in particular is useful for addressing dietary differences and/or changes among and within groups. Although the etiology of caries involves a complex interplay among the oral microbiota, dietary elements, dental microstructure, and saliva, the study of earlier human populations indicates that diet played a critical role in the formation of carious lesions. When the term diet is used in the context of caries, specific reference should be to carbohydrate consumption, in particular the ingestion of simple sugars (e.g., sucrose, glucose, fructose), which serve as the substrate for acidogenic bacteria. Fats and proteins do not promote carious lesions. [*See* Dental Caries.]

As the emergence of food production (domestication of plants and animals) did not occur until the Holocene, an economy based on hunting and gathering wild foods characterized the first several million years of human evolution. As the constituents of a hunter-gatherer diet did not generally promote the formation of carious lesions, these groups are characterized by low caries frequencies. In fact, populations of the far north subsisting on high protein–high fat diets often had no caries at all.

Despite the positive effects associated with the rise of agriculture in the Old and New Worlds, it also had its adverse consequences. With an increased reliance on plant foods and food preparation techniques, which broke down complex carbohydrates into simpler sugars, caries rates increased. But, despite the fact that carious lesions increased in earlier agricultural populations, this increase was modest compared with the extremely high caries rates in modern populations.

Today, the widespread usage of refined sugars processed from sugar cane and sugar beets has led to a major increase in caries frequencies.

The analysis of caries rates, like crown wear, is most informative when studied in the context of circumscribed geographic populations. Temporal comparisons among British populations, for example, show how caries rates can indicate the introduction of specific dietary components during particular historical periods. Comparisons between prehistoric and modern populations of the far north also show a dramatic rise in caries rates following the introduction of refined carbohydrates into native diets that had hitherto consisted primarily of animal products (protein and fat).

B. Implemental Behavior

Modern technology provides us with a tool for almost any mechanical task. Lacking such advantages, earlier human populations with relatively simple tool kits were often forced to rely on their own biological equipment. This is particularly true for teeth, which can serve functionally as pliers, vises, strippers, gravers, etc. In this sense, teeth can literally be used as a third hand with unique properties. While the use of teeth as tools is most commonly associated with populations of the far north, particularly Eskimos, this behavior is not limited in either time or space. Humans throughout history have taken advantage of the strength, form, and ready availability of their teeth to perform a variety of functions from carding wool to holding bobby pins.

Teeth do not record all instances of tool use, but they can reflect repetitive behaviors and traumatic episodes. In many hunter-gatherer populations, for example, the anterior teeth were used for skinworking (Fig. 8). The ultimate effect of this usage was to generate a distinctive pattern of labial rounding especially evident on the anterior teeth (Fig. 9). When manipulation centered on the manufacture of thread-like items from sinew or grasses, the pattern generated on the anterior teeth would take the form of notches or grooves rather than uniform surface wear.

In addition to patterns of uniform wear generated by attrition and abrasion, enamel and dentine can also be removed through traumatic fracturing. This process, called dental chipping, occurs when teeth are subjected to forces that exceed their load-bearing capacity. Commonly, such chipping takes the form of small enamel flakes removed around the margins of the teeth, much like the pressure-chipping evident

FIGURE 8 St. Lawrence Island Eskimo female using anterior teeth to soften and crimp walrus or bearded seal hide for mukluk sole. [Used with permission from the Denver Museum of Natural History, photo archives, BA21-419K; all rights reserved.]

along the margins of stone tools. Although chipping can be caused by such things as grit accidentally introduced into food, it is frequently attributed to using the teeth as tools, especially among Eskimos.

C. Incidental Cultural Behavior

Several patterned behaviors, which do not reflect either implemental use or intentional modification, leave an imprint on teeth. One such behavior is pipe smoking. Habitual pipe smokers commonly hold a pipe on either or both sides of the mouth in the region of the left or right canines. Pipes with highly abrasive clay stems are among the worst offenders for generating deep ovate notches that extend over several teeth. A more hygienic habit, the use of probes, or toothpicks, to remove food debris lodged between the teeth, may leave grooves on the interstitial surfaces of the

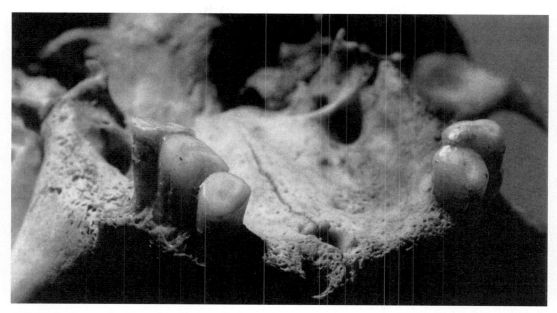

FIGURE 9 Upper dentition of prehistoric Alaskan Eskimo. Incisors lost antemortem, but canines and premolars exhibit distinct pattern of labial rounding indicative of implemental use (e.g., skin-working).

tooth crowns. Another cultural practice that leaves unintended wear is labret usage. Labrets, which are inserted through the cheeks or lips, come in different shapes and sizes and are made from a variety of raw materials, including wood, ivory, bone, and stone (Fig. 10). The wear pattern produced by labret use is very distinctive; it is manifest as a polished facet on the labial or buccal surfaces of the anterior or posterior teeth, respectively (Fig. 11). A thorough perusal of the ethnographic literature would probably reveal many other cultural practices that leave unintentional marks on the teeth.

D. Intentional Cultural Modification

Although the primary function of the oral cavity relates to the ingestion of food and water, the mouth also serves as a major social organ for many animals, including humans. Unlike other animals, however, which make do with the biological equipment they are provided with, humans can modify the appearance of their mouths in a variety of ways. In some cases, these modifications are external in nature, e.g., beards, tatoos, lip plugs, labrets, or lipstick. In others, groups directly modify the appearance of their teeth, especially the more visible incisors and canines.

Culturally prescribed dental mutilation takes several forms. For example, individuals can chip or file their incisors and canines to produce notched or

FIGURE 10 Early 19th century Eskimo male from Kotzebue Sound, Alaska, with paired composite labrets. The internal aspect of such labrets would contact the buccal surfaces of the lower anterior teeth, resulting in polished wear facets. [Sketch by Tim Sczawinski, courtesy of Ernest S. Burch, Jr.; all rights reserved.]

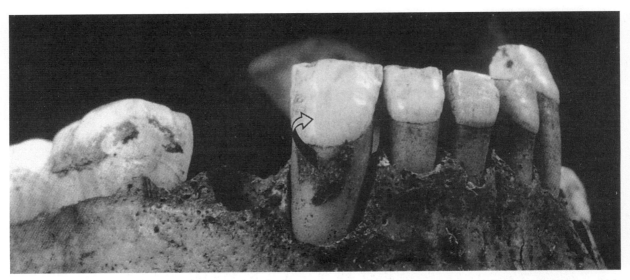

FIGURE 11 Labret facet on the lower right canine of a prehistoric Alaskan Eskimo. Arrow points to distal margin of facet, which extends from crown–root junction to occlusal surface.

pointed teeth. Incising tools can be used to engrave patterned lines on the labial surfaces of the anterior teeth, or small holes can be drilled to serve as settings for precious metals (e.g., gold) or gemstones (e.g., jade, turquoise). Precious metals can also be inlayed as bands on the labial surface or around the entire crown. In addition to modifications in form, entire teeth can be removed tramatically through the practice of ablation.

The reasons for dental mutilation may be idiosyncratic or culturally prescribed. In the first instance, individuals may choose to modify their teeth to achieve a desired cosmetic effect (e.g., to enhance beauty or fierceness). On the other hand, some populations require some form of dental mutilation as a symbol of group membership. In some cases, mutilation is involved in rites of passage, especially those rites involving a transition in status (e.g., adolescence to adulthood, unmarried to married). An interesting and not yet fully exploited anthropological usage of dental mutilation would be to assess the diffusion of specific practices from one region to another.

IV. DENTAL INDICATORS OF ENVIRONMENTAL STRESS

Anthropologists have long sought methods to estimate relative levels of environmental stress on earlier human populations. Growth arrest lines in long bones (i.e., Harris or transverse lines) provide one measure

of this phenomenon. In addition, events that disrupt crown and root formation may also mirror episodes of environmentally induced stress. Although the course of dental development is, to a large extent, stable and predictable in terms of matrix formation, calcification, and eruption, the dentition is not immune from external influences. This is apparent in size asymmetry between antimeres and micro- and macrostructural defects in the enamel and dentine.

Antimeres exhibit mirror imagery because dental development is moderated by a common genetic control system for the teeth in the two sides of the jaw. When differences are evident among antimeres in size, morphology, and number, they are environmental in origin. At one time, the varying levels of fluctuating asymmetry in tooth size among different populations were thought to provide some indication of relative levels of environmental stress. Although the logic of the argument remains sound, methodological problems pertaining to measurement error and small sample size have recently curtailed the use of dental asymmetry as a relative measure of stress.

As dental asymmetry appears to have certain limitations as a broad scale indicator of comparative stress levels, dental anthropologists have shifted their attention to the analysis of irregularities in the tooth crown that arise during amelogenesis (enamel formation) and dentinogenesis (dentine formation). The most readily observed manifestation of such growth irregularities is linear enamel hypoplasia (LEH), which takes the form of horizontal circumferential bands and/or

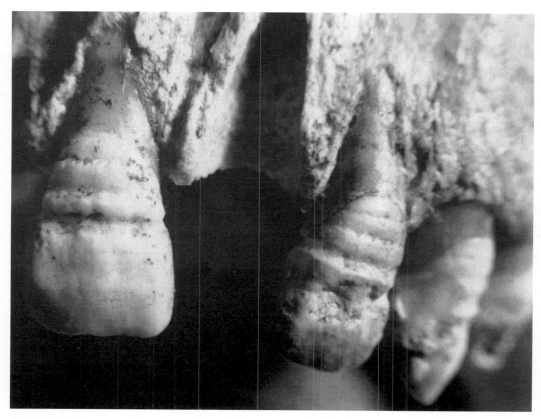

FIGURE 12 Pronounced linear enamel hypoplasia on anterior teeth of prehistoric St. Lawrence Island Eskimo. Three distinct bands can be observed on right central incisor, while four bands are visible on left lateral incisor.

pits on the tooth crown (Fig. 12). LEH can be observed on any tooth, although it is more common and pronounced on the anterior teeth. Experimental and clinical evidence shows that a wide range of phenomenon can disrupt amelogenesis and stimulate hypoplastic banding/pitting. However, the key stimulus in earlier human populations probably involved some combination of nutritional deficiency and disease morbidity.

Because matrix formation and calcification for specific teeth occur during predictable time intervals, it is possible to estimate the ages at which specific LEH bands developed and the approximate duration of a particular episode of stress. For this reason, enamel defects are used to address a number of problems in the analysis of human skeletal remains. For example, in some early agricultural groups in North America, LEH banding appears to concentrate in 2- to 4-year-old age category, possibly marking the shift from mother's milk to a weaning diet of cereal gruel deficient in certain essential amino acids. The distances between bands, in some cases, may indicate seasonal patterns of stress, a common phenomenon in earlier populations. The numbers of bands and their degree of expression may also provide insights into the differential treatment of male and female children or differences in status within a population. Further experimental and comparative work on surface irregularities and dental histological indicators of growth disturbance (e.g., Wilson bands) provide a promising avenue of research for scholars interested in discerning differential stress levels among and within earlier human populations.

BIBLIOGRAPHY

Brace, C. L., Rosenberg, K. R., and Hunt, K. D. (1987). Gradual change in human tooth size in the late Pleistocene and post-Pleistocene. *Evolution* **41**, 705–720.

Hillson, S. (1986). "Teeth." Cambridge University Press, Cambridge.

Hillson, S. (1996). "Dental Anthropology." Cambridge University Press, Cambridge.

Kelley, M. A., and Larsen, C. S. (eds.) (1991). "Recent Advances in Dental Anthropology." A. R. Liss, New York.

Kieser, J. A. (1991). "Human Adult Odontometrics." Cambridge University Press, Cambridge.

Lukacs, J. R. (1989). Dental paleopathology: Methods for reconstructing dietary patterns. *In* "Reconstruction of Life from the Skeleton" (M. Y. Iscan and K. A. R. Kennedy, eds.). A. R. Liss, New York.

Lukacs, J. R. (ed.) (1992). Culture, ecology and dental anthropology. *J. Hum. Ecol.* Special Issue, Vol. 2. Kamla-Raj Enterprises, Delhi.

Rose, J. C., Condon, K. W., and Goodman, A. H. (1985). Diet and dentition: Development disturbances. *In* "The Analysis of Prehistoric Diets" (R. I. Gilbert, Jr., and J. H. Meilke, eds.). Academic Press, New York.

Scott, G. R., and Turner, C. G., II (1988). Dental anthropology. *Ann. Rev. Anthrop.* **17**, 99–126.

Scott, G. R., and Turner, C. G., II (1997). "The Anthropology of Modern Human Teeth." Cambridge University Press, Cambridge.

Turner, C. G., II (1986). The first Americans: The dental evidence. *Nat. Geogr. Res.* **2**, 37–46.

Dental Caries

RENATA J. HENNEBERG

University of Adelaide Medical School

GLOSSARY

Dental calculus Mineralized microbial plaque covering the enamel or root surfaces firmly attached to them; supragingival and subgingival calculus types are distinguished with respect to where they form on the root relative to gingival margin

Gingivitis Inflammation of the gingiva

Periodontal disease General term denoting diseases of tissues (periodontium) that support teeth in jaws; periodontium includes gingiva, cementum, the periodontal ligament, and the alveolar bone

Plaque Layer of microorganisms of various species and strains embedded in an extracellular matrix and covering tooth surfaces; structure of plaque layer is highly variable

TOOTH DECAY, OR DENTAL CARIES (FROM LATIN *caries,* rottenness), is a pathological process of external origin and multifactorial nature in which localized destruction of the hard tissues of a tooth occurs. The disease begins with dissolution of the inorganic structures of the tooth surface (softening of the surface, decalcification, demineralization) and progresses to disintegration of the organic matrix of the tooth tissues, ultimately producing cavitation.

Several theories attempt to explain caries initiation and development. More prominent among them are the chemicoparasitic, or acidogenic, theory, the proteolytic theory, the proteolysis-chelation theory, the sucrose-chelation theory, and, most recent, the autoimmunity theory. Postulated in the 19th century, the chemicoparasitic theory is the oldest, and available evidence seems to provide it with the most support. According to this theory, microorganisms present in the oral cavity produce acids from food carbohydrates at or near the tooth surface. The acids dissolve apatite crystals in the tooth enamel.

I. SCORING TECHNIQUES USED FOR STUDYING CARIES

A. Background

Several classifications of dental caries were used in dental practice until 1961; the same situation existed among authors describing archeological dental material. This resulted in difficulties for comparative studies. In 1961, the International Dental Federation [FDI (Federation dentaire internationale)] made the first attempt to develop standards and a unitary system for compiling statistics on caries. The results of the FDI Special Commission on Oral and Dental Statistics were later adopted by the World Health Organization (WHO) and, with further improvements, were used in epidemiological studies of caries worldwide. In archeological and anthropological studies, the WHO-approved standards are also used but with some modifications due to specificity of the human prehistoric material (Table I).

From the diagnostic point of view, caries may be classified as initial caries and clinical caries. Initial caries (synonyms: microscopic carious lesion, radiographic lesion, questionable caries) is described as a white, chalky, or discolored rough spot on the enamel without visible surface breakdown. Clinical caries (synonyms: macroscopic carious lesion, untreated carious defect, cavity) is described as a visible cavity, which can be diagnosed by physical (clinical) exami-

ENCYCLOPEDIA OF HUMAN BIOLOGY, Second Edition, VOLUME 3.

TABLE I
Methods Used to Describe Occurrence of Caries in Living Human Populations and in Skeletal Samples

Living human populations	Skeletal samples

1. Caries frequency

$$\frac{\text{No. of individuals affected by caries}}{\text{total No. of individuals examined}}$$

or

$$\% = \frac{\text{No. of individuals affected by caries}}{\text{total No. of individuals examined}} \times 100$$

2. Caries intensity[a]

Living human populations	Skeletal samples
a. $\dfrac{\text{No. of carious teeth}}{\text{No. of individuals with caries}}$	a. $\dfrac{\text{No. of carious teeth}}{\text{No. of individuals with caries}}$
b. $\dfrac{\text{No. of cavities}}{\text{No. of individuals with caries}}$	b. $\dfrac{\text{No. of carious teeth}}{\text{No. of teeth examined}}$
c. $\dfrac{\text{No. of carious surfaces}}{\text{No. of individuals with caries}}$	
d. $\dfrac{\text{No. of carious surfaces}}{\text{No. of carious teeth}}$	
e. $\dfrac{\text{No. of cavities}}{\text{No. of carious surfaces}}$	

3. DMF index (DMFT or DMFS)[b] recommended by WHO, average number of decayed (D), missing (M) (due to caries), and filled (F) permanent teeth (T) or tooth surfaces (S) per person.

3. DM index (dmt or DMS)[b] Average number of decayed (D) and missing (M) (due to caries) permanent teeth (T) or tooth surfaces (S) per individual.

4. dmf index (dmft or dmfs) Recommended by WHO, sum of decayed (d), missing (m), and filled (f) primary (deciduous) teeth, which should be present in the mouth at the time of examination, or tooth surfaces (s) per person.

4. dm index (dmt or dms) Sum of decayed (d) and missing (m) primary (dciduous) teeth (t) or surfaces (s) per person.

5. def index Sum of primary teeth decayed (d), beyond repair (e), and filled (f).

5. —

6. Root Caries Index (RCI)

$$\text{a.} \quad \frac{\text{No. of root caries lesions} \times 100}{\text{No. of teeth or surfaces} \times \text{No. of individuals with gingival recession}}$$

$$\text{b.} \quad \frac{\text{No. of root caries lesions} \times 100}{\text{No. of carious teeth or surfaces}}$$

[a]Describes severity of caries experience.

[b]DMF or DM indices can be made age specific, sex specific, or population specific. WHO recommends epidemiologic surveys to be made when children and youths are 5, 12, 15, and 18-years old. DM or dm indices for skeletal material are very rough estimates due to errors resulting from incompleteness of archeological samples, thus their numerical values should be treated with caution.

nation. The main interest of dental clinicians, epidemiologists, and paleopathologists is concentrated on clinical caries. According to the WHO recommendations, the examination for dental caries should be conducted with a plane mouth mirror and a sharp, sickle-shaped dental explorer in natural light. Carious lesions found during such macroscopic examination may be classified as follows: (1) according to anatomical location on a tooth (Fig. 1), (2) according to speed with which the lesion progresses, and (3) according to the tissue attacked (Fig. 2).

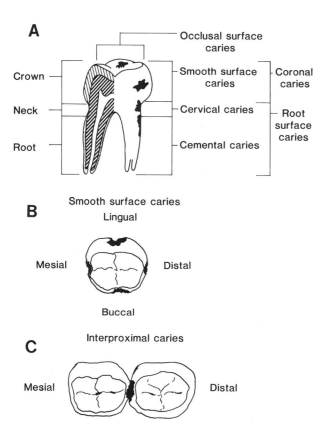

FIGURE I Classification of carious lesions according to anatomical location on a tooth. Coronal caries is a lesion localized on the crown of a tooth. Coronal caries may occur on the occlusal surface of the tooth (A), where it starts in pits or fissures on the surface, or on the smooth surface of the crown, where it can be localized on the buccal (or labial), lingual, mesial, or distal surface (B); a special case of smooth-surface caries is an interproximal lesion (C), in which adjacent surfaces of two neighboring teeth (proximal surfaces) are involved. Root surface caries (A), in which cementum covering the root of a tooth is affected, is a lesion that may be localized on the cemento enamel junction of the tooth (cervical caries, caries of the tooth neck) or lower down on the root (cemental caries). A separate category of carious lesions is "circular caries," which occurs in hypoplastic enamel on the buccal surface of teeth.

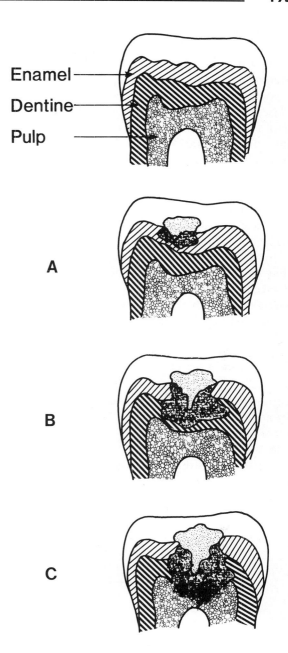

FIGURE 2 Classification of carious lesions according to the tissue affected. (A) Enamel caries (first-degree caries, caries prima, caries superficialis). (B) Caries of the dentine (second-degree caries, caries media). (C) Caries of the pulp (third-degree caries, caries profunda). This classification also describes the depth of the lesion's penetration in the coronal part of the tooth. Cemental caries constitutes a separate category of the classification, but categories of root caries have not been fully established yet. The term fourth-degree caries is sometimes used to denote complete tooth destruction by the carious process (missing tooth). Another classification is that based on the progression of the disease: nonpenetrating caries, in which only enamel and dentine are affected, and penetrating caries, in which the pulp chamber is opened. This classification is most often used by paleopathologists rather than dental practitioners.

B. Scoring of Caries in Living People

The total amount of dental destruction by caries, including tooth loss due to decay, is an individual's or a population's life caries experience. The rate at which caries disease occurs in a population is described as caries prevalence (i.e., a fraction of a population having any experience of caries). The caries prevalence is usually calculated separately for various age groups (age-specific prevalence). Caries incidence is the rate per unit of time (usually 1 year) at which new carious lesions develop in an individual or a population. Occurrence of caries in a population is expressed by means of various indices (Table I).

In epidemiologic studies, assessment of dental caries may be desired for the entire dentition, in which case all teeth are examined, or by using the half-mouth method, in which observations are made and recorded for one-half of the upper dental arch and one-half of the lower arch of the opposite side (e.g., upper-left and lower-right quadrants).

The latter method gives a reasonably accurate assessment of dental caries in living populations because caries is considered largely a bilateral phenomenon. In archeological material that is often fragmentary, recording of caries on all teeth available for examination is logical. Observations may be recorded in various ways. Recording is usually done using a simple code. The WHO recommended Basic Oral Health Assessment Form and the Treatment Assessment Form, which contain sections on caries, are widely used in both clinical practice and epidemiologic surveys. The individual's dentition is divided into quadrants denoted numerically (1, upper right; 2, upper left; 3, lower left; 4, lower right) and teeth are numbered 1–8 from the central incisor to the third molar in each quadrant. On a chart, each tooth is given a box into which information is written with a code (Table II). The chart also shows how many surfaces were restored on the tooth or the reason for eventual extraction.

C. Scoring of Caries in Archeological Samples

An example of a scoring chart used for human skeletal remains is presented in Table III.

D. Radiographic Techniques in Caries Identification

Modern radiographic techniques are a very useful tool for the identification of carious lesions in dental practice and in epidemiologic studies. Caries appears on a radiograph as a radiolucent (dark) area surrounded by more radiopaque (lighter) tissues of a tooth. Depending on its anatomical location, caries may be visible on a radiograph more or less readily. Some carious lesions are visible at such an early stage that

TABLE II
Codes Used for Tooth Assessment

	Dentition	
	Primary	Permanent
Sound tooth (no treated or untreated caries)	A	0
Decayed tooth (macroscopically visible caries)	B	1
Filled tooth with no active caries	C	2
Filled tooth with primary decay	D	3
Filled tooth with secondary decay	E	4
Deciduous teeth missing due to caries	M	—
Permanent teeth missing due to caries (only in individuals <30 years old)	—	5
Permanent teeth missing for any reason other than caries (<30 years old only)	—	6
Permanent teeth missing for any reason (>30 years old)	—	7
Unerupted tooth	—	8
Excluded tooth	X	9

TABLE III
Scoring Chart for Caries in Skeletal Material[a]

Right side		Tooth[b]							
		8	7	6	5	4	3	2	1
Maxilla (upper)	Degree of penetration	u		2	3	4		al	pl
	Surface attacked			o	o				
Mandible (lower)	Degree of penetration		1, 2	2		1		1	1
	Surface attacked		o, m	d		l		n, b	n, d
Degree of penetration (tissue attacked)		1, enamel caries; 2, dentinal caries; 3, pulpal caries; 4, crown completely decayed; al, antemortem tooth loss; pl, postmortem tooth loss; u, impacted (unerupted) tooth (if visible)							
Surface attacked		o, occlusal; b, buccal; l, lingual; d, distal; m, mesial (d + m = i, interproximal); n, neck (cervical caries, root surface caries)							

[a]Only one-half of the chart (for the right side) is shown.

[b]Examples given on the chart: Right upper third molar (8) is impacted. Right upper first molar (6) shows occlusal surface dentinal caries. Right upper second premolar (5) shows occlusal surface caries that penetrate into the pulp. Right upper first premolar (4) displays completely decayed crown due to caries. Right upper lateral incisor (2) was lost during life, most probably due to caries. Right upper first incisor (1) was lost after death, most probably during excavations. On the lower second molar (7) there are two cavities: one in the enamel only (first degree) on the occlusal surface and the other of the second degree (dentinal caries), interproximal with the cavity on the distal surface of the first molar (6). Right lower first premolar (4) shows enamel caries on the lingual surface. Both right lower incisors (2, 1) show enamel caries located on buccal (2) and distal (1) surfaces of the neck of the tooth.

they are undiagnosable by physical examination (with a dental probe).

Intraoral radiographs have proved to be extremely helpful as an aid in clinical examination. For instance, detection of interproximal caries with the use of X-rays improves more than twofold over detection by the mirror-probe method. Occlusal caries can be observed radiographically only after the disease has progressed to the dentinoenamel junction. At this stage the occlusal lesion is already clearly identifiable macroscopically. Detection of buccal, lingual, and cemental lesions can be done radiographically, but differentiation of these types with X-ray pictures is unnecessary because such lesions are clearly visible clinically. Panoramic radiographs (panorex) are an excellent survey tool. They are taken by various types of apparatus based on the principle of tomography (scanning). Orthopantomography produces one continuous image from the right to the left temporomandibular joint (Fig. 3). As a result of technological developments and education in radiology, use of radiographic techniques for caries diagnosis is increasing, especially in developed countries such as Sweden, Japan, Finland, and Denmark; it is also popular in the United States.

E. Caries in Forensic Medicine

Because of the progressive nature of the disease, caries as such has little value in the process of forensic identification. The pattern of treated carious lesions, however, and materials and methods used by various dental practitioners may play an important role in the elimination of false identifications (negative identification). The reliability of results from such examinations increases substantially when radiographs of the dentition of the person in question exist in dentist's files.

II. CARIES AS A MULTIFACTORIAL DISEASE

A. Antiquity of Caries

Dental caries is a disease of the human species and is one of the most common microbe-caused diseases in

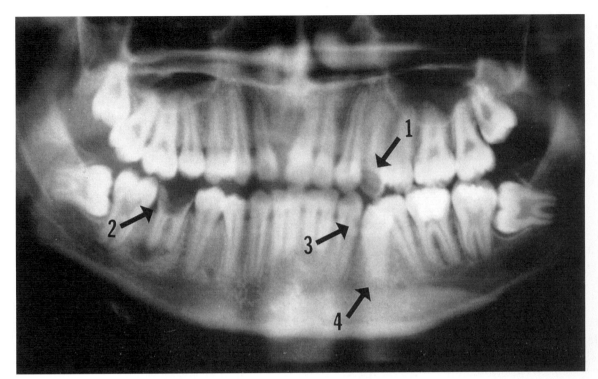

FIGURE 3 Orthopantomogram of jaws indicating presence of caries. (1) Caries of occlusal and buccal (labial) surface of the upper left first (third) premolar. (2) Penetrating caries of the right lower first molar. The crown is almost entirely destroyed by the disease. Radiolucency around the tip of the root of the tooth is due to the inflammation of surrounding tissues (periodontitis), which is a sequel to pulp involvement following penetration of the pulp chamber. (3) Interproximal caries on the cementoenamel junction of the distal surface of the left lower canine (cervical caries). (4) Rotated left lower second (fourth) premolar with periapical root resorption probably following trauma. [Courtesy of R. Hendricks, University of Western Cape.]

modern man. Although some carious lesions have been described in animals, they are uncommon and differ from those found in humans. Caries has been found in fossil reptiles, camel, mastodont, and European cave bears. Caries occurs in extant wild animals, especially among aged individuals of primates, horses, rabbits, rats, dogs, sheep, and cats. It can be induced in laboratory animals such as hamsters and rats.

Dental caries accompanied hominids during their evolution. The earliest examples of caries are found in South African australopithecines. The evidence of the first authentic dental caries of human type is found in the dentition of the skull from Kabwe in Zambia ("Rhodesian Man," Broken Hill). Of the 13 teeth preserved in this specimen, 11 showed carious lesions. This is an unusually high caries intensity for this period (Middle Paleolithic; Table IV). Caries prevalence was very low in hunters and gatherers. It increased substantially during the first major economic transi-

tion of neolithization, during which time dietary conversion occurred due to the introduction of agriculture as a major food source, and further increased dramatically with the second dietary conversion when refined food, rich in carbohydrates, became the basic source of nutrients. Caries is an undoubted disease of civilization.

B. Heritability and Prevalence of Caries

Evidence that genetic factors influence the prevalence of caries is derived from (1) animal experiments, (2) human family and twin studies, and (3) studies of metabolic diseases and defects, which themselves are heritable. Selective breeding of rats has demonstrated that genetically caries-resistant and caries-susceptible strains of animals can be developed. Under the same conditions of a carbohydrate diet, the caries-resistant rats produced caries at a rate several times slower

TABLE IV
Intensity and Prevalence of Caries in Various Periods of Prehistory and History

Period	Mode of subsistence	% carious teeth	% individuals with caries
Middle Paleolithic (100,000–40,000 BC)	Hunting and gathering	0–?[a]	0–?
Upper Paleolithic (35,000–10,000 BC)	Mostly large game hunting and gathering	1.0–?	—
Mesolithic (10,000–4000 BC)	Small game hunting, fishing, and gathering	0.4–7.7	?–41.6
Neolithic (8000–2000 BC)	Agriculture	1.4–12.0	0–60.0
Classic antiquity (1000 BC–500 AD)	Agriculture	5.0–14.6	?–67.0
Middle ages (500–1500 AD)	Agriculture	2.8–40.0	24.8–65.5
Modern times[b]	Industrialized agriculture	5.9–45.0	36.3–100.0

[a]?, no data.
[b]Cane sugar became commercially available at the beginning of this period.

than rats from the caries-susceptible strain. In the investigation of a large number of families, caries frequency among siblings of caries-susceptible school children was approximately double the frequency among siblings of caries-resistant individuals. Children of highly caries-resistant parents had lower DMF scores than children of caries-susceptible parents, despite the same exposure to water containing fluoride.

Twin studies showed greater variation of caries experience among dizygotic (DZ) twins than among monozygotic (MZ; i.e., identical) ones. Recent studies of twins reared apart confirm earlier findings that MZ twins show a strong resemblance of caries status despite different dietary patterns, oral hygiene practices, professional dental care, etc. In DZ twins, resemblance in caries status is weaker.

Individuals who could not taste the phenylthiocarbamide had 25% lower def scores than those who could. It is believed that individuals with Down's syndrome, whether institutionalized or living at home, have lower than average caries experience. The complexity of caries indicates that its hereditary determination is polygenic.

Genetic factors, however obviously present, seem to be less important for the development of caries than environmental factors such as food, fluoride content of water, or oral hygiene. Further detailed studies of the genetic–environmental interaction in the development of caries are needed.

C. Factors Influencing Prevalence of Caries (Fig. 4)

Etiologic factors responsible for dental caries can be divided into two major groups: essential and modifying. Factors essential for the development of caries are host susceptibility (i.e., natural susceptibility of the tooth), plaque, and food taken orally. Modifying factors are composition of the saliva, soil and water mineral contents, systemic diseases, sex and age of individuals, tooth anatomy, and also variations in all three essential factors, such as composition of the plaque, variations in susceptibility or resistance, variations in food composition, and the frequency of food intake. These modifying factors may affect distribution of caries (site on a tooth, type of tooth), speed of the lesion's progression, and remineralization of the lesion.

I. Host Susceptibility

Caries-resistant teeth have less complicated morphology and better enamel integrity. The simple evidence of this fact may be that molars and premolars, having a more complex structure and being generally bigger, are more prone to caries than anterior teeth. The first permanent molar is known to be the most carious tooth in the entire human dentition. The pattern of caries distribution by tooth type is usually as follows (in decreasing order of intensity): molars, premolars,

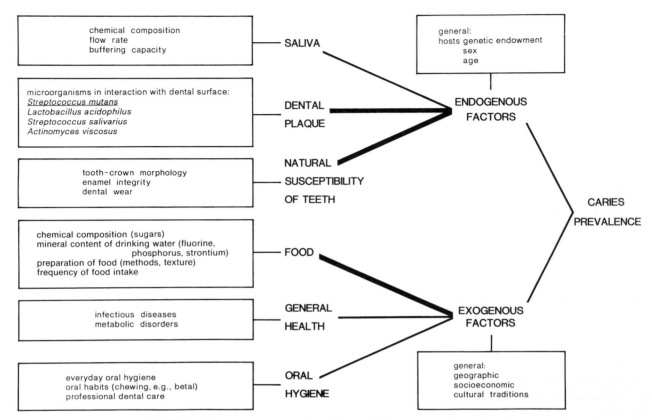

FIGURE 4   Schematic representation of factors influencing prevalence of caries.

incisors, and canines. This pattern has not changed much from ancient to modern populations. Teeth that do not erupt and are impacted in the jaws do not develop caries.

2. Plaque

Plaque formation begins with coating of the tooth surface with saliva. Salivary film is colonized by species of *Streptococcus, Actinomyces, Staphylococcus, Veillonella, Lactobacillus,* and a few others. There is a difference between the bacterial composition of the plaque from caries-active and caries-resistant individuals. Proof that microorganisms embedded in plaque are essential for the occurrence of carious lesions comes from animal experiments. Susceptible rodents do not develop caries on a highly cariogenic diet when kept under sterile, germ-free conditions. The same animals develop caries after infection with bacteria.

People who develop caries have a higher level of *Streptococcus mutans* and Lactobacilli in their plaque. Distribution of these organisms inside the mouth correlates to some extent with the distribution of carious lesions. *S. mutans* is considered to be the most active etiologic agent in caries disease. Acids produced by organisms living in plaque can decrease pH at the tooth surface from neutral to <5.0, which seems to be a critical point for initiation of the carious process.

3. Food

Food has the greatest impact on the initiation and development of caries. The concept of food carbohydrates as a major cariogenic factor has been proven in studies of human and experimental animal diets.

The cariogenicity of various carbohydrates depends on the speed of their disintegration in the process of fermentation and on their ability to reduce the pH of plaque and saliva by production of acids. Simple sugars, mono- and disaccharides, ferment more rapidly than complex polysaccharides. Sucrose is the greatest cariogenic agent among simple sugars; it is closely followed by glucose. Fats and proteins are not metabolized by cariogenic bacteria; they do create rather high pH conditions and their cariostatic properties are well documented. Cariogenicity of the diet is determined by (1) content of simple sugars, (2) texture of foodstuffs, and (3) frequency of food intake.

Soft, sticky substances are more cariogenic than foods of rough, hard texture. Hard and rough food requires more chewing and stimulates flow of the saliva, thus providing better cleansing of teeth. Sticky foods remain on the tooth surface longer and constitute a continuous source of substrates for fermentation. It has been proven that snacking increases the incidence of caries. Food intake seems to be the most flexible among factors influencing the multifactorial phenomenon of carious disease.

4. Saliva

Saliva contains proteins (enzymes: amylase, lactoperoxidase, lysozyme, lactoferrin, cationic glycoprotein, esterases, ribonucleases, and secretory IgA) and minerals (with ions of calcium, phosphate, magnesium, bicarbonate, chloride, sodium, potassium). The proportion of contents varies and depends on the type of salivary gland (parotid, submandibular, sublingual, minor glands), on intensity of stimulation, time of the day, diet, age, sex, a number of diseases, and pharmacological agents. Salivary pH is extremely variable, ranging from 5.8 to 7.7 under normal circumstances. By changing its contents and the rate of flow, saliva can affect caries by way of mechanical cleansing of the tooth surface; reducing enamel solubility by providing calcium, phosphate, and fluoride; buffering and neutralizing acids produced by cariogenic organisms and foods; and by antibacterial activity. [*See Salivary Glands and Saliva.*]

5. Mineral Content of Drinking Water

a. Fluoride

People who live in areas where drinking water contains a high concentration of fluoride (F^-) develop fewer carious lesions than people whose water does not contain fluoride salts. The inhibitory effect of fluoride depends on the time of exposure to the mineral and its concentration. The best results are found in populations exposed to F^- from birth and where F^- concentration in drinking water is 1.0–1.5 ppm (parts per million).

Higher fluoride concentration in water and food results in increased concentration of the mineral in the tooth enamel and dentine. The surface of the enamel and outer layers of dentine contain more fluoride than inner layers. It is suggested that F^- incorporated chemically in the hard tissues of the tooth as fluorohydroxyapatite or fluoroapatite either during tooth formation or after its eruption reduces the solubility of the enamel in acids. A constant supply of F^- in water, food, and dental care treatment (topical application of fluoride, fluoride tablets, toothpastes, mouthwashes, etc.) provides a reservoir of F^- when demineralization of the enamel in low pH occurs. F^- can substitute for carbonate in the apatite structure of dental hard tissues and, thus, increases their resistance to dissolution of minerals. Recent studies support the view that fluoride is particularly effective in remineralization of the tooth. It inhibits the progression of caries and stimulates the flow of minerals back into the enamel structure by binding them in chemical forms more resistant to dissolution. F^- has little effect on initiation of the carious lesion.

b. Strontium

Another trace element found in soil and water of which cariostatic properties were proved is strontium (Sr^{+2}). Similar to F^-, strontium is incorporated in the dental hard tissues. The element can substitute for calcium in the structure of the apatite in the tooth. A positive correlation has been found between concentrations of strontium in drinking water and concentrations of the mineral in the enamel. Unlike F^-, which has a greater posteruptive effect on the resistance to caries, strontium is most effective against caries when the element is supplied in high concentrations during tooth development. No posteruptive effect of strontium on caries prevalence has been observed.

Other trace elements such as molybdenum, manganese, vanadium, lithium, and boron also show anticariogenic properties. On the other hand, the presence of selenium, lead, cadmium, and silicon in drinking water increases the incidence of caries.

6. Systemic Diseases

Certain systemic diseases are accompanied by an increase in dental caries; in others caries activity is lower than average. High dental caries experience in neurologic and renal diseases may be explained mainly by difficulties in carrying out dental treatment. In Sjörgen's disease with xerostomia (dry mouth symptom due to reduced or absent flow of saliva), caries develops rapidly unless special precautions are taken. Decrease of dental caries incidence was observed in hypothyroidism related directly to thyroxine deficiency and indirectly to a deficient secretion of thyroid-stimulating hormone. Diabetics develop extensive caries despite diets containing very small amounts of carbohydrates. Individuals with congenital syphilis usually have increased caries levels. Their hypoplastic teeth can be more susceptible to caries because of thinner enamel and an increased number of pits and grooves on the crown.

7. Sexual Dimorphism in Caries Experience

Sex differences in the prevalence of caries were documented in almost every study of permanent teeth. Usually females exhibit more caries than males of the same age. No difference between sexes was found with respect to caries in deciduous teeth. The most popular theory emphasizing the earlier tooth eruption in females and their relatively longer exposure to cariogenic environment does not fully explain the difference. Although the difference may be partly explained by diet differences between males and females (in some historical populations or in pregnancy) or by gender-specific chewing habits, the differential susceptibility to caries in males and females remains unexplained.

8. Age and Caries

In prehistoric and historic times, caries seemed to be associated with old age rather than with childhood. Caries prevalence was observed to increase in adults from the age of 20 years upwards. In modern humans, caries is typically a childhood disease. It develops rapidly up to 18 years of age and then stabilizes. Almost 100% of adults of Western economies experience caries at least once in their lives (life caries experience). Distribution of various types of carious lesions changes with the age of an individual. Older people develop more root surface caries because of gingival recession causing root exposure.

D. Caries and Socioeconomic Status

Many aspects of the socioeconomic status (SES) such as income per capita, education, age of parents, dietary customs, oral hygiene habits, availability of dental services, dental attendance patterns, and others were studied in relation to the incidence and prevalence of caries. Because of the complexity of the socioeconomic status as a study variable, results of various studies are inconsistent and vary with the country in which they were conducted.

In developed countries, low SES groups have DMF (or dmf) scores higher than high SES groups. This is generally due to poor availability of dental services and poor oral hygiene habits, while sugar consumption is very high. In contrast, in underdeveloped countries, higher DMF scores are found among higher SES people. These are associated with increased sugar consumption as compared with low SES groups, while other SES factors play a minor role.

Socioeconomic status, because of its complexity, has only an indirect influence on dental caries prevalence. Thus the incidence of the disease cannot be treated as a simple and universal indicator of living conditions.

III. Conclusion

The reasons for which dental practitioners and anthropologists study caries are different. The practitioners are interested in combating the disease, whereas anthropologists use it as an indicator of varying living conditions.

Since the early 1970s, the prevalence of dental caries in developed countries declined rapidly by about 50%. This was due to improvements in prevention and dental care and to changing lifestyles of people. A further decline in DMF scores by another 60% is expected to occur by the year 2050. Developing countries will—at the same time—face the problem of increases in caries due to lack of efficient preventive action and increased sugar consumption. The changing patterns of caries throughout the world and a constant need for better preventive methods will remain the major subjects of investigation in the next decades.

Anthropologists have found the phenomenon of caries a useful tool in studying the living conditions of prehistoric and historic populations. Observations of caries experience and frequency can shed light on major changes in lifestyles such as subsistence strategies, techniques of food preparation, and dietary customs. Reconstruction of prehistoric diets helps explain patterns of human evolution and microevolution.

BIBLIOGRAPHY

Bowen, W. H., and Tabak, L. A. (eds.) (1993). "Cariology for the Nineties." University of Rochester Press, Rochester, NY.

Brothwell, D. R. (1963). The macroscopic dental pathology of some earlier human populations. *In* "Dental Anthropology" (D. R. Brothwell, ed.). Pergamon Press, London.

Driessens, F. C. M., and Woltgens, J. H. M. (eds.) (1986). "Tooth Development and Caries." CRC Press, Boca Raton, FL.

Hefferren, J. J., and Koehler, H. M. (eds.) (1981). "Foods, Nutrition and Dental Health," Vol. 1. Pathotox Publishers, Park Forest South, IL.

Keene, H. J. (1981). History of dental caries in human populations: The first million years. *In* "Animal Models in Cariology," (J. M. Tanzer, ed.). Information Retrieval, Washington, D.C.

Larsen, C. S., Shavit, R., and Griffin, M. C. (1991). Dental caries evidence for dietary change: An archaeological context. *In* "Advances in Dental Anthropology" (M. A. Kelley and C. S. Larsen, eds.), pp. 179–202. Wiley-Liss, New York.

Nikiforuk, G. (1985). "Understanding Dental Caries," Vols. 1 and 2. Karger, Basel.

Ortner, D. J., and Putschar, W. G. J. (1985). "Identification of Pathological Conditions in Human Skeletal Remains." Smithsonian Institute Press, Washington, D.C. [see especially chapter "Lesions of Jaws and Teeth"].

Powell, M. L. (1985). The analysis of dental wear and caries for dietary reconstruction. *In* "The Analysis of Prehistoric Diets"

(R. I. Gilbert, Jr. and J. H. Mielke, eds.). Academic Press, Orlando, FL.

Turner, C. G., II (1979). Dental anthropological indications of agriculture among the Jomon people of central Japan. *Am. J. Phys. Anthropol.* 51, 619.

World Health Organization (1977). "Oral Health Surveys. Basic Methods." World Health Organization. Geneva.

Wuehermann, A. H., and Manson-Hing, L. R. (1981). "Dental Radiology," 5th Ed. C. V. Mosby Co., St. Louis [see especially p. 305].

Dental Embryology and Histology

GEORGE W. BERNARD

University of California, Los Angeles

Primary epithelial band Thickened horseshoe-shaped band of oral epithelium seen at the future site of the mandible and the maxilla

Successional lamina Lingual extensions from the dental lamina, which is the site where 20 permanent incisor, canine, and bicuspid teeth will develop

Tooth bud/germ Structure from which the tooth and adjacent structures develop; includes the dental organ, dental papilla, and dental follicle (sac)

GLOSSARY

Alveolar bone Supporting bone immediately surrounding the tooth

Amelogenesis Development of tooth enamel, which is the outer cover of the tooth crown

Cementogenesis Development of cementum, which is the outer cover of the tooth root

Dental (enamel) organ Crown-shaped epithelial masses growing from the dental, successional, and molar laminae from which the enamel develops

Dental lamina Localized epithelial ingrowths from the epithelial band where each of the 20 deciduous teeth will develop

Dental pulp Innermost soft tissue of the tooth in which the lymphatic, nervous, and vascular structures are found; develops from the dental papilla

Dentinogenesis Dentin formation, which is the inner hard structure of the tooth

Gingiva Outer coat of epithelium and adjacent connective tissue (mucosa) that surrounds the tooth in the mouth

Molar lamina Rear extensions of the dental lamina from which the 12 molar teeth will develop, three for each quadrant of the dental arches

Periodontal ligament Oriented connective tissue fibers (collagen) that connect the cementum of the tooth to the alveolar bone

FROM THE 27TH TO THE 37TH DAY OF INTERuterine life two ectodermal epithelial bands are formed: one in the presumptive upper jaw (the maxillary process) and the other in the presumptive lower jaw (the mandibular process). From this band, a downgrowth of epithelial cells into the deeper layer of mesenchyme becomes the dental lamina from which 20 deciduous and 32 permanent tooth germs will develop. They become the baby or deciduous and the permanent teeth. Each tooth, whether deciduous or permanent, is made up of four different parts. Three are mineralized: the enamel, the dentin, and the cementum. The fourth is the soft tissue dental pulp, which occupies the central portion of the tooth.

Concomitant with the development of the teeth is the synchronous growth and development of adjacent tissues, which become an intrinsic part of the functioning tooth. The adjacent tissues are the periodontal ligament, the gingiva, and the alveolar bone. When the crowns and roots of the roots are mineralized, each tooth emerges into the mouth by breaking through the mucosa, ultimately meeting and articulating with its corresponding tooth in the opposite jaw.

Each tooth, whether deciduous or permanent, essentially follows the same pattern of development. Therefore the accompanying description of the development of *a* tooth is useful as a description of *all* teeth.

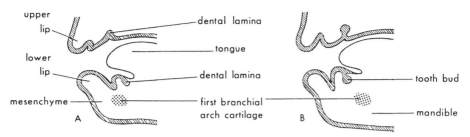

FIGURE I Sagittal sections through the developing jaws, illustrating early development of the teeth. (A) Early in the sixth week, showing the dental lamina. (B) Later in the sixth week, showing tooth buds arising from the dental laminae.

I. DEVELOPMENT OF THE PRIMITIVE MOUTH

The earliest indication that a presumptive mouth is forming is the appearance of the stomatodeum, which is formed by the anterior end of the embryo folding downward, the head fold. This creates a space bounded by the neural plate above and the cardiogenic area below. The bucco-pharyngeal membrane lies between the stomatodeum and the foregut. On either side (i.e., laterally), the stomatodeum becomes bounded by the first branchial arch, in which will ultimately develop the bony maxilla and the mandible. The boundaries of the stomatodeum consist of ectoderm on the outside and mesenchyme inside derived from the inner areas of the first bronchial arch and the head fold. Thus, the primitive mouth is a cavity lined by ectodermal epithelium, supported by mesenchymal connective tissue, and delineated internally by a temporary buccopharyngeal membrane. After the disappearance of this membrane, the mouth and the foregut are connected, presaging a continuous gastrointestinal tract.

A. Primary Epithelial Band

The teeth begin when two bands of epithelial tissue thicken in the upper and lower arches of the stomatodeum at 37 days postfertilization. This is the primary epithelial band. From it, two subbands will form, the vestibular lamina, which will become the cleft or sulcus between the teeth on one side and the cheek and lips on the other, and the dental lamina, from which all the teeth will develop (Fig. 1). At 20 distinct areas, there will be downgrowth of the epithelium at the sites where the 20 deciduous (baby) teeth will grow. Surrounding these sites is increased mitosis of the ecto-mesenchyme, primitive connective tissue that is derived from neural crest ectoderm. It is the expansion, growth, and refinement of these two interacting tissues (epithelium and ecto-mesenchyme) that will result in the formation of 20 deciduous teeth followed by 32 permanent teeth.

II. DEVELOPMENT OF TEETH

A. Dental Lamina

There are 20 areas of growth of epithelium into the ectomesenchyme along the epithelial band, 10 in each presumptive jaw. These become string-like, remaining attached at one end to the oral ectoderm. On the end farthest away from the mouth, the epithelial cells of the dental lamina begin to divide at a greater rate, heralding the development of the deciduous tooth germ or bud (Fig. 2).

B. Successional Lamina

On the lingual side of each of the 20 outgrowths of the dental lamina, an outcropping of epithelial cells indicates where permanent teeth will begin to develop. All the deciduous incisors and canine teeth will be replaced by permanent teeth of the same name from successional laminae, branches of the deciduous dental laminae. The successional laminae of the eight deciduous molars give rise to eight permanent bicuspids, two in each quadrant of the dental arches.

C. Molar Lamina

Posterior to the last or second deciduous molar on each side of each jaw, there is a backward extension of the dental lamina. These epithelial outgrowths on each side are the beginning sites of permanent molar tooth bud formation.

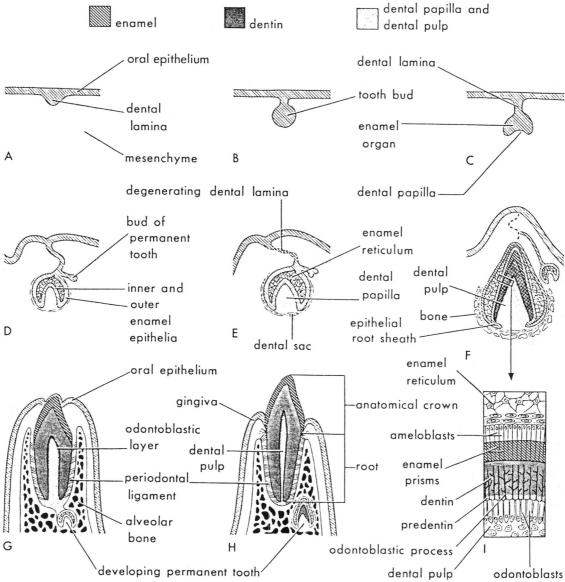

FIGURE 2 Sagittal sections showing successive stages in the development and eruption of an incisor tooth. (A) Six weeks, showing the dental lamina. (B) Seven weeks, showing the tooth bud developing from the dental lamina. (C) Eight weeks, showing the cap stage of teeth development. (D) Ten weeks, showing the early bell stage of the deciduous tooth and the bud stage of the developing permanent tooth. (E) Fourteen weeks, showing the advanced bell stage of the enamel organ. Note that the connection (dental lamina) of the tooth to the oral epithelium is degenerating. (F) The enamel and dentin layers at 28 weeks. (G) Six months postnatal, showing early tooth eruption. (H) A fully erupted deciduous incisor tooth at 18 months postnatal. The permanent incisor tooth now has a well-developed crown. (I) Section through a developing tooth, showing the ameloblasts (enamel producers) and the odontoblasts (dentin producers).

All the deciduous teeth begin to form between the 6th and 8th week of intrauterine life. The successional permanent teeth begin in the 20th week and end in the 10th month after birth. The permanent molars are formed between the 20th week of fetal life and 5 years into childhood.

D. Dental or Enamel Organ

As the cells at the free ends of the dental laminae divide, they form knots of epithelial cells, which begin a pattern of morpho- and histodifferentiation for the 20 deciduous teeth.

1. Bud Stage

This is the first indication that the epithelia are forming into a defined structure. The surrounding ectomesenchyme also begins to condense around this knot of cells, an indication of increased mitosis.

2. Cap Stage

The name of this stage is descriptive of the shape that the dividing epithelia take as they multiply. At this stage, differentiation has established the primitive dental (enamel) organ. It is made of an internal dental (enamel) epithelium, an external (enamel) epithelium, and an intermediate zone of less differentiated epithelial cells. The ectomesenchyme, in the meantime, has condensed around the cap as the dental follicle (sac) and under the cap as the dental papilla. The cells of the dental follicle will differentiate into the periodontal ligament, the cementum, and the inner portion of the alveolar bone. The cells of the dental papilla will become the dental pulp as the dentin is mineralized around it. The dental follicle, dental papilla, and the dental organ comprise the tissues of the tooth germ from which all the tooth and its adjacent structures will develop.

3. Bell Stage

The tooth germ continues to grow into the shape of a bell. Mitosis expands the size of the tooth germ. A layer of flattened differentiated cells has condensed in the dental organ side of the cuboidal inner enamel epithelia. This is the stratum intermedium, which is a source of new cells for the internal enamel epithelium. Water inhibition into the center of the dental organ causes intercellular spaces to appear. Macula adherens attachments between cells at limited spots on the plasma membranes now cause stretching of the cells into star shapes. This middle zone of the dental organ is now called the *stellate (star-shaped) reticulum*. The linear area at the periphery of the bell, where the inner enamel epithelium meets the outer enamel epithelium, is called the *cervical loop*. All the cells of the dental organ are attached to each other by maculae adherens. The whole dental organ is separated from the ectomesenchyme of the dental papilla and the dental follicle by a basement membrane.

The original dental lamina begins to break up into small clusters of cells, which disappear leaving the deciduous tooth germ and the successional lamina as isolates. When this occurs, the dental organ gradually takes the shape of the future crown. The stellate reticulum is squeezed to the sides of the dental organ and is sparse at the tip of the presumptive tooth crown.

The dental papilla is now the site of ingrowth of blood vessels and nerves. In contrast, the dental organ is avascular and must receive its nutrition by diffusion from blood vessels in the dental follicle. During the late bell stage, cells of the internal enamel epithelium change shape from cuboidal to tall columnar. Internally these cells develop organelles and the nucleus moves to the basal pole of the cell closer to the cells of the stratum intermedium. This cellular differentiation results in presecretory ameloblasts. Ectomesenchymal cells of the dental papilla close to the apical pole of the presecretory ameloblasts differentiate into elongate cells and line up along the underlying basement membrane. These cells are now called *odontoblasts*. They begin to secrete organic matrix, which will become calcified as dentin. The basement membrane disappears at this time. The ameloblasts will become secretory only after 2–3 μm of calcified dentin has been laid down, but odontoblasts will not be differentiated without the inductive activity of the ameloblasts. This epithelial–mesenchymal interaction is called *reciprocal induction*.

4. Appositional Stage

The shape of the crown has been determined by the design of the dental organ. Dentin has formed at the cusp tip. Enamel begins to calcify immediately after organic enamel matrix is secreted by the secretory ameloblasts (Fig. 3). During this stage, the differentiation of ameloblasts and odontoblasts continues down the sides of the crown, and dentin, followed by enamel, is mineralized immediately. In the older areas, both dentin and enamel thicken as they mineralize in opposite directions from each other, the dentin toward the papilla, the enamel toward the external enamel epithelium. The stellate reticulum collapses at this stage, allowing the ameloblasts to come closer to their nutrient vascular supply from the dental follicle.

5. Root Formation

The cervical loop of the appositional stage is where the inner enamel epithelium ends. It is also where the enamel ends its formation, as the last cells of the inner enamel epithelium are differentiated into ameloblasts. At this time cells of the inner and outer enamel epithelium continue dividing and form a two cell-layered sheet of epithelial cells, which grow away from the crown. This is Hertwig's epithelial root sheath, which will provide a scaffolding for the development of the tooth root. Cells of Hertwig's sheath are similar in origin to the ameloblasts, which have induced odontoblasts to differentiate from the ectomesenchyme of

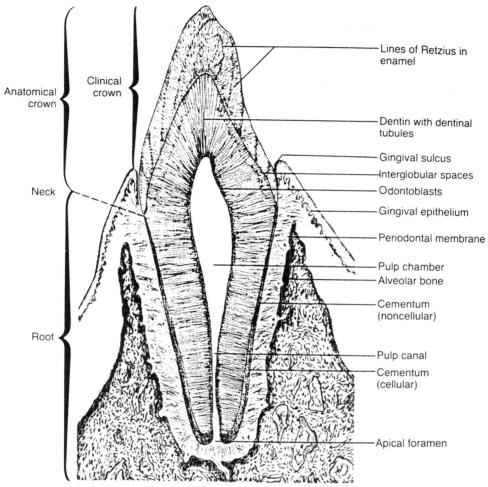

FIGURE 3   The fully formed functional tooth and adjoining structures. [Source: A. A. Paparo, T. S. Leeson, and C. R. Leeson (1988). "Text/Atlas of Histology," p. 402. Saunders, Philadelphia.]

the papilla. They repeat this process (i.e., papilla cells differentiate to form odontoblasts adjacent to Hertwig's sheath). These odontoblasts will now secrete matrix that will calcify as root dentin. Continued growth of Hertwig's sheath and subsequent dentin formation is how the root is formed. Multirooted teeth such as molars are developed from ingrowths of Hertwig's sheath, which allows each root to grow as a separate structure but attached to the crown of the tooth.

III. DEVELOPMENT AND STRUCTURE OF PARTS OF TEETH

Odontogenesis means tooth embryogenesis or development. Amelogenesis is enamel development, dentin-

ogenesis is dentin development, and cementogenesis is cementum development. Concomitant with tissue embryogenesis is the initiation and growth of the mineral within the tissues (mineralogenesis). As the mineralized tissues develop they surround the growing soft tissue in the dental pulp.

A. Amelogenesis

Enamel formation is composed of two processes: the growth and development of the dental organ, and the differentiation of its quintessential substructure, the inner enamel epithelium. As the dental organ matures from the cap through the bell, appositional, and root formative stages, the inner enamel epithelium matures from a flattened, squamous cell to a cuboidal cell, to a columnar presecretory cell, to a specialized secretory ameloblast, to a mature columnar cell, and finally to a cuboidal protective covering cell, which ultimately will be lost from the mineralized enamel surface after the tooth erupts into the mouth.

As discussed earlier, the presecretory ameloblasts induce proximal ectomesenchymal cells of the dental papilla to become a layer of juxtaposed odontoblasts. These cells produce extracellular matrix, which will calcify in a process discussed in Section III,B. After 2–3 μm of dentin is mineralized, presecretory ameloblasts differentiate into secretory ameloblasts and begin to produce enamel matrix. Secretory ameloblasts differentiate into secretory ameloblasts and begin to produce enamel matrix. Secretory ameloblasts are distinguished by the formation of Tomes apical processes, which are instrumental in the formation of rod and interrod matrix (see Section III,A,1). This matrix calcifies almost immediately with mineralized crystals of calcium, phosphate, and hydroxyl moieties, called *hydroxyapatite* (HA). These grow from the HA crystals already mineralized in the dentin. As enamel continues to thicken, the thin plate-like crystals of HA grow in length, width, and height at the expense of a pH labile enamel protein, which is denatured and resorbed. Eventually less than 1% of organic matrix remains in mature enamel. Enamel protein is exquisitely sensitive to small changes in pH. In the presence of potassium, an abundant extracellular cation, the crystal surface pH is elevated, significantly influencing protein depolymerization. There are one-half as many crystals of HA at the surface of enamel as compared with the smaller circumference at the dentino–enamel interface or junction (DEJ), indicating fusion of crystals as they grow and mature. Enamel matrix has two major constituent proteins: amelogenin, which is 19 times more abundant than the second protein, enamelin. Amelogenin, with a molecular mass of approximately 4000 Da, is unbound to HA and therefore is the protein that is resorbed during amelogenesis. Enamelin has a high molecular mass (67,000 Da) and because it is bound to HA, it mostly is retained in mature enamel. [*See* Extracellular Matrix.]

I. Structure of Enamel

The first enamel to form is relatively homogeneous. After 2 or 3 μm of this early enamel has been laid down, Tomes processes develop on the apical surfaces of the ameloblasts. Almost immediately the enamel becomes divided into rod and interrod structures, mediated by Tomes processes. The enamel rod or prism is the basic unit of enamel and is roughly hexagonal to round with a variable keyhole shape in cross section. It

is filled with innumerable crystals of HA, oriented more or less perpendicular to the DEJ. Theoretically, HA crystals of enamel can extend the whole 2- to 3-mm width of enamel from the DEJ to the tooth surface. Rod diameter averages 4 μm, varying between 2.5 μm at the DEJ to 5 μm at the surface. There are 5 million rods in a lower incisor and 12 million rods in an upper molar. They run a wavy course and therefore are longer than the 2- to 3-mm thickness of the enamel of the cusp tip.

The interrod substance is filled with small HA crystals and contains a greater amount of organic material than does the rod itself. Each rod has 4-μm cross striations, indicating a diurnal rhythmicity of formation. Periodically, there are larger, more distinct rhythmic bands called *incremental lines* or the *stria of Retzius*, caused by long-acting metabolic alterations or minute changes of rod direction. Sometimes, when there are even more prolonged metabolic changes, the incremental lines are increased, as either in febrile diseases of early childhood or the neonatal line of deciduous teeth or in permanent teeth that are in the process of mineralization at the time of birth. Just as is the case in the first enamel formed without Tomes processes of the ameloblasts, surface enamel shows no rod outlines because the maturing ameloblasts, which secrete the final enamel matrix, also are devoid of Tomes processes.

2. Manifestations of Developmental Faults in Enamel

Enamel lamellae are faults or cracks extending from the DEJ to the enamel surface.

Enamel tufts are interwoven areas of poor mineralization that look like tufts of grass extending out from the DEJ up to one-fifth to one-third of the enamel thickness.

Enamel spindles are poorly mineralized beginnings of odontoblastic processes (see later) trapped in the early enamel.

B. Dentinogenesis

Dentin formation begins with the differentiation of secretory odontoblasts from undifferentiated ectomesenchymal cells. These cells line up as a single layer on the basement membrane, which was previously secreted by presecretory ameloblasts. The secretory odontoblasts have a basally placed nucleus, which is at the pulpal pole of the tall columnar cell. The large number of intracellular membranous organelles is indicative of a secretory cell. The odontoblasts now secrete extracellular matrix (predentin). They migrate away from the DEJ, leaving behind cellular extensions called *odontoblastic processes,* which remain as part of the odontoblasts throughout life. The precalcified extracellular matrix contains the primary protein of connective tissues (i.e., collagen), as well as glycoproteins, phosphoproteins, glycosaminoglycans, lipoproteins, and, most importantly, matrix vesicles. The matrix vesicles are the primary calcification loci of dentin. It is within the matrix vesicle that the first crystals of HA are formed. Calcification that originates in matrix vesicles is called *primary* (initial calcification). As crystals continue to grow either related to subunits of the matrix vesicle membrane or by secondary crystallization from already crystallized HA, the matrix vesicle membrane breaks down and the crystals continue to grow in, on, and within ubiquitous collagen fibers. This in time separates dentin into mineralized dentin and unmineralized predentin. Predentin lies between the mineralized dentin and the secretory odontoblast. The odontoblastic process continues to grow in length as the odontoblast migrates toward the pulp. A highly calcified structure surrounds it in the mineralized dentin (*peritubular dentin*), forming a canal, the dentinal tubule, in which the odontoblastic process occurs.

The first dentin that is formed is *mantle dentin,* which represents the coalescence of calcification nodules. These are outgrowths of matrix vesicles. After several microns of this first dentin has been formed, the odontoblasts stop secreting matrix vesicles and begin to secrete collagen in sheets, more or less oriented radially from the mantle dentin toward the pulp. When this is mineralized, it is called *circumpulpal dentin.*

In ameloblasts do not develop, dentin will not form. If dentin does not form, enamel will not develop because each is reciprocally related to the other in development.

1. Structure of Dentin

Dentin is 70% inorganic, essentially HA, and 30% organic, which is primarily collagen with glycoproteins, phosphoproteins, lipoproteins, and glycosaminoglycans. Dentin underlies the enamel of the crown and the cementum of the root. As such, it comprises the bulk of the calcified structure of the tooth. It has a slightly yellowish color, which gives the tooth much of the background shade as light is reflected through the translucent enamel. All dentin, including the root, has mantle dentin beginning at the junctions with enamel or cementum and extending pulpward for about 10–20 μm. The rest of the dentin

consists of circumpulpal dentin, which is lined by a continuously viable layer of cells (i.e., the odontoblasts). This is the boundary for the dental pulp. Another way to subdivide the mineralized structure of dentin is by peritubular dentin, the heavily calcified zone surrounding the odontoblastic processes, which lie in the dentinal tubules, and by intertubular dentin, the areas between the processes. Still another way is temporal. Primary dentin forms before the root has been completed. Secondary dentin forms afterward and is represented by a sharp change of direction of the dentinal tubules. Tertiary dentin (reparative dentin) results from a response to insult or injury, and the tubules are irregular in their arrangement. Dentinal tubules are straight in the cusps and the roots and S-shaped in the crowns. In cross section, there are 20,000 tubules mm^2 at the DEJ and 45,000 at the pulp, indicating crowding of the odontoblasts as the dentine grows. Whenever dentinal tubules continue to calcify and completely occlude, it is called *sclerotic dentin.* Daily growth lines are called *incremental lines of von Ebner.* These are 6 μm in the crown and 3.5 μm in the roots.

The contour lines of Owen are analogous to the striae of Retzius of the enamel, i.e., they are structural bands resulting from long-lasting metabolic changes during dentinogenesis.

C. Cementogenesis or Cementum Formation

The extension of the inner enamel epithelial cells into Hertwig's epithelial root sheath forms the cellular basis for inducing root papillae ectomesenchymal cells to differentiate into root odontoblasts. Odontoblasts form several microns of mantle dentin of the root. The Hertwig's sheath now disintegrates. Mesenchymal cells from the dental sac (follicle) invade through the spaces caused by the breakdown of Hertwig's sheath. When these mesenchymal cells reach the mantle dentin of the root, they differentiate into cementoblasts. Cementoblasts, which are similar to osteoblasts, secrete collagen, glycoproteins, and glycosaminoglycans. This matrix calcifies by *secondary (subsequential) calcification* as crystals of HA grow in, on, and between collagen fibers by epitaxy from the HA of the mantle dentin. Groups of sheath cells, which do not degenerate, may last into adult life as epithelial rests of Malassez. These sometimes become cystic.

Collagen oriented perpendicularly to the surface of the mantle dentine is secreted by fibroblasts lying on the outside of the cementoblasts, and this collagen forms between the cementoblasts. Matrix secreted by the cementoblasts engulfs this collagen. All these structures are calcified from the mantle dentin by secondary calcification. The oriented collagen bundles produced by the fibroblasts when they are embedded within calcified cementum are called *Sharpey's fibers* and are the tooth terminal end of the periodontal ligament fibers, which terminate at the other end in alveolar bone as Sharpey's fiber as well. [*See* Collagen, Structure and Function.]

I. Structure of Cementum

This tissue is less hard than dentin because it is only 50% inorganic. Like other mammalian mineralized tissues, the basic inorganic moiety is HA. The organic substructure is primarily the fibrous protein, collagen with smaller amounts of glycoproteins, and glycosaminoglycans.

There are two types of cementum: (1) acellular, in which there is no incorporation of cells in the mineralized structure and which is sometimes missing in the apical one-third of the tooth; and (2) cellular, in which cellular cementum is studded with spider-like cementocytes with canaliculi, mainly directed toward the outer root surface, which are mineralized into the cementum. It is found in the apical one-third and the *furcations* of teeth. Furcations are the areas where roots join at the body of the tooth.

Collagen is found as intrinsic fibers, which are secreted by cementoblasts, and extrinsic fibers, which are secreted by periodontal ligament fibroblasts as Sharpey's fibers. Cementum is thinnest at the cemento–enamel junction (20–50 μm) and thickest at the apex (150–200 μm).

Incremental lines are parallel to the surface of the tooth and are found in both cellular and a cellular cementum.

The cemento–enamel junction is a butt joint 30% of the time, does not meet at all 10% of the time, and exhibits an overlap of the cementum over the enamel 60% of the time. When overlapping, the cementum is *afibrillar*.

Cementum is the tooth tissue to which the collagen bundles of the periodontal ligament attach. Cementum is incremental throughout life. Resorption is seen in deciduous teeth but is less frequent in the adult. Sharpey's fibers attach only to the outside layer of cementum. Each incremental layer has a new attachment of fibers. The deposition of apical cementum compensates for occlusal wear, to maintain the vertical height of the tooth.

D. Dental Pulp

The origin of dental pulp is in the dental papilla, which begins as a condensation of ectomesenchymal cells close to the dental organ when it reaches the cap stage of development. As dentin grows inward from the crown and downward to become the root of the tooth, nerves and vascular and lymphatic vessels grow in from the root side. The pulp is primarily a connective tissue with certain primitive qualities. The cells include odontoblasts, which have a peripheral location, fibroblasts, mesenchymal cells, macrophages, nerve cells, vascular cells, and lymphocytes. There are four zones in the pulp: the odontoblastic cell zone, the cell-free zone of Weil, the cell-rich zone, and the pulp core.

The extracellular matrix contains glycoproteins and glycosaminoglycans, but unusually, it has a high concentration of Type III embryonic collagen as well as the more common Type I collagen. When the pulp is completely contained by the crown and the root, it is said to reside in the pulp chamber. Access into the pulp chamber occurs only through the variably sized apertures at the root tip.

E. Vasculature and Lymphatics

One hundred fifty micron arterioles enter the pulp at the root tip, remain centrally located, shed some musculature, and give off lateral branches in the root that extend to the pulp periphery. Here branching develops into a subodontoblastic capillary network, which continues into the coronal area where the network is the most extensive. Venules of equivalent size accompany the arterial supply. Lymphatics function to relieve fluid pressure in the pulp.

F. Nerves

Nerves of different sizes follow the arborization of blood vessels. They form (similar to the vasculature) a subodontoblastic plexus of nerves called the *plexus of Raschkow* in the cell-free zone of Weil. Axons are primarily myelinated sensory afferents of the trigeminal nerve and nonmyelinated sympathetic branches from the superior cervical ganglion, which innervate blood vessels. Myelin is lost by the time the nerves reach the subodontoblastic plexus. Some nonmyelinated fibers enter dentinal tubules for a short distance after passing between the cell bodies of the odontoblasts.

G. Dental Pain

Pain is stimulated by diverse stimuli (e.g., hot, cold, mechanical stimulation, dessication). The tooth is most sensitive at the dentino–enamel junction. There is increased sensitivity with an inflamed pulp. Possible mechanisms of pain include the stimulation of free endings in the dentinal tubules, odontoblasts and their processes may serve as pain receptors, and/or fluid movement in the dentinal tubules may cause the stimulation of free nerve endings in the dentin and pulp. [See Pain.]

H. Pulp Stones

Pulp stones are calcified masses in the pulp chamber that are either free or attached. True pulp stones are surrounded by odontoblasts. False pulp stones are dystrophic calcifications without odontoblasts.

I. Age Changes

There is a gradual reduction in volume of the pulp chamber caused by the continual growth of secondary and tertiary dentin. Some dentinal tubules become sclerotic dentin as they calcify.

IV. DEVELOPMENT AND STRUCTURE OF DENTAL SUPPORTING TISSUES

All the supporting tissues of the teeth develop from the dental follicle or sac. As described earlier, during cementogenesis, Hertwig's epithelial root sheath disintegrates and cementoblasts differentiate from follicular mesenchymal cells to secrete cementum matrix. Some mesenchymal cells become fibroblasts, which secrete periodontal ligament collagen fibers, whereas others differentiate into osteoblasts and produce the adjacent alveolar bone.

A. Alveolar Bone

Alveolar bone develops by intramembranous osteogenesis in both the maxilla and the mandible. By the third fetal month there is a bony groove that contains the tooth germs, alveolar nerves, and blood vessels. In time, bony septae separate the tooth buds and the alveolar bone (which surrounds the teeth). Alveolar bone now becomes firmly attached to the body of the maxilla and the mandible without any distinct boundary between them.

The alveolar process is that part of the maxilla and mandible that forms the sockets and supports the teeth. Strictly speaking, it is defined only after the teeth have erupted and disappears when teeth are lost.

There are two parts to the alveolar process: (1) the alveolar bone proper, which gives attachment to the periodontal ligament fibers and therefore surrounds the tooth; and (2) the supporting alveolar bone, which is composed of two cortical plates on either side of spongy bone. This is attached to the outside of the alveolar bone proper.

In the anterior regions of both jaws, the supporting bone is thin so that the cortical plates merge with the alveolar bone proper.

In oral radiographs of the teeth and jaws, the alveolar bone proper is called the lamina dura. Because this bone is perforated by nutrient canals, it is sometimes called the cribiform plate.

Marrow spaces contain hemopoeitic bone marrow in the young but mostly only fatty marrow in the adult.

Alveolar bone begins as woven bone, and because there are no calcified structures present in the vicinity, calcification is initiated in matrix vesicles, produced by osteoblasts. It is only when several microns of woven bone have developed that matrix vesicles disappear and lamellar bone appears, with calcification proceeding by secondary nucleation and growth of HA crystals.

A peculiar and unique type of bone is found only in the alveolar process and is called bundle bone. Bundle bone has incremental areas of lamellar bone with incorporated Sharpey's fibers, even in the deeper layers where no attachment to the periodontal ligament exists.

Alveolar bone, like all bone, is 65% inorganic, primarily calcium HA, and 35% organic, primarily collagen and glycoprotein.

B. Gingival Tissues

Gingival tissues are the pink mucosal soft tissues that surround the teeth in the mouth. They originate from ectodermal cells and underlying ectomesenchymal cells of the stomatodeum. Early in life, for example, at birth, there is only the primitive oral mucosa devoid of both gingiva and teeth. Gingiva can only be demonstrated after teeth have erupted into the mouth because its structure and function are dependent on the presence of teeth. Its primary function is to seal the oral cavity from the connective tissue surrounding the tooth. It resists abrasive, hot, cold, and irritating foods. It forms a first line of defense against the invasion of microorganisms. The importance of this function is underscored by the fact that bacteria are normal inhabitants of the oral cavity and obviously must be contained.

Gingival tissue is made up of epithelium and underlying connective tissue. There are three gingival epithelial subdivisions.

1. The outer gingival epithelium covers the outer surface of the gingiva. Its surface is stratified, squamous, and keratinized as the horny outer layer. It has predominant epithelial ingrowths, the rete pegs, between which are connective tissue outgrowths.

2. The oral sulcular epithelium lines the superficial half of the gingival crevice. It is stratified, squamous, and nonkeratinized with prominent rete pegs. The gingival crevice or sulcus is the space formed around the tooth between the epithelium and the tooth. It is bounded below by the epithelium attachment to the tooth and is open to the mouth above.

3. The junctional epithelium lines the deep half of the gingiva below the crevice with stratified squamous nonkeratinized epithelium. It has no rete pegs and is only one to five cells thick.

Between the junctional epithelial cells and the tooth, a glycoproteinacious cement substance attaches the cells to the tooth. This is the epithelial attachment. It is secreted by the epithelial cells, which in turn are attached to it by hemidesosomes. These cells are attached to each other with tight junctions, which act as a further seal against bacterial invasion.

The attached connective tissue is organized as dense collagen fibers mainly perpendicular to the surface. There are no elastic fibers and no muscle fibers. The collagen fibers form a variety of structures that run from the alveolar bone to the gingiva, from the cementum to the gingiva, and around the tooth below the gingival crest.

C. Periodontal Ligament

The periodontal ligament is really a group of precisely oriented collagen fibers, which surround the tooth, cradle it, and attach it through the cementum with the alveolar bone (Fig. 4). At one time it was called the *periodontal membrane* because in early light microscopy it appeared homogeneous rather than, as it turned out, fibrillar. The *periodontium* is the term used to describe the tissues that surround and attach the tooth in its bony socket. It consists of cementum, periodontal ligament, inner alveolar bone, and junctional epithelial gingiva. The periodontal ligament originates in the ectomesenchymal cells of the dental follicle. These cells differentiate into fibroblasts, which begin to secrete collagen and associated noncollagenous proteins as well as glycosaminoglycans. The first collagen fibers are seen after the initiation of root formation. The first fibers are oriented in an oblique direction passing upward from the cementum because the tooth develops inferior to the crest of alveolar bone. As the tooth erupts, it moves past the alveolar crest and the fibers become horizontally oriented. When the tooth is completely erupted, the most coronal fibers again become oblique, but now with fibers directed from the cementum downward to the alveolar crest.

After the root has formed, the periodontal ligament becomes denser and the fibers assume particular orientations depending on the position in the tooth socket. The fibers are categorized into the following groups: alveolar crest, horizontal, oblique, transitional, apical, interdental, circular, and tangential. The alveolar crest group are oblique fibers that run downward between the cementum, at the cemental–enamel junction (CEJ) and the crest of alveolar bone. Just apical to them are horizontal fibers followed by oblique fibers that run upward from the cementum to the alveolar bone. Next is a group of horizontal fibers that are transitional to the oblique and vertical apical fibers surrounding the apex of the tooth. Interdental (or transseptal) fibers traverse the area above the gingival crest, attaching the cementum of one tooth with the cementum of an adjoining tooth. Circular fibers surround the tooth, attaching, in part, the gingiva to the bone, and tangential fibers run from one side of the alveolar bone to the other. The collagen of periodontal ligament fibers is attached on either side by being calcified into bone and cementum. These partially calcified fibers are called Sharpey's fibers. The diversity of orientation of the fibers suspends the tooth in the alveolar socket. This permits a slight limited movement during mastication while holding the tooth firmly in place. In addition to collagen, there are some elastic fibers associated with blood vessels and a unique fiber, the *oxytalin* fiber, which could be an elastin precursor. The periodontal ligament is highly vascularized, the blood supply being derived from the inferior alveolar artery for the mandibular periodontium and teeth and from the superior alveolar artery for the maxillary structures. Collagen of the periodon-

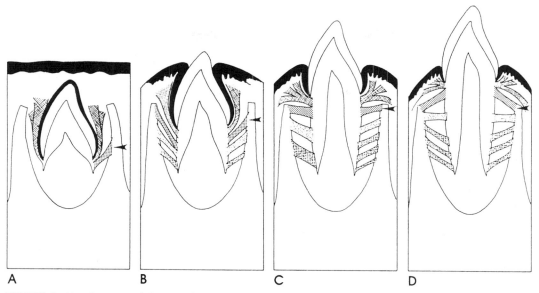

FIGURE 4 Development of the periodontal ligament. (A) The first fibers are oblique. (B) With eruption of the tooth, new fibers are added. (C) With continued eruption, the first fibers become horizontal. (D) Later still, they become oblique again but in the opposite direction. New fibers are added as the root continues development.

tal ligament is secreted and resorbed rapidly by the same cells, the fibroblasts. The ligament increases in thickness with increased force on the tooth.

Because all the components of the periodontium have the same origin, it is not surprising that the growth of bone, ligament, and cementum stems from the same cells. These precursors to the periodontal ligament fibroblast subsequently can differentiate into osteoblasts, cementoblasts, and fibroblasts. All these cells and the derived tissues function in unison under varying physiologic conditions. For example, if a tooth is moved orthodontically, the periodontal ligament is compressed against the alveolar bone. Osteoclasts are activated, causing the bone to resorb. On the other side of the tooth, where the periodontal ligament is under tension, the width of the ligament is increased, fibroblasts secrete oriented collagen, and new bone is deposited. Sharpey's fibers in both cementum and bone are either destroyed or remade on both sides of the tooth.

V. ERUPTION OF TEETH INTO THE MOUTH

At the time that the enamel of any tooth, deciduous or permanent, has been mineralized and the crown is

complete, the dental organ is compressed on the enamel surface as the reduced enamel epithelium. This two-layered structure is the remains of the dental organ. The layer in contact with the tooth consists of the nonsecretory protective phase ameloblasts and is firmly attached to the enamel. The outer layer is the remnant of all the other dental organ components. As the tooth erupts, the connective tissue that lies between the cusp of the tooth and the oral epithelium is resorbed. The outer cells of the reduced enamel epithelium and the inner cells of the oral epithelium merge. The cells now are induced to proliferate, particularly on the sides of the crown as the cusp tip pushes through the merged epithelium. The merged epithelial cells along the side of the tooth become the gingiva of the dentogingival junction. By continued growth of the tooth into the mouth and proliferation of the adjacent cells, the gingiva becomes the functional tissue described earlier. With age there is an apical migration of the dentogingival junction. In all this growth and development, the epithelium maintains an absolute separation of the gingiva and the crown from the underlying connective tissue.

1. Tooth Movement During Development

The deciduous teeth all begin development from the 6th to the 8th week postfertilization. The permanent teeth begin development from the successional laminae of the incisors and the molar laminae of the first molars during the 20th week. During this phase and the subsequent time period when the rest of the permanent teeth develop from the successional laminae, the jaws are enormously crowded. Fortunately, the jaws continue growth in all dimensions, permitting the teeth to continue their orchestrated growth. The permanent incisor and canine teeth begin growth lingual to (i.e., on the tongue side of) the deciduous teeth. As simultaneous growth of both deciduous and permanent teeth continues, each lies within its own bony crypt. The permanent premolars (bicuspids) grow between the roots of the deciduous molars. The permanent molars develop beind the deciduous molars from the molar lamina, at first in tight quarters. As the jaw gets larger, the tooth germs move into position in the greater space that growth has provided. As the tooth germs move forward and toward the midline, remodeling takes place in the bony crypts. There is resorption ahead of movement and deposition of bone behind the movement of the tooth germ. After root formation has begun and the tooth has begun to erupt, the overlying bone is resorbed, allowing the tooth to migrate through the epithelium into the mouth. The deciduous tooth, when erupted into the mouth, continues its root development until the apex is calcified. Meanwhile, the successional permanent tooth continues its growth. This pressures the deciduous tooth and starts the process of resorption of the roots. Osteoclast-like cells, the odontoclasts, mediate the resorption process. In time, the deciduous tooth roots are resorbed, and the successional permanent tooth crown is fully formed. The roots are partially complete. The deciduous tooth, which is now only a disembodied crown, is exfoliated into the mouth. Shortly afterward, the permanent tooth erupts into the mouth through a canal in the bone called the *gubernacular canal*. The remains of the successional lamina are in the canal.

Most evidence points to the pulling characteristics of the periodontal ligament fibers as the major force that causes the eruption process to continue. This is true for both deciduous and permanent teeth, although the deciduous teeth are eventually exfoliated because of the joint forces of root resorption and masticatory pressure. Exfoliation of deciduous teeth begins with the lower incisors at 6–7 years of age, then the upper incisors. The lower canines are followed by the upper canines, and all the deciduous molars erupt almost simultaneously at about 8–10 years of age. The successional permanent teeth erupt shortly after the deciduous teeth are exfoliated. The permanent first molars erupt at 6–7 years, the second molars at about 12 years, and the third molars between 17 and 22 years.

VI. MINERALOGENESIS: PRIMARY AND SECONDARY CALCIFICATION

Mineralization of tissues *in vivo* can be divided into two distinct phases, each giving rise to distinct anatomic entities. (Phase I is primary or initial calcification whereas phase II is secondary or subsequential calcification.) Initial calcification involves the matrix vesicle, which provides the locus for the nucleation of HA in woven bone, calcified cartilage, and mantle dentin. Subsequential calcification is the process by which lamellar bone, circumpulpal dentin, cementum, and enamel develops. It does not depend on matrix vesicles for the initiation of mineralization; rather HA continues growing by direct extension and secondary nucleation of the preexistent crystalline structure formed during initial calcification. Although each adult calcified tissue develops from a different cell lineage, all share the matrix vesicle as the initial calcification locus.

Matrix vesicles are only associated with mineralization of the primary dentin (mantle dentin) of the tooth. The tissue is unique in that mineralization develops segmentally, beginning at the top of the crown and extending to the apex of the root. Therefore, during development, initial mineralization may occur in one segment of the root while several segments of the crown are completely mineralized. In any one segment, mineralization of the tooth begins in matrix vesicles close to the presumptive interface with the enamel of the crown or the cementum of the root. As is true in woven bone and calcified cartilage, random growth of HA crystals from the initial nucleation site in the matrix vesicles of dentin determines the spheroidal structure of calcification nodules. These nodules coalesce as the HA crystals continue to grow into, onto, and between the collagen fibrils of the dentin extracellular matrix. When several micrometers of mantle dentin have calcified, HA crystals grow in two directions simultaneously. The first is directed toward the pulp of the tooth, as constantly receding odontoblasts secrete matrix glycoproteins and collagen precursors. The collagen fibers, particularly, are oriented in parallel array, more so than is apparent in the random orientation seen in the mantle dentin. This orientation of collagen signals the formation of circumpulpal dentin, which constitutes the major bulk of dentin. Structurally, HA crystals appear to grow from already formed crystals in the mantle dentin as the crystals infiltrate the collagen. The second direction in which the HA crystals grow begins at the enamel–mantle dentin or the cementum–mantle dentin interfaces. This growth is directed to the outside of the tooth, exactly opposite to the growth in the circumpulpal dentin. In cementum, HA crystals are related to collagen in an analogous pattern to the relation in circumpulpal dentin. Initiating enamel formation, ameloblasts secrete enamel matrix (enamel protein and glycoprotein) directly onto the surface of the mantle dentin HA. Characteristic long, thin crystals of enamel HA are initiated from the HA crystals in the mantle dentin, maturing and growing in all dimensions at the expense of labile enamel protein.

From these observations, a pattern of growth and development of the different calcified tissues has emerged. There are two subdivisions of mineralogenesis: (1) primary or initial, and (2) secondary or subsequential. The criteria for the inclusion of a mineralizing tissue as primary calcification are that calcification begins in the matrix vesicle and that there is no preexisting calcification in the tissue before matrix vesicle emergence in the matrix. The criteria for inclusion as secondary calcification are that growth of HA crystals is a continuation by direct extension and secondary nucleation from a previously formed seam of calcified tissue and not from matrix vesicles. There are three tissues that have HA nucleation sites in matrix vesicles and therefore are in the primary calcification subdivision: woven bone, calcified cartilage, and dentin (see Table I). There are four tissues that are in the secondary calcification subdivision: lamellar bone, which develops subsequent to woven bone, and

TABLE I
Lineage of Mammalian Calcification

Initial or primary	Subsequential or secondary
Woven bone	Lamellar bone
Calcified cartilage	
Mantle dentin	Circumpulpal dentin
	Cementum
	Enamel

enamel, cementum, and circumpulpal dentin, all three of which develop subsequent to mantle dentin.

In endochondral osteogenesis, bone formation begins *de novo* on the surface of calcified cartilage. That is, matrix vesicles of woven bone initiate mineralogenesis as if there were no seam of calcified tissue in the region. This may be because direct extension of HA crystals is impeded by a carbohydrate-rich coating on the surface of calcified cartilaginous spicules.

BIBLIOGRAPHY

Avery, J. K. (1994). "Oral Development and Histology," 2nd Ed. Thieme Medical Publishers, New York.

Berkovitz, BK. B., Holland, G. R., and Moxham, B. J. (1992). "Oral Anatomy, Histology and Embryology," 2nd Ed. Mosby Year Book, St. Louis, MO.

Bernard, G. W., and Marvaso, V. (1982). Matrix vesicles as an assay for primary calcification *in vivo* and *in vitro*. *In* "Matrix Vesicles" (A. Ascenzi, E. Bounuci, and B. de-Bernard, eds.), pp. 5–11. Wichtig Editore, Milane.

Moss-Salentijn, L., and Hendricks-Klyvert, M. (1990). "Dental and Oral Tissues." Lea and Febiger, Philadelphia.

Provenza, D. V. (1988). "Fundamentals of Oral Histology and Embryology." Lea and Febiger, Philadelphia.

Ten Cate, A. R. (1994) "Oral Histology, Development Structure and Function," 4th Ed. C. V. Mosby, St. Louis, MO.

Depression

PETER J. LAUTZ

San Diego

GLOSSARY

Clinical depression Distinguished from the mood changes of everyday life by severity, pervasiveness, and duration

Cognitive therapy Method of psychotherapy useful in treating depression which focuses on how thoughts and perceptions influence moods and behaviors

Dysthymia Chronic, low-level to moderate depression lasting a minimum of 2 years in adults

Neurotransmitters Brain chemicals acting as messengers between nerve cells which can influence moods

DEPRESSION, AS A HUMAN AND CLINICAL PHE-nomenon, is both multifaceted and complex. Clinical depression is to be distinguished from the more common, depressing difficulties of everyday life by its intensity, pervasiveness, and duration. Characteristics of the clinical disorder include a sad, discouraged, and/or empty emotional state; a significant reduction of pleasure in usually enjoyable activities; a diminished interest in one's life; a thinking process oriented toward helplessness, hopelessness, and personal guilt; changes in normal sleeping and eating patterns; fatigue; diminished concentration; irritability; social isolation; decreased libido; and thoughts of suicide and/or suicide attempts. All of these characteristics may not be present in an individual who is clinically depressed. However, many depressed people will experience the full range of these symptoms. The disor-der may be acute and relatively brief in duration or long standing and chronic.

I. SYMPTOMATOLOGY AND PHENOMENOLOGY

A. Mood Symptoms

People who are clinically depressed frequently experience varying degrees of sadness and will report feeling "blue" or "down." They may have a lack of normal emotion, as if consumed by an internal emptiness. Hopelessness and an anguished despair often are experienced by severely depressed individuals. Annoyance and irritability, particularly in relation to others, are common features of the mood dimension of depression. Loneliness, often despite the presence of supportive people, and apathy are other common alterations in mood. Many people experience the depressed moods as happening to them irrespective of their current external life situation. Finally, a sense of overwhelming personal guilt may exist in depression.

B. Anhedonia

A reduction in the capacity to enjoy normally pleasurable pursuits is a hallmark of depression. Some persons lose all sense of pleasure. Others may be able to enjoy very intensive, positive experiences but find they have a sense of day-to-day activities as lacking color and vividness.

C. Cognitive Symptoms

Concentration is frequently impaired and people may have difficulty attending to a conversation or comprehending what they read. Sometimes people become quite indecisive and may have much difficulty in making decisions. Thoughts and views about oneself, the world, and of people are pessimistic and cynical. Often negative events are amplified and positive or neutral events are minimized in an automatic, nonconscious manner.

D. Sleep Disturbances

Depressed people often experience an involuntary reduction in sleep. They may have difficulty falling asleep initially and/or awaken frequently (as often as every 1–3 hr sometimes) throughout the night with great difficulty returning to sleep. The sleep pattern often changes to include an abbreviated rapid eye movement (REM) latency period, increased REM sleep, and diminished deep sleep in stages three and four. Less frequently, hypersomnia occurs and some persons will sleep several hours more daily than is their usual pattern. [*See* Sleep Disorders.]

E. Appetite Disturbances

Appetite for food is often significantly lessened. This reduction in interest can result in substantial weight loss and may become a serious medical concern. At times, the sense of taste is even modified, which results in restricted enjoyment of food. A minority of individuals experience an increased appetite and may eat in a somewhat compulsive manner. They may gain weight during a depressive episode. [*See* Eating Disorders.]

F. Fatigue

Frequently, depressed persons experience a sense of physical and emotional depletion. They report that normal activities are much more difficult to do, leading to feelings of being overwhelmed by demands.

G. Psychomotor Changes

There may be a marked retardation in movement, talking, and taking action in severe depressions. Conversely, some people exhibit a significant agitation and may tremble, pace nervously, wring their hands, or pull at their hair or clothing. Some individuals will experience either psychomotor retardation or agitation alone, whereas others may fluctuate between these states.

H. Libidinal Symptoms

Many individuals experience a marked diminishment in sexual interest and desire while depressed. Others report an increase in sexual longing and/or sexual activity, which may represent an effort to transform a dysphoric mood into a more euthymic state.

I. Interpersonal Symptoms

Social withdrawal and isolation often occur during the course of depression. People paradoxically avoid human connection at a time when support and caring are very much needed. The experience of trying to socialize while depressed is often felt to be futile and thus motivation to relate to others is reduced.

J. Suicidal Thoughts and Actions

When suicidal thoughts are present they may be vague and without any actual intention to harm oneself, but simply a representation of easing the emotional pain. However, completed suicides do occur as a result of serious clinical depression. Elderly white males are at the highest risk for suicide, but suicide rates for young white males (ages 15–24 years) have significantly increased since the early 1990s. Males outnumber females by a ratio of 3:1 in completed suicides, although females make significantly more attempts at suicide than do males.

II. TYPES OF DISORDERS

A. Unipolar Depressions

This category of depressive disorders includes major depression, dysthymia, and depression not otherwise specified (e.g., disorders that do not meet full criteria for either major depression or dysthymia but have significant clinical signs of depression).

1. Major Depressive Disorder

This disorder is characterized by one or more major depressive episodes in an individual's life but without prior manic, mixed, or hypomanic episodes. Additionally, the depressive symptoms may not be caused by the ingestion of a substance (e.g., alcohol) or by a medical condition (e.g., hypothyroidism) to qualify for this diagnosis. The primary component of a major depressive episode is a period of at least 2 weeks during which there is either depressed mood or the loss of enjoyment or interest in most of life on a daily basis. Additionally, the person must have a minimum of four other symptoms from the following: changes in sleep, appetite, or psychomotor activity; fatigue; sense of guilt or deep inadequacy; problems in thinking clearly, in concentrating, and in decision-making; and/or repetitive thoughts of death or suicide or of suicidal plans or attempts. These symptoms are accompanied by marked distress and anguish and usually result in impaired functioning in the person's usual roles.

Sleep electroencephalograph (EEG) studies show abnormalities in 40–90% of persons with a major depressive episode. Brain neurotransmitters that appear to be involved in major depression include serotonin, norepinephrine, acetylcholine, dopamine, and γ-aminobutyric acid.

Children with a major depressive disorder typically exhibit irritability, physical complaints, and social isolation more often than the sad, hopeless emotions of depression. Some variance in the proportion of the adult population with this disorder has been reported by different studies. Approximately 10 to 25% of women and 5 to 12% of men can expect to experience a major depressive disorder during their lifetimes.

The average age of onset is in the mid-20s; however, data now suggest that the age of onset is decreasing for people born more recently. In about two-thirds of people the episode will end completely and in about one-third of individuals there is either no or a partial remission of symptoms.

2. Dysthymia

Dysthymia is characterized by a chronic disorder of depressed mood for a minimum of 2 years for most of the day on most days during the 2-year period. In children and adolescents, the mood can be one of irritability and may be for a minimum of 1 year. Additionally, two or more of these symptoms must be present: poor appetite or overeating; insomnia or hypersomnia; fatigue; low self-esteem; poor concentration or indecisiveness; or feelings of hopelessness. There may not be a remission of these symptoms for more than 2 months at a time to meet the criteria for dysthymia. The symptoms cannot be caused by a medical condition or by the effects of a substance (e.g., drug of abuse or prescription medication). The individual who has this disorder will feel quite distressed by it and/or experience impairment in life functioning (e.g., at work, in relationships, etc.). The lifetime prevalence of dysthymic disorder is approximately 6%.

3. Depressive Disorder Not Otherwise Specified

This disorder includes depressions that do not meet the diagnostic criteria for major depressive disorder, dysthymic disorder, or for depressive reactions to life stressors known as adjustment disorders. Examples of this disorder would include premenstrual dysphoric disorder and a depression of at least 2-weeks' duration, but lacking sufficient symptoms for a major depressive disorder.

4. Postpartum Depression

Postpartum depression is characterized by symptoms of a major depressive episode occurring within 4 weeks after childbirth. Lability of mood and a more changeable course are often more common than in a major depressive disorder. Psychomotor agitation, angry feelings toward the infant, guilt about the depression, anger, and suicidal ideation may be present. Postpartum mood episodes with psychotic features appear to occur in from 1 in 500 to 1 in 1000 deliveries. This psychotic component is particularly dangerous in that the mother may hear auditory hallucinations to kill the child. Women with these depressive episodes often experience disinterest in the baby, insomnia, and intense anxiety. Dysphoric feelings persist long after the more common "baby blues," usually about 2- to 7-days' duration.

5. Seasonal Affective Disorder (SAD)

SAD has a temporal pattern in both onset and remission of major depressive episodes. Usually the episodes commence in fall or winter and resolve in spring. Some people experience recurrent summertime depressions with remission in the fall or winter. Symptoms often include fatigue, hypersomnia, overeating, and weight gain, as well as depressed mood. SAD treatment now may include the regular use of special lighting to influence the restabilization of mood. This is quite effective for some persons with this disorder.

B. Bipolar Depressions

Another category of mood illnesses is known as bipolar disorder and is characterized by periods of depression and of mania. In general, strong evidence shows a genetic influence for Bipolar Disorder (previously known as manic-depressive illness) with first-degree biological relatives of persons having this disorder experiencing elevated rates of both bipolar disorder and major depressive disorder.

Clinical depression is experienced by persons with this disorder and then alternates with a manic episode period. The latter is a distinct period during which there is an unusually intense elevated, expansive, or angry mood for a minimum of 1 week. Additionally, a manic episode includes at least three other symptoms from the following: grandiosity; decreased need for sleep; pressured speech; flight of ideas; high distractibility; or increased involvement in enjoyable actions with probable painful consequences (e.g., overspending, sexual promiscuity). The prevalence of bipolar I disorder derived from community samples ranges from 0.4 to 1.6%. It is a recurrent disorder with approximately 60–70% of manic episodes occurring immediately before or after a major depressive episode. Studies show that greater than 90% of people who have a single manic episode will have future episodes. To date there have been no laboratory findings for biochemical features which might distinguish major depressive episodes found in major depressive disorder from those in bipolar I disorder. [See Mental Disorders; Mood Disorders.]

III. ETIOLOGY

The causation of clinical depression is complex and, in many cases, is probably determined by multiple factors: biological, psychological, and interpersonal.

Research in brain neurochemistry has led to hypotheses about the biology underlying the depressive disorders. The biogenic–amine theory holds that mood is regulated in the brain by complex chemicals functioning as transmitters in the synapse between cells. Messages between brain cells occur via the release of chemical neurotransmitters across the synapse; the messages are then completed by means of a two-stage process in which the biogenic amines are taken back into the transmitting cell and rendered inactive by specific enzymes. It is believed that major depressive disorder, and possibly other forms of depression, follow from a deficiency in the availability of certain neurotransmitters to the postsynaptic nerve cell.

Another hypothesis views depression to be a result in abnormalities of the cells receiving the neurotransmitter's message. These postsynaptic cells, it is speculated, have developed an overabundance of specific receptors which causes a chronic underproduciton of biogenic amines and leaves the person vulnerable to depressed moods.

The neurotransmitters norepinephrine, serotonin, and acetylcholine are among the amines conceptualized to be central to depression. It is probable that no one theory of the biochemistry of depressive disorders can accurately depict this complex phenomenon. Research continues to explore the intricate relationships among neurochemicals, mood, and behavior. [See Neurotransmitter and Neuropeptide Receptors in the Brain.]

Research in the studies of both identical and fraternal twins has demonstrated an apparent genetic predisposition to depression in some people. Identical twins have a concordance rate for depressive disorders from 40 to 70% whether they were raised together or apart from each other. The decisive factor in their vulnerability to depression was found to be whether a biological parent also experienced depression. Fraternal or nonidentical twins have a concordance rate from 0 to 13%. This is the same rate as for non-twin siblings.

Both clinical and research data have shown a very strong correlation between losses in early life and subsequent depressive disorders. For example, adults whose parent(s) died during childhood are much more likely to experience both dysthymia and major depressive disorder.

The frustration of natural developmental needs of children for support and contact on a chronic basis may increase a person's vulnerability to clinical depression. Children and adults who have been subject to ongoing experiences in which they are helpless to influence the outcome in important situations often develop a belief in their own generalized helplessness. This learned helplessness may become a coping style and lead to depression. Traumatic and stressful experiences suffered due to the actions of others (e.g., child abuse, violence) or from the environment (e.g., unrelenting poverty, natural disasters) can lead to vulnerabilities for various emotional difficulties, including depression.

External stressful life experiences frequently precipitate depression in persons susceptible to the disorder. For example, significant disruptions such as divorce, unemployment, death of a family member or friend, major geographic moves, or serious illness can trigger depression. The likelihood of these types of stressors resulting in a serious depression appears to be related to whether there is a past history of depression, how the person interprets and responds to the difficulty, the strength of their support system, and the genetic predisposition to depression. Additionally, substance abuse (e.g., overuse of alcohol) and some medical problems (e.g., untreated thyroid dysfunctions) can amplify and prolong depressive symptoms until they are also treated.

IV. TREATMENT

Depression, although not always curable, is able to be significantly ameliorated in 65–70% of cases with the appropriate, effective treatment. Therapies include psychotherapy, pharmacotherapy, and electroconvulsive shock therapy. These therapies are utilized both individually and cooperatively to relieve the symptoms of the depressive disorder.

Psychotherapy for depression generally will be targeted to reduce suffering by helping people develop more effective emotional, cognitive, and behavioral patterns. Both cognitive therapy and interpersonal therapy have been researched in detail regarding their efficacy in treating depression. Other psychotherapies, including psychodynamic therapy, experiential therapy, and behavior therapy, may be useful when practiced competently and when they are appropriately tailored to the patient's needs.

Cognitive therapy focuses on learning more adaptive and optimistic ways to view the self, others, and the world. It is oriented in the present and toward the future while helping people discover how their automatic thinking styles create or maintain depressive symptoms. New patterns of interpreting experience are developed so that more accurate thinking about one's life becomes predominant.

Depressed people typically see problems in life as pervasive, permanent, and personal. Cognitive therapy aims to develop specific skills to view problems as usually only part of life, temporary, and not due to a personal failure. As these skills are integrated, depressed mood and related symptoms often abate. This therapy is effective in bringing substantial relief to about 70% of depressed persons. Therapy is often brief in duration: 8 to 20 sessions over a 2- to 5-month period. It is not, however, a panacea and is not effective for all people. In cases of chronic, moderate depression or for more severe major depressive episodes, a combination of cognitive therapy and antidepressant medication has been found to be more helpful.

Interpersonal therapy often occurs in the context of group therapy and explores how the person and his world of relationships interact. The National Institute of Mental Health conducted a major outcome study of the treatment of depression and found this form of therapy to be as effective as tricyclic antidepressant medications and perhaps slightly more effective than cognitive therapy in relieving depression. It consists of 12 to 16 sessions, usually over 3 to 4 months.

Interpersonal therapy explores four dimensions in the present life of the depressed person: grief, conflicts with others, role transitions and deficits in social skills. Emotional expression around unresolved past losses helps to complete the grieving process. New relationships are developed to begin to create positive experiences in the person's life. Various skills are taught to aid in conflict resolution and strengthening social networks. Transitions in living such as marriage, retirement, or divorce are directly dealt with to enable the person to develop new skills, outlooks, and resources. People may experience more mastery in their lives and more hopeful anticipation for the future when this therapy is effective. [See Psychotherapy.]

Antidepressant medications are utilized in treating major depressive disorders and also may be helpful in ameliorating the symptoms of dysthymic disorder when psychotherapy alone has not been effective. At present there are three major classes of these medications: tricyclics, monoamine oxidase inhibitors (MAO-I), and serotonin-specific reuptake inhibitors (SSRI). The tricyclic antidepressants essentially seem to affect the transmission of norepinephrine and/or serotonin, thereby stabilizing mood. The MAO-I class prolong the effective life of certain neurotransmitters by inhibiting an enzyme which oxidizes, or inactivates, these amines. SSRI are of more recent development. They work by specifically reducing the reuptake of serotonin to the transmitter cell, thus making more serotonin available to the receptor cell. [See Antidepressants.]

When taken daily, these medications will typically relieve symptoms significantly within 2 to 4 weeks. The patient's physician monitors the response to medication to assess its effectiveness and any side effects. Generally, if there has been an insufficient therapeutic response within 3 to 4 weeks, either the dosage will be increased or another antidepressant medication will be utilized (e.g., if a tricyclic has been ineffective, an SSRI may be used).

Research assessing their efficacy in treating depression shows these drugs to significantly help from 60 to 80% of people within weeks. Generally, people take these medications for a minimum of 6 months, and in some cases will continue to use them to prevent relapse throughout their lifetime. The majority of people treated with antidepressant medications alone will have a return of symptoms after the medication is stopped. Outcome research data show that a combination of medication and psychotherapy predicts a better long-term prognosis than medications used alone.

Side effects are not experienced by all people using these medications. However, various side effects are fairly common, including dry mouth, hypotension, constipation, nausea, sexual difficulties, and overstimulation. These side effects may remit on their own in a brief time or they may continue, and it is up to the patient whether to adjust to them or to stop using the medication. MAO-I can cause seriously elevated blood pressure if certain foods are eaten while on this drug. These foods include, but are not limited to, cheese, bananas, red wine, ripe figs, and fava beans. Patients on MAO-I must be instructed to abstain from these foods and should be in compliance with the restrictions for their safety.

Lithium, which is a naturally occurring salt, is often a very useful medication in treating the mood swings of bipolar disorder. People treated with lithium often experience a cessation of manic episodes, as well as a reduction in depressive episodes. If the depressive component of bipolar disorder does not respond to lithium alone, an antidepressant medication is often used in conjunction with it.

Electroconvulsive therapy (ECT) can be effective in treating severe, intractable depressions that have not responded to other treatments. It relieves depression about 75% of the time, usually in a series of electrical shocks to the brain over several days. Side effects of confusion and memory loss can result. For deeply depressed persons who do not obtain relief from medications and/or psychotherapy, ECT can be helpful in providing acute relief.

BIBLIOGRAPHY

Hales, D., and Hales, R. E. (1995). "Caring for the Mind: The Comprehensive Guide to Mental Health." Bantam Books, New York.

Karp, D. A. (1996). "Speaking of Sadness: Depression, Disconnection, and the Meanings of Illness." Oxford University Press, New York.

Mondimo, F. (1990). "Depression: The Mood Disease." Johns Hopkins University Press, Baltimore, MD.

Schucter, S., Downs, N., and Zisook, S. (1996). "Biologically Informed Psychotherapy for Depression." The Guilford Press, New York.

Seligman, M. (1993). "What You Can Change and What You Can't." Ballantine Books, New York.

Task Force on DSM-IV. (1994). "Diagnostic and Statistical Manual of Mental Disorders." American Psychiatric Association, Washington, D.C.

Depression, Neurotransmitters and Receptors

ROGER W. HORTON

St. George's Hospital Medical School

GLOSSARY

Neurotransmitter receptors Specialized membrane macromolecules that detect neurotransmitters and mediate their cellular effects

Neurotransmitters Chemicals that mediate communication between nerves and between nerves and other organs

Receptor agonists Drugs that mimic the effects of a neurotransmitter

Receptor antagonists Drugs that prevent the effects of a neurotransmitter

DEPRESSION AND MANIA ARE TERMED AFFECTIVE disorders; the primary abnormality is in mood or "affect." Affective disorders are often recurrent, may require repeated periods in hospital, and diminish the quality of life for the sufferers and their families. Various forms of treatment are available, but the development of more effective treatments requires a better understanding of the chemical changes in the brain in affective disorders.

I. CLINICAL FEATURES OF DEPRESSION

A. Symptoms

The clinical concept of depression embraces a wide range of symptoms. A patient with a clinically depressed mood will describe themselves as feeling sad, miserable, hopeless, pessimistic, or some similar colloquial equivalent. Loss of interest in usual and pleasurable activities is always present. Guilt and a sense of worthlessness are exaggerated and often inappropriate. A decrease in energy level is invariably present and is experienced as sustained fatigue in the absence of physical effort. Difficulty in concentrating, slowed thinking, and indecisiveness are common. Delusions and hallucinations are sometimes present. Patients often feel they would be better off dead and suicidal thoughts and acts are common. Depressed moods often lessens in severity as the day goes on (diurnal variation). Physical symptoms are also usually present. Appetite is frequently reduced with a consequent weight loss; a minority of patients eat more and gain weight. Sleep is nearly always disturbed, often with early morning wakening or sometimes with difficulty getting to sleep. Some patients are unable to be still, constantly pacing or wringing their hands (psychomotor agitation); others may show slowed body movements and slow and monotonous speech (psychomotor retardation). The presence or absence and severity of particular symptoms vary widely between patients. [*See* Depression.]

The distinction between depressive illness and the downward mood swings that we all experience as part of daily living lies in the severity and persistence of symptoms throughout nearly every day for at least

ENCYCLOPEDIA OF HUMAN BIOLOGY, Second Edition, VOLUME 3.

2 weeks of duration and by changes from previous social and occupational functioning.

Some patients experience only a depressed mood (termed unipolar depression). Others experience periods of euphoric mood (mania), intermingled with depressive episodes (termed bipolar disorder or manic depressive illness). Manic episodes are characterized by inflated self-esteem, ranging from uncritical self-confidence to delusional grandiosity, often resulting in reckless behaviors. Energy is increased with a decreased need for sleep and food and an increased sexual interest. Thought and speech are rapid and copious, with abrupt and frequent changes from topic to topic. Frequently, manic patients have little insight into their illness and resist treatment.

Symptoms of depression usually develop over days to weeks and, if left untreated, an episode may last 6 months or longer. Manic episodes usually last from a few days to months and begin and end more abruptly than depressive episodes.

B. Frequency and Epidemiology

Depressive illness is among the most prevalent psychiatric disorders in adults. Estimates of point prevalence (the proportion of the population with the diagnosis at a point in time) for unipolar depression is 2–3% in adult males and 5–9% in adult females, with a lifetime risk of 8–12% for men and 20–26% for women. Bipolar disorder is much less common, with a lifetime risk for men and women of about 1%. Risk factors for unipolar depression include being female between the ages of 35 and 45 years, being of low socioeconomic class, having a family history of depression or alcoholism, experiencing a parental loss before the age of 11 years, living in a negative home environment, lacking a confiding relationship, and having a baby in the previous 6 months. Bipolar disorder is more prevalent in higher socioeconomic classes and mean age of onset is earlier than unipolar depression. Although negative life events play a similar role to unipolar depression, being female or lacking a confiding relationship does not increase the risk for a bipolar disorder. Bipolar disorder is more strongly associated with a family history of affective disorder than unipolar depression, indicating a stronger genetic component.

C. Classification and Measurement

The division into bipolar and unipolar depression is generally accepted. Further subdivisions of unipolar depression have been proposed, but the terminology is confusing. Endogenous depression refers to a symptom cluster, including marked physical signs (early morning wakening, diurnal variation, weight loss), with a relative lack of external precipitant and a well-adjusted premorbid personality. Melancholia is an approximate equivalent term. Nonendogenous depression refers to a symptom cluster where physical signs are not prominent, mood varies with external circumstances, and delusions and hallucinations never occur. Other terms used include depressive neurosis and reactive depression. However, there is considerable disagreement as to whether these are distinct illnesses, which it is important to differentiate, or extremes of a continuous spectrum.

The inadequacies of these classifications are important for research investigations because of the need to identify relatively homogeneous populations. Consequently, a number of classification systems have been developed based on standardized structured interviews and operational definitions of symptoms. High interrater reliability can be achieved. The Research Diagnostic Criteria and the Diagnostic and Statistical Manual of Mental Disorders (DSM III-R) are widely used for research purposes.

Standardized rating scales exist for the numerical assessment of the severity of depression. The observer-rated Hamilton depression rating scale and the self-rating Beck scale are widely used for research purposes. These measures are quite sensitive to changes in the intensity of the illness and eliminate subjective bias. [See Mental Disorders.]

D. Treatment

Monoamine oxidase (MAO) inhibitors were the first class of drugs used for the treatment of depression. However, because of the need for dietary restrictions (foods containing high concentrations of tyramine, such as cheese, induce marked increases in blood pressure in patients receiving MAO inhibitors) and other adverse effects, their use declined rapidly following the introduction of tricyclic antidepressants. The development of isoenzyme selective MAO inhibitors and subsequently of reversible MAO$_A$ inhibitors, such as moclobemide that eliminate the need for rigid dietary restrictions, have rekindled interest in this area. The tricyclic antidepressants, which include imipramine, amitriptyline, and dothepin, became the first line of drug treatment for depression; a position they maintained for about 20 years, despite their adverse effects. These include atropine-like (antimuscarinic) effects such as dry mouth, blurred vision, constipation, and urinary retention, and cardiovascular effects including postural hypotension, arrhythmias, and tachycardia. "Second generation" or atypical antidepressants, such as mianserin, iprindole, trazodone, maprotiline, and viloxazine, have reduced antimuscarinic effects and less cardiotoxicity than the tricyclics. More recently, the major change has been the introduction and widespread use of the selective serotonin (5-hydroxytryptamine, 5-HT) reuptake inhibitors (SSRIs), which include fluoxetine, fluvoxamine, paroxetine, and sertraline. These drugs are largely devoid of antimuscarinic effects and cardiotoxicity, although they can induce adverse gastrointestinal effects, including diarrhoea, nausea, and vomiting. A consistent feature of all classes of drugs used to treat depression is that there is a delay of about 3 weeks before beneficial effects become apparent.

Electroconvulsive therapy (every 2–3 days for 10 treatments) works quicker than drugs, is often the treatment of choice in suicidal patients, and is more effective than tricyclics if delusions are present. Psychological treatments (interpersonal therapy and cognitive therapy) appear to be as effective as drugs in mild depressions, particularly depressive neurosis. Psychosurgery (subcortical tractotomy) is occasionally performed in severe long-standing cases of depression where other treatments have failed.

Acute mania is often treated initially with neuroleptics (most often haloperidol) and lithium carbonate. Neuroleptics are withdrawn as soon as the effect of lithium is established. Long-term lithium treatment is effective in preventing further episodes. Carbamazepine and sodium valproate have been advocated as effective treatments in rapid cycling bipolar patients (who experience four or more episodes of affective illness per year).

II. THEORIES UNDERLYING THE BIOLOGICAL BASIS OF DEPRESSION

A. Historical Background

Attempts to explain depressive illness in terms of altered brain chemistry have been dominated since the mid-1960s by theories involving abnormalities in the functioning of the monoamine neurotransmitters, noradrenaline (NA) and 5-HT. The link between these neurotransmitters and the affective state in man arose from a combination of clinical observations and animal experimentation. In the

1950s reserpine was widely used as an antihypertensive drug and a proportion of patients developed symptoms of depression and attempted suicide. Clinical trials of the antituberulosis drug, iproniazid, produced mood elevation and reversal of depressive symptoms associated with chronic physical illness. Reserpine administered to animals caused a profound depletion of brain monoamines by preventing their intraneuronal storage, allowing them to leak into the cytoplasm and be metabolized by the enzyme MAO. Iproniazid reversed the behavioral effects of reserpine in animals by inhibiting MAO and allowed leakage of monoamines, rather than their inactive metabolites, from nerve terminals. Imipramine, a tricyclic antidepressant, prevented the behavioral effects of reserpine, not by inhibiting MAO, but by blocking the reuptake mechanism (the main mechanism of inactivation of monoamine neurotransmitters) and thus increasing their synaptic lifetime.

B. Formulation and Evolution of the Monoamine Hypothesis of Depression

This and other evidence led to the formulation of the catecholamine hypothesis of depression which stated that ". . . some, if not all, depressions are associated with an absolute or relative deficiency of catecholamines, particularly NA, at functionally important receptor sites in the brain. Elation conversely may be associated with an excess of such amines. . . ." Because much of the evidence was unable to distinguish between an involvement of NA and 5-HT, an analogous indoleamine hypothesis of depression quickly followed. The therapeutic benefits of the MAO inhibitors (that prevent the breakdown of monoamines) and of the tricyclic antidepressants (that increase the synaptic concentrations of monoamines by blocking their reuptake) added considerable support to the monoamine hypotheses.

Although the monoamine hypotheses have undergone refinement over the years, further developments have stimulated a major reexamination. The introduction of "second generation" or atypical antidepressants, such as mianserin and iprindole, that lack both MAO inhibition and amine reuptake blocking properties and the lack of antidepressant action of established amine reuptake blockers, such as cocaine and amphetamine, cast doubts on the relationship between these biochemical events and antidepressant drug action. This is further emphasized by the rapid onset of uptake or MAO inhibition compared to the therapeutic response in patients which is delayed for up to 3 weeks despite daily medication. This temporal mismatch suggests that longer-term adaptive changes, which may be initiated by the acute effects, are necessary for the development of therapeutic antidepressant activity. This has prompted a large number of studies to examine the effects of repeated administration of antidepressant drugs to animals. A range of techniques have been used, including behavioral studies and electrophysiological responses of single nerve cells to direct application of neurotransmitters. The quantitation of neurotransmitter receptors *in vitro* by labeling with radioactive drugs (radioligand binding) has made an important contribution.

An important finding was that antidepressants reduce the number of β-adrenoceptors and diminish the response of NA-sensitive adenylate cyclase (an intracellular enzyme which catalyzes the conversion of ATP to cyclic AMP and which acts as a second messenger system to mediate the effects of β-adrenoceptor stimulation) in rat cortex. This effect was not apparent after acute administration but was found to develop over a 3-week period of daily antidepressant administration. This property is shared by most antidepressant drugs (including tricyclics, MAO inhibitors, and atypical antidepressants), but not other psychoactive drugs, and by repeated but not single application, of electroconvulsive shocks [(ECS) a procedure in animals designed to mimic electroconvulsive therapy]. The importance of this finding is that the time course of these effects is similar to that of clinical improvement in depressed patients and that the effects are shared by most antidepressant treatments (although the SSRIs appear to be an exception).

However, effects of repeated administration of antidepressants are not limited to the β-adrenoceptor system. Reductions in α_2-adrenoceptor number and function have been reported, but are more variable depending on the individual drug, the dosage, and the duration of administration. Decreases in the number of 5-HT$_2$ receptors have also been reported following the antidepressant drugs of many chemical classes. A similar time course is seen to the effects on the β-adrenoceptor system. However, repeated ECS, unlike drug treatment, increase 5-HT$_2$ receptor number and function.

Although the detailed mechanisms involved in these adaptive receptor changes in animals and their relationship to antidepressant action remain unclear, these findings raise the possibility that depression per se may be related to abnormalities in the number or function of neurotransmitter receptors. Such effects may be secondary to primary abnormalities in neurotransmitter availability or turnover.

III. METHODS USED TO STUDY THE BIOLOGY OF DEPRESSION

Animal studies of antidepressant drugs have strongly influenced the thinking about depression and the mechanisms involved in the therapeutic benefits of treatment. However, critical testing of the hypotheses generated in animals requires experiments in depressed patients. Such studies are limited by practical and ethical issues. If the objective is to identify biological abnormalities associated with depression, a relatively homogeneous patient group usually based on operational classification systems, such as DSM III or RDC (see Section I,C), is compared with healthy volunteers. To exclude the confounding effects of recent or current drug treatment, depressed patients need to have never previously been treated with drugs (rarely achievable) or to be drug free for a period (ideally 4–6 weeks, in practice more often 1–2 weeks) prior to measures being performed. To minimize other confounding variables, patients and controls need to be matched as closely as possible for other factors, such as age, sex, race, or socioeconomic group, and in females, menstrual status. Depressed patients are often studied when drug free and throughout a standardized course of treatment. Such studies not only provide information on the effects of treatment, but allow distinctions to be made between biological abnormalities that are related to mood (so-called state-dependent markers) and those unrelated to current mood (so-called trait markers) which may reflect a predisposing abnormality.

The following section describes some of the methods used, together with their advantages and disadvantages.

A. Measurements in Blood, Urine, and Cerebrospinal Fluid (CSF)

Measurements of neurotransmitters and their metabolites in readily accessible body fluids, such as blood or urine, are poor substitutes for our inability to perform such measures in living human brain. Measurements in blood and urine are technically possible; the difficulty is in trying to distinguish that component derived from the brain from that derived from other body organs. For most neurotransmitters this distinction is not possible. Measures in CSF are a more useful, but still imperfect, measure of neurotransmitter turnover in the whole of the central neural axis; in the case of some neurotransmitters, there may be a significant contribution from the spinal cord. Concentration gradients within CSF necessitate matching patients and controls for height.

B. Peripheral Model Receptor Systems

Blood platelets possess a number of features in common with nerve cells, including a common embryonic origin. Platelets are able to take up 5-HT by a carrier-mediated mechanism and store it in intracellular organelles in a manner analogous to nerve cells. Platelets also possess a number of cell surface receptors for neurotransmitters, including 5-HT and NA. Thus platelets have been widely studied as readily accessible models of neurones. However, the validity of such models has not been convincingly established. [*See* Platelet Receptors.]

C. Neuroendocrine Challenge Tests

The basis of such tests lies in the ability of drugs to interact with central neurotransmitter systems to influence the circulating concentrations of pituitary hormones. Such tests have the advantage over many procedures in that they provide an index of transmitter function instead of a static measure of receptor numbers. Some well-established neuroendocrine challenge tests, e.g., stimulation of growth hormone secretion by the α_2-noradrenergic agonist clonidine, exist, but the number of tests has been limited by the availability of selective pharmacological agents suitable for administration in man. [*See* Neuroendocrinology.]

D. Postmortem Studies

The opportunity to study brain tissue from depressed patients is obviously a great advantage over the more indirect approaches outlined earlier. However, it is rare that such tissue is available from living subjects. An alternative is to study brain tissue obtained after death, i.e., at postmortem examination. However, great care is needed to ensure that apparent biological differences between depressed subjects and controls are due to depression and do not arise from some other cause. The greatest difficulty is in being able to identify suitable subjects. Two approaches have been used. One is to study subjects with an antemortem diagnosis of depression (based on established criteria)

who die by natural causes in hospitals. Such subjects are invariably elderly. The second approach is to study suicide victims. The rationale for this approach is that people who commit suicide almost always are suffering from psychiatric illness, of which depression is the most common. A limited number of studies have been performed in selected groups of suicides in whom there was sufficient documentary medical evidence to allow a firm retrospective diagnosis of depression. Such studies are likely to identify biological abnormalities more specifically associated with depression than studies performed in more heterogeneous groups of suicides which are likely to include subjects with other psychiatric diagnoses (such as schizophrenia, personality disorder, or alcoholism). However, it is rarely possible to achieve the detailed subclassification of depression that is possible in living subjects. The choice of control subjects in postmortem studies is also important. Control subjects need to be free of psychiatric illness and to be closely matched to the index subjects, particularly for age, sex, and postmortem delay (the time from death to freezing of tissue). Subjects dying after a protracted physical illness, particularly if drug treatment is involved, are clearly less than ideal. There is also a clear need in postmortem studies to establish that biological differences do not arise from artifactual causes related to instability of substances after death or upon storage. [*See* Suicide.]

IV. INVOLVEMENT OF 5-HYDROXYTRYPTAMINE

A. CSF Studies

Measurement of CSF 5-hydroxyindoleacetic acid (5-HIAA) concentration, the major metabolite of 5-HT, has been extensively studied as an index of central 5-HT function in depression. Although many studies have reported lower CSF 5-HIAA concentrations in drug-free depressed patients than controls, other studies have found no differences. Thus an association between low CSF 5-HIAA concentration and depression does not appear to be very consistent. Stronger evidence favors an association between low CSF 5-HIAA concentration and suicidal behavior, particularly in subjects who make violent suicide attempts. This relationship is not restricted to depressed patients as it is also seen in subjects with personality disorder, alcoholism, and schizophrenia. A low CSF 5-HIAA concentration has also been associated with various forms of impulsive and aggressive behavior. The important relationship appears to be between low CSF 5-HIAA concentration and impulsivity and aggression, which may be inwardly directed in the case of violent suicide rather than with depression per se.

B. Peripheral Models

Most studies that have employed adequate methodology have reported that the ability of platelets to take up 5-HT (measured *in vitro*) is lower in drug-free depressed patients than in controls, although a considerable overlap in values is evident in all studies. The reduction in uptake is due to a decrease in the maximum velocity of uptake (which is directly related to the number of uptake sites) rather than to a difference in the affinity of the uptake carrier for 5-HT.

The 5-HT uptake sites in platelets and brain can be quantitated by radiolabeling with the antidepressant drug [3H]imipramine. A practical advantage of this approach is that the platelet [3H]imipramine-binding sites are stable to repeated washing and freezing whereas active platelet 5-HT uptake is unable to withstand such treatments and requires that experiments are completed with a few hours of blood sampling. Initial studies reported that the number of platelet [3H]imipramine-binding sites was lower, by as much as 50%, in drug-free depressed patients compared to controls, whereas binding affinity was unaltered. Many studies have replicated this finding, although the magnitude of the difference has generally been less than that initially reported. However, a significant number of studies have found no differences in the number of [3H]imipramine-binding sites between depressed patients and controls. A number of factors have been proposed to explain these discrepant findings, including differences in assay methodology and patient selection. However, no single factor to date has been convincingly demonstrated to account for these differences. It has been argued that reduced platelet [3H]imipramine binding may be restricted to certain subgroups of depressed patients, such as those with a family history of depression or dexamethasone test nonsuppressors (see Section VII), or with specific clinical symptoms, such as retardation or psychosis. However, no such associations have been consistently replicated.

A problem with [3H]imipramine is that it can also label sites that are not related to 5-HT uptake. More recent studies have used [3H]paroxetine, a radiolabeled SSRI, that has high affinity and high selectivity for 5-HT uptake sites and, unlike [3H]imip-ramine, does not label additional sites unrelated to 5-HT uptake. Several studies have reported no differences in the number of platelet [3H]paroxetine-binding sites between drug-free depressed patients and controls.

Platelet 5-HT receptors, which are of the 5-HT$_2$ type (see Section III,D), do not differ in number or affinity between drug-free depressed patients and controls. However, treatment of depressed patients with tricyclic antidepressants or administration of desmethylimipramine to healthy volunteers resulted in a marked increase in the number of platelet 5-HT$_2$ receptors. This is in sharp contrast to the reduction in 5-HT$_2$ receptors in rat cortex following repeated antidepressant administration. Accumulating evidence shows that the numbers of platelet 5-HT$_2$ receptors are higher in patients who attempt suicide; however, this finding is not restricted to depressed patients. Platelet 5-HT-mediated aggregation has been reported to be reduced and unaltered in drug-free depressed patients compared to controls.

C. Neuroendocrine Studies

Activation of the 5-HT system by precursor loading (with tryptophan or 5-hydroxytryptophan), fenfluramine (which releases 5-HT and inhibits its reuptake), and clorimipramine (5-HT reuptake inhibitor) stimulates the release of prolactin and growth hormone. The prolactin response to these agents has generally been found to be lower in depressed patients than in controls. These findings strongly suggest an underactivity of the 5-HT system in depression, but are unable to unequivocally distinguish among transmitter synthesis, release, and receptor-mediated effects. More selective agents are now being used in such studies, e.g., agonists at 5-HT$_{1A}$ receptors such as ipsapirone and buspirone. These drugs increase the circulating concentrations of prolactin, growth hormone, adrenocorticotrophic hormone, and cortisol, and produce hypothermia. Hormonal responses to 5-HT$_{1A}$ agonists in depressed patients have not produced consistent findings, but the hypothermic response appears to be attenuated compared to controls, suggesting a decreased response of at least some 5-HT$_{1A}$ receptors in depression.

D. Postmortem Studies

There are several studies of 5-HT and 5-HIAA concentrations in brain tissue from psychiatrically undefined suicide victims. Although isolated differences have been reported, most studies found that concentrations were not significantly different from controls in cortical areas. The most consistent finding has been a reduction in 5-HT and or 5-HIAA concentration in the hindbrains of suicides. Taken together, the evidence suggests that the lower CSF 5-HIAA concentrations seen in violent suicide attempters are likely to be a reflection of a reduced turnover of 5-HT in hindbrain rather than in higher brain centers. The 5-HIAA concentration in patients with an antemortem diagnosis of depression who died by natural causes did not differ from controls in the frontal or occipital cortex or the hippocampus.

Studies of [^{3}H]imipramine-binding sites in suicide victims have yielded inconsistent results. In the frontal cortex, increased, decreased, and unaltered binding has been reported. These inconsistent findings may result from [^{3}H]imipramine-labeling sites that are unrelated to 5-HT uptake. Such sites are far more abundant in human cortex than in rat cortex or human platelets. A further factor may be the marked differences in [^{3}H]imipramine binding in frontal cortical samples from left and right hemispheres. In control subjects the number of binding sites was twofold higher in the right hemisphere than in the left hemisphere, whereas the converse was true in subjects with a variety of psychiatric disorders. The significance of this finding is not clear because when [^{3}H]paroxetine was used to label 5-HT uptake sites, no hemispheric asymmetry was found in control subjects or depressed suicides. One possibility is that the hemispheric asymmetry relates to sites labeled with [^{3}H]imipramine that are unrelated to 5-HT uptake, although the exact nature of such binding sites has not been established. At least three studies have examined [^{3}H]paroxetine binding in depressed suicides; no differences in the number of sites were found compared to controls in several cortical and subcortical areas. In patients with an antemortem diagnosis of depression who died by natural causes, [^{3}H]imipramine has been reported to be lower in the occipital cortex and hippocampus, whereas [^{3}H]paroxetine binding in the frontal cortex did not differ. Thus the most recent studies that have used the selective radioligand [^{3}H]paroxetine provide no evidence for an alteration in the number of 5-HT uptake sites in depression, either in blood platelets or in brain. This does not preclude other abnormalities in the uptake system that influence the efficiency of the process.

It is now generally accepted that 5-HT interacts with at least four distinct classes of receptors, termed 5-HT$_{1-4}$. Further heterogeneity exists within the 5-HT$_1$ and 5-HT$_2$ classes. At least three studies have examined 5-HT$_{1A}$ receptor binding, and no differences

were found between suicides or suicides with a retrospective diagnosis of depression and controls. Similarly, no differences in 5-HT$_{1D}$ receptor binding were found in cortical samples, although higher numbers of 5-HT$_{1D}$ sites have been reported in globus pallidus of suicides who died by violent methods than in controls.

5-HT$_2$ receptors (now more correctly termed 5-HT$_{2A}$ receptors) have been extensively studied, but again the findings are conflicting, with some studies reporting higher numbers in the frontal cortex of suicides, whereas other studies found no differences between suicides and controls. A further study provided evidence that higher numbers of frontal cortical 5-HT$_2$ receptors were restricted to those suicides who died by violent means; the number of frontal cortical 5-HT$_2$ receptors in suicides who died by ingestion of toxic agents did not differ from controls. This provided an opportunity to reconcile the previous findings since those studies that had found higher numbers of cortical 5-HT$_2$ receptors had included largely or exclusively suicides who had died violently, whereas those studies that had not found higher numbers of 5-HT$_2$ receptors had included a higher proportion of suicides who had died nonviolently. However, this proposed relationship between violence of death and cortical 5-HT$_2$ receptors in suicide was not replicated in a subsequent large-scale study. No significant difference in 5-HT$_2$ binding was found in the frontal cortex of subjects with an antemortem diagnosis of depression who died by natural causes.

V. INVOLVEMENT OF NORADRENALINE

A. CSF Studies

Studies of CSF NA and 3-methoxy-4-hydroxyphenylglycol (MHPG), the major central metabolite of NA, in affective disorder patients have not produced dramatic or consistent findings. Most studies report a greater variation in both NA and MHPG concentrations in patients than in controls. In general, manic patients and some unipolar depressed patients have increased concentrations of NA and MHPG, whereas bipolar patients when depressed tend to have lower concentrations than controls.

B. Peripheral Models

Blood platelets possess cell surface α-adrenoceptors of the α_2 type, which have a similar pharmacological profile to their central counterparts. Activation of these receptors *in vitro* induces platelet aggregation. This has been used as an *ex vivo* test of platelet α_2-adrenoceptor function. Quantitation of human platelet α_2-adrenoceptors by radioligand binding is dependent on the pharmacological nature of the radioligand used. The α_2-adrenoceptor antagonists, such as yohimbine and rauwolscine, appear to label the total population of α_2-adrenoceptors with uniform affinity. In contrast, the α_2-adrenoceptor agonists, such as clonidine and UK 14,304, label a proportion of sites with high affinity and the remainder of sites with lower affinity. The relative affinities for these two classes of sites vary from one agonist to another. High- and low-affinity conformation sites are in dynamic equilibrium and the proportion of sites can be reversibly altered *in vitro* by changes in the ionic composition of the medium and the presence of guanine nucleotides.

Several studies have examined platelet α_2-adrenoceptors in drug-free depressed patients and controls using antagonist radioligands. None of the studies found differences in the number of binding sites. In contrast, higher numbers of α_2-adrenoceptor-binding sites have been reported in drug-free depressed patients compared to controls using the agonist ligand [^{3}H]clonidine. The increase was initally reported to be confined to the high-affinity binding sites. One might conclude that while the total number of binding sites are unaltered in depression, there is a selective increase in the proportion of sites that exist in high-affinity agonist conformation. However, not all the evidence is in favor of this conclusion. A subsequent study with the same ligand found the number of both high- and low-affinity sites to be increased in depressed subjects, and other studies using restricted concentrations of other agonists, selected to label only the high-affinity conformation, have found the number of sites to be reduced or unaltered in depressed patients. Studies showing increased sensitivity to the aggregatory effects of adrenaline in platelets from depressed patients tend to support the findings of increased high-affinity binding sites. Other studies have shown that the aggregatory responses do not differ.

Although blood platelets possess a small number of β-adrenoceptors, these receptors have been most extensively studied in white blood cells. Two early studies provided evidence that lymphocyte β-adrenoceptor function was reduced in depression. Stimulation of adenylate cyclase with the β-adrenoceptor agonist isoprenaline resulted in lower cyclic AMP formation in lymphocytes from drug-free depressed and manic patients than in control subjects. The β-adrenoceptor binding was also lower in patients than in controls. A specific reduction in β-adrenoceptor-mediated cyclic AMP production was advocated since prostaglandin E_1 (PGE_1)-stimulated cyclic AMP production (which is via specific receptors distinct from β-adrenoceptors) did not differ between the groups of subjects. Similar independent findings of lower isoprenaline, but unaltered PGE_1-stimulated cyclic AMP production, were reported in unipolar depressed and bipolar manic patients. A subsequent study has confirmed the reducing isoprenaline-stimulated cyclic AMP production, but in this case in the absence of any alteration in β-adrenoceptor binding. A further interesting approach has been the establishment of lymphoblastoid cell lines in culture by transforming lymphocytes with Epstein–Barr virus from manic depressed patients, their unaffected relatives, and controls. The β-adrenoceptor binding was reduced to less than half the control values in 4 out of 6 cell lines from manic depressed patients but in only 1 out of 18 cell lines from unaffected relatives. Other cell surface markers did not differ between cell lines derived from manic depressive and control subjects.

The further classification of β-adrenoceptors into β_1 and β_2 subtypes is well established. In rodent and human brain, β_1-adrenoceptors are the predominant form in most brain regions (except cerebellum) and the reduction in β-adrenoceptors in rat brain following repeated antidepressant administration is limited to the β_1 subtype. Although the evidence is not convincing, lymphocytes β-adrenoceptors appear to be predominantly of the β_2 subtype, casting doubts on the suitability of lymphocyte β-adrenoceptors as a suitable model of central β-adrenoceptors.

C. Neuroendocrine Studies

The intravenous infusion of clonidine in man induces a release of pituitary growth hormone. This is mediated by α_2-adrenoceptors thought to be located in the arcuate nucleus of the hypothalamus. The growth hormone release in response to clonidine has been reported by several groups to be lower in patients with endogenous depression than in sex- and age-matched controls. Since growth hormone release induced by dopamine agonists is unaltered, this suggests a selective reduction in the sensitivity of the α_2-adrenoceptors that mediate this response rather than a generalized defect in growth hormone secretion in depressed patients. Most studies have found that other effects induced by clonidine, such as reduced blood pressure and increased sedation, which are also mediated by α_2-adrenoceptors, do not differ between endogenously depressed patients and controls. This suggests a selective reduction in the sensitivity of certain α_2-adrenoceptors in depression rather than a global change in all α_2-adrenoceptors. Preliminary experiments in drug-free recovered depressed patients indicate that the growth hormone response remains reduced, suggesting that α_2-adrenoceptor subsensitivity represents a state-independent biological abnormality in endogenous depression.

There are no well-established neuroendocrine challenge tests of β-adrenoreceptor function in man. One area of study is the release of melatonin from the pineal gland, which is stimulated by β_1-adrenoceptor activation via a sympathetic innervation from the superior cervical ganglion. Unstimulated melatonin secretion appears to be reduced in depressed patients, suggesting that net noradrenergic activity within this system is reduced in depression. Treatment of depressed patients for up to 3 weeks with desmethylimipramine consistently increased nighttime plasma melatonin concentrations, suggesting that down-regulation of β-adrenoceptors is not the predominant effect of antidepressant treatment, at least within this system.

D. Postmortem Studies

Although rather few in number, studies in postmortem brain from suicides or depressed patients have generally found no differences in NA or MHPG concentrations compared to controls.

Two studies have reported markedly higher β-adrenoceptor binding in the frontal cortex of suicide victims compared to controls. One study reported a mean increase of 50% in the number of sites in the suicide group, with increases in individual subjects ranging from 30 to 110% of the values in individually matched controls. The increase in β-adrenoceptor binding was largely restricted to the β_1-adrenoceptor subtype. The other study found a mean increase of 73% in suicides compared to controls. The suicides in both studies died largely by violent means. In contrast to these relatively large differences, at least two other studies found no differences in β-adrenoceptor binding to the frontal cortex between suicides and controls. As with other studies in psychiatrically unclassified suicide victims, it is difficult to attribute the biological differences, or the lack of, specifically to depression. The only study restricted to suicide victims with a retrospective diagnosis of depression who had

not been treated with antidepressant drugs recently found modestly lower rather than higher β-adrenoceptor binding in cortical areas. In subjects with an antemortem diagnosis of depression who died by natural causes, the β-adrenoceptor binding in the frontal cortex did not differ from controls.

No differences in the number of α_1-adrenoceptor-binding sites have been reported between suicides and controls. Two studies have reported higher numbers of α_2-adrenoceptor-binding sites in suicides with a retrospective diagnosis of depression compared to controls; in one study the increase was found in the frontal cortex, in the other the temporal cortex.

VI. INVOLVEMENT OF OTHER NEUROTRANSMITTERS

A. Dopamine (DA)

Homovanillic acid (HVA) concentration, the major metabolite of DA, has been reported to be lower in drug-free depressed patients than in controls in several studies, whereas patients with delusional/psychotic depression and bipolar manic patients have higher HVA concentrations than nondelusional depressed patients and controls. Although the number of studies is rather limited, neuroendocrine responses to DA agonists are not consistently different from controls in depressed or manic patients. Antidepressant effects have been reported with DA agonists, particularly in bipolar depressed patients. Numerous reports also show that high doses of neuroleptics, which block DA receptors, induce depressive symptoms and that a rebound improvement in mood and even hypomania are seen following neuroleptic withdrawal. Low doses of neuroleptics in combination with antidepressants are a clinically acceptable treatment for delusional/psychotic depression and there is substantial evidence that the efficacy of neuroleptics in treating mania is due to the blockade of DA receptors.

B. Acetylcholine

Evidence suggests that drugs which mimic or antagonize acetylcholine have effects on mood in man. Muscarinic agonists and inhibitors of acetylcholine esterase (the enzyme which catalyzes the inactivation of acetylcholine) can intensify depressive symptoms in unipolar patients, cause a depressed mood in euthymic bipolar patients, and reduce symptoms in manic patients. Anecdotal evidence also shows that muscarinic antagonists alleviate depression. Some tricyclic antidepressants are potent muscarinic antagonists. This property does not appear necessary for antidepressant activity since many clinically effective "second generation" drugs lack this action. Antimuscarinic action is an important source of troublesome side effects, such as dry mouth and blurred vision. While there are isolated reports of increased muscarinic receptor binding in cultured fibroblasts from affective disorder patients and in the frontal cortex of suicides compared with controls, other studies have not replicated these findings.

C. γ-Aminobutyric Acid (GABA)

Several lines of evidence suggest an involvement of GABA-mediated neurotransmission in antidepressant drug action. GABA is a major inhibitory neurotransmitter in the mammalian central nervous system and exerts its action through two separate classes of receptors: bicuculline-sensitive $GABA_A$ receptors, which include binding sites for benzodiazepines and barbiturates, and are coupled to chloride ion channels, and bicuculline-insensitive $GABA_B$ receptors, which are linked to potassium and calcium ion channels. While selective agonists at $GABA_A$ or $GABA_B$ receptors are not antidepressant, mixed agonists at $GABA_A$ and $GABA_B$ receptors, such as progabide and fengabine, are active in animal behavioral models of antidepressant drug activity and are clinically effective antidepressants in open and double blind trials. The repeated administration of antidepressant drugs to rats has been reported in some, but not all, studies to increase the number of $GABA_B$-binding sites in the frontal cortex and hippocampus.

However, there are relatively few studies of indices of GABA neurochemistry in depressed patients. CSF GABA and plasma GABA (which is thought to be largely of central origin) concentrations have generally been reported to be lower in drug-free unipolar depressed patients and, in some studies, higher in mania than in controls. The number of benzodiazepine binding sites (a marker for $GABA_A$ receptors) has been reported to be higher in the frontal cortex of depressed suicides than in controls, but did not differ in temporal cortex, amygdala, or hippocampus. $GABA_B$-binding sites did not differ in frontal or temporal cortex or hippocampus between drug-free depressed suicides and controls. Glutamic acid decarboxylase activity (the enzyme that catalyzes the synthesis of GABA from glutamic acid)

does not differ between depressed suicides and controls (when subjects who had died by carbon monoxide poisoning were excluded).

VII. ENDOCRINE ABNORMALITIES IN DEPRESSION

Hyperactivity of the hypothalamic–pituitary–adrenal (HPA) axis is one of the best established biological findings in depression. HPA activity is manifest as increased plasma cortisol, increased urinary excretion of free cortisol and cortisol metabolites, and failure of the synthetic steroid dexamethasone to reduce plasma cortisol [the dexamethasone suppression test (DST)]. The DST clearly distinguishes groups of depressed patients from controls; about half the patients with a major depressive disorder were nonsuppressors in the DST (i.e., had midafternoon plasma cortisol concentrations >5 μg/dl on the day following dexamethasone) compared to only 7–8% of controls. However, comparable rates of nonsuppression to those in major depressive disorder are also found in acutely psychotic and demented patients.

It has been proposed that the hypercortisolemia of depression is due to a defect at or above the hypothalamus that results in hypersecretion of corticotropin-releasing factor (CRF), leading to an increased release of adrenocorticotrophic hormone (ACTH) from the pituitary and an increased release of cortisol from the adrenals. The corticotrophic cells subsequently become less sensitive to CRF, ACTH secretion returns to normal, but the hypersensitivity of the adrenal cortex to ACTH continues to result in hypersecretion of cortisol. The neurosecretory cells of the hypothalamus receive numerous neuronal inputs from higher brain centers, acting via many different neurotransmitter systems. The involvement of these systems in the proposed hypersecretion of CRF remains unclear. It has also been suggested that alterations in the neurotransmitter receptor number and function in depression may be the result of sustained increases in circulating cortisol.

Increased secretion of CRF may not be limited to the hypothalamus. In addition to its endocrine role, strong evidence shows that CRF acts as a neurotransmitter within the brain. CRF administered into the brain ventricles of animals activates the locus coeruleus (the nucleus containing the cell bodies of the ascending NA pathways) and decreases feeding and sexual behavior. CRF concentrations have been reported to be higher in CSF of drug-free depressed patients compared with controls. CRF receptor-binding sites have also been reported to be lower in the frontal cortex of suicides compared to controls. This finding is compatible with an adaptive decrease in CRF receptors in response to chronic hypersecretion of CRF.

is now only dependent on the development of suitable radioligands and should make a vital contribution to our understanding of the illness. Although affective disorders demonstrate complex inheritance, the powerful techniques of molecular biology have begun to make significant impacts, particularly in bipolar disorder.

VIII. FUTURE PROSPECTS

Despite considerable advances in recent years, the biological basis of depression still remains in doubt. Heterogeneity in the clinical presentation of the illness may indicate heterogeneity within the underlying chemical pathology and may account for the disappointing lack of replication of findings between studies. However, developing techniques provide exciting prospects for the future. The ability to image neurotransmitter receptors in the brains of living subjects

BIBLIOGRAPHY

Bloom, F. E., *et al.* (eds.) (1994). "Psychopharmacology: The Fourth Generation of Progress." Raven Press, New York.

Caldecott-Hazard, S., *et al.* (1991). Clinical and biochemical aspects of depressive disorders. *Synapse* **9**, 251–301.

Horton, R. W., and Katona, C. L. E. (eds.) (1991). "Biological Aspects of Affective Disorders." Academic Press, London.

Tipton, K. F., and Youdim, M. B. H. (eds.) (1989). "Biochemical and Pharmacological Aspects of Depression." Taylor & Francis, London.

Willner, P. (1985). "Depression: A Psychobiological Synthesis." Wiley, Chichester, UK.

Developmental Neuropsychology

JOHN E. OBRZUT
ANNE UECKER
University of Arizona

I. Theoretical Issues and Research
II. Common Disorders
III. Perspectives and Issues in Assessment

GLOSSARY

Affective disorders Group of disorders characterized by a disturbance of mood and not caused by any other physical or mental disorder: unipolar depression, disorder of individuals who have experienced episodes of depression but not of mania; bipolar disorder, disorder of people who have experienced episodes of both mania and depression or mania alone

Aphasia Absence or impairment of the ability to communicate through speech, writing, or signs, due to dysfunction of brain centers; it is considered to be complete or total when both sensory and motor areas are involved

Broca's area Area located in the inferior frontal gyrus of the brain that is responsible for expressive language; a lesion to this area results in expressive aphasia

Cerebral asymmetries Specialization of the two cerebral hemispheres: The left hemisphere has superiority for functions such as language, whereas the right hemisphere specializes in behaviors such as music and the holistic perception of patterns and faces

Corpus callosum Great commissure of the brain connecting the two cerebral hemispheres

Down's syndrome A form of moderate to severe mental retardation most generally caused by an extra 21st chromosome

Equipotentiality All areas of the cerebrum possessing an equal ability to assume behavioral functions; specific areas in the brain do not govern specific aspects of behavior

Hemidecortication Removal of one-half of the cortex; a hemidecorticate is an individual with only one hemisphere

Intelligence quotient Standardized measure indicating how far an individual's raw score on an intelligence test falls away from the average raw score of his or her age group; abbreviated IQ

Lateralization Localization doctrine in respect to hemispheric specialization for behaviors such as language, spatial tasks, and visual perceptions, e.g., in the majority of the population, language is assumed to be lateralized to the left hemisphere

Localization Doctrine which suggests that certain brain areas are dedicated to specific behavioral or psychological functions

Plasticity Undamaged areas of the brain taking over the function of damaged areas

Schizophrenia Group of psychotic disorders characterized by disturbances in thought, emotion, and behavior; individuals are usually impaired in daily functions such as work, social relations, and physical care

Unilateral hemidecortication Removal of the cerebral cortex from one side of the brain; removal of the left or right cerebral hemisphere has permitted further investigation into equipotentiality, lateralization, and localization

Wernicke's area Area found in the posterior portion of the left superior temporal gyrus that is responsible for the comprehension of written or spoken words; damage to this area results in receptive aphasia

DEVELOPMENTAL NEUROPSYCHOLOGY IS THE study of brain–behavior relationships as they apply to the developing human organism. It can be said that developmental neuropsychology attempts to discover and understand the intricate neurological mechanisms involved in learning. Although child and adult neuropsychologies are similar in research and clinical issues, they differ in emphasis and conceptualization. However, a problem in developmental neuropsychology is a lack of theoretical direction and data-based studies.

ENCYCLOPEDIA OF HUMAN BIOLOGY, Second Edition, VOLUME 3.   Copyright © 1997 by Academic Press.   All rights of reproduction in any form reserved.

Scientific and clinical goals cannot be met without a clearly established direction. A common goal of neuropsychology in laboratories throughout the United States, Canada, and other countries is to learn more, in a descriptive way, about brain–behavior relationships. Once validity has been established, the ultimate goal is the development of remediation and rehabilitation procedures.

To attain an adequate description of the patient's problems and to learn more about brain–behavior relationships in children, the developmental neuropsychologist is faced with a variety of problems. A foremost issue in the study of neuropsychological dysfunction in children is the many developmental factors which must be considered. The age of the child at the time of injury, the type and size of lesion, the extent and location of damage, and the specific mental activity involved and its cognitive complexity are just a few of the factors which must be considered. The consideration of these factors affects the reliability and validity of research in developmental neuropsychology.

Although developmental neuropsychology has developed out of the larger field of clinical psychology, it is imperative that one does not generalize adult data to a child population. Data from a child population follow a set of rules not necessarily similar to that of the adult population. As suggested earlier, age of onset of the injury is a main factor involved in determining patterns of behavioral deficit. Also, the effects of children's injuries seem to be more generalized, whereas damage in adults tends to be more localized and specific.

There are several obstacles in interpreting data from a child population: (1) adult brain injuries are usually more apt to undergo diagnostic and surgical procedures, and therefore receive more attention than childhood cerebral dysfunction; (2) assessments of children are more difficult because of the complex nature of interaction between brain injury and the natural progressive changes due to development; (3) there is a limited range of dependent variables that are appropriate for use with children; and (4) it is difficult to obtain a representative sample of brain-damaged children to study.

I. THEORETICAL ISSUES AND RESEARCH

Developmental neuropsychology has a long history. In the 1800s Gall postulated the brain to consist of numerous individual organs. Each organ, according to Gall, had a specific psychological function. An organ of the brain might be responsible for reading, writing, arithmetic, walking, talking, or friendliness, among other things. In addition, it was the size of the organ which determined the amount of function. A gifted reader would be said to possess a large reading organ, while a learning disabled reader would be said to possess a smaller reading organ. It was this belief about organ size that led to the study of skull configuration, or phrenology. [*See* Brain.]

Gall's theories serve as the basis for the localizationist doctrine, which, however, did not meet with unconditional acceptance. Another scientist, Flourens, found little support for it in his experiments. When he selectively removed portions of pigeon and chicken brain, he found that the area removed had little to do with the nature of the symptoms shown by the hen or chicken. Flourens' observation that the mass of the lesion was the cause of the symptoms led to the equipotential theory of brain functioning. He stated that all areas of the brain are equipotential; there is no differentiation of brain tissue for psychological behavior, as was suggested by the localizationists. It is from this assumption that the name "equipotentialism" comes, the name indicating that all brain tissue is equivalent in terms of what it does or can do. A second, related, assumption is the postulate of mass action. Since all brain tissue is equal, the effects of brain injury are determined by the size of the injury rather than its location.

Localizationist doctrine and equipotential theory have essentially remained the same since the time of the first studies and have been the two approaches that have generally dominated American psychology and education. Theories of brain function, rehabilitation, and assessment use the assumptions of one of these approaches in their formulations, although these underlying theoretical beliefs are not always recognized. For example, the classic description of the brain-damaged child is an equipotential explanation. The classic symptoms include attentional deficits, emotional lability, coordination difficulties, and poor academic functioning and are characteristic of such children. Although never stated, such a description implies that all brain-damaged children are alike, regardless of the localization of their injury, and that the brain is homogeneous in terms of function; the description is thus a reflection of equipotential thinking.

A. Developmental Neurolinguistics

Lenneberg is a well-known proponent of equipotentiality in the study of language. His three now well-known tenets include: (1) Hemispheric equipotentiality exists at birth for language mediation. (2) Lateralization of language processes (generally to the left hemisphere) is realized gradually. Influenced by both maturational and environmental factors (exposure to and use of language), the development of lateralization (a subsequent decrease in equipotentiality) proceeds most rapidly between 2 and 3 years of age and more slowly after that until puberty, when the process is felt to be complete. (3) Interhemispheric plasticity exists for language development. This plasticity (actually denoting the combined effects of tenets 1 and 2 above) is necessary for learning language naturally and completely (at the critical age for language learning) and also enables the right hemisphere to assume language mediation in cases of damage to the left, language-lateralized, hemisphere. [See Hemispheric Interactions.]

These ideas generated much neurolinguistic research on the topics of hemispheric specialization and interhemispheric plasticity within the maturing child. Evidence to the contrary, however, suggests a left-hemisphere specialization for language, present from birth, which could limit the degree of interhemispheric plasticity available to the brain-damaged child.

B. Development and Measurement of Cerebral Lateralization

By 1977 enough data suggested that newborns exhibit rudimentary hemisphere specialization by at least 2–3 months of age, and possibly even before full-term birth, analogous to the adult pattern of left-hemisphere superiority for language-related functions and right-hemisphere superiority for music and the holistic perception of patterns and faces. [See Cerebral Specialization.]

Lateralization is an important issue in developmental neuropsychological research. For example, it has been suggested that deficiencies in cognitive tasks such as reading and language are due to an abnormal or weak pattern of lateralization. The study of lateralization, though, presents many problems. For instance, the validity of the method is often suspect. The wide applicability of these methods, however, as well as the lack of any clearly superior alternatives, provides incentives to investigators to continue using laterality methods while attempting to increase their validity. However, other evidence, such as that derived from computerized axial tomography (CAT) scans, magnetic resonance imaging scans, neurological examination, and medical history, is used to confirm or support signs of lateralized dysfunction. In

time other noninvasive measures (e.g., the recording of cortical evoked potentials) or minimally invasive procedures (e.g., positron emission tomography scanning) could prove to be far superior.

Lateralized brain damage or dysfunction is determined by comparing performance of the right and left hemispheres. Because the two hemispheres are contralaterally organized, the right side of the body is primarily controlled by the left hemisphere and the left side of the body is primarily regulated by the right hemisphere. While somatosensory, motor, and auditory systems are almost completely crossed, other pathways send impulses from the same side of the body to the same hemisphere (e.g., right ear to right hemisphere). The visual system is more complex than the other systems because the visual fields (not the eyes) are crossed in the hemispheres. Thus, part of the left visual field projects to the right visual cortex, and part of the right visual field projects to the left visual cortex. Several noninvasive methods exist with which to measure brain lateralization; these are reviewed below.

I. Dichotic Listening

This noninvasive procedure can be used on all subjects. In this test two lists of numbers are presented simultaneously, one to each ear, arranged in such a way that one number arrives at the left ear at the same time a different number arrives at the right ear. Subjects are asked to listen to the numbers and then to report as many as they could, in any order. Normal subjects are more accurate on the right ear than on the left, and clinical patients with known language lateralization are better on the ear contralateral to their language-dominant hemisphere. The results suggest that the contralateral pathway from each ear to the cerebral cortex is more efficient than the ipsilateral pathway, and with competitive dichotic stimulation there is a suppression of the ipsilateral input by the contralateral input. This right ear advantage is thought to reflect the left-hemisphere representation for language. Most right-handed children report right ear stimuli more accurately than left. However, children with learning problems do not show this same pattern of performance.

2. Manual Preference

A gross estimate of handedness is obtained through questionnaires. There is a significant relationship between handedness and speech lateralization. However, handedness is not an adequate basis for identifying speech lateralization because it leads too frequently to misclassifications. In fact, although the majority of the population are left-hemispheric for both language and production and language perception, there are many exceptions, especially those who are left handed.

3. Dichhaptic Stimulation

In this procedure stimuli are presented through the sense of touch. Subjects are given two different shapes to palpate simultaneously, one with each hand. Because the ascending somatosensory systems are crossed, information from the right hand is transmitted first to the left hemisphere, while the reverse is true for left-hand information. This gives the dichhaptic procedure a superficial advantage over the dichotic procedure, in that there are no ascending ipsilateral pathways from the hands.

4. Tachistoscopic Procedures

Visual asymmetry studies commonly used tachistoscopic procedures. Verbal or spatial stimuli are presented briefly, usually less than 180 msec, to the right and/or left visual field. The unilateral procedure involves random presentation to the right or left of a central fixation point; bilateral presentation, in contrast, involves different stimuli being presented simultaneously to the left and right of fixation. Stimuli perceived in the left visual half-field are processed in the right cerebral hemisphere whereas stimuli perceived in the right visual half-field are processed in the left cerebral hemisphere. It is common for children to perform poorly at brief exposure durations, and it is often necessary to increase the exposure duration into the range in which eye movements are possible in order to raise accuracy to acceptable levels. Also, to identify verbal material, the child must be able to read it; consequently, there is an inevitable confound between reading ability and accuracy that could influence observed laterality measures.

5. Dual-Task Performance

Dual-task performance can also be referred to as verbal–manual time sharing. Time sharing is a type of experiment that contrasts the subject's ability to perform concurrent activities when they are performed in the same hemisphere (e.g., speaking and right-hand manual activities) and when they are programmed in separate hemispheres (e.g., speaking and left-hand manual activities). The consequence of this effect of "hemisphere sharing" seems to be competition and "crosstalk" between incompatible timing mechanisms hierarchically organized in the brain. Basically, the idea is to occupy one hemisphere with a particular

task and to see how this affects the performance of some other task. Thus, for example, reciting animal names or a familiar nursery rhyme should involve left-hemisphere activity and disrupt right-hand tapping more than left-hand tapping. Such an effect is found, but is not related to age between 3 and 12 years.

Although the methods of measurement of cerebral lateralization and asymmetry are imperfect, much has been learned from their studies. A pattern emerges that lateralization effects do not change significantly after 3 years. There might be meaningful changes in the first few years of life, but better techniques and more careful longitudinal studies are needed to establish their significance.

6. Mental Rotation

Mental rotation is a nonverbal spatial cognitive paradigm which requires the ability of the subject to mentally rotate the visual image of an object from one position to another. In one such tachistoscopic task, subjects are asked to rotate mentally a stick-figure stimulus and then press a response button that corresponds to the side on which the stick figure is holding a ball. The stick figure is presented at 0° to 360° orientations in 45° increments. Accuracy and reaction time of responses are recorded. A number of studies suggest a left visual field or right hemisphere superiority for tasks involving some form of mental rotation. Mental rotation functions characteristically show a linear relation between angular disparity and reaction times. Reaction times that increase with angular disorientation indicate that analog processing is occurring. That is, mental rotation, as physical rotation, takes time: the greater the distance, the greater the amount of time.

C. Anatomical Evidence of Lateralization

Anatomical studies could present a more solid case for the early lateralization of language because the left temporal planum, a region of the brain important for receptive language, is larger than the right planum in about 88% of infants. The temporal planum is part of Wernicke's area and therefore is in the language territory. The existence of this asymmetry in newborns lends to the interpretation that language lateralization begins very early.

Another language-specialized area in the left hemisphere is Broca's area in the frontal lobe. The size of this crucial speech production area is paradoxically smaller in the left hemisphere than in the right in the majority of both adult and fetal brains, when measured as the visible surface area. However, it is more deeply fissurated in the left hemisphere. If the cortical surface is measured, including the cortex buried inside the folds, it is larger on the left than on the right in three-quarters of the cases.

D. Plasticity and Recovery of Function in the Central Nervous System

The issue of plasticity also presents an ongoing debate in the field of developmental neuropsychology, between structuralists, who believe that early brain damage is deleterious to the developmental potential, and those advocating plasticity, who believe that the young brain shows a greater restoration of function in comparison to an adult brain. Still other theorists suggested that there are critical periods for the successful transfer of function. Although there is agreement that the earlier the damage, the better the chances for transfer of function to occur, there is no agreement as to the optimal times when damage can be minimized. A majority of children between the ages of 6 and 15 years show substantial improvement in language functions 1 year after injury to the left hemisphere, and language impairment is less severe when left-hemispheric damage occurs in infancy as opposed to later in life. [*See* Plasticity, Nervous System.]

As to the period beyond which restoration cannot occur, some investigations put it at about age 14, whereas others put it at about 2 years. In general, the earlier the damage, the better the chance for transfer of function.

When unilateral hemidecortications are performed prior to the development of language (i.e., in the first year of life), transfer of function is possible. If the left hemisphere is removed, simple language tasks can be performed by the right hemisphere without a decrease in visuospatial abilities. If the right hemisphere is removed, simple visuospatial abilities can be mediated by the left hemisphere without impairment to language functions. While simple tasks can be performed by the remaining hemisphere, more complex tasks cannot be. Thus, left hemidecortication results in a loss of complex language functions, whereas right hemidecortication is followed by a loss of complex visuospatial abilities. It seems that each hemisphere has the ability to mediate functions of the opposite hemisphere, but neither hemisphere is able to assume all of the functions of the other. These studies support the idea of plasticity in the young brain, but not equipotentiality.

II. COMMON DISORDERS

The most common childhood neuropsychological disorders include mental retardation, learning disabilities, epilepsy, closed-head injuries, and psychiatric disorders. Although each of these areas has been researched, the majority of neuropsychological research with exceptional children has focused primarily on the learning-disabled child. Neuropsychological research in the other areas has not progressed as rapidly due to technological and psychometric limitations in evaluating such children. It is hoped that in future years continued research into the spectrum of neuropsychological disorders will help to reveal aspects of brain–behavior relationships of interest to the field of developmental neuropsychology.

A. Mental Retardation

The concept of mental retardation is, like the concepts of brain damage and epilepsy, clouded by attempts to identify a homogenous or unitary entity from what is, in reality, a widely varying assortment of neuropsychological conditions, often with little more in common than a poor intelligence quotient (IQ) on formal IQ tests. Brain dysfunction is currently viewed as being concomitant to mental retardation; in fact, neurostructural damage was found in necropsies on a large sample of institutionalized retarded individuals. A tendency was also noted for the less severe cases of retardation to have the less severe brain anomalies. Further evidence can be seen with electroencephalographic (EEG) procedures; specifically, with increasing severity of mental retardation in children and adolescents, there is an increase in EEG abnormalities. Down's syndrome subjects seem to be an exception, with significantly fewer EEG abnormalities.

Neuropsychological aspects of mental retardation have not received attention for several reasons: (1) many difficulties exist when administering neuropsychological tests to this population, (2) there are few specific neuroanatomical and neurophysiological correlates of behavior in this population, and (3) there is much uncertainty with regard to the normal neuropsychological data.

Little laterality research has been completed with severely language-impaired retarded children. The usual dichotic listening procedure requires that subjects comprehend and reliably follow verbal instructions, attend to and discriminate the dichotic stimuli, remember them, and report back what has been heard. Children with severely limited language lack some of the skills necessary to perform the task adequately. Recent research looking at ear advantage in Down's syndrome children using dichotic listening tests has indicated that these children have a left ear advantage for linguistic serially processed auditory stimuli (e.g., digits and common objects). This is in contrast to the right ear advantage commonly found. This left ear advantage seems to be related to the syndrome itself and not to retardation in general since a control group of non-Down's syndrome, but retarded, children were shown to have the usual right ear advantage. There is also an increase in the incidence of dichotic listening left ear advantage for autistic compared to normal children, suggesting that severe and pervasive language disabilities might be associated with an increased incidence of right-hemisphere specialization for language functioning.

B. Learning Disabilities

"Learning disability" is a generic term that refers to a heterogenous group of disorders manifested by significant difficulties in the acquisition and use of listening, speaking, reading, writing, reasoning, or mathematical abilities. Little is known about the etiology of learning disabilities. Neuropsychological studies have shown that children who have large discrepancies between their verbal and nonverbal abilities are often the most severely learning impaired.

It has been proposed that learning-disabled children have deficits in their ability to transfer information from one hemisphere to the other through the corpus callosum. The result of this deficit is that the two cerebral hemispheres in these children might function somewhat independently and without the contralateral interaction found in normal children.

The difficulty in studying learning disabilities is that most childhood disorders are rare and are also diffuse or indeterminate, making it difficult to state with any assurance that something is wrong in any particular area of the brain. For learning disabilities there is rarely any neurological evidence for characteristic lesions, and the evidence obtained from many children is contradictory.

C. Epilepsy in Children

Epilepsy is the most predominant neurological disorder of childhood. Epileptic children attending ordinary schools are at a greater risk of developing learning problems than other children. In one study of 85 school children with recurrent seizures, it was reported that 16% were regarded as falling seriously behind and 53% were functioning at a below-average educational level; 42% of the children were described as inattentive by their teachers and had poorer school performance. Several factors affect intelligence score outcomes in epilepsy. Low seizure frequency is associated with higher intelligence scores, whereas early-onset seizures tend to be indicative of lower intelligence scores (IQ). [See Epilepsy.]

Children with minor motor and atypical absence (petit mal) seizure tend to have the lowest IQ, whereas those with generalized tonic–clonic and classic absence have the highest average full-scale IQs. With respect to other areas of cognitive functioning, no specific pattern of impairment has been identified.

D. Closed-Head Injury in Children

A head injury can range from a simple bump on the head to a more complex penetrating injury. Infants and young children are at a high risk for head injury, especially for nonpenetrating closed-head injuries. Closed-head injury can have no observable consequences or, in a very severe case, can result in death. Brain injury from trauma to the skull is one of the more common neurological disorders in children. The effects vary considerably with age because the brain of the child is still developing postnatally.

Trauma at an early age affects an incompletely developed brain and an incomplete repertoire of behaviors. Early brain lesions have a greater effect, perhaps because of the smaller repertoire of skills and knowledge the young individual has to rely on. Around the age of 5 years, damage to the right hemisphere no longer has much of a disruptive effect on language. Between the ages of 5 and 12 years, a left-hemisphere injury produces aphasia, although it is generally milder and more transitory than that found with similar injury later in life. It is only after about age 16–18 that adult-like aphasia is seen with left-hemisphere injuries. In a group of children whose injuries in the left hemisphere occurred at various times from infancy through the preschool years, deficits were evident in both verbal and nonverbal skills relative to healthy children. However, dysphasia, which is typically seen in adults with left-hemisphere damage, was not prevalent.

Perhaps even more important, the infant brain often shows a reduction in size following brain damage, suggesting that, from a structural point of view, early damage might be more disastrous than later damage. When the mature brain of the adolescent or adult is damaged, loss of function tends to be more highly localized and specific.

Intellectual impairment several years after head trauma is more frequent in children younger than 8 years of age at the time of injury than in those over 10 years of age. Thus, although younger children recover more rapidly, they do so less completely. A lesion occurring during the first year of life tends to be associated with intellectual deficits involving both verbal and nonverbal skills; lesions occurring after the first year have effects more dependent on the side of the lesion. Later left-hemisphere lesions were found to be related to decreased verbal and nonverbal test scores, whereas right-hemisphere lesions were found to be associated with impaired nonverbal skills.

E. Psychiatric Disorders

Psychiatric disorders present a continuing puzzle to developmental neuropsychologists. Defects in the lateralization of functions to the hemispheres of the brain have frequently been cited as an etiological factor in a number of unadjusted behaviors (e.g., learning disorders and emotional disturbance). Confused lateralization seems to be especially prominent in those with schizophrenia, a disorder that is more clearly related to an overall level of neuropsychological dysfunction than specific patterns of impairment. As a group, patients with affective disorders, both unipolar and bipolar, were relatively more deficient on right-hemisphere tasks than schizophrenics. Inconsistency in the patterns of peripheral activities in children might accompany emotional instability and compromised frustration tolerance. [*See* Mental Disorders; Mood Disorders; Schizophrenic Disorders.]

Left-hemisphere dysfunction has been repeatedly suggested to be related to schizophrenia; such a dysfunction is suggested by flat affect, speech disorders, and paranoia. In neuropsychological studies a reduced anterior left-hemisphere activation was found when ipsilateral skin conductance responsiveness and cerebral blood flow were used as measures of activation. In studies of the cognitive capacity of the left and right hemispheres, schizophrenics were found to be much less able to process both verbal and nonverbal information tachistoscopically presented to the left hemisphere.

Head injury is also likely to produce psychiatric symptoms, different in left- or right-hemisphere damage. Right-hemisphere damage is likely to result in clinical levels of anxiety, general denial, and inappropriate indifference to the medical condition. The emotional response to lesions of the left hemisphere seems to be most often expressed as a depression–catastrophic reaction.

III. PERSPECTIVES AND ISSUES IN ASSESSMENT

Much remains to be learned regarding neuropsychological developmental disorders. Lateralization studies provide one avenue of investigation, but an important area not to be neglected is the standard neuropsychological test battery. The use of these tests can aid in further defining brain–behavior relationships in neuropsychological disorders. In addition, use of the tests can help provide answers as to how best to rehabilitate the patient in need of neuropsychological assistance.

A. Standard Neuropsychological Batteries for Children

Two common neuropsychological test batteries include the Halstead–Reitan Neuropsychological test batteries for children and the Luria–Nebraska Test Battery: Children's Revision. The Halstead–Reitan test is said to epitomize North American standardized assessment procedures. The focus in this test is on quantitative norms. The Luria approach, in contrast, is more qualitative in nature and uses a functional approach. The Luria–Nebraska Battery is an attempt to wed Luria's techniques with American clinical neuropsychology.

The Halstead–Reitan Neuropsychological Test Battery for Older Children (i.e., 9–14) and the Reitan–Indiana Neuropsychological Test Battery for Younger Children (i.e., 5–8) are two of the most commonly used neuropsychological test batteries for children. These batteries were developed by Ralph Reitan based on the adult version of the Halstead–Reitan Neuropsychological Test Battery. The major theoretical basis of the Halstead–Reitan and Reitan–Indiana tests is the proposition that behavior has an organic basis and that performance on behavioral measures can be used to assess brain functioning.

The Luria–Nebraska Neuropsychological Test Battery was developed by Charles Golden, who studied with A. Luria in the Soviet Union. Luria's test procedures have several advantages over techniques traditionally used in the United States. First, they comprehensively assess impaired neuropsychological functions from a clinical perspective. Functions are broken down into their most basic components. As a result, a qualitative evaluation of the patient is performed, allowing for the identification of the basic neuropsychological processes underlying overt behavior. A second major advantage is that the Luria–Nebraska Test Battery is somewhat speedier.

Finally, the Luria–Nebraska Test Battery was also designed for the development of rehabilitation programs directed toward the individual's deficits, using techniques that optimize recovery and minimize staff time. Considering the limitation of rehabilitation sources available for the brain-injured patient, this is an important and potentially powerful advantage.

Luria repeatedly stressed that variability and flexibility in the administration of neuropsychological measures were crucial. The emphasis here is on how problems are solved by the patient, what types of compensations are possible, and areas of deficit and strength. Within such an intraindividual model, the psychologist is encouraged to adapt neuropsychological assessment procedures in such a way that the maximum amount of information is gleaned. However, lack of standardization has limited the use of Luria's test procedures in the United States.

In summary, actuarial and clinical approaches to assessment in neuropsychology are adequately illustrated in the Halstead–Reitan and Luria–Nebraska test batteries. Reitan's test hallmark is the strict scientific standardization of the clinical approach. Luria's method, on the other hand, is actuarial and encourages the clinician to exercise his professional expertise. The clinical approach is more readily accepted in scientific circles, having conformed to the rigors of experimental investigation; actuarial approaches are not usually so well accepted. Nevertheless, it is acknowledged that both approaches are necessary in pediatric neuropsychological practice.

B. Prospectives of Developmental Neuropsychology

Clinical developmental neuropsychology is more than a search for tests of brain damage. Diagnostic information as an end result of neuropsychological assessment limits the effectiveness of the testing situation. The recognition by many clinical neuropsychologists of the need to complete assessments with more direct treatment implications is an important step forward. While the procedures have shown clinical validity for differentiating behavioral deficiencies resulting from

brain dysfunction, the main objective in neuropsychological evaluation is to provide descriptive information about the behavioral consequences of neuropathology, which can be used to design relevant rehabilitation or remedial programs for individuals with brain-related disorders.

The therapeutic role of the neuropsychologist is only beginning to be defined. Traditionally, the neuropsychologist has worked as a diagnostician, with little emphasis on devising treatment programs. It is now a common argument in favor of a neuropsychological perspective that goes beyond the diagnosis of impaired neurological processes to the structuring of educational programs that maximize a child's assessed strengths.

Until recently, the brain and its relationship to behavior have been largely unrecognized in attempts to rehabilitate and educate the individual. Developmental neuropsychology can signify a great step forward in its promise to teach humans about humans. It is important to move away from diagnosis as a sole criterion of neuropsychological assessment and move toward ways in which the individual can best be helped. The study of children, as developing individuals, has the potential to teach us much about a subject we are most interested in: the human being.

BIBLIOGRAPHY

Bryden, M. P. (1982). "Laterality: Functional Asymmetry in the Intact Brain." Academic Press, New York.

Davidson, R. J., and Hugdahl, K. (eds.) (1995). "Brain Asymmetry." MIT Press, Cambridge, MA.

Hynd, G. W., and Obrzut, J. E. (eds.) (1981). "Neuropsychological Assessment and the School-Age Child: Issues and Procedures." Grune & Stratton, New York.

Hynd, G. W., and Willis, W. G. (1988). Pediatric Neuropsychology." Grune & Stratton, Philadelphia.

Kolb, B., and Whishaw, I. Q. (1980). "Fundamentals of Human Neuropsychology." Freeman, San Francisco.

Lenneberg, E. H. (1967). "Biological Foundations of Language." Wiley, New York.

Obrzut, J. E., and Hynd, G. W. (eds.) (1986). "Child Neuropsychology," Vol. 1. Academic Press, Orlando, FL.

Obrzut, J. E., and Hynd, G. W. (eds.) (1986). "Child Neuropsychology," Vol. 2. Academic Press, Orlando, FL.

Obrzut, J. E., and Hynd, G. W. (eds.) (1991). "Neuropsychological Foundations of Learning Disabilities: A Handbook of Issues, Methods, and Practice." Academic Press, San Diego.

Orton, S. T. (1937). "Reading, Writing, and Speech Problems in Children." Norton, New York.

Rourke, B. P., Bakker, D. J., Fisk, J. L., and Strang, J. D. (1983). "Child Neuropsychology." Guildford, New York.

Segalowitz, S. J., and Gruber, F. A. (eds.) (1977). "Language Development and Neurological Theory." Academic Press, New York.

Development of the Self

ROBERT L. LEAHY

American Institute for Cognitive Therapy and Cornell University Medical School

I. Psychoanalytic Theories of Self
II. Cognitive Theories of Self-Development
III. Cognitive–Developmental Models of the Self
IV. Conclusions

GLOSSARY

Object–relations Process by which the individual forms attachment and subsequently individuates self–other

Schematic processing Bias in information processing resulting in greater attention and recall of information consistent with a prototype

Self-schema Representation of the self that forms the early prototype of self-understanding

Structural model Psychoanalytic model of the id–ego–superego of the self

ONE OF THE MOST IMPORTANT QUESTIONS TO BE answered by any psychological theory of human behavior is how the self develops. The self is not distinctly "human," if we include in our definition of the self the ability to recognize one's own physical attributes, since monkeys and cats are capable of this. If we limit our definition to self-recognition, i.e., the ability to recognize that a spot or cap on one's head does not belong there when seen in a mirror image, then we can say that this is clearly established in infancy. These clever experiments on nonverbal subjects, however, do not investigate what we generally refer to as "self." Any review of theory and research on self-development leads to a recognition of the often incompatible variation with which the self is defined and explained. Does the self control anxiety and impulses, are we always conscious of the self, does the self change with age, is the self primariliy emotional or cognitive, are there different contents to the self, can we describe psychopathology by reference to self, and what accounts for individual differences in the self? Each of these intriguing questions has special relevance to different theories of the self.

This article attempts a description of the self's development by examining how different theoretical systems approach this issue. It begins with a discussion of psychoanalytic theory which reflects the importance of how the understanding of the self has led to revolutionary developments in the theory itself such that contemporary object–relations theory often seems far removed from the original work of Freud. Next it turns to cognitive models of the self, which, in a sense, are more "general process" approaches than "developmental" in that they do not describe qualitative stages of functioning which are related to age or development. Finally, it examines "cognitive-developmental" theories which attempt to relate self-functioning to qualitative changes in cognitive and social functioning associated with age.

I. PSYCHOANALYTIC THEORIES OF SELF

A. Freudian Theory of the Self

Freud's theory of the self is composed of two models: the *topographic* and the *structural*. According to the topographic model, consciousness of the self's emotions and experiences may be blocked by repression and denial. Freud distinguished among conscious, preconscious, and unconscious knowledge and motivation, such that preconscious thought could be accessed with some effort, but unconscious thought is generally unaccessible. According to Freud, thought remains unconscious because of its capacity to arouse anxiety. These unconscious qualities of the self are expressed through ego defenses, such as projection, introjection, displacement, and reaction formation.

According to the structural model, the self is composed of three often competing elements: id, ego, and superego. The id refers to innate libidinal energy, such as hunger, sex, and aggression, which seeks release and satisfaction. According to Freud's theory of anxiety, frustration during infancy is first experienced with the infant's separation from the mother. In order to reduce this frustration, Freud postulated, the infant forms a mental representation of the mother ("hallucinatory image") which allows the infant to internalize the mother's presence in her absence. [See Psychoanalytic Theory.]

This early representation of the external world is the first source of the infant's differentiation of self from external world (i.e., mother) and marks the origin of the ego which attemps to control the id and negotiate adaptation to reality: "Where id was, ego shall be." This "libidinized cognition" was criticized by Heinz Hartmann, who proposed that perception and cognition are "preadapted" to reality and are not derived through frustration of drive. Hartmann's theory of ego psychology placed considerable emphasis on analyzing the functions of the ego which are independent of early infantile or oedipal conflict.

Freud outlined psychosexual stages of development—oral, anal, genital, latency, and phallic—which correspond to body modalities in which the self attaches libidinal energy ("cathects"). Thus, at the oral stage, during the first 2 years of life, the self experiences the world largely through oral preoccupations such as dependency, narcissism, and "oral rage" (when frustrated). Freud viewed the sequence of these stages as invariant due to biological maturation of body zones. Failure to resolve conflicts at any stage, or experience of excessive gratification at a stage, would result in negative and positive fixation in later development. For example, strict and punitive toilet training would be expected to result in fastidious and retentive patterns in later life.

The development of the superego is important during the genital stage. The superego represents the internalization of the values of parents and others. According to Freud, the male child fears castration by the father for his sexual desires for the mother. In the Oedipal complex, he resolves this fear by "identifying with the aggressor," i.e., he renounces his desires for his mother, internalizes the father's identity (as his superego), and displaces his sexual and aggressive energy in more culturally acceptable behavior through sublimation during the latency period. Classic Freudian theory reflects a Victorian male bias in that it proposes that the female, lacking the threat of castration, has a less severe superego and, thus, is less capable of cultural achievement. Feminist critics have indicated that, to the contrary, females are equal and, in some cases, more advanced on moral and social–cognitive indices of development.

Freud's theory of development places the origins of the most important qualities of the self in the "pre-Oedipal" phase, i.e., between 4 and 6 years of age. Except for his discussion of the origins of the ego during the oral phase, there is little that is interpersonal in the self. The self, in Freud's theory, seems to be an interiorized self with conflicts raging between different layers and different structural unities.

B. Non-Freudian Psychoanalytic Theories of Self-Development

Harry Stack Sullivan advanced an *interpersonal theory* of the self, which he proposed as a psychoanalytical alternative to Freud's emphasis on sexuality in self-development and the emphasis on the "privacy" of Freud's ego. In Sullivan's view, the earliest source of anxiety for the child is *empathy* by the infant of the mother's emotions. The self arises as a means of reducing this anxiety. The infant splits the self into three components—good-me, bad-me, and not-me—where the not-me corresponds to those traumatic experiences that cannot be integrated into consciousness. Sullivan's terminology—dramatization, consensual validation, and egocentricity—all point to the interpersonal construction of the self in his model.

Erik Erikson has outlined eight psychosocial stages which mark changes in self-development. In Erikson's model, each stage reflects a social, not just sexual, quality of self-construction. These "eight ages of man" reflect competing issues: trust vs mistrust, autonomy vs shame and doubt, initiative vs guilt, industry vs inferiority, identity vs role confusion, intimacy vs isolation, generativity vs stagnation, and ego integrity vs despair. In Erikson's system, failure to resolve an earlier conflict results in persistence of that conflict at later stages. For example, failure to positively resolve the conflict of trust vs mistrust may result in mistrust as an impediment in identity achievement during late adolescence or early adulthood.

Of specific interest to us is Erikson's discussion of "ego identity," which he defines as the awareness of self-sameness and continuity of one's "style of individuality" where this sense corresponds to the meaning for significant others. Ego identity is defined interpersonally, just as all other stages in Erikson's system have interpersonal implications. Erikson claims that commitment and choice (e.g., career, religion, marriage) contribute to the formation of ego identity. According to Erikson, many individuals experience considerable anxiety (i.e., an "identity crisis") in the formation of their identity, resulting in uncertainty about the self's past, present, and future, as well as intense questioning of one's own values. Some individuals, however, who experience "identity foreclosure" never question their conventional values or identity, simply accepting the traditional stereotypes of their parents.

Jane Loevinger has proposed that development is characterized by successive levels in the functioning and structure of the ego which parallel intellectual, moral, and social functioning. Loevinger's levels span the life span, from birth to maturity: presocial, symbiotic, impulsive, self-protective, conformist, self-aware, conscientious, individualistic, autonomous, and integrated. These levels are characterized by differences in impulse control, interpersonal style, conscious preoccupations, and cognitive style. Ego-development theory originates with Sullivan's interpersonal theory. Loevinger's view of the ego goes beyond Freud's model of id–ego–superego; in Loevinger's system, the self not only controls anxiety, but it also implies the manner by which others are viewed and how relationships are judged. Because of the clear interpersonal parallels of self–other conceptualization that develop with each level, Loevinger's theory has had more direct impact than Freud's theory on developmentalists interested in social cognitive functioning.

Object–relations theory attempts to trace interpersonal developments of the self. It posits several levels of self-development, beginning in the first months of life as "symbiotic" relations, such that there is no distinction between infant and mother. This symbiotic stage gives way to phases of differentiation of self and mother, to "rapprochement" with (return to) mother, and final independence. An extension of the theory proposes that the representation of self and other follows parallel development in that the self is largely an internal representation, or mirror, of the image of the other. Primitive, and sometimes unresolved, images of self and other involve idealization and splitting, with idealization referring to a projected positive omnipotence and nurturance, often as a result of a defense against a projected mother image which is negative, punitive, and destructive. Thus, during infancy the other is viewed as either meeting all needs or frustrating all needs. This "splitting" of the image of other is mirrored in the image of self, either all good or all bad. Normal development beyond infancy results in the modification and integration of positive and negative images, allowing for the direction and control of anxiety.

Freud's emphasis on repression of sexuality and his focus on hysterics and obsessive–compulsives reflected the typical Victorian patient of his day. However, many of the psychiatric disturbances seen today are better conceptualized in an object–relations model than in the Oedipal model of Freud's time. Object–relations theorists have been especially helpful in elucidating the complexities of serious personality disorders, such as the borderline and narcissistic personalities. These disorders are viewed as failures to resolve the earlier splitting which results in the inability to integrate the complex, competing emotions in the self. [*See* Personality Disorders.]

II. COGNITIVE THEORIES OF SELF-DEVELOPMENT

Cognitive theories of the self place emphasis on the role of thought processes in the construction of the self. Unlikely psychoanalytic theory which emphasizes regulation of anxiety in the development of the self, resulting in an emphasis on defense mechanisms in regulating the self, cognitive theories view the self as an information-processing sysetm. There are a variety of cognitive theories of self-development, with no single unifying approach.

A. Information-Processing Models of the Self

According to Aaron Beck's cognitive model of depression, individuals differ in their underlying *self-schemas*. These self-schemas refer to labels of the self, established during early childhood, which direct information processing. For example, the individual who begins with a self-schema of being unworthy will differentially focus on information consistent with the idea that he is unworthy and filter out information inconsistent with that image. This process of *schematic processing* results in cognitive distortions which are referred to as *automatic thoughts*. Typical automatic thought distortions include mind-reading ("He thinks I'm a failure"), discounting the positive ("That test was easy"), all or nothing thinking ("I fail at everything"), mislabeling ("I'm a loser"), and magnification ("That quiz was really important"). In addition, the negative self-schema is reinforced by *maladaptive assumptions* which are the "formulas" or rules by which information is evaluated. These include "if–then" statements ("If I fail one thing, then I'm a failure" or "If someone criticizes me, then I'm wrong") and "should statements" ("I should be perfect at everything," "I should understand things immediately," "I should be better than everyone else"). The simultaneous operation of automatic thoughts, maladaptive assumptions, and negative self-schemes is schematic processing.

Although these concepts do not provide a truly developmental model of the self (i.e., a model suggesting qualitative stages of development), they indicate that early experiences of loss, rejection, or punishment contribute to the formation and persistence of these self-schemas. Moving to personality disorders, cognitive therapists have attempted to speculate about the origins of early self-schemas.

Another information-processing model proposes that not everyone who encounters experiences of failure develops depression. The central factor affecting depression, according to this model, is the set of *attributions* or explanations the individual gives for his failure. For example, if one explains his failure by claiming that it was caused by an internal-stable cause (e.g., lack of ability), that others would have done well ("personal helplessness"), and that this failure will generalize to many other situations ("global helplessness"), then self-critical depression will result. There are no "developmental" stages of the self in the attribution model. In fact, many of the attribution patterns characteristic of depressed adults are also characteristic of young children who demonstrate helplessness on difficult tasks. The depressive attribution pattern is related to early childhood experiences of exposure to punitive and rejecting parents and to early separation or loss of a parent.

In a somewhat different vein, *self-perception theorists* have proposed that the self-concept is developed by the individual "observing" his own behavior and, subsequently, drawing inferences about his motives, thoughts, or traits, e.g., "I must like pasta, because I'm always eating it." An assumption of this model is that we make inferences about ourselves in a manner similar to the way we make inferences about others. If children are rewarded for behavior that was already intrinsically interesting to them, there is a subsequent decrement in their interest in that activity. The inferred information processing that "turned play into work" was that the child would say to himself, "It must not be that interesting if I have to be bribed to do it." Too great an emphasis on reinforcements in early education may cause children to lose interest in intrinsically interesting behavior.

III. COGNITIVE–DEVELOPMENTAL MODELS OF THE SELF

The foregoing models of self are not truly "developmental" in that there is no clear description of qualitative changes in self associated with age. This section reviews cognitive–developmental models of the self which propose that there are either parallels or structural similarities in self-cognition with other forms of cognitive development.

A. Social–Cognitive Developments Affecting Self-Image

The self is an object of cognition similar to other social objects. Consequently, it is reasonable to assume that there may be parallel developments in social and self-cognition. With increasing age, there is an increasing emphasis on intention, distal causes (i.e., causes that are not immediately present in the situation), complexity, qualification, past time perspective as well as future time perspective, and social competence in evaluating self and other. Between the ages of 5 and 10 there is a dramatic increase in the emphasis of "social comparison" information in evaluating the self's performance, i.e., older children use the norms of other children's performance to evaluate their own performance. In addition, motivational systems also undergo change during this age period, with older children placing a greater emphasis on competence or mastery, with less reliance on either social approval or extrinsic reinforcement.

One model of the self emphasizes the *self-regulatory* aspects in self-development. It draws on the work of the Soviet psychologists Luria and Vygotski, who proposed that speech begins as social, or external, speech, but with development, speech becomes internal, or private. This internal speech serves a self-regulating function, such that the older child may rely on these internal representations to control and guide behavior. Children are taught to stay on task, examine all alternatives, and engage in self-reward for completion of the task. This cognitive movel was first applied to children with behavioral deficits (e.g., hyperactive children), teaching these children to use self-instruction to control their impulsive behavior. Although successful in reducing impulsive behavior and in increasing "on-task" behavior, self-control seldom generalizes to situations other than the initial training situation. There is little that is "developmental" in the model and little that allows us to determine the content or developmental course of self-regulatory processes. The only development appears to be the overcoming of the *deficit* in self-regulation.

John Bowlby's model of attachment has been extended to include self-development. Contrary to traditional psychoanalytic theory, which proposes that attachment is a result of the learned association of the mother with reduction of drive (e.g., hunger or anxiety), this model advances an *ethological* theory according to which attachment is an innate behavior pattern which seeks completion. Self–other development reflects disturbances in these attachment patterns. According to the model the child develops *object–representations* of self–other interactions, where the self may be viewed as unlovable, alone, or helpless. The model also refers to the self as an information-processing system in which information threatening to the self is excluded from memory. This theory places considerable emphasis on the priority of early experience in affecting all future object representations. Thus, early loss of a parent (through death or divorce) or threats to attachment bonds result in lifelong vulnerabilities to depression, agoraphobia, or anxiety disorders.

This theory does not propose any qualitative stages in self-development and does not specify the content of the self at any level. It appears to have replaced Freud's structural model (id, ego, superego) with an ethological model emphasizing innate tendencies (rather than the id) and information processing (rather than ego and superego). It explicitly excludes "psychic energies" or "libido" in favor of information processing and innate attachment tendencies. Its consideration of the self is primarily focused on self–other interactions rather than on the private experiences of the self.

George Herbert Mead proposed that the self is a phenomenal object distinct from all other objects in that it is reflexive. The problem for the self would be how the individual is able to stand outside the self to view the self. The proposal is that the self becomes an object of experience by taking the role of others toward the self, e.g., "How does the teacher (or parent) see me?" This theory of the self emphasizes the cognitive (e.g., role taking) rather than the emotional aspects of the self and proposes that the self is a social construction, i.e., the self is known only from the perspective of other selves.

In this theory, the epistemology of the self is based on an analogy to games. According to this view, the child comes to understand who he is and what individual differences are through playing social games with other children. The theory distinguishes several stages in this "self-understanding." The first stage is the *play stage,* during which the child plays at reciprocal roles (e.g., teacher–student) in which the child acts like another toward himself (e.g., praising himself). At the *game stage* the child coordinates the views of others in a game through which the "team perspective" is organized and constructed. Further, games entail cooperation by which the child internalizes the standards and expectations of others ("the generalized other").

This theory has been extended by Robert Selman who proposes that self-understanding changes with qualitative changes in role-taking ability. This model identifies a series of stages of role taking, which develop between childhood and adolescence. At the egocentric stage, the child is unable to recognize differences in knowledge and perspective between self and other, whereas at the subjective stage the child understands that these differences do exist. With self-reflective role taking the child or adolescent understands that he may be the object of someone else's thoughts. At the level of mutual role taking the individual understands that the self and other may be constructed from a third person's point of view. Finally, with the conventional system of role taking, the individual recognizes that the views, needs, and dilemmas of self and other may be reconciled and unified through conventions.

Another proposal is that increased development is characterized by increasing differentiation and internalization of values. Numerous studies support this model, showing that higher self-image disparity (between the way one sees oneself and the way one wishes to be) is associated with greater chronological and mental age, IQ, and social competence, with emotional maladjustment unrelated to disparity.

An attempt to specify the cognitive factors accounting for this greater disparity tries to identify social cognitive factors accounting for greater self-image disparity. As already discussed, it has been proposed that the self becomes an object of experience through the process of role taking and that through role taking the child internalizes the values of his peers and parents. Moreover, levels of moral judgment may be conceptualized as increasingly more general and abstract forms of role taking. Consistent with this structural model, it was found that greater self-image is associated with higher levels of moral judgment and higher levels of role taking.

It has been suggested that the foregoing findings imply that there are costs of development: Increasing social cognitive development results in an increased capacity for self-critical depression and an increased probability of identity diffusion. This is implied by the fact that development results in greater internalization (higher ideal self-image), greater self-image disparity, increased uncertainty and qualification in describing the self, increased uncertainty regarding the "true" perspectives of others, and greater future-time perspectives.

B. Cognitive-Development and Self-Understanding

A comprehensive model of the development of self-understanding that is the individual's conceptualization of different schemes of the self (e.g., physical, activity, social, and moral) has been advanced. Four *levels* of self-understanding of the "me" component are identified. This model stands in contrast to other developmental models which have proposed that self-understanding begins with descriptions of physical attributes and actions and later develops into concepts of social, psychological, and moral phenomena (which correspond to *self-schemes*). These self-schemes are operative at every level of development, but what changes is the manner in which they are conceptualized.

IV. CONCLUSIONS

The self is described and explained in a variety of ways. Psychoanalytic theory has undergone considerable modification from Freud's early description of a self with different layers of consciousness and competition among id–ego–superego. With the advent of objective–relations theory, the self has gained primary importance in psychoanalysis, with its essential developmental milestones established in the first few years of life. The information-processing models of self, especially the cognitive theory and the ethological theory, have been important in advancing our understanding of affective disorders, such as depression, which result from negative cognitive distortions in processing information and from disturbances in early attachment. Finally, the cognitive–developmental approaches to self have demonstrated that the self-concept undergoes qualitative change with increasing age.

BIBLIOGRAPHY

Beck, A. T., Rush, A. J., Shaw, B. F., and Emery, G. (1979). "Cognitive Therapy of Depression." Guilford, New York.

Bowlby, J. (1980). "Attachment and Loss (Vol. 3): Loss, Sadness and Depression." Hogarth Press, London.

Damon, W., and Hart, D. (1988). "Self-Understanding in Childhood and Adolescence." Cambridge Univ. Press, New York.

Guidano, V. (1988). "The Complexity of the Self: A Developmental Approach to Psychopathology and Therapy." Guilford, New York.

Guidano, V., and Liotti, G. (1983). "Cognitive Processes and the Emotional Disorders." Guilford, New York.

Kernberg, O. (1975). "Borderline Conditions and Pathological Narcissism." Jason Aronson, New York.

Kohut, H. (1977). "The Restoration of the Self." International Univ. Press, New York.

Leahy, R. L. (ed.) (1985). "The Development of the Self." Academic Press, San Diego.

Leahy, R. L. (1996). "Cognitive Therapy: Basic Principles and Applications." Jason Aronson Publishers, Northvale, New Jersey.

Masterson, J. F. (1983). "The Narcissistic and Borderline Disorders: An Integrated Developmental Approach." Bruner Mazel, New York.

Development, Psychobiology

MYRON A. HOFER
Columbia University

I. History
II. Evolution and Development
III. Nature and Nurture
IV. Principles
V. Events and Processes
VI. Range and Prospects of the Field

GLOSSARY

Critical or sensitive periods Limited time during development when a system is particularly open to modification of its characteristics in response to some external influence

Epigenesis Historically, the idea that complex structures and functions originate from formless material within the egg. Now used to refer to the influences on development that do not arise from the actions of genes. In particular, denotes the stepwise nature of behavior development, whereby new characteristics emerge serially as a result of the repeated interactions of the organism with its environment

Heterochrony Evolutionary change in the onset or timing of development. Rates of development of features are not synchronized within each individual but are capable of separate variation. Particularly, the rate of development of a feature in descendants may be either accelerated or retarded in relation to its schedule in ancestors

Ontogenetic adaptation Specific adaptive mechanism, peculiar to a given stage in development and its special environmental conditions, which is not present in the adult

Stage Period during the development of a system or behavior during which changes are taking place relatively slowly but which differ from earlier or later stages in important characteristics. Stages are bounded by transitions in characteristics of the organism (e.g., infant, juvenile) or its environment (e.g., prenatal, postnatal)

THE FIELD OF DEVELOPMENTAL PSYCHOBIOLOGY brings together research efforts aimed at understanding the development of behavior and of the physiological systems related to it. The goal is to discover the underlying processes that determine the course of development through experimental and observational studies of animals, including humans. An effort is made to integrate psychological with biological frames of reference, through an approach to behavior that is based on the principles of evolution and ecology. The development of physiological systems and of behavior are treated as parts of an integrated unit organized to promote adaptation of the developing organism to its changing environment. Research in this field has shown that these adaptive interactions provide complex neural and hormonal feedback to the developing nervous system that regulates the course of behavior development and ultimately determines the nature of the organism. This approach promises to resolve the ancient "nature vs nurture" riddle by shifting the level of analysis to the psychological, physiological, cellular, and genetic mechanisms involved in the processes of development.

I. HISTORY

The relationship among the processes underlying inner experience, behavior, and bodily functions has puzzled humans since they first began to think. Once Darwin established the evolutionary continuity between animals and humans, the stage was set for a vast broadening and enrichment of scientific approaches to the problems of psychology and of development. The word *psychobiology* has been used since the end of the 19th century to denote an integrative approach that cuts across traditional academic disciplines to emphasize the fundamental unity of the processes underlying these various manifestations of human nature. Comparative psychology in North America and ethology in Europe emerged as early forerunners of contemporary psychobiology, and in the area of medicine, Adolph Meyer is generally credited with establishing the term in his "psychobiological life history" approach.

In the first half of the 20th century, however, these trends were almost buried in waves of extreme genetic determinism (e.g., eugenics) and of a polar opposite, extreme environmentalism (e.g., behaviorism). Furthermore, for most of this century, developmental psychology and embryology (or "developmental biology") were fields that maintained almost complete isolation from each other, turning away from the synthesis that had been emerging at the close of the 19th century. Finally, in the years after the second world war, research on displaced war refugees, followed by more controlled experimental work with animals, produced dramatic examples of the impact that early life experience can have on the development of behavior and of underlying neural structures. Embryologists (e.g., C. F. Waddington) as well as ethologists (e.g., Konrad Lorenz) were writing about critical or sensitive periods in development when experience could have unique effects. Developmental psychologists (e.g., Piaget) were discovering stages in the acquisition of knowledge by children during which information was processed in ways that were different from the adult. Moreover, research in endocrine influences on behavior (e.g., Frank Beach and Daniel Lehrman) demonstrated how biological regulatory systems, behavior, and experience interacted in orderly sequences to produce complex adaptive patterns such as courtship, mating, and nursing.

These research findings in a number of different fields appeared to have much to contribute to one another and created the need for an organization to foster the study of developmental processes that would draw not only from psychology and embryology but also from physiology, genetics, biochemistry, endocrinology, and neuroanatomy. The potential for novel applications of these approaches to clinical medicine attracted psychiatrists, pediatricians, and neurologists. In 1967, The International Society for Developmental Psychobiology was formed, and the first issue of the journal *Developmental Psychobiology* was published in March of 1968. Since that time, the Society for Neuroscience (in 1971) and the International Society for Developmental Neuroscience (in 1981) have come into being as forums for research

on the nervous system, primarily at the cellular and molecular levels.

Currently, the study of development is being carried out in fields that are defined by levels of biological organization and the different methods that are applicable at each level: cellular and molecular neuroscience at one extreme, cognitive/developmental psychology at the other, and developmental psychobiology operating in the area between these two, at the level of integrative physiology and behavior.

II. EVOLUTION AND DEVELOPMENT

Ultimately, the understanding of development derives from evolutionary theory in that developmental patterns are products of evolution. But attempts to integrate the two forms of biological change within a single explanatory principle have not yet been successful. In Darwin's day, adaptations that were acquired during the lifetime of an individual, like particular habits of behavior, were thought to gradually become part of the inherited characteristics of individuals after a number of successful generational repetitions. In this way, natural selection was thought to be capable of building a developmental schedule of adaptive behaviors from the history of experiences that occurred over a number of generations. For example, Douglas Spalding, probably the first developmental psychobiologist, discovered in 1872 that chicks raised without visual experience would peck at moving insects with "infallible accuracy" on their first experience after removal of the hoods he had fitted on them at hatching. He concluded that such highly complex behavior patterns ("instincts") were the result of the "accumulated experiences of past generations . . . and . . . may be conceived to be, like memory, a turning on of the nerve currents on already established tracks."

The use of memory as an analogy for the mechanisms of heredity and development was central also to the immensely influential "biogenetic law" of Ernst Haeckel, put forward in 1874. He portrayed the stages in the development of a human as "recapitulating" the adult forms of its ancestors progressively from single cells through cellular aggregates, invertebrate, vertebrate, and lower mammals in a stepwise march through our evolutionary history. This portrayal of human development as an accelerated version of the developmental patterns of all our ancestors linked together was captured in the phrase still taught in high school biology: "ontogeny recapitulates phylogeny."

The rediscovery of Mendel's experimental work on the mechanisms of inheritance in 1900 and the work of experimental embryologists in the first half of the century provided a mass of evidence that forced the abandonment of these appealing theories that so neatly linked together development and evolution. They are mentioned here because they have not yet disappeared from the psychological literature and clinical psychiatric works on development, and the student should be aware of their historical place.

Our best present evidence leads us to believe that the germ cells containing the genes are isolated and are not affected by the consequences of events in the life of the individual. We have learned that developmental rates of different features may be delayed as well as accelerated in evolution. Thus, early features of an ancestor's development may be retained and appear later in the development of descendants (neoteny) as frequently as the accelerated appearance of ancestor's adult features described by Haeckel. Thus, developmental processes are clearly not offering us a "time capsule" in which we can trace the history of evolution in any direct manner. The ontogeny of a species is what has been retained *and subsequently modified* in the process of evolution. It embodies transformations that have come about through mutations in structural genes and through heritable alterations in regulatory genes that control the timing of developmental events in different cell lines within the organism. New species can originate through the addition of new traits early in development (insertion) as well as at the end of a developmental sequence (terminal addition), and alterations in the rate of development of one system or characteristic in relation to others (heterochrony) can have profound long-term effects, as will be described at the end of this section.

Developmental processes are usually viewed as products of evolution. But the recent discovery of genes that regulate the timing and occurrence of developmental events has suggested a mechanism by which development may actually drive evolutionary change. New species appear to have evolved through relatively simple modifications in regulatory genes controlling the timing of development in different systems. A particularly intriguing example is the divergence of humans and chimpanzees from their common ancestor; the two species have an excess of 99.9% identical DNA. The retention of certain early bodily features into adulthood ("neoteny") and, in particular, the

large, slowly developing brain of the human are likely to be due to changes in a few genes regulating the relative timing of developmental events within different cellular systems. The prolonged period of responsiveness to experience in humans, resulting from this genetic mechanism, is thought to underlie our acquisition of complex language, abstract thought, and even civilization itself. [*See* Evolution, Human.]

But evolution has not created new forms with developmental patterns that are entirely new designs. Evolution has conserved many basic developmental processes in the human by which the long line of our ancestors developed, going back to single-cell organisms and beyond. The general similarity of early stages in development across widely divergent species is the result of our common ancestry. In different species, similar structures and functions may develop on different schedules and through different developmental processes. Likewise, similar developmental processes can have different outcomes in different species. Within closely related groups such as the mammals, however, the conservative nature of evolution is predominant, with many developmental processes being closely similar and principles derived from one mammalian species often being found to generalize to others. This provides us with the basis for the use of animal models to understand human development.

Thus, in the patterns of human development we see the footprints of our evolution. The diversity of the complex patterns underlying our behavioral development appears to have the same origin as the diversity of successful life forms. For these patterns embody residues from phases of evolutionary history when the units of natural selection shifted from the simplest replicating nucleotides to more complex molecules, organelles, then to cells, and finally to multicellular organisms. The many different paths and processes by which the nervous system of even the simplest animals is assembled appear to represent the end result of the operation of a multitude of selection pressures in changing environments over a prodigious time scale.

III. NATURE AND NURTURE

We humans have great difficulty sorting out our feelings about the many influences that make us what we are. This difficulty is particularly acute in the areas of our behavior and our psychological makeup. These feelings get in the way of our understanding development and, on a larger scale, periodically erupt into violent societal disagreements over the primacy of nature or of nurture in determining differences between people.

But we should know better. There can be no traits that are solely the result of environmental influences on development nor are there any that are solely "programmed" by the genes. No "blueprint" for bodily structure or for behavior exists. Genes are as influenced by their location in the developing embryo as children are by their location among the cultures of the world. Environments have become as important a part of what is inherited as the information carried in the genome. The intrauterine and early postnatal mother–infant interactions are good examples of heritable environments in which experience and behavior have a powerful shaping force on subsequent outcome (see Section IV).

Since regulation by feedback from the immediate environment is present even in relatively simple biochemical processes (e.g., end-product inhibition of enzymes), the role of environmental interactions has been an integral part of developmental processes from the beginning. Development has made use of both genetic and environmental sources of information at every stage in the transformation of germ cells into an adult organism in all species (see Section IV,C and V).

Thus, the answer to the nature–nurture debate is to be found in understanding development. But this is not a simple matter, for we do not have a single principle by which the diversity of processes and schedules found in human development can be said to operate, something similar to the explanatory power that the principle of natural selection provides for evolution. And yet, if our present understanding is correct, such a principle does not exist. Instead, developmental processes are the *result* of natural selection, and as such they represent diverse solutions to the problem of building a successful organism.

The contributions of genetic and environmental influences can only be understood in terms of the interactive processes underlying a given developmental event. They cannot be predicted by generalizations such as that genes determine form while experience determines function. (For example, the shape of a lobster's claw and the markings of a Siamese cat's fur are strongly influenced by experience at sensitive stages of development, while the ability of some nestling birds to fly does not require learning.)

One of the major discoveries of the past two decades in this field is the extent to which biological as well as psychological systems are shaped by environmental influences during development. Conversely, the extent to which psychological processes are influenced by genetic differences is rapidly becoming appreciated. These trends have led to the abandonment of notions that associated biology with nature and psychology with nurture. This unification has greatly expanded the horizons of biology while giving psychology a firm foundation upon which to launch comparative and developmental branches of the field.

IV. PRINCIPLES

A. Stages and Transformations

Development is often defined in dictionaries as "a gradual unfolding, a fuller working out of things" (Oxford English Dictionary). But it has a very different character, which is the result of the forces considered in the previous section. Evolution did not invent wholly new developmental plans at each stage in the history of a species but utilized existing developmental processes, modifying them through terminal addition, insertion, and heterochrony as described earlier. The inheritance of these modifications, involving changes in sequences and timing of growth, proliferation, and differentiation in different cellular systems, gives development its complex programmatic character in which reasonably well-defined stages of relative stability alternate with periodic rapid transformations in multiple systems. The temporal organization of development in current species thus reflects a long evolutionary history of successful strategies, pieced together and modified in novel ways.

Stages are periods during which changes are taking place relatively slowly but which differ from subsequent or past stages in important characteristics of structure or function. Transitions mark the limits of stages. Structures and functions are generally continuous within stages and may be continuous or discontinuous across transition periods, depending on the extent of reorganization taking place during the transition. Thus embryos, fetuses, newborns, children, adolescents, adults, and the elderly are remarkably different and inhabit different environments. They differ more from each other than do adults of different species.

B. Adaptations and Precursors

One of the tasks of a developmental approach is to attempt to understand organisms at a particular developmental stage and their adaptation to the unique environment of that stage because environments change as much during development as do the young themselves (for example, the transition from prenatal to postnatal life in mammals). Special characteristics that appear transiently and suit developing forms to a given stage-specific environment are known as "ontogenetic adaptations."

Two forms of selection pressure have acted together to determine the nature of the developing organism at any particular stage. One has selected features that prepare for more complex and more adaptive functioning in subsequent stages (e.g., capabilities for perception and for action are necessary antecedents for learning to occur). The results of this form of selection give development its "progressive" character. The other has selected features that favor adaptation to the unique environment of a given stage (for example, the following response of newly hatched birds that favors attachment to the parent). Features that maximize adaptation to such early environments as the uterus or egg have resulted in greater numbers of individuals surviving to a reproductive age.

Thus, traits of the young are usually understood both in terms of their contribution to future development, as precursors, and in terms of their role within the ecology of their stage in development, as adaptive traits. The interesting issue then becomes: how do these two processes interact? Variations in the nature of adaptive behavior at a given stage often feed back upon developing systems (through the neurochemical and hormonal effects of these experiences) to alter the schedule and even the direction of their development in subsequent stages. For example, different patterns of mother–infant interaction have been shown to alter the behavioral and physiological responses of young as they develop into juveniles and adults.

A great deal of current research activity in the field is directed at understanding more precisely how these interactive effects take place, both in terms of what later changes are produced as a result of the early experiences and also in terms of the mechanisms by which earlier adaptive interactions alter developmental trajectories in specific systems. Sensitive or "critical" periods when certain experiences (e.g., sensory stimulation, learning, drug exposure) have maximal influence on subsequent development are of particular interest, along with the developmental events that form the boundaries of these sensitive periods. For example, the period for imprinting of attachment in certain species of birds is ended by the development of "fearful" avoidance responses to novel moving objects.

C. Dynamic Forces

The foregoing considerations help us to understand the stepwise form that development takes but do not account for its dynamic quality. What are the forces that propel the developing organism through this series of stages and transformations? How are these forces regulated?

It is difficult to single out certain elements as primary in a complex process like development that operates through a series of interactions between different kinds of events occurring at different levels of organization. But to state that everything depends on everything else does not provide a good basis for further understanding. Table I provides a conceptual schema which proposes three classes of events as the most dynamic and pervasive ("driving forces") and others which act to modulate, guide, and shape the interplay of these forces ("regulatory processes"). This schema is designed to provide a basis for discussing mechanisms of development at the level of behavior and integrative physiology, and is congruent with a similar schema proposed by Gerald Edelman for developmental processes at the level of groups of cells (brain cells in particular). In Edelman's classification, the driving forces are cell division, migration, and death whereas regulatory processes are cell adhesion, differentiation and induction. These cellular events result in the formation of a diversity of tissues and organs (morphogenesis) from cells with identical genomes. The natural extension of morphogenesis is maturation, which is listed as the first of three driving forces of development at the level of integrative physiology and behavior. Thus maturation can be considered to be a link across levels of biological organization in development.

Maturation denotes the growth and functional changes in the brain and other organs that result from the interactions of structural and regulatory genes as controlled by intracellular "messenger" proteins linking the genome with the environment of the organism in a series of steps involving neural and hor-

TABLE I
Components of Psychobiological Development

Driving forces	Regulatory processes
Maturation	Nutrient
Behavior	Thermal
Environmental change	Hormonal
	Sensori-motor
	Integrative

monal pathways. The second driving force, behavior, begins to exert its effects when maturing nerve cells begin to generate and conduct impulses in early fetal life. Activity in motor and sensory nerve cells plays a major role in shaping the musculoskeletal system and in the organization of brain networks for processing information. Motor activity propels the developing organism into new interactions with the environment after birth, and sensory function allows these interactions to be organized into novel patterns (e.g., through learning and memory). The third major force is environmental change. As development proceeds, the environment that the organism encounters changes radically. Some of these changes are thrust upon the young (e.g., birth), some are the result of its own increasing range of activity, and many are due to both operating together.

These three forces, operating together, propel the development of behavior along its course. But in order for development to follow a relatively predictable route and for the organism to adapt, within limits, to variation in its ecology, a number of regulatory influences exist which initiate and control the operation of these forces and act to integrate them in adaptive patterns. These regulatory processes exist in an extraordinary variety, and the analysis of specific instances of such regulation is the subject of much current research in the field. In outline, these regulatory processes can be categorized into five main areas: nutrient, thermal–metabolic, endocrine, sensori-motor, and higher integrative functions (see Table I).

This article gives one or two brief examples from the wide range of regulatory influences in the five major categories. Nutrient intake regulates feeding behavior in different ways at different ages; early deprivation of nutrient can have irreversible long-term effects on brain structure and cognition in adulthood; the nutrient has unexpected actions in regulating early cardiovascular function; and young raised on diets of different composition show characteristic behavioral differences. Core body temperature, together with oxygen-dependent metabolic processes, controls rates of maturation of cells and other physiological processes. Ambient temperature regulates the expression of a variety of behaviors in infants and influences the nature of the mother–infant interaction. Endocrine influences have their most profound effects on reproductive development. For example, low levels of testosterone in the late fetal period of male primates appear to be essential for masculine sexual behaviors to develop in puberty when activated by the greatly increased rates of secretion of testosterone in males

at that later stage of development. Sensori-motor regulation is perhaps the largest category, involving influences in all sensory modalities as well as feedback from all behavioral activity. Motor activity takes place in human fetuses from the seventh week on, and sensory function develops at about the same time. These early sensori-motor activities constitute a major shaping influence on maturational processes. For example, the shapes of joints between bones and the growth of supporting cartilage structure are dependent on the existence of the jerky, uncoordinated activity of the early fetus. Later, the development of coordinated and adaptive movements is dependent on shaping by sensory feedback and by the temporal relationship between motor and sensory functioning. The fifth category, higher integrative functions, includes learning, memory, emotion, motivation, cognitive plans, and inner consciousness of these mental events. Much of developmental psychology is concerned with this last category, which nevertheless rests upon the operation of many less-apparent developmental processes within the other four categories.

D. Levels of Organization

The driving and regulatory forces just described clearly operate at a number of different levels of organization, ranging literally from molecules to mind. This is inevitable when considering human development which begins with single cells, the sperm and the egg, and progressees to complex psychological events. Bridging these different worlds are a number of properties which are of central interest. *Environments* exist for cells and molecules as well as for the adult organisms and provide similar opportunities for interactions or *experience*. Cells as well as organisms are changed as a result of that interaction, through molecular mechanisms. *Behavior* is a word that we use to describe molecules and cells as well as complex physiological systems and the actions of the whole organism. The use of this word in these different settings is not simply metaphorical. The three basic properties of behavior as we know it (activity, receptivity, and integration) are found even within the chemical systems of cells. Some of the simplest of life processes have an organized and adaptive character. This principle underlying the organization of behavior across many levels is understandable in terms of evolutionary history: behavior has been extraordinarily useful.

These continuities across levels of organization help the developmental psychobiologist make sense out of the diverse events that characterize different stages of development. In the adult human, a given behavior or psychological process can be analyzed at a number of different levels of organization using similar principles, although the methods of analysis may differ enormously between the levels of molecular genetics and cognitive psychology.

E. Emergent Properties

It is evident that as one advances into more complex levels of organization, new properties emerge. Psychological terms such as *learning* and *emotion* are used to deal with properties of complex levels of organization for which biological terms do not exist. Much of current research in developmental psychobiology centers on the general area of the transition from complex physiological systems to psychological processes. Behavioral terms which apply to both systems provide a conceptual bridge for moving back and forth between the historically (and academically) separate realms of psychology and biology.

Throughout the course of human development, new properties emerge at each stage. In its first emergence, we have an example of the property in a simple form when we are most likely to be able to analyze and understand it. Since evolution and development build up more complex organization out of earlier simpler structures, these are not abandoned but are used as building blocks. Thus this field hopes to understand such complex mental functions as motivation, emotion, and cognition by finding out how they are put together.

V. EVENTS AND PROCESSES

It would be well beyond the scope of this article to summarize current knowledge in the field. It is possible, however, to select some of the major events in early human development and use them to illustrate the operation of the principles outlined above.

The first phase of human development takes place in the form of germ cells within individuals of the previous generation, prospective parents. The ova have a long existence, being formed during the fetal period of the mother. The sperm are formed 60–90 days before fertilization in the prospective father and become motile when ejaculated with hundreds of millions of others into the fluid environment provided by the male and female at the time of sexual intercourse. The swimming behavior of the sperm is the first critically important behavior in a human life, for

one of the 300–400 million sperm will be the first to make its way through the cervical canal, the cavity of the uterus, and the fallopian tube to fertilize the egg, blocking all other sperm from penetrating the egg membrane. Individual sperm vary greatly in shape, motility, and surface proteins, as well as in the genetic material within their nuclei. The competitive balance between different sperm can be shifted by their interaction with characteristics of the female genital tract and the egg. The mother's behavior and emotional state, through changing hormonal and autonomic balance, may alter the environment through which the sperm must move (antibodies, mucous flow, and uterine contractions) and thereby act to select which sperm will fertilize her egg. The genetic content of the sperm also may influence the selection process by affecting sperm surface proteins and motility patterns. (For example, more girl babies are born to women who have intercourse soon after ovulation.) The gene–environment interaction is nowhere more directly represented than in this prefertilization stage of human development. [*See* Sperm.]

Fertilization represents the first major transformation in human development and brings to an end the bicellular phase of human life, one that is barely recognized by us, although it is both long and, in the end, eventful. The next phase lasts for only 9 months, but in that short time a whole human being is built from a single cell. Until rcently, it has been viewed as a stage in our life devoted to the development of structure, while behavioral development was considered to begin at birth. The invention of ultrasound imaging, the ability to support life in younger premature babies, and advances in developmental neurobiology in the past decade have revealed for the first time the extent and nature of prenatal behavior. [*See* Fertilization.]

The fetus begins to move early in gestation, during the 7th week after fertilization in the human. By 9 or 10 weeks, 15 different kinds of movements can be described, including startles, hiccups, breathing movements, and sucking, as well as movements of limbs, trunk, and head separately and together. Sensory function begins, independently, at about the same time with the first response occurring to stimulation in the area of the mouth, at a stage when the growing sensory nerves have not yet reached the basement membrane of the skin. Early axonal outgrowths conduct impulses and transport trophic substances even before synapses are fully formed and before primitive muscle tissue becomes differentiated. [*See* Fetus.]

What is the "purpose" of this early behavior? It

has become clear that neural activity and the transport of trophic substances are major components of the means by which the nervous system is constructed and by which its fine structure is organized. For example, it is the pattern of early neural activity that determines whether muscles will develop into "fast twitch" or "slow twitch" types, with their different biochemical structure and performance. Activity in sensory systems is needed for the formation of cerebral architecture such as "cortical columns." The early, frequent, and persistent movements of fetal limbs are critical for the shaping of joint cavities and the formation of intraarticular ligaments. This use of early behavior as a sculptor of the developing brain and limb is a good example of an ontogenetic adaptation. Early behaviors are not fully explained as precursors of later, more complex behaviors; they serve an adaptive function unique to the particular requirements of that stage in ontogeny. The dynamic forces of maturation and activity are clearly illustrated here as well.

The intrauterine space provides a controlled, predictable environment that is reliably heritable. High rates of nutrient and oxygen flow can be supplied, as well as other substances such as steroid hormones and neuroactive peptides, which can serve a regulatory function for the developing fetus. We are just beginning to learn about maternal influences on behavior development during this period. For example, it has been discovered that the mother sets the biological clock of her fetus in the last trimester to her own circadian rhythm and day length. Also, recent evidence suggests that the fetus learns the special characteristics of its own mother's voice so that it can recognize her as distinct from others within hours of birth. These functions clearly serve an adaptive role as precursors of traits useful in a later stage after birth.

As the fetal period progresses, we can observe a series of transformations that represent the maturation of progressively higher neural systems which come to exert a modulating and organizing effect on the activity of earlier developing units. Motor activity of the fetus reaches a peak between the 15th and 17th weeks, declining gradually thereafter as movements become more patterned in time and are coordinated with each other into sequences. This gathering together of units from within the broad matrix of early diffuse activity tends to organize behavior into a hierarchical structure with different levels of organization. New properties emerge. For example, a cycling between activity and rest characterizes the behavior of a single neuron, but then assemblages of neurons begin to function together with their own particular patterns, and when cardiovascular and respiratory systems are joined with neuromuscular patterns, an integrated behavioral state is formed that can be specially suited to carrying out certain highly specialized functions. The appearance of a precursor of dreaming sleep [rapid eye movement (REM) sleep] appears at 32 weeks of gestation. In this state there is a profound inhibition of muscle tone and movements while diffuse activating volleys project from the brain stem to higher cortical centers as well as to eye and ear muscles, blood vessels, and heart. This intense internal activation in an otherwise passive organism appears to represent an evolutionary solution to the paradoxical requirements for neural activation in preparation for life after birth and behavioral quiescence in the confined space of the uterus.

Birth represents the most abrupt regularly imposed environmental change in our normal development. It is clearly a most powerful force in the development of human behavior since it provides a whole new range of stimuli with which the infant can interact. This greatly increases the complexity of sensori-motor interactions, sets the stage for learning, and initiates the possibility of higher cognitive functioning. The range of regulatory or shaping forces is thus expanded to a great degree over those present during the prenatal period.

The behavioral capabilities of the newborn human, in contrast to its environment, change very little in the period immediately after birth. Continuity with the sleep–wake state organization of the late fetal period is the most striking characteristic, although episodes of clear-cut wakeful state continue to become more frequent and quiet sleep gradually begins to appear, alternating with the REM sleep previously described. When held in its mother's arms, the newborn's behavior reveals that its visual and auditory systems have become predisposed to respond to the human face and voice, and its sensori-motor systems are predisposed to respond to the breast with a complex sequence of behaviors enabling it to find the nipple, attach, and suck.

The competence of the newborn and its ability to discriminate and learn have only become apparent as investigators have found ways of studying its behavioral capacities in its natural environment. This is because early adaptive behaviors have evolved in relation to the heritable aspects of the environment and not to other modalities, intensities, and patterns of stimuli, such as those present in laboratory tests of learning. The distinction that until recently was drawn between instinctual behaviors and those that are learned has now been abandoned since it has become clear that all animals are predisposed to learn certain things more easily than others and that even the learning of novel tasks is guided by preexisting hierarchies and patterns of response tendencies. In turn, predispositions and complex response sequences, like suckling, that appear to occur in newborns without any opportunity for learning have been found to depend on nursing experience in order to be maintained by the infant. With careful analysis, their first appearance can be found to depend on a number of prior interactions between activity and local environmental cues within developing sensory and motor systems in the prenatal period.

Of the many behavioral and physiological systems that have been studied experimentally in animal model systems by developmental psychobiologists, those mediating feeding have been the most extensively analyzed. One of the major surprises to come from this work is the finding that, in all species studied, independent feeding on solid food and fluid does not develop out of nursing as a precursor behavior but instead has its own developmental course that is for the most part independent of the nursing experience. An infant's suckling thus is another example of ontogenetic adaptation.

Experimental studies of the mechanism underlying nursing behavior in rats have revealed an interweaving of response predisposition, sensori-motor regulation through olfaction and touch, more complex associative learning, and dependence on feedback from active initiation of behavior, all of which illustrate the dynamic forces described in the previous section. Pups are born with sensory-guided response tendencies (present also prenatally) which bring them to their mother's ventrum. The scent of amniotic fluid deposited by the mother on her nipple lines by licking after birth guides them closer; then highly developed touch and sensory hairs of their snouts provide the most proximal sensory cues. The first few nursing bouts are the result of these response tendencies, but if some of the pups are artificially fed by stomach tube, these behaviors will gradually cease to be elicited by their dam, and nursing by the pups will cease in 2–3 days. Even if the pup is placed on the nipple each day, nursing will fade away at the same rate. But if the pups are merely placed near the ventrum and are allowed to search out and locate nipples for as little as 15 min each day, their capacity to suckle will be maintained. This maintenance, through active initiation, does not depend on reinforcement by milk.

The next phase in nursing, its decline at weaning, is assumed in humans to be under control by the mother. But some infants seem to wean themselves, and little is known of the processes underlying this developmental transformation which does so much to define the stage of infancy from that of childhood. Some understanding has been gained by experimental analysis in laboratory rats. The decline in nursing by pups between days 15 and 21 has been shown to be mediated, at least in part, by maturation of a serotonergic receptor forebrain system. This is not to suggest that weaning takes place simply due to processes intrinsic to the pup. Dams of 21-day-old litters given to 14-day-old pups will hasten weaning, and dams of 7-day-old pups will retard it. In fact, placing a single pup of 15–17 days age with a dam and litter of 14-day-old pups and then replacing the dam and litter with a new 14-day-old "family" each week will result in rats continuing to nurse into young adulthood, well after sexual maturity (50–70 days), when they are often as large as the mother. This phenomenon has been shown to be the result of continued milk availability plus a lesser component contributed by the social facilitation of the younger litter mates.

It is clear from these analyses of the processes underlying the development of early feeding that the infant mammal and its parent participate in a range of interactions in which they are linked by their sensitivity to subtle cues originating from the other. The course of nursing is initiated, maintained, and regulated by these interactions. But nursing is only one of a number of transactions taking place between the parent and the infant. Parents also keep infants warm, clean them, and engage in a wide range of communicative actions and reciprocal sequences referred to as "play." These activities have roles in the health of the infant, the development of its emotional life, its ability to communicate, and a wide variety of motor skills.

Recent experimental research in laboratory rats has revealed a number of inapparent processes, built into the many mother–infant interactions, by which the developing physiological and behavioral systems of the infant are regulated. Growth hormone and corticosterone levels, autonomic cardiovascular function, sleep–wake cycles, central catecholamine neurotransmitter levels, and behavioral reactivity have all been shown to be maintained at their characteristic levels in infants by various aspects of the intensity, frequency, and patterning of different interactions between the dam and her pups. Each system is controlled by a particular modality or aspect of the interaction. For example, behavioral reactivity is regulated by intermittent tactile and olfactory stimulation, cardiac rate by gastric interocepters sensitive to nutrient composition, and sleep–wake states by the periodicity and timing of nutrient and behavioral interaction.

These regulatory systems within early social relationships are an extension of the principles of regulatory forces in development that were present in prenatal life. Their existence is made possible by the predictable inheritance of the early mother–infant interaction secondary to the mammalian characteristic of nursing young. The behavioral attachment system provides the young with a powerful motivational system for maintaining close contact with early social companions. This system makes developmental regulation by the parents and feeding by nursing possible.

As a result of the existence of these systems, prolonged separation of young infants from their mothers can have widespread and even lethal consequences. In addition to the loss of nutrition from nursing and mobilization of the attachment system in prolonged but unsuccessful efforts to search out and reunite with the mother, all maternal regulatory systems are withdrawn at once, producing a pattern of widespread physiological changes. These separation responses give dramatic evidence of the existence of these several developmental processes that characterize this phase of our development.

As development proceeds and new functions mature, the infant gradually comes to increasing independence not only in feeding (as described earlier) but also in homeostatic regulation of its internal physiology and expansion of its range of behaviors beyond the confines of interaction with its immediate caretaker. But there continue to be dependences of development in a number of systems on social interactions. For example, children born congenitally deaf show similar vocalizations to hearing children during the first 3 months of life. They cry and murmur normally and show normal "gooing." But they do not, beginning in the 4th month, produce normal, resonant vowel-like sounds, and they lag months behind hearing infants in the production of mature syllable strings like "ba ba ba." Without special training, deaf babies' vocalizations do not progress beyond the early stages. This represents the importance of auditory experience in maintaining the normal development of speech.

VI. RANGE AND PROSPECTS OF THE FIELD

These few examples only hint at the range of topics currently under study in developmental psychobiology and focus on very early stages, although development continues throughout life into old age. There is a good deal of activity in the area of abnormal experiences, environmental hazards, and how these influences contribute to the production of pathological deviations in development. Animal model systems useful for the study of clinical conditions in psychiatry (such as anxiety disorders) and pediatrics (e.g., prenatal drug abuse, sudden infant death syndrome) have been developed. Altered susceptibility to stress as a result of differential early experience has been found in studies demonstrating marked influences on corticosteroid responses, the production of gastric erosions, and hypertension. The course of experimental diseases such as cancer, infections, and diabetes in animals has been found to be significantly influenced by the modification of early experience such as premature separation, handling by the experimenter, or crowding of the pregnant dams. The physiological regulatory processes that originate in the early environmental interactions described in Sections IV and V account for these complex and wide-ranging early experience effects.

Future progress in the field should result in a deepening understanding of the behavioral and physiological mechanisms of these effects, operating both in normal and disturbed development. The advent of genetically engineered strains with specifiable discrete changes in DNA should lead to a new level of analysis of environmental and other epigenetic processes in gene expression.

BIBLIOGRAPHY

Aslin, R. N., Alberts, J. R., and Petersen, M. R. (eds.) (1981). "Development of Perception. Psychobiological Perspectives," Vol. 1. Academic Press, New York.

Blass, E. M. (ed.) (1988). "Handbook of Behavioral Neurobiology," Vol. 9. Plenum Press, New York.

Edelman, G. M. (1988). "Topobiology." An Introduction to Molecular Embryology." Basic Books, New York.

Gould, S. J. (1977). "Ontogeny and Phylogeny." Belknap Press of Harvard University Press, Cambridge, MA.

Greenough, W. T., and Juraska, J. M. (1986). "Developmental Neuropsychobiology." Academic Press, New York.

Hofer, M. A. (1981). "The Roots of Human Behavior: An Introduction to the Psychobiology of Early Development." Freeman, New York.

Krasnegor, N. A., Blass, E. M., Hofer, M. A., and Smotherman, W. P. (eds.) (1987). "Perinatal Development: A Psychobiological Perspective." Academic Press, New York.

Prechtl, H. R. F. (1984). "Continuity of Neural Functions from Prenatal to Postnatal Life." Lippincott, Philadelphia.

Raff, R. A., and Kaufman, T. C. (1983). "Embryos, Genes and Evolution." Macmillan, New York.

Shair, H. S., Barr, G. A., and Hofer, M. A. (eds.) (1991). "Developmental Psychobiology: New Methods and Changing Concepts." Oxford University Press, New York.

Smotherman, W. P., and Robinson, S. R. (eds.) (1988). "Behavior of the Fetus." Telford Press, Caldwell, NJ.

Diabetes Mellitus

NIR BARZILAI
HARRY SHAMOON
Albert Einstein College of Medicine

GLOSSARY

Glycation Nonenzymatic glycosylation of proteins such as hemoglobin associated with the elevated blood glucose level of diabetes

Insulin-dependent diabetes mellitus Type 1 or autoimmune diabetes, formerly called juvenile diabetes

Keto acids Organic acids that are the by-product of oxidation of fatty acids accumulate in the bloodstream in uncontrolled diabetes

Nephropathy Specific form of renal disease associated with renal glomerulosclerosis due to diabetes

Neuropathy Symptoms and nerve fiber damage associated with diabetes

Noninsulin-dependent diabetes mellitus Type 2 diabetes, formerly called adult- or maturity-onset diabetes

Prandial Meal associated

Retinopathy Specific lesion of the vascular and neural retina associated with long-standing diabetes

Sorbitol Glucose-derived polyol that accumulates in tissues under conditions of hyperglycemia

DIABETES MELLITUS SO CALLED BECAUSE OF THE Greek term for "sweet urine," is a common disease characterized by elevated levels of blood glucose. These disorders cluster in phenotypes that are classified by clinical criteria, but the molecular basis for most forms of diabetes is not known. This article summarizes the major causes of diabetes and describes the abnormalities of the immune system, metabolic pathways, environmental factors, and the genetic background leading to diabetes.

I. CLINICAL MANIFESTATIONS OF DIABETES

Diabetes mellitus is characterized by elevated levels of blood glucose (hyperglycemia). Even though there are different genetic and phenotypic characteristics of the multiple forms of diabetes, hyperglycemia is the common biochemical denominator. In addition to the metabolic derangement and symptoms attributable directly to hyperglycemia, diabetes is distinguished by several specific vascular and neurological complications which appear after many years of uncontrolled hyperglycemia.

Based on the epidemiological data relating the development of complications to hyperglycemia, the World Health Organization has defined the level of blood glucose that indicates a diagnosis of diabetes. Individuals whose morning fasting plasma glucose concentration (that achieved following an evening meal by 8 to 14 hr earlier) is equal to or exceeds 140 mg/dl have diabetes. These diagnostic criteria are currently under review, and fasting plasma glucose levels of 125 mg/dl (6.9 mmol/L) may be appropriate for diagnosis. Milder degrees of fasting hyperglycemia can be associated with excessive increases in blood glucose following an oral glucose challenge; thus diabetes can be diagnosed at this less overt state with a glucose tolerance test.

ENCYCLOPEDIA OF HUMAN BIOLOGY, Second Edition, VOLUME 3.   Copyright © 1997 by Academic Press.   All rights of reproduction in any form reserved.

At plasma glucose concentrations sufficiently high to exceed the renal tubular maximal reabsorptive capacity, glucose excretion results in calorie, water, and electrolyte loss. This explains the symptoms of excessive thirst, urination, and weight loss that are classically associated with the onset of hyperglycemia. Less commonly, diabetes is discovered because of altered responses to infection and injury. All of these "acute" symptoms may be more severe if the level of plasma glucose is dramatically high (>600 mg/dl) with or without a concomitant accumulation of keto acids which can result in "diabetic coma." The latter is usually associated with the extreme insulin deficiency of insulin-dependent diabetes (see later); however, both insulin-dependent (Type 1 diabetes) and noninsulin-dependent diabetes mellitus (Type 2 diabetes) are now known to develop insidiously after many years of subtle abnormalities in insulin secretion and/or action.

Although beyond the scope of this discussion, it is important to note that the overall impact of diabetes is not limited to the metabolic consequences per se but stems from its long-term vascular and neurological complications. These latter disorders explain why diabetes is a leading cause of mortality by way of its association with coronary atherosclerosis and why it is among the leading causes of blindness, kidney failure, and nerve damage due to microvascular complications. Most evidence points to hyperglycemia per se being involved in the etiology of these latter diseases either by its metabolic effects (e.g., intracellular accumulation of sorbitol and depletion of *myo*-inositol) or by the production of long-lived nonenzymatically glycated proteins which form cross-links and become pathogenic for the complications. None of the presently available treatments of diabetes readily normalize the blood glucose level; hence, some degree of hyperglycemia is present in most patients. The prevalence of diabetes in the United States is ~16 million persons, and the estimated annual direct costs of medical care may be as great as \$102 billion. Most of these costs are associated with the treatment and other consequences of the long-term vascular complications. Nevertheless, the results of the Diabetes Control and Complications Trial clearly implicate the chronic average level of blood glucose as a major determinant of many vascular complications. In this trial, over 1400 persons were followed for a median of 6.5 years, and it was determined that measures which keep blood glucose levels close to normal could prevent the development and progression of retinopathy, neuropathy, and nephropathy, and that any improvement

in hyperglycemia is associated with a reduced risk of complications. Efforts at treatment of complications, the treatment of diabetes, and the prevention of diabetes continue to be the focus of a number of trials in progress.

II. PHYSIOLOGY OF THE ENDOCRINE PANCREAS

The pancreas of higher species is both an *exocrine* gland—secreting digestive enzymes into the gastrointestinal tract—as well as an *endocrine* gland—secreting hormones into the blood that regulate energy metabolism. The hormone-producing cells are all located within microscopic clusters called islets of Langerhans scattered throughout the pancreas which comprise only $\sim2\%$ of the total mass of the pancreas. Within these islets is a specific architecture of cell types, characterized by their content of secretory granules, their staining pattern as cells originating from the neural crest, and specific patterns to immunostaining with antisera to the hormone product of each cell. Insulin-producing cells (β cells) constitute $\sim80\%$ of the total islet cell mass, with A cells (glucagon), D cells (somatostatin), and F cells (pancreatic polypeptide) comprising the remainder. Each islet contains about a thousand β cells and each pancreas about a million islets. The significance of the close association of all these cell types within an islet is not entirely clear, but a modulatory role for insulin secretion has certainly been demonstrated. Within the islets, cell-to-cell interactions may take place via tight junctions or by the actions of locally released hormones (so-called paracrine effects). Potential candidates that may play a role in insulin secretion include pancreastatin (a fragment of chromogranin A present in insulin secretory granules); islet amyloid polypeptide (IAPP) which is $\sim50\%$ homologous with calcitonin gene-related peptide (CGRP) and also produced by β cells; galanin (a neuropeptide present in many endocrine cells, including β cells); NPY, a neuropeptide present within the islets which may play a role as a neurotransmitter; vasoactive intestinal peptide and CGRP, both found within islets which may regulate islet blood flow; and somatostatin. [*See* Peptide Hormones of the Gut.]

The highly vascular pancreas receives its major blood supply from vessels which drain from the gut and, in turn, the blood exiting the pancreas is directly delivered via the portal vein to the liver. These relationships position the islets to receive both metabolic substrates (such as sugars and amino acids) as well as insulin secretory "incretin" hormones (e.g., GLP1, gastric inhibitory peptide, secretin, and somatostatin) from the gut. Thus, the normal route for stimulation of insulin secretion involves multiple inputs to the β cells (Fig. 1).

A variety of known stimuli regulate insulin secretion, including the level of blood glucose and amino acids such as arginine and leucine. The mechanism of glucose-mediated insulin secretion is thought to involve the actual metabolism of glucose within the β cell via the enzyme glucokinase. In the presence of hyperglycemia, the rate of glucose entry into β cells and its phosphorylation to glucose-6-phosphate are increased by glucokinase. It is this latter step that probably provides the sensing mechanism of the β cell for the level of blood glucose and which, in turn, results in insulin secretion. While the specific glucose transporter of the β cell (GLUT2) is necessary for glucose entry, it is believed that the activity of the phosphorylating enzyme—glucokinase—is key to insulin secretion in normal islets. Experimental agents that interfere with glucose metabolism interfere with insulin release, and in situations where glucokinase is deficient, insulin secretion is impaired. At least one genetically distinct form of early onset diabetes is associated in ~50% of families with mutations of one allele of the glucokinase gene. In these patients, insulin secretion is dependent on the presence of high plasma glucose concentrations.

Insulin release is known to involve the action of intracellular calcium and of cAMP formation and depolarization of the voltage-dependent potassium channel. Glucose metabolism, through production or flux of ATP, or an increased ATP/ADP ratio, inhibits ATP-sensitive K^+ channels in the β cell. The resulting plasma membrane depolarization opens voltage-gated Ca^{2+} channels and induces an influx of Ca^{2+}, triggering the fusion of preformed insulin-containing vesicles with the plasma membrane and their subsequent release into the blood. The stimulatory effects of glucose *in vivo* are complemented by circulating substances (e.g., epinephrine) and local nerve endings (both sympathetic and parasympathetic), as well as by the gut-derived hormones.

The role played by the gut hormones to enhance insulin secretion, termed the "enteroinsular axis," was first noted for GIP (glucose-dependent insulinotropic hormone) released from endocrine cells within the upper gastrointestinal tract when stimulated by carbohydrates and fatty acids. In addition, mammalian proglucagon contains another peptide known to enhance insulin secretion. These glucagon-like peptides (mainly GLP1) are the product of posttranslational processing of gut-derived glucagon. GLP1 enhances insulin secretion with feeding. It is important to note that insulin is transported to the liver after release into the portal vein; insulin concentrations at the portally supplied hepatocyte are two- to threefold higher than in peripheral blood. This gradient is due to the receptor-mediated uptake of insulin by liver cells, both as a pathway for clearance of the hormone as well as a reflection of the primary role played by the liver in overall fuel economy. [*See* Endocrine System; Pancreas, Physiology and Disease.]

III. REGULATION OF GLUCOSE TRANSPORT

Glucose is the ubiquitous six-carbon molecule which serves as the primary energy source in aerobic metabolism. Its intracellular fate has been detailed elsewhere, but prior to its first metabolic transformation into glucose-6-phosphate by the enzyme hexokinase, it must be transported across membranes. Transport of glucose across cell membranes is facilitated by two classes of molecules, each derived from a distinct gene family. The facilitated diffusion glucose transporters are expressed throughout all cells, whereas the Na^+/ glucose cotransporters are limited to the cells of the kidney and intestine which transport glucose molecules from a lumen into the blood. The facilitated glucose transporters are structurally similar, but distinct types have been cloned in different tissues and represent the unique requirements of the tissue in

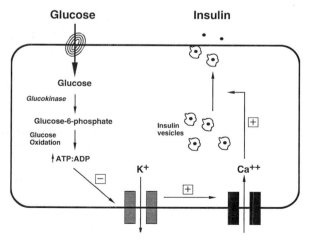

FIGURE I Major pathways which mediate glucose-induced insulin secretion by the β cell.

TABLE I

Glucose Transporter Molecules

Glucose transport	Location of tissue expression	K_m (mM)	Function
GLUT1	Erythrocytes, brain, placenta	~1–2	Basal uptake
GLUT2	Liver, β-cell intestine, kidney	~15–40	Uptake and release Glucose sensing Glucose efflux
GLUT3	Brain	1	Basal uptake
GLUT4	Muscle, fat	~2–5	Insulin-stimulated glucose uptake
GLUT5	Jejunum	?	Unknown

which they are expressed. Because the number of glucose transporters in any cell appears to be limited, the facilitated diffusion they augment shows saturation kinetics that can be characterized by the half-maximal transport coefficient (K_m). Insulin appears to enhance the activity of the transport system by increasing the translocation of the insulin-sensitive carrier molecules (GLUT4) from the interior of the cell to the plasma membrane (Table I and Fig. 2).

IV. INSULIN-DEPENDENT DIABETES MELLITUS (TYPE 1 DIABETES)

Type 1 diabetes, formerly known as "juvenile diabetes," is a chronic disorder that affects humans and other mammals and can occur at any stage in life. Insulin deficiency in Type 1 is now known to be the result of autoimmune destruction of the islet β cells, ultimately requiring that patients receive insulin to maintain life. Type 1 occurs in varying prevalence in many different populations, although it is most common among individuals of Northern European descent. Since there may be a variety of Type 1 susceptibility genes, however, other distinct ethnic groups (such as Sardinian Italians) also have high prevalence rates. The lifetime prevalence of Type 1 in the United States is ~0.4%; thus, 1 million persons have this form of diabetes.

An important component of Type 1 susceptibility resides in a gene(s) within the major histocompatibility complex (MHC) located on chromosome 6. Various other susceptibility loci have been identified in the human genome, but this region, which involves immune recognition molecules [or histocompatibility

leukocyte antigens (HLA)], most certainly accounts for the preponderance of the genetic susceptibility in most persons studied. As described elsewhere, autoimmune diseases are associated with genes in this area since the interaction between cells expressing an HLA–antigen complex and T lymphocytes, which possess receptors capable of recognizing the complex, is responsible for proliferation of this subset of lymphocytes. In addition to this region, Type 1 may be associated with another gene located on the short arm of chromosome 11 near the insulin and IGF-II genes, as well as genes associated with the intracellular protein that transports antigenic peptides to the endoplasmic reticulum.

Susceptibility to Type 1 is closely associated with HLA-DR3 or DR4 and DQ genotypes. Conversely, other HLA haplotypes protect against Type 1. The interaction between genetic susceptibility and environmental factors is best exemplified by the fact that monozygotic twins who are 100% genetically identical display only ~35% concordance for Type 1. HLA-associated susceptibility modulates either the intensity of the immune response to a pancreatic β-cell autoantigen or an altered presentation of β-cell antigens such that they are not appropriately tolerated as self-antigens in that person.

The characteristically specific loss of β cells in Type 1 is due to an inflammatory infiltrate of CD8, CD4, B lymphocytes, macrophages, and natural killer cells (Fig. 3). This "insulitis" is associated with the appearance of autoantibodies in the blood to islet cell proteins and products. Islet cell antibodies and autoantibodies to insulin are found in patients at the early stages of Type 1. It is hypothesized that environmental factors trigger the immune destruction of the β cells by a cellular immune mechanism, and that these autoantibodies are markers for the ongoing destruction. Potential triggers include viruses such as mumps, rubella, and Coxsackie B4, possibly acting by the production of viral antigen-specific cytotoxic T or T effector cells which may cross-react with a β-cell-specific autoantigen. Other environmental factors that have been implicated are dietary, either via direct toxic effects on β cells or via autoimmune mechanisms. For example, the occasional association between Type 1 and a history of early exposure to cow's milk in infancy has been ascribed to the homology of a milk albumin peptide ABBOS and a 69-kDa islet cell protein that is its autoantigen. It is also possible that chemical agents present in the environment (such as the rodenticide Vacor or hydrogen cyanide from spoiled tapioca or cassava root) may play a role. Vari-

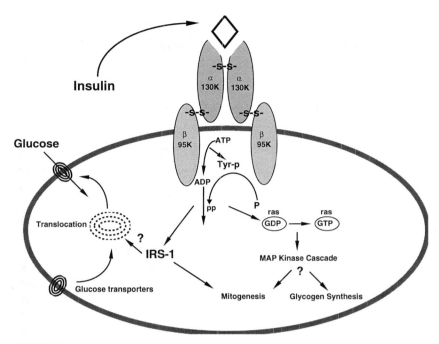

FIGURE 2   Schematic overview of insulin action and putative pathways mediating insulin signal transduction.

eties of Type 1 are phenotypically similar to the typical pattern noted earlier, but without associated autoantibodies (such as in nonwhite populations), suggesting that β-cell loss need not always be on an autoimmune basis.

In animal models of Type 1 (e.g., the NOD mouse or the BB rat), islet tissue transplantation at an early stage reverses all the changes due to diabetes (Table II). Limited evidence with transplanted pancreas tissue or islet cells tends to corroborate these observations, but most of the advanced complications in the eyes, kidneys, nerves, and vasculature do not regress. In humans, current treatment of Type 1 requires a complete or near-complete replacement of insulin deficiency, either by injections of insulin or, in special cases, by the transplantation of islet cells or pancreatic

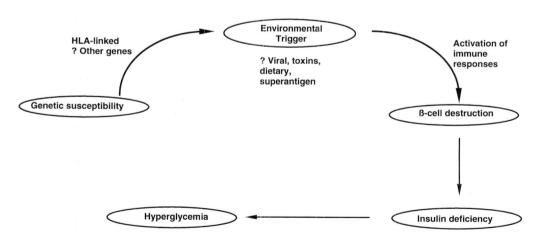

FIGURE 3   Interactions among genetic susceptibility and immune response to the development of Type 1 diabetes.

TABLE II
Antigens Eliciting Autoantibody Formation in Type I Diabetes

Sialoglycolipid	38-kDa islet cell protein
Insulin	52-kDa islet cell protein
Glutamic acid decarboxylase	69-kDa islet cell protein
Bovine serum albumin	hsp 65
Carboxypeptidase H	ICA 12/ICA 512
Pancreatic sulfatide	150-kDa islet cell protein
β-cell GLUT2	RIN polar
37- or 40-kDa islet cell protein	Insulin receptor

grafts. The major limitation of insulin use is that replacement is imperfect due to a lack of feedback regulation as occurs in the β cell. Inadequate insulin levels result in hyperglycemia, and excessive insulin levels in hypoglycemia. This latter problem may be compounded by defects in the counterregulatory system that normally antagonizes and reverses these effects via the action of hormones such as glucagon and epinephrine. At present, the transplantation of allograft islet tissue is hampered by the availability of donor cells and by the problems of immunologic rejection. In addition, the autoimmune destruction of β cells has been reported after the successful transplantation of HLA-identical tissue in individuals who have not been immunosuppressed. Strategies to improve the treatment of Type 1 include the development of insulin or its delivery systems that allow insulin to act more physiologically, methods of preventing immune access to or recognition of transplanted islet tissue, agents that mimic insulin action, and earlier diagnosis to prevent the loss of normal β cells. The focus of prevention of Type 1 is the modulation of antigen presentation on the β-cell surface in its early stages. It has been demonstrated, for example, that exposure to exogenous insulin in animal models and in humans with a high susceptibility to diabetes significantly retards the development of insulin deficiency.

V. NONINSULIN-DEPENDENT DIABETES MELLITUS (TYPE 2 DIABETES)

While insulin deficiency is the only characteristic of Type 1 diabetes, this is not the case in Type 2 diabetes. In fact, with the development of a radioimmunoassay for the measurement of insulin, it became clear that subjects with mild hyperglycemia have an increased fasting and postprandial insulin level compared to normal. Thus, it is believed that peripheral and hepatic insulin resistance, as well as a decrease in insulin secretion, are both implicated in the pathogenesis of Type 2. Additionally, Type 2 is different from Type 1 because of features such as adult age of onset (usually >40 years old), a genetic influence (>90% concordance in identical twins), obesity in most patients (60–80%), and no association with HLA alloantigens.

Understanding the pathogenesis of Type 2 is complicated by several factors. First, it has become apparent that this is a disease with a multitude of causes and presentations. A few single gene defects have been characterized both clinically and genetically. These include mutations in the glucokinase gene associated with the disease "maturity onset diabetes of the young" (MODY). Mutations of the insulin receptor that render it poorly responsive to insulin or mutations of the insulin molecule also occur. These defects are relatively rare and do not represent the cause of diabetes in the majority of Type 2. Second, it has been suggested that Type 2 is a consequence of genetic selection of "thrifty genes." In this hypothesis, the survival value of factors which promote caloric storage and more efficient energy metabolism favor the survival of organisms in times of food restriction. The Pima Indians in the southwest United States are an example of a genetically homogeneous group in which diabetes was virtually unknown in the 19th century when the tribe existed on a subsistence diet, derived at a considerable cost of energy expenditure. In the present day, the increased availability of calories in the tribal reservations and decreased exercise have resulted in dramatic increases in the proportion of adults who are obese and approximately 50% ultimately develop Type 2. This is not attributed to a single genetic defect but rather to a number of genes involved in metabolism, not only of glucose but likely also of other fuels such as fats. Whole body metabolism, which was effective in conserving energy and ensuring survival of the individual during famine, results in metabolic diseases when food is abundant. Thus, it is not likely that a single gene will be found defective. Finally, hyperglycemia per se causes further abnormalities in muscle, liver, and β-cell glucose metabolism, thus making it difficult to characterize primary vs secondary defects in human and animal models of diabetes due to "glucose toxicity" (see later).

Diabetes has been linked with a cluster of disorders

which seem to have resistance to insulin action in common. Hypertension, abdominal obesity, and dyslipidemia are linked in epidemiological studies with either impaired glucose tolerance or frank Type 2. Termed syndrome X, these disorders may share a common genetic basis and/or pathogenesis, suggested by the fact that exercise and weight loss, which improve insulin resistance, may lower blood pressure, improve lipid levels, and lower circulating plasma insulin concentrations.

A. Peripheral Insulin Resistance

Fasting and postprandial hyperinsulinemia suggest an insulin resistance state. Indeed, with an euglycemic hyperinsulin-emic clamp technique, in which a high dose of insulin is infused at various "clamped" levels of plasma glucose, the uptake of glucose can be used as an index of tissue sensitivity to insulin. In such studies, both diabetic patients and their nondiabetic (i.e., nonhyperglycemic) siblings have been shown to have severe insulin resistance when compared with controls.

The first event in insulin action is its binding to its receptor. Most Type 2 patients have genetically normal insulin receptors, but decreased insulin binding has been described in lean and obese Type 2 patients, as well as in animal models of diabetes. This decrease is attributed to changes in receptor number with no change in receptor affinity. This phenomenon may be the consequence of down-regulation of the receptor number by hyperinsulinemia and is likely secondary to diabetes rather than a primary defect. In rare patients with very severe insulin resistance, antibodies to the insulin receptor have been identified, explaining their insulin resistance. Whether such diabetes variants are also associated with some degree of insulin deficiency is not known. Other rare genetic abnormalities in the insulin receptor have been characterized, but these mutations do not explain the "garden variety" of Type 2.

The next event in insulin action is generation of a transmembrane signal by activation of the receptor tyrosine kinase and initiating a phosphorylation cascade involving multiple molecules such as the insulin receptor substrate (IRS-I), mitogen-activated protein (MAP) kinase, phosphatidylinositol-3-kinase, and other intermediaries. This phosphorylation cascade results in activation of the glucose transport system and the glycolytic and glycogen synthetic pathways. Although there are as yet no reports of genetic abnormalities in this cascade in Type 2 patients, it is likely that defects in insulin signaling mechanisms will be found to underlie some forms of diabetes.

Insulin signaling results in the translocation of the muscle/fat glucose transporter (GLUT4) from the intracellular pool to the plasma membrane. In human and animal models of diabetes, the mRNA and protein for GLUT4 are generally normal; however, translocation of GLUT4 may be impaired. While this phenomenon was best described in fat cells, muscle cells (representing the major insulin responsive tissue) are more difficult to fractionate, and it remains to be shown that insulin-mediated GLUT4 translocation is defective in Type 2. This is further complicated by the fact that GLUT4 seems to have intrinsic activity, and therefore glucose transport is not necessarily proportionate to the number of GLUT4 molecules. Again, this defect may be secondary to hyperinsulinemia and glucose toxicity.

Other postreceptor events in insulin signaling result in enzymatic activation (such as increase activity of glycogen synthase) or an increase in mRNA (such as in pyruvate kinase) in the pathways of glycogen storage or glycolysis in muscle. While oxidative glucose metabolism into CO_2, water, and lactate are not impaired, patients with Type 2 exhibit a major metabolic block in glycogen synthesis (nonoxidative metabolism) *in vivo*. Although some linkage has been found between Type 2 and an abnormal glycogen synthase gene, an abnormal enzyme has not been demonstrated in Type 2.

Other factors may be involved in the development of insulin resistance. Amylin, a 37 amino acid β-cell peptide, is stored and cosecreted with insulin and has been shown to induce insulin resistance in animal models. Protein and mRNA for tumor necrosis factor-α is elevated in proportion to fat mass and may also induce insulin resistance in animal models. Peptide RP-1 has been shown to be increased in Type 2 and may be associated with defective insulin action. Finally, as the mechanism of obesity becomes elucidated, it is plausible that defective genes leading to obesity in the general population in some way accentuate the (inherited) susceptibility to develop diabetes in others.

B. Hepatic Glucose Production

Insulin resistance in Type 2 is also present in the liver. Insulin action in the liver inhibits hepatic glucose production, and the decreased sensitivity of the liver to insulin thus promotes the overproduction of glucose. Indeed, in Type 2, the liver loses its sensitivity to the suppressive effects of both insulin and glucose. In particular, increased gluconeogenesis in the face of hepatic insulin resistance causes alterations in hepatic glycogen metabolism and results in an increase in hepatic glucose production. When plasma glucose levels are low in Type 2 patients, hyperinsulinemia may still suppress hepatic glucose production to normal rates (~2 mg/kg/min), but when glucose levels are greater than 180 mg/dl, glucose production increases. With marked hyperglycemia, hepatic glucose production may be doubled to rates of 4 mg/kg/min, which will result in over 50 g of glucose released from the liver into the bloodstream every night in a patient with decompensated Type 2. To date, however, no primary defects in the hepatocyte have been identified in diabetes, and the contribution of the liver to hyperglycemia can be entirely explained on the basis of defects in insulin availability or action.

C. Insulin Secretion

Although fasting plasma insulin levels are relatively elevated for the extent of hyperglycemia in the early stages of Type 2, the extent of such elevation depends on the degree of glucose intolerance. Insulin secretion is maximal at postprandial plasma glucose levels of ~200 mg/dl (so-called "prediabetes"). While plasma insulin concentrations are still increased in early stages of diabetes (plasma glucose >140 mg/dl) when compared to normal, they are decreased when compared with the prediabetic stage. With longer duration of Type 2, the plasma insulin concentration tends to continue to decline, although rarely to the absolute insulin-deficient state of Type 1. A possible exception to this scenario is the recent recognition that Type 1 presenting in older persons may be a slowly evolving disease; hence autoimmune markers for Type 1 have been reported in some individuals presumed to have "Type 2." For example, among Caucasians with phenotypic Type 2, 10–30% prevalence rates of positive anti-ICA or -GAD have been reported in Scandinavian populations.

Even during the early stages of Type 2 when the insulin response to an oral glucose load is relatively well preserved, there is a loss of the early, first-phase insulin response. It has been suggested that this is an early event in Type 2, preceding even the development of insulin resistance. Stimuli for insulin secretion such as β-adrenergic agonists and arginine are normally preserved at this stage, reflecting the specific loss of the glucose stimulus as well as the potential roles for GLP-1 and other gastrointestinal hormones in the development of Type 2. Accumulation of amylin (see earlier discussion) in the secretory granules may contribute not only to insulin resistance but also to late failure of insulin secretion. Families of patients with MODY (see earlier discussion) are heterozygous with respect to a defective glucokinase gene and therefore have impaired glucose sensing for insulin secretion. This results in a rightward shift to their response to glucose stimuli, and usually a mild form of diabetes.

Finally, a rare form of diabetes has been associated with mutations in the insulin gene; patients who are heterozygous for such mutant insulins also display a relatively mild form of Type 2. In a few families, there is impaired cleavage of the proinsulin molecule, whereas the remainder of families have amino acid substitutions within the insulin molecule. Because the mutant insulin binds relatively weakly to the insulin receptor, its plasma concentrations tend to be elevated. However, there is little or no insulin resistance in these individuals (Fig. 4).

Although beyond the scope of this article, treatment of Type 2 is an implicit aspect of the definition of this heterogeneous groups of disorders. Because most patients are overweight and/or sedentary, Type 2 is regarded as a disease of the "Westernization" of mod-

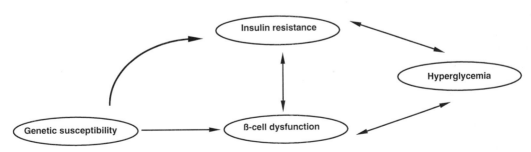

FIGURE 4   Schema relating defects in insulin secretion and action to the development of Type 2 diabetes.

ern life. This may account for the worldwide explosion in the rates of Type 2. Sulfonylurea drugs and biguanides are oral medications that increase insulin secretion and enhance insulin action, respectively. In the case of sulfonylureas, the activation of receptors linked to β-cell K^+ channels is responsible for their augmentation of glucose-induced insulin secretion. Biguanides may act directly to increase the sensitivity of hepatic glucose production to insulin, and the effect of other compounds which enhance insulin action and lower plasma glucose confirms that improvement of this defect in Type 2 has important biologic effects. Orally active inhibitors of intestinal glucosidases have also been used to lower postprandial glucose levels. Finally, a new class of oral insulin sensitizers—thiazolidinediones—have become available for treating patients with Type 2 diabetes. However, Type 2 is phenotypically a progressive disorder in terms of the severity of the metabolic defects and the resultant hyperglycemia. Most patients ultimately fail to be controlled with any of the oral medications alone or in combination, and exogenous insulin is required for the prevention of symptoms and/or severe hyperglycemia. Thus, the paradox of Type 2 is that it frequently requires the use of insulin for treatment.

VI. OTHER FORMS OF DIABETES

The most common form of diabetes is probably the temporary form of diabetes accompanying third trimester gestations in 2–3% of all pregnancies. So-called gestational diabetes (GDM) may result from either insulin resistance exacerbated by hormones secreted during pregnancy and/or insulin deficiency. The diabetes is usually mild, but delay in its detection can result in fetal macrosomia and complications related to delivery and placental development. A fraction of women with GDM have diabetes persisting postpartum, while most have some temporary remission for months or years. The risk of development of diabetes in later life is related to factors such as obesity, prior family history, and the severity of diabetes during the pregnancy. Most women with GDM who develop permanent diabetes develop Type 2, whereas some develop a Type 1-like picture.

Not surprisingly, destruction of significant amounts of the pancreas results in diabetes due to the deficiency of insulin. This form of diabetes is complicated by the severe absence of glucagon which makes the treatment with exogenous insulin therapy somewhat more prone to induce hypoglycemia. Pancreatic diabetes can result from trauma, surgery, and chronic alcoholism, and is always associated with pancreatic enzyme deficiency and nutrient malabsorption. [*See* Insulin and Glucagon.]

Drugs that interfere with insulin release and/or action may cause glucose intolerance or diabetes. Thiazide diuretics, phenytoin, α-adrenergic blockers, and β-adrenergic blockers have all been implicated. Furthermore, drugs that cause β-cell destruction such as pentamidine can also produce diabetes.

Diseases of other endocrine glands may also cause glucose intolerance. Acromegaly with hypersecretion of growth hormone, Cushing's syndrome with glucocorticoid excess, pheochromocytoma with catecholamine hypersecretion, glucagonoma due to an A-cell islet tumor, or somatostatinomas have all been associated with abnormal carbohydrate tolerance. Cure of these diseases frequently ameliorates, and sometimes cures, the coexisting diabetes. There is some doubt as to whether these disorders may simply reveal an underlying genetic propensity for diabetes.

A large number of rare congenital or inherited disorders are also associated with diabetes. Maternally acquired rubella results in diabetes in a large portion of neonates who survive, with a Type 1-like islet disease presenting in childhood. Prader–Willi syndrome, Downs' syndrome, and some forms of myotonic dystrophy are associated with obesity and Type 2.

A. "Glucose Toxicity"

While hyperglycemia may lead to long-term complications, the term glucose toxicity is often used to describe the effects of hyperglycemia on further deterioration of the impairment of insulin secretion and action. Several clinical observation have led to the characterization of this phenomenon. First, while nondiabetic patients produce approximately 20 units per day of endogenous insulin to maintain normal glucose levels, Type 1 patients require at least twice as much on average, and despite this greater availability of insulin, their plasma glucose levels are frequently increased. While Type 1 patients are characterized by decreased insulin secretion, this suggests that with hyperglycemia they become resistant to insulin action. Indeed, in well-controlled Type 1 patients whose plasma glucose levels were experimentally increased for 24 hr, peripheral insulin resistance has been induced. Conversely, it has become apparent from the treatment of newly diagnosed Type 2 patients that any mode of therapy (diet, insulin, or sulfonylurea drugs) improves insulin resistance once plasma glucose levels are decreased. Indeed, insulin secretion is also improved under the same circumstances. Subsequent studies have examined insulin resistance in diabetic patients prior to and several weeks following intensive insulin therapy and have uniformly shown improvements in insulin resistance and insulin secretion. Studies in the rat established that insulin resistance and insulin secretion were improved due to the reduction of hyperglycemia, not to the direct effects of insulin. A serendipitous finding that glucose can induce insulin resistance in primary cell cultures of adipocytes only in the presence of the amino acid glutamine has pointed to the hexosamine pathway as a potential mechanism by which glucose can induce insulin resistance. This pathway involves the formation of glucosamine-6-phosphate from fructose-6-phosphate and glutamine by an amido transferase enzyme. While only 1–2% of glucose normally fluxes through this pathway, an increase in this flux during hyperglycemia may result in insulin resistance, possibly by impairing the translocation of glucose transporters from an intracellular pool.

Thus, hyperglycemia is not only a consequence of diabetes, but may also cause a further deterioration in insulin action and secretion.

VII. ANIMAL AND TRANSGENIC MOUSE MODELS IN DIABETES

Animal models have been widely used as models for diabetes. The most commonly employed models are the pancreatectomized and the streptozotozin-injected rats. These models have reproduced many of the metabolic changes associated with hyperglycemia and/or hypoinsulinemia. In addition, mice and rat models for naturally occurring diabetes have been investigated (e.g., the ob/ob mouse, the Zucker fatty rat). These models may have specific genetic defects not related to defects in human Type 2; however, although these animal models have not generally contributed to the understanding of the genetics of human Type 2, they have yielded significant insights regarding the metabolic alterations associated with diabetes.

The complex interaction between both genetic and acquired metabolic defects in the development of Type 2 has encouraged the application of transgenic technology. With this technology, genes evolved in the regulation of glucose metabolism may be either "overexpressed" or "knocked out." Using these interventions, diabetic models may be created and/or alternative compensatory mechanisms may be studied. Transgenic animal models have been created for genes involved in β-cell, liver, muscle, and fat glucose metabolism.

A. Modulation of Insulin Secretion

Overexpression of the insulin gene has resulted in hypoglycemic mice, although older mice also develop insulin resistance and hyperlipidemia. These findings support a hypothesis that hyperinsulinemia might be a primary event leading to Type 2. "Knocking out" the β-cell glucose transporter (GLUT2) has not resulted in decreased insulin secretion. However, deleting the glucose sensing element of the β-cell glucokinase has resulted in impaired insulin secretion and hepatic glucose metabolism. This animal model is the first to be genetically engineered that is similar to an established form of diabetes in human (MODY).

B. Modulation of Hepatic Glucose Production

A critical step in hepatic gluconeogenesis involves the conversion of oxaloacetate to phosphoenolpyruvate by the enzyme PEPCK. A transgenic model overexpressing PEPCK has resulted in an insulin resistant and diabetic mouse, indicating that the primary increase in hepatic glucose production is sufficient to cause a form of NIDDM. The hepatic defect of the glucokinase knock out is also suggestive of the importance of the liver in diabetes.

C. Modulation of Insulin Sensitivity and Action

Overexpression of the insulin receptor, the insulin-mediated glucose transporter in muscle and fat (GLUT4), and the basal glucose transporter (GLUT1) has resulted in an increase in insulin sensitivity and hypoglycemia. Knock out transgenic mice have provided some of the most surprising findings. IRS-1, thought to play a central role in the signaling pathway of the insulin receptor, does not result in a phenotype with a major abnormality. It was subsequently found that the lack of IRS-1 resulted in overexpression of IRS-2, pointing to some powerful compensatory mechanisms which may prevent diabetes. Complete knock out of GLUT4 was compatible with life and has resulted in diabetes and hyperinsulinemia in such animals. It did not result in compensation by the expression of other glucose transporters, but severe defects in heart function as well as in other organs resulted in early death.

ACKNOWLEDGMENT

The authors are indebted to Dr. Norman Fleischer for his critical review of the manuscript.

BIBLIOGRAPHY

Atkinson, M. A., and Maclaren, N. K. (1994). The pathogenesis of insulin-dependent diabetes mellitus. *N. Engl. J. Med.* **331**, 1428–1436.

DeFronzo, R. A., Bonadonna, R. C., and Ferrannini, E. (1992). Pathogenesis of NIDDM: A balanced overview. *Diabetes Care* **15**, 318–368.

Moller, D. E. (1994). Transgenic approach to the pathogenesis of NIDDM. *Diabetes* **43**, 1394–1401.

Porte, Jr., D., and Sherwin, R. S. (1997). "Ellenberg and Rifkin's Diabetes Mellitus" 5th Ed. Appleton & Lange, Stamford, CT.

Diagnostic Radiology

ALEXANDER R. MARGULIS
RUEDI F. THOENI
University of California, San Francisco

GLOSSARY

Angiography Radiographic visualization of blood vessels following the introduction of contrast material into the vasculature

Computed tomography Two-dimensional representation of the linear X-ray attenuation coefficient distribution through a narrow planar cross-section of the body

Contrast medium Substance that is introduced into the body because of the difference in absorption of X-rays by the contrast medium and the surrounding tissues and which permits radiographic visualization of that structure

Interventional radiology Roentgenographic technique using the ability to observe and control manipulations within the body via a large variety of catheters, stents, and guide wires

Magnetic resonance imaging Imaging of the body which involves the interaction of nuclei of a selected atom with an external magnetic field and an external oscillating (i.e., radiofrequency) electromagnetic field that is changing as a function of time at a particular frequency

Mammography Roentgenography of the breast

PACS Picture archiving and communication system which reduces the dependence on film for the acquisition, storage, and display of medical images

Roentgenography, or radiography The making of film records (i.e., radiographs) of internal structures of the body by passage of X-rays or rays through the body to act on specially sensitized film. Roentgen rays are produced when electrons generated by high voltage and moving at high velocity impinge on various substances, particularly heavy metal

Ultrasonography Visualization of deep structures of the body by recording the reflection of ultrasonic waves directed into the tissues

DIAGNOSTIC RADIOLOGY EMBODIES MULTIPLE IMaging approaches to render diagnoses on the basis of altered morphology. It consists of conventional radiography and fluoroscopy, ultrasonography, computed X-ray tomography, and magnetic resonance imaging (MRI). Nuclear medicine, which involves imaging with radioactive isotopes, is not included here.

I. PLAIN FILM RADIOGRAPHY

X-rays were discovered in 1895 by Wilhelm Conrad Röntgen, who published his first results in 1896. X-rays are produced when cathode rays in a vacuum tube strike the anode, producing photons of energy up to 140 kV in modern diagnostic X-ray machines. After filtration, the useful range in diagnostic procedures is between 80 and 120 kV.

Fluoroscopic rooms permit the direct visualization of structures inside the body, including bones and soft tissue, via image intensifiers and television surveying systems, whereas simple radiographic rooms are used only for obtaining radiographs. Modern fluoroscopic equipment consists of a generator, an X-ray tube, an image-intensifying tube, and a fluoroscopic viewing system. An up-to-date radiographic room does not need fluoroscopic viewing or image-amplifying systems. Some modern radiographic rooms, however, are equipped with digital systems. These digital systems are used to minimize radiation exposure to the patient, to reduce the amount of contrast material administered, and to permit image manipulation (e.g., instant subtraction angiography, freeze action, and image enhancement).

In addition to the standard radiographic and fluoroscopic equipment, specialized rooms for angiography and interventional radiography also have digital subtraction devices. These devices depict the background before the injection of contrast medium in a positive or negative mode and display the same background after the injection of contrast medium into the vessel in the opposite mode. The subtraction results in deletion of all of the anatomic structures, except the vessels. Angiographic and interventional suites are frequently equipped with biplane fluoroscopic equipment, which has two X-ray tubes and two image intensifiers, providing the simultaneous viewing of images at right angles to each other. The image intensifier and the X-ray tubes are often attached to semicircular mobile devices (i.e., C arms) which hold the X-ray tube mount on one end and the image intensifier device on the other. This permits rapid movement of the X-ray tube and the image intensifier without the need for additional alignment of the two.

Some angiographic rooms are also equipped with serial film changers, devices which permit multiple X-ray exposures on films that are quickly changed and synchronized to X-ray exposures. Such equipment allows the recording of rapidly recurring events, such as the flow of radiopaque contrast medium through blood vessels or heart chambers. The majority, however, use cine filming devices and/or videotape. In more recent years, magnetic resonance and computed tomography (CT) angiography (MRA and CTA) often are employed in place of conventional angiography. Mammographic rooms have equipment that allows the display of fine soft tissue detail of the breast.

Portable mobile X-ray equipment is being used with increasing frequency in hospital rooms and hospital operating rooms. It is flexible and capable of providing superb images, and its use reflects the greater orientation of the hospitals toward surgical procedures.

Most of the portable machines are equipped with a high-energy rechargeable battery to render the unit independent of electric power fluctuation in the hospital.

II. PLAIN FILM FINDINGS

A. Skull

Films of the skull are today generally used for screening: to rule out skull fractures in emergency rooms and to diagnose scalp abnormalities, including tumors or inflammatory processes of the cranium. Today, the brain itself is best examined with MRI or CT. In newborns and small infants, ultrasound may be employed, particularly for assessment of the ventricles.

B. Spine

Plain film radiography of the spine is used for the screening examination of fractures of the cervical, thoracic, or lumbar spine; for the evaluation of osteoarthritis associated with osteophytes, which often represent bone spurs resulting from even mild injuries; for the evaluation of the height of the intervertebral disks; for the diagnosis of primary and metastatic tumors; and for the study of congenital anomalies (e.g., fused vertebrae, hemivertebrae, and defects in the neural arch). In the cervical region, traumatic dislocations or fractures are of particular importance, and plain film radiography is essential. In the lumbosacral spine, slipping of vertebral bodies on each other either because of congenital defects in the intervertebral articulating processes or because of trauma is another pathological change in which plain film radiography is helpful. Since the introduction of helical CT, a fast CT scanner that uses slip-ring technology and a continuously rotating X-ray tube, a CT with lateral localizer views and axial slices often is used for the rapid screening of trauma patients.

C. Chest

Plain film evaluation of heart size (Figs. 1 and 2) and shape and the detection of coronary calcifications are still of great clinical value. A diagnosis of enlargement of the atrium and/or ventricle on the left or the right side of the heart can be made in this way. Plain film radiography of the lungs is the primary method of evaluating lung disease and is still the most frequent radiographic procedure performed. It can demon-

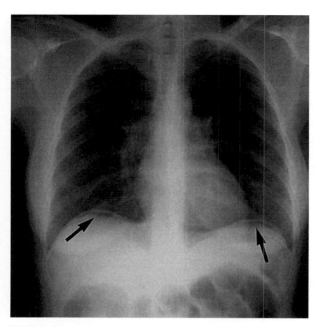

FIGURE 1 Conventional chest radiograph (posterior–anterior projection) of a 24-year-old patient with mild upper abdominal pain. The lungs are clear and the heart is normal, but a small amount of air (arrows) is seen under each hemidiaphragm.

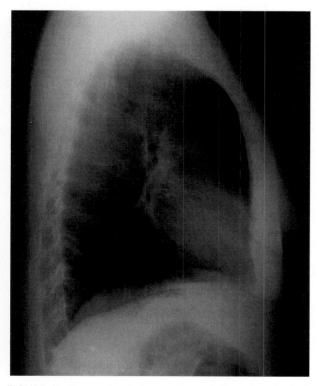

FIGURE 2 Conventional chest radiograph (lateral projection). No definite abnormality is seen on this lateral projection of the chest. A normal-sized heart is well outlined.

strate diffuse or lobar processes, primary and metastatic neoplasms, and a whole range of alveolar or interstitial disease.

In the examination of the mediastinum (i.e., the space between the lungs), plain film radiography is of less use than methods that provide cross-sectional anatomy (CT or MRI). Nevertheless, widening or change of the shape of the mediastinum can be well evaluated, and plain film radiography is often used as a screening method for abnormalities in this region.

D. Abdomen

Plain film radiography of the abdomen is still one of the most important examinations performed in a radiology department, particularly in emergency situations. Perforations of viscera are diagnosed based on the detection of free air caught under the diaphragm in the upright position or between the edge of the liver and the abdominal wall, with the patient lying on the left side. With this approach, even a few milliliters of free air can be accurately detected. Obstructions of the intestinal tube can also be properly visualized by obtaining upright and recumbent films which demonstrate dilated loops of bowel with air–fluid levels. However, these examinations often must be supplemented by administering radiopaque contrast material either by mouth or by rectum to determine the exact location and course of the obstruction. On radiographs, calcium appears white and fat dark, but less dark than air. Adynamic ileus, which shows multiple radiolucent (i.e., dark) bowel loops, represents decreased or absent motility, and dilatation of the small bowel can usually be differentiated from mechanical obstruction by the presence of gas distension throughout the small and large bowels, particularly in the rectum.

This finding is in contradistinction to obstruction that is diagnosed based on distension of the bowel proximal (i.e., toward the mouth) to the obstruction and collapse of the bowel distal (i.e., toward the anus) to it. Masses in the abdomen are best seen on plain radiographs if they contain calcification or fat. In the absence of these features, masses are detected by displacement of the gas-filled bowel. Occasionally, masses are diagnosed because they obliterate fat planes enclosed in the retroperitoneum. Calcifications in tumors, in the peritoneal cavity, or in arteries or veins and stones in the kidneys, gallbladder, or urinary bladder are also clearly seen on plain films.

E. Extremities

Plain radiographs of extremities, including wrists, hands, ankles, and feet, are obtained with great frequency to detect fractures, dislocations, inflammations, and/or tumors of the soft tissues or bones. Today, many inflammatory or neoplastic processes of the soft tissue or bone are better assessed by magnetic resonance or CT.

III. ANGIOGRAPHIC PRINCIPLES

A. Contrast Media

Contrast media are used in roentgenography to increase the contrast between specific organs and surrounding tissues. They easily demonstrate vessels injected by either the intravenous or the arterial route. Modern contrast media are based on the radiopacity of iodine incorporated in organic compounds. Current ionic intravascular contrast media have been used since the mid-1960s, but have been challenged by nonionic intravascular contrast agents, which appear to be safer and have fewer side effects. All of these contrast agents are excreted by the kidneys and are used for excretory urography, angiography, and CT. If intravenous contrast agents are used for CT, tissue contrast is increased, in addition to the demonstration of vessels because parenchymal organs and inflamed tissue are enhanced by the contrast material. With digital subtraction angiography, smaller amounts of contrast media can be used, particularly if arterial injections are employed.

B. Angiography

Angiography uses rapid film sequences during a bolus of contrast material, which consists of rapid injection of contrast medium (e.g., 60 ml at a rate of 2–5 ml/sec). This bolus is administered with a mechanical adjustable injection. For this purpose catheters are placed through arteries or veins, using a femoral, cubital, axillary, jugular, or carotid approach. In recent years digital angiography has permitted the use of smaller doses of contrast material, due to computer manipulation of the data obtained, which includes subtraction techniques and edge enhancement.

IV. GASTROINTESTINAL TRACT

Fine-particle barium suspensions are used orally or by enema to examine the gastrointestinal tube. For the upper gastrointestinal examination the best detail is obtained by adding carbon dioxide-forming crystals together with surface tension-reducing agents (e.g., simethicone) before administering barium. For examinations of the colon, a small amount of barium is administered per rectum. This part of the examination is followed by rectal insufflation of air for complete distention of the colon. Both of these procedures are called double-contrast examinations because air and barium combine to permit a "see-through" effect. For the single-contrast barium enema, barium alone is infused.

Double-contrast examination of the small bowel is obtained by enteroclysis (Fig. 3). The examination consists of passing a tube beyond the duodenojejunal junction and injecting barium followed by methylcellulose. This method permits better control of distention and allows the demonstration of superb mucosal detail. In many instances, enteroclysis provides more information than that obtainable by the oral administration of barium and assessment of the small bowel by means of spot films and serial overhead films.

For spot films, the X-ray tube is below the patient and the film is above. Spot films are obtained during fluoroscopy, with the examiner recording any anatomic detail or abnormality seen during screening on film. Overhead films are obtained by the technologist following fluoroscopy, with the patient lying on the table, the film in the cassette located underneath, and the X-ray tube above the patient.

Water-soluble iodine-containing contrast media are

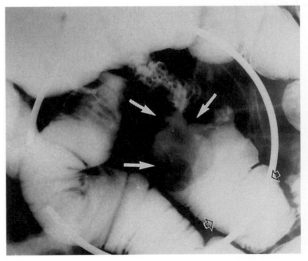

FIGURE 3 Single-contrast enteroclysis. A large mass (white arrows) is seen. Note dilatation of the small bowel (open arrows) proximal to the lesion.

used if perforation of the intestinal tube is suspected because they are absorbed from the soft tissues and do not produce an inflammatory reaction, resulting in scarring, such as is caused by barium. The same substances are used for cleansing a colon impacted with viscous fecal material since they act as laxatives, similar to milk of magnesia, due to their high salt concentration (i.e., high osmolarity). If an aspiration (i.e., inadvertent passage of food or liquid into the pulmonary system) or esophagopulmonary fistula is suspected, water-soluble iodine-containing contrast media should never be used orally because pulmonary edema could ensue, due to its high osmolarity.

In the performance of gastrointestinal examinations, the esophagus, stomach, colon, and small bowel are all examined first fluoroscopically and then with overhead radiographic films. Fluoroscopy is considered to be important since it directs the subsequent conduct of the examination. Barium examinations of the gastrointestinal tract are predominantly for diagnosing ulcers, inflammatory conditions, tumors, congenital abnormalities, and obstruction.

V. ENDOSCOPIC RETROGRADE CHOLANGIOGRAPHY AND PANCREATOGRAPHY

Endoscopic retrograde cholangiography and pancreatography is an examination that combines endoscopic and radiographic techniques and consists of introducing a fiber-optic endoscope by mouth into the duodenum and a catheter via scope through the papilla of Vater into the biliary and pancreatic ducts. Stones, inflammatory processes, or tumors can thus be detected. By combining the endoscopic diagnostic technique with cauterization for sphincterotomy (i.e., cutting of the sphincter of Oddi, which controls the opening of the common duct into the duodenum), sweeping of the duct with a balloon for stones, or introduction of a stent to bridge an obstruction and permit biliary drainage, surgery can frequently be avoided. A stent is a tube, usually made of plastic, which has a configuration (i.e., curled or flared ends) that prevents it from getting dislodged.

VI. MAMMOGRAPHY

After a period of controversy about the dose of radiation administered by radiography to breast tissues,

new low-dose techniques have been introduced for the screening of women at risk. High-resolution techniques are capable of demonstrating nonpalpable cancers in addition to inflammatory disease and benign tumors. Screening mammography has significantly decreased mortality from breast cancer in countries where it is widely used. In a newer development, magnetic resonance has been used for the evaluation of possible neoplasms in the breast and has been successful in demonstrating small cancers or infiltrating tumors that were not detected by conventional mammography.

VII. COMPUTED TOMOGRAPHY

CT is an X-ray-based cross-sectional imaging modality using computer assistance. Images are reconstructed in a process converting detector readings from hundreds of thousands of data samples into an electronic picture that represents the scan section. The picture is composed of a matrix of picture elements (i.e., pixels), each of which has a density value represented by its CT number. CT devices offer high resolution (6 to 20 line pairs per centimeter), high speed (50 msec to 2 sec per scan), fast patient throughput, and a wide range of software options. Helical or spiral CT permits obtaining scans of the entire liver in one breath hold. This new development uses slip-ring technology and continuous rotation of the X-ray tube to achieve a rapid set of data acquired as a volume which can be used for three-dimensional reconstruction or reformation in planes other than the axial sclices. With thin slices and large and rapidly delivered contrast boluses, three-dimensional reconstruction of vascular structures has become possible. Also, electron beam CT, which uses an electron gun instead of an X-ray tube to create images, can be used in the evaluations of cardiac function and in pediatric cases in which patient cooperation without sedation is difficult to obtain.

Because the price of CT devices has precipitously declined, and because of their versatility, they have become widespread in the Western world, and their use has encroached on conventional radiography. It is the examination of choice for the lungs and the abdomen, particularly the peritoneal cavity and the kidneys. Because of its wide field of view, it is an excellent screening method, but does require the use of intravenous and oral iodine-containing contrast media. Although MRI is superior to CT in many areas of the body and the head, CT is still widely used for

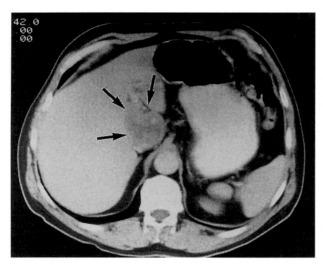

FIGURE 4 CT of the upper abdomen. A low-density lesion (arrows) is seen in the caudate lobe of the liver of a patient with metastatic colon carcinoma.

examinations of the head, spine, heart, abdomen, and extremities (Figs. 4 and 5). This is largely due to the relative scarcity of MRI devices, the high cost of magnetic resonance, and the long examination times.

VIII. MAGNETIC RESONANCE IMAGING

MRI requires a large magnet into which the patient can fit. Systems with permanent, resistive, or superconducting magnets are available. All of these systems are connected to computer image-displaying monitors and use radiofrequency-emitting and -receiving coils. This method is based on the fact that nuclei with an uneven number of particles in a strong external magnetic field have a magnetic moment and become oriented by the field. In MRI, protons (i.e., hydrogen nuclei) are used predominantly because of their ubiquity in the body and their great sensitivity. By bombarding the aligned protons with specific (Larmor) frequency, their direction of spin is changed. They emit signals of the same radiofrequency as the one absorbed, and after cessation of the pulse they return to the original direction of spin. By introducing a known gradient of radiofrequencies and with the use of computer-assisted reconstruction approaches, images are produced in any plane, with superb soft tissue contrast resolution. More recently, low-field magnetic resonance units with an open design were introduced which permit performance of biopsies and other interventional procedures under magnetic resonance guidance. [*See* Magnetic Resonance Imaging.]

MRI is the method of choice for the examination of pathological processes involving the brain, spinal cord, and musculoskeletal system, including the spine and the pelvis. It is also an excellent method for demonstrating abnormalities of the large vessels and is the preferred method for the diagnosis of dissecting aneurysms of the aorta. MRI often is as good as CT for examining the neck, mediastinum, heart, retroperitoneum, liver, spleen, and large vessels (Figs. 6 and 7). MRI is superior to CT in the area of the pelvis, particularly in evaluation of the male and female organs. Its use in these areas is getting more frequent as the number of MRI devices and experience with them increase.

IX. ULTRASONOGRAPHY

B-mode ultrasonography uses a transducer that converts a short electric energizing pulse into an ultrasonic pulse, which propagates along the line of the sight of the transducer. Part of the energy in this ultrasonic pulse is reflected. Tissues vary in the echoes reflected to the transducer, where they are reconverted into electric pulses and amplified. In modern devices a cross-sectional image is obtained in gray scale and is digitized. Ultrasonography is the leading screening method for examination of the abdomen and for the detection and localization of masses. It is also used for screening neonates for suspected congenital anomalies of the brain. It is used intraoperatively with sterile wrapped transducers to gauge the success of tumor

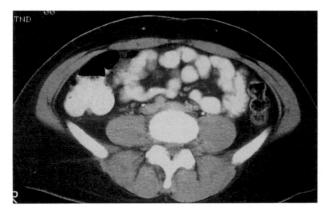

FIGURE 5 CT of the lower abdomen. Multiple small bowel loops are well demonstrated due to good filling of the bowel with oral contrast material.

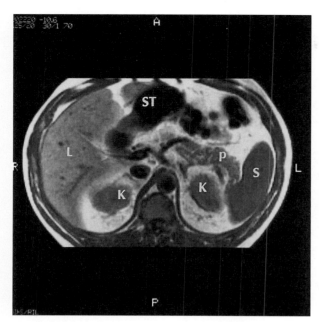

FIGURE 6 Breath-hold MRI of the upper abdomen (transverse plane). Due to the absence of respiratory motion, anatomical detail is shown clearly. P, pancreas; S, spleen; L, liver; ST, stomach filled with air; K, kidneys.

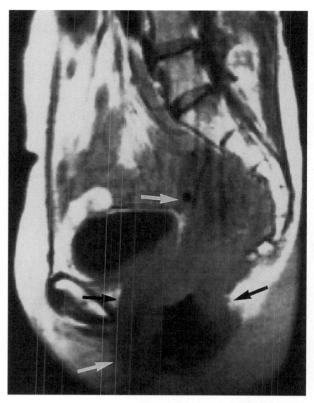

FIGURE 7 MRI of the pelvis (sagittal plane). A large mass (arrows) is seen in the rectum, with invasion of the bladder and a large ulceration near the anus.

removal. Ultrasonography is the standard method for examination of the gallbladder and detection of gallstones (Fig. 8). [*See* Ultrasound Propagation in Tissue.]

The most frequent use of ultrasonography, and possibly its greatest value, is in obstetrics, in which it is employed for diagnosis as well as for the guidance of intervention procedures. High-risk obstetrics depends heavily on the expert use of ultrasonography, which has greatly reduced morbidity and mortality in this branch of medicine. Newer ultrasound devices are also equipped with a Doppler apparatus, which can depict arteries and veins and their patency, direction, amount, and speed of flow. The Doppler apparatus is a noninvasive means of assessing blood flow. This is an excellent method for screening the patency of major blood vessels and is presently heavily used in transplant surgery.

X. INTERVENTIONAL RADIOLOGY

Using fluoroscopy, ultrasonography, or CT for guidance, a new approach to therapy has been developed called interventional radiology. This method permits diagnostic biopsies, drainage of abscesses, introduc-

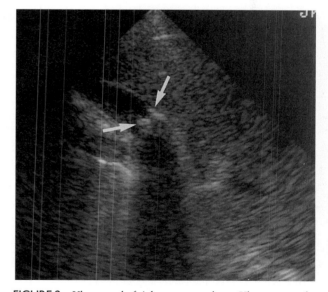

FIGURE 8 Ultrasound of right upper quadrant. Ultrasonography demonstrates the gallbladder with stones (arrows).

tion of stents which bypass obstructions, and balloon dilatations of stenosis of the gastrointestinal tube or blood vessels (i.e., angioplasty). Angioplasty is used with increasing frequency for the dilatation of stenoses of coronary and peripheral arteries. Laser techniques as well as "roto-rooter" devices under fluoroscopic guidance have also been used, but are presently experimental. In the biliary and vascular systems of the liver, interventional radiology has facilitated the treatment of benign and malignant strictures as well as portal hypertension with a transjugular intraportosystemic shunt.

Interventional procedures are of particular importance in deep areas of the brain and the spinal cord, where surgery can produce adverse results. They are used either instead of surgery or to facilitate surgical approaches in the occlusion of aneurysms (i.e., dilatations of blood vessels which result from weakness of the wall and are prone to rupture) with detachable balloons or in the occlusion of arteriovenous malformations.

Much of the success of neurointerventional techniques is due to the development of microinstruments. Tracking techniques have also been helpful. These are based on the fact that it is possible to freeze an image of the background and the contrast-filled vessels on the screen without further irradiation and superimpose on this background the active image of the advancing catheter. As the instruments become smaller, it is of the utmost importance to improve the resolution of the image intensifiers.

XI. ELECTRONIC HIGHWAY IN RADIOLOGY

The use of the electronic highway for actual diagnostic imaging, storage of the digital studies, and connection of the radiology information system with information systems that serve the hospitals, clinics, and private physicians' offices is a newer concept which leads to better and faster communication between the referring physicians and the health care systems and can expand diagnostic and therapeutic procedures. These information systems, often referred to as PACS (picture archiving and communication system) or MDIS (medical diagnostic imaging support system), are expected to reduce the dependence on film for the acqui-

sition, storage, and display of medical images. Telemedicine or teleradiology is achieved with computer resources whose capabilities become increasingly more powerful. The delivery of telemedicine is based on workstations that are challenged in spatial and contrast resolution, patient and image data access, image display speed, and image manipulation and processing. Presently, difficulties still exist for microcalcifications in mammography, subtle bone lesions, or minimal pulmonary changes.

Not only can the electronic highway be used for quicker access to and manipulation of imaging data, it can serve as a means for obtaining current scientific medical information at sites of care delivery as well as a tool for teaching and learning. Currently, the usefulness of the concept of telemedicine is limited by the overall lack of information infrastructure that might be available at advanced centers but not in rural, remote, or financially disadvantaged sites.

BIBLIOGRAPHY

Callen, P. W. (1988). "Ultrasonography in Obstetrics and Gynecology." Saunders, Philadelphia.

Elliot, K., Fishman, R., and Brooke, J., Jr. (1995) "Spiral CT: Principles, Techniques, and Clinical Applications." Raven Press, New York.

Feig, S. A., and McLelland, R. (1983). "Breast Carcinoma: Current Diagnosis and Treatment." Masson, New York.

Fraser, R. G., and Pare, J. A. (1988). "Diagnosis of Diseases of the Chest," 3rd Ed. Saunders, Philadelphia.

Freeny, P. C., and Stevenson, G. W. (1994). "Margulis and Burhenne's Alimentary Tract Radiology," 5th Ed. Mosby, St. Louis, MO.

Higgins, C. G., Hricak, H., and Helms, C. A. (1992). "Magnetic Resonance Imaging of the Body." 2nd Ed. Raven, New York.

Hricak, H., and Carrington B. M. (1991) "MRI of the Pelvis: A Text Atlas." Bernadette: Appleton & Lange, East Norwalk, CT.

Moss, A. A., Gamsu, G., and Genant, H. K. (1992). "Computed Tomography of the Body." 2nd Ed., Saunders, Philadelphia.

Newton, T. H., Hasso, A. N., and Dillon, W. P. (1988). "Computed Tomography of the Head and Neck," Vol. 3. Raven, New York.

Ney, C., and Friedenberg, R. M. (1981). "Radiographic Atlas of the Genitourinary Systems," Vols. I and II. Lippincott, Philadelphia.

Resnick, D., and Niwayama, G. (1988). "Diagnosis of Bone and Joint Disorders," 2nd Ed., Vol. 1–6. Saunders, Philadelphia.

Ring, E. J., and McLean, G. K. (1981). "Interventional Radiology: Principles and Techniques." Little, Brown, Boston, MA.

Stark, D. D., and Bradley, W. G. (1988). "Magnetic Resonance Imaging." Mosby, St. Louis, MO.

Wong, A. W. K., Huang, H. K., Arenson, R. L., and Less, J. K. (1994). Digital archive system for radiologic images. *Radiographics* **14**, 1119–1126.

Dielectric Properties of Tissue

RONALD PETHIG

University of Wales, Bangor

GLOSSARY

Dielectric Material having a relatively low electrical conductivity and capable of supporting electrostatic stress

Dielectric loss Amount of energy dissipated as heat in a dielectric when it is subjected to a time-varying electric field; this energy loss is associated with the frictional work done in the orientation of polar molecules along the field direction

Dielectric relaxation When an electric field is removed from a dielectric, the orientation of polar molecules is lost by thermal agitation; for a given polar molecule, this relaxation of field-induced orientation occurs at a characteristic relaxation time, the reciprocal of which defines the relaxation frequency of the corresponding dielectric loss

Electric capacitance Measure of the amount of electrical charge that must be given to a material to raise its electrical potential by 1 unit; capacitance of 1 farad requires 1 coulomb of charge to raise its potential by 1 volt

Electrical conductivity Measure of the ease with which electrical charges can move through a material under the influence of an applied electric field; it is quantified by the amount of charge transferred across 1 unit area per unit voltage gradient per second; reciprocal of resistivity

Impedance Measure of the opposition of a material to the passage of electric current; for direct current this is just the resistance of the material, but for alternating current the effect of the electrical capacitance of the material must also be taken into account

Relative permittivity Principal property of a dielectric, also known as dielectric constant; it is the factor by which,

for a given charge distribution, the electric field in a vacuum exceeds that in the dielectric

THE PASSIVE ELECTRICAL PROPERTIES OF A MATErial can be completely characterized by measurement of its electrical conductance [G; ohm^{-1} or siemens (S)] and electrical capacitance (C; farads). Taken together, these two parameters define the dielectric properties of the test material. If measurements are made on the same sample using the same electrode arrangement, then the conductance and capacitance can be defined by the equations

$$G = k\sigma \text{ and } C = k\varepsilon_0\varepsilon, \qquad (1)$$

where k is a geometry factor (in m) often referred to as the measurement cell constant, and the conductivity (σ; siemens/m) is the proportionality factor between the induced electric current density and the applied electric field, and is a measure of the ease with which delocalized electrical charges can move through the material under the influence of the field. For aqueous biological materials, the conductivity is mostly associated with mobile protons and hydrated ions. The factor ε_0 is the dielectric permittivity of free space (of value 8.854×10^{-12} farad/m), while ε is the permittivity of the material relative to that of free space. For historical reasons, ε is often referred to as the dielectric constant, but because it is a material property that varies (e.g., with electrical frequency and temperature), the term relative permittivity is preferable. The relative permittivity is a measure of the extent to which localized charge distributions can be distorted or polarized by the applied field. For biological materials, such effects are mainly associated with electrical double layers occurring at membrane surfaces or

ENCYCLOPEDIA OF HUMAN BIOLOGY, Second Edition, VOLUME 3.

around solvated macromolecules, or with molecules that possess a permanent electric dipole moment.

I. UNDERLYING THEORY

A. Background

Dielectric measurements have played a significant role in the biological sciences. For example, in 1910, the electrical impedance of suspensions of erythrocytes was measured and found to decrease with increasing frequency, a result that implied that the cells were composed of a poorly conducting membrane surrounding a cytoplasm of relatively low resistivity. In later years, a more detailed analysis of such electrical measurements led to the first indications of the ultra-thin molecular thickness of this membrane. Later observations of a negative capacitance effect in the electrical properties of squid axons led directly to the concept of voltage-gated membrane pores as the vital mechanism of neurotransmission, and dielectric measurements on protein solutions have provided quantitative details of the molecular size, shape, and extent of hydration of protein molecules. More recently, dielectric studies have provided insights into the diffusional motions of proteins and lipids in cell membranes, as well as the influence of hydration on enzyme activity and on protonic and ionic transport processes in protein structures. Studies of the ways that electric fields interact with tissues are also of importance to the development of radiofrequency and microwave diathermy and clinical hyperthermia, in impedance plethysmography and tomography, in the gentle thawing of cryogenically preserved tissue, and in the use of pulsed electric fields to aid tissue and bone regeneration and healing.

B. Theory

Figure 1 shows in a highly schematic form the basic concept of a molecule possessing a dipole moment and examples of electrical double layers around a globular protein and at the surface of a biological membrane. The simplest molecular dipole consists of a pair of opposite electric charges, of magnitude $+q$ and $-q$, separated by a distance (d), and, in this case, the molecular dipole moment m is given as $m = qd$ and has units of coulomb.m. For a protein molecule, the presence of ionizable acidic and basic amino acid residues in its structure gives rise to a comparatively large permanent dipole moment whose value varies

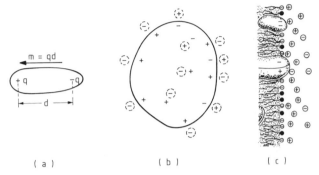

FIGURE I (a) The basic form of a molecule possessing a dipole moment (m), consisting of a pair of opposite electrical charges ($+q$ and $-q$) separated by a distance (d). (b) A schematic illustration of an electrical double layer formed around a charged solvated protein and (c) at the surface of a charged biological membrane.

with pH and protein conformation. Counterions in the ionic aqueous solution around the protein form an electrical double layer, an effect that also occurs at the surface of all charged biological membranes. When an electric field is applied to such molecular dipoles or electrical double layers, the dipoles tend to align with the field and charge displacements occur in the double layers, that is, they become polarized. This polarization does not occur instantly on application of the field, and each polarizable entity will respond in its own characteristic way. To describe this, the relative permittivity is written as a complex function of the form

$$\varepsilon(\omega) = \varepsilon_\infty + (\varepsilon_S - \varepsilon_\infty)/(1 + j\omega\tau). \qquad (2)$$

In this equation, ε_∞ is the relative permittivity measured at a frequency so high that the polarizable entity cannot respond quickly enough to the electric field, and ε_S is the limiting low-frequency permittivity where the polarization is fully manifest. The characteristic response time or relaxation time is given as τ, and ω is the angular frequency (radians/sec) of the sinusoidal electric field. The real and imaginary components of the complex permittivity are given as

$$\varepsilon = \varepsilon' - j\varepsilon'', \qquad (3)$$

where, as in Eq. (2), j is $\sqrt{-1}$. The real part ε', corresponding to the permittivity giving rise to the capacitance described in Eq. (1), is given by

$$\varepsilon'(\omega) = \varepsilon_\infty + (\varepsilon_S - \varepsilon_\infty)/(1 + \omega^2\tau^2). \qquad (4)$$

The imaginary component ε'' is given by

$$\varepsilon'' = (\varepsilon_S - \varepsilon_\infty)\omega\tau/(1 + \omega^2\tau^2), \qquad (5)$$

and corresponds to the dissipative loss associated with the movement of polarizable charges in phase with the electric field. The frequency variations of ε' and ε'' for pure water are shown in Fig. 2. Water is a relatively simple molecule having one effective relaxation time (τ). More complicated polarizable systems have a distribution of relaxation times, and the plots of ε' and ε'' extend over a wider frequency range than is shown in Fig. 2. Because the dielectric loss factor ε'' reflects the extent to which the polarizable charges move in phase with the field, ε contributes to the overall conductivity, which can be written as

$$\sigma = \sigma_0 + \sigma(\omega) = \sigma_0 + \omega\varepsilon_0\varepsilon''. \qquad (6)$$

In this equation, σ_0 is the steady-state conductivity arising from mobile charges, and $\sigma(\omega)$ is the frequency-dependent conductivity arising from dielectric polarization. Also, because the total energy associated with the electric field is constant and is either stored [as reflected in $\varepsilon'(\omega)$] or dissipated [as reflected in $\varepsilon''(\omega)$] by the material, the change in conductivity is directly proportional to the change in permittivity as a dielectric loss process is traversed through its frequency range. In terms of the parameters $\Delta\varepsilon'$ and $\Delta\sigma$ shown in Fig. 2, the relationship is

$$\Delta\sigma = 2\pi f_c\varepsilon_0\Delta\varepsilon', \qquad (7)$$

where f_c is the relaxation frequency ($f_c = 1/2\pi\tau$).

II. TISSUES

The permittivity of biological tissues typically decreases with increasing frequency in three major steps (i.e., dispersions), which are usually designated the α, β, and γ dispersions. This is shown for muscle tissue in Fig. 3, together with the corresponding increments in the conductivity. For frequencies below around 100 MHz, the conductivity of most tissues is relatively insensitive to the measurement frequency, whereas at higher frequencies, and especially in the GHz range, the conductivity rapidly increases with increasing frequency.

Although the various controlling factors are not fully understood, the α dispersion is generally associated with the tangential and radial displacement of counterions at membrane surfaces. Membranes, largely composed of lipid and protein molecules, are

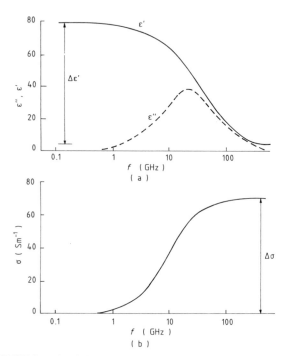

FIGURE 2 The dielectric dispersion exhibited by pure water at 20°C as a function of frequency (f), illustrated in terms of the changes (a) in the dielectric permittivity parameters ε' and ε'' and (b) of the conductivity σ.

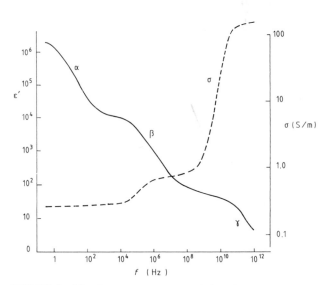

FIGURE 3 The frequency variation of the relative permittivity (ε) and conductivity (σ) for skeletal muscle. Many body tissues exhibit the three distinctive dielectric dispersions α, β, and γ.

poor conductors. At low frequencies, this high resistance of cell membranes insulates the cell interior (cytoplasm) from external electric fields, and no current is induced within the cell interior. Effectively, the membrane acts rather like a capacitor and becomes electrically "charged-up." At higher frequencies, the short-circuiting effect of the membrane capacitance allows the electric field to penetrate into the cell until at a sufficiently high frequency the effective membrane resistance becomes vanishingly small and the cell appears dielectrically as a globule of cytoplasm dispersed in an electrolyte. Thus, the effective permittivity and conductivity of cellular structures will fall and rise, respectively, with increasing frequency, leading to the existence of the β dispersion shown in Fig. 3. As the frequency is increased to the GHz range, dielectric relaxation of free-water molecules begins to occur and gives rise to the γ dispersion. [*See* Membranes, Biological.]

As a result of complicating experimental factors such as electrode polarization effects, comparatively few reliable dielectric data exist for tissues in the frequency range up to a few kilohertz. A good example of the α dispersion is given in Fig. 4 for normal and malignant breast tissue, where it may be seen that malignancy appears to markedly influence the dielectric properties. After excision of tissue, the α dispersion steadily decreases with time, and this effect is consistent with the loss of integrity and physiological viability of the cell membranes. Another example of low-frequency data is that shown in Fig. 5 for human knee ligament and human cancellous bone. The dielectric properties of bone are direction-dependent, and

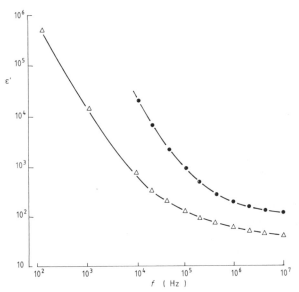

FIGURE 5 Frequency variation of the relative permittivity for human cancellous bone and knee ligament. △, bone; ●, ligament.

for cancellous bone this is considered to be related to the orientation of the trabeculae.

A good example of the β dispersion is shown in Fig. 6 for an aqueous suspension of eye lens fibers (bovine). In this figure, the effect of lysing the cell membranes using digitonin is shown, and clearly the

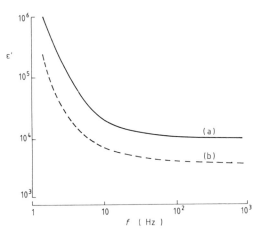

FIGURE 4 Frequency variation of the relative permittivity for (a) normal breast tissue and (b) breast tissue with a malignant tumor.

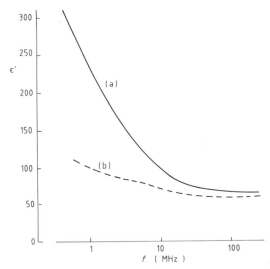

FIGURE 6 The relative permittivity of a suspension of eye lens fibers (a) before and (b) after lysis of the cell membranes using digitonin.

β dispersion depends on the integrity of cellular membranes for its existence. After excision of tissue, the β dispersion, like the α dispersion, steadily decreases with time. Therefore, dielectric measurements can be used to monitor the "freshness" of tissue. Sometimes, as for the cases of ligament and bone shown in Fig. 5, the α and β dispersions tend to merge into one another.

For frequencies >100 MHz, where electrical charging effects of the cell membranes become negligible, the dielectric characteristics of tissue can be expected to reflect the properties of the inter- and intracellular electrolytes. As such we will expect to see evidence of the dielectric dispersion associated with the relaxation of water molecules, as can be seen in Fig. 7. Here, the dielectric properties of brain, fat, and muscle >100 MHz are shown alongside that for saline solution of normal physiological strength. Apparently, the microwave dielectric properties of tissues are influenced by their water contents. Muscle, with a water content ranging from 73 to 78 wt%, has a much greater permittivity and conductivity than fat, for example, which has a water content ranging from 5 to 15 wt%.

The dielectric properties of various tissues at 37°C are given in Tables I and II for several frequencies commonly used for clinical therapeutic and diagnostic purposes. Because of the variability of water content values, the data presented in these tables should be regarded as average values. Also, when reviewing the literature, it is important to note that different measurement methods can provide large differences in data. An example of this can be seen by comparing the results for excised brain tissue given in Tables I and II with the *in vivo* data shown in Fig. 7. Table III gives the normal ranges of water contents for various tissues and organs; these can be used to estimate the likely spread of dielectric properties expected at microwave frequencies for *in vivo* measurements.

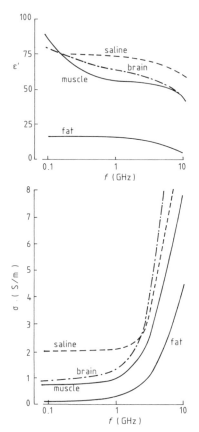

FIGURE 7 The relative permittivity and conductivity at microwave frequencies for muscle, brain tissue, fat, and physiological-strength (0.9%) saline solution.

III. SKIN

The average *in vivo* dielectric properties of skin over the frequency range 1 Hz to 10 GHz are shown in Fig. 8. These properties exhibit considerable regional variability over the body, with permittivity and conductivity greatest in areas such as the palms, where sweat ducts are in abundance. The dielectric properties of skin can be understood in terms of its inhomogeneous structure and composition and the way in which these features vary from the skin surface into the underlying dermis and subcutaneous tissue. [*See* Skin.]

Close study of the dielectric dispersion exhibited by normal skin over the frequency range 0.5 Hz to 10 kHz shows that it is characterized by two separate relaxation processes, centered around 80 Hz and 2 kHz, respectively. These relaxation processes are considered to be located within the stratum corneum and to be associated with relaxation of counterions surrounding the corneal cells. Psoriatic skin exhibits significantly different dielectric properties from those of normal skin.

IV. BLOOD

The relationship between the conductivity of whole blood and hematocrit is shown in Fig. 9. At a frequency of 100 kHz and at 37°C, the numerical rela-

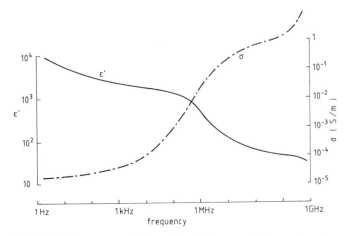

FIGURE 8   Frequency variation of the relative permittivity and conductivity of skin at 37°C.

TABLE I

Relative Permittivity of Biological Tissues at 37°C

Material	13.56 MHz	27.12 MHz	433 MHz	915 MHz	2.45 GHz
Artery	—	—	—	—	43
Blood	155	110	66	62	60
Bone					
With marrow	11	9	5.2	4.9	4.8
In Hank's solution	28	24	—	—	—
Bowel (plus contents)	73	49	—	—	—
Brain					
White matter	182	123	48	41	35.5
Grey matter	310	186	57	50	43
Fat	38	22	15	15	12
Kidney	402	229	60	55	50
Liver	288	182	47	46	44
Lung					
Inflated	42	29	—	—	—
Deflated	94	57	35	33	—
Muscle	152	112	57	55.4	49.6
Ocular tissues					
Choroid	240	144	60	55	52
Cornea	132	100	55	51.5	49
Iris	240	150	59	55	52
Lens cortex	175	107	55	52	48
Lens nucleus	50.5	48.5	31.5	30.8	26
Retina	464	250	61	57	56
Skin	120	98	47	45	44
Spleen	269	170	—	—	—

TABLE II
Conductivity (SM⁻¹) of Biological Tissues at 37°C

Material	13.56 MHz	27.12 MHz	433 MH	915 MHz	2.45 GHz
Artery	—	—	—	—	1.85
Blood	1.16	1.19	1.27	1.41	2.04
Bone					
With marrow	0.03	0.04	0.11	0.15	0.21
In Hank's solution	0.021	0.024	—	—	—
Brain					
White matter	0.27	0.33	0.63	0.77	1.04
Grey matter	0.40	0.45	0.83	1.0	1.43
Fat	0.21	0.21	0.26	0.35	0.82
Kidney	0.72	0.83	1.22	1.41	2.63
Liver	0.49	0.58	0.89	1.06	1.79
Lung					
Inflated	0.11	0.13	—	—	—
Deflated	0.29	0.32	0.71	0.78	—
Muscle	0.74	0.76	1.12	1.45	2.56
Ocular tissues					
Choroid	0.97	1.0	1.32	1.40	2.30
Cornea	1.55	1.57	1.73	1.90	2.50
Iris	0.90	0.95	1.18	1.18	2.10
Lens cortex	0.53	0.58	0.80	0.97	1.75
Lens nucleus	0.13	0.15	0.29	0.50	1.40
Retina	0.90	1.0	1.50	1.55	2.50
Skin	0.25	0.40	0.84	0.97	—
Spleen	0.86	0.93	—	—	—

TABLE III
Water Content Values for Various Tissues and Organs

Tissue	Wt% (water content)
Bone	44–55
Bone marrow	8–16
Bowel	60–82
Brain	
White matter	68–73
Grey matter	82–85
Fat	5–15
Kidney	78–79
Liver	73–77
Lung	80–83
Muscle	73–78
Ocular tissues	
Choroid	78
Cornea	75
Iris	77
Lens	65
Retina	89
Skin	60–76
Spleen	76–81

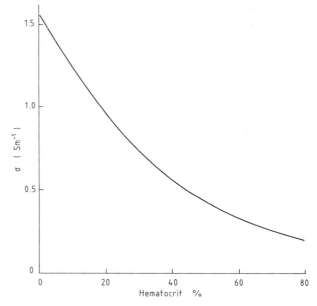

FIGURE 9 Variation of the conductivity of whole blood with hematocrit (H%) at 100 kHz and 37°C (normal H is around 42%).

tionship between conductivity (σ) and hematocrit (H) is given by

$$\sigma = 1.5 \exp(-0.025 \, H\%) \text{ S/m.} \qquad (8)$$

No difference is found between the conductivity and hematocrit relationship for adult normal, neonatal, and placental blood. Reconstituted, time-expired bank blood exhibits a different relationship of the form

$$\sigma = 1.9 \exp(-0.22 \, H\%) \text{ S/m.} \qquad (9)$$

For patients on hemodialysis, the blood conductivity varies according to the linear relationship

$$\sigma = 4.8 + 8(H\%)^{-1} \text{ S/m.} \qquad (10)$$

This markedly higher conductivity probably is related to the elevated levels of urea, sodium, and potassium found in the blood predialysis. The relative permittivity of whole blood falls markedly in the frequency range from 1 to 100 MHz, typically from around 350 down to 70, whereas the conductivity increases by no more than around 10%. [*See* Hemoglobin.]

BIBLIOGRAPHY

Pethig, R. (1987). Dielectric properties of body tissues. *Clin. Phys. Physiol. Meas.* **8A**, 5–12.

Pethig, R., and Kell, D. B. (1987). The passive electrical properties of biological systems: Their significance in physiology, biophysics and biotechnology. *Phys. Med. Biol.* **32**, 933–970.

Schwan, H. P. (1985). Dielectric properties of the cell surface and biological systems. *Stud. Biophys.* **110**, 13–18.

Diet

E. F. PATRICE JELLIFFE
DERRICK B. JELLIFFE[1]
University of California, Los Angeles

GLOSSARY

Anorexia nervosa Severe eating disorder, characterized by refusal to eat, commonest in young women, seeking exaggerated thinness

Anthropometry Measurement of the human body, especially weight, height, subcutaneous fat

Atherosclerosis The major type of arteriosclerosis (hardening of the arteries) most influenced by lifestyle factors

Bioavailability Digestibility and absorption from the alimentary canal

Bulimia Severe eating disorder with excessive food intake followed by food restriction and "purging" (induced vomiting, laxatives, and/or strenuous exercise), usually in young females, seeking exaggerated thinness

Carrying capacity The largest number of any given species that a habitat can support indefinitely

Cartesian Belief that the universe (including human biology) follows mathematical rules (from René Descartes, 1696–1750)

Food security All people at all times have access to the food they need for an active and healthy life

Hyperkinesis (Attention deficit syndrome) A purposeless and impulsive overactivity often associated with a short attention span and distractability

Kwashiorkor Severe form of protein-energy malnutrition (PEM), usually in weanling children (1–2 years of age), with water retention (edema), caused by low-protein diet containing carbohydrate calories, combined with frequent infections

Malting Village-level or commercial treatment of cereal grains to increase nutrient concentration and digestibility (via increased availability of starch-splitting enzyme, amylase)

Marasmus Severe form of PEM, usually in infants, caused by marked lack of calories and protein in the diet, often with infective diarrhea

Multifactorial Condition caused by several interacting factors or causes

Multimix Mixture of foods that are nutritionally complementary

Osteoporosis Condition characterized by poorly calcified bones, commonest in older women, related to long-term low-calcium intake and postmenopausal hormonal changes, characterized by fractures and bent back (kyphosis)

Protein-energy malnutrition (PEM) Various conditions caused by lack of calories and protein in the diet, usually with added infections, commonly in young children in poor communities

Reductionism Analysis or scientific consideration focused on one or very limited factors in what are actually highly complicated circumstances

Soil solarization An agricultural technique to disinfect infested soils, using solar radiation

---

[1]Deceased.

DIET USUALLY REFERS TO THE CUSTOMARY MIXture of foods consumed. It is influenced by adaptations dating from gatherer-hunter times in human evo-

ENCYCLOPEDIA OF HUMAN BIOLOGY, Second Edition, VOLUME 3.   Copyright © 1997 by Academic Press.   All rights of reproduction in any form reserved.

lution, modified by historical and technological changes in life-style from agriculture to traditional cities to modern cities.

Present-day diets consumed vary greatly in different ecologies and cultures, but all have to supply nutrients needed for all stages of life and levels of activity. Modification of diet by various means can best be emphasized for two contrasting populations: the often poorly fed majority in less technically developed countries, and the frequently overnourished in more technically developed regions.

I. CUSTOMARY DIETS

A. Selection

The customary or general diet of a population or an individual comprises the usual mixture of foods consumed. This is influenced by availability (agricultural practices, seasonal variation, transport, preservation, and storage facilities), existing cultural attitudes (food classification, culinary methods, meal patterns), and accessibility (cultivable land and income). Customary diets need to contain sufficient amounts of energy (calories), essential nutrients [protein (amino acids), carbohydrates, fats (fatty acids), vitamins, minerals (calcium, phosphorus, magnesium), trace elements (iron, zinc, iodine, copper, fluoride etc.), electrolytes (sodium, potassium), and water] and certain nonnutrients (e.g., fiber) (1) to supply the body's requirements with foods of appropriate consistency for basic metabolic maintenance at different biological stages of life (including the pregnant woman and her fetus, and the elderly), (2) to sustain "good health," avoid nutrient deficiency, and protect against infections, (3) to provide sufficient nutrients for varying levels of physical activity and body build at different phases of life and climatic conditions, (4) to provide for growth, especially during pregnancy, early childhood, and adolescence, and (5) to minimize the risk of some multifactorial chronic, so-called degenerative diseases that can develop in adult life (e.g., obesity, diabetes, coronary heart disease, cirrhosis of the liver, certain types of cancer) in which prolonged overnutrition or excess or imbalance of some nutrients are considered "risk factors." Sometimes an oversimplistic view of single nutrients is made. Although there are individual functions and need for each, complex interactions occur between different nutrients in the alimentary canal and after

absorption into the body. Some are undoubtedly still unrecognized or ill-defined. [See Nutrients, Energy Content.]

The role of saturated fats and cholesterol have dominated the dietary scene in recent years. In fact, cholesterol is necessary for the formation of various essential body tissues (e.g., the brain in infancy, cell membranes, and some hormones). Normally, it is synthesized in the liver. Excessive dietary intakes may be deposited as plaques in blood vessels, notably the coronary arteries of the heart and brain. This can lead to obstruction and coronary heart disease or stroke. However, the deposition of cholesterol in arteries is increased by the presence in the blood of high density lipoproteins (HDL) found in saturated fat, mostly in animal fats, palm oil, and coconut oil. Conversely, unsaturated fats present in all other plant oils make available low density lipoproteins (LDL), which assist in the removal of arterial cholesterol. The situation is further complicated by variation in individual susceptibility to cholesterol intake—in extreme instances a basic genetic hypercholesteremia occurs independent of dietary intake. [See Cholesterol; Nutrition, Fats and Oils.]

During some stages of social evolution and in different cultures in the past and present, varied diets, usually centered on the local staple, have been adopted, which successfully supply all nutrient needs. Without this process, these communities could not have survived. Originally, selection must have been the result of long-term trial. This included recognizing and avoiding poisonous items (e.g., polar bear liver by Eskimos [toxic levels of vitamin A]) or making them harmless, as by removing hydrocyanic acid from cassava (in many cultures). Also, some inherent biological drives must have contributed to dietary selection via uncertain mechanisms, using edible items made possible by the climate and geography of the particular area.

Recognizable physiological effects (satiation, [feeling of repletion, largely related to the fat content and retention in the stomach], "mouth-feel," digestibility) and genetic variation in ability to metabolize certain nutrients have influenced the foods included in diets in different communities. More recently, economic affordability, agricultural and food technology, rapid forms of transportation, persuasion by commercial marketing practices, and fast-paced urban life-styles have become increasingly significant.

In all cultures (including so-called Westernized societies), the development of often little appreciated classifications of foods, restricts or encourages the use of some of the potentially available edible items for all the population, particularly for vulnerable groups (i.e., pregnant women and young children). Common categories in such classifications include (Table I) foods/nonfoods; "cultural superfoods" (usually the cereal grain or root crop staple); indigenous concepts of body physiology and disease (e.g., "hot–cold" classification of foods and illnesses); prestige/celebration foods; foods desirable or otherwise for some groups (e.g., women, young children); "sympathetic magic" foods, food items forbidden, restricted, or accepted for religious reasons (e.g., kosher foods); and/or for moral concerns, as with vegans (strict vegetarians). Examples of cultural influences include religious fasts (e.g., the Muslim month of Ramadan) or feasts (e.g., Thanksgiving in the United States) and the currently cultural ideal "body-image" (e.g., *gordito* or "little fat one" for infants in some Latin American circumstances; the emaciated, over-slim ballerina-fashion model look for women in some Western countries; or the preferred large majestic bulk in Polynesia).

Also, in all cultures, foods, the types of food, and composition of meals eaten together have psychosocial symbolism and bonding significance, as indicated by sayings and proverbs in many languages. This is especially so on culturally important occasions, including such religious commemorations as Hanukkah and Christmas.

Currently, and increasingly in the future, high-level technology and the vagaries of international economics, competitive forces in trade, and the food industry will influence the range and decisions, positively and negatively, concerning dietary choices all over the world, in both poor and wealthy communities.

B. Evolutionary Perspectives

1. General

During the past 2 million years, humankind has evolved into increasingly complex forms of social organization, levels of technology, and dietary patterns from preindustrial—gatherer-hunters (GH), agricul-

TABLE I
Cultural Classification of Foods Commonly Used in Many Parts of the World

	Definition	Examples More technically developed countries	Less technically developed countries
Nonfood*	Edible items not usually eaten	Rats, cats, dogs, insects (Europe, N. America)	Beef (Hindus, India)
"Cultural superfood"	Main food (staple) source of calories and other nutrients, often historical, mythical, and religious significance	Wheat (Europe—now shared with the potato)	Rice (S. China) Maize (corn) (Central America) Potato (Andes)
Body physiology	Related to local concepts of body physiology (e.g., "hot–cold," yin–yang)	Bodily humors (milk = melancholic—not food for young men) (UK)	Animal milk = garam (hot); not for infants with diarrhea ("hot" illness) (parts of India)
Age group	Suitable or unsuitable, often for women (especially pregnant) and young children		Fish not appropriate for young children as can produce worms (some communities in rural Malaysia)
Sympathetic magic	Conveying some similar property to consumer	Underdone steak to give strength to atheletes (UK)	Walnut (similar appearance to human brain) (brain power) (Gujerat, India)
Prestige/**celebration	Rare, expensive	"Game" (venison, boat salmon, honorific, forbidden to population at large in Medieval times)	Special milk-dessert (India)

*Nonfoods: only eaten if severe shortage, as in famines.
**Usually animal product even in vegetarian societies (i.e., milk-based desserts with Hindus in India).

turalists, pastoralists, traditional city dwellers—to industrial—early and modern techno-cities.

2. Gatherer-Hunters (99.9% of Human Existence)

During this predominant period of time, an extremely mixed diet mainly consisted of wild roots, berries, nuts (including acorns in California), fruits, seeds, grains (including wild rice), insects, birds' eggs, etc., eaten on the move ("grazing") and as available, together with varying amounts of relatively lean game meat, hunted, trapped, or scavenged from predators.

Probably the only communal "meal" that could take place was in the evening, after the discovery of fire and its protection against nocturnal carnivores. Meat and roots would be barbecued, resulting in improved masticability and taste and a limited degree of food preservation. Usually, the diet was mainly plant foods, unless climatic and geographic circumstances made game abundant.

Such diets, currently still found in the few, rapidly diminishing GH peoples (e.g., the Kung of the Kalahari and the Hadza of Tanzania) are high in fiber, vitamin C, potassium, and calcium (from vegetable sources and bones), usually adequate in calories (mainly derived from complex plant carbohydrates) even for their highly active way of life. Also, they are low in salt (sodium), fat, and sugars. Vitamin D is both a nutrient and a hormone synthesized in the skin after exposure to the ultraviolet light in sunshine. [See Vitamin D.]

Nevertheless, sweet foods [breast milk (7% lactose), ripe fruits, and especially honey] and fat, calorie-rich portions of animals (abdominal omentum, bone marrow) are prized. Physiologically, this may be correlated with the location of the taste buds on the tongue (anterior "acceptors": sweet, salt; posterior "rejectors": bitter, sour) and with the satiation effect of fatty foods.

In GH peoples, so-called prolonged, on-request breastfeeding for several years is the norm, followed by the introduction in later infancy of various softer foods, including bone marrow, and meat and roots prechewed by the mother.

Dietary studies in present-day GH and paleolontological investigations of types of teeth of prehistoric humans (incisors, canines, molars) indicate an omnivorous diet (with the use of all kinds of foods) but suggest a special emphasis on plant foods. However, a basic characteristic is to eat as much as available at any time, especially prized fatty and sweet foods. This helps to develop a reserve for possible less abundant food supplies later.

As a generalization, GH probably had and currently have little overt malnutrition because of their very varied food collection over large areas. A short life span is likely as a result of infections and especially trauma (accidents, animal attacks) of special importance for survival because of the need for rapid mobility to keep up with the group.

3. Agriculturalists and Pastoralists

Revolutionary changes in the diet occurred when "food obtaining" was succeeded by "food production" (early agriculture, domestication of animals). These were dependent on the development of essential agricultural knowledge and village technology (metallurgy, pottery-making), including storage, preservation [as with the complex freeze-dried potato (chuno) in the Andes], and food preparation (as with various techniques for the removal of hydrocyanic acid from cassava) and the recognition of the special properties of the gluten in wheat flour for bread-making. Pastoralist and nomad groups developed, centered on different milk-flesh-producing animals, including sheep, cattle, camels, horses, and reindeer (Lapland). In these communities, the use of animal milk and its products was dominant, as opposed to some other cultures, such as those in much of Polynesia (including Hawaii) where no animal milk was available.

Agriculturalists developed a necessarily more static way of life around the cultivation of a particular cereal grain or root crop staple. The range of different foods consumed was much more limited than with the GH but would be more abundant, provided growing conditions were favorable, without drought, floods, or pests and storage feasible and successful. Often, fatness would be an advantage as a reserve for the annual "lean" or "hungry" season, before the crops were harvested. In some African societies, a degree of obesity was purposely produced in women by overfeeding before the nutritional drain of marriage and repeated reproductive cycles. Sometimes, particularly early in this evolutionary change, a mixed, "transitional" life-style occurred (and still occurs) with limited agriculture and some hunting and food gathering or with a settled village, but with mobile flocks herded by men during the dry season in search of pasture and water.

In all these circumstances, successful traditional diets moved toward nutritionally complementary food mixtures. Food selection and recipes were influenced by cultural and physiological factors including the "mouth-feel," taste, texture, and other sensations perceived during eating (organolepsis). Such diets were normally centered on a staple cereal grain plus a legume [Mexican corn tortillas and red bean frijoles, rice and dhal (lentil) in India] or a root crop and animal product [poi (cocoyam) and shellfish in Polynesian Hawaii]. In some diets, excessive fiber can be consumed, with the phytates present limiting absorption of iron and/or zinc.

Genetic intolerance to certain items (lactose in animal milk) or apparent physiological adaptation (the traditional, almost exclusively sweet potato diet of the New Guinea Highlands) also played a part in evolving dietary patterns, either as cause or effect. Sometimes, such indigenous diets necessitated consuming known potentially harmful items in times of shortage. These include cereal infected with the fungus aflatoxin (with possible liver damage) or use of the grain *Lathyrus sativus* (with potential damage to the spinal cord).

Very extensive food mixtures have developed in many cultures. For example, Chinese cuisine is characterized by an exceptionally wide range and mixture of foods used, with small quantities of meat and fat, fresh ingredients, and rapid stir-cooking using minimal fuel. All are nutritionally and ecologically desirable practices.

4. Traditional Cities

Traditional cities could only develop when sufficient food was available and transported from the agricultural countryside or grown nearby. Most written historical accounts of diets, including meal patterns and eating utensils, come from the possibly laudatory records of banquets of the aristocracy, where often exotic meat dishes figured prominently and vegetables were little used. All over the world, the least nutritionally desirable, most "white" overmilled cereal grain staple was preferred as being more expensive and exclusive, including white wheat bread and highly polished rice. The usually little mentioned basic diet of the general population would have consisted of less refined staple, fortunately often combined with legumes (such as soy bean preparations in some Asian countries, or other pulses elsewhere) or less expensive animal products, [e.g., "white meat" (cheese)].

In Europe, ill-preserved meat became tainted in the winter, and the search for "masking" spices was one of the motives for the explorations commencing in the 15th century. This trend also resulted in important widespread changes in foods in many areas, with the importation of the potato and corn (maize), with

sugar moving from an expensive luxury item to an increasingly common food, and with the limited availability of mainly "flavor foods," such as cocoa and vanilla. Conversely, the diet in what is now Latin America was altered by the introduction of cattle, pork, and wheat.

In most preindustrial societies, problems usually relate to bacterial contamination of food (associated with environmental factors such as poor household hygiene, limited contaminated water), to general food shortage, in the annual "lean" or "hungry" season, especially in drought, and to supplying the high dietary needs of physiologically and culturally vulnerable groups 'in the weaning period of' infancy and in pregnancy, especially if restrictive customs limit foods permitted. In some circumstances, specific nutrients may be generally borderline and sometimes inadequate in the usual diet (e.g., iodine, vitamin A, thiamin). As an unusual example, iron deficiency anemia was common in older children in the Bahima people in Uganda, because of an overemphasis on widely available (but iron-poor) cow's milk as their main food. Geographical nutritional factors may be principally responsible in some multifactorial diseases, as with the severe cardiac illness (Keshan disease) seen in parts of China, where the soil has a low selenium content.

5. Early Industrial Cities

The Industrial Revolution commencing 150 years ago led to a massive urban migration in Europe and elsewhere in search of employment. The diet of the poor, dependent on a cash economy, often deteriorated. This was especially so in infants for whom breastfeeding was often supplanted by inadequate "hand-feeding" (artificial feeding) with paps and dilute contaminated animal milk, while mothers worked in factories for low wages. This practice leads to marasmus and infective diarrhea, as it does now under similar slum circumstances in less technically developed countries. Rickets, caused by deficiency of vitamin D, also increased, as little exposure to sunlight was possible, as did scurvy (lack of vitamin C), especially in young children, as a result of limited availability of fresh fruit and vegetables.

The latter part of the 19th century was characterized by the development of an overconfident (but actually very limited) nutritional science, mainly concerned with the gross composition of foods and the needs of the working poor. It was also paralleled by early examples of mechanical food technology (e.g., canned condensed milk, corned beef) and, as a good early example of "techno-food," (e.g., margarine, originally manufactured in 1869 from beef fat as a cheap butter substitute).

The social impact on the family and its "foodways" was considerable and is currently still increasing. Traditional family meal patterns and foods consumed changed with the advent of earlier versions of store-bought foods. The level and type of food production at village level became changed everywhere, as did the role of family members, especially women. Often in countries under colonial rule, large scale, usually foreign-owned plantation agriculture was developed to produce the raw materials for food products exported and processed in industrialized countries. These included plantations for palm and coconut oil, cane sugar, cocoa, and bananas. Social and nutritional disruptive effects also occurred in the agricultural food-producing countries with major emphasis moving away from indigenous foods for local consumption to cash crops and the use of slaves or imported indentured labor to work the plantations.

6. Modern Techno-Cities

Changes in the food chain, beneficial and otherwise, have accelerated geometrically since World War II, with increasing emphasis on agribusiness. Influences include advances in mass production (chemical agriculture, hybrid seeds, etc.), modern preservation (canning, dehydration, irradiation, freeze-drying), marketing [including rapid transport, large-scale food outlets (supermarkets, fast-food establishments)], and home technology (refrigerator, deep-freeze, microwave oven).

Modern high technology has lead to the availability of the year-round supply of foods needed for increasingly large nonfood-producing urban populations and, in more technically developed wealthy communities, to a bewildering array of highly advertised, blended processed products containing numbers of unfamiliar preservatives, emulsifiers, solvents, stabilizers, and artificial colors. Another innovation has been the technological development of novel food items. These include *mechanically modified* fish protein concentrate and textured meat analogues from spun soy fiber, *surimi* (mock crab meat formulated from trash fish); *underused primary foods* antarctic prawns (krill *Euphasia superba*), yeast, plankton, leaf protein, seaweed, algae *(Spirulina)*, *fungus; edible waste products whey* protein (from the cheese industry), and even processed feathers and wool; and *substitute foods* carob bean flour for chocolate. These are being used in blended processed products in industrialized communities in various ways but also have formed part of the food-aid mixtures employed in emergency famine and refugee situations.

Modern processed foods have also often posed problems (1) from thousands of inadequately tested chemicals that can be used in the agribusiness chain, including pesticide residues, antibiotics and hormones, additives, and preservatives of little-known long-term, cumulative significance, and (2) from items with high levels of calories, fat (particularly if saturated animal fat rich in cholesterol) and salt and a low "nutrient density" (amounts of other essential nutrients compared with calories), including an overuse of so-called junk foods. The widespread availability of many high-calorie modern processed foods has coincided with the emergence of mechanical energy-reducing devices, for work, convenience, and pleasure, including automobiles, television, power tools, and central heating. This has resulted in an imbalance between calorie intake from energy-rich foods and energy output from physical activity.

The nutritional results in areas where techno-food is abundant has been the virtual elimination of old-style malnutrition, except among the impoverished and disadvantaged, including some minority communities, unskilled immigrants, the homeless, and the elderly. Instead, the major issues (risk factors) are related in varying degrees directly or partly to dietary excess or to chemical additives in the food chain. These conditions include obesity (also involving adolescents and the elderly) and such multifactorial conditions as coronary heart disease, stroke, diabetes, hypertension, dental caries, some intestinal problems (e.g., constipation), cirrhosis of the liver (overconsumption of alcohol), possibly certain forms of cancer (large intestine, breast, prostate), and osteoporosis.

As a reaction to overprocessed, highly advertised foods, a "health food movement" emerged in more technically developed countries in recent decades concerned with emphasizing the use of more natural, traditional foods produced without chemicals. Indeed, this movement has lead the way for changes in some general and scientific dietary concepts and practices. Examples include an increased emphasis on better soluble and insoluble fiber in the diet, awareness of the unknown risks of chemical agriculture, and the value of exercise. This trend can be seen in products sold in some supermarkets. For example, there has been an awakening of interest in soluble fiber in the diet, as in oat bran or in bran from the Indian grain psyllium, because of its cholesterol-lowering potential, as well as the anti-constipation effect of the insoluble fiber. Nutritious, abandoned "new-old" cereal grains, such as amaranth and Andean quinoa (*Chenopodium quinoa)*, have claimed attention. At the same time, the health food movement has become riddled with financially exploitive enterprises ("pseudo-nutrition") with bizarre items promoted as having incredible medieval magical properties. These include, for example, royal bee jelly, pineapple capsules, "negative calorie" meals, and unnecessary high, even dangerous, intakes of vitamins for "stress." A recent example of a harmful alleged panacea has been L-tryptophan, now shown to cause a serious blood disorder in susceptibles.

Scientific research has also continued at a rapid pace but still tends to be overconfident concerning the mechanical mathematical certainty of nutrition. For example, the intake of different nutrients needed for health often have become regarded as precise numerical truths.

Fortunately, it is now more appreciated (1) that there is much inherent genetic interpersonal variation between individuals and communities long accustomed to certain diets, (2) the bioavailability of nutrients, such as iron and calcium, depends on the mixture and interaction of different foods and some medicines when actually present in the intestine. Complications arise more commonly among the elderly as a group as sensitivity to drugs change with the aging process (increase in body fat, lower metabolic rate, higher requirements in water, etc.). Many older people consume a number of medications "poly-pharmacy" whilst being in the care of several physicians. The ready availability of non-prescription drugs (antihistamines, vitamins, laxatives, etc.) may cause accidents due to dizziness and curtail their intake of food, dehydration and malnutrition may ensue leading possibly to hospitalization [*See* Gerontology, Physiology.], (3) the mathematical tools used for nutritional assessment are all useful approximations, and (4) sensational, alarming scientific reports of limited studies (Cartesian reductionism) and subsequent media publicity give rise to a unwarranted overattention, information overload and public confusion ("national nutritional neurosis"), with a specific "nutrient or food of the month or year." Currently, the important micronutrient selenium has been touted as an unproven preventive against a wide range of maladies ("selenophilia"). Olive oil (justifiably) and garlic (less certainly) seem to be highly prized foods of the present. More recently both chocolate and red wine, which contain flavinoid phenolics, are being scientifically investigated as antioxidants for low density lipoprotein (LDL) oxidation and their possible beneficial role in the prevention of coronary heart disease (Lancet, September 1996). These findings should not promote excess consumption of these products.

Optimal health and nutrition cannot be defined, and even the widely quoted RDA or RDI (recommended dietary allowances or intakes) are extremely approximate figures, based on imperfect facts derived from limited studies from humans and animals. In the United States, they are intended for use with groups, not individuals ("needs of the majority of the healthy population"), with a "safety factor" of 25% added. The necessity for caution in their use is emphasized by comparison of RDI from one American compilation (1980) to the next (1989) and between various countries. Different figures are sometimes given for certain nutrients; the age groupings selected vary; some countries give two levels, one for individuals and the other for planning for population groups.

II. DIETARY MODIFICATION

The aim for an individual, a family, or a nation should be to consume an enjoyable, affordable, and culturally acceptable diet, largely based on local foods, with a sufficient intake of all nutrients, leading to the best possible level of health, both currently and in later life. National policy and advice on dietary details will need to vary greatly in different cultures, but everywhere have to be approximations based on presently available knowledge. Much effort has been given by governments, international agencies, and nutritionists as to how to modify those aspects of diets that appear to be dangerous or in need of improvement in particular community. This can be made more difficult by inadequate training in nutrition for physicians and nurses whose advice is often asked. As food habits are learned early in life and often difficult to change, demonstration and motivation are needed, as well as information. Also, the role of nondietary causes (e.g., genetics, infections, parasites, and psychologic and social stresses) in the causation of different forms of malnutrition and in long-term effects of dietary intakes have to be taken into account.

A. Assessment of the Diet, Its Causes and Nutritional Problems

Such an assessment is logically needed but is difficult to carry out. The techniques and tools used in dietary assessment are all approximate, qualitative, yet useful guides. They need, when possible, to comprise several validating methods of nutritional appraisal, such as a combination of clinical signs, anthropometry, laboratory tests and food consumption). Dietary intake may be measured by questionnaires (often 24-hr dietary recall, especially with the limited range of foods in Third World countries), by observation and measuring food consumption and preparation (including, as far as possible, foods eaten out of the home), and, in some literate communities, by dietary records. Errors can result from an atypical, too short period, or from deliberate over- or underreporting of some items, including alcohol consumption.

Biologically, individual (or community) variation in nutrient needs have to be recognized. Also, the bioavailability of nutrients from foods depends on physiological modifications (e.g., increased iron absorption in pregnancy) and on the variable digestion, absorption, and utilization at the cellular level resulting from the actual mixtures of foods in the alimentary canal (e.g., vitamin C-containing foods' increase in absorption of iron; phytates interfere with the uptake of calcium and zinc, and tea with iron; a range of factors in breast milk facilitate absorption of many nutrients, such as iron). Account must be taken of cooking methods used, including the length of cooking time and amount of water as these factors may result in the loss of vitamin content present in some foods.

However, as in all community studies, investigations into the causes of interrelated dietary inadequacy and infection (or other factors), it is difficult to differentiate those that are "associated" from those that have a predominant or minor "causative" effect. Mathematical analysis (including multiple regression analysis) may be helpful, but a sensible practical view is to regard highly suspect influences as potential "risk factors."

Lastly, dietary analysis in a country is usually complicated by the need to consider many socioeconomic and/or cultural groups, as is the case in America. Also, all over the world, urbanization, the persuasive influence of commercial marketing, and a continually increasing number of technologically modified foods are changing and tending to Westernize dietary habits, especially in cities. This has occurred, for example, in Japan.

B. Nutritional Dietary Modification

This is often termed *nutrition education*. It comprises information, advice, and probably more importantly, persuasion and motivation to change, often related to local cultural considerations (preferred body image; concepts of body physiology, such as "hot-cold", greater attention given to male children). Such approaches include the following overlapping methods: (1) mathematical, (2) food groups, (3) key nutrients, and (4) general advice.

I. Mathematical Method

Mathematical nutrition education is usually self-defeating or unattainable, although appearing to Western Cartesian culture. First, the numbers suggested are often debatable and vary between individuals. Also, no normal person knows—still less customarily calculates—the amount in grams, calories, etc., or the percentages of even the most obvious nutrients consumed daily. Indeed, in modern urban circumstances, these calculations are impossible, with the unsuspected "invisible" fat, sugar, and salt in many processed products (e.g., pastries, canned soups, etc . . .). The only way this could be done would be by a trained dietitian calculating the intake of major nutrients (e.g., fat) and advising on reduction or increase in practical domestic measures (e.g., using specified spoonfuls less or more oil) and/or in customary foods units (e.g., locally standard tortillas). In some modern Westernized circumstances, some individuals become expert at calorie or cholesterol counting by memorizing (or having lists) of the calorie or cholesterol contents of commonly used foods. Equally when advice regarding fiber consumption is considered there exists no RDA's for a daily intake. Studies in experimental animals or human volunteers have yielded controversial results regarding the possible prevention of chronic diseases e.g. colon cancer, diabetes and promotion of weight loss etc. Further data is required as sources of fiber are many and methods to estimate intake are imprecise. It has been postulated that a high fiber diet for mothers in pregnancy and lactation could exert a protective effect on the baby's colon and the changed composition of the mother's milk may affect gut elasticity and flora in the breastfed baby. A low fiber intake however could be a risk factor for colonic diverticulosis.

More recently nutrition pyramids have been devised which illustrate choices in six categories of foods eaten by consumers in the USA and also traditional diets in the Mediterranean region as well as in Asia and Latin American countries. The pyramids can be used as a teaching aid to promote "nutrition literacy", but the assistance of a dietitian would be beneficial to help consumers distinguish between foods with a modest fat content and those which should be eaten occasionally, those which have antioxidant properties etc. Also no indication of portion size is given. For example, some scientists have suggested an intake of 12–30 grams of fiber a day would be a reasonable amount. The most important message should stress that a diet which include a variety of foods with emphasis on fruits, vegetables and whole grains which provide necessary micronutrients and calories should be the aim in nutrition planning (Fig. 1).

A major role for an approximate mathematical approach to modification is with therapeutic or preventive diets worked out by dietitians for use in institutions (e.g., schools or hospitals) and for nutrition-related diseases (e.g., diabetes). Specialized guidance will be essential for recently developed hospital "nutrition support" services, particularly with parenteral feeding, using the intravenous route in seriously ill persons.

Major dietary issues include (1) obesity, eating disorders (anorexia nervosa and bulimia), diabetes, gout, renal and cardiac illnesses, food allergies, (2) the management of various "inborn errors of metabolism" [such as phenylketonuria (PKU), in which individuals cannot metabolize the amino acid phenylalanine from birth, with serious consequences, including mental retardation, and (3) the role of attention to diet in the supportive treatment of cancer and AIDS.

2. The "Food Group" Method

This approach has been used in the United States and other countries for some decades, usually based on four (or sometimes more) so-called "groups"—meat and meat substitutes, milk and milk products, fruits and vegetables, grains (bread and cereal products)—often directed toward a difficult-to-define "balanced diet" in which all needed nutrients are present and with a mutually complementary effect.

In fact, this concept has only limited usefulness in modern America, especially with the increasing number of commercial processed products. Also, this group method ignores food quality (e.g., nutrient density, fiber content, method of preparation). It gives rise to misconceptions, for example, that protein is only present in animal products. The system is culturally biased; for example, the majority of the world

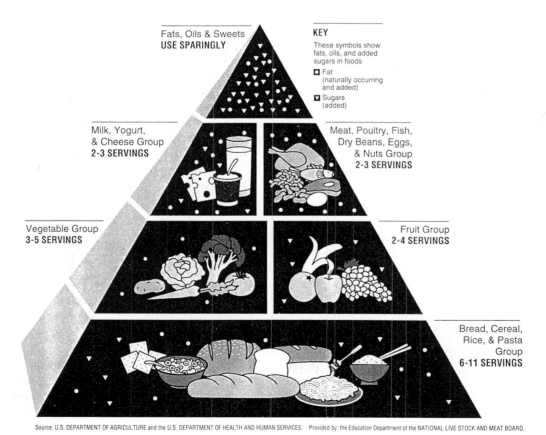

Source: U.S. DEPARTMENT OF AGRICULTURE and the U.S. DEPARTMENT OF HEALTH AND HUMAN SERVICES.   Provided by: the Education Department of the NATIONAL LIVE STOCK AND MEAT BOARD.

FIGURE I   The food guide pyramid.

obtains most of its proteins from cereal grains and legumes.

In technically less developed countries, the U.S. "food group" classification is still less relevant, and advice and information should be "staple-centered." The diet will always have to be envisaged as being the staple (usually a cereal grain such as rice, or a root crop such as potato or yam) together with other nutritionally complementary foods.

3. Key Nutrient Approach

This approach may be practicable and useful in different circumstances, based on knowledge of foods rich or low in particular nutrients. In parts of Indonesia, the main nutrient lacking is vitamin A, or its precursor, beta carotene. In such areas, available beta carotene or vitamin A would be key nutrients needing special attention, either in the diet or as a supplement. Foods advised would include yellow-orange pigment vegetables and fruits, as well as red palm oil and dark green leafy vegetables, especially for pregnant and lactating women and young children.

Conversely in overfed communities, the key nutrient is often saturated fat, so that major emphasis should be directed toward discouraging its consumption by substituting other foods and/or (more realistically), limiting or modifying various popular items. For example, in the United States, foods high in animal fat include "hot dogs," regular hamburgers, french fries, ice cream, and fattier bacon.

At the same time, current consensus in pediatric nutrition is that restriction of fat and cholesterol is not indicated in young, rapidly growing children in the first 2 years. After this, moderation seems prudent, partly because of the need to influence lifelong food habits.

4. General Advice

Usually, nutrition education and advice have to be mainly in general terms, at best quasimathematical, within the economic, hygienic, culinary, and cultural restraints indicated earlier, and channeled through schools, the press and mass media, and individual counselling. They must be geared to minimizing locally important nutritional problems ("key nutrient approach"), and be directed to the physiologically and socially at-risk, including pregnant women, young children, teenagers, and the elderly. Three universal generalizations are (1) that the more mixed and varied any diet the better, especially if based on unprocessed foods, (2) that available animal products

should be well-cooked, and (3) plant foods should be as well-washed as possible.

Traditional time-periods at which nutrition behavior is implanted or modified is by the family diet and as a result of school meals. The success of a school meal program depends on numerous factors such as cooperation of parents, ready acceptance of culturally appropriate foods by the children, their improved health status, and increasing school attendance, when possible. At times, school gardens and small animal production units may be included if conditions are suitable. Whether schools are situated in urban or rural areas, school meals should include a nutrition education component.

C. Nonnutritional Dietary Modification

For better and worse, urban influences and industrialization increasingly affect the diet all over the world, especially in more technically developed countries. These include marketing (advertising, promotion) (including less desirable foods [e.g., infant formulas, "fast foods"] and processed products containing many ingredients unknown to the consumer), the introduction of artificial substitutes or "fake foods" [e.g., (such as saccharine, aspartame, etc.), nonabsorbable fat substitutes such as "Simplesse" and "Olestra". Labeling is required to alert consumers of some serious side effects which may arise in unusually sensitive individuals including children (vitamins) dizziness and black outs recently reported by airline pilots (aspartame) diarrhea etc. in other consumers (olestra). Influence of special interest groups (e.g., the milk, meat, egg, and dairy industries)] can affect the items given emphasis in large-scale programs (e.g., school meals in the United States or as part of international food aid, especially for emergency situations). Recently in California, some well-known non-profit health foundations are endorsing brand name commonly used over-the-counter pain killers and Florida orange juice ("Cause marketing"). They in return are financially recompensed by the companies. This represents a conflict of interest and will reflect negatively on their former neutral reputation. However, in recent years, in more educated technically developed communities, some marketing practices have changed to increase sales appeal to more educated consumers (i.e., low fat, pesticide-free foods). Food categories particularly likely to expand in sales are those labeled "natural," organic, "diet," "convenience," and "light," despite difficulties in definition.

Dietary choices also for better and worse have also been much influenced by the "health food industry." This movement originated as a reaction to overmechanized, overchemicalized food production, preparation and consumption. In many ways, it has publicized and headed the present-day moves toward a "prudent diet," with an emphasis on natural foods. At the same time, it has lead to a completely unwarranted quackery, with mystical values attributed to high dosage vitamin supplements and to certain foods, especially many unusual, often exotic, and usually high-priced items, particularly including those directed to psychologically vulnerable groups (e.g., athletes, teenagers, and the elderly).

Promotion of "high potency" vitamins or excessive use of single doses of a specific vitamin, is not only unnecessary and expensive but can be dangerous because of toxicity; for example oil soluble vitamins A, D and vitamin E (alpha-tocopherol) and water soluble vitamins C, B_6 and folic acid. If a "prudent diet" is consumed there is usually no need for such vitamins. Excessive consumption of foods containing beta carotene may result in yellow staining of hands and feet (carotenemia) which disappears when dosage is reduced.

Medical advice should be sought prior to purchasing any nonprescription drugs, because of possible side effects. Controversy persists on the possible role of antioxidants such as beta carotene, vitamins C and E in the prevention of epithelial cancer (? caused by free radical damage) and lung cancer also if vitamins C and E may decrease the susceptibility of LDL cholesterol (Low density lipoproteins) to oxidation and have a preventive effect against atherosclerosis. Queries are currently being voiced on values which could represent individual "normal" levels of LDL and HDL (High density lipoproteins) instead of group values.

Vitamin therapy may need to be implemented in certain cases. Recently in 1995 the Center for Disease Control and Prevention has recommended that women who have had an infant with a neural tube defect should take 4 mg a day of folic acid from at least 4 weeks before conception through the first 3 months of pregnancy; all women capable of becoming pregnant should consume 0.4 mg a day.

It is possible to obtain this dosage by eating according to the US dietary pyramid. Cigarette smoking, alcohol, some disease and drugs may affect folate status [See Food Toxicology.]. Factors which may influence acceptance of this recommendation include lack of compliance, knowledge, unplanned pregnancies, etc. [See Embryo, Body Pattern Formation.].

Dietary frauds are also obvious in the continuing flow of books and regimens offering miraculous "guaranteed" methods of slimming or weight reduction, sometimes combined with "diet pills" and apparatus for "passive exercise." Their value is probably related to their usual high cost and temporary novelty. Their lack of success ("yo-yo weight regain") is indicated by their continual change. Two diet pills, which are appetite suppressants Redux and Fen/Phen have mild side effects, however, very rarely Primary Pulmonary Hypertension which can be fatal, may arise with the use of Redux. Both these drugs, which are expensive, should preferably only be used by morbidly obese individuals under medical supervision. The Western obsession with overweight is indicated by the fact that the word *dieting* usually refers to food restriction. This is exemplified by serious eating problems that are probably on the increase, especially in young women. These are anorexia nervosa and bulimia, both rooted in the desire for an ultra-thin, almost emaciated figure. [See Eating Disorders.]

III. WORLD PERSPECTIVES

Satisfactory mixtures of a wide range of foods make up the present-day diet of the world's many cultures, including those in the United States. It is then only possible to highlight two main dietary themes and changes currently occurring in them.

A. Non-Western Diets

This encompasses an extremely wide array of different food combinations and most societies usually include an often small "Westernized" urban minority. On the whole, the basic food mixtures of such traditional diets are usually satisfactory. The main deficiency is often in quantity, particularly the staple. This may itself be changing, as with the increasing spread of the consumption of potato and cassava around the world. In many areas, this problem has been made worse by increasing cost and decreasing production of foods for local use for various reasons. These have included overdependence on a limited range of cash crops e.g., coffee, cocoa or cane sugar, (whose world price varies yearly depending on yields) and consumer preferences (sugar substitutes). Diminished returns from the sugar cane industry have led some farmers to invest into more lucrative export crops such as flowers (carnations, (Guatemala) exotic tropical flowers, (West Indies). Recent financial economic adjustments (1980s) were also made to try to rectify huge accumulated national debts. Such changes often led to greatly increased food prices and a move to cheaper, less preferred items [e.g., from imported "salt-fish" (cod) to shark in the Caribbean].

A marked change in non-Western style diets has been in the feeding practices for young children. In many countries, breastfeeding has declined or become accompanied by falsely prestigious apparently "modern" bottle feeding (mixed milk feeding), with locally available animal milk or, increasingly, highly expensive infant formulas, often imported. This practice has spread for a variable mixture of reasons perceived status, unethical commercial advertising and promotion, unhelpful health services (with professionals untrained in breastfeeding management), and, especially in some urban areas, by women employed in salaried work far away from home. As a result, there has been an increase in avoidable marasmus as a result of over-dilute feeds, often made worse by infective diarrhea.

In addition, dietary problems still occur commonly in the traditionally and physiologically dangerous periods, especially when the young, rapidly growing child is in transition from sufficient breast milk to the full adult diet. At this time many infections occur and add to the nutritional burden in various ways. It is at this stage of life (1–2 or 3 years) that kwashiorkor (and similar less severe forms of PEM) occur more commonly. A major consideration in nutrition in such circumstances is to protect against infections and to institute easy inexpensive treatment (by improved hygiene, immunization, oral rehydration etc.) and to devise home-prepared weaning "multimixes" of local foods, which are nutritious, digestible, culturally acceptable, affordable and culinarily practicable.

Pregnant and lactating women are also especially vulnerable, particularly after repeated reproductive cycles of pregnancy and lactation, often made worse by specific culturally defined food avoidances and limitations. This can lead to various forms of maternal depletion, most commonly anemia due to deficiency of iron or folate. In areas of special risk, other deficiencies may occur, for example, osteomalacia [adult rickets from vitamin D (actually mainly a hormone produced by exposure of the skin to sunlight) and calcium deficiency], where strict purdan (veiling and seclusion of women) is customary, or beriberi (severe thiamin deficiency), when the diet consists almost entirely of polished rice]. Under these special circumstances, the issue of vitamin supplements may be indicated.

B. Westernized Diets

Until recently, Western diets and those of Westernized minorities in the urban elite in many countries have been characterized by (1) usually abundant food supplies, with often hundreds or thousands of new processed items in supermarkets annually, (2) low intakes of fiber and complex carbohydrates (e.g., cereal grains), (3) increasing fat consumption (especially saturated animal fats, such as meat, eggs, and dairy products, including cholesterol), sugar (sucrose), sodium (mainly sodium chloride or salt), and nutritionally unnecessary amounts of protein (sometimes 2–3 times real needs), all of which are present in many purchased, processed food products and fast-food items (e.g., pizza), (4) recognition, diagnosis and overdiagnosis of food allergies, if no other aetiology can be ascertained, (5) overemphasis on vitamin supplements as being universally needed (over-the-counter vitamin sales in the US ± $1.5 billion or more per year), (6) increased intake of up to 50,000 chemicals used in agriculture and food processing and preservation (e.g., sulfite in preventing wilting in lettuce, dioxin present in bleached-paper milk cartons, lead and aluminum in canning, and (7) an unexpected rise in illness and unproductive sick, even deaths, caused by food poisoning (microbial infections: salmonella, listeria and campylobacter organisms, viruses, botulism toxin, etc.), most frequently from animal products (e.g., chicken [contaminated disemboweling machines, water-pooling of chicken carcasses], undercooked eggs, "soft" cheese, seafood, and meat infected during mass production, inadequately inspected importation from abroad and distribution, and by unsatisfactory home culinary practices, including storage, hand-washing and undercooking). Especially dangerous items include raw steak *tartare and sushi* (particularly the potentially toxic *fugu* fish). Other syndromes sometimes related to modern diet include a range of allergies and possibly hyperkinesis (hyperactivity) in young children believed by some to be possibly due to an allergy to additives in food, especially certain artificial colorings, e.g. yellow dye No. 5 (tartrazime). Although the aetiology is not certain it is important to rule out mental illness or mental retardation if hyperactivity continues for 6 months or more in children up to 7 years old.

Apart from recent mass-production induced infections, the risks of such diets are not in early onset malnutrition, but rather as risk factors in the development of multifactorial so-called degenerative illnesses (mentioned previously) occurring as chronic burdens in later adult life, and thereby limiting active longevity (sometimes termed the "epidemiological transition"). Awareness of these ill effects have become increasingly known to the literate public through informed newspapers and many newsletters from consumer groups and universities and also magazines. Such educational information seems to have had some effect. It has caused changes (1) in some restaurants (use of vegetable oil, french fries cooked in vegetable oil, rather than a mixture of animal fat (lard) and vegetable oil as before; grilled items rather than fried, opening of salad bars, turkey meat hot dogs, availability of nutritional menu cards emphasizing fish and lean meat); (2) in supermarkets (conspicuous labeling as "pesticide free," etc.); (3) in advice on home food preparation (increased use of vegetable oil and fish, decreasing, use of items high in animal fat, e.g., poultry skin, "marbled" beef, eggs), methods of minimizing the chances of "food poisoning" (actual bacterial infection), including avoidance of raw or "underdone" fish and meat, and care with refrigerated foods, and (4) increased emphasis on physical exercise as part of a nutritional/fitness regimen.

As a result of aroused awareness recognition of the dangers of Westernized diets during the past 20 years,

at least 15 countries, including Norway, UK, and the United States have produced national nutritional guidelines. All are similar in their general emphasis: decrease calories to suit energy needs (especially animal fat and sugar); limit salt intake; increase fiber (especially soluble fiber) and possibly, calcium for girls from childhood (as they may be at risk of postmenopausal osteoporosis, largely caused by hormonal changes). A high emphasis on sugar (especially more sticky preparations) as the main cause of dental caries has been reduced somewhat because of the clearly proven protective effects of adequate intakes of the nutrient fluoride nowadays present in many items, including toothpaste, and in water, where safe fluoridation [1 part per million (ppm)] has been introduced. Recent scientific debate also suggests that advice concerning limiting cholesterol intake remains, but with less *direct* emphasis than before. This can, in fact, be taken care of by limitation of dietary consumption of saturated, mainly animal fat and some tropical vegetable oils (palm, coconut). (Table II Dietary Recommendations).

In the United States, "Dietary Recommendations" have been published by the National Research Council (1989). The main general points are given in Table II. However, in their application there are obvious problems in translating mathematical figures for daily intake (e.g., 6 g salt, 1.6 g/kg protein) into practical terms. This is particularly difficult when considering populations of different ages, size, and cultural dietary

TABLE II
Dietary Recommendations

- Reduce fat intake to 30% or less calories. Reduce saturated fat intake to less than 10% of calories and the intake of cholesterol to less than 300 mg daily.
- Every day eat five or more one-half cup servings of a combination of vegetables and fruits, especially green and yellow vegetables and citrus fruits. Also increase intake of starches and other complex carbohydrates by eating six or more daily servings of a combination of breads, cereals, and legumes. Carbohydrates should total more than 55% of calories.
- Maintain protein intake at moderate levels, i.e., approximately the currently Recommended Dietary Allowance (RDA) for protein, but not exceeding twice that amount for 1.6 g/kg of body weight for adults.
- Balance food intake and physical activity to maintain appropriate body weight.
- We do not recommend alcoholic beverages. If you do drink, limit yourself to less than 1 ounce of pure alcohol daily. This is equivalent of two cans of beer, two small glasses of wine, or two average cocktails. Pregnant women should avoid alcoholic beverages altogether.
- Limit total daily intake of salt to 6 g or less. Limit the use of salt in cooking and avoid adding it to food at the table. Salty, salt preserved, and salt-pickled foods should be consumed sparingly.
- Maintain adequate calcium intake.
- Avoid taking dietary supplements in excess of the RDA for 1 day.
- Maintain an optimal intake of fluoride, particularly during the years of primary and secondary tooth formation and growth.

Reproduced with Permission from National Academy of Sciences. "Diet and Health: Implications for Reducing Chronic Disease Risk" (1989). National Academy Press, Washington, D.C.

practices anywhere in the world, including America. Also, no mention is made in these recommendations concerning food safety regarding either chemicals present or bacterial contamination. A simplified guide is given in Table III.

IV. SUGGESTIONS FOR THE FUTURE

Essentially, a mismatch has developed in more technically developed countries between diets supplying genetically determined physiological nutritional *"needs"* evolved over thousands of years and the recent availability of abundant, highly advertised, largely processed, often fatty, salty or sugary foods catering to human *"wants."* Unfortunately, an opposite situation has occurred for the majority of populations in many poor, less technically developed countries, so that dietary advice has to be distinctly different with the dangers of advocating elements of a Westernized diet that are currently being warned against. For example, the general population of such countries need more calories and other nutrients in their diet. These programs should be based, when possible, on improved environmental sanitation, on the prevention of contributary infections, and on "appropriate technology" to increase local production, preservation and storage of foods making up the traditional diet including small scale urban agriculture. The last include breast milk for infants and the use of homemade, village-level multimix "weaning foods," with special attention to ensuring a high calorie in-take, via frequent feeds, added oils or fats, and malting. Usually, the need is to retain the traditional diet but to try to increase its availability and to reinforce when really indicated (e.g., iodine fortification in areas with goiter and fetal iodine-deficiency syndromes; vitamins A or D in regions where these nutrients are lacking). Unfortunately, imitation of "Western" practices and unaffordable status seeking modern advertising is tending increasingly to modify dietary patterns (e.g., infant formulas, calorie dense fatty fast foods), involving cities and both the poor and the relatively well-to-do.

In Westernized societies, the mathematical specifics of nutritional recommendations (Table II) always have to be simplified to useful, practical generalizations, such as lists of foods that should be "limited" or "consumed in greater amounts," unless more definite guidance can be obtained from a dietitian. This will rarely be the case, except for sick individuals or for institutional catering.

Legislation in some countries appears to be moving slowly, usually with delaying tactics from industrial concerns involved, toward limiting the use of chemicals at the many different links in the food chain, including soil pollutants. Stronger regulations are required to ensure all new products include environmental considerations prior to development and a levy of taxes if necessary may be required. Many new technologies in bioengineering (rice, wheat, etc.) are already in progress, but will not be available in the immediate future. Easy understandable, enforceable labeling of products must be mandatory. Selective

---

TABLE III

Central Guide to an Optimal Diet Major Points

General	Some specifics	Practical examples
Decrease		
Overall intake of fat	Especially of animal origin, rich in saturated fatty acids and cholesterol	Use vegetable cooking oil and low-fat margarine
		Use lean red meat, fish, poultry
Sodium intake		Use least amount of salt in cooking
Sugar intake	Especially sweet sticky cariogenic candy	
Increase		
Fiber		Whole grain bread
Complex carbohydrates	Vegetables	Particularly potatoes
Vitamin C		Fruits
		Vegetables (especially fresh)
Exercise		
Variable emphasis		
Vitamin supplements	Usually do not need if mixed diet taken	
	Megadoses to be avoided	

use of chemicals is important; many pesticides are neurotoxins and may lead to health hazards (psychiatric problems, irritability, abherent behavior possibly Parkinsonism), latex paints contain mercury and many leads as well, as do diagnostic X-rays, leaded automobile fuels, etc. Side effects are dangerous for all age groups and higher rates of cancer occur among farmers and agricultural workers.

A major need is for clearly visible, easily interpretable information on the three commonest dietary excesses (saturated fats, calories, and salt) and on chemical additives or residues, as well as small print details of the percentage of RDA nutrients present. Regulations are needed for false, or more usually misleading, labels giving, for example, information on one item only (e.g., "cholesterol free") and ignoring others.

Recently a much improved new label "NUTRITION FACTS" provides information on serving size, calories per serving, total fat (% value) (saturated, poly-unsaturated, monosaturated) cholesterol, sodium, total carbohydrates, dietary fiber and sugars and percentage of vitamins present based on a 2,000 calorie diet (See Fig. 2). Also unsafe food products such as raw milk cheeses, apple and fruit juices which are unpasteurized should carry a label to this effect as should categories of milk (low fat, half and half), margarine (light, low fat, etc.) so that consumers can make informed choices. The choice of margarine sold in tubs is preferable as it will contain less transfatty acids than do solid sticks.

Factors which affect food production and consumption are numerous and still being identified. The use of a new environmentalism approach, the aim of which is the preservation of the whole ecosystems is much needed when a food crisis could be imminent. Multiple factors such as ecological damage, and the rapid rise in population growth, climatic changes, ozone depletion in Antarctica still continue to deplete food supplies, the safety of which is debatable.

Apart from an universal need to decrease the chemical pollution of foods, future dietary needs in different parts of the world are simple to state in general, although extremely complex, variable, and uncertain in detail. In poor, less technically developed countries, more food is needed, particularly in rural areas, especially for young children and pregnant women, including paradoxically, "compact calories" (oils and fats) during the weaning period. In urbanized, more technically developed countries, a decreased intake is often the main priority, with special reference to foods

Nutrition Facts		
Serving Size 1 cup (228g)		
Servings Per Container 2		
Amount Per Serving		
Calories 260 Calories from Fat 120		
		% Daily Value*
Total Fat 13g		20%
Saturated Fat 5g		25%
Cholesterol 30mg		10%
Sodium 660mg		28%
Total Carbohydrate 31g		10%
Dietary Fiber 0g		0%
Sugars 5g		
Protein 5g		
Vitamin A 4%	•	Vitamin C 2%
Calcium 15%	•	Iron 4%

*Percent Daily Values are based on a 2,000 calorie diet. Your daily values may be higher or lower depending on your calorie needs:

	Calories:	2000	2500
Total Fat	Less than	65g	80g
Sat Fat	Less than	20g	25g
Cholesterol	Less than	300mg	300mg
Sodium	Less than	2400mg	2400mg
Total Carbohydrate		300g	375g
Dietary Fiber		25g	30g

Calories per gram:
Fat 9 • Carbohydrate 4 • Protein 4

FIGURE 2   "Nutrition Facts" on new food labels.

mentioned earlier. Among some poor, as well as a rising more affluent middle class, obesity, an independent "risk factor" for the development of diabetes mellitus and hypertension is a major problem and attention particularly needs to be given to the use of smaller portions and to limiting foods rich in animal fats (as concentrated sources of calories, saturated fats, and cholesterol) as well as pesticides and other chemicals, which are mostly lipid-soluble. The consumption of less fatty varieties of fish and new forms of lean pork (50% less fat) and beef is receiving attention, and ostrich and bison meat are now being sold, but their popularity will depend on cultural acceptance, availability and cost, as is the use of nontropical vegetable oils (e.g., corn oil, safflower), which are composed of polyunsaturated fats with no cholesterol, and fiber-rich foods. More recently, mono-unsaturated vegetable oils—olive oil, canola or rape seed oil, and peanut oil—seem preferable as, containing no cholesterol (as with all plant foods), having no saturated fat and also not having the risk of being related to some forms of cancer, especially of the colon, for which polyunsaturated vegetable oils are considered by some scientists to be risk factors.

The future of the world's diet is impossible to foresee. Factors that influence it in different directions include the geometric global increase in population size (especially in cities in less technically developed countries), including rising numbers of elderly not looked after by the extended family paralleled by the impact of worsening ecological changes such as, land degradation, soil compaction and erosion, air pollution, acid rain, flooding, increased ultraviolet radiation (affecting humans and crops), severe drought, global warming as well as current fears of food shortages world-wide. This critical situation has arisen because of a false assumption that food production was adequate following the Green Revolution and that the response of the genetically altered grains to fertilizers would prevail. In 1990 national food production was deficient in Third World countries (who were major food exporters in the past) and their own needs of cereal imports is estimated to increase by 50% in the next 15 years or so. No investment was allocated to agriculture and grain stocks are now at a low ebb globally. Feed grain allotted to cattle, pigs, poultry, fish, etc. will have to be reduced and feeding with bagasse (rice, straw, sugar cane waste, etc.) instituted, also consumer education linked with advice of the superiority of the prudent diet and its health benefits may reduce consumption of fatty meat among affluent consumers. Concerns regarding availability and management of grain distribution to refugees and to famine areas prevail.

Although growth of populations and economies are exponential, the resources which support them are not e.g. land, water, forests, etc (STATE OF THE WORLD 1994). Since the 1970's warnings from ecologists and other groups regarding diminishing availability of grain stocks have been frequent. In 1992 the Union of Concerned Scientists produced a document entitled "The World Scientists Warning To Humanity" which was signed by 1,600 of its members in order to stimulate action and prevent further damage.

The future of the world's diet is impossible to foresee. Factors that influence it in different directions include the geometric global increase in population size (especially in cities in less technically developed countries), including rising numbers of elderly not looked after by the extended family.

V. FOOD PRODUCTION

In order to improve and maintain food production a revolution in approaches must be established. Numerous interacting factors must be taken into account as rapid population growth requires an increasing food supply.

Urgent attention must be given to the following areas: 1) The new role of the farming industry. More efficiency in the use of available resources must include dynamic changes such as the recent adoption by some farmers of integrated pest management and sustainable agricultural techniques with careful selection of chemicals for the protection of the environment. In addition soil solarization can be employed should heavily infested soils be present. 2) Land availability, protection and careful management. The worsening of ecological damage such as land degradation, compacting, erosion, loss of top soil etc. is linked to air pollution, acid rain, ultraviolet radiation, ozone depletion which affects the growth of a number of crops (soybeans, rice, wheat, etc.). Also changing soil chemistry, waterlogging and salinity and global warming due to increased emissions from the use of fossil fuels worldwide, exacerbate the problem. Overgrazing or browsing by livestock sheep and goats, and periods of drought have caused major deterioration of grasslands and desertification. Cattle losses due to starvation in the 1990's has resulted in foreclosure of farms and increased poverty in many countries worldwide. The availability of lands which can sustain intensive crop production (e.g., rain-fed, waterlogged lowlands and irrigated land) is diminishing in some areas.

Despite the challenge of increasing population pressure only 2/5 of the world's food needs can be grown on 17% of land which is irrigated. It is estimated that food demand will increase by 75% in the next three decades to feed a population of nine billion people and 80% of the food will be grown on this land. In the past two decades food production worldwide (excluding Africa due to internal strife and wars) has only grown by 50% whilst population rate rose by 40% and is still increasing. As an example China is adding 12 million people a year to the world population and projected purchases of grain years 1995–2030 will be 200 million tons, equal to the world's entire grain exports in 1994. The USA is the second nation with large grain consumption (160 million tons in 1991). Estimates of average food consumption per capita vary e.g. 323 kg/year (data in 15 selected countries in the world (1991) or much higher e.g. 800 kg/

year from other sources. In 1996 plant scourges are also devastating crops i.e. Pierce's disease in which bacterium *Xylella fastidiosa*, carried by a leafhopper insect prevents absorption of water by plants e.g. grape vines, alfalfa, peach, almond and ornamental trees in California and coffee and citrus crops in Brazil have all been affected. The use of other predator insects, birds, etc. may prove to be useful in combatting this disease.

A. Water

Shortages of water supplies are well documented, the most severely affected region is the Middle East where seven nations compete for water from the Nile river. This may lead to serious conflicts unless agreements between nations can be reached. Replenishing of aquifers from rainfall has been low in drought periods and the seeping of polluted seawater will also contaminate drinking water as will runoff of pesticides from farms and forests in which agriculture is maintained and heavy metals such as copper and lead etc. are now pollutants in water supplies.

There is a pressing need for pollution control laws, for water conservation guidelines at government level worldwide, and new technologies for water conservation for example rich economies such as some Asian countries and China have high levels of consumption which are incompatible with the present situation. Recycled water should be used in agriculture and rainwater stored for domestic purposes.

Pollution of oceans and rivers and coastlines are increasing, dumping of waste material, sewage, etc. have rendered beaches unsafe as well as drinking water.

B. Fisheries and Aquaculture

The neglect of the environment is well exemplified by unlimited access to fishing grounds by fishing fleets and especially commercial enterprises. Overfishing has become a global concern as demand for fish is rising in developed countries whilst world catches are decreasing. Between 1991 and 1993 fish catches declined by 50% and many of species such as cod, flounders and herrings are diminishing in numbers because of disruption of their marine habitats. The collapse of the Peruvian fisheries of the 1970s was a warning sign of future events. Many factors have contributed to the decline of available marine catches 1) air pollution (already described) which has also increased pollution of oceans and rivers 2) loss of

mangroves and coral reefs which were plentiful sources of fish and mollusks 3) decimation of plankton which provided habitat and food for fish species and 4) in great part human activities. The latter include, increasing populations along coastal areas, causing ecological pollution (sewage, refuse, oil spills); and housing projects etc. The demands of the tourism industry, (silting from dams, deforestation) and rising sea levels from global warming which can flood coastal environments. A major impact includes types of fishing methods and equipment (explosives, trawlers, etc.) used by commercial fishermen, some of whom are still using drifnets which capture also marine mammals and seabirds as well as "trash fish" which is discarded (as much as 27 million tons) as it has little commercial value but it would provide a source of revenue for local fishermen and an excellent food source for the poor (FAO 1996). "Biomass" fishing is also practiced, catches include algaes, immature fish, etc. and can be sold to inland fisheries. Aquaculture is a rapidly developing industry worldwide, but its environmental impact needs constant monitoring. This industry provides marketable products (salmon, talapia, catfish, shrimp, mussels, etc.) which are more likely to be consumed by rich rather than poor consumers. Many suggestions have been made over the years to prevent mismanagement of fisheries, such as limiting catches (e.g. krill) quotas, levying of rents (Australia) but many are difficult to enforce. Action to prevent further depletion of marine catches should include a review of past international agreements to protect coastal environments with concerted support of grassroot organizations and the wide network of non-governmental organizations (NGO).

C. Deforestation

Deforestation worldwide has continued for many decades at an accelerating pace. In the recent past, forests covered 30% of the land mass, less of 1/3 of them remained in 1991. This desecration has brought about global warming loss of top soil, flooding, sedimentation of rivers decline of fisheries and other related problems. Forests have provided recreational activities, food for indigenous forest dwellers, fruits, nuts, fuel (Biomass), plant fibers (cloth making) and herbal remedies for a number of illnesses and income has been derived from the sale of these materials. Forests also are the reservoir of gene pools from numerous species as yet unexplored which could benefit future generations. Already "green" products are available such as pulp required in paper making, which is made from bamboo, hemp, and a fast growing Kenaf plant and not wood. In 1992, 9% of the world's paper supply was treeless (State of the World 1994). Carbon taxes are already levied in some countries to stem damaging activities as deforestation.

Advocacy is needed at all levels to maintain the existence of forests, to recognize the important role of indigenous forest inhabitants in protecting their habitat and hopefully heads of governments may recognize that the value of forests fast transcends the economic returns from the mining, logging ranching and paper industries and with selective "logging", adoption of innovative strategies, establishment of sustainably managed forests should prevail.

D. Energy

Protection of the environment requires investment in energy efficient technologies. Over the years sources of renewable energy solar, wind, geothermal have been utilized in both third world and developed countries e.g. photovoltaric cells installed on roofs of houses which can be used to power radios, television sets, solar hot water heaters, etc. Biomass "wastes" could possibly supply 30% of global electricity. Wind power turbines are also used in California and parts of Europe. Goals for reducing global warming are being implemented in transport systems, advocating walking, bicycling, ride sharing, bus and train riding. In recent years fuel obtained from sugar cane wastes have been used in the Brazilian car industry and safer "green" cars using hydrogen fuel cells are also being produced, however, the cost of manufactured hydrogen would be excessively high at present.

VI. FOOD SAFETY AND HANDLING

Mechanized mass production of animal meats (factory foods) such as chicken, beef, eggs opposed by "animal rights" groups have required consumers to take precautionary measures in handling and cooking these products. In the recent 1990's, food borne diseases have been reported in newspapers and medical journals such as "Mad Cow Disease" (Bovine Spongiform Encephalopathy) with possible links with the incurable Creutzfeld-Facob disease in humans.

Microbial diseases which cause severe diarrhea and may lead to death have been highlighted recently such as strains of *Echerichia coli: 0157: H7* in South Japan affected 9,000 people and caused 10 fatalities and mustard and cress sprouts were possibly the cause in school meals. In the U.S.A. fruit juices and other products containing unpasteurized apple juice which were recalled recently were also contaminated with *E. coli*. Many infants and young people became ill and one infant died from kidney failure. Cyclospora infections recently afflicted several hundreds of individuals in the USA and Canada who consumed unsafe foods. Many other foods such as ice cream, salami, uncooked hamburgers, etc. have also been implicated.

In a recent report issued from the US Department of Agriculture (November 1996) regarding the safety of the supply, it was estimated that well over 6.5 million people are made seriously ill from tainted foods from numerous sources and as many as 7,000 people will succumb from the illness. Every year the ingestion of poisonous mushrooms claims many victims e.g. the Ukraine (1996).

VII. FUTURE AND CURRENT AVAILABILITY OF FOOD SUPPLIES

The present dilemma facing food producers globally is the need to meet the demand of increasing consumers in an overpopulated world without future deterioration of the environment.

The majority of world governments and their political advisers have not taken into account the carrying capacity of the land and their attitudes towards providing family planning services has varied widely. Although the world rate of population growth declined to below 1.6% in 1995, projections indicate that an additional yearly addition of 900 million people is probable.

It is now anticipated that a 64% expected increase in food demand will be required over the next 25 years, at the same time as a shortage of available grain harvested area and a reduced world grain production (which, in 1995, was 5% below the harvest of 1.78 billion tons in 1990) are now evident. In 1996, the world carry over stocks of grain have dropped precipitously to an estimated 49 days of world consumption, the lowest ever recorded, and rising food prices confirm existing food scarcity. A recent survey by the United Nations indicated that 1 billion individuals are malnourished, however famines recently reported in parts of Africa are no longer solely due to crop failure only, but to other events such as landless labourers migrating to cities, increasing urbanization affecting the poor, both in urban and rural areas. The deeply entrenched governmental policies worldwide to increase meat production using intensive rearing of cattle, pig, chicken and fish, has deviated large amounts of land to the production of grain for animals, which in the past would have been devoted to growing hopefully accessible, culturally appropriate inexpensive foods for all groups of consumers. Thirty eight percent of total world grain is used in feed lots (7 kg of grain is required to produce 1 kg of beef, 4 kg for pork and 2 kg for poultry and fish).

It is also difficult to assess the amount of food losses between production and retailing of grains, estimates have varied from 20% to as much as 1/3 due to pests and crop failures and storage problems (1991–1993). It is evident that catastrophes such as famines may continue unless immediate remedial action can be taken at the global level. It is necessary to sensitize governments and their political advisers to carry out capacity assessment.

Newer technologies to increase grain production are few and depend on funding for research by agricultural scientists although it is believed that if achievable increases can be engendered, it may not be sufficient to meet global needs and maintain food security for all. The survival of the poor and displaced, and homeless individuals is threatened further as funding for food aid programs has been drastically reduced by 50% in the past three years. The one food resource which has continued to be available in both good and lean times is breast-milk which provides a unique and complete food for infants up to 6 months and an excellent supplement for a couple of years, should mothers wish to continue lactation. The efficiency of conversion of nutrients by nursing women is known to be high, and this milk needs to be included in both food composition tables and food balance sheets. Its economic significance should be viewed at three overlapping levels 1) large scale (national and community, 2) family and 3) commercial levels. Its benefits in the prevention of many illnesses in mothers and children is well known, and its use in refugee camps has been actively promoted by different agencies. It was calculated in 1968 that to supply cow's milk for all women with young children in India would require the immediate development of an additional herd of 114 million lactating cows, however the environmental damage from discarded feeding bottles, nipples, cans, etc. was not included (Jelliffe & Jelliffe 1968).

Universal concerns with balance of payments and expenditure of foreign currency may lead governments to promote breast feeding more energetically rather than importing excessive amounts of expensive substitutes, as well as promoting facilities, lactation breaks for working women who have elected to breastfeed.

In general, the future scenario, if it persists indicates that great changes in lifestyles, food consumption and reproductive behaviour, and limited use of non-renewable resources etc. may be required over time to prevent social disintegration at the global level. It has been suggested that rather than basing projections for the future on past events, analysis of the current interactions between multiple factors affecting food availability might be more pertinent, accurate and effective, if it is also linked with a fiscal policy (State of the World 1996).

Hopefully, these changes may eventually engender popular support by world consumer groups as this revolutionary and dynamic movement would sensitize them to the need to limit their family size in the present, thus ensuring reasonable and stable food supply for future generations.

BIBLIOGRAPHY

American Dietetic Association (1989). "The Dietary Guidelines: Seven Ways to Help Yourself to Good Health and Nutrition." American Dietetic Association, Chicago, Illinois.

Brown, L. R. et al. (1993). "State of the World." W. W. Norton & Company.

Brown, L. R. et al. (1994). "State of the World." W. W. Norton & Company.

Brown, L. R. et al. (1995). "State of the World." W. W. Norton & Company.

Brown, L. R. et al. (1996). "State of the World." W. W. Norton & Company.

Brown, L. R. (1995). "Who Will Feed China. Wake Up Call For A Small Planet." W. W. Norton & Company.

Cunningham, A. S., Jelliffe, D. B., and Jelliffe, E. F. P. "Breastfeeding, Growth, and Illness. An Annotated Bibliography." UNICEF. New York, N. Y., USA.

Eaton, S. B., Shostak, M., and Konneer, M. (1988). "The Paleolithic Prescription." Harper and Row, New York.

Food and Agriculture Organization. (FAO 1996). "Food Production and Environmental Impact." World Food Summit (Rome). 13–17 November.

Food and Agriculture Organization. (FAO 1996). "Synthesis." Vol. 1, 2, 3. World Food Summit (Rome). 13–17 November.

Food and Agriculture Organization. (FAO 1996). "Rome Declaration on World Food and World Food Summit Plan of Action." World Food Summit (Rome). 13–17 November.

International Planned Parenthood Federation. Family Planning for Doctors IPPF. Medical Publications.

James, W. P. T. (1988). "Healthy Nutrition." WHO Regional Office for Europe, Copenhagen.

Jelliffe, D. B., and Jelliffe, E. F. P. (1979). "Human Milk in the Modern World." (2nd edition). The English Language Book Society and Oxford University Press, Oxford.

Jelliffe, D. B., and Jelliffe, E. F. P. (1988). Programs to Promote Breastfeeding. Oxford University Press, Oxford.

Jelliffe, D. B., and Jelliffe, E. F. P. (1990). "Community Nutritional Assessment." Oxford University Press, Oxford.

Jelliffe, D. B., and Jelliffe, E. F. P. (1991). "Dietary Management of Young Children With Diarrhoea. A Manual for Managers of Health Programs." (2nd edition). World Health Organization and United Nations Children's Fund (UNICEF).

Krause's Food Nutrition and Diet Therapy. (1996). 9th edition. W. B. Saunders Company.

Latham, M., and Van Veen, M. S., eds. (1989). "Dietary Guidelines: Proceedings of an International Conference, Toronto, Canada." Cornell International Nutrition Monograph Series No. 21, Ithaca, New York.

National Research Council. (1989). "Diet and Health: Implications for Reducing Chronic Disease Risk." U.S. National Academy Press, Washington, D.C.

National Research Council. (1989). "Alternative Agriculture." National Academy of Sciences, Washington, D.C.

National Research Council. (1989). "Recommended Dietary Allowances." National Academy Press, Washington, D.C.

The Surgeon General's Report on Nutrition and Health. (1988). US Department of Health and Human Services Public. Nos. 88–50211, U.S. Government Printing Office, Washington, D.C.

The Hunger Project. (1989). "Ending Hunger." Praeger Publishers, Sparks, Nevada.

Truswell, A. S., ed. (1983). Recommended dietary intakes around the world. *Nutr. Abstr. Rev.* **53**, 11.

U.S. Department of Agriculture and U.S. Department of Health and Human Services (1995). Nutrition and Your Health Dietary Guidelines for Americans. 4th Edition.

World Health Organization. (1989). "Health Surveillance and Management Procedures for Food-handling Personnel." Tech. Rept. Ser. No. 785, WHO Geneva.

Dietary Factors of the Cancer Process

LAWRENCE H. KUSHI

University of Minnesota School of Public Health

GLOSSARY

Case–control study A retrospective study design in which cases with disease and a comparable series of controls without disease are enrolled and information about past exposures is collected

Cohort study A prospective study design in which individuals without disease are enrolled and information about current or past exposures is collected. The cohort of individuals is then followed forward in time, and differences in exposures between those who develop disease and those who do not are compared

Epidemiology Study of the causes and prevention of disease in human populations

THE ROLE THAT DIETARY FACTORS MAY PLAY IN the development of cancer is being studied with great interest. This is exemplified by estimates that approximately one-third of all cancers in the United States may be attributable to dietary causes. Thus, only tobacco approaches the impact of diet as a causal factor in the development of cancer. Studies to elucidate the role of diet in the cancer process have largely been either epidemiologic, i.e., determining associations of dietary patterns or specific dietary factors with cancer incidence and mortality in human populations, or focused on biological mechanisms. The latter studies, usually conducted in animal models or tissue culture systems, are often focused on examining the effects of isolated compounds that are found in the diet and

that may either increase or decrease the likelihood of developing cancer. Agreement between the two types of studies increases the likelihood that the specific dietary factors of interest play an important role in cancer causation. On this basis, it is generally recognized that an increased consumption of vegetables and fruits is likely to decrease the risk of developing cancer, whereas an increased consumption of animal foods, particularly red meat, is likely to increase the risk of cancer. The impact of these dietary effects is likely to differ for cancers of different sites. These observations form the basis for most dietary recommendations for the prevention of cancer.

I. EPIDEMIOLOGIC BASIS

A. Ecologic and Cross-Cultural Studies

It is well known that cancer incidence and mortality rates vary substantially across different regions and nations. For example, rates of a common cancer such as breast cancer vary sevenfold between areas with relatively low rates (such as many regions in Africa or Asia, with national rates of about 10 to 30 cases per 100,000 persons per year) and areas with relatively high rates (such as the United States, with rates ranging from about 70 to 100 cases per 100,000 persons per year). Cancers of other sites can vary even more substantially; e.g., the rates for esophageal cancer vary from about 4 cases per 100,000 persons per year in southern Ireland to between 165 and 195 cases per 100,000 persons per year in the Caspian Sea region of Iran.

That the substantial variation in cancer rates are likely to be due to environmental factors is demonstrated by studies that have examined the cancer experience of migrants from areas with relatively low cancer rates to areas with relatively high cancer rates.

Generally speaking, migrants adopt the disease experience of their host country rather than maintaining the disease experience of their country of origin. If genetic factors were the predominant factors influencing geographic differences in cancer rates, the disease experience of migrant populations would be similar to that of their population of origin, a situation that is rarely observed. For example, colon cancer rates in Japan are substantially lower in Japan than in the United States; however, colon cancer rates among Japanese Americans are higher than in Japan, and within two generations approach those of the United States.

Among the many factors that have been hypothesized to explain observations of substantial differences in cancer rates are dietary factors. For example, it has been observed that international differences in dietary fat availability correlate strongly with international differences in breast cancer incidence. Figure 1 shows this correlation for 21 countries with good cancer registries. In addition, as migrant populations adapt to their new environment, it is well known that dietary habits are among the many things that are altered. Although the countries in Fig. 1 differ in many ways other than the availability of fat in the diet, these observations have provided a basis for other investigations of diet and cancer using more rigorous study designs. There is little dispute that cancer rates vary

substantially across countries, and that these variations are associated with substantial differences in food intake and dietary patterns.

B. Analytic Epidemiologic Studies

Among epidemiologic study designs, cross-sectional studies such as those investigating international comparisons of diet and cancer are considered to be among the weakest for establishing causality. Analytic epidemiologic studies, which are designed explicitly to examine associations between exposures, e.g., dietary factors, and disease, e.g., cancer, are generally better suited to the investigation of diet–cancer relationships. This is in large part because such studies are able to take into account other confounding factors that may differ between people who develop cancer and people who do not. For example, people who consume diets that are relatively high in fat also tend to be less physically active and to smoke cigarettes more often. Analytic epidemiologic studies provide the ability to examine diet–cancer associations while taking into account these potential confounders.

The two basic study designs in analytic epidemiology are the case–control study and the cohort study. In a case–control study, subjects with a disease such as cancer, the cases, are enrolled in the study. A comparable series of people who do not have the disease,

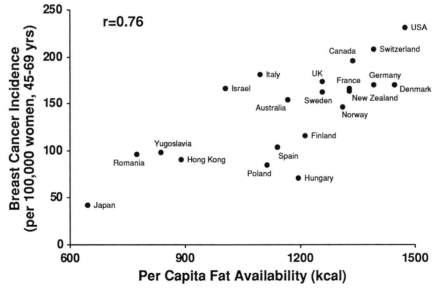

FIGURE I Correlation of dietary fat availability and breast cancer incidence among women aged 45–69 years, among 21 countries with good cancer registration. Adapted from Prentice, R. L., Kakar, F., Hursting, S., Sheppard, L., Klein, R., and Kushi, L. H. (1988). Aspects of the rationale for the Women's Health Trial. *J. Natl. Cancer Inst.* 80, 802–814.

the controls, are also enrolled. Because age is a major determinant of cancer risk, controls are often selected to be of similar age to the cases. In studies of diet, the cases and controls are asked to recall their usual diet at some point in the past; often, this is 1 or 2 years prior to diagnosis for the cases, and a comparable period of time in the past for the controls. The diets of the cases and controls are then compared to determine whether they differed in any way. Because dietary habits and other factors that occurred in the past are queried of study subjects after they developed cancer, case–control studies are regarded as retrospective studies.

In a cohort study, a study population in which none of the individuals have the disease of interest is enrolled. In studies of diet and cancer, information on the dietary habits of these individuals is collected as they are enrolled in the study. The study population, or cohort, is then followed forward in time; some members of the cohort will eventually develop cancer. The dietary habits of cohort members who develop cancer are compared to those who did not, to identify factors that may have resulted in the development of cancer. Because dietary habits and other factors are ascertained before the study participants have developed cancer, and the participants are then followed forward in time, cohort studies are considered to be prospective in nature.

Because of the differing nature of case–control versus cohort studies, they provide somewhat different perspectives on the question of whether dietary factors are related to risk of cancer. One principal difference is the necessity in case–control studies that study participants are asked to recall what they consumed at some point in the past. Because participants are likely to recall dietary habits with a relatively high degree of inaccuracy, assessment of associations of dietary factors with cancer will be compromised by measurement error. Measurement error will also exist in cohort studies in which participants report their current dietary habits, although this may be somewhat less problematic. A more important distinction between case–control studies and cohort studies, however, is that in the case–control study, subjects will have knowledge of whether or not they have cancer as they recall their dietary habits. This may result in what is known as recall bias, in which the cases are more likely to remember factors that they think may have resulted in their cancer; this would not occur among the controls, as they do not have cancer. This bias, in which dietary habits are recalled in a differential manner between cases and controls, does not exist in

a cohort study, as none of the participants have the disease at the time that dietary habits are ascertained. Thus, if it is generally thought that fat intake increases risk of cancer, a case–control study may be more likely to detect a positive association of fat intake with cancer than a cohort study would. For this reason, it is generally recognized that cohort studies are more likely to provide an estimate closer to the truth regarding the association of diet and cancer. Ultimately, the consistency of findings both among and between case–control and cohort studies strengthens the likelihood that dietary factors are associated with cancer.

In the past two decades, there have been an increasing number of analytic epidemiologic studies conducted to examine the relationship of dietary patterns or specific dietary factors with cancers of specific sites. The vast majority of these studies have been case–control studies. This is generally because cohort studies require large populations and a relatively long follow-up period in order for adequate numbers of cancer cases to occur to allow informative investigations of diet and cancer to be conducted. For example, most of the cohort studies of diet and cancer that are being conducted consist of tens of thousands or hundreds of thousands of individuals. In contrast, it is possible to conduct case–control studies of cancer in a relatively time-efficient manner, in which a few hundred individuals with cancer and a comparable number of controls may be enrolled in the study. For rare cancers, case–control studies are the only practical study design for the investigation of possible factors that may increase the risk of developing that cancer.

Several hundred case–control studies of diet and cancer have been published. These studies have focused on a wide variety of cancers, including the common cancers of economically developed countries, such as lung, colon, breast, and prostate, as well as cancers that are more common in other parts of the world, such as stomach and cancers of the upper gastrointestinal tract. Studies have been conducted in populations around the world, using different age groups, methods of dietary assessment, and variations in study design. Although these studies are therefore quite varied, there is a remarkable consistency in their findings relating to diet. Table I summarizes the consistency of association for certain dietary factors for some of the more common cancers based on case–control and cohort studies.

The most consistent finding in these studies is that an increased consumption of vegetables and fruits is associated with a decreased risk of development of

TABLE I

Risk Implications for Major Cancers by Consumption of Selected Foods and Nutrients[a,b]

Cancer site	Vegetables	Fruit	Meat	Protein	Fiber	Saturated fat	Alcohol
Oral cavity	Reduce	**Reduce**					Increase
Esophagus	**Reduce**	**Reduce**					Increase
Stomach	**Reduce**	Reduce					
Large bowel	Reduce	Reduce	**Increase**	Increase?	Reduce?	Increase?	Increase
Liver	Reduce?	Reduce?					
Pancreas	Reduce?	Reduce?	Increase?	Increase?	Reduce?		**Increase**
Larynx	**Reduce**	**Reduce**					
Lung	**Reduce**	Reduce				Increase?	
Breast	Reduce?	Reduce?	Increase?			Neutral	**Increase**
Endometrium	Reduce?	Reduce?		Increase?		Increase?	Increase
Cervix uteri	Reduce?	Reduce?					
Ovary	Reduce?	Reduce?					
Prostate	Reduce?					Increase	
Urinary bladder	**Reduce**	Reduce					
Kidney	Reduce	Reduce?					

[a] A question mark indicates current data are only suggestive; bold type indicates data are convincing.

[b] Adapted from Willett and Trichopoulos (1996). Summary of the evidence: Nutrition and cancer. *Cancer Causes Contr.* 7, 178–180.

cancer. That is, in the vast majority of these studies, cases with cancer reported that they consumed fewer vegetables or fruits, or with less frequency, than the controls without cancer. For certain cancer sites, this evidence is extremely consistent with virtually no studies suggesting contrary evidence. The cancers that are most consistently related to decreased consumption of vegetables and fruits include stomach, esophagus, larynx, lung, and bladder cancer. Evidence for an inverse association of vegetables and fruits with cancer of the colon and rectum is also consistent. Evidence for other cancer sites is less striking, either because studies are less consistent in their findings (e.g., breast) or because there have been relatively few studies that have examined these associations. These cancers include those of the endometrium, cervix, ovary, prostate, pancreas, and liver. Even so, the preponderance of evidence suggests that the risk of these cancers may also be decreased with increasing consumption of vegetables and fruits.

A second consistent finding from case–control studies is the association of animal food, principally red meat from domesticated mammals such as cows, sheep, and pigs, with an increased risk of cancers of the colon and rectum. Unlike the findings for vegetables and fruits, however, this association appears to be relatively site specific. That is, with the exceptions

of cancers of the pancreas, prostate, and possibly breast, there is relatively little evidence that red meat intake increases the risk of other cancers. In the case of prostate cancer, this may be a result principally of associations with animal fat intake, whereas with large bowel, breast, and possibly pancreatic cancer, associations with red meat intake are likely to be a consequence of other factors in red meat.

A third consistent finding from case–control studies is the effect of alcohol consumption on increasing risk of several cancers. These include cancers of the oral cavity, esophagus, colon, rectum, pancreas, breast, and cervix. [*See* Alcohol, Impact on Health.]

Generally, these consistent observations are paralleled by similar findings for nutrients that are associated with these foods. For example, a relatively consistent observation is an increased risk of cancer associated with total fat or saturated fat intake. This holds particularly for prostate cancer, but has also been observed with some consistency for cancers of the colon and rectum, lung, and endometrium. Case–control studies also provide some evidence of a positive association between fat intake and increased risk of breast cancer, although this is an inconsistent finding. Dietary fiber is generally found to be inversely associated with the risk of colon and rectum cancer, and various cancers appear to be associated with a decreased risk of vitamins that are found in abundance in vegetables and fruits, such as vitamins C and E. [See Ascorbic Acid; Nutrition, Dietary Fiber; Vitamin C.]

In contrast to the large number of case–control studies, there have been relatively few cohort studies published on the associations of dietary factors and cancer; those that have been published have resulted from perhaps a dozen different cohort studies. With few exceptions, associations of diet with cancer observed in case–control studies have also been seen in cohort studies. For example, several cohort studies suggest that vegetable or fruit intake, or dietary factors in vegetables or fruits, is associated with a decreased risk of lung cancer and colon cancer. Most cohort studies of colon cancer also observe an association of some aspect of red meat intake and an increased risk of these cancers; studies of breast cancer have found either an increased risk or no association with red meat intake. Alcohol intake is seen consistently to increase the risk of breast cancer in the several cohort studies that have examined this relationship.

Virtually all cohort studies of dietary fat and breast cancer have failed to find an association between these factors. Thus, the suggestion from some case–control studies and international correlation studies that dietary fat may increase risk of breast cancer is not well supported. This is a principal exception to the general observation that associations of dietary factors with cancer are seen with relative consistency in both case–control and cohort studies.

There have been few cohort studies of cancers of the oral cavity, esophagus, larynx, bladder, or pancreas as these are relatively rare cancers. Thus, findings from case–control studies for these cancers have generally not been replicated in enough settings in cohort studies to conclude that there is a consistency of findings in cohort studies related to diet for these cancers.

C. Intervention Studies

Another type of study in which the association of dietary factors with cancer can be investigated is the clinical trial or intervention study. In such studies, study participants are randomized into an intervention and a control group; because of randomization, the two groups are usually comparable in all respects except for the intervention. Intervention studies are also prospective in nature, with study participants followed forward in time until a reasonable number of cancers have occurred. Differences in cancer rates between the intervention and the control group can be attributable to the intervention.

For studies of dietary factors and cancer, there have been only a few studies completed; almost all of these studies have used dietary supplements (such as β-carotene or vitamin E) as the intervention. This is despite the fact that the principal associations that have been consistently observed in case–control and cohort studies have been of foods and food groups rather than of single nutrients. The primary reasons for this are that it is possible to conduct double-blind studies of nutrient supplements, and that it is thought that an intervention focused on dietary change is much more difficult to achieve and maintain over long periods of time.

The results of these supplement trials have generally been disappointing. A prime example are clinical trials that have examined the effect of β-carotene on lung cancer. As noted previously, many case–control and cohort studies have noted that a relatively high intake of vegetables and fruits is associated with a decreased risk of lung cancer. Some investigators suggested that this may be due to the relatively high intake of β-carotene, a strong antioxidant and precursor of vitamin A that is found in many vegetables and certain fruits. This hypothesis formed the basis of several randomized intervention trials of β-carotene supplementation and risk of lung cancer. The results of three of these trials have been published. In two of the trials, lung cancer rates were actually higher among those who received β-carotene than among those who received placebo. In the other trial, there was no difference between the two groups. Other studies of single or combination supplements have generally not demonstrated a substantial effect of the intervention on decreasing the risk of cancer.

Although clinical trials of isolated dietary factors and cancer have been disappointing, clinical trials aimed at dietary change are underway and promise to provide some insight into whether changes in dietary patterns may result in a decreased risk of certain cancers. One of the largest of these is the Women's Health Initiative, in which about 75,000 women are being randomized to either a low-fat diet or no dietary intervention. This study is coordinated by the National Institutes of Health and involves numerous clinical centers around the United States in order to enroll the requisite number of women. Primary endpoints for this study include coronary heart disease and breast cancer, despite the fact that there is little support from cohort studies for the hypothesis that dietary fat intake is associated with a risk of breast cancer. This study is designed to follow the women through the year 2005, so it will be several years before it is known whether a low-fat dietary intervention will lower the risk of breast cancer. Another dietary intervention study, conducted by the National Cancer Institute, is the Polyp Prevention Trial, which is using a high fiber, high vegetable and fruit dietary intervention for the prevention of polyps, a precursor of colon cancer. Although the endpoint of this study is not colon cancer, this study should be completed in the near future and will provide some insight into whether such a dietary intervention may decrease the risk of colon cancer.

D. Summary of Epidemiologic Studies

Overall, epidemiologic studies indicate that an increased consumption of vegetables and fruits is associated with decreased risk of many cancers. This has been observed in case–control and cohort studies in many different populations and using many different dietary assessment methods. The association with vegetables and fruits appears strongest and most consistent for epithelial cancers of the gastrointestinal and respiratory tracts, including lung, larynx, esophagus, and stomach, and is also consistent for cancers of the large bowel and bladder, whereas it is somewhat weaker for cancers that are thought to be hormone dependent, including breast, prostate, and endometrial cancer. Epidemiologic studies also indicate that an increased consumption of red meat is associated with an increased risk of colon cancer; red meat intake may also increase the risk of pancreatic cancer, prostate cancer, and breast cancer, although the findings are less consistent for these cancers. Alcohol intake has also been shown to increase the risk of breast cancer and cancers of the gastrointestinal tract. Although few of these associations have been confirmed in randomized clinical trials, the consistency of these findings across many studies in many different populations suggests that these associations are causal. [*See* Breast Cancer Biology; Colon Cancer Biology; Gastrointestinal Cancer; Prostate Cancer.]

II. MECHANISMS

Although epidemiologic studies demonstrate that vegetables and fruits likely decrease the risk of developing cancer, *in vitro* studies and animal studies provide a basis for examining potential mechanisms for the cancer-preventive effects of specific compounds found in these foods. A listing of some of these compounds are provided in Table II.

TABLE II

Selected Compounds Found in Plant Foods that May Decrease Cancer Risk

Type of compound	Example	Food source examples	Example of role in cancer prevention
Allium compounds	Diallyl sulfide	Onions, garlic	Induction of detoxification enzymes
Carotenoids	Lycopene	Tomatoes	Antioxidant
	β-Carotene	Carrots, green leafy vegetables	Antioxidant
Dietary fibers	Wheat bran	Whole grains, vegetables	Bile acid metabolism
Dithiolthiones		Cruciferous vegetables	Induction of detoxification enzymes
Flavonoids	Quercetin	Apples, kale, onions	Antioxidant
Glucosinolates	Glucobrassicin	Cruciferous vegetables	Induction of detoxification enzymes
Indoles	Indole-3-carbinol	Cruciferous vegetables	Induction of detoxification enzymes
Isothiocyanates	Benzyl isothiocyanate	Cruciferous vegetables	Induction of detoxification enzymes
Isoflavones	Genistein	Soy foods	Phytoestrogen
Lignans	Matairesinol	Whole grains, seeds	Phytoestrogen
Phenols	Caffeic acid	Many plant foods	Antioxidant
Protease inhibitors		Legumes	Protease inhibition
Saponins		Soy foods	Bile acid metabolism
Selenium		Whole grains, vegetables	Antioxidant
Thiocyanates	Benzyl thiocyanate	Cruciferous vegetables	Induction of detoxification enzymes
Vitamin E	α-Tocopherol	Vegetable oils, seeds	Antioxidant
Vitamin C		Citrus fruit	Antioxidant

315

A. Antioxidants

As noted previously, β-carotene was thought to be a likely candidate for a specific compound that may decrease lung cancer risk because it is a precursor of vitamin A, which is known to enhance cellular differentiation, and because it is known to be an antioxidant. Antioxidants may prevent oxidative DNA damage by helping to quench singlet oxygen, therefore preventing the formation of free radicals. β-Carotene is abundant in certain vegetables and fruits, such as carrots, green leafy vegetables, sweet potatoes, and winter squashes, and therefore might partly explain the decreased risk of cancer associated with vegetable and fruit intake.

Although β-carotene was shown in clinical trials to increase and not decrease the risk of lung cancer, the possibility that antioxidants that occur in vegetables and fruits play an important role in cancer prevention remains. In addition to β-carotene, there are several hundred other carotenoids that exist in nature. Several of these occur in significant quantities in the food supply. For example, lycopene, which is a more potent antioxidant than β-carotene, is abundant in tomatoes and tomato products. At least one prospective cohort study suggests that an increased intake of tomato products is associated with a decreased risk of prostate cancer. Lutein, another carotenoid with antioxidant properties, is rich in dark-green leafy vegetables.

Aside from carotenoids, other compounds that are found in vegetables and fruits have antioxidant properties. These include vitamin C, vitamin E, flavonoids such as quercetin, kaempferol, and myricetin, and the mineral selenium. These compounds may act as antioxidants because of their ability to interrupt free radical damage (such as vitamin E) or because of their role as cofactors in enzymatic reactions that reduce oxidized metabolic products (such as selenium as a cofactor for the enzyme glutathione peroxidase). These compounds and their antioxidant properties may in part be responsible for the anticancer effects observed for plant foods in many epidemiologic studies.

Vitamin C may also prevent cancer through mechanisms other than its role as a water-soluble antioxidant. It is required for the hydroxylation of proline and lysine in the formation of collagen, and therefore plays a central role in maintenance of the extracellular matrix. It is also known to enhance immune function and, of relevance to stomach cancer, can prevent the formation of nitrosamines by scavenging and reducing nitrite precursors.

B. Induction of Detoxification Enzymes

Several compounds with possible anticancer properties are particularly abundant in cruciferous vegetables such as broccoli, kale, cauliflower, cabbage, Brussels sprouts, mustard greens, rutabaga, and turnips. These include dithiolthiones, glucosinolates, indoles, isothiocyanates, and thiocyanates. These compounds have been shown to induce enzymes of the mixed function oxidase or cytochrome P450 system or to induce enzymes such as glutathione-S-transferase or glutathione reductase. These enzymes may be involved in the detoxification of carcinogens, thereby decreasing the risk of cancer.

Other compounds that are found in vegetables and fruits and may also induce glutathione-S-transferase include coumarins, found in many vegetables and citrus fruits; allium compounds, found in allium vegetables such as onions, garlic, chives, and leeks; D-limonene, found in citrus fruit oils; and phenolic compounds such as caffeic, ferulic, and ellagic acids, ubiquitous in many plant foods. The phenolic compounds may also inhibit the formation of nitrosamines, similar to the action of vitamin C. The extent to which these compounds are consumed in usual diets is not known as most food composition tables do not have information on these compounds. Some are largely contained in parts of the vegetable or fruit that are usually not eaten; an example is D-limonene, found principally in the peel of citrus fruits.

C. Hormonal Effects

Isoflavones, including genistein and daidzein, occur in high concentrations in soybeans and are also found in many other plant foods. Isoflavones are known to have weak estrogenic activity. Because of their structural similarity to estrogens, they may bind to estrogen receptors and prevent more potent endogenous estrogens such as estradiol from exerting their cell proliferation effects. Isoflavones may also enhance the production of sex hormone binding globulin in the liver, thereby resulting in lower concentrations of free, biologically active estrogens. Other phytoestrogens may have similar effects to the isoflavones. These include lignans such as secoisolariciresinol and matairesinol, which are found in relatively high concentrations in whole grains and seeds.

D. Protease Inhibition

In addition to isoflavones, soybeans also contain protease inhibitors. Other foods that contain various protease inhibitors include other legumes, such as kidney beans and chickpeas, and whole grains, potatoes, and certain vegetables. Protease inhibitors have been suggested to prevent the action of proteases that are produced by neoplastic cells. These proteases result in destruction of the extracellular matrix, allowing the local invasion and growth of neoplastic cells.

E. Effects on Bile Acid Metabolism

Of particular relevance to cancers of the large bowel, various plant compounds have been postulated to influence bile acid metabolism in ways that decrease carcinogenesis. Dietary fibers are among the most well studied of these compounds and are thought to bind bile acids and cholesterol in the gut, reducing the production of secondary bile acids that may act as promoters of carcinogenesis. Fermentation of fibers by colonic bacteria results in a more acidic environment in the colon through the production of short-chain fatty acids, which may in turn reduce the solubility of free bile acids, and therefore their availability for cocarcinogenic activity. Butyrate, one of these short-chain fatty acids, may also have anticarcinogenic effects. Insoluble fibers, such as those that predominate in wheat bran, tend to increase fecal bulk and decrease the concentration of fecal bile acids. High fecal bile acid concentration may increase the risk of colon carcinogenesis.

Dietary fibers are not the only compounds that may influence bile acid metabolism. Saponins, found in soybeans and other plant foods, also have the ability to bind bile acids and cholesterol, thereby reducing the production of secondary bile acids. Plant sterols such as β-sitosterol occur in many plant foods and have been shown to decrease the development of carcinogen-induced colon tumors in rats. Whether this is due to effects on bile acid metabolism is uncertain. [*See* Bile Acids.]

F. Effects of High-Fat or High-Meat Diets

In addition to the large number of compounds present in plant foods that may decrease the risk of cancer, plant-rich diets also tend to be lower in fat intake and animal foods.

Dietary fat is known to be more calorically dense than other components of diet. One gram of dietary fat provides approximately 9 kcal of energy, whereas 1 g of dietary protein or carbohydrate provides ap-

proximately 4 kcal. Thus, for a given amount of food, diets high in fat will tend to be higher in total energy. Numerous studies have established that animals that consume energy-restricted diets will develop fewer tumors and have longer life spans than animals allowed to consume as much food as they desire. Although the relevance of this observation to human populations that have an adequate food supply is not well established, the increased energy density of high-fat diets may be one of the mechanisms by which high-fat diets may promote carcinogenesis. Energy-dense diets may also promote the development of obesity, a risk factor for several cancers, including endometrial, postmenopausal breast, renal cell, and possibly colon cancer.

Animal studies also demonstrate that, given isoenergetic diets, diets that have a higher fat content will promote the tumorigenesis relative to diets with a lower fat content. For hormone-dependent cancers, dietary fat, especially saturated fat, may enhance the availability of endogenous estrogens. Epidemiologic studies also suggest that dietary fat may influence the risk of cancers that are not primarily hormone dependent as well, including lung and colon cancers. The specific mechanisms by which dietary fat may enhance the development of these cancers may relate to effects on bile acid and cholesterol metabolism, as high-fat diets increase the need for bile acid secretion in the gut. [*See* Nutrition, Fats and Oils.]

Although dietary fat, especially saturated fat from animal sources, may be responsible for the increased risk of prostate cancer due to meat intake, the association of meat intake with colon and possibly other cancers may be due to additional factors related to meat intake. These factors include the availability of readily absorbed heme iron in red meat, a possible pro-oxidant that may enhance oxidative DNA damage; the production of carcinogenic heterocyclic aromatic amines through the cooking of proteins; the presence of nitrosation precursors in cured and processed meats; and the relatively high proportion of sulfur-containing amino acids that may alter the availability of folic acid for normal DNA replication.

G. Summary

It is clear that there are numerous compounds in plant foods that may have anticancer effects; this overview covers only some of the compounds that have been hypothesized to decrease the risk of cancer and should not be considered exhaustive. The presence of these compounds in diets high in vegetables and fruits provides an underlying biological basis for the strong and consistent epidemiologic observations that diets high in these foods are associated with a decreased risk of a wide variety of cancers. There are also several plausible mechanisms by which saturated fat or red meat intake may increase the development of colon and other cancers. These mechanisms correspond with observations from epidemiologic studies that diets high in red meat increase the risk of colon and perhaps other cancers.

III. DIETARY RECOMMENDATIONS FOR CANCER PREVENTION

The consistency of findings across epidemiologic studies regarding diet and cancer and the corresponding animal and tissue culture studies that provide a biological basis for these relationships have led to the promulgation of dietary recommendations for the prevention of cancer by several organizations, including the National Academy of Sciences and the National Cancer Institute. One of the more recent sets of dietary guidelines for the prevention of cancer are those revised in 1996 by the American Cancer Society (ACS). These guidelines consist of four principal recommendations, with implementation steps as part of these guidelines; these are summarized in Table III.

The principal finding from epidemiologic studies of

TABLE III

American Cancer Society 1996 Guidelines on Diet, Nutrition, and Cancer Prevention[a]

1. Choose most of the foods you eat from plant sources
   Eat five or more servings of fruits and vegetables each day
   Eat other foods from plant sources, such as breads, cereals, grain products, rice, pasta, or beans several times each day
2. Limit your intake of high-fat foods, particularly from animal sources
   Choose foods low in fat
   Limit consumption of meats, especially high-fat meats
3. Be physically active: Achieve and maintain a healthy weight
   Be at least moderately active for 30 min or more on most days of the week
   Stay within your healthy weight range
4. Limit consumption of alcoholic beverages, if you drink at all

[a]Used with permission from the American Cancer Society 1996 Advisory Committee on Diet, Nutrition, and Cancer Prevention (1996).

diet and cancer is the consistent and strong association of increased vegetable and fruit intake with the decreased risk of cancer. In agreement with these observations, the first ACS dietary guideline for the prevention of cancer is to "choose most of the foods you eat from plant sources." It is specifically suggested that people consume "five or more servings of fruits and vegetables each day," and that "other foods from plant sources, such as breads, cereals, grain products, rice, pasta, or beans," should also be consumed several times daily. It is further suggested that whole grains are preferred to processed or refined grains and that beans provide a reasonable protein alternative to meat.

The second ACS dietary guideline for cancer prevention is to "limit your intake of high-fat foods, particularly from animal sources." This guideline is supported by the evidence that red meat and high-fat diets increase the risk of colon and prostate cancers, and may play a role in other cancers. It is suggested under this guideline that foods low in fat should be selected and that meat intake, particularly high-fat meats, should be limited.

The third ACS guideline has not explicitly been addressed in this article. It is to "be physically active: achieve and maintain a healthy weight." This guideline is supported not only by the evidence that obesity increases the risk of certain cancers, but also by the observation that regular physical activity may decrease the risk of certain cancers independent of its effects on body size. This effect is seen most consistently and strongly for colon cancer, but may also play a role in breast, endometrial, and prostate cancers.

The fourth and last ACS guideline is to "limit consumption of alcoholic beverages, if you drink at all."

The evidence supporting these and similar guidelines for the dietary prevention of cancer from other organizations are strong and consistent. They are based on a large and growing body of studies examining the role of dietary factors and cancer. The consistency of the evidence supporting these guidelines has led to programs such as the "Five A Day" program, first initiated by the California Department of Health Services and subsequently adopted by the National Cancer Institute. The purpose of this program is to encourage the daily consumption of at least five servings of fruits and vegetables; recent national food consumption surveys suggest that on average, the U.S. population consumes these foods about half as frequently.

IV. SUMMARY

The evidence linking dietary factors with cancer comes from a wide variety of epidemiologic studies, in conjunction with more focused experimental studies in animals and in tissue culture. The wide variation in cancer rates internationally suggests that the widespread adoption of dietary patterns such as recommended in the American Cancer Society's 1996 guidelines for nutrition and cancer prevention may eventually result in substantially lower cancer rates in the United States and other economically developed nations. Although the specific mechanisms by which these dietary patterns may decrease cancer risk are speculative—witness the sobering experience from clinical trials that examined β-carotene as a possible cancer prevention agent—there are a wide variety of compounds and associated biological mechanisms that may in part explain the decreased cancer rates with increased vegetable and fruit intake and the increased colon cancer rates with increased red meat intake. Overall, the evidence relating dietary factors to cancer is a cause for optimism and suggests that substantial benefit may accrue from the adoption of dietary patterns suggested by these studies.

BIBLIOGRAPHY

American Cancer Society 1996 Advisory Committee on Diet, Nutrition, and Cancer Prevention (1996). American Cancer Society guidelines on diet, nutrition, and cancer prevention: Reducing the risk of cancer with healthy food choices and physical activity. *CA Cancer J. Clin.* **46,** 325–341.

Ames, B. N., Gold, L. S., and Willett, W. C. (1995). The causes and prevention of cancer. *Proc. Natl. Acad. Sci. USA* **92,** 5258–5265.

Schottenfeld, D., and Fraumeni, J. F., Jr. (eds.) (1996). "Cancer Epidemiology and Prevention," 2nd Ed. Oxford University Press, New York.

Steinmetz, K. A., and Potter, J. D. (1991). Vegetables, fruit, and cancer. I. Epidemiology. *Cancer Causes Contr.* **2,** 325–357.

Steinmetz, K. A., and Potter, J. D. (1991). Vegetables, fruit, and cancer. II. Mechanisms. *Cancer Causes Contr.* **2,** 427–442.

Trichopoulos, D., and Willett, W. C. (eds.) (1996). *Cancer Causes Contr.* **7,** 3–180.

Digestion and Absorption of Human Milk Lipids

OLLE HERNELL
LARS BLÄCKBERG
University of Umeå, Sweden

GLOSSARY

Bile salt Physiological detergent supplied with the bile into the intestinal lumen, for example, cholate ($3\alpha,7\alpha,12\alpha$-trihydroxycholanoate)

Emulsion Droplets of oil (mainly triacylglycerol) in water stabilized with a surface coat of amphiphilic substances (detergents)

Lipase Enzyme that hydrolyzes ester bonds of emulsified acylglycerols (triacylglycerol hydrolases, E.C. 3.1.1.3)

Micelle Aggregate of bile salt spontaneously formed above a certain concentration (critical micellar concentration, CMC) that solubilizes products of lipolysis (mainly monoacylglycerol and free fatty acid) to form mixed micelle

Milk fat Globules of mainly triacylglycerol (>98%) covered with a surface coat (membrane) of mainly phospholipid and protein

Vesicle Bilayered structure of lipolysis products saturated with bile salt that coexist with mixed micelles in the aqueous portion of postprandial upper small intestinal contents

DURING THE LAST DECADE THERE HAS BEEN AN increasing awareness that not only is the nutritional composition of human milk ideal to meet the requirements of the newborn infant, but the bioavailability of many nutrients is also better from milk than from formulae with a composition made as similar to milk as possible. For instance, it is well known that the coefficient of fat absorption is higher from raw human milk than from such formulae. To achieve the same high coefficient of absorption, manufacturers have used a high percentage of either medium-chain fatty acids or long-chain polyunsaturated fatty acids, for example, linoleic acid. Recently, there have been calls for caution regarding both types of formulae because of observed, or expected, side effects. [*See* Fatty Acids; Lipids.]

For scientific as well as for practical reasons it is therefore important to fully understand why the coefficient of fat absorption is so unexpectedly high in breast-fed infants. Particularly since, when the accepted model of fat digestion and absorption in healthy adults was applied to the breast-fed infant, it became evident that it was an oversimplification. In this article we review our current understanding of the mechanisms of lipid digestion and absorption in breast-fed infants, and why these mechanisms in some respects may be different, and therefore less efficient, in formula-fed infants. In doing so we will also discuss why regarding human milk merely as a source of

ENCYCLOPEDIA OF HUMAN BIOLOGY, Second Edition, VOLUME 3.

nutrients and energy, albeit of optimal composition to the newborn, is no longer valid. Rather, breast-feeding could be considered as an extension of the umbilical cord through which immunological, biochemical, and endocrine support and information are also communicated.

I. INTRODUCTION

An ever-increasing demand for energy is one of the major characteristics of the dynamic and vulnerable perinatal period. The energy requirements of the fetus are mainly for growth and tissue maintenance; during the last trimester the fetus more than triples its weight. Through placental transport of nutrients the mother provides the fetus with the required energy substrates. Although carbohydrate (i.e., glucose) is the dominating fetal energy substrate, it seems clear that lipids [i.e., free fatty acids (FFA) and ketone bodies] may be used as alternative fuels. Via placental transfer and/or synthesis the fetus is also supplied with essential lipid nutrients, that is, fat-soluble vitamins and essential fatty acids, including long-chain polyenoic derivatives of both the *n*-3 and *n*-6 series required for membrane synthesis, particularly of the rapidly developing brain and retina.

Yet birth represents a radical transition from more than a nutritional point of view. Because of its relative concentration in human milk, and in milk formulae patterned after human milk, lipid now becomes the major energy substrate. In fact, newborn infants ingest three- to fivefold more lipid per kilogram body weight per day than do adults consuming a typical Western diet. Therefore, the human neonate immediately becomes dependent on effective mechanisms for digestion and absorption of dietary lipids, and on dietary sources of essential nutrients including lipids.

II. COMPOSITION OF HUMAN MILK FAT

Energy-rich triacylglycerol (triglyceride) accounts for at least 98% of the lipids of human milk. The balance is made up by phospholipids, cholesterol, fat-soluble vitamins, and other minor constituents. The triacylglycerols are synthesized in the endoplasmic reticulum and cytosol of the epithelial cells of lactating mammary gland. They form the cores of the milk fat globules, which, during synthesis and intracellular storage,

are stabilized by a surrounding monolayer of phospholipid. When extruded from the epithelial cell, the globules become enveloped by the apical part of the plasma membrane. Hence, this so-called milk fat globule membrane is a trilayer composed mainly of phospholipid, other amphiphilic lipids, and membrane proteins. Some of the proteins are high-molecular-weight glycoproteins that form a remarkable array of filaments, approximately 5×10^2 nm in length, sticking out from the surface. These filaments, absent from bovine milk globules, are extracted by heat treatment (pasteurization) of the milk. It has been suggested that the filaments may enhance the digestibility of the globules, and therefore could be one reason for the observed reduced fat absorption from pasteurized milk.

Although the size of the globules varies, they typically have a diameter of $1-4 \times 10^3$ nm. Thus, composed milk fat globules represent a very stable physiological lipid, or oil-in-water emulsion. However, the stabilizing surface coat, independent of heat treatment, also functions as a barrier with regard to accessibility of the milk triacylglycerol for the necessary gastrointestinal digestion. [*See* Nutrition, Fats and Oils.]

III. THE LIPASES OF GASTROINTESTINAL FAT DIGESTION

To be utilized efficiently the nonabsorbable dietary triacylglycerol must be hydrolyzed to absorbable products. This digestion is accomplished by certain triacylglycerol hydrolases (E.C. 3.1.1.3), or lipases, operating in the lumen of the upper gastrointestinal tract. Figure 1 represents a schematic presentation of the main lipases involved in the digestion of human milk triacylglycerol, as well as the tissues of secretion for the respective lipase. [*See* Digestive System, Physiology and Biochemistry.]

A. Colipase-Dependent Pancreatic Lipase

Pancreata of all studied mammalian species secretes a potent lipase, an approximately 45-kDa single-chain glycoprotein. The amino acid sequence and the complete gene structure are known for several species. This pancreatic lipase is designed to act preferentially at lipid–water interfaces of the chyme entering the

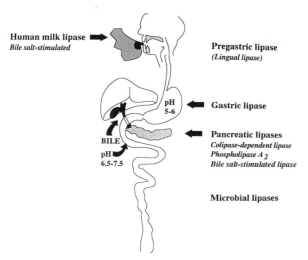

FIGURE I Schematic presentation of the main sources of lipases relevant to milk triacylglycerol digestion.

a lipid substrate. The elucidation of the three-dimensional structure of the human enzyme, by X-ray crystallography, revealed that it contains two distinct domains; one larger amino-terminal domain including the active site structure and another carboxy-terminal domain to which colipase binds. The active-site groove is, however, covered by a lid structure. Binding of colipase in the presence of micellar phospholipid alters the conformation of the enzyme, leading to an opening of the lid that gives easy access to the active site. Moreover, the open lid structure together with bound colipase forms a hydrophobic surface to which lipids bind. This is the mechanism behind the so-called interfacial activation typical for colipase-dependent lipase. Colipase may also affect the catalytic efficiency of the bound lipase. Hence, this pancreatic lipase is now often designated colipase-dependent (Table I).

B. Gastric Lipase

Gastric contents of several species, including humans, contain lipase activity (Table I). Until recently it has been uncertain from which tissue(s) the responsible lipase(s) are secreted, the explanation being that lipase is secreted from different tissues in different species. From carefully conducted activity determinations of different sections of the upper gastrointestinal tract, it is now believed that in humans the enzyme is almost exclusively secreted by the gastric mucosa. By immunocytochemistry it was further shown that lipase originates in the chief cells of the gastric fundus. In contrast, in the rat it is secreted from the von Ebner glands localized beneath the circumvallate papillae of the tongue, and in the calf it is secreted from pharyngeal tissue. Because of these differences in tissue of origin, the lipase(s) have been given several names, for example, lingual lipase (rat), pregastric lipase (calf), or gastric lipase (humans). The properties of the purified enzymes are very similar, implying that regardless of origin they have a common physiological function.

upper small intestine from the stomach. Although it has some activity on water-soluble ester substrates (esterase activity, E.C. 3.1.1.1), its activity is enhanced manyfold when acting at the surface of aggregated substrates such as emulsion particles. This so-called "interfacial activation" is a property unique to this lipase when compared to the two other gastrointestinal lipases to be discussed. For optimal function, pancreatic lipase depends on a protein cofactor colipase. This 9-kDa protein is secreted from the pancreas as a procolipase in approximately equimolar amounts. Trypsin cleaves off an amino-terminal tetrapeptide, enterostatin, from procolipase to yield colipase. Evidence suggests that enterostatin induces satiety, particularly when fat-rich meals are fed.

One important function of colipase is to anchor the lipase to lipid substrates covered with amphiphilic components, for example, bile salts, phospholipids, and proteins. As mentioned earlier, the milk fat globule is a physiologically relevant representative of such

TABLE I

Lipases in the Gastrointestinal Tract of Breast-Fed Infants

Lipase	Origin	Approximate molecular mass (kDa)	Cofactor	Site of action
Colipase-dependent lipase	Pancreas	45	Colipase	Duodenal contents
Gastric lipase	Gastric mucosa	42	None	Gastric contents, duodenal contents?
BSSL	Milk/pancreas	107/100	Primary bile salt	Duodenal contents

From the amino acid sequences of the rat and human enzymes, obtained from cloning of their respective cDNA, it became obvious that they represent not only a functional entity but are also very similar on the molecular level; the sequence homology between the two is more than 75%. Here we will use the name gastric lipase.

The properties of gastric lipase make it ideally suited to function in gastric contents. It has an acidic pH optimum, is very stable in the acidic environment of the stomach, and is also resistant to digestion by pepsin. In contrast, when treated *in vitro* with pancreatic proteases, that is, trypsin and chymotrypsin, the enzyme activity is rapidly lost, particularly in the presence of bile salts. If this is also true *in vivo*, gastric lipase will not remain active in intestinal contents and hence would not be expected to make a major contribution to intraduodenal lipolysis in healthy human adults.

C. Bile Salt-Stimulated Lipase

During the last decade the bile salt-stimulated lipase (BSSL) has become one of the most thoroughly studied of all human milk enzymes. Its concentration in milk (0.1 mg/ml) is, compared to other milk enzymes, remarkably little influenced by length of gestation, duration of lactation, or diurnal rhythm. BSSL represents a striking example of how the mother, via her milk, provides the newborn with a biologically active protein important for its growth and development. This interplay is further illustrated by the enzyme's activation mechanism. The lipase is inactive in the milk but becomes activated when, after passage through the stomach into the intestinal lumen, the partially digested milk mixes with bile salts. Only free or conjugated primary bile salts, that is, cholate and chenodeoxycholate, are effective activators. The exact mechanism of activation is as yet unknown. The current model states that bile salts interact with BSSL at two sites: one lipid-binding promoting site and one activation site. In duodenal contents, bile salts not only activate but also protect the enzyme from inactivation by proteases.

BSSL is a constituent of milk from only a limited number of species, for example, higher primates, African green monkey, cat, dog, ferret, and mouse, and is lacking in many others, for example, cow, horse, pig, goat, rat, and rhesus monkey. There is no explanation yet as to why this lipase is secreted into milk in some species but not in others. A correlation to the milk content of long-chain triacylglycerol was recently suggested; seal milk, which has an extremely high fat content, also has one of the highest BSSL activities found in any species. BSSL is indeed synthesized by lactating human mammary gland, and the primary structure as well as the gene structure of human BSSL is known. The same gene product is secreted from human exocrine pancreas. The pancreatic enzyme has been given many different names, for example, carboxyl ester lipase, carboxyl ester hydrolase, bile salt-dependent lipase, cholesterol esterase, and lysophospholipase. We will use the name BSSL regardless of source.

Human BSSL consists of 722 amino acid residues and shows very little homology to gastric lipase or representatives of the super lipase family, for example, colipase-dependent lipase and lipoprotein lipase. However, there is a marked homology in the amino-terminal part to typical esterases, for example, acetylcholine esterase. This agrees well with the unusually broad substrate specificity of BSSL. BSSL, in contrast to colipase-dependent lipase, is a nonspecific lipase inasmuch as it hydrolyzes not only tri-, di-, and monoacylglycerols (see the following) but also cholesteryl esters and esters of fat-soluble vitamins. Recent observations suggest that BSSL may be of particular importance for utilization of dietary essential fatty acids, in particular the long-chain polyunsaturated derivatives of linoleic and α-linolenic acids. Furthermore, its activity is not restricted to emulsified substrates. Micellar and soluble substrates are also effectively hydrolyzed. It has been suggested that BSSL is not only an enzyme but also facilitates the transport of cholesterol from the intestinal lumen to the enterocytes. However, whether or not BSSL should be regarded as a transport protein facilitating product absorption is still an open question.

A characteristic structural feature of BSSL is a region in the carboxy-terminal end consisting of 16 near-identical, proline-rich, highly O-glycosylated repeats of 11 amino acid residues. The functional significance of these repeats is as yet unknown, however, they are not important for catalysis or enzyme stability. There is a slight difference in apparent molecular mass between BSSL found in milk and that secreted from the pancreas; 107 kDa for milk BSSL and 100 kDa for pancreatic BSSL. This most likely reflects differences in glycosylation.

In contrast to BSSL in milk, which is absent in many species, the pancreatic enzyme has been found in all species so far tested. In fact, pancreatic BSSL is phylogenetically older than the colipase-dependent pancreatic lipase. Probably the opposite could not have been possible. The apparent molecular mass of BSSL from various mammalian species ranges from around 70-kDa (rat) to above 100 kDa (humans). These size differences are due to differences in the number of repeats, which range from 3 to 16 in the species so far sequenced. An extreme example is BSSL from salmon, which completely lacks repeats.

Measurements of intraduodenal lipase activities in preterm infants one hour after consecutive feeds of raw or pasteurized human milk showed that about two-thirds of the BSSL activity found was from the milk, indicating that milk is the major source of enzyme in breast-fed infants. Interestingly, in the suckling rat, which does not receive any BSSL via milk, BSSL is the main lipolytic enzyme secreted from the pancreas and it is not until weaning that the colipase-dependent lipase becomes dominating. This observation further substantiates a physiological relevance of BSSL during the neonatal period.

D. Relative Physiological Importance of the Lipases

The exact relative quantitative contributions of each of the three lipases to net fat digestion are difficult to assess. Some conclusions may be drawn from observations of triacylglycerol utilization under pathological conditions represented by low levels of at least one of the lipases. In the healthy adult, the capacity to digest and absorb dietary triacylglycerol is high. For instance, it has been calculated that, theoretically, the amount of colipase-dependent lipase secreted into intestinal lumen is in 1000-fold excess of what is required. However, the use of more physiologically relevant conditions gave considerably lower figures. However, the importance of other lipases is illustrated by the fact that patients suffering from isolated deficiency of colipase-dependent lipase and/or colipase may still absorb 50% or more of dietary triacylglycerol. Likewise, cystic fibrosis patients suffering from such deficiency as part of a complete pancreatic insufficiency, but known to have normal or close to normal activities of gastric lipase, often digest and absorb at least half of dietary triacylglycerol, also without supplementation with pancreatic extracts. Isolated gastric lipase or pancreatic BSSL deficiencies have not yet been described. From indirect evidence, however, it has been argued that when gastric lipolysis is circumvented, fat absorption may be reduced by perhaps 30%. This figure is close to those in a recent report on duodenal intubation studies in healthy adults. It was calculated that 10% of triacylglycerols are hydrolyzed to diacylglycerol in the stomach. Gastric lipase continues hydrolysis in intestinal contents and accounts for 25% of total triacylglycerol hydrolysis to absorbable monoacylglycerol and free fatty acids.

The physiological significance of milk BSSL is of course restricted to milk lipid digestion. Its importance was first suggested merely from its characteristics, and from the frequent observation that fat malabsorption is more common in newborn infants fed cow's milk-based formulae (devoid of BSSL activity) than in breast-fed infants. More direct evidence came from fat balance studies comparing the coefficient of absorption from raw and pasteurized human milk. Pasteurization inactivates the milk enzyme and reduces fat absorption by as much as 30% in preterm infants. The recent finding of a BSSL activity in cat milk made it possible to directly study the importance of BSSL in this species. The effect of BSSL on the weight gain of kittens was studied by use of three groups of kittens fed either raw cat milk, a kitten formula supplemented with purified human BSSL, or the formula alone. Based on recordings of the daily weight gain during a 5-day period, it was concluded that kittens fed formula supplemented with BSSL had a weight gain comparable to that of milk-fed kittens but twice that of kittens fed the unsupplemented formula, strongly implying that BSSL was the causative factor.

IV. THE TWO-PHASE MODEL OF FAT DIGESTION

Based on the now classic studies of Hofmann and B. Borgström, lipid digestion in general, and triacylglycerol digestion in particular, was until recently regarded as a two-phase process occurring almost exclusively in the upper part of the small intestinal contents. In short, the triacylglycerol enters the duodenum as an emulsion of oil-in-water, that is, as large water-insoluble lipid droplets, with a hydrodynamic radius (R_h) of 125–250 nm, dispersed in the aqueous phase of intestinal contents. At the oil/water interphase, colipase-dependent pancreatic lipase hydrolyzes the triacylglycerols. Since this lipase cannot release the fatty acid esterified to the sn-2-position of the glycerol molecule, each triacylglycerol gives rise to one sn-2-monoacylglycerol and two FFA. The bile lipids (mainly bile salts and phospholipids) distribute between the aqueous phase and the oil/water interphase, where they displace amphiphilic substances, including lipolysis products, from the interphase into the water. In the aqueous phase, bile salts above their critical micellar concentration (CMC) spontaneously form aggregates called micelles, which dramatically increase the solubility of the products of lipolysis. Mixed micelles of bile salts and lipolysis products are spherical particles ($R_h \leq 4$ nm) from which absorption occurs.

According to the described two-phase model of fat digestion and absorption, low intraluminal concentrations of colipase-dependent lipase and/or of bile salts should be obvious reasons for fat malabsorption. Considering their high lipid intake, one would therefore expect newborn infants to have high intraluminal concentrations of lipase and bile salts. However, because the pancreatic and liver functions at birth are not fully developed, this is not the situation. Intraluminal fasting levels of digestive enzymes secreted from the pancreas are considerably lower than in adults during the first 2–12 months, and adult levels of response to pancreozymin may not be seen until the end of the first year of life. Consistent with this, we found levels of colipase-dependent lipase in duodenal contents after a meal that were on average only 4% in preterm infants when compared with adults. Similarly, others have reported a 5- to 10-fold lower postprandial intraluminal bile salt concentration in preterm infants than in adults. In fact, such low concentrations, that is, 1–2 mM, are close to or even below that required for micelle formation. Hence they would not, according to theory, be compatible with efficient solubilization, transport, and ultimately absorption of lipolytic products. Nonetheless, digestion and absorption of triacylglycerol are often remarkably efficient even in preterm breast-fed infants.

In contrast to colipase-dependent lipase, the capacity for gastric lipase secretion is fully developed in preterm newborns. Since the major part of BSSL is supplied with the milk, intraduodenal levels are much less dependent on the developmental stage of the infant.

V. ACCESSIBILITY OF HUMAN MILK TRIACYLGLYCEROL FOR HYDROLYSIS

To fully understand the exact function of the three lipases for milk triacylglycerol utilization it was necessary to make model experiments by use of purified

enzymes and the natural dietary lipid substrate for the breast-fed newborn, that is, human milk fat globule triacylglycerol. Table II shows the lipolysis rate obtained with each of the three lipases with human milk fat as substrate as compared to that obtained with a synthetic triacylglycerol emulsion under conditions chosen to be suitable for the respective lipase. Even in the presence of colipase, pancreatic colipase-dependent lipase by itself was unable to hydrolyze human milk triacylglycerol. The reason for this is that the lipase does not bind to the milk fat globule membrane, and thus it does not get access to the substrate. Obviously BSSL is devoid of activity against milk fat in milk as secreted since no bile salts are present. However, even when bile salts were added in concentrations sufficient for activation with the artificial substrate, no lipolysis of milk fat occurred (Table II). Consequently, neither colipase-dependent lipase nor BSSL activities *in vitro* are, even in the presence of their respective cofactors, by themselves or together, sufficient to account for the efficient hydrolysis of human milk fat found *in vivo*. In sharp contrast, gastric lipase hydrolyzed human milk triacylglycerol at the same rate as the synthetic substrate. Thus, gastric lipase is unique among the gastrointestinal lipases in that it can by itself attack human milk fat globule triacylglycerol. Model experiments have shown that the surface pressure, which depends on the nature and packing of the constituents of the emulsion surface coat (such as the milk fat globule membrane), affects gastric lipase and colipase-dependent lipase activities differently. The conclusion to be drawn is that it is an oversimplification to consider the relative

TABLE II

Accessibility of Human Milk Triacylglycerol for Gastrointestinal Lipases[a]

Lipase	Substrate	
	Triolein/gum arabic	Milk fat
Colipase-dependent lipase[b]	500	<0.5
BSSL[c]	100	<0.5
Gastric lipase[c]	70	80

[a]Conditions were chosen to obtain optimal activity for the respective lipase.
[b]Human pancreatic juice was used as source of enzyme and cofactor; values are expressed as μmol/min $\times$ ml juice.
[c]Values are expressed as μmol/min $\times$ mg enzyme protein.

contribution of each lipase to gastrointestinal lipolysis in merely quantitative terms.

VI. SEQUENTIAL AND CONCERTED ACTION OF THE LIPASES

To reveal the physiological function of the different lipases from a qualitative viewpoint it was important to study their concerted effects. If human milk fat globules were pretreated with gastric lipase, resulting in hydrolysis of only 5–10% of the triacylglycerol ester bonds, this dramatically altered the accessibility of the remaining triacylglycerol for subsequent hydrolysis. Unlike the intact milk fat globules, these partially digested globules were readily cleaved by colipase-dependent lipase, or BSSL. This further emphasizes a unique function of gastric lipase. For hydrolysis by colipase-dependent lipase this triggering effect is, at least partly, caused by the released FFA. FFA enables the lipase/colipase complex to bind efficiently to the substrate surface. Thus, *in vitro* this effect of gastric lipase can be replaced by addition of exogenous FFA, particularly long-chain unsaturated FFA. The synergistic action of gastric lipase and colipase-dependent lipase results in hydrolysis of about two-thirds of total triacylglycerol ester bonds giving rise to *sn*-2-monoacylglycerols and FFA. As discussed earlier for the two-phase model, these are considered the relevant steps of triacylglycerol digestion in adults.

In breast-fed newborns, BSSL also contributes to net triacylglycerol digestion. The exact nature of the triggering effect of gastric lipase on BSSL-catalyzed hydrolysis is unclear; the essential causative factor is not the released fatty acids. Possibly the partial acylglycerols, also generated by gastric lipase, may be of importance. The route of lipid digestion for encountering gastric lipase and BSSL, but circumventing colipase-dependent lipase, is unique to the breast-fed infant. It may be of particular importance in infants born prematurely, that is, with particularly low levels of colipase-dependent lipase and/or bile salts, or in breast-fed infants with pancreatic insufficiency for pathological reasons.

In the physiological situation, milk lipid digestion results from the concerted action of all three lipases. Based on *in vitro* experiments with purified human enzymes and cofactors and human milk as substrate, a model for triacylglycerol digestion in the breast-fed newborn can be proposed. The results of such experiments are shown in Fig. 2. Each lipase contributes its specific properties. Under conditions resembling the environment of gastric contents, gastric lipase initiates lipolysis, which is, however, limited. The change of conditions to resemble the environment of upper small intestinal contents, that is, raised pH and the addition of colipase-dependent lipase, colipase, and bile salts (Fig. 2A, point a), results in hydrolysis of more than 60% of the total ester bonds, giving rise to FFA and monoacylglycerol as end products. BSSL, being a nonspecific lipase, lacks positional specificity. Therefore, it can also hydrolyze *sn*-2-monoacylglycerol. Hence, when BSSL was added to the incubation after the milk fat had been extensively hydrolyzed by gastric lipase and colipase-dependent lipase (Fig. 2A, point b), hydrolysis recommenced (Fig. 2A, point c) owing to hydrolysis of the monoacylglycerol by BSSL. Although the released FFA was not removed by absorption as it would have been *in vivo*, the concerted action of all three lipases resulted in hydrolysis of more than 90% of the total triacylglycerol ester bonds. Hydrolysis of monoacylglycerol was confirmed by product compositional analyses after the various steps of hydrolysis (Fig. 2B). Together with FFA, free glycerol became the main product formed. In fact, more than 80 mol% of the glycerol initially bound as triacylglycerol was released as free glycerol.

However, the sequential addition of the lipases is not a true representation of the *in vivo* situation. Obviously, colipase-dependent lipase and BSSL will act simultaneously. When this was mimicked *in vitro* (Fig. 2C), BSSL affected not only the product pattern, but also the initial rate of lipolysis. The rate was doubled as compared to that obtained with colipase-dependent lipase alone (cf. Figs. 2A and 2C), illustrating its contribution to tri- and diacylglycerol digestion. *In vivo*, monoacylglycerol and FFA are expected to be removed from the site of lipolysis by micellar solubilization and transport. However, since the capacity of micellar transport depends on the intraluminal bile salt concentration, this will be a rapid and efficient process only when this concentration is high. Most monoacylglycerol will not be available for hydrolysis, and BSSL will contribute to lipolysis chiefly by supporting colipase-dependent lipase in hydrolysis of tri- and diacylglycerol. Yet if the bile salt concentration is low, which may be the normal state in many newborn preterm infants, the monoacylglycerol will remain in the lumen for a longer time and BSSL will complete lipolysis by also hydrolyzing the monoacylglycerol.

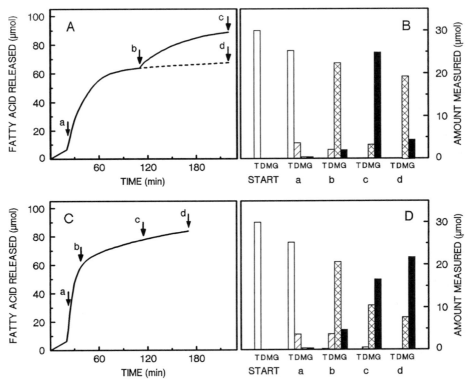

FIGURE 2 Hydrolysis of heat-treated human milk fat *in vitro*. Conditions and concentrations of purified lipases and cofactors were chosen to resemble the physiological situation. Panels A and C show the continuous recording of released fatty acids. At 0 min, gastric lipase was added (A and C), and after 20 min (a) the pH was raised from 6.0 to 7.5 and bile salts, colipase-dependent lipase, and colipase were added. BSSL was added either after 115 min (A, solid line) or after 20 min (C). At the start and at the times indicated (a–d) samples were removed and the product pattern was determined. Panels B and D show the decrease in relative concentration of triacylglycerol (T) and the increase in diacylglycerol (D), monoacylglycerol (M), and glycerol (G) in the incubations of panels A and C, respectively. [For further details see S. Bernbäck, L. Bläckberg, and O. Hernell (1990). *J. Clin. Invest.* **85**, 221. Reproduced by permission.]

VII. THE PHYSICAL-CHEMICAL STATE OF DIETARY LIPIDS IN POSTPRANDIAL INTESTINAL CONTENTS

Because the aqueous solubilities of partially ionized long-chain FFA and monoacylglycerol are extremely low but increase considerably (1,000,000-fold) in the presence of bile salts above their CMCs, mixed micelles have, according to the two-phase model, been considered the sole vehicles for solubilization and transport of lipolysis products from the emulsion surfaces to the enterocytes for subsequent absorption. Furthermore, lipid uptake is believed to be monomeric and therefore mixed micelles provide a favorable concentration gradient across unstirred layers and muco-

sal surfaces. In fact, solubilization of lipolysis products within bile salt micelles has been considered a rate-limiting step in absorption of dietary lipids. However, in the complete absence of intraluminal bile salts, for example, in patients with bile fistulas, 50–75% of dietary triacylglycerol is absorbed. Moreover, it has been reported that in breast-fed infants in contrast to formula-fed infants, the coefficient of fat absorption does not correlate to the intraluminal bile salt concentration. These observations suggest that non-micellar mechanisms may be of importance for dispersion of lipolysis products in aqueous duodenal contents, and hence for product absorption.

Recently, we gave direct evidence for the coexistence of small (R_h = 20–60 nm) unilamellar vesicles and mixed micelles in the aqueous-rich portion of postprandial duodenal contents in healthy adults. It seems that during lipolysis products will first locate mainly at the emulsion surface, presumably as multilamellar liquid-crystalline bilayers (liposomes). As the core of the emulsion droplet shrinks during the progress of lipolysis, parts of the excess surface coat will pinch off as large liquid-crystalline structures (Fig. 3) due to forced phase separation resulting from increased surface pressure. Provided sufficient bile salt is present, bile salt catalyzes formation of small unilamellar vesicles from these multilamellar liposomes and, via the continued presence of bile, a two-phase system of vesicles and mixed micelles will form. Phase separations showed that at equilibrium the mixed micelles are composed of bile salt micelles saturated with lipolysis products, whereas the vesicles are composed of lipolysis products saturated with bile salts. The relative amount of lipolysis products solubilized in

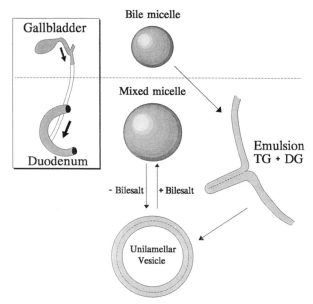

FIGURE 3 Proposed model of physical-chemical states of lipids in postprandial upper small intestinal contents. During progress of lipolysis the core of emulsion particles shrinks while the products formed locate mainly at the interphase. Increased surface pressure forces a phase separate and excess material buds off, presumably as multilamellar liquid-crystalline structures. Provided sufficient bile salt is present, bile salt will catalyze formation of small unilamellar vesicles from these multilamellar liposomes, and via the continued presence of bile salts a physiological two-phase system of vesicles and mixed micelles will form. At low bile salt/lipid ratios, vesicles will constitute the major product phase from which absorption occurs. However, as long as the micellar phase is not saturated with lipolysis products there will be a spontaneous dissolution of vesicles into mixed micelles.

each phase depends on the bile salt concentration relative to the total lipid (product) concentration (Fig. 3). At low bile salt/lipid ratios the vesicle phase will be favored. We also found evidence for spontaneous dissolution of vesicles into micelles, as long as the micellar phase was unsaturated with lipolysis products.

Saturation of micelles with lipolysis products provides the ideal thermodynamic environment for maximum rates of intestinal absorption. The driving force down a concentration gradient results in lipid product absorption, in healthy adults probably preferentially from micelles, owing to their larger number, smaller sizes ($R_h \leq 4$ nm), and thereby their more rapid diffusive access to the mucosal surface. However, absorption could also take place from unilamellar and perhaps even multilamellar vesicles. This would explain the relatively unimpaired fat absorption seen at low or absent intraluminal bile salt concentrations, for example, in patients with bile fistula and in infants during the neonatal period. It seems that partially ionized fatty acids have a higher solubility in unilamellar vesicles than do monoacylglycerols. If so, this could be the major explanation as to why a complete hydrolysis of triacylglycerol to water-soluble glycerol and free fatty acids catalyzed by the milk lipase promotes efficient product absorption in breast-fed infants, and why this absorption is less dependent on the intraluminal bile salt concentration. [*See* Bile Acids.]

VIII. CONCLUDING REMARKS

Our present view on human milk triacylglycerol digestion involves the concerted action of three lipolytic enzymes: gastric lipase, pancreatic colipase-dependent lipase, and BSSL of human milk. Figure 4 is a schematic presentation of the sequential steps catalyzed by the respective lipase(s). Each enzyme has a unique function, but also overlapping functions with the other two. When the intraluminal bile salt concentration is comparatively high, BSSL will aid colipase-dependent lipase in generating monoacylglycerol and FFA, which, using mixed micelles as vehicles, are absorbed. In infants with low intraluminal bile salt levels, which is normal for many preterm infants, the concerted action of the three lipases will result in complete hydrolysis with FFA and free glycerol as end products. Under these conditions, substantial product absorption may occur from unilamellar vesicles.

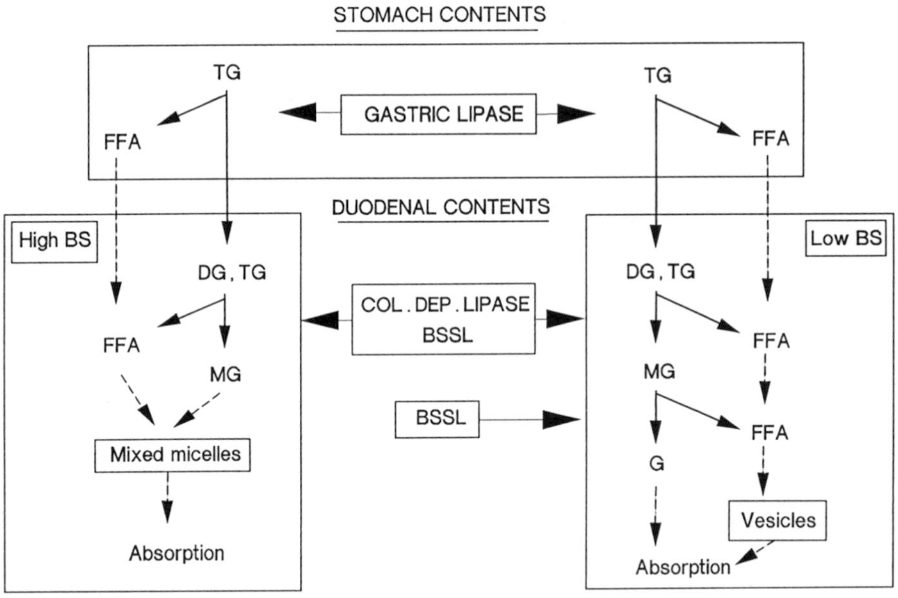

FIGURE 4 Schematic presentation of the sequential steps involved in digestion of human milk triacylglycerol. TG, triacylglycerol; DG, diacylglycerol; MG, monoacylglycerol; G, glycerol; FFA, free fatty acids; BS, intraluminal bile salt concentration. Solid lines represent enzymatic and dashed lines nonenzymatic processes, for example, transport.

BIBLIOGRAPHY

Bernbäck, S., Bläckberg, L., and Hernell, O. (1990). Complete digestion of human milk triacylglycerol *in vitro* requires gastric lipase, colipase-dependent lipase, and bile salt-stimulated lipase. *J. Clin. Invest.* **85**, 221.

Borgström, B., and Brockman, H. L. (1984). "Lipases." Elsevier, Amsterdam.

Carey, M. C., and Hernell, O. (1992). Digestion and absorption of fat. *Sem. Gastrointest. Dis.* **3**, 189.

Carriere, F., Barrowman, J. A., Verger, R., and Laugier, R. (1993). Secretion and contribution of lipolysis of gastric and pancreatic lipases during a test meal in humans. *Gastroenterology* **105**, 876.

Hamosh, M. (1989). Enzymes in human milk: Their role in nutrient digestion, gastrointestinal function, and nutrient delivery to the newborn infant. *In* "Textbook of Gastroenterology and Nutrition in Infancy" (E. Lebenthal, ed.), 2nd Ed., pp. 121–134. Raven, New York.

Hamosh, M., Iverson, S. J., Kirk, C. L., and Hamosh, P. (1994). Milk lipids and neonatal fat digestion: Relationship between fatty acid composition, endogenous and exogenous digestive enzymes and digestion of milk fat. *World Rev. Nutr. Diet.* **75**, 86.

Hansson, L., Bläckberg, L., Edlund, M., Lundberg, L., Stömqvist, M., and Hernell, O. (1993). Recombinant human milk bile salt-stimulated lipase. Catalytic activity is retained in the absence of glycosylation and the unique proline-rich repeats. *J. Biol. Chem.* **268**, 26692.

Hernell, O., Bläckberg, L., and Lindberg, T. (1989). Human milk enzymes with emphasis on the lipases. *In* "Textbook of Gastroenterology and Nutrition in Infancy" (E. Lebenthal, ed.), 2nd Ed., pp. 209–217. Raven, New York.

Hernell, O., Staggers, J. E., and Carey, M. C. (1990). Physical-chemical behavior of dietary and biliary lipids during intestinal digestion and absorption. II. Phase analysis and aggregation states of luminal lipids during duodenal fat digestion in healthy adult human beings. *Biochemistry* **209**, 2041.

Lidberg, U., Nilsson, J., Strömberg, K., Steinman, G., Sahlin, P., Enerbäck, S., and Bjursell, G. (1992). Genomic organization, sequence analysis and chromosomal localization of the human carboxyl ester lipase (CEL) gene and for a CEL-like (CELL) gene. *Genomics* **13**, 630.

Lowe, M. E. (1994). Pancreatic triglyceride lipase and colipase: Insights into fat digestion. *Gastroenterology* **107**, 1524.

van Tilbeurgh, H., Egloff, M-P., Martinez, C., Rugani, N., Verger, R., and Cambillau, C. (1993). Interfacial activation of the lipase–procolipase complex by mixed micelles revealed by X-ray crystallography. *Nature* **362**, 814.

Wang, C.-S., and Hartsuck, J. A. (1993). Bile salt-activated lipase. A multiple function lipolytic enzyme. *Biochim. Biophys. Acta* **1166**, 1.

Digestive System, Anatomy

DAVID J. CHIVERS
University of Cambridge

I. Background
II. Foregut
III. Midgut
IV. Hindgut
V. Overview

GLOSSARY

Attachment, primary Original body wall attachment

Attachment, secondary Subsequent attachment of mesentery of organ, to body wall or to mesentery of another organ

Faunivory Consumption of animal matter, vertebrate or invertebrate

Folivory Consumption of vegetative parts of plants—leaves (blades, midrib, buds, young, mature), stems, exudates (gum/saps), bark

Foregut In abdomen, stomach and first part of duodenum

Frugivory Consumption of reproductive plant parts (primary production)—fruits (pulp and/or seeds, mature/immature), flowers

Hindgut Left colon, rectum

Mesentery Two-layered membrane attaching organ to body wall

Midgut End of duodenum, jejunum, ileum, caecum, right colon, transverse colon

Peritoneum, parietal Membrane lining body cavity

Peritoneum, visceral Membrane covering organ

Retroperitoneal Behind parietal peritoneum, organ with no mesentery

ALTHOUGH DIGESTION MAY START IN THE ORAL cavity or mouth, through the secretion of salivary enzymes, which continues to act as the food passes through the pharynx (the crossroads of air and food pathways) and esophagus, it is the gastrointestinal tract in the abdomen and pelvic cavities that is considered here as the digestive system. These biochemical processes of digestion and absorption contrast with the mechanical processes of food degradation that occur in the mouth, through movements of the jaws and occlusion of the teeth in regular cycles. Thus, the anatomy of the human digestive system concerns the structure and arrangement of the stomach, small intestine, caecum, colon, and rectum in relation to function and evolution. In contrast to the gastrointestinal tract of most other mammalian orders, which are specialized either for meat-eating or for leaf-eating, those of primates, especially humans, are relatively unspecialized.

I. BACKGROUND

A. Development

When, in early development, the embryonic disc, made up of three cellular layers, folds up to assume fetal form, one of the layers overlying ectoderm forms the epidermis of the skin, whereas the underlying layer, the endoderm, forms the mucosa (inner lining) of the digestive system. Initially they are separated cranially and caudally from the degenerating yolk sac (Fig. 1), the intervening layer, the mesoderm, forms the dermis of the skin, the muscles and bones of the body wall, various other organ systems, and the smooth muscle of the digestive system. Thus, the digestive system becomes tubular, divisible into foregut (future pharynx, esophagus, stomach, and start of duodenum), midgut [still in connection with the yolk sac (rest of duodenum, jejunum, ileum, caecum, and right and transverse colon)], and hindgut (left colon, rectum, and urogenital sinus); these three parts of the

ENCYCLOPEDIA OF HUMAN BIOLOGY, Second Edition, VOLUME 3.

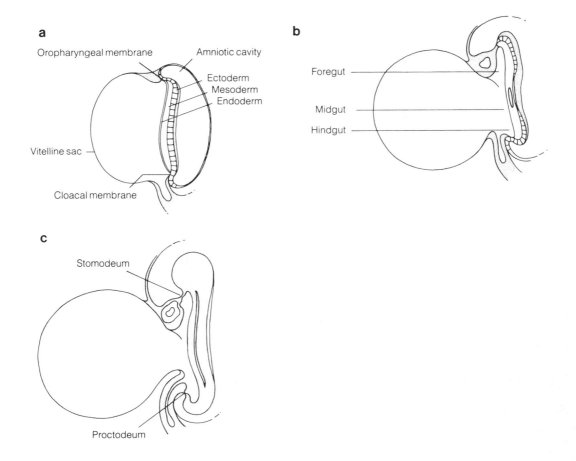

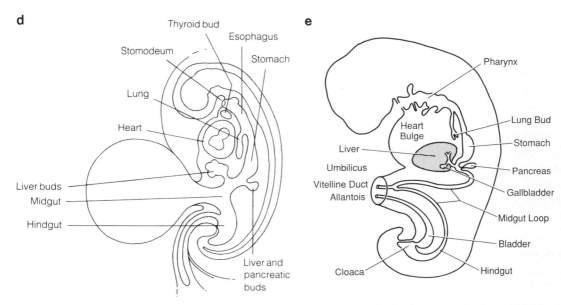

FIGURE 1   Development of gut tube in fetus: (a) beginnings late in third week, (b) elongation of embryo, (c) folding in fourth week, (d) tubular form and organ buds by fifth week, and (e) key features of fetal gut. [Source: A. J. Gaudin and K. C. Jones (1989). "Human Anatomy and Physiology," p. 442. Harcourt Brace Jovanovich, San Diego. Reproduced with permission.]

gut maintain distinctive blood supplies and innervation through development (see Color Plate 1).

The abdominal foregut expands into the stomach, which rotates to the left acquiring a greater curvature, and becomes U-shaped, in part because of anchorage of the start of the duodenum to the septum transversum [i.e., the mass of mesoderm ventral to the stomach, between the pericardial and peritoneal cavities, within which the liver develops and the cranial vestige of which becomes the main part of the diaphragm (the partition between thoracic and abdominal cavities)]. The anchorage comes about because of a duodenal diverticulum (outgrowth) that develops into the bile duct, gallbladder, and biliary system, ramifying through the liver tissue. Two other diverticula from the foregut part of the duodenum develop into the dorsal and ventral lobes of the pancreas, which subsequently fuse in part.

The midgut loop, retaining a connection with the yolk sac, herniates through the umbilicus as it elongates. It is subsequently withdrawn and laid down within the abdomen more or less simultaneously from both ends in an anticlockwise direction (viewed from ventrally). Thus, the duodenum ends up to the right and dorsally, and the transverse colon crosses from left to right to the ascending or right colon (and caecum). The coils of small intestine (jejunum and ileum) are bundled in last, and a vestige of the yolk sac (Meckel's diverticulum) may persist in a few individuals. By this time the umbilicus is closed down to allow passage only of the two umbilical arteries and the one umbilical vein (Fig. 2). The hindgut remains in all species as a simple tube, the left colon leading into the rectum (and thence anus) in the pelvic cavity.

None of these structures is actually in the peritoneal cavity; they are all supported by (and contained within) a double peritoneal membrane (or mesentery) attached initially to the dorsal midline. Subsequent differential growth and movements of the gut tube mean that these original dorsal attachments are often distorted or modified (for the fore- and midgut parts) (Fig. 3). Furthermore, the caudal migration of the liver out of the septum transversum and the cranial migration of parts of the urogenital system and umbilical arteries out of the caudal pelvic wall lead to the acquisition of "ventral" mesenteries. Humans are unique, even among primates, in the secondary reattachment of alternate parts of the gut back behind the peritoneum of the dorsal abdominal wall; this must relate to the acquisition of bipedalism and the need to move back the center of gravity as close as possible to the vertebral column (spine).

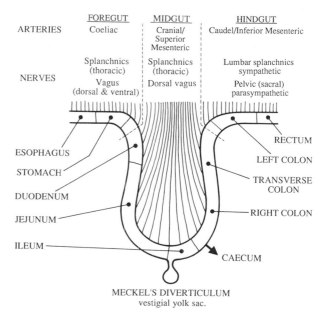

FIGURE 2 Basic design of gastrointestinal tract, with blood supply and innervation.

B. Gross Structure and Diet

Various measures can be obtained. The functional significance of surface area is for absorption of volume, for fermentation, and, of weight, for muscular activity in digestion.

The stomach and first part of the large intestine are associated with fermentation among mammals in general, whereas the small intestine is the main, but not the only, site of absorption; the amount of muscle reflects the degree of activity in each region, whether it be to pass the food through more rapidly, to mix it more thoroughly, or to reverse its passage to prolong digestion.

Most mammals are consumers either of animal matter [invertebrate or vertebrate (faunivores)] or of the vegetative parts of the primary production [foliage of some kind (folivores)]. It is anatomically and physiologically impossible to digest significant amounts of both kinds of foods, hence the inappropriateness of the term *omnivore*. In faunivores (e.g., insectivores, carnivores, cetaceans, pholidotes), whose food is rare but readily digested, the gut is simple and dominated by small intestine. Folivores are exploiting a common but indigestible food; they depend on the fermenting action of bacteria and have to expand either the stomach and/or the caecum and right colon (midgut derivatives, hence the error of hindgut fermentation common in the literature). Body size tends to be larger,

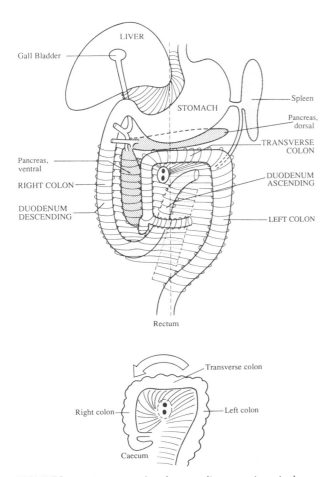

FIGURE 3 Basic topography of mammalian gastrointestinal tract, showing rotations and mesenteries, ventral view.

to house a more voluminous gut. The more specialized foregut fermenters (e.g., ruminants and colobine monkeys) contrast with the caeco-colic fermenters [e.g., rodents, equids, some strepsirhine (Prosiman) primates, and some monkeys and apes] (Table I).

Apart from bats, some rodents, and pigs, only primates have really exploited food of intermediate availability and digestibility (i.e., the reproductive parts of plants, fruit). It is this flexibility of diet, along with the capacity for year-round breeding afforded by the menstrual cycle, that has made primates so successful and led to the evolution of humans. Such frugivores, have guts of more even proportions, dominated by small intestine but with some expansion of the stomach and/or, more often, the caecum and first part of the colon. Not all amino acids occur in fruit, hence the smaller frugivores supplement their fruit diet with animal matter and the larger ones with foliage, with the appropriate modifications of gut dimensions (Table I).

From Table I the main contrasts can be seen. In carnivores, the gut is dominated by small intestine, and the stomach is large to accommodate the infrequent meals; the guts are relatively light to facilitate the chase, and digestion is aided by prolonged retention time. The guts of ungulates are bulky, especially if they subsist on leaves. Pigs have a comparable proportion of small intestine to carnivores but a smaller stomach and large caecum and colon. Ruminants have a massive stomach for fermentation, whereas the horse has a relatively larger caecum and colon.

TABLE I
Gut Dimensions in Various Mammals

| Dietary staple | Species | Body weight (kg) | % gut surface area | | | Total surface area (cm²) |
			Stomach	Small intestine	Caecum colon	
Meat	Dog	9.7	20	67	14	1,320
Meat	Cat	2.5	21	57	22	580
Fruit	Pig	58	3	65	32	18,860
Leaves	Goat	56	67	28	14	35,930
Leaves	Horse	202	2	22	76	48,950
	Monkey					
Leaves	langur	5.8	35	40	25	3,518
Fruit	macaque	3.2	15	48	37	2,129
	Ape					
Fruit	orangutan		7	49	44	13,373
Fruit	chimpanzee	34	9	49	42	7,662
Leaves	gorilla	51	10	38	52	10,508
Meat	Human	61	8	71	21	6,320

The frugivorous (ancestrally) monkeys and apes have reduced areas of small intestine and increased areas of caecum and colon (for fermentation of the leaf components); leaf-eating colobine monkeys (e.g., langurs) have much enlarged stomachs and reduction of caecum and colon; in all cases, just under half the gut area is small intestine. Humans have an even greater proportion of small intestine than carnivores, but smaller stomachs and slightly enlarged caecum and colon, reflecting their long meat-eating ancestry. Modern techniques of food preparation and processing allow humans to be omnivorous; data on long-term vegetarians are still needed to confirm the expectation that their gut proportions are more similar to those of apes. The full range of parameters of the human gastrointestinal tract is given in Table II.

C. Basic Anatomy

The gastrointestinal tract is basically a tube of smooth muscle (mesodermal), lined with a mucous membrane (endodermal) and covered by peritoneum (mesodermal). The muscular wall is divided into two layers: an inner circular layer and an outer longitudinal layer [separated into discrete bundles (taenia coli) in the large intestine]; there is an innermost oblique layer reinforcing the body of the stomach. Striated muscle extends some way down the esophagus, and the initial rapid propulsion initially is transformed into the slow, rhythmic contractions of peristalsis lower down, which continue through the rest of the tract.

The mucous membrane is thick but loose, allowing considerable dilation to occur. The lining is divided into two layers (mucosa and submucosa) by a thin sheet of smooth muscle (muscularis mucosae). The mucosa, a single layer of columnar cells, is thrown into folds or crypts, into which cells secrete, forming glands, that vary through the tract. The surface of the small intestine is projected into villi that increase the surface area significantly and contain capillaries and lacteals (lymphatic vessels). The submucosa contains blood vessels, Meissner nerve plexus (Auerbach's plexus is myenteric, between longitudinal and circular muscles), and lymphatic follicles (Fig. 4).

II. FOREGUT

A. Stomach

This is the most dilated part of the gastrointestinal tract, with a short lesser curvature running from the esophageal orifice (cardia) just to the left of the midline behind the 7th costal cartilage to the pyloric orifice, which is located just to the right of the midline at the level of the 1st/2nd lumbar vertebrae (behind the 9th costal cartilage); thus it faces right, then cranially, and finally to the left. The muscular bag is created by the great expansion of the greater curvature (primitive dorsal surface), which faces left, then caudally, and finally, near the pylorus, to the right; to it is attached the equally expanded dorsal mesentery (dorsal mesogastrium, greater omentum) to which it is attached to the dorsal body wall. The cavity created therein (the omental bursa or lesser sac) is a diverticulum of the main peritoneal cavity. This shape has come about by differential growth and rotation of the primitive gastric spindle to create a large sac mainly to the left of the midline (see Color Plate 2).

TABLE II
Human Gut Dimensions[a]

	Stomach		Small intestine		Caecum + colon			
	$\overline{x}$	%	$\overline{x}$	%	$\overline{x}$	%	$\overline{x}$	%
Surface area (cm^2)	498		4,520		66		1,234	
		8		7		1		20
Weight (g)	150		587		22		348	
		14		53		2		31
Volume (cm^3)	1,069		2,470		48		854	
		24		56		1		18
Length (cm)	—		662		8		146	
				81		1		18

[a]n, 6; body weight, 60.8 kg; body length (crown–rump), 84.6 cm; % gut total.

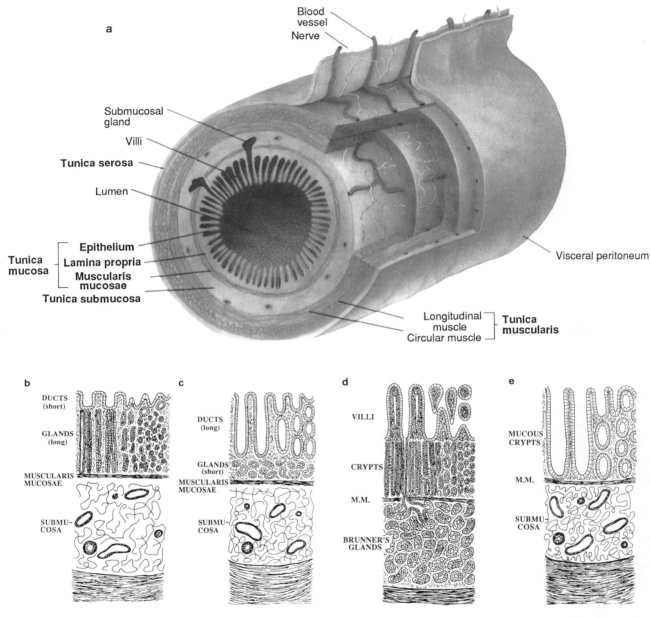

FIGURE 4 Histology of the gastrointestinal tract: (a) general; mucosa and submucosa in (b) body of stomach, (c) pylorus of stomach, (d) duodenum, and (e) colon. [From R. J. Last (1978). "Anatomy: Regional and Applied," 6th Ed. Churchill Livingstone, New York. Source: A. J. Gaudin and K. C. Jones (1989). "Human Anatomy and Physiology," p. 443. Harcourt Brace Jovanovich, San Diego. Reproduced with permission.]

The fundus is the cranial projection to the left of the esophagus, up against the dome of the diaphragm. As in the body of the stomach, the mucosa is thrown into folds or rugae containing fundic glands, secreting acid and gastric enzymes. These give way to mucus-secreting glands in the pylorus. The lesser curvature is attached to the liver by the lesser omentum (or hepato-gastric ligament); it extends a short way along the duodenum, containing the bile duct from the gallbladder in its free edge. This "ventral" mesentery of the stomach is produced by the caudal migration/expansion of the gastric spindle from over the septum transversum (mesodermal block in embryo between thorax and abdomen) and by the growth of the liver out of the septum transversum.

Hence the stomach is related to the diaphragm and liver cranially, the abdominal wall ventrally, the kidneys and vertebral column dorsally, the transverse colon caudally (and intestinal mass), the spleen to the left, and the liver and duodenum to the right.

B. Duodenum

The forgut extends as far as the entrance of bile and pancreatic ducts, near the end of the second part of the duodenum. The first part, or duodenal cap, loops 5 cm upward and then downward, in front of the right crus of the diaphragm, the right psoas muscle, and the head of the pancreas (Fig. 5). The next 7 cm curves caudally over the hilus of the right kidney, behind the right colon and its mesentery. The common bile duct and ventral pancreatic duct open together at the duodenal papilla 2 cm from the end of this part of the duodenum. The duodenum is retroperitoneal; it is not suspended in a mesentery in humans but has been reattached to the dorsal body wall (see Section V,A). Glands secreting digestive enzymes are concentrated at the start of the duodenum, extending through the muscularis mucosa into the submucosa as Brunner's glands, opening into the crypts of Lieberkuhn that extend into the mucosa throughout the intestine. The villi are largest and most numerous

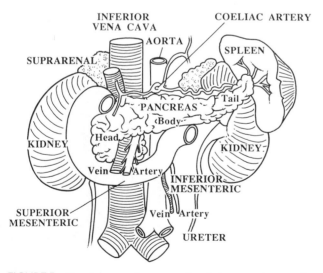

FIGURE 5 Duodenum and pancreas, in relation to major vessels, spleen, kidney, and suprarenal glands. [From R. J. Last (1978). "Anatomy: Regional and Applied," 6th Ed. Churchill Livingstone, New York.]

at the start of the small intestine; mucus-secreting cells, which are scarce, are represented by the large Paneth cells.

C. Accessory Glands

Reference has been made to the development of the liver in the septum transversum from a meshwork of fetal blood vessels (umbilical veins, increasingly just the left, and vitelline veins from the gut, developing into the hepatic portal vein, bringing nutrients and waste products of digestion) infiltrated by bile canaliculi from the bile/gallbladder diverticulum of the duodenum.

The liver is wedge-shaped, with a small left lobe and large right one (divisible into right, quadrate, and caudate processes) (see Color Plate 3). The parietal surface, covered by peritoneum, except for the left and right triangular and coronary ligaments attaching it to the diaphragm, relates closely to the dome of the diaphragm around the inferior vena cava, which is connected to the heart. The caudally facing visceral surface relates to the stomach on the left and the duodenum, right kidney, and colon on the right, receiving the hepatic portal vein, hepatic artery, and proper bile ducts at the porta at the lower limit of the attachment of the lesser omentum. The left and right lobes are separated by the ligamentum teres, the peritoneal fold from the umbilicus, and abdominal floor, containing the vestige of the left umbilical vein.

The gallbladder is situated between the quadrate and right lobe (see Color Plate 3), receiving bile from the liver lobules, which contain cells in intimate relation with bile canaliculi and with capillaries from hepatic artery and vein and hepatic portal vein. The common bile duct then runs down to the duodenal papilla in the free edge of the lesser omentum.

The pancreas develops as two buds: the ventral bud in common with the bile duct, and the dorsal bud in the dorsal mesentery of the duodenum. The latter enlarges toward the left in relation to the greater curvature of the stomach and transverse colon. The ventral bud escapes from the confines of the ventral mesentery, when the duodenum rotates in response to the rotations of the stomach; it moves under the peritoneal covering of the duodenum into the greater spaces of the dorsal mesentery, extending caudally within the duodenal loop, which is large in most mammals. This rotation brings the ventral bud into contact with the dorsal bud, and they fuse at their bases to form the body of the pancreas, with dorsal (or left) and ventral (or right) lobes around the superior mesenteric arteries and veins (the latter a main branch of the hepatic portal vein).

The pancreas is compact in humans because of the small, C-shaped duodenal loop (Fig. 5), and the body is moulded to the concavity of the duodenum, in front of the inferior vena cava and renal veins. The right lobe is reduced to an uncinate process, in front of the aorta and behind the mesenteric vessels; the left lobe, or tail, extends right across to the left kidney.

The main pancreatic duct (originally the ventral one) opens into the common bile duct; the accessory duct (originally dorsal) opens into the duodenum opposite and nearer the pylorus. These two ducts are only separate in 9% of humans; in 61% the dorsal duct loses contact with the duodenum, so that drainage is through a link into the ventral duct. In the remaining 30% of cases, the dorsal duct is reduced from its original shape and the connecting duct increased.

Mention should finally be made of the spleen, which, although it serves no digestive function, develops from the mesoderm of the stomach wall to migrate out into the greater omentum and into one leaf of the two-layered mesentery, remaining close to the greater curvature to the left of the stomach against the diaphragm at the level of ribs 9–11. The left leaf of greater omentum comes together at the hilus on the visceral side of the spleen through which pass the vessels and nerves of the organ.

D. Blood Vessels and Nerves

Three unpaired arteries leave the aorta [that to the foregut is the coeliac artery (Fig. 4), the most cranial of the three]. It has three main branches: (1) the splenic, to the spleen and greater curvature of the stomach, left gastro-epiploic, proximally; (2) the common hepatic, to the liver, pyloric end of the stomach, greater and lesser curvatures (right gastro-epiploic and right gastric, respectively), and (3) proximal duodenum (superior pancreatico-duodenal, anastomosing with the inferior artery from the midgut artery) (see Section III,F).

As with the rest of the gut, the veins coming from the wall of the gut do not closely follow the arteries; gastric, gastro-epiploic, splenic, and pancreatico-duodenal veins converge on the mesenteric veins to form the hepatic portal vein, which runs into the porta of the liver (Fig. 6).

The motor innervation of the foregut muscular wall comes from the thoracic sympathetic chain and from the vagus (10th cranial) nerve, parasympathetic

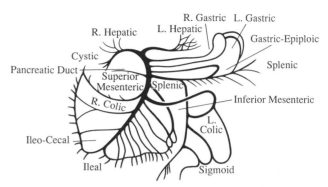

FIGURE 6 Hepatic portal vein and gastrointestinal branches. [From R. J. Last (1978). "Anatomy: Regional and Applied," 6th Ed. Churchill Livingstone, New York.]

(Fig. 7). Together they form plexuses along the walls of the arteries and carry sensory fibers from the mucosa back to the central nervous system. The sympathetic fibers, if they have not synapsed in one of the last six thoracic chain ganglia through which they pass, synapse in the coeliac ganglion having passed through the crura of the diaphragm and/or the aortic hiatus at the base of the coeliac artery, thus the coeliac plexus contains only postganglionic neurons. The parasympathetic neurons synapse in the gut wall and have short postganglionic fibers; the vagal fibers enter the abdomen through the esophageal hiatus of the diaphragm, dorsal and ventral to the esophagus (each being a mixture of left and right vagal fibers in the thorax). The ventral vagus, predictably, runs along the lesser curvature of the stomach and right side of

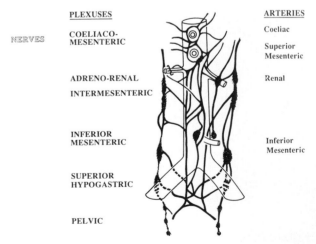

FIGURE 7 Sympathetic and parasympathetic nerves in abdomen, in relation to arterial trunks. [From R. J. Last (1978). "Anatomy: Regional and Applied," 6th Ed. Churchill Livingstone, New York.]

the start of the duodenum and up into the liver (i.e., within the lesser omentum). By contrast, the dorsal vagus, in dorsal mesentery, has access to the midgut, as well as along the coeliac artery to the greater curvature of the stomach and start of duodenum, as well as spleen and part of pancreas. [*See* Gastric Circulation.]

III. MIDGUT

A. Duodenum

The third part of the C-shaped retroperitoneal duodenum passes back toward the midline for 8 cm over the psoas muscle, gonadal vessels, and ureter to cross the inferior vena cava and aorta (Fig. 5). The final 5 cm ascend to the left of the aorta to reach the lower border (caudal) of pancreas. This first part of the midgut leads into the midgut loop that is anchored cranially at this duodeno-jejunal flexure by the ligament of Treitz, which descends from the right crus of the diaphragm in front of the aorta and behind the pancreas, to be anchored to the left psoas muscle by connective tissue.

B. Jejunum

The remainder of the small intestine retains its mesentery and is liberally coiled below and in front of the transverse colon. It is the main site of digestion and absorption, with villi getting slowly fewer and smaller, digestive juice secretion decreasing, and the frequency of mucous cells (lubricative) increasing; the wall is thick. Meckel's diverticulum, vestige of the yolk sac (with which the gut lumen was originally confluent), persists in 2% of humans, about 60 cm from the end of the small intestine (Fig. 2).

C. Ileum

The ileum is the terminal part of the small intestine, crossing caudally (inferiorly) from left to right, to the ileo-caeco-colic junction in the lower right part of the abdomen. It is thin-walled and narrower, with Peyer's patches (aggregated lymphatic follicles) sometimes visible externally.

D. Caecum

The caecum is a blind pouch of the large intestine that hangs free in the right iliac fossa, secondarily attached by a mesentery to the ileum; the terminal

appendix is a concentration of lymphatic tissue (Fig. 8).

E. Colon

The first part of the colon [right, ascending (15 cm)] is retroperitoneal crossing the iliac crest and ascending in front of the hypaxial muscles; the longitudinal muscle has separated into three bands (taenia coli), which remain distinctive until the rectum. In front of the lower pole of the right kidney, it forms the right colic flexure into the transverse colon, which hangs down in a mesentery as it crosses to the left colic flexure across the descending duodenum and pancreas to the lower pole of the left kidney (see Color Plate 4). This terminal 45 cm of the midgut loop is attached to the diaphragm below the spleen at the level of the 11th rib by a peritoneal fold (the phrenico-colic ligament). The mucosa has a honeycomb appearance under the lower-power microscope, reflecting the profusion of crypts into which mucous cells secrete and through which absorption occurs.

F. Blood Vessels and Nerves

The cranial or superior mesenteric artery is the artery of the midgut. Its first branches (inferior pancreatico-duodenal and common colic) go to the opposite, dor-

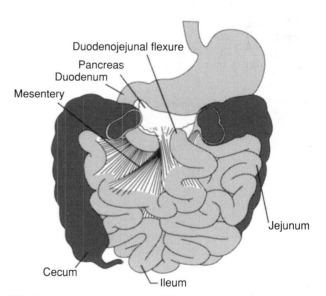

FIGURE 8 Arrangement of small intestine, in relation to stomach and colon. [Source: A. J. Gaudin and K. C. Jones (1989). "Human Anatomy and Physiology," p. 455. Harcourt Brace Jovanovich, San Diego. Reproduced with permission.]

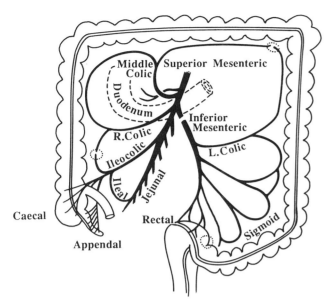

FIGURE 9 Superior and inferior mesenteric veins, in relation to colon. [From R. J. Last (1978). "Anatomy: Regional and Applied," 6th Ed. Churchill, Livingstone, New York.]

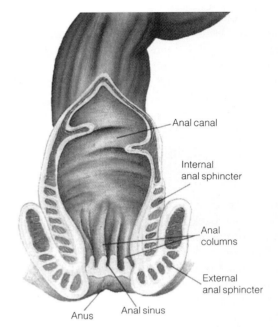

FIGURE 10 Rectum and anal canal. [Source: A. J. Gaudin and K. C. Jones (1989). "Human Anatomy and Physiology," p. 456. Harcourt Brace Jovanovich, San Diego. Reproduced with permission.]

sal ends of the midgut loop; the terminal branches (jejunal) pass to all parts of the main loop, forming anastomosing arcades on the intestinal wall. The common colic artery gives branches to the ileum, caecum, and ascending and transverse colons. Corresponding veins form the superior mesenteric vein, which heads past the pancreas into the hepatic portal vein (Fig. 9).

As mentioned earlier, parasympathetic innervation comes from the dorsal vagus, with preganglionic fibers that ramify along the artery, being joined by postganglionic sympathetic fibers from the superior mesenteric ganglion at the base of the artery (Fig. 7). [*See* Intestinal Blood Flow Regulation.]

IV. HINDGUT

A. Colon

The left or descending colon descends for 30 cm behind the peritoneum, crossing the iliac crest and iliac fossa to the pelvic brim, where it runs into the 45 cm of sigmoid colon, which loops up in a mesentery against the psoas muscle and then down in front of the sacrum into the rectum (see Color Plate 4).

B. Rectum

The rectum is retroperitoneal, curving against the sacrum to pass into the anus, with, once again, a continuous sleeve of longitudinal muscle; the upper part is usually empty but the lower part is expanded into the ampulla, containing flatus and feces (Fig. 10).

C. Blood Vessels and Nerves

The inferior mesenteric artery is the artery of the hindgut, giving left colic, sigmoid, and superior rectal branches; caudally the rectum is supplied by branches of the internal iliac artery. Veins form the inferior

FIGURE 11 Rotations of midgut loop: (a) in relation to main arteries, lateral views; (b) origin and movement of posterior body wall attachments; (c) final adherence of primitive mesenteries of gastrointestinal tract, resulting in parts of tract becoming retroperitoneal. [From R. J. Last (1978). "Anatomy: Regional and Applied," 6th Ed. Churchill Livingstone, New York].

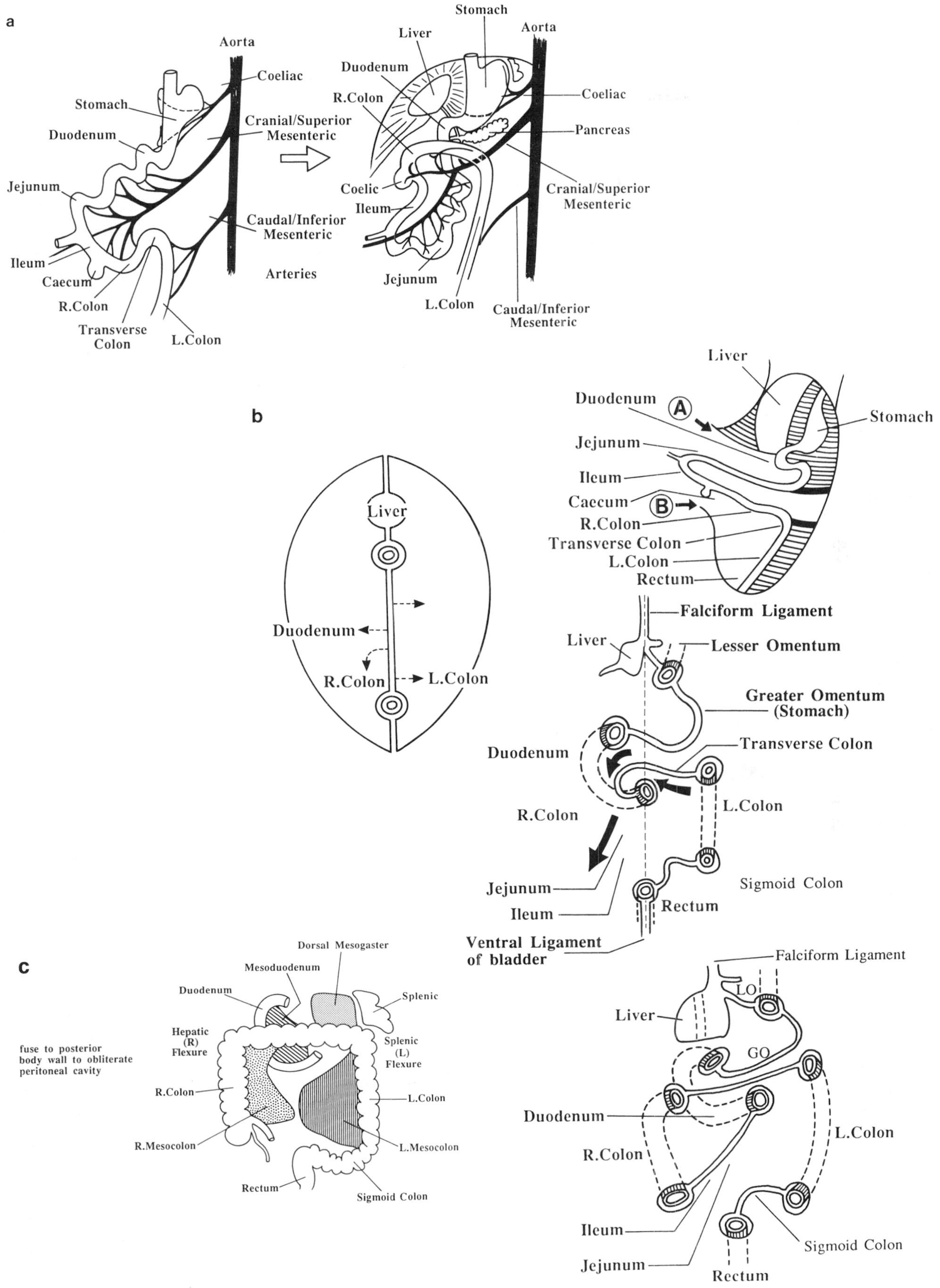
a
Aorta
Coeliac
Stomach
Duodenum
Cranial/Superior Mesenteric
Jejunum
Caudal/Inferior Mesenteric
Ileum
Caecum
R.Colon
Transverse Colon
L.Colon
Arteries
Stomach
Liver
Aorta
Duodenum
R.Colon
Coeliac
Pancreas
Coelic
Ileum
Cranial/Superior Mesenteric
Jejunum
L.Colon
Caudal/Inferior Mesenteric
b
Liver
Duodenum
R.Colon
L.Colon
Liver
Duodenum
A
Stomach
Jejunum
Ileum
Caecum
B
R.Colon
Transverse Colon
L.Colon
Rectum
Falciform Ligament
Liver
Lesser Omentum
Greater Omentum (Stomach)
Duodenum
Transverse Colon
R.Colon
L.Colon
Jejunum
Sigmoid Colon
Ileum
Rectum
Ventral Ligament of bladder
c
Dorsal Mesogaster
Mesoduodenum
Duodenum
Splenic
Hepatic (R) Flexure
Splenic (L) Flexure
fuse to posterior body wall to obliterate peritoneal cavity
R.Colon
L.Colon
R.Mesocolon
L.Mesocolon
Rectum
Sigmoid Colon
Falciform Ligament
LO
Liver
GO
Duodenum
L.Colon
R.Colon
Ileum
Sigmoid Colon
Jejunum
Rectum

mesenteric vein, which joins the superior one near its base (Fig. 9).

The sympathetic nerves come from the lumbar outflow, which pass to and have all synapsed by the inferior mesenteric artery at the base of the artery of the same name; postganglionic fibers form a plexus on the left colic artery and run down into the pelvis in the two hypogastric nerves to form a plexus along the visceral branches of the internal iliac artery. Here they are joined by parasympathetic fibers from the sacral outflow to reach the pelvic parts of the gut; this sacral outflow reaches the left colon either by running up the wall of the gut or by running up the hypogastric nerve to the left colic artery. Sensory fibers return to the spinal cord and brain by any of these routes but more often by the sympathetic routes back into the thoracic-lumbar spinal cord.

Throughout the gastrointestinal tract, parasympathetic fibers stimulate muscular contraction and glandular secretions, whereas sympathetic fibers are inhibitory to the gut muscle and stimulate the closure of sphincter and blood vessels.

V. OVERVIEW

A. Attachments

The anticlockwise rotations (viewed from ventrally) of the midgut loop (Fig. 11a) pull the original dorsal midline body wall attachment (Fig. 11b) to the right cranially (and then caudally) and to the left caudally, the stomach also having pulled its attachment over to the left. This results in the mesenteries "applying, adhering, and absorbing" (the three A's) to varying degrees to the peritoneum of the roof of the abdominal cavity (Fig. 11c), so that alternate parts are retroperitoneal (duodenum, right colon, and left colon) and alternate parts retain their mesentery (stomach, jejunum/ileum, transverse colon, and sigmoid colon), albeit displaced to the left or right of the midline in some cases (Fig. 11b). As mentioned earlier, this uniquely human phenomenon seems to be related to assisting in upright posture, by shifting the center of gravity of the abdominal viscera as close as possible to the vertebral column over the pelvis and axis of the legs.

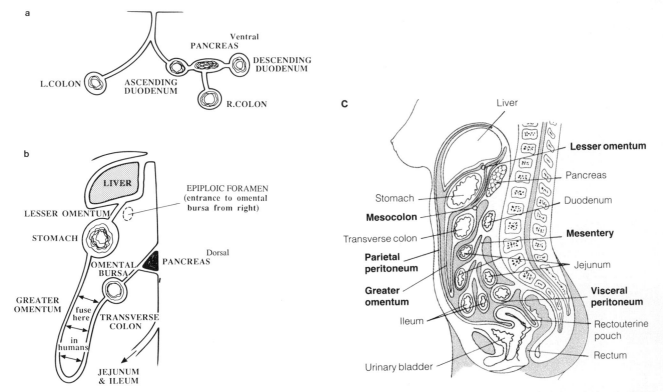

FIGURE 12   Secondary attachments of mesenteries of gastrointestinal tract: (a) transverse section in mammal such as dog, (b) longitudinal section in humans, and (c) sagittal section through human abdomen and pelvis. [Source: A. J. Gaudin and K. C. Jones (1989). "Human Anatomy and Physiology," p. 472. Harcourt Brace Jovanovich, San Diego. Reproduced with permission.]

Inevitably, the cranial or caudal movement of parts of the intestine result in the mesentery of one being secondarily attached to another; this is more obvious in other mammals, where no parts of the intestine are retroperitoneal—where the whole tract is suspended in mesentery. Hence, the jejunal mass (most ventral) may be attached to the base of the colon mesentery, which in turn is attached to the mesenteries of stomach and duodenum (most dorsal—retracted from the umbilical hernia first, the jejunal coils last) (Fig. 11a).

B. Relations

So we return to our basic layout (see Color Plate 1 and Fig. 3). In mammals the right colon can be traced back to the dorsal body wall via the mesentery of the duodenum and thence the mesentery of the left colon, and the transverse colon can be traced back via the greater omentum containing the dorsal (left) limb of the pancreas (Fig. 12a). In humans, with the duodenum and dorsal pancreas retroperitoneal, the mesentery of the transverse colon is fused to the peritoneal covering of the peritoneum and the greater omentum (Fig. 12b).

A distinctive feature in all mammals is the voluminous ballooning of the greater omentum from the caudally facing part of the greater curvature of the stomach (the spleen restricts such expansion more proximally from the left-facing part) (Fig. 12c). Thus, the omental sac comes to lie ventral to the intestinal mass, providing a protective cushion in quadrupeds and acting in all species as a site of fat storage and supposedly having a physical and immunological role in blocking and healing any damage to the intestinal wall.

Thus, we have an enlarged stomach to the left cranially, with the bulk of the liver displaced to the right, a short duodenum in humans to the right, a large loop of colon arching from caudally to the left, across cranially close to the stomach and down on the right, with coils of jejunum centrally and ventrally, all obscured from view (ventrally) by the expansion of the greater omentum caudally.

Although the primate gut is relatively unspecialized for fruit-eating, in contrast to the meat- or leaf-eating specializations of most other mammals, the human gut conforms to this pattern, although assuming more the proportions of a meat-eater, with unique adaptations in attachments for bipedalism.

BIBLIOGRAPHY

Chivers, D. J., and Hladik, C. M. (1980). Morphology of the gastrointestinal tract in primates: Comparisons with other mammals in relation to diet. *J. Morphol.* **166,** 337–386.

Hill, W. C. O. (1958). Pharynx, oesophagus, stomach, small and large intestine: Form and position. *Primatologia* **3,** 139–207.

Langer, P. (1987). Evolutionary patterns of Perissodactyla and Artiodactyla (Mammalia) with different types of digestion. *Z. Zool. Syst. Evolut-forsch.* **25,** 212–236.

Last, R. J. (1978). "Anatomy: Regional and Applied," 6th Ed. Churchill Livingstone, New York.

MacLarnon, A. M., Chivers, D. J., and Martin, R. D. (1986). Gastro-intestinal allometry in primates and other mammals including new species. *In* "Primate Ecology and Conservation" (J. G. Else and P. C. Lee, eds.), Vol. 2, pp. 75–85. Cambridge Univ. Press, Cambridge, England.

Martin, R. D., Chivers, D. J., MacLarnon, A. M., and Hladik, C. M. (1985). Gastro-intestinal allometry in primates and other mammals. *In* "Size and Scaling in Primate Biology" (W. L. Jungers, ed.), pp. 61–89. Plenum, New York.

Digestive System, Physiology and Biochemistry

ELDON A. SHAFFER

Calgary Regional Health Authority and University of Calgary

GLOSSARY

Alimentary Relevant to food or nutrition

Assimilate To absorb and incorporate digested food products into the body

Carbohydrates Saccharides or sugars composed of carbon, hydrogen, and oxygen, with the empirical formula $(CH_2O)_n$

Chyme A creamy, semifluid, or gruel-like material produced by partial digestion in the stomach

Digestive system System of structures which process and absorb food; synonymous with the alimentary tract (or canal) or the gastrointestinal tract

Electrolyte Compound that dissolves in water to form charged ions, such as sodium (Na^+), hydrogen (H^+), bicarbonate (HCO_3^-), or chloride (Cl^-). Such an electrolyte solution conducts an electrical current. Acids, bases, or salts (inorganic or organic) are electrolytes if the substance forms ions in solution.

Endocrine Internal secretions (e.g., hormones) of a gland usually enter the blood. Although typical endocrine glands are the thyroid, adrenal, and reproductive glands, the gastrointestinal tract is the largest endocrine gland in the body.

Enzyme Specialized protein molecules which act as catalysts to enhance the rate of a specific chemical reaction. The substance on which the enzyme works is the substrate; each substrate usually is changed by a specific enzyme. Many enzymes are named by adding the suffix -ase to the name of the substrate on which they act. For example, the enzyme lactase acts on lactose. Other enzymes have less informative names (e.g., pepsin, which means to cook or digest).

Epithelia Cells which cover a surface

Exocrine External, or outward, secretion of a gland, usually via a collecting or duct system (e.g., pancreatic enzymes and bile). Some glands, such as the pancreas, have both exocrine (e.g., enzymes) and endocrine (e.g., the hormone insulin) functions.

Gastrointestinal Referring to both the stomach and the intestine; gastroenteric

Hormone An organic substance, produced in one organ or part of the body, which regulates another organ or cell type. Some hormones reach distant sites of action by circulating in the blood; others are released locally and diffuse toward their target cells. A number of hormones are formed by ductless glands; other hormones, such as cholecystokinin and secretin, are produced by the gastrointestinal tract. Such gut hormones are polypeptides, composed of short chains of 10–100 amino acids.

Lumen The cavity in a tubular structure

Mucosa Mucous membrane, composed primarily of epithelial cells, which lines the inner surface of the digestive system

Mucus A clear viscid secretion of mucous membranes

Secretion Production of a substance by a cell or group of cells (a gland), usually in liquid form

Transport To move a substance from one site to another

Triglycerides The major form of fat present in the diet of humans. This lipid is a neutral fat consisting of glycerol, to which are attached three fatty acids.

HUMANS REQUIRE FOOD, WATER, AND ELECTRO-
lytes for sustenance, growth, and survival. The digestive system (also known as the gastrointestinal tract
or alimentary canal) is a continuous hollow tube with
specialized parts of distinct size, shape, and function.
Its purpose is to make nutrients useful. Food enters
the mouth, passes down the esophagus to the stomach, and is propelled through the small and large
intestines before emptying out the anus (Fig. 1). This
long winding canal and its accessory organs (i.e., the
salivary glands, pancreas, and liver) are responsible
for digestion, the simplification or chemical breakdown of complex molecules. Digestion is facilitated
physically by contractions of the stomach and the
intestine and biochemically by enzymes and bile secreted from the exocrine organs. Digestive enzymes
are secreted by the salivary glands, the pancreas, and
the epithelial cells lining the stomach and the small
intestine; bile is secreted by the liver, stored in the
gallbladder, and emptied into the upper small intes-

tine. Once food has been processed to a sufficiently
simple state, absorption occurs across the mucosal
lining of the gastrointestinal tract. The absorbed nutrients are then carried by the bloodstream for further
metabolic processing in the liver and other body tissues. Fluid and electrolyte absorption also occurs in
the intestine. Nondigestible and unabsorbed materials
are eliminated via the large intestine or colon.

I. ANATOMIC AND FUNCTIONAL ORGANIZATION OF THE DIGESTIVE SYSTEM

The digestive system is the normal avenue by which
nutrition enters the body and is therefore essential for
life. Humans require food (commonly divided into
carbohydrates, proteins, and fats), water, inorganic
salts, minerals, and vitamins. The food we eat consists
of large molecules, too complex for direct use by the
body; food must first be digested. These large, sometimes water-insoluble, substances are mechanically
and chemically broken down into simpler water-
soluble molecules. Digestion takes place within the
hollow cavity, or lumen, of the alimentary canal.
Absorption is the movement or transport of the digested and dissolved nutrients across the lining of
the small intestine. The digestive system is selective,
keeping out microorganisms and eliminating waste
and toxic material.

The major functions of this system, then, are digestion and absorption, which are accomplished by muscular contractions of the canal and the secretion of
digestive juices. Food enters the body from the external environment through the oral cavity and is then
propelled through different regions of the digestive
system. A number of digestive juices are secreted en
route. The resultant smaller molecules can then be
absorbed into the transporting fluids of the body:
blood and lymph.

The organs of digestion consist of more than the
stomach and the intestinal tract. Collectively, the digestive tract includes the mouth, esophagus, stomach,
small and large intestines, plus the following glandular outpouchings, whose secretions provide important
digestive function: salivary glands, liver and biliary
system (bile ducts and gallbladder), and pancreas
(Fig. 1). The inner surface of this canal is lined by a
continuous sheet of cells (i.e., epithelial cells), some
of which secrete digestive juices, while others are in-

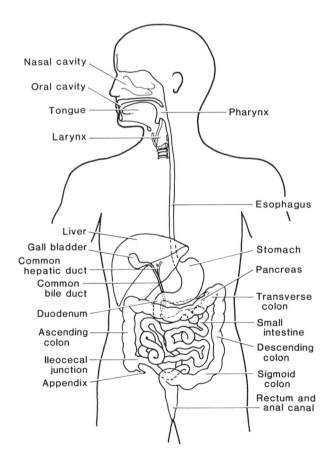

FIGURE I   Digestive system from the mouth to the anus.

volved in absorption. This important lining is termed the mucosa.

The gastrointestinal tract structurally is well suited for its functions: a long coiled tube, into which a variety of glands secrete digestive juices and around which a muscular wall contracts and relaxes to churn and propel the contents. From the upper end of the esophagus to the lower end of the rectum, this wall contains an inner layer of circular muscle (which constricts the lumen) and an outer layer of longitudinal muscle (which shortens the gut).

Two basic types of muscular activity are involved in digestion and absorption: propulsive movements and mixing movements. A repetitive pattern of altered muscle tension either facilitates or impedes progress through the gut. For propulsion, increased pressure in one region squeezes the contents toward a lower region, in which there is reduced pressure within the lumen. Peristaltic contractions are coordinated, sequential, ring-like waves of muscular constriction, which pushes liquids and solids forward.

The rate at which the contents traverse the digestive tract varies between regions, depending on the different functions occurring in that specific organ. Entry of food, for example, through the mouth, pharynx, and esophagus is aided by gravity and is rapid: a conduit to the stomach. The stomach and the small intestine, which are responsible for the majority of digestion and absorption, involve the second process, mixing. Increased tone in the absence of forward progression allows the stomach to grind solids into finer products and mixes the digestive juices with the ingested nutrients. Thus, transit from the stomach through the intestines is slow. Mixing in the small and large intestines increases the exposure of chyme to the surface lining, enhancing absorption.

The circular muscle layer becomes thickened at certain points along the alimentary tract, creating areas of constriction, waist-like sphincters (see the pyloric sphincter depicted later in Fig. 2). Sphincters control the forward and backward fluxes of material; they remain closed until it is appropriate to allow forward flux to proceed or initiate egress (say from the bile ducts into the duodenum), yet to limit regurgitation (backward fluxes) at all times. They generally function like one-way muscular valves, but are not simple flap valves. Rather, sphincters are high-pressure zones which tighten to retard backward passage, yet relax to allow forward passage when contents arrive from above. A variety of nervous, hormonal, and intrinsic muscular controls regulate the tension and, hence,

resistance in these zones. Control of traffic is found in the following sites: at the entrance to the esophagus (*the upper esophageal sphincter*); at the outlet of the esophagus (*the lower esophageal sphincter*); between the stomach and the first portion of the small intestine, the duodenum (*the pylorus*); at the entrance of the pancreatic and biliary ducts into the duodenum (*Oddi's sphincter*); between the small and large intestines (*ileocecal sphincter*); and at the outlet (the *internal* and *external anal sphincters*). Although sphincters do not necessarily occlude the lumen completely or continuously, they do tend to create "closed" systems.

The musculature of the digestive tract is composed of smooth (involuntary) muscle, except for the entrance and exit of the canal. The striated (voluntary)-muscled organs are the mouth, the pharynx, the upper one-third of the esophagus, and the external anal sphincter. These striated muscles are dependent on nerves for voluntary muscular contraction. The smooth muscle cells differ; they exhibit involuntary spontaneous activity. Nerves and hormones modify, rather than initiate, these contractions. [*See* Smooth Muscle.]

The gastrointestinal tract is innervated by the autonomic nervous system, which reaches out from many levels of the brain and the spinal cord to connect with local nerves contained in the wall of the gut. The autonomic nervous system consists of two divisions: the sympathetic and the parasympathetic nervous systems. Either can transmit inhibitory or excitatory impulses to smooth-muscled organs or secretory glands. These autonomic nerves also contain sensory fibers involved in the reflex regulation of gut function and nociception (including pain, nausea, and vomiting). The local nerves in the wall are termed the **enteric nervous system**. These nerves are unique. Although situated outside the central nervous system, they can regulate gut function on their own, even when separated from the central nervous system. That is, the enteric nervous system is truly "autonomous," yet is modulated in turn by impulses from the central nervous system reaching these local nerves via the autonomic nerves. [*See* Autonomic Nervous System; Brain Regulation of Gastrointestinal Function.]

Smooth muscle contraction/relaxation is also controlled by hormones, the most important of which are the gut hormones synthesized by endocrine cells within the gastrointestinal tract. The autonomic nervous system also influences gut hormone production and release. Activation of the vagus nerve of the parasympathetic nervous system promotes the release of hormones such as gastrin, cholecystokinin, and secretin which enhance digestion; the sympathetic nervous system has the opposite effect.

II. MOUTH

Although the preparation and cooking of meals may initiate some alteration of complex foods, the digestive process truly begins when food enters the mouth.

A. Mastication

The mechanical chewing of food dramatically reduces the size of food particles, so that swallowing is then possible through the relatively narrow esophagus. Smaller pieces also have a relatively larger surface area on which digestive processes can work. Chewing is both a voluntary and an involuntary process; food in the mouth stimulates a chewing reflex. The presence of food in the mouth and its smell also trigger nervous reflexes which stimulate digestive function in anticipation of the arrival of food. For example, the vagus nerve increases gastric and pancreatic secretions and initiates gallbladder emptying.

B. Salivation

The salivary glands consist of three paired structures: the large parotid glands near the angle of the jaw, the submandibular glands under the middle of the jaw, and sublingual glands under the tongue. [*See* Salivary Glands and Saliva.]

Saliva has two important functions: digestion and protection. For digestion, saliva lubricates the food as it is chewed into smaller pieces, facilitating swallowing. Saliva also possesses enzymes which initiate the breakdown (i.e., hydrolysis) of starch and fat. As a protective secretion, saliva dissolves and washes away retained food particles, preventing dental caries, possesses some activity against bacteria, and physically buffers ingested substances which could be noxious (from hot beverages to acidic drinks). The alkalinity of saliva when swallowed also protects the lower esophagus from acid refluxing up from the stomach.

III. ESOPHAGUS

A. Anatomy

The esophagus in adults is a hollow tube about 25 cm long, extending from the pharynx in the lower part of the neck through the chest (i.e., thorax) and via an opening in the diaphragm into the abdomen for 2 cm before joining the stomach. In its passage from the thoracic cavity to the abdominal cavity, the esophagus moves from a low-pressure region in the chest (especially when one inhales) to a high-pressure area. To maintain this pressure differential, the lower esophageal sphincter acts as a one-way "gatekeeper," keeping corrosive gastric acid from rising up the esophagus from below, while allowing ingested food to pass unhindered from above. The esophagus, like the mouth, is lined with a stratified squamous epithelium, layers of cells like the skin but without the superficial coat of keratin (Color Plate 5).

B. Swallowing

Swallowing involves the mouth, pharynx, and esophagus. It begins when a food bolus is pushed backward by the tongue into the pharynx (a funnel-shaped tube between the mouth and the esophagus). This solid ball of food sets up nerve reflexes to the brain, which automatically protects the airways: The soft portion of the roof of the mouth (soft palate) moves upward to close off the communication between the nose and the mouth, the vocal cords in the larynx close together, and a roof-like structure (the epiglottis) swings backward to help seal off the entrance to the lungs (the tracheal tube). Breathing momentarily ceases. Movements of the tongue and the pharynx propel the food back and down to the esophagus.

As food enters the esophagus, involuntary but coordinated contraction waves spread down the esophagus, progressively narrowing the lumen in a propulsive manner. The lower esophageal sphincter relaxes to allow food to be pushed into the stomach. Normally, food passes from the pharynx to the stomach in 5–10 sec.

IV. STOMACH

A. Anatomy

The stomach is the most dilated region of an otherwise narrow digestive tract. It is connected above with the esophagus and ends in the first portion of the duodenum. Its anatomy has been divided into regions (see Fig. 2): (1) the cardia (surrounding the entrance of the esophagus), (2) the fundus (the dome of the stomach to the left and above the cardia), (3) the body (the major portion—the long stem of the J-shaped

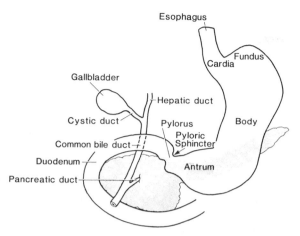

FIGURE 2 Anatomic relationships of the lower esophagus, stomach (including its regions), duodenum, pancreas (stippled gland), and biliary tract.

stomach, (4) the antrum (the more horizontal lower end from the right angulation to the pylorus), and (5) the pylorus (the distal, narrowed, tubular part, whose thickened muscular wall forms the pyloric sphincter).

The surface lining of the human stomach is composed of a simple columnar epithelium (simple meaning a single cell thick; columnar meaning long cells). Numerous pits on the surface lead into branched tubular glands which extend deep into the wall of the stomach (Color Plate 6). Functionally, there are two major gland areas: (1) an acid-secreting portion located in the upper 80% of the stomach and (2) the antrum, which produces the hormone gastrin. Gastrin is released into the bloodstream and circulates to stimulate the cells in the body of the stomach to secrete acid into the lumen of the stomach. Other surface cells produce a slimy protective coat of mucus.

B. Gastric Motility

The motor function of the stomach represents its major contribution to nutrition. Among mammals, its structure and function are tailored to the feeding habits of the animal. Ruminants (such as the cow) have several separate cavities; rodents, an upper stomach clearly defined into a fundus and body; and humans, a single chamber with less apparent differences between parts.

Gastric motility consists of three functions: (1) accommodating and storing the large volumes of material ingested, (2) mixing this food with gastric secretions and grinding solids into small particles, and (3) emptying the partially digested gastric contents in a controlled manner into the duodenum.

Two anatomic divisions of the stomach coordinate these functions. The upper stomach (the fundus and the upper part of the body) is involved in storage. The antrum grinds particulate matter, while the pylorus sieves and retains solid masses until reduced to an appropriate size (less than about 1 mm in diameter).

1. Storage Function of the Stomach

When fasting, the cavity of the stomach is quite empty, except for small volumes of secretions. Though empty and supposedly at rest, periodic antral contractions occur and may be responsible for the "hunger pains" we normally experience. When food enters, distending the stomach, muscle tone relaxes, enabling the average stomach to accommodate 1 liter with little increase in intragastric pressure.

2. Mixing and Grinding of Food

Weak contractions in the body of the stomach gently mix and blend food with the gastric secretions. Antral contraction waves are propulsive and more vigorous, churning and grinding the solid content of food before the small particles are allowed to exit through the narrowed pylorus. The pyloric sphincter controls delivery of the gastric contents into the duodenum.

3. Gastric Emptying

Emptying occurs concurrently with mixing and grinding and can take several hours. Liquids are squeezed out by the fundus, which contracts down like a piston. Emptying into the duodenum quite rapidly, a liquid meal will leave the stomach in 90 min or less, depending on the total volume imbibed. Solids empty only after being ground down to particles less than 1–2 mm in diameter. Nondigestible solids, which cannot be broken down, are eliminated later—by contraction waves which sweep residual contents out of the stomach every 90 min—a true "housekeeper," cleaning the gastrointestinal tract during fasting.

C. Gastric Secretion

The important constituents of gastric juice are gastric acid in the form of hydrochloric acid (HCl), a protein-digesting enzyme (pepsinogen), and a factor essential to vitamin B_{12} absorption (intrinsic factor).

The human stomach normally contains about 1 billion specialized cells to produce acid at high concentrations (≈ 150 mM) and large volumes (1–2 liters $\neq$

day). This acid is HCl. The resultant fluid has a pH of 1.0 during fasting, compared to blood at a pH of 7.4. Under basal or fasting conditions, acid secretion is less than 10% of maximum. Gastric acid secretion is stimulated by three substances which have separate receptors on the back surface (i.e., basolateral membrane) of the acid-secreting cell: (1) gastrin, the hormone synthesized in the antrum; (2) acetylcholine, a chemical substance liberated by nerve endings; and (3) histamine, from cells usually associated with inflammation, mast cells, or some other specialized cells containing histamine. The final event is entrusion of H^+ by the proton pump (H^+, K^+-ATPase) which exchanges H^+ for K^+, yielding a net secretion of H^+ and Cl^- at concentrations of about 160 mM (pH 0.8). ATP provides the energy for the active pumping of protons out of the parietal cell into gastric juice. With meals, gastric secretion is stimulated in a classical sequence:

1. In the cephalic phase, the sight, smell, taste, or chewing of food initiates nervous impulses via the vagus nerve to release acetylcholine. This response can easily be "conditioned" psychologically.

2. In the gastric phase, food in the stomach further increases gastric secretion through nerve reflexes stimulated by the mechanical distension of the stomach and from an increased release of gastrin into the bloodstream. Different cell types in the stomach wall sense distension, an increase in the pH (less acidity, with food-buffering HCl), and the presence of partially digested protein. This results in increased gastrin.

3. In the intestinal phase, chyme in the upper small intestine also stimulates acid secretion. The intestine is also a site for powerful inhibitory mechanisms which suppress acid secretion in response to fat, hyperosmolar contents, or acid.

To prevent the stomach from digesting itself under the onslaught of acid and digestive enzymes, several defense mechanisms are available. Mucus is secreted by superficial epithelial cells throughout the stomach (Color Plate 5). This viscous gel adheres to the surface, providing an important lubricating and mechanical protective function. Bicarbonate is also secreted, creating an alkaline environment at the base of the mucus coat, acting as a neutralizing buffer to protect the underlining epithelial cells. Gastric acid is not only important in protein digestion, but also functions to

reduce the entry of microorganisms ingested from the environment.

V. SMALL INTESTINE

Most digestion and virtually all absorption of the major dietary products take place in the small intestine, making it and its associated structures (i.e., the pancreas and the liver) the most important parts of the alimentary tract.

A. Anatomy

This 300-cm tube begins at the pylorus and ends at the ileocecal junction (Fig. 1). It is divided into three parts: the duodenum, the jejunum, and the ileum.

The duodenum, about 25 cm in length, forms a C-shaped curve, which cradles the head of the pancreas (Fig. 2). The pancreatic and bile ducts open into its central part. The remainder of the small intestine consists of the jejunum (arbitrarily, the upper two-fifths) and the ileum (the distal three-fifths). The jejunum is generally located in the upper left portion of the abdomen; the ileum is lower and to the right.

Three modifications dramatically increase the surface area exposed for absorption: circular folds involving the lining, mucosal villi, and absorptive cells microvilli. First are the transverse folds, which are quite obvious to the naked eye. Next, the mucosal villi, evident at a microscopic level, are numerous, 0.5- to 1-mm, finger-like projections of the luminal surface (Color Plate 7). Each contains a small artery, vein, and lymph vessel and is covered by columnar epithelial cells. Millions of these villi are present throughout the small bowel, like the surface of a rug. Villi increase the total surface area for absortion to approximately 10 m². They are also capable of swaying movements, which stir the food lying adjacent to the intestinal mucosa, aiding absorption. Pit-like crypts encircle the villi around their bases (Fig. 3). The epithelial cells divide in the base of the crypts and mature as they are pushed up and out of the crypts onto the sides of the villi, to be extruded within 2–3 days at the tips. A third feature increases the surface area of the small intestine 30-fold: multiple, tiny, brush-like projections of the luminal surface of each intestinal epithelial cell, forming microvilli, termed the "brush border" membrane because of their appearance under a high-powered microscope. This

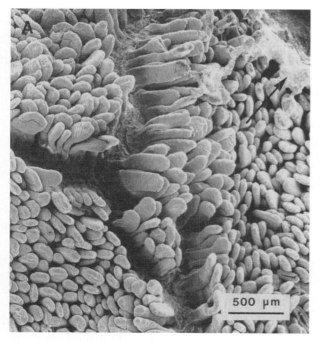
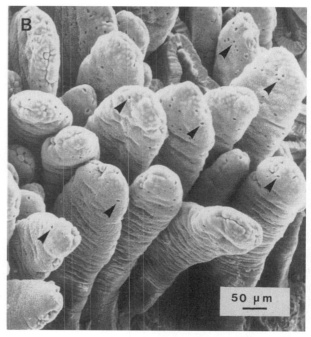

FIGURE 3 Scanning electron micrographs of the small intestine mucosa, demonstrating the many frond-like villi protruding into the lumen. (A) At the top right corner (arrow) is a patch of mucus. (B) This higher magnification shows the tips of the villi. The tiny holes (arrowheads) are the openings to mucus-secreting cells. (Courtesy of André Buret.)

microvillous membrane also contains digestive enzymes.

B. Motility

The motor action of the small intestine is organized to optimize the processes of digestion and absorption of nutrients. The small intestine receives a stream of chyme from the stomach. The chyme, which is partially digested solids and liquids, almost immediately meets secretions from the pancreas and the biliary tract in the duodenum. These endogenous (i.e., from the body) secretions contain important enzymes and solubilizers to complete the digestive process. Thorough mixing is necessary for chemical digestion and for absorption through contact with the mucosa. Failure of digestion and absorption within the small intestine allows the nutrients to pass into the colon, where bacterial metabolism produces osmotically active end products, gas, and certain fatty acids which can result in diarrhea. Finally, small-bowel contractions must move along any indigestible solids; if left, mechanical obstruction ensues.

Such mixing and movement of contents through the small intestine are controlled by an intrinsic activity of individual smooth muscles modified by the autonomic nervous system and certain gut hormones.

Three general sequences of contractions are recognized: (1) *peristalsis*, one or more ringed contractions of the circular muscle, propel the contents in an oral-to-anal direction; (2) *segmentation*, two or more isolated contraction rings, form short occluded segments, which chop the luminal contents into sausage-like portions; and (3) *pendular movements*, periodic contractions of longitudinal muscle, cause a to-and-fro motion of intestinal contents which contribute to mixing. Peristalsis moves intestinal contents along as if encircling one's fingers around a long thin tube of paste and pulling the tube and squeezing the paste forward. Segmentation and pendular movements provide mixing.

The gastrointestinal tract does not rest during fasting. About every 90 min in humans, electrical and motor activities migrate from the lower esophagus or antrum through the small intestine to the terminal ileum. This interdigestive contractile ring complex moves down the small bowel at 6–8 cm/min. Its periodicity is likely governed by the enteric nervous system plexus, but can be modified by other neural or hormonal factors. This motor complex is believed to act as a "housekeeper" during fasting, clearing the small intestine of residual food, secretions, and sloughed epithelial cells into the colon. During the active phase of these complexes, the epithelium switches from absorption to a secretory state, aiding this cleaning function. Ridding the small intestine of such debris prevents any ingested or resident bacteria from proliferating and overgrowing the region. Excessive bacteria in the upper small bowel impairs digestion and absorption.

VI. EXOCRINE ORGANS

Two gut-related organs provide important digestive juices to the duodenum. The pancreas and liver are termed exocrine organs because they produce an aqueous solution (containing pancreatic enzymes and bile, respectively) and deliver it via a conduit or ductal system directly to the lumen of the duodenum.

A. Pancreas

1. Anatomy
This large gland, 12–15 cm long, lies transversely across the upper abdomen, behind the stomach, with its head cradled to the right in the C-shaped sweep of the duodenum (Fig. 2). The body and the tail of the pancreas continue from the head and the neck to the left and upward.

The pancreas has a dual function, with two important glandular cell types. Endocrine cells (about 20%) produce important hormones, such as insulin, which circulate in the blood. Exocrine cells (80%) make digestive enzymes, which are secreted via a duct system into the duodenum.

2. Pancreatic Secretions
The pancreas secretes approximately 1 liter of fluid per day, consisting of two components: digestive enzymes and a high sodium bicarbonate fluid. The enzyme-rich juice originates in the exocrine (i.e., glandular) cells, while the bicarbonate component is produced by the duct cells. Two gut hormones regulate pancreatic secretion: cholecystokinin results in a concentrated enzyme output and secretin causes the duct cells to produce a large volume of bicarbonate-rich fluid. Secretin is released from the first few centimeters of the duodenum when the acidic chyme enters from the stomach. Cholecystokinin is released from the duodenum in response to the digested products of dietary fat (i.e., fatty acids) and proteins (i.e., amino

acids). Both hormones enter the bloodstream and circulate to act to another site. Secretin got its name because it causes the pancreas to secrete. (Historically, it was the first hormone described.) Cholecystokinin causes the gallbladder (which stores bile) to empty; its name means "to move bile." Other gut hormones and autonomic nerves also regulate pancreatic secretion.

B. Liver and Biliary System

1. Anatomy
The liver, the largest solid organ in the body, weighs 1500 g in adults. Its strategic location in the right upper quadrant of the abdomen (Fig. 1) accommodates a double blood supply: the portal vein brings blood from the intestine (commencing in the villi), and the hepatic artery conveys well-oxygenated blood from the aorta and the heart. Venous drainage is into the inferior vena cava. Bile, the yellow–green fluid produced by the liver, flows along a ductal system and exits by the common bile duct into the duodenum (Fig. 2). The gallbladder, a distensable sac, can accommodate up to 25–50 ml of bile.

2. Bile Secretion
The liver transports a variety of substances from blood to bile. The liver cell, or hepatocytes, has one surface bathed in plasma and another forming a groove with an adjacent liver cell. Bile is secreted into these tiny canals between liver cells, then empties into small ducts and eventually into the large hepatic ducts. During fasting, most bile enters the gallbladder to be stored rather than exiting into the duodenum (Fig. 2). Eating causes neural (i.e., vagus nerve) and hormonal (i.e., primarily cholecystokinin) stimulation of the gallbladder smooth muscle, which contracts, expelling its contents. Cholecystokinin also causes the sphincter at the outlet of the common bile duct (i.e., Oddi's sphincter) to relax, facilitating bile output into the duodenum.

Bile, a complex solution of organic and inorganic components, provides the main excretory pathway for toxic metabolites, cholesterol, lipid waste products, and bilirubin (a pigment which is the breakdown product of hemoglobin in red blood cells). The liver is the most important site of cholesterol synthesis (the small bowel being the second most important) and the means by which this lipid is excreted from the body in bile. Bile is 90% water, yet cholesterol is water insoluble. The liver transforms cholesterol into a biological detergent: bile salts. Bile salts solubilize the lipids in bile and also aid in the digestion and the absorption of dietary fat in the upper small intestine. Thus, bile is a fluid "waste" from the liver, yet provides an important function for the assimilation of the fat. [*See* Bile Acids.]

3. Other Functions

The liver also plays a central role in the handling of carbohydrates, fats, and proteins. It stores glucose as glycogen and mobilizes this sugar when necessary. The liver takes up and stores fat and cholesterol and also synthesizes lipids, all in a well-controlled balance. It also synthesizes most proteins that circulate in the plasma. The liver is also involved in the metabolism of drugs and chemicals, excreting some in bile and altering others into products which can be safely stored or eliminated from the body through the kidneys.

VII. DIGESTION AND ABSORPTION

Digestion is the chemical decomposition of foodstuff into simpler components to facilitate absorption. Although digestion of certain nutrients may begin in the mouth and the stomach, the process is completed in the upper small intestine. Complex molecules are hydrolyzed. Hydrolysis introduces a water molecule to split the bonds joining the two component molecules: Part of the water molecule goes with one component; part, with another. This process occurs with all three major foods—carbohydrates and proteins, which are water soluble, and fat, which is insoluble in water. The resultant products are smaller, more water soluble, and more readily absorbed.

Absorption is the transport across the mucosal lining of the gastrointestinal tract from the lumen into epithelial cells. For substances readily absorbed, transport is a simple matter of diffusion, passive movement down concentration gradients. For others, active transport requiring energy is necessary. The rate and the efficiency of absorption vary in different parts of the digestive tract. For most digested nutrients, this takes place in the upper jejunum, with some notable exceptions: ethyl alcohol (i.e., the drinkable kind) in the stomach, iron in the duodenum, and bile salts and vitamin B_{12} in the terminal ileum.

A. Carbohydrates

1. Digestion

Carbohydrates or saccharides (sugars) are organic compounds consisting of carbon, hydrogen, and oxy-

gen. The ratio of hydrogen to oxygen atoms is always $2:1$, as can be seen in the formulas for glucose ($C_6H_{12}O_6$) and sucrose ($C_{12}H_{22}O_{11}$). Carbohydrates can be divided into three major groups: (1) monosaccharides, the single units (e.g., glucose); (2) disaccharides, consisting of two joined monosaccharides (e.g., sucrose); and (3) polysaccharides, containing very long chains (eight or more) of monosaccharides (e.g., starch and cellulose) and molecular weights of several million. Unlike the simpler sugars, polysaccharides are not sweet. Glucose, the most important component, fuels our energy requirements which are "glucose dependent." [*See* Carbohydrates (Nutrition).]

Our daily intake of carbohydrates is about 350–450 g, consisting of 50% starch, 30% sucrose, and less than 10% lactose (milk sugar). Starch, a complex polymer of many simple sugars, must be hydrolyzed to smaller simpler units before absorption can occur. Both salivary and especially pancreatic amylase attack dietary starch, eventually producing disaccharides. Further hydrolysis depends on disaccharidases, specific enzymes bound to the surface (microvillous) membrane of the intestinal epithelial cell. The resultant single sugar, usually glucose, can then be absorbed.

2. Transport

Active transport occurs against a concentration gradient and requires energy. The active transport of glucose is indirect, being linked to sodium uptake. Glucose and sodium move into the cell together. Once inside, sodium is actively pumped out the other side of the cell into the interstitial space. This basal surface of the epithelial cell possesses an energy-requiring sodium pump. Pumping Na^+ out the bottom of the cell into the high Na^+ concentration of the interstitial fluid is active transport, which maintains a low intracellular sodium concentration. The low concentration of sodium in the cell relative to that in the lumen encourages further uptake of sodium (and with it, glucose). Meanwhile, the entry of glucose raises its intracellular concentration, favoring its passive movement out of the base of the epithelial cell toward the blood. Other sugars, including some glucose, traverse the luminal cell membrane by passive diffusion. This simple transport process is controlled primarily by physicochemical laws, the difference in concentration gradients across cell membranes. The driving force is the random movements of the molecules, which are dependent on the body temperature.

The absorptive process appears not to be an easy one, yet, for sugars, the capacity of the small intestine is enormous. The equivalent of more than 10 kg of table sugar (i.e., sucrose), equivalent to 40,000 kcal, can be absorbed daily. Ironically, sugar substitutes, such as sorbitol, are not well absorbed and, if consumed in excess, reach the colon, where they act as osmotically active particles to retain water and cause diarrhea.

B. Proteins

These complex organic compounds contain carbon, hydrogen, oxygen, and nitrogen. Amino acids are the building units of proteins, linked head to tail by peptide bonds; the carbon end (i.e., the carboxy terminus) of each amino acid is joined to the nitrogen (i.e., the amino terminus) of another amino acid. Such polymers are called polypeptide chains, with molecular weights of 4000 to 1 million or more. [*See* Proteins (Nutrition).]

I. Digestion

Hydrolysis breaks the peptide bond between amino acids. This begins in the lumen of the stomach. The stomach secretes pepsinogen, which is activated in the presence of acid to pepsin. Pepsin cleaves amino acids from large protein molecules, resulting in smaller combinations of amino acids (i.e., polypeptides). Gastric acid also denatures protein. The pancreas, in response to cholecystokinin, secretes proteolytic enzymes, which further reduce the size of the polypeptides inside the lumen of the upper small intestine. Enzymes within the microvillous membrane of the intestinal epithelial cells complete the job, producing free amino acids. Membrane digestion and absorption are closely related.

2. Absorption

Several different systems exist for amino acid absorption. Some require sodium, much like the active transport of glucose, whereas others are transferred by simple diffusion. A small percentage of short-chain peptides (with two or three amino acids) are absorbed intact. Some are further hydrolyzed inside the cell by cellular enzymes; the rest, a small percentage of peptides, enter the blood intact. In some people, these represent a potential source of allergic reactions.

Dietary protein is efficiently absorbed, less than 1% being eliminated through the colon each day. About 60% is absorbed in the jejunum.

C. Fats

Fats or lipids are water-insoluble organic substances. The Western diet contains upward of 80–160 g of fat, 95% of which is triglycerides. This represents half of the daily calorie intake. Triglycerides are also the major storage forms of fat in animal and plant cells. Triglycerides, like carbohydrates, consist of carbon, hydrogen, and oxygen in two components: a glycerol molecule and three fatty acid molecules. Each fatty acid is attached to a separate carbon atom of the glycerol molecule. By definition, fat is hydrophobic, or water hating. This presents a dilemma. The luminal contents of the gastrointestinal tract and its secretions are predominantly in a water medium. Further, the primary enzyme that breaks down these bulky lipids, pancreatic lipase, acts only at surfaces. Like a salad dressing of oil (i.e., fat) and vinegar (i.e., water) left standing, the oil rises to the top, while the water remains a separate phase at the bottom. The surface area between these two phases is relatively small compared to the droplets that can be produced by shaking. This dispersion of fine oil droplets through a water phase produces an emulsion. Thus, within the lumen of the small intestine, the dilemma is resolved by emulsification. Inside the body, any lipid must also be compatible with biological fluids such as blood, which is primarily water. [*See* Fats and Oils (Nutrition).]

I. Digestion

The digestion of lipids involves progressive hydrolysis of the bonds linking the three fatty acids to the glycerol backbone of triglycerides. This is carried out by lipase enzymes. Several are present in human breast milk, saliva, the stomach, and, most importantly, pancreatic juice. Lipase from the mouth and (for infants) in breast milk initiate lipid digestion in the stomach. This is aided by the churning action of the antrum, which disperses fat as an emulsion of six triglycerides. As fat is retained in the stomach for up to several hours, significant (10–40%) hydrolysis can occur. Lipolysis in the stomach is critical to the newborn, whose diet is primarily milk, yet it has an immature pancreatic secretory function.

The duodenum is the major site of fat digestion in adults. There is a well-orchestrated entry of bile from the biliary system to further assist in the dispersion of fat into fine oil particles as a suspension and pancreatic juice containing enzymes and bicarbonate. Remember that fatty acids release cholecystokinin from the duodenal mucosa, and this hormone causes the gallbladder to expel its bilious contents of detergents and emulsifiers into the duodenum. Cholecystokinin also stimulates the pancreas to secrete pancreatic lipase. Pancreatic lipase acts at the oil droplet surface, splitting off one fatty acid at a time from the triglycerides, leaving a di- or monoglyceride (with two or one fatty acid still attached, respectively). The products of lipid digestion—fatty acids, monoglycerides, and glycerol—are carried from the surface of the oil droplet by a detergent mixture of bile salts. Bile salts themselves are not absorbed in the upper jejunum, but seem to shuttle back and forth, transferring the lipid digestive products from the oil droplet toward the mucosal surface. Eventually, bile salts are actively absorbed farther on, in the terminal ileum.

2. Absorption

The first barrier through which these lipid products must pass is a fine unstirred water layer adjacent to the epithelial cell lining. Passage across the cell membrane is then comparatively easy. Absorption is passive. For some lipids, the whole process, however, is relatively inefficient: Less than half of ingested cholesterol is absorbed. The inner fluid of the cell (i.e., cytosol) is mainly water, requiring some carrier or mechanism to get these water-hating compounds from the plasma membrane through the cell.

Once inside the intestinal epithelial cell, the components are resynthesized into triglycerides. Covered with a protein coat and some cholesterol, the resulting chylomicrons (i.e., emulsified fat) are assembled and then expelled into the space between the epithelial cells to be transported in lymph. The relatively large chylomicrons, about 1 μm in diameter, cause the lymph (and, in some individuals, the blood) to become milky. With minor changes, chylomicrons soon reach the liver for further processing.

VIII. FLUIDS AND ELECTROLYTES

Over 9–10 liters of fluid enters the gastrointestinal tract each day. We imbibe about 2 or more liters, and another 7–8 liters enters from internal secretions: 1 liter as saliva, 2–3 liters from gastric secretions, 0.5 liter as bile, 1–2 liters from pancreatic juice, plus a smaller contribution by the small intestine. Thus, large volumes of fluid must be absorbed, or we would soon become dehydrated. The absorptive capacity of the small intestine is quite significant, up to and perhaps exceeding 12 liters per day. Of the 1–2 liters

presented to the large intestine each day, all but 100–200 ml is absorbed, representing a 90% efficiency.

The gut thus absorbs most of the salt and water that enters. There is no active transport process for water: fluid movement occurs secondary to the transport of either an electrolyte (mainly sodium) or a nonelectrolyte (e.g., glucose). Both active and passive transport mechanisms are responsible for solute movement in the intestine, although their relative contributions differ in the jejunum, ileum, and colon. Passive diffusion involves the movement of solutes from an area of high concentration (i.e., overcrowding) to one of lower concentration. To maintain osmotic equilibrium, water accompanies solute transport. The jejunum, whose epithelium is relatively leaky, has a low resistance to the passage of water and electrolytes and readily allows passive transfer. Active transport, being energy dependent and one way, requires an epithelium more resistant to movement in order to retain the transported particles. In general terms, absorption of sodium and water occurs passively in the jejunum, whereas, in the ileum, the active transport of sodium and glucose draws in water. At a microscopic level, the villi predominantly absorb, whereas the crypts secrete. At any moment, net transport is a balance between these two opposing movements of electrolytes and fluid.

Other important electrolytes, such as potassium, bicarbonate, chloride, and hydrogen, also have one or more transport mechanisms. Iron is primarily absorbed in the duodenum whereas calcium is absorbed in the jejunum.

IX. COLON AND RECTUM

A. Anatomy

The large intestine is continuous with the ileum (Fig. 1). It begins with the cecum in the lower right quadrant of the abdomen. The appendix, about 10 cm in length, protrudes from the back surface of the cecum. This worm-like tube ends blindly.

The colon consists of the following parts: the ascending colon, which runs from the cecum on the lower right-hand side of the abdomen up to below the liver to become the transverse colon; the transverse colon, which crosses the upper abdomen beneath the rib cage to the left upper quadrant; here, it makes another acute turn to become the descending colon, which passes down the left side to enter the pelvis, where, as the sigmoid colon, it is joined to the rectum. The rectum is not completely straight, despite its name, but has a gentle concavity, sloping forward in front of the sacrum and coccyx (i.e., tailbone). The anal canal is the termination of the digestive tract and is surrounded by sphincter muscles. The entire length of the large intestine in adults is about 150 cm in length. Its diameter, though somewhat larger than that of the small intestine, diminishes from the cecum to the anus. The wall of the large intestine consists of layers similar to that of the small intestine, but the arrangement of columnar cells lining the lumen differs (Color Plate 8). There are no true villi in the colon. Instead, the absorptive surface is flat, with many straight tubular crypts.

B. Motility

The colon contains smooth muscle throughout its length, except for the external anal sphincter, which is composed of skeletal (i.e., voluntary) muscle. The gross pattern of flow and motility through the colon is not completely understood. "Mass movement" is a sudden translocation of significant fecal mass over a long segment of the colon, reaching one-third to one-half of the colonic length. Such shifts do not occur regularly, perhaps only a few times per day, and are fleeting, lasting a few seconds. They certainly have a temporal relationship to eating and defecation. The flow of luminal contents through the colon is not steadily progressive, unlike that in the small intestine. Nonpropulsive segmental contractions may move luminal contents in both directions, forward and back. Solid contents in general move much more slowly than does liquid or gas.

Continence is maintained by the internal and external anal sphincters. Large amounts of material in the rectum can initiate reflex relaxation of the tonically contracted internal sphincter. Once this happens, material passes spontaneously into the anal canal. Volitional contractions of the external sphincter may prevent defecation, but once the urge to stool occurs, this skeletal muscle relaxes and elimination follows.

C. Absorption

The liquid residue which remains once the chyme has passed through the small intestine enters the cecum. This residue consists of undigested matter plus fluid taken in with food or secreted as digestive juices, but not resorbed by the small intestine. The colon is presented with about 1 liter of fluid from the ileum each day, but releases less than 200 ml. The absorption of water which accompanies sodium and chloride transport takes place continuously as the residue passes through the large intestine—a journey of 12–14 hr. Some bacterial hydrolysis of colonic matter produces nutrients, such as short-chain fatty acids (with less than 12 carbon atoms); when absorbed, these can provide energy. As water is absorbed, the material in the colon becomes increasingly more solid. The solid (though normally soft) which collects in the rectum is the feces, or stool. No nutrients are absorbed in the colon. Hence, sodium absorption is not linked to that of glucose or other nonelectrolytes, as in the small intestine. The colon must rely on active sodium transport and a small element of passive absorption. Its normal absorptive capacity is 4 liters per day. The net result is feces, consisting of water and solid materials—undigested food residue, microorganisms, and mucus.

X. CONCLUSIONS

The digestive system constitutes a simple, yet elegant, winding tube about 5 m in length with specialized regions extending from the mouth to the anus. The food we eat enters from the external environment in a rough and complex form. Propelled and chopped up by the digestive system, this food is mixed with digestive juices, which convert complex molecules to simpler forms. These small digestive products are absorbed into the portal venous blood and lymph for subsequent distribution as vital nutrients to the cells of the body. Unabsorbed residues and other waste products are moved to the end of the tract and are eliminated from the body. The total time from ingestion to elimination is normally 40–60 hr. Nutrient digestion and absorption take 3–6 hr, a relatively short time for processes essential to life.

BIBLIOGRAPHY

Aries, I. M., Boyer, J. L., Fausto, W. B., Jakoby, W. B., Popper, H., Schachter, D., and Shafritz, D. A. (1994). "The Liver: Biology and Pathobiology," 3rd Ed. Raven, New York.

Davenport, H. W. (1988). "Digest of Digestion." Year Book, Chicago.

Davison, J. S., ed. (1989). "Gastrointestinal Secretion." Wright, London.

Johnson, L. R. (ed.). (1994). "Physiology of the Gastrointestinal Tract," 3rd Ed. Raven, New York.

Digital Dermatoglyphics (Fingerprints)

ROBERT J. MEIER
Indiana University

GLOSSARY

Admixture Evolutionary process by which individuals from one population mate with those from another and thereby transfer genes between groups

Friction skin Epidermal skin surface of the palms and the soles that bears ridge and furrow systems

Mendelian inheritance Inheritance that is governed by a single genetic locus

Microevolution Genetic change that occurs as a result of one or more of the following processes: mutation, random genetic drift, admixture, and natural selection

Natural selection Evolutionary process based on differential survival and/or differential reproduction among individuals who possess variable traits

Polygenic inheritance Inheritance that is dependent on the interaction of several genes

Sexual dimorphism Observable and measurable differences between males and females

DERMATOGLYPHICS IS THE STUDY OF THE EPI-dermal ridges as found on the skin surface of the palms, fingers, soles, and toes. This study encompasses the embryological formation of ridges, their minute structure, and their pattern configurations. It can deal at any one of several levels, from individual digits or separate hands to comparisons between the sexes, among different human populations, or even among different primate species. A primary interest is to learn about the development and variation of dermato-glyphic features both in normal persons and in known cases of birth defects. Each of these topics is discussed in the following sections.

I. DEVELOPMENT OF DERMAL RIDGES

A. Volar Pads

During the first trimester of gestation, at approximately the sixth week, volar pads appear on the fingertips, as well as on other areas of the palm and the foot (see Fig. 1). Volar pads consist of mesenchymal tissue, forming raised mound-like structures which are thought to directly influence the patterning of epidermal ridges. Volar pads are most prominent in the third to fourth fetal months before regressing, and it is during this time that ridges begin to form on the pads.

B. Primary and Secondary Ridge Formation

Primary ridge formation begins at the dermal or basal layer of the skin, initially as undulating folds surrounding paired rows of dermal papillae, from which newly formed skin cells arise (see Fig. 2). These folds underlie the primary ridges, and through them pass the ducts from sweat glands that exit on the top of the ridge as a pore. Numerous nerve endings are also exposed along the ridge. Adjacent to the primary ridges are secondary ridges, which are actually furrows extending deeper into the dermis. Commencing about the sixth fetal month, the ridge-and-furrow system is visible on the skin surface. It is after this time that no change will occur in the epidermal ridge patterns themselves; they will simply increase in size in

355

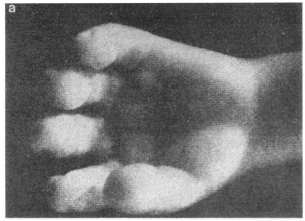

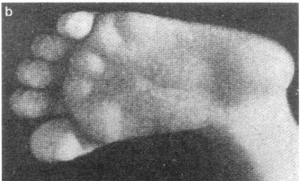

FIGURE I Volar pads (a) on the hand of a 10-week human fetus and (b) on the foot of a fetus about 2 weeks older. [From H. Cummins and C. Midlo (1961). "Finger Prints, Palms and Soles: An Introduction to Dermatoglyphics," p. 179. Dover, New York.]

response to the growing fetus. These same patterns will persist as well throughout the remainder of a persons' lifetime.

C. Functions of Epidermal Ridge Systems

Given the rather unique appearance of the epidermal ridged skin surface, there is good reason to expect that it performs some important functions. Indeed, several have been proposed, and all would seem probable to a degree. First, and perhaps foremost, the ridge-and-furrow system serves to retard slipping in a manner similar to automobile tire treads. Accordingly, "friction skin" is sometimes used to denote the anti-slipping nature of epidermal ridges. Second, since sweat pores open only on the primary ridges and not in the furrows, it is thought that a small amount of sweat bathing the ridges might improve their level of friction and further aid against slipping. Then, since nerve endings are extensive along the ridges, the ridges

themselves might well serve to enhance the sense of touch or to improve tactile stimulation. Sir Francis Galton, a pioneer in dermatoglyphic study, performed an experiment which showed that blindfolded people moved their fingertips around in circles when asked to identify objects, as if to actively engage the nerve endings. Finally, there is the possibility that the ridge-and-furrow system provides some added structural stability analogous to corrugated cardboard or metal.

D. Evolutionary and Hereditary Considerations

These many functions probably point to an evolutionary history within primates, all of whom possess "fingerprints," that placed a premium on the ability to carry out secure grasping behaviors, to perceive valuable information about the environment through the sense of touch, and perhaps to withstand extensive use of the volar surfaces in moving about and feeding in trees. In short, primates, in part, adapted to their arboreal niche by means of epidermal ridges and dermatoglyphic patterns. We humans simply continue to reflect some of that ancestral adaptive response, since it is rather unlikely that our current survival strongly depends on functioning epidermal ridge patterns.

Various models of inheritance have been proposed to account for the expression of dermatoglyphic features. Some of these provide evidence of Mendelian inheritance or single gene control for the pattern configurations, while others treat the development of the ridges under a polygenic mode of inheritance subject to a certain amount of environmental interaction. Very likely, genic control is effected at various levels, in the formation and regression of the volar pads, as well as in the expression and patterning of the primary and secondary ridges.

II. PATTERN CONFIGURATIONS AND CLASSIFICATION

A. Pattern Landmarks: Triradius and Core

The two main features found in a pattern are its triradius and core. Figure 3 depicts a pattern on which these landmarks are indicated. The triradius is a place on the pattern from which the ridge system extends along in three directions. The details of the triradius might vary somewhat, but the resultant three radiants are generally easily observed. The core is the cen-

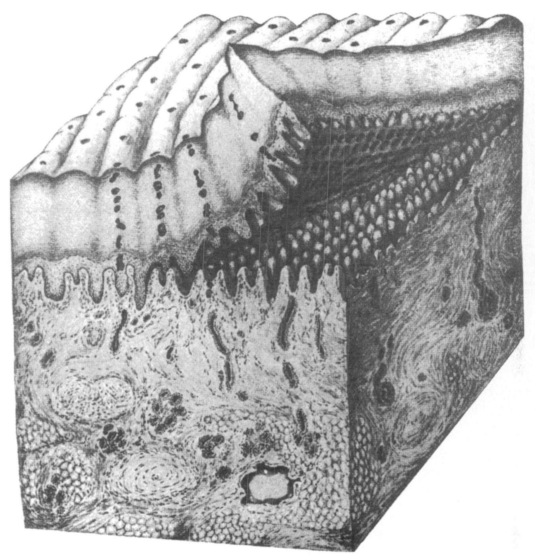

FIGURE 2   The histology of ridged skin. The epidermis is partially lifted away from the dermis to show the underlying dermal papillae. [From H. Cummins and C. Midlo (1961). "Finger Prints, Palms and Soles: An Introduction to Dermatoglyphics," p. 38. Dover, New York.]

termost ridge or portion of ridge within the pattern. Again, there is variation in this feature, but it is likewise readily determined.

B. Ridge Counting

Many studies utilize the triradius and the core in analyzing fingerprints, initially to classify them into patterns and then to count the number of ridges that make up the patterns. Ridge counting is done usually with the aid of a magnifier (to 10× at least) and a pointed marker. Ridges are counted between the core

and the triradius, but these two landmarks are not included in the count.

C. Arch–Loop–Whorl Classification

Ridge systems form recognizable patterns that can be classified according to the number of triradii and ridge counts they possess. The most basic classification consists of three types: arches, loops, and whorls (see Fig. 4). Arches are the simplest patterns, generally not having a triradius or a core, and hence having a zero ridge count. Loops usually have one triradius, a single

FIGURE 3 A fingerprint pattern indicating the triradius and the core.

core, and one variable ridge count. Loops can be further classified as ulnar if the ridges on the side opposite the triradius open to the ulnar side of the hand (i.e., toward the little finger) or radial if the ridges open to the radial side (i.e., toward the thumb). Whorls have usually two triradii and one or two cores, with two variable ridge counts. There are many subtypes of whorls, which justifies their being considered the most complex of patterns. Ridge counts might range from 1 to 2 ridges in small loops to more than 25 ridges in large whorl patterns. A total finger ridge count (TFRC) is calculated as the sum of the ridges across all fingers; only the larger of the two counts is used in the case of whorl patterns. An alternative is to record the ridge count of each digit separately in terms of both an ulnar and a radial count. Hence, each person would be represented by a 20-count matrix or array. Another calculated variable is the pattern intensity index (PII), which is the sum of the triradii for all digits. This can range from zero (10 arches) to 20 (10 whorls). There is a strong correlation between TFRC and PII (see Table I).

D. Unusual Dermatoglyphic Features

On the digits, sometimes a pattern configuration is observed that consists of a combination of two patterns, such as a loop and a whorl. These "accidentals" occur in less than 1% of individuals, and then most likely on the index finger (see Fig. 4). Another rare condition involves an unusual appearance of ridges composed of a series of short segments referred to as a "string of pearls." They are more generally classified as ridge dissociations. Frequently, it is possible to see pattern configurations so large that the triradius is not present and is said to be "extralimital," or beyond the ridged skin surface.

III. POPULATION COMPARISONS

A. Distribution in Human Groups

It is of considerable interest that dermatoglyphic features express so much variability. There is, of course, the forensic and legal acceptance of the uniqueness of each individual's fingerprints. That is, no two persons have exactly the same fingerprints, and this even applies to identical twins (see Table II). These unique differences actually are found in the detailed makeup of the ridges. Ridges are segmented to form fragments of varying length; they often bifurcate to make fork-like structures or rejoin in forming enclosures (see Fig. 5). It is these details (i.e., minutiae) that uniquely identify the individual, but this identification is generally aided by means of pattern configurations.

I. By Digit

Pattern types do not occur randomly across the fingers (see Table III). To illustrate, the thumb (digit I) and the ring finger (digit IV) of each hand tend to have the highest frequency of whorl patterns, while the little finger (digit V) and the middle finger (digit III) bear most of the ulnar loops. The index finger (digit II) has nearly all of the radial loops, and arches are more likely to occur on the index, middle, and little fingers. Developmental factors may underlie this differential distribution, which appears to be firmly based, since these tendencies occur throughout nearly all human populations.

2. By Hand

There are also left–right differences in the frequencies of pattern types, usually involving whorls versus arches. Accordingly, whorls are more often found on the right-hand digits, while arches appear more frequently on the left side. Studies of bilateral asymmetry between homologous digits have been done to test notions of canalization of growth. Work has also been done in associating fingertip patterns and hand preference, revealing inconsistent findings for right- versus left-handers.

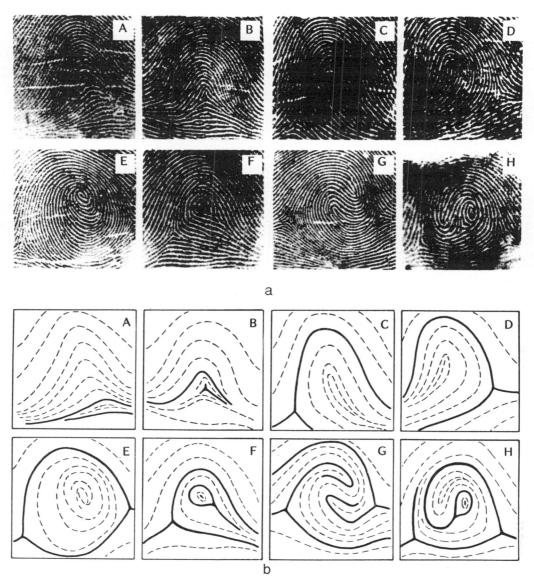

FIGURE 4 Digital pattern types: actual prints (a) with (A) simple arch, (B) tented arch, (C and D) loops, (E) simple whorl, (F and G) variant whorls, (H) accidental pattern. (b) The "type ridges" of each of the patterns. [From M. Alter (1966). Dermatoglyphic analysis as a diagnostic tool. *Medicine* **46**, 35–56, © by Williams & Wilkins.]

3. By Sex

Sexual dimorphism with respect to fingertip pattern frequencies has received considerable attention. The general tendencies show that females have more arches and ulnar loops, whereas males have more whorls and radial loops. Again, there could be minor developmental differences, such as in the timing of embryonic growth and differentiation in males versus females, that might account for the tendencies. Because of the sexual dimorphism in pattern configurations, males generally have higher TFRC and PII values than females due to their larger percentage of whorls and fewer arches (see Table I).

4. By Population

Strong research interest in population variation has likewise characterized dermatoglyphic study. Indeed, much of the earlier work provided descriptive and comparative analyses of human groups from around the world. While it is not possible to summarize all studies of this nature without obscuring some of the important findings, it might be of interest to bring

TABLE I

Correlations between Total Finger Ridge Count (TFRC) and Pattern Intensity Index (PII)[a]

Population	TFRC			PII		Correlation[b]
	No.	Mean	SD	Mean	SD	
Males						
Point Hope	51	157.7	52.6	16.0	2.8	0.60
Anaktuvuk Pass	35	133.8	40.5	13.5	2.8	0.84
Barter Island	23	134.0	42.0	13.5	2.9	0.74
Barrow	100	133.0	45.9	13.3	3.2	0.77
Wainwright	113	115.9	45.8	12.6	2.8	0.74
Females						
Point Hope	66	136.5	58.6	14.4	3.8	0.69
Anaktuvuk Pass	37	124.8	38.0	13.1	2.7	0.86
Barter Island	16	120.4	52.4	13.0	3.0	0.83
Barrow	118	129.9	47.2	13.4	3.3	0.81
Wainwright	122	104.6	41.9	11.7	3.0	0.81

[a]Modified from R. J. Meier (1978). Dermatoglyphic variation. *In* "Eskimos of Northwestern Alaska: A Biological Perspective" (P. L. Jamison, S. L. Zegura, and F. A. Milan, eds.), p. 83. Dowden, Hutchinson & Ross, Stroudsburg, PA.

[b]All correlations are significant beyond 0.001.

out once more some general tendencies with regard to the major pattern type distributions (see Table IV). Accordingly, arch patterns have their highest frequencies in sub-Saharan African groups (particularly the

TABLE II

Pattern Types (PT) and Ridge Counts (RC) in a Set of Monozygotic (Identical) Twins[a]

Left hand		Digit				
		I	II	III	IV	V
Twin						
PT	A[b]	W	RL	UL	UL	UL
	B[c]	UL	RL	UL	UL	UL
RC	A	21–4	3	17	16	12
	B	22	4	18	18	17
Right hand						
PT	A	UL	UL	UL	UL	UL
	B	UL	UL	UL	UL	UL
RC	A	22	20	16	13	10
	B	23	16	16	12	13

[a]Code: digit 1, thumb to digit V, little finger. PT, pattern type; W, whorl; UL, ulnar loop; RL, radial loop; RC, ridge count [note that there are two ridge counts for the whorl pattern on the left thumb (digit I) of twin A, while all of the loop patterns have a single ridge count].

[b]TFRC in twin A, 150; PII in twin A, 11.

[c]TFRC in twin B, 159; PII in twin B, 10.

FIGURE 5 Ridge details, or Galton's minutiae. [From H. Cummins and C. Midlo (1961). "Finger Prints, Palms and Soles: An Introduction to Dermatoglyphics," p. 31. Dover, New York.]

Pygmy), while Asian groups show the lowest percentages, and European populations fall in between. With respect to whorls, Asians generally have the highest frequency (along with Pacific Island groups and Australian Aborigines), Africans have the lowest frequency, and Europeans occupy the middle position. Last, Europeans usually have more radial loops on their digits than any of the other populations.

TABLE III

Percentages of Frequency of Digital Patterns in 5000 People from the Scotland Yard Files[a]

Digit	Side[b]	Pattern type			
		Whorls	Ulnar loops	Radial loops	Arches
I	R	41.43	55.89	0.22	2.47
	L	29.39	65.90	0.20	4.51
	R + L	35.41	60.89	0.21	3.49
II	R	30.79	32.30	26.03	10.87
	L	28.15	38.10	23.37	10.36
	R + L	29.47	35.20	24.70	10.62
III	R	16.56	74.81	2.53	6.09
	L	16.18	73.32	2.51	7.98
	R + L	16.37	74.07	2.52	7.03
IV	R	41.07	55.61	1.47	1.85
	L	27.82	68.92	0.50	2.75
	R + L	34.44	62.27	0.98	2.30
V	R	13.80	85.46	0.20	0.54
	L	9.03	89.79	0.02	1.17
	R + L	11.41	87.62	0.11	0.85
All	R	28.74	60.83	6.08	4.36
	L	22.12	67.21	5.31	5.35
	R + L	25.42	64.02	5.69	4.85

[a]Modified from H. Cummins and C. Midlo (1961). "Finger Prints, Palms and Soles: An Introduction to Dermatoglyphics," p. 67. Dover, New York.

[b]R, right; L, left.

TABLE IV

Percentage Ranges and Means of Fingertip Pattern Frequencies in Different Populations[a]

		Fingertip pattern type								
		Whorls		Ulnar L.		Radial L.		Arches	
Population	N[b]	Range	Mean	Range	Mean	Range	Mean	Range	Mean
Caucasians	112	26–49	35.4	50–66	55.6	4–7	4.3	2–9	4.3
Negroes	88	15–42	27.4	53–74	61.4	1–4	2.6	2–18	8.8
Amerindians	76	32–58	42.6	37–58	49.4	1–5	3.1	1–9	5.0
Orientals	55	36–56	46.7	39–57	48.1	1–4	3.0	1–5	1.8
Australasians	60	33–78	52.7	22–64	44.9	0–3	1.1	0–4	1.4
Asian Indians	7	33–56	42.6	33–59	51.8	1–4	2.2	1–7	3.4

[a]Modified from B. Schaumann and M. Alter (1976). "Dermatoglyphics in Medical Disorders," p. 83. Springer-Verlag, New York; and based on C. Plato (1973). "Variation and Distribution of the Dermatoglyphic Features in Different Populations." Penrose Memorial Symposium, Berlin.

[b]Number of population samples summarized.

Given the partial genetic control over the expression of dermatoglyphic features, it is to be expected that a certain amount of historical interpopulation differentiation would come about as a result of microevolution, perhaps more through random processes than predominantly through natural selection. Population contacts and any resultant admixture of interbreeding between the groups also can be expected to influence the distribution of digital pattern types and ridge counts. For example, Table V shows the likely consequence of European admixture in Easter Islanders and in northwestern Alaskan natives by lowering the PII, which is mainly the result of whorl pattern reduction.

A large amount of current research in digital dermatoglyphics is devoted to examining both microevolution and population structure. Dermatoglyphic variables derived from ridge counting, especially the 20-count array, have been found to be highly suitable for carrying out multivariate analyses. For example, Table VI shows an application of principal component analysis which has reduced the 20 counts to only four components that adequately characterize the nature of dermatoglyphic development according to either the radial or the ulnar side of particular sets of digits. Hence, component 1 for males combines the ulnar counts of digits II, III, IV, and V. For females, all five digits fall within this first component. Component 1 obviously relates to a developmental correspondence among ulnar counts. Likewise, the other three components sort out radial counts according to certain digits. For example, component 4 represents the radial count on digit V for both sexes. Very close

TABLE V

Pattern Intensity Index in Admixed and Unadmixed Groups from Easter Island and Alaska, by Sex

Group	Unadmixed	Admixed
Easter Island		
Males	15.8	14.9
Females	15.0	14.6
Point Hope		
Males	16.0	14.1
Females	14.4	12.0

TABLE VI

Principal Components Based on a Digital 20-Count Analysis of Northwestern Alaskan Inupiat (Eskimos)[a]

	Component			
Group	1	2	3	4
Males	Ulnar II, III IV, V	Radial II, III IV	Ulnar/radial I	Radial V
Females	Ulnar I, II, III IV, V	Radial II, III IV	Radial I	Radial V

[a]Left and right sides showed identical components.

agreement can be seen between the sexes for the components. This is not always the case and, in addition, comparable studies have shown population variation in the digits and counts that make up the components. These differences could be reflecting aspects of growth and development along with particular evolutionary histories that characterize human populations.

B. Nonhuman Primate Studies

Although nonhuman primates (e.g., prosimians, monkeys, and apes) have not been studied thoroughly in terms of their dermatoglyphics, enough research has been done to clearly demonstrate a significance in dermatoglyphic variation comparable to that found in humans. However, it appears to be a lesser amount of variation that distinguishes nonhuman primates from humans. Descriptive series have been collected for both New and Old World species, and there is a tentative observation that, with respect to their digital dermatoglyphics, nonhuman primates show less intraindividual variation than is found in humans, with reduced bilateral differences as well as a lesser degree of sexual dimorphism. Additionally, nonhuman primates show more uniform expression of the whorl-type pattern across all digits.

At the species and population levels, basic comparisons have been made and applied to questions of phylogenetic relationships and taxonomy, with the microevolutionary mechanisms of population differentiation, and with discerning the functional and adaptive significance of dermatoglyphic features. Areas that have received little attention so far, but would be most appropriately examined in nonhuman primates, include investigations of inheritance patterns, the role of the environment in dermatoglyphic expression, and the details surrounding the morphogenesis of volar pads and primary and secondary ridges. Volar pads do persist as prominent structures in some nonhuman primates, which might account for their tendency to express more complex whorl-type patterns. These patterns, in turn, might represent a morphological adaptation involving extensive grasping function during locomotion and other manipulatory activities.

IV. MEDICAL–CLINICAL APPLICATIONS

During the prenatal period, when dermatoglyphic features are being formed, there can be disturbances that originate from genetic and/or environmental sources and alter the course of development. An understanding of this interaction assists in applying dermatoglyphic analysis to two medical situations. First, there is the potential of using dermatoglyphics in screening newborns for known birth defects. Furthermore, an analysis of dermatoglyphics of newborns could also be useful in detecting abnormal morphogenesis in general, and thus alert the clinician to closely examine for developmental defects. This application has been moderately successful as, for example, in profiling Down syndrome patients and patients with aberrant numbers of chromosomes, such as Trisomy 18. Importantly, it is not only the digital dermatoglyphic features that are informative. Very often the palmar and plantar dermatoglyphics are similarly diagnostic and should always be included in the analysis.

The second application involves attempts to discover associations between a given birth defect or medical disorder and unusual dermatoglyphic findings. If such an association is found, this would signify that the defect, or a predisposition toward developing the defect, had possibly originated as early as the end of the first trimester, when dermatoglyphic features are under formation. This kind of study requires an adequate control sample in order to specify just how unusual the patient's dermatoglyphics are, for the differences are rarely any more than statistical peculiarities of already known and described traits. A fairly common finding for digits is that affected cases have an increased frequency of arches or loops and a corresponding reduction in TFRC and PII. Associations have been found between dermatoglyphics and numerous congenital malformations, leukemia, cytomegalic inclusion disease, and rubella embryopathy. Environmental agents having teratogenic effects are also known to alter dermatoglyphic development.

ACKNOWLEDGMENT

Comments from Blanka Schaumann were greatly appreciated.

BIBLIOGRAPHY

Cummins, H., and Midlo, C. (1961). "Finger Prints, Palms and Soles: An Introduction to Dermatoglyphics." Dover, New York.

Durham, N. M., and Plato, C. C. (1990). "Trends in Dermatoglyphic Research." Kluwer Academic Publishers, Dordrecht, The Netherlands.

Jantz, R. L. (1987). Anthropological dermatoglyphic research. *Annu. Rev. Anthropol.* **16**, 161–177.

Loesch, D. Z. (1983). "Quantitative Dermatoglyphics: Classification, Genetics, and Pathology." Oxford Univ. Press, Oxford.

Mavalwala, J. (1977). "Dermatoglyphics: An International Bibliography." Mouton, The Hague.

Meier, R. J. (1975). Dermatoglyphics of Easter Islanders analyzed by pattern type, admixture effect, and ridge count variation. *Am. J. Phys. Anthropol.* **42**, 269–276.

Meier, R. J. (1978). Dermatoglyphics variation. *In* "Eskimos of Northwestern Alaska: A Biological Perspective" (P. L. Jamison, S. L. Zegura, and F. Milan, eds.). Dowden, Hutchinson & Ross, Stroudsburg, PA.

Meier, R. J. (1980). Anthropological dermatoglyphics: A review. *Yearbook Phys. Anthropol.* **23**, 147–178.

Meier, R. J. (1981). Sequential developmental components of digital dermatoglyphics. *Hum. Biol.* **53**, 557–573.

Plato, C. C., Garruto, R. M., and Schaumann, B. A. (1991). "Dermatoglyphics: Science in Transition." Wiley-Liss, New York.

Schaumann, B. A., and Alter, M. (1976). "Dermatoglyphics in Medical Disorders." Springer-Verlag, New York.

Diuretics

EDWARD J. CAFRUNY

University of Medicine & Dentistry of New Jersey

I. Physiology of Renal Salt and Water Excretion
II. Classification and Primary Renal Actions of Diuretics
III. Pharmacology and Therapeutics

GLOSSARY

Adverse reaction Undesirable or toxic reaction induced by a drug

Edema Surplus of extracellular fluid. The fluid may be confined to a part or spread throughout the body (systemic)

Hypokalemia, hypochloremia, hyponatremia Abnormally low plasma levels of K^+, Cl^-, or Na^+, respectively

Lumen (pl. lumina or lumens) Hollow space or cavity within a duct or tubule. Urine flows through the lumens of the kidney

Natriuretic Substance that increases the renal excretion of sodium salts and fluid

Nephron Functional unit of a vertebrate kidney comprising a glomerulus and attached tubule. More than 1 million nephrons are in each human kidney

Side effect Effect of a drug on an extraneous organ system (i.e., a system the therapist does not intend to influence). Side effects are not necessarily adverse

Tubular reabsorption Movement of any substance from the tubular lumen to the surrounding capillary network

Tubular secretion Movement of any substance from the surrounding capillary network to the tubular lumen

Visceral epithelium Single layer of flattened cells covering glomerular capillaries

Water diuretic Osmotic agent that increases urine flow but has little effect on the excretion of sodium salts

DIURETICS ARE DRUGS THAT ACT DIRECTLY ON the kidneys to increase the rate of excretion of urine and the sodium salts dissolved in urine. Because the direct source of the fluid and salt excreted is the circulating plasma, blood volume decreases. This leads to the formation of a gradient that favors the passive movement of interstitial fluid (extracellular fluid surrounding cells) into the bloodstream. The volume of the plasma reexpands as the interstitial fluid compartment contracts.

In past years, physicians used diuretics primarily to eliminate the disturbing surplus of extracellular fluid and salt (edema) that accumulates in patients with diseases such as chronic congestive heart failure, severe liver damage, or renal insufficiency; use of the drugs has been extended in recent years to ailments in which edema is not present or is not a conspicuous feature. Table I lists some of the common indications for dispensing diuretics. Physicians write more than 100 million prescriptions annually in the United States alone, a large fraction of the total for the treatment of high blood pressure.

I. PHYSIOLOGY OF RENAL SALT AND WATER EXCRETION

Animals that live on land regulate the volume and composition of their body fluids with precision to avoid sizable deviations that can interfere with the normal activity of cells. Because intake is variable, output also must vary. The vertebrate kidney is the primary organ for controlling output, although other organs assist in important ways. The kidney maintains salt and water balance by adjusting the volume and composition of the urine it makes. The renal output of water and dissolved solutes, added to output via other routes (e.g., lung, skin), then matches dietary intake.

The complexity of this task is immense. Table II lists the compartments that sequester body fluids. Average total body water content of a 70-kg adult is about 42

TABLE I
Some Indications for Diuretic Therapy

Condition	Effect of diuretics	Diuretic class
Congestive heart failure	Mobilize and excrete ECF[a]	All natriuretics except CAI[b]
Cirrhosis of the liver	Mobilize and excrete ECF	All natriuretics except CAI
Hypertension	Lower blood pressure	All natriuretics except CAI
Acute pulmonary edema	Improve lung and heart function	Loop agents
Impending renal failure	Arrest progression	Osmotic or loop agents
Glaucoma	Lower intraocular pressure	CAI only
Renal calcium stones	Decrease urinary calcium levels	Thiazides only
Elevated blood calcium	Increase urinary calcium excretion	Loop diuretics
Chemical intoxication	Increase urine flow to eliminate toxin	Loop diuretics
Hypokalemia	Increase blood potassium	Potassium-sparing diuretics

[a]ECF, extracellular fluid.
[b]CAI, carbonic anhydrase inhibitors.

liters, 60% of the body weight. More than one-half of this volume is virtually imprisoned within healthy cells, kept there by the osmotic forces of dissolved ions and molecules. Active transport systems control ionic movements into and out of many types of cells (discussed in the following). Intracellular water com-

TABLE II
Distribution of Water in an Adult Weighing 70 kg

Compartment	Liters	% of total body water
Extracellular H$_2$O		
Circulating Plasma	4	9
Interstitial	11	26
Total	15	35
Intracellular H$_2$O	23	55
Bone and transcellular H$_2$O	4+	10
Total body H$_2$O	42+	100

prises about 55% of the total fluid content of the body. Bone and the ducts and lumens of organs contain an additional 10%. These compartments, containing 65% of the total body water, resist rapid changes in volume. The kidney accomplishes its task by communicating with the circulating plasma, the most accessible component of the extracellular fluid compartment. Physicochemical and hormonal messages transmitted between the kidney and plasma compartments invoke the renal mechanisms that regulate fluid and electrolyte balance.

Figure 1 is a diagram of the structure of a single nephron, the basic unit of vertebrate kidneys. It shows the path that fluid filtered out of the blood-stream follows before leaving the body as urine. Some of the tubular segments of the diagram are convoluted; the

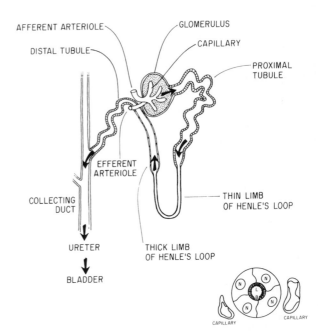

FIGURE I Diagram of the structure of a mammalian nephron. The afferent arteriole divides to form a capillary network within the glomerulus. An ultrafiltrate of capillary plasma begins its journey through the tubular lumens (top arrow), passing progressively through the proximal tubule, loop of Henle, and distal tubule. Incipient urine then enters the collecting duct and finally exits through the ureter before reaching the bladder. Glomerular capillaries merge and subsequently exit as the efferent arteriole. This arteriole then splits into a tubular capillary network (not shown in the large diagram). The cross section below illustrates the relation between a proximal tubule and the tubular capillary network. L, lumen; B, brush border of proximal cells; N, nucleus of cell. Luminal fluid and solutes traverse the cell to return to the blood via capillaries.

parts are not drawn to scale. More than a million nephrons are packed in a single human kidney. The first part of each unit, the glomerulus, comprises a central region of supporting cells and a capillary network through which blood flows. An outer capsule encloses these structures (see Fig. 1). A fraction of plasma water and low-molecular-weight solutes passes into the tubular lumens. This process is called *glomerular filtration*. The large sizes and shapes of protein molecules hinder their passage through the walls of the glomerular capillaries. Contractions of the heart transmit the energy needed to maintain the filtration process, the first step in the formation of urine. Because glomeruli filter about 180 liters every 24 hr (an amount 45 times greater than the plasma volume), almost all the fluid and solutes, other than waste products, must traverse the tubular cells and subsequently diffuse into tubular capillaries. The rest, usually 1.0–1.5 liters/24 hr, is urine. [*See* Urinary system, Anatomy.]

The crucial second step, tubular reabsorption, is energized by active transport systems in the cellular membranes of transporting segments.

Primary active transport (PAT) describes the movement of an ion or molecule between two compartments in a direction opposite to that predicted from its existing electrochemical gradient, which depends on the salt concentrations in the two compartments. Because the flux is "uphill," from a compartment of low concentration to one of high concentration, the transport is active (i.e., requires expenditure of energy). The sodium pump of the proximal tubule of Fig. 2 illustrates this type of active transport. The energy required to drive sodium out of the cell comes from the enzymatic breakdown of adenosine triphosphate (ATP) in the basolateral membrane. Potassium simultaneously enters the cells through the same pump but exits again because its concentration within the cell is higher than in the surrounding interstitial space. Sodium enters cells passively from the lumen because the pump ejects it so rapidly into the interstitial space that cellular concentration of the ion is always less than luminal concentration. In the illustration, a negatively charged chloride or bicarbonate anion enters passively, thus preventing the electrical charge separation that would supervene if only positively charged sodium ions moved out of the cell. Water follows the ions. The sodium pump is present in all transporting cells of the renal tubule. [*See* Adenosine Triphosphate (ATP); Ion Pumps.]

Secondary active transport (SAT) is "uphill" movement of an ion across a membrane, together with an

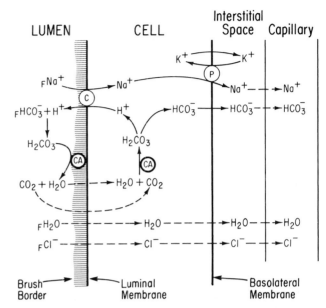

FIGURE 2 Diagram of ion transport systems in the proximal tubule. Filtered sodium (FNa^+) moves into the cell as H^+ moves out. Ions are countertransported across the membrane (C). Na^+ and K^+ are pumped across the basolateral membrane (P), the first entering the capillary, the second moving back into the cell. Within the cell, carbonic anhydrase (CA) catalyzes the hydration of CO_2 to form H_2CO_3, the source of H^+. Both HCO_3^- and Cl^- can follow Na^+ into the capillary. CA in the brush border catalyzes the dehydration of H_2CO_3 and the products of this reaction enter the cell to reform H_2CO_3. See text.

ionic partner traveling "downhill." A PAT system drives the partner across a second membrane. The Na–K–2Cl cotransport mechanism located in the thick limb of Henle's loop (see Fig. 1) is an example of SAT. Na^+ and K^+ in the lumens of this tubular segment serve as ionic partners, permitting Cl^- to enter the cell against an opposing electrochemical gradient. The ions are carried by a symporter, a transporter protein that helps an ion (in this instance, Cl^-) pass through a membrane with other ions. PAT across the basolateral membrane (the second membrane) establishes a favorable gradient for the admission of luminal Na^+ into the cell.

The proportions of water and sodium reabsorbed in each segment of the renal tubule are listed in Table III. Values listed are averages for healthy adults. These segments are potential sites of action for diuretics; the efficacy of a given diuretic will depend, in part at least, on the location and reabsorptive capcity of the part(s) of the tubule on which it acts. Note that a 1.0% reduction in tubular reabsorption of water will double the urine flow rate.

TABLE III

Reabsorption of Filtered Water and Sodium[a]

| Tubular segment | H$_2$O reabsorbed | Na$^+$ reabsorbed |
	(listed as a percentage of the amount filtered)	
Proximal tubule	70.0	70.0
Loop of Henle	15.0	22.0
Distal tubule and collecting duct	14.0	7.5
Total	99.0	99.5

[a]Glomerular filtration rate = 180 liters/24 hr; urine volume = 1.8 liters/24 hr; Na$^+$ filtered = 580 g/24 hr; Na$^+$ excreted = 3 g/24 hr.

II. CLASSIFICATION AND PRIMARY RENAL ACTIONS OF DIURETICS

The chemical structures and mechanisms of action of diuretic drugs vary considerably. For this reason, the classification is not rigid or systematic. Table IV illustrates this point. Thiazides are classified according to chemical structure; loop diuretics by site of action in

TABLE IV

Classification and Primary Action of Diuretics

Class	Primary action	Examples[a]
Carbonic anhydrase inhibitors	Prevent reabsorption of NaHCO$_3$ in proximal and late distal tubules	Diamox
Thiazides and thiazide-related	Inhibit reabsorption of NaCl in early part of the distal tubules	Hydrodiuril Hygroton
Loop agents	Inhibit Na–K–2Cl cotransport in loops of Henle	Lasix Edecrin
Potassium-sparing	Block receptors for Na$^+$–K$^+$ exchange in cells of late distal tubules	Aldactone
	Retard the entry of luminal Na$^+$ into late distal cells	Dyrenium Midamor
Osmotic	Hold fluid in lumens of all tubular segments	Mannitol (USP)

[a]Examples listed are brand names of the most frequently used preparations. Generic names appear in the text.

the kidney; the rest by the primary action they exert. Although each group includes many drugs, the table lists prototypes only.

Except for the osmotic group and spironolactone, all the diuretics prescribed in the United States are organic acids or bases. The glomeruli filter and the proximal tubular cells secrete these organic electrolytes. The secretory systems require the expenditure of metabolic energy. Proximal tubular secretion of diuretics enhances both the rate of delivery to and the concentrations achieved in the kidney.

III. PHARMACOLOGY AND THERAPEUTICS

A. Carbonic Anhydrase Inhibitors

Carbonic anhydrase (CA) is a zinc-containing enzyme that catalyses the reversible reaction CO$_2$ + H$_2$O = HCO$_3^-$ + H$^+$. The pH of the medium in which the reaction occurs establishes the prevailing direction of the reaction: a rise of pH increases the hydration of CO$_2$ to form HCO$_3^-$ (bicarbonate), whereas a fall increases the release of CO$_2$. In the proximal tubule, filtered sodium ions enter the cells passively, exchanging for hydrogen ions that replace luminal sodium. The mechanism of exchange, countertransport, is a form of cotransport and involves a membranal transporter. Figure 2 illustrates the process. Because cellular pH rises when H$^+$ leaves the cells, CA within the cell accelerates the formation of bicarbonate. The bicarbonate anion then crosses the basolateral membrane with Na$^+$ and diffuses into surrounding capillaries. CA in brush border membranes, where extruded H$^+$ ions acidify the environment, converts filtered bicarbonate to CO$_2$ and H$_2$O. The intracellular enzyme effectively replaces the filtered bicarbonate ions that are lost by this conversion. The upshot of this seemingly frivolous process is impressive. Replication keeps the plasma concentration of bicarbonate, an important acid–base buffer, from falling after the ion is filtered and subsequently converted to CO$_2$ and H$_2$O. In addition, a significant fraction of filtered sodium, accompanied by H$_2$O and cellular bicarbonate, undergoes reabsorption when H$^+$ exchanges for Na$^+$ (see Fig. 2).

Drugs that inhibit CA impede the countertransport process in the proximal tubule, and probably in the distal segment as well, by limiting the supply of H$^+$. The unreabsorbed fraction of Na$^+$ inevitably expands. Larger quantities are excreted with HCO$_3^-$. Urine flow rate increases because the ions hold water in what may be described as an "osmotic embrace."

Figure 3 shows the chemical structure of acetazolamide (Diamox), the prototypical CA inhibitor. It is a sulfonamide relative of antibacterial sulfa drugs. Its free sulfamyl group ($-SO_2NH_2$), however, bestows CA inhibitory activity. Acetazolamide is an effective diuretic but is not often prescribed for several reasons: (1) tolerance to its renal action develops within 3–4 days so that it becomes less effective; (2) acidosis (increased acidity of plasma) occurs when excessive excretion causes plasma bicarbonate levels to fall; and (3) newer diuretics that do not deplete plasma bicarbonate are available. A more frequent indication for CA inhibitors is the treatment of some types of glaucoma (elevated intraocular pressure). Inhibition of CA present in ocular tissues suppresses the production of ocular fluid and thus lowers intraocular pressure. Additional therapeutic indications include the prevention of epileptiform seizures and high-altitude sickness. Acetazolamide alkalinizes the urine, and elevated urinary pH accelerates the excretion of some, but not all, acidic drugs or toxins.

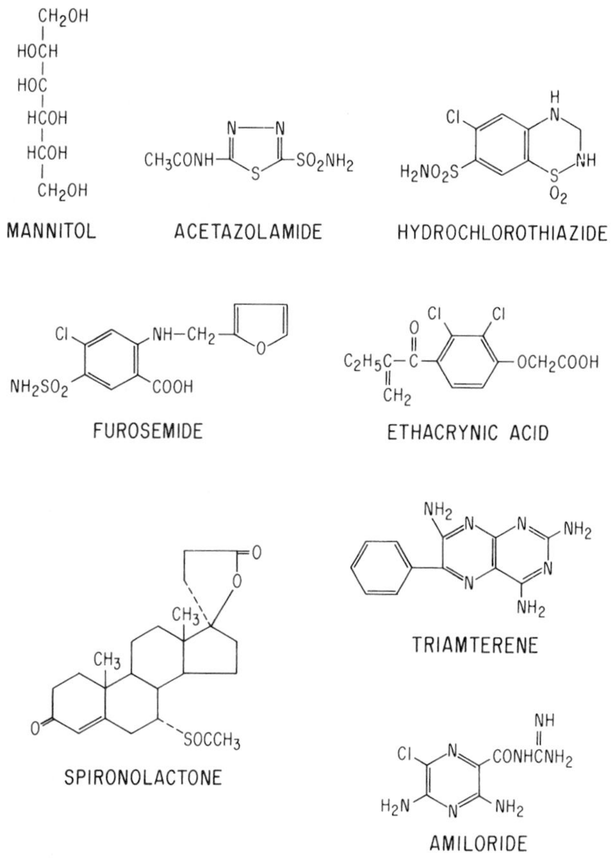

Adverse reactions to CA inhibitors occur less frequently when the dosage is not excessive and drug administration is intermittent or is interrupted periodically. Acidosis and potassium depletion are inevitable consequences of overuse. Depression, malaise, and gastrointestinal intolerance occur less often. Like other sulfonamides, CA inhibitors may rarely elicit signs of hypersensitivity: rash, drug fever, or disordered blood-cell formation.

B. Thiazides and Related Drugs

Successors to CA inhibitors, thiazides, appeared in 1960 and soon became the diuretics of choice because they were continuously effective in the treatment of edema (tolerance did not develop). Use of diuretics increased enormously when clinical researchers reported that thiazides lowered the blood pressure of hypertensive patients without provoking the unpleasant side effects of the poor drugs of the time that were, if anything, only marginally effective in controlling hypertension. Physicians soon discovered that they could reduce the dosage of antihypertensives that were likely to cause side reactions simply by prescribing a thiazide concurrently. Combination therapy not only maintained or improved the response but also reduced the incidence and intensity of adverse reactions.

Chlorothiazide, the progenitor of the thiazide series, promoted the urinary excretion of both bicarbonate and chloride salts. It contained a benzothiadiazide ring flanked by a sulfamyl group ($-SO_2NH_2$), the moiety that imparts CA inhibitory activity. The simple addition of two hydrogen atoms (for the synthesis of hydrochlorothiazide) had an important impact. Hydrochlorothiazide in clinically effective doses increased chloride excretion but had little effect on bicarbonate excretion. The addition of hydrogen atoms had weakened the ability to inhibit the enzyme. The chemical structure of hydrochlorothiazide, the most popular member of the series, appears in Fig. 3. The sulfamyl group is not replaceable but the benzothiadiazide ring structure may be altered without loss of diuretic activity. Absence of significant effect of hydrochlorothiazide and newer members of the thiazide group on bicarbonate excretion solved the problems of acidosis and refractoriness.

The primary site of action of thiazides is the early portion of the distal tubule, where they interfere with the electroneutral passage of NaCl across the luminal

FIGURE 3 Chemical structures of the diuretics discussed in the text.

membrane, retarding entry of the salt into the cells. If the major site of action were in the proximal tubule, a large part of the effect probably would be annulled by a compensatory rise in the activity of the Na–K–2Cl cotransport system in the loop of Henle. Because the entire distal tubule and collecting duct system reabsorbs only about 7–8% of filtered sodium (see Table III), an action in the distal segment cannot be expected to be as pronounced as a similar action in the loop of Henle, which controls the reabsorption of as much as 25% of filtered NaCl. However, it is not often desirable or necessary to reduce the reabsorption of sodium more than 2–3%. A change of this magnitude will usually double or triple the flow of urine.

All the useful thiazides share the same mechanism of action and are equally effective. Differences with respect to oral absorption, distribution and metabolic fate in the body, and rapidity of excretion ("pharmacokinetic properties") account for obvious differences in potency and duration of action. As a general rule, potency (i.e., the dose required to achieve a given response) is not important because efficacy (i.e., the maximal response achievable) is the same for all. [See Pharmacokinetics.]

Most of the complications of thiazide therapy are manifestations of their pharmacologic effects on the kidney. Depletion of potassium is the chief problem, but other ions also may be depleted. The plasma levels of uric acid and of calcium may rise when the extracellular fluid volume declines. Elevated serum uric acid can cause gouty attacks in susceptible people. Several studies have shown that serum cholesterol may increase during a period of 2–4 months when patients take thiazides daily. The elevation is small, 5–6%, and is not evident in several clinical trials that lasted a year or more. Thiazides also lessen sensitivity to insulin, an effect that prompts readjustment of the latter drug's dosage in diabetics. The clinical relevance of this finding in nondiabetics is not established at this time. [See Cholesterol.]

Before 1960, the only drugs that could substantially increase the renal elimination of sodium were the mercurial diuretics (not orally effective) and the CA inhibitors (not continuously effective). The arrival of thiazides resolved these two problems. Similarly, there were no safe and effective drugs for lowering blood pressure. Thiazides filled this void and soon became first-line agents in the pharmacotherapy of hypertension. Although newer drugs can lower pressure more effectively or evoke diuresis in patients who are refractory to the older drugs, thiazides continue to be maintained stays. In the United States alone, more than 15 million patients use them.

C. Loop Diuretics

The diuretic response to thiazides is usually hindered or inadequate when the glomerular filtration rate is depressed, heart failure is severe, or the liver is damaged. In such cases, the therapist uses an inhibitor of the Na–K–2Cl cotransport, the SAT system that controls the entry of monovalent ions into cells of the ascending limbs of Henle's loop. These inhibitors are often called "high-ceiling diuretics" because they can induce the renal elimination of enormous quantities of salt and water. Cells of the loop reabsorb 20–25% of the filtered sodium (see Table III).

Two chemically unrelated types of loop agent are available in the United States (see Fig. 3): furosemide, a sulfamylated derivative of anthranilic acid, and ethacrynic acid, an unsaturated ketone. A third, bumetanide, is a sulfamylated compound pharmacologically akin to but more potent than furosemide.

Most investigators discount, but do not disprove, evidence that ancillary sites of action of the high-ceiling diuretics exist in parts of the nephron other than the loop. The argument against actions in the proximal and distal tubules is based on the assumption that a reduction of NaCl reabsorption proximal to the loop of Henle can be compensated in the loop, whereas a more distal action would be quantitatively trivial in any case. This notion neglects the well-established fact that high-ceiling diuretics limit the capacity of the loop to compensate for an action that takes place upstream. Response to loop diuretics depends on dosage. At low levels, NaCl elimination is equivalent to that elicited by thiazides; at high levels, NaCl and water losses exceed by far any losses that a thiazide can muster. Response, however, depends on the delivery of drug to site of action. When renal function is impaired, less drug gets to the kidney. In such cases, large doses of loop diuretics can be both safe and effective whereas no quantity of a thiazide will suffice. Furosemide is an agent of choice in acute pulmonary edema. It increases pulmonary and venous compliance, affording rapid relief, and then maintains the beneficial effects by reducing plasma volume.

Adverse reactions to loop diuretics include all the depletion phenomena and metabolic derangements listed for the thiazides. In addition, vertigo and deafness may develop in patients who require large intravenous doses. It is prudent to avoid the concurrent administration of certain (potentially ototoxic) antibiotics.

D. Potassium-Sparing Diuretics

Depletion of body potassium with or without hypokalemia (significant reduction of potassium in the blood) is the most common adverse reaction attributable to diuretics. Depletion can initiate serious disorders in the rhythm of the heart, impair the function of nerves and muscles, and cause intestinal disturbances. Hypokalemia is especially troublesome in patients with congestive heart failure who take both a digitalis preparation (a cardiac glycoside) and a diuretic. Potassium wasting sensitizes the heart to the toxic effects of the cardiac glycoside. Hypokalemia can sometimes be avoided if the patient eats foods that contain large amounts of potassium or, alternatively, takes supplements containing potassium chloride. If these measures fail, oral supplements should be stopped and a potassium-sparing agent started. Because this category of drug influences the tubular transport of only small quantities of Na^+, the diuretic response is weak.

Drugs of this group comprise two pharmacologically distinct groups. The steroidal antagonists of aldosterone resemble the hormone aldosterone, which is synthesized in the adrenal cortex. Aldosterone attaches to a cytoplasmic receptor in the late portion of the distal tubule. The aldosterone–receptor complex subsequently migrates to the nucleus, where it stimulates the production of messenger RNA (mRNA) molecules. The mRNA molecules then move into the cytoplasm. Here, they direct the formation of a protein that increases the permeability of the luminal membrane to Na^+, facilitating the entry of larger quantities of luminal Na^+ into the cell. The sodium pump spurs egress from the cell. Cellular potassium ions move into urine to replace some of the sodium ions that left, thus increasing the excretion of potassium. Because the potassium-sparing drug, spironolactone (see Fig. 3), is structurally similar to the natural hormone, it can attach to the same cytoplasmic receptor. However, the two steroids differ enough so that spironolactone cannot initiate formation of the mRNA responsible for generating the permeability-enhancing protein. When the drug is provided, fewer receptor sites are free to react with the natural hormone. For this reason, the drug behaves as a competitive inhibitor.

The second group of potassium-sparing drugs includes two organic bases: (1) triamterene, an aminopteridine chemically related to folic acid, and (2) amiloride, a pyrazinoylguanidine. Both drugs slow the entry of luminal sodium into cells of the late portion of the distal tubule. The molecular mechanism of this action is unknown. Neither drug competes directly with aldosterone. Researchers suggest that both may cause closure of channels through which Na^+ ions gain entry. Like spironolactone, the drugs are weak diuretics but do strongly promote the conservation of potassium.

Triamterene and amiloride are usually prescribed in combination with a thiazide or loop diuretic when maintenance of a normal serum potassuim level is essential (e.g., patients with abnormal cardiac rhythms, low levels of serum potassium, or taking a digitalis preparation). They are sometimes used together with another diuretic to amplify the response of individuals who are refractory to a single drug.

All the potassium-sparing diuretics may cause hyperkalemia, even when a potassium-wasting diuretic is prescribed concurrently. This adverse reaction occurs more often when renal function is impaired or renal reserve is limited (e.g., in diabetic or elderly patients). The drugs should not be prescribed together with potassium supplements, when serum potassium exceeds the normal value, or when renal function is poor.

Because of its steroidal structure, spironolactone can also attach to nonrenal hormonal receptors and provoke unpleasant reactions. Examples are menstrual irregularities, hirsutism in females, and gynecomastia in females or males.

E. Osmotic Diuretics

Osmotic agents do not react with receptors or directly inhibit any renal transporting system. Their activity depends, instead, on the osmotic force their molecules exert in solution. When present in interstitial fluid, the molecules attract cellular fluid to the extracellular compartment; in the tubular lumens of the kidney, they oppose the movement of water into renal cells. The amount of water held back depends on the number of molecules of the osmotic diuretic present in the lumens, not on their molecular size. For this reason, the molecular weights of all useful agents are low. Large molecules would increase the weight of the required dose. In addition, the ideal osmotic diuretic should not undergo reabsorption from the tubular fluid or metabolism (chemical alteration) in the body. These events would reduce its concentration in the renal lumens, the sites of action.

Mannitol, a six-carbon sugar (see Fig. 3) found in certain plants, approaches this ideal and is the drug of choice. It is not effective when given orally because it does not cross gastrointestinal membranes. Isosorbide and glycerol are effective by the oral route but are less efficient because they penetrate cellular membranes in the kidney (are reabsorbed). When injected intravenously, mannitol is distributed throughout the extracellular fluid compartment but does not penetrate cells. The extracellular fluid volume increases rapidly as the osmotic agent draws fluid from the cells. In the proximal tubule, mannitol slows the passive reabsorption of water that normally follows the active transport of sodium. In effect, the osmotic force of a nonreabsorbable solute opposes the osmotic force of reabsorbable sodium. Urine flow rises in proportion to the dose of the osmotic agent. As a general rule, the fractional excretion of sodium increases but to a much smaller extent than the fractional excretion of water. For this reason, osmotic drugs are not first-line diuretics. Loss of fluid unaccompanied by salt induces thirst. A patient soon reaccumulates fluid. More commonly, mannitol is used to reduce the elevated intracranial pressure of cerebral edema, to prevent the development of renal failure associated with severe trauma or lengthy surgical procedures, or to promote renal "washout" of drugs or toxins.

Initially, the rapid expansion of extracellular fluid brought about by osmotic drugs increases the work of the heart, and some patients suffering from congestive heart failure may develop pulmonary edema. Later, especially when administration of osmotic diuretics is prolonged and the renal response is large, severe volume depletion and electrolyte imbalances supersede.

BIBLIOGRAPHY

Cafruny, E. J. (1991). Diuretics: Drugs that increase the excretion of water and electrolytes. *In* "Human Pharmacology" (L. B. Wingard, T. M. Brody, J. Larner, and A. Schwartz, eds.). Mosby, St. Louis.

Cafruny, E. J., and Itskovitz, H. (1982). Sites of action of loop diuretics. *J. Pharmacol. Exp. Ther.* **223**, 105–109.

Goodman, L. S., Limbird, L. E., and Milinoff, P. B. (eds.) (1996). "Goodman & Gilman's the Pharmacological Basis of Therapeutics," 9th Ed. McGraw–Hill, New York.

Lant, A. (1985). Clinical pharmacology and therapeutic use of diuretics. *Drugs* **29**, 57–87 (Part I), 162–188 (Part II).

Maren, T. H. (1967). Carbonic anhydrase: Chemistry, physiology, and inhibition. *Physiol. Rev.* **47**, 597–781.

Shackleton, R., Wong, N. L. M., and Sutton, R. A. L. (1986). Distal potassium-sparing diuretics. *In* "Diuretics: Physiology, Pharmacology, and Clinical Use" (J. H. Dirks and R. A. L. Sutton, eds.), pp. 117–134. Saunders, Philadelphia.

Vander, A. J. (1995). "Renal Physiology," 5th Ed. McGraw–Hill, New York.

Velazquez, H. (1988). Thiazide diuretics. *Renal Physiol. (Basel)* **10**, 184–197.

DNA and Gene Transcription

MASAMI MURAMATSU
Saitama Medical School

GLOSSARY

Amphipathic α-helix α-helical portions of a polypeptide that have affinity to both polar and nonpolar solvents

***cis*-acting element** When a DNA sequence regulates a gene on the same DNA molecule, the sequence is said to act in *cis*; in this article it refers to any DNA sequence element that participates in the regulation of the gene on the same DNA

Consensus sequence Most common or frequent sequence chosen from available data when variable sequences exist for a *cis*-acting element (a regulatory signal on DNA)

Domain Portion of a protein molecule having a defined function

***trans*-acting factor** Any protein factor that acts on the regulation of a gene but is encoded on a separate DNA molecule; in this article it is a regulatory protein of a gene that works by interacting with one of the *cis*-acting elements

DEOXYRIBONUCLEIC ACID (DNA) IS THE GENETIC material and the central molecule of life. Its major function is to express genetic information via messenger RNA to achieve the phenotype of an organism. In this article, the first step of gene expression—transcription—will be detailed together with some fundamental aspects of the molecular structure of DNA.

I. WHAT IS DNA?: STRUCTURE, PROPERTIES, AND FUNCTION

A. Definition and Structure

DNA is an informational macromolecule contained in virtually all living organisms[1] on earth, and it has all of the genetic information of each species to which it specifically pertains. It is a long, filamentous molecule consisting of a backbone of D-2-deoxyribose moieties connected by 3′-5′-phosphodiester linkages, each having one of the four nitrogenous bases—adenine (A), guanine (G), thymine (T), or cytosine (C)—at the 1′-C atom of the deoxyribose. DNA is actually a polymer of deoxyribonucleotides and is synthesized in the cell from four kinds of deoxyribonucleoside triphosphates. [*See* Enzymology of DNA Replication.] DNA is made up of two filaments, or strands, connected to each other by hydrogen bonds between A and T, and G and C, forming the Watson–Crick double helix (see Color Plate 9). The two strands are said to be complementary.

The complementary double helix is of prime importance in every aspect of DNA function such as replication, transcription, repair, and recombination. A DNA double helix usually takes one of three different conformations, designated the A-, B-, or Z-form, depending on its environment and sequence. The main features of each form are presented in Table I. The A- and B-forms are right-handed and wind rather smoothly, whereas the Z-form is left-handed and zigzag shaped (for which it is named). Under physiological conditions, most DNAs appear to exist as the B-form. Some DNAs, which have a special nucleotide sequence such as a stretch of alternate CG or TG, assume the Z-form *in vivo*. This is especially true

---

[1]Some viruses have RNA genomes and no DNA.

TABLE I

Conformation Comparison of Three Forms of DNA

	A-form	B-form	Z-form
Overall shape	Short and thick	In between	Long and thin
Winding direction	Right-handed	Right-handed	Left-handed
Pitch per turn	24.6 Å	33.2 Å	45.6 Å
Base pairs per turn	10.7	10.0	12
Diameter of helix	25.5 Å	23.7 Å	18.4 Å
Tilt of base to helix axis	$+19°$	$-1°$	$-9°$
Major groove	Narrow and deep	Wide and medium deep	Flat and exposed on surface
Minor groove	Broad and shallow	Narrow and medium deep	Narrow and deep

when some of the C residues are methylated. Negative supercoiling of DNA that gives a stress of anticlockwise rotation to the long axis of DNA double helix also favors the formation of the Z-form. The specific effects of DNA nucleotide sequence on its higher-order structure are not yet fully understood, particularly when there are DNA–protein interactions, as in chromatin. These aspects of higher-order structure in relation to the interaction with proteins will be one of the most important targets of future research.

B. Properties of DNA

Because double-stranded DNA is a very long, filamentous molecule, DNA solutions have a high viscosity, which disappears by denaturation (separation of the two strands) or shearing. Due to their great length, DNA molecules are rather fragile and are easily broken into smaller fragments by just pipetting through a narrow tube. When DNA molecules are centrifuged in a high concentration of CsCl, the latter forms a density gradient in the solution and the DNA molecules form a band where the density of the solution is equal to that of the DNA molecule. Because the DNA molecule is much denser than a protein molecule, this is an excellent procedure for isolating DNA from biological materials.

The DNA double helix may be separated by a variety of means, including high temperature and high pH, to break hydrogen bonds, which causes transition from a double helix to a single strand (denaturation). The temperature at which half of the DNA molecules denature is called the melting temperature (T_m). T_m elevates as the GC content of DNA increases due to the increased amount of hydrogen bonding; a GC pair has three hydrogen bonds, whereas an AT pair has only two. Denatured single-stranded DNA can reform native double-stranded DNA when incubated at a

proper temperature if the two strands have sufficient complementarity (renaturation). Single-stranded RNA having a complementary sequence can also form a (hetero)duplex with a single-stranded DNA by renaturation (hybridization). The ability of a DNA strand to form homo- and heteroduplexes with a complementary nucleic acid (DNA or RNA) strand is the central function of DNA *in vivo* and *in vitro*. These features form the basis of DNA analysis methodology.

Chemically, DNA is much more stable than RNA because of the absence of the 2′-OH group in the ribose and by the protection provided by hydrogen bonding in the double helix. Nevertheless, DNA may be attacked by a number of enzymes that act in various ways. DNases (endonucleases) randomly cut a DNA molecule within its nucleotide sequence; exonucleases cleave it from either the 3′ or 5′ end. Specific cleavage at a defined nucleotide sequence may be obtained by an enzyme family designated restriction endonuclease (or restriction enzyme). The existence of a large number of restriction enzymes that can cut a variety of possible combinations of nucleotide sequence makes them an important tool of gene engineering. DNA ligases, which can splice DNA; reverse transcriptase, which can synthesize DNA on RNA templates; nuclease S1 (or P1), which digests only single-stranded DNAs; and RNase H, which digests only RNA in DNA–RNA hybrids are among the other DNA-related enzymes frequently used in gene technology.

C. Function of DNA

DNA is the central molecule of life in that it stores genetic information, replicates itself, and then expresses the information as the phenotype of a living organism. The flow of genetic information is drawn schematically as follows:

$$\text{Replication}\left(\text{DNA} \underset{\substack{\text{Reverse}\\\text{transcription}}}{\overset{\text{Transcription}}{\rightleftharpoons}} \text{RNA} \xrightarrow{\text{Translation}} \text{Protein}\right.$$

The major functions of DNA are replication and transcription, or, more precisely, gene expression. Other important aspects of DNA function are repair (which is accompanied by DNA synthesis) and recombination (which includes transposition). In this article, only one major aspect of DNA function—gene expression (transcription)—will be discussed in molecular detail. The focus will be on mammalian systems because they are the most relevant to human biology.

II. MECHANISM OF TRANSCRIPTION

A. Chemistry of Transcription or RNA Synthesis

Cellular RNA, including messenger RNA (mRNA), ribosomal RNA (rRNA), transfer RNA (tRNA), and small nuclear RNA (snRNA), are synthesized on respective DNA templates. The overall stoichiometry may be written as follows:

$$\left.\begin{cases} n_1 \text{ ATP} \\ n_2 \text{ GTP} \\ n_3 \text{ CTP} \\ n_4 \text{ UTP} \end{cases}\right\} + \text{template} \xrightarrow[\text{Mg}^{2+}, (\text{Mn}^{2+})]{\overset{\text{RNA polymerase}}{\downarrow}}$$

(mononucleotides)

$$\begin{array}{c} \text{AMPn}_1 \\ | \\ \text{GMPn}_2 \\ | \\ \text{CMPn}_3 \\ | \\ \text{UMPn}_4 \end{array}$$

(a polynucleotide)

$$+ (n_1 + n_2 + n_3 + n_4)\text{PP}_i + \text{template}$$

where ATP and AMP are adenosine tri- and monophosphate, respectively; GTP and GMP are guanosine tri- and monophosphate, respectively; CTP and CMP are cytidine tri- and monophosphate, respectively; UTP and UMP are uridine tri- and monophosphate, respectively; PP_i is inorganic pyrophosphate; and n_2–n_4 are variables defined by the base composition of the template DNA. The essential reaction of the polymerization of an RNA chain is the nucleophilic attack of the α-phosphate group of a nucleoside triphosphate by

the 3′-hydroxyl group of the ribose moiety at the 3′ end of RNA. Phosphodiester linkages are formed by the energy released from the hydrolysis of the nucleoside triphosphates into component monophosphates. Template DNA assists in the alignment of the incoming nucleotide to the proper position by hydrogen bonding, thus determining the complementary nucleotide sequence. Note that the RNA chain elongates from 5′ to 3′ direction by this mechanism.

B. Initiation of Transcription

For correct readout of DNA, transcription must start from a fixed point, i.e., the top of the gene. In prokaryotes such as *Escherichia coli*, the signal that directs transcription initiation, the promoter, is composed of three regions near and upstream of the transcription start site. These are the Pribnow box at −10 (nucleotide positions are numbered with minus for upstream and with plus for downstream from the transcription start site [+1]), consisting of 5–6 bp (consensus: TATAAT), the "−35 sequence," having another specific sequence of 5–6 bp (consensus: TTGAC), and a region consisting of several nucleotides around the actual transcription initiation site.

For transcription to begin, RNA polymerase must bind to the initiation site. RNA polymerase is composed of five subunits; four subunits (two α's, β, β') constitute the core of the enzyme, while the fifth subunit (σ) is required for promoter recognition and transcription initiation. Once transcription begins, the σ factor is released from the transcribing complex and the core polymerase continues RNA chain elongation. Prokaryotic gene transcription is regulated mainly by a sequence called the operator, located near the promoter. The operator binds repressor molecules, which in turn interact with RNA polymerase and inhibit transcription initiation. Prokaryotes also have positive regulation; for instance, the binding of the catabolite gene activator protein (CAP)–cyclic adenosine monophosphate complex to the promoter of the *lac* operon can stimulate transcription.

III. TRANSCRIPTION INITIATION IN EUKARYOTIC CELLS

A. Three Polymerase Systems Are in Operation in Eukaryotes

All eukaryotes, from yeast to humans, have three RNA polymerase systems, which transcribe different categories of the gene (Table II). RNA polymerase

TABLE II
Classes of Eukaryotic RNA Polymerase

RNA polymerase	Localization	α-Amanitin sensitivity	Product
I (A)	Nucleolus	—	rRNA
II (B)	Nucleus (chromatin)	$+++$ $(0.0025\ \mu g/ml)^a$	mRNA, snRNA
III (C)	Nucleus (nucleoplasm)	$+$ $(25\ \mu g/ml)^a$	5SRNA, tRNA, snRNA, VA RNA[b]

[a]ID_{50}.
[b]Virus-associated RNA.

I transcribes solely ribosomal RNA genes that are repeated several hundred times in most eukaryotic genomes. This polymerase is characterized by complete insensitivity to the fungal toxin α-amanitin but extreme sensitivity to the antibiotic actinomycin D. Ribosomal RNA synthesis is inhibited >95% when 0.05 μg/ml of actinomycin D is added to the culture medium of animal cells. More than 20 times this concentration is required to inhibit the transcription of genes by RNA polymerase II.

RNA polymerase II transcribes protein-coding genes and most of the snRNA genes. It is very sensitive to α-amanitin, as shown in Table II. RNA polymerase III transcribes transfer RNA, ribosomal 5S RNA, and some snRNA genes, as well as virus-encoded low molecular weight RNAs. This enzyme is characterized by a moderate sensitivity to α-amanitin (Table II). All three polymerases consist of 2 large subunits (resembling E. coli β and β') and nearly 10 smaller subunits. About half of the components are unique among these three enzyme classes, with the remaining being shared by two or three classes.

B. Promoter Sequences and General Transcription Factors

Each category of the eukaryotic gene has its own promoter sequence and is specifically transcribed by the cognate polymerase. For precise initiation and control of gene transcription, eukaryotes have developed multiple transcription factors that recognize and bind the promoter sequence. These protein factors are required for transcription in each polymerase system and are called general transcription factors. Other transcription factors are gene- or category-specific (see Section IV,B). Promoter sequences and the general transcription factors for each class of polymerase are described below.

I. RNA Polymerase I System

There is no consensus promoter sequence for RNA polymerase I common to different species, even among vertebrates. This is reflected in the species-specificity of this system among mammalian species such that a crucial general transcription factor cannot be exchanged between the human and mouse. Apparently, this evolutionary divergence occurred only in the RNA polymerase I system. Because it transcribes only ribosomal RNA, RNA polymerase I and its factors may have coevolved with its cognate promoter sequences much faster than the RNA polymerase II system where the polymerase must interact with factors specific for many different genes.

The promoter region of the eukaryotic ribosomal RNA gene can be divided into two parts: one is the core promoter encompassing the transcription start site and part of the upstream sequence (up to -50 depending on species); the other is the upstream control element that is present near -100 (and varies among species). These sequences are recognized and bound by a protein complex (designated TFID, TIF-IB, or SL-1) with the assistance of UBF1, another factor. SL-1 is composed of TATA-binding protein (TBP) and at least three TBP-associated proteins (TAFIs). RNA polymerase I then recognizes and binds this complex with the possible assistance of another factor, TFIA. When the four kinds of nucleoside triphosphates are supplied to this preinitiation complex, RNA polymerase I begins RNA chain-elongation; the other factors are left behind on the promoter. A new polymerase I molecule can now join the promoter complex, allowing transcription to be initiated many times. Transcription by RNA polymerase I stops at a termination site having a specific sequence. In the case of the mouse, it is called the *Sal*I box, to which a termination factor is known to bind.

2. RNA Polymerase II System

There are two types of genes in this category: one has a TATA box (the comsensus sequence is TATA, but it is subject to slight changes) at approximately -30 and the other has no TATA box. In many cases, multiple GC boxes (GGGCGG) are found in the upstream region instead. The latter type of promoter is seen frequently in so-called housekeeping genes, which code for general proteins or enzymes required for basic functions of the cell such as energy production. In genes with a TATA box, the first step of transcription initiation is the binding of TFIID to the TATA box. TFIID is composed of the TBP and more than seven (depending on species) TBP-associated factors (TAFII).

RNA polymerase II, with at least six different factors (TFIIA, TFIIB, TFIIE, TFIIF, TFIIH, and TFIIJ), can then form the initiation complex and begin transcription. As the purifications of these factors progress, the precise order of the binding and the function of each factor is now emerging. TFIIA appears to stabilize the TATA box–TFIID complex, and TFIIF binds directly to RNA polymerase II and helps its entry in the initiation complex. TFIIB appears to bind with the promoter from the opposite side of TBP and also interacts with upstream transcription factors. TFIIH has three enzyme activities (CTD kinase, DNA helicase, and ATP ase) and promotes transcription initiation. It should be mentioned that many of these factors are composed of multiple subunits. For TATA-less promoters, TBP is also required, as well as other factors, but precise interaction of these factors with DNA remains to be determined. To understand the functions of all these factors, more knowledge about the function of each subunit of RNA polymerase II, as well as that of transcription factors interacting with them, is required. Termination of transcription of RNA polymerase II is not well defined but appears to occur far downstream from the poly(A) addition site and may occur at multiple sites, causing transcription to cease gradually.

3. RNA Polymerase III System

A remarkable feature of the promoters for RNA polymerase III is that they are usually present inside the gene, i.e., in the transcribed region of the gene. In the 5S ribosomal RNA gene, the internal promoter is divided into three regions known as Box A, the I region, and Box C. The sequences and the spacing of these regions are rigorously determined. Transfer RNA genes also have two regions—the A- and B-blocks—as the internal promoter. The A-block has a sequence similar to Box A of the 5S RNA gene. Both the 5S ribosomal RNA gene and the tRNA gene, having the internal promoter in common, appear to have an upstream sequence, which enhances the promoter activity.

RNA polymerase III utilizes at least three transcription factors: TFIIIA, TFIIIB, and TFIIIC. In 5S RNA gene transcription, TFIIIA first binds with the internal promoter followed by TFIIIC and TFIIIB binding and complex stabilization. RNA polymerase III can then bind and transcribe the gene. TFIIIC, but not TFIIIA, is required for tRNA gene transcription. TFIIIC first binds with the gene recognizing both A- and B-blocks, followed by TFIIIB and RNA polymerase III binding, resulting in transcription. Recently, TFIIIB was found to contain TBP, thus making the latter the central molecule for all classes of transcription machinery. Termination occurs either at a stretch of more than four straight Ts (TTTT) or at a sufficiently long repeat of ATs.

IV. HOW IS GENE TRANSCRIPTION REGULATED?

A. *cis*-Acting Elements

Eukaryotic genes are exquisitely regulated to achieve the developmental program of the genome and to maintain homeostasis of the cell and the whole body against external stimuli and perturbation. Regulation is carried out by a number of sequences that are engaged in the control of gene transcription (Fig. 1). These can be divided into several categories. The first

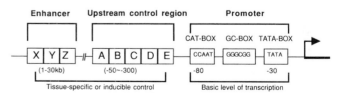

FIGURE I *cis*-acting elements in gene control region. Not all promoters have all these elements. Both the upstream control region and the enhancer consist of more than one element and may share some of the elements (A ~ E, X ~ Z). Only the enhancer can work at a distance.

category, the promoter, specifies the transcription start site of every gene. It is composed of the TATA box and, in many cases, some sequence at the transcription start site. In genes without a TATA box, GC boxes with certain unidentified sequences near the transcription start site appear to determine transcription initiation. General transcription factors previously described also bind to these structures, forming basic transcription apparatus.

The second category is the upstream regulatory region, or the enhancer. Enhancers, when first found in some viruses such as SV40, were defined as activating DNA elements that exerted influence from distant places in an orientation-independent manner, either upstream or downstream of the promoter. Some enhancers, such as the immunoglobulin heavy-chain enhancer, were further found to be tissue specific. (Tissue specificity may provide a basis for specific gene activation during differentiation.) However, it was later found that some other cell- and/or induction-specific regulatory sequences work only at a certain upstream proximity to the promoter and in an orientation-dependent manner. These seem to be intermediate forms of promoters and enhancers and are regarded either as a part of the promoter (together with the core promoter represented by TATA box) or a type of enhancer working at short range. They are sometimes referred to as upstream control regions (or sequences).

Some sequences with positive effects are frequently found near and upstream of the TATA box (e.g., CAT box [CCAATG] and GC-box [GGGCGG]). They are thought to strengthen the promoter activity and are sometimes included in the promoter structure. Most upstream control sequences are found within several hundred base pairs of the promoter; however, other enhancers, many of which are tissue specific, are located at several kilobases and sometimes as far as 30 kb upstream. Both upstream control sequences and enhancers are generally composed of a few elements clustered within several dozen base pairs. Because many of these control regions share some of the elements with homologous sequence, the cell or gene specificity of an enhancer may be determined by the combinatorial interaction of these elements with different trans-acting factors of a particular cell type. Alternatively, one cis-acting element sometimes partially overlaps or superimposes another element, allowing competition as well as synergism between different trans-acting factors. Negative control elements, called silencers or dehancers, which cancel out enhancer effects, have also been described. Some of the cis-acting elements that have been relatively well studied are shown in Table III together with their cognate transacting factors.

B. trans-Acting Factors

The previously mentioned cis-acting elements exert their gene-activating or -repressing effects by binding specific protein factors. Several examples of this are presented in Table III. These protein factors, generally called trans-acting factors, are also transcription factors. They are generally low in abundance in the cell. By using gene engineering methodology, various deletion and substitution mutants of these factors have been synthesized and functionally tested in cultured cells. The results indicate that these transacting molecules have separable "domain" structures, with each domain having a different function, such as DNA binding, protein–protein interaction, and trans-activating capacities.

1. DNA-Binding Domains

For most of the trans-acting factors, a DNA-binding domain consisting of several dozen amino acids is required, although not sufficient, for activity. There are at least five, and probably more, categories by which a trans-acting factor could interact with DNA (Table III; Fig. 2). The first one, known as a zinc finger, was originally found in a general transcription factor TFIIIA (see Section III,B,3) and in the steroid hormone receptor family. Zinc fingers are characterized by about 30 amino acids that include two cysteine and two histidine (or again cysteine) residues forming a fingertip-like structure surrounding a Zn^{2+} ion. The trans-activator SP1, which is found rather ubiquitously in mammalian cells and binds with the GC box, has three zinc fingers near its carboxy terminus. These proteins bind with DNA by inserting the finger structure into the major groove of DNA. The sequence specificity appears to be determined by some amino acids near the finger (e.g., at the base of the stem of the finger).

The second category by which a trans-acting factor could bind DNA is the helix-turn-helix-type structure. This structure has been well analyzed in prokaryotic repressor molecules and, in eukaryotes, is represented by a so-called homeodomain. This domain, containing about 60 amino acids, was first identified in Drosophila homeotic genes that regulate different aspects of embryonic development. It has since been found in a number of mammalian transcription factors, including OCT-1 (or OTF-1, NF-A1, and NFIII),

TABLE III
Mammalian *trans*-Acting Factors and Binding Sequences[a]

trans-Acting factor	DNA-binding site (5′ → 3′)	Class	Some characteristics
SP1	GGGCGG[b]	Zinc finger	Glutamine-rich domain, O-glycosylated
GR[c]	GGTACAN₃TGTTCT[d]	Zinc finger	Act as dimer
C/EBP	TGTGGAAAG	Leucine zipper	Rat liver nuclear protein, dimerization
c-*Jun*	TGA ^G_C TCA[e]	Leucine zipper	Form heterodimer with c-*fos*
CREB	TGACGTCA[f]	Leucine zipper	Weakly cross-reactive with c-*Jun*-binding site
OCT-1	ATTTGCAT[g]	Helix-turn-helix[h]	Ubiquitous
OCT-2	ATTTGCAT[g]	Helix-turn-helix[h]	Lymphoid cell specific
OCT-3	ATTTGCAT[g]	Helix-turn-helix[h]	Embryonic cell-specific, proline-rich domain
Pit-1 (GHF1)	$^{TT}_{AA}$ TATNCAT	Helix-turn-helix	Pituitary specific, serine, threonine (tyrosin)-rich domain
CTF/NF-1	GCCAAT		Proline-rich domain by alternative splicing
SRF	GATGTCCATA TTAGGACATC[i]		Dimerization
E12/E47	CANNTG[j]	Helix-loop-helix	Ubiquitous
MyoD	CANNTG[j]	Helix-loop-helix	Directs muscle differentiation, regulated by Id having a similar sequence

[a]Only a part of the known transcription factors of mammals.
[b]GC box.
[c]Glucocorticoid receptor.
[d]Glucocorticoid-responsive element.
[e]TPA-responsive element.
[f]cAMP-responsive element.
[g]Octamer.
[h]This is more specifically a "POU domain," which consists of a homeodomain and a POU box.
[i]Serum-responsive element.
[j]Core consensus sequence.

OCT-2 (or OTF-2 and NFA2), Pit-1 (or GHF1), and OCT-3 (see Table III). Some of these proteins have a POU domain, which consists of a homeodomain and a POU box comprising about 160 amino acids. OCT-1 is found almost ubiquitously and can activate the histone H2B gene *in vitro*. In contrast, OCT-2 is lymphoid cell specific, and OCT-3 is confined to undifferentiated embryonic cells.

The third category of *trans*-acting factors consists of proteins that have a so-called leucine zipper structure. The leucine zipper consists of four to five consecutive leucines that appear every seven amino acids. In this way, they are all aligned when the polypeptide chain forms an α-helix. The series of leucines are thought to interact with another similar series on another molecule, facilitating the dimerization of these proteins to form a hetero- or homo-dimer. In these proteins, a DNA-binding domain of about 30 amino acids with a high net basic charge is present immediately upstream (N-terminal side) of the leucine zipper.

Dimer formation is a prerequisite for the specific binding of these proteins to the cognate DNA sequences and is thought to have regulatory significance. The palindromic nature of the target sequences suggests the interaction of each monomer with the half-sequence of the dyad symmetry. These proteins include C/EBP, Jun, Fos, and CREB. An oncogene c-*jun* product is known to bind to the AP-1 site (or TRE) as a heterodimer with the oncogene c-*fos* product and activates the transcription of a gene having this site (Table III).

The fourth category of the DNA-binding domain is found in CTF/NF-1. It has a high content of basic amino acids and can form an α-helix; no similarity is apparent with any of the previously described motifs. The fifth and most recently identified category of the DNA-binding domain is the helix-loop-helix structure. It is characterized by two amphipathic helices separated by an intervening loop. Several hydrophobic amino acid residues in the helices are strictly con-

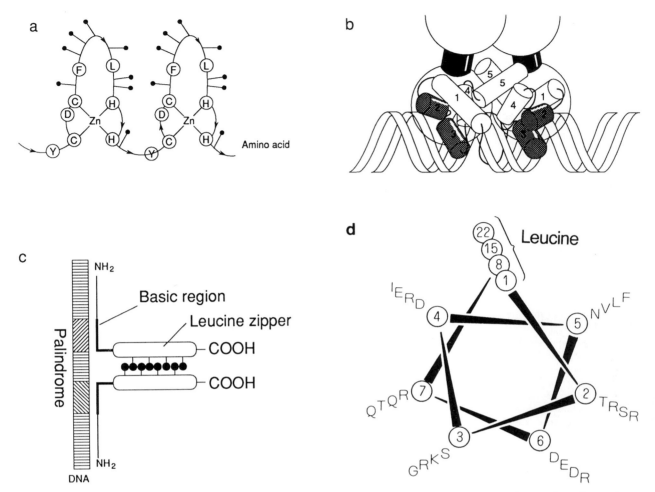

FIGURE 2 Three modes of DNA binding of a *trans*-acting factor. (a) Zinc finger. Amino acids in the circles are shown by one-letter symbols. (b) Helix-turn-helix. λ repressor molecules (as a dimer) bound to an operator site. Helices are shown as rods. Two helices (Nos. 2 and 3) interact with DNA in the major groove. [Reproduced, with permission, from M. Ptashne, 1980, "A Genetic Switch. Gene Control and Phage," Cell Press, Cambridge Massachusetts, and Blackwell Scientific Publications, Palo Alto, California.] (c) Leucine zipper. Two proteins with a leucine zipper such as Jun and Fos form a dimer and interact with a palindromic sequence such as TRE of DNA via the basic region adjacent to the leucine zipper. (d) Formation of a zipper. An α-helix with four leucines at every seven amino acids aligns the leucine residues in a row, which can interact with a similar structure as shown in (c). Amino acid residues are represented by one-letter symbols.

served. Many of these proteins have a basic region just N-terminal side of the first helix. The helices are supposed to interact to form a heterodimer of different classes of these proteins, and the basic domain is required for the interaction with DNA. This category of *trans*-acting proteins include immunoglobulin enhancer-binding proteins E12/E47, Myo-D, c-Myc, *da* (daughterless) protein, etc. Intriguingly, some of them may have specific regulatory proteins that have a similar helix-loop-helix structure but without the basic DNA-binding domain. They appear to form a superfamily of proteins that are engaged in the regulation of gene expression during development and differentiation. DNA-binding proteins lacking any of the known domain structures are increasing as the number of structurally defined *trans*-acting factors increases. Therefore, more different families of *trans*-activating protein with different modes of interaction will be found in the future.

2. *trans*-Activating Domains

The *trans*-activating domain of a transcription factor has the size of 30–100 amino acids and apparently works independently of its DNA-binding domain.

(This has been demonstrated by making chimeric molecules in which a *trans*-activating domain is exchanged between unrelated transcription factors and hooked to the DNA-binding domain of another factor.) Some factors have more than one *trans*-activation domain. These domains are located at different regions of the molecule and work independently of each other. Several different types of *trans*-activating domains are known. One *trans*-activating domain, an acidic domain, is found in yeast GAL4 and GCN4 and is characterized by containing a substantial negative charge (a high content of glutamic and aspartic acids) and the ability to form amphipathic α-helix. The α-helix is sometimes called an "acidic noodle," based on the postulated shape of this domain. In this structure, there appears to be no definite amino acid sequence requirements, only a net negative charge requirement. Jun proteins are one class of proteins that have negatively charged α-helical regions, which may act as a *trans*-activating domain. These acidic noodles are postulated to interact with one of the general transcription factors, possibly TFIID, TFIIB, or with a subunit of RNA polymerase II.

A second type of *trans*-acting domain is the glutamin-rich domain, which is found in SP1 and characterized by a high content (about 25%) of glutamine and very few charged amino acids. Regions with a high glutamine content are found in some of the *Drosophila* homeotic genes, yeast genes, mammalian OCT-1, OCT-2, Jun, and other genes, although no sequence homology is apparent among the different groups. A third type of transcriptional activating domain is the proline-rich domain found in the carboxyl terminus of CTF/NF-1. Proline-rich regions are also noted in a number of mammalian *trans*-activating proteins including Jun, OCT-2, OCT-3, AP-2, and serum responsive factor. How these different motifs in the transactivating domain interact with the basic promoter apparatus, however, is not yet known. It is tempting to speculate that different motifs interact with different general transcription factors and/or different subunits of RNA polymerase II.

3. Modification of *trans*-Acting Factors

Transcription factors are subject to modifications such as phosphorylation and glycosylation. When signal transduction-dependent gene activation is insensitive to protein synthesis inhibitors such as cycloheximide, it is deduced that activation does not involve new transcription factor synthesis but rather is regulated by some modification of preexisting factors. Although relatively few instances are known at present,

yeast heat-shock transcription factor and mammalian CREB factor appear to be positively regulated by phosphorylation. The latter factor is known to dimerize as a result of phosphorylation. Glycosylation may also be a posttranslational regulatory mechanism of *trans*-acting factors since SP1, which is highly glycosylated, could be inhibited *in vitro* by wheat germ agglutinin, which binds to the sugar components.

4. Adaptor or Coactivator Molecules

Recent studies have shown that a number of enhancer-binding proteins exert their effect on the primary initiation complex not directly but indirectly via another protein which is called an adaptor or a coactivator. This type of protein was found in yeast *Saccharomyes cerevisiae* first (ADA1-3), but was soon identified in mammalian systems. For example, adenovirus early gene product EIA is known to bind with TBP and work as an coactivator of Oct-3 which is present in embryonic stem cells. Retinoic acid receptor α also transmits its effect on transcription through EIA and then TBP. Another example is CBP (CREB-binding protein). When CREB is phosphorylated by cAMP-dependent kinase (A kinase), it can bind with the CBP and then transmit an activation effect to the initiation complex on the promoter. Examples such as these must increase as the quest for transcriptional modulators expands.

V. CONCLUDING REMARKS

The mechanism of regulation of gene transcription is not yet completely clarified. Recent advances in molecular biology with combined gene, protein, and cell technology have established a fundamental approach to this problem. A number of genes are now being analyzed with respect to their DNA signals and the protein factors interacting with them. As a result, some useful pictures are emerging as testable models of macromolecular interaction between DNA and protein as well as protein and protein. One important point that is already apparent is that both the *cis*-acting elements and the *trans*-acting factors are probably limited in number as compared with the complexity of the genome and its regulation. Not only are enhancers and other regulatory regions composed of a limited number of *cis*-acting elements by combinatorial arrangements, but the *trans*-acting factors also appear to be composed of a limited number of DNA-binding and *trans*-activating domains, which are combined during the evolution of the gene. Nevertheless, many more genes must be studied before a comprehensive understanding of the *cis*-acting elements and the *trans*-acting factors is obtained. When sufficient numbers of these components are identified and analyzed, we will then be able to define the network of multicomponent and interdependent gene regulation, which is the basis for development, differentiation, and homeostasis of organisms such as humans. Other closely related subjects not discussed in this article include chromatin structure, nuclear scaffold or matrix, and DNA methylation, for which separate articles are available. [*See* Chromatin Folding; DNA Methylation in Mammalian Genomes: Promoter Activity and Genetic Imprinting; DNA in the Nucleosome.]

BIBLIOGRAPHY

Chrivia, J. C., Kwok, R. P. S., Lamb, N., Hagiwara, M., Montiminy, M., and Goodman, R. H. (1993). Phosphorylated CREB binds specifically to the nuclear protein CBP. *Nature* 365, 855.

Comai, L., Tanese, N., and Tjian, R. (1992). The TATA-binding protein and associated factors are integral components of the RNA polymerase I transcription factor, SL1. *Cell* 68, 965.

Evans, R. M. (1988). The steroid and thyroid hormone receptor superfamily. *Science* 240, 889.

Gehring, W. J. (1987). Homeo boxes in the study of development. *Science* 236, 1245.

Hernandez, N. (1993). TBP, a universal eukaryotic transcription factor? *Genes Dev.* 7, 1291.

Johnson, P. F., and McKnight, S. L. (1989). Eukaryotic transcriptional regulatory proteins. *Annu. Rev. Biochem.* 58, 799.

Latchman, D. S. (1995). Eukaryotic Transcription Factors. 2nd Ed. Academic Press, London.

Lewin, B. (1994). Genes V. Oxford Univ. Press, New York.

Maniatis, T., Goodbourn, S., and Fischer, J. A. (1987). Regulation of inducible and tissue-specific gene expression. *Science* 236, 1237.

McKnight, S., and Tjian, R. (1986). Transcriptional selectivity of viral genes in mammalian cells. *Cell* 46, 795.

Mitchell, P. J., and Tjian, R. (1989). Transcriptional regulation in mammalian cells by sequence-specific DNA binding proteins. *Science* 245, 371.

Ptashne, M. (1986). "A Genetic Switch. Gene Control and Phage." Cell Press, Cambridge, MA.

Ptashne, M. (1988). How eucaryotic transcriptional enhancers work. *Nature* 336, 683.

Saltzman, A. G., and Weinmann, R. (1989). Promoter specificity and modulation of RNA polymerase II transcription. *FASEB J.* 3, 1723.

Watson, J. D., Hopkins, N. H., Roberts, J. W., Steitz, J. A., and Weiner, A. M. (1987). "Molecular Biology of the Gene," 4th Ed. Benjamin/Cummings, Menlo Park, CA.

DNA-Binding Sites

OTTO G. BERG
Uppsala University Biomedical Center

GLOSSARY

Nonspecific Refers to properties shared by all DNA sequences

Protein–DNA specificity Ability of a protein molecule to bind preferentially to one particular sequence of DNA base pairs, but not to others

Pseudosites DNA sequences in the genome that resemble specific sites, but have appeared by random chance and do not serve any function

IN ALL ORGANISMS A LARGE FRACTION OF THE DNA is covered by bound protein of various kinds. Most of these proteins serve to organize—or roll up—the linear DNA molecule into a more compact structure. These structural proteins are bound mostly in a nonspecific way, i.e., they bind anywhere along the DNA. Other regulatory proteins require for their proper function that they bind predominantly to specific sites on the DNA. Such specific sites are defined by a certain sequence of DNA base pairs that the protein can recognize and bind to. The biological function and molecular design of these specific DNA-binding sites are the subjects of this article.

I. PURPOSE OF DNA BINDING

The genetic information is stored in the huge linear DNA molecule. For this information to be disseminated in a systematic and regulated way, regulatory proteins must be directed to the appropriate regions of the DNA and retained there while their action is needed. Thus, the main purpose of DNA-binding sites is to serve as attachment points for gene regulatory proteins. As a first step in the process of reading the genetic information, genes are transcribed into RNA by the enzyme RNA polymerase. Starting from its attachment site—the promoter—RNA polymerase follows the linear DNA and sequentially copies a gene—or a cluster of genes—into RNA. The promoter not only signals the starting point, but can also determine how often a gene is transcribed. [*See* DNA and Gene Transcription.]

The need for different gene products depends strongly on the developmental stage of the cell and on its environment. Therefore, a sensitive control is required so that genes can be turned on and off by internal or external signals. This is achieved by a whole range of regulatory proteins that can bind to specific sites in or around the promoters of their particular target genes and interact with the polymerase to either block or stimulate its activity at the promoter site. Genes can be turned off by repressor proteins that bind to operator sites in or near the promoter and thereby block the access of RNA polymerase. Sites for the binding of activator proteins, which stimulate RNA polymerase, are sometimes also located in the promoter region, but often such sites (e.g., the enhancer sites) can be found at a considerable distance.

Almost all of the known specific DNA-binding sites for protein are involved in gene regulation. Other such sites are found, for example, at the origin of DNA replication. Some structural proteins could also require specific DNA sequences to direct their binding; often such proteins that wrap DNA around them use the innate tendency of some DNA sequences to bend. It can be expected that the molecular design of these other types of binding sites follows the same principles as discerned for the gene regulatory sites discussed here.

The basic physical principles for protein–DNA recognition are the same in all organisms, but there are important differences in the functional arrangements of binding sequences. For example, prokaryotic genes have distinct promoter sites, while the promoter sites in eukaryotic genes are much more complex in terms of DNA sequence. These differences in molecular design might be connected with the much greater complexity and needs of development for eukaryotes.

The restriction sites offer an example of recognition sites of a somewhat different nature. These sites, where the DNA chain can be cut by special enzymes (i.e., restriction enzymes), are spread throughout the genome instead of being situated at specific locations. These sites, therefore, do not have the same functional requirements as gene regulatory sites, although in the two cases the recognition mechanism is based on the same kind of physical protein–DNA interactions.

II. PROTEIN–DNA RECOGNITION IS BASED ON PHYSICAL INTERACTIONS

The target for a DNA-binding protein—the DNA molecule—constitutes a long regular helical structure. The outer surfaces of this helix look much the same along the entire length. A specific DNA sequence is defined from a certain combination of the four base pairs (i.e., $A \cdot T$, $T \cdot A$, $C \cdot G$, and $G \cdot C$). The genome of a bacterial cell consists typically of a DNA molecule with a few million base pairs ordered in a linear array; the genome of a higher organism can be 1000-fold larger. Thus, a protein molecule that should bind specifically to only a few sites must be able to distinguish a specific DNA sequence among a vast excess of all other structurally similar parts of the DNA molecule. The resolution of this problem requires both adequate physical interactions and proper base pair coding of individual sites. [See Human Genome and Its Evolutionary Origin.]

A. Structural Fit

The primary requirement for binding is a structural fit among the molecules that allows a sufficient number of weak interactions to be established. Such a structural fit often involves a cleft on the protein where the DNA helix can be lodged; also, chemical groups on the protein can be fitted into the helical grooves of the DNA for more intimate contacts. To achieve sufficient binding strength, this structural complementarity—called "lock-and-key" fit—must be combined with an interactional complementarity such that the groups that are brought into contact also will be held together by physical interactions.

B. Electrostatic Interactions

The binding of protein to DNA has a strong electrostatic component mediated by the attraction between the negative charges on the phosphate groups along the backbone on the outside of the DNA helix and a collection of positive charges on the DNA-binding surface of the protein. This binding is predominantly nonspecific as the backbone is essentially the same along the DNA molecule and independent of the particular sequence of DNA base pairs from which it is built. The electrostatic interactions provide the main contribution to binding for nonspecific proteins. The binding of proteins to specific sites also has a strong electrostatic component; as a consequence, these proteins bind not only to their specific sites, but also, although much more weakly, to all other sequences in the genome.

C. Hydrogen Bonds

A hydrogen bond is formed when a hydrogen atom is shared by two other atoms. It occurs when a hydrogen that is covalently linked to a particular atom (i.e., the donor, most often oxygen or nitrogen) is brought into contact with another atom (i.e., the acceptor, also most often oxygen or nitrogen) with which it can be shared.

Both protein and DNA molecules carry many molecular groups that can function as hydrogen bond donors or acceptors. Of particular importance are those that are exposed in the major groove of the DNA helix; these are most easily accessible and also provide the best possibility to distinguish different base pairs, i.e., they are sequence specific in the sense that different DNA sequences will expose different patterns of donors and acceptors. Specific binding of a protein to DNA is achieved when the pattern of hydrogen bond donors and acceptors on the protein is complementary to that of a particular DNA sequence, such that the juxtaposition of the molecules can correctly align individual donors with acceptors and vice versa. Even if the juxtaposition leaves only a few donors or acceptors unpaired, the loss of potential binding energy is too large and the complex cannot hold together. Thus, the hydrogen bond complementarity constitutes an extremely sensitive determinant for the specificity of protein—DNA binding and is the main criterion for the molecular design of DNA binding sites.

Many hydrogen bonds also exist between the protein and the DNA backbone. These interactions are mainly nonspecific but could contribute to specificity in an indirect way. Since the detailed structure of the backbone helix depends to some extent on the base pair sequence, the hydrogen bonds between backbone and protein could have a better structural fit for some DNA sequences than others. Furthermore, the nonspecific interactions with the backbone can be important by holding the protein in a fixed position where it can sense the presence or absence of the specific interactions in a more efficient way.

D. Other Interactions

Other effects influence both binding and specificity. For example, hydrophobic interactions, particularly with the methyl group of thymine, can also contribute to specific binding.

Futhermore, the binding of a protein is likely to distort the structure of the DNA somewhat in order to achieve the best fit among interacting groups; in some cases bends or kinks are introduced in the DNA. Such structural distortions are energetically costly and therefore reduce the binding energy. Because the flexibility of DNA depends on the base pair sequence, the energetic cost is also sequence dependent. Thus, while they can only decrease the overall binding strength, structural DNA distortions can nevertheless contribute to specificity by decreasing the binding to different extents on different sequences.

These effects should be considered mostly as a modulating influence on the basic contributions from electrostatic and hydrogen bond interactions since they are not sufficiently limiting by themselves to define a specific binding site for a protein.

Some particular sequences of DNA base pairs can take on structures that differ greatly from the usual DNA helix (e.g., "left-handed" DNA or cruciform DNA), and proteins might exist that, through a complementary fit, specifically recognize such alternate structures. This could provide an effective way of defining specific binding sites, but no such case has been unequivocally identified as yet.

III. FUNCTIONAL REQUIREMENTS

The most important requirements for the biological function of DNA-binding sites are that the protein be able to find its target site(s) within a reasonable time and, once there, be able to stay long enough to execute its function. The physical protein–DNA interactions and the base pair coding of the specific sites are arranged to meet these requirements.

A. Binding Strength

For many gene regulatory proteins (e.g., repressors and activators) the biological activity is determined primarily by the probability of occupancy at individual binding sites; the extent to which a gene is repressed, for example, is proportional to the fraction of time that a repressor is bound at the operator. The occupancy, in turn, is determined by the binding strength and the availability of the required protein. A weak binding can to some extent be compensated for by an increased amount of protein to provide a sufficient level of occupancy.

Function is not always determined by binding alone, and in these cases specific activity would be a more important property; for example, RNA polymerase requires a series of activation events in which the DNA helix is opened before transcription initiation can take place. However, binding to the specific recognition sequence is the first step in function. To the extent that signals for the activity are carried by the DNA sequence, for simplicity of discussion in this article, they are lumped together with the binding specificity.

B. Functional Specificity

Apart from their strong binding to specific sites, most gene regulatory proteins also have a weak—mostly electrostatic—nonspecific affinity for DNA in general. Because of the large amounts of DNA in the cell, even a weak affinity can lead to a strong competitive effect when large fractions of the gene regulatory proteins are bound at nonspecific and nonfunctional DNA sites. While the specificity of a particular protein is large, exhibiting a large difference in binding strength between a specific and a nonspecific site, the specificity in the cell is effectively much smaller.

In the living cell a large fraction of the genome is covered by structural proteins. Even if this fraction is not available and is therefore not contributing to the competitive binding of the gene regulatory protein, nonspecific competition is expected to be appreciable. Although the exact organization of the DNA in the living cell is not sufficiently well established for any definite conclusion, it can be surmised that an increased availability of the genetic control regions relative to other parts of the DNA could serve to increase effective specificity.

Apart from the purely nonspecific competition, also expected are a whole range of randomly occurring pseudosites that more or less resemble the specific ones in sequence (see Section IV,A). This follows from the limited site sizes used and the sometimes slow decrease in binding strength as more "wrong" base pairs are allowed in a site.

Both a weak binding strength and the binding competition from nonspecific and pseudospecific sites can be compensated for by increasing the amount of protein in the cell. The functional specificity can be defined from the amount of protein that is "lost" on nonfunctional sites. Thus, specificity must be balanced against the physiological cost of investing in more protein.

C. Functional Variability

Specific sites for the same protein require different binding strengths, depending on the physiological need for the particular gene(s) they control. To achieve sufficient fine-tuning of the binding strength, some of the base pairs that contribute to specific binding do so only to a smaller degree. In this way a precise adjustment of binding at individual sites can be achieved by replacing weakly interacting base pairs. However, to achieve sufficient specificity within the limited binding site size, other base pairs must be discriminatory and therefore contribute strongly to binding. Thus, the specificity and variability requirements lead to a whole range of base pair interaction levels. As discussed in Section IV,C, these different interaction levels are to some extent reflected in the statistics of base pair choice.

D. Kinetics

Molecular complexes can be formed after random diffusional motions have brought the molecules into contact. The specific binding of protein to DNA requires a precise alignment of the molecules before the

interacting groups can form their bonds. This precise alignment has a low probability of occurring by chance when the molecules are tossed around by the random thermal motions. As a consequence the time it would take to form such a specific complex might become exceedingly long.

A nonspecific complex requires much less precision and can therefore be formed much more rapidly; it is sufficient that the positive charges on the DNA-binding surface of the protein are brought into the neighborhood of the negative charges on the outside of the DNA. Once nonspecifically bound, the protein can explore the grooves of the DNA helix for specific interaction possibilities. Such a two-step binding can significantly increase the rate of specific complex formation.

It has been found that some gene regulatory proteins can make even more efficient use of the nonspecific binding intermediate in that they can actually "slide" in a one-dimensional diffusion along the nonspecific DNA in search of specific binding interactions. While this sliding mechanism can lead to a fast association in experimental systems, its efficiency is expected to be much smaller in the living cell; even a weak nonspecific binding to the large amounts of DNA present leads to competition effects that decrease the efficiency. Consequently, there is a delicate balance between the rate-enhancing and rate-decreasing effects of the nonspecific DNA binding. Nevertheless, the kinetic requirements for a sufficiently fast association could be a major reason that the gene regulatory protein has a nonspecific binding at all. This could therefore provide an important limitation on the usefulness of too large a specificity.

E. Protein–Protein Interactions

Specific binding sites are often arranged in tandem (i.e., next to each other) on the DNA, such that a regulatory protein bound at one site can interact with proteins bound at neighboring sites. In this way the specific DNA binding of a particular protein can require the presence of several other protein molecules on nearby sites. This use of protein–protein interactions is one way of effectively increasing specificity without unduly increasing the size and specificity requirements of the individual proteins. This arrangement also affords more sensitive and intricate control of gene expression by requiring the simultaneous presence of sometimes many different protein factors.

As only a limited number of sites can be arranged in the regions surrounding a promoter, packing problems can arise when the control mechanisms require more binding sites. One way of avoiding such problems is to make use of the flexibility of the linear DNA molecule, since bending and looping can bring distant DNA segments into close contact. This seems to be the strategy used by the enhancer sites, which can be placed at distances of 1000 bp or more away from the promoter site. Proteins bound at such sites can be brought into contact with RNA polymerase at the promoter by looping of the DNA. This kind of arrangement allows much more freedom in the specification and location of specific binding sites when the linear constraints of the DNA molecule are effectively removed.

IV. CODING CONSTRAINTS AND THE STATISTICS OF BASE PAIR CHOICE

The pattern of interaction possibilities exposed on the DNA-binding surface of the protein has evolved to interact more favorably with some DNA sequences than others. Simultaneously, the specific DNA sequences have been chosen to provide a physiologically adequate binding at the required sites in the DNA molecule. The result of this mutual evolutionary adaptation can be seen in the DNA sequences used to define the functional sites.

A. Site Size

The size of the protein imposes a limit on how many DNA base pairs it can interact with simultaneously; for a typical gene regulatory protein this would be about 15–30 bp. This physical limitation on the site size provides an upper bound on the possible number specific interactions, although not all of the base pairs in a site contribute to specificity.

There is also a statistical limitation on the site size: A specific site cannot be too small either. If the number of base pairs that contribute specific interactions to the binding of a gene regulatory protein is too small, a large number of sites that are identical to a specific one could appear by random chance. The probability that a specified sequence of n base pairs will appear by random chance is $(\frac{1}{4})^n$, if each of the four possible base pairs is equally likely to be chosen. For $n = 11$, this corresponds approximately to one chance in 4 million. Thus, in a bacterial genome with a few million base pairs, unique specific sites would have to be de-

fined by sequences of about 11 base pairs or more. In higher organisms, with their much larger genomes, this limitation could be much more severe.

This statistical minimal site size estimate is based on an absolute discrimination such that sites with one wrong base pair will not bind at all. In actuality, in real sites many of the specific base pairs contribute relatively weakly to binding discrimination, and a site can carry a number of wrong base pairs and still be a fairly effective binding site. This is necessary for the functional variability of different sites. As a consequence, however, the site size must be larger than the minimal estimate to avoid the competitive presence of too many nonfunctional random sites in the genome. In fact, the site sizes actually used seem barely sufficient to keep the expected number of such randomly occurring pseudosites within reasonable limits.

B. Sequence Variability

Most gene regulatory proteins can bind to several functional sites in the genome. On one extreme there is RNA polymerase, which recognizes 1000 or more promoter sites in the bacterial genome, whereas some repressor proteins bind only a few operator sites. Although similar in sequence, the different binding sites for the same protein are not identical.

Sometimes this sequence variability is functional: Genes that are regulated by the same protein are needed to different extents and therefore require different binding of the protein. There is also a possibility that a binding site overlaps with other distinct sequences—either a binding site for a different protein factor or part of a gene—such that other base pair requirements must be simultaneously satisfied.

If the same binding and function can be achieved by several somewhat different sequence combinations, there is also a statistical variability. In this case the actual base pair sequence used among the functionally equivalent possibilities might be a matter of evolutionary history and random choice.

The pattern of interaction possibilities on the DNA-binding surface of the protein is expected to define a distinct DNA sequence to which it has the best complementarity and binding. Because of the large variability, however, this best binding sequence is not likely to be in wide use as a functional site in the genome. Nevertheless, binding sites for a particular protein often share sufficient similarities in the sequence choice that potential sites in a stretch of sequenced DNA can be identified by the human eye.

C. Consensus Sequences

By examining the base pair sequences of a set of functional specific sites for a particular protein, one can derive information on the importance of individual base pair choices at the various positions in a site. Some base pairs are conserved and occur almost always at particular positions in all sites of the set. Such base pairs are likely to contribute strong binding interactions for the protein. Others are only weakly preferred and are therefore expected to contribute less to the binding or activity.

By listing the most commonly occurring base pair at each position, one finds the consensus sequence. This can be expected to be similar to the sequence that would exhibit the best complementary fit with the binding surface of the protein. In this way the statistics of base pair choice carry information about the molecular design of the protein.

One can also use the variability in the base pair choices to quantitatively correlate the degree of preference for a certain base pair with the strength of its contribution to protein binding. This makes it possible to predict the relative binding strengths for arbitrary sequences. Although such predictions agree fairly well with experimental data (when available), they are fraught with uncertainties, both conceptual and statistical. Care must also be taken that conserved base pairs are not required for other purposes, perhaps contributing to the binding of some other factor at overlapping binding sites. When only a few binding sites are known, there is also a large statistical uncertainty in the estimate of the degree of preference for an individual base pair. Consequently, such a prediction can only provide a rough idea of the strength of binding and activity before these quantities have been experimentally determined.

V. EXAMPLES

The molecular design of DNA-binding sites is based on the physical protein–DNA interactions and on an adequate base pair coding to satisfy the functional requirements discussed earlier. To see how real systems have developed to cope with these demands, it is helpful to look at some examples.

A. Restriction Sites

As a defense against invading foreign DNA in a prokaryotic cell, restriction enzymes can cut DNA at well-defined sites. Over 100 enzymes with different DNA specificities are known. The recognition sequences are defined by 4 to 10 bp. In contrast to the other types of sites discussed here, the function does not require specified location of the sites, and they are found anywhere. This is one reason that their sequence length can be so much smaller than that of the gene regulatory sites. A specific sequence of only 4 bp is expected to occur by random chance every few hundred base pairs, while one with 10 occurs once in every million base pairs.

Furthermore, the function does not require a graded or varied response at different sites; a site is either cut or not. Thus, there is no requirement for a functional sequence variability and consequently no need for base pairs that contribute only weakly to specific recognition.

This example shows that the physical protein–DNA interactions can be sufficiently discriminatory to distinguish DNA sequences of only a few base pairs. However, the gene regulatory sites must involve a large number of less discriminatory base pair interactions to satisfy their requirements for a unique location and a functional variability.

B. Operators

The design of bacterial DNA-binding sites is best understood for operator sites, which were the first to be studied in systematic detail, both functionally and by physicochemical means. More recently, several repressor–operator complexes have also been examined by X-ray crystallography.

Most repressor proteins are built up from two or more identical subunits, providing a twofold symmetry. Similarly, the operator sites have a matching symmetry in their sequence. This is one way of increasing specificity (i.e., twice the interactions with DNA) without requiring an expanded protein design, since the same protein unit is used twice. Furthermore, the twofold symmetry allows the protein to bind in either orientation, thereby doubling its likelihood of finding a specific site.

Table I lists a set of six recognition sites for two repressors, the cro protein and the λ repressor in phage λ. The first entry is the consensus sequence for which the twofold symmetry is most evident, as the lower strand read from right to left is almost identical to the upper one read from left to right. The other entries are the six real sites, labeled with their names. The elements of conservation, variability, and symmetry are obvious in these sequences. Similar properties

TABLE I

Recognition Sequences for cro Protein and
λ Repressor

Consensus[a]	TTACCACCGGCGGTGATAA
	AATGGTGGCCGCCACTATT
O_{R3}	CTATCACCGCAAGGGATAA
	GATAGTGGCGTTCCCTATT
O_{R2}	CTAACACCGTGCGTGTTGA
	GATTGTGGCACGCACAACT
O_{R1}	TTACCTCTGGCGGTGATAA
	AATGGAGACCGCCACTATT
O_{L1}	ATACCACTGGCGGTGATAC
	TATGGTGACCGCCACTATG
O_{L2}	TTATCTCTGGCGGTGTTGA
	AATAGAGACCGCCACAACT
O_{L3}	TAACCATCTGCGGTGATAA
	ATTGGTAGACGCCACTATT

[a] The consensus sequence is formed by taking the most commonly occurring base pair at each position in the six recognition sequences listed.

are found in most, if not all, other lists of both operator and activator sites in prokaryotes.

C. Prokaryotic Promoters

The recognition sites for RNA polymerase (i.e., promoters) are placed at the beginning of a gene and serve to set the polymerase at the correct position for transcription initiation. Since the gene is to be transcribed on one strand only and in one direction, the promoter sequence must also bind the polymerase in a correct orientation. Therefore, in contrast to the operator sites, promoters cannot be symmetrical in sequence.

RNA polymerase is a large protein that can interact with a long stretch of the DNA. The bacterial promoters are approximately 50 bp or more in length, but not all of these base pairs are important for recognition and activity. Two regions of 6 bp each seem to contribute most to specificity. These are located at around 10 and 35 bp upstream from the point where the transcription is initiated. In the −10 region the consensus sequence is TAtaaT (small letters symbolizing less conserved base pairs), and in the −35 region it is TTGaca.

The importance of these two regions for RNA polymerase activity have been demonstrated in several ways: They contain the most conserved base pairs, most of the deleterious mutations identified are local-

ized there, and chemical probes show that RNA polymerase or its associated factors make important physical contacts with the DNA in and around them.

Clearly, 12 bp—of which perhaps only 6 or fewer are really strongly required—are not sufficient to define unique sites. In fact, based on these criteria, promoter-like elements can be found scattered everywhere in the genome. Although there are other regions with weakly conserved base pairs in the promoter sequences that also contribute to specificity, it has proved difficult to define unequivocally the necessary and sufficient sequence requirements for a functioning promoter. Possibly, regions outside the promoter also contribute to the specificity of transcription initiation by RNA polymerase, as seems to be the case in eukaryotes.

D. Promoters in Higher Eukaryotes

In eukaryotes there are three different forms of RNA polymerase that recognize different kinds of promoters, with varied and sometimes confusing DNA sequence requirements.

RNA polymerase II transcribes those genes that code for protein, and it is the polymerase that must recognize the largest number of sites. Its recognition sequences share some regions that are partially conserved also among different species. One of the most important ones is the so-called TATA box, which is located approximately 25 bp upstream from the start site. This seems to serve a role similar to the −10 region in the bacterial promoter: to position the polymerase accurately for transcription initiation. Further upstream from the start site is quite often a CCAAT sequence, but its exact location does not seem to be crucial, and it can even occur on the opposite DNA strand. Also, other upstream sequences seem to be required; some of these are present only for certain kinds of genes and might serve as specific signals to identify such genes.

These upstream sequences seem to be binding sites for the auxiliary protein factors needed before RNA polymerase II can recognize the start site. Thus, the physical recognition of the DNA sequence by the polymerase is partially indirect through other bound protein factors. This could create a much more efficient recognition surface for the polymerase by removing the requirement for a linear search along the DNA. Exactly what factors are needed and where they bind is currently under intense experimental study.

The logic of this design could be connected to the large size and complexity of the eukaryotic genome;

the probability that these various binding elements would occur at random in the same region is quite small, even in a large amount of DNA. At the same time the seeminly lax sequence requirements allow a large measure of freedom in both the construction and the control of promoter sites.

E. Enhancer Elements

The activity of polymerase II at promoter sites can also be strongly influenced by DNA sequences far outside, sometimes several thousand base pairs away from the start site, either upstream or downstream inside the gene. Of particular interest are the enhancer elements, first identified in some DNA tumor viruses. These consist of individual sequences—enhancer motifs—that are organized into fairly long stretches of DNA of up to a few hundred base pairs. Their function is largely independent of position and orientation with respect to the gene itself. The individual sequences are thought to be binding sites for activator (or sometimes inhibitor) proteins, which can be brought into direct contact with the RNA polymerase at the promoter by bending of the DNA into a loop. Other mechanisms are also possible, however; for example, protein binding at the enhancer could change the DNA structure and thereby make the promoter more accessible.

One example of enhancer motif is the hormone-responsive elements present in hormone-sensitive enhancers. These sequences are about 15 bp in length and serve as binding sites for hormone-receptor proteins. Steroid hormones work as signal molecules that, by binding to their receptor proteins, change the expression of particular genes. Binding of the hormone activates the receptor protein, and from its recognition sites in the hormone-sensitive enhancers, it can stimulate polymerase activity.

Because of the complex structure of the enhancer elements comprising recognition sequences for a large number of different proteins, their exact function has been difficult to resolve. Clearly, the modular struc-

ture of the enhancers, where different binding sequences can be introduced and shuffled around, is similar to the structure of the promoters themselves and allows for complex control mechanisms. The modular structure also allows for a relatively facile reorganization of such control circuits; it is likely that such changes constitute one of the major evolutionary pathways for higher organisms.

VI. CONCLUDING REMARKS

Because of recent advances in experimental techniques, notably genetic engineering and DNA sequencing methods, there has been explosive development in the study of gene control and the role of DNA-binding sites. However, this is still the beginning and many important questions remain. Although the basic principles of protein–DNA binding and the fundamental design of binding sites are fairly well understood, the organization and the function of these binding sites in their biological and physiological contexts remain to be elucidated, particularly for higher organisms. This is of immense importance for further advances in both biology and medicine.

BIBLIOGRAPHY

Berg, O. G., and von Hippel, P. H. (1988). Selection of DNA binding sites by regulatory proteins. *Trends Biochem. Sci.* **13**, 207.

Pabo, C. O., and Sauer, R. T. (1992). Transcription factors: Structural families and principles of DNA recognition. *Annu. Rev. Biochem.* **61**, 1053.

Ptashne, M. (1988). How eucaryotic transcriptional enhancers work. *Nature (London)* **335**, 683.

Steitz, T. A. (1990). Structural studies of protein–nucleic acid interactions: The sources of sequence-specific binding. *Q. Rev. Biophys.* **23**, 205.

von Hippel, P. H., and Berg, O. G. (1986). On the specificity of DNA–protein interactions. *Proc. Natl. Acad. Sci. USA* **83**, 1608.

Watson, J. D., Hopkins, N. H., Roberts, J. W., Steitz, J. A., and Weiner, A. M. (1987). "Molecular Biology of the Gene," 4th Ed. Benjamin/Cummings, Menlo Park, CA.

DNA in the Nucleosome

RANDALL H. MORSE

New York State Department of Health and SUNY School of Public Health

GLOSSARY

Chromatin Ordered complex of protein and DNA, including both nucleosomal and nonnucleosomal proteins, found in eukaryotic cells

Gel electrophoresis Method for separating DNA molecules according to size by allowing them to migrate through a gel matrix under the influence of an electric field

Major and minor grooves of DNA Larger and smaller of the two grooves that follow the path of the helix in double-stranded DNA (see Fig. 1, bottom)

Nuclease Enzyme that cleaves DNA or RNA

IN EUKARYOTIC ORGANISMS (ORGANISMS WHOSE cells have nuclei, such as yeast, plants, and animals), DNA is complexed with proteins in an orderly array. The first level of packaging occurs in the nucleosome, a structure consisting of approximately 160 bp of DNA wrapped around a disk-shaped complex of proteins called histones. Nucleosomes have been crystallized and, using physical and biochemical techniques, much has been learned about their structure. The packaging of DNA into nucleosomes presents a potential problem inside the cell, where vital processes such as transcription and replication depend on the recognition of particular DNA sequences by protein factors. Whether and how such factors can recognize DNA sequences that are both structurally deformed and sterically obscured by their incorporation into nucleosomes is currently under investigation.

I. INTRODUCTION

The DNA content of a cell of the bacterium *Escherichia coli* is about 5 million bp, whereas human cells have a content of about 3 billion bp. These amounts of DNA, if stretched out in the form of solitary double helixes, would measure 1.7 mm and 1 m in length, respectively. Because these lengths are many times greater than the diameter of the vessels that contain them (bacterial cells or eukaryotic cell nuclei), some form of compaction or packaging of the DNA must take place. Although it is conceivable that the DNA might simply be squashed into the cell like string, this is not the case. In both prokaryotes (bacteria) and eukaryotes (higher organisms), proteins associate with and compact the DNA. In prokaryotes, this function is fulfilled by the histone-like, or HU, proteins, which are not discussed here. In eukaryotes, the packaging of DNA takes place at several levels, giving an overall compaction in length relative to the extended double helix of about 10^5 (Fig. 1). The highest degree of compaction can be visualized as the metaphase chromosome, a complicated structure that allows the orderly segregation of the genetic material to progeny cells during cell division (mitosis). This structure arises from the ordered folding of what is already an organized structure, the 30-nm fiber (1 nm = 10^{-9} m, or a millionth of a millimeter). The 30-nm fiber can be further unfolded to yield a 10-nm fiber of chromatin. The 10-nm fiber, seen under the electron microscope, has the appearance of beads on a string (Fig. 2). The string is DNA; the beads represent nucleosomes, the basic units of eukaryotic chromatin. [*See* Chromosomes; DNA and Gene Transcription; Histones.]

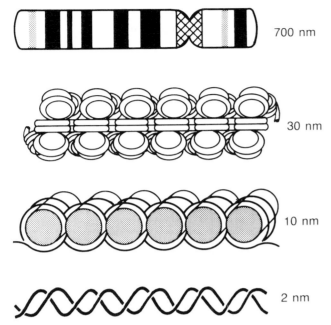

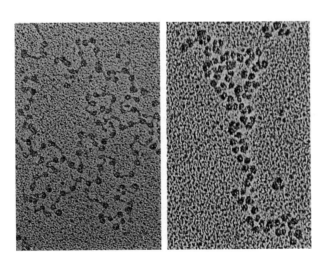

FIGURE 1 Hierarchies of packaging of DNA in eukaryotic cell nuclei. From top: metaphase chromosome; 30-nm chromatin fiber; 10-nm chromatin fiber; naked duplex DNA. The diameters of the various structures are given at the right. Various structures have been proposed for the folding of nucleosomes in the 30-nm fiber; for simplicity, only one is shown. (Redrawn, with permission, from Alberts *et al.*, 1989, "Molecular Biology of the Cell," Garland Publishing, Inc., New York.)

II. STRUCTURE AND PROPERTIES OF THE NUCLEOSOME

A. The General Picture

The nucleosome diagrammed in the 10-nm fiber in Fig. 1 (see also Fig. 3) is a simple structure: roughly two turns of DNA are wrapped around the outside of a disk-shaped object like a length of hose around a hockey puck. DNA, however, is in some respects a more complex structure than a length of hose, and the hockey puck of Fig. 1 is not so featureless either. What the disk is meant to represent is the protein core of the nucleosome. This core is made up of two copies each of the four core histone proteins: H2A, H2B, H3, and H4.

The details of how DNA and histones are arranged in the nucleosomes have been deduced primarily from X-ray diffraction and chemical cross-linking studies. The cross-linking work, by establishing particular sites of contact between the histone proteins and DNA, helped in deducing the orientation of the polypeptide chains of the histones when the crystal structure was being solved. A rough picture of the accepted structure is shown in Fig. 3. It consists of 146 bp of DNA wrapped around a disk-shaped core, which is

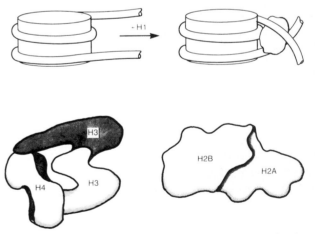

FIGURE 3 The nucleosome. The DNA helix is shown narrower than it would be in proper scale, so that the disk-shaped histone octamer is not obscured. The addition of H1 to the nucleosome core particle gives rise to the chromatosome. The arrangement of the histone proteins in the octamer core is schematized at the bottom, as they would be viewed from the top of the disk in the upper part of the figure. Histones H3 and H4 combine in a tetrameric structure, as shown; the second copy of H4 is represented by the shaded region below the upper copy and is mostly obscured. This tetramer is sandwiched in the nucleosome between two symmetrically disposed dimers of histones H2A and H2B.

FIGURE 2 Electron micrographs of chromatin. Chromatin as viewed by electron microscopy ($10^5\times$ enlargement) reveals "beads-on-a-string" structure of nucleosomes on the DNA fiber. In the left panel, histone H1 is absent; in the right panel, histone H1 is present, and the chromatin fiber is condensed (see Fig. 1). (Reproduced from Thoma, Koller, and Klug, 1979, *J. Cell. Biol.* **83**, 403, courtesy of Dr. Fritz Thoma; and with permission from The Rockefeller University Press, New York.)

7.3 nm in diameter and 4.0 nm in height. The entire structure is 11.0 nm in diameter and 5.7 nm along the axis of the disk and possesses near dyad symmetry. (This means that the structure is largely symmetric with respect to a 180° rotation around the central axis; as a consequence, if a person could sit on the exact center of the DNA sequence wrapped around the outside, he or she could travel in either direction along the DNA, and the view of the protein core could be nearly the same.) The DNA wraps around the outside for about 1.8 turns, with about 80 bp of DNA in a single turn. The histones make contacts all along the inside of the DNA helix but have little if any contact with the outside surface of the DNA. Histones H3 and H4 are assembled in a tetrameric (four-part) structure, which forms the most central part of the nucleosome. Two pairs formed by one histone H2A and one H2B sit on opposite sides of this tetramer, but still inside the DNA loop. These dimers do not sit exactly symmetrically in the crystallized nucleosome. It is not known if this reflects asymmetry also present in nucleosomes in solution (or in the cell) or if it is due to crystal-packing forces.

An additional protein, histone H1, can bind to the nucleosome at the exit and entry points of the DNA (Fig. 3) to yield a structure termed the chromatosome. In most cell types, approximately one molecule of histone H1 is present for each nucleosome. The interaction of histone H1 with the nucleosome core stabilizes the interaction of an additional 20 bp of DNA so that two full turns, comprising 166 bp of DNA, are stabilized in the chromatosome. Histone H1 (and related forms) is important in higher-order folding of nucleosomes in chromatin. Despite this important role for H1, the precise nature of its interactions with the nucleosome is largely unknown, and the chromatosome has yet to be crystallized.

B. Histones

The core histones are small proteins, ranging from 11 to 16 kDa in molecular weight in most eukaryotes (Table I). All four are rich in basic amino acid residues, particularly lysine and arginine, which account for 20–25% of the amino acid content of the histones. This property might be expected of proteins designed to bind tightly to an acidic polymer such as DNA. All of the histones, as they exist in the nucleosome, can be divided into a central structured region and a more extended, amino-terminal tail. In some cases (histones H2A and H3), the carboxyl terminus is also present in an extended configuration. This is evident

TABLE I
Properties of the Histone Proteins

	Histone				
	H1	H2A	H2B	H3	H4
Molecular weight (kilodaltons)	23	14–15	14–16	15	11
Lysine content	29%	11%	16–18%	10%	11%
Arginine content	1.5%	7–9%	2–5%	14%	14%

because the histone tails in the nucleosome are more accessible to proteases (enzymes that cleave the polypeptide chain of proteins) than are the central domains and from nuclear magnetic resonance studies, which show the amino acid residues in these same regions to be much more mobile than those in the central domains. The histone tails are the site of a number of chemical modifications (e.g., acetylation, phosphorylation, and methylation) that take place within the cell. The central structured regions, on the other hand, are subject to virtually no modifications, probably reflecting their relative inaccessibility in the nucleosome.

The core histones are remarkably well conserved throughout eukaryotes. For example, the major form of H2A from humans differs from that of chickens in only 5 out of 129 amino acid residues, from that of trout in 10 residues, and from that of yeast in 31 residues. Histones H3 and H4 are even more highly conserved: differences between human and yeast H4 occur in 8 out of 102 residues, and the trout and chicken proteins are identical to human H4. Some sort of evolutionary pressure seems to be operating to keep the histone proteins so highly conserved in their primary structures, but the nature of the damage that an organism would suffer if, for example, some of the amino acids in histone H3 were altered is unknown. (Surprisingly, yeast cells lacking all or part of the N-terminal regions from histones H2A, H2B or H4, including some highly conserved amino acids, are able to survive quite well.)

Compared with the core histones, histone H1 shows substantial variation among species. Moreover, the types of H1 found in a single organism may vary according to cell type and stage of development and within the cell cycle. Variants of the core histones also can be found, but these represent smaller contributions to the total pool of core histones than the variants of H1 do to its total pool. Some progress has been made toward understanding the role that certain specific histone subtypes play in the cell, but the function of most histone variants, like the function of many of the chemical modifications that the histones undergo, remains enigmatic.

C. Properties of the Nucleosome

Some of the most widely exploited properties of the nucleosome concern the way in which the histone proteins mask nucleosomal DNA from particular chemical and enzymatic agents. Enzymes known as nucleases, for example, will cut naked DNA in solution into very small pieces so that if the exposure is sufficiently long, the DNA will be degraded to pieces less than 10 bp (or nucleotides) long. One such enzyme is micrococcal nuclease. When chromatin is treated with this enzyme, only the DNA between nucleosomes—the linker DNA—is rapidly cut. This results in about 180–200 bp of DNA per nucleosome being protected from digestion under mild conditions and 146 bp showing strong resistance to nuclease cutting even during protracted digestion. Because not all of the linker DNA is immediately cut by the enzyme under mild conditions, when the DNA digestion products are analyzed by gel electrophoresis, a ladder of fragments is seen corresponding to DNA associated with 1, 2, 3, . . . nucleosomes. This allows the length of the linker DNA to be measured. The core DNA—the 146 bp that are strongly protected against micrococcal nuclease digestion in the nucleosome—is virtually invariant in all species studied, again demonstrating the extreme evolutionary conservation of nucleosome structure. The length of the linker DNA, on the other hand, varies among species, among tissues, and even along the DNA within a single cell. Evidence shows that histone H1, the linker histone, plays a part in determining the length of the linker DNA, but how this is accomplished and the significance of variations in linker DNA length remain to be discovered.

Another nuclease used to probe nucleosome structure is DNase I. This enzyme interacts with chromatin in a very different way from micrococcal nuclease. It is capable of cutting DNA in the nucleosome core, but only every 10 bp, so that the digestion products, when visualized by gel electrophoresis, form a ladder of fragments of 10, 20, 30, . . . 140 (and possibly as large as 160 or 170) bp in length. This is because DNase I cuts DNA only in its minor groove and can apparently only gain access to the minor groove where it faces outward on the nucleosomal surface, which occurs once every helical repeat, or about every 10 bp.

Why does DNase I cut DNA in the nucleosome, whereas micrococcal nuclease is much less able to do so? A clue may lie in the crystal structure of DNase I complexed with a small piece of DNA. The structure shows that the DNA fragment is bent away from the enzyme. Because nucleosomal DNA bends the same way relative to the enzyme (i.e., when DNase I or any other enzyme approaches DNA in a nucleosome, the DNA is bent away from the enzyme), the enzyme may have no difficulty in inducing the correct DNA conformation to cut the phosphodiester bond (the bond connecting adjacent nucleotides in one strand of the helix). Indeed, the correct conformation is already present. Micrococcal nuclease, on the other hand, may only cut DNA well when it is straight or it may even need to bend the DNA toward itself, either of which would be difficult with nucleosomal DNA. This is only a conjecture, however, because the structure of micrococcal nuclease complexed with DNA has not been determined.

In addition to protecting DNA in different ways from digestion by micrococcal nuclease and DNase I, the nucleosome also protects DNA against digestion by restriction endonucleases (enzymes that cut at specific DNA sequences) and hydroxyl radicals. Nucleosomes also stabilize DNA so that it denatures (i.e., separates into the two component strands) at higher temperatures than does naked DNA and prevents DNA from undergoing thermal untwisting (a change corresponding to a slight increase in the separation between adjacent bases, which DNA undergoes with increasing temperature). Other characteristics of the nucleosome include a well-defined sedimentation coefficient, which is greatly different from that of naked DNA, and a change in certain physical characteristics of the DNA, such as the circular dichroic spectrum.

D. DNA in the Nucleosome

Although the packaging of DNA into nucleosomes is only the first level of compaction in the cell, the double helix must undergo substantial deformation to be wrapped around the histone core. If the 80 bp of DNA that form one turn were smoothly bent, the edge-to-edge separation of adjacent base pairs would increase from 3.4 (on average) to 3.9 Å at the outside edge (a 14% increase) and decrease to 2.9 Å at the inside. Some of the details of the crystal structure confirm this kind of deformation: the minor groove of the DNA double helix varies from 7 Å on the inner face of the helix (facing the histone core) to 13 Å on the outside, whereas the major groove varies from 11 to 20 Å. These changes in the conformation of DNA when it is incorporated into nucleosomes may be important in the cell and will be discussed later.

Two other kinds of perturbations that the DNA structure undergoes when it is incorporated into the nucleosome are kinks, which are sharp bends in the helix, and a change in the twist, or helical period. Sharp bends or kinks are found in the DNA helix at one and four helical turns (one turn is about 10 bp) on either side of the dyad (or center) of the DNA in the nucleosome. These bends are apparent in the crystal structure and have also been detected by chemical means. The twist of the DNA helix, which can be defined as the number of base pairs to complete one turn of the helix, changes from an average value in solution of 10.5–10.6 to 10.0 at the edges and 10.7 near the center of the nucleosome. This may be due in part to the altered ionic environment seen by DNA in the immediate neighborhood of the nucleosome, caused by the high local concentration of positive charges contributed by the histones.

The wrapping of DNA around the outside of the histone octamer results in another profound change in a fundamental property of the DNA: its writhe, or supercoiling. These terms refer to the three-dimensional shape of DNA in space. Supercoiling might be described as how twisted up the helix is on itself. This is a different kind of twisting from that described earlier; the two strands of the helix can maintain their same helical period, or twist, but the helix as a whole can still be twisted, like a rubber band or the cord to a telephone receiver. Each nucleosome introduces a single supercoil into the DNA, as depicted in Fig. 4. Both the path that the DNA follows around the outside of the histone octamer and the slight change in

twist that occurs upon incorporation of DNA into the nucleosome contribute to the exact magnitude of supercoiling that is observed. Supercoils can be put into DNA in the absence of nucleosome formation (e.g., by the prokaryotic enzyme DNA gyrase), but this process requires energy. It has been speculated that removal of a nucleosome or nucleosomes in a local domain of DNA *in vivo* (for instance, in the region of a gene's promoter) could produce a locally supercoiled region of DNA, which might be recognized by factors involved in transcriptional activation. However, to date no direct evidence has been found for this kind of mechanism being involved in transcriptional regulation in eukaryotes.

E. Nucleosome Assembly

Nucleosomes can be assembled *in vitro* from DNA and purified histones. This may seem fantastic—that such a complex structure, involving four different proteins and an intertwined pair of DNA strands, can be faithfully put together from its solubilized constituents—but in fact, the nucleosome is virtually a self-assembling structure. The most commonly used method for reconstituting nucleosomes is simplicity itself: purified histones and DNA are mixed in a buffered solution having sodium chloride at a concentration of 1–2 M and dialyzed against solutions of successively lower ionic strength. The assembly process is strongly aided by the affinities the component histones have for each other. Histones H3 and H4 readily bind together to form a tetramer (H3 · H4)$_2$, resembling the inner core of the histone octamer, and H2A and H2B can form dimers with a structure similar to that

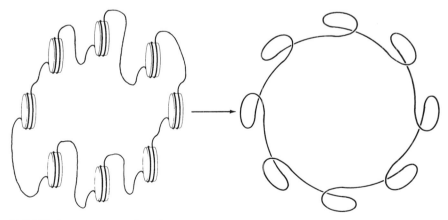

FIGURE 4 Disruption of nucleosomes on a closed, circular DNA molecule leads to the formation of loops, or supercoils, in the DNA. One supercoil arises from each nucleosome. These supercoils are said to be left-handed, or negative. If the strands crossed in the opposite sense the supercoils would be right-handed, or positive.

in the nucleosome (Fig. 3). Nucleosomes resulting from *in vitro* reconstitution are identical to those isolated from cells in histone content, sedimentation coefficient, supercoiling of DNA, appearance under the electron microscope, protecting DNA against nuclease digestion, and so on.

One important property of chromatin that is not faithfully mimicked by *in vitro* reconstitution is the spacing between nucleosomes. This spacing, as mentioned earlier, is a regular feature in living cells that depends on species, cell type, and so forth. In contrast, when nucleosomes are assembled *in vitro* by salt dialysis onto DNA sufficiently long to accommodate many nucleosomes, the spacing between particles is fairly random. At low nucleosome densities, individual particles are usually spread far apart, with occasional pairs being close-packed (meaning that there is almost no linker DNA between them). At high densities, most of the nucleosomes become close-packed. Certain cell extracts, however (e.g., from frog eggs or oocytes), are capable of assembling chromatin in which the individual nucleosomes are faithfully reconstituted and correct spacing is generated as well. These extracts contain histones as well as assembly factors, such as the protein nucleoplasmin and a factor called N1, which appear to maintain histones in a form that allows for their efficient assembly into chromatin. Investigators are employing such extracts to create templates that can be used in studies of transcription and replication of chromatin. One extract, made from cultured mammalian cells, has the interesting property of causing exogenously added DNA to be replicated and then preferentially assembling the replicated DNA into nucleosomes. Use of this extract should eventually allow dissection of the process by which newly replicated DNA is packaged into chromatin *in vivo*. [*See* Chromatin Folding; Chromatin Structure and Gene Expression in the Mammalian Brain.]

F. Nucleosome Positioning

Some DNA sequences are incorporated into nucleosomes, *in vivo* and/or *in vitro*, with a preferred rotational and translational orientation with respect to the histone octamer. In such positioned nucleosomes, if the DNA sequence that is protected against micrococcal nuclease digestion in the nucleosome were numbered from base pair 1 to 146, base pair 5 (for example) would always be found at the same place in the nucleosome—5 bp from one end, base pair 73 in the center, and so forth. But of course the DNA sequence is effectively numbered by its precise nucleo-

tide sequence, so that investigators have been able to show that some sequences do indeed give rise to positioned nucleosomes. Other sequences, however, when reconstituted *in vitro* into nucleosomes, are found to be situated randomly with respect to the histone octamer. Similarly, some but not all DNA sequences are associated with positioned nucleosomes *in vivo*.

What determines whether a particular DNA sequence is capable of forming a positioned nucleosome? One determinant is the ease with which a sequence can be bent. Certain DNA sequences form bends in the helix, so that if they are repeated with the same periodicity as the double helix, the DNA is curved overall. Such molecules can even form small circles visible as such under the electron microscope. Given the tight turning of DNA in the nucleosome, it is not surprising that such a sequence would be incorporated in a nucleosome with a preferred rotational orientation. What structural features lead to an exact translation orientation (i.e., why the DNA sequence at, say, +30 from one end is not located at +40, +50, etc.) are not yet clear, although the kinks found near the center of nucleosomal DNA may serve as some kind of positioning signal.

Bending (or perhaps bendability) can allow nucleosome positioning both *in vitro* and *in vivo*. Another mechanism that can position nucleosomes *in vivo* is the presence of other nonnucleosomal proteins that associate with particular DNA sequences. Such proteins may act by either a passive or an active mechanism. In passive nucleosome positioning, as occurs with the GRF2 protein in yeast, protein binding causes the adjacent nucleosomes to be restricted in the way the histone octamer can be positioned on the DNA, and this effect can in turn ripple out for several nucleosomes. In active nucleosome positioning, as occurs with the yeast α2/MCM1 complex, the protein interacts with the histones in a way that leads to precise nucleosome positioning. Regardless of the mechanism, nucleosome positioning may sometimes have a functional role, as discussed later.

III. NUCLEOSOME FUNCTION

A. General Considerations

Processes that use DNA as a substrate in eukaryotic cells (e.g., transcription, replication, and repair) depend on the recognition of specific DNA sequences by various regulatory factors. However, for the DNA to be packaged inside the nucleus, it must become tightly associated with histone proteins in the form of nucleosomes, and these in turn are arranged into higher-order structures. The structure of the nucleosome discussed earlier is likely to hinder the association of DNA with regulatory proteins which recognize specific DNA sequences. These proteins must recognize the molecular surface and charge presented by the base pairs (as seen in the minor or major groove), the exact conformation of the phosphate backbone of the double helix, or both. When DNA is organized into nucleosomes, these features are greatly perturbed. Kinks are introduced into the helix, the DNA is strongly bent, and the twist is altered, albeit subtly. In addition, the intimate association of the histone proteins with the DNA, as well as the close proximity of the (nearly) two turns of the DNA helix to each other (Fig. 3), may provide steric and ionic barriers sufficient to render the DNA inaccessible to many proteins.

These changes impose strong constraints on protein–DNA recognition, requiring either that those structural features recognized by a particular protein be preserved in nucleosomal DNA or that the relevant DNA sequence somehow be freed of the histones long enough to be recognized. Recent work suggests that regulatory proteins cover a whole spectrum, from recognizing only protein-free DNA to recognizing nucleosomal DNA in a particular orientation to being completely indifferent as to whether DNA is packaged into nucleosomes. In other words, from the perspective of proteins that act on DNA, chromatin may appear transparent, completely opaque, or something in between.

B. Transcription, Replication, and Nucleosomes

Binding of RNA polymerase II to promoter DNA and initiation of transcription require the association of several proteins with the promoter. In general, two kinds of interaction are necessary: proteins of the preinitiation complex must bind to sequences near the start site of transcription, and transcriptional activator proteins must bind to sites further upstream, termed enhancers in higher eukaryotes and upstream activation sequences (UASs) in yeast. Experiments done *in vitro* and *in vivo* have resulted in substantial progress toward understanding how these interactions are affected by packaging of DNA into chromatin.

Many eukaryotic genes contain a TATA box, a sequence motif found 25–30 bp upstream of the transcription start site. Such genes can be transcribed accurately *in vitro* in the absence of enhancers or UASs at a low or basal level. Two pieces of evidence suggest that such basal transcription is repressed by nucleosomes *in vivo*. The central component of the preinitiation complex formed on TATA-containing genes is the transcription factor TFIID, a multiprotein complex built around the TATA-binding protein (TBP). If nucleosomes are assembled *in vitro* onto a gene with a TATA box, transcription by RNA polymerase II is greatly inhibited. If, however, TFIID is allowed to associate with the TATA box prior to nucleosome assembly, transcription is able to proceed when RNA polymerase II and other appropriate factors are added. This suggests that basal transcription could be inhibited by nucleosomes *in vivo* unless something happened to allow TFIID to bind to its recognition sequence before it was covered up by the histones.

The second series of experiments that suggest that nucleosomes repress basal transcription *in vivo* has been performed in yeast cells. Yeast (*Saccharomyces cerevisiae*) have only two copies of the genes for each histone protein, and gene replacement strategies have allowed workers to eliminate one of these (one H3 gene, for example) and mutate or inactivate its partner. The effect of these manipulations on both overall cell viability and transcription of particular genes has been studied and correlated with effects on chromatin structure. Little effect was seen upon deletion of one copy of any of the individual histone genes. Deletion of both copies of any of the histone genes is lethal. However, when only a single copy of a histone gene is present, temporary inactivation of the gene causes cells to stop growing because they are unable to progress through part of the cell cycle (G2). Chromatin structure is perturbed under these conditions, and transcription of a number of genes artificially removed from their UASs is derepressed; i.e., basal transcription, which is repressed in wild-type cells, is allowed by perturbation of chromatin structure.

If basal transcription is repressed by nucleosomes *in vivo*, what overcomes this repression to allow transcription? The obvious solution is that proteins that bind to enhancers or UASs somehow overcome this repression. Such an antirepressive effect has been observed for derivatives of the transcriptional activator GAL4 *in vitro*, but it has not yet been directly demonstrated *in vivo*. Nevertheless, a picture is beginning to emerge that upstream activator proteins in conjunction with auxiliary, non-DNA-binding proteins such as the SWI/SNF complex in yeast act to overcome the repressive effect of chromatin in transcriptional activation.

If upstream activators are responsible for overcoming repressive effects of chromatin in living cells, they must themselves be able to bind to DNA that is packaged into chromatin. *In vitro* studies have shown that some transcription factors can recognize DNA in nucleosomes whereas others cannot. Of those factors that can recognize their binding sites in a nucleosome, some can bind without much effect on nucleosome structure, whereas others destabilize histone–DNA interactions. TFIIIA, whose binding to the internal promoter of the 5S rRNA gene is required for that gene to be transcribed by RNA polymerase III, can bind to a site in a particular positioned nucleosome *in vitro* only when the histone amino-terminal regions are acetylated or removed by proteolysis. Other examples of histone modification affecting the ability of chromatin to serve as a transcriptional template may remain to be discovered; it is certainly noteworthy that large chromosomal regions which are functionally homogeneous, such as the inactive X chromosome in humans, have been shown to be preferentially associated with particular histone modifications.

The effect of chromatin on the ability of activator proteins to bind to DNA *in vivo* is only beginning to be explored. Studies in yeast indicate that some factors can successfully compete with histones for occupancy of their binding sites whereas others cannot. For example, a replication origin in yeast, which is normally found outside a nucleosome, is inactivated when incorporated into a nucleosome. The most likely explanation for this result is that the histone proteins prevent recognition of specific DNA sequences by proteins required for replication. On the other hand, placement of a GAL4-binding site or a tRNA start site into a nucleosomal location does not affect binding, and the nucleosome in question is lost. Parameters which could affect whether a given protein can outcompete histones for its site include its abundance in the cell, the way it interacts with DNA, and the point in the cell cycle when it is able to bind to its recognition sequence.

In some instances, properly positioned nucleosomes may actually enhance transcriptional activation compared to naked DNA. Activation sometimes requires proteins bound at different sites on DNA to interact; this is often accomplished by bending of the DNA between the two sites. One way this DNA could be bent is by being incorporated into a nucleosome. If the intervening length is just right, the two sites may be brought into proximity and thereby help the proteins interact. This appears to be the case for the frog vitellogenin gene *in vitro* and the *Drosophila melanogaster* HSP26 gene *in vivo*.

An interesting example in which histones appear to play an active role in repression is provided by the α2/MCM1 operator in yeast. The α2/MCM1 complex binds upstream of genes that are expressed in yeast haploid **a** cells, which lack the α2 protein, but are repressed in haploid α or diploid **a**/α cells, which contain α2. Tight regulation of these genes is essential to proper regulation of mating type. The α2/MCM1 complex appears to repress in part by interacting with the transcription machinery, but it also precisely places a nucleosome directly adjacent to its binding site. This nucleosome occupies the TATA box upstream of the repressed genes. In mutant yeast cells which lack the amino-terminal region of histone H4, this nucleosome is no longer precisely positioned and a slight but significant derepression of transcription is observed. This strongly suggests that the α2/MCM1 complex represses transcription partly by using the repressive capability of the histone proteins.

Nucleosomes can also inhibit transcription by creating or allowing formation of higher-order structures. Histone H1, for example, which has been implicated in the folding of chromatin into higher-order structures, has long been known to be associated with inactive genes. A body of evidence now indicates that histone H1 is involved in the developmental regulation of the oocyte 5S rRNA gene family in the frog *Xenopus laevis*. First, when the repressed 5S rRNA gene is isolated as chromatin, it is not transcribed; if histone H1 is removed, the gene is transcribed, and readdition of H1 again abolishes transcription. Second, under appropriate conditions, expression of histone H1 *in vivo* can enhance repression of this gene, and conversely, removal of H1 can increase its expression *in vivo*. Whether these effects actually involve higher-order chromatin structures or more local effects of H1 binding is not yet known.

Nucleosomes or higher-order chromatin structures can also affect elongation once polymerase has recognized its promoter sequences. *In vitro* studies show that eukaryotic RNA polymerase II can transcribe through nucleosomes, but RNA polymerase III does not. The short genes transcribed *in vivo* by RNA polymerase III, however, may not be incorporated into nucleosomes, whereas those transcribed by RNA polymerase II are associated with nucleosomes. How does transcription proceed through nucleosomes? *In vitro* work suggests that nucleosomes may be transferred from sites in front of transcribing polymerase to sites behind without gross disruption of histone–histone contacts. *In vivo* studies of chromatin structure of transcribed sequences suggest some transient disruption or modification, but the precise nature of these changes remains obscure.

Nature must accommodate the packaging of DNA into chromatin in eukaryotic cells with the need for regulation of transcription and replication. In a few isolated cases, primarily from studies done with genetically manipulable yeast cells, chromatin appears to have been coopted to serve a regulatory function. Do higher eukaryotes use chromatin as a regulator as well? Some hints come from the aforementioned finding of modified histones associated with inactive genomic regions, and also from the complete conservation of the histone H4 amino-terminal region between humans and yeast. As mentioned earlier, this region is required for complete repression by the α2/MCM1 complex in yeast; it is also needed for repression of the silent mating type loci. It is not, however, needed for cell viability or for construction of a nucleosome. Why then it is so well conserved among eukaryotes? We have much yet to learn.

BIBLIOGRAPHY

Grunstein, M. (1992). Histones as regulators of transcription. *Sci. Am.* 68–74B.

Morse, R. H. (1992). Transcribed chromatin. *Trends Biochem. Sci.* **17**, 23–26.

Paranjape, S. M., Kamakaka, R. T., and Kadonaga, J. T. (1994). Role of chromatin structure in the regulation of transcription by RNA polymerase II. *Annu. Rev. Biochem.* **63**, 265–297.

Simpson, R. T. (1991). Nucleosome positioning: Occurrence, mechanisms, and functional consequences. *Prog. Nucleic Acid Res. Mol. Biol.* **40**, 143–184.

Wolffe, A. P. (1995). "Chromatin: Structure and Function." Academic Press, San Diego.

Wolffe, A. P. (ed.) (1995). The nucleosome. *Adv. Mol. Cell. Biol.*, **8**. JAI Press, Inc., Greenwich, CT.

Workman, J. R., and Buchman, A. R. (1993). Multiple functions of nucleosomes and regulatory factors in transcription. *Trends Biochem. Sci.* **18**, 90–95.

DNA Markers as Diagnostic Tools

ANNE BOWCOCK

Stanford University and The University of Texas Southwestern Medical Center at Dallas

GLOSSARY

Allele One of two or more alternate forms of a DNA sequence or a gene occupying the same locus on a particular chromosome; an individual with two similar alleles is said to be homozygous; one with two unlike alleles is said to be heterozygous

Autosomal Chromosome other than a sex chromosome

Carrier Individual heterozygous for a single recessive gene

Dominant Allele manifesting its phenotypic effect also in the heterozygotes; a trait determined by a dominant allele

Genotype Sum total of the genetic information of an organism

Haplotype Combination of individual alleles at linked loci

Karyotype Somatic chromosomal complement of an individual, analyzed according to size and banding patterns of each chromosome

Locus Position that a gene or a specific sequence of DNA occupies on a genetic map; alleles are situated at identical loci in homologous chromosomes

Phenotype Observable properties (structural and functional) of an organism; for many DNA polymorphic systems, both alleles are observed and the system is said to be codominant; when only one allele is observed it is considered a dominant trait; the other allele is a recessive trait because it can only be observed when it is on both chromosomes

Polymorphism Existence of two or more different genes or nucleotide sequences in a population; a locus is considered to be polymorphic when the second most common allele is present at a frequency of $>1\%$ in the population

Probe Laboratory definition for a recombinant DNA clone that can be used to detect its specific sequence within complex (e.g., genomic) DNA; can be inserts (of approximately 0.5–5 kb) in plasmid (an extrachromosomal element) vectors, inserts (of approximately 15 kb) in bacteriophage vectors, or inserts of approximately 40 kb in cosmid vectors

Recombination Can occur during reduction division (meiosis), which results in the generation of sperm and eggs that are haploid; refers to the process that gives rise to new combinations of linked genes due to the physical exchange of material between homologous chromosomes; the probability that recombination will occur between two loci is proportional to the physical distance between them

Restriction endonuclease Bacterial enzyme that cuts double-stranded DNA at or near a specific nucleotide sequence

Restriction fragment-length polymorphism Polymorphism in DNA sequence that is recognized by a restriction endonuclease; the loss or gain of the restriction endonuclease site results in a DNA fragment (allele) of different length; the DNA fragment is detected with a specific DNA probe. Because both alleles can be observed, the system is said to be codominant; alleles are inherited as Mendelian traits

Vector DNA molecule that can accept a piece of foreign DNA; the resultant recombinant DNA molecule can be taken up by and amplified in a host cell such as a bacterium

RESTRICTION FRAGMENT-LENGTH POLYMORPHISM (RFLPs) are dispersed throughout human chromo-

ENCYCLOPEDIA OF HUMAN BIOLOGY, Second Edition, VOLUME 3.

somes. The specificity of the restriction endonucleases combined with DNA probes can provide markers that are genetically linked to disease loci. If a marker locus is closely linked to a disease locus, one can follow the segregation of a particular allele in a family that is also segregating a disease allele. The alleles at a marker locus are usually independent of the disease mutation, and in different families different alleles may cosegregate with the disease locus; however, within one family, a particular allele will track a disease gene. For a number of human genetic diseases, prenatal and preclinical diagnoses are now possible by determining which marker alleles have been inherited in the individual at risk. In addition, it is often possible to predict the genotype of relatives. This allows the detection of asymptomatic-affected individuals in the case of dominant diseases, carrier females in the case of sex-linked diseases, and the discrimination of carriers versus normals in the case of autosomal recessive diseases. The goal in the detection of disease is to directly detect the mutation giving rise to the disease without relying on linked RFLPs where a low probability of recombination exists between the marker locus and the disease locus. For several diseases, direct detection is now possible.

I. INTRODUCTION

The first example of RFLPs used as a tool for antenatal diagnosis, but using a well-characterized gene, was with sickle-cell anemia, which is due to a mutation at codon 6 of the β-globin gene. A restriction endonuclease, HpaI usually cuts the DNA bearing the β-globin gene into a 7.6-kb fragment in American Blacks. It was found that the β-globin gene with the sickle-cell mutation (β^S) was localized on a 13-kb HpaI fragment >60% of the time. Loss of the first HpaI site and the sickle-cell mutation are unrelated base-pair substitutions, but recombination between the two mutations is very unlikely because they are so closely linked. Thus, in this population, the presence of the 13-kb HpaI fragment also indicated the presence of the sickle-cell mutation, and it was possible to track the inheritance of the β^S allele by tracking the inheritance of the 13-kb HpaI fragment. This development allowed fetal blood sampling to be replaced by DNA analysis of cells obtained from amniocentesis in fetuses at risk of having sickle-cell disease. Many RFLPs within the β-globin gene cluster were later found, resulting in a high probability that genetic diseases at this locus could be diagnosed. It is now

possible to directly detect many mutations at this locus, avoiding the use of linked RFLPs. [See Sickle Cell Hemoglobin.]

Because DNA polymorphic sites are scattered throughout the genome, it follows that some are physically close to genes that can be mutated and result in disease. If the DNA polymorphic site and the disease gene region are so close that they are often inherited together, they are said to be linked; i.e., very little recombination occurs between them. The advantage of using linked RFLPs for the diagnosis of genetic diseases is that it is not necessary to have isolated the disease gene. In many genetic diseases, the disease gene is not known or isolated, and prenatal and preclinical diagnoses are only possible with linked RFLPs. [See Genetic Diseases.]

The alleles generated by restriction endonuclease site polymorphisms can be detected as fragments of different lengths on autoradiographs after Southern blotting (see below). The development of the polymerase chain reaction (PCR; see below) subsequently allowed the direct detection of site polymorphisms. In addition, PCR allows the direct detection of mutations giving rise to disease, once they are identified.

There are now over 2000 probes detecting RFLPs in the human genome. Some of these have been shown to be linked to disease loci and are used in the diagnosis of their respective genetic diseases.

II. CHARACTERIZATION OF RESTRICTION FRAGMENT-LENGTH POLYMORPHISMS

A. Detection by Southern Blotting

The detection of RFLPs was first achieved by cleavage of total human DNA with a restriction endonuclease, agarose gel electrophoresis, Southern blotting, hybridization with a radioactive probe, and autoradiography to detect the region of interest (Fig. 1). In the simplest case, where a RFLP is due to the presence or absence of a restriction endonuclease site contained within a region detected by a DNA probe, there will be three possible fragment patterns on the autoradiograph (Fig. 2).

B. Types of RFLPs

RFLPs can be due to an alteration of 1 bp, which abolishes the restriction endonuclease site; alternatively, a DNA rearrangement such as an insertion or

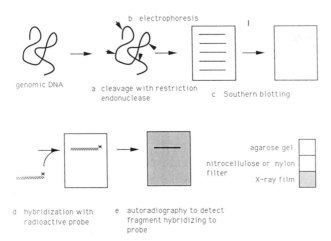

FIGURE 1 Detection of RFLPs. (a) Cleavage of human DNA by a 6-bp cutter (a restriction enzyme that recognizes 6 bp and cleaves the DNA at that sequence) generates approximately 1 million DNA fragments. (b) DNA fragments are fractionated according to their size by being subjected to agarose gel electrophoresis in an electric field. This forces the DNA, which is negatively charged, to migrate to the positive pole. For a particular DNA fragment, its rate of migration is inversely proportional to its size. (c) Southern blotting involves the denaturation of DNA fragments and their transfer to a solid support (such as nitrocellulose, or a more resilient nylon membrane). (d) The addition of a radioactively labeled DNA probe in the appropriate buffer results in hybridization, or annealing, of the probe and the specific DNA fragment to which it is homologous, or complementary. Unbound probe is washed off the membrane. (e) The fragment, which is bound to the probe, is visualized by autoradiography.

deletion can occur, which results in alleles of different fragment lengths that can be detected with a variety of enzymes (Fig. 3).

For RFLPs due to loss of a restriction site, there are usually only two alleles. The frequency of these alleles in the population determines how useful the RFLP is in tracking a disease gene. Some of the most useful RFLPs for diagnostic purposes have many alleles at a single locus. Several workers have identified DNA probes that identify such highly polymorphic loci. Many of these are due to variable numbers of a tandemly repeated sequence (VNTRs). The repeated sequence can be a single nucleotide, dinucleotide (e.g., CA) tri- or tetranucleotide, or a "minisatellite" sequence or longer. The minisatellite regions were originally detected because they shared a 10- to 15-bp "core" sequence. Some minisatellite probes detect several loci at once, and, thus, typing an individual with only a few of these multilocus probes is likely to yield a genetic fingerprint specific to that individual. This has been particularly useful in paternity testing, immigration disputes, and forensic testing.

VNTR probes are likely to be extremely useful in genetic fingerprinting samples, which enter the laboratory to be analyzed, protecting the laboratory from errors that can occur from sample mix-ups.

C. Types of Probes

Probes need to detect unique restriction fragments in human DNA; therefore, a prerequisite that they do not contain DNA sequences that are repeated elsewhere in the genome is required. One of the most common forms of repeat is the "Alu" sequence, a

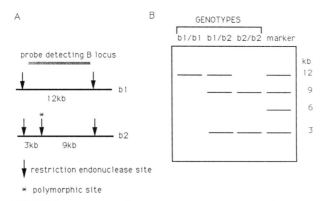

FIGURE 2 Detection of a RFLP. (A) The generation of a restriction site in human DNA yields fragments of 3 and 9 kb (b2 allele) instead of 12 kb (b1 allele). (B) The different fragment lengths can be visualized at different positions on the autoradiograph, according to where they migrated in the original gel before Southern blotting. DNA markers included in the electrophoresis step allow one to size them accurately. The patterns illustrated are for a homozygote for the b1 allele (b1/b1), a heterozygote for the b1 and b2 alleles (b1/b2), and a homozygote for the b2 allele (b2/b2).

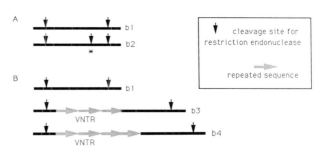

FIGURE 3 Different types of RFLPs. (A) Base-pair substitutions at restriction endonuclease sites generate two alleles: those with the site (b2) and those without the site (b1). (B) DNA rearrangements such as those due to insertions of variable numbers of tandem repeats (VNTRs) can generate several alleles. Whereas RFLPs due to base-pair substitution can usually only be detected with one restriction endonuclease. RFLPs due to DNA rearrangements can be detected with several enzymes.

300-bp sequence, occurring in approximately 300,000 copies in the haploid genome. They exist singly and in clusters, and on average it is estimated that an Alu sequence occurs every 1800 bp.

Hybridizing a Southern blot of human DNA with repeated sequences results in a smear instead of discrete fragments. The larger the insert, the greater the probability that it contains repetitive sequences. To eliminate this problem, it is necessary either to include a "cold" (not radioactivity labeled) repeated sequence in the hybridization mix to competitively hybridize with the repetitive DNA in the probe or to subclone the probe (so that only a portion of the insert, lacking the repeated DNA, is contained within the probe). Exceptions to this are the multilocus probes, which hybridize to many different unlinked fragments; however, these probes still allow for discrete fragments to be typed.

Probes can be derived from genomic DNA or from complementary DNA [cDNA; cDNA is complementary to messenger RNA (mRNA) and is synthesized with an enzyme known as reverse transcriptase]. A hypothetical region of the genome showing the origin of different types of probes is shown in Fig. 4. [See Human Genome.]

D. Applications of Linked RFLPs

Routinely, DNA samples from a family at risk and requesting diagnosis are digested with restriction enzymes that produce RFLPs revealed by DNA probes identifying loci close to the disease locus. The restric-

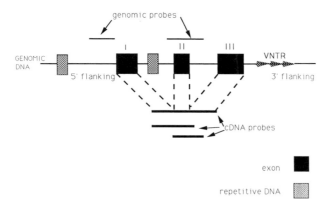

FIGURE 4 Hypothetical region of the genome containing a gene with three exons (I, II, and III), repetitive DNA, and a VNTR in the 3'-flanking region of the gene. The relationship of cDNA probes (obtained by reverse transcriptase off the mRNA template) to genomic probes is shown.

tion fragments (alleles) are then scored for each family member (e.g., the affected individual, fetus, sibs of unknown genotype that wish to know if they are carriers or are asymptomatic but affected, parents, grandparents). Because the disease and the alleles are known to cosegregate, the inheritance of the alleles will track the inheritance of the disease. When polymorphic loci, defined by RFLPs, are a distance from the disease locus so that recombination can occur between the two loci, the reliability of the test is lowered, and this is included in the final risk assessment.

The RFLP probes themselves have also allowed the mapping of a number of human diseases to human chromosomes. The first was Huntington's disease (HD) to chromosome 4p with a probe known as G8. The localization of cystic fibrosis (CF) to chromosome 7 was achieved with the demonstration that a random probe, pLAM-917, segregated 85% of the time with the CF defect. The localization of these and other genetic diseases allowed the further identification of other closer DNA markers. These markers can be used in prenatal and preclinical diagnoses, and their reliability depends on the distance they are from the disease locus. The information gained from each marker depends on the frequency of the alleles at the loci in question. When many alleles exist at one locus, such as in the case of the VNTR-containing loci, far more information can be gained than with simple two-allele systems, especially when the frequency of the one allele is low. In a fully informative mating, the inheritance of the parental chromosomes can be determined in the children. Figures 5 and 6 demonstrate the possible diagnoses that can be obtained in sibs of an individual with an autosomal dominant versus an autosomal recessive disease when a two-allele marker locus is linked to a disease locus.

The identification of markers extremely close to a disease locus eventually aids in the identification of the gene and the mutation(s) giving rise to the disease. The aim at this stage is to develop means of detecting the mutation(s) directly. However, for many genetic diseases, the identification of the mutation(s) has not yet been achieved, or the mutations are so numerous that the use of linked RFLPs in diagnosis is more reliable and efficient.

DNA diagnosis is now available for a number of inherited diseases; however, in most cases, because the diagnosis relies on knowing which allele is linked to the disease locus, and either allele may be linked, depending on the family, this type of diagnosis can only be performed in families where at least one child is already affected. Before 1984, DNA analysis was

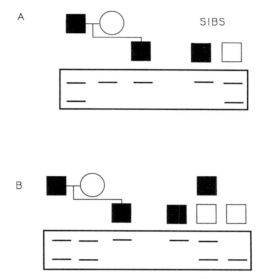

FIGURE 5 Informative (A)- and partially informative (B)-linked markers in a family segregating for an autosomal dominant trait. Open symbols are unaffected individuals; closed symbols are affected individuals. The fragments (alleles) seen in sibs with and without the trait are shown. In B, sibs that are heterozygotes for the marker alleles have a 50% chance of having the trait.

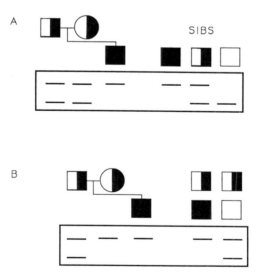

FIGURE 6 Informative (A)- and partially informative (B)-linked markers in a family segregating for an autosomal recessive trait. Open symbols are unaffected individuals; closed symbols are affected individuals; and half-filled symbols are carriers. In B, sibs homozygous for the upper fragment (allele) have a 50% chance of being affected and a 50% chance of being a carrier. Sibs heterozygous for the marker alleles have a 50% chance of being normal and a 50% chance of being carriers.

carried out on amniocytes obtained at 15–17 weeks of gestation. Since then, chorionic villus sampling at 9–11 weeks has been available as an alternative way of obtaining fetal cells.

III. THE POLYMERASE CHAIN REACTION AND ALLELE-SPECIFIC OLIGONUCLEOTIDE PROBES

With PCR technology, RFLPs can be detected in <1 day, in contrast to the time taken to obtain results from Southern blotting, which is usually several weeks. The principle is to use small stretches of single-stranded DNA (oligonucleotide primers) that flank a region of double-stranded DNA of interest (one from one strand, the other from the other strand). The complex (genomic) DNA template, containing the region to be amplified, is denatured to single strands by heating at 94°C, the oligonucleotides are annealed to the DNA substrate at an appropriate temperature (usually between 50 and 60°C), and DNA polymerase extends the oligonucleotides with the DNA substrate as a template. With successive cycles of denaturation, annealing, and extension, it is possible to amplify the stretch of DNA flanked by the oligonucleotide primers. This permits a 220,000-fold amplification of the target sequence with as few as 100 cells as starting material (Fig. 7). More recently, the use of a polymerase derived from a thermostable bacterium has facilitated this procedure because it is active at 72°C and is not denatured at 94°C when the DNA substrate is being denatured. Thus, the same *Taq* polymerase can be cycled repeatedly through the two or three temperatures required for amplification.

Such rapid amplification of a region of interest allows rapid typing of that region for a restriction endonuclease site polymorphism (Figs. 8B1 and 8B2). It is also possible to ask directly if the restriction endonuclease site or a particular DNA sequence is present. This is achieved by hybridization with short probes specific for a particular DNA sequence termed allele-specific oligonucleotides (ASOs). The DNA to be tested is spotted onto a membrane and hybridized with an ASO probe. The ASO only binds if there is a perfect match between it and the amplified DNA sequence (Figs. 8A1 and 8A2). The presence of the ASO is detected by labeling the ASO with radioactivity or a substance that can be detected with a chemical reaction that converts a colorless substrate to a colored precipitate. More recently, the ASO has been

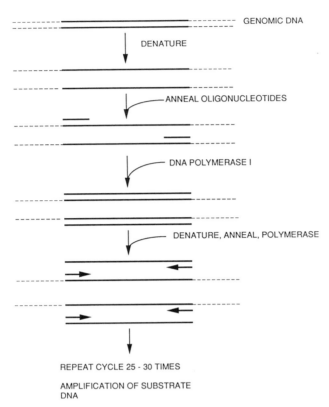

FIGURE 7 The polymerase chain reaction, consisting of successive rounds of denaturation of a complex DNA template, annealing with oligonucleotide primer sequences, and extension with DNA polymerase I.

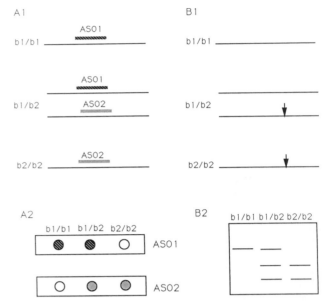

FIGURE 8 Detection of DNA sequence differences after polymerase chain reaction. (A1) The amplified product can be hybridized with allele-specific oligonucleotides (ASOs); in this case ASO1 binds to DNA from the b1 allele and ASO2 binds to DNA from the b2 allele. (A2) Amplified product is dot-blotted onto a solid support such as nitrocellulose, and hybridization with a specific ASO determines which allele is present. (B1) The alleles can sometimes be discriminated by cleavage with a specific restriction endonuclease. (B2) Agarose gel electrophoresis of amplified DNA digested with the appropriate enzyme will indicate which alleles are present.

immobilized on a solid support, and the amplified product has been hybridized to it.

The first application of PCR in DNA diagnosis was to enhance the sensitivity of the prenatal diagnosis of sickle-cell anemia. PCR is now being used for prenatal diagnosis and carrier detection of sickle-cell anemia. CF, the thalassemias, Duchenne muscular dystrophy (DMD), hemophilia, and a number of other genetic diseases. In conjunction with linked RFLPs, this is a highly reliable, informative diagnostic procedure. In certain cases, when the gene giving rise to the disease is known, but where different mutations are known to give rise to the disease, it is sometimes desirable to determine the mutation in a particular disease family. This can be carried out efficiently by amplification with PCR followed by direct sequencing of the amplified product directly (this determines the base-pair composition of the amplified DNA). This is sometimes carried out in the case of the thalassemias when the battery of ASOs do not detect the mutation. PCR can also be used to detect deletions (as in the case of

multiplex amplification of the DMD gene; see below). These direct approaches do not rely on the availability of linked RFLP probes.

The great advantage of PCR is that it can be performed on partially degraded DNA or where the supply of DNA is limiting. This is illustrated by the demonstration that it is possible to diagnose both CF and phenylketonuria (PKU) from old Guthrie cards (newborn blood spots stored on filters).

IV. DNA MARKERS FOR COMMON HUMAN GENETIC DISEASES

A. Autosomal Dominant

Table I lists a few autosomal dominant diseases where diagnosis with DNA markers is offered. In addition to prenatal diagnosis, it is often possible to detect heterozygotes who will develop the disease but who are presymptomatic at the time of the analysis.

TABLE I

Description of Some Autosomal Dominant Diseases
Detectable at the DNA Level

Disease (location; incidence in Caucasians)	Mode of detection
Huntington's disease (4p16.2–4p16.3; 1/40,000)	Closely linked RFLPs
Von Recklinghausen neurofibromatosis (17q11.2; 1/4000)	Flanking RFLPs
Adult polycystic kidney disease (16p13.11–16pter; 1/1000)	Flanking RFLPs
Retinoblastoma (RB1) (13q14.2; 1/3400–1/15,000)	RFLPs within and flanking the RB1 gene. PCR to detect DNA sequence polymorphisms
Myotonic dystrophy (19q13.2; 1/5000–1/15,000)	Closely linked RFLPs

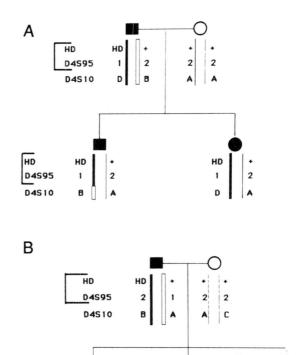

FIGURE 9 Segregation of D4S95 in Huntington's disease (HD) families with recombinants between D4S10 and HD. Open symbols are unaffected individuals; closed symbols are individuals diagnosed as having HD. Squares represent male individuals; circles represent female individuals. The letters listed for D4S10 represent haplotypes A–D. The D haplotype for D4S10 is associated with the disease gene for the family shown in *a*; the B haplotype for D4S10 travels with the disease gene for the family shown in *b*. The first sib in family *a* and the second sib in family *b* each inherited the nondisease haplotype for D4S10, yet they both have HD. In neither family has a crossover occurred between D4S95 and HD. [Reproduced, with permission, from Wasmuth *et al.* (1988).]

I. Huntington's Disease

HD develops late in life, usually after the affected individuals have had children and the opportunity to transmit the HD gene to them. It is a progressive neurodegenerative disorder and results in numerous psychiatric symptoms and a characteristic gait. The age of onset of the disease is usually between 30 and 50 years and the clinical symptoms vary.

This disease was mapped to the terminal cytogenetic band of chromosome 4p with a probe known as G8, which identifies the locus D4S10. The genetic distance between D4S10 and HD is 3–4 centiMorgans (cM). A number of closer probes have now been found. Some of these are now used for first trimester exclusion diagnosis. It is now also possible to offer predictive testing for individuals with a family history of HD; in other words, in a large proportion of cases, one can distinguish between carriers without symptoms (because they are too young to have yet developed the disease) and their siblings who have not inherited the mutant gene. Before this, children of individuals affected by HD, who were at 50% risk of the disease, had to wait until they were 65 years or older to be sure they had escaped HD. A successful predictive test can modify the risk from 50% to as low as 2% or as high as 98% in individuals at risk. Several centers now offer predictive testing to those who spontaneously request such tests.

Figure 9 demonstrates the segregation of alleles at two loci (D4S10 and D4S95) that are closely linked to the HD locus in a family also segregating this disease. In these families, rare recombination events have occurred between HD and D4S10, but the D4S95 alleles still segregate with the HD allele.

Although no therapy can be offered at present, knowledge that an individual is likely to have HD allows them to plan their life-style and future with this in mind.

Predictive testing for HD is likely to be a model for the presymptomatic testing of other disorders such as Alzheimer's disease, schizophrenia, manic depression, and familial cancer once the genetics for these diseases is elucidated and markers are isolated. [*See* Huntington's Disease.]

2. Neurofibromatosis

Von Recklinghausen neurofibromatosis (NF1) is an autosomal dominant disease characterized by cafe-au-lait spots, multiple neurofibromas (benign tumors of peripheral nerves) that increase in size and number with age, and an increased risk of malignancy, especially glioma and neurofibrosarcoma. The disease occurs with variable manifestations and severity. Approximately 30–40% are new mutations. The mutation giving rise to NF1 was shown to be located near the centromere on chromosome 17 with the probe pA10–41. Figure 10 shows the DNA fragments seen with this probe in a family with NF. NF has now been mapped to the long arm of chromosome 17 at 17q11.2. At present, presymptomatic and prenatal diagnoses are performed with many probes closely linked to this locus.

3. Polycystic Kidney Disease

Adult polycystic kidney disease (APCKD) is a common and often lethal disease affecting many organs. The most significant defect is the development and enlargement of cysts in several organs, including the liver, pancreas, spleen, and kidneys. Cyst development cannot be prevented by any known therapy and leads

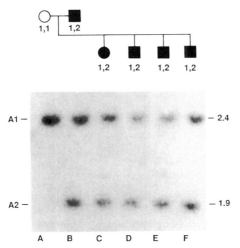

FIGURE 10 Hybridization of probe pA10–41 to *Msp*I-digested DNAs from individuals in a neurofibromatosis kindred. The alleles are a 2.4-kb (A1) or a 1.9-kb (A2) band. Squares represent male individuals, circles represent females. Open pedigree symbols are unaffected individuals; closed symbols are affected individuals. [Reproduced, with permission, from Barker *et al.* (1987). Copyright 1987 by the AAAS.]

to irreversible renal failure at an average age of 51 years unless dialysis or transplantation is used. The first symptoms tend to occur after the age of 40 years, after individuals have reproduced; however, there is marked variability in onset. In 1985, researchers showed that the APCKD locus (PKD1) is closely linked to the α-globin locus on the short arm of chromosome 16. They used a 3'HVR probe that consists of a tandem repeat sequence from the 3' end of the α-globin cluster. Subsequently, many probes have been isolated that detect loci flanking the disease locus. These allow an improved method of presymptomatic diagnosis of the disease.

One problem with the diagnosis of APCKD that is likely to be a paradigm for the diagnosis of some other genetic disorders is that in 2–3% of cases, the disease is caused by a mutation at another locus (i.e., APCKD is genetically heterogeneous). This form of the disease, which is also autosomal dominant, has been described for approximately six families and is clinically indistinguishable from the linked form of APCKD.

4. Retinoblastoma

Retinoblastoma is a malignant tumor that arises in the eyes of children, usually before the age of 4 years. Thirty to 40% of patients with retinoblastoma have inherited the predisposition to the tumor, in addition to several other cancers, particularly osteosarcoma. The cancer develops in 80–90% of individuals who carry at this locus any one of a variety of mutant alleles associated with a predisposition to the tumor. Much evidence supports the hypothesis that retinoblastomas that develop in patients without the hereditary form of retinoblastoma arise from retinal cells that have acquired somatic mutations at the same genetic locus.

Retinoblastoma and osteosarcoma arise from cells that have lost both functional copies of the retinoblastoma gene. This can be due to a loss of one copy of the gene in somatic tissue in an individual who has inherited an initial mutation on the other allele. Alternatively, mutations on both chromosomes can occur somatically.

Numerous advances have been made in the study of retinoblastoma at the molecular level, and it is one of the few familial cancers where prenatal and preclinical diagnoses are available.

The human retinoblastoma gene (RB1) contains 27 exons (protein-coding regions) and spans a region of approximately 200 kb. The gene has been cloned, and several RFLPs have been described. The locations of

Kilobases of DNA

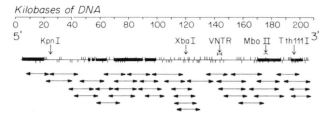

FIGURE 11 Restriction enzyme map of the genomic locus of the retinoblastoma gene. Vertical markers above the map represent the locations of *Hin*dIII sites; those below the map represent the locations of *Eco*RI sites. The boxed areas represent *Hin*dIII fragments that contain sequences found in the cDNA (exons). Each double-headed arrow beneath the map represents a distinct recombinant bacteriophage clone. The positions of the polymorphic DNA sites are indicated with arrows. The location of the polymorphic *Mbo*II site or sites has not been determined precisely but is shown at approximately 175 kb on this map. [Reprinted, by permission from Wiggs *et al.* (1988).]

some of these are shown in Fig. 11. Approximately 90% of families referred for DNA-based diagnoses are informative for these RFLPs; however, RFLP screening may have detected only 5% of the potentially useful polymorphic sites. It has been shown that additional DNA sequence polymorphisms can be detected with PCR.

In 10–20% of all families, it is sometimes possible with Southern blotting to determine that a parent has an altered copy of the gene (deleted or rearranged). The identification of children who also have that altered chromosome predicts which ones have a tendency to develop retinoblastoma.

The diagnostic test identifies individuals who have inherited a tumor-predisposing mutation at the retinoblastoma locus and allows eye examinations to be focused on children who carry a mutant allele, reducing the need for such examinations in children thought to have a normal genotype. [*See* Tumor Suppressor Genes, Retinoblastoma.]

B. Autosomal Recessive Diseases

Table II contains a description of some of the more common recessive disorders detected with linked RFLPs. In the case of autosomal recessive disorders where the appropriate DNA markers are available, prenatal diagnosis and the detection of heterozygotes (carriers) are possible.

1. Cystic Fibrosis

CF, the most common autosomal recessive disease in Caucasians, was mapped to chromosome 7 in 1985.

TABLE II

Description of Some Autosomal Recessive Diseases Detectable at the DNA Level

Disease (location; incidence in Caucasians)	Mode of detection
Cystic fibrosis (7q31; 1/2000)	Closely linked RFLPs PCR and ASOs
Phenylketonuria (12q22–q24.2; 1/4500–1/16,000)	RFLPs detected with PAH[a] cDNA probe
Sickle-cell anemia (16p13.3; 1/3000 in United States)	Direct detection with restriction enzyme (*Mst*II, *Cvn*I, *Oxa*NI) or PCR and ASOs
β-Thalassemia (16p13.3)	Direct detection—battery of ASOs for different ethnic groups RFLPs within β-globin cluster
α_1-Antitrypsin deficiency (14q32.1)	RFLPs within and flanking the gene PCR and ASOs for the Z allele
Friedreich's ataxia (9p22-cen; 1/50,000)	Linked RFLPs
Wilson's disease (13q14.3–q21; 1/33,500)	Flanking RFLPs

[a]Phenylalanine hydroxylase.

This was with a probe known as LAM917, which was shown to recombine with the CF locus at an incidence of 15%. Soon after that two other probes, the oncogene MET, and a probe identifying a piece of DNA of unknown function, J3.11 (locus D7S8), were shown to be extremely close to the CF locus and to probably flank it. At this stage, it was possible to offer prenatal diagnosis for CF with a reasonable amount of confidence. Many other RFLP probes closer to CF have been isolated subsequently. These have increased the probability of a fully informative and reliable prenatal diagnosis. It is possible to directly type many of the restriction site polymorphisms giving rise to the closely linked RFLPs with PCR.

In 1989, the CF gene was isolated and shown to span a distance of 250 kb. Twenty-four exons code for an mRNA 6.5 kb long. Analysis of CF versus normal DNA has shown that the defect on 68% of CF chromosomes is due to a deletion of 3 bp within exon 10. This results in deletion of the amino acid phenylalanine. This mutation occurs most commonly on haplotypes that could be considered part of a related group; however, this deletion has been observed

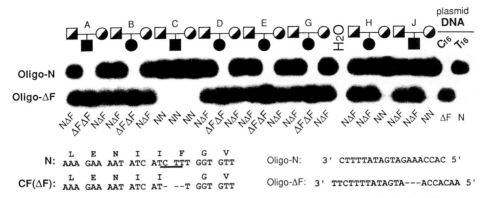

N: L   E   N   I   I   F   G   V
    AAA GAA AAT ATC ATC TTT GGT GTT

CF(ΔF): L   E   N   I   I   G   V
         AAA GAA AAT ATC AT- --T GGT GTT

Oligo-N: 3' CTTTTATAGTAGAAACCAC 5'

Oligo-ΔF: 3' TTCTTTTATAGTA---ACCACAA 5'

FIGURE 12 Detection of the ΔF508 mutation in cystic fibrosis (CF) by oligonucleotide hybridization. Autoradiographs show the hybridization results of genomic DNA from representative CF families with the two specific oligonucleotide probes as indicated. Oligo-N detects the normal DNA sequence, and oligo-ΔF detects the mutant sequence. Genomic DNA samples from each family member were separated by electrophoresis and transferred to a nylon membrane. The membrane was hybridized with radioactively labeled oligonucleotide probes, washed, and exposed to X-ray film. Samples without DNA (H2O) and plasmid DNA, T16 (N cDNA), and C16 (cDNA with the ΔF508 deletion) were included as controls. The illustration is based on the assumption that the triplet CTT was deleted; the sequencing data do not allow distinction between deletion of these nucleotides or other combinations. [Reproduced, with permission, from Kerem *et al.* (1989).]

on at least one other haplotype, suggesting that this mutation has arisen independently at least twice. The phenylalanine deletion is associated with pancreatic insufficiency, and it is suggested that other mutations may be less severe and sometimes result in the pancreatic-sufficient form of CF. Direct detection of this deletion with PCR and ASOs is shown in Fig. 12.

At least 40 other mutations have been found that can give rise to CF. A large number such as this is not ideal for the nationwide carrier screening of this disease. [*See* Cystic Fibrosis, Molecular Genetics.]

2. Phenylketonuria

PKU is the most common inborn error of amino acid metabolism in Caucasoid populations and is due to a deficiency of hepatic phenylalanine hydroxylase (PAH). In the normal human liver, PAH catalyzes the rate-limiting step in the hydroxylation of phenylalanine to tyrosine. If PAH is defective, serum phenylalanine accumulates, which results in hyperphenylalaninemia and abnormalities in the metabolism of many compounds derived from phenylalanine. Approximately 1 in 50 individuals is a carrier of the disease. If not detected early, affected children develop severe mental retardation. The consequences can be avoided with a rigid implementation of a restricted diet low in phenylalanine during the first decade of life. Neonatal screening for PKU has been routinely carried out in Western countries, but until recently there was no

way of detecting affected fetuses. This is now possible with RFLPs detected with a cDNA probe for the PAH gene. Although genetic analysis has defined two mutations, both associated with different haplotypes, until the identification of all PKU mutations, direct detection is not practical. [*See* Phenylketonuria, Molecular Genetics.]

3. Sickle-Cell Anemia and β-Thalassemia

The detection of sickle-cell anemia with the linked *Hpa*I RFLP has evolved into the direct analysis of the single base-pair change within the β-globin gene causing sickle-cell disease. The specific mutation within the β-globin gene, which is due to an amino acid substitution of glutamic acid for valine, also abolishes the site for the restriction enzymes *Dde*I and *Mst*II or its isoschizomers *Cvn*I or *Oxa*NI (isoschizomers are restriction enzymes that recognize the same nucleotide sequence). This meant that prenatal diagnosis could be done without family studies and more reliably than with a linked RFLP. With the advent of PCR and rapid amplification of the region in question, it was possible to ask directly if the restriction site was present and Southern blotting could be avoided entirely. Subsequently, the application of allele-specific oligonucleotide probes allows one to do without a restriction endonuclease, as the presence or absence of the mutant site can be determined directly by dot blotting and hybridization with ASO probes.

Developments such as hybridization of the PCR product to an immobilized array of oligonucleotide probes further improve the efficiency of the test. The progression of the diagnosis of sickle-cell anemia is shown in Fig. 13. Direct detection with ASOs is preferable to cleavage by a restriction enzyme such as *Mst*II, as a second common mutant allele at codon 6 exists known as beta-C (B^C). This mutation does not alter the endonuclease recognition site of either *Mst*II or *Dde*I.

The thalassemias are a heterogeneous group of hereditary anemias characterized by defective β-globin synthesis. Ethnic groups at risk for β-thalassemia are Mediterranean, North African, Middle Eastern, Asian Indian, Chinese, Southeast Asian, and Black African. At the end of 1988, >60 β-thalassemia mutations had been discovered, and it is estimated that alleles giving rise to 99% of β-thalassemia genes in the world have been identified. Direct detection of β-thalassemia mutations with PCR has only been possible since these different lesions have been characterized. Before that, indirect detection with RFLPs was carried out. This carried a theoretical error rate of between 1 in 300 and 1 in 500 due to meiotic recombination between

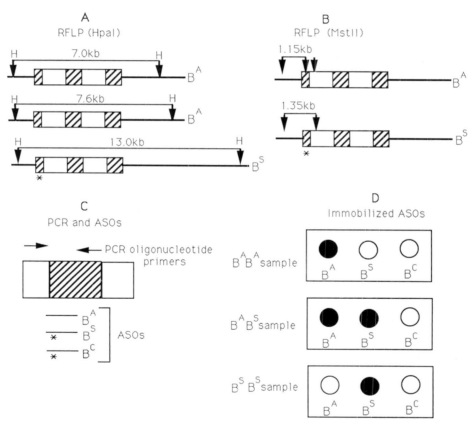

FIGURE 13 Progression of sickle-cell anemia diagnosis. (A) Linkage association of the 13.0-kb *Hpa*I fragment with the β-globin gene containing the sickle-cell mutation (B^S). The 7.0- and 7.6-kb fragments are associated with the normal β-globin gene (B^A). The cleavage sites for *Hpa*I (H) are shown. The position of the sickle-cell mutation is marked with an asterisk. Hatched boxes are β-globin exons; open boxes are introns. (B) The restriction endonuclease *Mst*II cleaves at codon 6 of B^A but not B^S, allowing direct detection of the mutation. The arrows show the *Mst*II sites. (C) Allele-specific oligonucleotide (ASO) recognition of B^A, B^S, and B^C. In conjunction with amplification of the DNA containing this sequence with polymerase chain reactions (PCR), this provides a highly specific approach to allele typing. (D) Detection of the B^A, B^S, and B^C alleles with ASOs immobilized on a solid support. The amplified DNA, which has incorporated a biotinylated primer, hybridizes to the ASO probes and is detected with a reaction that converts a colorless dye into a colored precipitate. Results are shown after hybridizing DNA from individuals with the genotypes B^A/B^A, B^A/B^S, and B^S/B^S. [Adapted, with permission, from Caskey (1987).]

a site used in tracking the β-thalassemia mutation and the mutation itself.

Each ethnic group has been found to have its own set of β-thalassemia mutations, and in any particular group, four to six alleles make up 90% of the β-thalassemia genes. This has simplified the detection of the disease-producing mutations. Prenatal diagnosis is now usually done by direct detection of the mutation, which involves a combination of dot-blot hybridization and restriction analysis. Approximately half of the β-thalassemia mutations can be detected with restriction enzyme analysis of the PCR product. PCR can also be used to detect deletion in these genes.

If difficulty is encountered in the direct detection of sickle-cell anemia or β-thalassemia, haplotype analysis of various family members by PCR will allow rapid prenatal diagnosis by indirect detection. This demonstrates the use that linked RFLPs play in prenatal and preclinical diagnoses, even when methods for the direct detection of the mutations are available. Linked RFLPs also provide a way of checking conclusions drawn from direct detection.

4. α_1-Antitrypsin Deficiency

This disease results in predisposition to childhood liver cirrhosis and pulmonary emphysema in early adult life. This occurs earlier in smokers than in nonsmokers (the third decade for smokers, and the fifth decade for nonsmokers).

A number of alleles exist for α_1-antitrypsin. M is the most common, normal allele. There are also a number of alleles that give rise to deficiency, the most common of which is the Z allele, which occurs in 1 out of 7000 North Americans. This allele can be distinguished by isoelectric focusing and is due to a base-pair substitution in codon 342 of exon V, which results in conversion of Glu to Lys. Many different null alleles also exist. RFLPs exist within and flanking the AAT gene. The haplotypes derived from these RFLPs have been shown to be strongly associated with different α_1-antitrypsin alleles. Thus, the determination of haplotypes present in diseased individuals can give an indication of the mutant alleles that are present.

It is now possible to detect the Z allele with PCR and ASO typing, and direct detection systems have also been developed for a number of other alleles. However, RFLPs still play an important role in the diagnosis of this disease due to the number of other mutant alleles that exist that are not detectable with ASOs.

C. Sex-Linked Diseases

Table III describes some sex-linked diseases detectable at the DNA level. In families affected with sex-linked diseases, where the appropriate DNA markers are available, prenatal diagnosis is offered to determine the affected males. The detection of female carriers is also often possible.

1. Duchenne Muscular Dystrophy

Affected males are often confined to a wheelchair by the age of 12 years and die in their late teens. One-third of all cases appear to arise via a new mutation. This disease has been shown to be due to defects in the X-linked dystrophin gene, which normally codes for an mRNA of 14 kb. Becker muscular dystrophy (BMD), which is a clinically milder form of the disease, has been shown to also be due to mutations in the dystrophin gene. No cure or effective therapy exists; however, it is now possible to perform highly accurate prenatal diagnosis and carrier detection of DMD and BMD via Southern analysis using cDNA and genomic clones for dystrophin.

The DMD gene spans more than 2 Mb of DNA and is split into at least 60 exons. Partial gene deletions have been shown to cause 50–60% of DMD lesions. These deletions usually span several hundred kilobases of the DMD locus and generally overlap two specific regions located approximately 0.5 and 1.2 Mb from the promoter. Fifteen percent of muta-

TABLE III

Description of Some Sex-Linked Diseases Detectable at the DNA Level

Disease (location; incidence in Caucasian males)	Method of detection
Duchenne and Becker muscular dystrophy (xp21.3–p21.1)	cDNA probes to detect deletions Multiplex amplification to detect deletions Linked RFLPs within and flanking the dystrophin gene
Hemophilia A (Xq28; 1/10,000)	RFLPs within and flanking the factor VIII gene
Hemophilia B (Xq26.3–q27.1; 1/50,000)	RFLPs linked to factor IX gene
X-linked hydrohidrotic ectodermal dysplasia (Xq11–q21.1)	Linked RFLPs

tions are due to duplications. This gene is too complex to be analyzed simultaneously with the complete cDNA probe: a *Hin*dIII digest generates at least 65 restriction fragments, and eight cDNA subcloned probes are required to detect all deletions in this gene. For the detection of carrier females, because of the presence of one normal gene, a dosage analysis is necessary in the region of the deletion. Alternatively, pulsed-field gel electrophoresis (which resolves large DNA fragments, i.e., between 50,000 and 5 million bp) can be used to detect an abnormal breakpoint fragment. In the one-third of cases that have no visible deletion, analysis has to be performed with intragenic RFLP markers, and several informative DNA sequence polymorphisms have been identified within the dystrophin gene. These allow diagnosis in 37% of samples. One problem with this approach is that the accuracy of diagnosis with linked RFLPs is reduced. This may be due to the large size of the dystrophin gene because recombination between a linked marker and the mutation occurs in 5% of meioses.

In some laboratories, PCR is now being used for the rapid detection of 80–90% of all dystrophin gene deletions. Currently, 18 separate oligonucleotide primers are being combined in a single reaction to simultaneously amplify nine deletion-prone exons of the gene. This approach, known as multiplex amplification, can be used in both the prenatal and postnatal diagnoses of DMD (Fig. 14); however, it is not applicable to nondeletion cases or for carrier diagnosis.

If one of the bands is missing after multiplex amplification, the extent of the deletion is determined with cDNA probes. Even with multiplex amplification, RFLP analysis with intragenic or extragenic probes is useful to determine if nonpaternity or sample switching has occurred.

Due to the high rate of new mutations in the dystrophin gene, genetic counseling and prenatal diagnosis will only have a limited effect on the incidence of the disease. [*See* Muscular Dystrophies.]

2. Hemophilias

Hemophilias are disorders of blood coagulation due to a deficiency of clotting factors VIII and IX. Hemophilia A, the classic type of hemophilia, is one of the most common severe, inherited bleeding disorders in humans. It is due to a deficiency in a protein involved in coagulation. This protein, known as factor VIII, which normally circulates in the plasma with von Willebrand factor, is an essential protein cofactor in the intrinsic coagulation pathway. The lesion giving

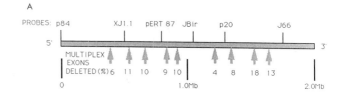

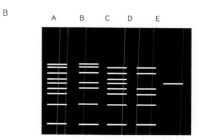

FIGURE 14 DMD mutations. (A) Schematic illustration of the DMD gene illustrating the relative locations of the nine currently amplified regions (arrows) relative to six genomic probe markers. The percentage of DMD patients carrying deletions of each of the nine regions is also shown. (B) Detection of DNA deletions at the DMD locus by multiplex amplification. Lane A shows the typical pattern of amplification products seen in normal DNA. Lanes B–E show the patterns that can be seen in DMD males with deletions of the dystrophin gene. [Adapted, with permission, from R. A. Gibbs, J. S. Chamberlain, and C. Thomas Caskey (1989). Diagnosis of new mutation diseases using the polymerase chain reaction. *In* "PCR Technology" (H. A. Erlich, ed.), Stockton Press, New York.]

rise to hemophilia A is due to a Glu-Gly substitution in exon 7 of factor VIII. Like DMD, the incidence of sporadic (noninherited) cases is high. The human gene for factor VIII has been cloned and sequenced and shown to be 186 kb with 26 exons and 25 introns. Until the identification of the gene, prenatal diagnosis was only possible with fetal blood sampling that was performed in the second trimester. This presented a 5% risk to the fetus. In addition, carrier detection was extremely unreliable due to the wide range of clotting found in normal and heterozygous females.

Since the discovery of several RFLPs within and linked to the factor VIII gene, carrier identification and prenatal diagnosis have been achieved with chorionic villus sampling in the first trimester.

Many mutations have been detected in the factor VIII gene. These mutations are diverse; in a total of 240 patients, 28 molecular defects have been found: 5.8% are deletions, 5.0% are base-pair substitutions, and 0.8% are insertions. The existence of many unique mutations underscores the necessity of using linked RFLPs for DNA-based diagnosis where the gene can be tracked in families at risk, while remaining independent of the point mutation causing the hemophilia.

Hemophilia B, or Christmas disease, is due to a defect in the factor IX gene. Diagnosis has previously relied on the familial segregation of linked RFLPs. This is successful in 65% of Caucasoid cases but fails where the marker is uninformative (15% of cases) of where there is homozygosity for closely linked alleles in key family members (20% of cases). Effort is being made to identify the mutations causing the disease so that they can be detected directly. [See Hemophilia, Molecular Genetics.]

V. OTHER USES OF DNA MARKERS AND FUTURE PROJECTIONS

A. Polygenic Diseases

One of the most challenging applications of RFLP research and association of genetic diseases is the study of polygenic diseases such as cardiovascular disease and atherosclerosis. Coronary heart disease is the leading cause of morbidity and mortality in the United States. While it is evident that many different genes are likely to contribute to these diseases, the role and interactions of the different genes are not understood. By using RFLPs to study the inheritance of candidate genes in high-risk families, it is likely that their interaction will slowly be unraveled. [See Coronary Heart Disease, Molecular Genetics.]

B. Autoimmune Diseases

Similarly, RFLPs linked to genes predisposing to familial cancers and autoimmune diseases are likely to be determined, allowing individual risk assessment. A variety of autoimmune diseases such as insulin-dependent diabetes mellitus, rheumatoid arthritis, multiple sclerosis, and myasthenia gravis already have been associated with different serological types of the human leukocyte antigen (HLA) class II antigens; however, it is not known if the HLA association with human autoimmune diseases is due to a specific immune response or is due to linkage disequilibrium with disease-susceptible alleles. Nevertheless, genes at other loci are also likely involved in these autoimmune diseases. [See Autoimmune Disease; Multiple Sclerosis; Myasthenia Gravis.]

C. Psychiatric Diseases

There is also hope that RFLP probes will be useful in the study of the inheritance of psychiatric diseases.

The extent to which genetic heterogeneity exists is still not known (how many independent genes at different loci give rise to these diseases). Familial Alzheimer's disease (FAD) is a neurodegenerative disorder that occurs late in life; most cases occur after the age of 70 years. There are several large families where the disease appears to be inherited as an autosomal dominant trait. In these families, FAD has been shown to be linked to loci on the long arm of chromosome 21 in four large FAD families where the age at onset was younger than usually observed (40+ years). This suggests that early onset and late onset may have different genetic causes. Studies of RFLPs linked to manic depression, schizophrenia, and autism are also likely to provide information and, possibly, the potential for prenatal and preclinical diagnoses. [See Alzheimer's Disease.]

D. DNA Markers in Tissue Typing

The major histocompatibility complex (MHC) of mammals encodes many different cell-surface glycoproteins and serum complement components that mediate a variety of immunological functions. The human MHC, known as HLA, region lies on the short arm of chromosome 6 and is highly polymorphic. The class I molecules are the classical transplantation antigens and are encoded by HLA A, B, and C loci. The value of HLA typing for transplantation is demonstrated by the increased graft survival observed in HLA-matched donor–recipient pairs. Polymorphisms within class I have been defined serologically. These can be used for tissue typing for transplantation. RFLPs are also detected with HLA genes, although entirely correlating specific serologic types with specific RFLP patterns has not been possible. [See Immunobiology of Transplantation.]

PCR followed by hybridization with ASOs has been used with great success to type HLA polymorphisms, particularly HLA-DQA, HLA-DQB, and HLA-DRB. This provides the opportunity for much more refined HLA typing than that afforded by current serologic methods and should prove useful in tissue typing for transplantation.

E. Paternity Determination and Forensics

Because the combination of alleles an individual possesses serves as a "genetic fingerprint," DNA markers are being used in paternity testing and forensic analysis. Currently, DNA markers detected with PCR, and highly polymorphic DNA probes such as the minisatellite probes and other VNTR probes, are being used for this purpose.

F. Uses of RFLP Probes for Karyotype Detection

As the genome becomes saturated with DNA markers, they likely will be used to determine molecular karyotypes. In particular, subtle chromosomal rearrangements, which are not visible cytogenetically, would be detectable. This would also involve the identification of microdeletions in certain congenital disorders. This approach has been applied in the case of cri-du chat, which is commonly due to a deletion of 5p15.1–pter. A translocation involving this region of chromosome 5, yet undetectable cytogenetically, could be detected with DNA probes, and prenatal diagnosis and carrier testing could be provided.

VI. CONCLUSION

By providing direct or indirect determination of disease state or individual identity, DNA diagnosis is having a major impact on the diagnosis of genetic disease, tissue typing, paternity testing, and possibly karyotype detection in the future.

ACKNOWLEDGMENTS

I thank A. J. Cooper (The University of Texas Southwestern Medical Center) and R. T. Taggart (Wayne State University) for helpful comments on this article.

BIBLIOGRAPHY

Antonarakis, S. E., and Kazazian, H. H. (1988). The molecular basis of hemophilia A in man. *Trends Genet.* **4**, 233.

Barker, D., Wright, E., Nguyen, K., Cannon, L., Fain, P., Goldgar, D., Bishop, D. T., Carey, J., Baty, B., Kivlin, J., Willard, H., Waye, J. S., Greig, G., Leinwand, L., Nakamura, Y., O'Connell, P., Leppert, M., Lalouel, J.-M., White, R., and Skolnik, M. (1987). Gene for von Recklinghausen neurofibromatosis is in the pericentric region of chromosome 17. *Science* **236**, 1100.

Beaudet, A., Bowcock, A., Buchwald, M., Cavalli-Sforza, L., Farrall, M., King, M.-C., Klinger, K., Lalouel, J.-M., Lathrop, G., Naylor, S., Ott, J., Tsui, L.-C., Wainwright, B., Watkins, P., White, R., and Williamson, R. (1986). Linkage of cystic fibrosis to two tightly linked DNA markers: Joint report from a collaborative study. *Am. J. Hum. Genet.* **39**, 581.

Botstein, D., White, R. L., Skolnik, M., and Davis, R. W. (1980).

Construction of a genetic linkage map in man using restriction fragment length polymorphisms. *Am. J. Hum. Genet.* **32**, 314.

Bowcock, A. M., Farrer, L. A., Hebert, J. M., Agger, M., Sternlieb, I., Scheinberg, I. H., Buys, C. H. C. M., Scheffer, H., Frydman, M., Chajek-Saul, T., Bonne-Tamir, B., and Cavalli-Sforza, L. L. (1988). Eight closely linked loci place the Wilson disease locus within 13q14–q21. *Am. J. Hum. Genet.* **43**, 664.

Caskey, C. T. (1987). Disease diagnosis by recombinant DNA methods. *Science* **236**, 1223.

Chamberlain, J. S., Gibbs, R. A., Ranier, J. E., Nguyen, P. N., and Caskey, C. T. (1988). Deletion screening of the Duchenne muscular dystrophy locus via multiplex DNA amplification. *Nucleic Acids Res.* **16**, 11141.

Chamberlain, S., Shaw, J., Rowland, A., Wallis, J., South, S., Nakamura, Y., von Gabain, A., Farrall, M., and Williamson, R. (1988). Mapping of mutation causing Friedreich's ataxia to human chromosome 9. *Nature* **334**, 248.

DiLella, A. G., Marvit, J., Brayton, K., and Woo, S. L. C. (1987). An amino-acid substitution involved in phenylketonuria is in linkage disequilibrium with DNA haplotype 2. *Nature* **327**, 333.

DiLella, A. G., Marvit, J., Lidsky, A. S., Guttler, F., and Woo, S. L. C. (1986). Tight linkage between a splicing mutation and a specific DNA haplotype in phenylketonuria. *Nature* **322**, 799.

Erlich, H. A., Sheldon, E. L., and Horn, G. (1986). HLA typing using DNA probes. *Biotechnology* **4**, 975.

Gitschier, J., Drayna, D., Tuddenham, E. G. D., White, R. L., and Lawn, R. M. (1985). Genetic mapping and diagnosis of haemophilia A achieved through a BclI polymorphism in the factor VII gene. *Nature* **314**, 738.

Glick, B. R., and Pasternak, J. J. (1994). "Molecular Biotechnology: Principles and Applications of Recombinant DNA." ASM Press, Washington, D.C.

Goldgar, D. E., Green, P., Parry, D. M., and Mulvihill, J. J. (1989). Multipoint linkage analysis in neurofibromatosis type I: An international collaboration. *Am. J. Hum. Genet.* **44**, 6.

Innis, M. A., Gelfand, D. H., and Sninsky, J. J. (1995). "PCR Strategies." Academic Press, San Diego.

Jeffreys, A. J., Wilson, V., and Thein, S. L. (1985). Hypervariable "minisatellite" regions in human DNA. *Nature* **314**, 6006.

Kan, Y. W., and Dozy, A. M. (1978). Polymorphism of DNA sequence adjacent to human beta-globulin structural gene: Relationship to sickle mutation. *Proc. Natl. Acad. Sci. USA* **75**, 5631.

Kazazian, H. H. (1989). Use of PCR in the diagnosis of monogenic disease. *In* "PCR Technology" (Henry A. Erlich, ed.), p. 153. Stockton Press, New York.

Kazazian, H. H., and Boehm, C. D. (1988). Molecular basis and prenatal diagnosis of beta-thalassemia. *Blood* **72**, 1107.

Kerem, N., Rommens, J. M., Buchanan, J. A., Markiewicz, D., Cox, T. K., Chakravarti, A., Buchwald, M., and Tsui, L.-C. (1989). Identification of the cystic fibrosis gene: Genetic analysis. *Science* **245**, 1073.

Kogan, S. C., Doherty, M., and Gitschier, J. (1987). An improved method for prenatal diagnosis of genetic diseases by analysis of amplified DNA sequences: Application to hemophilia A. *N. Engl. J. Med.* **317**, 985.

Michaelis, A., Green, M. M., and Rieger, R. (1991). "Glossary of Genetics: Classical and Molecular," 5th Ed. Springer-Verlag, New York.

Montandon, A. J., Green, P. M., Gianelli, F., and Bentley, D. R. (1989). Direct detection of point mutations by mismatch analysis: Application to haemophilia B. *Nucleic Acids Res.* **17**, 3347.

Nakamura, Y., Leppert, M., O'Connell, P., Wolff, R., Holm, T., Culver, M., Martin, C., Fujimoto, E., Joff, M., Kumlin, E., and White, R. (1987). Variable number of tandem repeat (VNTR) markers for human gene mapping. *Science* **235**, 1616.

Overhauser, J., Bengtsson, U., McMahon, J., Ulm, J., Butler, M. G., Santiago, L., and Wasmuth, J. J. (1989). Prenatal diagnosis and carrier detection of a cryptic translocation by using DNA markers from the short arm of chromosome 5. *Am. J. Hum. Genet.* **45**, 296.

Reeders, S. T., Keith, T., Green, P., Germino, G. G., Barton, N. J., Lehmann, O. J., Brown, V. A., Phipps, P., Morgan, J., Bear, J. C., and Parfrey, P. (1988). Regional localization of the autosomal dominant polycystic kidney disease locus. *Genomics* **3**, 150.

Riordan, J. R., Rommens, J. M., Kerem, B., Alon, N., Rozmahel, R., Grzelczak, Z., Zielenski, J., Lok, S., Plavsic, N., Chou, J.-L., Drumm, M. L., Iannuzzi, M. C., Collins, F. S., and Tsui, L.-C. (1989). Identification of the cystic fibrosis gene: Cloning and characterization of complementary DNA. *Science* **245**, 1066.

Rommens, J. M., Iannuzzi, M. C., Kerem, B., Drumm, M. L., Melmer, G., Dean, M., Rozmahel, R., Cole, J. L., Kennedy, D., Hidaka, N., Zsiga, M., Buchwald, M., Riordan, J. R., Tsui, L.-C., and Collins, F. S. (1989). Identification of the cystic fibrosis gene: chromosome walking and jumping. *Science* **245**, 1059.

Rubin, C. M., Houck, C. M., Deininger, P. L., Friedman, T., and Schmid, C. W. (1980). Partial nucleotide sequence of the 300-nucleotide interspersed repeated human DNA sequences. *Nature* **284**, 372.

Saiki, R. K., Walsh, P. S., Levenson, C. H., and Erlich, H. A. (1989). Genetic analysis of amplified DNA with immobilized sequence-specific oligonucleotide probes. *Proc. Natl. Acad. Sci. USA* **86**, 6230.

Shaw, D. J., and Harper, P. S. (1989). Myotonic dystrophy: Developments in molecular genetics. *Br. Med. Bull.* **45**, 745.

St. George-Hyslop, P. H., Tanzi, R. E., Polinsky, R. J., Haines, J. L., Nee, L., Watkins, P. C., Myers, R. H., Feldman, R. G., Pollen, D., Drachman, D., Growdon, J., Bruni, A., Foncin, J.-F., Salmon, D., Frommelt, P., Amaducci, L., Sorbi, S., Piacentini, S., Stewart, G. D., Hobbs, W. J., Conneally, P. M., and Gusella, J. (1987). The genetic defect causing familial Alzheimer's disease maps on chromosome 21. *Science* **235**, 885.

Tsui, L.-C., Buchwald, M., Barker, D., *et al.* (1985). Cystic fibrosis locus defined by a genetically linked polymorphic DNA marker. *Science* **230**, 104.

Wasmuth, J. J., Hewitt, J., Smith, B., Allard, D., Haines, J., Skarecky, D., Partlow, E., and Hayden, M. R. (1988). A highly polymorphic locus very tightly linked to the Huntington's disease gene. *Nature* **332**, 734.

Weber, J., and May, P. E. (1989). Abundant class of human DNA polymorphisms which can be typed using the polymerase chain reaction. *Am. J. Hum. Genet.* **44**, 388.

Wiggs, J., Nordenskjold, M., Yandell, D., Rapaport, J., Grondin, V., Janson, M., Werelius, B., Petersen, R., Craft, A., Riedel, K., Liberfarb, R., Walton, D., Wilson, W., and Dryja, T. (1988). Prediction of the risk of hereditary retinoblastoma, using DNA polymorphisms within the retinoblastoma gene. *N. Engl. J. Med.* **318**, 151.

Yandell, D. W., and Dryja, T. P. (1989). Detection of DNA sequence polymorphisms by enzymatic amplification and direct genomic sequencing. *Am. J. Hum. Genet.* **45**, 547.

Zonana, J., Sarfarazi, M., Thomas, N. S. T., Clarke, A., Marymee, K., and Harper, P. S. (1989). Improved definition of carrier status in X-linked hypohidrotic ectodermal dysplasia by use of restriction fragment length polymorphism-based linkage analysis. *J. Pediatr.* **114**, 392.

DNA Methodologies in Disease Diagnosis

RAFFAELE PALMIROTTA
ALLESSANDRO CAMA
RENATO MARIANI-COSTANTINI
PASQUALE BATTISTA
Università "Gabriele D'Annunzio"

LUIGI FRATI
Università "La Sapienza"

GLOSSARY

Exons Coding sequences of genes that, after transcription and excision of introns (splicing), form mature mRNA

Genetic marker Detectable variant of the DNA sequence that can be linked to disease transmission

Germ-line mutation Mutation that was carried by sperm or egg and that is present in all somatic cells and in 50% of the germ cells

Introns DNA sequences that interrupt the coding sequence of genes. After transcription, introns are excised from RNA (splicing)

Linkage Relationship between two genetic markers, or between a phenotypic characteristic and a genetic marker. Genes or genetic markers that are close to each other on the same chromosome are likely to be coinherited

Locus Site of chromosomal location of a gene or of a genetic marker

Mutation Any structural alteration of the sequence of DNA or RNA that alters the function of the gene product. Mutations may range from single nucleotide substitutions (point mutations) to extensive deletions or rearrangements and may affect the number, organization, and sequence of genes

Oligonucleotide Short, single-stranded DNA molecules that can be synthesized *in vitro*

Polymorphism Inherited variation in the sequence of DNA that does not result in alterations of gene or protein function(s)

Primer Oligonucleotide necessary for DNA polymerases to start synthesis of new strands

Probe Specific DNA or RNA sequence, labeled with radioactive or nonradioactive tracers, that is hybridized to the target sequence, allowing its visualization

Restriction endonuclease Enzyme that recognizes and cleaves a specific deoxyribonucleotide sequence in a double-stranded DNA molecule

Restriction fragment Fragment of DNA generated by cleavage with restriction enzymes

Somatic mutation Mutation occurring in a cell clone originating in tissues or organs forming parts of the body, other than germ cells (sperm or egg)

IN RECENT YEARS, REMARKABLE ADVANCES IN THE field of molecular genetics allowed the development of a powerful technology that uses recombinant DNA methods, reverse genetic strategies, and *in vitro* synthesis of nucleic acids for the isolation, characterization, and manipulation of genes. This rapidly evolving technology, as well as the accumulating information on the structure and function of many genes, has relevant practical applications in the fields of agriculture, veterinary medicine, drug production, and human clinical medicine. The science of molecular genetics is rapidly becoming one of the most important

ENCYCLOPEDIA OF HUMAN BIOLOGY, Second Edition, VOLUME 3.

fields of human biology and of modern medicine. In disease diagnosis, molecular genetic techniques are now successfully applied to the diagnosis of genetic disorders, to the diagnostic and prognostic evaluation of cancer, and to the identification of pathogens in infectious diseases. Nucleic acids are relatively stable following fixation or desiccation of tissues, and polymerase chain reaction (PCR) amplification can be performed on nucleic acids that are partially degraded. Thus, a wide range of biological samples, including whole blood or serum, stored or fresh biopsy or autopsy tissues, swabs, sputa, urine, fecal materials, crusts, cerebrospinal, pleural and peritoneal fluids, bronchial washings, eye secretions, and so on, can be used to obtain precise diagnostic information. PCR-based diagnostic techniques have also been successfully adapted to the analysis of nucleic acids from paraffin-embedded tissue, which permits the retrieval of retrospective genetic diagnosis when fresh tissue is not available, or even when the affected patient has died. This article describes the diagnostic use of DNA technology in the clinical investigation of diseases at the molecular level.

I. METHODS

Genetic diagnosis can be achieved by indirect or direct methods. Indirect diagnostic methods are utilized to define patterns of familial inheritance of hereditary diseases that have not yet been assigned to cloned genes and to localize the responsible genes. Direct diagnostic methods allow the unambiguous genetic identification of mutations responsible for hereditary diseases or cancer development, and of biological pathogens responsible for infectious diseases.

A. Indirect Genetic Diagnosis

In indirect analysis, genetic markers closely linked or within the gene that causes the disease are identified by Southern blot or, more conveniently, by PCR amplification of microsatellite sequences (see Section I,A,2). A carrier of these markers has a high probability to have inherited the disease. Methods for indirect genetic diagnosis are relatively simple and rapid compared to some of the methods for direct diagnosis. The limitation of indirect methods is that they can be applied only when biological samples from several members of a family are available. In addition, they

leave a certain degree of uncertainty about the diagnosis.

The genomes of two individuals are estimated to differ in 1 out of 100–300 base pairs (bp). Most changes in DNA sequence do not affect gene function (DNA polymorphisms). DNA polymorphisms provide markers that can be used to identify different allelic forms of a gene or of a DNA sequence and can be exploited for indirect genetic diagnosis. Genetic markers utilized in indirect genetic diagnosis include restriction fragment-length polymorphisms (RFLPs) and length polymorphisms in repetitive sequences.

1. Restriction Fragment-Length Polymorphisms

Restriction enzymes cut at specific sites in the DNA molecule and generate DNA fragments whose length corresponds to the distance between restriction sites in the DNA molecule. Sequence polymorphisms may result in the loss or gain of specific restriction sites. Accordingly, restriction enzyme digestion may generate DNA fragments of different sizes that can be resolved by gel electrophoresis, followed by either Southern blot analysis or, in the case of PCR-amplified products, ethidium bromide staining.

2. Analysis of Repetitive Sequences

Multiple/tandem sequence repeats are common in the human genome. Repetitive sequences are particularly prone to errors during DNA replication. As a result, the number of repeated units at a given locus is highly polymorphic in the population and each individual is likely to inherit alleles of different sizes that can be readily identified. Thus, sequence repeats represent useful markers for indirect genetic diagnosis. In addition, sequence repeats are utilized for gene localization, for the identification of carriers of diseases associated to expansion of trinucleotide repeats (see Section II), and for the diagnosis of hereditary and acquired defects in mismatch repair systems (see Section IV). Moreover, the combination of allele sizes at multiple loci, which has been compared to fingerprints (DNA fingerprinting), is used for individual identification in forensic medicine. Repetitive sequences used in genetic diagnosis include minisatellites, variable number of tandem repeats (VNTRs), and microsatellites. Minisatellites and VNTRs are repetitive sequences consisting of units within the size range studied by Southern blot analysis. Microsatellites are small DNA sequences (50–200 bp) consisting of 2-

to 4-bp tandem repeats that can be rapidly analyzed after amplification by PCR.

B. Direct Genetic Diagnosis

In the last few years, genes responsible for a broad spectrum of human genetic diseases have been molecularly characterized, and mutations resulting in the expression of altered proteins that cause specific functional defects have been identified. Thus, genetic diagnosis based on direct analysis of the gene responsible for the disease is currently available for several hereditary disorders. The detection of genetic alterations may also contribute to the diagnosis of tumors and to the molecular analysis of the neoplastic progression. Furthermore, direct diagnostic methods, based on the analysis of nucleic acid sequences of microorganisms, are becoming an important complement of the conventional diagnosis of infectious agents in the clinical pathology setting. The following sections briefly describe some of the methods most commonly used for direct genetic diagnosis of diseases.

1. General Methods for Direct Genetic Analysis

a. In Situ Hybridization

The fluorescent *in situ* hybridization (FISH) technique utilizes synthetic nucleic acid probes, labeled with fluorescent reagents, to localize complementary sequences in cytological or histological preparations. By this technique, unique sequences, chromosomal subregions, and entire chromosomes or genomes can be specifically highlighted in metaphase or interphase cells. FISH is routinely used to identify chromosomes, detect chromosomal abnormalities, and determine the chromosomal location of specific sequences. In addition, FISH is used to evaluate tissue sections and cells for the presence of mRNA transcripts and of pathogen nucleic acid sequences.

b. Southern Blot

For Southern blot analysis, genomic DNA is cleaved using restriction enzymes, size-fractionated by electrophoresis, and immobilized on a solid support. The structure of the gene or DNA sequence of interest is then analyzed using radiolabeled complementary probes. This method, now largely substituted by simpler PCR-based analyses, is still used to detect extensive gene deletions and recombinations.

c. Northern Blot

For Northern blot analysis, RNA is size-fractionated by denaturing electrophoresis and immobilized on a solid support. The structure and the abundance of a specific mRNA are then analyzed using radiolabeled probes. This method has been largely substituted by simpler PCR-based analyses [e.g., reverse transcriptase (RT) PCR] or by more accurate methods of RNA quantitation (see Section I,B,2,c,iii).

d. Methods for Amplification of Nucleic Acid Sequences

Most techniques used for genetic analyses require the purification of relatively large amounts of target sequence. The approaches that are most commonly used to amplify target sequences are DNA cloning and *in vitro* amplification using PCR.

i. *DNA Cloning* For DNA cloning, nucleic acids (DNA, or reverse transcribed RNA = cDNA) are introduced into host cells (usually bacteria or yeasts) using appropriate vectors. Host cells are then rapidly grown *in vitro,* yielding a doubling of the copies of foreign DNA at each cell division. Colonies of host cells containing the appropriate target sequence are identified by means of hybridization with specific probes and isolated by further cycles of purification and amplification in host cells. This technique is powerful but time-consuming, therefore it is often substituted by PCR.

ii. *PCR Amplification* PCR amplification is obtained through cycles of target-specific DNA synthesis, using a thermostable DNA polymerase and primer pairs flanking the sequence to be amplified. At each cycle, the number of copies of DNA can be doubled, achieving approximately a millionfold amplification in 20 cycles. By this approach, one or more target sequences are amplified simultaneously in 1 to 3 hr. PCR is very powerful, sensitive, fast, and specific. Several improvements to the basic PCR protocols have been introduced in the past few years. Amplification protocols involving the sequential use of external and internal primer sets (nested PCR) have been developed to increase the specificity and sensitivity of the technique. Protocols for the simultaneous amplification of several target sequences in a single reaction tube have also been developed (multiplex PCR). In addition, protocols for the reverse transcription of RNA followed by cDNA amplification have been developed

(RT-PCR). One of the limits of PCR was that only segments of DNA up to 2000 bp could be efficiently amplified. This limit has been recently overcome by the development of thermostable DNA polymerases and reagents that amplify efficiently up to 30,000 bp. Special PCR protocols that allow quantitative evaluation of the original template have also been developed (quantitative PCR). Because of the great sensitivity of the technique and its ability to amplify target sequences up to a million- or billionfold, extreme care has to be taken to avoid contaminations with amplified sequences of the sample to be studied.

e. Sequencing

The sequence of cloned or PCR-amplified DNA can be determined by a number of different techniques. The most frequently used involves DNA or cDNA sequencig using dideoxyterminators. Sequence determination allows definitive genetic diagnosis, but the approach of sequencing entire genes or their coding sequences is often too expensive and time-consuming

for diagnostic uses. The development of rapid and relatively simple screening techniques that identify the area of a gene containing a mutation has allowed sequence analysis to be restricted to small regions of the gene. Coupling screening techniques to sequencing expedites the task of characterizing mutations and allows the simultaneous analysis of a relatively large number of samples.

2. Screening Methods for Mutational Analysis

An ideal screening method for the detection of gene mutations should be simple, fast, and sensitive and should minimize the use of toxic or radioactive reagents. Furthermore, it should analyze long sequence stretches and provide the exact localization of the mutation. Several screening techniques applicable to routine mutational analysis of relatively large numbers of samples are now in use. Some of the most common screening techniques, outlined in Fig. 1, will be briefly described.

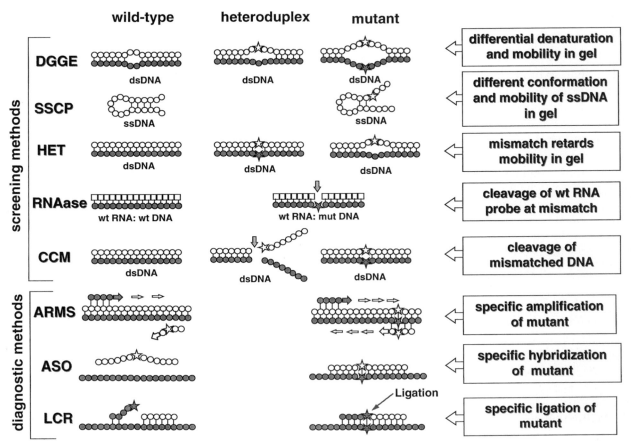

FIGURE I   Schematic representation of general methods for direct genetic diagnosis.

a. Single Strand Conformation Polymorphism

Single strand conformation polymorphism (SSCP) analysis of PCR products (PCR-SSCP) is a rapid and relatively simple method for the detection of base changes in DNA sequences and is widely used for diagnostic screening. This technique is based on the different secondary structures of DNA strands that can result even from single-nucleotide substitutions (Fig. 2). The target sequence is first amplified and labeled by PCR. The labeled PCR product is then denatured, quickly cooled, resolved by nondenaturing polyacrylamide gel electrophoresis, and autoradiographed. Strands containing insertions, deletions, or point mutations are identified on the basis of their differential mobility during nondenaturing electrophoresis (Fig. 2). Bands displaying abnormal electrophoretic mobility can be eluted from the gel for sequence analysis. Nonisotopic SSCP has also been developed. The limitation of SSCP is that its sensitivity does not achieve 100% and drops considerably when DNA fragments longer than 300 bp are analyzed. Fragments longer than 450 bp should not be analyzed by SSCP because >50% of mutations may escape detection.

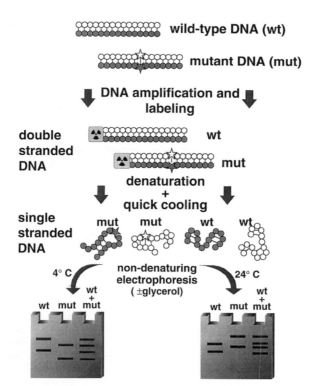

FIGURE 2 Schematic representation of the single strand conformation polymorphism analysis.

b. Dideoxyfingerprinting

Dideoxyfingerprinting (ddF) consists of the SSCP analysis of DNA fragments generated by sequencing reactions that use a single dideoxyterminator nucleotide. ddF is more sensitive than SSCP and requires only one termination reaction compared to the four required for DNA sequencing. However, this method is more complex than SSCP and therefore is less applicable to routine diagnostic screening.

c. Methods Based on Heteroduplex Formation

At least four screening methods exploit the formation of heteroduplexes between wild-type and mutant strands of nucleic acids. These include direct electrophoretic analysis of heteroduplexes (Fig. 3), denaturing gradient gel electrophoresis (DGGE) (Fig. 3), chemical modification (Fig. 3), and RNase cleavage (Fig. 4). In the first three methods, heteroduplexes are formed during DNA amplification by PCR (DNA is extracted from heterozygotes, or mixtures of wild-type and mutant DNA), and in the fourth method, heteroduplexes are formed by hybridization of target DNA or RNA to a labeled RNA probe. Additional methods employing mismatch repair enzymes have been recently used to detect heteroduplex formation.

i. *Direct Electrophoretic Analysis of Heteroduplexes* Heteroduplexes can be detected by nondenaturing gel electrophoresis because they migrate slower than the corresponding homoduplex. The slower migration is due to a loop formed at the site of mismatch. Electrophoresis on agarose and polyacrylamide minigels has been successfully used to detect the heteroduplex (Fig. 5). The size of the DNA fragments analyzed is generally on the order of 300 bp, but fragments greater than 1000 bp may be used. This method is fast and simple and does not require the use of radioisotopes. Its sensitivity may be enhanced by the use of special gel matrices. Direct electrophoretic analysis of heteroduplexes may be used as one-step screening for mutations, or to complement SSCP analysis.

ii. *Denaturing Gradient Gel Electrophoresis* Heteroduplexes can be detected by denaturing gradient gel electrophoresis (DGGE) because their strands dissociate sooner and consequently migrate slower than the corresponding homoduplexes (Fig. 3). Furthermore, depending on their sequence, the dissociation of wild-type and mutant homoduplexes may also occur at different rates in a denaturing gradient. Denaturing

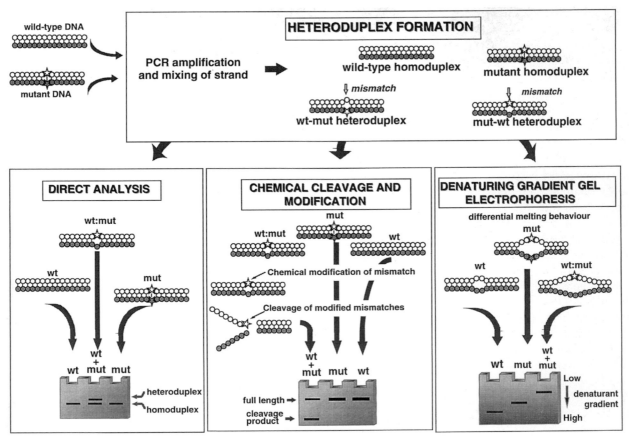

FIGURE 3 Schematic representation of screening methods for direct genetic analysis based on heteroduplex formation.

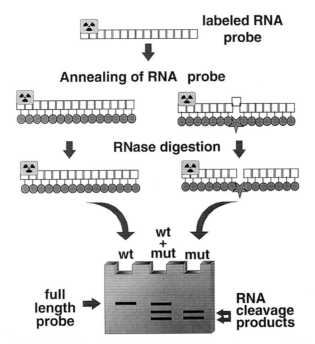

FIGURE 4 Schematic principle of the RNase cleavage method.

FIGURE 5 Screening of the *APC* gene by heteroduplex analysis on agarose minigel in 11 unrelated, familial adenomatous polyposis patients. Two fragments of PCR-amplified DNA, spanning codons 998–1141 (top) and 1260–1410 (bottom) of the *APC* gene, were tested for heteroduplex formation by electrophoresis on a 4% agarose minigel stained with ethidium bromide. Lanes 1 and 2 show representative samples that resulted negative by this screening technique. Samples in lanes 3–11 show the presence of heteroduplex bands that correspond to *APC* gene deletions at codons 1068 (lane 4), 1061 (lanes 8 and 9), and 1309 (lanes 3, 5, 6, 7, 10, and 11). [From A. Cama *et al.* (1995). Multiplex PCR analysis and genotype–phenotype correlations of frequent *APC* mutations. *Hum Mutat.* **5**, 144–150. Reprinted by permission of John Wiley & Sons.]

agents (formamide and urea) are usually employed to create such a gradient. Alternatively, temperature gradients can be employed to denature the DNA template during electrophoretic fractionation (temperature gradient gel electrophoresis or TGGE). Because of the high sensitivity and the ability to detect mutations in DNA fragments up to 600 bp, DGGE and TGGE are widely used in mutation detection. However, these methods require the use of specially designed primer pairs, with a 30 to 60-bp high-melting-temperature "GC-clamp" attached to one of the primers used for PCR amplification. In the absence of such a clamp, the melting profile of the target sequences may affect the sensitivity of the method. Moreover, special care and/or equipment has to be used to ensure uniform denaturing gradients.

iii. *Chemical Modification and/or Cleavage*   Chemical modification and cleavage techniques are based on modification of heteroduplexes at the site(s) of mismatch(es), followed by detection of the modified nucleotide(s) using different systems (Fig. 3). One system uses a relatively nontoxic reagent, carbodiimide, to modify mismatched nucleotides. The technique is very sensitive when coupled to primer extension protocols. In this case, DNA synthesis will terminate at the site of carbodiimide modification. The size of DNA fragments generated by primer extension indicates the site of mutation. The advantage of carbodiimide modification is that it allows one to estimate the location of a mismatch within target DNA sequences spanning over 1500 bp. In the second system, heteroduplexes are chemically modified by osmium tetroxide and hydroxylamine, and then cleaved by piperidine, generating DNA fragments whose size indicates the site of mismatch. The sensitivity of the chemical cleavage method approaches 100%, and target sequences of up to 1700 bp have been analyzed successfully. The main disadvantages of this method are that it uses toxic chemicals and requires several manipulations of the DNA samples.

iv. *RNase Cleavage*   RNase cleavage is a very sensitive method that can be used without prior PCR amplification of the target sequence. However, its sensitivity and reliability in mutation detection are increased by the use of amplified DNA templates. The method requires the use of a wild-type RNA probe that is annealed to the target DNA (or RNA) sequence (Fig. 4). As in other heteroduplex-based methods, homoduplexes and heteroduplexes will be formed between the RNA probe and the wild-type or the mutant target sequences, respectively. RNA mismatches in the heteroduplexes are cleaved by RNase A, and the labeled RNA fragments thus generated are analyzed by denaturing gel electrophoresis. The advantage of RNase cleavage is that it allows one to estimate the location of a mismatch within a target DNA sequence of several hundred base pairs. However, this method has the disadvantage of a relatively low sensitivity. Furthermore, it requires the synthesis of a special RNA probe. Despite these disadvantages, RNase cleavage has been extensively used to study variations in viral strains and to detect gene mutations.

d. In Vitro Transcription-Translation of Amplified DNA

The transcription-translation of amplified DNA detects mutations that cause premature termination of protein synthesis (Fig. 6). Such mutations occur frequently in some hereditary diseases and in cancers. The method consists of the *in vitro* transcription and translation of amplified DNA (PCR products or cloned DNA), corresponding to the open reading

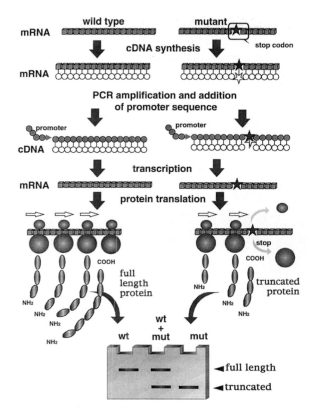

FIGURE 6   Principles of the *in vitro* transcription-translation of amplified DNA for the detection of mutations that cause premature termination of protein synthesis.

frame of the gene examined. The upstream primer used for PCR is engineered to contain the sequence of a bacterial promoter and a translation-initiation signal. A bacterial RNA polymerase will drive *in vitro* RNA transcription. Reticulocyte extracts, contained in the same reaction tube, will translate the RNA molecules into protein. The *in vitro* translated protein is fractionated by polyacrylamide gel electrophoresis and revealed by either radioactive or nonradioactive methods. This system, also designated protein truncation test (PTT) or *in vitro* synthesized protein (IVSP), is simple and fast and does not require the purification of PCR products. Furthermore, relatively large target sequences (>2500 bp) can be analyzed in a single test, and the size of the truncated protein indicates the location of the stop codon in the DNA sequence (Fig. 7).

3. Methods for the Rapid Identification of Known Nucleic Acid Variants

Rapid identification methods have been developed to identify known nucleic acid sequences and represent an effective alternative to sequence analysis. They are based on the use of PCR or RT-PCR to amplify the target DNA or RNA sequence and may be employed to diagnose recurrent common mutations responsible for some hereditary disorders or for specific forms of cancer, to identify disease carriers in pedigrees where the mutation has already been determined in one affected patient, and to detect nucleic acid sequences that allow the identification and differential diagnosis of biological pathogens. Because of their simple and reliable design, these methods can be adapted for routine use in diagnostic laboratories.

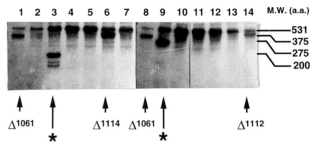

FIGURE 7 Example of screening for truncating mutations of the *APC* gene in familial adenomatous polyposis patients by *in vitro* transcription-translation of PCR-amplified DNA products of the *APC* gene. The normal peptide product (531 amino acids) is present in all lanes. Truncated proteins corresponding to premature termination codons, introduced by germ-line mutations, are detected in lanes 1, 3, 6, 8, 9, and 14.

a. Oligonucleotide Hybridization

Oligonucleotide hybridization is based on the differential hybridization of amplified target DNA to a specific oligonucleotide probe (approximately 20 bp). The specificity of short oligonucleotide hybridization allows a reliable identification of sequences differing even by a single nucleotide. Either the amplified DNA or the probe (reverse hybridization) can be fixed to a matrix or to a multiwell plate, and the hybridization can be revealed using radioisotopic or nonradioisotopic detection. Oligonucleotide hybridization is widely used for differential allele identification (allele-specific oligonucleotide hybridization or ASO) and for strain typing of pathogenic microorganisms.

b. Restriction Analysis

In mutational analysis, restriction analysis is applicable when the mutation causes the loss or gain of a restriction site and can be identified by differential digestion of wild-type and mutant DNA with restriction enzymes. Restriction analysis may also allow strain typing or species identification of biological pathogens whose genomic sequence contains diagnostic restriction sites. DNA can be analyzed by restriction enzyme digestion either directly (in association with Southern blot analysis) or after PCR amplification. In the latter case, digested DNA products can be visualized by nonradioactive staining following electrophoresis. Restriction enzyme digestion of PCR-amplified DNA is a very simple, rapid, and reliable diagnostic method.

c. Allele-Specific PCR

Allele-specific PCR is based on the selective amplification of DNA corresponding to a specific allele (wild type or mutant) by placing one of the PCR primers at the point where the sequence of the alleles differs. Allele-specific PCR is particularly helpful for single-step detection of a mutant allele and can be used in conjunction with multiplex allele-specific PCR for the screening of multiple mutant alleles that recur frequently in a given disease (Fig. 8).

d. Ligase Chain Reaction

The ligase chain reaction (LCR) is based on the selective ligation of two primers designed on the same strand of a specific allele (wild type or mutant). The primer on the 5′ side has the 3′ end falling on the region where the sequence of the alleles differs (Fig. 1). The second primer is designed so that the first 5′

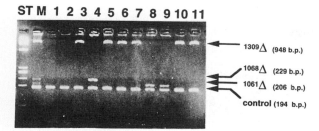

ST M 1 2 3 4 5 6 7 8 9 10 11

1309Δ (948 b.p.)

1068Δ (229 b.p.)

1061Δ (206 b.p.)

control (194 b.p.)

FIGURE 8 Single-step screening and diagnosis of frequent *APC* gene mutations in unrelated familial adenomatous polyposis patients using multiplex PCR. Three downstream primers specific for the ^{1061}Δ, ^{1068}Δ, and ^{1309}Δ mutant alleles and a common upstream primer were included in the same reaction mixture. In addition, a set of control primers was included in each amplification. Lanes 1–11 correspond to samples in lanes 1–11 of Fig. 5. Lane ST contains a molecular weight standard and lane M contains a mixture of amplified products corresponding to the three allele-specific amplifications and the control. The number in parentheses indicates the size of the bands in base pairs. [From A. Cama *et al.* (1995). Multiplex PCR analysis and genotype–phenotype correlations of frequent *APC* mutations. *Hum Mutat.* 5, 144–150. Reprinted by permission of John Wiley & Sons.]

base falls immediately downstream to the 3′ end of the other primer. When the sequence of the primers matches the DNA template, the 3′ end of the upstream primer is contiguous to the 5′ end of the downstream primer, and the two primers can be joined by a ligase enzyme. When the sequence of the primers does not match the template DNA, the two primers cannot be joined by the ligase. The use of a thermostable ligase allows multiple cycles of primer annealing and ligation that result in increased signals. This assay does not require the use of radioisotopes. To increase its sensitivity, LCR is generally coupled to PCR amplification of DNA templates.

e. Determination of the Relative Levels of Allele Expression

Some mutations that cause a decrease in the expression of mRNA are difficult to identify because they fall outside the coding sequence of a gene. If no mutation can be identified, sequence polymorphisms can be used as markers to measure the relative levels of allele expression. RNA of the gene to be examined is amplified by RT-PCR. The relative levels of allele expression can be estimated by allele-specific signals obtained using ASO hybridization, LCR, or restriction enzyme digestion of the RT-PCR product.

II. MOLECULAR DIAGNOSIS OF GENETIC DISEASES

In the last few years, chromosomal and mitochondrial genes responsible for a broad spectrum of human genetic diseases have been molecularly characterized, and mutations, resulting in the expression of altered proteins that cause specific functional defects, have been identified. Thus, genetic screening of carrier status, based on direct diagnostic analysis of the gene responsible for the disease, is available for several hereditary pathologies. This permits rigorous genetic counseling for affected families in whom mutations are identified and may allow targeted preventive therapies. [*See* Genetic Counseling.] In addition, other measures designed to retard or reverse the pathologic manifestation of the disease can be adopted to treat carriers of mutations that are diagnosed before the manifestation of the disease. Because inherited genomic or mitochondrial DNA alterations are present in each nucleated cell of the affected individual, it is sufficient to analyze DNA extracted from easily available cells, such as peripheral blood lymphocytes or skin fibroblasts or, in case of prenatal diagnosis, fetal aminocytes or trophoblasts from chorionic villi.

The pattern of familial inheritance of genetic diseases that have not yet been assigned to cloned genes can be indirectly traced in affected and unaffected members of a family by analyzing DNA markers that are in strict linkage with disease transmission. Simple sequence repeats, or microsatellites, that are scattered throughout the genome and are highly polymorphic are ideally suited for linkage analysis by PCR of many hereditary diseases when the gene(s) responsible have not been identified.

Genetic diagnosis represents a great improvement compared to diagnosis based on functional or biochemical assays and on family history, methods that often yield ambiguous results. Direct or indirect genetic diagnosis is crucial for heterozygote detection in X-linked or autosomal recessive diseases, and for the early detection of carrier state in late onset diseases. Direct genetic analysis may be the only diagnostic option in families with no prior history of genetic disease. In genetic diseases with heterogeneous or variable clinical manifestations, knowledge of the exact type of molecular defect allows genotype/phenotype correlations that may improve the accuracy of prognostic evaluations and may affect genetic counseling.

The techniques used for direct genetic diagnosis vary according to type, frequency, and localization of the mutations known to be responsible for the hereditary disease. In diseases known to be caused by recurrent common mutations, targeted diagnostic methods, including allele-specific PCR, multiplex allele-specific PCR, or mulitiplex PCR coupled with allele-specific oligonucleotide hybridization, may allow rapid single-step screening and diagnosis. For example, common mutations of the β-glucocerebrosidase gene account for the majority of the alleles associated with Gaucher's disease and can be rapidly screened and diagnosed in a single step utilizing allele-specific PCR. Similarly, common mutations represent about 90% of the total known mutations of the cystic fibrosis gene and can be rapidly diagnosed by multiplex amplification and allele-specific oligonucleotide hybridization.

A rapid and simple procedure for the diagnosis of common mutations abolishing or creating restriction endonuclease recognition sites is based on DNA amplification by PCR, followed by analysis of the restriction enzyme profile. For example, the common β-globin gene mutation responsible for sickle-cell anemia abolishes an *Mst* II restriction site in the β-globin gene (Fig. 9). Thus, homozygosity and heterozygosity for sickle-cell disease may be easily detected by Southern blot analysis, after restriction enzyme cleavage of PCR products, Similarly, a rapid and simple diagnostic procedure, based on DNA amplification by PCR and cleavage by the *Nla* III endonuclease, allows detection of the mutation Arg 3500 → Gln of the human apolipoprotein B-100 gene. This mutation abolishes an *Nla* III recognition site and is frequently responsible for a form of familial hypercholesterolemia.

Recent studies demonstrated that unstable mutations, caused by expansion of trinucleotide repeats, are implicated in various hereditary neurological disorders. These disorders include fragile X mental retardation syndrome, associated with expansion of CGG/CCG repeats at the 3′ end of the *FRAXA* locus, myotonic dystrophy, associated with expansion of CTG repeats at the *DM* locus, and four neuronal diseases due to expansion of CAG repeats: Huntington's disease (*HD* gene), X-linked spinobulbar muscular atrophy (androgen receptor locus), spinocerebellar ataxia type I (*SCA1* locus), and dentatorubral-pallidoluysian atrophy (*DRPLA* locus). PCR analysis or Southern blot of the polymorphic repeat alleles of the gene or chromosomal locus implicated in the disease may provide accurate diagnostic tests for this group of hereditary disorders.

In hereditary diseases caused by molecularly heterogeneous genetic alterations, the identification of the

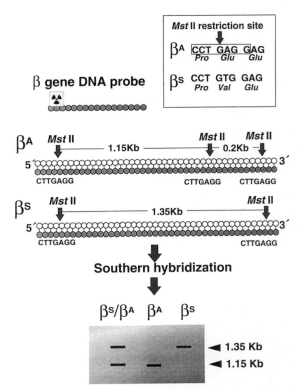

FIGURE 9 Schematic representation of rapid identification of known nucleic variants by restriction analysis. The common β-globin gene mutation responsible for sickle-cell anemia abolishes the intermediate *Mst* II restriction site. Cleavage with *Mst* II followed by Southern blot hybridization using a β-globin gene probe complementary to the sequence between the 5′ and the intermediate *Mst* II restriction site will result in a single hybridization band at 1.15 Kb for the normal β^A sequence in the homozygous state, a single hybridization band at 1.35 Kb for the mutant β^S in the homozygous state, and both the 1.15- and 1.35-Kb bands for the heterozygous β^A/β^S state.

mutation responsible for disease transmission may be complex and time-consuming in the first disease-affected member of the family analyzed, depending on the size and structure of the gene. In these cases, mutational analysis may be conducted using one of the screening methods described earlier, such as *in vitro* transcription-translation assay, SSCP, DGGE, or heteroduplex analysis. After the identification of the mutation, targeted diagnostic methods can be employed for the screening of family members.

III. MOLECULAR DIAGNOSIS OF INFECTIOUS DISEASES

Conventional microbiological, biochemical, and immunological techniques used to characterize biological pathogens are often time-consuming and may lack in sensitivity. In contrast, the genetic diagnosis of infectious agents using molecular techniques is rapid, accurate, and sensitive, does not require the viability of the infectious agent, is independent from its antigenic properties and from the immune response of patients, and can be performed in both prospective and retrospective fashion. Moreover, for strain typing purposes, the genetic analysis of uncultured microorganisms from patient samples is preferable to the analysis of *in vitro* cultures that may not correspond to the *in vivo* dominant strain.

For the aforementioned reasons, diagnostic methods based on DNA technology, and particularly on *in vitro* PCR amplification, are rapidly becoming an important complement to the conventional diagnosis of infectious agents. However, caution is needed in the application of PCR to the diagnosis of common infections, because infections unrelated to the main pathology of the patient may lead to incorrect diagnostic associations. Moreover, the extreme sensitivity of molecular tests to environmental contamination may result in false positivities, particularly when amplified DNA fragments from previous PCRs (amplicons) have been manipulated in the same laboratory where sample preparation is performed. These risks can be avoided if the techniques used for DNA extraction and purification are appropriately selected and accurately performed, under environmental conditions that minimize the risk of sample contamination. In general, sample preparation techniques that require minimum manipulations reduce the risk of contaminations. In most cases, when dealing with unfixed tissues other than blood, the simple microwaving or 20 min of boiling of the sample may free DNA in amounts sufficient for PCR analysis of pathogens. Test samples should be collected, stored, and processed in a laboratory completely independent from that in which amplicons are manipulated and analyzed. Nested amplification techniques should be used to increase the sensitivity and specificity of the molecular diagnosis.

DNA amplification with PCR is particularly effective in the diagnosis of latent infections, associated with low numbers of infectious agents, and in the identification of pathogens that elude conventional methods of detection, or that are difficult or hazardous to grow in a clinical microbiology laboratory. Molecular techniques may be used to detect genes encoding bacterial toxins and may also contribute to define drug susceptibilities in bacteria and viruses. PCR amplification followed by image analysis to quantify band intensities relative to a known dilution series, may estimate the load of pathogens and contribute to the monitoring of specific therapies. Amplification by PCR, using primers targeting conserved genomic sequences, is also a powerful tool for the identification of infectious microorganisms responsible for diseases of unknown etiology. Universal primers to conserved regions of the 16S and 18S rRNA bacterial genes have been applied to the search for bacteria associated with infections of unknown origin. The bacterial species is recognized following sequence analysis of amplification products. In the case of previously unknown microorganisms, their evolutionary relationships with known microorganisms can be calculated through maximum parsimony analysis of sequence data. Likewise, the amplification of conserved viral gene sequences from cDNAs prepared from infected tissues has led to the cloning and identification of viral pathogens, as exemplified by the identification of hepatitis C and hepatitis E. [*See* DNA Amplification.]

At the epidemiological level, the application of molecular techniques to the diagnosis of infectious diseases and to the genetic typing of microorganisms allows a more rapid, precise, and sensitive evaluation of the impact of a given pathogen in human populations. PCR-based methods are rapidly contributing to the characterization of strain diversity of microbial pathogens obtained from different outbreaks and to the determination of relationships between different strains or species. Thus, molecular diagnosis is critical for the control of disease outbreaks that cannot be rapidly and reliably identified using standard diagnostic methods.

A. Diagnostic Analysis of Viral Diseases

Oligonucleotide primers that generate products of different sizes, or differentiated by restriction enzyme cleavage sites, are efficiently used to distinguish related viral types with conserved regions in their genomes. This is exemplified in the case of the human papilloma viruses and of the herpes-group viruses, including cytomegalovirus, Epstein-Barr virus, varicella zoster virus, and herpes simplex viruses. On the other hand multiplex PCR may allow the single-step screening and detection of unrelated viral genomes in blood products or in subjects at risk for different viral diseases, such as drug addicts and dialysis or immunocompromised patients. Reverse transcription, coupled to PCR amplification or to nested or seminested PCR, has been used for the detection of hepatitis C virus RNA in serum, plasma, or blood samples, hepatitis A virus in stool, and influenza A virus in nasal swabs, and for the detection and rapid identification of picornaviruses, dengue virus, and other flaviviruses.

PCR-based diagnostic methods represent an important complement to conventional techniques of viral identification. PCR amplification of viral sequences may contribute to the early recognition of relapses or chronic evolution of viral diseases, to the discrimination of different virus carrier states, and to the follow-up of antiviral therapies. DNA technology is also required for the production of recombinant viral antigens, obtained by cloning PCR-amplified open reading frames of antigenic viral proteins in appropriate vectors, that are expressed in bacterial and/or eukaryotic cells. Homologous recombinant antigens allow sensitive and specific "second generation" immunoblot assays, even when conventional serological diagnosis is not available, because of difficulties in virus isolation.

An important area of PCR application concerns the diagnosis of those viral infections that may escape the detection by less sensitive techniques. PCR amplification, using oligonucleotide primers derived from regions of the viral genome conserved among different viral isolates, is much more rapid and sensitive than viral culture and is of critical importance in the management of viral diseases, particularly when specific and effective therapies depend on early diagnosis. DNA technology can reliably detect the presence of congenitally acquired viral infections during the first 2 months of life, a period in which the presence of maternal antibodies and the absence of immune response of the infant do not allow the use of serological methods. Another critically important area of application of PCR-based diagnostic technology concerns the screening of blood products for blood-borne viruses, such as the hepatitis viruses, the human immunodeficiency virus type 1 (HIV-1), and the human T-cell leukemia virus type 1 (HTLV-1). Finally, molecular techniques are extremely useful for epidemiological and environmental monitoring of viruses.

B. Diagnostic Analysis of Bacterial, Rickettsial, Chlamydial, Protozoal, and Fungal Diseases

Simplified PCR assays represent very promising alternatives to standard methods for definitive diagnosis of a variety of important bacterial pathogens, including *Helicobacter pylori, Legionella, Bordetella, Neisseria,*

Haemophilus, Streptococcus, Pseudomonas, Escherichia, Listeria, Staphylococcus, Leptospira, and *Treponema* species. The relevance of PCR in the diagnosis of active tuberculosis has been recently questioned in view of the diffusion of the infection and of the extreme sensitivity of the technique. However, a distinct advantage of PCR-based methods in the diagnosis of mycobacterial diseases consists in the ability to discriminate mycobacterial species, including *M. avium, M. intracellulare, M. kansasii,* or others, as well as strains within *M. tuberculosis.* This can be done by hybridization of PCR products with species-specific oligoprobes or by other PCR-based techniques (see Section I,B,3). Thus, PCR-based diagnosis may contribute to a more precise definition of mixed mycobacterial infections.

Progress in the understanding of the molecular phylogeny of bacteria is profoundly influencing the design of diagnostic tests. Phylogenetic analysis of 16S rDNA sequences, obtained after amplification with universal eubacterial primers, has been applied to the diagnosis of bacterial diseases and has allowed the identification of novel species of bacterial pathogens. PCR-based randomly amplified polymorphic DNA patterns can be obtained directly from boiled supernatants of bacterial colonies and provide highly discriminatory DNA fingerprints for the identification and epidemiological characterization of bacterial strains implicated in outbreaks of infection.

PCR-based diagnostic systems, combined with hybridization on DNA-precoated, enzyme-linked immunosorbent assay plates, have also been successfully used for the diagnosis of rickettsial and chlamydial infections. These new methods appear to be more sensitive than culture and allow the prompt differential identification of related pathogenic species.

Recombinant DNA technology offers distinct advantages for the epidemiological monitoring of protozoal and fungal parasites and for the management and control of diseases caused by these pathogens, particularly in cases with atypical clinical presentation or those occurring in nonendemic areas. In fact, many of these organisms are relatively difficult to identify using conventional parasitological, cultural, or immunological methods, and molecular confirmation of the diagnosis is particularly desirable prior to the initiation of prolonged and potentially toxic therapies. The amplification of specific regions of ribosomal DNA genes, followed by Southern blot hybridization with specific internal oligonucleotide probes, may lead to the rapid and precise identification of pathogenic *Plasmodium, Toxoplasma, Babesia, Trypanosoma,* and

Acanthamoeba species. In the case of trypanosomiases, amplification of the hypervariable region of trypanosomal kinetoplast minicircle DNA allows sensitive and specific parasite detection. PCR-based techniques could prospectively be applied to the identification of most protozoal pathogens.

IV. MOLECULAR DIAGNOSIS OF CANCER

Mutations in key genes controlling normal cell proliferation and cell–cell interactions are responsible for the development and progression of malignancies. The recent advances in the understanding of the genetic bases of neoplasia are producing relevant clinical applications that are progressively increasing the sensitivity and specificity of cancer diagnosis. The detection of genetic alterations is an important element for the diagnosis, staging, and prediction of the clinical evolution of several neoplasms.

Chromosomal translocations, gene rearrangements, and other tumor-associated DNA mutations may generate modifications in the restriction endonuclease cleavage pattern of the involved genes. If present in a sizable fraction of the tumor-cell population from which genomic DNA is extracted, these alterations can be easily revealed by comparative Southern blot analysis of matched normal and tumor DNAs. This simple analysis may detect novel tumor-associated hybridization bands, indicating the presence of specific mutations, or show the comigration in the same restriction fragment of DNA sequences from different genes, which is diagnostic for chromosomal translocations and gene rearrangements that juxtapose previously separated DNA sequences. Southern blot analysis may not be sensitive enough to identify genetic lesions present in a restricted number of tumor cells, or in tumor cells diluted with normal cells that were present in the biopsy sample. Thus, PCR techniques that selectively amplify altered DNA sequences are now largely substituting for Southern blot analysis. For the detection of DNA mutations, the amplified PCR product may be analyzed by restriction endonuclease cleavage or differential ASO hybridization when the mutated sequence is known, and by *in vitro* transcription-translation assay, SSCP, DGGE, or RNase A cleavage to screen for the presence of unknown modifications in the nucleotide sequence (see Section I). Chromosomal translocations and gene rearrangements can be revealed using primers complementary to DNA sequences on opposite strands of each genomic region involved in the phenomenon. Obviously, such primer pairs will yield an amplification product when brought within the range of PCR performance, which occurs if one of the target sequences moved to an abnormal position next to the other, following the occurrence of reciprocal translocation or rearrangement in tumor DNA.

Genetic amplification of protooncogenes and of genes conferring multidrug resistance (*mdr* genes) represents alterations that may have important prognostic significance. The presence and extent of a protooncogene amplification is usually determined on autoradiograms of Southern blots or dot blots, hybridized with the specific cDNA-labeled probe. The amplification level is evaluated by comparing the hybridization signal of the amplified gene with that of another gene ("reporter gene") that is not affected by cancer-associated alterations. The comparison should be made using appropriate analytical imagery systems. As an alternative to genomic analysis by Southern blotting, special protocols for quantitative PCR may be used to estimate the level of amplification of the gene of interest, in comparison to a "reporter gene." The degree of genetic amplification is reflected in the relative ratio between the levels of the two PCR products, which is determined after incorporation of a radiolabeled deoxyribonucleotide during PCR, or by simple ethidium bromide staining, after gel electrophoresis. As a caveat, the detection of gene amplification is influenced by the ratio between tumor and normal cells in the biopsy sample from which DNA is extracted. A significant contamination by normal cell DNA may in fact dilute the signal of the amplified target gene.

Other genetic events associated with neoplastic progression, such as allelic losses, may provide important prognostic markers for tumors. Allelic losses can be determined by standard Southern blot analysis of paired normal tissue and tumor DNA or, more conveniently, by PCR amplification of microsatellite markers, also from paired tumor and normal tissue DNA samples. In the latter case, the test can be performed on formalin-fixed, paraffin-embedded archival biopsy samples.

A. Diagnosis and Staging of Leukemias and Lymphomas

The impact of DNA technology is particularly perceived in the diagnosis and management of tumors of white blood cells. This follows the accumulation of a vast amount of knowledge on the molecular basis of normal growth and differentiation of hemopoietic stem cells, and the ever-increasing accuracy in the characterization of genetic alterations associated with specific forms of neoplasms of the blood and lymph nodes.

Examples of chromosomal translocations associated with leukemias or lymphomas are: the t(8;14)(q24;q32) translocation of Burkitt's lymphoma; the t(9;22)(q34;q11) translocation, which results in an abnormally truncated chromosome 22 (Philadelphia chromosome), detected in >90% of chronic myelogenous leukemias and in approximately 20% of acute lymphoblastic leukemias; and the t(15;17) (q22;q12 or q21.1) translocation of acute promyelocytic leukemia. In most cases of typical Burkitt's lymphoma, the c-*myc* protooncogene, located on the long arm of chromosome 8, is involved in reciprocal translocations with the immunoglobulin heavy-chain gene on the long arm of chromosome 14. In chronic myelogenous leukemias, most of the coding sequence of the c-*abl* protooncogene, located on chromosome 9, is fused to the 5′ region of the *bcr* gene, the breakpoint cluster region on the long arm of chromosome 22. The chromosome breakpoints of the t(15;17) of acute promyelocytic leukemia involve the retinoic acid receptor α (*RAR-α*) gene and the recently described promyelocytic leukemia (*PML*) gene on chromosomes 17 and 15, respectively. Rearrangements of these loci are found in virtually 100% of acute promyelocytic leukemias. Acute myelomonocytic leukemia with bone marrow eosinophilia is associated with pericentric inversion of chromosome 16, inv(16)(p13q22), which results in the fusion of two genes, *CBFB* on 16q and *MYH11* on 16p, and in the expression of *CBFB/MYH11* fusion mRNAs. In leukemias or lymphomas associated with specific chromosomal translocations, PCR amplification with primers complementary to sequences nearby breakpoints may be used to amplify hybrid genes that serve as tumor-specific markers. This represents a sensitive tool for the detection of minimal residual disease, with a detection limit of 1–10 malignant cells in $10^{(6)}$ normal cells. [*See* Leukemia; Lymphoma.]

Molecular markers of differentiation provide means to characterize the lineage and stage of maturational arrest of leukemic cells that cannot be reliably identified using cytochemical and immunological techniques. In acute leukemias with early maturational arrest, the identification via RT-PCR of transcripts of the myeloperoxidase gene may help to determine or exclude the myeloid lineage. In acute lymphoid neoplasms, the cloning and sequencing of PCR-amplified fragments of immunoglobulin or T-cell receptor genes provide clonospecific DNA markers, formed by the junction of rearranged variable (*V*), diversity (*D*), and joining (*J*) regions of antigen receptor genes. The molecular analysis of *VDJ* junctions allows the assignment of lymphoid neoplasms to the B- or the T-cell lineage, and the definition of the stage of maturational arrest of the neoplastic lymphocytes. Lymphoid leukemia and lymphoma cells can be reliably identified among normal cells, following PCR amplification of clonospecific antigen receptor regions. Quantitative PCR techniques, developed to detect low levels of leukemic cells, can be used to monitor the response to chemotherapy and to detect minimal residual disease.

Genetic alterations of protooncogenes and tumor suppressor genes may contribute to define the biological aggressiveness of leukemic and preleukemic cells. In particular, mutations of *ras* genes (c-H-*ras*, c-K-*ras*, c-N-*ras*) represent predictive markers of the malignant evolution of myelodysplasia, a disorder of hemopoiesis, characterized by the alteration of blood cell differentiation, which progresses to acute myelogenous leukemia in 10–30% of cases. Mutations of *ras* genes and of p53 are indicative of a poor prognosis in chronic myelogenous and in T-cell leukemias, respectively.

B. Diagnosis and Staging of Solid Tumors

PCR-based analysis of clonal mutations responsible for the malignant conversion of protooncogenes or for the inactivation of tumor suppressor genes may contribute to the diagnosis and staging of solid tumors. This is well documented for mutations of the *ras* family of protooncogenes and of the p53 tumor suppressor gene that represent critical events in the pathogenesis of a variety of neoplasms. For example, in pancreatic adenoncarcinoma, the search for mutations of the Ki-*ras* oncogene and of p53 may contribute to early tumor detection in pancreatic secretions obtained at routine endoscopies or in fine-needle biopsy samples. Similarly, the analysis of p53 and *ras* gene mutations in sputum and bronchial washing specimens provides a novel approach for the early detection of lung cancer. Somatic mutations of the adenomatous polyposis coli (*APC*) gene occur in the vast majority of gastrointestinal tumors, and their detection may allow the early identification of gastric and colorectal carcinomas and their precursor lesions.

The cloning of genes involved in translocations asociated with specific subtypes of solid tumors provides DNA probes for diagnostic tests, including Southern blot analysis and fluorescence *in situ* hybridization, and allows the pathognomonic amplification of the fusion gene or of the fusion transcript by PCR or RT-PCR. Examples of translocations characteristic of specific soft tissue tumor subtypes include t(11;22)(q24;q12) in Ewing's sarcoma, t(9,22)(q22;q12) in extraskeletal myxoid chondrosarcoma, t(X;18)(p11;q11) in synovial sarcoma, and t(12;16)(q13;p11) in myxoid liposarcoma.

The genetic amplification of protooncogenes, particularly of those coding for growth factors or growth-factor receptors, and of *mdr* genes may have a prognostic value in defining tumors that have high biological aggressiveness and metastatic potential or are resistant to chemotherapy. This is the case for the amplification of the c-*erb*-B2 protooncogene (also known as *HER*-2 or *neu*) that encodes a cell-membrane receptor-like protein, related to the epidermal growth-factor receptor (EGFr). Significant increases in c-*erb*-B2 copy number (more than five copies per diploid genome) have reportedly been associated with aggressive clinical behavior in carcinomas. The genetic amplification of the c-*erb*-B2 protooncogene may identify those breast cancers that, despite the absence of apparent metastases, will have poor clinical evolution and require chemotherapy to prevent disease relapse after surgery. Likewise, amplifications of the N-*myc* protooncogene in neuroblastomas and of the c-*erb*-B1 (EGFr) gene in gliomas correlate with rapid tumor progression, even in patients at early stages of the disease.

C. Molecular Diagnosis of Cancer Predisposition

Some human tumors are caused by germ-line mutations that are transmitted as hereditary traits. Recent discoveries allowed the identification of genes responsible for various forms of hereditary predisposition to tumors, including colorectal, skin, mammary, ovarian, and thyroid cancers and some pediatric neoplasms. Thus, in families with such hereditary forms of cancer, direct genetic analyses may now determine the specific mutation responsible for disease transmission (Fig. 10). These inherited genetic alterations are present in all tissues of the carrier, including peripheral blood lymphocytes or skin fibroblasts, that are easily available for analysis.

Based on current knowledge, genetic analysis allows a precise determination of cancer susceptibility only in subjects from families in which the specific germ-line

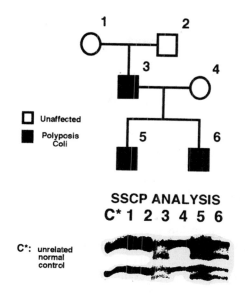

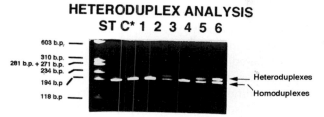

FIGURE 10 Example of PCR-SSCP and heteroduplex analysis of the *APC* gene in a familial adenomatous polyposis kindred. The numbers in the pedigree (upper panel) correspond to the numbering of the lanes in the SSCP (middle panel) and heteroduplex (lower panel) analyses. Lanes ST and C represent the molecular weight DNA standard and an unrelated normal control, respectively. The abnormal SSCP conformers and heteroduplex bands correspond to a 3-bp deletions of the *APC* gene and are detected in disease-affected members of the kindred. [From A. Cama *et al.*. (1994). A novel mutation at the splice junction of exon 9 of the *APC* gene in familial adenomatous polyposis. *Hum Mutat.* **3**, 305–308. Reprinted by permission of John Wiley & Sons.]

mutation that causes the disease has been identified. The identification of the mutation in the first disease-affected member of the kindred analyzed may be costly and time-consuming. However, once this is achieved, by using less expensive tests targeted at the specific mutation it is possible to screen rapidly other disease-affected or at-risk members of the family, thus possibly identifying disease carriers prior to the development of clinical signs. [*See* Genetic Testing.]

In families affected by hereditary forms of cancer, the identification of the mutation responsible for tumor predisposition carries several distinct advantages. The first is that it may free noncarrier members from the anxiety of the disease and from the necessity of continuous clinical surveillance. Obviously, this situation can be rapidly extended to branches of the family that may include several subjects. A second advantage lies in the fact that it is possible to give specific indications of the modalities and timing of clinical controls to carriers of the mutation that are identified prior to the manifestation of the disease. In these individuals, careful follow-up minimizes the risks of malignant evolution of the disease and may allow optimum timing of eventual surgical therapies. A third advantage is related to the possibility of administering specific forms of preventive treatment, devised to retard or inhibit tumorigenesis, to carriers of mutations identified before disease development. In this respect, knowledge of the specific type of molecular alteration responsible for tumorigenesis and the possibility of initiating preventive therapies before the development of the disease will probably allow further progress in pharmacological prevention.

Based on present knowledge, the immediate impact of direct genetic diagnosis primarily concerns two important hereditary forms of predisposition to colorectal cancer, familial adenomatous polyposis (FAP) and hereditary nonpolyposis colorectal cancer (HNPCC, also designated "Lynch syndrome"), and at least one form of hereditary predisposition to breast and ovarian cancer. FAP represents an ideal model for the rapid transfer of molecular diagnosis to the clinical practice. Germ-line mutations of the *APC* gene are responsible for this disease, characterized by the early development of innumerable colorectal adenomas, colorectal cancer, and extracolorectal neoplastic lesions, including aggressive fibromatoses (desmoids). The *APC* gene encodes a cytoplasmic protein of about 300 kDa. The observation that the majority of FAP-associated mutations result in stop codons allowed the development of assays that rapidly identify mutations resulting in frameshifts that cause the expression of truncated *APC* proteins. These assays permit the detection of mutations responsible for the disease in about 90% of the cases.

The observation that over 80% of colorectal tumors from HNPCC patients manifest replication errors at microsatellites (RER⁺ phenotype) marked a turning point in the understanding of the molecular bases of HNPCC. The RER⁺ phenotype is due to functional defects of genes that are homologous to prokaryotic genes involved in mismatch repair (*mutator* genes). At least four human genes involved in the pathogenesis of HNPCC have been cloned: *hMSH2, hMLH1, hPMS1,*

and *hPMS2*. In some families, HNPCC is almost exclusively associated with increased risk of colorectal tumors (Lynch type 1), whereas in other families the spectrum of neoplasms includes colorectal and extracolorectal malignancies, such as carcinomas of the uterus, ovary, stomach, pancreas, small intestine, larynx, and ureter (Lynch type 2). Several issues that are relevant to the genetic diagnosis of HNPCC need to be clarified. However, given the rapid pace of discoveries, it is probable that simplified and unambiguous tests for genetic detection of HNPCC will soon be available.

Hereditary forms of breast cancer account for about 5% of the cases. The recently identified *BRCA 1* gene, mapped to chromosome 17q21, is thought to be responsible for about 50% of hereditary breast tumors. The *BCRA 1* protein does not appear to be related to previously described proteins and contains in the amino terminus a zinc finger domain capable of interacting with DNA. *BRCA 1* mutations described so far provide compelling evidence of the role of this gene in the pathogenesis of early-onset (before age 50) hereditary breast cancer associated to ovarian tumors. At least one other gene, designated *BRCA 2*, plays a role in early-onset hereditary breast cancer. To date, the genes(s) involved in late-onset hereditary breast cancer remain to be identified. Familial breast cancers with *BRCA 1* mutations are phenotypically heterogeneous. Preliminary mutational analysis data on *BRCA 1* suggest that the site and type of mutation may have an effect on the age of onset of neoplasia and/or the incidence of ovarian cancer in the family.

An additional hereditary cancer syndrome that can be effectively diagnosed and controlled by direct genetic analysis is multiple endocrine neoplasia type 2A (MEN-2A) characterized by familial medullary thyroid carcinoma, associated with pheochromocytoma and, sometimes, parathyroid adenoma. Germ-line mutations in the *RET* protooncogene are responsible for the familial transmission of this syndrome. Mutational analysis of the *RET* protooncogene may allow the presymptomatic identification of MEN-2A carriers, in whom risk of medullary thyroid carcinoma approximates 100%. Careful follow-up of these subjects may allow an effective control of the disease, which can be prevented by thyroidectomy at the appropriate time.

V. SUMMARY

DNA technology provides powerful tools that result in an extraordinary improvement in the capability to diagnose and prevent human diseases. By introducing higher levels of sensitivity and specificity in the identification of infectious agents, in the diagnosis and staging of cancer, and in the detection of carriers of genetic disorders, these new techniques, so far restricted to the investigation of selected cases in research laboratories, are rapidly becoming the backbone of diagnostic services for the general population.

ACKNOWLEDGMENT

The work of the authors is supported by the Associazone Italiana Ricerca Cancro.

BIBLIOGRAPHY

Banfi, S., and Zoghbi, H. Y. (1994). Molecular genetics of hereditary ataxias. *Baillieres Clin. Neurol.* **3**, 281.

Cama, A., Palmirotta, R., Curia, M. C., Esposito, D., Ficari, F., Valanzano, R., Ranieri, A., Tonelli, F., Battista, P., and Mariani-Costantini, R. (1995). Multiplex PCR analysis and genotype–phenotype correlations of frequent *APC* mutations. *Hum. Mutat.* **5**, 144.

Cherian, T., Bobo, L., Steinhoff, M. C., Karron, R. A., and Yolkaen, R. H. (1994). Use of PCR-enzyme immunoassay for identification of influenza A virus matrix RNA in clinical samples negative for cultivable virus. *J. Clin. Microbiol.* **32**, 623.

Cinque, P., Brytting, M., Vago, L., Castagna, A., D'Arminio, M. A., Lazzarin, A., and Linde, A. (1994). Diagnosis of virus-associated opportunist diseases of the central nervous system in patients with HIV infection by polymerase chain reaction on cerebrospinal fluid. *Ann. N.Y. Acad. Sci.* **724**, 170.

Cone, R. V., Hobson, A. C., Brown, Z., Ashley, R., Berry, S., Winter, C., and Corey, L. (1994). Frequent detection of genital herpes simplex virus DNA by polymerase chain reaction among pregnant women. *JAMA* **272**, 792.

Connelly, B. L., Stanberry, L. R., and Bernstein, D. E. (1993). Detection of varicella-zoster virus DNA in nasopharyngeal secretions of immune household contacts of varicella. *J. Infect. Dis.* **168**, 1253.

Cotton, R. G. H. (1993). Current methods of mutation detection. *Mutat. Res.* **285**, 125.

Crum, C. P. (1994). Genital papillomaviruses and related neoplasms: Causation, diagnosis and classification (Bethesda). *Mod. Pathol.* **7**, 138.

Diccianni, M. B., Yu, J., Hsiao, M., Mukherjee, S., Shao, L. E., and Yu, A. L. (1994). Clinical significance of p53 mutations in relapsed T-cell acute lymphoblastic leukemia. *Blood* **84**, 3105.

Echevarria, J. E., Tenorio, A., Courouce, A. M., Leon, P., and Echevarria, J. M. (1994). Polymerase chain reaction can resolve some undefined cases of hepatitis B virus antigenic subtyping. *J. Med. Virol.* **42**, 217.

Feldmann, H., Sanchez, A., Morsunov, S., Spiropoulou, C. F., Rollin, P. E., Ksiazek, T. G., Peters, C. J., and Nichol, S. T. (1993). Utilization of autopsy RNA for the synthesis of the nucleocapsid antigen of a newly recognized virus associated with hantavirus pulmonary syndrome. *Virus Res.* **30**, 351.

Fulop, L., Barrett, A. D., Phillpotts, R., Martin, K., Leslie, D., and Titball, R. V. (1993). Rapid identification of flaviviruses based on conserved *NS5* gene sequences. *J. Virol. Methods* **44**, 179.

Futreal, P. A., *et al.* (1994). *BRCA 1* mutations in primary breast and ovarian carcinomas. *Science* **266**, 120.

Gaydos, C. A., Eiden, J. J., Oldach, D., Mundy, L. M., Auwerter, P., Warner, M. L., Vance, E., Burton, A. A., and Quinn, T. C. (1994). Diagnosis of *Chlamyida pneumoniae* infection in patients with community-acquired pneumonia by polymerase chain reaction enzyme immunoassay. *Clin. Infect. Dis.* **19**, 157.

Gill, J. E., and Gulley, M. L. (1994). Immunoglobulin and T-cell receptor gene rearrangement. *Hematol. Oncol. Clin. North Am.* **8**, 751

Grompe, M. (1993). The rapid detection of unknown mutations in nucleic acids. *Nature Genet.* **5**, 111.

Hayashi, K. (1991). PCR-SSCP: A simple and sensitive method for detection of mutations in the genomic DNA. *Cold Spring Harbor Lab. Press* **1**, 34.

Kammerer, U., Kunkel, B., and Kom, K. (1994). A nested PCR for specific detection and rapid identification of human picornaviruses. *J. Clin. Microbiol.* **32**, 285.

Loeb, L. A. (1994). Microsatellite instability: Marker of a mutator phenotype in cancer. *Cancer Res.* **54**, 5059.

Mandel, J. L. (1994). Trinucleotide diseases on the rise. *Nature Genet.* **7**, 453.

Miki, Y., *et al.* (1994). A strong candidate for the breast and ovarian cancer susceptibility gene *BRCA 1*. *Science* **266**, 66.

Powell, S. M., Petersen, G. M., Krush, A. J., Booker, S., Jen, J., Giardiello, F. M., Hamilton, R., Vogelstein, B., and Kinzler, K. W. (1993). Molecular diagnosis of familial adenomatous polyposis. *New Eng. J. Med.* **329**, 1982.

Rustgi, A. K. (1994). Hereditary gastrointestinal polyposis and nonpolyposis syndromes. *New Eng. J. Med.* **331**, 1694.

Schiffman, M. H., and Schatzkin, A. (194). Test reliability is critically important to molecular epidemiology: An example from studies of human papillomavirus infection and cervical neoplasia. *Cancer Res.* **54** (7 Suppl.).

Zeviani, M., and Taroni, F. (1994). Mitochondrial diseases. *Baillieres Clin. Neurol.* **3**, 315.

DNA Methylation in Mammalian Genomes: Promoter Activity and Genetic Imprinting

MARC MUNNES
WALTER DOERFLER
Universität zu Köln

STUDIES ON THE BIOLOGICAL SIGNIFICANCE OF DNA methylation in mammalian genomes continue to present challenging problems for basic and medically relevant research. In what way could patterns of DNA methylation, which appear to be unique for each cell type and which can be interindividually conserved in the human population, contribute to the overall organization of mammalian genomes? Foreign DNA integrated into the host genomes upon virus infection or by artificial transfection becomes frequently *de novo* methylated in specific patterns. This modification contributes to the long-term silencing of integrated foreign genes in mammalian cells. The insertion of foreign DNA into established mammalian genomes elicits marked changes in the patterns of DNA methylation in cellular genes far remote from the integration site, even on different chromosomes. Sequence-specific promoter methylation can lead to promoter inactivation in many genes in different organisms. As a consequence, the implications of DNA methylation for the mechanism of genetic imprinting have been intensely investigated; for example, in human disease imprinting plays a role in explaining different clinical phenotypes after the deletion of specific regions on the maternally or the paternally inherited chromosome.

I. THE BIOLOGICAL SIGNIFICANCE OF DNA METHYLATION IN MAMMALIAN GENOMES

The nucleotide 5-methyldeoxycytidine (5-mC) in eukaryotic DNA has been considered a fifth nucleotide, although the methyl group is attached postreplicationally to deoxycytidine (C) residues in DNA. The amount of 5-mC in eukaryotic DNA varies with species, developmental stage, site in the genome, and perhaps other parameters. The 5-mC content probably constitutes between 2 and maximally 10% of the C residues in mammalian genomes. It has been shown that 5-mC is distributed in highly specific patterns in mammalian genomes. These patterns are specific for different parts of the genome and appear to vary among cell types. At least in certain segments of the human genome, these patterns are identical among different individuals even of different ethnic origins. This high degree of specificity and the uniqueness of patterns imply that they have significance for the organization of the mammalian genome and its function. An increasing number of researchers has studied interrelationships between patterns of DNA methylation and their functional significance.

The sequence-specific methylation of mammalian promoter sequences can serve as a long-term signal

ENCYCLOPEDIA OF HUMAN BIOLOGY, Second Edition, VOLUME 3.

for promoter inactivation. At present, a general rule as to which sequences in a given promoter will lead to promoter inactivation has not been established. The decisive nucleotide sequences, whose methylation will entail promoter inactivation, have to be experimentally determined for each promoter. Two alternative but mutually not exclusive mechanisms have been proposed to explain promoter silencing by a specific pattern of DNA methylation: (1) the binding of specific transcription factors or cofactors in the promoter or enhancer segment of a gene can be counteracted by specific promoter methylation; and (2) the methylation of certain motifs in a promoter sequence can prepare these motifs for the binding of proteins whose specificity in the interaction with sequence motifs is strictly dependent on the presence of 5-mC.

In general, the presence or introduction of 5-mC residues into specific promoter sequences is associated with or can lead to promoter inactivation, respectively. However, as an example of the complex interdependencies between promoter methylation and gene inactivation, frog virus 3 (FV3) promoters have provided challenging insights. The DNA in this iridovirion is completely methylated in all 5′-CG-3′ dinucleotides. At least the late viral promoters appear to be completely methylated when they are actively transcribed. In accordance with these findings in the FV3 promoters, the *in vitro* methylation of all 5′-CG-3′ sequences in an isolated, late FV3 promoter, which has been fused to a reporter gene, leaves this promoter active. However, when only the eight 5′-CCGG-3′ (HpaII) sequences in the same promoter and in its upstream elements are *in vitro* premethylated, the promoter is silenced. Apparently, a certain methylation pattern, possibly different for each promoter, is required to inactivate the specific promoter. This notion is compatible with the modulating effect that DNA-motif methylation can have on the interaction of this motif with specific proteins required for promoter activity. Since the requirements in transcription factor and cofactor binding are different for each promoter and cannot *a priori* be assessed for any promoter on the basis of reliable rules, predictions about the methylation-sensitive sites in mammalian promoters will have to await the in-depth elucidation of promoter structure and function in mammalian systems. In this context, it is worth mentioning that the *in vitro* methylation of viral and mammalian promoters has been shown to lead to structural changes even of the naked promoter DNA sequences, for example, with respect to DNA bending.

The long-term silencing effect of sequence-specific DNA methylation has thus far become one of the few well-documented biological functions of DNA methylation in mammalian and other eukaryotic genomes. This recognized interrelationship is also the basis for medically important studies on the role of DNA methylation, in particular in tumor biology and medical genetics. The inactivation of one of the human X chromosomes and the role of DNA methylation in this process or in the genetic imprinting of one of the alleles in specific segments of mouse or human chromosomes are interesting and mechanistically challenging examples for further research. Investigations on the detailed mechanism of genetic imprinting in mammalian chromosomes promise exciting insights into fundamental concepts of genome organization and function. Patterns of DNA methylation will be an important part of, but most likely not the ultimate solution to, these unorthodox genetic mechanisms. [*See* Human Genome.]

There is a considerable body of experimental evidence supporting the notion that the methylated variant of a 5′-CG-3′ dinucleotide pair in DNA, perhaps not in any sequence context, is prone to deamination and hence can lead to C → T transitions, that is, to specific mutations. Though this mutational mechanism due to a mutagen intrinsic in the nucleotide sequence of many DNA segments has been unequivocally documented, it remains unknown to what extent these mutations can be swiftly repaired or counterselected during the replication of a nucleotide sequence in which only one strand has been altered by the transition. Nevertheless, many examples in tumor tissues have been described in which hot spots of mutations have been observed, for example, in the tumor suppressor genes p53, RB (retinoblastoma), and others, whose locations coincide with the presence of a 5-mCG dinucleotide in the premutation nucleotide sequence. However, in many of these studies the critical controls of sequencing randomly selected genes in the same tumor tissue in order to determine the overall mutation rate due to 5-mC to T transitions have unfortunately not yet been carried out. Therefore, it remains to be seen what relevance the described mutations in the p53 or RB genes might have for the tumorigenic phenotype. [*See* Tumor Suppressor Genes; p53; Tumor Suppressor Genes, Retinoblastoma.]

In the process of DNA methylation, the DNA methyltransferase becomes transiently linked to C residues in DNA by a covalent bond. Depending on the concentration of *S*-adenosylmethionine (SAM)—the methyl-group donor in the methylation reaction—in the cell nucleus during the time of DNA methyltransferase action, C residues in DNA are methylated (high levels of SAM) or deaminated to thymidine (low levels or absence of SAM). In the latter case, a mutation or transition can be generated, unless the generated mismatch is subsequently repaired. This reaction mechanism is still somewhat hypothetical, but is supported by experimental evidence.

When the gene for the known DNA methyltransferase has been "knocked out" in mice, development of the embryos has been impaired. It is not certain whether the thus manipulated embryonal stem cells or embryos are completely devoid of DNA methyltransferase activity.

II. DOES DNA METHYLATION CONTRIBUTE TO GLOBAL GENOME ORGANIZATION?

Sequence-specific DNA methylation can affect the interaction of specific proteins with sequence motifs in a functionally relevant sense. The clearest example for this modulator function of 5-mC in a specific sequence context is the abrogation of restriction endonuclease functions on specifically methylated DNA sequences in prokaryotic systems. For the modified nucleotide N^6-methyldeoxyadenosine, which occurs only in prokaryotes, there are examples for restriction endonuclease inhibitory effects, but also those instances in which the presence of this modified nucleotide is a precondition for the activity of the restriction endonuclease. The restriction enzyme DpnI, for example, can cleave at the nucleotide sequence 5'-GATC-3' only when the A residue is methylated. On the basis of these well-documented enzymatic mechanisms, one would expect that 5-mC could affect the binding capacity of specific proteins to nucleotide sequence motifs in either a positive or a negative way depending on the nature of the protein–DNA interactions and on the specific nucleotide sequences involved.

The effect of 5-mC in specific sequence contexts is unlikely to be restricted to a regulatory role in eukaryotic promotors. Moreover, the very specific and apparently interindividually conserved nature of methylation patterns in the human and other eukaryotic genomes suggests that the biological significance of DNA methylation transcends the function in a single biochemical mechanism, however important transcriptional regulation has turned out to be in developmental biology, tumorigenesis, and the coordination of functions in an organism. For these reasons, it appears mandatory to study these patterns of DNA methylation and their generation, conservation, alterations, abrogation, and reinstitution during development in considerable detail. How would one ever be able to understand function unless the structure of these patterns was elucidated? For these same reasons, current "genome programs," which aim at determining the organization and nucleotide sequence of the four traditional nucleotides in a number of eukaryotic genomes, might fall short of their proclaimed destinations as long as these programs will not include the localization of the fifth nucleotide in the genomes under investigations. Since this determination is a technically difficult and often still frustrating task, this desirable goal will not quickly be reached.

Patterns of DNA methylation will have to be investigated in different parts of the human and other mammalian genomes. Without this information, one will be incapable of completely understanding the biological function of 5-mC residues in the mammalian genome during development and later life. The organization of these complex genomes with a large, but hitherto unknown, number of proteins interacting in a highly pliable way with DNA sequence motifs could be based in part on patterns of DNA modification. The failure of the *Drosophila melanogaster* genome, whose importance for studies of eukaryotic genome organization and function is beyond doubt, to contain detectable amounts of 5-mC cannot be accepted as an argument against the biological importance of the nucleotide 5-mC in mammalian systems. Of course, the successful application of highly sensitive techniques to the analysis of the *D. melanogaster* genome may still bring forth results that will permit one to include it in the group of 5-mC-bearing genomes.

III. CONSEQUENCES OF FOREIGN DNA INSERTION INTO ESTABLISHED MAMMALIAN GENOMES

A. Changes in Cellular DNA Methylation upon the Insertion of Foreign DNA

During the lifetime of an organism, the organization of its genome is thought to be stable. Of course, it is unknown to what extent rearrangements, deletions, or insertions of foreign DNA can alter the architecture of a genome. Experimentally, particularly with DNA or DNA tumor viruses, insertions of foreign DNA into established mammalian genomes have been investigated in detail. The mode and mechanism of the insertion of adenovirus DNA into hamster genomes have been described; it is able to insert at many different sites. At present, there is no evidence for the notion that specific insertion sites would exist, although in individual instances cell lines have been characterized that carried the viral genome in apparently identical or very similar locations. It would be surprising if the insertion of multiple copies of adenovirus DNA with a genome length of 30 to 35 kilobase pairs, depending on the type of adenovirus, could not have far-reaching consequences for the stability and organization of the entire target cell genome. We have demonstrated that the insertion of foreign DNA into mammalian genomes can be associated with a previously not recognized type of insertional mutagenesis. Upon the insertion of adenovirus, plasmid, or bacteriophage lambda DNA, striking changes have been observed in different parts of the cellular genome, even at sites far remote from the site of insertion and on different chromosomes. We have not yet shown that these changes are directly linked to altered patterns of cellular DNA transcription. An increase or decrease in DNA methylation in segments of the cellular genome as a consequence of foreign DNA insertion could lead to the silencing or to the activation, respectively, of cellular genes that had been in the opposite state of activity prior to the insertion event. Considering these experimental findings, the concept of insertional mutagenesis could now be understood in a much wider sense than the traditional one in that the overall activity profile of a cell could be significantly altered by the insertion of foreign DNA into an established genome owing to alterations of DNA methylation. Thus the mutagenic effect would not be limited to the destruction or alteration of sequences at the immediate insertion site. It will be interesting to investigate whether this long-range insertional mutagenesis effect could play a role in viral oncogenesis.

B. *De Novo* Methylation of Inserted Foreign DNA

When foreign DNA is inserted into an intact mammalian genome, the integrated foreign DNA is *de novo* methylated in specific patterns. Such patterns have been observed in integrated adenovirus DNA, integrated retroviral DNA, integrated plasmid constructs or bacteriophage λ DNA in mammalian genomes, and in foreign DNA integrated into plant genomes. The *de novo* methylation of integrated foreign DNA thus appears to be a general consequence of foreign DNA insertion in eukaryotic organisms. It is conceivable

that *de novo* methylation serves as a protective or defense mechanism of the recipient cell or organism against the activity of foreign genes that have intruded into the cell's genome and must then be permanently silenced by sequence-specific DNA methylation. *De novo* methylation is perhaps an evolutionarily old defense mechanism against the introduction of ubiquitously present foreign DNA.

The mechanism of foreign DNA methylation and its regulation are not well understood. From the results of studies on integrated adenovirus genomes, it is likely that the site of initiation of *de novo* methylation of foreign DNA is not primarily dependent on a specific nucleotide sequence, since the same adenovirus DNA sequence can become *de novo* methylated or remain un- or hypomethylated, depending probably on the site of integration into the cellular genome. Other authors have suggested that a specific nucleotide sequence could have an important influence on the initiation of *de novo* methylation. However, though a role of DNA sequence may exist in site selection, some of these data interpreted to document sequence contributions in the selection process do not allow one to rule out the decisive effect of the site of foreign DNA integration on the initiation of *de novo* methylation, since in different experiments the experimentally introduced foreign DNA resided in different locations in the target genome.

Many of the cited investigations have been carried out with cells growing in culture. Foreign DNA inserted into the mouse genome in transgenic organisms can also be subjected to *de novo* methylation, depending on the genetics of the mouse strain, the site of insertion in the genome, and possibly other factors. Undoubtedly, the nature and interdependencies of different factors affecting the establishment and generation of *de novo* patterns of DNA methylation are poorly understood, and much more research will be required to clarify this important mechanism. In transgenic mice, the newly generated patterns of foreign DNA methylation have been found to be identical or very similar in all organ systems investigated, except for the testis. Many lines of evidence in different eukaryotic systems suggest that in the germ line and in early evolution, drastic changes in patterns of DNA methylation do occur. Patterns are apparently reset at this early developmental stage, and specific, previously existing patterns can be reinstalled *de novo* with surprising precision. We do not understand what structures and/or enzymatic mechanisms preserve the memory for cell type-specific and developmental stage-specific patterns of DNA methylation.

The solution of these biochemical, genetic, and epigenetic problems not only constitutes an intellectual challenge for researchers in the field of DNA methylation, but will have repercussions on the interpretation of results gleaned from work with transgenic organisms and for the field of gene therapy. How can anyone hope to formulate meaningful research on gene therapy without a thorough understanding of the fate of foreign DNA in mammalian organisms and its organizational and transcriptional regulation after the fixation in an established genome?

IV. SEQUENCE-SPECIFIC PROMOTER METHYLATION CAN LEAD TO GENE INACTIVATION

The notion that sequence-specific promoter methylation can lead to gene inactivation is mainly based on results derived from experiments performed in cell culture systems. It remains a matter of debate whether, in organisms, promoter methylation is the cause or a consequence of promoter inactivation or whether both possibilities exist. The results of extensive experimental work (Table I) indicate that promoter methylation exerts a causal role in promoter inactivation.

In our own work, we have initially pursued the question of why in integrated adenovirus genomes some of the viral genes are transcribed, whereas others are apparently permanently silenced. Early investigations have shown that an inverse correlation exists between the transcriptional activity of integrated viral genes and their level of DNA methylation. This observation has been subsequently extended to many other viral and nonviral eukaryotic systems (Table I). The most convincing example of an inverse correlation in our line of investigations has come from studies on the promoter region of the late E2A gene of integrated adenovirus type 2 (Ad2) DNA, which is completely methylated in all 5'-CCGG-3' sequences in a cell line that carries an inactive E2A gene. In a different cell line containing the same integrated viral promoter in an actively transcribed form, this promoter lacked DNA methylation in all 5'-CCGG-3' sequences. These conclusions were initially based on promoter analyses with the methylation-sensitive restriction endonuclease HpaII from *Hemophilus parainfluenzae*, which allows one to investigate only the 5'-CCGG-3' sequences for their state of methylation. Later on, these results were confirmed and extended by applying the genomic sequencing technique to the analyses of the promoter in the two different cell lines. In the inactive promoter, all 5'-CG-3' dinucleotides were found to be methylated, whereas in the active promoter, none of them was methylated. [*See* Adenoviruses.]

The E2A region of Ad2 DNA codes for a DNA-binding protein that is required for viral DNA replication and also plays a role in viral DNA transcription. This early viral gene is controlled by two promoters, the early and the late E2A promoters. For much of our experimental work, we have concentrated on the late E2A promoter of Ad2, since we reasoned that the results would be more complete and convincingly interpretable when the same promoter was tested under a variety of experimental conditions and in different biological systems. The results adduced with this viral system were later confirmed by many investigators using a large number of different cellular or viral promoters (Table I).

Since the early results could necessarily provide only correlative data, we designed experiments in which the late E2A promoter of Ad2 DNA was fused to a reporter gene, like the prokaryotic gene for chloramphenicol acetyltransferase or for luciferase, which were absent from eukaryotic cells. The unmethylated construct or the methylated construct was then investigated for transcriptional activity in various biological systems:

(1) Upon microinjection into *Xenopus laevis* oocytes, the unmethylated construct remained active and the 5'-CCGG-3' methylated construct was not transcribed. The results of some of these experiments demonstrated that the promoter region was sensitive to DNA methylation, whereas the methylation of the gene proper with the promoter region left unmethylated did not lead to the loss of transcriptional activity of the late E2A gene. Similar results, however, with a more pronounced and reliably silencing effect, were obtained when the promoter was methylated in all 5'-CG-3' dinucleotide sequences. This type of *in vitro* premethylation of a construct became feasible with the advent of the 5'-CG-3' DNA methyltransferase M.SssI from *Spiroplasma* species.

(2) Methylation of the late E2A promoter of Ad2 DNA led to transcriptional inactivation also in short-term transcription experiments after transfection of the constructs into mammalian cells growing in culture. However, the E1A transactivator encoded in the E1A region of adenovirus DNA could at least partly cancel the inactivating effect of E2A promoter methylation. Similarly, the strong enhancer from an early

TABLE I

Summary of Experiments Investigating Promoter (Gene) Methylation
and Transcriptional Inactivation of Promoters[a]

Species	Gene	Location	Effect[b]	Type of evidence[c]
Human	Alpha-fetoprotein	5′-region	Inactivated	A1
	Alu	Promoter	Inactivated	A1
	Apolipoprotein-E	Gene	No effect	A1
	Beta-amyloid protein precursor	5′-region	Inactivated	A1, C
	B-cell tyrosine kinase	Gene	Inactivated	A1
	c-myc	Gene	Questionable	A1, A2
	Cyclin D2	Gene	Inactivated	C
	Cytochrome *P*-450 2E1	Gene	Questionable	A1
	E-selectin	Promoter	Inactivated	B
	Estrogen receptor	Gene	Inactivated	A1, C
	Fragile X mental retardation 1	Promoter	Inactivated	A2, C
	Gamma globin	Promoter	Inactivated	A1
	Glucose-6-phosphate dehydrogenase	5′-region	Questionable	A1
	HLA class I loci A, B, G, E	Gene	Inactivated	A1
	Hypoxanthine phosphoribosyltransferase	Gene	Inactivated	A2
	IFN-gamma	Promoter	Inactivated	A1, C
	IGF2 and H19	Gene	Questionable	A1
	Lactate dehydrogenase	5′-region	Inactivated	A1
	Leukosalin	5′-region	Inactivated	A1
	Major breakpoint cluster region	Gene	Inactivated	A1
	mb-1	5′-region	Inactivated	A1
	Multiple tumor suppressor 1	Gene	Inactivated	A1, C
	Myeloperoxidase	5′-region	Inactivated	A1
	myoD	5′-region	Inactivated	A1
	O^6-Methylguanine-DNA methyltransferase	Promoter	Inactivated	A1
	Parathyroid hormone-related peptide	Promoter	Inactivated	A1, C
	Phosphoglycerate kinase	5′-region	Inactivated	C
	pi-class glutathion *S*-transferase	Promoter	Inactivated	A1
	Platelet-derived growth factor A-chain	Promoter	Inactivated	B, C
	Pro alpha 1(I) collagen	Promoter	Inactivated	B, C
	Proenkephalin	5′-region	Inactivated	B
	Proliferating cell nuclear antigen	Gene	No effect	A1, C
	Retinoblastoma tumor suppressor	5′-region	Inactivated	A1
	β-actin	Gene	Inactivated	B
	Sea urchin retroposon family -1/-2	Promoter	Inactivated	B, C
	Thymidine kinase	Promoter	Inactivated	A1
	Tumor necrosis factor-alpha	5′-region	Inactivated	B
	von Hippel-Lindau	5′-region	Inactivated	A1
Mouse	Adenine phosphoribosyltransferase	Gene	Inactivated	A1, C
	Alpha 1(I) collagen	Promoter	Inactivated	B
	c-alb thymidine kinase	Promoter	Inactivated	B
	c-Ha-rasVal 12 oncogene	5′,3′-region	No effect	A1
	H19	Promoter	Inactivated	A1, B
	HRD-transgene	Gene	Inactivated	A1
	Hypoxanthine phosphoribosyltransferase	Gene	Inactivated	A2
	IFN-gamma	Promoter	Inactivated	A1, C
	Insulin-like growth factor 2	5′-region	No effect	A1
	Insulin-like growth factor 2-receptor	5′-region	Questionable	A1
	Metallothionein I promoter	Promoter	Inactivated	A1, B
	myoD	Gene	Inactivated	C
	Ovine growth hormone	Gene	Inactivated	A1
	Phosphoglycerate kinase	5′-region	Inactivated	A2
	Sex-limited protein	5′-region	Inactivated	B
	Sea urchin retroposon family -1/-2	Promoter	Inactivated	B, C
	U2AF1-rs1	Promoter	No effect	A1
	Xist	Promoter	Inactivated	A2

continues

TABLE I (*Continued*)

Species	Gene	Location	Effect[b]	Type of evidence[c]
Rat	Apolipoprotein-E	Gene	No effect	A1
	Gamma-crystallin	5'-region	Inactivated	B
	Gamma-glutamyl transpeptidase	5'-region	Inactivated	C
	Glial fibrillary acidic protein	5'-region	Inactivated	A1
	Pro alpha 2(I) collagen	Promoter	Inactivated	B
	Prolactin	Promoter	Inactivated	B
	Testis-specific H2B histone	Promoter	Inactivated	A2
Dog	Thyroglobulin	Promoter	Inactivated	B
Chicken	Pepsinogen	Promoter	Inactivated	A1
Xenopus	Ribosomal genes	Promoter	Questionable	A1
Viruses[d]				
Ad2	E2A late	5'-region	Inactivated	A1, A2, B
Cau MV	35s	Gene	Inactivated	C
CRPV	Different	5'-region	Inactivated	A1
EBV	BCR2	5'-region	Inactivated	B
EBV	BNLF1	5'-region	Inactivated	B
EBV	BamHI W promoter	Promoter	Inactivated	C
FV3	L1140	5'-region	Inactivated	B
H CMV	Early promoter	Coding	Inactivated	B
HIV-1	LTR	Complete	Inactivated	B
HPV-18	URR	Complete	Inactivated	B
HSV	Thymidine kinase	5',3'-regions	Inactivated	B
SFV-3	Different	Complete	Inactivated	C
SV40	Early promoter	Promoter	Inactivated	B
Ciliate				
Colpoda	General	Complete	Questionable	C

[a]The data shown in this compilation concern genes investigated after 1989.

[b]The term "inactivated" refers to the effect of promoter inactivation due to DNA methylation. In some cases (see footnote c), treatment of cells with 5-aza-2'-deoxycytidine has led to the reactivation of the gene. See evidence C.

[c]These interpretations are based on inverse correlations between promoter methylation and gene inactivation analyzed with (**A1**) restriction enzymes or (**A2**) the method of genomic sequencing, or (**B**) on the premethylation of promoter-indicator gene constructs whose activity has subsequently been assessed upon transfection into mammalian cells. In some experiments, previously silent and methylated genes have been reactivated by 5-aza-2'-deoxycytidine (**C**).

[d]Ad2, human adenovirus type 2; Cau MV, cauliflower mosaic virus; CRPV, cottontail rabbit papillomavirus; EBV, Epstein–Barr virus; FV3, frog virus 3; H CMV, human cytomegalovirus; HIV-1, human immunodeficiency virus-1; HPV-18, human papillomavirus-18; HSV, herpes simplex virus; SFV-3, simian foamy virus type 3; SV40, simian virus 40.

cytomegalovirus (CMV) promoter could overcome promoter silencing by methylation when this enhancer was spared methylation in the construct used for activity testing. Differential methylation experiments in the CMV enhancer were feasible, since it did not contain 5'-CCGG-3' sites and hence escaped *in vitro* methylation by the HpaII DNA methyltransferase. However, when the CMV enhancer was methylated in all 5'-CG-3' sequences with the DNA methyltransferase from *Spiroplasma* species, the construct was completely silenced. These data indicated that promoter methylation was not a stringently irreversible signal of gene inactivation but could be overcome by transactivators or by the presence of a nonmethylated strong enhancer in the vicinity of the methylated promoter.

(3) Very similar results were obtained when the methylated or unmethylated promoter construct was genomically fixed by integration in the genome of mammalian cells in culture or in transgenic mice. In the experiments with transgenic mice, stability of a preimposed methylation pattern or of the nonmethylated state of the integrated construct turned out to be dependent on a number of parameters, like the site of foreign DNA integration and the genetics of the mouse strain. In transgenic mice, the activity of the late E2A promoter, which was a relatively weak promoter requiring transactivation, was dependent on

the organ system in which the activity of the construct was tested. In some of the mouse organs, the late E2A promoter construct failed to exhibit any activity, perhaps because of the lack of required transactivators, hence the effect of sequence-specific methylation could not be ascertained in these organs. In organs in which the late E2A promoter could attain activity, its methylation led to transcriptional inactivation.

(4) The late E2A promoter as well as the much stronger major late promoter of Ad2 DNA could also be shown to be inactivated by sequence-specific methylation in *in vitro* transcription experiments using a cell-free system from HeLa cell nuclear extracts.

(5) By using the RNA polymerase III-transcribed VAI region of Ad2 DNA as a test system, it was demonstrated that the control region of this viral gene was also silenced by sequence-specific DNA methylation. The VAI region of Ad2 DNA encodes an ≈ 160-nucleotide RNA that is termed virus-associated (VA) RNA, although it is not really virion-associated but is synthesized in large quantities in Ad2-infected cells. VAI RNA is a translational activator and counteracts the action of interferon, thus rendering adenovirus infections largely insensitive to interferon inhibition.

(6) The adenovirion DNA inside the mature adenovirus particle is not detectably methylated in strong contrast to frog virion 3 DNA. We have investigated adenovirion DNA or the free intracellular adenoviral DNA in infected cells with chemical methods, with restriction endonucleases, or, in certain segments of the Ad2 genome, by the genomic sequencing technique. We failed to adduce any evidence for the occurrence of 5-methyldeoxycytidine in virion DNA or in the free intracellular adenoviral DNA early or late after infection. Therefore, we conclude that adenoviral DNA methylation most likely plays no role whatsoever in the adenovirus infection cycle. Adenovirus DNA has been found to become *de novo* methylated only after the integration into an established cellular genome. Adenovirus DNA probably shares this fate with most (any?) foreign DNA and with retroviral genomes integrated into eukaryotic chromosomes.

The viral model systems employed in our experiments have provided a frame for studies on the biological effects of DNA methylation. The results deduced with the same promoter in different biological contexts lend credence to the interpretation that sequence-specific promoter methylation can lead to promoter inactivation. Each promoter will have to be investigated in detail to determine its methylation-sensitive sequences and the parameters enhancing or

counteracting the promoter-inactivating effect due to sequence-specific methylation. Such modifying factors can be transactivators, enhancers, and possibly other regulatory elements in or close to the promoter sequence. These conclusions have been derived from several different experimental systems. Of course, it is still difficult to ascertain and to prove that, in a living cell, promoter methylation plays the same long-term inactivating role, although the results of correlative studies for many different promoters in different eukaryotic systems support exactly that interpretation. Nevertheless, further investigations will be needed to strengthen the present notion about the role of promoter methylation in the long-term silencing of eukaryotic genes. The regulatory signal of promoter methylation serves a long-term inactivating role. In systems that require activation and silencing cycles within a short time span, promoter methylation would obviously not be the signal of choice.

(7) The mechanism of promoter inactivation has not yet been elucidated. It is likely that the interaction of essential transcription factors with sequence motifs in the promoter can be interfered with by promoter methylation. Early studies on promoter constructs revealed that the inactivating effect of promoter methylation was somehow dependent on the presence of chromatin-like structures in the transfected constructs. There are a number of reports that have demonstrated that specific transcription factors cannot bind anymore to decisive promoter sequences when these regions are methylated. Moreover, some promoter motifs exhibit a propensity to bind proteins only when the motifs are methylated. These methylation-dependent protein–DNA interactions may also play an important function in the inactivation of promoters by sequence-specific methylation.

By choosing the adenovirus system as one of the models to introduce the reader to problems of promoter methylation and long-term gene inactivation, we did not intend to detract from the importance of the incisive work in many other laboratories that came to very similar results and conclusions. In Table I, results from many nonviral and viral eukaryotic systems have been compiled. The table includes information on the organism and the gene(s) with which the studies have been carried out. Moreover, the sites of DNA methylation studied have been indicated, as well as the type of experiments performed and the results obtained. In most instances, promoter methylation leads to inactivation. Only in rare cases did the promoter remain active, in spite of DNA methylation, perhaps because the right, that is, methylation-sensitive, sequence was not modified. Some of the genes analyzed were genetically imprinted, and this topic will be discussed in more detail in the following section. Analyses of the effects of promoter methylation on gene activity can be rendered experimentally more decisive when *in vitro* systems are included in the experimental approach, and this aspect has also been considered in Table I. A considerable number of references presented in the Bibliography may help the reader access the ever-growing literature on DNA methylation.

(8) The nucleotide analog 5-azadeoxycytidine (5-azaC) can be incorporated into DNA and serves as an attractant for the DNA methyltransferase, which can bind covalently to 5-azaC. In contrast to the transient covalent bond formed between DNA methyltransferases and the nucleotide C, the binding to 5-azaC is thought to be irreversible, hence the available DNA methyltransferase levels in a 5-azaC-treated cell are quickly and strikingly decreased such that in growing cells the maintenance of the cell's methylation pattern will be compromised. Consequently, many segments of the cell's genome can eventually become demethylated due to the lack of the availability of free DNA methyltransferase activity. It has been demonstrated that the treatment of growing cells in culture with 5-azaC causes the activation of previously inactive genes. In fact, developmental programs have been shown to be switched on, or endogenous retroviral genes have become reactivated, by 5-azaC treatment. The activation of previously silenced cellular or endogenous retroviral genes by the 5-azaC treatment of cells has been one of the arguments in favor of the long-term silencing function of promoter methylation. This reasoning is somewhat weakened by the fact that 5-azaC has a number of less well known side effects that might be involved in promoter activation by mechanisms independent of promoter methylation and the loss of specific 5-mC residues in the promoters.

(9) Since regions in a gene other than the bona fide promoter and 5′-upstream segments have been shown to contain regulatory elements, it is not surprising that changes in DNA methylation in these noncanonical, because less well understood, regulatory regions can also affect gene transcription. One of the earliest described examples of such an effect of DNA methylation has been reported for the thymidine kinase gene, which was inactivated when its 3′-located structural region was methylated.

V. GENETIC IMPRINTING AND DNA METHYLATION

The term genetic imprinting designates the transcriptionally nonequivalent state of a gene, a genetic region, or more extensive parts of the X chromosome, for example in the human genome, between the two allelic chromosomes of which in general one has been inherited via the paternal and the other via the maternal gamete. As far as is known at present, in most regions of the human or the mouse chromosome both alleles of the coding regions of the chromosomes can be transcribed. It is not certain whether both alleles can be transcribed at all times and to the same extent in all cell types. In contrast, in some regions of the genome, one estimates about 10%, for example, for the mouse genome, only one chromosomal allele, the paternally or the maternally inherited one, is transcribed. The other allele apparently has been permanently silenced at some time during the individual's development. It is completely unknown for what biological or evolutionary reasons some genomic segments become permanently inactivated. The nature of the gene products, the specific localizations, or genomic organization at distinct loci may hold clues to this fascinating genetic phenomenon, but a definite explanation cannot yet be offered. [*See* Genetic Imprinting.]

Regarding the molecular mechanism underlying genetic imprinting, sequence-specific DNA methylation was an obvious signal to investigate, as it had the well-documented capacity to inactivate a gene for a long time. Imprinting models were derived from experiments with transgenic mice, and experimental evidence was adduced for an important, perhaps not exclusive, role that DNA methylation would play in this process. It is likely that additional parameters render an important contribution to the imprinting mechanism. In what way does the organization of genetically imprinted regions differ from that of most other parts of the mammalian genome? What is the biological significance of genetic imprinting? Have the imprinted regions resided in an established genome as long as the nonimprinted segments?

Table II presents a compilation of some of the currently recognized imprinted sections of the human and mouse genomes. The names of the regions or genes and their chromosomal locations are listed, as well as the parental imprint and the possible involvement of DNA methylation in the imprinted region. We have also included possible consequences of im-

TABLE II

Imprinted Genes and DNA Methylation

Imprinted gene/locus	Chromosome	Parental imprint	Region[a] methylated	Involvement in diseases[b]	Tissue specificity
Mouse					
H19	mu7F	Paternal	+	Embryonic lethality	Placenta, kidney
Insulin-like growth factor 2	mu7F	Maternal	+	Embryonic lethality	Placenta, kidney
Insulin-like growth factor 2-receptor	mu17A	Paternal	+	Embryonic lethality	Placenta, liver
Small nuclear riboprotein N	mu7C	Maternal	+	Embryonic lethality	Brain, heart
U2af-rs1 (SP2)	mu11	Maternal	+		Brain, liver
Ins1-Ins2	mu7F	Maternal	ND		Yolk sac
p57 *KIP2*	mu7F	Paternal	+		
Mas	mu17F	Maternal	ND		
Mash2	mu7	Paternal	ND	Embryonic lethality	Trophoblast
Mo2-macrosatellite			+		
Sex-limited protein		Maternal	+		
Xist	muX	Paternal	+		Embryogenesis up to morula
Human					
H19	hu11	Paternal	+		Placenta, kidney
Insulin-like growth factor 2	hu11	Maternal	+	Beckwith–Wiedemann	Placenta, kidney
Small nuclear riboprotein N	hu15	Maternal	+	Prader–Labhart–Willi/Angelman	Brain, heart
Wilms' tumor suppressor	hu11	Paternal	ND	DDS/WAGR syndrome	Placenta, brain
ZNF127	hu15	Paternal	+	Prader–Labhart–Willi/Angelman	

[a] + stands for extensive DNA methylation.

[b] DDS stands for Denys–Drash syndrome, a rare human condition in which severe urogenital aberrations result in renal failure, pseudohermaphroditism, and Wilms' tumor. WAGR stands for aniridia, ambiguous genitalia, and mental retardation, a syndrome of hemihypertrophy and other congenital anomalies with Wilms' tumor.

printing for pathogenesis and development, as well as information on tissue specificities of the genetically imprinted genome sections. Since imprinted regions are functionally represented only once in a genome, deletions or mutations in imprinted regions on one of the chromosomes can have catastrophic sequelae for the organism. Genetic imprinting has been recognized to be of considerable medical importance in the causation of some important genetic diseases.

VI. IMPRINTING AND GENETIC DISEASE

Depending on the location of a major deletion on the long arm of human chromosome 15 (15q) on the paternally or the maternally inherited chromosome, totally different symptoms are elicited in the patient presenting with this type of deletion. The phenomenon cannot be explained by classic Mendelian genetics. The two chromosomes are imprinted differently at least in the segment 15q11-13. It has been shown

that the methylation patterns on the two chromosomes differ decisively in this region.

A deletion on the paternal chromosome in 15q11-13 is the cause for the Prader–Labhart–Willi syndrome (PWS), whereas deletions in a similar region on the maternally imprinted chromosome lead to the Angelman syndrome (AS). Both genetic diseases are also known as microdeletion syndromes, although megabase pairs can be deleted in these regions. By clinical phenotype, PWS and AS are completely different diseases. In PWS, developmental and mental retardation, psychological difficulties, obesity due to dysregulation of eating habits, small hands and feet, hypogenitalism and others are the most prominent symptoms. The patient with AS is characterized by developmental and severe mental retardation with absence of speech, difficulties in the ability to establish contact with others, cerebral seizures, a characteristic spastic gait, emotional incontinence ("laughing spells," which in reality have nothing to do with laughter), and other symptoms. Since within the same general region on 15q11-13 different segments are imprinted in opposite directions and, hence, exhibit opposite transcriptional patterns on the paternally or on the maternally inherited chromosome, deletions on one or the other chromosome will cause very different phenotypes. It has been recognized that PWS or AS can also be due to maternal or paternal uniparental disomy, respectively, a rare disturbance in chromosomal distribution thought to occur in meiosis. In such individuals, two copies of the maternal or the paternal chromosome are inherited with a concomitant lack of the opposite parental allele. As a consequence, all or part (mosaicism) of the cells in the embryo carry two maternally or two paternally inherited chromosomes 15. Maternal disomy of chromosome 15 is functionally equivalent to a deletion in the imprinted region on the paternal chromosome with PWS as the phenotypic consequence. *Mutatis mutandis,* paternal disomy of chromosome 15 causes AS.

Data collected in Table II present some of the diseases caused by so-called microdeletions in other imprinted parts of the human chromosome. Another well-investigated region comprises segments on the short arm of chromosome 11 (11p). In the context of this review, we cannot elaborate in more detail on the complex phenomena associated with genetic imprinting in this chromosomal segment. The general concept that DNA methylation may be causally related to the mechanism of imprinting should not be overemphasized, since the complete explanation may well turn out to be more complex. The involvement of alterations in DNA methylation in specific parts of the human chromosomes and their relations to human genetic diseases have given studies on DNA methylation continued impetus.

ACKNOWLEDGMENTS

Research in the authors' laboratory was supported by the Deutsche Forschungsgemeinschaft, the Federal Ministry of Education, Science Research and Technology, Bonn, and the Wilhelm Sander-Foundation, Munich, Germany.

BIBLIOGRAPHY

Bartolomei, M. S., Zemel, S., and Tilghman, S. M. (1991). Parental imprinting of the mouse H19 gene. *Nature* **351**, 153.

Beard, C., Li, E., and Jaenisch, R. (1995). Loss of methylation activates Xist in somatic but not in embryonic cells. *Genes Dev.* **9**, 2325.

Bestor, T. H., Laudano, A., Mattaliano, R., and Ingram, V. (1988). Cloning and sequencing of a cDNA encoding DNA methyltransferase of mouse cells. The carboxyl-terminal domain of the mammalian enzymes is related to bacterial restriction methyltransferases. *J. Mol. Biol.* **203**, 971.

Brown, T. C., and Jiricny, J. (1987). A specific mismatch repair event protects mammalian cells from loss of 5-methylcytosine. *Cell* **50**, 945.

Cattanach, B. M., and Jones, J. (1994). Genetic imprinting in the mouse: Implications for gene regulation. *J. Inher. Dis.* **17**, 403.

Cedar, H., and Razin, A. (1990). DNA methylation and development. *Biochim. Biophys. Acta* **1049**, 1.

Church, G. M., and Gilbert, W. (1984). Genomic sequencing. *Proc. Natl. Acad. Sci. USA* **81**, 1991.

Constantinides, P. G., Jones, P. A., and Gevers, W. (1977). Functional striated muscle cells from non-myoblast precursors following 5-azacytidine treatment. *Nature* **267**, 364.

Deobagkar, D. D., Liebler, M., Graessmann, M., and Graessmann, A. (1990). Hemimethylation of DNA prevents chromatin expression. *Proc. Natl. Acad. Sci. USA* **87**, 1691.

Dittrich, B., Robinson, W. P., Knoblauch, H., Buiting, K., Schmidt, K., Gillessen-Kaesbach, G., and Horsthemke, B. (1992). Molecular diagnosis of the Prader-Willi and Angelman syndromes by detection of parent-of-origin specific DNA methylation in 15q11-13. *Hum. Genet.* **90**, 313.

Doerfler, W. (1983). DNA methylation and gene activity. *Annu. Rev. Biochem.* **52**, 93.

Doerfler, W. (1995). The insertion of foreign DNA into mammalian genomes and its consequences: A concept in oncogenesis. *Adv. Cancer Res.* **66**, 313.

Doerfler, W., and Böhm, P. (eds.) (1995). The molecular repertoire of adenoviruses. *Curr. Top. Microbiol. Immunol.* **199**, I–III.

Engler, P., Haasch, D., Pinkert, C. A., Doglio, L., Glymour, M., Brister, R., and Strob, U. (1991). A strain-specific modifier on mouse chromosome 4 controls the methylation of independent transgene loci. *Cell* **65**, 939.

Frommer, M., McDonald, L. E., Millar, D. S., Collins, C. M., Watt, F., Grigg, G. W., Molloy, P. L., and Paul, C. L. (1992). A genomic sequencing protocol that yields a positive display of 5 methylcytosine residues in individual DNA strands. *Proc. Natl. Acad. Sci. USA* **89**, 1827.

Hansen, R. S., Gartler, S. M., Scott, C. R., Chen, S. H., and Laird, C. D. (1992). Methylation analysis of CGG sites in the CpG island of the human FMR1 gene. *Hum. Mol. Genet.* **1**, 571.

Heller, H., Kämmer, C., Wilgenbus, P., and Doerfler, W. (1995). Chromosomal insertion of foreign (adenovirus type 12, plasmid, or bacteriophage lambda) DNA is associated with enhanced methylation of cellular DNA segments. *Proc. Natl. Acad. Sci. USA* **92**, 5515.

Hermann, R., Hoeveler, A., and Doerfler, W. (1989). Sequence-specific methylation in a downstream region of the late E2A promoter of adenovirus type 2 DNA prevents protein binding. *J. Mol. Biol.* **210**, 411.

Holliday, R. (1987). The inheritance of epigenetic defects. *Science* **238**, 163.

Iguchi-Ariga, S. M., and Schaffner, W. (1989). CpG methylation of the cAMP-responsive enhancer/promoter sequence TAGCGTCA abolishes specific factor binding as well as transcriptional activation. *Genes Dev.* **3**, 612.

Jones, P. A., and Taylor, S. M. (1980). Cellular differentiation, cytidine analogs and DNA methylation. *Cell* **20**, 85.

Kochanek, S., Toth, M., Dehmel, A., Renz, D., and Doerfler, W. (1990). Interindividual concordance of methylation profiles in human genes for tumor necrosis factors alpha and beta. *Proc. Natl. Acad. Sci. USA* **87**, 8830.

Koetsier, P. A., Mangel, L., Schmitz, B., and Doerfler, W. (1996). Stability of transgene methylation patterns in mice: position effects, strain specificity and cellular mosaicism. *Transgenic Res.* 5, 235.

Kruczek, I., and Doerfler, W. (1983). Expression of the chloramphenicol acetyltransferase gene in mammalian cells under the control of adenovirus type 12 promoters: effect of promoter methylation on gene expression. *Proc. Natl. Acad. Sci. USA* 80, 7586.

Langner, K. D., Vardimon, L., Renz, D. and Doerfler, W. (1984). DNA methylation of three 5' C-C-G-G 3' sites in the promoter and 5' region inactivate the E2a gene of adenovirus type 2. *Proc. Natl. Acad. Sci. USA* 81, 2950.

Levine, A., Cantoni, G. L., and Razin, A. (1991). Inhibition of promoter activity by methylation: possible involvement of protein mediators. *Proc. Natl. Acad. Sci. USA* 88, 6515.

Li, E., Bestor, T. H., and Jaenisch, R. (1992). Targeted mutation of the DNA methyltransferase gene results in embryonic lethality. *Cell* 69, 915.

Meehan, R. R., Lewis, J. D., McKay, S., Kleiner, E. L., and Bird, A. P. (1989). Identification of a mammalian protein that binds specifically to DNA containing methylated CpGs. *Cell* 58, 499.

Muiznieks, I., and Doerfler, W. (1994). The topology of the promoter of RNA polymerase II- and III-transcribed genes is modified by the methylation of 5'-CG-3' dinucleotides. *Nucl. Acids Res.* 22, 2568.

Munnes, M., Schetter, C., Hölker, I., and Doerfler, W. (1995). A fully 5'-CG-3' but not a 5'-CCGG-3' methylated late frog virus 3 promoter retains activity. *J. Virol.* 69, 2240.

Orend, G., Knoblauch, M., Kämmer, C., Tjia, S. T., Schmitz, B., Linkwitz, A., Meyer zu Altenschildesche, G., Maas, J., and Doerfler, W. (1995). The initiation of de novo methylation of foreign DNA integrated into a mammalian genome is not exclusively targeted by nucleotide sequence. *J. Virol.* 69, 1226.

Reik, W., Collick, A., Norris, M. L., Barton, S. C., and Surani, M. A. (1987). Genomic imprinting determines methylation of parental alleles in transgenic mice. *Nature* 328, 248.

Saluz, H. P., Jiricny, J., and Jost, J. P. (1986). Genomic sequencing reveals a positive correlation between the kinetics of strand-specific DNA demethylation of the overlapping estradiol/glucocorticoid-receptor binding sites and rate of avian vitellogenin mRNA synthesis. *Proc. Natl. Acad. Sci. USA* 83, 7167.

Santi, D. V., Garrett, C. E., and Barr, P. J. (1983). On the mechanism of inhibition of DNA-cytosine methyltransferases by cytosine analogs. *Cell* 33, 9.

Stuhlmann, H., Jähner, D., and Jaenisch, R. (1981). Infectivity and methylation of retroviral genomes is correlated with expression in the animal. *Cell* 26, 221.

Sutter, D., and Doerfler, W. (1980). Methylation of integrated adenovirus type 12 DNA sequences in transformed cells is inversely correlated with viral gene expression. *Proc. Natl. Acad. Sci. USA* 77, 253.

Sutter, D., Westphal, M., and Doerfler, W. (1978). Patterns of integration of viral DNA sequences in the genomes of adenovirus type 12-transformed hamster cells. *Cell* 14, 569.

Toth, M., Lichtenberg, U., and Doerfler, W. (1989). Genomic sequencing reveals a 5-methylcytosine-free domain in active promoters and the spreading of preimposed methylation patterns. *Proc. Natl. Acad. Sci. USA* 86, 3728.

Vardimon, L., Neumann, R., Kuhlmann, I., Sutter, D., and Doerfler, W. (1980). DNA methylation and viral expression in adenovirus-transformed and -infected cells. *Nucl. Acids Res.* 8, 2461.

Wang, R. Y., Zhang, X. Y., and Ehrlich, M. (1986). A human DNA binding protein is methylation-specific and sequence-specific. *Nucl. Acids Res.* 14, 1599.

Weisshaar, B., Langner, K. D., Jüttermann, R., Müller, U., Zock, C., Klimkait, T., and Doerfler, W. (1988). Reactivation of the methylation-inactivated late E2A promoter of adenovirus type 2 by E1A (13S) functions. *J. Mol. Biol.* 202, 255.

Willis, D. B., and Granoff, A. (1980). Frog virus 3 DNA is heavily methylated at CpG sequences. *Virology* 107, 250.

DNA Repair

LAWRENCE GROSSMAN
The Johns Hopkins University

GLOSSARY

Endonuclease Nuclease that hydrolyzes internal phosphodiester bonds

Excision Removal of damaged nucleotides from incised nucleic acids

Exonuclease Nuclease that hydrolyzes terminal phosphodiester bonds

Glycosylase Enzymes that hydrolyze N-glycosyl bonds linking purines and pyrimidines to carbohydrate components of nucleic acids

Incision Endonucleolytic break in damaged nucleic acids

Ligation Phosphodiester bond formation as the final stage in repair

Nuclease Enzyme that hydrolyzes the internucleotide phosphodiester bonds in nucleic acids

Resynthesis Polymerization of nucleotides into excised regions of damaged nucleic acids

THE ABILITY OF CELLS TO SURVIVE HOSTILE ENVIronments is due in part to surveillance systems that recognize damaged sites in DNA and are capable of either reversing the damage or removing damaged bases or nucleotides, generating sites that lead to a cascade of events restoring DNA to its original structural and biological integrity.

Both endogenous and exogenous environmental agents can damage DNA. A number of repair systems are regulated by the stressful effects of such damage, affecting the levels of responsible enzymes, or by modifying their specificity. Repair enzymes appear to be the most highly conserved proteins, thus demonstrating their important role throughout evolution.

Either the enzyme systems can directly reverse the damage to form the normal purine or pyrimidine bases or the modified bases can be removed together with surrounding bases through a succession of events involving nucleases, DNA polymerizing enzymes, and polynucleotide ligases, which assist in restoring the biological and genetic integrity to DNA.

I. DAMAGE

As a target for damage, DNA possesses a multitude of sites that differ in their receptiveness to modification. On a stereochemical level, nucleotides in the major groove are more receptive to modification than those in the minor groove, the termini of DNA chains expose reactive groups, and some atoms of a purine or pyrimidine are more susceptible than others. As a consequence, the structure of DNA represents a heterogeneous target in which certain nucleotide sequences also contribute to the susceptibility of DNA to genotoxic agents. [See DNA and Gene Transcription.]

A. Endogenous Damage

Even at physiological pHs and temperatures in the absence of extraneous agents, the primary structure of DNA undergoes alterations. A number of specific reactions directly influence the informational content as well as the integrity of DNA. Although the rate constants for many reactions are inherently low, because of the enormous size of DNA and its persistence

ENCYCLOPEDIA OF HUMAN BIOLOGY, Second Edition, VOLUME 3.   Copyright © 1997 by Academic Press.   All rights of reproduction in any form reserved.

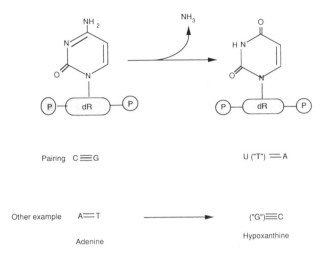

Pairing   C ≡ G   U ("T") = A

Other example   A = T   ⟶   ("G") ≡ C
                  Adenine   Hypoxanthine

FIGURE I   Deamination reactions have mutagenic consequences because the deaminated based cause false recognition.

in cellular life cycles, the accumulation of these changes can have significant long-term effects.

1. Deamination

The hydrolytic conversion of adenine to hypoxanthine (Fig. 1), guanine to xanthine, and cytosine to uracil-containing nucleotides is of sufficient magnitude to affect the informational content of DNA.

2. Depurination

The glycosylic bonds linking guanine in nucleotides are more sensitive to hydrolysis than the adenine and pyrimidine glycosylic links. The resulting apurinic (AP), or apyrimidinic, site is recognized by surveillance systems and, as a consequence, is repaired.

3. Mismatched Bases

During the course of DNA replication, there are those noncomplementary nucleotides that are incorrectly incorporated into DNA and manage to escape the editing functions of the DNA polymerases. The proper strand as well as the mismatched base is recognized and repaired.

4. Metabolic Damage

When thymine incorporation into DNA is limited through either restricted precursor deoxyuridine triphosphate (dUTP) availability or inhibition of the thymidylate synthetase system, dUTP is utilized as a substitute for thymidine triphosphate. The presence of uracil is identified as a damaged site and acted upon by repair processes.

5. Oxygen Damage

The production of oxygen radicals (superoxide or hydroxyl radicals) as a metabolic consequence as well as at inflammatory sites causes sugar destruction, which eventually leads to strand breakage.

B. Exogenous Damage

The concept of DNA repair in biological systems arose from studies by photobiologists and radiobiologists studying the viability and mutagenicity in biological systems exposed to either ionizing or ultraviolet irradiation. Target theories, derived from the random statistical nature of photon bombardment, led to the identification of DNA as the primary target for the cytotoxicity and mutagenicity of ultraviolet light. The photoproducts in DNA responsible for the effects of such irradiation have been attributed to 5,6-pyrimidine cyclobutane cis,syn dimers, 6,4-pyrimidine-pyrimidone dimers, and 5,6-water-addition products of cytosine (hydrates). In addition, most of the structural and regulatory genes controlling DNA repair in *Escherichia coli* were identified, facilitating the molecular characterization of the relevant enzymes.

1. Ionizing Radiation

The primary cellular effect of ionizing radiation is the radiolysis of water, which generates mainly hydroxyl radicals (HO·). HO· is capable of abstracting protons from the C-4' position of the deoxyribose moiety of DNA, thereby labilizing the phosphodiester bonds and generating single- and double-strand breaks. The pyrimidine bases are also subject to HO· addition reactions.

2. Ultraviolet Irradiation

Most ultraviolet photoproducts are chemically stable; their recognition provides direct biochemical evidence for DNA repair. The major photoproducts are 5,6-cyclobutane dimers of neighboring pyrimidines (intrastrand dimers), 6,4-pyrimidine-pyrimidone dimers (6–4 adducts), and 5,6-water-addition products of cytosine (cytosine hydrates).

3. Alkylation

Alkylation takes place on purine-ring nitrogens (cytotoxic adducts), the O^6 position on guanine, the O^4 positions of the pyrimidines (mutagenic lesions), and the oxygen residues of the phosphodiester bonds of the DNA backbone (biologically silent). Alkylating agents are environmentally pervasive, arising indirectly from

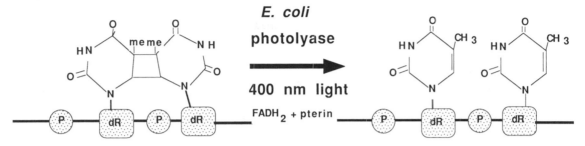

FIGURE 2 The direct photoreversal of pyrimidine dimers in the presence of visible light. FADH₂, reduced flavin adenine dinucleotide.

many foodstuffs and from automobile exhaust, in which internal combustion of atmospheric nitrogen results in the formation of nitrate and nitrites.

4. Bulky Adducts

Large, bulky, polycyclic, aromatic hydrocarbon modification occurs primarily on the N-2 and C-8 position of guanines, invariably from the metabolic activation of these large, hydrophobic, uncharged macromolecules to their epoxide analogues. The major source of these substances is from the combustion of tobacco, petroleum products, and foodstuffs.

II. DIRECT REMOVAL MECHANISMS

The simplest repair mechanisms involve the direct photoreversal of pyrimidine dimers to their normal homologues and the removal of O-alkyl groups from the O^6-methylguanine and from the phosphotriester backbone as a consequence of alkylation damage to DNA.

A. Photolyases (Photoreversal)

The direct reversal of pyrimidine dimers to the monomeric pyrimidines is the simplest mechanism (Fig. 2), and parenthetically it is chronologically the first mechanism described for the repair of photochemically damaged DNA. It is a unique mechanism characterized by a requirement for visible light as the sole source of energy for breaking two carbon–carbon bonds.

The enzyme protein has two associated light-absorbing molecules (chromophores), which can form an active light-dependent enzyme. One of the chromophores is reduced flavin adenine dinucleotide (FADH₂), and the other is either a pterin or a deazaflavin, capable of absorbing the 365- to 400-nm wavelengths required for photoreactivation of pyrimi-

dine dimers. It is suggested that photoreversal involves energy transfer from the pterin molecule to FADH₂ with electron transfer to the pyrimidine dimer resulting in nonsynchronous cleavage of the C-5 and C-6 cyclobutane bonds.

Enzymes that carry out photoreactivation have been identified in both prokaryotes and eukaryotes.

B. Alkyl Group Removal (Methyl Transferases)

Cells pretreated with less than cytotoxic or genotoxic levels of alkylating agents before a lethal mutagenic dose are more resistant. This is an adaptive phenomenon with antimutagenic and anticytotoxic significance. During this adaptive period, a 39-kDa *Ada protein* is synthesized, which specifically removes a methyl group from a phosphotriester bond and from an O^6-methyl group of guanine (or from O^4-methyl thymine) (Fig. 3). The O^6-methyl group of guanine is not liberated as free O^6-methyl guanine during this process but is transferred directly from the alkylated DNA to this protein; the Ada protein (methyl transferase) and an unmodified guanine are simultaneously generated. These alkyl groups specifically methylate cysteine 69 and cysteine 321, respectively, in the protein.

The methyl transferase is used stoichiometrically in the process (i.e., does not turnover) and is permanently inactivated. Nascent enzyme is, however, generated because the mono- or dimethylated transferase activates transcription of its own "regulon," which includes, in addition to the *ada* gene, the *alk*B gene of undefined activity and the *alk*A gene (which controls a DNA glycosylase). The latter enzyme acts on 3-methyl adenine, 3-methyl guanine, O^2-methyl cytosine, and O^2-methyl thymine. The methylated Ada protein can specifically bind to the operator of the *ada* gene acting

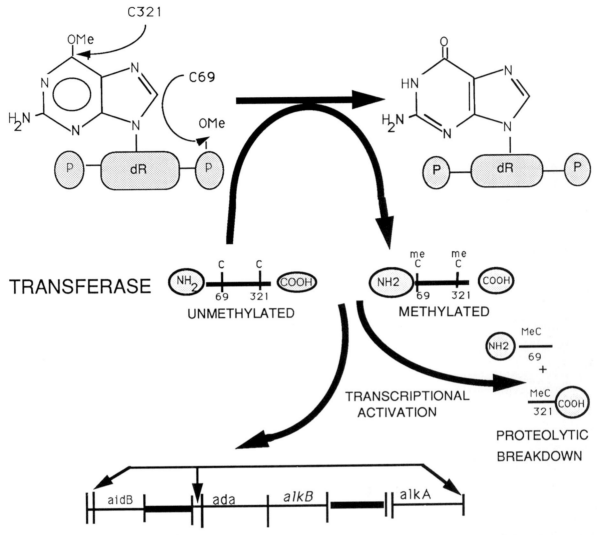

FIGURE 3 The direct reversal of alkylation damage removes such groups from the DNA backbone and the O^6 position of guanine. Such alkyl groups are transferred directly to specific cysteine residues on the transferase, the levels of which are influenced adaptively by levels of the alkylating agents. The methylation of the transferase inactivates the enzyme, which is used up stoichiometrically in the reaction. The alkylated transferase acts as a positive transcriptive signal turning on the synthesis of unique mRNA.

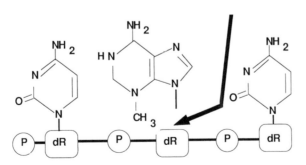

FIGURE 4 DNA glycosylases hydrolyze the N-glycosyl bond between damaged bases and deoxyribose, generating an apurinic (AP), or apyrimidinic, site.

as a positive regulator. Down-regulation may be controlled by proteases acting at two hinge sites in the Ada protein.

III. BASE-SPECIFIC RESPONSES

A. Base Excision Repair by Glycosylases and AP Endonucleases

Bases modified by deamination can be repaired by a group of enzymes called DNA glycosylases, which specifically break the N-glycosyl bond of that base and the deoxyribose of the DNA backbone generating

AP endonuclease

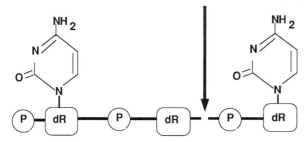

FIGURE 5 Endonucleases recognize AP sites and hydrolyze the phosphodiester bonds at either 3', 5', or both sides of the deoxyribose moiety in damaged DNA.

an AP site (Fig. 4). These are rather small, highly specific enzymes, which require no cofactor for functioning. They are the most highly conserved proteins, attesting to the evolutionary unity both structurally and mechanistically from bacteria to humans.

As a consequence of DNA glycosylase action, the AP sites generated in the DNA are acted upon by a phosphodiesterase specific for such sites, which can nick the DNA either 5' and/or 3' to such damaged sites (Fig. 5). If there is a sequential action of a 5'-acting and 3'-acting AP endonuclease, the AP site is excised, generating a gap in the DNA strand.

B. Glycosylase-Associated AP Endonucleases

An enzyme from bacteria and phage-infected bacteria, encoded in the latter case by a single gene (*den*V), breaks the *N*-glycosyl bond of the 5'-thymine moiety of a pyrimidine dimer followed by hydrolysis of the phosphodiester bond between the two thymine residues of the dimer (Fig. 6). This enzyme, referred to as the pyrimidine dimer DNA glycosylase, is found in *Micrococcus luteus* and phage T4-infected *E. coli*. This small, uncomplicated enzyme does not require cofactors and is presumed to act by a series of linked beta-elimination reactions.

An enzyme behaving in a similar glycosylase-endonuclease fashion but acting on the radiolysis product of thymine, thymine glycol, has been isolated from *E. coli* and is referred to as endonuclease III.

IV. NUCLEOTIDE EXCISION REPAIR

A. Prokaryotes

The ideal repair system is one that is somewhat indiscriminate and can respond to virtually any kind of

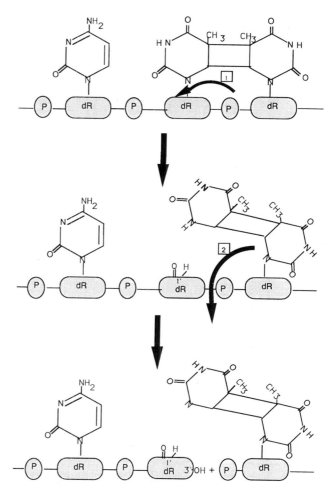

FIGURE 6 The same enzyme that can hydrolyze the *N*-glycosyl residue of a damaged nucleotide also hydrolyzes the phosphodiester bond linking the AP site generated in the first *N*-glycosylase reaction. UV, ultraviolet.

damage. Such a repair system has been characterized in *E. coli*, where it consists of at least six gene products of the *uvr* system. These proteins consist of the UvrA protein, which binds as a dimer to DNA in the presence of adenosine triphosphate (ATP), followed by the UvrB protein, which cannot bind DNA by itself. Translocation of the UvrAB complex from initial undamaged DNA sites to damaged sites is driven by a cryptic ATPase associated with UvrB, which is activated by the formation of the UvrAB-undamaged DNA complex. This complex is now poised for endonucleolytic activity catalyzed by the interaction of the UvrAB-damaged DNA complex with UvrC to generate two nicks in the DNA, seven nucleotides 5' to the damaged site and three to four nucleotides 3' to the same site (Fig. 7). These sites of breakage are invariant regardless of the nature of the damage. In the presence

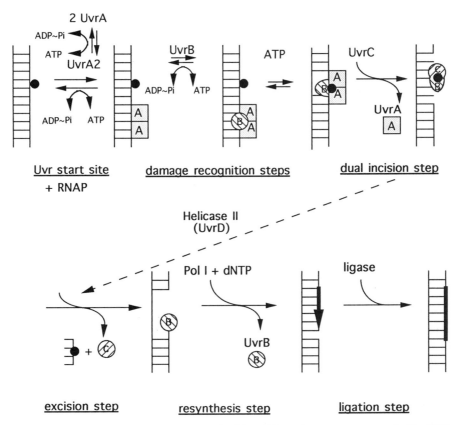

FIGURE 7 Nucleotide excision reactions. In this multiprotein enzyme system, the UvrABC proteins catalyze a dual incision reaction seven nucleotides 5′ and three to four nucleotides 3′ to a damaged site. The UvrA protein, as a dimer, binds to undamaged sites initially and in the presence of UvrB, whose cryptic ATPase provides the energy, is able to translocate to a damaged site. This preincision complex interacts with UvrC, leading to the dual incision reaction. The incised DNA–UvrABC requires the coordinated participation of the UvrD and DNA polymerase reactions for damaged fragment release and turnover of the UvrABC proteins. Ligation, the final reaction, restores integrity to the DNA strands. +RNAP-RNA polymerase.

of the UvrD (helicase III) and DNA polymerase I, the damaged fragment is released, accompanied by the turnover of the UvrA, UvrB, and UvrC proteins. The continuity of the DNA helix is maintained based on the sequence of the opposite strand. The final integrity of the interrupted strands is restored by the action of DNA polymerase I, which copies the other strand, and by polynucleotide ligase, which seals the gap.

The levels of the Uvr proteins are regulated in *E. coli* by an "SOS" regulon monitoring a large number of genes, which includes the *uvrA, uvrB*, possibly *uvrC*, and *uvrD* as part of the excision repair system; it also includes the regulators of the SOS system, the *lexA* and *recA* proteins; cell-division genes *sulA* and *sulB*; recombination genes *recA, recN, recQ, uvrD*,

and *ruv*; mutagenic bypass mechanisms (*umuDC* and *recA*); damage-inducible genes; and the lysogenic phage lambda. The lexA protein negatively regulates these genes as a repressor by binding to unique operator regions. When the DNA is damaged, for example, by ultraviolet light, a signal in the form of a DNA repair intermediate induces the synthesis of the recA protein. When induced, the recA protein acts as a protease assisting the lexA protein to degrade itself, activating its own synthesis and that of the recA protein as well as some 20 other different genes. These genes permit the survival of the cell in the face of life-threatening environmental damages. Upon repair of the damaged DNA, the level of the signal subsides, reducing the level of recA and stabilizing the integrity of the intact lexA protein and its repressive properties

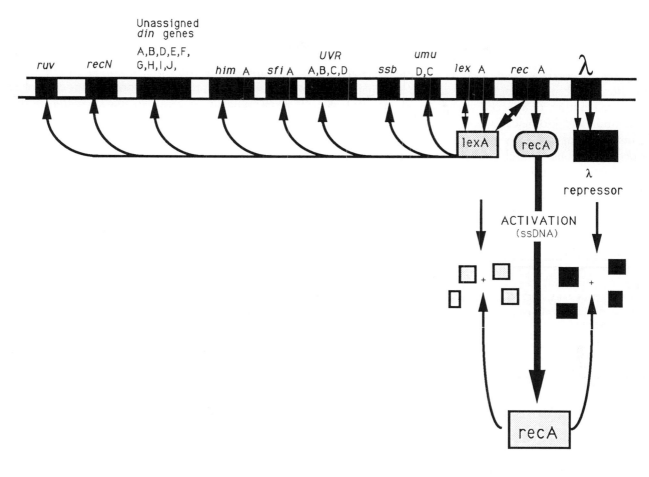

FIGURE 8 Regulation of the nucleotide excision pathway by the "SOS" system. The lexA and lambda repressors negatively control a multitude of genes, which are turned on when cells are damaged, leading to the overproduction of the *recA* protein, which assists in the proteolysis of the *lexA* and lambda proteins, thereby derepressing the controlled gene systems. When DNA is fully repaired, the level of *recA* declines, restoring the "SOS" system to negative control.

on all the other genes (Fig. 8); then the cell returns to its normal state.

B. Eukaryotes

Nucleotide excision repair in eukaryotes, although considerably more elaborate than that in prokaryotes, is mechanistically similar. Some 30 different repair and transcription proteins are required to excise damaged oligonucleotides. Human defects in nucleotide excision repair, such as xeroderma pigmentosum, trichothiodystrophy, and Cockayne syndrome, result in chronic diseases such as cancer and neurodegenerative disorders. Although the molecular mechanisms of nu-

cleotide excision repair in eukaryotes have not been completely documented, it is assumed that there are evolutionary similarities to the prokaryotic ones for the following two reasons. Human cells with mutations affecting DNA repair show a sensitivity to the same spectrum of damaging agents as *E. coli uvr* mutants. As a consequence, the mechanism proposed for the *E. coli uvr* system is probably similar to that of the human DNA nucleotide excision repair system. Bacterial and human uracil DNA glycosylases have been shown to have as much as a 73% amino acid sequence homology and may be the most highly conserved group of proteins, pointing to the evolutionary unity of repair mechanisms. However, inherent structural and physiological differences exist between bacterial and eukaryotic cells that must be mechanistically accommodated. The chromosomal structure imposes a steric challenge not only to a complicated multiprotein repair complex but also to the highly regulated nature of the cell cycle. Apparently, the structural and biological specificity associated with transcriptional processes limits DNA repair to those damaged regions of the chromosome undergoing transcription, and that damage in those quiescent regions is persistent.

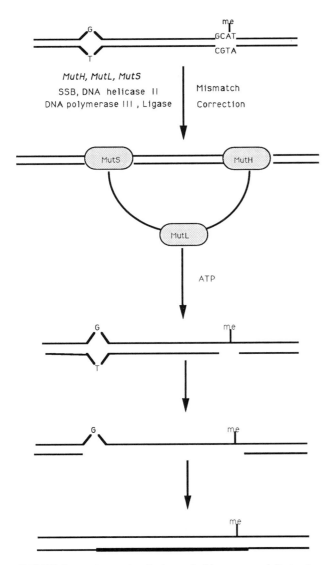

FIGURE 9 In the repair of mismatched bases, strand distinction can be achieved by the delay in adenine methylation during replication. It is the nascent, unmethylated strand that serves as a template for the incision reactions catalyzed by a number of proteins specifically engaged in mismatch repair processes.

V. MISMATCH REPAIR

A number of mechanisms recognize not damage, but rather mispairing errors that occur in all biological systems. In *E. coli*, mismatch correction is controlled by seven mutator genes: *dam, mutD, mutH, mutL, mutS, mutU (uvrD),* and *mutY*. In mismatch correction, one of the two strands of the mismatches is corrected to conform with the other strand (Fig. 9). Strand selection is one of the intrinsic problems in mismatch repair, and the selection is operated in bacterial systems by adenine methylation, which takes place at d(GATC) sequences. Because such methylation occurs after DNA has replicated, only the template strand of the nascent duplex is methylated. In mismatch repair, only the unmethylated strand is repaired, thus retaining the original nucleotide sequence. The MutH, MutL, and MutS proteins are involved in the incision reaction on this strand with the remainder of the proteins plus DNA polymerase III and polynucleotide ligase participating in the excision–resynthesis reactions.

Some sporadic colorectal tumors and most tumors developing in hereditary nonpolyposis colorectal cancer patients display a replication error phenotype (RER$^+$) in simple repeated sequences in microsatellite DNA. It has been demonstrated that the mutation rate of dinucleotide repeats in RER$^+$ tumor cells is 100-fold that in RER$^-$ tumor cells and that in *in vitro* assays the increased mutability of the RER$^+$ cells is associated with a significant defect in strand-specific mismatch repair. There is evidence that this is a true mutator phenotype existing in a subset of tumor cells, which is likely to lead to transitions and transversions in addition to microsatellite alterations.

BIBLIOGRAPHY

Ahn, B., and Grossman, L. (1996). "RNA Polymerase Signals UvrAB Landing Sites." *J. Biol. Chem.* **271**, 21453–21461.

Friedberg, E. C., Walker, G. C., and Seide, W. (1995). "DNA Repair and Mutagenesis." ASM Press, Washington, D.C.

Grossman, L., and Thiagalingam, S. (1993). Nucleotide excision repair, a tracking mechanism in search of damage. *J. Biol. Chem.* **268**, 16871–16874.

Lloyd, R. S., and Linn, S. (1993). "Nucleases" (S. Linn, R. S. Lloyd, and R. J. Roberts, eds.). Cold Spring Harbor Laboratory Press, Cold Spring Harbor, NY.

Modrich, P. (1991). Mechanism and biological effects of mismatch repair. *Annu. Rev. Gen.* **25**, 229–253.

Parsons, R., Li, G-M., Longley, M. J., Fang, W-H., Papadopoulos, N., Jen, J., de la Chapelle, A., Kinzler, K. W., Vogelstein, B., and Modrich, P. (1993). Hypermutability and mismatch repair in RER$^+$ tumor cells. *Cell* **75**, 1227–1236.

DNA Replication

ULRICH HÜBSCHER

Universität Zürich-Irchel

GLOSSARY

Deoxyribonucleic acid (DNA) Carrier of the genetic information, which is encoded in the sequence of bases. It is present in chromosomes and chromosomal material of cell organelles (e.g., mitochondria and chloroplasts) and also in some viruses. DNA is a polymeric molecule made up of deoxyribonucleotide repeating units (composed of the sugar 2-deoxyribose, phosphate, and a purine or pyridine base) linked by the phosphate group joining the $3'$ position of one sugar to $5'$ position of the next

DNA polymerase Enzyme that catalyzes the addition of deoxyribonucleotide residues to one end of a primer hybridized to a template strand. The new nucleotide has to fit to the template according to the Watson–Crick base pairing rule so that the template is faithfully copied.

DNA replication Duplication of DNA before cell division. Both mother templates must be copied

Double helix Double-stranded DNA held together by hydrogen bonds. Because of the antiparallel arrangement of the two strands, the DNA usually forms a right-handed helix structure.

Enzymes Catalytic proteins that mediate and promote the chemical processes of cell functions without being altered or destroyed

Okazaki fragments Short DNA fragments that are synthesized on one DNA strand during replication

Plasmids Extrachromosomal DNA in some bacteria. They may harbor antibiotic resistance genes and are important tools for recombinant DNA technology

Processivity Defines the numbers of monomers that are polymerized in one binding event by a polymerizing enzyme (e.g., by a DNA polymerase)

Semiconservative replication Duplication of DNA in a fashion in which the template strand is conserved and is used for an exact copy of a new complementary strand

Watson–Crick base pairing rule Double helix formed by hydrogen bonds between the bases guanine and cytosine or adenine and thymine

ANY LIVING ORGANISM HARBORS GENETIC information coding for the protein molecules that are responsible for structure and function of a cell. Most organisms bear this information (comparable to the "software" of a computer) in the form of a long thread called deoxyribonucleic acid (DNA). The components of a DNA molecule are phosphoric acid, the pentose sugar deoxyribose, and the four bases adenine (A), guanine (G), cytosine (C), and thymine (T). It is the linear order of the four bases that encodes for the genetic information of the DNA. A human cell contains about 3×10^9 bp in its nucleus corresponding to a total length of 2 m. Before cell division, which generates two daughter cells from a single mother cell, the DNA must be duplicated.

I. DNA REPLICATION IS A VITAL, UNIVERSAL BIOLOGICAL TASK

DNA replication is the process that guarantees a faithful duplication of the genetic information. This vital event is crucial for the correct maintenance of the genetic information. Let us assume a cell nucleus corresponds to a tennis ball; then the DNA would be a thread of 20 km (13 miles). The process of DNA replication enables duplication of the 20-km thread in a tennis ball at an enormous speed and with an extremely high accuracy. Finally, the complete distribution of these two DNA threads to the two daughter nuclei during mitosis is another extremely challenging task of DNA replication.

The mechanism of DNA replication depends on the chemistry of DNA. The DNA backbone is built by a series of phosphodiester bridges between the carbon 3 of one deoxyribose and the carbon 5 of the next. Every deoxyribose itself is bound to one of the four bases via its carbon 1 residue. Millions of these phosphodiester bridges form a long molecule. At one (called the 3′ end), the deoxyribose contains a free hydroxyl group at the carbon 3 and a free hydroxyl group at the carbon 5 residue on the other (called the 5′ end). The thermodynamically more stable form of the DNA is the double helix. The stability is guaranteed by the simple rule of base pairing between adenine and thymine or guanine and cytosine. In the helix the two strands have different free hydroxyl groups at their ends: at one end of the helix, one of the strands has a free carbon 3 whereas the other strand has a free carbon 5. The arrangement of the two strands is therefore antiparallel and has, according to these free hydroxyl groups, a direction of 5′ → 3′ on one strand and 3′ → 5′ on the other. [See DNA and Gene Transcription.]

Basic mechanistic features of DNA replication have emerged since the mid-1960s. Replication of double-stranded DNA is semiconservative. The two strands (called *mother strands*) serve each as templates for the synthesis of a new strand (called *daughter strand*). For this purpose, the DNA double helix has to be brought into single-stranded form before DNA synthesis. Because of the basic reaction mechanism of any DNA polymerase known (for details see Section III), the direction of synthesis of a DNA strand is always 5′ → 3′. The difference derives from the antiparallel arrangement of the DNA and the fact that the overall process of replication proceeds in the same direction on both strands. Of the two mother strands, one is continuously copied as a whole and is called the *leading strand* whereas the other is discontinuously copied in small pieces and is called the *lagging strand*. During DNA synthesis, the discontinuously copied lagging strand consists of many pieces of DNA with still attached short ribonucleic acid (RNA) segments, which are called *Okazaki fragments*. These primers serve the DNA polymerases as starting points, as no

DNA polymerase can start DNA synthesis *de novo* on a template, unless it is provided with a 3′-hydroxyl group of a previously synthesized primer with a complementary sequence to the template. Most primers in nature are short pieces of RNA. During replication the RNA primers are later removed and replaced by DNA; the Okazaki fragments are then joined one to another. As a result, both new strands must be synthesized with the same accuracy, copied entirely, and lead to two completely separated double strands (Fig. 1). [See DNA Synthesis.]

Considering these facts, it is not astonishing that DNA replication is a complex event. The logical consequence appears to be that a cell must possess many enzymes that can carry out all these steps in a concerted action.

II. DNA REPLICATION REQUIRES MANY DIFFERENT ENZYMES

A complex network of interacting proteins and enzymes is required for DNA replication. Much of our present understanding is derived from studies of the bacterium *Escherichia coli* and its viruses (called *bacteriophages*). The DNA replication system was also studied *in vitro* using genomes of bacteriophages (e.g., T7, T4, and φX174) or plasmids carrying the origin of bacterial replication as models. By genetical analysis, 25–30 proteins were isolated, and their functional tasks are understood today at an elementary level. These proteins work together in a coordinated order as multisubunit assemblies. These results served as a guideline for the search and the purification of similar proteins in eukaryotes. Here again, model systems for replication lead the way, especially with viral DNAs from simian virus 40, adenovirus, and herpesvirus.

For simplicity, this article concentrates exclusively on enzymes acting on a DNA in the process of replication (Table I). The structure that the DNA has at that time is designated as the replication fork. DNA replication *in vivo* likely occurs in an ordered and highly organized way, in which all the enzymes and proteins mentioned in Table I have their exact role in a replication complex called the replisome. Models have been proposed on how the enzymatic machinery might be spatially arranged at a replication fork. They were based on the idea that DNA polymerases dimerize and that the DNA loops on the lagging strand in a way that the "directionality" for the DNA polymerases is the same (Fig. 1). If one postulates that the replisome is fixed to structures in the nucleus, such as

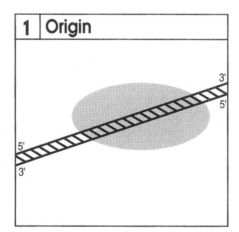

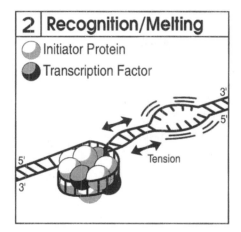

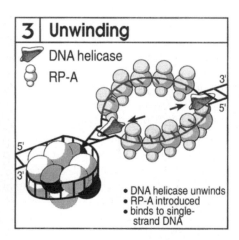

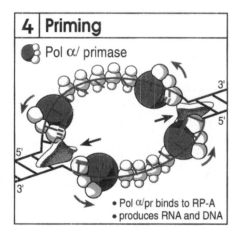

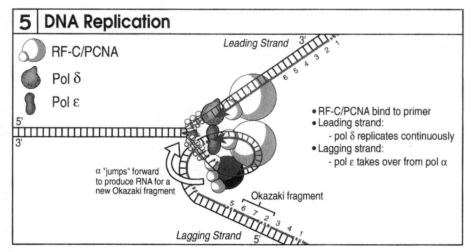

FIGURE 1 Current view of the eukaryotic replication fork. An origin of DNA replication (1) is recognized by specific proteins called initiator proteins and transcription factors (2) inducing a tension so that a neighboring region such as an AT-rich sequence preferentially opens, thus allowing a DNA helicase to unwind the DNA and replication protein A (RP-A) to cover single-stranded regions immediately (3). DNA polymerase α/primase (pol α) binds to RP-A and starts to produce RNA and short stretches of DNA (4). The arrows outside indicate the direction of movement of pol α whereas those inside indicate the direction of replication fork movement. The free 3′ OH groups are covered by a primer recognition complex made up by proliferating cell nuclear antigen (PCNA), replication factor C (RF-C), and ATP, thus preventing pol α to bind, but allowing pol δ and pol ε to form DNA polymerase holoenzyme complexes (5). Pol α "jumps" forward to produce RNA for a new Okazaki fragment (open arrow). Pol δ replicates the leading strand continuously and pol ε might take over gap filling from pol α. The final processing of the Okazaki fragment is not shown. For further details see text. From Hübscher and Spadari (1994).

TABLE I

Enzymes and Proteins at the Eukaryotic Replication Fork and
Their Likely Functional Roles[a]

Enzyme/protein	Role
Sequence-specific DNA-binding protein	Recognition and alteration of DNA at the origin of replication
DNA helicase	Transient melting of DNA
Single-stranded DNA-binding protein	Covering and protecting single-stranded DNA, interaction with DNA primase and DNA helicases
DNA primase	Synthesis of RNA primers
DNA polymerase	DNA synthesis
$3' \rightarrow 5'$-exonuclease	Proofreading for the DNA polymerase
DNA polymerase auxiliary proteins (=proliferating cell nuclear antigen and replication factor C)	Processivity, efficient primer recognition, clamp for DNA polymerase, recycling for DNA polymerase, selectivity for a particular DNA polymerase, interaction with proteins that regulate the cell cycle
RNase H, $5' \rightarrow 3'$-exonuclease (=maturation factor I)	Removal of RNA primers on Okazaki fragments
DNA ligase	Ligation of Okazaki fragments
DNA topoisomerase	Conversion of topological DNA isomers, segregation of DNA strands

[a]In many cases (e.g., DNA helicase, DNA polymerase, DNA ligase, DNA topoisomerase), several enzmyes are found, which have particular functional tasks in replication and other DNA metabolic events (e.g., repair and recombination).

the nuclear matrix, it would thread the DNA through itself. The assembly and the actions of a replisome might occur as follows (see Fig. 1):

1. An initiator protein binds to an origin of replication. The protein has to be in an activated form, e.g., properly phosphorylated, and it might act together with transcription factors. This leads to the formation of an *initial complex* that is able to alter DNA structures locally in its vicinity.

2. DNA helicases are thus attracted and *open the DNA* to produce a single-stranded DNA substrate.

3. The single-stranded DNA has to be immediately *protected* and *stabilized* by the single-stranded DNA-binding protein, called replication protein A (RP-A), which can help unwind the DNA by its unwinding

activity and, possibly, through its interactions with DNA helicase and DNA polymerase α/primase.

4. With its tightly associated DNA primase, DNA polymerase α initiates DNA synthesis at the leading strand, thus acting as an *initiating DNA polymerase* (Table II).

5. After initiation and very limited DNA synthesis, a *DNA polymerase switch* occurs most likely by the two DNA polymerase auxiliary proteins (Table III): proliferating cell nuclear antigen (PCNA) and replication factor C (RF-C), which cover the primer termini and thus "kick out" the DNA polymerase α/primase complex, which advances at the lagging strand of the replication fork to initiate another Okazaki fragment and then acts as a *lagging strand DNA polymerase*.

6. The DNA polymerase δ holoenzyme (DNA polymerase δ, PCNA, and RF-C) is then formed and acts on the leading strand as a very processive *leading strand DNA polymerase*.

7. At the same time on the lagging strand, a DNA polymerase ε holoenzyme (DNA polymerase ε, PCNA, and RF-C) may be formed and act as the *second lagging strand DNA polymerase*. Alternatively, it is conceivable that this step could be performed by DNA polymerase δ as well.

8. In both cases the two DNA polymerases (either the δ/ε dimer or the δ/δ dimer) would guarantee *proofreading* with their inherent $3' \rightarrow 5'$ exonuclease activities.

9. The initiator RNA at the lagging strand is removed by the combined action of $5' \rightarrow 3'$ exonuclease (=maturation factor 1) and RNase H1.

TABLE II

The Five Eukaryotic DNA Polymerases

DNA polymerase	Possible biological function
α	Initiating DNA polymerase: Initiation of the leading strand; initiation of the lagging strand
β	Repair DNA polymerase: Base excision repair Translesion DNA polymerase
γ	Replicative DNA polymerase of mitochondrial DNA
δ	Leading strand DNA polymerase: Replication of the leading strand; recombination
ε	Gap-filling DNA polymerase: Second lagging strand enzyme; nucleotide excision repair; recombination

TABLE III

Eukaryotic and Prokaryotic DNA Replication Accessory Proteins[a]

Functional component	Organism				
	Escherichia coli	Bacteriophage T4	Eukaryote (yeast to human)	Herpes simplex virus	Adenovirus
Processivity factor or sliding clamp	β subunit	gp45	PCNA	UL 42	AdDBP[b]
Clamp loader, brace protein, or matchmaker	γ-complex	gp44–gp62 complex	RF-C	?	?
Single-strand DNA-binding protein	SSB	SSB (*E. coli*)	RP-A	ICP 8	AdDBP

[a]Other DNA replication accessory proteins include $3' \to 5'$ exonuclease, DNA primase, RNase H, $5' \to 3'$ exonuclease, DNA helicases, and DNA ligases.

[b]The adDBP increases the strand displacement activity of the adenovirus-encoded DNA polymerase.

10. Okazaki fragment synthesis is possibly completed by DNA polymerase ε holoenzyme and the fragment sealed by DNA ligase I.

11. Topological constraints are released by DNA topoisomerase I upon movement of the replication fork.

12. Finally, the replicated DNA can be separated by DNA topoisomerase II.

III. DNA POLYMERASES ARE KEY ENZYMES FOR DNA REPLICATION

DNA polymerases are the main actors for polymerization of deoxyribonucleoside triphosphates (dNTP) at the replication fork. These enzymes use the two strands of the mother DNA as templates to guide their DNA polymerization according to the Watson–Crick base pairing rule. In addition, in order to start DNA synthesis the enzyme needs a $3'$-hydroxyl primer of a short nucleic acid strand (RNA or DNA) bound to the mother template strand. The four bases (A, T, G, and C) are required in the dNTP form. First, DNA polymerase binds to a template–primer junction. Then the corresponding dNTP binds to the active center of the enzyme, and the base is covalently attached as a deoxynucleoside monophosphate to the $3'$-hydroxyl group of the primer, releasing a pyrophosphate molecule. A divalent cation (usually magnesium) is required for the enzymatic activities of all known DNA polymerases.

Five different DNA polymerases are found in all eukaryotic cells (Table II).

A. DNA Polymerase α

DNA polymerase α is responsible for important roles in nuclear DNA replication. After initiation of DNA replication at the origin of replication (see Fig. 1), the DNA polymerase α primase complex initiates leading strand synthesis. Subsequently, Okazaki fragments are initiated at the lagging strand of the replication fork.

B. DNA Polymerase β

DNA polymerase β is the main repair DNA polymerase in the nucleus and appears to act in base excision repair. It was also identified to act in synthesis through DNA lesions.

C. DNA Polymerase γ

DNA polymerase γ is the replicase for replication of the mitochondrial DNA.

D. DNA Polymerase δ

DNA polymerase δ is involved in nuclear DNA replication, with a likely role at the leading strand of the replication fork. This DNA polymerase was first described in 1976 and was originally distinguished from DNA polymerase α by its $3' \to 5'$ exonuclease activity. Subsequently, all $3' \to 5'$ exonucleases containing DNA polymerases were called DNA polymerase δ. The "disadvantage" of this enzyme at this early time was its inability to replicate naturally occurring DNA templates. Today we know that the two auxiliary proteins PCNA and RF-C, are required for DNA polymerase δ to form a functional holoenzyme (see Table III).

E. DNA Polymerase ε

DNA polymerase ε is the most recently discovered enzyme. A function of DNA polymerase ε has originally been proposed in nuclear DNA repair, but it also appears to have an important role in nuclear DNA replication as well since its gene is essential in yeast. Experimental evidence suggested that DNA polymerase ε might be required as a second DNA polymerase at the lagging strand of the replication fork. DNA polymerase ε has also been identified in a protein complex that repairs double-stranded breaks and deletions by recombination, suggesting a role in DNA recombination as well. Furthermore, DNA polymerase ε appears to be involved in nucleotide excision repair. Thus it appears that DNA polymerase ε is involved in various gap-filling reactions such as Okazaki fragment DNA synthesis on the lagging strand of the replication fork, nucleotide excision repair DNA synthesis, and recombinational DNA repair synthesis.

The genes for the three replicative DNA polymerases (α, δ, and ε) have been isolated from various tissues, including human, and details at the DNA level are known. All three enzymes contain at least six conserved amino acid boxes that are required for important enzyme functions such as DNA binding, dNTP binding, $3' \rightarrow 5'$ exonuclease, polymerization, and pyrophosphate (ppi) binding. The conservation within a DNA polymerase species is the highest for DNA polymerase δ. The overexpression of functional DNA polymerases in the appropriate expression vectors and site-directed mutagenesis will tell us more about the structural properties and the functional anatomy of DNA polymerases as well as their interactions with DNA polymerase auxiliary proteins and other proteins of the replication apparatus.

As already indicated, DNA polymerases δ and ε contain a second enzymatic activity called a $3' \rightarrow 5'$ exonuclease. This enzyme can remove nucleotides that were wrongly incorporated by the DNA polymerase; this is performed through exonucleolytic removal of a nonbase-paired deoxyribonucleoside 5'-monophosphate. The accuracy of the DNA polymerase is thus increased about 100 times and provides to the overall accuracy of DNA replication. The $3' \rightarrow 5'$ exonuclease of DNA polymerase δ appears to be regulated

by the single-stranded DNA-binding protein RP-A and by the two auxiliary proteins PCNA and RF-C.

IV. DNA REPLICATION NEEDS THREE DNA POLYMERASES WITH A COMPLEX ARCHITECTURE

The function of a replication-specific DNA polymerase is to produce an accurate copy of the genetic material. The replicative complex, in which the DNA polymerase is active, must perform several functions other than that of the polymerization of dNTP. For example, it must take part in the unwinding of the two mother strands so that these can be used as templates. On the leading strand the DNA must be synthesized in a highly processive way, whereas on the lagging strand, cooperation with DNA primases in the synthesis of oligomeric RNA is required to prime the discontinuous DNA synthesis. Furthermore, on the lagging strand, the DNA polymerase and its associated proteins must be frequently recycled. The fidelity of DNA replication is also achieved by enzymatic control (e.g., by proofreading with the aid of a $3' \rightarrow 5'$ exonuclease). Finally, the entire replication machinery must be able to cooperate with other enzymes, proteins, and factors involved in the control of cell cycle and DNA replication.

During DNA replication, both strands have to be replicated with the same speed and accuracy. The antiparallelity of the mother DNA strands raises the fundamental biological question of how nature has untangled the difficult problem of continuous DNA synthesis on one strand and a discontinuous one on the other. A model has been proposed in which two DNA polymerases act coordinately as a dimeric complex at the replication fork (Fig. 1). One DNA polymerase replicates the leading strand continuously, whereas another replicates the lagging strand discontinuously. The lagging strand would bend back on itself to form a loop at the replication fork. This allows a dimeric DNA polymerase to move along both template strands in the "same" direction without violating the $5' \rightarrow 3'$ directionality rule. When the lagging strand DNA polymerase abuts on the Okazaki fragment synthesized during the previous round, the freshly synthesized DNA is threaded through the enzyme, allowing the DNA polymerase to recycle to a newly exposed single-stranded region more 3' to the previous priming site for initiation of a new Okazaki fragment. In this way, a dimeric DNA polymerase would guarantee an efficient and coordinated progression of the replication fork (see Fig. 1).

Studies on DNA polymerase III *Escherichia coli*, the main replicative enzyme, clearly indicate that forms of the multiprotein complex with several degrees of complexity and functional capacity can be isolated (Table IV). The simplest form is the DNA polymerase III core, which possesses three subunits, and the most complicated form is the DNA polymerase III holoenzyme dimer, which contains 22 polypeptides. The enzyme has an asymmetric form. The leading part is less complex than the lagging one. This is likely because the lagging part of the complex has to recycle frequently.

In eukaryotes it is also likely that DNA polymerases have a similar complexity. In addition, as already discussed, it appears that the three different DNA polymerases (α, δ, and ε) share the task of continuous and discontinuous DNA replication. All three enzymes have different degrees of complexities in analogy to *E. coli* (Table IV). An attractive hypothesis is that these DNA polymerases form DNA polymerase complexes, in which they can perform different functions, resulting in a functional replisome.

In summary, the replicative DNA polymerases have an extremely complex architecture and we are far from understanding how this machinery works *in vivo*.

TABLE IV
Complexities of Replicative DNA Polymerases

Species		Polypeptides
Escherichia coli		
DNA polymerase III core		3
DNA polymerase III′		9
DNA polymerase III*		18
DNA polymerase III holoenzyme (asymmetric dimer)		22
Eukaryotes, including human		
DNA polymerase α		
Core	1	
α/primase	4	4
DNA polymerase δ		
Core	2	
δ/PCNA/RF-C (holoenzyme)	10	10
DNA polymerase ε		
Core	2	
ε/PCNA/RF-C (holoenzyme)	10	10
Minimum of polypeptides required for the DNA synthesis machinery		24

V. CONCLUSIONS AND FUTURE PERSPECTIVES

As one of the most important events in any living cell, DNA replication is only understood at an elementary level. For unscrambling the puzzle of important biological events we need to understand the details. This should be achieved at several levels: (1) the complete inventory of components involved (actors as well as regulators) is required and (2) the details of the reaction intermediates must be elucidated. This can then at the third stage lead to the detailed rates of all transitions and to the component structures at the atomic resolution.

So far only the principal parts of the replication components are known and how they are assembled or regulated. Thus we have to learn now how they interact to become an efficient and regulatable engine. One can expect that information about functional details eventually leads to a better understanding of two very crucial unsolved problems in modern medicine: (1) we will comprehend better the small bandwidth between a normal and a malignant cell and (2) an understanding of differences between cellular and viral replication mechanisms (e.g., human immunodeficiency virus, herpes simplex virus, adenovirus, hepatitisivirus) will help to develop more potent and less toxic drugs.

The practical human and veterinary medicine already profits from drugs against replication enzymes from cells, bacteria, and viruses. Examples are inhibitors against DNA polymerases (human immunodeficiency virus reverse transcriptase, herpes simplex virus DNA polymerase), DNA topoisomerases (antibacterial gyrase inhibitors, anticancer drugs), and DNA ligase (anticancer drugs). By exploring more mechanistic details in basic sciences we can be optimistic about having more potent drugs within the next decade.

ACKNOWLEDGMENTS

The work in the author's laboratory has been supported by the Swiss National Science Foundation and the Kanton of Zürich.

BIBLIOGRAPHY

Alberts, B., and Miake, L. R. (1992). Unscrambling the puzzle of biological machines: The importance of the details. *Cell* **68**, 415–420.

Baker, T. A., and Kornberg, A. (1992). "DNA Replication," 2nd Ed. Freeman, San Francisco.

De Pamphilis, M. L. (ed.) (1996). "DNA Replication in Eukaryotic Cells: Concepts, Enzymes, Systems." Cold Spring Harbor Monograph, Cold Spring Harbor Press.

Fanning, E., Knippers, P., and Winnacker, E.-L. (1992). "DNA Replication and the Cell Cycle." Springer-Verlag, Berlin.

Hübscher, U., and Spadari, S. (1994). DNA replication and chemotherapy. *Physiol. Rev.* **74**, 259–304.

Hübscher, U., and Thömmes, P. (1992). DNA polymerase ε: In search for a function. *Trends Biochem. Sci.* **17**, 55–58.

Podust, L. M., Podust, V. N., Sogo, J., and Hübscher, U. (1995). Mammalian DNA polymerase auxiliary proteins: Analysis of replication factors C-catalyzed proliferating cell nucleus antigen loading onto circular double-stranded DNA *Mol. Cell Biol.* **15**, 3072–3081.

So, A. G., and Downey, K. (1992). Eukaryotic DNA replication. *Crit. Rev. Biochem. Mol. Biol.* **27**, 129–155.

Stillman, B. (1994). Smart machines at the DNA replication fork. *Cell* **78**, 725–728.

Stuckenberg, P. T., Turner, J., and O'Donnell, M. (1994). An explanation for lagging strand replication polymerase hopping among DNA sliding clamps. *Cell* **78**, 877–887.

Down Syndrome, Molecular Genetics

YORAM GRONER
ARI ELSON
The Weizmann Institute of Science

I. Introduction
II. Theoretical Aspects of Gene Dosage
III. Mapping of Chromosome 21
IV. Model Systems for Investigation of the Molecular Biology of Down Syndrome
V. Future Prospects

GLOSSARY

Cloned DNA Large numbers of DNA molecules identical to each other and to an ancestral molecule

Genomic clone DNA sequence derived from the genome and carried by a cloning vector

Karyotype Entire chromosome complement of a cell or an individual as determined by the spreading and staining of mitotic metaphase chromosomes

Nod-disjunction Failure of duplicated chromosomes to migrate to opposite poles during cell division

Phenotype Appearance and other characteristics of an individual, resulting mainly from its genetic constitution

Restriction fragment length polymorphism Subtle variations in the precise DNA sequence exist between individuals, some of which create or destroy sites of restriction endonuclease cleavage, thereby changing the length of restriction fragments

Robertsonian translocation Also called centric fusion, a special type of reciprocal translocation occurring between two acrocentric chromosomes that are joined in such a way that the two long arms are essentially preserved

Somatic cell hybrids Cell lines that are the result of fusion of human cells in culture to cells obtained from rodents; cells of this type gradually lose their human chromosomes and thus serve as an invaluable tool for correlating phenomena with the presence of specific human chromosomes

Transfection of eukaryotic cells Introduction of exogenous DNA into the cells

Transgenic mice Created by introduction of exogenous cloned DNA into the germ line of the animal

DOWN SYNDROME (DS OR TRISOMY 21) IS A SEVERE genetic disorder caused by the triplication of the distal part of the long arm of chromosome 21. It is assumed that the presence of extra copies of genes that reside in the triplicated region results in the synthesis of added amounts of gene products, which upset the normal biochemical pattern of existence; this in turn causes the physiological symptoms of the disease. One of the main thrusts in research into the molecular biology of DS is the precise definition of the minimal length of the triplicated fragment that causes the disease and the identification of the genes that reside in it. Other efforts are directed toward cloning of these genes and linking the results of their enhanced expression to the physiological symptoms of the syndrome through the use of several types of model systems.

I. INTRODUCTION

Down syndrome is the most common human genetic disease, occurring approximately once in every 700–1000 live births. It is thought to be the single most frequent cause of mental retardation in the industrialized countries. DS patients suffer from a wide range of symptoms. Most obvious among these are morphological defects (e.g., muscular flaccidity, short stature, and the epicanthic eye folds, which give rise to the eye shape characteristic to the syndrome). Patients are mentally retarded, and those who survive past their

ENCYCLOPEDIA OF HUMAN BIOLOGY, Second Edition, VOLUME 3.

mid-30s usually develop Alzheimer's disease. Premature aging and an increased incidence of leukemia and other hematological disorders are common in affected individuals, as are cardiac defects, a high susceptibility to infections, and several types of endocrinological disorders. The risk of a child being born with trisomy 21 sharply increases as maternal age progresses into the fourth decade of life, and because presently many couples in Western societies postpone parenthood until these age groups, the incidence of DS is expected to increase. [See Alzheimer's Disease; Genetic Diseases.]

The presently available techniques for prenatal screening for DS (amniocentesis and chorion villi biopsy) are costly and not without risk; they are therefore applied routinely only to at-risk pregnancies, with the result of being that most pregnancies are not screened at all for DS. Moreover, because of continuous improvements in all aspects of clinical treatment, the life expectancy of DS patients has tripled during the past two decades; middle-aged DS patients are no longer a rare occurrence. Thus, despite the medical and technological advances of recent years, the prevalence of DS individuals in society is not likely to be significantly decreased in the near future.

The disease was first described by the English physician John Langdon Down in 1866. Early descriptions of the patients used the term *mongol* in reference to their facial appearance; hence, the disease is sometimes referred to as *mongolism*. In 1959, Jerome Lejeune, Marthe Gautier, and Raymond Turpin of the Institute de Progenese in Paris recognized that a strong linkage exists between the presence of an extra copy of chromosome 21 and the occurrence of DS. It has since been found that this is true in about 95% of the cases of DS. The source of the extra chromosome in the great majority of such cases is a maternal nondisjunction event that occurs during the formation of germs cells at the time of the first meiotic division. In the remaining 5% of the cases, only part of chromosome 21 is present in triplicate, translocated onto another chromosome. Analysis of such cases gradually led to the conclusion in the mid-1970s that the presence of an extra copy of the entire chromosome was not an absolute prerequisite for DS: rather, trisomy of the distal half of the long arm, cytologically known as band 21q22, was sufficient. Band 21q22 comprises approximately a third of chromosome 21 and is estimated to contain several hundred genes.

Reports of DS patients whose chromosomes appear to be normal exist in the literature. Such cases are suggested to have arisen by an aberration too fine to be detected by karyotyping, the technique commonly used to verify a clinical diagnosis of DS. The normal appearance of the chromosomes in a DS patient can be explained by assuming that a duplication of a small segment of band 21q22 has taken place; alternatively, one or more recombination events may have occurred, resulting in very limited translocation of part of band 21q22 to another chromosome. Elucidation of such cases may serve to further narrow down the size of the pathological segment. However, the number of cases of this type that have been reported is small, and it is not clear whether these patients exhibit all the symptoms of "classical" DS.

Moving to the molecular level, the mechanism by which the presence of the extra chromosomal material causes the symptoms of DS is unknown. However, a direct proportionality may exist between the number of gene copies in the genomes of DS patients and the amount of gene products synthesized; stated somewhat bluntly, 50% more gene copies results in 50% more gene products. These products are thought to be perfectly normal, but their increased amounts are believed to upset the balance of biochemical reactions that ensure normal existence.

Down syndrome is set apart from most other genetic diseases. The latter are, for the most part, the result of a problem affecting a single gene. Usually, the resulting gene product is damaged; its capabilities to carry out its function are greatly reduced or even eliminated. In contrast, in DS the wide range of symptoms is caused by small (~50%) increases in the amounts of a large number of normal gene products, the exact number and identity of which are presently unknown.

However, as will become evident from the description of the nature of gene dosage effects, it is safe to assume that not all the genes that reside in band 21q22 contribute to the same extent in causing the DS phenotype; some should bear a more prominent role in this than others. Elucidating the molecular basis of DS is then an effort to identify the gene products whose augmentation by 50% or more causes the symptoms of DS; to explain the precise mechanisms by which the suspected gene products cause the symptoms of the disease; and to explain the possible interrelations between the gene products that might serve to amplify or ameliorate the severity of the symptoms.

II. THEORETICAL ASPECTS OF GENE DOSAGE

A primary gene dosage effect (GDE) is defined as the existence of a direct relation between the number of copies of a gene and the amounts of its product. The hypothesis that DS is caused by GDEs encompasses two levels of reasoning: that GDEs exist in DS and that they have physiological manifestations. As the main emphasis in DS is on GDEs of added DNA, there will be no mention of GDEs of decreased DNA content, as in monosomies or deletions.

The process by which a gene gives rise to its product includes many steps and is regulated by a complex system of checks and balances. In general, not all DNA encoding for products is used at all times or in all tissues; the expression of many gene products is specific for certain tissues or developmental stages in the life of an organism. An extra copy of a gene may exist without being expressed in some cells because of mechanisms that prevent the expression of the gene in inappropriate contexts. As the extra DNA includes the normal regulatory signals of the genes, the pattern of expression of the added DNA should be similar to that of the normal DNA complement.

A GDE resulting from the presence of an extra copy of a chromosome might be expected to result in increased synthesis of a large number of gene products. However, attempts to demonstrate such an increase by the technique of two-dimensional electrophoresis of total protein extracts from cells known to be trisomic have not generally succeeded. Some studies have shown that some gene products are present in enhanced amounts and in a manner that can be related to the existence of added chromosomal material. However, the extent of this phenomenon was far below what was expected, and the amount of some of the other gene products was, in fact, reduced. In contrast, an examination of messenger RNA in skin fibroblasts known to be monosomic, disomic, or trisomic for chromosome 21 revealed that transcription of chromosome 21-specific sequences was proportional to the copy number of chromosome 21. In addition, there have been several successful attempts to demonstrate enhanced synthesis of the products of several specific genes from band 21q22 in cells from DS patients (Table I) as well as in other types of cells. The failure to demonstrate gross changes in trisomic material by two-dimensional gel electrophoresis might stem from the relative insensitivity of this technique. An additional parameter that hinders detection of global GDEs is the timing of expression. It is possible that some genes (e.g., those responsible for defects already evident at birth) are expressed at high levels during embryonal development; analysis of postnatal material might not reveal any abnormalities in their expression.

At the physiological level, the presence of a gene dosage effect does not automatically mean that the normal biochemical pattern of the organism will be disturbed because the extra amount of a gene product can still be compensated for at the level of protein function. In certain cases the organism can adjust the amounts and activities of other parts of its metabolism to counteract these effects, creating what are known as *secondary GDEs*. An often-quoted example of a secondary GDE in the context of DS is the case of glutathione peroxidase, an enzyme known to be part of the mechanisms that defend cells from the damages of oxidative stress. The gene for this enzyme is located on chromosome 3; yet, its activity is increased in erythrocytes of DS patients, presumably as part of the defense mechanisms against the damage caused by the overproduction of hydrogen peroxide by copper-zinc superoxide dismutase (CuZnSOD), a chromosome 21-encoded enzyme for which a primary GDE has been demonstrated in DS. The reverse is also true: The physiological result of a GDE might be amplified, producing an end result much larger than that expected from an increase by a mere 50% in product synthesis. Again, using an example from chromosome 21, it has been demonstrated that some of the functions triggered in cells by the interferon-α receptor, for which a gene dosage effect of 50% has been demonstrated, are increased three- to eightfold when compared with cells devoid of such a GDE.

The physiological effects a GDE can have are intimately linked to the identity of the gene product itself, irrespective of whether the GDE is of primary or of secondary nature.

A. Enzymes

Despite the large body of information that has been accumulated on genes that code for enzymes, there is relatively little experimental information regarding the physiological effects of small changes in enzymatic activity. A theoretical explanation backed by some experimental evidence suggests that this is due to the very nature of enzymatic systems. Enzymes do not usually function alone, rather they are part of metabolic pathways and are linked to other enzymes through common substrates, products, or effectors. Multienzymatic pathways in which control of the pathway's flux is not concentrated at a single step have an inherent capacity to absorb small changes in the activity of one of their constituent enzymes so that the total flux through the pathway remains more or less unaffected. One result of this is that the sizes of pools of intermediate metabolites in the pathway are changed. As the number of enzymes in a pathway of the type described increases, the relative share of each

TABLE I

Genes Assigned to Chromosome 21[a]

Gene symbol	Name	E.C. number	Chromosomal location	Remarks
AABT	β-Amino acid transport system		Chromosome 21	Tentative
AD1	Familial Alzheimer's disease		21q21-21q22.1	
AML1	Acute myloid leukemia		21q22.1	
APP	β-Amyloid precursor protein		21q21	
ASNSL2	Asparagine synthetase (?)		Chromosome 21	Tentative
BCEI	Estrogen-inducible protein from breast cancer cell line		21q22.3	
CBS	Cystathionine β-synthetase	4.2.1.21	21q22	GDE in fibroblasts trisomic for chromosome 21
CD 18	β subunit of LFA antigen and related proteins		21q22.3, distal to ETS-2	
COL6A1 and COL6A2	α-1 and α-2 chains of type 6 collagen		21q22.3	
CRYA1	α-1 crystallin		21q22.3	
ERG	ETS-related gene		21q22.3	
ETS-2	Cellular homolog of the E26 avian transforming retrovirus		21q22.3	
GPXP1	Glutathione peroxidase (?)		Chromosome 21	Tentative
HSPA3	HSP70 heat shock protein		Chromosome 21	Tentative
HTOR	Hydroxytryptophan oxygenase regulator (?)		Chromosome 21	Tentative
IFNAR and IFNBR	Receptor(s) for interferon type α/β		21q21-21qter	GDE in fibroblasts aneuploid for various parts of chromosome 21
IFNGT1	Transducer (?) of interferon-γ receptor		Chromosome 21	
MX1, MX2	Myxovirus (influenza) resistance loci		Chromosome 21	
PAIS	Phosphoribosyl aminoimidazole synthetase	6.3.3.1	Chromosome 21	
PFKL	Liver-type subunit of phosphofructokinase	2.7.1.11	21q22.3	GDE in erythrocytes and fibroblasts
PGFT	Phosphoribosyl glycinamide formyl transferase	2.1.2.2	21q11.2-21q22.2	
PNY1	Anonymous polypeptides		Chromosome 21	PNY1, M_r 82 kDa, pI = 7.0
PNY2				PNY2, M_r 65 kDa, pI = 6.0
PNY3				PNY3, M_r 33 kDa, pI = 5.1
PNY4				PNY4, M_r 72 kDa, pI = 6.3
PNY5				PNY5, M_r 40 kDa, pI = 5.2
PRGS	Phosphoribosyl glycinamide synthetase	6.3.4.13	21q22.1	GDE in fibroblasts aneuploid for regions of chromosome 21
RNR4	Ribosomal RNA		21p12	
S14	Unidentified surface antigen		Chromosome 21	
S100-B	β subunit of S100 protein		21q22	
SOD1	Cu/Zn superoxide dismutase	1.15.1.1	21q22.1	GDE in fibroblasts, various types of blood cells

[a]GDE, gene dosage effect; M_r, molecular mass; pI, isoelectric point; E.C., enzyme commission classification number.

of its constituent enzymes in controlling the flux rate decreases. A pathway's buffering capacity is then expected to increase with the number of enzymes that make it up. One can therefore predict that enzymatic GDEs will succeed in provoking a physiological response only in special cases. A prime example of such a case are enzymes that do not abide by the just-mentioned generalization (e.g., those that catalyze rate-limiting reactions). As the control of flux through the pathway is determined in such cases by the activity of the extra-active enzyme, the degree of freedom left for the remaining enzymes to compensate for this is limited and cannot prevent the flux through the pathway from changing. Another exception to the rule is when changing an enzyme's activity results in the accumulation of a harmful metabolite. In such cases, the physiological effect can be noticed even when the total flux through the pathway is unchanged.

B. Receptors and Ligands

Many cellular processes are initiated by the process of a ligand binding to a receptor molecule located on the outside of the cell. The rules that describe receptor–ligand interactions can be used to show that the fraction of the receptors that bind ligand molecules at a given ligand concentration is independent of the amount of receptors available. Therefore, any increase in the number of receptor molecules on the cell as a result of a GDE will not change the fraction of receptors that actively bind ligand molecules. The absolute number of bound receptors will, however, increase by definition under such circumstances, and it is precisely this parameter that determines the strength of the signal that the cell receives. A cell with more receptors on its surface will receive a stronger signal at a given ligand concentration than a similar cell with less surface receptors; the two types of cells might then be expected to react differently to similar environmental signals. As the degree of binding of receptors to ligands is dependent on the concentration of ligand molecules, GDEs that increase ligand concentrations are expected to have physiological effects similar to those of increased receptor concentrations.

C. Assembly of Macromolecules

Some of the enzymes and structural components of cells are multimers constructed from several smaller protein molecules. Usually the identity and amounts of these subunits are strictly defined. In such cases, a GDE leading to an excess of a subunit could be expected to change the total amount of the multimer only if its amount was the limiting factor for the formation of the multimer in the first place. Cases exist, however, in which the precise composition of the multimer is not strictly defined, rather it is the result of the random association of whatever types of subunits are present. In such a case, increasing the amount of one type of subunit could result in its increased incorporation into the multimer with possible impacts on its properties and function. An example of the latter case is the glycolytic enzyme phosphofructokinase (PFK). PFK functions as an association of four subunits of which three different types exist. The subunits come together to form active PFK molecules in a random manner; the precise composition and properties of PFK tetramers in different tissues have been shown to reflect the levels of expression of the three types of subunits in these tissues. Because each of the subunit types is known to possess different biochemical properties, enhanced expression of any type of subunit could be expected to sway the composition and properties of the PFK tetramers in its direction. We should add at this point that the gene for the liver-type subunit of PFK has been mapped to the pathological segment of chromosome 21, and the composition of PFK tetramers in erythrocytes of DS patients has been shown to include a higher than normal fraction of L-type subunits. Similar effects are expected to occur in organized structural components of cells. Disruption of the equilibrium of subunit concentrations might change the overall geometry of the multimer and influence the spatial organization of cells and the organ in which they reside.

D. Regulatory Systems

Some of the functions that regulate gene expression are encoded by the genome itself. It is therefore not inconceivable that a GDE in a gene whose product regulates the expression of other genes could result in their incorrect expression. The expression of these genes could be either enhanced or repressed, as determined by the function of the amplified regulatory gene product. The finding mentioned earlier that the amounts of some proteins from total protein extracts from trisomic sources were reduced in comparison to normal controls might loosely be interpreted as indirect demonstration of the increased activity of a regulatory gene product that decreases the expression of these proteins. No direct evidence suggests that the activity of a regulatory gene product is affected by trisomy 21. However, the implications of such a finding would be wide as it could suggest a mechanism by which seemingly unrelated and unexplained secondary GDEs could be justified.

From all that has been described, it is clear that the results of the presence of extra copies of genes may not be noticeable. However, in many of these cases, although the organism might seem normal, its capabilities to maintain its biochemical balance under unfavorable conditions might be decreased. Stressful situations can therefore result in the realization of the potential harm that GDEs can have.

III. MAPPING OF CHROMOSOME 21

A. Theory

The human genome is estimated to contain 3×10^9 bp of DNA and in the order of magnitude of 100,000 genes. Chromosome 21 contains about 1.5% of this (i.e., around 45–50 million bp). Assuming that genes are equally distributed within the genome, chromosome 21 is expected to contain spine 1500 genes. Band 21q22, the so-called pathological segment of the chromosome, comprises about a third of the chromosome; its 15 million bp are therefore estimated to harbor several hundred genes. The task of identifying the genes that reside on chromosome 21 and mapping their precise location and order is formidable. At the present time, relatively few (i.e., only about 25) of the expected number of genes have been shown to reside on the chromosome (Fig. 1); most of these have been regionally mapped to band 21q22. However, the fine mapping of chromosome 21 is of prime importance to enable a better understanding of the causes of DS and of other diseases whose loci have been mapped to this chromosome (e.g., Alzheimer's disease). A precise map of the chromosome would allow a reduction in what is perceived to be the size of the fragment whose presence in triplicate causes DS; determining which genes reside in this segment is absolutely vital to the process of understanding the molecular mechanisms of the disease. [See Genes; Genome, Human.]

A prerequisite for using methods of molecular biology to map particular areas of chromosome 21 (and chromosomes in general) is the existence of specific, unique probes (i.e., DNA fragments that can be used to hybridize to, in this case, chromosome 21). Such a probe must be unique (i.e., it must hybridize only to the gene it was meant to locate and possibly also

to closely related sequences and not to repetitive sequences that exist in the genome). For the purpose of mapping, it is not necessary to know the biological function of the DNA used as a probe. Hence, probes can be either parts of known genes or they can be anonymous, in the sense that their identity, functions, or even their precise nucleotide sequence is unknown. Of the several dozen unique DNA probes for chromosome 21 that exist today, the vast majority are anonymous. This bias reflects the fact that anonymous probes are generally easier to isolate because probes that are fragments of known genes are usually the by-product of the time-consuming task of the cloning and identification of genes in which they reside. It is conceivable that some of the anonymous probes will be recognized in the future as parts of genes that have yet to be identified. [See DNA Markers as Diagnostic Tools.]

Once a gene is assigned to a chromosome, attempts are usually made to localize it to a particular subregion of the chromosome, thus producing a map of the chromosome. Two basic types of chromosome maps exist. Physical maps depict the actual physical distance between points on chromosomes; distances are measured in physical maps in base pairs of DNA or in multiples thereof. Genetic linkage maps, on the other hand, depict the degree of recombination between loci; the greater the degree of recombination between two points, the farther apart they will be placed on this type of map. Distances in genetic linkage maps are measured in crossing-over units, which are essentially a measure of the probability of a recombination event occurring between points on the map. When comparing a genetic linkage map with a physical map of the same region, we find that the order of the points on both maps is similar, although the distances between points are usually not. Because the recombination rate along a chromosome is not constant, there is no constant linear factor for comparing distances measured in both types of maps. [See Genetic Maps.]

A probe used for genetic linkage mapping must, aside from being unique and located on chromosome 21, recognize a site of restriction fragment length polymorphism (RFLP); some 40% of probes used for physical mapping fulfill this additional criterion. In such cases, the probe, in conjunction with the specific restriction endonuclease that generated the RFLP, can be used to mark the fragment and to measure its size. Each of the different lengths of fragments represents a different sequence version at a certain point in the genome and is essentially an allele at this locus. When

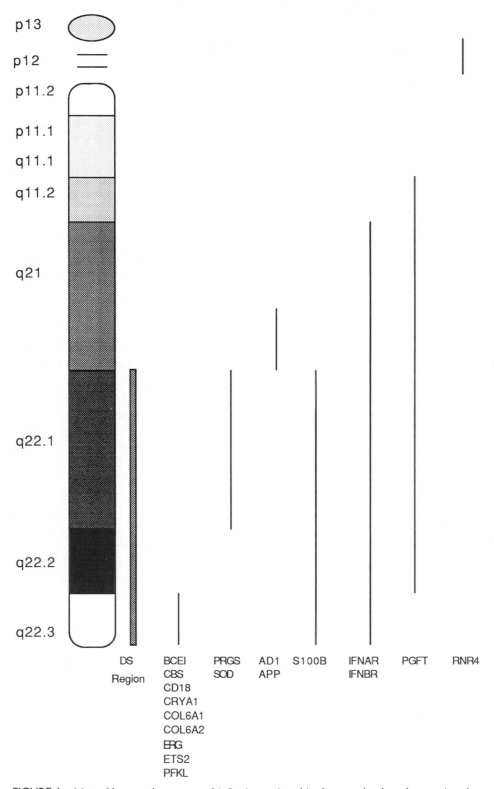

FIGURE I   Map of human chromosome 21. Loci, mentioned in the text, that have been assigned to but not mapped on the chromosome are HSPA3, IFNGT1, MX1, PAIS, PNY1-PNY5, and S14.

used to screen the RFLP patterns of populations, probes can be used to measure the frequency of each allele and to determine which allele exists in each individual. Following the identity of alleles along a pedigree can reveal the frequency of recombinations, which is the raw data from which genetic maps are constructed.

B. Brief Overview of Genes Mapped to Chromosome 21

Speculations abound as to the possible connections between the genes known to reside on chromosome 21 and particular symptoms of DS. Some of the suggested connections are reasonable; most, however, remain unproven. Moreover, ideas relating to one gene may not necessarily be applicable in a system beset by many GDEs whose combination may produce, cancel, or change the intensity of symptoms in an as yet unpredictable manner. Theoretical linkages between overexpression of specific genes and symptoms of DS are therefore generally not discussed here; connections demonstrated experimentally are described in the following sections. Table I lists the genes that have been assigned to chromosome 21; here we present a brief overview of some of them.

1. AD1—The Locus for Early-Onset Familial Alzheimer's Disease

Not all cases of Alzheimer's disease are thought to be caused by a genetic defect. However, in some large families the disease is transmitted in autosomal dominant fashion, indicating that at least in these cases a genetic factor is involved. Cases of familial Alzheimer's disease (FAD) are indistinguishable from "classical" Alzheimer's disease except for a somewhat younger age of onset. The genetic locus for FAD has been mapped to the region 21q21-21q22.1 by virtue of its linkage to anonymous DNA probes whose location on chromosome 231 is known. The FAD locus is thought not to be included in the DS pathological region of the chromosome.

2. APP—The Precursor of the β-Amyloid Protein

The β-amyloid protein is a 4200-DA polypeptide, which has been isolated from the amyloid cores of neuritic plaques that are one of the most prominent pathological findings in the brains of humans who have died of Alzheimer's disease. The β protein is derived from a much larger protein, the amyloid precursor protein (APP), which is expressed in many tis-

sues, including the brain, of both healthy and sick subjects. Recently, a meaningful degree of similarity was found between APP and nexin II, a molecule that seems to inhibit proteases that cleave and thereby activate several types of growth factor molecules. It is plausible that the biological function of APP is to participate in the regulation of the activity of certain growth factors. The APP gene is in close proximity to the locus for FAD and was mapped to the proximal edge of the DS region. Reports of recombination occurring between these two loci indicate that they are distinct from one another with the FAD locus farther away from the pathological segment. The finding that both the APP and FAD loci map so close to the DS pathological region is intriguing, as DS patients who survive into their fourth decade invariably develop symptoms and pathology identical to those of Alzheimer's disease patients.

3. BCEI—The Gene for an Estrogen-Inducible mRNA That Was Isolated from a Human Breast-Cancer Cell Line

The product of the gene is a small 84-amino-acid-long protein, which has been shown to be expressed by stomach mucosa cells. The exact function of this protein is unknown, although a similar porcine protein inhibits gastrointestinal motility and gastric secretion.

4. CBS—Cystathionine β-Synthase

This is an enzyme of sulfur amino acid metabolism, which catalyzes the condensation of homocysteine and serine to form cystathionine, along the pathway converting methionine to cysteine. Comparison of the physical symptoms of patients of homocystinuria, an inborn error of metabolism that is most commonly caused by deficient CBS activity, with those of DS patients led to the conclusion that some of the physical symptoms characteristic of both diseases could be viewed as opposites of one another. This led to the suggestion by Jerome Lejeune in 1975, some 10 years before the assignment of the CBS gene to chromosome 21, that abnormalities in CBS activity could play a part in causing the symptoms of DS.

5. CD18—The Gene for the β Subunit of Lymphocyte Function–Associated Antigen (LFA-1) and Related Proteins

The LFA-1 molecule is intimately involved in mediating adhesive interactions between several species of cells involved in the immune reaction. Humans suffering from a deficiency in LFA-1 and its related antigens suffer from serious defects in the adhesive-dependent functions of B and T lymphocytes, monocytes, and granulocytes and from recurrent life-threatening infections.

6. COL6A1, COL6A2—Two Genes Coding for the α-1 and the α-2 Polypeptide Chains of Collagen Type 6

Collagens are a large family of trimeric, triple-helical proteins that form fibers in the extracellular matrix. Type 6 collagen is a ubiquitous structural protein, yet its precise functions are unknown. It is also a somewhat unique member of the collagen family, as more than two-thirds of each of its three constituent polypeptides are globular domains. [See Collagen, Structure and Function.]

7. CRYA1—The Gene for α-1 Crystallin

Crystallins are major structural proteins found in lenses of vertebrates. Overexpression of this gene may be connected to the early cataracts that DS patients suffer from.

8. ETS2—The Gene for One of the Two Human Cellular Homologs of the Transforming Avian Erythroblastosis Retrovirus E26

Children affected with DS are prone to develop acute leukemia or to undergo a transient leukemia-like reaction. In acute myelogenous leukemia of subtype M2, a translocation of sequences from band 21q22 to chromosome 8 exists in many of the cases. However, although mapping studies have determined that apparently the EST2 locus is contained in the segment that is translocated, the precise point of chromosome breakage is far removed, on a genetic scale, from the ETS2 gene. This gene is therefore not rearranged in this translocation.

9. ERG—The ETS-Related Gene

This gene codes for at least two proteins that share limited domains of similarity with the ETS-2-encoded protein. Like the ETS-2 gene, the ERG gene has been shown to be translocated to chromosome 8 in many patients with acute myelogenous leukemia of subtype M2. Although genetic linkage studies have mapped the ERG gene to a point closer to the precise point of chromosome breakage than the ETS-2 gene, it too has been shown not to be rearranged in this translocation.

10. HSPA3—A Possible Gene for the HSP70 Heat-Shock Response Protein

Cells in culture respond to heat shock and other stimuli by the induced synthesis of a small number of specific proteins and by repression of other genes that are normally active. The precise function of heat-shock proteins is unknown, but it is held that they are part of the mechanism that defends the cell in situations of stress. The human genome contains several copies of the HSP70 gene. The assignment of the HSPA3 gene to chromosome 21 is somewhat uncertain, and it seems that although chromosome 21 is involved in the expression of heat shock proteins, the manner by which this is done is unclear. [See Heat Shock.]

11. IFNAR, IFNBR—The Receptor for Interferon α and β, Respectively

Interferons are proteins that are produced by many different types of cells in response to various stimuli, including viruses and polynucleotides. Binding experiments suggest that interferons α and β bind to the same receptor, but it has not been definitely proven that the IFNAR and IFNBR loci are one and the same. GDEs for the receptor have been observed in fibroblasts with varying copy numbers of chromosome 21. The cDNA for the interferon α/β receptor has been cloned. Its expression in mouse cells conferred interferon α/β sensitivity on them, although at levels lower than expected. It is therefore not inconceivable that another locus involved in response to these interferons is encoded on chromosome 21. [See Interferons.]

12. IFNGT1—A Locus Required for the Ability of Cells to Respond to Interferon γ

This locus is distinct from the interferon γ receptor itself, which has been convincingly mapped to chromosome 6. Experiments using hamster–human hybrid cell lines show that the presence of chromosome 6 is sufficient for the binding of interferon γ to cells. However, to confer a biologically measurable response to interferon γ (the induction of the major histocompatibility complex antigens), the presence of chromosome 21 is also required. It has been therefore suggested that chromosome 21 encodes a component of the mechanism for cellular response to interferon γ that is distinct from the receptor itself, possibly a transducer.

13. MX1—Myxovirus (Influenza) Resistance Locus

This locus encodes a protein that is induced by interferons of type α and β. Its function is unknown, although it has been shown that a homologous interferon-induced protein in mice is responsible for protection against influenza virus. [*See* Influenza Virus Infections.]

14. PRGS, PGFT, and PAIS—Genes for the Enzymes That Catalyze Three of the Reactions of the Purine Biosynthetic Pathway

Several lines of evidence suggest that all three enzymatic activities are encoded by one gene, probably located in the DS pathological segment of chromosome 21. A GDE has been reported for PRGS in human fibroblasts trisomic for chromosome 21, and DS patients have been known for a long time to suffer from impaired purine metabolism. However, as none of these enzymatic activities is considered rate-limiting in purine biosynthesis, the linkage between their possible overproduction in DS and the disease's symptoms remains to be demonstrated.

15. PKFL—The Liver-Type Subunit of Phosphofructokinase

Phosphofructokinase catalyzes one of the major rate-limiting reactions in glycolysis, an essential pathway in the use of sugar energy in cells. GDEs for PKFL have been demonstrated in erythrocytes and in fibroblasts obtained from DS patients, and the composition of PFK from such erythrocytes shows an increased proportion of the liver-type subunit.

16. PNY1, PNY2, PNY3, PNY4, PNY5—Five Polypeptides of Known Size and Isoelectric Points but of Unknown Functions, Which Have Been Assigned to Chromosome 21

These are based on two-dimensional electrophoresis of extracts of a human—mouse hybrid cell line that contains chromosome 21 as its only human complement.

17. RNR4—A Locus for Ribosomal RNA Mapping to Band 21p12

This probably is not involved in causing DS.

18. S14—A Surface Antigen, the Identity or Function of Which is Unknown

The precise subregion of the chromosome in which it resides is also unknown.

19. S100B—The β Subunit of the S100 Protein, a Calcium-Binding Protein of Unknown Function

It is found mostly in the nervous system of vertebrates, predominantly in brain glial cells.

20. SOD1—Copper-Zinc Superoxide Dismutase

This enzyme catalyzes the reaction that transforms superoxide anion radicals into hydrogen peroxide as part of the organism's defense mechanisms against damage by free radicals. The hydrogen peroxide produced, which is in itself toxic, is disposed of by other enzymatic and nonenzymatic reactions. GDEs have been demonstrated for this enzyme in various types of blood cells, fibroblasts, and brain samples obtained from patients with DS. This gene was the first chromosome 21–encoded gene that was cloned and used in studies of GDEs in transfected cells and transgenic mice.

21. Several Other Genes Have Been Assigned to Chromosome 21

These, however, have not been regionally mapped.

Comparison of the findings obtained from physical maps to those of genetic maps of chromosome 21 reveals that as in other chromosomes, a large degree of recombination takes place near the distal part of the chromosome in band 21q22.3. Consequently, this region, which spans about 10% of the physical-cytogenetic length of the chromosome, takes up some 40% of its genetic length. It is of interest to note that the majority of chromosomal breaks in chromosome 21 that are associated with translocation of segments to other chromosomes cluster in band 21q22, especially in subband 22.3.

The distal region of band 21q22.3 seems to contain more genes than expected. Some two-thirds of all unique probes assigned to and regionally localized on chromsome 21 map to band 21q22; of these, half map to band 22q22.3, as exemplified by the data in Table I. The clustering of genes to this region correlates well with other findings. Genes are generally thought to cluster in parts of the genome that are GC rich (i.e., that contain a higher than average proportion of guanidine and cytidine residues); actively transcribed genes are believed to be found near regions of DNA that are rich in the dinucleotide CG and that are undermethylated. Band 21q22 fulfills both of these criteria, further attesting to its "gene-rich" character. It is therefore possible that the DS pathological segment contains more genes than can be estimated by assuming an even distribution of genes throughout the human genome.

As mentioned previously, there are several examples of DS patients whose karyotype seems normal. This result is attributed to chromosomal aberrations too fine to be detected by karyotyping and is therefore limited in size to 2000–3000 kb of DNA. As this length is about a fifth of the size of band 21q22, a clear definition of the boundaries of such fine aberrations can serve to trim down the size of the pathological segment and make its analysis more manageable. Research into this question has so far suggested that triplication of the part of band 21q22 that is composed of bands 21q22.2 and 21q22.3 is linked to the occurrence of DS, although the precise borders of this region are ill-defined at present. A major drawback to this approach is the precise clinical diagnosis of DS. The disease manifests itself in a large possible number of clinical and morphological symptoms, not all of which are expressed in each and every patient. In fact, the karyotypic finding of an extra copy of part of or all chromosome 21 is generally accepted as the ultimate proof that the patient in question does indeed suffer from DS. In the absence of this aid to the clinical diagnosis, it is feared that different patients with varying karyotypically undetectable triplications and somewhat different symptoms would all be similarly categorized as suffering from DS. Correcting for the variability in the disease's symptoms is extremely difficult but is necessary when interpreting the results of this type of studies.

IV. MODEL SYSTEMS FOR INVESTIGATION OF THE MOLECULAR BIOLOGY OF DOWN SYNDROME

Two main reasons motivated efforts to develop a suitable system for studying the molecular genetics of the syndrome. The first one stems from difficulties attendant in research on humans. Most of the pathological consequences of trisomy 21 are manifested during fetal development; research on human subjects, especially *in utero*, is ethically complicated and practically impossible. Therefore, attention was turned toward the development of model systems. A second reason has to do with the large number of genes residing in the pathological segment and the consequent need to identify and sort out the quintes-

sence. It is not clear how many genes are involved in determining the characteristic DS phenotype and which one is doing what. Is overexpression of individual genes responsible for certain features associated with the syndrome (e.g., the high incidence of leukemia, the large protruding tongue, the low blood serotonin)? Or does imbalance in the expression of several genes act in a nonspecific fashion to produce the DS phenotype? A model system should be of help in answering these questions.

To date, two types of model systems have been developed: (1) a cellular system, consisting of cultured cells overexpressing candidate genes from chromosome 21, and (2) an animal model, employing either mice with trisomy 16 (animals in which many of the genes homologous to human genes from the region 21q22 are triplicated) or transgenic mice, carrying a gene from chromosome 21 in their genome and producing increased quantities of the gene product.

A. Cellular Model System: Transformed Cultured Cells Expressing Human Genes from the DS Locus

Recombinant DNA technology has made possible the expression of individual genes in established cell lines. This methodology offers a relatively simple approach for investigating GDEs at the cellular level and has already led to some understanding of how imbalanced expression of the genes may contribute directly or indirectly to the Down phenotype. [*See* DNA Technology in Disease Diagnosis.]

When a particular gene derived from the DS segment is introduced into cultured animal cells, the recipient cells resemble trisomy 21 cells except for one important difference: The imbalance is limited to one particular gene rather than the whole chromosome. A cellular system of this type permits the study of the biochemical effects of the altered dosage of a particular gene in a defined background, irrespective of the overexpression of other genes from chromosome 21.

Although studies concerning the GDEs of a single gene have already been conducted on several candidate genes, including ETS2, CBS, APP, and PFKL, the first and most detailed ongoing study is concerned with the gene encoding CuZnSOD, a key enzyme in the metabolism of oxygen free radicals. Overexpression of the CuZnSOD gene, because of gene dosage, may disturb the steady-state equilibrium of active oxygen species within the cell, resulting in oxidative damage to biologically important molecules. In particular, the polyunsaturated fatty acids of membranes may be affected in a process known as lipid peroxidation. Because brain function is highly dependent on membrane interactions and because the brain might be particularly susceptible to lipoperoxidation damage, it was suggested that such a mechanism may in part be responsible for the mental retardation, hypotonia, and Alzheimer's disease pathology associated with the DS phenotype. Experimental studies were conducted by introducing a cloned human CuZnSOD gene into the rat PC12 cell line, which possesses characteristics of neuronal cells grown in culture. The exogenously introduced human gene was stably integrated into the host chromosome, giving rise to increased amounts of enzymatically active CuZnSOD. While outwardly maintaining their response to nerve growth factor and their typical appearance of cultured neurons, the cells expressing the extra gene had a greatly reduced capacity to take up neurotransmitters. Neurotransmitters transfer the signals from one neuron to the next at the junction between them—the synase. Following detailed analysis of the phenomenon, it was discovered that the vacuoles (called chromaffin granules) responsible for accumulating neurotransmitters in the transformant-CuZnSOD cells have a lesion in their membrane, possibly caused by lipid peroxidation, which prevents them from taking up the neurotransmitters at the normal rate. This deficiency could have important consequences for neurons in the brain that use a similar organelle (called the synaptic vesicle) for accumulation of neurotransmitters. If a released transmitter substance persists for an abnormally extended period, new signals cannot get through at the proper rate. This observation demonstrates that even at the cellular level, an imbalance in the expression of the CuZnSOD gene has a deleterious effect, which, if it occurs in the central nervous system, would produce alterations in neuron function, which would impair the transduction of signals and mimic the deficiencies apparent in DS.

A cellular model system of this type was also used in studies on the possible involvement of the APP gene in the neurodegenerative process characterizing the pathology of Alzheimer's disease, which constitutes a clinial symptom in DS patients older than the age of 40. In this case, PC12 cells were transfected with portions of the gene for the human APP, and colonies of cells containing the human DNA integrated into their genome were obtained. It was found that the PC12 cells expressing the APP gradually degenerated when induced to differentiate into neuronal cells by treatment with nerve growth factor. This observation indicates that a peptide derived from the APP gene product is neurotoxic and, therefore, overexpression of this gene in DS may result in the accumulation of a neurotoxic peptide in the brains of affected individuals.

B. Animal Models of DS

The examples described earlier illustrate the usefulness of the cellular model for studies of the biochemical events resulting from gene dosage. Exploitation of this system will further increase as more candidate genes assigned to the 21q22 segment are cloned. However, this type of model suffers from a serious drawback in that it does not address the more complex issues of development and morphogenesis (i.e., the physiological consequences of gene dosage at the level of the whole animal). The appropriate model for that purpose is obviously an animal model, preferably one that shares a large number of biological similarities with humans. For practical reasons the mouse has been the animal of choice.

1. Trisomy 16 and Chimeric Mice

Genetic homology exists between mouse chromosome 16 and human chromosome 21 (i.e., many of the genes residing at the DS region 21q22 of human chromosome 21 are found on the distal segment of mouse chromosome 16). Mice trisomic for this chromosome were therefore considered as an animal model system for studies of DS. Trisomy 16 mice are produced by mating normal females with males carrying two different Robertsonian translocations. Trisomy 16 mice exhibit a number of phenotypic characteristics observed in DS (e.g., the endocardial cushion defect, the aortic arch defect, and a shortened neck), and many informative studies are being performed on these animals. However, one of the major limitations of this model is that only few trisomic 16 fetuses survive to birth; they usually begin to die at day 14 of gestation, whereas those that are born alive survive but a few days. Mouse chromosome 16 constitutes a much larger portion of the genome than does human chromosome 21, thus the degree of genetic imbalance is considerably more extensive and results in fetus lethality. Therefore, developmental studies as well as attempts to identify putative ameliorative therapies are precluded, and the major research emphasis involving trisomy 16 mice has been placed in studies confined to the cellular level. Cell cultures of various types have been established from trisomic fetal or neonatal mice, and studies on growth kinetics, life expectancy, and the sensitivity of various receptor systems have been carried out. In general, cells explanted from trisomy 16 tissues grow poorly in primary culture; trisomy 16 dorsal root ganglion cells degenerate and exhibit exaggerated electrical membrane properties.

To circumvent the early lethality of trisomy 16 fetuses and to obtain both survival and postnatal development, an alternative procedure was developed in which chimeric mice are produced having both trisomic and normal cells. Such chimeras are formed by fusing together an early embryo exhibiting trisomy 16 and a normal embryo. Chimeras of this type could be considered analogous to humans that are mosaic for trisomy 21 because they have both trisomic and normal cells within their tissues. In some chimeric mice, the proportion of trisomy 16 cells in the brain was as high as 50–70%. Studies on these animals revealed neurochemical abnormalities, as well as altered behavior (i.e., increased activity during the dark part of the light cycle with greater distance traveled, increased speed of movement, and excessive grooming activity).

Although the trisomy 16 and the chimeric 16 mice provide the opportunity for anatomical and physiological studies during gestation that cannot be performed in humans, this animal model leaves much to be desired, not only because of the early lethality mentioned earlier, but also because of the excess in genetic imbalance. Mouse chromosome 16 is considerably larger than human chromosome 21 and contains many genes, the human analogs of which reside on chromosomes other than 21. Mouse chromosome 16 represents 3.9% of the haploid autosomal complement of the mouse genome, whereas chromosome 21 constitutes only 1.9% of the human complement. In addition, several genes mapped at the DS region of chromosome 21 have been localized to chromosomes other than the mouse 16. Therefore genes that might be contributing to the etiology of DS are not implicated in the phenotype of the mouse trisomy 16. For these reasons, a better model for DS-related studies would be trisomic mice, which contain only a portion of chromosome 16, the segment homologous to the DS region 21q22, or, even better, a triplication of one or several candidate genes from the DS locus. Trisomy mice of this kind will certainly have a better survival capacity and will also permit investigating the participation of individual genes in the DS phenotype. Such trisomy mice were recently obtained by introducing specific cloned genes into the mouse germ line.

2. Transgenic Mice Carrying Chromosome 21–Encoded Human Genes

Gene transfer into mice, leading to the creation of so-called transgenic mice, can be achieved by microinjecting a foreign gene into one of the pronuclei present in every fertilized egg. The exogenously introduced DNA becomes stably integrated into the mouse chromosome, and the resultant embryo develops into a mouse that carries an extra gene and transfers it to subsequent generations in a Mendelian fashion.

Advances in recombinant DNA techniques facilitated the development of transgenic mice as in *in vivo* model for genetic diseases. Transgenic mice harboring candidate genes from the DS region have an advantage over cultured cells with transfected genes in that they are closer to the natural situation; the inserted gene is present in every cell of the animal, and its influence is manifested throughout its entire developmental history. By overexpressing individual genes in transgenic mice, it might be possible to dissect the trisomy 21 phenotype, gene by gene. The first strains of transgenic mice harboring a candidate gene from the DS region were constructed in parallel with the CuZnSOD-cellular system, using the human CuZnSOD as the transgene.

3. Transgenic-CuZnSOD Mice

Following the interesting observation of diminished uptake of neurotransmitters by cultured neuronal cells overexpressing the CuZnSOD gene, a cloned DNA segment containing this gene was microinjected into the male pronucleus of fertilized mouse eggs, and several strains of transgenic mice that carry the gene were obtained. These animals expressed the transgene in a manner similar to that of humans and showed an increased activity of the enzyme, from 1.6- to 6-fold in the brain and to an equal or lesser extent in several other tissues. Outwardly, the transgenic mice appeared normal, without any obvious deformities. This is not surprising because there is no reason to expect that elevation of CuZnSOD activity alone will cause the major dysmorphic features of DS. Rather, we anticipate that overexpression of CuZnSOD will affect more subtle aspects of tissue function and integrity, particularly in those tissues that might be affected by the altered metabolism of oxygen-free radicals. Bearing in mind the effect observed in the CuZnSOD-cellular system, the concentration of the neurotransmitter serotonin was measured in the blood of transgenic-CuZnSOD mice and was found to be significantly lower than the corresponding value in non-transgenic littermate mice. This observation generated much interest because reduced concentration of blood serotonin is a well-known clinical symptom in DS patients. When the deficiency was first noticed in the 1960s, it aroused considerable attention because of the possible relevance of serotonin uptake by blood platelets to neurotransmitter function in the central nervous system, and hence its involvement in the hypotonia and mental retardation of DS. At that time, attempts were made to raise the levels of blood serotonin in DS infants by administration of its precursor, 5-hydroxytryptophan; muscular tone, motor activity, and sleep abnormalities were reported to improve concomitantly with its administration. However, the development of infantile spasms, a severe seizure syndrome, in 17% of the patients receiving the drug brought these studies to a halt. Serotonin is an important neurotransmitter in the central nervous system, both in the embryonic state and in infants. It usually does not appear free in the blood circulation because of its efficient uptake by platelets, where it is accumulated and stored in the dense granules. Detailed analysis of platelets isolated from the transgenic mice bearing the extra CuZnSOD gene revealed that the uptake process in these granules is impaired, and this constitutes the cause for the reduced concentrations of blood serotonin in these mice. The dense granules of the platelets are in many respects similar to the vacuoles in the PC12 cells, which, as described in the previous section, were damaged by the elevated activity of CuZnSOD. It is intriguing that this same lesion appears both in the cellular-CuZnSOD system and in the transgenic-CuZnSOD mice and that the consequent defect is a well-known deficiency diagnosed in DS. This observation in the first example in which a direct link between a clinical symptom of the syndrome and a GDE of an individual gene has been established. The transgenic-CuZnSOD mice also have abnormalities in the connections between nerve terminals and the muscles (the so-called neuromuscular junctions), which are similar to the defects observed in neuromuscular junctions of patients with DS. This is an additional indication of the connection between gene dosage of CuZnSOD and defects characterizing the syndrome.

4. Transgenic Mice and Alzheimer's Pathology

Since the construction of transgenic-CuZnSOD mice, additional strains of mice bearing other human chromosome 21 genes have been developed and studied.

The relation between overexpression of the gene encoding APP and the Alzheimer's disease–type neuropathology found in adult DS patients is currently being investigated in mice carrying and overexpressing the gene for the amyloid precursor. Overproduction of APP was detected in neurons, particularly in the hippocampus, the deeper layer of the cortex, and in the Purkinje cells in the cerebellum. The effects of this elevated level of the amyloid precursor are presently being studied.

In summary, the etiology of DS is a complex process, involving many genetic factors that produce the profound disturbances of development and morphogenesis seen in DS. Progress toward understanding the molecular basis of the deficiencies is slowed down by our inability to conduct studies of GDEs in human patients. An animal model system consisting of transgenic mice, with only the DS genes triplicated, is therefore essential for meaningful exploration of this complex disease in the context of a living animal.

V. FUTURE PROSPECTS

To delineate, at the molecular level, the relative importance of different genes on chromosome 21 in determining the DS phenotype, the minimal chromosomal regions whose triplication produces the syndrome has first to be determined and the genes it contains identified. Despite rapid progress in the molecular analysis of the genetic structure of chromosome 21, the minimal size and precise localization of the DS region are still undefined. It is also not known whether the DS region is contiguous or exists in patches scattered over the 21q22 segment. Molecular and clinical investigation of DS individuals possessing unbalanced translocations of parts of chromosome 21 should permit the precise definition of the DS region. These clinicogenetic studies should be extensive because some of the DS symptoms are variable and difficult to identify on the basis of an individual case.

The recent advances in human molecular genetics should provide the means to construct a detailed genetic and physical map of genes along chromosome 21. A physical map of the entire chromosome, consisting of ordered clones containing overlapping sets of DNA fragments in cosmid or yeast vectors, is also a prerequisite for determining the complete nucleotide sequence of chromosome 21. This undertaking is within the framework of the international effort to map and sequence the whole human genome. Knowing the nucleotide sequence will facilitate the identification of all the genes residing on this chromosome and will eventually lead to the isolation of those mapped to the minimal DS region.

Availability of cloned genes will allow researchers to construct other cellular and animal model systems of the type described in this article and to investigate the biochemical and morphological consequences of imbalance of these genes. The ability to generate transgenic mice carrying several genes from defined regions of the DS segment will further facilitate the exploration of GDEs on development and morphogenesis. The powerful technique of gene transfer to the mouse germ line, followed by a mating program between strains of transgenic mice, each carrying one candidate gene, permits the development of a mouse strain that overexpresses several transgenes. Eventually a battery of transgenic strains with a full complement of the DS region triplicated will become available. Such an animal model will not only lead to the identification of the genes participating in the syndrome and enable detailed study of the Down abnormalities during the entire life span from embryogenesis to the fully developed animal, but it will also provide a test system for therapeutic or ameliorative procedures.

Although the application of therapy for DS is still a long way off, the animal models provide systems in which various therapeutic strategies that may prevent or ameliorate some of the pathologies associated with the syndrome can be tested. The molecular biological approach may include gene therapy, which, in the case of DS, will consist of genetic means to eliminate the overexpression of the genes that have been identified as causing clinical symptoms. This approach requires gene targeting (i.e., homologous recombination between DNA sequences residing in the chromosome and newly introduced DNA sequences). This technique, if sufficiently developed, may permit one day the selective inactivation of those extra genes that contribute to the Down phenotype. This technology is currently being used to introduce insertion mutations that will silence the human CuZnSOD gene carried by the transgenic-CuZnSOD mice.

BIBLIOGRAPHY

Antonarakis, S. E. (1993). Human chromosome 21: Genome mapping and exploration, circa 1993. *Trends Genet.* **9,** 142–148.

Carritt, B., and Litt, M. (1989). Report of the committee on the genetic constitution of chromosomes 20 and 21. Human Gene Mapping 10 (1989); Tenth International Workshop on Human Gene Mapping. *Cytogenet. Cell Genet.* **51,** 351–371.

Cooper, D. N., and Hall, C. (1988). Down syndrome and the molecular biology of chromosome 21. *Prog. Neurobiol.* **30,** 507–530.

Epstein, C. J. (1986). The consequences of chromosomal imbalance: Principles, mechanisms and models. *In* "Developmental and Cell Biology" (P. W. Barlow, P. B. Green, and C. C. Wylie, eds.), Vol. 18. Cambridge Univ. Press, Cambridge.

Epstein, C., and Patterson, D. (eds.) (1990). "21st Chromosome and Down Syndrome." A. R. Liss, New York.

Groner, Y. (1995). Transgenic models for chromosome 21 gene-dosage effects. *In* "Etiology and Pathogenesis of Down Syndrome," pp. 193–212. Wiley-Liss, New York.

Stewart, G. D., Van Keuren, M. L., Galt, J., Kurachi, S., Buraczynska, M. J., and Kurnit, D. M. (1989). Molecular structure of human chromosome 21. *Annu. Rev. Genet.* **23,** 409–423.

Dreaming

ERNEST HARTMANN

Tufts University School of Medicine and Sleep Disorders Center, Newton Wellesley Hospital

---

GLOSSARY

Delta activity Brain wave activity characterized by relatively low frequency of 0.5–4 cycles per second

Electroencephalogram Record of brain waves (electrical activity of the brain) obtained from electrodes placed on the scalp; the instrument used to make these recordings consists of oscillographs and amplifiers and is known as an electroencephalograph

Neurotransmitter Chemical substance that transmits messages between neurons (nerve cells); the substance is usually released by one neuron, called presynaptic, into the space between cells, called the synapse, where it has an excitatory or inhibitory effect on a second neuron, called the postsynaptic neuron

DREAMING IS MENTAL ACTIVITY THAT OCCURS during sleep. Aside from this, there is no universally agreed upon definition of dreaming; however, it is generally accepted that dreaming involves consciousness, albeit a different kind of consciousness from that of the waking state. Dreaming tends to be perceptual rather than conceptual, with a great deal of direct sensory experience and relatively little thinking. The sensory experience is most often visual, but 20–40% of dream reports mention auditory experience, and smaller percentages include touch, pain, smell, and taste, in that order. In a typical dream, the dreamer lacks the experience of free will, which is so characteristic of the waking state. In addition, dreams usually appear in isolated pieces, with sharp discontinuities between dreams and sometimes within parts of a single dream. Thus, our dream lives are not tied together by the threads of memory, which provide a sense of continuity in our waking lives. This difference in continuity allows us to distinguish our own waking states from our dreaming states, and lets us answer the ancient Chinese conundrum: How do I know I am a man who dreams (at night) he is a butterfly, and not a butterfly who dreams (by day) that it is a man?

I. HISTORICAL IMPORTANCE OF DREAMING

Almost all human cultures have given significance to dreaming, whether the dream is considered a voyage of the soul, a message from the gods, a prophecy of the future, a guide to the direction of one's life, or an aid in healing the body or mind.

A dream as a voyage taken by the soul or spirit is a very widespread idea and is not at all unreasonable, even though the words "soul" and "spirit" grate on the Western scientific ear. Obviously, in our dreams we are able to see and to converse with people who actually live far away, or even with people who are no longer living. If we wish to concretize and give a name to the part of us that does this seeing and conversing, it cannot be the body, peacefully sleeping in bed, nor the ordinary mind, which guides us in our daily routines, but something else. A word such as "soul" or "spirit" fills this place.

In fact, this aspect of dreaming can explain one of the most widespread beliefs about the soul: in numerous traditional cultures, it is believed that the soul of a person who has died remains on earth, among the living, for a number of months or years, visits its relatives, and so on. This view is not unknown even in modern Western culture, where mediums attempt to make contact with the souls of the recently departed. Surely this view can be explained by the fact that relatives and friends are dreaming of the recently dead person, and, of course, this dreaming activity occurs most in the years immediately after the person's death, when he or she is well remembered. After 30–60 years, the chances are that the living are dreaming little or not at all about that particular person, whose soul is therefore believed to have finally departed.

A belief in prophetic dreams is found in many ancient cultures. The Chester Beatty Papyrus documents dream interpretations—frequently involving predictions of future events—from the twelfth dynasty of ancient Egypt (1991–627 B.C.). The Old Testament contains many prophetic dream interpretations; perhaps the most famous is the Pharaoh's dream of seven fat cows and seven lean cows, interpreted by Joseph as foretelling seven years of plentiful crops followed by seven lean years. In some cultures, skepticism is expressed about prophetic dreams: in the *Odyssey*, dreams are divided into false prophecies ("passing through the gate of ivory") and true prophecies ("passing through the gate of horn").

In classical Greece and in many cultures of the Middle East, dreams also played a part in healing. Patients or supplicants seeking help with a physical or mental disease, or seeking for guidance in making plans for their lives, were treated at specially designated dream temples. Usually after various preparations and rituals, the patients went to sleep in the temple and carefully noted any dreams they experienced, which were then interpreted either by priests and priestesses or by the patients themselves in an effort to find a cure or solution. The cure or solution might involve a direct message or prescription from the gods but could also involve psychological insight in the dreamer. In the latter sense, this use of dreams continues to the present: dreams are used in psychoanalysis (see the following section) and a variety of psychotherapies to increase a patient's self-understanding. There has been a recent resurgence of interest in using dream incubation (without a temple, but simply by asking the dreamer to concentrate on a particular problem before sleep) as a means of solving personal problems.

Western science, however, has historically shown little interest in dreaming and, in fact, not much interest in sleep. By the nineteenth century, dreaming was considered a by-product of disordered bodily functioning during sleep, of so little consequence that it was barely mentioned in textbooks of biology or psychology and was studied only by an occasional eccentric. Since 1990, scientific interest in dreaming has revived and flourished, nourished by two very different sources: Sigmund Freud and psychoanalysis, starting around 1900, and Eugene Aserinsky and Nathaniel Kleitman's discoveries leading to the biology of dreaming, which began in the 1950s.

A. Freud and Psychoanalysis

Freud actually proposed not one but several interconnected theories about dreaming. First, he insisted that the dream is a meaningful mental product, created by the dreamer from recent waking events (called day residues) and from unconscious wishes. He felt that, when analyzed in detail, every dream represents the fulfillment of a wish; however, the wish is by no means always obvious. Freud believed that the direct expression of a wish or other drive material would be disruptive and awaken the sleeper; therefore, the "dreamwork" transforms and disguises the wish and other "latent dream thoughts," using mechanisms such as condensation (i.e., joining several thought elements to produce one dream image) and displacement (i.e., moving the emphasis and emotion from one element to another). [*See* Psychoanalytic Theory.]

By asking the patient (the dreamer) to associate to each element of the dream, the analyst tries to interpret the dream by determining the wishes and other unconscious material that formed it. Freud considered the interpretation of dreams to be the most direct path to such unconsciousness material and, thus, called dreams the "royal road to the unconscious."

Attempts to test rigorously the various hypotheses embedded in Freud's theories have met with mixed results. For instance, evidence supports a weak version of wish fulfillment: subjects dream more of drinking when they are thirsty (artificially deprived of water) than when they are not. However, it is certainly not established that every dream represents the fulfillment of a wish; nightmares do not easily fit, and Freud himself was not satisfied with his attempts to explain them. Also, laboratory evidence tends to contradict Freud's view that the dream functions to protect sleep.

Despite the lack of proof, Freud's views have been extremely influential. Most forms of psychoanalysis and psychotherapy practiced throughout the world are based directly or indirectly on Freud's work. Not all therapists subscribe to the totality of Freud's view on dreams, but insofar as dreams are used at all in therapy, they are seen as meaningful mental products and as the royal road, or at least one good road, to the patient's unconscious.

A number of Freud's followers accepted his main tenets but suggested significant amendments. Carl Gustav Jung, a prominent early psychoanalyst and ardent student of dreams, proposed that dreams are not only a road to the person's individual unconscious but also lead to the "collective unconscious," a postulated tendency of the entire species to dream in certain patterns and to dream of certain basic images or symbols, which Jung called archetypes. At first these views seem far-fetched and mystical; however, "collective unconscious" and "archetypes" can be taken as referring to aspects of the organization of the human cortex. These aspects impose constraints upon the individual's construction of his or her own dream, making dreams of different individuals similar in some basic ways—much as myths from very different cultures reveal striking similarities, as emphasized repeatedly by Jung himself and by his follower Joseph Campbell.

Other researchers have emphasized that dreams often appear to be attempts at solving current problems of the dreamer. Many analysts and others have recently emphasized the importance of the manifest dream, insisting that the dream exactly as dreamt can provide important information and insight, without the need for deep interpretation.

In any case, dreams have been alive and well in the world of psychotherapy and self-understanding since 1900, yet they received scant attention from biologists and nonclinical psychologists until the mid-1950s.

B. Aserinsky and Kleitman: Rapid Eye Movement Sleep

Kleitman, a professor of physiology at the University of Chicago, had no particular interest in dreaming, but he was one of the few scientists to study human and animal sleep in great detail. He performed many research studies, often using himself as a subject; for instance, he spent weeks living underground in a cave to examine his sleep–wake patterns without the usual influence of light and dark. Aserinsky, his student, was interested in blinking and eye movements in children and decided to record these throughout a night of sleep. He found that on a number of occasions each night, the eyes moved in a conjugate pattern, similar to the movements involved in watching something while awake. Aserinsky and Kleitman hypothesized that the child was dreaming during these periods, which rapidly led to studies in adults demonstrating that awakenings during these periods of rapid eye movements (REMS) usually produced reports of dreams. These findings initiated a whole era of laboratory research on the biology of dreaming. The highlights of this research will be summarized in the following sections.

II. A NIGHT OF SLEEP IN THE HUMAN ADULT: WHAT IS RECALLED WHEN?

Many thousands of nights of sleep recorded in the laboratory demonstrated the following basic pattern: as a person falls asleep, the muscles gradually relax, pulse and respiratory rates decrease slightly, and the electroencephalogram (EEG) shows a decrease of waking alpha activity, which is replaced by random low-voltage activity without any clear rhythmic pattern. This is known as stage 1 sleep, sometimes called drowsiness or sleep onset rather than true sleep. In the next few minutes, the EEG begins to demonstrate specific sleep rhythms called sleep spindles—activity at 13–15 cycles per second in bursts 0.5–2 sec long. The onset of sleep spindles marks the beginning of stage 2 sleep. The spindles continue, but the low-voltage background is gradually replaced by delta activity (0.5–4 cycles per second). Arbitrarily, a record in which 20% of each 30-sec epoch consists of delta activity is assigned to stage 3 sleep, and when 50% of each epoch consists of delta activity it is called stage 4 sleep. [See Sleep.]

In a healthy young adult, the transition from stage 1 to stage 4 sleep occurs within 15–20 min. The next hour of sleep is typically stage 3 and stage 4 sleep. Then, rather suddenly, after 80–120 min, the EEG reverts to what looks like very light sleep, or stage 1 sleep, which continues for 5–15 min. This episode of stage 1 is accompanied by rapid, conjugated eye movements and, in terms of awakening threshold, the episode is not light but is approximately as deep as stage 4 sleep. This is the first REM period of the night. The remainder of the night consists of 90-min alternations between non-REM sleep (stages 2, 3, and 4) and REM sleep. Non-REM sleep gradually lightens, and mostly stage 2 sleep occurs toward the end of the night. The REM periods gradually lengthen, and the last one of the night typically lasts 20–30 min.

REM sleep is also known as D-sleep (desynchronized, or dreaming, sleep), whereas non-REM sleep is called S-sleep (synchronized sleep). Typically in a young adult, four or five REM periods last a total of 100–120 min, or 25% of total sleep time.

Numerous studies have examined mental activity at different times in a night of laboratory-recorded sleep. Awakenings during REM sleep result in a report identified as a dream by the sleeper, and by independent raters, 60–90% of the time. Awakenings within a few minutes of the onset of a REM period result in short dreams, and awakenings further into a REM period produce longer dreams. Awakenings during the later REM periods of the night result in more emotional dreams, and dreams containing more material from earlier in the dreamer's life. Awakenings from non-REM sleep do not necessarily produce reports of no content. About 50% of such awakenings result in a report of something going on—a thought or image or fragment of a dream. Reports scored as dreams by raters occur in 10–40% of such awakenings in different studies. Reports scoreable as dreams also occur at other times during sleep, especially at sleep onset, or stage 1 sleep; at these times, even though the person is barely asleep, awakenings result in very vivid, although usually short, dream-like reports.

Overall, therefore, dreaming should not be equated simply with REM sleep. REM sleep is certainly the time when most typical dreams occur, and in many subjects every REM awakening produces a dream or at least a fragment, but the converse is not true—non-REM sleep cannot be considered a time when no dreaming occurs.

Different dreams during the same night generally show some relationship to one another. If a person is awakened in the laboratory during each REM period and asked to report a dream each time, the dream reports will not be identical, nor will they involve exactly the same dream setting or characters; however, they will usually show some similarity in theme or content, leading the experimenter in several studies to conclude that the dreams were dealing with the same issues or problems in the dreamer's life.

III. BRAIN BIOLOGY IN REM SLEEP

REM sleep occurs in almost all mammals (see Section III,B). Hundreds of studies, chiefly in the cat and the rat, have elucidated the brain physiology underlying REM sleep. In brief, REM sleep and the REM–non-REM cycle described earlier depend on centers in the pons, the midportion of the brain stem. An animal with a transsection above the pons, or with the entire brain above the pons removed, continues to have cycles of REM and non-REM sleep.

Specifically, REM and non-REM sleep are regulated by the interaction of several small cell groups within the brain stem. A group of large cholinergic cells, sometimes referred to as the giant cell nucleus, appears to have a dominant role in turning REM sleep on, whereas two groups of cells—the serotonergic cells of the raphe nuclei and the noradrenergic cells of the locus coeruleus—act to turn REM sleep off and non-REM sleep on.

These three neurotransmitters—acetylcholine, serotonin, and norepinephrine—among others, not only play roles in regulating the states of waking and sleep, but their release in the forebrain helps to determine the characteristics of the states. The "dreamlike" qualities of mental experience during REM sleep, so different from the qualities of wakefulness, probably result from the almost complete lack of norepinephrine and serotonin activity in the forebrain during REM sleep.

A. Ontogeny of Dreaming: Dreams in Childhood

Dreams are recalled more or less throughout an individual's life, although with different frequencies in different individuals. Do they begin at a particular time in childhood? And do they develop in a particular way?

REM sleep is definitely present in children. In fact, the amount of REM sleep is highest at birth and decreases gradually during childhood; a newborn child spends 16 hr asleep, and approximately half of that time is spent in REM sleep. Studies have also demonstrated the presence of REM sleep in the fetus. But what about actual dream reports?

Dreams have been recorded in laboratory settings from children as young as 4–5 yr old and in clinical settings from children as young as 2.5 yr. Reports of nightmares are quite frequent at 3–5 yr old. Apparently dreams, including nightmares, can be described by children as soon as their verbal skills are sufficient to describe any experience—usually around 2–3 yr old. However, some sort of dream probably occurs even before that time. In one case, the parents of a 1-yr-old child noticed certain movements during sleep in their son, who would then awaken crying, apparently upset. He continued to have these episodes occasionally for a number of months while learning to talk, and finally he was able to describe a frightening dream to his parents. Since the movements and upset crying were the same over this period, the parents concluded that their child had probably had the same dream, or a similar dream, as early as 1 yr old.

However, these early dreams and nightmares quite likely did not involve the full visual-spatial reality of adult dreams, which appears to develop only gradually with age. Thus dreams may develop much as do other mental products—images, thoughts, stories— suggesting that dreams are indeed one sort of production of the human mind and that they can be examined and analyzed, like the others, by the methods of cognitive psychology.

B. Phylogeny of Dreaming: Do Animals Dream?

Physiological recordings of sleep have been performed in a wide variety of animal species. Both REM and non-REM sleep occur in the sleep of all mammals studied so far, with only one exception: the spiny anteater, or echidna. Birds have non-REM sleep and occasional short episodes of REM sleep; the latter are usually seen only in young (immature) birds and disappear or almost disappear in the adult. Reptiles definitely show non-REM sleep, and a few appear to have a rudimentary form of REM sleep. Amphibians and fish do not appear to have REM sleep, but they do show electrical changes suggesting a form of non-REM sleep. Invertebrates have been less methodically studied. Many invertebrate species demonstrate regular rest and activity cycles, with periods of inactivity that can be called behavioral sleep, but without all the electrophysiological characteristics of sleep seen in vertebrates.

However, even in mammals who definitely have REM sleep, does this mean that they experience dreams? Leaving aside the philosophical question of whether or not we can ever be sure that another human being dreams, or thinks, we generally accept the report of another person who tells us that he or she was dreaming. We are unable to ask animals whether or not they were dreaming, and they are unable to answer us in a language we understand. So perhaps the question cannot be answered; however, there are at least suggestive hints. First, we have all seen pets, especially young cats and dogs, running or playing in their sleep. We usually conclude that they must be dreaming. The Roman poet and philosopher Lucretius, in his great work "De Rerum Naturae," written

2000 years ago, noted horses running in their sleep and concluded that they were dreaming. Such observations are suggestive, but not proof.

Second, physiological recordings indicate that the pattern of brain activity in other mammals during REM sleep closely resembles that found in humans. For instance, there is intense activation in the visual cortex in many mammals, similar to that found during active waking or during search activity. This certainly suggests that the animal is seeing something.

Third, in one study, monkeys who were trained for other reasons to press a bar whenever any picture was projected onto a screen in front of them were found to be making bar-pressing movements at times during their sleep. Recordings were not made to determine times of REM sleep, and there were other methodological problems, but this sort of study is perhaps the closest we can come to an answer at present. Perhaps the researchers who are teaching speech ("signaling") to chimpanzees will eventually be able to ask them directly about dreams. Meanwhile, indirect evidence suggests a strong probability that animals other than humans do experience at least some form of dreams.

IV. NIGHTMARES

When someone says "I have terrible nightmares," she or he usually means one of two things, which have only recently been studied and clearly distinguished in sleep laboratory studies. The first phenomenon is the true nightmare: a long, vivid, frightening dream with increasingly scary content that awakens the sleeper. True nightmares occur during REM sleep, like most dreams, usually from very long REM periods in the second half of the night. The second phenomenon consists of waking in absolute terror, with a scream but generally without a dream; the sleeper recalls either no content or a single frightening image: "something was sitting on me" or "I was being crushed." This second phenomenon occurs during an arousal from deep non-REM sleep during the first hours of sleep and is best called "night terror." It is physiologically as well as psychologically different from the nightmare.

Nightmares are not rare. Questionnaire surveys suggest that the average adult experiences one to two definite nightmares per year. Although almost half of adults claim to have no nightmares, they usually mean no recent nightmares; on careful questioning, they often will remember one or two nightmares in childhood. Nightmares appear to be most common in children 3–6 yr old; at least 75% of children report at least some nightmares at this age. Most surveys suggest that nightmares are somewhat more common in women than in men, but interviews suggest that women may be more willing to admit to having nightmares rather than necessarily experiencing more.

A nightmare almost always consists of something harmful or dangerous happening to the dreamer; she or he is being chased, threatened, wounded, tortured, or killed. We ourselves are the victims in our nightmares. The only consistent exception occurs in mothers of young children who often dream that something is happening to their child rather than to themselves.

Nightmares can sometimes be initiated by acute trauma. In this case, the nightmare often involves a repetitive playback of the traumatic event with small variations. Without trauma, nightmares tend to occur at stressful times, especially if the stress reminds the person of childhood vulnerability.

Some people tend to have nightmares all the time—every few days for many years, often since childhood. These people have been studied intensively. They are not necessarily very anxious people, nor do they necessarily have mental illness or serious problems; rather, they are people who are unusually open, appear defenseless or vulnerable, and are often highly creative (many artists, including some of the most famous, suffered from frequent nightmares). They are described as having "thin boundaries" in many different senses.

V. OTHER SPECIAL DREAMS

The lucid dream is a special kind of dream that has recently aroused considerable interest. This is a dream in which one becomes aware that one is dreaming while the dream is happening. If one does not awaken, one can sometimes direct the dream and make changes in it. Lucid dreams sound like a state in between dreaming and waking with features of both: laboratory studies have demonstrated that lucid dreams almost always occur in REM sleep. Lucid dreamers have been able to signal to the experimenter, using prearranged eye-movement signals, to indicate that they know they are dreaming. The EEG record shows the continuation of REM sleep before and after the signal.

Both dreams-within-dreams and false awakenings refer to a situation in which the dreamer dreams of having awakened from a dream but still finds himself within a dream from which he later truly awakens.

Little is known about these dreams, except that they appear to occur more frequently in persons who also report nightmares or lucid dreams.

VI. REMEMBERING AND FORGETTING DREAMS: WHO REMEMBERS DREAMS?

One of the most intriguing features of dreams is how easily they are forgotten. One recalls a dream on awakening and finds that after breakfast it is totally gone. This is probably related to the nature of the chemical environment at the forebrain during REM sleep, which is very different from the environment during active waking (see earlier).

Some people remember dreams very well, whereas others remember nothing. We all have four or five REM periods per night, each of which usually yields a dream if interrupted by an awakening in the laboratory. Theoretically, we should remember 30 dreams per week from our REM periods and a few others from sleep onset or non-REM sleep. When people are asked on questionnaires how many dreams per week they recall, the answers range from 0 to 21; the mean is about 2.5 per week. Obviously we forget a lot of dreams. But what accounts for the huge interpersonal differences? First, physiological factors are involved. Several studies show that good dream recallers have slightly more REM sleep than poor recallers and also tend to awaken more often during the night, thus having a greater chance of awakening during a REM period. One study examined complicated measures of EEG frequency in both hemispheres during REM sleep and during waking and demonstrated that good dream recallers showed less difference on these measures between REM and waking than did poor dream recallers. In other words, good dream recallers "didn't have as far to go" when they switched from dreaming to waking.

Psychological factors are involved as well. Personality studies demonstrate that good dream recallers become more absorbed in daydreams and fantasies, have greater "tolerance for ambiguity," and have thinner boundaries (as discussed earlier), meaning they are more open, flexible, artistic, and vulnerable. Indeed, frequent nightmare sufferers and lucid dreamers, both groups with thin boundaries, turn out to have a very high rate of dream recall. The dimension "thin versus thick boundaries" may actually encompass the physiological differences as well. Persons with relatively thick boundaries keep things in separate compartments and maintain very separate states, and their REM sleep state differs greatly from their waking state. This may account for their remembering few dreams.

VII. THE BASIC NATURE AND FUNCTIONS OF DREAMING

We have discussed how dreaming, initially defined as mental activity occurring during sleep, can be thought of as mental activity occurring during certain portions of sleep—mostly during REM sleep—and perhaps further defined as activity occurring when the cerebral cortex is under certain biochemical conditions, especially when norepinephrine and serotonin are at very low levels.

But what exactly is this mental functioning? What differentiates it from waking mental functioning? A contemporary theory of dreaming is developing along the following lines. The mind, or the cortex (the physical basis of mind), can be thought of as consisting of a net—a broad network of simple units. Such a network, sometimes called a neural net, can to a limited extent be modeled by "connectionist nets" on the computer. In such a net, all that exists are simple units with varying connection strengths. The flow of excitation in the net is determined by connection strengths between units, and in turn the use of the net changes the connection strengths so that each time the net has been used the connection strengths differ slightly from what they were before.

In such a net all that can happen during waking, or during sleep, is the activation of certain patterns and the strengthening or weakening of connections. We make connections all the time. However, the connections are more constrained during waking. Under the influence of waking brain chemistry, with its high levels of norepinephrine for instance, which is thought to produce "increased signal-to-noise ratio," the net is constrained into acting as a relatively direct input-to-output feed-forward net. In dreaming, connections are made more broadly. The net, under different chemical conditions, is not so constrained and acts more as an "auto-associative net."

Thus in dreaming the connections are made more broadly or widely than in waking. However, the connections are not made randomly, but appear to be guided by the dominant emotion or emotional concerns of the dreamer. This can be seen most clearly when there is a single obvious emotional concern—for instance, when someone has just experienced a trauma. Someone who has just escaped from a fire in which others have died may dream about the fire but also dreams about being in a tidal wave and about being chased by a vicious gang. Clearly what is dreamt of is predominantly the emotion or emotional concern of the dreamer. According to this contemporary theory, dreaming "contextualizes emotion."

Of course this contextualizing does not occur in words or in mathematical symbols. In the dream it occurs in a visual space very similar to that of our waking world. The dream is pictured in visual-spatial reality and the dream, or at least the most striking portions of it, can be seen as explanatory metaphor, as metaphoric pictures of the dreamer's emotional concerns or state of mind. Sometimes, such as after trauma, this is obvious. When emotional concerns are less or are multiple, the dream is less clear. In such cases, some form of "dream interpretation" is necessary to arrive at the emotional concern.

Finally, does this connecting and making of metaphoric pictures that contextualize emotion have a function or is it simply something that happens while we are in the odd state of sleep or REM sleep? A number of researchers have concluded that dreaming does have a kind of adaptive function, although this is hard to prove. In the clearest case, after trauma, a dream appears to make connections between the recent very disturbing event and similar emotionally related material elsewhere in the nets. The dream appears to weave in the new material and produce increased complexity of connections. This may constitute a basic function that allows both the disturbance in the net produced by new material to subside (thus leading to an emotional calming) and new and richer connections to be made, which presumably will be useful later in dealing with somewhat similar or new traumatic material.

Aside from this suggested basic function of dreaming, it can be useful in a number of ways. We can make use of the broader connections made during dreaming in psychotherapy, in problem solving, in scientific or artistic creativity, and generally in advancing our self-knowledge.

BIBLIOGRAPHY

Aserinsky, E., and Kleitman, N. (1953). Regularly occurring periods of eye motility and concomitant phenomena during sleep. *Science* **118**, 273–274.

Foulkes, D. (1985). "Dreaming: A Cognitive Psychology Approach." Lawrence Erlbaum Associates, Hillsdale, New Jersey.

Freud, S. (1900). "Die Traumdeutung (The Interpretation of Dreams)." Hogarth, London.

Hartmann, E. (1973). "The Functions of Sleep." Yale Univ. Press, New Haven, Connecticut.

Hartmann, E. (1984). "The Nightmare." Basic Books, New York.

Hartmann, E. (1991). "Boundaries in the Mind." Basic Books, New York.

Hartmann, E. (1996). Outline for a theory on the nature and functions of dreaming. *Dreaming* **6**(2), 147–70.

Hobson, J. (1988). "The Dreaming Brain." Basic Books, New York.

Hunt, H. (1989). "The Multiplicity of Dreams." Yale Univ. Press, New Haven, Connecticut.

Kleitman, N. (1963). "Sleep and Wakefulness." Univ. of Chicago Press, Chicago.

Drug Analysis

DAVID B. JACK
M W E & C, Glasgow

GLOSSARY

Chromatography Any system that produces a separation by partition between a stationary phase and a mobile one

Drug Any medicinal compound of animal, vegetable, mineral, or synthetic nature

Retention factor In thin-layer chromatography, the distance traveled by the compound relative to the mobile phase

Retention time Time taken for a compound to pass through a chromatographic system from injection to detection

DRUG ANALYSIS INCLUDES ALL METHODS THAT are generally used to determine the concentration of a drug in a matrix. This matrix may be relatively simple (e.g., a tablet) or complex (e.g., urine). This covers a wide range of techniques from the old and inexpensive to the new and costly.

I. INTRODUCTION

Drugs are substances used to treat or prevent disease, and they may be used alone or in combination. Drugs are administered in a variety of ways, depending on the nature of the drug, the disease, or the speed with which it is to act. Most drugs are given by mouth and rapidly enter the bloodstream; other routes include intravenous, intramuscular, and rectal administration as well as through the skin or by inhalation. They may only need be taken for a short time to cure a headache, for 1 or 2 weeks to treat an acute infection, or for life to control a chronic disease like high blood pressure.

To use drugs as safely as possible, it is important that the amount of active agent in each formulation (tablet, capsule, ointment, etc.) is carefully controlled and that the way it is given produces the required action with a minimum of side effects. Drugs are not just measured for therapeutic reasons: drug screening in sport, employment, and overdose is now common and likely to remain so. A wide variety of techniques are available, and the one chosen largely depends on the end in view. The techniques can be classified as physical, chemical, or biological, and Table I lists the most important. One method may be used in preference to another because of the nature of the material containing the drug (e.g., tablet, ointment, blood, urine) or because of the amount of drug likely to be present (1 part in 10, 1 in 10^3, 1 in 10^6, and so on).

Drugs may often be converted in the body to more polar, water-soluble metabolites, which may exert no pharmacological effect or may exhibit a different effect, so it is important that any method measuring drugs in body fluids such as blood or urine can distinguish between the unchanged drug and its metabolites or, at least, can determine the unchanged drug without interference from metabolites.

II. SAMPLE PREPARATION

Every analytical technique requires some form of sample preparation. This may range from the simple dilution of a cough mixture or plasma sample to a complex series of extractions. Some methods require little in the way of pretreatment of the sample, especially if the drug is present in high concentration in a relatively simple matrix (e.g., an active ingredient in a tablet

TABLE I

Techniques Used for Drug Analysis

Physical	Chemical	Biological
Radioisotope counting	Volumetric analysis	Immunoassay
Gravimetry (including thermal gravimetry)	Chromatography	Radioimmunoassay
Polarography	Gas–liquid	Enzyme immunoassay
Ion-selective electrodes	High-performance liquid	Fluoroimmunoassay
Spectroscopy (UV, VIS, IR, FLUOR)	Thin-layer	Luminescence immunoassay
Atomic absorption	Hyphenated techniques (GC-MS, etc.)	Microbiological
Spectrography		
Nuclear magnetic resonance		

formulation). However, much drug analysis is carried out using complex biological fluids such as plasma or urine where there will be large amounts of interfering substances and where the drug may be present in only nanogram quantities. Techniques requiring little in the way of sample preparation are radioisotope counting, atomic absorption, volumetric analysis, some electrochemical methods, infrared (IR), ultraviolet (UV), and visible (VIS) spectroscopy, fluorescence, and many immunoassays. Techniques requiring considerable preparation are gravimetry and most types of chromatography, although column switching in liquid chromatography can sometimes allow analysis without pretreatment. When drugs are present in relatively low concentration in complex media (e.g., plasma or urine), some means of separating the drug from interfering compounds and concentrating it is needed. The most frequently used methods are liquid–liquid and liquid–solid extraction.

A. Liquid–Liquid Extraction

This is based on the principle that the drug will selectively distribute between one of two immiscible liquids under a particular set of experimental conditions. By altering the pH, the ionization of the drug can be suppressed and its solubility in organic solvents increased. As a general rule, the least polar solvent possible should be chosen to extract the drug because this will ensure that much of the polar material is left behind in the aqueous phase. Some common solvents arranged in order of polarity are given in Table II. If acidic, neutral, and basic drugs are present, together these can all be separated by a sequential change in pH as is shown in Fig. 1. Generally this type of extraction procedure works well for drugs that are very lipid soluble, and high recoveries (>90%) can often be obtained. However, some drugs may pose particular problems. For example, some are amphoteric, having both acidic and basic groups, and here the relative strengths of each group will determine the pH chosen for extraction. A number of drugs are ionized at all pH values and are too polar to be extracted. However, they can often be paired with a suitable ion of opposite charge to give an ion pair with an overall charge of zero and hence soluble in organic solvents. Careful choice of this counter-ion can lead to a very selective type of extraction indeed; the counter-ion can also be chosen because it possesses a particularly favorable property (e.g., fluorescence), and in this way the original drug can be measured at low concentrations even if it has no fluorescence itself. Liquid–liquid extraction is widely used, but it does have some disadvantages: it is costly in terms of organic solvent used, large quantities of solvent constitute a potential fire hazard, large volumes of waste solvent are generated

TABLE II

Common Organic Solvents Arranged in Increasing Polarity

Heptane	Dioxane
Hexane	Ethyl acetate
Carbon tetrachloride	Acetonitrile
Toluene	Pyridine
Diethyl ether	Ethanol
Chloroform	Methanol
Dichloromethane	Acetic acid
Acetone	(Water)

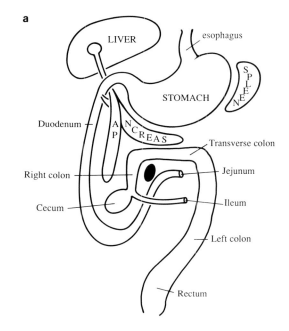

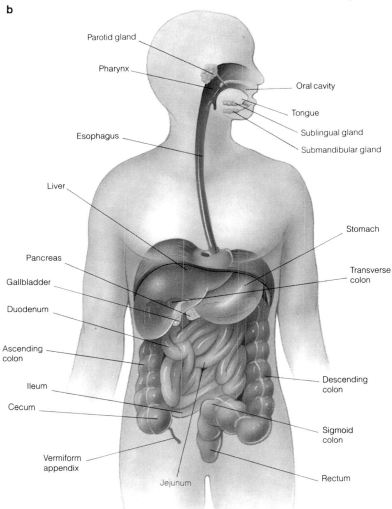

COLOR PLATE I   Basic topography of (a) mammalian gastrointestinal tract and (b) layout in humans. [Source: Gaudin, A. J., and Jones, K. C. (1989). "Human Anatomy and Physiology." Harcourt Brace Jovanovich, San Diego. p. 441. Reproduced with permission.] [*See* Digestive System, Anatomy.]

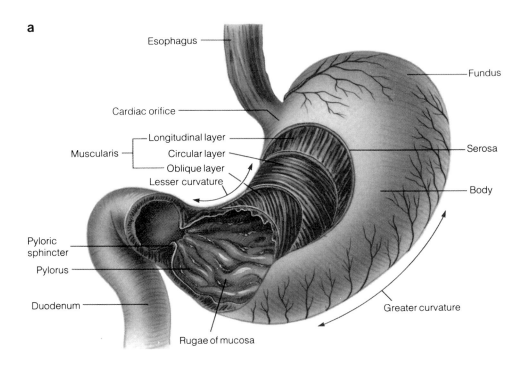

a

Esophagus

Cardiac orifice

Muscularis
— Longitudinal layer
— Circular layer
— Oblique layer
Lesser curvature

Pyloric
sphincter

Pylorus

Duodenum

Rugae of mucosa

Fundus

Serosa

Body

Greater curvature

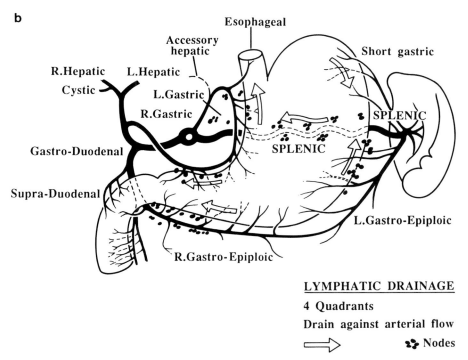

b

Esophageal

Accessory
hepatic

R.Hepatic   L.Hepatic

Cystic

L.Gastric

R.Gastric

Gastro-Duodenal

Supra-Duodenal

Short gastric

SPLENIC

SPLENIC

R.Gastro-Epiploic

L.Gastro-Epiploic

LYMPHATIC DRAINAGE

4 Quadrants

Drain against arterial flow

⇨   ❧ **Nodes**

COLOR PLATE 2 Structure of stomach, showing (a) muscle layers and (b) blood supply and arterial drainage. [Source: Gaudin, A. J., and Jones, K. C. (1989). "Human Anatomy and Physiology." Harcourt Brace Jovanovich, San Diego. p. 451. Reproduced with permission.] [*See* Digestive System, Anatomy.]

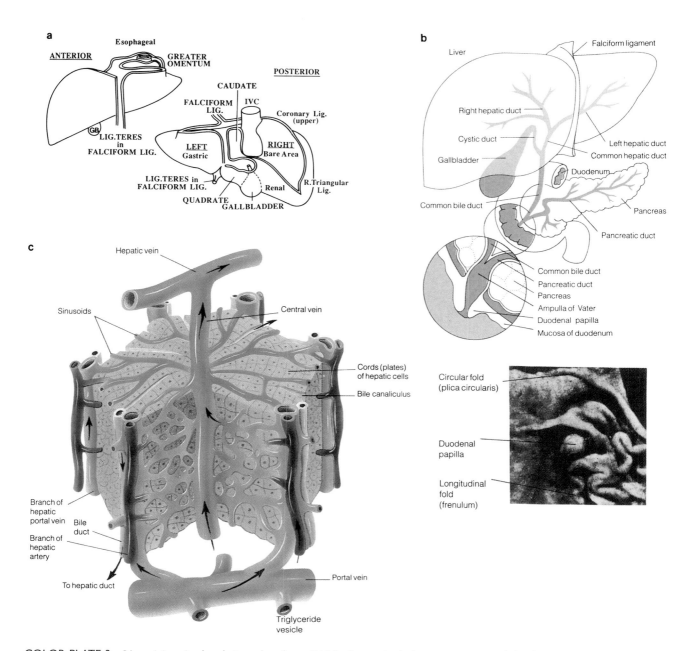

a

ANTERIOR

Esophageal

GREATER OMENTUM

POSTERIOR

CAUDATE

FALCIFORM LIG.

IVC

Coronary Lig. (upper)

GB

LIG.TERES in FALCIFORM LIG.

LEFT Gastric

RIGHT Bare Area

LIG.TERES in FALCIFORM LIG.

QUADRATE

Renal

GALLBLADDER

R.Triangular Lig.

b

Liver

Falciform ligament

Right hepatic duct

Cystic duct

Gallbladder

Left hepatic duct

Common hepatic duct

Duodenum

Pancreas

Common bile duct

Pancreatic duct

Common bile duct

Pancreatic duct

Pancreas

Ampulla of Vater

Duodenal papilla

Mucosa of duodenum

c

Hepatic vein

Sinusoids

Central vein

Cords (plates) of hepatic cells

Bile canaliculus

Branch of hepatic portal vein

Bile duct

Branch of hepatic artery

To hepatic duct

Portal vein

Triglyceride vesicle

Circular fold (plica circularis)

Duodenal papilla

Longitudinal fold (frenulum)

COLOR PLATE 3 Liver: (a) parietal and visceral surfaces, (b) bile duct and relations to pancreas and duodenum, and (c) internal organization of liver lobule. [Source: Gaudin, A. J., and Jones, K. C. (1989). "Human Anatomy and Physiology." Harcourt Brace Jovanovich, San Diego. p. 458, 469. Reproduced with permission.] [*See* Digestive System, Anatomy.]

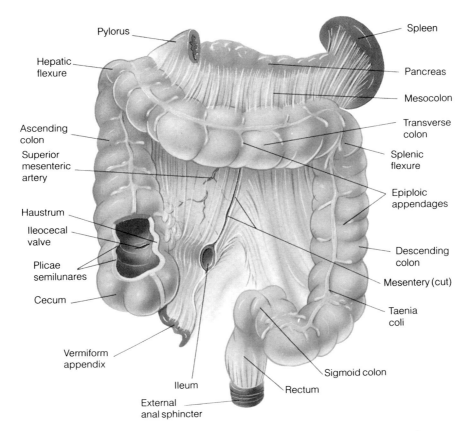

COLOR PLATE 4   Cecum and colon in relation to pancreas, spleen, and ileum. [Source: Gaudin, A. J., and Jones, K. C. (1989). "Human Anatomy and Physiology." Harcourt Brace Jovanovich, San Diego. p. 464. Reproduced with permission.] [*See* Digestive System, Anatomy.]

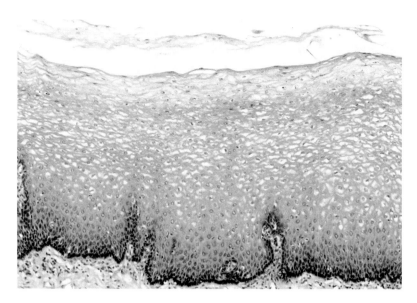

COLOR PLATE 5 Esophageal mucosa. The esophageal lining is shown with the luminal surface at the top. The stratified squamous epithelium has many scale-like layers of epithelium. The dark blue nuclei become dense near the base where cells are generated and then move upward toward the surface where they slough off. Similar epithelium lines the mouth and pharynx. [Photomicrograph courtesy of Dr. J. Kelly.] (*See* Digestive System, Physiology and Biochemistry.]

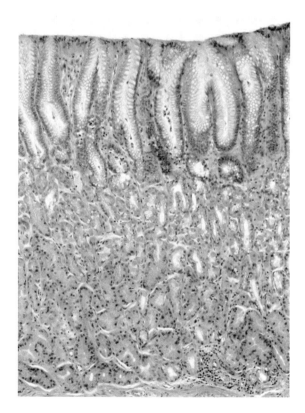

COLOR PLATE 6 Gastric mucosa with two layers. The upper third is composed of tubules lined by mucus-secreting columnar epithelium. The mucus appears paler and slightly foamy compared to the dark nucleus at the base of each cell. The lower two-thirds consists of glands which secrete acid and the enzyme, pepsinogen. The cells are arranged in circular configurations. As in Color Plate 5, the open space at the top represents the luminal cavity. [Photomicrograph courtesy of Dr. J. Kelly.] [*See* Digestive System, Physiology and Biochemistry.]

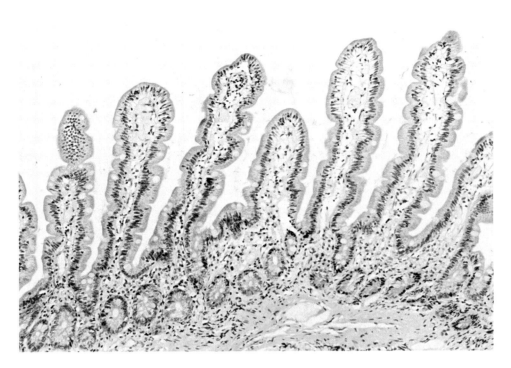

COLOR PLATE 7   The small intestine is lined by columnar cells. Multiple fingerlike villi protrude into the lumen and increase the surface area available for absorption. [Photomicrograph courtesy of Dr. J. Kelly.] (*See* Digestive System, Physiology and Biochemistry.]

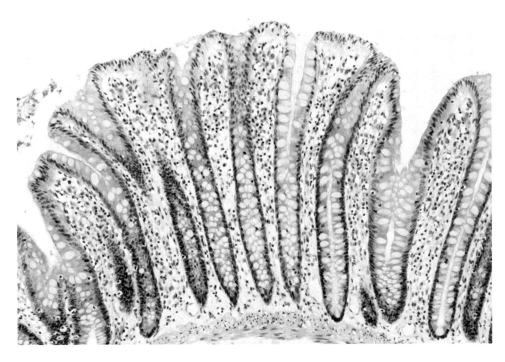

COLOR PLATE 8   The large bowel (colon) mucosa is composed of tubular glands lined by absorptive cells or mucus-secreting goblet cells. No villi are present. [*See* Digestive System, Physiology and Biochemistry.]

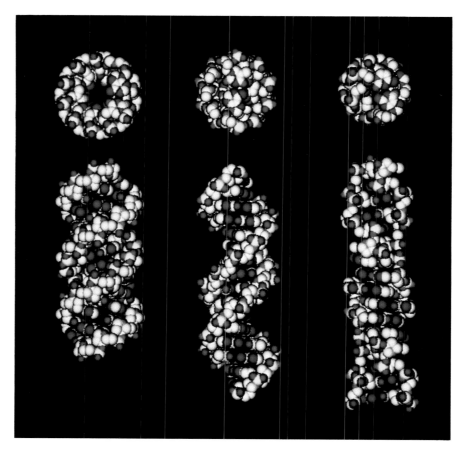

COLOR PLATE 9   Three-dimensional models of DNA double helix. Three different forms (A-, B-, and Z-forms) are shown. Their characteristics are summarized in Table I. [*See* DNA and Gene Transcription.]

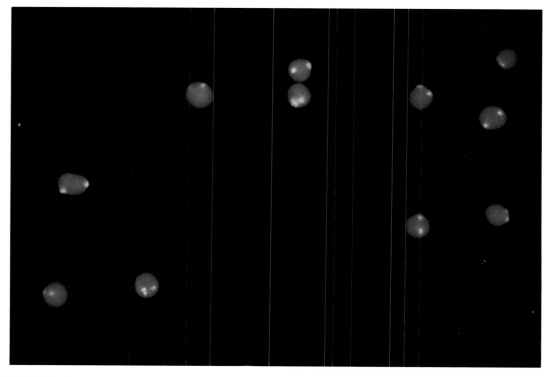

COLOR PLATE 10   Fluorescent *in situ* hybridization performed on interphase nuclei using an X chromosome-specific probe. Two hybridization signals are present indicating the presence of two X chromosomes. [*See* Embryofetopathology.]

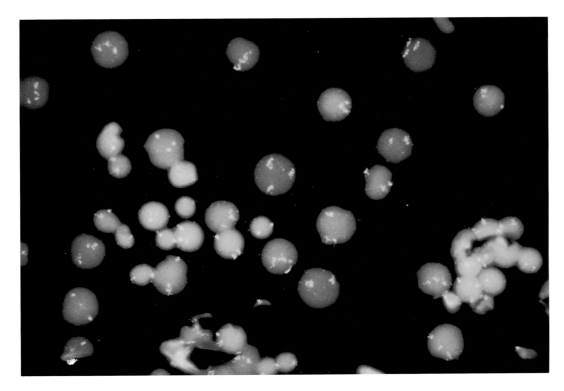

COLOR PLATE 11   Fluorescent *in situ* hybridization performed on interphase nuclei using an X chromosome-specific probe. Three hybridization signals are present indicating the presence of an extra X chromosome. [*See* Embryofetopathology.]

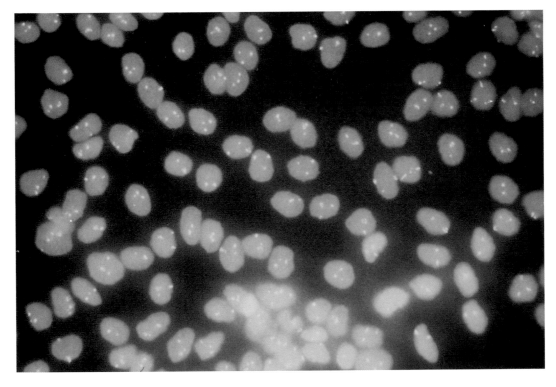

COLOR PLATE 12   Fluorescent *in situ* hybridization performed on interphase nuclei using a chromosome 18-specific probe. Three hybridization signals are present indicating trisomy 18, also known as Edward's syndrome. [*See* Embryofetopathology.]

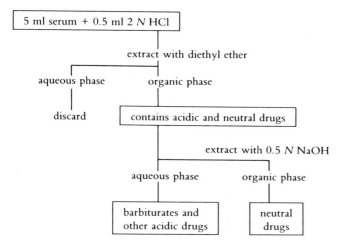

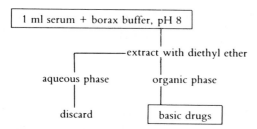

FIGURE I Separation of acidic, neutral, and basic drugs. From Jack (1984).

and disposal can be expensive, and finally, this type of extraction is difficult to automate.

B. Liquid–Solid Extraction

This approach was developed to solve the above problems and involves passing the material containing the drug over a matrix that will adsorb it while letting most of the other material through. The adsorbed drug can then be washed with a suitable buffer and then eluted with an organic solvent. This procedure is simple, economical, and much easier to automate. A range of solid phase cartridges is commercially available, and their properties and use are outlined in Table III. Sample recoveries of 80–100% can now be achieved for many drugs, and completely automated robotic systems can analyze about 100 samples per day. No matter which type of extraction is chosen, care should be taken to ensure that all glassware is scrupulously clean, and it should be borne in mind that some drugs, especially those with tertiary amine groups, bind strongly to glass in their unionized forms.

TABLE III

Comparison of Normal and Reverse-Phase Cartridges for Liquid-Solid Extraction

Cartridge	Silica	C_{18}
Polarity	Polar	Nonpolar
Sample applied in	Low-polarity solvent (e.g., hexane)	High-polarity solvent (e.g., water)
Wash solvent	Low polarity	High polarity
Eluting solvent	Increased polarity (e.g., buffer)	Decreased polarity (e.g., methanol, acetonitrile)
Elution order	Nonpolar first, most polar last	Most polar first, least polar last

III. ANALYTICAL TECHNIQUES

As already seen, the choice of technique is determined by the amount of drug likely to be present and the nature of the matrix. The physical techniques will be considered first, followed by chemical and biological.

A. Physical Techniques

I. Radioisotope Counting

Administering a drug in a radioactively labeled form to an animal is an excellent way of gaining information on how a drug distributes itself in the body tissues. This approach obviously is not applicable to studies in humans, and unless some means of separating the unchanged drug from any metabolites is used, this approach will only give information about the "total" drug and will be of limited value in pharmacokinetic studies. Another use of the technique is to estimate recovery from complex extractions by adding labeled drug in a known amount at the beginning and counting what remains at the end.

2. Gravimetry

This method has been in use for hundreds of years, ever since accurate chemical balances were developed. It can be used when drugs are present in a high concentration (e.g., the major component of a dosage form). Some simple initial extraction is usually applied to separate the drug, and it is then weighed.

3. Thermogravimetry

The change in weight of a sample on heating in a carefully controlled way is measured. It is widely applied to bulk drugs to investigate water content and stability.

4. Polarography

This can be applied to metals and many drugs and, in certain circumstances, can be applied directly to body fluids. The current is measured with changing potential between two electrodes. A dropping mercury electrode is the cathode (negative electrode) and constantly renews itself, whereas a pool of mercury is the positive anode. A characteristic half-wave potential is produced for the compound or group undergoing reduction. A mathematical relation links current, mercury flow, and concentration.

5. Ion-Selective Electrodes

These can be used with solutions containing drugs and require little or no preparation. The basic design consists of an electrical sensing device separated from the sample by a selective barrier of conducting material. New liquid ion-exchange membranes, enzyme electrodes, and membranes containing macrocyclic compounds have extended the use of this technique, and a number of drugs (e.g., antibiotics, alkaloids, and nonsteroidal antiinflammatory agents) have been measured in the range of 10–40 μM.

6. Spectroscopic Techniques

These can be used when the drug is present in relatively high concentration. They are usually applied after some form of extraction procedure, although, in the case of many dosage forms, this may not always be necessary. It uses light in the UV, VIS, and IR spectrum. The relatively high energy of the UV and VIS raises the electrons forming bonds in drug molecules to excited states, whereas lower energy absorption (IR) increases the vibrational energy of the bonds. These energy changes can be measured using appropriate equipment.

a. Ultraviolet and Visible Spectroscopy

These techniques are widely used for the quantitative analysis of compounds known to be present in formulations and body fluids. Measurement depends on the absorption of energy when the radiation traverses the sample. By preparing a series of standards containing known amounts of drug and by comparing their absorption with that of the sample, the amount of drug present can be calculated. The use of the technique has been extended by employing difference spectroscopy, in which different ionized forms of the drug are measured simultaneously, or derivative spec-

troscopy, in which slight changes in slope can be more easily detected or interference suppressed.

b. Infrared Spectroscopy

When drugs absorb in this region, they show changes in their vibrational and rotational bond energies, which means that IR is a useful tool in studying structure. However, it can also be used quantitatively if the drug possesses any strongly absorbing groups (e.g., double or triple bonds). In principle it is more versatile than either the UV or VIS range, in which drugs will often only display one or two absorption peaks. Sample preparation is simple because the technique is generally applied to dosage forms. A number of tablets are ground, and a sample is compressed into a disc with a halide such as potassium bromide or is made into a mull with liquid paraffin. Standards are prepared in the normal way, and the absorption of the chosen band allows calculation of the amount of drug present.

7. Fluorimetry

This is more sensitive than either UV or VIS and is more selective because the sample can be irradiated at one wavelength and emit light at a higher wavelength. Fluorescence is sensitive to a number of factors (e.g., pH, temperature, and solvent). Glassware should be scrupulously clean and all materials used checked for interfering fluorophores. The linear range is narrower than with either UV or VIS, but its greater sensitivity has allowed it to be used successfully for a number of common drugs including catecholamines, phenothiazines, β-adrenoceptor antagonists, and some antibiotics. After suitable chemical derivatization, it can also be used for drugs that do not possess any native fluorescence.

8. Miscellaneous Techniques

Atomic absorption is a sensitive and selective means of measuring metals and can be applied to biological fluids after only a simple dilution of the sample. This is because the energy sources (lamps) used are specific for the element in question. It is used frequently to monitor lithium concentrations in patients receiving lithium carbonate for manic depression. It can also be used for the trace analysis of zinc, copper, and aluminum. Spectrography can also be used in drug analysis, usually to determine trace contaminants, and X-ray diffraction has been used to provide structural information, but its use is hindered by the difficulty in obtaining high-quality crystals. Nuclear magnetic resonance is a powerful technique that has been used for quantitative and qualitative analysis. It is theoretically possible to measure several different drugs simultaneously, but the expense of the instrument and its demanding operating conditions have restricted its use. Finally, chemiluminescence is becoming more popular because appropriate labeling can result in very sensitive methods of detection for many steroid hormones.

B. Chemical Techniques

1. Volumetric Analysis

This is an old technique still used today. Here a reagent is chosen that will react chemically with the drug to be determined. Both are dissolved in a solvent and one is titrated against the other using the appropriate indicator of the end point. The volume of reagent consumed is used to calculate the amount of drug present. In its simplest form, this is an inexpensive technique, but its sensitivity is limited and it is restricted to the analysis of drugs in dosage forms. Its range and sensitivity can, however, be extended by using electrochemical methods to detect the end point, and titrations can also be carried out in nonaqueous media.

2. Chromatography

Chromatography has expanded rapidly, particularly since the mid-1950s, and it is now the most important technique used for drug analysis. An extensive literature is available on all aspects of its use. Essentially, the technique separates complex mixtures by allowing the individual components to partition between a stationary and a mobile phase. The stationary phase is usually contained within a column, and the mobile phase can be a gas (gas–liquid chromatography, GLC), a liquid (high-performance liquid chromatography, HPLC) or a supercritical fluid—a liquefied gas such as CO_2 (supercritical fluid chromatography, SFC). If the stationary phase is spread two-dimensionally, the process is called thin-layer chromatography (TLC), and when the particle size of this phase is carefully graded, it is called high-performance thin-layer chromatography (HPTLC).

The most important advantage of chromatography is that it can be made selective by (1) choosing the phases from a wide range of polarities and (2) using a number of sensitive and selective detectors. Under a given set of instrumental conditions, the time taken for a compound to move through the system from the start (injection) to the finish (detector) will be constant. This is called the *retention time* (t_R) and is usually measured in minutes. In HPLC, in which the mobile phase is a liquid, it may be measured either in minutes or volume (e.g., milliliters). In TLC the distance moved by the compound to be measured (e.g., drug) relative to the solvent front is called the *retention factor* (R_f). Column efficiency is defined by the number of theoretical plates, and this terminology derives from distillation theory. The larger the number of theoretical plates, the more efficient the column. The theory of chromatography is well documented in a number of standard works.

Most modern analytical laboratories will possess both GLC and HPLC equipment, and their use is largely complementary, although HPLC is now probably more popular than GLC. A brief comparison is given in Table IV.

a. Gas Chromatography

No matter whether GLC or HPLC is chosen, the choice of column is critical because an unsuitable column will give poor results even with the most sophisticated equipment. Two types of column are used in GLC: packed and capillary. The former are relatively short (1–4 m) and wide bore (2–4 mm), whereas the latter are much longer (20–50 m) with a narrower bore (0.2–0.7 mm). The gas used as the mobile phase depends on the detector system but can be nitrogen, helium, or mixtures (e.g., argon:methane. [*See* Gas Chromatography, Analytical.]

i. *Packed Columns* The column packaging consists of a stationary phase coated on a relatively inert support to give a large surface area. The general efficiency of such columns is relatively low, about 400–2000 theoretical plates; this may not be a disadvantage where the components of relatively simple extracts have to be separated [e.g., the active ingredients of a cough mixture or an extract of plasma in which the drug is present in high concentration (μg/ml)].

For analytical drug work, the percentage coating of the stationary phase on the support is low (1–3%).

ii. *Support* This is usually made from diatomaceous earth, which is composed almost entirely of silica in a pure form, which is carefully graded to give a narrow particle size distribution: this is important for an efficient column.

iii. *Stationary Phase* A good phase should be thermally stable and manufacturers' catalogues will indicate the recommended operating range for each. More than 200 phases of a wide range of polarity are avail-

TABLE IV
Comparison of GLC and HPLC

	GLC	HPLC
Stationary phase	Range of polarity	Range of Polarity
Mobile phase	Gas	Liquid (organic/aqueous)
Operating temperature	Subambient—350°C	Subambient—50°C
Programming available	Yes, temperature	Yes, gradient elution
Sample volume (μl)	Small <5	Large <500
Range of detectors	Wide, FID, N/PD, ECD	Wide, UV, FLUOR, EC
Detector sensitivity	High, especially after derivatizing	High, with EC or derivatizing
Detector selectivity	Good, N/PD and ECD	Poor–moderate
Suitable for high-MW drugs	No	Yes
Suitable for polar drugs	Not without derivatizing	Yes
Suitable for heat-labile drugs	No	Yes
Automation available	Yes	Yes

able, but most drug analysis can be carried out with about a dozen. To achieve an efficient separation, it is important that the components of the mixture partition between the stationary phase and the gaseous mobile phase (note that no separation takes place in the mobile phase itself).

The simple principle that "like dissolves like" can be applied in chosing a phase. For example, if a relatively nonpolar mixture of drugs is to be separated, a nonpolar phase should be chosen so that the components will truly dissolve and take a finite time to pass through the column. If, instead, a polar phase were chosen, the compounds would pass through quickly and emerge from the column without being separated.

In the early days of chromatography, columns were classified as being simply nonpolar, polar, or very polar, but columns can now be compared on quantitative terms. This is done by chromatographing a test mixture of a series of compounds containing a range of chemical groups on different stationary phases and comparing their retention times with those obtained on squalane, the least polar of the stationary phases. This gives a series of retention index differences that characterize each phase. The higher the retention index difference the more polar the phase. Tables of these *McReynolds constants*, as they are called, can be found in some suppliers' catalogues and are useful.

iv. Capillary Columns

These are much longer and narrower than packed colunms. The stationary phase can be bonded directly to the etched inner wall of the column, and an inert support is unnecessary. When this is done, the column is referred to as wall-coated open tubular (WCOT).

Supports such as microcrystalline sodium chloride can also be used, and these are first chemically bonded to the inner wall before coating the stationary phase to give support-coated open tubular (SCOT) columns. Most laboratories buy their capillary columns directly from commercial suppliers, although some still prefer to prepare their own. The thickness of the stationary phase can have an important effect on the separation because thin films (0.1–0.5 μm) give the most efficient separations with the shortest analysis times. However, greater amounts of sample can be handled by thicker films (1–1.5 μm). Column efficiency is high, and 10,000–50,000 theoretical plates can be obtained. The polarity of the stationary phase is less important with capillary columns, and much drug analysis is carried out on nonpolar phases such as OV 1. A great deal of packed column work is carried out at constant temperature. This is not always satisfactory, especially when mixtures containing high and low boiling components are present. In such cases, temperature programming is used. This option is now available on most modern instruments and is essential for capillary operation.

v. GLC Injection Systems

Generally drug extracts from formulations or body fluids are dissolved in a volatile organic solvent such as hexane or toluene, and for injection onto packed columns, a suitable volume is taken up in a capillary syringe (1–10 μl) and injected through a silicone septum onto the column. The injection port temperature is always kept at a temperature higher than the column so that the injected material is efficiently swept onto the column. Because of the narrow bore, injection onto capillary columns is more complicated. Syringes with fine silica needles are available, and injection directly onto the column is possible. Such on-column injection is efficient. Variations such as split injection are available for concentrated solutions, in which only a preselected portion of the extract reaches the column while splitless injection is available for very dilute solutions, in which the entire injected volume enters the column; modern instruments usually allow both systems to be used. Headspace analysis is useful for volatile materials (e.g., the determination of ethyl alcohol in blood).

vi. *Detectors*   A range of detectors are available, and flame ionization or electron capture are most frequently used for routine analytical work.

(a) Flame Ionization Detection

Here the detection at the end of the column consists of a hydrogen flame burning in air. A potential difference is maintained between the flame tip and a collector electrode above. When an organic compound emerges from the column, it burns in the flame, giving an increased flow of electrons detected as a peak. This detector is robust, with a large linear range, and is relatively sensitive (down to about 0.1 μg injected) for organic compounds that do not contain large numbers of oxygen or halogen atoms.

Nitrogen/phosphorus detection (N/PD) is a modification of flame ionization detection (FID), which can be tuned to give a selective response to compounds containing nitrogen (most drugs) or phosphorus (many insecticides). This is particularly useful because when drugs are being measured in biological fluids, even after a complex extraction, many other substances will also be present. It is possible to measure some nitrogen-containing drugs down to 1 ng/ml plasma or less with capillary column instruments equipped with this detector.

(b) Electron-Capture Detection

In this system a radioactive source (usually Ni^{63}) produces a stream of slow electrons by interacting with the carrier gas. Any drug containing halogen atoms, nitro-groups, or a highly conjugated system of double bonds will efficiently capture electrons. In practice this does not include many drugs, but chemi-

cal derivatization can easily produce electron-capturing compounds (see Section III,vii). The linear range is less than with FID, but subnanogram quantities of suitable drugs can be detected while reduced responses are produced by material that does not capture electrons. Thermal conductivity, photoionization, and flame photometric detection are also available, but for a number of reasons they are not as useful as the ones discussed earlier. For some purposes, FID and electron-capture detection (ECD) can be used together to gain important structural information about drugs and their metabolites.

vii. *Derivatization for GLC*   Although many drugs are volatile enough to be chromatographed directly, a number are too polar and have first to be converted to a less polar derivative. The conversion of carboxylic acids and amines to esters and amides is a good example. Silylation reactions are also widely used to give volatile trimethylsilyl (TMS) ethers. These latter types of derivative are generally more sensitive to water and are less stable than their nonsilylated counterparts. Derivatization can also be carried out to make the drug more electron-capturing. There is a wealth of chemical literature on chemical derivatization techniques.

b. High-Performance Liquid Chromatography

This is now the most widely used technique for the determination of drugs in dosage forms and body fluids. This is due to its more general applicability (e.g., it can be applied to polar and nonpolar drugs and high molecular weight compounds without derivatization, and it is less demanding in terms of gases, injection systems, and detectors). The different forms of HPLC are normal phase, reversed phase, ion-pair, and size exclusion chromatography. Table V gives an indication of the differences and how the right technique can be chosen. [*See* High Performance Liquid Chromatography.]

i. *Columns*   These are generally made from stainless steel and are 10–25 cm in length and $\frac{1}{8} - \frac{1}{4}$ inch in internal diameter. Good separations can be obtained with normal phase columns such as silica gel, particularly when closely related compounds (e.g., drugs and their metabolites) are being investigated. Reversed-phase chromatography involves a nonpolar column with a more polar mobile phase, which usually contains water. Selective separations can be obtained because the degree of polarity can be carefully chosen, the mobile phase buffered at a selected pH, and the

TABLE V
Choosing an HPLC Column

Drug	Chromatography	Column	Mobile phase	Example
Water insoluble				
Nonpolar	Partition, reversed phase	Bonded C_2-C_{18}	Polar	Antibiotics
Weakly polar	Adsorption, solid phase	Silica, alumina	Weakly polar	Alkaloids
Polar	Partition, normal phase	Bonded $-CN$, $-NO_2-NH_2$	Nonpolar	Alkaloids
Water soluble				
Anionic	Ion exchange	Anion exchanger	Polar	Sulfonamides
Both	Ion pairing	Bonded C_{18}	Polar	Many drugs
Cationic	Ion exchange	Cation exchangers	Polar	Catecholamines
Molecular weight >2000	Size exclusion	LiChrospher Microgel	Polar	Proteins, polymers

ionic strength controlled. Ion-exchange chromatography is now much less frequently used because it has largely been replaced by reversed-phase operation, which is more versatile. Size exclusion (gel permeation) chromatography depends on the different molecular sizes of the components of a mixture and is usually only used for compounds of MW >2000 and, hence, is not applicable to most drugs. It does, however, play an important role in the separation of peptides and polymers.

ii. *Injection Systems* These are much simpler than for GLC and usually involve a loop system where the extract, often dissolved in mobile phase, is injected into a loop isolated from the column. Rotation of a valve allows the extract to be efficiently swept onto the column for separation. Loops are usually of a fixed volume (20–500 μl) and hence larger volumes can be handled than by GLC. Again it is much easier to automate such a system.

iii. *Detectors* A range of different detectors are available and some are now approaching the sensitivity of GLC detectors.

(a) UV Detection

This is the nearest we can get to a universal detection system for drug analysis with HPLC because most drugs exhibit some UV absorption even if only at low wavelengths. Detectors can be variable or fixed wavelength, the latter being obviously more versatile if a range of different drugs is to be measured. Under suitable circumstances, these detectors can measure drugs in blood down to about 50 ng/ml. Diode array systems are a recent development and allow the recording of a complete UV spectrum of a compound as it emerges from the column. This is particularly useful where identification is important (e.g., in drug metabolism or screening). It can also be used to monitor the column effluent at different wavelengths simultaneously and is useful to confirm whether a peak consists of a single component.

(b) Fluorescence (FLUOR) Detection

This is more sensitive than UV, but fewer drugs possess a natural fluorescence. However, chemical derivatization can allow conversion to strongly fluorescent compounds (see Section III,iv). The ability to vary the excitation and emission wavelengths also makes the technique more selective, and low drug concentrations in plasma can be measured (5 ng/ml or less).

(c) Electrochemical (EC) Detection

This can offer a sensitive detection system, and it has become increasingly popular in the past 10 years or so. Most detectors are electrolytic and depend on the oxidation or reduction of the drug to generate a current. Because of the electrochemical nature of the system, mobile phases have to be capable of dissolving suitable electrolytes, and hence their use is largely restricted to aqueous systems and reversed-phase operation.

iv. *HPLC Derivatization* In HPLC, volatility is not important, and here derivatization is aimed at coupling strongly UV-absorbing or fluorescent groups to drug molecules to allow them to be determined at a low concentration in body fluids. Reaction conditions of temperature, time, solvent, and concentration all have to be carefully optimized. Reactions may be carried out before chromatography or as the compound emerges from the column. The latter approach is called *postcolumn derivatization,* and although this reduces the chances of decomposition, it restricts the choice of derivative.

c. Supercritical Fluid Chromatography

This is a relatively recent development that is claimed to combine some of the best features of GLC and HPLC. The main advantage is that a supercritical fluid possesses the solvating properties of a true liquid but is much less viscous, and more efficient separations can be obtained using open tubular columns or conventional HPLC columns. The polarity of the mobile phase can be altered simply by changing the pressure, and the technique is highly compatible with FID. At present it works better with nonpolar than polar compounds because of the limited choice of mobile phases (e.g., carbon dioxide and nitrous oxide).

d. Thin-Layer Chromatography

This is a versatile technique and, even in its most inexpensive form, can provide a great deal of useful information regarding the presence of contaminants in dosage forms and the structure of metabolites and as a simple means of drug screening. TLC can be regarded as two-dimensional HPLC, and TLC systems are often used in the early stages to develop conditions for a new HPLC separation.

The plates themselves are made of glass, plastic, or aluminum and can be normal phase (silica gel), cellulose, or reversed-phase using hydrocarbons 8–18 carbon atoms long. Plates can be obtained with a fluorophore added, which allows UV-absorbing spots to be readily detected. A wide range of spray reagents are available to allow different chemical groups to be detected (e.g., phenols, amines, and carboxylic acids).

TLC has the advantage that spots can be scraped off the plate and the compounds eluted for study by some other technique.

e. High-Performance Thin-Layer Chromatography

This can be a useful means of drug analysis. For the best results, the sample "spot" size must be as small as possible, and this can rarely be achieved by hand. Automated sample application is not expensive, and volumes of as much as 50 μl can be applied.

Development is carried out usually in a glass tank,

and the choice of stationary and mobile phase is carried out as in HPLC. The volume of the vapor phase can be reduced by using specially designed tanks, and in overpressure layer chromatography (OPC), linear development takes place with the minimum of mobile phase. Circular development (CTLC) is also posible, and gradient elution has also been used. Once the plate has been developed, the drugs can usually be quantitated by scanning densitometry in the UV or fluorescence mode.

f. Capillary Zone Electrophoresis

The use of this technique has increased very rapidly. It depends on the separation of charged species due to differences in their rate of migration in an electric field and its main advantage is that it offers very rapid separations of acidic, basic, and neutral drugs with low or high molecular weights.

g. Combination Techniques

A number of so-called hyphenated techniques are used to provide powerful analytical instruments such as GC-MS (mass spectrometry), GC-IR, GC-IR-MS, LC-MS, and LC-FID. GC-MS is used mainly to obtain structural information on a drug or its metabolites or as a selective analytical instrument when set to detect a single mass ion. The most useful combination with HPLC is designed to allow FID to be used. Not unexpectedly, the combination of a technique that uses a liquid mobile phase with a flame detector has proved difficult, but instruments are now available and, as we have seen, SFC provides another approach to the problem. The combination of HPLC with the most sensitive form of IR, Fourier transform, is able to provide good structural information, but mobile phases containing water or other polar solvents cannot be used, and hence reversed-phase chromatography is impossible.

3. Separation of Optical Isomers

Many drugs contain at least one optically active center and can, therefore, give rise to compounds of the same composition (isomer) that rotate the plane of polarization of light either to the right or to the left. Most synthetic methods produce racemic mixtures containing equal amounts of both isomers. Usually only one of the isomers possessed the desired pharmacological activity, but little attempt is made to separate the two because the other isomer is usually regarded as inert. However, it is now appreciated that in some cases the "inert" isomer may contribute substantially to unwanted side effects and its kinetics may be different from the other isomer. Resolution and analysis of optical isomers can be achieved by HPLC, TLC, and even GLC, and this is an active area of research.

C. Biological Techniques

I. Immunoassays

These are used for the rapid determination of drugs in biological fluids, and the principle behind their application is simple. Many foreign substances are antigenic and, when introduced into animals, will generate an antibody (Ab) response, and the antibody produced will bind the foreign material. In practice, most drugs have molecular weights too low to generate a significant antibody response and first have to be coupled with larger molecules such as albumin. The albumin used has to be from a species other than that in which the antibody is to be raised. The way in which the drug is coupled to the albumin (i.e., the three-dimensional stereochemistry of the protruding drug molecule) will determine the specificity of the antibody produced (i.e., whether closely related compounds will also bind to the antibody). The size of the animal chosen will generally govern the amount of antibody produced, and rabbits and horses are frequently used. Once the antibody has been isolated and purified, a dilution curve is prepared to find the antibody dilution that will respond most sensitively to the drug concentrations likely to be present in the specimens to be assayed.

How immunoassays work can be seen from the following simple example. If it is desired to measure drug A in patients, a small amount of the drug is prepared in a modified form, A*, which may be radio-labeled, fluorescence labeled, etc. A known amount of A* is added to the patient plasma sample and, after mixing, the antibody to A is added, when the following reaction takes place:

$$A + A^* + Ab \rightleftharpoons \underset{free}{A\text{-}Ab} + \underset{bound}{A^*\text{---}Ab}$$

in which the unmodified drug (A) competes with A* for binding to the antibody. By separating the free from the bound fraction and estimating the amount of A* free, the amount of A in the original sample can be calculated. If the drug is labeled with an enzyme, fluorophore, or chemiluminescent substrate and binding to the antibody produces a change in property that can be detected without separation of bound from free, the assay is called homogeneous. This is obviously useful because little sample treatment is needed. If the label is in a form that produces no change in property (e.g., radiolabeling), a separation of bound from free is necessary, and the assay is called heterogeneous. Separation can be carried out in various ways. [See Immunoassays, Nonradionucleotide.]

a. Radioimmunoassay

A good example of the usefulness of immunoassays can be illustrated by the cardio-active agent, digoxin, a powerful drug given in low doses. The clinical signs of under- and overdosing are not easy to distinguish, and therapeutic monitoring is necessary. Digoxin has a high molecular weight and is unsuitable for GLC. It also does not have a strong enough UV absorption for HPLC. It is routinely and rapidly monitored by radioimmunoassay. Three isotopes are commonly used for radiolabeling. Their properties are compared in Table VI. Chemical synthesis can supply ^3H and ^{14}C labeling of most drugs, but few contain iodine. Techniques exist for iodination of such drugs. These techniques, however, may modify the properties of the drug because of the introduction of such a large atom. [See Radioimmunoassays.]

b. Optical Immunoassays

In this approach the reagents used are stable, and most assays are homogeneous. In enzyme-linked immunoassays, the drug to be determined is linked to an enzyme such as peroxidase or glucose-6-phosphate dehydrogenase. The sample volume required is gener-

TABLE VI
Properties of Isotopes Used for Immunoassays

Isotope	Symbol	Half-life	Radiation	Energy (KeV)	Counter	Advantages	Disadvantages
Tritium	^3H	12.3 years	β	18	Liquid scintillation	Long half-life, low hazard	Counting expensive
Carbon	^{14}C	5730 years	β	155	Liquid scintillation	Long half-life, low hazard	Counting expensive
Iodine	^{125}I	60 days	X- and γ	27	Crystal scintillation	No extraction, high specific activity	Iodination may alter binding

ally small (10–50 μl), and all the materials needed for the assay are supplied commercially in kit form. Fluoroimmunoassays are useful and their sensitivity can be controlled by the choice of fluorophore. Fluorescein is the most popular, although others have also been used. Luminescence requires the drug to be linked to a substance that emits light and, because light is not introduced into the system, measurement is carried out in a luminometer in the dark, and increased sensitivity can be obtained. Luminol is frequently used as the label, and because luminescence is generally short lived (about 500 nsec), rapid mixing and signal integration are essential.

c. Advantages and Disadvantages of Immunoassays

Immunoassays are rapid, and many hundreds if not thousands of patient specimens can be processed daily. This is far in excess of what can be achieved by any of the chromatographic methods, even if fully automated. They require relatively little equipment, especially the optical immunoassays, and can be set up in outpatient clinics or adjacent to hospital wards to provide immediate results. Their main disadvantage is that they are designed to provide information on only one drug or a single group of drugs, unlike the chromatographic or spectroscopic systems that can easily be modified to look at completely different compounds. Price is also important, and the kits are expensive but bulk buying can bring down unit costs. However, because immunoassays are generally supplied commercially, kits are only available for drugs for which there is a demand, and new or experimental compounds have to be determined by other, usually chromatographic means.

2. Microbiological Assay

This is a simple and useful method of determining antibiotics in body fluids. In essence, agar plates containing a selected microorganism are prepared, and diluted specimens of patient plasma containing the antibiotic to be determined are introduced into wells cut in the plates. After a suitable incubation period, the area of microbial growth inhibition around each well is measured and compared with the areas around wells on the same plate containing standard concentrations of antibiotic. Much of the early work on antibiotic pharmacokinetics was carried out using this technique, but its main disadvantage is that only microbial activity is measured. If the antibiotic is metabolized and any of the metabolites possess similar activity, falsely high results will be obtained. If patients are also receiving other antibiotics, as frequently happens, these may also produce an inhibition in growth of the chosen organism.

For these reasons, although the method is still used in certain circumstances, it has been superseded largely by HPLC.

IV. QUALITY CONTROL

No matter which method is chosen, it is important to ensure that it functions properly, with satisfactory precision and accuracy, and gives the "correct" result. Every method involves the preparation of some type of calibration curve with standards of known concentration, and in each batch processed, several other samples can be incorporated containing known amounts of drug covering the anticipated concentration range. These are "quality control" specimens and are not used in the construction of the calibration curve. Monitoring the day-to-day results for these samples can give important information on how a method is performing. Samples may also be reanalyzed at frequent intervals to give repeat analysis checks. Important as it is, intralaboratory control is not enough. Interlaboratory comparison is more demanding but ultimately more rewarding in terms of raised standards and increased operator confidence. A number of national and international schemes exist for drugs that are monitored regularly for therapeutic purposes (e.g., digoxin, diphenylhydantoin, theophylline). At regular intervals, laboratories are sent freeze-dried plasma samples containing relevant drugs and these are analyzed in the same way as normal patient samples. The results are collected by a central body, and each laboratory receives a document showing how it performed in relation to the other participating laboratories. These data are often supplied in histogram form, and the identities of the other laboratories are not revealed. This type of approach is useful as an independent check on performance and it rapidly identifies laboratories using unsatisfactory methods. For example, the determination of many anticonvulsant drugs by GLC was clearly shown to be inferior to HPLC by external quality control.

BIBLIOGRAPHY

Aszalos, A. (ed.) (1986). "Modern Analysis of Antibiotics." Dekker, New York.

Gennaro, A. R. (ed.) (1990). "Remington's Pharmaceutical Sciences." Mack Publishing, Easton, PA.

Grob, R. L. (1985). "Modern Practice of Gas Chromatography." Wiley, Chichester, United Kingdom.

Jack, D. B. (1984). "Drug Analysis by Gas Chromatography." Academic Press, Orlando, FL.

Jack, D. B. (1990). Chemical analysis. *In* "Comprehensive Medicinal Chemistry" (J. Taylor, ed.), Vol. 5. Pergamon Press, Oxford.

McDowall, R. D., Pearce, J. C., and Murkitt, G. S. (1986). Liquid–solid sample preparation in drug analysis. *J. Pharm. Biomed. Anal.* **4,** 3–21.

Moffat, A. C. (ed.) (1986). "Clarke's Isolation and Identification of Drugs," 2nd Ed. Pharmaceutical Press, London.

Munson, J. W. (ed.) (1981). "Pharmaceutical Analysis Modern Methods." Dekker, New York.

Poole, C. F., and Schuette, S. A. (1984). "Contemporary Practice of Chromatography." Elsevier, Amsterdam.

Souter, R. W. (1985). "Chromatographic Separation of Stereoisomers." CRC Press, Boca Raton, FL.

Wong, S. H. Y. (ed.) (1985). "Therapeutic Drug Monitoring and Toxicology by Liquid Chromatography." Dekker, New York.

Drugs of Abuse and Alcohol Interactions

MONIQUE C. BRAUDE
BRACOL
(previously with the National Institute on Drug Abuse)

GLOSSARY

Antagonism Mechanism by which a drug inhibits the action of another drug

Metabolism Drug metabolism is a general term applied to the chemical processes taking place in living tissues after the administration of a drug that alter its original chemical composition

Synergism Quality of two drugs whereby their combined effects are greater than the algebraic sum of their individual effects

INTERACTIONS BETWEEN ALCOHOL AND OTHER drugs can result in synergism or antagonism. A direct interaction is the result of the pharmacologic effects of the drugs. Indirect effects include pharmacokinetic interactions (in which a drug alters the absorption, distribution, metabolism, or excretion of another) and tolerance phenomena, in which a drug alters the response of the target tissue to itself or another drug. Tolerance to alcohol has been reported after repeated use. Cross-tolerance between alcohol and other drugs occurs when the physiologic changes induced by alcohol, for instance, carry over to another drug (e.g., a barbiturate) and the observed effect of the second drug is diminished. Interactions between alcohol and other drugs can usually be understood in terms of one or more of these phenomena.

The important consequences of psychotropic drug interactions for an individual and society include the following:

1. Unexpected degree of impairment in performing daily tasks associated with a serious potential for injury (driving a car, operating machinery).
2. Accidental death from overdose.
3. Increase in behavioral toxicity (cognitive functions, memory impairment, etc.).

The concomitant oral ingestion of alcohol and other drugs could theoretically affect the absorption of these drugs, but information is mostly lacking in this area and the existing evidence does not point to clinically significant effects. Alcohol in the body is primarily metabolized by the liver alcohol dehydrogenase to yield acetaldehyde. Research since the early 1980s also indicates that acute and chronic administration of alcohol can inhibit the biotransformation of many drugs that are normally degraded by liver enzymes. The most prominent result of this alcohol-induced inhibition of enzymes *in vivo* appears to be a prolongation of the plasma life of drugs such as pentobarbital, meprobamate, and amphetamines, but not phencyclidine (PCP). The amount of alcohol use and the direction of its use during a long or short period are also important factors in drug interactions.

Whereas acute alcohol intoxication usually inhibits the biotransformation of drugs, the long-term ingestion of alcohol can lead to hepatic microsomal enzyme induction and can produce enhancement of the formation of intermediate metabolites that may be toxic to the liver, thus increasing the hepatotoxic effect of alcohol. Such appears to be the case with cocaine. Finally, alcohol can influence the elimination of other drugs indirectly by causing hepatic dysfunction and/or nutritional deficiencies.

ENCYCLOPEDIA OF HUMAN BIOLOGY, Second Edition, VOLUME 3. Copyright © 1997 by Academic Press. All rights of reproduction in any form reserved.

I. INTERACTIONS OF ALCOHOL WITH CENTRAL NERVOUS SYSTEM DEPRESSANTS AND BENZODIAZEPINES

Interactive effects between alcohol and central nervous system (CNS) depressants such as barbiturates, benzodiazepines (e.g., valium), and methaqualone are frequent. In terms of CNS depression, use of these drugs with alcohol is usually additive. With drugs that are metabolized by the liver enzymes (e.g., phenobarbital and pentobarbital), synergism has been reported, leading to increased toxicity and death in humans.

A. Alcohol and Opiates

Concomitant alcohol and narcotic abuse is common. More than half of the "overdose" cases with heroin or methadone are, in fact, cases in which concomitant abuse of alcohol has played a prominent role.

Several studies have shown that alcohol may have a biphasic effect on the disposition of another opioid. During chronic use of large quantities of alcohol and at times when high blood levels of alcohol are present, alcohol may inhibit the metabolism of other drugs such as methadone. But when alcohol is no longer present in the body, drug metabolism may be accelerated because, in the chronic alcoholic, alcohol has produced an enhancement of the liver detoxifying enzyme system. The presence of chronic alcoholic-induced liver disease may result in significant alterations in methadone disposition and prevent achievement of the steady state during methadone treatment.

Interestingly, the opiate antagonist naltrexone has been approved by the Food and Drug Administration for use in the treatment of alcoholism in the United States and abroad. This is probably due to a genetic factor in individuals whose endogenous opioid system may be susceptible to alcohol effects.

B. Alcohol and Cannabinoids

Marijuana and alcohol are ubiquitous. The combined use of alcohol and marijuana in variable amounts, frequencies, and settings has become a well-established fact. In a study of fatally injured truck drivers, analyses of blood specimens showed that cannabinoids and alcohol were found in 13% of the drivers. Impairment due to marijuana use was a factor in all cases where the $\Delta 9$-tetrahydrocannabinol (THC) concentration exceeded 1.0 ng/ml, and alcohol impairment contributed to all accidents where the blood alcohol concentration was 0.04% (w/v) or greater. Studies in rats showed that simultaneous administration of THC and alcohol resulted in an increased rate and magnitude of alcohol tolerance and physical dependence. Complete alcohol tolerance was established within 12–16 days in animals receiving both drugs, whereas only minimal alcohol tolerance or cross-tolerance was detected in alcohol- or THC-treated rats, respectively, at this time. [*See* Marijuana and Cannabinoids.]

A few studies in humans have indicated that the acute administration of THC, the primary psychoactive component of marijuana, in combination with alcohol results in enhanced impairment of physical and mental performance, more marked subjective effects such as pupil size, and a greater increase in pulse rate and conjunctival congestion compared with ingestion of either drug alone. However, the magnitude of the enhancement of the drug effects was not great.

More recent studies, using batteries of cognitive, perceptual, and motor function tests, also found that both THC and alcohol produce significant decrements in the general performance factors, but there was no evidence of any interaction between THC and alcohol, and the effects of a combination of both compounds were no more than additive.

A study of the effects of alcohol and marijuana on mood and behavior found that moderate doses of alcohol and marijuana, alone or in combination, produced acute behavioral and subjective impairment, but few residual effects were seen on the following day.

Cannabidiol (CBD), the other major cannabinoid of the marijuana plant, did not produce any demonstrable effects either alone or in combination with alcohol. However, as CBD is one of the major components of some marijuana plants, there is still the possibility that there may be a THC–CBD–alcohol interaction. The combination of alcohol with CBD in humans may result in significantly lower blood alcohol levels compared with alcohol given alone. However, there are few differences between the pharmacological effects of alcohol given alone or given with CBD.

In conclusion, it seems that there is no more than an additive effect between cannabinoids and alcohol (both acting as CNS depressants). However, this may be dangerous in pregnancy, as it was shown that alcohol produces a 100% enhancement of marijuana-induced fetotoxicity in rodents. This is especially significant in view of a study on concordant alcohol and marijuana use in women, showing strong correlations between alcohol and marijuana use. The heavy marijuana smokers reported drinking significantly more alcohol than the light marijuana smokers. For pregnant women who combine the use of alcohol and marijuana, this suggests a potential danger that may be far greater than that associated with using either drug alone.

C. Benzodiazepines and Alcohol

Benzodiazepines have now significantly replaced barbiturates as anxiolytics and sedative hypnotics. The most extensively studied drug interactions have been those of benzodiazepines and alcohol. Alcohol and benzodiazepines interact at both pharmacokinetic and pharmacodynamic levels in a predictable manner qualitatively or quantitatively by enhancing each other's effects. However, benzodiazepines do not seem to potentiate the effects of alcohol as much as other sedatives do.

Kinetically, acute doses of alcohol impair the disposition of benzodiazepines that are metabolized by demethylation or hydroxylation. For instance, it increases the blood levels of orally administered diazepam but not of oxazepam, which undergoes glucuronide conjugation. Chronic alcohol ingestion, however, increases the clearance of benzodiazepines that are demethylated or hydroxylated.

In general, alcohol and benzodiazepines, when studied alone, produce sedation and impair motor coordination, reaction time, memory acquisition, retention, and recall in a dose-related manner. Many studies indicate that the acute combination of both drugs produces the same impairment at a lower dose than given separately or reveals an impairment not apparent with the control dose of the drug alone. In this enhancing effect, alcohol appears to be the dominant partner. Although alcohol–benzodiazepine interactions may be less important than those involving alcohol and other psychotropic drugs (e.g., cannabinoids, neuroleptics, stimulants, and antidepressants), Chan points out that this combination is associated with drug-induced deaths, drug overdoses, traffic accidents, and fatalities.

These interactions may in part be explained by the effects of these compounds on the γ-aminobutyric (GABA) receptor. GABA is a major inhibitory neurotransmitter in the mammalian CNS. The GABAergic synapse is also an important site of action for a variety of centrally acting drugs, including alcohol, benzodiazepines, and barbiturates. In drug combination studies, subeffective doses of alcohol, in combination with subeffective doses of GABAmimetics, potentiate each other's effects. The potentiating effect of alcohol is blocked by GABA antagonists and the inverse agonists of the benzodiazepine receptor site.

Interestingly, a major clinical use of benzodiazepines has been in the short- and long-term treatment of alcoholics as benzodiazepines were found effective in alleviating alcohol-withdrawal symptoms. This is now controversial, as some studies have pointed out that alcoholics may be more at risk of developing benzodiazepine or alcohol–benzodiazepine dependence than the general population. There is therefore a dire need for large-scale controlled studies concerning the efficacy of benzodiazepines in the long-term treatment of alcoholics.

II. COMBINED ABUSE OF ALCOHOL AND STIMULANTS: AMPHETAMINES AND COCAINE

Surprisingly, few studies have dealt with the interaction of alcohol and amphetamines in humans despite the widespread use of both. Most of the studies that have investigated the effects of combined alcohol and amphetamine use have used either animal or volunteer subjects rather than members of the subculture directly involved in such abuse. A novel and neurotoxic metabolite of amphetamine, a tetrahydroisoquinoline derivative, was found in the brain of chronic alcohol-intoxicated rats subjected to repeated amphetamine administration. These rats exhibited behavioral abnormalities and repeated seizures. This compound was not found in the brains of rats receiving only amphetamine administration but no alcohol. Although synergism between alcohol and amphetamines was demonstrated in some preclinical studies, a large majority have reported antagonism between the two. The clinical evaluation of patients for treatment of alcohol abuse shows that their clinical histories indicate the existence of two types of abuse patterns, one primarily concerned with the effects of amphetamine and the other with the effects of alcohol; the latter type of patients use amphetamine only to help them maintain a wakeful state.

Cocaine, a naturally occurring alkaloid, is a powerful and rapid CNS stimulant of short duration, which has become a prominent and favored drug of abuse. Cocaine is rapidly degraded and metabolized in the body by various enzymes. Alcohol pretreatment produces inhibition of the esteratic metabolism of cocaine to benzoyl-ecgonine, resulting in higher cocaine levels and metabolism through alternative pathways to compounds such as norcocaine and cocaethylene that are more toxic than cocaine.

Cocaethylene is a psychoactive metabolite of cocaine that is formed exclusively during the coadministration of cocaine and alcohol. Not a natural alkaloid of the coca leaf, cocaethylene can be identified in urine, in blood, in hair, and in neurological and liver tissue samples of individuals who have consumed both cocaine and alcohol.

To investigate the pharmacologic effects of the interaction between cocaine and alcohol, human volunteers were given these two compounds successively. *Depending on the order of administration,* some of the resulting subjective and cardiovascular effects were different. When alcohol was given prior to cocaine, significant increases in cocaine plasma concentrations, subjective ratings of cocaine "high," and increased heart rate were observed. This did not happen when alcohol was administered after cocaine. These results underline again the importance of the mode of administration in interaction studies.

Both clinical experience and epidemiological studies in community and specialized (e.g., treatment) populations indicate that the prevalence of the co-use of alcohol and cocaine and the comorbidity of alcoholism and cocaine addiction are greater than would be expected from the chance occurrence of two independent conditions. Thus, the prevalent co-use of alcohol and cocaine has important implications for drug abuse treatment.

Finally, one should not forget that alcohol and other drugs of abuse such as morphine and barbiturates alter the hemodynamic and cerebral response to cocaine in the newborn and may contribute to central nervous system abnormalities seen in "crack" babies.

III. CONCLUSION

In summary, most of the abused compounds interact with alcohol in a predictable manner. What is important to remember is that the pharmacokinetic and pharmacodynamic interactions of a given drug with alcohol are in part dependent on the length of the individual exposure to alcohol. The end response may be enhanced in the acutely intoxicated individual who does not have a history of alcohol abuse. Conversely, it may be diminished in the otherwise physically healthy but alcohol-dependent individual. As mentioned with marijuana, the combined use of alcohol and abused substances in pregnancy requires further attention as it may be hazardous to the fetus and the offspring.

In general, appreciation of alcohol–drug interactions should provide a useful basis for the management of alcohol patients and the prevention of accidental overdoses in the elderly.

BIBLIOGRAPHY

Abel, E. (1985). Alcohol enhancement of marijuana induced fetotoxicity. *Teratology* **31**, 35–40.

Albuquerque, M. L., *et al.* (1995). Ethanol, barbiturate and morphine alter the hemodynamic and cerebral response to cocaine in newborn pigs. *Biol. Neonate* **67**(6), 432–440.

Belgrave, B. E., Bird, K. D., *et al.* (1979). The effect of delta-9-tetrahydrocannabinol, alone and in combination with ethanol on human performance. *Psychopharmacology* **62**, 53–60.

Braude, M. C. (1986). Interactions alcohol and drugs of abuse. *Psychopharmacol. Bull.* **22**(3), 717–721.

Braude, M. C., and Ginzburg, H. M. (eds.) (1986). Strategies for research on the interactions of drugs of abuse. *Natl. Inst. Drug Abuse Res. Monogr. Ser.* **68**, DHHS, Washington, D.C.

Chait, L. D., and Perry, J. L. (1994). Acute and residual effects of alcohol and marihuana, alone and in combination, on mood and performance. *Psychopharmacology* **115**(3), 340–349.

Chan, A. W. K. (1984). Effects of combined alcohol and benzodiazepines: A review. *Drug Alcohol Depend.* **13**, 315–341.

Closser, M. H., and Kosten, T. R. (1992). Alcohol and cocaine abuse: A comparison and epidemiological characteristics. *Rec. Dev. Alcohol* **10**, 115–128.

Crouch, D. J., *et al.* (1993). The prevalence of drugs and alcohol in fatally injured truck drivers. *J. Forens. Sci.* **38**(6), 1342–1353.

Forster, L. E., *et al.* (1993). Alcohol use and potential risk for alcohol-related adverse drug effects among community-based elderly. *J. Commun. Health* **18**(4), 225–239.

Gorelick, D. A. (1992). Alcohol and cocaine: Clinical and pharmacological interactions. *Rec. Dev. Alcohol.* **10**, 37–56.

Kipperman, A., and Fine, E. W. (1974). The combined use of alcohol and amphetamines. *Am. J. Psychiat.* **131**(11), 1277–1280.

Kreek, M. J., and Stimmel, B. (1984). Dual addiction: Pharmacological issues in the treatment of concomitant alcoholism and drug abuse. *Adv. Alcohol. Subst. Abuse* **3**(4), 1–6.

Landry, M. J. (1992). An overview of cocaethylene, an alcohol-derived, psychoactive cocaine metabolite. *J. Psychoact. Drugs* **24**(3), 273–276.

Lex, B. W., Griffin, M. L., *et al.* (1986). Concordant alcohol and marijuana use in women. *Alcohol* **3**(3), 193–200.

Makino *et al.* (1990). A novel and neurotoxic tetrahydroisoquinoline derivative in vivo: Formation of 1,3-dimethyl-1,2,3,4-tetrahydroisoquinoline, a condensation product of amphetamines, in brains of rats under chronic ethanol treatment. *J. Neurochem.* **55**(3), 963–969.

Mello, N. K., Mendelson, J. H., *et al.* (1978). Human polydrug use: Marijuana and alcohol. *J. Pharmacol. Exp. Ther.* **207**(3), 922–935.

Muhoberac, B. B., Roberts, R. K., *et al.* (1984). Mechanism(s) of ethanol-drug interactions. *Alcohol. Clin. Exp. Res.* 8(6), 583–593.

O'Brien, C., and Volpicelli, J. (1996). Naltrexone in the treatment of alcoholism: A clinical review. *Alcohol* 31, 35–39.

Perez-Reyes, M., and Jeffcoat, A. R. (1992). Ethanol/cocaine interaction: Cocaine and cocaethylene plasma concentrations and their relationship to subjective and cardiovascular effects. *Life Sci.* 51(8), 553–563.

Perez-Reyes, M. (1994). The order of drug administration: Its effects on the interaction between cocaine and ethanol. *Life Sci.* 55(7), 541–550.

Roberts *et al.* (1993). Inhibition by ethanol of the metabolism of cocaine to benzoylecgonine and ecgonine ethyl ester in mouse and human liver. *Drug Metab. Dispos.* 21(3), 537–541.

Sellers, E. M., and Busto, V. (1982). Benzodiazepines and ethanol: Assessment of the effects and consequences of psychotropic drug interactions. *J. Clin. Psychopharmacol.* 2(4), 249–262.

Shen *et al.* (1995). The effects of chronic amphetamine treatment on prenatal ethanol-induced changes in dopamine receptor function: Electro-physiological findings. *J. Pharmacol. Exp. Ther.* 274(3), 1054–1060.

Ticku, M. K. (1990). Alcohol and GABA-benzodiazepine receptor function. *Ann. Med.* 22(4), 241–246.

Drug Testing

Massachusetts General Hospital

GLOSSARY

Bioavailability Ratio of the amount of a drug which reaches the circulation to the amount administered

Chain of custody An account of who had a specimen, when they had it, and where they had it, from time of collection to time of testing

Cloned enzyme donor immunoassays Immunoassay technique in which the activity of two coupled enzymes correlates with the amount of drug in a specimen

Cytochrome P450 mixed function oxidase system An inducible network of enzymes in hepatic microsomes involved in the oxidative inactivation of many drugs

Drug Chemical taken internally or applied topically for its desirable physiological or psychological effects, but having the potential for toxicity

Enzyme-multiplied immunoassay technique Immunoassay technique in which the activity of an enzyme correlates with the amount of drug in a specimen

Fluorescence polarization immunoassay Immunoassay technique in which the polarization of fluorescent light correlates with the amount of drug in a specimen

Gas chromatography (CG) A separation method with high sensitivity and good resolution that is best suited for volatile compounds

High-pressure liquid chromatography A separation method with high sensitivity and good resolution; applicable to a broader range of compounds than GC

Plasma The liquid component of blood, separated from the cellular components of blood by centrifugation in the presence of an anticoagulant

Serum The liquid component of blood after blood has been allowed to clot. Similar, but not identical in composition to plasma, except for the loss of coagulation factors

Therapeutic index (TI) Ratio of drug dosage producing toxicity in 50% of subjects (LD_{50}) to that producing a therapeutic effect in 50% of subjects (ED_{50}):

$$TI = LD_{50}/ED_{50}$$

Thin-layer chromatography A separation method with moderate sensitivity and fair resolution

Volume of distribution Ratio of the total amount of drug in the body to the plasma concentration of the drug

TOXICOLOGY CAN BE BROADLY DEFINED AS THE assay of exogenous chemicals in tissues and body fluids. Drug testing is a subset of toxicology. The boundaries of this topic are somewhat arbitrary, depending on the definition of a "drug." The definition just given entails current prescription and over-the-counter therapies and their legitimate and illicit applications. It also includes exogenous sources of substances normally found in the body, such as hormonal supplements, but excludes food. There are three principal indications for drug testing: therapeutic drug monitoring (TDM); identifying overdosage in the acute care setting; and forensic, workplace, and athletic drug screening. This article briefly addresses the relevant pharmacology and methodology. Examples from each category of drug testing are discussed.

I. IMPORTANT CONCEPTS FROM PHARMACOLOGY

A general property of medically useful drugs is that the therapeutic dose is less than the toxic dose. This is not a universal property, however, depending on how one defines toxicity. Certain cancer chemotherapeutic regimens, for example, produce hair loss and nausea in a majority of patients.

The therapeutic index (TI) is defined as the ratio of drug dosage producing toxicity in 50% of subjects (LD_{50}) to that producing a therapeutic effect in 50% of subjects (ED_{50}). The TI is a measure of the margin of safety in administering a drug: the larger the TI, the safer the drug. However, a single number may not suffice in predicting toxicity, especially if the LD range overlaps the ED range. Other parameters can be more clinically relevant, such as the ratio of LD_1/ED_{99}.

Given a specific dose of a drug with a known TI, why is further testing needed? To answer this question, it is helpful to draw on some basic principles from pharmacology. [See Pharmacokinetics.] A simplified model of pharmacokinetics entails three phases: absorption, distribution, and elimination.

Absorption depends primarily on the chemical nature of the drug and the route of administration. Bioavailability is a measure of absorption, defined as the ratio of the amount of drug which reaches the circulation to the amount administered. Intravenous drugs are assumed to have a bioavailability of one. Oral, nasal, rectal, and even intraperitoneal routes yield lower bioavailabilities for several reasons. The drug may be partially inactivated by the method of administration, such as pyrolysis of drugs which are smoked or precipitation of drugs in solution. Drugs may be inactivated by enzymes at the mucosal interface or be activated from a prodrug form. Because charged drugs will not readily diffuse across the epithelium, acidic drugs preferentially cross the gastric mucosa, where pH is low and they are neutrally charged. Similarly, basic drugs may pass through the oral or duodenal mucosa, where pH is high and they are neutrally charged. Processes which modulate transit time or pH in these areas of the gastrointestinal tract may facilitate or hinder absorption. If facilitated or active transport is required for drug uptake, other drugs or an underlying illness may inhibit the process.

Once absorbed, the drug can be distributed across many body "compartments." Lipophilic molecules generally concentrate in cell membranes, whereas charged molecules tend to remain in the circulation and lymphatics. Even within the plasma, multiple compartments exist, albeit in rapid equilibrium. Many positively charged drugs are highly bound to albumin, whereas many negatively charged drugs are bound to

α_1-acid glycoprotein. Generally, it is the concentration of unbound ("free") drug in the plasma that correlates with activity. Yet it is usually total plasma drug concentration which is measured.

Because many drug targets are membrane receptors, lipophilic drugs may demonstrate activity, despite a low plasma concentration. One important underlying assumption in TDM is that of kinetic homogeneity: that the plasma drug concentration is in fixed proportion to the concentration in the various body compartments. Certain areas of the body, such as the central nervous system or testicles, have a functional permeability barrier that slows the accumulation of drugs there. For drugs active at such sites, steady-state equilibrium may never be obtained, and the assumption of kinetic homogeneity may never be valid.

Another helpful theoretical construct is the volume of distribution (V_d), defined as the total amount of drug in the body divided by the plasma concentration. For lipophilic drugs, the V_d can be several times the actual plasma volume. Drugs with a large V_d may be slow to clear from the system.

Partitioning can be extremely rapid between bound and free plasma compartments. Displacement of one bound drug by another agent can rapidly increase the concentration of the free drug, which increases both activity and clearance. This is an important form of drug interaction. Such potentiation can also result from the accumulation of metabolites as in uremia.

Elimination of most drugs occurs via the liver or kidneys. Some drugs are excreted unchanged in the urine or bile. Often, drugs are metabolized to a more polar and less active form by hepatic microsomal enzymes, then filtered and excreted by the kidneys. The cytochrome P450 mixed function oxidase system (CP450) is an inducible network of enzymes in hepatic microsomes. [See Cytochrome P450.] The oxidation step catalyzed by this system inactivates many drugs, but can actually form toxic metabolites from other drugs. The induction of CP450 enzymes by one drug, with a consequent potentiation of toxicity of a second drug, represents another important drug interaction. Another elimination reaction in hepatic microsomes is drug conjugation, which renders a drug more hydrophilic, facilitating renal filtration. Elimination mechanisms that are enzyme mediated, either through chemical modification or transport, can operate in two kinetic regimes. At lower drug concentrations the clearance is "first-order," meaning that the change in plasma concentration with time is proportional to the prevailing concentration. However, at higher drug concentrations, saturation can occur and only a fixed amount of drug is cleared per unit time: the "zero-order" regime. In the zero-order regime, small increments in dose can result in significant changes in plasma drug concentration.

II. METHODOLOGY

A. Obtaining the Specimen

Drug testing is most commonly performed on blood or urine, although less intrusive sampling of saliva or hair is gaining popularity. The most appropriate specimen to test depends largely on the pharmacokinetics of the drug under investigation and the method employed to assay for it. Serum concentrations are preferred for monitoring therapeutic or toxic levels. Urine serves as a reservoir for excreted drugs and their metabolites which may test positive long after the blood has been cleared. However, no reliable extrapolation can be made to estimate total dose consumed or plasma concentration. Postmortem forensic toxicology permits more thorough sampling: heart and peripheral blood, vitreous humor, urine, bile, gastric contents, and various tissues such as liver and brain.

When penalties are contingent on the results of drug testing, verifying the authenticity of a specimen assumes paramount importance. The laboratory must have a paper trail to document the disposition of the specimen since the time of collection to certify that no substitution or adulteration of the specimen could have happened. This is known as the "chain of custody." The collection itself must be directly observed. This can be humiliating in urine collection, but it markedly increases the positive test rate, as substitution or adulteration of urine is common. Even direct observation can be confounded, as illustrated by three woman athletes who sequestered intravaginal tubes of urine, but were detected only because their urine chemistries were absolutely identical. In light of the potential harm from reporting a false positive result, all positive tests should be confirmed by an independent technique.

B. Chromatographic Methods

Thin-layer chromatography (TLC) provides a rapid and inexpensive technique for low-resolution separation of body fluid components. Specimen prepro-

cessing may consist of an extraction step. High-pressure liquid chromatography (HPLC) affords substantially greater resolution. The use of "chiral" columns permits even the resolution of enantiomeric compounds. Again moderate preprocessing is required, and by both methods the purified drug can be isolated for subsequent testing. Gas chromatography (GC) also provides high resolution, but requires specimen derivatization for many compounds to be made volatile. Mobilities by each of these chromatographic methods can be used to characterize a compound. Both HPLC and GC can be directly coupled with ultraviolet (UV) detectors or mass spectrometers (MS), providing an independent, real-time characterization of eluted drugs. The combination of GC/MS is regarded as the gold standard for most drug testing.

C. Immunologic Methods

Immunoassays provide rapid detection of specific drugs. [See Immunoassays, Nonradionucleotide.] A drug-specific antibody is combined with a known concentration of that drug, linked covalently to some molecular reporter. Upon addition of the patient's sample, the drug present in his bodily fluid will displace the labeled drug from the antibody-binding site. The amount of labeled drug displaced correlates with the amount of drug present in the specimen. The antibodies are typically monoclonal and highly specific for the drug or drug class in question, although the potential for cross reactivity has been well documented. The result can be quantitative or qualitative, depending on whether the antibody is drug or drug-class specific. The process can be readily automated. Several immunoassay techniques are currently utilized, differentiated by their drug-bound labels. In the enzyme-multiplied immunoassay technique (EMIT), the reporter is an enzyme whose activity is suppressed by antibody binding. The fluorescence polarization immunoassay (FPIA) depends on the rapid fluorescence of the reporter group after illumination by a polarized light source. If the drug–reporter complex is free, it tumbles rapidly and the polarization of illumination is lost in the fluorescent emission. When antibody bound, the complex tumbles slowly, and much of the fluorescence has the same polarization as the illumination source. Cloned enzyme donor immunoassays (CEDIA) rely on the coupled activity of two enzymes, one of which is drug bound. When the labeled drug is antibody bound, the second enzyme cannot interact with the bound enzyme.

III. THERAPEUTIC DRUG MONITORING

A. History

"The right dose distinguishes a poison from a remedy"
Paracelsus

The history of iatrogenic poisoning is probably as long and colorful as the history of medicine itself. Many substances that were formerly considered poisons are now considered medicines, and vise versa. Hippocrates's writings betray a familiarity with overdosage. The *Lex Cornelia* of the Roman Emperor Sulla was probably the first law to proscribe poisoning and careless drug dispensing. Benvenuto Cellini, the Renaissance artist, included the goddess of venereal disease in his sculpture of Perseus in honor of his enemies. Those enemies attempted to poison him with small amounts of mercury added to his food, but succeeded only in curing him of syphilis.

B. Indications

Assessing whether a patient is appropriately medicated can be complicated for a number of reasons. TDM may be indicated for any combination of the following.

1. The Drug Has a Low Therapeutic Index

This is the most common indication for therapeutic drug monitoring. Many important agents, including some antiarrythmics, antibiotics, anticonvulsants, antidepressants, antineoplastic agents, and immunosuppressive agents, have relatively low TIs. The mechanism of efficacy and toxicity can be physiologically linked or mediated through different pathways.

2. The Treatment Goal Is To Maintain a Specific Plasma Concentration of Drug

The therapeutic efficacy of a drug may be difficult to assess due to subtle clinical endpoints or a slow onset of effects. When hormone replacement therapy is indicated, a specific plasma concentration of the hormone defines treatment success. Many antibiotics, particularly of the aminoglycoside class, require peak and trough plasma concentration measurements. The peak level should exceed the minimum inhibitory concentration (MIC) determined by sensitivity testing of the isolated pathogen. A sufficiently low trough level is necessary to ensure that side effects are minimized.

3. The Physiology of the Patient Complicates Dosing

Illness, as well as genetically determined physiologic variability, can significantly impact drug absorption, distribution, and elimination. A gastrointestinal disease or surgery can impede oral uptake. Uremia can displace drugs bound to plasma carrier proteins, speeding their clearance. Pregnant women can have elevated plasma protein and total drug levels, while maintaining therapeutic free drug levels. Patients often have impaired renal function, sometimes as a direct nephrotoxic consequence of the drugs being monitored. This can slow or hasten elimination, depending on the glomerular filtration rate and whether the drug undergoes active reuptake in renal tubules. Liver function is frequently impaired in the critically ill. Hepatic microsomal function can be impaired, and bile excretion can be blocked. Individuals may also have dose-dependent idiosyncratic reactions to specific drugs.

4. The Patient Is on Multiple Medications

Elimination pathways can be inhibited or induced by concurrent drug use. Drugs bound to plasma carrier proteins can displace each other. Acetaminophen is normally eliminated via five alternative pathways. One minor pathway is mediated by CP450 and involves toxic intermediates that are inactivated by glutathione. Chronic or binge ethanol consumption significantly induces CP450 enzymes, shunting more acetaminophen toward toxic metabolites. Furthermore, glutathione is consumed by ethanol metabolism, reducing the ability of the liver to neutralize those metabolites.

5. The Drug Exhibits Zero-Order Kinetics

Theophylline is a bronchodilator and pulmonary vasodilator used in asthma treatment. It is eliminated according to zero-order kinetics. Overdosage is a constant concern and can result in seizures or death. Phenytoin is an anticonvulsant that is widely used for psychomotor and tonic–clonic seizure prophylaxis. It is also eliminated according to zero-order kinetics. Some side effects are common even at therapeutic levels. Interactions with coadministered anticonvulsants are a frequent problem. Phenobarbital induces the CP450 system, which enhances phenytoin clearance. Valproate displaces phenytoin from plasma proteins, potentiating both phenytoin's activity and clearance.

Two additional indications for TDM may arise in special clinical circumstances.

a. Patient Compliance Is an Issue

Noncompliant patients can confound treatment, posing a risk to themselves and others. The alarming increase in multidrug-resistant tuberculosis is largely a consequence of poor compliance with antimicrobial regimens.

b. The Drug Is under Investigation

Correlating drug plasma concentrations with side effects and efficacy is an important part of the FDA approval process for new drugs.

C. Drugs Requiring Monitoring

A partial list of drugs commonly monitored is given in Table I.

IV. TOXICOLOGY IN THE ACUTE CARE SETTING

Drug testing in the acute care setting can be much more complex than TDM. The number of drugs to be screened for includes those requiring TDM, other clinical agents which in proper usage do not require TDM, and a large number of illicit drugs. The process may be further complicated by a lack of reliable history. As the results are usually intended for treatment purposes rather than legal proceedings, chain of custody protocols are not followed. Testing typically starts with a chromatographic analysis of urine as a broad spectrum screen followed by immunoassays on urine or blood for confirmation or quantification as needed. Toward more cost-effective screening, limited toxicology panels based on clinical presentation are gaining popularity. Lists of drugs typically assayed in urine and serum are provided in Tables II and III, respectively.

V. DRUG SCREENING

A. Workplace Drug Screening

The annual cost of drug abuse to the American economy has been estimated as over $100 billion in terms of lost productivity and health care costs. Employees who test positive for illicit drugs on pre-employment screening have more than a 60% greater risk of accidents, injuries, absenteeism, and turnover than employees who test clean. More than 90% of companies with 5000 or more employees have some form of testing in place. This testing can be pre-employment, in instances of probable cause, and at random. Probable cause requires that certain criteria are satisfied, such as the suspicion that a person is intoxicated or when an accident has occurred. The incidence of positive pre-employment testing is diminishing. As the consequences have significant legal ramifications, formal chain of custody protocols should be observed. Typical drugs tested for in the workplace are listed in Table IV.

B. Athletic Drug Screening

The unfortunate confluence of sports and performance-enhancing drugs may well predate the classical

TABLE I
Commonly Monitored Therapeutic Drugs

Antiarrythmics	Anticonvulsants
Amiodarone	Carbamazepine
Digoxin	Clonazepam
Disopyramide	Ethosuximide
Lidocaine	Gabapentin
Procainamide	Lamotrigine
Quinidine	Mephenytion
Antibiotics	Phenobarbital
Amikacin	Phenytoin
Gentamycin	Primidone
Tobramycin	Vigabatrin
Vancomycin	Valproate
Antineoplastic agents	Antidepressants
Methotrexate	Amitriptyline
Brochodilators	Desipramine
Theophylline	Imipramine
Immunosuppressants	Lithium
Cyclosporin	Nortriptyline
FK506	Trazodone

TABLE II
Drugs Typically Detectable on a Urine Toxicology Panel

Amphetamines	Methaqualone
Benzodiazepines	Opiates and their metabolites
Barbiturates	Phencyclidine
Cannabinoids	Propoxyphene
Cocaine metabolites	

TABLE III
Drugs Typically Detectable on a Serum Toxicology Panel

Alcohols	Benzodiazepines
Ethanol	Alprazolam
Methanol	Chlordiazepoxide
Isopropanol	Clonazepam
Analgesics	Demoxepam
Acetaminophen	Diazepam
Salicylates	Flurazepam
Ibuprofen	Lorazepam
Naproxen	Oxazepam
Anticonvulsants	Temazepam
Carbamazepine	Other sedatives
Ethosuximide	Meprobamate
Phenytoin	Carisoprodol
Phenobarbital	Ethchlorvynol
Primidone	Glutethimide
Barbiturates	Methaqualone
Amobarbital	Methyprylon
Butabarbital	
Pentobarbital	
Phenobarbital	
Secobarbital	

Olympic games. Probably the first drug-related sports fatality in modern times involved an English cyclist, who overdosed on "trimethyl" in 1886. A cocktail of strychnine and alcohol was popular among athletes in the first half of this century, but was supplanted by amphetamines in the World War II era. Amphetamine overdose claimed the lives of several cyclists, including an Olympian in 1960 and a participant in the 1967 Tour de France. Testosterone supplementation was pioneered by Soviet athletes in the 1950s and quickly spread to the United States and other countries. A survey of drug use by athletes in the 1964 Olympic games prompted bylaws prohibiting "doping." Rudimentary drug testing occurred in the 1968 Olympic games, but it was not until 1972 that full-scale drug testing began at the Olympics. Drug abuse in athletes is not merely limited to performance-enhancing agents, as evidenced by the long list of professional athletes suspended for cocaine or marijuana use.

The list of prohibited substances of the United States Olympic Committee is too long to reproduce here, but is listed by category in Table V. Some of these categories deserve more explanation. The general heading of restricted substances includes substances that are not specifically prohibited in most sports, but which in certain contexts may lead to sanctions. Prohibited hormones and analogs include chorionic gonadotropin, corticotrophin, growth hormone, and erythropoietin. Also prohibited are probenecid and related compounds, which inhibit renal excretion, and epitestosterone, which masks testosterone administration.

C. Drug Rehabilitation

Patients in drug rehabilitation programs often participate by court order in lieu of stricter sentencing. Enforcing abstinence from drugs of abuse is a continuing struggle. Drug screening plays an important role in ensuring compliance. The drugs tested for are typically the same as in workplace drug testing (Table IV) or may be limited to the drug of previous addiction.

If a patient tests positive, one common excuse is that of second-hand exposure from drugs abused by family or friends. The toxicologist may be consulted as to the likelihood that such exposures could fully explain a positive test result. One interesting case brought to the toxicologists at the Massachusetts General Hospital involved a young woman who tested positive for PCP. She denied such use, but attributed the result to her boyfriend. He was an active user, and she may have been exposed by performing fellatio on him. This mechanism is under active investigation.

TABLE IV
Drugs Assayed in Workplace Drug Testing

Ethanol	Phencyclidine
Marijuana metabolites	Amphetamine
Morphine	Methamphetamine
Codeine	Cocaine metabolites
6-Monoacetylmorphine	

TABLE V
USOC Categories of Prohibited and Restricted Substances

Prohibited substances	Restricted substances
Stimulants	Alcohol
Narcotics	Marijuana
Anabolic agents	Local anesthetics
Diuretics	Corticosteroids
Hormones and analogs	β-blockers

DRUG TESTING

D. Other Forensic Applications

In forensic drug testing, the issue is not simply whether a drug is present, but also whether the subject was under the influence of that drug at a specific time. Drunk-driving cases underscore this distinction, as the blood alcohol level determines whether a suspect was legally intoxicated. This testing is not only performed on suspects, but also on victims who may have been surreptitiously drugged. Postmortem testing can evaluate intoxication as well as noncompliance with prescribed medications as contributors to the cause of death. Strict chain of custody protocol must be observed.

Estimating antemortem ethanol levels is a difficult issue in postmortem toxicology. Postmortem ethanol levels can be falsely elevated due to microbial fermentation. There can also be large regional variations in ethanol levels once the circulation stops. For these reasons, vitreous humor and urine ethanol levels are important to correlate with blood levels. If ethanol is present in the blood, but absent in the urine and vitreous humor, the ethanol is probably a by-product of postmortem fermentation.

VI. CONCLUSION

Drug testing is an essential clinical tool, which plays an increasing role in the workplace, athletic field, and courts of law.

ACKNOWLEDGMENTS

The author is indebted to the following individuals for critical reading of this manuscript: James G. Flood, Ph.D., Director, Chemistry Laboratory, Massachusetts General Hospital; Anna Toy-Palmer, Ph.D., Applications Scientist, Tripos, Inc.; and Jane Yang, M.D., Assistant Director, Chemistry Laboratory, Massachusetts General Hospital.

BIBLIOGRAPHY

Christophersen, A. S., and Morland, J. (1994). Drug analysis for control purposes in forensic toxicology, workplace testing, sports medicine and related areas. *Pharmacol. Toxicol.* **74**, 202.

Kwong, T. C. (1994). Toxicology. *In* "Clinical Laboratory Medicine" (K. D. McClatchey, ed.). Williams and Wilkins, Baltimore.

Landry, G. L., and Kokotailo, P. K. (1994). Drug screening in the athletic setting *Curr. Prob. Pediatr.* **24**, 344.

Levine, B. S., Smith, M. L., and Froede, R. D. (1990). Postmortem forensic Toxicology. *Clin. Lab. Med.* **10**, 571.

Moyer, T. P., and Lawson, G. M. (1994). Therapeutic drug monitoring. *In* "Clinical Laboratory Medicine" (K. D. McClatchey, ed.). Williams and Wilkins, Baltimore.

Pincus, M. R., and Abraham, N. Z., Jr. (1996). Toxicology and therapeutic drug monitoring. *In* "Clinical Diagnosis and Management by Laboratory Methods" (J. B. Henry, ed.). Saunders, New York.

Puopolo, P. R., Volpicelli, S. A., Johnson, D. M., and Flood, J. G. (1991). Emergency toxicology testing (detection, confirmation, and quantification) of basic drugs in serum by liquid chromatography with photodiode array detection. *Clin. Chem.* **37**, 2124.

United States Olympic Committee Drug Education Program. (1996). "National Anti-Doping Program: Policies and Procedures." U.S. Olympic Committee, Colorado Springs, CO.

Wolf, P. (1994). If clinical chemistry had existed then . . . , *Clin. Chem.* **40**, 328.

Dyslexia–Dysgraphia

ANNELIESE A. PONTIUS
Harvard Medical School

I. Hypotheses on Etiological Factors in "Developmental" and/or Brain Damage-Related "Acquired" Dyslexia–Dysgraphia

II. Anatomico-Functional Components of Dyslexia–Dysgraphia

III. New Approaches, Suggesting Correlations between Written Language and Brain Systems

GLOSSARY

Agnosias Disorders of recognition, associated with inaccurate representation in memory and faulty processing of accurately perceived sense impressions

Alpha rhythm Type of electroencephalographic recording that reflects brain activity of fast rhythmical oscillations in the electrical potential at the rate of 8 to 12 cycles per second, at a voltage of about 40–50 microvolts on average. It occurs in the normally inattentive brain at rest, as in drowsiness, light or hypnagogic sleep, narcosis, or when the eyes are closed, or even when open, if there is no intention to perceive, as in the initial stages of some meditation

Dyslexia "A disorder in children, who, despite conventional classroom experience, fail to attain the language skills of reading, writing, and spelling commensurate with their intellectual abilities." (World Federation of Neurology)

Ecological dyslexia In addition to essential criteria of dyslexia, large-scale, ecologically determined severe under-exposure and underuse of certain precursors to written language are factors

Grapheme Product of the recoded phoneme into a visual sign, shaped with regard to the spatial-relational properties of the elements of a word

Kimura figures-instruction Task requiring one to remember a set of six abstract figures presented on index cards

Phoneme Distinct, discrete unit of speech sounds, composing the spoken word

Recognition Cognitive function viewed as a second step following encoding in memory (representation)

Representation Cognitive function with high-level cross- and supramodal integrative processing (including memory); distinct from the primary level of modality-specific perception

Theta rhythm Type of electroencephalographic recording reflecting brain activity, consisting of slow waves (4 to 8 cycles per second) at about 60 to 100 microvolts. It is produced, for example, during a progressed stage of meditation and dreaming (REM sleep)

THE COGNITIVE COMPLEXITY OF WRITTEN language, its late acquisition (not yet attained by one-third of the world's population), and the range of individual distinctiveness of its disorders have remained a challenge for over a century. The here-employed classic categorization, mainly meant as a heuristic device, is based on anatomico-clinical findings and is selected out of many other classificatory systems.

The operational definition of dyslexia (see Glossary) excludes the following as possible causal factors: psycho-socioeducational influences, autism, mental retardation, and severe sensory loss. Thus, by strict neurological definition, dyslexics have normal visual and auditory acuity. Presently, about 3.5–15% of otherwise well-developed North American children are labeled "dyslexic." At least 10–15% of adults are said to be mildly or functionally dyslexic.

This overview aims to alert researchers, clinicians, and educators to the wide range of potential deficits implicating specific brain system dysfunctioning (see Section II). This can be due to lesions, immaturity, or severe ecologically determined lack of exposure and

ENCYCLOPEDIA OF HUMAN BIOLOGY, Second Edition, VOLUME 3.

practice (see Sections II,C,3,b and III,B,7). Thus, the emphasis is not on statistics, addressing frequency and degree of dysfunction, nor on specificity within a given modality. Rather, the goal is to present a neuro-functionally based ordering system to differentiate among the large diversity of specific symptoms and their typical grouping in syndromes. The awareness of such association among symptoms is of heuristic value, inviting the search for specific, potentially linked additional dysfunctions. Such diagnostic differentiation is necessary for theoretical reasons, as well as for specific, finely honed remediation.

I. HYPOTHESES ON ETIOLOGICAL FACTORS IN "DEVELOPMENTAL" AND/OR BRAIN DAMAGE-RELATED "ACQUIRED" DYSLEXIA–DYSGRAPHIA

A. Familial Predisposition and Sex Difference

Dyslexia tends to be familial and is claimed to affect males three to four times as often as females. There is also a higher incidence in left-handers. A hypothesized sex-linked recessive genetic transmission, however, is not supported by available evidence, as is also the case in other forms of genetic inheritance (polygenetic or autosomal). Thus far, no single genetic model fits the data.

A hypothesized relationship between autoimmune disorders and developmental dyslexia has found support by an animal model (the New Zealand Black mouse), which shows cortical malformations similar to those found in the brains of dyslexics as well as learning problems.

In addition, hormonal hypotheses have remained unconfirmed, positing testosterone interference with the usual lateralization of language functions in the left hemisphere, thereby contributing to dyslexia. [See Cerebral Specialization.]

B. Subtle Brain Dysfunction (Pre-, Peri-, and/or Postnatal Onset)

1. Anatomico-Biological Factors

A recent review by A. M. Galaburda implicates three abnormal anatomical factors in dyslexia that appear to be interrelated in an as yet undetermined way:

a. There is a lack of the expression of cerebral asymmetry of the temporal lobes planum temporale, which is larger in the left hemisphere in about 75% of the general population (whereas only about 5% of the latter are dyslexic).

b. Developmental focal microabnormalities of various language-related areas of the cerebral cortex were found in eight out of nine dyslexics studied at autopsy.

c. Abnormal visual pathways were recently found. For example, the transient, magnocellular visual pathway was shown to be impaired in developmental dyslexia. This pathway is implicated in the processing of rapid visual transitions at low contrast and in motion perception, stereopsis, and saccade suppression. Such abnormal dysfunctions in the visual system of dyslexics raise the question of whether they represent independent factors or are related to (a) and (b).

To address such a question, two main sets of interrelated research are indicated: (1) specific brain tests, such as positron emission tomography (PET) studies (see Section III,A) or echoplanar functional magnetic resonance imaging, as well as (2) the neuroecological correlation method (see Section III,B).

2. Neurochemical Factors

Thus far, neurochemical factors, such as a suggested reduction in certain neurotransmitters (dopamine and serotonin), have not been found to be specific for dyslexia but can pertain to learning disabilities in general.

3. Electrophysiological Factors

a. Auditory Event-Related Potentials

Some evidence indicates that dyslexics may utilize the left hemisphere for tasks that normal readers accomplish with their right hemisphere. Activation of the left hemisphere occurs in some "disabled readers" during ongoing, relatively low-level reading-related cognitive tasks that employ "visual" components: letter-sound and letter-shape processing. Such a different use of neuronal systems is speculated to reflect a deficit either in right (rather than left) hemispheric functioning or in right–left hemispheric integration.

b. Brain Electrical Activity Mapping

Recently, interest has switched from studying right–left differences against within-hemispheric (anterior–posterior) differences. Using the brain electrical activity mapping (BEAM) technique, one finds in dyslexics (1) increased alpha activity in both frontal lobes (left somewhat more than right). (Behaviorally, such findings can be viewed as a lowering of the level of frontal lobe-mediated attention in dyslexia, which in turn is frequently found linked with attention deficit disorder, also implicating frontal lobe system immaturity or other dysfunctioning.) (2) Further, there is a slight difference in theta activity in two tests. (i) The Kimura figures-instruction task showed difference in theta activity in the left antero-(lateral)-frontal regions. (Note, however, that this task may tap especially interobject spatial relations rather than the intraobject relations, preferentially used for written language processes.) (ii) In a silent reading test, a difference in theta activity was found in the left midtemporal region and bilateral medial frontal region (left side more than right side). With eye closing, such a small difference was also evident in the right parietal area.

C. Maturational Lag

Children with a developmental delay in attaining any of Luria's following phases of reading and/or writing skills (e.g., those with attention deficit disorder) (see Section I,B,3,b) can be expected to outgrow their dyslexic difficulties more readily than children with a certain implied brain damage, or subject to severe and long-lasting underuse in extreme ecological context (see Section III,B).

Step 1: Voluntary organized activity requires the conscious analysis (and synthesis) of the component sounds to be produced in a smooth sequence. This process requires the analysis of the phonetic complex to detect its components, the phonemes.

A maturational lag at this level is analogous to disturbances of the linguistic ("acoustic," "phonetic") components (see Section II,B) of written language.

Step 2: Graphemes, the optic structures corresponding to phonemes, have visuospatial properties, defined by visuospatial representation. These are specific for each grapheme and are required for encoding (writing) and for decoding (reading). This visuospatial representation requires complex spatial analysis and synthesis, considering the components' orientation (up–down, right–left), proportions, and ratios. If such intrapattern spatial representation (including memory representation) is inaccurate (e.g., too globalized), mistakes occur: such as erroneously reading and/or writing as if p = q = d = b; was = saw; from = form; N = Z; e = l; O = Q. Developmental delay at this level is analogous to disturbances of mainly spatial ("visual") components of written language (see Section II,C).

Step 3: Graphemes are recoded into a system of motor acts step by step, until the process becomes increasingly more automatic with improved writing skills. Delays at this phase are analogous to disturbances of the motor components at various levels of integration (see Section II,A), particularly those mediated by the prefrontal lobe system: there are difficulties in sentence sequencing and in flexibility of planning strategies and of applying the rules of syntax. (Given that the prefrontal system is a supramodel one, its functioning also overlaps with that of other such systems (e.g., with spatial and "quasi-spatial" processing by the angular gyrus) (see Section II,C,3).

D. Developmental and Ecological Hypotheses

The visual system carries specific subfunctions that appear to remain segregated from the retina through the lateral geniculate nucleus and possibly up to the higher cortical association areas, including the prestriate visual cortex. At the level of the lateral geniculate nucleus, there is a time difference in the maturation of two subsystems, of which the magnocellular system matures prior to the parvocellular one. Consequently, there is a segregation in functioning of the two subsystems—both of which mature only postnatally.

Such segregation of visual functions includes color selectivity, contrast sensitivity, temporal resolution, and acuity. Of particular interest here is the *magno*cellular system that *generally* mediates fast and low-contrast visual information and *specifically* visuospatial functions such as spatial localization, depth perception, localization, stereoacuity, and figure/ground segregation. Based on the segregated magnocellular functioning, two interrelated hypotheses can be proposed with regard to linguistic (auditory) and to the spatial (visual) subtype of dyslexia concerning mainly the just listed general or specific visuospatial functions, respectively.

1. Linguistic (Auditory) Dyslexia

Recent studies by M. S. Livingstone *et al.* have specifically addressed segregated functions related to verbal-auditory processing at the level of the lateral geniculate nucleus. In five dyslexic brains, defects were found exclusively in the magnocellular system, which processes mainly fast (and transient) low-contrast visual

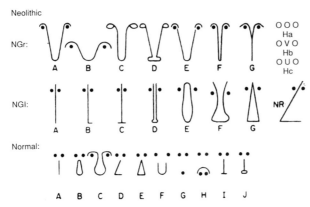

FIGURE I Frontal view of forehead, nose, and eyes—sector of face patterns, measured with a ruler. "Neolithic face" patterns NGr and NGl: continuation of forehead into nose without any indication of a narrowing, indentation, or discontinuity at the bridge (root) of the nose, that is, at the area between the eyes. Thus, a ruler placed at the upper and lower borders of the eyes measures the presence or absence of a Neolithic face configuration. Subtype NGr is present if at least at the lower border of the eyes (if not, even above it) the nose is as wide there as it is at its tip (NGr: A–Ha, b, c), discounting a bulbous enlargement of the cartilaginous parts around the tip of the nose (NGr: D). Note that patterns Ha, b, and c depict the same principle of obliteration of the bridge of the nose area: even at the *upper* border of the eyes, the nose is at least as wide here as it is at its "tip" area. Subtype NGl is present if the nose extends above the upper border of the eyes into the forehead (NGl: A–G). A half-profile type "NR" is not rated, as it is intermediate between Neolithic and normal faces. Normal face patterns (A–J) are characterized by a discontinuity between forehead and nose at the bridge of the nose area through (a) narrowing or (b) indentation, or (c) a beginning of the nose design only below the lower border of the eyes. In all face patterns, dots and circles can be interchanged without affecting the rating. Note that the DAPF is best performed automatically, without looking at a face, from memory, in a (playful) mode of relaxed attention. Thereby, face representation from memory is automatically depicted. The most telling results can be obtained from subjects naive in face drawing; otherwise, some false negative test results may ensue, masking a subtle spatial relational deficiency. The latter is likely to be severe if a person trained in face drawing nonetheless depicts Neolithic face patterns. Note three neglected but essential aspects in the determination of subtle spatial representation, as necessary for reading and writing: (1) Face representation is not identical with face recognition (see Glossary), (2) nor with visual perception, which requires focused attention. By definition, perception is intact in dyslexia. Thus, attention-demanding tests, such as those using unfamiliar and/or complex geometrical figures or signs, are too inappropriate and/or insensitive to rule out visuospatial deficits in dyslexia. (3) Furthermore, there are two subtypes of spatial representation: interobject (used for maps) and intraobject representation of the spatial relations within the constraints of a pattern (face pattern or lexical items). [Reprinted with permission of the publisher from A. A. Pontius (1983). Links between literacy skills and accurate spatial relations in representations of the face: Comparisons of preschoolers, school children, dyslexics, and mentally retarded. *Perceptual and Motor Skills* 57, 659–666.]

Suggestions to increase quantification of the Draw-A-Person-With-Face-In-Front (DAPF) Test: (1) Measurement of the length

information. Based on these findings, a hypothesis has been proposed that mainly addresses difficulties in the auditory processing as also demonstrated by physiological tests in the linguistic subtype of dyslexia, such as slow processing of various sensory modalities, particularly of the auditory kind.

2. Spatial (Visual) Dyslexia

The just cited abnormalities in the magnocellular layer of the lateral geniculate nucleus of five dyslexic brains might also be viewed as related to the spatial (visual) subtype of dyslexia, which is also implicated in visuospatial functions, such as spatial localization, depth perception, stereoacuity, figural groupings, illusory border perception, and figure/ground segregation.

Visual processing proceeds developmentally from a globalized type (sufficient for interobject representation) to an increasingly more subtle kind, particularly with regard to intraobject spatial relational representation, which is a prerequisite for written language, and which has also turned out to be required for accurate intrapattern depiction of the simple human facial features (Fig. 1) (see Section III,B,1–8).

II. ANATOMICO-FUNCTIONAL COMPONENTS OF DYSLEXIA–DYSGRAPHIA

Cognitive functions can be disturbed in various ways. Presently, the focus will be on certain probable sites

of deviation from the interrelations within the face pattern regarding lines, directions, and ratios between the facial features of the eye–nose–forehead area. (2) Measurement of eye-scanning path to determine whether there is a deviation from the norm (a) as to perception or (b) as to the motoric scanning path. When designing a new method of scanning path, certain requirements about the object's size and distance from the viewer's eyes should be noted. Rule out any additional factors that would implicate the frontal lobe system wherever there is a shift of attention, rather than of eye movements per se. (3) (a) Request a drawing of a human face performed with the left hand, as well as with the right hand; (b) observe whether there are any age or sex differences, independent of literacy skills. (4) Test other sensory modalities, especially touch and kinesthesia, by blind-folded finger-tracing (e.g., on sand paper) or by walking along various stick-figure face patterns to distinguish between an exclusively visual misrepresentation in the form of Neolithic face patterns (NGr and NGl) as against a preserved mental representation of the face pattern, in case of accurate mediation by sensory modalities other than vision. (5) Test recognition of subtle spatial relational differences: instruct subjects by drawing attention to the distinct facial features characterizing normal as against Neolithic face patterns. Then present a mixture of the two main types of faces, requesting subjects to sort them out.

of brain lesions or dysfunctions (implying anatomical and/or electrochemical changes).

Neuropathological constraints (based on lesion-acquired dyslexia and dysphasia) are here utilized as a structuring device to implicate specific brain areas or systems potentially dysfunctional to a milder degree in developmental dyslexia and dysgraphia.

In such structuring, the following principles pertain: (1) A lesion may not result in a complete, but only partial, loss of function, and a (compensatory) adaptation of the rest of the brain may occur, especially in young persons. (2) Each site contributes a specific quality to the ensuing functional disturbance. (3) Typically, entire specific brain systems are involved, and dysfunctions are overlapping, forming (specific) groupings ("syndromes"). Principles 2 and 3 have a heuristic diagnostic value in focusing on the site(s) of brain lesion and, at times, suggest a certain etiology (vascular accident, trauma, tumor, metabolic, etc.). Frequently, a diagnosis becomes even more focused when additional specific nonlanguage-related symptoms of a known syndrome exist. (4) For each set of the main components contributing to a given "higher" (i.e., late-emerging) cognitive function, such as written language, three levels, or zones, of interrelating brain systems exist. (i) Primary (projection) cortex for sensory input and motor output. (Most language-related functions, however, require higher levels of mediation.) (ii) Secondary zones, or "association cortices" (including their connecting white matter fiber tracts). Here, cross-modal input is processed, and programs are prepared. (iii) Tertiary zones, characterized by their late anatomical maturation (not before the seventh year of life; note that prefrontal and certain parieto-occipital brain systems require several decades). There is overlapping participation and supramodal integration of many cortical (and subcortical) areas, with an increasingly abstract, more general representation of what is perceived and mediated. Thus, language can be used by the blind and the deaf. Writing can occur with any medium (e.g., body parts other than the hand, sky-writing by airplanes, skating, computers, making knots in strings, coded representations in various art forms). In the tertiary zones, information is coded and functionally organized. In the left hemisphere (language-dominant in right-handers), such organization occurs especially with the help of language, whereas the right hemisphere (nondominant, or minor, in right-handers) mediates especially non-language-related spatial aspects.

A. Motor Components and Related Dysfunctions

1. Primary Motor Cortex

The primary motor cortex includes the motor cortex (left lower third of the prerolandic gyrus), adjacent areas, and connections to the motor association cortex. (i) With lesions of the monosynaptic connections from the motor cortex to the brain stem motor neurons (cortico-bulbar tract), impaired speech movements ensue: dysarthria, with potential secondary spelling difficulties [compare the following dysfunction (v)]. (ii) Motor dysgraphia shows impaired elements of writing movements. (iii) Efferent (kinetic) dysgraphia (left inferior part of the premotor cortex) is characterized by disturbed phonetic analysis and synthesis of words, which may be combined with the following dysfunction. (iv) Impaired kinesthetic tracing, a necessary feedback from the speech muscles, is found in afferent motor dysphasia as well as concomitant disturbance of "silent writing" (left posterior part of the premotor cortex). (v) Imprecise articulation (left posterior sensorimotor area) can lead to secondary inaccurate spelling of consonants. (Note that such articulatory imprecision may impair spelling in conjunction with certain street dialects, or with certain Pidgin versions of English, such as "tok pisin" of Papua New Guinea). [See Speech and Language Pathology.]

2. Secondary Zone: Motor Association Cortex

a. Broca's Area

Broca's area is in the inferior prerolandic region of the frontal operculum (with projections from the primary motor cortex and to the same area in the contralateral hemisphere), involving both speech and language.

b. Frontal Operculum

The frontal operculum or its outflow, when lesioned, initially leads to nonfluent speech disturbance with a transient buccofacial apraxia and, at times, with a secondary mild writing disturbance.

c. Corpus Callosum

The corpus callosum (and other transcortical tracts) connects the right-sided area homologous to the left motor association cortex. In turn, the motor association cortex receives a projection from the posterior sensory association cortex (also see Sections II,B,2 and II,D,1,a).

d. Cerebellum

When the cerebellum is lesioned, one finds (i) dysgraphia and/or (ii) dysarthria in the context of ataxia also involving other body parts (poor coordination of muscles, such as used for writing or speaking, respectively) [compare dysfunction (iv) in Section II,A,1).]

3. Tertiary Cortical Zone for Motor Integration
Prefrontal Lobe System

The prefrontal lobe system mediates the "highest" (most integrated) level of the regulating speech and language. This level includes Luria's conception of "inner language," which is required for the organization of any mature action behavior. When lesioned, or relatively immature (see Section I,C), for a certain task one finds the following dysfunctions.

a. General Dysfunctions

In general, one finds (i) frontal dysgraphia in the context of abulia and overall disorganization of action; (ii) in left- or right-sided lesions, aspontaneity, "pathological inertia" (tendency not to apply oneself to a task) in the language-related realm, impulsive guessing when reading, and inadequate verification of the results of one's actions; (iii) impairment of various reading strategies requiring prefrontal mediation for anticipation and flexibly appropriate switching between various strategies and rules of action, narrated in writing and/or regarding regular or irregular spelling; (iv) deficient "chunking" [phonological segmentation is necessary especially for the decoding of continuous written language and for noncontent (context-dependent) words that emphasize structure (prepositions, conjunctions, auxiliary verbs, articles, and the copula); such words, which are mastered last and are absent in certain near stone-age native languages within their ecological context, pose the greatest difficulties for developmental as well as for acquired dyslexics; (v) poor search by eye movements for active analysis of most significant elements of complex sentences during reading; (vi) loss of order of the elements in spoken and written language; (vii) reduced retention of required sequences; and (viii) perseveration of concrete units or of the principle of activity.

b. Specific Dysfunctions

Specifically, when the left prefrontal convexity is impaired, its cortico-subcortical connections or the most caudal (backward placed) part of the second frontal convolution, dysorthographia (spelling and graphic disturbances) ensues.

B. Linguistic (Verbal, Acoustic, Phonetic) Components

In general, disturbances of written language can occur secondary to those of spoken language. Therefore, only certain specific difficulties will be listed.

1. Primary Cortical Region for Auditory Input

This region connects to Heschl's gyrus in the left transverse temporal gyrus. Here, lesions are not primarily essential in dyslexia and/or dysgraphia (both dysfunctions by strict neurological definition include no perceptual deficits, auditory or otherwise; see Glossary).

2. Secondary Zone: Left Auditory Association Cortex

This zone includes posterior planum temporale (see Section I,B,1) and adjacent regions in the left superior temporal gyrus, and the caudal part of the left, first temporal convolution (Wernicke's area), projecting to the frontal motor association cortex.

a. General Dysfunctions

In general, lesions involve the linguistic (acoustic, phonetic) components of reading and writing, typically constituting mild forms of the neuropathological syndrome of sensory dysphasia (type I Wernicke's dysphasia) with fluent but paraphasic speech: (i) poor phonetic synthesis and analysis with omission of sounds and (ii) substitution of sounds with closely related ones, which may have different phonetic properties. (Such disturbances become especially apparent when trying to write unfamiliar words on the basis of their phonetic analysis or when writing a series of words or phrases.)

b. Specific Dysfunctions

Specifically, the following associations between lesions and deficits pertain: (iii) Wernicke's or adjacent areas: occasionally dissociation between auditory and written comprehension; (iv) posterior to Wernicke's area: dysphasia with relatively more impairment of reading; (v) posterior parts of the superior temporal gyrus (here, type I Wernicke's sensory dysphasia can be associated with relatively more severely disturbed auditory or reading comprehension, depending on more anterior or posterior lesion extension, respectively); and (vi) left temporo-occipital junction: impairment of word retrieval and of short-term memory.

In midtemporal lesions, the naming deficits involve audioverbal "speech memory." Compare deficits in corpus callosum lesions (see Sections II,A,2,c and II,D,1,a) isolating the right hemisphere from the speech area. Further deficits in left ventrolateral and pulvinar of the thalamus functioning can occur (see Section II,D,2,b).

3. Tertiary Cortical Zone for Linguistic (Verbal, Acoustic, Phonetic) Integration

The inferior parietal lobe system comprises Brodmann areas 39 and 40 (same as Section II,C,3) and its surrounding structures, or the parieto-(temporo)-occipital system (Brodmann areas 37 and 21; same as Section II,C,3), especially the angular gyrus (highest level of supramodal integration). Thus there exists an overlap with other tertiary zones and between the subregions, here outlined mainly for emphasis, not implying absolute distinction between them.

a. Left Parieto-Occipital System

In lesions one finds amnestic dysphasia (compare Section II,C,3,h,v) with deficient nominative functions as well as impaired visuospatial representation of corresponding objects to be named upon confrontation. The following abilities may be deficient: classificatory semantic schemes; distinction of the characteristic features of a shape (e.g., letters, words), of a real object, or of one depicted in a stylized manner; and description of the details of the drawing of a shape or of its completion when the drawing was started by someone else. Naming with correct pronunciation is made immediately possible by prompting with the first sound or first syllable of the forgotten word. This attests to an intact audioverbal memory with maintained acousticoverbal traces (in contrast to lesions of the (mid)temporal region; see Sections II,B,2,b,v and II,C,3,h,i).

b. Left Inferior Parietal Lobe or Its Surrounding Structures

When impaired, one finds transcortical sensory dysphasia (type II Wernicke's dysphasia). Lesions also interfere with the contact between the angular region (see Sections II,B,3,d and II,C,3,d) and the rest of the brain, isolating the speech area, and resulting in dyslexia with dysgraphia, an impairment of comprehension of oral and written language. By contrast, the Broca area, arcuate fasciculus, and Wernicke area

proper are spared, enabling intact audiophonatory transposition and repetition.

c. Left Inferior Parietal or Parieto-(Temporo)-Occipital System

Impairment is linked with (i) poor assembling of consecutive series of sounds into a word to be read simultaneously aloud and (ii) with loss of prepositions, there is impaired comprehension of *spoken* and written communication, which is reduced to that of events and of individual words as in ecological-evolutionary restrictions, reflected in certain near stone-age native languages (also see Sections II,C,2,e,ii and II,C,3,g).

d. Left Angular Gyrus

When partially lesioned, one finds type III Wernicke's dysphasia characterized by predominant disturbances in the expression and comprehension of written language (see Section II,C,3,d).

C. Spatial (Visual) Representational Components

1. Primary Cortical Zone for Visual Input

This zone (Brodmann area 17) connects to the calcarine cortex of the occipital lobes, which has no primarily essential role in dyslexia or dysgraphia, in which visual perception is intact, by strict neurological definition (see Glossary).

2. Secondary Cortical Zone: Visual Association Cortex

This zone (Brodmann areas 18 and 19) comprises the occipital lobes and the adjacent parieto-occipital-temporal junction. In this system, the significance of direct peripheral input decreases, and the contact with other related cortical regions increases. (It appears that here perception loses its importance at the expense of the more integrated process of "representation"; see Glossary.)

This cortex uses a method of specific stepwise feature discrimination, and at a higher level (compare Section II,C,3) there is high-level discrimination by modality-specific regions (e.g., concerning color and possibly specific spatial relations).

In visuospatial representation, a general, essential distinction is necessary: intraobject (within a pattern) versus interobject spatial relations. Thus, in preliterate populations with highly developed interobject spatial-relational abilities (required for rapid global assessment), there is typically underevolved intrapattern spatial relational discrimination, which only becomes essential for such skills as literacy. Supporting this distinction are experiments with monkeys that delineated two visual cortical subsystems, apparently analogous to those subserving humans' visual abilities.

a. Inferior Temporal Cortex

This cortex is connected to the occipital visual cortex by the inferior longitudinal fasciculus (see Section II,D,1,b,iii). When lesioned in the posterior part of the inferior temporal cortex in monkeys or further posteriorly in humans, there is impairment of the intricate (intrapattern) visual dicrimination found in a subtype of visuospatial dyslexia, whereas damage to the anterior part interferes with visual memory.

By contrast, interobject spatial relations with distance discrimination are mediated by occipital-parietal mechanisms, including the superior longitudinal fasciculus (see Section II,D,1,b,ii). Typically, in dyslexia and/or dysgraphia, interobject visuospatial discrimination is intact. Thus, when determining only this ability, the presence of intrapattern visuospatial dyslexia tends to be overlooked.

The following subsections add further specification, although more by emphasis than by strictly distinct localization, in that there is much overlap.

b. Parieto-Occipital System (Bilateral or Right Side)

(i) Spatial (visual, optic) dyslexia and/or dysgraphia with poor representation of spatial relations, requiring accurate subtle interrelations, ratios, and orientation of the features within a given configuration (e.g., letter, word, face); (ii) poor position analysis of letters or words; (iii) defective visual analysis and synthesis and disturbed general integration of graphemes and their visual(-spatial) distinctiveness; (iv) poor simultaneous representation of letters or words, impaired survey of the whole system of presented sounds and/ or signs in a whole word or in a series of sounds or signs as a single letter or word normally to be "perceived" simultaneously (literal dyslexia or verbal dyslexia, respectively). (Thus, when a word is perceived only by its first one or two letters, reading proceeds letter by letter.) Specifically, the findings of D. N. Levine and R. Calvanio are as follows.

c. Left Temporo-Occipital

In these lesions performance depends on the span of letters: as letter naming is limited to single letters only, so is reading. Horizontal three-letter arrays cannot be read.

d. Occipital Pole (Mainly Right Hemisphere) and/or Posterior Splenium

In hemidyslexia, horizontal (as opposed to vertical) arrays of letters can be read only on one side of the visual field, and writing is performed only in half of the space, all in the absence of visual field defects.

e. Temporo-Parietal-Occipital System (also see Section II,C,3,c)

(i) Poor recoding of identified phonemes into graphemes with unstable optic-acoustic connections; (ii) poor assembling of consecutive series of sounds and/ or written signs into a word to be read simultaneously (see Section II,B,3,c,i).

3. Tertiary Cortical Zone of Visuospatial Representation

This zone includes the inferior parietal system (Brodmann areas 39 and 40, same as Section II,B,3) and/ or the parieto-(temporo)-occipital system (Brodmann areas 37 and 21; same as Section II,B,3), especially including the angular gyrus. Such overlapping between these and other tertiary (at times also secondary) zones is due to supramodel integration. The following subparts also overlap and are listed only for emphasis, not as indicative of strict delineation.

In general, the parieto-occipital system mediates two kinds of spatial synthesis necessary to process written language: concrete spatial relations (pertaining to any object, shape, or depiction), as well as symbolic, quasi-spatial (Luria) relations (pertaining especially to language processes). In principle, one must consider that even in this tertiary zone of visuospatial processing, the left hemisphere (with the language area) is the most important one, because most visuospatial aspects of percepts in humans are actively influenced by language: Precise spatial organization of a shape (including letters and words) or of any complex object or its depiction makes it possible to code them into precise categories, not merely in a generic way, but as individual members of categories. Such a coding is necessary for later retrieval (recognition) of even ambiguous individual features.

a. Left (or Right) Parieto-Temporal-Occipital System

In general, despite lesions, there are intact acoustic and articulary bases of writing, as well as phonetic analysis and synthesis. Impaired writing occurs in the recoding of the identified phonemes into graphemes. Note that despite left temporal lesions, there are no such writing disturbances with hieroglyphic writing systems, in which the characters tend to denote actual concepts, not requiring acoustic and literal word analysis. Thus, in Japanese Kanji the verbal-linguistic subtype of dyslexia–dysgraphia occurs very rarely, distinguishing it from the visuospatial subtype of dyslexia–dysgraphia that occurs in lesions of the occipital-parietal cortex for visuospatial analysis and synthesis.

Thus, there appears to be a difference in visuospatial representation between the processing of any object, shape, or depiction as against signs denoting language.

Specifically, in left (or right) parieto-temporal-occipital lesions, there is deficient mediation of concrete spatial relations in graphic representation: (i) constructional dyspraxia, deficient reproduction of a constructed shape, even when asymmetrical (e.g., letter, word) from its component elements, requiring an exact position in space (before–behind, right–left, up–down, as in b = d = p = q; Z = N); from becomes form and was becomes saw. (ii) "Optic or visual-spatial dysgraphia": mirror writing and/or disorganized copying of letters or drawings. Milder cases have difficulties in producing a shape from memory or to mentally rotate a shape. All of these impairments are based on a disturbed ability to retain the required spatial relations, ratios, and positions of the lines forming the letters, despite clear distinction between the sounds or phonemes to be written.

b. Left and/or Right Occipital-Parietal Area, Particularly Inferior Parietal Lobule

When lesioned, one can find the Gerstmann syndrome, with several (though not necessarily all) of the following dysfunctions: finger agnosia, dyscalculia, spatial and particularly right–left disorientation, dyslexia with dysgraphia, and autotopagnosia (nonrecognition of own body, despite normal perception). Also frequently present are constructional disturbances of match stick or block designs (e.g., Kohs or other block design test) and of drawing [e.g., of schematized faces and/or topographic amnesia, even when tested in a purely visual (nonverbal) manner]. It is important to distinguish between the mostly intact interpattern spatial representation (e.g., orientation maps between objects in the environment) and a disturbed intraobject spatial representation (such as

within written words or within depicted face patterns (Fig. 1).

In support of the specificity of several essential symptoms found in dysfunctioning systems due to brain pathology, it is of note that Pontius delineated *ecological syndromes*. These are analogous to syndromes of Gerstmann, of prosopagnosia (see Sections II,C,3,e, II,C,3,h,iv, and III,B), and of color dysnomia, all with disturbances of written language functions in not brain-damaged, contemporary, near stone-age adult populations. Such findings corroborate a shared underlying brain "circuitry" given that similar linkage between specific dysfunctions pertains, whether in brain pathology or merely in situations of severe ecological underexposure and underuse (see Section III,B,7).

c. Parieto-Temporo-Occipital Junction and Angular Gyrus

Small lesions lead to relatively isolated deficits in written language.

d. Angular Gyrus

Lesions produce dyslexia with dysgraphia—loss of inner speech, analogous to illiteracy, as Geschwind referred to it (Section II,B,3,d).

e. Right Parieto-Occipital (Temporal) System

When lesioned, in some cases the language system can remain intact, as can symbolic (including mathematical) operations and the understanding of complex logical grammatical structures. Instead, one finds disturbance of visual recognition and of the sense of familiarity. As active visual searching is maintained in an attempt to compensate for the visuospatial deficits in representation, there is guessing (e.g., during reading). In particular, while preserving generic recognition, there is disturbed recognition of the visuospatial representation of individual members of one or more categories with ambiguous shapes (e.g., letters, words, cars, chairs, cows, or human faces). The latter kind of nonrecognition, prosopagnosia, is found mostly in bilateral occipital lesions and is associated with reading disturbances in half of the cases (see Section II,C,3,b).

f. Left Inferior Parieto-Occipital System

In general, damage here brings about impairment of the representation of visuospatial components of writing while speech comprehension is relatively preserved.

g. Right Occipito-Parietal-(Temporal) System

Lesions impair graphic representation (e.g., in writing and reading) of concrete spatial relations; in extreme cases, persons are unaware of the left half of the visual field (see Section II,C,3) during reading, writing, drawing, or copying complex configurations (e.g., letters and words). Further, there is impairment of those aspects impinging on written language, which A. R. Luria termed quasi-spatial: symbolic, logico-grammatical, syntactic relationships (including prepositions, adverbial clauses, etc.). With loss of prepositions there is impaired comprehension of spoken and written communication, which becomes reduced to that of events and of individual words (see Section II,B,3,c,ii).

h. Right (Nondominant) Parieto-Occipital System

Lesions are associated with impaired spatial aspects including these unrelated to language: (i) poor visual recognition of objects; (ii) impaired directionality with regard to the body and/or in external space: constructional apraxia, which can impinge on the writing process; (iii) unawareness of the left half of the visual field (see Section II,C,3,g) when reading, writing, or drawing or during spontaneous constructive activities (unilateral spatial agnosia); (iv) impaired sense of familiarity (e.g., of the face in prosopagnosia (see Sections II,C,3,b and II,C,3,e) (nonrecognition of the familiar face, linked with reading disturbances in half of the cases); (v) a specific kind of "amnestic dysphasia" (compare Section II,B,3,a), here associated with loss of sense of familiarity in recognizing written communication. There is not merely an impairment of speech memory (see Sections II,B,2,b and II,B,3,a), but especially of visuospatial representation and its classificatory spatial (and semantic) schemes, poor ability to distinguish the systems of characteristic features embodied in a shape (e.g., letter, word), to draw an object, to identify an object when depicted in a stylized manner, to complete such a drawing of an object started by someone else, or to describe its details. There is deficient visual representation of an object required to be named on confrontation, leading to a disturbed visuospatial basis of naming.

D. Contributory Subcortical Components

I. Fiber Tracts (White Matter)

a. Interhemispheric Connections

Interhemispheric connections are established by the corpus callosum. Its posterior part, the splenium (also see Section II,A,2,c), performs intercortical connections between homologous areas in left and right hemispheres. In lesions, N. Geschwind delineated various "disconnection syndromes" disrupting the functions of brain areas that are only unilaterally represented: (i) (quasi) pure word deafness disrupts the connection between the left and right primary auditory cortices on one hand and Wernicke's area on the other; (ii) pure alexia (very rare) without (or only with minimal) agraphia and/or aphasia, also called agnosic alexia (see Section II,D,1,b): specifically, there is interruption between the splenium and the left geniculocalcarine tract, disconnecting an otherwise intact left hemisphere from the visual input; (iii) in related disconnection symptomatology, it is controversial whether the result is a deficit to "see" letters, numbers, or colors (Luria) or to respond verbally to such stimuli [i.e., to name them on confrontation (N. Geschwind); see Section II,C,3,b], or whether both kinds of deficits exist, especially in adults' left visual field. Apparently, isolation of the left hemisphere from direct and transcallosal visual input is associated with impairment of that kind of visual discrimination learned in conjunction with speech, whereas other visual discrimination is only mildly affected.

b. Intrahemispheric Fiber Tracts

Intrahemispheric fiber tracts transfer impulses between specific association cortices: (i) In lesions of tracts linking motor and auditory association cortices of the left hemisphere, conduction dysphasia is claimed to ensue. This "associationistic" interpretation is, however, controversial, inasmuch as the ability to repeat and to read aloud can be preserved, although spontaneous language may include phonemic paraphasias (transformations, additions, and/or omissions). Relatively speaking, when repetition is disturbed, it is usually more severe than reading disturbance. The phonological code remains intact. (ii) In disconnection between left intrahemispheric occipito-frontal fiber tracts [also belonging to the superior longitudinal fasciculus (see Section II,C,2,a), which links the specific visual and motor association cortices], poor oral reading results despite intact comprehension of written language. (In most cases, the lesion also involves the cortex of the supramarginal gyrus and affects the arcuate fasciculus in its most caudal part of the left parietal operculum). (iii) The specific tract of the inferior longitudinal fasciculus is the major intracerebral direct pathway for the transfer of "visual" impulses from the occipital visual cortex (areas 18 and 19; see Sections II,C,2 and II,C,3) to the memory mediation in the temporal lobe and limbic systems (see Sections II,E,1 and II,E,2). If interrupted, nonrecognition of perceived objects, such as written material, ensues in various forms of visual agnosias (e.g., agnosic subtype of dyslexia; see Section II,D,1,a,ii).

2. Specific Subcortical Nuclei (Gray Matter)

It remains controversial whether such nuclei contribute directly to reading and writing or indirectly through the tracts (white matter) that lead to and from the following nuclei.

a. Basal Ganglia

When basal ganglia are lesioned, such as in Parkinsonism, poor articulation is claimed to result (potentially contributing to secondary spelling errors).

b. Thalamus

The thalamic nuclei maintain mutual axonal exchanges with the language zone. In addition, lesions here are typically accompanied by a distinct kind of thalamic pain (without any other source) (compare Section III,A). [See Thalamus.]

E. Contributory Limbic System Components

The limbic system (especially its hippocampal formation and amygdaloid nuclei), in conjunction with the temporal lobes (Section II,B,2) and the inferior longitudinal fasciculus (Section II,D,1,b,iii), mediates access to visual memory necessary for visual recognition. In bilateral lesions, stimuli (e.g., written material, faces) can be perceived but have lost their meaning: visual agnosias, including the agnosic subtype of dyslexia (Section II,D,1,b,iii). In unilateral lesions, dyslexia occurs only if the corpus callosum (Section II,D,1,a) is also dysfunctional.

I. Hippocampal Formation

This archicortex and its intralimbic connections are involved in the detection of novelty aspects and in the memory of learned, predictable regularities of various domains, including linguistic and individual visuospatial aspects of written signs (see Section III,A). When dysfunctional, memory for such detailed facts is more impaired than memory for procedures, or motor skills. [See Hippocampal Formation.]

2. Amygdaloid Nuclei

Amygdaloid nuclei also contribute to the visual recognition of individual words and of other configurations. In general, contribution to visual recognition concerns more the identification of an object located directly in front rather than one in the periphery.

III. NEW APPROACHES, SUGGESTING CORRELATIONS BETWEEN WRITTEN LANGUAGE AND BRAIN SYSTEMS

Some new approaches will be added to the already listed clinico-anatomical correlations (see Section II) and to the anatomico-biological and electrophysiological methods discussed in the context of etiological hypotheses (see Section I). The following approaches can address large populations of not brain-damaged, generally well-functioning populations in an attempt to suggest new aspects of the brain's mediation of reading and writing. Some clues about the roots and certain precursory components of these abilities might thereby be obtained.

A. Cerebral Blood Flow with Positron Emission Tomographic Studies

This new method holds great promise for the ultimate visualization (on a screen) of the neuronal substrate of cognitive functions. So far, it has been applied to the processing of a single common noun or of a syllable-like nonword (which follows gross rules of English spelling), and mostly by skilled readers at that. Experienced readers, however, typically process such overlearned stimuli as if they were processing a picture. Thus, such a holistic processing does not require semantic, orthographic, phonological, or visuospatial analysis and synthesis of component parts, as is necessary for neophyte readers. Thus, a certain caveat applies to the interpretation of the results as if they were fully applicable to the complex function of reading (with the components suggested in Section II). Rather, thus far, aspects of the single-word tasks overlap with main aspects of picture recognition (well developed by 4 years of age). Another factor to consider when determining neuronal activity (e.g., on the basis of measurements of increased blood flow) is that not only excitation requires energy, but so does inhibition.

Recent positron emission tomographic (PET) studies use water with oxygen-15 (half-life 123 sec) as the tracer for blood flow in the human brain during single-word processing. Measurement of regional cerebral blood flow is a marker of local neuronal activity. Seventeen normal subjects (intelligent and "highly skilled readers") performed several tasks. Each one took less than 1 min. During all tasks, the subjects fixated on a small crosshair presented on a color television monitor. There were four levels of complexity: (1) simple fixation; (2) passive observation (recognition) of a word presented either visually or auditorily; (3) verbal repetition of sounds (an output task, highlighting the pattern associated with verbal production; these single-word repetition tasks were found to bypass the association cortex); and (4) generation of a semantically appropriate verb in response to each presented noun (an association task). In addition, there were two monitoring tasks: a semantic one, in which members of a given semantic category (e.g., dangerous animals) had to be noted, and a rhyme-monitoring task, in which subjects had to judge whether visually presented pairs of words rhymed.

The PET studies are reported as delineating the localization of certain component mental operations, rather than the localization of an entire cognitive task such as reading.

With each task, few brain regions were activated. Auditory single-word representation activated primary auditory cortex, inferior anterior cingulate cortex, and superior temporal cortex. (Activation of the latter two regions occurred specifically upon word presentation.) In contrast, visual single-word presentation did not activate Wernicke's area (posterior temporal cortex). Visual presentation activated the primary visual cortex (the striate cortex in the occipital lobes), and the prestriate areas near the temporo-occipital border. The extrastriate areas and temporal cortex were specific to word presentation. An overlap occurred between activation by the auditory and visual word stimuli.

These PET studies suggested independence in the phonological input and output codes. Even though there was a common activation near the classic Broca area for word output on both visual and auditory single-word presentations, the Broca area was not activated by auditory single-word input.

The results showed that no auditory reading task (a single word) activated the left temporo-parietal cortex (Wernicke's area is traditionally implicated in phonological coding, although for more than one word or word-like syllable).

Among other findings, this led the researchers to a tentative interpretation in support of a "multiple-route" model in which visual input does not necessarily have to be phonologically recoded, as proposed by certain "single-route" models of cognitive functioning.

Whichever cognitive ability (reading or holistic single-word or picture-like recognition) was actually assessed by such PET studies, both tasks can share certain aspects of memory and of attention: various PET studies can distinguish between automatic activation and controlled processing by means of attention. Furthermore, there are two kinds of (interrelated) attention systems: (1) a more general system, largely mediated by the prefrontal system (such attention for action is required, e.g., when attending to only one meaning of a word, while suppressing alternative meanings), and (2) the visual spatial attention system, mediated by the posterior parietal lobe, a portion of the thalamus (part of the pulvinar) (see Section II,D,2,b), and areas of the midbrain related to eye movements.

Memory is another aspect of language. Once an item has left current attention, "the hippocampus performs a computation needed for storage," which enables conscious retrieval, as the PET studies argue (see Section II,E,1).

In general, the results from the PET studies discussed are congruent with those from brain lesion studies. Specific surprise findings in the PET studies include the greater tendency toward bilateral brain activation and the repetitive task bypassing the association cortex, which itself has access to the output system.

B. Neuroecological Correlation Method

A novel field of inquiry into dyslexia can implicate likely precursors to specific graphic representation of lexical signs: in general, humankind's early graphic representation in sacred art. Such art is often the only available record of virtually all of the human population's cognitive functioning, dating back over millenia including the Neolithic period, by definition the last completely illiterate period. Of specific interest in the context of written language is the study of Neolithic "artists" depicting the human face pattern (Fig. 1). Both configurations of face and of written signs are generally considered as sharing essential aspects, given that both can present an unlimited variety by subtle rearrangement among a few basic spatial features. A specific, largely overlooked subtype in testing visuo-spatial representation in both face and written signs concerns the spatial relations, orientation, and size ratios among the parts within their configurations. (This spatial ability must be distinguished from interpattern spatial relations, as used in mental maps, e.g., finding one's way among objects.) It remains to be determined whether this common denominator of intrapattern variation, linking face and lexical signs, pertains exclusively to a correlational or even to a neurologically based causal relationship between face pattern and written signs (see Sections II,C,3,b, II,C,3,e, and II,C,3,h,iv).

A perusal of art of the Neolithic (i.e., illiterate) period worldwide shows a prevalence (at least 3 of 4) of a specific misrepresentation within the face (Fig. 1) configuration, labeled "Neolithic" face patterns (NGr, NGl). The same patterns prevailed in contemporary illiterates still living in near stone-age situations prior to acquiring literacy and rules of spelling. These patterns are characterized by an inaccurate continuity between forehead and nose whether in a roundish type (NGr) or in a longish type (NGl). This continuity obliterates the subtle spatial relations among the features of the upper part of the face around the bridge of the nose area. (The mouth is not important and is frequently omitted or is tiny and deep set in Neolithic art.)

Underlying neurophysiological factors contributing to this Neolithic inaccuracy were reasonable to postulate when it was discovered that such spatial inaccuracy begins to disappear in art as well as in contemporary people's graphic representations with the advent of literacy (though it may resurface under the impact of alcohol or other drugs). About one-third of dyslexics (as opposed to 6–11% of normal controls) produced the Neolithic face patterns when asked to "Draw-A-Person-With-Face-In-Front" (DAPF) (see Fig. 1 legend). Two interrelated sets of factors must be considered: (1) A lack of exposure to specific external input, which promotes literacy. Probably, populations that live in severe ecological situations that neglect graphic-pictorial material (e.g., of small patterns) have survival needs that promote different skills, unrelated to those tapped by literacy. Such an ecological context selects out the fostering of certain cognitive abilities (e.g., interobject, map-like spatial representation, and rapid global shape assessment of attacker/prey/food). By contrast, other abilities are neglected (such as intraobject spatial representation required to process face and lexical patterns). (2) Subtle brain dysfunction or underuse. The following brain areas were tentatively implicated (likely through lack of exposure and of usage): the mediation of intrapattern spatial relations implicates especially the parietal-occipital system (see Sections II,B,3 and II,C,3). (This suggestion gains further support by an apparently analogous link found in a neuropathological counterpart to the here proposed ecological syndrome: pros-

opagnosia, linked with reading difficulties in half of the cases; see Sections II,C,3,b, II,C,3,e, and II,c,3,h,iv.) Such patients with occipital brain damage say they no longer "recognize" the familiar individual face, because they experience it as "flattened out, without any relief," and most of them also depict the face that way, which is of the Neolithic face type (Fig. 1, NGr, NGl). Other possibly involved brain regions may be the hippocampus (see Sections II,E,1, and III,A) concerning certain spatial representational globalization in memory (implicated in chronic alcoholism), as well as certain prefrontal lobe mediation of flexibility (see Section II,A,3,a,iii).

A working hypothesis, initially based on such consistently repeated observations of Neolithic face depictions in diverse situations, was engendered by such inaccurate face representation being linked with illiteracy and later with the visual spatial subtype of dyslexia (discussed in the following). Inasmuch as these two reflections of apparently evolutionarily early or ecologically modified visuospatial cognitive functioning may share a common denominator, Neolithic face depiction may serve as an indicator of certain potentially poor literacy skills, meriting further scrutiny.

A fairly wide array of evidence has been gathered within the context of the neuroecological paradigm in support of this hypothesis.

1. *Contemporary near stone-age people* in remote areas of New Guinea, Indonesia, and South America, whose spontaneous sacred art (where produced) is characterized by Neolithic face patterns, also preferred the same in their elicited face drawings. Surprisingly, the percentage of such globalized Neolithic faces (in >1000 subjects, collected over two decades) showed a significant positive correlation with the results from their direct reading and writing tests (UNESCO).

2. *Colonial North American stone masons,* whose rate of illiteracy (20.5%) is on record, and who left human face engravings on about 1500 tombstones, showed a positive correlation with their Neolithic face depictions (19.6%).

3. *"Before and after" studies* with such near stone-age peoples showed a significant increase in accurate face depictions with the availability of literacy instruction, although not necessarily producing full literacy skills.

4. Schooling alone is apparently not sufficient for the emergence of literacy when ecological situations continue to lack specific stimuli with subtle intrapat-

tern variety (e.g., in small geometrical patterns). In contrast to 4% of their classmates of European descent who drew Neolithic faces, about a third of 269 *Australian Aboriginal schoolchildren* did so. All of these Aboriginals were at least of average intelligence, functioning within regular class settings. They fulfilled the requirements to be classified as dyslexics (see Glossary), or here as "ecological dyslexics" (distinguishing this subgroup from that listed under 6). The Aboriginals' overall bleak Northern Territory desert and artificial settlement surrounds were presumably not fostering the development and utilization of those intraobject spatial skills necessary for reading.

5. In 407 *"Western" preschoolers,* the face depictions (collected by five different "blind" researchers in Europe and the United States) showed a consistent percentage of about one-quarter of the Neolithic face pattern. It is likely that most of these children had already been exposed to graphic material seen on TV and small geometrical patterns depicted on fabrics and on paper.

6. *Out of 297 European dyslexics,* about one-third of them drew Neolithic face configurations (Fig. 1, NGr, NGl) (as opposed to 6–11% of normal readers). These "Westerners" were naive as to face-drawing instruction, but otherwise grew up with exposure to TV and to other graphic pictorial materials, for example, small geometric patterns. In distinction from "ecological dyslexics" (see 4) in general, in "Western" dyslexics internal factors are more likely to be implicated (e.g., subtle brain dysfunctioning) (see Sections I,A and I,B).

7. Further support from ontogeny for the hypothesis is suggested by experiments with young *infants* (obviously illiterates). Their automatic, indiscriminate "smiling response" with wide-open mouth was elicited significantly faster ($p < 0.001$) by masks presenting Neolithic faces (Fig. 1, NGr, NGl) compared with control masks stressing the mouth, and even with the natural face up to 18 weeks of age, when the latter was preferred.

8. Finally, *prosopagnosia* can occur in occipito-(parietal) brain damage (see Sections II,C,3,e and II,C,3,h,iv). This neuropathological disturbance is characterized by experiencing (and drawing) the face configuration as "flattened out," analogous to typical Neolithic face patterns (Fig. 1, NGr, NGl). Furthermore, half of such patients also have reading disturbances. Based on certain preliminary findings, it is hypothesized that such brain damage may also occur

in certain persons suffering from early phases of AIDS or Alzheimer presenility. If so, then Neolithic face depiction in the DAPF test (Fig. 1) may constitute a sensitive early diagnostic indicator.

In summary, the neuroecological method implicates the following correlation in not brain-damaged illiterate (preliterate) populations worldwide: (1) When ecologically determined daily activities typically lack exposure to and practice with subtle spatial relations within (small) geometrical configurations (as in graphic–pictorial material), such a lack is likely to be reflected in (2) specific spatial-relational, globalized misrepresentation of the features within the face configuration, called Neolithic face patterns (see Fig. 1, NGr, NGl).

Thus, when such Neolithic face depictions are also found in drawings by "Western" dyslexics, such misrepresentation can be viewed as an indicator for the spatial (visual) subtype of dyslexia. Specifically, such a subtype of spatial (visual) dyslexia can be subclassified as "ecological dyslexia" when such Neolithic face patterns are depicted by populations in the process of emerging from preliteracy. On closer scrutiny, however, one finds that if "ecological dyslexics" continue their daily activities within their original ecological setting (poor in picto-graphic material) literacy instruction by itself, then, does not appear to be sufficient to attain full literacy. Such a persistent spatial underevolving also remains then typically correlated with the persistent drawing of Neolithic face patterns (Fig. 1, NGr, NGl), underlining the importance of daily exposure to picto-graphic material to develop the subtle spatial-relational abilities necessary for efficient literacy skills.

Neuroecological correlation studies—as a supplement to objective brain tests (see Section III,A)—appear to be of particular relevance to the developmental aspects of dyslexia. Such aspects are strongly implicated in the anatomical abnormalities and dysfunctions of developmentally transient visual pathways (see Section I,B).

It is generally acknowledged that usage and practice play an essential role in the evolving of initially transient pathways, hence the important impact of eco-cultural factors on postnatally late developing cognitive functions, so far almost exclusively studied in "Westerners."

S. Zeki's anatomical delineation of segregated single visual subfunctions is congruent with certain experimental ecocultural findings of a segregated evolving of specific visuospatial subfunctions in contemporary preliterate hunter-gatherers.

BIBLIOGRAPHY

Arbib, M. A., Caplan, D., and Marshall, J. C. (eds.) (1982). "Neural Models of Language Processes." Academic Press, New York.

Duffy, F. H., and Geschwind, N. (eds.) (1985). "Dyslexia. A Neuroscientific Approach to Clinical Evaluation." Little Brown, Boston.

Galaburda, A. M. (1993). Developmental dyslexia. *Rev. Neurol.* (Paris) **149**, 1–3.

Ingle, D. J., Mansfield, R. W., and Goodale, M. A. (eds.) (1982). "The Analysis of Visual Behavior." M.I.T. Press, Cambridge, MA.

Lecours, A. R., Lhermitte, F., and Bryans, B. (1983). "Aphasiology." Bailleire Tindall, London.

Livingstone, M. S., Rosen, G. D., Drislane, F. W., and Galaburda, A. M. (1991). Physiological and anatomical evidence for a magnocellular defect in developmental dyslexia. *Proc. Natl. Acad. Sci. USA* **88**, 7943.

Luria, A. R. (1980). "Higher Cortical Functions in Man." Basic Books, New York.

Petersen, S. E., Fox, P. T., Posner, M. I., Mintun, M., and Raichle, M. E. (1988). Positron emission tomographic studies of the cortical anatomy of single-word processing. *Nature* **331** (6157), 585.

Pirozzolo, F. S., and Wittrock, M. C. (1981). "Neuropsychology and Cognitive Processes in Reading." Academic Press, New York.

Pontius, A. A. (1980). Pre-literate people in New Guinea and Indonesia draw specifically distorted faces, as do "Western" dyslexics, using a paleo-visual-representational mode. *Experientia* **36**, 83.

Pontius, A. A. (1981). Geometric figure-rotation task and face representation in dyslexia: Role of spatial relations and orientation. *Percept. Mot. Skills* **53**, 607.

Pontius, A. A. (1989). Color and spatial error in block design in stone-age Auca Indians: Ecological underuse of occipital-parietal system in men and of frontal lobes in women. *Brain & Cognition* **10**, 54.

Pontius, A. A. (1993). Spatial representation modified by ecology. From hunter-gatherers to city dwellers in Indonesia. *J. Cross-Cultural Psychol.* **24**, 399–413.

Seidenberg, M. S. (1988). Cognitive neuropsychology and language. The state of the art. *Cog. Neuropsychol.* **5**, 403.

Zeki, S. (1990). Functional specialization in the visual cortex: The generation of separate constructs and their multistage integration. *In* "Signal and Sense" (G. M. Edelman, W. E. Gall, and W. E. Cowan, eds.), pp. 85–130. Wiley–Liss, New York.

Ears and Hearing

JAMES O. PICKLES

University of Queensland, Australia

GLOSSARY

Decibel (dB) Relative measure of sound pressure, derived from the formula: number of decibels $= 20 \log_{10}$ (measured sound pressure/reference sound pressure). If the reference sound pressure is taken as 2×10^{-5} N/m^2, the measure becomes an absolute measure of sound pressure known as dB sound pressure level (dB SPL)

Impedance In acoustics, the pressure needed to produce unit velocity of vibrations in the medium; the impedance for a plane wave in an infinite volume of the free medium (known as the specific impedance) is the pressure required to induce a unit velocity of vibration of the medium (units N sec/m^3)

Tuning curve (frequency-threshold curve) Curve of the stimulus intensity necessary to give a constant output of the system as the stimulus frequency is varied; it can be applied equally well, for example, to vibration of the basilar membrane (for constant amplitude of vibration), and responses of auditory neurons (for constant evoked firing rate); an advantage of the tuning curve is that it permits comparison of tuning of different stages of the auditory system

AUDITORY PERFORMANCE IS DETERMINED IN A number of important ways by the performance of the cochlea. Thus, for instance, both frequency resolution and the perception of distortion tones can be traced to the fundamental properties of cochlear mechanics and hair cells. Similarly, the absolute detection threshold can be traced to the cochlea, in conjunction with the outer and middle ears. Thus, understanding of the auditory system is closely allied to our understanding of cochlear physiology. The limitations to auditory performance imposed by the auditory central nervous system are much less clear.

I. THE OUTER EAR

The outer ear conveys the sound wave for the external field to the tympanic membrane (eardrum) and modifies the wave in two important ways. First, by virtue of its resonant cavities, the outer ear increases the sound pressure in the middle range of frequencies, enhancing sensitivity in this frequency range. Second, because the enhancement is directionally selective, the outer ear gives cues, which help in sound localization.

The outer ear has multiple resonant cavities. One important cavity is formed by the concha (Fig. 1), situated at the entrance to the external auditory meatus (ear canal). The other most important cavity is the meatus itself. Since the tympanic membrane offers a higher impedance to the passage of sound waves than does free air, the membrane reflects some of the sound energy that reaches the end of the canal. This sets up a quarter-wave resonance in the canal. The resonant wave has its node (low pressure) at the entrance to the canal, and its antinode (high pressure) at the tympanic membrane. In humans, this ear-canal resonance increases the

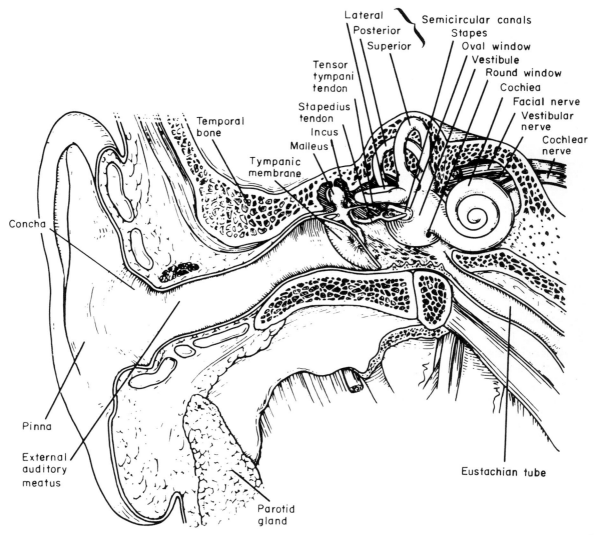

Lateral
Posterior
Superior
Semicircular canals
Stapes
Oval window
Vestibule
Round window
Cochlea
Facial nerve
Vestibular nerve
Cochlear nerve
Tensor tympani tendon
Stapedius tendon
Incus
Malleus
Temporal bone
Tympanic membrane
Concha
Pinna
External auditory meatus
Parotid gland
Eustachian tube

FIGURE I   The outer, middle, and inner ears in humans. [Adapted from R. G. Kessel and R. H. Kardon (1979). "Tissues and Organs: A Text-Atlas of Scanning Electron Microscopy," W. H. Freeman and Company, New York.]

pressure at the tympanic membrane by some 10–15 dB in a broad range around the resonant frequency of 2.5 kHz. The second most important resonance, in the concha, adds some 10 dB in the 5-kHz region. In humans, the sum of these two resonances, together with resonances in the pinna and reflections from the head, increases the sound pressure at the tympanic membrane by 10–20 dB above that in the free field, over the frequency range of 1.5–8 kHz.

The resonant amplification of sound pressure by the external ear depends on the direction of the sound source, in addition to the factors mentioned earlier. Although the most powerful cues for sound localization in humans are given by binaural (i.e., two-ear)

listening (see below), the outer ear adds cues, which can be used for monaural localization and for distinguishing the elevation of the source. Sound waves are scattered from the raised rim at the edge of the pinna and from the rear wall of the concha. When a sound source is moved behind the ear, the scattered waves interfere destructively with the wave transmitted directly, reducing the input for sound frequencies >3 kHz. Because the convolutions of the pinna and concha are asymmetrical above and below the entrance of the ear canal, the interaction is also dependent on the elevation of the source. Thus, when a sound source is raised above the horizontal, the intensity at the ear canal increases in the 5- to 10-kHz range.

II. THE MIDDLE EAR

A. The Middle Ear as an Impedance Transformer

The acoustic vibrations are transferred from the tympanic membrane via three small bones [the ossicles; namely the malleus, incus, and stapes (Fig. 2)] to the round window of the cochlea. Here, vibration of the stapes induces flow of the cochlear fluids. The cochlear fluids offer a much higher impedance to acoustic vibrations than does air. When sound waves meet a boundary with such an impedance jump, it can be calculated that much of the sound energy will be reflected, with only a small proportion transmitted. In the ear, we expect that only 1% of the incident energy to be transmitted to the cochlea if the sound waves met the oval window directly. The middle ear apparatus reduces this power loss substantially.

The middle ear reduces the power loss by acting as an impedance transformer. It changes the relatively large-amplitude, low-pressure vibrations of the tympanic membrane to low-amplitude, high-pressure vibrations at the round window. This is accomplished by three mechanisms: (1) the stapes footplate in the round window has a smaller effective area than the tympanic membrane; (2) the middle ear bones act as a lever, decreasing the amplitude of the movement at

the round window but increasing its force; and (3) the tympanic membrane buckles as it moves, similarly decreasing the amplitude of the movement and increasing its force. At 1 kHz, the three factors together increase the ratio of force/displacement of the vibrations by a factor of 185 in the cat, the species for which we have the most information (by far the greatest contribution, a factor of 35 times, arises from the first mechanism). This produces a substantial reduction of the effective impedance of the cochlea as seen at the tympanic membrane. Nevertheless, the measured impedance at the tympanic membrane is rather higher than the impedance of air by a factor of about four. This is due to (1) the middle ear transformer ratio being smaller than the ideal value by a factor of two and (2) friction and other losses in the middle ear.

B. Transfer Function of the Middle Ear

Losses due to friction in the middle ear reduce transmission evenly over the whole frequency range. Further factors are more frequency specific in their effects. The *stiffness* of the ligaments supporting the bones and membranes of the middle ear reduces transmission primarily at low frequencies (<1 kHz). At higher frequencies, the *mass* of the middle ear bones becomes dominant. In this frequency range, transmission also becomes less efficient because the stimulus is less effective at moving the tympanic membrane: The membrane vibrates as a number of separate zones, only some of which couple vibration to the malleus. Therefore, transmission is most efficient in the midrange of frequencies (around 1 kHz), falling off at higher and lower frequencies. The combined effects of both the outer and middle ears in enhancing certain frequencies substantially account for the variation of our acoustic sensitivity with frequeny, except at the very top and bottom of the frequency range, where cochlear factors are dominant.

C. Middle Ear Muscles

Transmission through the middle ear can be controlled by the middle ear muscles. The tensor tympani attaches to the malleus near the tympanic membrane, and the other—the stapedius muscle—attaches to the stapes. Contraction of the muscles increases the stiffness of the ossicular chain. As described earlier, an increase in stiffness reduces transmission at low frequencies (<1 kHz). In this range, middle ear muscles can reduce transmission by up to 20 dB. The middle

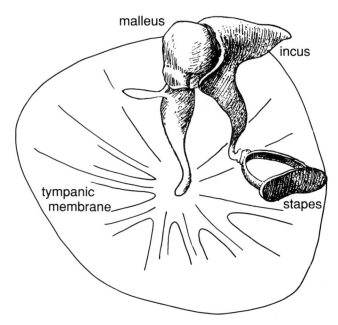

FIGURE 2 The tympanic membrane and middle ear ossicles in humans, as seen from within. The malleus and incus are suspended at their upper ends by ligaments (not shown).

A

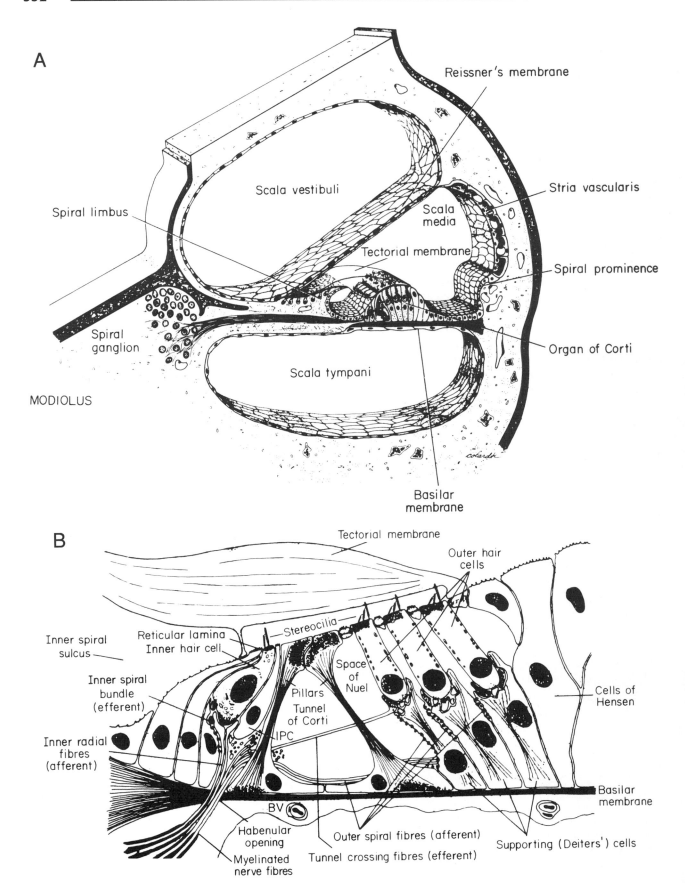

Reissner's membrane

Spiral limbus

Scala vestibuli

Stria vascularis

Scala media

Tectorial membrane

Spiral prominence

Spiral ganglion

Organ of Corti

Scala tympani

MODIOLUS

Basilar membrane

B

Tectorial membrane

Outer hair cells

Inner spiral sulcus

Reticular lamina
Inner hair cell

Stereocilia

Inner spiral bundle (efferent)

Pillars

Space of Nuel

Cells of Hensen

Tunnel of Corti

Inner radial fibres (afferent)

IPC

Basilar membrane

BV

Habenular opening

Outer spiral fibres (afferent)

Supporting (Deiters') cells

Myelinated nerve fibres

Tunnel crossing fibres (efferent)

ear muscles also seem to affect the vibration of the middle ear bones in further ways not understood because the response at higher frequencies can be influenced, with certain irregularities in middle ear transmission being reduced.

Contraction of the middle ear muscles can be induced by loud sound (the acoustic reflex), by vocalizations and bodily movements, by tactile stimulation of the head, and, in some people, voluntarily. Above the intensity required to elicit the acoustic reflex (some 75 dB above the absolute threshold), the muscles help protect the ear from acoustic trauma, although the acoustically driven reflex is too slow to protect the ear from impulsive sounds such as gunshots. In addition, at high intensities, low-frequency sounds are particularly effective at masking high-frequency sounds. Contraction of the muscles, by selectively reducing the transmission of low-frequency sounds, may also be beneficial for the perception of complex sounds with low-frequency components, such as speech, at high intensities.

III. THE COCHLEA

A. General Anatomy

The cochlea is a spiral fluid-filled tunnel within the petrous (stony) part of the temporal bone (Fig. 1). It forms part of the labyrinth, the other division being the vestibular system, with which it is in continuity. The cochlear duct is divided into three compartments by two membranes known as the basilar membrane and Reissner's membrane (Fig. 3A). The compartments are called the scala vestibuli, scala media, and scala tympani. The cochlear duct with its three compartments spiral together around a central core called the modiolus, which contains the auditory nerve and the cochlear blood supply. In humans, the cochlear duct is about 35 mm long and is packed into a spiral of two and a half turns in a space some 10 mm wide and 5 mm high. Lengths of duct in different mammals vary from 7 mm in the mouse to 57 mm in the elephant. The number of turns varies from half a turn in the relatively primitive monotremes to four turns in the guinea pig.

The organ of Corti (Fig. 3B), which contains the receptor apparatus, is situated on the basilar membrane. It is covered by a gelatinous and fibrous flap called the tectorial membrane. The position of the organ, on the interface between the scala media and the scala tympani with their fluids of very different composition, is considered critical for cochlear function.

B. Cochlear Fluids: Production and Electrochemistry

The scala media contains endolymph, which is unique for an extracellular fluid in that it has a high K^+ composition (≈ 150 mM) and a low Na^+ concentration (≈ 1 mM). It also has a high positive potential ($+80$ to $+100$ mV) with respect to the surrounding bone; the higher values are found in the more basal turns of the cochlea. In contrast, the perilymph, which is contained within the scala vestibuli and the scala tympani, has the ionic composition expected of most extracellular fluids and an electric potential similar to that of the surrounding bone.

The high positive potential and high K^+ concentration of the endolymph are directly dependent on active, ion-pumping processes within the stria vascularis, a structure situated on the wall of the scala media farthest away from the modiolus (Fig. 3A). As its name implies, the stria vascularis has a rich blood supply. The stria contains two types of cells: the more superficially located marginal cells and the more deeply located basal cells. The marginal cells have deep infoldings on the basolateral surface facing away from the scala media (i.e., on the surface facing the blood supply); the infoldings contain many mitochondria, indicative of high levels of energy-consuming activity. The marginal cells have high positive intracellular potentials, a few millivolts more positive than the endolymph, and it is suggested that they maintain the chemical and electrical composition of the endolymph. Although the details are still controversial, it is generally agreed that a Na^+–K^+-activated adenosine triphosphatase in the marginal cells is involved. Other processes, such as Na^+–K^+–Cl^- cotransport, also may play a part. [*See* Cochlear Chemical Neurotransmission.]

FIGURE 3 (A) Cross section of a single turn of the cochlear duct. [Reproduced from D. W. Fawcett (1986). "A Textbook of Histology," Saunders, Philadelphia.] (B) Cross section of the organ of Corti, as it appears in the basal turn. The modiolus is to the left of the figure. BV, blood vessel; IPC, inner phalangeal cell. [Slightly modified from A. F. Ryan and P. Dallos (1984). "Hearing Disorders" (J. L. Northern, ed.), pp. 253–266. Little, Brown and Company, Boston.]

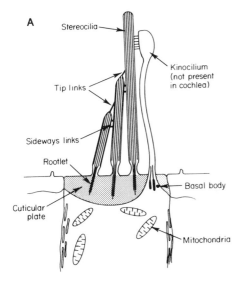

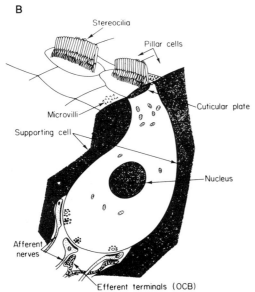

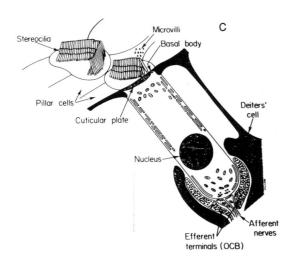

C. The Organ of Corti

Mechanical vibrations resulting from the incoming sound are transformed into neural activation in the organ of Corti (Fig. 3B). The hair cells form the mechanotransducing elements. The cells are situated with their apical ends in the reticular lamina on the top surface of the organ, with their sensory hairs, or stereocilia, projecting to touch or nearly touch the tectorial membrane. Figure 4A shows a schematic representation of the stereocilia and the other structures at the upper end of the hair cell. The hair cells are divided into two types, with complementary roles in cochlear function. The inner hair cells (Fig. 4B) are situated in a single row on the side of the organ of Corti nearest to the modiolus. They signal the vibrations of the cochlear partition to the central nervous system. The outer hair cells (Fig. 4C) are situated in three to five rows on the outer side of the organ of Corti. Their role is rather more problematic; however, they contribute to the sharp frequency selectivity and high degree of sensitivity of normal cochlear function. To analyze the mechanisms of hair cell function in the organ of Corti it has been necessary to bring together the results from a number of different types of experiments, including the basic biophysical analysis of the operations of single hair cells. Such experiments are rather difficult in the mammalian cochlea, and our most detailed information has come instead from the analysis of vestibular hair cells in other species.

D. Mechanotransduction in Cochlear Hair Cells

The structures in Fig. 4A, with minor variations, are common to all hair cells of the acousticolateral system, i.e., to hair cells of the cochlea and vestibular system and, in addition, to hair cells of the lateral line organ. The stereocilia are situated in several rows of graded height, with between 50 and 250 stereocilia per hair cell, depending on organ and species. The stereocilia are also richly interconnected by different sets of linkages. Linkages of one set connect the stereocilia later-

FIGURE 4 (A) Schematic diagram of the common structures on the apical (transducing) pole of acousticolateral hair cells. In the cochlea, the kinocilium is present only in immature hair cells. (B) Inner hair cells are shaped like a flask. (C) Outer hair cells are shaped like a cylinder. In all parts of the figure, the modiolus would be situated to the left. [Copyright J. O. Pickles, 1987.]

ally, so that all stereocilia move when some are deflected.

Mechanotransduction depends on deflection of the stereocilia. When the bundle is pushed in the direction of the tallest stereocilia, the tip links, which interconnect the stereocilia near their tips (Fig. 4A), are stretched (Fig. 5). This pulls open the mechanotransducer channels in the surface membrane of the stereocilia, by a direct mechanical action. Deflection of the bundle in the opposite direction relieves the tension on the links and allows the channels to close.

The role of the tip links in mechanotransduction has been shown by experiments in which the tip links were selectively broken by the removal of Ca^{2+} from around the bundle, which also has the effect of abolishing mechanotransduction permanently. While there is no direct evidence as to the actual sites of the mechanotransducer channels, it appears that the channels are at the upper ends of the stereocilia. This is the region in which the greatest current flows are produced when the channels open and in which channel blockers, applied locally through a micropipette, have their greatest effect.

The relation between bundle deflection and intracellular voltage is an asymmetric and saturating sigmoid, with the changes in the depolarizing direction being larger than the changes in the hyperpolarizing direction (Fig. 6). In the formulation put forward by A. J. Hudspeth and colleagues, the form of this function can be directly traced to the kinetics of the transducer channels and is determined by the Boltzmann distribution.

In the mammalian cochlea, it is expected that the stereocilia are stimulated by the vibration of the basilar membrane, which produces a shear movement between the tectorial membrane and the reticular lamina. Comparisons of the thresholds for activation of hair cells and auditory nerve fibers suggest that depolarizations of 2 mV raise the firing rate of the auditory nerve fibers to the hair cells by about 20 action potentials per second.

E. Cochlear Mechanics and the Frequency Selectivity of the Cochlea

In the mammal, the frequency selectivity of the hair cells, the frequency selectivity of the auditory central nervous system, and the frequency selectivity of much auditory performance are all a direct reflection of the selectivity of the mechanical vibrations of the basilar membrane. The early measurements of basilar membrane vibration by von Békésy showed the phenomenon of the traveling wave (Fig. 7A). This can be visualized as a series of ripples traveling up the cochlear duct from base to apex. The ripples grow in size as they travel apically, come to a peak, and then die away sharply. If the frequency of the input stimulus is constant, the envelope of the waveform is constant, and so the point of maximum vibration stays constant. The point of greatest response depends on the

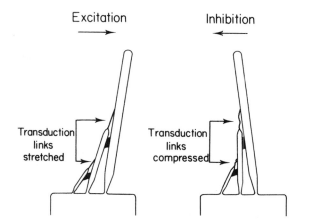

FIGURE 5 Hypothesis for transduction in hair cells, in which stretch of the tip links opens transducer channels at the tips of the stereocilia. [Reproduced from J. O. Pickles, S. D. Comis, and M. P. Osborne (1984). Cross-links between stereocilia in the guinea pig organ of Corti, and their possible relation to sensory transduction. *Hearing Res.* **15**, 103–112, Elsevier Science Publishers, Amsterdam.]

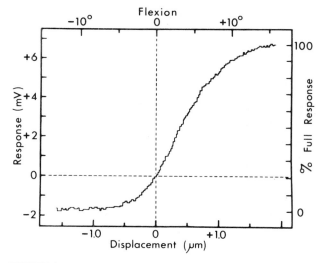

FIGURE 6 Relation between deflection of the stereocilia and intracellular voltage change in a hair cell of the bullfrog sacculus. [Reproduced from A. J. Hudspeth and D. P. Corey (1977). *Proc. Natl. Acad. Sci. USA* **74**, 2407–2411.]

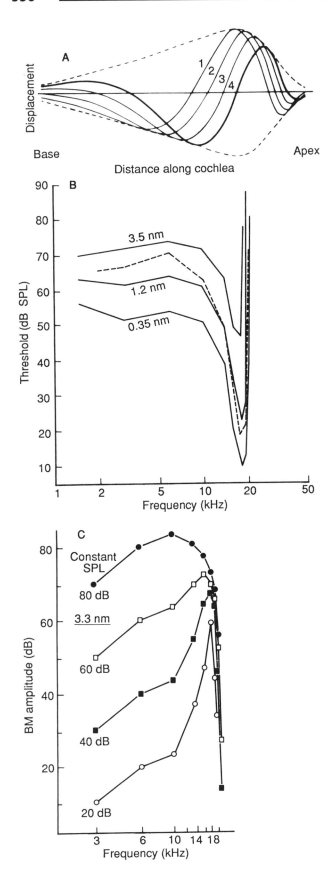

frequency of stimulation, with low frequencies producing the greatest response at the apex of the cochlea and high frequencies at the base.

The measurements of von Békésy were made in dead specimens; however, it has since been shown that the detailed pattern of the traveling wave is highly dependent on the cochlea being in a good physiological state. To ensure this, recent investigators have confined their measurements to single points along the cochlear duct, with the stimulus frequency being varied as the independent variable. The recent results can be shown in two ways. One form is the tuning curve, which relates the sound intensity necessary to produce a certain, fixed level of response for different frequencies of stimulation (Fig. 7B). If the same data are instead used to plot the amplitude of response for stimuli of constant intensity as the stimulus frequency is varied, we have the amplitude functions of Fig. 7C. The results differ from those of von Békésy in that the curves have steep slopes near the peak, indicating sharp frequency selectivity. In addition, the curve of Fig. 7B also has a very low threshold at its minimum, and the curve of Fig. 7C has a relatively large amplitude at its peak, indicating great sensitivity in the responses.

The origin of the large amplitude and sharp frequency selectivity of the cochlear mechanics is controversial. Simple mathematical analyses of cochlear models show only low-sensitivity broadly tuned vibrations, similar to those found by von Békésy. The models that are most successful in showing vibrations comparable to those found experimentally are those in which the cochlear partition contains a source of mechanical energy. It is suggested that this is provided by the outer hair cells.

FIGURE 7 (A) Traveling waves in the cochlea, according to the measurements of von Békésy. Frequency of stimulation: 200 Hz. [Reproduced from G. von Békésy (1960). "Experiments in Hearing," McGraw-Hill, New York.] (B) Tuning curves for basilar membrane vibration at the 18-kHz point on the guinea pig cochlea. The numbers on the curves show the vibration amplitudes at which the curves were made. The dotted line shows a tuning curve for the firing of an auditory nerve afferent, indicating similar tuning characteristics. [Reproduced from P. M. Sellick, R. Patuzzi, and B. M. Johnstone (1982). Measurement of basilar membrane motion in the guinea pig using the Mössbauer technique, *J. Acoust. Soc. Am.* **72**, 131–141.] (C) Amplitude functions for basilar membrane vibration at the 18-kHz point in the cochlea, measured at different intensities of stimulation. For a stimulus at 20 dB SPL, the vibration is sharply peaked, becoming more and more broadly peaked as the stimulus intensity is raised. [Reproduced from B. M. Johnstone, R. Patuzzi, and G. K. Yates (1986). Basilar membrane measurements and the travelling wave. *Hearing Res.* **22**, 147–153.]

Outer hair cells have evolved a unique mechanism for generating movements up to very high frequencies. The bodies of the outer hair cells are long and cylindrical; their cell membranes contain a large number of protein molecules packed tightly together. When an electrical field is applied across the membrane, the particles, which are likely to be electrically polarized, change their packing, altering the surface area of the membrane. The cell wall is underlain by filaments, wound spirally around the cylinder of the outer hair cell body. This stops the cell changing in width and means that the change in area is constrained into a change in the length of the cell body. The length change causes a deformation of the organ of Corti in the outer hair cell region and enhances the movement of the basilar membrane. The details of the way that the organ of Corti deforms and the way that it is translated into an enhancement of the traveling wave have not yet been worked out. However, unequivocal evidence indicates that some sort of movement-generating apparatus exists in the cochlea and that the apparatus is closely allied to mechanotransduction. This evidence comes from the observations that the cochlea can emit sound under some circumstances.

To detect acoustic emissions, a small speaker and a sensitive microphone are sealed into the ear canal. If the ear is stimulated with a click, a long series of waves are found, following the direct pressure pulse produced by the stimulus. The waves can be easily seen in human subjects; the form of the waves is different for each subject. If the click is presented at a very low intensity (e.g., within some 10 dB of the subject's absolute threshold), then sometimes more energy is returned to the ear canal than was originally introduced. Indeed, in some subjects the stimulus triggers a long series of oscillations, which can continue for seconds, and in others continuous vibrations are detectable without any input. The latter observations show that the ear can generate movement rather than just passively reflect a proportion of the incident energy back into the ear canal. A number of experiments (e.g., the effect of toxic agents, stimulation of the olivocochlear bundle) further suggest that the motile mechanism is closely allied to the outer hair cells. Because the emissions need at least some normal hair cells (although they seem to reflect discontinuities in the properties of the cochlea along the duct), the emissions can be used as a basis for objective audiometry.

The degree of outer hair cell motility is under the influence of the olivocochlear bundle, the pathway from the brain stem to the cochlea, some fibers of which give rise to synapses on the outer hair cells.

Therefore, this forms a way in which the central nervous system can control the amplitude of vibration of the mechanical traveling wave. Evidence indicates that the bundle helps protect the ear from acoustic trauma, and it may be involved in selective auditory attention.

Since hair cells produce a nonlinear response to deflection of the stereocilia (Fig. 6), it can be expected that any motile response from the outer hair cells will also be nonlinear. In other words, there will be slow shifts in the position of the basilar membrane, together with harmonics of the stimulus frequency and, for multiple input frequencies, intermodulation products. Evidence indicates that such distortion components exist in the movement of the basilar membrane. The distortion components are produced at the point at which outer hair cell feedback is maximal (i.e., around the peak of the traveling wave). If the components are lower in frequency than the input frequency, they can give rise to traveling waves of their own because they can then propagate away from the site of generation in the normal direction of the traveling wave (i.e., toward the apex of the cochlea). Two such distortion components are found at frequencies $f_2 - f_1$ and $2f_1 - f_2$ (where $f_2 > f_1$). These components can be detected both physiologically and psychophysically and may underlie some complex perceptual phenomena (e.g., the detection of musical consonance).

F. Gross Electrical Responses of the Cochlea

Correlates of hair cell and auditory nerve fiber activation can be measured by gross electrodes in and near the cochlea. Stimulation of the outer hair cells produces massed currents through the cells, giving rise to the cochlear microphonic, a potential which follows the waveform of the stimulus (Fig. 8). Nonlinear-

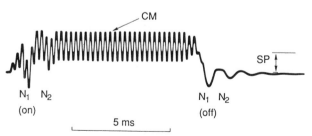

FIGURE 8 Diagram of the responses to a tone burst measured by an electrode on the round window. CM, cochlear microphonic; N_1 and N_2, compound action potentials; SP, summating potential (d.c., displacement of the microphonic).

ity in outer hair cell responses gives rise to d.c. components in the microphonic, producing the summating potential. Synchronized firing of auditory nerve fibers at the onset of a stimulus produces summed potentials called the compound action potential, or the N_1 and N_2 potentials. These potentials form important indicators of cochlear function in cases where it is not possible to measure the more detailed cochlear potentials.

IV. THE AUDITORY NERVE

A. Anatomy

In the mammalian auditory nerve, 95% of fibers innervate the inner hair cells and the remainder innervate the outer hair cells. This indicates that the job of inner hair cells is to detect the movement of the cochlear partition and to transmit the information to the central nervous system. It must be presumed that the records that we have of auditory nerve fiber activity arise from fibers innervating inner hair cells. In only one case do we have a proven recording from a fiber innervating an outer hair cell, and in this case the fiber showed no activity.

B. Tuning and Influence of Intensity

The firing of auditory nerve fibers as a function of frequency is primarily determined by the vibration of the cochlear partition. The tuning curves of auditory

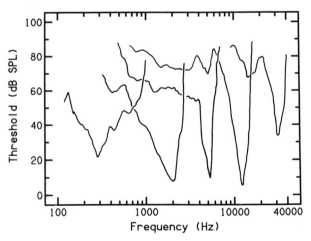

FIGURE 9 Tuning curves of five auditory fibers, all from the same cat. [Modified, with permission, from E. Javel, 1986, Basic response properties of auditory nerve fibers, *in* "Neurobiology of Hearing: The Cochlea" (R. A. Altschuler, R. P. Bobbin, and D. W. Hoffmann, eds.), pp. 213–245, Raven Press, New York.]

nerve fibers (Fig. 9) are similar to the tuning curves for vibrations of the basilar membrane (Fig. 7B). As far as current knowledge goes, any differences can probably be accounted for by the errors in measuring basilar membrane vibration.

The firing rate of auditory nerve fibers shows a sigmoidal relation with stimulus intensity; the firing rate in the majority of fibers reaches its maximum at a sound pressure level of 50 dB SPL and stays constant for further increases in stimulus intensity. The intensity at which the firing rate reaches a just-noticeable increment above the background or spontaneous firing rate gives the threshold of the fiber; most of the fibers recorded in an animal have thresholds within 20 dB of the animal's absolute threshold. However, a minority of fibers have higher thresholds, and the firing rate in these fibers also tends to increase relatively slowly with further increases in stimulus intensity. These fibers, which are driven by relatively high-threshold synapses on the inner hair cells, are used to code the responses to high-intensity stimuli.

C. Temporal Relations

The inner hair cell intracellular potential undergoes cyclic fluctuations in synchrony with the vibration of the basilar membrane. This evokes cyclic release of neurotransmitter (glutamate is the favored candidate), producing phasic firing in auditory nerve fibers; however, such behavior is found only at low frequencies. As the stimulus frequency is raised above about 600 Hz, the number of phase-locked action potentials decreases. This occurs because at high frequencies the transducer current flows through the capacitance of the hair cell external membranes, decreasing the intracellular a.c. voltage response to the stimulus. How then do high-frequency stimuli evoke action potentials? The asymmetric relation between stereociliar deflection and intracellular depolarization (Fig. 6) means that for sinusoidal stimuli a net depolarizing potential is superimposed on the a.c. potential in the hair cell. The sustained depolarization contributes to the release of transmitter, but in a nonphase-locked manner. The balance between the two mechanisms changes with increases in frequency, so that above about 5 kHz no action potentials are phase locked; therefore, temporal coding of auditory information is possible only below this frequency limit.

D. Response to Complex Stimuli

The cochlea functions nonlinearly, with the result that interactions are produced between the components of spectrally complex stimuli. One such interaction involves the generation of intermodulation products at frequencies $2f_1 - f_2$ (where $f_2 > f_1$) and $f_2 - f_1$, as described earlier. The products can be seen in auditory nerve activity in two ways. First, for frequencies of stimulation below the 5-kHz limit for phase locking, the firing patterns of auditory nerve fibers contain intervals, which correspond to the periods of the distortion tones. Second, distortion products can activate auditory nerve fibers tuned to the frequency of the distortion products, even though the fibers may give no response to the primaries (i.e., to f_1 and f_2) when presented alone. This can be explained by a mechanism in which the distortion products are generated in the cochlea at the point at which the traveling waves to f_1 and f_2 overlap. The active mechanical process at this point then produces new traveling waves to the distortion tones, which peak at their own characteristic place.

A second nonlinear interaction can be seen in the operation of two-tone suppression. This is also thought, predominantly or in part, to arise from the nonlinear input–output functions of outer hair cells (Fig. 6). If one stimulus produces sufficient deflections to drive a cell to the flat part of the input–output function, a superimposed stimulus will not be able to produce any greater response. Therefore, the outer hair cells will not be able to contribute to the active amplification of the traveling wave to the second stimulus, and the peak of its traveling wave will not reach its normal size. Thus, the presence of one stimulus will reduce the response to the other, to the extent that the traveling wave to one overlaps the active process to the other. Two-tone suppression is particularly powerful when the suppressor lies just above the suppressed tone in frequency.

V. THE AUDITORY CENTRAL NERVOUS SYSTEM

A. Information Flow

Auditory nerve fibers branch and innervate the three divisions of the cochlear nucleus (Fig. 10). The cells in the anteroventral cochlear nucleus have response properties similar to those of auditory nerve fibers. This nucleus therefore transmits auditory signals to the next stage, the superior olivary complex, with relatively little transformation. At the other extreme, the dorsal cochlear nucleus has very complex neuronal responses, the cells showing tuning curves with many

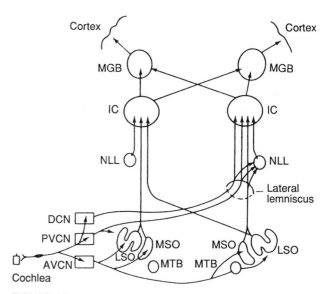

FIGURE 10 The main ascending pathways of the brainstem. Many minor tracts are not shown. AVCN, anteroventral cochlear nucleus; DCN, dorsal cochlear nucleus; IC, inferior colliculus; LSO, lateral superior olive; MGB, medial geniculate body; MSO, medial superior olive; MTB, medial nucleus of the trapezoid body; NLL, nucleus of the lateral lemniscus; PVCN, posteroventral cochlear nucleus.

inhibitory areas, and complex temporal patterns of firing. These can cause substantial enhancement of responses to the temporal and spectral contrasts in the auditory stimulus. The output of this nucleus bypasses the superior olivary complex to reach the inferior colliculus directly.

Significant binaural interaction in the auditory system first occurs in the superior olivary complex. The interaction is the basis of sound localization, with the relevant cues being differences in the intensity and time of arrival of sounds at the two ears. Diffraction around the head produces significant interaural intensity differences for high-frequency stimuli. The lateral, or S-shaped, superior olivary nucleus responds to the intensity differences, with sounds in one ear inhibiting the response to sounds in the other. The cells are also sensitive to differences in the time of arrival of transients in the envelopes of high frequency stimuli. In contrast, the medial superior olive represents low-frequency stimuli. Such stimuli are in the range in which phase locking is preserved, and here the sensitivity to differences in the times of arrival means that the cells are responsive to interaural phase.

Other subnuclei of the superior olivary complex relay information from one side of the brain stem to the other and also give rise to the fibers of the olivocochlear bundle (see earlier), the efferent pathway to the cochlea.

Activity from the superior olive is transmitted via the lateral lemniscus (some fibers synapsing in the nuclei of the lateral lemniscus) to the inferior colliculus, a substantial integrative and relay center of the brain stem. Here the location-specific information from the superior olive is combined with information from the complex pattern analyses of the dorsal cochlear nucleus. Thence, the information is transmitted via the specific thalamic relay for audition, the medial geniculate nucleus, to the auditory cortex.

Audition is represented in multiple areas in the cerebral cortex. In cats, the primarily auditory cortex is surrounded by fields called the secondary auditory (AII), anterior auditory, posterior ectosylvian, posterior, ventral, and ventral posterior fields. The secondary somatosensory area SII and the insulotemporal areas are also auditory fields. In primates, the auditory cortex is situated on the superior temporal plane in the lateral fissure; the primary field is surrounded by subsidiary cortical areas called the lateral, rostrolateral, and caudomedian fields, together with further auditory fields.

Cellular responses in the auditory cortex show tuning curves with single or multiple peaks. Some cells may respond only to complex auditory stimuli, such as frequency- or amplitude-modulated tones, or simulated speech sounds.

B. Principles of Organization

A number of principles of organization can be discerned in the central auditory system. First, the component nuclei are tonotopically organized, i.e., the cells are arranged in order of characteristic frequency. This facilitates interaction between neurons of adjacent characteristic frequency and may also represent economical packing of the afferent innervation. In the nuclei of the auditory brain stem and the thalamus, cells of the same characteristic frequency are situated in two-dimensional isofrequency planes across the nucleus. In the cortex, cells of the same characteristic frequency are situated in one-dimensional isofrequency strips across the cortical surface; separate tonotopic maps are found in several of the auditory cortical subareas.

Second, at and beyond the superior olivary complex, responses depend on the location of the sound source; cells in the higher nuclei are driven most strongly by stimuli on the contralateral side. Searches have also been made for location-specific information within each isofrequency plane or strip. Thus, in the primary auditory cortex, cells of differing binaural interaction (e.g., cells excited by both ears, as against cells excited from one ear and inhibited from the other) are situated in discrete strips running at right angles to the isofrequency strips. Evidence also indicates that there is a crude map of auditory space within each isofrequency plane in the inferior colliculus. However, no correlate has been found in the mammal of the precise map of auditory space found in the owl, in its homolog of the inferior colliculus, the lateral dorsal mesencephalic nucleus.

Third, there is a general progression in the complexity of responses as one ascends the auditory system. For instance, tuning curves are exclusively simple in the auditory nerve and anteroventral cochlear nucleus, while both simple and complex tuning curves are found in the auditory cortex. However, the progression is far less clear than for instance in the visual system, and, except for sound localization, it has not been possible to be sure of the functional implications of any progression in complexity. Evidently, there are multiple parallel pathways for the extraction of much auditory information, and the hierarchical relation between the different stages and systems is not clear. Here it must not be forgotten that the auditory system also has a rich centrifugal innervation, with fibers running from the higher levels of the system to the lower. The olivocochlear bundle is one example of such a pathway. Therefore, the opportunity exists for the responses at the lower levels of the auditory system to reflect the complex processing of the highest.

VI. BIOLOGICAL BASIS OF AUDITORY PERCEPTION

A. Threshold

Measurement of the displacement of the basilar membrane suggests that the membrane vibrates by about 0.3 nm at an animal's measured behavioral threshold. If the organ of Corti and tectorial membrane move in a simple geometrical manner (by no means a reliable assumption), then at threshold the stereocilia would be deflected by about 10^{-2} degrees, with the movement transmitted to the transducer channel being about 0.4 nm. This seems not an unreasonable value for the threshold movement at a channel. Because it is hypothesized that each of the hundred or so channels on a single hair cell is continually opening and closing under the influence of thermal energy, deflection merely causing a redistribution of the open and closed probabilities, the definition of threshold becomes a statistical one and dependent on the amount of averaging employed. An animal's absolute threshold lies just below the best thresholds of the most sensitive of its auditory nerve fibers, probably a reflection of the averaging possible in an intact auditory nerve.

B. Frequency Resolution

Psychophysical frequency resolution is likely to be a direct reflection of the frequency resolution shown in basilar membrane vibration (Fig. 7B), as also reflected in the tuning of auditory nerve fibers (Fig. 9). While it is possible to measure similar tuning functions psychophysically in humans, care in interpretation is necessary if a detailed comparison is to be made. Functions in humans are plotted using more complex paradigms such as masking, where the threshold to one stimulus (the probe) is measured in the presence of another. Because the cochlea behaves nonlinearly, tuning to one stimulus may be affected by the presence of another, in experiments where the masker and probe are present simultaneously. Experiments where the masker and probe are present nonsimultaneously, such as forward masking, are thought to give more reliable measures of peripheral frequency resolution. Using such techniques, it is also possible to measure correlates of further phenomena found electrophysiologically in animals, such as two-tone suppression.

In experimental cases of cochlear pathology, the tuning curves of basilar membrane vibration rise in best threshold and increase in bandwidth. This most likely reflects loss of the active mechanism from the cochlear vibration. Psychophysical results in humans show similar changes, with an increase in threshold in the audiogram, and a loss of frequency resolution.

C. The Perception of Speech

Responses to speech sounds have been measured extensively in the auditory nerve and have shown how the different aspects of the stimulus are represented in the firing of auditory nerve fibers. The formants (spectral peaks) in vowels are, for instance, represented by increased levels of activity in the auditory nerve fibers tuned to the formant frequencies. At a higher level, the critical involvement of the auditory cortex in speech processing is demonstrated by disabilities in the production and understanding of speech after lesions of the speech areas, namely Wernicke's area in the dominant (nomally left) cerebral hemisphere and Broca's area. The involvement of these areas is also shown through selective local increases in blood flow or by magnetic resonance imaging while listening to speech. It is also possible to use the asymmetry of the involvement of the two cerebral hemispheres in speech to tag anatomical specializations related to speech analysis: for instance, area 22, which contains Wernicke's area, is larger on the dominant side and contains certain neuronal specializations. However, caution is necessary because not all hemispheric asymmetries in the auditory areas may be related to speech. In addition, recordings made in the auditory cortex of unanesthetized human patients have shown that cells in the superior temporal gyrus respond efficiently to speech sounds and less effectively to other auditory stimuli. Wernicke's area was not explored in such operations, lest speech function be damaged. However, it is probably true to say that we have little or no information on the physiological mechanisms underlying the linguistic, as against the acoustic, aspects of speech.

BIBLIOGRAPHY

Altschuler, R. A., Bobbin, R. P., and Hoffman, D. W. (eds.) (1986). "Neurobiology of Hearing: The Cochlea." Raven Press, New York.

Altschuler, R. A., Bobbin, R. P., Clopton, B. M., and Hoffman, D. W. (eds.) (1991). "Neurobiology of Hearing: The Central Auditory System." Raven Press, New York.

Pickles, J. O. (1988). "An Introduction to the Physiology of Hearing." Academic Press, London.

Pickles, J. O., and Corey, D. P. (1992). Mechanoelectrical transduction by hair cells. *Trends Neurosci.* **15,** 254–259.

Popper, A. N., and Fay, R. R. (1992). "The Mammalian Auditory Pathway: Neurophysiology," Vol. 2. Springer, New York.

Romand, R. (ed.) (1992). "Development of Auditory and Vestibular Systems," Vol. 2. Elsevier, Amsterdam.

Webster, D. B., Popper, A. N., and Fay, R. R. (1992). "The Mammalian Auditory Pathway: Neuroanatomy," Vol. 1. Springer, New York.

Eating Disorders

University of Washington

GLOSSARY

Anorectic Pertaining to or affected with anorexia

Anorexia nervosa Nervous disorder in which a person eats very little food

Binge/gorge Episode of eating an unreasonably large amount of food

Bulimia nervosa Nervous disorder in which a person eats much more food than necessary and then voluntarily vomits or purges

Bulimic Pertaining to or affected with bulimia

Calorie Unit of measuring heat energy, commonly used to measure food energy

Developmental obesity Having an increase in weight beyond reasonable limits due to accumulation of fat throughout development

Endocrine Organs and structures of the body that produce hormones to regulate body processes

Hormones Chemical produced by an organ of the body that specifically regulates another organ and, therefore, body processes

Hyperphagia Eating larger than reasonable amounts of food

Purge To make the bowels empty out

Regurgitate To bring up undigested food

Rumination Disorder in which a person regularly brings up food to be chewed again after meals

Semistarvation Being nearly deprived of food for a long period

PERSONS WITH EATING DISORDERS (ANOREXIA nervosa, bulimia nervosa, rumination, and developmental obesity) fail to develop rhythms, or develop distorted rhythms, of eating and exercising. Their habits turn into obsessive abuses, taking food out of its normal function, namely that of maintaining life. The energy stored in their bodies as fat increases, decreases, or swings in wide variation, jeopardizing overall health. The disordered behaviors are linked to psychological inadequacies, as both causes and effects. People with eating disorders improve when the psychological basis of their habits is altered and when the physical symptoms are treated by a team of health care specialists.

I. THE SPECTRUM OF EATING DISORDERS

When the physical symptoms of eating disorders are illustrated on a spectrum, as in Fig. 1, anorexia will be at one end and developmental obesity at the opposite end. In the middle of the spectrum, the figures represent individuals of normal or close to normal weight. Examples are those with bulimia nervosa and rumination, recently described disorders.

A. Anorexia Nervosa

In anorexia nervosa, excessive restriction of food energy (measured in calories) and strenuous exercise

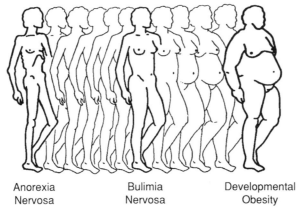

Anorexia Nervosa   Bulimia Nervosa   Developmental Obesity

FIGURE I Spectrum of eating disorders. Underlying psychosocial characteristics are held in common while physical conditions vary across the spectrum. [Adapted, with permission; originally published in J. M. Rees (1984). Eating disorders. *In* "Nutrition in Adolescence" (L. K. Mahan and J. M. Rees, eds.), p. 105. C. V. Mosby, St. Louis. © L. K. Mahan and J. M. Rees, Seattle, WA, (1989).]

depletes the energy pool of the body to the degree that the individual could die if not treated. The observable motivator of the affected person is a fear of fatness.

B. Bulimia Nervosa

In bulimia nervosa, the person who is trying to control body weight eats more food than he or she needs, or thinks is needed, and gets rid of it by vomiting and/or purging. Exaggerated feelings of inadequacy generally focus on the physique; the affected person is hypercritical of his or her anatomical form.

C. Rumination

The ruminator chews, swallows, and then regurgitates food and swallows it in a repetitive cycle after meals, apparently as a response to stress. Rumination is reportedly practiced most frequently by people who push themselves to be high achievers.

D. Developmental Obesity

The developmentally obese person grows up storing excess energy as fat, eating more than is necessary, and exercising less than normal. This condition could be called hyperphagia nervosa to parallel the nomenclature of other eating disorders and to distinguish it from overfatness not associated with psychological dependency on abnormal eating and exercise habits. There are sufficient other causes of storing excess en-

ergy as fat that the conditions are called "the obesities" by experts, even though the term obesity is still usually used to denote anyone weighing 20% more than normal body weight. [*See* Obesity.]

II. PHYSICAL VERSUS PSYCHOLOGICAL PROBLEMS

Most people focus on the physical characteristics of eating disorders because they are the most obvious. Average citizens can identify dancers who are skeleton-like rather than lithe and people who are so heavy they cannot move easily. Less commonly understood are the important underlying psychological problems associated with the disorders along the spectrum, problems that interfere with the ability of those affected to function normally and productively.

Most prominently, people with eating disorders use food inappropriately and obsessively in response to life stresses. Being preoccupied with weight and food, they have difficulty meeting developmental milestones and becoming independent adults. They have some degree of problem with experiencing bodily sensations arising within themselves as normal and valid, formulating their own value system instead of relying on social opinion, and realistically perceiving their body size as well as setting expectations for achievements and roles. Thinking flexibly and in the abstract, especially about themselves, may be difficult. They do not conceive of food as nourishment, know the value of nourishing themselves, or the ways to do it.

III. THEORIES REGARDING ETIOLOGY

Early psychoanalytic theories suggested that eating disorders grew out of an inability to deal with sexual drives. In a later period, they were ascribed to abnormalities in the endocrine system. Disturbances in the mother–child relationship and, more completely, interaction of the whole family are observed to foster the development of eating disorders and other psychosomatic diseases. A biopsychosocial explanation of the origins of disordered eating incorporates the previous theories, suggesting that a complex mix of characteristics leads to eating disorders. Thus, an individual born into a family whose interaction does not allow normal development of autonomy feels inadequate to cope with life and adopts abnormal eating and exercise behaviors. Patients may do this for the express purpose of changing their bodies or because they find gratification in the habits themselves. They may be fed inappropriately or taught aberrant eating and exercise behaviors by family members. Their bodies may also harbor a propensity for a biochemical imbalance.

However they originated, the abnormal behaviors and psychological disturbances often alter body function, including endocrine patterns. Thus, sexuality is indeed influenced through related changes in sex hormone production, an effect if not a cause of severe eating disorders, validating the original observations. The processes vary in each of the disorders, but family, individual psychological, and biochemical characteristics are all involved at different stages.

IV. FAMILY CHARACTERISTICS

The family that engenders eating disorders often tends to be overprotective and rigid rather than flexible. Conflicts are avoided and go unresolved because they are not faced openly. Children are enmeshed in the family, a characteristic that stands in the way of normal development. Because parents do not communicate well with each other, an affected child is pulled into their interactions as a kind of go between, a situation termed triangulation. This role pressures the child, who is not free to be child-like, and, in turn, is distracted from concentrating on his or her own development. While not all families of people developing eating disorders follow the same patterns, disturbances leading to unreal expectations are usually a contributing factor.

V. CULTURAL INFLUENCES

Several cultural influences are partly responsible for the recent proliferation of eating disorders in the United States. Valuing a slim physique together with fat- and sugar-laden foods is a dichotomy that is played out in many lives, either casually, by restricting food for a while after eating rich meals, or in the extreme, as in eating disorders. Psychological stress levels and demands for achievement in modern life are also strong influences. Women are more vulnerable to eating disorders than in former eras because they are faced with greater role expectations combined with tightly circumscribed standards for physique. These cultural features exist to some degree in all affluent nations but have developed to the extreme in the United States.

Collectively, society does not have a reasonable concept of nourishment, the realization that human beings eat food to stay alive, and that materials they get in food are built into their bodies. They also do not realize that about three-fourths of the energy in the food they eat is used in the basic processes that keep them alive at rest, with the other used in moving and working. The constant misuse of the word calorie encourages misunderstanding. To many people, calories are equated with mass or volume (chunks or lumps), or simply fat. When they realize that the term is a unit of measure and that energy is the thing being measured, they can understand more easily that it is necessary, positive, and dynamic and can be controlled and used. They begin to see food energy as like oxygen, a basic element of life; that it should not be restricted or overconsumed as influenced by fashion or used frivolously; and, further, that their bodies are *themselves*, needing to be cared for, not spoiled by overeating or punished with "dieting."

VI. ANOREXIA NERVOSA

Because often families of anorectics do not allow conflicts to surface, the symptoms seem suddenly to emerge in the children, usually female, of model families. It is as if the patient, feeling inadequate to cope with the complex life of adolescence, regresses to the earliest ways of showing independence—regulating food—to feel she is in control of at least one facet of her life. The resistance of many patients and their families to treatment stems from their desire at any cost to keep up the appearance of having no problems.

Quite a number of anorectics are overweight before starting to "diet." They take their expectations for slimness from models, dancers, and media stars, carrying them to extremes. A parent who is interested in food production or is obsessive about exercise will have a strong influence. The affected child tends to be more passive than siblings and becomes enmeshed. No predisposing biochemical vulnerability has been found so far, although some experts believe physical factors contributing to the onset of the disorder will be found. The body changes measurably throughout the course of the disease, in direct relationship to the amount and length of starvation, and recovers when nourishment is replenished.

According to surveys, about 1 in 100 affluent, highly educated and 1 in 200 of the general population of young women have anorexia nervosa. Only about 10% of patients in treatment are male. Throughout the anorexia and bulimia sections of this article the female pronoun will be used because most anorectics and bulimics are female.

A. Physical Manifestation

The person with anorexia nervosa will initially try to reduce to an unreasonably low weight by eating less and less and exercising obsessively. She also may force herself to vomit or take laxatives to purge herself. At the same time, she isolates herself from peers and is argumentative with her family. The menstrual cycle ceases, so patients are infertile; the psychological disturbance may be responsible initially. When weight is sufficiently low for any reason, the menstrual cycle will stop due to decreased endocrine function, which is also responsible for the decrease in sex drive. Growth, sexual development, and organ (including brain) makeup are interrupted during starvation. The potential harm to patients who are not physically mature is therefore great and the consequences may be long lasting.

Patients need comprehensive treatment with psychological, nutritional, and medical components. The focus will be the appropriate role of parents in setting limits while allowing the patient to develop without abnormal intrusion or pressures rather than on the bizarre eating and exercise rituals alone. As the person with anorexia and the family respond to treatment, the anorectic will be able to accept new ideas and resume healthy eating and exercise behaviors as advised by nutrition and medical professionals. Progress toward physical health is monitored throughout treatment.

B. Crisis

If the anorectic, and her family, is untreated or does not respond well, her physical condition will often deteriorate to a state of outright starvation. With inadequate energy supply, the body fat stores are depleted, and muscles begin to be used up for immediate nourishment. Breathing and heart rate slow and the skin is dry and scaly; downy hair grows over the face, trunk, and appendages. Meanwhile, hair on the head is thin, dry, brittle, and falls out. The slow heart beat causes low blood pressure with fainting spells. Sleep, adjustment to temperature change, and bowel regularity are disturbed. Cracks may appear at the corner of the lips, fingernails become brittle, and tips of fingers,

toes, and ear lobes turn blue. Body fluids and their contents are imbalanced and flow outside their usual compartments, causing the body to swell abnormally.

Thoughts of the starving person are centered on food while coherent, creative thinking is impaired. The person is apathetic, dull, exhausted, uninterested in sex, and depressed.

A supportive but firm hospital program replenishes energy stores so that the body systems, including the brain, begin to function again. With weight gain, the life-threatening phase is over and psychotherapy can continue.

C. Long-Term Recovery

Full recovery requires a long period—*years*—of counseling and education. Leaving the hospital at near normal weight is the real beginning of treatment. Individual psychotherapy and family therapy are key. The process is time-consuming because it is complex. The person will develop full potential as an autonomous adult after she and her family understand and can practice appropriate communication and interaction. Through information and changed attitudes and behavior gained in nutrition counseling, the anorectic will learn to control body weight at a reasonable level throughout the various stages and activity levels of the life cycle, without obsessive behaviors.

D. Outcome

Because of early recognition and modern treatment of anorexia nervosa, the death rate is down to 4%, or less, of treated patients as opposed to 10% in the 1970s. People can make a full physical, psychological, and social recovery if they have comprehensive treatment for a number of years. If patients and their families leave treatment after the physical crisis, they often continue to be obsessive and not able to educate themselves, work at full potential, or to establish healthy families; they often have repeated periods of starvation, bulimia, or become obese. People who have the disorders at a young age without proper treatment may fail to develop normally, both physically and psychologically.

VII. BULIMIA NERVOSA

Bulimia probably develops under similar circumstances as anorexia nervosa, although the goals and habits of many bulimics are not so self-destructive.

Families are often more hostile and chaotic than those of anorectics and symptoms may develop a little later in life. People with bulimia try to eat only small amounts of food to control body weight, but the body responds physically and psychologically to temporary starvation, triggering a binge. They vomit and/or purge to rid themselves of the food they have eaten.

Five to 10% of college females have bulimia nervosa while 20–25% practice some bulimic behaviors, although they do not have the complete disorder. Less than 1% of college males have bulimia nervosa. Persons with bulimia are usually a little older (late adolescent to young adult) than those with anorexia nervosa.

A. Physical Symptoms

Bulimics may arise each morning thinking they will follow their ideal semistarvation eating plan and do so for most of the day. When they are alone and unoccupied, however, the gorging begins. It will end when they are overfilled and they force themselves to vomit. Vomiting gets easier with time, and they can do it without great effort. They may take laxatives and diuretics to purge themselves. Unless the habits are drastic, they can practice them for many years without serious health problems. Most prominently, the stomach acid they vomit will erode the teeth, irritate the throat, and inflame the esophagus. Salivary glands swell, lips crack, and blood vessels in the face may be broken by the pressure of vomiting. Their hands may be calloused from resting their upper teeth on them as they vomit. They usually are at near normal weight and are well enough nourished that they remain fertile. The menstrual cycle is affected in as many as half, however, probably as a result of psychological stress. They may not experience the interruptions in sexual function that anorectics do.

They need hospital treatment if they also are starving themselves or if they take many medications and vomit or purge so frequently that the body fluids and its contents are imbalanced, holes develop in the esophagus, or the kidneys are damaged.

B. Psychological Characteristics

Overall, bulimics feel secretive and guilty. They fear rejection, abandonment, and failure. An all or nothing attitude rules their expectations. They lack confidence related to judgements they make about their bodies. As part of the maladaptive concept of nourishment, they also harbor distorted beliefs about food and find

it hard to accept their bodies and the way they function and look to others.

C. Treatment Goals

Psychological counseling is aimed at dispelling the negative feelings bulimic patients have about themselves and giving them the inner strength to cope with life as independent adults. Family counseling helps young persons still living at home to develop normally. With nutrition education and counseling, they nourish themselves adequately, set weight goals appropriately, and give up bulimic habits.

D. Outcome

Effective programs help people with bulimia to adjust psychosocially and improve their abnormal eating behaviors. Like anorectics, they need several *years* of therapy. Patients continue to improve while they are in treatment but unfortunately, also like anorectics, they often leave treatment before they have been fully rehabilitated.

VIII. RUMINATION

Some people ruminate as part of another eating disorder and others practice it as a sole disorder. The number and gender distribution of people who are affected is not well known and therapeutic programs are not common. It does not seem to cause nutritional problems or damaged teeth; one of the most common complaints of ruminators is that they have foul smelling breath. Helping people to deal with anxiety and monitor their regurgitation will most likely lead to success in giving up the habit.

IX. OBESITY

Obesity is one of the biggest health problems in the United States, affecting as many as one-third of Americans. The number of obese children alone rose by about 40% from the mid-1960s to the late 1970s and continues to rise. Debilitating conditions such as heart, blood vessel, and lung disease, diabetes, and some types of cancer are linked to obesity.

The prejudice of society, including health professionals in blaming the obese for causing their own problems by lack of will power, laziness, and gluttony, intensifies the dependency on simplistic treatments and hinders the search for reasonable solutions to obesity. In actuality, the obese need support and comprehensive long-term treatment to deal with difficult physical and psychological problems that arise from a multitude of factors, many outside their control. Many do not have psychological problems and as a group they do not demonstrate psychopathology. Treatment enables them to control the symptoms; there is no cure.

A. Categories of Obesity

1. Developmental Obesity

In parallel to anorexia nervosa, disturbances in families and individuals may lead to developmental obesity as depicted in the illustration (Fig. 1). The family interaction stunts psychosocial development and encourages overeating and underexercising. Growing obesity, in turn, interferes with normal development by hampering actual movement and making the obese child feel negative about himself and his or her experiences with others.

2. Reactionary Obesity and Connections with Other Psychological Disturbances

Some developmentally obese are not affected all the time but revert to abuse of food, moving and interacting with others less as a reaction to loss or other trauma. Obesity is also associated with serious psychological disorders in some individuals.

3. Culturally Based and Behaviorally Fostered

The popularity of rich foods and the emphasis on eating for gratification, celebration, and reward foster obesity in some people who are not otherwise predisposed as well as intensifying the problems of those who are. Culture fosters obesity in families and individuals who do not know how to take precautions against influences such as advertising and the constant availability of prepared food.

4. Physiological

Physically inherited body processes that store excess energy as body fat cause some obesities. However the obesity started, storage of excess energy activates other processes that work to sustain obesity much like a thermostat maintains heat. Both the inherited and acquired processes that conserve energy are survival traits, rooted in history when starvation was a great threat to humans.

B. Physical Factors

A person who is obese does not have to overeat to stay obese. If the person overeats, he or she will gain, and gain more easily, than a lean person. Compared with the lean body, the bodies of obese people save energy, for example, by producing less internal heat and pumping less sodium between cells. They store more fat because they have more energy to store and because the chemicals that allow fat into fat cells are more plentiful. Weight loss improves some, but not all, of these differences in the way energy is used and stored by the obese. It appears the size of fat cells can be reduced, for example, but the number cannot. The reduced fat cells tend to fill again with ease. None of the processes can be treated with drugs at this point. Investigation of energy conservation processes has led to the set point theory: the cumulative effect is to maintain an excessive weight—the set point—which the body regains easily if weight is ever reduced.

Carrying excess weight stresses and causes disease in the heart, blood vessels, lungs, spine, and weight-bearing joints of the obese. A disordered endocrine system can interrupt the menstrual cycle in women, who may be infertile as result.

C. Psychological Factors

The developmentally obese have trouble separating from parents and tend to live passively and to feel they cannot accomplish what they set out to do. Second, a person who has become obese through purely physical factors as well as the developmentally obese suffers psychological damage through societal prejudices in education, employment, and social settings. Coping with isolation, name-calling, and scapegoating throughout childhood and adolescence affects personality development profoundly, fostering many of the psychological characteristics of the developmentally obese. A result is that many obese people feel guilty each time they eat rather than understanding that they need a certain amount of food to stay alive no matter what they weigh.

Finally, even the psychologically healthy individual faces a set of psychological problems in trying to control weight. He or she will have to learn to break tasks down into accomplishable segments with rewards, have the patience to wait for long-term results, and to enjoy the process. The effort is comparable to that required by an average person who for some reason would need to train for running a marathon or prepare to be a concert-level musician and must be conceptualized as being comparably demanding.

D. Cultural Factors

Overfeeding associated with psychological problems or even merely social customs in a family where the physical predisposition to obesity is genetically passed on is especially detrimental. In a society where using body energy is made obsolete by mechanization and energy-rich foods abound, the results are devastating. A significant number of people in the United States are two to three times heavier than normal people.

E. Treatment Objectives

The first goal of obesity treatment is to slow down the rate of gain, if the person is still gaining, and the second is to stop gaining. Because of individual variation in the physical potential to lose weight, focusing on weight loss as a short-term goal is unreasonable. Weight loss has proved nearly impossible to maintain if it is rapidly accomplished by severely restricting food. Semistarvation diets, which have been the mainstay of treatment of obesity, appear to be self-defeating. If a person eats much less than the amount of energy used, the systems that defend body weight are challenged, and the tendency to store excess fat increases in response to the starvation. Severe energy-restricting "diets" are the root of the dismal experiences the obese have in dealing with their problem. An even higher set point weight, bringing an even greater problem of control, may be avoided by concentrating on the knowledge, attitudes, and behaviors that control weight.

Learning to eat only the needed amount of food (especially fatty food), but not to restrict it; increasing body movement; and improving psychosocial adjustment helps patients stay at a weight as close to normal as their physical characteristics will allow. Any person trying to learn to control weight needs psychological support, individualized if the obesity is severe, and counseling if there is preexisting or acquired maladjustment.

Acknowledging that weight management is multifaceted and requires complex treatment increases the chances for controlling symptoms in any of the obesities. Most successful programs are carried out by teams of nutritionists and medical and psychosocial health specialists. As with the other eating disorders, a number of *years* are required while people learn and prac-

tice behaviors they will need to continue throughout their lives.

F. Outcome

People who are not helped as children through effective prevention and treatment are most vulnerable to the consequences of obesity. Results of recorded studies show that about 95% of people fail to maintain a body weight they arrived at by a few months of drastic dietary treatment. These numbers do not include people who have learned to deal with obesity on their own or through individualized programs. Complete treatment strategies that facilitate more complete and sustainable long-term outcomes are beginning to be set up, thus increasing the number of people who can control the symptoms of obesity in the near future. A greater emphasis will need to be placed on prevention now that the complexity of changing a person's body weight is more fully understood.

X. CONCLUSION

A large portion of Americans are affected by eating disorders. Many others are overly concerned about weight and shape but do not have a complete disorder. Treatment helps disordered eaters put food and exercise in perspective and deal with psychosocial maladjustment. It is of primary importance that the complexity of the disorders be kept in mind by professionals and patients as they attempt to learn about, prevent, and deal with this spectrum of problems that impairs the function of so many in modern society.

BIBLIOGRAPHY

Bennett, W. (1987). Dietary treatments of obesity. *Ann. N.Y. Acad. Sci.* **499**, 250–265.

Blinder, B. J. (1986). Rumination: A benign disorder? *Int. J. Eat. Dis.* **5**, 385–386.

Bruch, H. (1973). "Eating Disorders." Basic Books, New York.

Comerci, G. (1990). Medical complications of anorexia nervosa and bulimia nervosa. *In* "Adolescent Medicine" (J. A. Farrow, ed.), Vol. 74, pp. 1293–1310. Saunders, Philadelphia.

Fisher, M., Golden, N., Katzman, D., Kreipe, R. E., Rees, J. M., Schebendach, J., Sigman, G., Ammerman, S., and Hoberman, H. (1995). Eating disorders in adolescents: A background paper. *J. Adol. Health* **16**, 420–437.

Garner, D. M., and Garfinkel, P. E. (eds.) (1985). "Handbook of

Psychotherapy for Treatment of Anorexia Nervosa and Bulimia." Guilford Press, New York.

Keys, A., *et al.* (1950). "The Biology of Human Starvation," Vols. I and II. University of Minnesota Press, Minneapolis.

Kreipe, R. E., Golden, N. H., Katzman, D. K., Fisher, M. D., Rees, J. M., Tonkin, R. S., Silber, T. J., Sigman, G., Schebendach, J., Ammerman, S., and Hoberman, H. M. (1995). Society for adolescent medicine position paper: Eating disorders in adolescents. *J. Adol. Health* **16**

Minuchin, S., Rosman, B. L., and Baker, L. (1978). "Psychosomatic Families: Anorexia Nervosa in Context." Harvard University Press, Cambridge, MA.

Rees, J. M. (1984). Eating disorders. *In* "Nutrition in Adolescence" (L. K. Mahan and J. M. Rees, eds.), pp. 104–137. Times/Mirror Mosby, St. Louis.

Rees, J. M. (1990). Management of obesity in adolescents. *In* "Adolescent Medicine" (J. A. Farrow, ed.), Vol. 74, pp. 1275–1292. Saunders, Philadelphia.

Rees, J. M. (1995). Rational nutrition component of comprehensive weight management. *In* "Obesity: New Directions in Assessment and Management" (A. P. Simopoulos and VanItallie, eds.). The Charles Press Publishers, Philadelphia.

Stunkard, A. J. (ed.) (1980). "Obesity." Saunders, Philadelphia.

Yager, J., Andersen, A., Devlin, M., Mitchell, J., Powers, P., and Yates, A. (1993). American Psychiatric Association practice guidelines for eating disorders. *Am. J. Psychiat.* **150**, 207–228.

Eating Styles and Food Choices Associated with Obesity and Dieting

BONNIE SPRING
REGINA PINGITORE
JUNE ZARAGOZA
Finch University of Health Science,
The Chicago Medical School, and
Hines VA Medical Center

GLOSSARY

Abstinence violation effect Behaviors ensuing from the perfectionistic cognitive appraisal that any lapse in self-discipline regarding eating, exercise, or other health behavior means that complete failure has occurred, justifying relapse to the problematic behavior

Basal metabolic rate Energy expended for the sustenance of life processes such as respiration, circulation, synthesis of organic constituents, pumping ions across membranes, and maintaining body temperature

Binge eating Consumption of a large amount of food in a short time accompanied by a feeling of lack of control over eating behavior during the binge

Doubly labeled water An indirect method to assess daily energy expenditure by calculating carbon dioxide production from the differential disappearance rates of two stable isotopes of water

Energy balance equation Quantifies the state of equilibrium between energy intake and all forms of energy expenditure, such that changes in energy stores result from imbalance between intake and expenditure

Lean body mass The fat-free mass of the body, which determines basal metabolic rate, including muscle, organs, bone marrow, the central nervous system, and the small amount of essential fat that is contained within these elements

Reactivity A phenomenon whereby the process of assessment alters the behavior being measured; as applied to dietary records, it suggests that the process of self-monitoring and recording food intake decreases the frequency of problematic eating behavior

Relapse prevention Behavior modification techniques used to sustain healthy behavior by developing coping strategies to manage and contain slips before full relapse can occur

Restraint A behavioral pattern characterized by prominent weight consciousness, recurrent efforts to restrict dietary intake, and proneness to periods of disinhibition during which large quantities of food are consumed

Weight cycling A pattern of weight loss followed by weight regain that is usually associated with "yo yo dieting," cycling on and off a diet; it is theorized that in successive cycles, weight losses occur more slowly and regains occur more rapidly

EATING STYLES AND FOOD CHOICES THAT ARE associated with obesity and dieting are the focus of this article. After characterizing ways in which excess energy intake relative to expenditure results in obesity, how energy balance can be altered to promote weight loss is described. Four styles of overeating are differentiated. These can be regarded as varying from mildest to most severe in terms of the number of domains of life function affected and the complexity of treatment that is warranted. The spectrum of aberrant eating patterns ranges from an uncomplicated pattern of excess intake relative to a sedentary life-style; through selective overeating in pursuit of self-nurturance, sensory experience, or self-medication of dysphoric moods; through restrained eating; to binge eating, a form of overeating that is complex and difficult to treat. This article describes general properties of interventions that successfully foster weight loss and illustrates how treatments can be tailored to address different aberrant eating patterns.

I. INTRODUCTION: THE PURSUIT OF SLIMNESS

Conventional wisdom in industrialized societies dictates that one can never be too rich or too thin. In contrast to affluence, which is almost universally prized, thinness has not always been positively valued. Intentional pursuit of slimness would have signified sheer folly in earlier centuries that were characterized by intermittent food shortages and famine. Similarly, in nonindustrialized societies where scarcity still prevails, plumpness and sometimes even morbid obesity are considered attractive, and thinness is disliked because it signifies poverty and undernutrition.

The pursuit of slimness has only become a meaningful personal aspiration against the background of nutritional abundance that characterizes most industrialized societies today. Whereas most of evolutionary history led organisms to develop biological mechanisms that cope with conditions of food scarcity, precious little evolutionary experience has equipped us to cope with nutritional surfeit. Because our defenses against excess nutrition work less well than those against undernutrition, many people become overweight as a result of a sedentary lifestyle and unlimited access to the palatable, high-calorie cafeteria that constitutes our daily diet. In the United States today, for example, 25–30% of adults can be characterized as obese, weighing at least 20% more than a desirable body weight. Since few defenses against overnutrition are inborn, except in those fortunate individuals who are genetically predisposed to leanness, most people must learn behavioral strategies to protect against overweight. There are ethnic differences in both the risk of obesity and the attitude toward overweight; for example, Hispanic and African-American women are at greater risk for obesity than upper-middle-class Anglo-American women, but may judge themselves less harshly for being overweight.

Although much of this article will address dieting strategies to alleviate obesity, it is important to bear in mind that dieting is not restricted to only the overweight or obese. On the contrary, a considerable amount of dieting, perhaps even the majority, is practiced by individuals who are not overweight, but who nonetheless desire to become slimmer. Their weight loss efforts bear testimony to the fact that weight dissatisfaction and dieting have become normative for an upwardly mobile sector of the industrialized world. On any given day, 40% of American women and 20% of American men are dieting, and the likelihood of dieting is greater among those who are more highly educated and affluent. In the United States alone, the weight loss industry nets nearly $29 billion annually. For young women, especially, the preoccupation with losing weight and achieving a slim physique begins early and lasts a lifetime. Even before the fat deposition that accompanies puberty, fear of fatness and the practice of dieting are alarmingly prevalent, characterizing 60–80% of 10-year-old girls. [*See* Obesity.]

II. WEIGHT LOSS STRATEGIES AND THE ENERGY BALANCE EQUATION

All weight loss strategies ultimately act upon the energy balance equation that maintains a stable body weight. Energy balance refers to the observation that as long as calorie intake is equal to calorie expenditure, no change (beyond normal day-to-day fluctuation) will occur in body weight. If energy intake regularly exceeds expenditure, however, excess calories will be stored in the form of body fat, and weight gain will be evident. Conversely, if energy expenditure constantly exceeds calorie intake, the body's tissue fuel stores will be depleted to meet its energy needs, and weight loss will occur. To give a very approximate prediction formula, intake of every 3500 kcal in excess of energy needs results in a weight gain of 1 lb (or 0.45 kg of adipose tissue). Conversely, each time that energy expenditure exceeds calorie intake by 3500 kcal, resulting in an energy deficit, 0.45 kg of adipose tissue will be oxidized, causing body weight to decrease by 1 lb.

Those who wish to lose weight must do so by creating an energy deficit. From an energy balance perspective, the weight loss consequences will be the same regardless of whether the deficit is achieved rapidly or slowly. A 500-kcal energy deficit attained on each of 7 days and a 35-kcal deficit achieved on each of 100 days will both yield the same weight change outcome.

From a behavioral perspective, however, the outcomes may differ importantly, as discussed later on.

An energy deficit can be created by decreasing calorie intake, increasing energy expenditure, or a combination of the two. Dieting, or intentionally restricting food energy intake, remains the most frequently practiced weight loss strategy. Different diets vary considerably in their recommended degrees of caloric restriction, the selectivity with which they advocate constraining intake of particular macronutrients or food categories, and in their format, duration, and integration of other life-style changes. Alternative dieting and food choice strategies will be discussed later. As noted elsewhere in this volume, however, a comprehensive portrayal of weight loss strategies must also acknowledge approaches that aim to modify the expenditure side of the energy balance equation. Such approaches aim to create an energy deficit by increasing calorie expenditure either alone or simultaneously with decreased energy intake.

There are several ways to achieve weight loss by increasing calorie expenditure. The greatest effect is achieved by interventions that enhance basal metabolism, the amount of energy the body requires simply to maintain its vital functions. This is because the greatest amount of the body's total energy expenditure (60–75%) is allocated to maintaining the body's "idling state." Energy is also used to support digestion (thermic effect of food) and to sustain movement (energy cost of physical activity), but the caloric demands of these latter functions (10% and 15–30%, respectively) are considerably less than the energy cost of sustaining basal metabolism. An average American male weighing 154 lb expends about 1680 kcal/day just to sustain his basal metabolic rate, and the average female weighing 128 lb uses about 1327 for the same purpose. The energy utilization added by exercise is relatively smaller. For example, after an hour of walking at 3 mph, the average man or woman has only increased energy expenditure by 150–265 kcal, not enough to offset snacking on a 300-kcal candy bar. [*See* Energy Metabolism.]

Great interest exists in identifying weight loss strategies that increase energy expenditure. For example, efforts are in progress to develop antiobesity drugs that have the potential to increase energy expenditure. Several pharmacologic agents produce weight loss at least partly by increasing resting metabolism, decreasing insulin resistance, or inhibiting the lipoprotein lipase enzyme that breaks down triglycerides so that they can be taken up and stored by adipocytes. On the other hand, many of these same pharmacological agents also inhibit food intake, making it difficult to determine the degree to which their antiobesity action occurs via decreased energy intake versus increased energy expenditure. Increasing exercise is a more widely practiced strategy that aims to promote weight loss by increasing energy expenditure. As noted earlier, a considerable amount of exercise is needed to directly create a large energy deficit, but the energy cost associated with heightened activity can be multiplied to the extent that exercise temporarily elevates the basal metabolic rate. Because small energy deficits do accumulate, however, regular exercise that is practiced over time as part of an active lifestyle can have a substantial effect on body weight, as long as energy intake remains constant or decreases.

III. DO THE OBESE OVEREAT?

The pendulum of professional opinion has swung back and forth several times on the question of whether the obese consume more calories than the lean. Long ago it went unquestioned that overweight resulted from a combination of gluttony (overeating) and sloth (lack of exercise). Shortly thereafter, however, consensus shifted to the viewpoint that some cases of obesity are "glandular" (derived from endocrine imbalance) or due to "slowed metabolism." It remains undisputed that endocrine and metabolic abnormalities play a role in the etiology of some forms of obesity, such as those associated with hypothyroidism or Prader–Willi syndrome. It is interesting to note, however, that even in these forms of obesity, biological constraints on energy expenditure often co-occur with overeating and excess calorie intake.

More controversial is the proposition that a decreased ability to expend calories because of a slowed basal metabolic rate *usually* plays a role in the pathophysiology of obesity. Although basal metabolism is normally increased rather than decreased among the obese (reflecting the heightened energy required to sustain a larger body mass), it is possible to find overweight individuals whose metabolic rate is slower than expected, given their large size. A plausible explanation in many such cases is that metabolic slowing is a direct consequence of caloric restriction, which can decrease the resting metabolic rate by 20–30% in both obese and lean dieters. As noted earlier, one residue of an evolutionary history characterized by repeated exposures to food scarcity is the existence of homeostatic mechanisms that preserve a stable body weight despite conditions of inadequate energy intake.

The body's propensity to slow metabolic expenditure in response to conditions of starvation is one such evolutionary vestige, and it is triggered by dieting in much the same way as it once was by famine.

Some experts believe that repeated cycles of diet-induced weight loss followed by weight regain result in a pattern of body *weight cycling* that is associated with chronic reductions in basal metabolic rate. The proposition is that "yo yo dieters," who go on and off diets repeatedly, lose both lean body mass and fat stores when restricting calories, and regain only fat after abandoning a diet. When people who are sedentary stop restricting calories after a diet, they regain weight disproportionately as fat stores because their life-styles create little demand to rebuild lean tissue or muscle. Weight gain, therefore, increases the sedentary person's percentage of body fat. Because metabolic expenditure is greater for lean tissue, especially muscle, than for fat mass, metabolic expenditure decreases. It seems plausible, therefore, that obesity might persist in weight cyclers because of decreased energy expenditure, despite their consumption of a normal or even an energy-deficient diet.

A considerable body of empirical research seemed to support the proposition that the calorie intake of obese individuals is equivalent or even less than that of normal weight individuals. Much of that research measured calorie intake by having people keep food diaries, in which they recorded everything that they ate throughout a 3- to 4-day period. Another assessment technique that was frequently used required recall of everything eaten during the previous 24 hr. In data collected from food diary and 24-hr recall studies, obese individuals often reported that they ate no more or even considerably less than their lean counterparts, implying that their overweight status could be attributed to decreased metabolic expenditure rather than to excess calorie intake.

In subsequent research, however, more sophisticated metabolic measures, such as the *doubly labeled water technique,* made is possible to directly test the assumption that obesity is associated with decreased metabolic expenditure in the presence of normal or subnormal energy intake. Results of these studies revealed that the dietary intake reports of many individuals cannot possibly be accurate because the energy intake they describe is insufficient to sustain life at their quantified level of energy expenditure. Moreover, the amount to which energy intake is underestimated appears to increase in proportion to the reporter's degree of obesity. Such findings suggest that reports of dietary intake are not only characterized by random error of measurement. Self-reports of food intake also seem to be influenced by directional or systematic response biases in reporting consumption that are associated with the respondent's weight status.

There is currently great interest in characterizing and explaining the causes of biased reporting of food intake by obese individuals. One suggestion has been that the social stigma associated with obesity leads overweight people to underreport their calorie intake because of fear that reports of high energy intake will exacerbate negative social appraisals. Another possible explanation proposes that the behavior of eating shows *reactivity,* such that food intake is altered by the very measurement techniques that are intended to assess consumption. The assignment to self-monitor and record problematic behavior is a common element of most habit modification interventions because it is well known that the undesirable behavior tends to decrease in frequency as its initiation becomes less automatic, more conscious, and more guided by conscious awareness and choice. Consequently, it may be that the requirement to record what is eaten increases self-awareness about food intake and decreases the likelihood of high-calorie eating occasions.

As we move next to a discussion of how to diagnose different overeating patterns and recommend diets that are matched to the overweight client's particular needs, the comments cited earlier create a context that explains why the task is difficult. First, after long and intense controversy, a consensus is just beginning to reemerge that excess calorie intake plays a significant role in the genesis of obesity. Second, the dietary record and recall techniques that are the major tools used to assess disturbed eating patterns are inaccurate and most likely elicit underreporting of intake from the very obese individuals who could benefit the most from dieting. A third difficulty that has not yet been addressed concerns the heterogeneity of disturbed eating patterns among the obese. The eating behaviors that contribute to overweight ultimately act via the final common pathway of creating excess energy intake relative to energy expenditure, but the specific inroads via which they reach that outcome vary considerably. If dietary interventions are to be well matched to target eating problems, it is useful to be able to correctly identify the kind of eating disturbance and intervene accordingly.

IV. THE DIVERSITY OF OVEREATING STYLES

In some sense every overweight person has a unique eating style; however, a few broad styles of overeating can be described. In what follows, four styles of overeating are differentiated that can be regarded as varying from mildest to most severe in terms of the number of domains of life function affected and the complexity of treatment that is indicated. The four overeating styles range from the relatively uncomplicated pattern of excess energy intake relative to a sedentary lifestyle, to selective overeating, to restrained eating, and finally to binge eating, a form of overeating that is complex and difficult to treat.

A. Excess Intake Relative to Sedentary Lifestyle

In industrialized societies, the modal pattern by which overweight develops involves a gradual accretion of modest amounts of weight across the life span. An increase in body weight at midlife is observed so commonly that age-related weight gain has come to be accepted as normal and natural. It has even been proposed that accepted criteria for defining degrees of overweight should be revised to set more liberal standards for older individuals so that a larger body mass could be considered a normal, healthy weight for people of middle age or older. Recent research findings associating heightened mortality with even modest age-related weight gains hve momentarily halted proposed revisions of ideal body weight standards. Results from the Nurses' Health Study show that the lowest mortality rates are found among women whose weight has remained steadily at least 15 percent less than the U.S. average. Those findings suggest that age-related weight gain may be not merely a cosmetic liability, but rather a potential health risk that warrants intervention.

Decreased energy expenditure from activity is a primary cause of age-related weight gain, although increased calorie intake also plays a role. In the school age years, fitness activities are institutionally required and appropriate nutrients and energy are supplied. Adults progress to a world of employment influenced by historic and socioeconomic forces that increasingly reward sedentary activities. In the gradual transition from an industrial- and service-based economy to the current information-based economy, the average daily energy expenditure required to do a day's work has fallen by an estimated 200 kcal/day. Given current reliance on computers and other convenience devices, the world of work requires sedentary activities that are pursued over extended working hours, usually preceded and followed by a long, sedentary commute. For most women and men, therefore, the energy cost of the physical activities associated with daily work

decreases dramatically from youth to adulthood. Few adults compensate by increasing their activity levels outside of the workplace. In fact, the Centers for Disease Control estimates that fewer than 20% of U.S. adults engage in frequent, vigorous physical activity, and more than half of the adult population lead inactive lives.

The increasingly sedentary adult could maintain energy balance and preserve a youthful body weight by reducing energy intake. There is ample temptation to do otherwise, however, when faced with the vast cafeteria of easily accessible, highly palatable, and energy-dense foods made available by commercial food manufacturers. What begins as the daily addition of coffee and a pastry eaten on the way to work shows up gradually in weight gains that can average 1–2 lb per month. Such small increases, if they are observed at all, can be dismissed as trivial weight fluctuations rather than understood as the outgrowth of energy imbalance. If sustained over time, though, the result can be a progression from normal weight to overweight and even to moderate obesity.

At first glance, the logical intervention for this pattern of overeating might seem to be the combination of a diet that restricts the amount and variety of treat foods plus a planned exercise program. Such an intervention, however, could actually foster more significant problems, while failing to achieve the desired weight loss. As will be noted in discussing restrained eating, diets that disallow certain categories of highly enjoyed foods often backfire. By triggering both heightened weight consciousness and feelings of deprivation, they initiate a self-perpetuating cycle of dietary restriction followed by overeating. Structured exercise programs that are externally imposed suffer a different kind of problem in that they do little to integrate physical activity into the daily routine and are unlikely to foster enjoyment of exercise. Although dieting alternatives will be discussed in the next section, it is suggested that the best initial intervention for this kind of age-related obesity is to monitor and decrease the time spent in sedentary activities.

B. Selective Overeaters

For selective overeaters, all foods are far from equal. Rather, certain foods are craved, sought out, and consumed in quantities that exceed energy requirements. The kinds of foods that are craved usually combine various characteristics, several of which have been highlighted in competing theories to explain the causes of selective overeating. Preferred foods tend to be highly palatable, associated with nurturance or pleasant social occasions, sweet, and high in carbohydrate and fat content.

1. Self-Nurturant Overeating

From early life onward, we are educated that food is a reward. That lesson is transmitted to the toddler who learns that she can have a cookie if she stops crying, the child who receives cake and ice cream to celebrate a birthday, and the laborer who returns from a hard day at work to be provided with a "man's meal" that will "stick to his ribs." As the grownups in our childhoods reinforced us with food, so do we as adults learn to use food to reward ourselves. Food supplies us with self-nurturance on many kinds of occasion. Consider the ice cream sundae that we self-administer to celebrate the completion of a project, the candy bar that we eat to restore morale during a tedious afternoon of studying, or the box of cookies that comforts us through the sorrows of a broken relationship. As these examples indicate, densely caloric sweet treat foods are commonly used to engender feelings of indulgence and self-nurturance, but other kinds of foods may serve as well. For many men, a hamburger or other red meat is self-administered as a favorite indulgence, in preference to sweets.

According to self-nurturance theory, risk of overweight occurs when preferred foods are self-administered to excess, both in relation to caloric needs and in preference to other forms of self-reward. The theory is psychological in suggesting that, via learning, any food can come to be associated with nurturance. Thus, if prevailing cultural conditions equate raw meat or butter with indulgent reward, then raw meat or butter will come to be a potent vehicle to supply self-nurturance. According to this theory, the essential property of a self-nurturant food is that it has been learned to be such, rather than that it has any particular sensory or macronutrient property.

Treatment of obesity that results from self-nurturant overeating is somewhat complex because the selective overeating serves an important psychological need. A prerequisite of successful treatment is, therefore, the need to identify and substitute alternative behaviors that are rewarding and that can be initiated in place of eating. If overeating is not replaced with some other means of gratification, dieting is likely to trigger a vicious cycle. Curtailment of food self-administration may exacerbate feelings of deprivation, which, in turn, create a greater need for the very feelings of self-indulgence that can be generated by selective overeating.

2. Overeating as a Sensory Pursuit

Alternative explanations of selective overeating attribute motivational power to the sensory properties of preferred foods. From this perspective, particular foods are liked, sought out, and overconsumed because of the way they taste, smell, or feel in the mouth. A corollary proposition is that eating stops when sensory changes produced by ingesting a food trigger satiety by rendering the food no longer palatable. An example of the manner in which sensory influences can promote overeating is the well-known "cafeteria effect," whereby energy intake can be enhanced above normal by exposure to a wide variety of foods of differing sensory properties. One explanation of the cafeteria effect invokes the hypothesis of "sensory specific satiety" which states that liking for a particular flavor decreases as a function of repetitive exposures to that flavor during consumption. Conversely, sequential exposure to a variety of different flavors makes it less likely that boredom will develop to any single flavor, and therefore increases the amount of eating that will occur before satiety develops.

Although crunchy, savory, and creamy foods all have their supporters, the "sweet tooth" is many dieters' Achilles heel. A liking for sweetness is already evident in neonates. The sweet tooth is further cultivated throughout childhood as commercial food manufacturers add vast quantities of sweeteners to many cereals, beverages, and snack foods. Individuals do differ reliably in their degree of preference for sweets, although it remains unclear whether the consumption of sugary foods can be fully explained by the degree of liking for them. Very rough parallels between consumption and liking can be found. For example, patients with anorexia, who eat very little, show a keen aversion toward sweets. In contrast, bulimia nervosa patients show diminished responsivity to sweetness, plausibly because their recurrent vomiting has caused a loss of sweet taste receptors on the tongue. Excessive consumption of treat foods during binges might, therefore, reflect an effort to trigger a sensory mechanism that is less sensitive because it has been damaged.

The sweet tooth can be conceptualized as described earlier: an appetitive force that initiates pursuit and consumption of treats. Alternatively, the sweet tooth can be thought of as a force that causes excessive persistence of the appetite for sweets. From the latter perspective, the sweet tooth involves a failure in the satiating mechanism that shuts off eating: a deficit that prevents a person from being able to eat just a single sweet instead of too many. A special case of sensory-specific satiety is the phenomenon called alliesthesia: a decline in preference and consumption of sweet foods after eating other sweets. It is alliesthesia (in addition to wisdom) that stops a weight-conscious person from eating an entire box of chocolates because the candies eventually begin to taste nauseating rather than appealing. For some overweight individuals, however, consumption of a sweet food fails to trigger the transition from liking to not liking to disliking, or triggers it only after a delay. In such cases, deficient alliesthesia fails to support the sensory shut-off mechanism that would ordinarily terminate consumption of a sweet dessert or snack. Lacking the sensory cues that would ordinarily motivate the cessation of eating dessert, the decision to put down the fork must be controlled consciously, effortfully, and often with great difficulty.

Another proposed mechanism to explain selective overeating is that foods that blend sugar and fat in combination comprise super-palatable mixtures that are difficult for any eater to resist. Sweet–fat combinations have been argued to have quasi-addictive properties because they may trigger the release of endogenous opioids. Taste panel reports indicate that sugary foods are liked for their sweetness, whereas moderate-to high-fat foods tend to be liked for their creaminess, viscosity, and how they feel in the mouth. Some findings do suggest that the heightened preference for foods that are both sweet and creamy exceeds the sum of the preferences for the two constituent sets of sensations. Unquestionably, foods like ice cream or chocolate that are both intensely sweet and creamy tend to be very high in calories, which usually derive even more from their fat than their carbohydrate content. Such densely caloric foods can wreak havoc with a moderate calorie diet because a single portion supplies a high proportion of the day's allowable calories. Moreover, some evidence suggests that the preference for sweet–fat mixtures is heightened above normal among recurrent dieters who have experienced multiple weight fluctuations. Thus, the foods whose ingestion can prove most problematic for maintenance of a diet are the very same ones that are most tempting to the dieter.

When an overweight person uses food to pursue palatable sensations, several considerations should influence weight-loss planning. Much as it is useful for the self-nurturant overeater to find alternative sources of gratification, it can be helpful for the sensory-oriented overeater to identify enjoyable alternative forms of sensory stimulation. The commercial market affords abundant options in the form of noncaloric sweeteners and other food flavorers, aromatic nasal sprays, and, most recently, fat substitutes. Interestingly, we lack clear understanding of techniques that could genuinely alter sensory preferences in such a way as to increase the palatability of low-calorie foods. For example, there is no known intervention that can cause a dieter to find carrots more hedonically appealing than ice cream. In dieting, though, sweets deprivation may ultimately provide its own reward, in the respect that sweets genuinely become less palatable after their prolonged elimination from the diet. Sustained elimination of sweet–fat mixtures also ultimately yields a similar hedonic shift, but it is much more transient because preference for these superpalatable combinations is almost immediately reacquired upon reexposure to such foods. Ironically, even though it is a nondietary form of intervention, exercise is one of the few treatments that has shown some ability to reduce the hedonic value of sweet–fat mixtures.

3. Carbohydrate Craving

Another theory about selective overeating describes the food intake patterns of people who preferentially consume carbohydrates, usually as late afternoon or evening snacks, or as desserts following meals. Their eating style has been described as carbohydrate craving because, when given equal access to foods high in carbohydrate and very low in protein versus foods high in protein, they reliably select the high-carbohydrate options. In many cases, their selectively high intake of carbohydrates generalizes across alternatives that are sweet or savory and high or low in fat content.

Carbohydrate preference has been documented not only in obesity, but also in other conditions that are associated with less severe degrees of overweight. Those conditions include seasonal affective disorder (winter depression), late luteal phase dysphoric disorder (premenstrual distress syndrome), and nicotine withdrawal. It has been suggested that in all of these syndromes, including obesity with carbohydrate craving, patients experience dysphoric moods that are triggered endogenously by states involving a functional deficiency of the brain neurotransmitter, serotonin. A further proposition is that a lifetime of experience in managing their personal moods and body chemistry has taught carbohydrate cravers that their dysphoric moods can be dissipated by eating carbohydrate-rich, protein-poor foods. The presumed mechanism is that such foods trigger an insulin secretion that causes most amino acids to be removed from the bloodstream and taken up into tissues. Tryptophan is an exception, however, because it binds to albumin as insulin strips away the fatty acids that are usually albumin bound. With clearance from plasma of the other large neutral amino acids that share the same carrier system for transport across the blood–brain barrier, tryptophan's competitive access to the carrier molecules increases. As brain influx of tryptophan, serotonin's precursor, increases, so do serotonin synthesis and release, ameliorating the presumed deficiency of brain serotonin.

Evidence suggests that carbohydrate cravers do experience positive mood states after consuming carbohydrate-rich, protein-poor foods, in contrast to more balanced eaters who tend to become sleepy. These empirical findings bear out patients' anecdotal reports that they feel better after eating carbohydrates and intentionally use such foods to self-medicate dysphoric moods. To the extent that carbohydrate self-administration is reinforced in such individuals by the induction of positive mood states, treatment requires the development of alternative techniques for reversing dysphoric moods. The treatment options are many, including the substitution of carbohydrate-rich snack foods that are low in fat and calories or the use of exercise or cognitive behavioral techniques to dispel dysphoric moods. In addition, a variety of serotoninergic agents, including dexfenfluramine and fluoxetine, have been shown to improve mood and selectively reduce carbohydrate consumption in individuals and at times when intake is disproportionately high.

C. Restrained Eaters

In one sense all of the eating styles discussed so far reduce to a single problem: excess energy intake. A more complicated eating pattern occurs in restrained eating, where calorie intake fluctuates between too little and too much. Individuals who exhibit the restrained eating pattern usually display several characteristics. First, in defining their self-concepts, they place great emphasis on weight and physical appearance. Their prominent weight concerns occur in conjunction with marked weight dissatisfaction, exacerbating an already low self-esteem. A second characteristic is a felt need to exercise a high degree of control over food intake. To sustain their constant efforts to diet, restrained eaters remain preoccupied with food. They consciously restrict their intake of "forbidden foods," palatable foods that they consider incompatable with a diet because of their high fat, carbohydrate, and calorie content. In fact, when they are successfully controlling food intake, restrained

eaters do consume fewer calories, carbohydrates, and fat–carbohydrate combinations than nonrestrained eaters. The difficulty is that these periods of dietary restriction are too extreme to be sustained and eventually give way to episodes of disinhibited eating during which great quantities of calorically dense foods are consumed. Disinhibited eating, the third characteristic of restrained eaters, may be triggered by emotional upset or by a small departure from extreme dietary restriction. During periods of disinhibition, the restrained eater's behavior reflects voracious hunger that is very easily triggered coupled with deficient satiety that is very difficult to induce.

Underlying many of the restrained eater's problems is a rigid and perfectionistic cognitive style. Foods are arbitrarily defined as "good" (permitted while on a diet) or "bad" (forbidden while on a diet). Similarly, the self is appraised as worthy or unworthy solely on the basis of success in maintaining calorie restriction and body weight. For example, self-appraisals like the following are commonplace: "I was bad; I ate a piece of cake" or "I was good; I lost 2 lb." The restrained eater's extremity of cognitition predisposes toward behavior that follows an "all of nothing" rule. Once a small amount of a forbidden food has been consumed, or even a food that is erroneously believed to be high calorie, there follow the appraisals that "all is lost" and "oh, what the hell?" Thus evaluated, small dietary infractions accelerate into very major eating occasions. That pattern illustrates the abstinence violation effect, whereby a small lapse from desired behavior is interpreted so pessimistically as to warrant a major behavioral transgression. The consequence is the restrained eater's pattern of "yo yo" dieting, whereby eating and body weight vacillate between phases of tight control and extreme lack of control.

Especially since many restrained eaters are not overweight at all, weight control is only a secondary target of treatment for restrained eating. More fundamental are the problematic cognitions and beliefs that sustain the aberrant behavior pattern: equation of self-worth with physical appearance, black versus white categorization of foods, all-or-none beliefs about dieting, and perfectionistic approach to eating and weight. If weight loss efforts are initiated before the aberrant thinking pattern is addressed, the clinician may collude with the client in perpetuating the cycle of vacillation between eating extremes. A similar problem can affect the introduction of an exercise plan. If the restrained eater applies the usual "all or none" thinking, she is likely to implement a rigid plan that cannot consistently meet her perfectionistic expectations. For these reasons, it is useful to address the extreme style of "black and white" thinking early in treatment.

D. Binge Eating

Binge eating can be regarded as the most severe disturbance on the continuum of aberrant eating styles. A subgroup of obese individuals has been called binge eaters or compulsive overeaters because their eating behavior is characterized by consuming large amounts of food in a short time period accompanied by a feeling of being out of control. During binges, which can last from minutes to hours or, more rarely, even days, patients may eat rapidly even when they feel uncomfortably full. Solitary eating is often followed by feelings of guilt, shame, depression, and disgust. Although bingeing bears some similarity to the disinhibited eating exhibited by restrained eaters, cycles of dietary restraint, even if present at some time in the patient's past, usually no longer intervene. In the quantity of food eaten and the sense of lack of control, binge eating resembles bulimia nervosa, except that purging is absent. In contrast to bulimics, however, compulsive overeaters' binges tend to last longer, contain a wider variety of foods, and often lack distinct beginnings and ends.

In one multi-site study of patients who sought treatment in hospital-based weight control programs, about 30% of participants were binge eaters. The condition is more prevalent in women than men and is associated with greater adiposity and less treatment-responsive obesity. In comparison with obese non-bingers, binge eaters are more likely to drop out of weight control treatment, are less likely to achieve weight loss if they remain in treatment, and are more likely to regain weight quickly.

The causes of binge eating remain in dispute. Because binge eaters exhibit more prominent depressive symptoms than comparably obese individuals who do not binge, and because binges are often triggered by emotional upset, it has been proposed that the condition involves a variant of an affective disorder. Because of the feelings of compulsiveness and lack of control that accompany eating episodes and because significant alcohol abuse is often present, binge eating has also been described as an "addiction," bearing similarities to compulsive gambling or substance abuse. Still another belief is that restrained eating often serves as the gateway to binge eating, which occurs when restrictive dieting conflicts with the body's biological "set point" for weight.

The treatment of binge eating poses an interesting challenge especially since this aberrant eating style has thus far been defined largely by its treatment refractoriness. Experimental trials of various pharmacotherapies are currently in progress. In the domain of behavioral intervention, there exists consensus that two differentiable problems require treatment, binge eating behavior and obesity, but there is no agreement about which problem should be treated first. Current behavioral treatments of bingeing follow the approach taken to treat other compulsive behaviors and include self-monitoring of surrounding cognitions and emotions, as well as the development of alternative coping skills and social supports. A complexity is that for food, unlike other problematic substances, abstinence is not a behavioral option. Consequently, binge eaters need to learn to selectively self-administer foods. Special difficulties may surround the handling of "trigger foods," which commonly include desserts, processed lunch meats, or high-fat condiments. Some patients find that certain foods are so certain to trigger a binge that it is best to eliminate them from the diet. Others find that trigger foods can safely be reintroduced after a period of stability. In general, binge eaters struggle against the same temptations as those who overeat for self-nurturance, sensory pursuit, or carbohydrate craving, but to a greater degree. The great challenge of treamtent is coping with the risk that the calorie restriction needed to produce weight loss will engender feelings of deprivation that exacerbate the urge to binge. [See Eating Disorders.]

V. CHOOSING AN APPROPRIATE DIET

Traditionally, the choice of an appropriate weight loss treatment has been based on two primary considerations: the medical risks associated with the patient's obesity and the amount of weight loss needed to normalize body weight. Figure 1 and Table I display the kind of risk classification algorithm that has been used and the corresponding treatment recommendations. As illustrated, calculations of obesity-associated health risks usually incorporate the patient's body mass index (BMI), computed as weight kg/height m^2, as well as the presence of other medical risk factors, such as noninsulin dependent diabetes, hypertension, or hyperlipidemia. Higher medical risks and more morbid degrees of obesity have been considered to warrant more intensive treatment, including more extreme restriction of calorie intake and more forms of

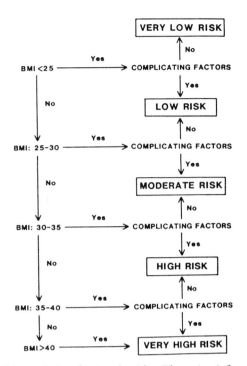

FIGURE I Risk classification algorithm. The patient is first placed into a category based on body mass index. The presence or absence of complicating factors determines the degree of health risk. Complicating factors include elevated abdominal–gluteal ratio (male, >0.95, female, >0.85), diabetes mellitus, hypertension, hyperlipidemia, male sex, and age less than 40 years. From Bjorntorp and Brodoff (1992).

medical treatment, including surgery and pharmacotherapy, that have been discussed elsewhere in this volume. Very low calorie diets (VLCDs) typically provide only 400 to 800 kcal/day in order to produce weight losses averaging 20 kg in 12 weeks. By providing most of their calories in the form of protein,

VLCDs attempt to produce large weight losses without jeopardizing lean body mass. VLCDs can consist of "real foods" (e.g., lean fish or fowl) or liquid formulas (Optifast, Medifast).

Contemporary thinking endorses VLCDs only for patients for whom acute weight loss is needed to correct a severe health risk or allow a necessary medical intervention, such as surgery. Some disadvantages associated with VLCDs are the risk of medical complications associated with too rapid a rate of weight loss, the likelihood of significant rebound weight gain when the diet ends, and the lack of opportunity for patients to learn how to eat moderately under normal circumstances. In addition, if patients regain weight, they have great difficulty sustaining the motivation to lose weight the same way a second time, because of the extreme deprivation associated with VLCDs.

A. Life-Style Modification

The approach to weight reduction currently advocated by the National Institutes of Health is that weight loss efforts should be incorporated into a more comprehensive initiative to establish a healthy life-style. The underlying rationale is that we reside in a cultural climate of excess that exposes us lifelong to a significant risk of obesity. Consequently, weight management is most likely to be achieved when it is the outgrowth of pervasive and sustained patterns of healthy eating, activity, and nutritional intake. From that perspective, it is preferable to make small to moderate behavioral changes that can be gradually incorporated into a healthy life-style and maintained over the long term rather than more extreme changes that cause abrupt disruptions in life-style and are unlikely to be sustained. Many of the recommended

TABLE I
A Classification of Obesity Relating Risk to Choices of Treatment[a]

Risk	Caloric intake (kcal/day)			Choice of treatments		
	<200	200–800	>800	Exercise	Drugs	Surgery
Low	NA	3	2	1	NA	NA
Moderate	NA	2	1–2	1	3	NA
High	NA	1	2	3	2	NA
Very high	2	1	1	3	2	1–2

Note. NA, not appropriate; 1, first choice; 2, second choice; 3, third choice.
[a]From Bjorntorp and Brodoff (1992).

behavioral changes presuppose self-regulatory skills such as planning, impulse control, frustration tolerance, and judgment. These skills may need to be taught as part of any weight loss strategy to prevent weight loss attempts from becoming another failure experience for the patient.

Several treatment components are shared by lifestyle modification programs appropriate to address all of the eating styles that have been discussed. These components include implementation of a moderate calorie balanced diet, a rate of weight loss not to exceed 1–1.5 lb/week, behavioral modification of eating and activity patterns, and ongoing support.

B. Dietary Intake

Moderate calorie restriction to a daily intake of 1200–1500 kcal typically results in a 1- to 1.5-lb/week weight loss, from a deficit of 500–750 kcal/day, depending on activity level. It should be noted that this current recommendation contrasts with earlier practices that endorsed a 22-kcal/kg actual weight, to produce a deficit of 1000 kcal/day, yielding a 2.2-lb weekly weight loss.

The most preferable food plan is a balanced, energy-controlled diet supplying high carbohydrate intake (55%), generous protein of high quality (0.8–1.2 g/kg ideal weight), and restricted fat (20–30%). Such plans can be followed for months without specific supplements, although over time the nutrients most likely to become deficient include iron, folacin, vitamin B_6, and zinc. Intakes of 1100 and below require vitamin and mineral supplements. The inclusion of carbohydrate prevents the ketogenic effects of low-calorie diets that lack sufficient starches and other carbohydrates. High-fiber starches, fruits, and vegetables supply food volume while reducing the caloric density of the plan, provide satiety by delaying gastric emptying, and slightly reduce the efficiency of intestinal absorption. Twenty to 30 g/day of dietary fiber from food is recommended. The protein amount allows some conversion of dietary protein to energy if necessary. Fat intake is reduced because of its lipogenic quality and high energy content (9 kcal/g vs 4 kcal/g for protein and carbohydrate), but moderate restriction ensures against essential fatty acid or fat-soluble vitamin deficiencies. Additional recommendations include limiting cholesterol intake to 200 mg/day or less and drinking at least 1 liter/day of water or 1 ml/cal/day, whichever is more.

For the average person who lacks nutritional train-ing, the prospect of understanding and implementing such a diet can be daunting. Many dieters find that they benefit from approaches that simplify and support them in food intake planning. A variety of community or commercial weight loss programs provide classes in nutrition and behavioral modification. These programs often structure food intake planning using the exchange system, which achieves both nutritional balance and calorie reduction by programming the required intake of foods from different nutritional categories. In the interests of simplifying dietary exchange planning and making the approach more accessible to consumers, the number of different food categories has recently been reduced from six to three: a starch category that includes milk, meat or meat substitutes, and fat. Exchange-based diets are usually well balanced and safe. They can be complicated, however, when they require clients to make all their own food choices from the outset, before acquiring nutritional training. A compromise approach that is sometimes adopted involves the use of prepackaged low fat meals. These plans are popular because they implement portion control and calorie reduction from the outset of a diet, while providing time to nutritionally educate the dieter to be able to make his or her own food selections. A related approach involves the use of formula diets that are used to partially replace meals. These supply approximately 900 kcal with 20% calories from protein, 30% from fat, and 50% from carbohydrate with vitamins and minerals to meet RDA standards. The advantage of meal-replacement approaches is that they begin the weight loss process while "buying time" to educate the patient nutritionally and to experiment with making gradual behavioral changes. To be effective in the long run, though, a critical requirement is that the transition to personal selection of grocery foods be made gradually and effectively. If not, patients may become dependent on the replacement plan and fail to develop transferable eating skills.

An alternative simplification strategy is for the dieter to focus on restricting one food group. An example is fat gram counting, a widely practiced technique, that is safe if practiced in the context of a nutritionally balanced eating plan. If followed too faithfully, though, an imbalance of macronutrients can result, which causes an imbalance of micronutrients. A diet that is very high in carbohydrate, low in protein (35 g), and low in fat (10% of calories) can lead to low intakes of salt, iron, essential fatty acids, and the fat-soluble vitamins.

C. Behavior Modification

Self-monitoring of eating behavior and activity are components of any effective weight-loss plan. By maintaining records that note the thoughts, feelings, and circumstances that surround problematic eating occasions, dieters bring formerly automatic behavior patterns into conscious awareness, gain insight into the triggers for aberrant eating, and create the possibility of behavioral choice and control. Daily records also highlight the amount and nature of the time spent in sedentary activities so that the amount of daily activity can be reengineered.

Self-monitoring is also a cornerstone of the assessment process needed to tailor a weight-loss plan to an individual's eating style. Four-day diet records that span both week and weekend time periods can help to establish the kinds of foods that are problematic and the eating pattern in which they are overconsumed. For example, food intake diaries can be informative about whether excess intake derives predominantly from overly large portions of meats that are eaten at meals versus desserts or between-meal snacks or social occasions. Diet records are especially useful when a selective pattern of overeating is suspected. They can help to establish whether preferred foods illustrate selectivity on the basis of macronutrient composition or taste, whether disinhibited eating or frank binges occur, and whether they follow periods of caloric deprivation, exposure to food triggers, or strong emotions. During the course of treatment, food records give feedback, show progress, and give clues about how to prevent eating lapses and weight gain.

As noted earlier, reactivity to the measurement tool can pose difficulties if the client fails to record certain foods or eating occasions out of embarrassment, or if self-monitoring artificially suppresses the problematic eating behavior. An empathic manner or the use of peer counselors who have experience with obesity can be helpful in overcoming patients' fears of being ridiculed. Also, because dieters are especially likely to be self-conscious about reporting consumption of "nonnutritious" snacks and treat foods, other measurement techniques can be useful. For example, weight-loss clients are often more willing to report their consumption of these kinds of foods on food frequency measures that inquire about their "usual intake" over an extended time period.

With regard to activity, simple behavioral changes that can be easily integrated into an overall life-style are preferred An effective first step is for patients to monitor and decrease the time that they spend in sedentary activities like television viewing. Not only does this approach lead to increased activity, it also results in increased liking of exercise, perhaps because the new activities are freely chosen rather than externally imposed. For those who have long been sedentary, it is best to gradually introduce moderate intensity activities, such as walking, gardening, golfing, dancing, or fly-fishing, that can be readily incorporated into daily life. Activities that can be practiced daily or almost daily activity are prefereable, even if, for convenience, they occur in brief 10-min bouts accumulating to about 30 min daily.

D. Social Support and Maintenance

The typical picture with weight loss maintenance is recidivism. Some reported success rates for maintenance of weight loss over a 5-year period are as low as 5%. After the early attrition due to program dropouts, many patients do lose weight successfully, but the vast majority later regain it. Concerns have been expressed that the weight losses produced chiefly by restricting energy intake include losses in body protein and fat. Increases in body fat tend to predominate when weight is regained, however, heightening the risk for hyperlipidemias, hypertension, and diabetes.

More encouraging long-term outcomes have been reported recently for programs that combine diet, exercise, and social support, and that continue these interventions into long-term maintenance of weight loss. The approach taken is that obesity is a chronic condition, treatment of which requires integrated life-style change and for which relapse prevention is continuously needed. For those patients who can be comfortable with regular moderate weight losses, weight reduction programs that integrate diet, exercise, and behavior modification hold the most promise for long-term maintenance of weight loss.

E. Tailoring the Approach to the Eating Style

We cannot overstate the need for an assessment adequate to diagnose the aberrant eating pattern. Once the eating style has been understood, treatment can begin, as outlined in Table II. For individuals whose overweight reflects an uncomplicated pattern of excess intake relative to activity level, treatment should commence with simple techniques that address both components of the energy balance equation. To increase energy expenditure, the time spent in sedentary

TABLE II
Treatment Recommendations Based on Eating Styles

Treatment	Eating style			
	Sedentary lifestyle	Selective overeater	Restrained eater	Binge eater
Energy intake	1200+ kcal	1200+ kcal	1200+ kcal	1500–1800+ kcal
Rate of weight loss	1–1.5 lb/week	1–1.5 lb/week	<1.5 lb/week	<1.0 lb/week
Diet composition	Balanced	May include high-carbohydrate, sweet, or creamy snacks	Balanced	Balanced
Trigger food restriction	None	Evaluate	Evaluate forbidden foods	Evaluate trigger foods
Therapy	Self-monitoring	Self-monitoring; substitute coping behaviors; consider pharmacotherapy	Self-monitoring; cognitive restructuring	Self-monitoring; social support; stress reduction; consider pharmacotherapy
Activity plan	Reduce sedentary time	Reduce sedentary time; moderately active life-style	Reduce sedentary time; after restrictive cognitions improve, include moderate activity	Reduce sedentary time; gradually include moderate activity
Relapse prevention	Short duration	Moderate duration	Moderate to long duration	Long duration
Social supports	Build social supports	Group support	Structured support (individual or group)	Structured group support

activities can be decreased, with encouragement to add enjoyable activities that fit the person's life-style. To produce a modest yet nondepriving reduction in calorie intake, fat gram counting can be encouraged. Given that a deficit of only 500–750 kcal/day is likely to result in weight loss of 1–1.5 lb/week, these two modest interventions may produce sufficiently powerful results. Adherence and maintenance will also be supported to the extent that these subtle changes can be incorporated into the life-style.

For those who selectively overconsume particular foods, the most important first step is to understand what foods are craved and why they are sought out. Those who overconsume foods to engender feelings of indulgence may be helped by understanding their style of using food to supply self-nurturance. An effort is usually needed to identify and increase the repertoire of alternative behaviors or activities that can be self-administered for self-gratification since food has usually been overutilized as a reward. It can prove especially useful to identify rewarding activities like exercise that are incompatible with eating since their implementation demonstrates to the patient that he or she can feel good without the overuse of food. If, on the other hand, the assessment shows that the difficulty in identifying alternative rewards stems from generalized anhedonia that is symptomatic of an un-

derlying depression or dysthymia, it may be appropriate to consider antidepressant treatment. Behavioral approaches such as thought monitoring, self-talk, progressive relaxation, and cognitive rehearsal can be attempted first, followed by evaluation of pharmacotherapy if these prove unsuccessful. Similar considerations surround the practice of carbohydrate self-administration as a way of self-medicating dysphoric moods. Because selective serotonin reuptake inhibitors, such as fluoxetine or the serotonin releaser, dexfenfluramine, have been shown to have positive effects on mood, overeating, and weight, their use can be considered.

Both carbohydrate cravers and overeaters who are motivated by the sensory properties of foods may be able to accommodate their preferences into a diet that promotes weight loss. The planned use of flavoring agents, sweeteners, fat substitutes, and/or high-carbohydrate, low-fat snacks within allowable calorie guidelines may preserve weight loss goals without introducing a sense of severe deprivation. Determination of whether this can be done depends on the assessment of whether the preferred treat foods or flavors serve as "triggers" that prime additional cravings and overconsumption. If so, strict portion controls may need to be implemented or these foods may need to be avoided, at least temporarily.

Treatment for both restrained and binge eaters needs to begin by addressing the negative cognitions that underly and precipitate the maladaptive eating behaviors. Because these negative cognitive sets are well developed and automatic, occurring largely without conscious awareness, they are highly resistant to change. Weight loss efforts should not begin too early in the treatment process because dieting will then re-initiate the cycle of perfectionistic efforts to achieve excessive control, followed by disinhibition and over-eating. A good beginning can involve supportive treatment or even feminist psychotherapy that addresses poor self-esteem. Cognitive restructuring should aim to achieve greater flexibility in understanding emotional states, conflicting feelings, interpersonal relationships, food, and eating. Eventually, treatment may even involve cognitive rehearsal and exposure to trigger situations and foods, with positive experiences leading to greater self-efficacy. For binge eaters, especially, only modest calorie restriction should be introduced and this should be done very gradually in the context of strong social support and continued relapse prevention.

ACKNOWLEDGMENTS

This work was supported in part by National Institutes of Health Grant HL52577 and a VA Merit Review award to Dr. Spring.

BIBLIOGRAPHY

Bjorntorp, P., and Brodoff, B. N. (1992). "Obesity." J. B. Lippincott, Philadelphia.

Drewnowski, A., Kurth, C., and Rahaim, J. (1991). Taste preferences in human obesity: Environmental and familial factors. *Am. J. Clin. Nutr.* **54,** 635–641.

Fairburn, C. G., and Wilson, G. T. (1993). "Binge Eating: Nature, Assessment, and Treatment." Guilford Press, New York.

Institute of Medicine (1995). "Weighing the Options." National Academy Press, Washington, D.C.

Mahan, L. K., and Arlin, M. T. (1992). "Krause's Food, Nutrition and Diet Therapy," 8th Ed. Saunders, Philadelphia.

Manson, J. E., Willett, W. C., Stampfer, M. J., Colditz, G. A., Hunter, D. J., Hankinson, S. E., Hennekens, C. H., and Speizer, F. E. (1995). Body weight and mortality among women. *N. Engl. J. Med.* **333,** 677–685.

McArdle, W. D., Katch, F. I., and Katch, V. L. (1994). "Essentials of Exercise Physiology." Lea and Febiger, Philadelphia.

Perri, M. G., Nezu, A. M., and Viegener, B. J. (1992). "Improving the Long-Term Management of Obesity." Wiley, New York.

Shils, M. E., Olson, J. A., and Shike, M. (1994). "Modern Nutrition in Health and Disease," 8th Ed. Lea and Febiger, Philadelphia.

Spring, B., Pingitore, R., Bruckner, E., and Penava, S. (1994). Obesity: Idealized or stigmatized? Sociocultural influences on the meaning and prevalence of obesity. *In* "Fitness and Fatness" (A. Hills and M. L. Wahlqvist, eds.), pp. 49–60. Smith-Gordon Publishers.

Stunkard, A. J., and Stellar, E. (1984). "Eating and Its Disorders." Raven Press, New York.

Wurtman, R. J., and Wurtman, J. (1987). "Human Obesity," Vol. 499. Annals of the New York Academy of Sciences, New York.

Ecotoxicology

F. MORIARTY
Consultant Ecotoxicologist

D. W. CONNELL
Griffith University, Australia

GLOSSARY

Bioaccumulation Increase of pollutant concentration in an organism from that in the ambient environment, taking into account intake from all sources

Bioconcentration Similar meaning to bioaccumulation, but applies to aquatic organisms and indicates intake from water alone

Chemical speciation Occurrence of well-defined chemical entities such as ions, molecules, and complexes; of particular relevance for heavy metals and some air pollutants

Ecosystem Assemblage of species that forms a distinct system (e.g., the species in a lake) and its habitat (e.g., the rocks, sediments, and water in a lake)

Environment All an organism's surroundings, both the habitat and the other plants and animals, including other members of its own species

Gene pool Collection of genes that is distributed among the individual members of a population

Habitat Inanimate, or abiotic, components of the environment for an individual, population, community, or ecosystem

Pollutant Substance that occurs in the environment at least in part as a result of human activities and that has a deleterious effect on living organisms

Population Group of individuals of the same species that are within a defined area or, alternatively, that have the possibility of mating with each other, provided the needs are met for opposite sexes and maturity

ECOTOXICOLOGY DEALS WITH THE EFFECTS OF pollutants on ecosystems. In many situations attention is often focused on the consequences to individual species of organisms from exposure to pollutants. However, apart from our own species and for cultivated and domesticated species, the important effect is on whole ecosystems, not on individual species. This shift of emphasis, from individual species to ecosystems, makes a sharp distinction from environmental toxicology, which is concerned with effects on individual species. Thus the fact that a pollutant kills, say, half of the individuals in a species population may have little ecological significance, whereas a pollutant that kills no organisms but retards development may have a considerable impact on the ecosystem.

There are many potential pollutants. It has been estimated that about 63,000 chemicals are in common use worldwide, with 3000 compounds accounting for almost 90% of the total weight of chemicals produced by industry. In addition, an estimated 200–1000 new synthetic chemicals come onto the market each year.

A considerable amount of legislation and regulation has been developed in many countries during the past three decades, aimed at avoiding or controlling pollution. However, the scientific basis for the development of effective control is often incomplete, which makes decisions on proposed control measures difficult. In addition to the potential adverse environmental consequences, considerable economic costs may be involved.

I. APPROACHES TO ENVIRONMENTAL POLLUTION

Many sections of human society are concerned about the effects of pollution but approach the problems from different viewpoints. First, administrators and regulators in governments tend to view the environment by habitat, and consequently describe pollution situations as pollution of the land, freshwater, marine, or atmospheric environments. This approach simply reflects the way in which administrators subdivide the world around them into management sectors. It is now becoming appreciated that sometimes efforts to reduce the degree of air pollution, for example, can possibly increase the degree of pollution in fresh water. There is therefore now a tendency to seek that combination of methods of waste control and disposal that minimizes the total extent of environmental damage rather than minimizing sectoral damage. The second approach occurs commonly in industry, where the immediate concern is for the risk of pollution from the chemicals produced by that particular industry. This means that attention is addressed toward pesticides, heavy metals, detergents, or some other class of pollutant. Both these approaches, by habitat and by chemical class of pollutant, are valid and have produced benefits in environment management. Also, most books and articles about pollution approach the subject in one or both of those ways.

In recent years the need for a third, more comprehensive, approach has been recognized. This approach attempts to place all habitats and chemicals into a general format where the effects can be seen on the ecosystem as a whole. This new science has been described as "ecotoxicology." In practice the simple principle of evaluating ecosystems can be difficult to apply. For example, ecosystems are often not stable; size and age distribution of organisms can change, sometimes rapidly, in response to natural environmental variables, including biological factors. It follows that field observations of correlations between a change in the amount of pollutant present in the environment and change in population size do not necessarily indicate an effect by the pollutant. Extrapolation from laboratory tests to field situations is also not straightforward because the rest of the environment can profoundly modify an organism's response to pollutants. In order to successfully apply the ecotoxicological approach, there needs to be an increase in our knowledge of the system itself and an enhanced ability to interpret the effects of pollutants on ecosystems.

Ecotoxicology can be considered to be based on a sequence of interactions and biological effects as illustrated in Fig. 1. First, the chemical originates from a source, then it distributes into the natural environment according to its properties. For example, if it is easily vaporized, it will evaporate into the atmosphere or, if it is readily soluble in water, it will concentrate in the ocean, lakes, rivers, and streams. At the same time natural chemical processes start degrading it to simpler substances. After a period, levels become established in air, water, soil, the atmosphere, and so on where organisms are then consistently exposed. The substance can then be taken up, sometimes with an increase in concentration referred to as bioaccumulation. The individual organisms can then respond by exhibiting sublethal effects, such as reduced growth, or occasionally lethal effects. The final outcome of all the effects on individuals results in an overall impact on the ecosystem. Many scientists include impacts on human societies, particularly human health as part of this approach.

II. EXAMPLES OF EFFECTS

Effective assessment and control of pollution requires a multidisciplinary approach. The complexities, and many of the relevant considerations, can best be illus-

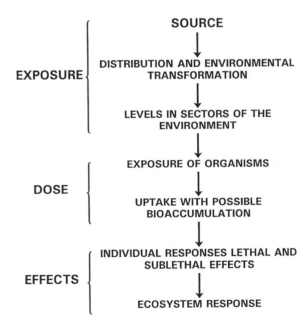

FIGURE I The sequence of actions and effects of a chemical on an ecosystem.

trated by real problems caused by pollutants with different effects on the ecosystem.

A. Effects on an Individual Species

Widespread concern about the possible ecological effects of chemicals developed during the 1950s and 1960s when some agricultural pesticides were found to affect wildlife. Much attention focused on the peregrine falcon (*Falco peregrinus*) in both North America and Great Britain. The size of the British breeding population appears to have been stable for several centuries, but it declined sharply and rapidly in the late 1950s. From 1947 onward, peregrines started to lay eggs with thinner, weaker shells, and the incidence of broken eggs within clutches increased from 4 to 39%. This thinning was not caused directly by DDT, but by DDE, a persistent metabolite of DDT. A similar situation had developed in North America.

From an ecotoxicological perspective the DDT originated from agricultural usage. It accumulated in soil and various seeds and was transferred to seed-eating birds subsequently consumed by the falcon. As a result, the falcon was not directly exposed but was exposed later through the consumption of contaminated foods.

The bioaccumulation of DDT, and its DDE metabolite, did not cause a lethal effect but reduced breeding success through the development of relatively thin fragile eggs. So a dramatic reduction in population occurred which, with the banning of DDT, is now recovering.

B. Effects of an Individual Substance: Lead

Lead is widely used in human society for many purposes. Lead-acid batteries for cars and lead containing octane enhancers are two particularly important applications. Of course many countries have introduced lead-free fuel, but large quantities of lead from previous usage have already accumulated, particularly in soils, of many cities.

Lead can enter plants through the roots and from deposits on leaves. Animals also take up lead from the atmosphere, from water, and from contaminated foods. As a result, residents of our cities now have significant lead concentrations in their bloodstreams. Exposure of many natural systems has occurred as well and both aquatic and terrestrial organisms can contain significant lead levels.

The high toxicity of lead to most living organisms, including many species of both plants and animals, is well established. This can be exerted so as to cause death or, at a sublethal level, reduced growth, reproduction, and life span. In many ecosystems lead has been identified as the agent exerting a detrimental effect. For example, the general decline of European spruce forests have been partly attributed to lead contamination. Some researchers have attributed a general decline in some aquatic ecosystems to lead contamination. The adverse effects on human health and well-being are well documented. Of particular concern is the observed decline in the reasoning ability of children exposed to sublethal levels of leads.

C. Effects on the Gene Pool

Acute toxicity in successive generations from a pollutant may alter the gene pool of the population. It will increase the incidence of those genes that confer resistance and therefore increase the proportion of the population that is resistant. Not all species contain the genetic variability to exhibit this effect but many do with a wide range of pollutants. For example, a considerable number of plant species can develop resistance to one or more heavy metals, and some plant species can develop resistance to sulfur dioxide. Probably the best-known example of these effects is the incidence of resistance to pesticides. Many populations of species of insects and mites are known to be resistant to one or more insecticides, and similar problems have developed with herbicides, fungicides, and rodenticides.

Many different mechanisms can increase resistance and sometimes resistance develops from more than one mechanism. In general terms, there are at least six possibilities: changes in structure or, for animals, behavior that reduces exposure; reduced rate of intake; increased excretion rate; increased storage capacity at inert sites; reduced sensitivity to the pollutant at the site of action; and an increased rate of metabolism of the pollutant. However, it should be noted that metabolism is not always a detoxifying mechanism.

III. PREDICTION OF EFFECTS

We are now much more aware of the potential risks from pollutants to both our own and other species. In consequence, in many countries, the government now has to give permission before significant quantities of any new synthetic chemical can be marketed. A potential problem also being considered is concerned with which of the chemicals already in use also need to be assessed for possible hazards. The government evaluation of chemicals usually follows an ecotoxicological approach as illustrated in Fig. 1. This usually attempts to follow the environmental history of the chemical through the sequence of its interactions and effects on the environment. The prediction process can be simply divided into three phases: an evaluation of the exposure of the biota, an evaluation of the dose that the biota take up, and the resultant effects of that dose.

A. Exposure

The first logical step is usually to attempt answers to a sequence of questions:

1. How much will be released into the environment?
2. How will the chemical be distributed within the environment?
3. How persistent will the chemical be?

Answers to these three questions indicate likely exposures, which is the first step for predicting effects (see Fig. 1). Release into the environment can be estimated from the amounts to be produced and the intended uses and methods of disposal. With volatile compounds (e.g., some organic solvents), we can often assume that all the amount produced will eventually reach the environment. With many industrial chemicals, previous experience has shown the possible levels in waste discharges. This can be used to estimate rates of discharge, quantity of discharge, and chemical forms of the hazardous chemical.

In predicting the distribution of a chemical in the environment, the environment itself must be first rationalized into a set of phases, e.g, air, soil, water bottom sediments, suspended sediments, and aquatic biota. The chemical can be considered to distribute between these phases as a result of a set of two phase partition processes between pairs of phases as in Fig. 2. These partition processes are characterized by constant partition coefficients (K values) leading to the various concentrations in the phases (C values). This process can be modeled successfully using physicochemical properties of the chemical such as vapor pressure, solubility in water, and so on. In this way estimations of likely concentrations in environmental phases can be made.

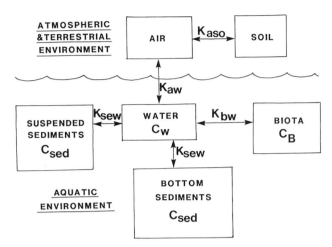

FIGURE 2 The distribution of a chemical between phases in the natural environment. The partition processes are characterized by the partition coefficients (K values) leading to the various concentrations in the phases (C values).

Persistence is a key characteristic, although it can have several related meanings. For present purposes, it means the time period a chemical remains in the environment before it is converted, either by organisms or in the physical environment, into other compounds. Often more restricted meanings are used where persistence means the time period a pollutant remains in one place, either in one part of the physical environment (e.g., the soil) or in an organism. Elemental pollutants, such as heavy metals, can persist in the environment indefinitely as a range of chemical species, although their persistence within organisms may be much less.

Persistence in the environment can be assessed from standard laboratory tests, which include degradation rates by microorganisms in both aerobic and anaerobic conditions. It is a property not only of the particular pollutant, but also of the organism or habitat in which it operates. The conventional measure of persistence is half-life, but this is usually a first approximation only since the rate at which a pollutant disappears (mass lost/unit mass/unit time) usually decreases with time. Mathematical models can then be used to predict amounts and concentrations of chemicals after different periods have elapsed. Models usually entail a considerable simplification of the processes involved and are complicated by the fact that concentrations of a pollutant are rarely in a steady state in the actual environment. [*See* Environmental and Occupational Toxicology.]

B. Dose

The term *dose*, taken from pharmacology, indicates the intake of a pollutant by an organism from its exposure. It is usually measured as the concentration of pollutant in an organ, tissue, or whole body and is expressed in a variety of units (e.g., wet weight, dry weight, and lipid weight). There is often the assumption that the magnitude of the dose indicates the probability or degree of effect.

In general terms, there are two routes by which animals can acquire pollutants: by ingestion with their food and from direct contact with their inanimate environment, across the body surfaces in general or the respiratory surfaces in particular. For most plants, direct contact is the only route—with the air or the soil. There are also two mechanisms by which organisms can lose pollutants (Fig. 3): by excretion and by metabolism, when the pollutant is converted into one or more other compounds within the organism. With a constant exposure, the amount of pollutant within the organism tends to approach a steady state when the rate of intake equals the sum of the rates of excretion and metabolism.

Pollutants are not distributed uniformly within organisms. Fairly sophisticated compartmental models have been developed to describe amounts of pollutant within organisms during and after exposure, but although such models often fit equations to the data well, their theoretical basis is not well established.

For aquatic animals exposed to organic compounds, a much simpler approach is commonly used. Evidence suggests, at least for many pollutants, that food is not a significant route of intake. For many organic compounds, the degree of bioconcentration increases with the *n*-octanol:water partition coefficient (commonly denoted as P or K_{OW}) (e.g., Fig. 4). The theory is that the degree of bioconcentration depends on the distribution ratio of the compound between the ambient water and the animal's fats, which is indicated by the bioconcentration factor (K_B concentration in organism/concentration in water). However, this use of K_{OW} is not completely reliable. Predicted bioconcentration factors usually fall within one order of magnitude of the experimentally measured value but some compounds occasionally exhibit greater deviations.

Terrestrial animals will often acquire most of their pollutant burden from their food. The relation between concentrations of pollutant in predator and prey depends on many factors, but it is untrue that persistent pollutants inevitably increase in concentration along food chains. Intake often depends on the specific details of the exposure: the distribution and chemical form of the pollutant within the air, soil water, or sediment. More attention to the pathways that pollutants follow in the environment could improve our ability to predict doses.

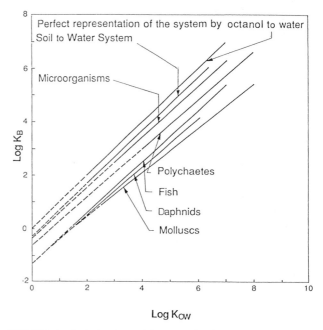

FIGURE 4 Relationship of log K_B (the bioconcentration factor = concentration in biota/concentration in water at equilibrium) for aquatic biota to log K_{ow} (the *n*-octanol/water partition coefficient). Also shown is the theoretically perfect relationship and the soil/water partition coefficient relationship. Solid lines indicate the range of values used to establish the relationship and the broken lines indicate extrapolation lines.

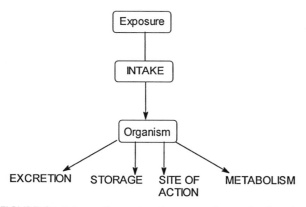

FIGURE 3 Scheme illustrating the routes that molecules of a pollutant can follow after uptake.

C. Effects

There is currently much interest in quantitative structure–activity relations (QSARs) which describe relationships between measures of the structural properties of chemicals and their biological activity. The presumption is that for a series of chemically related compounds, their relative degree of biological activity depends on the concentration at the site of action, indicated by the partition coefficient and by the ease of attachment of molecules to the receptor. Equations are now being developed in which the measure of biological activity (the concentration of chemical required to produce a standard biological response) is a function of a series of terms for partition coefficient, electron density, and other aspects of molecular structure. These equations are developed empirically and are at an early stage.

The common measures of dose indicate the amount of pollutant within the organism at one instant. Chemical laboratory tests usually give organisms a single acute exposure or a chronic constant exposure. We usually do not know what influence fluctuations of dose with concentration and duration have on the degree or likelihood of biological effect.

This means that the results of laboratory tests, no matter how comprehensive, cannot be applied directly to field situations. Also, ecosystems contain many species, and different species react differently to an input of pollutant for both genetic and environmental reasons. Moreover, the consequences of effects on individuals within one species on a population can be complex. The interactions between individuals within a population and between species within an ecosystem are often not well understood.

Studies are sometimes made on populations in the field or even on whole ecosystems, but most ecosystems are so complex that such studies inevitably entail a considerable amount of effort, and results can be difficult to interpret. Microcosms are sometimes used in an attempt to gain the advantages of both relative simplicity and representation of the natural ecosystem. They can range from fairly simple aquaria with sand, water, and a few selected plant and animal species to large marine enclosures with as much as 1700 m³ of seawater. The view is now developing that the chief value of these systems is not as simulations of nature but as experiments on selected components of the ecosystem. They can yield useful information on the pathways that pollutants follow in the habitat, but sometimes appear, for studies on populations, to suffer the worst of both worlds. They can be too artificial to represent natural ecosystems and too complicated for biological effects to be easily interpreted.

D. Assessment

The Organisation for Economic Co-operation and Development (OECD) has made considerable efforts to harmonize the tests required by different countries when assessing new chemicals. The details can vary considerably between countries, but there is widespread agreement on the rationale. This is to estimate amounts released, environmental distribution, and concentration and to compare the predicted environmental concentrations with those needed to affect organisms. This is applying the ecotoxicological approach previously described (see Fig. 1).

The assessment of new chemicals requires the application of a set of laboratory tests. These tests of biological effects usually include tests of the exposure needed to kill a range of species, which usually include fish and crustacea, and can include, particularly for pesticides, birds, bees, other beneficial insects, and earthworms. Tests on bioaccumulation are commonly required, usually with a species of fish, but the details vary with intended uses of the chemical, quantities involved, and the results of earlier tests. Testing is performed by the manufacturer or supplier and continues until either a decision about permitted uses, if any, is made or the manufacturer withdraws the application.

This procedure takes little account of possible interactions that may occur between the chemical being considered and other pollutants that may also be present in the environment. Sometimes the observed effect is greater than the sum of the individual effects of exposure to the contaminants alone or it can be less. The degree of effect of a pollutant often depends on the presence or absence of another pollutant. More information is required before we can decide whether pollutants affect each other's biological activity. To determine whether pollutants potentiate or antagonize each other's effects, one needs to determine the relation between the degree of effect and the proportion of the pollutants in toxic mixtures.

It is generally recognized that these assessments of effects are crude, and a safety factor is therefore used. This is a factor by which the dose needed to produce an effect in the laboratory exceeds the predicted environmental concentration. Sometimes, in cases of doubt for pesticides, permission for limited use is given, when safety is monitored during use in the field.

Despite safety factors, there is still the risk of unacceptable ecological effects from new chemicals. But it is difficult to determine how great this risk is. The present system of assessment may seem more effective than it really is because we do not know how many past or present ecological effects may have gone unnoticed. Moreover, apart from pesticides, which are released into the environment deliberately because of their biological activity, most new chemicals may not have serious ecological effects anyway. Both of these factors could make the system of assessment appear to be more effective than it really is.

IV. MONITORING

Present abilities to understand and to predict the pathways that pollutants follow in the environment and the effects that they exert are inadequate. We need, therefore, to make repeated surveys (surveillance), or what is usually called *monitoring*. The latter term sometimes has a more restricted meaning which is to check that the amounts of pollutants or that the effects being studied conform to an already stated standard.

Three types of information may be sought:

1. Rates of release of pollutants into the environment.
2. Degree and changes of environmental contamination.
3. Biological effects.

Few monitoring schemes measure both amounts of pollutant in the environment and the biological effects, if any. Whatever the purpose of a monitoring scheme, it is essential that objectives are determined at the start because the objectives dictate the details of the sampling program. It is also desirable that it should be known what specified degree of change can be detected with an appropriate degree of statistical significance. Many monitoring schemes do not meet these criteria.

Measurements of concentrations of a pollutant in nonliving samples or in organisms can be related to standards (i.e., acceptable limits of contamination). Organisms can have advantages over nonliving samples since they acquire much higher concentrations of some pollutants, making chemical analysis easier. They also give a measure of the pollutant's availability, which is more relevant for the probability of biological effects than is a measure of the amount in the environment. It is often also said that organisms integrate the amount of pollutant present in the environment during a period of time, but this is not strictly correct. Organisms do not retain pollutants indefinitely. They do give some sort of an average measure for fluctuating exposures, although the most recent exposures will contribute proportionately more than will earlier exposures.

A recent development is the establishment of environmental specimen banks, collections taken from both living and inanimate components of our environment for indefinite storage. These samples are being stored in anticipation of future, as yet unforeseen, needs in the expectation that this material will permit measurements of pollutants. Doubts stem from the difficulties of designing an effective sampling program when important details of the objectives are unknown. It must be accepted in relation to specimen banks as well as current chemical monitoring programs that although it is often easier to monitor amounts of pollutant rather than their biological effects, to rely on measurements of pollutant alone does ignore some potential problems, such as:

1. It is sometimes impracticable to measure all contaminants.
2. Routine analytical techniques for chemicals may be too insensitive.
3. The biological significance of pollutants, at the levels found, may not be fully appreciated.
4. Combinations of pollutants may interact.
5. Regular chemical measurements may miss occasional, significant, high values.

It is often not a straightforward matter to relate biological changes in the functioning of individual organisms or of ecosystems to the degree of environmental contamination. Many variables can affect the dose–response relationship since not all pollutants can readily be detected within organisms and since effects are sometimes indirect, mediated through other species or the abiotic environment. Moreover, one's concern is often with the whole ecosystem. As a consequence, numerous indices attempt to encapsulate in one number the abundance and presence of a range of species. These indices have many defects, of which perhaps the most fundamental is that it is often difficult to attach any biological meaning to the index. The monitoring of single named species can be more incisive. However, most species within a community are then likely to be ignored, but clear objectives can be stated from which practicable programs of monitoring can be devised, and any changes are relatively easy to interpret.

V. CONCLUSION

The investigation and prediction of the effects of chemicals on natural systems and human health has become a new branch of science described as ecotoxicology. A systematic sequential approach has been developed which utilizes models to describe the various stages in this process. But for the foreseeable future, prediction and detection of ecological effects by pollutants are often going to be difficult. Attitudes to pollution have changed, and it is beginning to be accepted that there is a need to consider the fate and effect of potential pollutants at all stages, from "cradle to grave." To a considerable extent, the political impetus for these new attitudes derives from public pressure, but what is often lacking is a consensus on the value judgements that underlie attempts to control effects by pollutants on wildlife.

BIBLIOGRAPHY

Carlow, P. (1993). "Handbook of Ecotoxicology," Vol. 1. Blackwell Scientific Publications, Oxford.
Carlow, P. (1994). "Handbook of Ecotoxicology," Vol. 2. Blackwell Scientific Publications, Oxford.
Connell, D. W., and Miller, G. J. (1984). "Chemistry and Ecotoxicology of Pollution." Wiley, New York.
Hoffman, D. J., Rather, B. A., Burton, G. A., and Cairns, J. (1995). "Handbook of Ecotoxicology," Lewis Publishers, Boca Raton, FL.
Kaiser, K. L. E. (ed.) (1987). "QSAR in Environmental Toxicology II." D. Reidel, Dordrecht, The Netherlands.
Moriarty, F. (1988). "Ecotoxicology: The Study of Pollutants in Ecosystems," 2nd Ed. Academic Press, London.
OECD (1981). "OECD Guidelines for Testing of Chemicals." Organisation for Economic Co-operation and Development, Paris.
Richardson, M. (ed.) (1986). "Toxic Hazard Assessment of Chemicals." The Royal Society of Chemistry, London.
Sheehan, P., Korte, F., Klein, W., and Bourdeau, P. (eds.) (1985). "Appraisal of Tests to Predict the Environmental Behaviour of Chemicals: SCOPE 25." Wiley, Chichester.

Elastin

ROBERT B. RUCKER
DONALD TINKER
University of California, Davis

GLOSSARY

Alu sequences Sequences of about 300 nucleotide pairs that are duplicated in human DNA so that they comprise 4–6% of the total DNA in the human genome

Coaccervation Process of temperature-dependent exclusion of hydrophobic polymers from polar environments; elastin coaccervate exists as a separate phase with the characteristics of an oil droplet

Ehlers–Danlos syndromes Syndromes representing a family of related connective tissue disorders, which are characterized by hyperextensible joints, hyperelasticity of skin, skin fragility, and tumor-like growths following trauma; both abnormal elastin and/or collagen fiber formation are components of the syndrome

Marfan's syndrome Major features are abnormalities of the heart and blood vessels as well as abnormal growth and development of bone and other connective tissues

Mesenchymal Mesenchymal tissue is the meshwork of connective tissue derived from the mesoderm, which is the middle layer of the three primary germ layers of the embryo; muscle, blood components, and epithelia are also derived from the mesoderm

Parenchyma The functional elements of an organ as distinguished from its framework or stroma

Promoter Regulates RNA synthesis from given genes; in the promoter, some sites are characterized by specific nucleic acid sequences (e.g., so-called TATA, CAAT, or CG-rich boxes), and other sites bind specific transcriptional factors; TATA boxes are recognition sites that are important to RNA polymerase II activity; AP-1 and SP-1 sites are sequences known to bind to specific regulatory proteins in response to signals important to cell replication and growth cycles

Pseudoxanthoma elasticum Skin disease that is characterized by swelling and degeneration of elastic fibers

ELASTIN IS AN IMPORTANT STRUCTURAL PROTEIN found in the extracellular matrix of tissues that are required to be highly compliant. Elastin is composed of hydrophobic amino acid sequences that are separated by sequences enriched in alanine and lysine. Lysyl residues in the alanyl and lysyl sequences can be modified to facilitate the cross-linking of individual polypeptide chains of elastin. This results in an insoluble protein as the final product with a structure analogous to the organization of many natural and synthetic rubbers. The production of the protein is regulated so that its synthesis coincides with specific periods in the development of given tissues and organs. Abnormal elastin metabolism can cause vascular, skin, and lung lesions, which are characterized by poor elasticity and laxity.

I. INTRODUCTION

Elastin is a unique protein that is important to the structure and development of many tissues. The synthesis of elastin is developmentally regulated by a gene that has unusual features. As a consequence, the regulation of elastin is of interest to both the developmental and the molecular biologist. Elastin is also of interest to the protein chemist because of its novel chemical and physical properties. Elastin is one of the few proteins in nature with the properties of a rubber-like elastomer. It possesses an unusual amino acid composition and undergoes complex posttranslational chemical modifications in the process of its assembly into mature fibers. Moreover, defects in elastin can also underlie important disease processes; thus the protein is also of interest to clinicians who deal with elastin-containing tissues such as skin, ligaments, arteries, and lung. Elastic fibers function for exceedingly long periods. Elastin metabolic turnover in some tissues is best estimated in years. Consequently, genetic conditions or agents that alter elastin biosynthesis, processing, and deposition can have long-lasting effects.

II. THE ELASTIC FIBER

Table I compares the collagen and elastin contents of selected human connective tissues. In such tissues, the arrangements of elastin fibers are diverse (e.g., extending from organized laminar sheets to structures that are interdispersed among other fibers). In tissues such as ligaments, elastin fibers are oriented in the direction of stress. In blood vessels, the fibers take on the laminar or sheet-like arrangements (see Fig. 1A). In skin, elastin fibers may be deposited as interlinked sheets of protein but with numerous fused and branched points. In lung parenchymal tissue, elastic fibers appear as laminar sheets that encapsulate individual alveoli.

Elastic fibers contain two components. Elastin is the predominant component and appears amorphous when examined at the electron micrograph level. In contrast, the second component is distinctly different and microfibrillar in appearance (see Fig. 1B). This component is composed of several proteins that vary in size from about 25,000 to 340,000 Da. The proteins comprising the microfibril include (1) the fibrillin family, (2) the family of so-called latent TGF-β-binding proteins, (3) the microfibrillar associated protein (MFAP) family, and (4) the microfibril associated glycoprotein (MAGP) family. The proteins in each of these families arise from unique genes localized on different chromosomes. Of interest are the observations that elastic fibers from different tissue locations differ in the distribution of microfibrillar proteins and elastin. For example, the distribution of fibrillins (currently three distinct fibrillins have been identified) differ markedly. The fibrillins have an average diameter of 10 nm. At high resolution, the fibrillins have a "beads on a string" appearance resulting from the head-to-tail polymerization of multiple fibrillin aggregates and contain repeated sequences homologous to the epidermal growth factor calcium-binding motif. Fibrillin-2 transcripts appear early in morphogenesis and accumulate for a shorter period of time than fibrillin-1 transcripts. Synthesis of fibrillin-1 correlates with late morphogenesis. Fibrillin-1 provides mostly force-bearing structural support whereas fibrillin-2 plays a more direct role in the early process of elastic fiber assembly. The microfibrillar component may also include lysyl oxidase, the enzyme that catalyzes elastin cross-linking. Also, proteins such as vitronectin and amyloid F may be absorbed onto aged elastic fibers. The microfibrillar proteins can exist in fibrillar arrangements or as aggregates independent of their association with elastin. Therefore, it is assumed that in elastic fibers they play roles in fiber organization. For example, it is speculated that a first step in elastin's assembly into fibers is by disulfide bond formation with one of the microfibrillar proteins. A chemical characteristic of the microfibrillar proteins is that they tend to be high in cysteine. The C-terminal amino acid sequence of elastin also contains several cysteine

TABLE I
Collagen and Elastin Content of Selected Connective Tissues[a]

Tissue	Collagen	Elastin
Bone, mineral-free	85–93	
Achilles tendon	80–90	4–5
Skin	70–75	0.5–1.5
Cornea	65–75	
Cartilage	45–65	1–2
Ligament	15–20	75
Aorta	12–24	25–35
Liver	2–4	
Lung parenchyma	10–20	20–25

[a]g/100 g of dry weight.

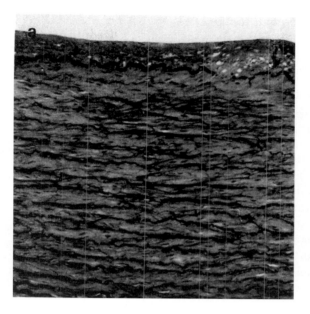

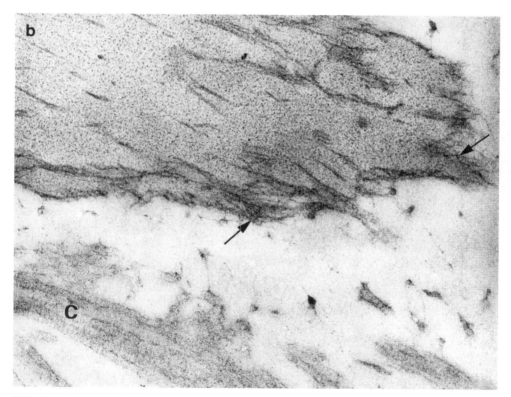

FIGURE 1 (a) A light microscopic transverse section of human thoracic aorta from a healthy 27-year-old male. The elastic fibers are darkly stained and are seen in a parallel arrangement. There is thickening of the intimal layer on the luminal surface within the thickened internal elastic lamella. Magnification ×350. (b) A transverse section of an elastic fiber showing a peripheral rind of electron-dense microfibrils (arrows) surrounding the amorphous elastin. Collagen fibers (c) in the transverse section are adjacent. Magnification ×70,000. [The light and electron micrographs were provided by Dr. E. G. Cleary and Dr. J. Kumaratilake, Department of Pathology, School of Medicine, University of Adelaide, Australia.]

residues that may undergo disulfide bond cross-linking.

III. CHEMICAL COMPOSITION AND PHYSICAL PROPERTIES OF ELASTIN AND ELASTIC FIBERS

Elastin is one of nature's most apolar proteins. It is also a very insoluble protein because of its extensive intermolecular (interpolypeptide chain) cross-linking. Elastin is resistant to solubilization by strong protein denaturants, autoclaving, treatment with hot alkali, or formic acid. Moreover, many of the novel amino acid sequences in the protein are resistant to mild hydrolytic conditions. The apolar sequences in elastin are enriched in glycine (Gly), valine (Val), and proline (Pro) (e.g., Gly-Gly-Val-Pro, Pro-Gly-Val-Gly-Val, or Pro-Gly-Val-Gly-Val-Ala. The Val-Pro imino bond is not cleaved as easily as typical peptide bonds. Elastin is also a basic protein. Many of the glutamyl and aspartyl residues in elastin are amidated; thus, cationic functions from lysyl and arginyl residues are in excess and not balanced by anionic functions, which normally arise from glutamyl or aspartyl residues. [See Proteins.]

However, long-range, reversible elasticity is the most important feature of elastin. The high degree of interchain cross-linking and the hydrophobic nature of the Gly-Val-Pro sequences result in a polymer that possesses properties analogous to rubber. The energy that results in elastic recoil comes about in part because of apolar–polar interactions between water and the hydrophobic portion of the elastin molecule. Interchain cross-linking also plays a role. The hydrophobic sequences in elastin are separated by short segments that are enriched in the amino acids alanine and lysine. Lysyl residues found in the alanyl- and lysyl-enriched sequences undergo specific chemical modifications that lead to interchain cross-links.

IV. RELATIONSHIP BETWEEN PHYSIOLOGICAL AND PHYSICAL PROPERTIES

All animals with pulsating blood vessels contain some type of elastin. Many animals also have valves and ligamental structures that are elastic and contain elastin.

The elastins that have been isolated from species

TABLE II
Lysine and Cross-Linking Amino Acid Content Commonly Found in Mature Elastin

Amino acid[a]	Residues/1000 total
Desmosine	1.5–2.0
Isodesmosine	0.5–1.3
Lysinonorleucine	0.8–1.5
Allysine	0.5–2.0
Aldol condensation products	0.8–1.5
Dehydrolysinonorleucine	0.1–0.3
Others	2.0–10
Lysine	6–12

[a]Structures for desmosine, lysinonorleucine, allysine, aldol condensation products, and dehydrolysinonorleucine are given in Figs. 3 and 4.

ranging from humans down to bony fishes contain lysine-derived cross-links. Elastins from chordates, however, do not contain cross-linking amino acids in easily measured quantities (Table II). Chordates have elastin-like proteins that are considerably less hydrophobic than the elastins found in higher animals.

In general, mammals and birds possess elastins that are highly cross-linked and hydrophobic. Reptiles and amphibians contain elastin with an intermediate percentage of hydrophobic amino acids. Bony fishes such as sharks contain lower amounts of hydrophobic amino acids. Such differences have evolved possibly because of the need to accommodate different fluid pressures as well as differences in body temperature. The hydrophobic portions of elasin amino acids are forced inward at lower temperatures, thus decreasing the potential for apolar–polar interactions. It may be predicted that, to accommodate high-fluid pressures or body temperatures, the more suitable forms of elastin should be relatively hydrophobic and highly cross-linked.

V. TROPOELASTIN AND CELLULAR SOURCES OF ELASTIC FIBERS

Tropoelastin is the soluble precursor to elastin. This protein has been isolated from a number of different tissues and from various species of animals. The major source of elastin is the smooth muscle cell; however, cartilage cells, myofibroblasts, ligamentum fibroblasts, endothelial cells, and lung interstitial cell fibroblasts also secrete elastins.

Tropoelastin is not a single entity and occurs in several molecular forms. These forms arise from deletion and transposition of selected exons in the elastin gene. The single deletions and transpositions, however, do little to alter the overall amino acid composition of tropoelastin or its relative size (about 70,000 Da). The deletions come about because of alternative processing of portions of tropoelastin mRNA molecules (Fig. 2). The deletions or transpositions may serve to produce insoluble elastin polymers that vary in their directional orientation or degree of branching.

VI. THE ELASTIN GENE

Information is now available regarding the structure of elastic genes, which has been useful in understanding why there is variability in elastic fiber structure. The elastin gene has many unusual features compared with other known genes. In humans, there is a single gene, which contains nearly 40 kb of a sequence that codes for an mRNA of about 3.5–3.7 kb. This gene is found on the long arms of chromosome 2. Chromosome 2 also contains genes for type II collagen, fibronectin, and type VI collagen.

The elastin gene has one of the highest intron–exon ratios (i.e., 19 to 1) of genes characterized to date. The exons that eventually code for the peptide sequences involved in cross-linking are interspersed among those exons that code for the hydrophobic regions. In addition to the human elastin gene, data are also available for the elastin genes that encode for bovine, chick, and rat elastins. The size of exons varies somewhat from one species to another. The spacing of introns to exons is also not well conserved among species. Alternative splicing is also a consistent and unusual feature that is associated with the processing of mRNA products from elastin genes. The splice borders are the same for each exon in elastin. Often a single cross-linking exon or hydrophobic exon is deleted or transposed. Alternative splicing gives rise to different mRNA molecules, which in turn translate into differing polypeptides. Consequently, the exons may be viewed as cassettes in which given segments of the molecule may be spliced out.

The human elastin gene is an Alu-enriched gene. At the 5′ promoter region, there are several CAT boxes but no TATA boxes in the promoter. The promoter also contains multiple initiation sites. Both ST-1 sites and AP-2 binding sites are present in the promoter. ST-1 and AP-2 are protein transcriptional factors. The AP-2 factor is responsive to changes in cellular cyclic adenosine monophosphate levels. Responsiveness to hormones and related physiological regulators (e.g., retinoids, hydroxylated sterols, or dexamethasone) are also features of gene regulation. These responses, however, appear sensitive to the developmental phenotype of the cells making elastin. For example, with exposure to dexamethasone, increased elastin expression is observed when cells with

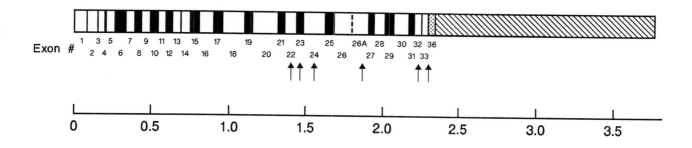

Nucleotides (kb)

FIGURE 2 Diagram of human elastin cDNA. The cDNA is divided into exons, which are numbered and drawn to scale. Characterization of the cDNA and the complete human elastin gene has evolved primarily from the work of Joel Rosenbloom and his colleagues at the University of Pennsylvania. All exons are multiples of the three nucleotides, and exon–intron borders always split codons in the same way. This can permit "cassette-like" alternative splicing. The exons encoding cross-link domains are represented by solid bars, whereas the hydrophobic domains are open boxes. The arrows indicate known sites for alternative splicing. The dashed line that separates exon 26 and 26A indicates that in this particular exon, splicing occurs primarily in the 3′ portion. Exon 36 is novel in that it encodes for cysteine residues, which may be involved in tropoelastin attachment to microfibrillar proteins. In the region that corresponds to a nontranslated segment (stippled), there are two polyadenylation sites.

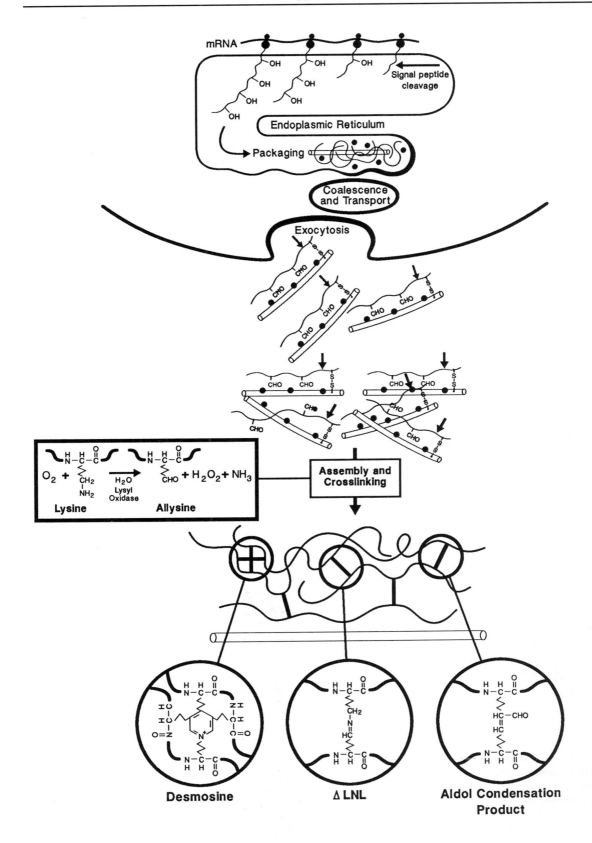

Desmosine

Δ LNL

Aldol Condensation Product

a fetal phenotype are used in studies; when adult cells are used, typically no change in elastin expression is observed.

VII. POSTTRANSLATIONAL STEPS IN ELASTIN FIBER FORMATION

Figure 3 outlines important posttranslational features of elastin fibrogenesis. First, about 3–8% of the total proline residues in elastin are hydroxylated. This step is analogous to the hydroxylation of prolyl residues in collagen and occurs intracellularly. The next phase in fibril formation appears to involve the organization of elastin around the microfibrillar components. Lysyl residues in elastin are then modified to give rise to a number of unusual cross-linking amino acids. The cross-linking amino acids arise following the oxidative deamination of lysyl residues in tropoelastin. Once lysyl residues are oxidized (to peptidyl α-amino acid-δ-adipic-semialdehyde, allysine), cross-linking of tropoelastin occurs by a number of unusual condensation reactions. It is now well established that the cross-links in elastin are extensive and significantly influence both the biophysical and the biochemical properties of the protein. For the most part, the cross-links are necessary to constrain the elastic fiber so that upon stretching, the individual polypeptide chains are allowed to realign and restore fibril organization upon release of tension. When specific lysyl residues are modified and there is appropriate ordering and alignment of the lysyl-enriched portions of elastin, the condensation reactions occur spontaneously. Once the condensation products are formed, subsequent redox reactions result in stable and irreversible cross-links. A possible mechanism of condensation and, ultimately, stable cross-link formation is presented in Fig. 4. The mechanism suggests that the Schiff-base product of allysine and lysine, dehydrolysinonorleucine (ΔLNL), serves as the agent for the oxidation of hydroxydesmosine derivatives to form the tetrafunctional cross-

links desmosine and its isomers. A pentafunctional variant of desmosine has also been isolated.

Much of the progress in our knowledge of the cross-linking amino acids in elastin and collagen resulted from studies that focused on the enzyme lysyl oxidase. Lysyl oxidase requires as cofactors copper and an orthoquinone component, derived from peptidyl 6-hydroxydopa. Severe dietary copper deficiency will cause defective cross-linking. Agents that react with carbonyl functions, such as β-aminopropionitrile, various hydrazines, and semicarbazides, also promote defective cross-linking and aneurysms.

VIII. ROLE OF ELASTIN IN ORGANOGENESIS AND DEVELOPMENT

The deposition of elastin is correlated with specific physiological and developmental events. For example, in the mammalian lung, elastic fibers are concentrated around the opening of alveoli. Alveolar shape is thought to result in part from the molding or scaffolding influences of lung elastic fiber. Elastin synthesis and deposition are coordinated with alveolarization.

In developing arteries, elastin deposition appears to be coordinated with changes in arterial pressure and mechanical activity. The transduction mechanisms that link mechanical activity to elastin expression undoubtedly involve cell surface receptors. Also, cell movement in ligaments and arteries is related to the orientation of elastin. A current hypothesis is that elastin-synthesizing cells are attached to elastin through cell surface receptors. Once attached, the synthesis of additional elastin and other matrix proteins and carbohydrates may be influenced by stretching or other factors that influence cellular shape.

One of the elastin receptors that has been studied most extensively is 67,000 Da and has features that are similar, if not identical, to receptors that bind to laminin. Laminin and other proteins (e.g., fibronectin

FIGURE 3 Steps in the posttranslational modification of elastin. The intracellular steps include signal peptide cleavage, prolyl hydroxylation, packaging, transport, and exocytosis by vesicles. The extracellular steps include association with microfibrillar protein components (rods) via sulfide linkages, conversion of selected lysyl residues to residues of allysine (the reaction defined in the box insert), cleavage of elastin associated with the microfibrils (arrows), and eventual coacervation and cross-link formation. The major cross-links in elastin (shown in the magnified circles) are desmosine, derived from the condensation of lysyl residues with three allysyl residues; dehydrolysinonorleucine (ΔLNL), a Schiff-base product that can be reduced in elastin to form lysinonorleucine (see Fig. 4); and the aldol condensation product, derived from allysyl aldol condensation reagents.

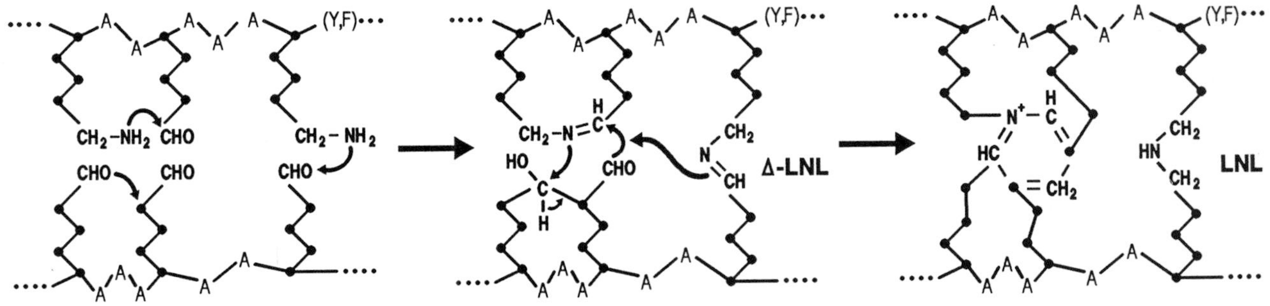

FIGURE 4 Oxidation of desmosine intermediates and reduction of dehydrolysinonorleucine (ΔLNL) to form stable cross-links in elastin. The cross-linking regions in elastin arise from exons that encode predominantly alanyl and lysyl residues. It has also been observed that when an aromatic amino acid residue appears adjacent to a lysyl residue, the lysyl residue is often not oxidized. As a consequence, this favors dehydrolysinonorleucine formation or dictates whether desmosine or a related isomer forms. LNL, lysinonorleucine.

and entactin) facilitate cell–extracellular matrix communication by binding to both cell receptor proteins and extracellular matrix components. Very active investigations currently relate to how such attachments mediate biomechanical responses in connective tissues. The linkage is of obvious importance to understanding extracellular matrix formation during development or during tissue repair following injury. [See Extracellular Matrix; Laminin in Neuronal Development.]

IX. OTHER FUNCTIONAL ROLES OF ELASTIN

Elastin receptors are also found on the surface of cells that are not involved in extracellular matrix protein production. Phagocytic cells, such as the macrophage, appear to chemotax (migrate in response to a signal) to sites that are enriched in hydrophobic elastin peptides (e.g., VGVAPG) as a result of elastin degradation. Elastin is a long-lived protein that is normally not degraded, except following tissue injury. Consequently, it can be speculated that elastin may play an important role in chemotaxis. Elastin peptide levels are usually very low; therefore, any rise in elastin peptides could serve as a localized signal of mesenchymal tissue damage that may require the aid of cells with phagocytic functions.

A knowledge of the receptor–ligand association process is important to the understanding of certain diseases. Some tumor cells that metastasize to highly vascularized tissues attach to elastin by receptor-mediated processes. Once attached, the tumor cell secretes proteinases, thereby allowing penetration into the interstitial tissue.

X. TURNOVER OF ELASTIN, ELASTOLYTIC PROCESSES, AND ELASTIN-RELATED DISEASES

As noted, once elastin is synthesized and stabilized by cross-links, it does not undergo rates of turnover common to other proteins. In some tissues, measurable turnover is difficult to demonstrate, particularly when corrections are made for new tissue growth. Under normal situations, elastin peptides are released only in minute quantities as the result of degradation. Concentrations in human serum are normally in the nanogram per deciliter range. When urinary desmosine or elastin peptides are measured, the amounts correspond to no more than milligram quantities of elastin degraded per day. Because the body pool of elastin in humans is in gram amounts, the small quantities of elastin products that are excreted under normal situations clearly suggest that turnover is very slow. Likewise, when very young animals are injected with radioactively labeled lysine at those developmental periods corresponding to maximal expression of elastin, the labeled lysine that is incorporated into chemically stable crosslinks, such as desmosine, may persist in elastic fibers from months to years; however, exceptions are evident, especially in tissues in which continual remodeling is required because of growth (e.g., the uterus).

For those diseases in which elastin destruction is of biological significance, the "inertness" of the protein takes on considerable importance. Focal disruption of the elastic lamina of a blood vessel or alveoli need not be extensive to alter mechanical or biochemical properties. One of the best examples of how disruption of elastic fibers leads to pathology is pulmonary emphysema, a process that involves the proteolytic destruction of alveolar elastin. Emphysema appears to involve inappropriate elastolytic digestion by proteinases derived from neutrophils, macrophages, and/or even some mesenchymal cells. Furthermore, if elastin cross-linking becomes defective, elastin can serve as a substrate for many proteinases that normally do not degrade elastin (e.g., kallikrein or proteinases associated with blood coagulation).

Considerable effort and study have gone into understanding the process of elastolysis. Clearly, nature goes to some effort to modulate and counteract the inappropriate proteolysis of extracellular proteins by the secretion of a wide variety of proteinase inhibitors. One such inhibitor is α_1-antiproteinase, an inhibitor made in the liver. Inactivation or genetic deficiencies of α_1-antiproteinase can promote obstructive pulmonary disease (e.g., emphysema) because of enhanced elastolysis.

Other diseases that involve elastin and elastic fibers include pseudoxanthoma elasticum, Buschke-Ollendorf syndrome, cutis laxa, several Ehlers–Danlos syndromes, Menkes' disease, Marfan's syndrome, endocardial fibroelastosis, elastoderma, and Williams' syndrome. A better understanding of these disorders has recently come about from the use of molecular genetics, specifically genetic linkage analysis, positional cloning, and mutational analyses. For example, Williams' syndrome is now known to result from mutations involving the elastin gene on chromosome 7q11.23. The mutations include intragenic deletions and translocations of the elastin gene that can result in a quantitative reduction in elastin during vascular development. Characterization of fibrillin mutations in Marfan syndrome patients suggests that, in addition to serving a structural protein in elastic fibers, fibrillin may be involved in regulating cellular activities and morphogenetic programs important to vascular tissue development. It is also better appreciated that the effects of some mutations on the elastic fiber are complex and indirect. Menkes' disease affects elasin fiber integrity, but the process is probably most related to defects in copper transport and the inability in Menkes' disease to produce functional lysyl oxidase. Indeed, many diverse disorders are usually expressed genetically as X-linked, autosomal recessive, or dominant.

The preceding description of the elastin gene suggests that several mechanisms may underlie the heritable diseases that involve elastin (e.g., mutation, insertion, or deletion of given exons, abnormal transcriptional or translational regulation, or mRNA instability). As a general rule, primary protein structural

defects are often autosomal dominant and lethal. The disorders that are recessive in inheritance are usually less lethal and involve a processing enzyme or a gene regulatory element.

XI. CONCLUDING COMMENTS

Elastin is an important component of connective tissue in that it facilitates tissue extensibility. It imparts rubber-like properties to tissues such as lung, blood vessels, ligaments, skin, and the uterus. It is one of the most apolar proteins known to be synthesized by mammalian cells. In some tissues, highly cross-linked elastic fibers persist for months to years with little turnover. Slow protein turnover takes on considerable importance because it can limit the possibility of appropriate repair following tissue injury.

Consequently, a knowledge of elastin's function is of obvious importance to human and veterinary medicine. Loss of elastin or elasticity contributes to a number of disease processes such as emphysema, atherosclerotic vessel wall disease, and cutaneous disorders.

BIBLIOGRAPHY

Bashir, M. M., Indik, Z., Yeh, H., Ornstein-Goldstein, N., Rosenbloom, J., Abrams, W., Fazio, M., Vitto, J., and Rosenbloom, J. (1989). Characterization of the complete human elastin gene. *J. Biol. Chem.* **264**, 8887–8891.

Dietz, H. C., Ramirez, F., and Sakai, L. Y. (1994). Marfan's syndrome and other microfibrillar diseases. *Adv. Hum. Genet.* **22**, 153–186.

Indik, Z., Yeh, H., Goldstein, N., Kucich, U., Abrams, W., Rosenbloom, J. C., and Rosenbloom, J. (1989). Structure of the elastin gene and alternative splicing of elastin mRNA: Implications for human disease *Am. J. Med. Genet.* **34**, 81–90.

Mecham, R. P., Hinek, A., Griffin, G. L., Senior, R. M., and Liotta, L. A. (1989). The elastin receptor shows structural and functional similarities to the 67-kDa tumor cell lamin receptor. *J. Biol. Chem.* **264**, 16652–16657.

Reiser, K., McCormick, R. J., and Rucker, R. B. (1992). The enzymatic and non-enzymatic crosslinking of collagen and elastin. *FASEB J.* **6**, 2439–2449.

Rosenbloom, J., Abrams, W. R., and Mecham, R. (1993). Extracellular matrix 4: The elastic fiber. *FASEB J.* **7**, 1208–1218.

Swee, M. H., Parks, W. C., and Pierce, R. A. (1995). Developmental regulation of elastin production: Expression of tropoelastin pre-mRNA persists after down-regulation of steady-state mRNA levels. *J. Biol. Chem.* **270**, 14899–14906.

Tinker, D., and Rucker, R. B. (1985). Role of selected nutrients in the synthesis accumulation and chemical modification of connective tissue proteins. *Physiol. Rev.* **65**, 607–657.

Electron Microscopy

PETER J. GOODHEW
The University of Liverpool

DAWN CHESCOE
University of Surrey

GLOSSARY

Contrast Difference in intensity that permits a feature of interest to be distinguished from its (presumably less interesting) background. It is commonly quantified in terms of the local intensity I and the background intensity I_b as $(I - I_b)/I_b$

Resolution, or resolving power Closest spacing of two point objects at which they can be distinguished to be two objects rather than a single feature. This is not necessarily the same as the size of the smallest isolated object that could be seen

ELECTRON MICROSCOPY IS A GENERAL TERM THAT embraces two major imaging techniques and a host of subsidiary activities such as the preparation of suitable samples and the use of the microscope for elemental analysis of small volumes. Scanning electron microscopy (SEM) is primarily used to reveal the external morphology of a small sample at magnifications typically between ×10 and ×10,000. Transmission electron microscopy (TEM) is used to reveal the internal microstructure (or ultrastructure) of a thin specimen at magnifications that are usually in the range of ×100 to ×1,000,000. Both microscopes can be used to perform elemental microanalysis using characteristic X-rays excited by the electron beam.

I. INTRODUCTION

The resolving power, d, of a light microscope is limited to some fraction (about one-third in the best case) of the wavelength of light, λ, via the relation

$$d = 0.6\lambda/\mu \sin \theta,$$

where μ is the refractive index of the medium between the objective lens and the specimen and θ is the half-acceptance angle of the lens. Using a green light of wavelength 500 nm, resolution is therefore limited to about 150 nm, and features closer than 150 nm apart in the specimen cannot be resolved as two objects but will appear as a single feature in the image. If electrons of wavelength 0.0037 nm are used, the attainable resolution would appear to be about 0.001 nm, or 1 pm, which is much smaller than a single atom. Although in practice such good resolution cannot actually be achieved, this logic has led to the development of a whole family of electron microscopes.

Because electrons carry a charge and are rather light, it is simple to construct electromagnetic lenses that act on a beam of electrons in the same way that a conventional thin glass lens acts on a beam of light. Using such lenses, the strength of which can be varied by varying the current through their windings, it is possible to make many different electron microscopes. Two configurations are in common use: (1) In SEM, a fine beam of electrons is used to probe the outer surface of a solid specimen revealing the external mor-

ENCYCLOPEDIA OF HUMAN BIOLOGY, Second Edition, VOLUME 3.

phology; and (2) in TEM, a higher-energy electron beam is passed through a thin specimen, revealing the internal ultrastructure in a manner exactly analogous to the light microscope used in transmission.

Even high-energy electrons cannot travel far in a solid without being strongly scattered and eventually stopped. Therefore, SEM using 20-keV electrons will reveal the detail of the outermost few tens of nanometers of the sample, whereas TEM using 100-keV electrons can only provide useful images of specimens less than about 1 μm thick. Nevertheless, these capabilities have been found to be extremely useful, and many thousands of microscopes of both types are now in use around the world.

II. SCANNING ELECTRON MICROSCOPY

A SEM works in a manner similar to a conventional television camera. A beam of electrons is scanned across a small, usually rectangular, region of the specimen. A cathode ray tube (CRT) display is scanned in synchronism with the specimen, and its brightness is modulated according to the intensity of a signal emitted by the specimen. Most frequently this signal is related to the intensity of secondary electrons emitted from the surface, in which case the microscope is said to be operating in the secondary mode. However, many other signals [e.g., X-rays or light (cathodoluminescence, CL)] can be used, in which cases the SEM is said to be operating in X-ray or CL mode. [*See* Scanning Electron Microscopy.]

Secondary mode SEM images are usually rather easy to interpret because the image contrast is similar to that experienced by the eye when looking at everyday objects. Thus, holes appear dark and hills usually have a bright side and a dark side in shadow.

III. TRANSMISSION ELECTRON MICROSCOPY

In a TEM the thin specimen is illuminated by a beam of energetic electrons. Electrons that are transmitted through the specimen with only a small loss of energy are used to form a projection image. The image is formed by a series of objective and projector lenses, which usually provide magnifications in the range of a few hundred to a million diameters.

Figure 1 is a ray diagram of a TEM. The essential components of any such microscope are an electron gun, two or more condenser lenses, an objective lens, several projector lenses, a viewing screen, and a camera. Additionally there are always three or more sets of apertures that permit the operator to limit the diameter of the electron beam at several points in the microscope column. Each of these components will be briefly considered.

The electron gun accelerates a beam of electrons through a potential, usually in the range of 80–400 keV, giving each electron an energy of 80–400 keV. Higher energy electrons penetrate farther and can thus be used to study thicker samples. However, the cost of a microscope rises almost linearly with electron energy, so much biological microscopy is performed with microscopes capable of maximum electron energies of 100 or 120 keV.

All lenses in a TEM are electromagnetic and consist of a coil wound around a specially shaped magnetic core. Optically they can be considered to be "thin lenses," and ray diagrams for a TEM are similar to those that describe the operation of a projection light

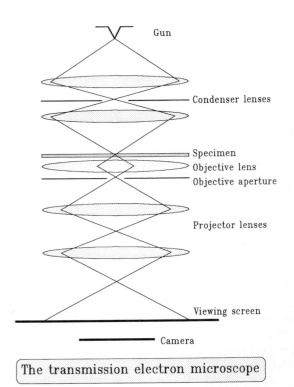

The transmission electron microscope

FIGURE I Outline ray diagram of a TEM illustrating the arrangement of the lenses and other major components with respect to the specimen.

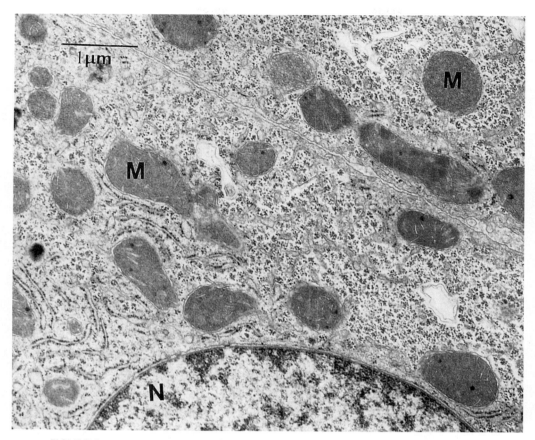

FIGURE 2   TEM micrograph of human liver showing a nucleus (N) and mitochondria (M).

microscope. Electron microscope lenses differ from lenses for light in two important ways: (1) Their focal length can be changed easily by altering the current flowing through their coils and (2) the image undergoes a rotation because the electrons travel a spiral path through each lens. The ability to vary the focal length of each lens makes it possible to use a TEM in many different ways and to change focus, illumination, and magnification without moving either the specimen or the lenses. Image rotations, which may occur as lens strengths are altered, may confuse the novice but are not serious and are no longer present in many modern microscopes. Normal viewing of the image is via a window in the vacuum system, revealing the projection image on a green fluorescent screen. Micrographs are permanently recorded on photographic film loaded within the vacuum system directly below the viewing screen or directly into a computer via a TV camera in the same position.

Apertures are primarily used to control the diameter of (and current in) the illuminating beam and to enhance the image contrast (discussed later).

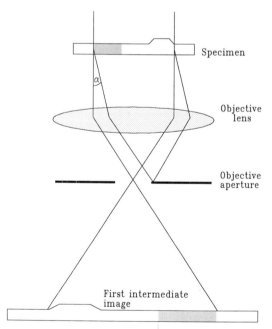

FIGURE 3   Ray diagram showing how the objective aperture is used to prevent electrons scattered through an angle greater than α contributing to the image. Without the objective aperture, there will be negligible contrast between different regions of the specimen.

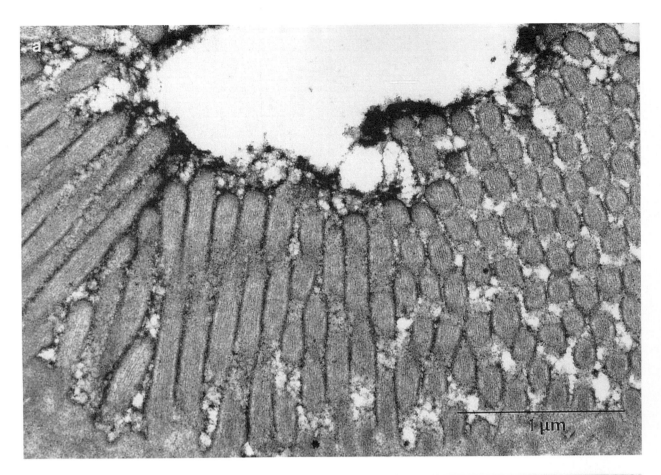

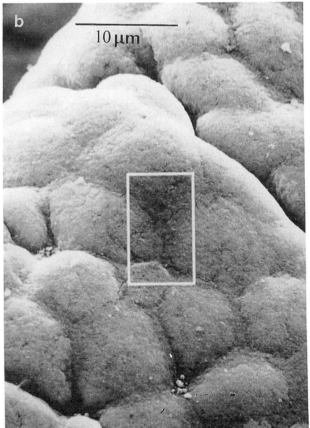

The micrograph of human liver shown in Fig. 2 illustrates the type of contrast seen in most biological material. Dark regions such as the nucleus and mitochondria in Fig. 2 are those that strongly scatter electrons.

Figure 3 shows how contrast is created in the TEM. Essentially all electrons incident on the specimen are transmitted. However, those that are scattered through an angle greater than α are stopped by the objective aperture and do not contribute to the image. Regions of the specimen that strongly scatter electrons (e.g., those that are much thicker or contain heavier elements) therefore lead to dark contrast in the image. Selection of a smaller objective aperture increases the number of scattered electrons that are stopped and therefore tends to increase contrast.

Because most tissue sections are intended to be parallel-sided (see Section VI), there should be no large changes in thickness across the specimen. Dark contrast therefore usually indicates the presence of a heavy element. Specimens are frequently stained with heavy elements to decorate and thus reveal particular features of the ultrastructure.

The resolution (resolving power) of a TEM is limited by aberrations in the electromagnetic lenses rather than by the electron wavelength. In a modern microscope it is possible to resolve two points that are about 0.3 nm (3 Å) apart in an ideal specimen. However, biological specimens are rarely ideal for this purpose, and the resolution achievable on a particular specimen is more commonly determined by the specimen itself or by the degradation of the specimen under the influence of the beam. The high-energy electrons rapidly destroy the fine structure of the specimen, leaving a carbonaceous but chemically altered replica of the original. Stained features will be visible because of the heavy-metal atoms that they contain, which remain in the same place in the specimen. Many topographical features of the original specimen are preserved, and the image will probably look like the original specimen. However, the most finely resolved features are likely to be the heavy-metal deposits, which may now be rather larger than the ultrastructural features that they once decorated. In these circumstances, a more realistic resolution limit, imposed by the specimen, is 1–2 nm.

If high resolution is required, special experimental strategies are required, which are likely to involve "low-dose" microscopy of unstained material using short exposures to reduce the electron beam damage. It is then usually necessary to enlist the aid of computer image processing to extract the information from rather noisy images.

Figure 4 illustrates the different types of information available using the SEM and the TEM. Figure 4a shows a section of human gut cut on a microtome: The microvilli are clearly outlined by the stain, and it is evident that they have been sectioned at a variety of angles. Some microvilli are seen in almost circular cross section, whereas others have, by chance, been sectioned almost longitudinally. Figure 4b shows a similar gut specimen in the SEM: On the left at low magnification, the whole villus can be seen, whereas the magnified image of the boxed region on the right shows the microvilli end on. It should be obvious that the TEM and SEM views give complementary information.

IV. ANALYTICAL MICROSCOPY

The TEM is an excellent tool for performing local chemical analysis. Many of the electrons that are scattered as they pass through the specimen lose a significant amount of energy (from a few electron volts to several thousand electron volts) and excite one or more atoms to high-energy states. In a fraction of cases, the excited atoms will relax by emitting a characteristic X-ray. Chemical analysis can be performed either by measuring the energy of the X-ray or by measuring the amount of energy lost by the transmitted electron. These principles form the basis of the techniques commonly known as energy dispersive X-ray analysis (EDX) and electron energy loss spectroscopy (EELS).

Energy dispersive X-ray analysis is widely available on both TEM and SEM instruments. Characteristic X-rays excited within the specimen are detected by a semiconductor detector, which is typically sensitive to X-rays from elements 11 (sodium) to 92 (uranium) in the periodic table. Special "windowless" detectors can extend this range to the lighter elements with atomic number down to 5 (boron). Thus virtually all the elements of potential interest to biological microscopists are accessible to analysis. The volume of the

---

FIGURE 4 (a) TEM micrograph of sectioned human gut. (b) SEM micrograph of a similar gut sample. (Right) A magnified image of the region within the white box.

specimen that can be analyzed can be as small as 10 nm × 10 nm × t. The thickness of the specimen, t, will typically lie in the range of 20–500 nm, giving an analyzed volume of only about 10^{-23}m^3, or about 10^{-17}g. Because the signal (i.e., the number of emitted X-rays) is small, the analytical accuracy is not high, but the ability to localize the analysis to a known region provides an extremely powerful microanalytical tool. Figure 5 shows a section of human thyroid and the X-ray spectrum collected from the region of colloid labeled C. The small iodine signal that arises from the naturally occurring iodine within the colloid can be seen. The prominent osmium peak results from the fixation and staining procedures, whereas the chlorine peak comes from the embedding resin. The sulfur and calcium peaks are real features of the structure.

EELS has been developed more recently than EDX and offers high sensitivity for light elements. Quantitative analysis is less well-developed using EELS than EDX, and the majority of analyses are carried out using the longer-established technique.

For analytical microscopy to be meaningful, it is obviously essential that the composition of the specimen is unchanged from that of the *in vivo* material. This means that special care has to be taken during specimen preparation (see Section VI). Clearly the electron beam itself can alter the chemical nature of the specimen while it is under observation. This is not of great concern if the changes involve carbon, hydrogen, and oxygen atoms because it is unlikely that these would be of analytical interest. However, the microscopist should be aware that many ions of great interest (e.g., Na, K, Ca) are likely to be removed during specimen preparation and, even if they remain, can be mobile under the influence of the beam. Analyses involving these elements can therefore be erroneously interpreted.

V. IMMUNOCYTOCHEMISTRY

The principle of this rapidly developing technique is to label a surface antigen site with a specific labeled antibody. Peroxidase can be used for labeling the antibody because it is easily highlighted by subsequent osmium fixation and can therefore be made visible in the TEM. However, the use of antibodies labeled with colloidal gold is increasing and is already the dominant technique. Colloidal gold particles have the advantage that they can be seen in the image without further staining. The effectiveness of the technique depends on the specific antigen. It is possible with many specimens to fix them lightly with, for example, a dilute mixture of glutaraldehyde and formaldehyde. Depending on the antigen, labeling can take place before or after embedding. It is also possible in favorable circumstances to label frozen hydrated or freeze-substituted specimens (see Section VI). A gold-labeled section of human granulocytes is shown in Fig. 6.

VI. PREPARATION OF SPECIMENS

Specimen preparation is absolutely central to biological electron microscopy. The treatment of the sample at this stage will determine the quality of the image and analytical information that can be deduced by TEM. Preparing the specimen usually takes an order of magnitude longer than performing the microscopy, and great importance therefore is attached to the development of a sound technique. The ideal TEM specimen would be thin, representative, and unaltered from its *in vivo* state. These ideals can rarely be attained, but there are three common approaches to preparing specimens that are at least interpretable. Briefly, these techniques involve cutting thin slices (sections) from the embedded bulk specimen (ultramicrotomy), mounting small discrete entities (e.g., viruses) on a thin support film, or replicating a carefully created internal surface (e.g., by freeze-fracture).

A. Section Cutting

The essential steps in the preparation of a thin section are fixation, staining, embedding, and ultramicrotomy. A small piece of the tissue material is typically fixed in a dilute 2–4% glutaraldehyde solution, washed in a buffer solution, and then impregnated with a 1–2% solution of osmium tetroxide. This acts both as a secondary fixative and as a heavy-metal stain. During fixation, efforts should be made to keep the pH, osmolarity, and ionic constitution of the material unchanged as much as possible.

At this stage it may be desirable to use a specific stain to render specific chemical (and thus biological) sites visible. However, for many samples the action of the osmium tetroxide in staining protein sites is adequate. There are many alternative staining methods.

The specimen must now be dehydrated by immersion in a series of water/ethanol mixtures with increasing ethanol content, finishing in a bath of pure ethanol. It may then be conditioned in propylene oxide

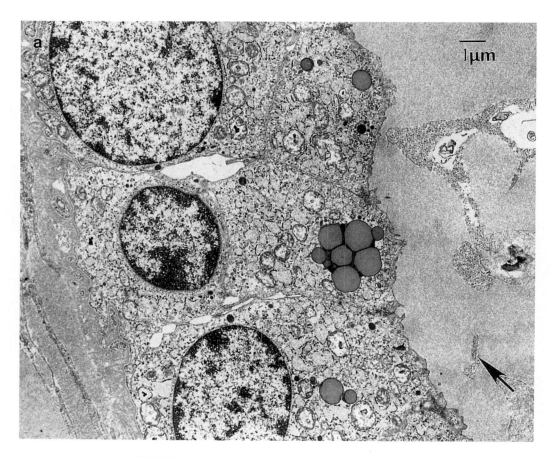

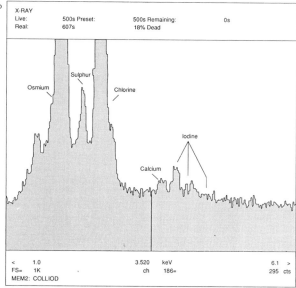

FIGURE 5   (a) Section of human thyroid. (b) X-ray spectrum was collected from the region of colloid.

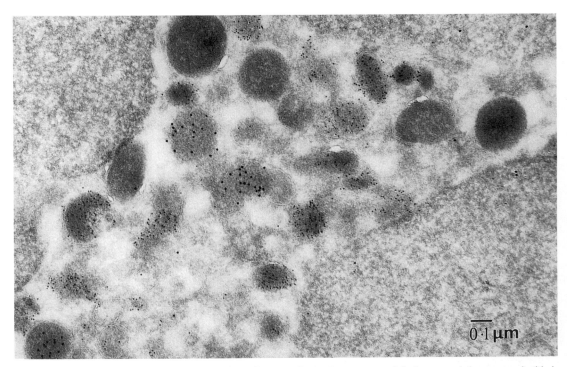

FIGURE 6 A gold-labeled section of human granulocytes. The section was cut while frozen and then postembedded. Two different antibodies (rabbit antibody raised against lactoferrin and mouse monoclonal antibody raised against elastase) were labeled using two sizes of gold spheres (15 nm and 5 nm), which show as large and small black dots in the micrograph. The large spheres were coated with goat anti-mouse Ig, whereas the small spheres were coated with goat anti-rabbit Ig. It is clear that the different antibodies accumulate in separate granules. (Courtesy of Dr. J. E. Beesley. Wellcome Research Lab, United Kingdom.)

before being embedded in resin. Once the resin is cured, the block (now about 8 mm diameter and 15 mm long) is gripped in the stage of an ultramicrotome, and sections of controlled thickness are shaved off using a glass or diamond knife.

Thin sections can now be collected and transferred to the microscope, and they will usually be stained again, using, for example, lead citrate and uranyl acetate. Staining at this stage is more efficient than at an earlier stage because the specimen is so thin that penetration into the tissue is rapid.

Tissue sections prepared in this way are ideal for the purposes of comparative morphology, and a vast amount of microscopy is performed on sections prepared by this standard method. However, such sections are not well-suited to analytical microscopy because virtually all the sodium and potassium will have been removed during preparation, together with much of the calcium. These specimens are not ideal for immunocytochemistry because surface antigens are likely to have been altered by cross-linking and are thus unavailable for a labeled antibody.

B. Negative Staining

Small particulate materials, especially viruses, are difficult to handle and prepare by sectioning. They are usually dispersed, via a liquid medium, onto a carbon film, which can be supported on an electron microscope grid. They are then "negatively" stained: A solution of the chosen stain is flooded across the carbon film and collects around the outside of the virus particles. On drying, it leaves a deposit of heavy metal, defining the external morphology of the particle. An example of a negatively stained influenza virus is shown in Fig. 7.

C. Cryofixation

Rapid cooling techniques are becoming widely used as a means of preventing elemental migration and thus preparing useful specimens for analytical TEM. Two common methods of rapidly cooling the specimen and thus avoiding cell damage by ice formation are (1) plunging the specimen into liquid propane or

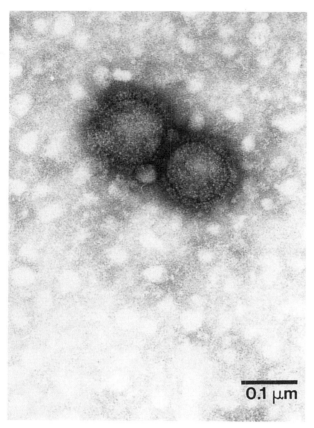

FIGURE 7   Pair of negatively stained influenza viruses.

chemical structure and thus the distribution of elements. However, because the specimen is not stained, image contrast can be low and morphological information is not easy to extract.

Several techniques have been developed that offer a compromise between frozen and room-temperature methods. For instance, it is possible to fix a sample with a very dilute (e.g., 0.5%) glutaraldehyde solution and/or stain it lightly before freezing. Alternatively, freeze-substitution can be attempted. After rapid freezing, the sample is held for several days in an organic solvent at about −90°C to allow substitution to take place and is then warmed to about −40°C and impregnated with a special resin that is still fluid at this temperature. The resin must be cured by irradiation with ultraviolet light while the sample is still cold, but the ultramicrotomy can then be performed at room temperature.

BIBLIOGRAPHY

Amelinckx, S., *et al.* (eds.) (1996). "Handbook of Microscopy." VCH, Weinheim.

Bozzola, J. J., and Russell, L. D. (1992). "Electron Microscopy: Principles and Techniques for Biologists" (Jones and Bartlett eds.), Boston, MA.

Chescoe, D., and Goodhew, P. J. (1989). "The Operation of Scanning and Transmission Electron Microscopes," Royal Microscopical Society Handbook No. 20. Oxford University Press, Oxford.

Elder, H. Y., and Goodhew, P. J. (eds.) (1989). "Proceedings of EMAG/MICRO 89," Institute of Physics Conference Series, Vol. 98. Institute of Physics, Bristol.

Glauert, A. M. (ed.) (1973–present). "Practical Methods in Electron Microscopy" series, Elsevier, Amsterdam.

Goodhew, P. J. (1987). "The Transmission Electron Microscope," a software package, Engineering Materials Software Series, Institute of Metals, London.

Goodhew, P. J., and Humphreys, F. J. (1988). "Electron Microscopy and Analysis." Taylor & Francis, London.

(2) "slamming" the specimen onto a cold block. The cold specimen can then either be cut into sections while still frozen using a cryo-ultramicrotome or held at a higher temperature (still less than 0°C) until it has freeze-dried.

The advantage of specimens prepared in these ways is that there tends to be good preservation of the

Embryo, Body Pattern Formation

JONATHAN COOKE

National Institute for Medical Research, London

GLOSSARY

Commitment Setting of the course of development in a cell or cell group so that it can no longer be modified by exposure to normal stimuli that cause specification or respecification in its embryo of origin

Germ layers Three fundamental superimposed cell layers in development of the embryo; organs are characteristically contributed to by two adjacent germ layers

Histodifferentiation Functional specialization of different cell types, involving massive synthesis of special products and arrangement into special tissue structures, visualized in microscopy by histological techniques

Homology Relationship of structures in two organisms by common descent from a structure in a shared ancestral organism; can be a DNA sequence, the protein it encodes, an anatomical structure, or an episode of cellular activity in development

Models (1) Organism with an embryo type accessible to experimentation and presumed to show useful homology of mechanism with, ultimately, human development; (2) quantitative (usually computer-run) simulation of a formally possible mechanism for pattern formation to assess performance characteristics

Morphogenesis Chiefly, the aspect of development in which tissue reveals its pattern of diverse specifications and commitments by specific form-building movements and differentiations

Morphogens Intercellularly acting molecules (gene products or smaller metabolites) whose dynamics of synthesis and interaction result in a pattern within embryo tissue that cells can use to become different from each other according to relative positions in the whole

Specification Cell state that will lead to development of a particular tissue type, or region from the normal body pattern, upon culture in isolation from any subsequent, respecifying stimuli such as may be experienced in the embryo or by experimental means (compare to Commitment)

BODY PATTERN FORMATION IS A DISTINCT, EARLY process occurring during development of embryos, in which small regions within the tissue are designated as giving rise to particular parts of the future body. It occupies only a brief time in relation to the remainder of embryonic and fetal life, during which differentiation and the onset of growth occur. Pattern formation largely coincides in time with the distinctive set of mechanical rearrangements known as gastrulation, which reorganize the cell population of the embryo into, in the case of complex animals such as the human, three layers. At their foundation, body patterns normally exhibit the correct spatial arrangement, and rather closely controlled proportions, among the precursor regions that have been designated. While still in formation, the pattern typically exhibits regulatory properties; that is, proportions and completeness of the sets of regions tend to attain normality despite earlier experimental rearrangement, subtraction or addition of tissue, and in face of natural variation among individual embryos in the amount of tissue available for the process. It has nevertheless been possible to begin investigation of the mechanism that accomplishes this regulated patterning, using a variety of other embryos as "models" for the human development. Such investigation employs surgical and biochemical interferences that leave permanently disturbed body pattern, as well as direct visualization by molecular technology of the newly synthesized products of the genes that are involved. We might expect these genes themselves to have been highly conserved in the process of evolution, and thus among embryos of different types, but the strategy of gene deployment—that is, the cell biology and anatomy of the development—to have changed much more. Our understanding of how the activities of specific genes control the beginning of mammalian development is still primitive but has recently begun to accelerate. Nevertheless, a complete understanding of biological pattern formation must ultimately incorporate knowledge not only of the sequence of gene products and their network of interactions, but also of "engineering" principles whereby these can have been welded into a mechanism with the regulatory properties and reliability that are observed. This article surveys theoretical requirements and possibilities and summarizes our cell and molecular biological knowledge to date.

I. BODY PATTERN FORMATION AS A DISTINCT PHASE IN THE HIGHER MULTICELLULAR LIFE CYCLE

Many people, including scientific specialists in other areas of biology, have surprisingly vague notions of the sequence of events in embryonic development. We are used to thinking of the nervous and endocrine systems as constituting the great integrative mechanisms of highly evolved multicellular life, which enable their possessors to live the distinctive lifestyle that would be unavailable to a heap of cells, a plant, or a primitive, relatively undifferentiated multicell such as a sponge. Yet these systems of organism-wide coordination have only reached their present pitch of complexity as aspects of the complexity and consistency of the anatomy achieved in the body of each species. Anatomy is a reflection of a precise spatial array of cellular activities in later morphogenesis, that is, in form-building movements and in histodifferentiation to give specialized types. Evidence indicates that certain cell types of the body, though generated at many widely scattered locations within it, retain long-term "memories" of belonging to the particular geographical regions of their origin. They maintain the tissue characteristics proper to one region (e.g., the locally characteristic pattern of hair growth on skin) when transplanted elsewhere. Thus body pattern formation, the system

of interactions that first regionalizes embryonic tissue to give a properly proportioned set of founder territories from which the anatomy is later constructed, must rank as the third great integrative mechanism of higher animal life-forms.

As we shall see in Section V, pattern formation largely uses the same categories of signals within and between cells as do the processes that integrate bodily function later in life. The bulk of what we need to learn about developmental biology is simply cell biology, with the crucial molecules of intercellular communication being relatively few, and some as yet unidentified. But in at least one way, this early phase of development *is* unique; only then are cells being instructed as to which parts of the body, as well as to which specialized cell types, their descendants will give rise. Thereafter, intercellular signals are used only to decide between preset options for differentiation or physiological function that are already set in the cells that are predesignated as the targets for those particular signals.

The integrative systems of multicellularity seem to have evolved from relatively loosely, indefinitely controlled networks of interactions to more precise, specialized ones. Body pattern formation has been accompanied by compression in time, relegating the events to the earliest part of development, and the following of a precise sequence of cell-specification steps within that brief time. Initial evolution of multicellular bodies probably occurred by opportunistic harnessing of physiological or environmentally imposed differences arising among regions of a growing cell mass. These differences were co-opted as stimuli causing diversification of cell types, thus optimizing the survival ability of the whole. Subsequently, the stimuli for diversification and their spatial pattern became regularized as a mechanism within the embryo, and largely separated from a prolonged, later process of growth within the committed tissues. Thus, in most evolutionarily advanced animal forms, including mammals and the experimental models described later, pattern formation for the body as a whole is integrated, stereotyped, and compressed into 100 hours or less. This is preceded by a period of cell multiplication, often (though not in mammals) of a specially rapid kind, that provides enough tissue space or a high enough cell number for the mechanism of body pattern formation to operate in. Then, at the other end of the crucial, brief period, the embryo is still largely undifferentiated histologically but is in a very different state. It has become cryptically parceled into a sequence of regions, some of which may contain very few cells. If such a region is subsequently removed or killed, or if the set of regions is incomplete or very wrongly proportioned owing to physiological stress at the crucial time or to genetic deficiency, the individual body concerned will be correspondingly incomplete or imbalanced in anatomy. Each region has become committed, containing the uniqe precursor cells for a crucial component of the structure of a particular organ system or district of the body. The final pattern of commitments is achieved stepwise, such that some regions go through transient states of specification during which, without receiving further information, they would form structures other than those that are fated to be their final commitment. This can be shown by explanting and culturing such regions from particular stages, in isolation from their normal environment within the embryo.

Each region of commitment seems to be programmed for its future pattern of growth and for provision of particular sequences of terminally differentiated cell types or stem cells for such types. The program may be simple or complex (e.g., within the various regions of the brain or the blood-forming tissue). Further instructions *within* each initial region of commitment will, of course, occur to increase the complexity finally attained; but by "body pattern formation," we refer to the episode after which the primary parcellation into noninterchangeable regions, giving the large-scale plan of the body, has been completed. This is dramatically early within development as a whole and, especially in our own animal group, the vertebrates, is somehow associated with the integrated set of cell movements called gastrulation. These movements reorganize the sheet of cells, which will form the embryo, into the definitive three-layered structure (the germ layers) on which the body is based. Thus the crucial events occur, relatively, so early that in human development they begin around 15 days after conception but are essentially over 7 days after that. It is worth remembering that rather few of the world's women are aware of being pregnant at these dates. A simple view would be that subsequent development is simply the playing out of a pattern of local schedules for growth and differentiation. Tissue could be surgically rearranged within the embryo with little effect on these schedules and, thus, with catastrophic effects on final anatomy. In experimental vertebrate embryos during or just after gastrulation, precursor tissue for the vertebrae or the brain can be reversed in polarity or exchanged between sites along the future body axis. This is followed by the final development of individuals with correspondingly misplaced

or back-to-front structures embedded within an otherwise normal anatomy.

II. EVOLUTIONARY CONSIDERATIONS

A. The Embryo Body Itself

The transition to highly regularized development, from that of some primitive and less tightly controlled body form, may have occurred only once in the ancestry that gave rise to all of those animals whose embryos are now used as experimental models. It has sometimes been proposed that certain rare evolutionary steps might have involved single "hopeful monster" genetic changes that altered body patterns in massive ways that turned out to be survivable and then adaptive. But generally, the earlier in development that a process was affected by any genetic change, the less likely can it have been that that change was advantageous and thus perpetuated in evolution. The assemblage of genes providing the basis for body pattern formation is thus expected to be largely the same, and relatively highly conserved in structure, among such diverse "higher" animals as fruit flies, worms, and various vertebrates. This expectation has been generally borne out so far (see Section V). The study of forms in which molecular genetic analysis is relatively easy, and thus advanced, has therefore greatly helped us to make sense of vertebrate development, in which the list of known gene products has grown only slowly while the anatomical structure accompanying pattern formation is nevertheless more complex. However, there are certain problems entailed in using those anatomically simpler model systems to arrive at a direct understanding of mammalian (human) development.

Embryos themselves are subject to evolutionary pressures. They therefore specialize in form and function, along with the bodies they give rise to. This means that, although the available menu of molecular signals is conserved, the selection made from that menu and the strategy of development at cell and anatomical levels are varied, often in quite a dramatic way, even among closely related forms. Within vertebrates, for instance, at the close of gastrulation (thus pattern formation), the primitive anatomy of the body passes through a stage that is easily recognizable across all types. The anatomical arrangements in earlier development leading up to this, however, vary much more, in ways that relate to the "life-style" of the organisms concerned but obscure the homology of the body patterning mechanism. Knowledge to date of the very initial phase of development comes mainly from the fruit fly, *Drosophila,* with significant additions from lower vertebrates but little directly from mammals. Certain idiosyncracies of these most common model systems, which limit the direct extrapolation from them to ourselves, must therefore be mentioned.

In *Drosophila's* specialized, rapid version of insect development, the early events occur in a layer of cell nuclei that are not separated by true cell membranes. A whole class of interactions, those whereby a messenger RNA or a protein resulting from gene activation in one nucleus can translocate in space directly to interact with another nucleus, can thus occur in this system. This is ruled out for vetebrates, however, in which pattern is formed in a truly cellularized tissue from the outset. Therefore, particular initial stages of spatial organization in the vertebrate must involve additional classes of genes, those encoding intercellular signals and their signal-transducing pathways. A further limitation of "model" systems concerns the varied strategies among vertebrates themselves. The frog egg, which may represent something like the ancestral vertebrate strategy, employs a structural localization of some kind, which is set up by mechanical events immediately after fertilization. This, then, is a reference point, organizing the much later intercellular signaling events so that patterning and its orientation occur in relation to those cells that have "inherited" it during cleavage. In fish, birds, and mammals, the embryo body itself forms in a sheet of cells—the blastoderm or blastodisc—deriving from only part of the cell population of the original, cleaving egg. We know that the signaling events that initiate and orientate this pattern occur only after much cell division and in response to subtle cues extrinsic to the egg's structure (see Section V). This is important in relation to adequate theories of pattern formation (see Section IV). It means that in embryos of our own type, pattern must be self-initiating or, rather, triggered by the amplification of some very small difference of physiological state between one locality and its surroundings, within a sheet of cells of initially homogeneous state. Initial localization of a gene product to certain cells is not required and cannot be assumed.

Despite these caveats, we believe that once it is under way, pattern formation by intercellular signaling in the primitive tissue at onset of development probably employs the same mechanism among all vertebrate embryos. Indeed, evidence indicates homology

of function of genes involved in slightly later aspects of pattern establishment, as between *Drosophila* and vertebrates. Products of these genes mediate the acquisition of permanent identities among the cells, as members of the regions that constitute the body pattern (see Section V).

B. The Problem of Apparently Redundant Gene Expression in Early Mammalian Development

Very recent experimental work has produced several instances of puzzling discrepancy, where observational and experimental evidence has implicated a gene in some vital early step in embryo organization—its encoded product is first made at particular sites that embryologists recognize as sources of developmental information, for instance, and can even be shown to simulate the vital step concerned when experimentally overexpressed in certain test situations—and yet the molecular genetic technique of effectively removing the activity of the gene from the mouse embryo ("knocking out" the gene concerned) does not result in the expected very early abnormality of development. The problem is not that such genes are functionless; one or more later, more localized failures of the developmental process reveal corresponding necessary functions for each one. Rather, it appears as if some genes also have early, specific-appearing aspects of their expression that are redundant, or at least represent duplications of the functioning of some other gene or genes with regard to certain developmental steps. Only extensive future work will tell us whether such apparent redundancy represents the "fossilized" expression by such genes of functions once held in evolutionary history, or whether instead our laboratory techniques of diagnosing a "normal" example of development up to a particular stage are much less sensitive than is the selection that operates in nature. In either case, the possibility of the parallel operation of several genes in particular well-defined embryonic processes is a fascinating one.

III. STRUCTURE OF THE EMBRYO DURING PATTERN FORMATION

A. Gastrulation and Secondary Inductions

Figure 1 represents, in general form, the cell movements that occur while the embryo tissue is undergo-

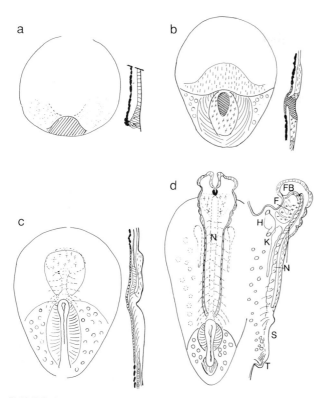

FIGURE I Sequence of morphological stages in body pattern formation of a generalized vertebrate. The figure is best studied after Section III has been read. The sheet of tissue in which the embryo forms (blastoderm, or blastodisc) is shown in plan view from above (left) and sectional view from the future body's left side (at right) in each figure part. In mammalian development, this sheet is not, in fact, planar but a closed-ended cylinder, but representation of that fact obscures rather than aids understanding. In (a), the future dorsal axial midline, whence mesoderm is to arise in gastrulation, is represented as a thickened region of the upper cell layer where the posterior end of the axis will finally lie (obliquely hatched). Cells from within the stippled area may be in course of being specified as mesoderm. The lower layer of cells or hypoblast (black) may be important in inducing pattern formation but does not contribute to the embryo. In (b), the first cells of the future embryonic endoderm (inner cell layer) have begun to leave the upper cell layer in gastrulation, and the processes of primary regionalization into future ectodermal, mesodermal, and endodermal derivatives is in train but has reached an unknown stage of refinement. Note that the embryonic endoderm (cross-hatched) is displacing the hypoblast, whereas subsequent regions will form a middle cell layer between ectodermal and endodermal layers. Codes for major regions in this and, where shown, in parts (c) and (d): head-mesoderm (crosses), dorsal axial mesoderm (notochord and somites) (oblique curved hatching), lateral-ventral mesoderm (circles), area destined for nervous system formation (not involuting) (vertical dashed). In (c), the region of active gastrulation has elongated to form the definitive, thickened primitive streak, or blastopore. The already gastrulated, anterior, dorsal axial mesoderm is also actively elongating and narrowing beneath the now morphologically distinct neuralized area, and its segmentation is beginning. It and the head mesoderm are shown by a dashed outline (i.e., below the surface) in plan view. Much posterior dorsal axial and (lateral-ventral) mesoderm remains to gastrulate and probably does so through different but adjacent regions of the streak as it shortens (see part d). In (d), the body axis has taken shape and major secondary inductions (interactions between cell layers) have occurred, even though gastrulation is still incomplete through the shortened streak in the tail region. Major structures and organ systems are labeled in addition to previous regional coding: H, heart; K, primitive kidney; N, notochord (midline skeletal rod between two rows of somite segments); FB, forebrain region ahead of notochord underlain by the head mesoderm; F, face and mouth structure, contributed by head-mesoderm and -endoderm, with additions from the neural tube (see text). Endoderm as a whole (the gut lining) is omitted from (d) for clarity. S, remainder of primitive streak; T, tailbud.

ing body pattern formation. At the beginning, the tissue is a sheet, one or a few cells thick, and breaks up into the obvious populations that will make the various structures only after gastrulation has produced the three fundamental germ layers. There is an arrangement whereby an orderly supply of cells leaves the original epiblast sheet at a rather restricted site, to migrate and pass beneath that cell layer, whereupon it generates two further tissue layers. Thus, the three-layer structure is progressively formed. The rudiments of the body take shape and begin to differentiate in sequence, from the head (first-immigrating mesoderm and the parts of the other two layers that are finally associated with this) toward more posterior regions.

Immigrating mesoderm at the outset may be of two different types defined by cell behavior, in addition to revealing regionalizations within those two overall components. Mesoderm of the dorsal axial region, immediately upon immigration, forms a rather cohesive tissue that also adheres rather strongly to the overlying midline of the epiblast, while generating a forceful shape change (convergent extension) that turns the whole into an elongated axial formation. This version of gastrulation behavior is termed involution of mesoderm, and it has the most highly organized sequence, whereby the newly involuted cells at successive time points are destined for the front, middle, or back of the body. More laterally and ventrally fated mesoderm behaves more as individually immigrating cells, with less capacity to produce force by mutual adhesions and more capacity to spread into available space between the two layers.

The appearances of gastrulation in various vertebrates differ as to how the orderly supply of mesoderm cells from the epiblast is arranged to occur, from a reservoir of uninvoluted tissue rather than in the structures formed after involution. The arrangement in mammals is the least understood at this level, but this need not dismay us unduly. Precise understanding of particular versions of their anatomical arrangement may not be relevant to understanding the underlying diversification of these cells, leading to their specification and commitment as members of different regions of body pattern.

The definitive endoderm, the innermost of the three layers of the embryo, is derived from the very first-immigrating cell population from the epiblast. This is, in turn, migrated upon by the succeeding mesoderm when the latter inserts itself between the two other "germ layers." The definitive germ layers and their ultimate derivatives in the body are ectoderm (future nervous system and epidermis of the skin), mesoderm (all primary connective tissue and skeletal, muscular, excretory, and blood systems), and endoderm (lining of the gut, including the epithelium of various digestive glands). One rather confusing term used by mammal embryologists should be addressed at this point. The upper cell layer of the blastoderm or blastodisc from which the whole embryo is probably derived is often called by them the "primitive ectoderm" before the start of gastrulation, even though, as we have just seen, the definitive endoderm and mesoderm as well as both epidermal and neural parts of the definitive ectoderm are to come from it.

Sequences of local interactions (best defined as secondary inductions and briefly referred to in Section IV) exist whereby short-range instructive signals pass between cells of the different layers to coordinate their patterns of differentiation. The best studied of these are involved in establishing the regional axial pattern within the area of ectoderm, overlying the midline of the involuted mesoderm, that has been specified to give rise to the nervous system. These signals, emitted from that mesoderm, are thus an expression of the body pattern of regionalization already emerging within it. Generally, mesoderm is the controlling cell layer in most of the body, in the system of local inductions that ensures appropriate contributions to the organ systems from the other two layers. The situation may be different in the head, however, where a second population of mesoderm-like cells, derived from the brain rudiment after its induction, enters the middle cell layer to collaborate in morphogenesis with the primary mesoderm. These cells may actively import the information back into the mesodermal and endodermal layers regarding which pieces of head structure are to be formed by their participation, even though they themselves originally acquired this regional infor-

mation at least partially from underlying mesoderm in the process of patterned induction of the neural plate (see the following subsection). This difference in the status of the germ layers in providing positional information may be one sign of a rather deep difference between an anterior and ventral domain of vertebrate body structure, devoted originally to a feeding apparatus, and the remaining elongating axial domain that provides the locomotory (primitively, swimming) apparatus.

B. Primary Embryonic Induction

The preceding section shows that secondary inductions are essentially a means whereby the pattern of regionalization in one cell layer is imprinted onto superimposed ones. They then respond in a spatially appropriate manner in their own differentiation. In large-scale regionalization to make up the vertebrate body plan, the primary role is thereby given to pattern formation within the mesoderm itself, or at least within the cell population comprising anterior (pharyngeal) endoderm and the entire mesoderm. We need to understand a mechanism that sets aside, and specifies as (endo +) mesoderm, a subpopulation of the original epiblast cells. At the same time, this mechanism gives a regional diversity to that subpopulation so that it can show the spatial and temporal pattern of adhesive and locomotory behaviors that orders gastrulation and hence orders its final positions in the body plan. The evidence is that the regional characters of this tissue, which are expressed in an orderly way during gastrulation, lead directly to commitments as regions of the body, although the activities of gastrulation movements themselves perhaps refine and finalize the regionalization. Local inductive signals are, in normal development, given out during this process.

Before the cell rearrangements of gastrulation actually start, signals within the epiblast or primitive ectoderm may result in the setting aside of an area that will form the neural plate (the entire future nervous system). This area, left in the outer cell layer and changing shape greatly as the mesoderm migrates under it, is then subregionalized into the axial plan of the nervous system by a process more akin to the patterned local inductive signals between germ layers, mentioned in the previous subsection.

By the onset of the precise sequence of gastrulation movements, then, specification of neural, endodermal, and mesodermal territories, and regional diversification within at least the latter, are well advanced. Diversity within mesoderm exists at two levels: segregation into dorsal axial, convergently extending, versus anteriormost and lateral sectors, with a different behavior, and the sequence of autonomous timing for cell behavior change *within* mesoderm of each sector, leading to the head-to-tail sequence of the body at gastrulation. In amphibian development, both of these types of diversity occur via mechanisms significantly related to original positions of cells, during the induction process but before the times of involution. We do not yet know whether this is true in the mammalian version, or whether the diversification of mesoderm cells is by a mechanism not involving their relative positions at the time. In Section IV, we shall see that this is of importance for theory about mechanisms of patterning.

C. Segmentation

Even adult human anatomy shows a repeated unit of structure, with regional variations of architecture imposed upon its derivatives, in the axial skeleton and musculature and the peripheral nervous system below head level. In more primitive aquatic vertebrates and especially in younger embryos or larval forms, this segmentation is much more prominent. It originates in segmentation of the dorsal axial mesoderm on either side of the midline rod (notochord) into a head-to-tail series of discrete units, the somites. Their cells contribute normally to provide the vertebrae, muscle blocks, and domains of dermal (skin) tissue and, secondarily, to organize the nervous system. They are of regular size at initial segregation, though this varies in different general regions of the axis and is later obscured by patterns of growth.

In the anterior part of the embryo head, the role of the mesodermal segmentation is less clear, because a repetitive pattern in the brain rudiment or even the endoderm (the gill bars) may control the pattern instead (see Section III,A). But behind this, somite pattern appears to be strongly controlled as to the constancy of segment numbers found in different individuals between the back of the skull and particular landmarks in the body, such as the position of articulation of the pelvic girdle with the vertebral column, or that of the kidney rudiment. Regulation must be involved (see Sections I and IV), because total axial tissue available is considerably variable in different individual embryos during pattern formation, so that sizes of the individual segmented cell groups must be adapted in relation to this to ensure the constancy of numbers. All of this might imply that the somites are a manifestation of fundamental, regular repeating domains of genetic activity, on which variation is

played to compose the body pattern. This has been strikingly demonstrated in insect development, where the body is essentially made up of a fixed number of segments, as opposed to just containing one segmenting cell layer. These segments bear a fixed relationship to the early domains of gene activity that are known to give regional characters to the insect embryo blastoderm.

In fact, there is no evidence for and some evidence against the idea that mesodermal segmentation in vertebrates is so fundamental in the mechanism of constructing the body. Division of the muscle and skeleton into repeated units may initially have been simply an adaptation to locomotion in water, in fish-like vertebrate ancestors and their larvae, which has become deeply imprinted in the developing axial mesoderm. Under experimental and natural conditions, somite numbers can be made to vary slightly in relation to the regions of the body by extreme ambient temperatures, or transitions of physiological conditions, during development. It cannot be said, for instance, that somite number N will *always* form the last lumbar vertebrae, whereas $N + 1$ forms the first component of the sacrum (where N is on the order of 25). In insects, by contrast, such physiological perturbations, or genetic malfunctions, leave the unaltered species-typical number of body segments. Instead, they replace the normal characters of particular members of the series with those appropriate to other members, thus duplicating and deleting complementary segment types in the phenomenon of "homeosis."

IV. THEORY OF BIOLOGICAL PATTERN FORMATION

A. The Leading Role of the Mesoderm

The local signals that act at later stages to coordinate the differentiation of the germ layers contributing to each organ system are of great importance but will not be considered much further in this article. The signaling molecules are probably secreted proteins, and their correct action seems to require the intimate contact between inducing and responding cell layers, involving the special extracellular structure known as basal lamina, that can only set in after the large-scale rearrangements of gastrulation are over. The commonest situation is that the instructive signals, those establishing the character of the organ that is to form, come from the mesoderm at this later stage of detailed patterning as they did while the overall plan was being set up in gastrulation. Subsequently, there may follow a complex reciprocal series of interactions between the (usually two) germ layer components of each emerging organ. This is the case, for instance, for fore- versus hindlimb character of limb rudiments, and for the character of the kidney and of various major glands. During gastrulation, however, an important principle may be that sequences of signals are received from mesoderm, particularly at positions within the ectodermal layer where the nervous sytem is being specified, because the gastrulation movements progressively bring later-generated (thus more posteriorly fated) mesoderm beneath them.

B. The Generation of Diversity

In view of all the foregoing, we can see that the real heart of body pattern formation is the process whereby the original cell sheet—the "primitive ectoderm" in the case of mammal embryos—diversifies into a progressively regionalizing endoderm plus mesoderm that will drive gastrulation by its movements, a neuralized area that is itself imprinted with more regionalization as gastrulation proceeds, and a remaining definitive ectoderm that will provide the body's epidermis. Up to a certain stage, these events have the additional property of regulation; that is, the *proportions* in which the available cells are allocated to the parts can be normalized in the face of natural or experimentally imposed variation in their total number. The more slowly developing of the amphibian embryo types, widely used in traditional experimental embryology, have these regulative properties well developed. For instance, an entire domain such as the mesoderm fated for the anterior part of the head, together with its distinctive (i.e., forebrain inducing) properties can be replaced by respecification of the surrounding cells if surgically excised from its known location in the pregastrula or early gastrula. The mammalian primitive ectoderm as a whole almost certainly passes through an even more profoundly regulative stage, as evidenced by the production of identical twins—two bodies from one blastula—or by the production of a rather normal fetal body after destruction of up to 80% of its cells around the outset of gastrulation in the case of experimental mice. What kind of mechanism might underlie such development and its regulative capacities?

First, some kind of diversification of the states of the early cells is needed. Such diversification could initially be only physiological, and lead later to differential gene transcription, or else could result almost

immediately in diverse gene activity. Second, a set of gene activities is required that locks in place the specification given by the generator of diversity so that they become "long-term memories" within the cells regarding which restricted regions of the future pattern they and their descendants can contribute to. Body patterning starts with a particular orientation within the tissue when, in mammals at least, no reference point for this has been inherited structurally in the egg. Correct ordering and proportions for the regions or groups of cells with the different specifications then follow. The reference point, necessary to determine orientation and thus the axis of bilateral symmetry, must involve one eccentric location becoming established as special within the tissue. However, it is often assumed that the other features of pattern, because they are finally *expressed* as a spatial layout, are necessarily generated by mechanisms of *position-dependent specification* of cells. But this is neither a formally necessary nor, for higher vertebrate embryos, an experimentally proven assumption, and may turn out to have been misleading when the real mechanism is understood. Some formal possibilities for patterning mechanisms are represented in diagrammatic form in Fig. 2, which should be studied in conjunction with what follows.

In the relatively well understood embryo of the fruit fly *Drosophila*, position-dependent specification by early diversification of gene activity is indeed the mechanism. Interactions of gene products in space form gradients and periodic patterns, which in turn are responsible for precisely localized domains of activation for further genes. In the frog's version of vertebrate development it appears that cells' initial positions, particularly in relation to localized sources of intercellular signals (see Section V), are at least partly responsible for their diversification in pattern formation. But in the smaller types of vertebrate embryo, and especially in the minute tissue space available during the early part of mammalian pattern specification, it is difficult to see how a pattern due to diffusing and interacting substances (i.e., morphogens—see the Glossary), containing sufficient levels of information to specify a diversity of cell states, might stably exist for long enough to produce the reliability of development that is observed.

The property of regulation does seem to imply intercellular communication, whether in formation of a gradient or other morphogen pattern (Fig. 2a), or by emission of specific signals by cells of each of the types that the system is generating, which act as feedback to influence the "choices" of less committed cells else-

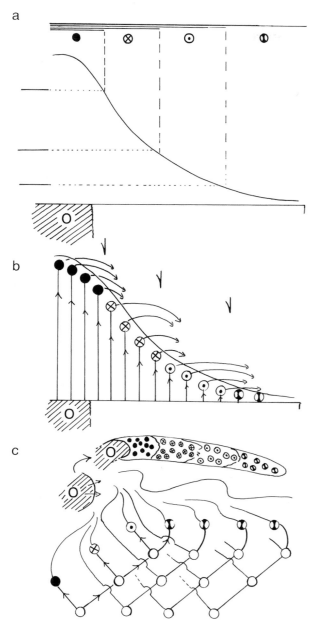

where in the tissue (Fig. 2b). But an extreme alternative possibility, perhaps implying less intercommunication, would be that in the dividing cell population of the early epithelium from which the embryo is to be generated, there is spontaneous and random generation of the diversely specified cells that are required, with the probabilities of entering each specificity set such that proper proportions of the cell states are produced. In this model, regulation is automatic at early stages, because random removal of cells from the tissue will remove representative numbers from the mixture of cell types and leave the random generator of diversity unaffected. Because what is finally manifested is a spatial pattern, a necessary capacity of cells specified in this way would be to migrate selectively early during the movements of gastrulation, that is, to sort out, so as to attain their correct relative positions. The special small cell group, constituting the reference point for overall orientation of the body pattern, would in this case emit signals orientating the movements that allowed sorting out of cells elsewhere (Fig. 2c).

Whereas only the pure forms of the alternative mechanisms can reasonably be illustrated in Fig. 2, it now seems very likely that embryo formation in higher vertebrates occurs by a composite mechanism involving elements of both position-dependent specification and then refinement of pattern by selective aggregation of the diversified cells. Such a relatively complex mechanism, involving elements of all three formalisms

FIGURE 2 Three principles that could operate in the spatial diversification required for body pattern formation. All ideas start from the notion that a relatively restricted sector of the tissue (probably the future dorsal and anterior mesoderm/endoderm) becomes set aside as being in a special state and emits signals or "morphogen" molecules as the organizer (O) of body pattern (oblique hatched). Pattern is arbitrarily shown as consisting of four regions or cell states (represented by the symbols), which correspond to neither of the initial two states but are organized as a result of interactions between them. In (a) and (b), a gradient is set up by the "source" and "sink" activities of the organizer region and its surroundings. In (a), this is interpreted directly to give the cell states, which result from combinations of the active versus inactive states of a small set of genes, each of which has its own threshold of morphogen signal level for switching between active and inactive states. Thus,

the morphogen could be a repressor *or* an activator of the genes, but the cell type specified at one end of the system has all the genes turned off, whereas that at the other has them all activated. In (b), the same final arrangement is controlled through the gradient but in a less direct way. The gradient rank orders the *rate* of development of cells; cells near the organizer are determined first, and those far away last, in a race for occupancy of the hierarchy of available cell states. Cells entering each state in the hierarchy emit a specific diffusing signal that diverts less-developed cells toward the next preferred state and prohibits further access to their own state. Half-arrowheads mark positions at which buildup of each successive signal *diverts* the character of the wave front of cell determination to give the next state. In (c), the required cell types are generated initially without respect to position and are intermixed. An asymmetrical stem-cell lineage is shown, tied to the cell cycle, but any cell-autonomous generator of diversity may prove to be operating. The role of the organizer region is here to act as a focal point for cell migration through its emitted signals, with cells selectively positioning themselves through adhesive preferences, etc., during migration. None of these three mechanisms is likely to be a complete one, but the real mechanism may embody elements of all.

in Fig. 2, certainly underlies development of one very well studied, primitive multicellular organism, the cellular slime mold. So in higher embryos the reference point for pattern orientation and symmetry—the "organizer" region that has a special role in setting up pattern as well as having its own fate—probably functions both as a focus for cell aggregation and as an initiating point in a system of morphogens. This role may in fact pass from one cell group to another during development, but it seems likely that at least for one period it resides with the cells that will become the structure known as Hensen's node, lying at the end of the primitive streak or region where gastrulation is occurring. These cells or their equivalents, in the different vertebrate embryo types in which the experiment has been performed, organize extra body patterns when transferred to sites in other, host embryos.

A fascinating idea, though a hard one to test, is as follows. Early embryo cells do seem to monitor elapsed "developmental time" in a striking way and may do this in relation to the progression of cell cycles while in an actively growing tissue. If, in a certain region of the blastodisc, conditions allowed cells to change their specification state systematically with time, but cell population pressure or aggregation toward the organizing center continuously removed cells from this special zone as development progressed, then a supply of tissue trapped into states of successively more posterior specification might be generated. The mammal embryo in particular is known to be laying down an axial pattern across a period of massive growth, and some evidence indicates a region specially active in cell division near the presumed site of origin of the axial tissues.

C. Positional Memory in Embryo Cells

The second-acting part of any pattern formation mechanism must be that which records the specifications that have been assigned to groups of founder cells, as long-term commitments in them and their descendants to form each of the body regions. It is becoming clear that particular genes, from several different families that encode proteins of the kind believed to control activity of specific sets of other "downstream" genes, make up this mechanism. These are not themselves genes that trigger differentiation of the various histologically defined cell types. Each such cell type characteristically occurs throughout much of the body, but tail tissue is coded differently from head tissue before, say, cartilage or dermal cells have been differentiated in either region. Constructing a body is not primarily a matter of consigning particu-

lar geographical patches to particular differentiated cell states, but rather of establishing a series of geographical units, within each of which a distinct variation on a "theme" pattern of cellular differentiations is then played out by more local interactions (see Section III,C). Direct visualization of the system first encoding body position as combinations of activated genes has been possible using molecular technology, first in the *Drosophila* embryo and more recently in mammal and bird embryos, and confirms this principle. These genes encode records of cells' relative positions, in the body as a whole, on a combinatorial basis along the lines illustrated in Fig. 2a, although there is no evidence that this is arranged by direct and simple interpretation of a morphogen gradient. Thus each gene is activated in an extensive, continuous domain along the future axis, suggesting a simple response to some variable, but this could equally well be, for instance, cellular "age" at which cells became involved in gastrulation at each position (see previous subsection). A number of uniquely encoded domains adequate to specify a body plan are built up because the sectors of activity for each gene overlap extensively, so that unique combinations of such genes characterize much more restricted regions. The "downstream" or executive genes, which such combinations of positional gene activity control in order to dictate precise arrangements of commitments to differentiation within each body unit, are still essentially unknown but would be expected to include those controlling aspects of cell-selective adhesion and motility as well as differentiation. As might be expected in view of the demands of the final anatomical structure, the most fine-grained partitioning by positional gene activities in vertebrate embryos seems to occur in the brain rudiment. Here, as in insect development, the domains of such genes' activities are integrated precisely with the segmented structure that seems to be set up, at least transiently, to control aspects of head pattern.

V. CURRENT CANDIDATES FOR MOLECULES INVOLVED IN VERTEBRATE BODY PATTERN FORMATION

A. Reference Point for Pattern Orientation and, Hence, Bilateral Symmetry

No vertebrate is thought to utilize structural prelocalization of a particular gene product in the egg as a mechanism of body axis initiation or orientation. In amphibians, however, symmetrization does occur by a physical reorganization of the massive egg cell to install a special state upon one narrow sector within it, during the one-cell stage after fertilization. The movements normally relate to gravity, but probably do not require a gravity vector. The mechanism seems to act by arranging that special inducing signals, giving rise to dorsal axial tissue specification, occur only within the narrow sector (see the next subsection). In higher (mammalian) vertebrate embryos, establishment of a dorsal midline may indeed be accomplished as a similar patterning of inducing signals, that is, one widespread around the blastodisc and one more restricted and eccentric in location. But this eccentric "organizer" site cannot be caused by special biochemical or physiological states inherited by cells from one position in the egg's structure, because determination of the site of axis formation and thus its orientation does not occur until multicellular tissue is present. An asymmetric relationship of the blastodisc to gravity or to local environmental variables (e.g., O_2 or CO_2 tensions) may trigger the breaking of symmetry, but the mechanism must be put in gear by a self-amplifying response that (a) "locks" the initially favored cells into a new state in which they produce dorsal axial-type inducers and perhaps cell aggregation signals and (b) stabilizes the symmetry by suppressing (through intercellular signaling) the attainment of this organizing state by cells elsewhere.

B. The Intercellular Signals Generating Diversity for Body Pattern

There are currently no clues to molecular mechanisms for spontaneous, or cell cycle-related, intracellular generation of diverse states of specification (Fig. 2c; Section IV). We have seen that, in vertebrates, position-dependent mechanisms of diversification must be by truly intercellular signals. Recently, exciting candidates for the first-acting members of such a signal cascade have been identified from amphibian development. These belong to families of small, secreted proteins whose first identified members were called "growth factors" because of effects on the cell-cycle status of defined mammalian target cells in tissue culture. The changes of specified state that they evoke in pregastrula-aged embryo cells correspond with the diversification that normally accompanies gastrulation, in the mammalian "primitive ectoderm" or its equivalent in other vertebrates. They thus correspond with the events that embryologists have defined as

primary inductions (see Section III,B). The several possible responses of the cells show evidence of progressive change with concentration of the "inducing" molecules, as would be expected for a position-related or morphogen-like patterning mechanism. Though several such secreted molecules that are well characterized from their later roles in the life of the vertebrate organism act experimentally in this way, it is as yet unclear whether the true endogenous, primary inducing members of these families have been identified (see Section II,B). Another "intercellular" molecule now implicated in patterning of the vertebrate body is retinoic acid or one of its relatives. In a concentration-dependent way, it can permit or prevent specification of particular sectors of axial body structure. The precise principle upon which such secreted molecules as these induce geographically patterned mesodermal and neural specification is currently unknown, but of particular interest is that among the first genes transcribed in response to them in development are members of the principal "homeobox"-containing class to be mentioned in the next and final subsection.

C. Molecules Encoding Cell Memories for the Regionality of the Body

In the fruit fly, which forms a pattern largely before advent of cell boundaries, the initial position-dependent regionalization, as well as the establishment of longer-term memories for regional identity, results from interactions among proteins that are sequence-specific DNA binders, principally of two structural families. The function of such proteins as part of a morphogen mechanism of first regionalization is clearly impossible in fully cellularized development, where this role may be mediated by the smaller, secreted molecules mentioned earlier. The evidence is now overwhelming, however, that DNA binding proteins of these two functional classes, referred to popularly as the homeobox-containing (helix-turn-helix) and "zinc-finger"-containing proteins, are central to intracellular recording of regional identity in vertebrates. Their genes may respond directly to morphogen concentrations and also have a network of cross-interactions as controllers of one another's transcriptional activity within individual cell nuclei. Restricted axial domains of synthesis for these proteins, reminiscent of those in the fruit fly, are seen in relatively newly induced mesoderms and nervous systems of all vertebrate-type embryos. Systematic interference with body pattern has resulted from experimental alteration of function of specific members. It is nevertheless

Embryofetopathology

DAGMAR K. KALOUSEK
GLENDA HENDSON
British Columbia Children's Hospital

GLOSSARY

Abortion or miscarriage Premature expulsion or removal of the conceptus from the uterus before it is viable to sustain life on its own; 18 developmental or 20 gestational weeks is considered the lower limit of viability. Older fetuses can be delivered either as stillborn or premature, mature, and postmature newborns

Amnion Innermost membrane of the sac enclosing the embryo or fetus

Body stalk Early primitive umbilical cord connecting vascular systems of the embryo and developing placenta

Chorion Outermost of the two membranes that completely envelop the embryo or fetus

Chorionic sac Precursor of the placenta

Conceptus or products of conception Includes all of the structures that develop from the zygote, i.e., the embryo–fetus and the placenta with its membranes

Decidua Maternal tissue derived from the mucosal lining of the uterus that peels off at childbirth or abortion

Developmental or conceptional age Extends from the day of fertilization to the day of intrauterine death or expulsion

Deoxyribonucleic acid (DNA) probe A nucleic acid fragment that is complementary to a sequence of nucleotides on a specific chromosome that will, by hydrogen bonding to the latter, locate or identify that sequence

Diploid Having two sets of chromosomes as normally found in the somatic cells of higher organisms; in humans the diploid number is 46

Embryo Developing human from conception until the end of the 8th week, by which time all organ systems have been formed

Embryonic growth disorganization Represents generalized abnormal embryonic development, which can be divided into four categories ranging from GD1, which is most severe with complete lack of embryonic development, to GD4 in which there is a severely abnormal embryo present

Gestational and menstrual age Extends from the first day of the last menstrual period to the expulsion or the removal of the conceptus; therefore, gestational or menstrual age is 2 weeks greater than developmental or conceptional age

Haploid Having a single set of chromosomes as normally carried by a gamete; in humans the haploid number of chromosomes is 23

Hybridization Association of single-stranded DNA to form double-stranded DNA which occurs when the nucleotide base sequences are complementary

Implantation Attachment of the blastocyst to the epithelial lining of the uterus occurring 6–7 days after fertilization of the ovum

Interphase The "resting phase" of the cell; the interval between two successive cell divisions during which the chromosomes in the nucleus are not individually distinguishable

Intrauterine retention period The time between the death of the embryo–fetus and its expulsion or removal

Karyotype Chromosomal characteristics of an individual or a cell line, usually presented in a systemized array of metaphase chromosomes of a single cell from a photomicrograph or computer-generated image. The chromosomes are arranged in pairs in decreasing order of size and according to the banding pattern and position of the centromere

Maceration Degenerative changes, discoloration, and softening of a fetus retained in the uterus after its death

Metaphase The second stage of cell division during which the contracted chromosomes, each consisting of two chromatids, are visible

Morphology Features comprised in the form and structure of an organism

Placenta Vascular organ within the uterus connected to the fetus by the umbilical cord and which serves as the structure through which the fetus receives nourishment from and eliminates waste into the circulatory system of the mother; the lay term is the "afterbirth"

Previable fetus Developing human from the beginning of the 9th week postconception until gestational viability is reached (20 gestational weeks)

Somite The paired, block-like mass of mesoderm arranged segmentally alongside the neural tube of the embryo, forming the vertebral column and segmental musculature

Uterus Hollow muscular organ of the mother in which the ovum is deposited and the conceptus develops; the lay term is the "womb"

Yolk sac Extraembryonic structure that serves temporarily for the nourishment of the early embryo

EMBRYOFETOPATHOLOGY IS A NEW DISCIPLINE of human pathology dealing with the evaluation of the development of human embryos (up to 8 developmental weeks) and previable fetuses (9–18 developmental weeks). It has evolved in response to clinical interest in the early monitoring of human pregnancy, intensive prenatal care, and increasing demands for genetic counseling. Embryofetopathologists investigate both spontaneously aborted embryos and fetuses and conceptions terminated due to prenatal diagnosis of an abnormality. The investigations consist of the evaluation of embryonic–fetal and placental development and the detection of any defects. The information is used in determining the risk of repeated spontaneous abortions, in genetic counseling of the parents regarding risks of having a normal or abnormal conceptus in future pregnancies, and for guidance to prenatal monitoring of future pregnancies.

I. INTRODUCTION

Pregnancy loss is surprisingly frequent. It has been estimated that about 50% of human conceptions spontaneously abort. The highest conceptus mortality exists in the early weeks of pregnancy, usually before the pregnancy is clinically recognized. Detailed hormonal studies at this very early stage suggest that approximately 25% of conceptuses fail to implant.

The rate of pregnancy loss after implantation and before clinical recognition is estimated to be around 15%. Another 15–20% of clinically recognized pregnancies spontaneously abort prior to 20 weeks gestation.

Among the causes of pregnancy loss are chromosomal aberrations, developmental defects, inborn errors of metabolism in the conceptus, maternal illness, uterine structural abnormalities, cervical incompetence, infection, and immunologic and hormonal imbalances. Depending on the particular stage of the pregnancy when fetal loss occurs, different etiologies are more common; the chromosomal defects are the most common among the early spontaneous abortions, whereas pregnancy loss caused by infection is commonly detected in late abortions, especially after 16 weeks of gestation. [*See* Abortion, Spontaneous; Implantation, Embryology.]

Early experience with the pathological examination of aborted human embryos has led to the description of four basic types of embryonic maldevelopment and to the realization that embryonic developmental anomalies not only differ morphologically from defects at birth, but also are much more common. Many abnormal embryos do not show specific malformations and represent generalized embryonic growth disorganization. The high incidence of abnormal embryos in abortion specimens has been confirmed by many investigators. In one of the largest morphologic studies of early spontaneous abortion, among 1226 specimens, 84% of the embryos were abnormal, with localized defects in 5% and generalized abnormal embryonic development in 79%. [*See* Birth Defects; Chromosome Anomalies.]

Findings among aborted fetuses are different from aborted embryos. The majority of aborted fetuses are morphologically normal and their intrauterine demise or expulsion is caused mainly by infection, placenta-related problems, or cervical incompetence. Less than 20% of spontaneously aborted fetuses show developmental defects.

II. TYPES OF EMBRYOFETOPATHOLOGY SPECIMENS

There are two types of abortion specimens: early (corresponding to embryonic) and late (corresponding to fetal development between 11 and 20 gestational weeks).

A. Early Abortion Specimens

For practical purposes, early abortion specimens may be divided into *complete* and *incomplete*. A complete specimen of early abortion usually consists of a chorionic sac with an embryo, decidual tissue, and blood clots. The chorionic sac may be intact, ruptured, or fragmented. If the chorionic sac is not intact and an embryo is not found, the specimen is categorized as an incomplete specimen. Incomplete specimens of early abortions are common, often only consisting of fragments of chorionic sac, decidua, and blood clots.

An aborted, intact chorionic sac is usually found embedded in maternal decidua, which resembles a cast of the uterine cavity (Fig. 1). Such specimens are carefully opened from the cervical end to release the intact chorionic sac (Fig. 2). The dimensions of the chorionic sac vary from 5 to 80 mm in diameter. The surface of a well-developed chorionic sac is like a "fuzzy tennis ball" and is completely covered by abundant chorionic villi which, if normal, are of uniform diameter with symmetrical branches and multiple buds along their length. Villi are best observed and examined floating in saline solution under the

FIGURE 2 Intact chorionic sac 1.3 cm in diameter covered with well-developed chorionic villi (arrows).

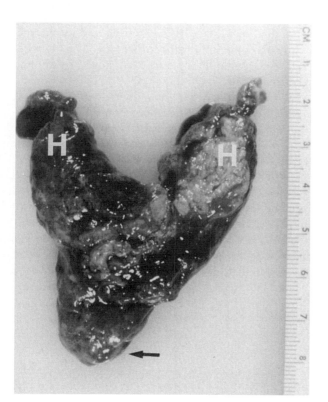

FIGURE 1 Decidual cast of uterine cavity with two labeled horns (H) and an arrow pointing to the cervical end.

dissecting microscope. Sparse villi with abnormal morphology are common in specimens of early abortions and usually indicate a chromosomally abnormal conceptus.

Rupturing the chorionic sac is usually associated with rupture of the amniotic sac and loss of amniotic fluid. The ruptured chorionic sac has a collapsed corrugated appearance. An embryo is frequently missing from such a sac. Even in the absence of the embryo, the evaluation of the amniotic sac, villi, body stalk, and yolk sac, both morphologically and histologically, allows categorization of such specimens as a normally or an abnormally developing conceptus. A fragmented chorionic sac without a detectable embryo can yield information about embryonic development by the histological examination of the chorionic villi.

B. Late Abortion Specimens

Simply due to their size, late abortion specimens are usually complete. They consist of an easily identifiable fetus (9–18 developmental weeks or 11–20 gestational weeks) and a placenta consisting of an umbilical cord, extra-embryonic membranes (amnion and chorion), and a disc of chorionic villi. Both fetus and placenta are examined to establish the cause of the spontaneous abortion (Fig. 3).

III. INVESTIGATIVE APPROACH

A. Examination of the Chorionic Sac

The size of the sac, the amount, and the morphology of villi are evaluated. After opening the intact chorionic sac, the presence or absence of the amniotic sac, yolk sac, body stalk/umbilical cord, and embryo is noted. When the amniotic sac is present, its size and relationship to the chorionic sac are recorded. An abnormally large amniotic sac and its premature fusion with chorion are the most frequent abnormalities (fusion of the amniotic membrane to the chorion only occurs after 11 weeks developmental age). The abnormal gross and histological morphology of the chorionic sac and its villi gives the embryopathologist insight into the sequence and timing of the insult during early embryogenesis; if the villi show no embryonic vessels, swelling of the stroma, and abnormal morphology of the epithelial lining, embryonic death occurred before the establishment of the embryonic circulation in the placental villi (i.e., before the end of the third week). However, if the embryonic circulation was established and then ceased because of embryonic death, the remnants of vascular channels and nucleated fetal red blood cells can be observed.

After intrauterine embryonic death, the chorionic sac may remain in the uterus for several weeks, and significant changes such as collapse of villous vessels or an increase or decrease in villous fluid content occur within the chorionic villi. Calcifications in villi, either focally or along the villus basement membrane, are a common finding. These degenerative changes must be distinguished from true abnormal development.

B. Examination of the Embryo

The development of human embryos from fertilization until the embryo has attained a crown–rump (CR) length of 30 mm is divided into 23 stages, or horizons. In the early stages, the embryo is characterized by a number of somites, whereas during the later stages the specific size of the embryo and its external features become important for staging. The main external embryonic features of each stage are summarized in Table I.

The complete external examination of embryos is done under the dissecting microscope. The examination consists of a detailed evaluation of characteristic facial, limb, and body features to provide data for determination of the embryonic developmental stage. The difference of 14 days between the developmental age thus derived and the gestational age calculated from the last menstrual period reflects the length of time of retention *in utero* after embryonic death.

Based on external examination, human embryos collected from spontaneously aborted tissues can be classified into four major categories:

1. Normal embryos, appropriately developed for their developmental stage (Fig. 4).
2. Growth disorganized embryos, with highly abnormal embryonic development (Fig. 5).
3. Embryos with specific developmental defects (Fig. 6).
4. Macerated, damaged, unclassifiable embryos (Fig. 7).

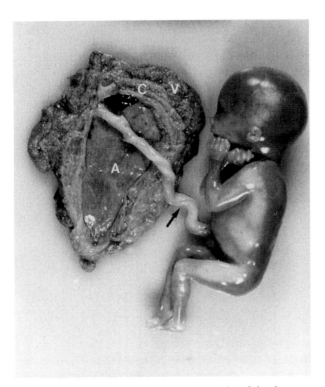

FIGURE 3 Fetus and its placenta at 16 weeks of development. Note umbilical cord (arrow) connecting the vascular systems of the fetus and placenta. In the placenta, amnion (A), chorion (C), and placental villi (V) are indicated.

TABLE I
Developmental Stages in Human Embryos[a]

Age (days)	Stage	No. of somites	Length (mm)	Main characteristics
20–21	9	1–3	1.5–3.0	Deep neural groove and first somites present. Head fold evident.
22–23	10	4–12	2.0–3.5	Embryo straight or slightly curved. Neural tube forming or formed opposite somites, but widely open at rostral and caudal neuropores. First and second pairs of branchial arches visible.
24–25	11	13–20	2.5–4.5	Embryo curved due to head and tail folds. Rostral neuropore closing. Otic placodes present. Optic vesicles formed.
26–27	12	21–29	3.0–5.0	Upper limb buds appear. Caudal neuropore closing or closed. Three pairs of branchial arches visible. Heart prominence distinct. Otic pits present.
28–30	13	30–35	4.0–6.0	Embryo has C-shaped curve. Upper limb buds are flipper-like. Four pairs of branchial arches visible. Lower limb buds appear. Otic vesicles present. Lens placodes distinct. Attenuated tail present.
31–32	14	[b]	5.0–7.0	Upper limbs are paddle-shaped. Lens pits and nasal pits visible. Optic cups present.
33–36	15		7.0–9.0	Hand plates formed. Lens vesicles present. Nasal pits prominent. Lower limbs are paddle-shaped. Cervical sinus visible.
37–40	16		8.0–11.0	Foot plates formed. Pigment visible in retina. Auricular hillocks developing.
41–43	17		11.0–14.0	Digital, or finger, rays appear. Auricular hillocks outline future auricle of external ear. Trunk beginning to straighten. Cerebral vesicles prominent.
44–46	18		13.0–17.0	Digital, or toe, rays appearing. Elbow region visible. Eyelids forming. Notches between finger rays. Nipples visible.
47–48	19		16.0–18.0	Limbs extend ventrally. Trunk elongating and straightening. Midgut herniation prominent.
49–51	20		18.0–22.0	Upper limbs longer and bent at elbows. Fingers distinct but webbed. Notches between toe rays. Scalp vascular plexus appears.
52–53	21		22.0–24.0	Hands and feet approach each other. Fingers are free and longer. Toes distinct but webbed. Stubby tail present.
54–55	22		23.0–28.0	Toes free and longer. Eyelids and auricles of external ears are more developed.
56	23		27.0–31.0	Head more rounded and shows human characteristics. External genitalia still have sexless appearance. Distinct bulge caused by herniation of intestines still present in umbilical cord. Tail has disappeared.

[a]Modified, with permission, from K. L. Moore (1982). Criteria for estimating developmental stages in human embryo. *In* "The Developing Human: Clinically Oriented Embryology," 4th Ed. Saunders, Philadelphia.
[b]At this and subsequent stages, the number of somites is difficult to determine and therefore is not a useful criterion.

Any discrepancy between the embryonic CR length and the expected development for that stage suggests the existence of a developmental defect. For example, if a well-preserved fresh embryo measures 9 mm and shows incomplete closure of the neural tube, this is diagnostic of an open neural tube defect, as complete closure takes place before the embryo is 5 mm CR length. Accurate evaluation of normal embryonic development is best done by comparison of an embryo with photographs of normally developed human embryos from anterior, posterior, and lateral views as well as comparison with actual preserved embryonic specimens illustrating the normal development of each embryonic stage (Fig. 8). Although the CR length is very important for staging normally developing em-bryos with focal defects, CR length cannot be used for embryos that are growth disorganized.

Occasionally, damaged and severely macerated embryos cannot be properly evaluated. Their developmental age can only be estimated based on specific structure or organ development [e.g., hand/eye (Fig. 7)]. A pitfall with macerated embryos is that maceration can mimic focal developmental defects such as an open neural tube defect or cleft lip.

C. Examination of the Fetus

Routine examination of the fetus consists of both external and internal examination, radiological examination, photographic documentation, and his-

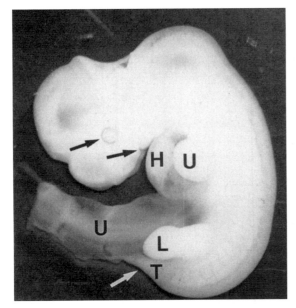

FIGURE 4 Well-developed embryo at stage 15, length 7 mm, showing developing eye (arrow), mouth (arrow), heart (H), upper limb (U), and lower limb (L). The presence of tail (T) is a normal finding at this stage. Umbilical cord (u) contains three vessels.

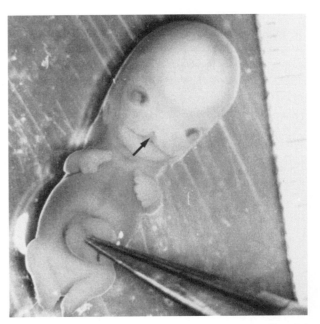

FIGURE 6 Embryo stage 21, length 23 mm, showing midline cleft of upper lip (arrow). No other developmental defects are present.

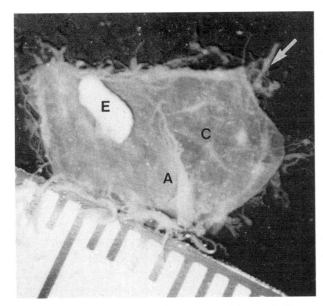

FIGURE 5 Opened abnormal chorionic sac with growth disorganized embryo (E). The embryo measures 3.5 mm in length but shows no evidence of differentiation. It appears as a cylinder of tissue with a barely recognizable head and tail end and a very short body stalk. In the chorionic sac, amnion (A), chorion (C), and villi (arrow) can be recognized. Very sparse and abnormal villous development is obvious.

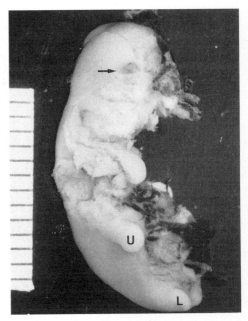

FIGURE 7 Embryo about stage 15 showing severe maceration and damage. Note complete loss of embryonic curvature and damage of face, neck, and abdomen. Eye is identified by an arrow. L, lower limb; U, upper limb.

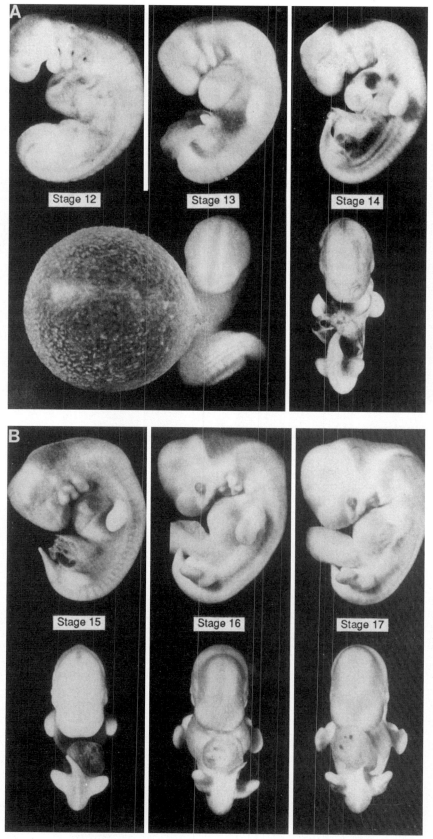

FIGURE 8 Illustration of embryonic stages 12–23. [Reproduced with permission from N. Exalto, R. Rolland, T. K. A. B. Eskes, and G. P. Vooijs (1982). "De jonge swangerschap." Boehringer, Ingelheim.]

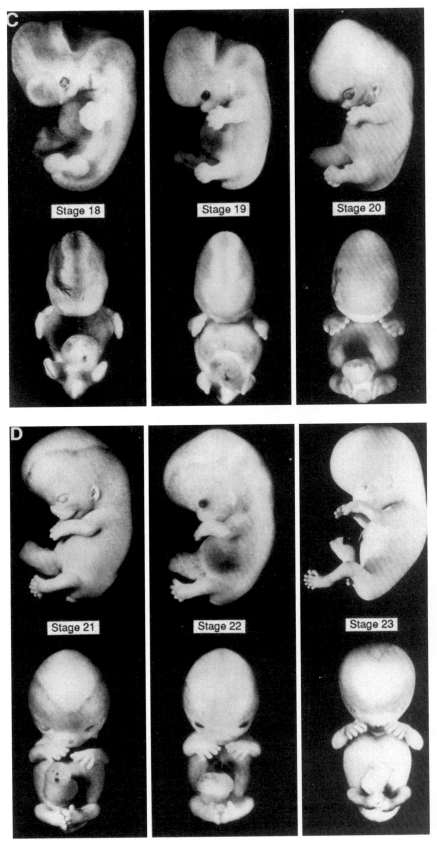

FIGURE 8 (*Continued*)

tological examination. Cytogenetic, biochemical, and microbiological studies are done on selected cases when warranted.

1. External Examination

The fetus is weighed and CR length, crown–heel length, and head circumference are recorded. CR length is the main criterion used for establishing the fetal developmental age according to tables based on average measurements of a large population of fetuses with similar criteria. Tables are also available for determining developmental age from foot and hand lengths. Hand and foot lengths can be used instead of CR length when the specimen is incomplete or fragmented or can be used in addition to CR length to verify the accuracy of the CR length measurement.

A detailed external examination of the fetus is an essential part of morphological studies. The head and facial features are noted, with attention being paid to the shape of the head, position of the ears, presence or absence of scalp hair, abnormalities in the shape, slant, and distance between the eyes, shape of the nose, patency of the nasal passages, mouth, size of the tongue, completeness of the palate, intactness of the lips, shape and size of the mandible, and the configuration of the ears. The limbs are examined, recording their length, number, and position of the digits, the position of the limbs, presence or absence of flexion deformities, or other abnormalities. The chest shape and defects, abdominal shape and defects, and external genitalia complete the external examination.

Macerated and damaged fetuses may be grossly distorted and hinder pathological investigation (Fig. 9); however, this does not negate investigation as they may provide clues to abnormality. Thus, despite cranial suture molding, distortion, and head collapse, the major external malformations, if present, such as a neural tube defect, facial clefting, and missing or extra digits, can be diagnosed. The documentation of the absence of any developmental defects is as important as the diagnosis of an abnormality.

2. Internal Examination

Dissection is usually done in the manner of a miniautopsy with the help of a dissecting microscope. The internal examination is directed toward the diagnosis of obvious abnormalities such as the absence of an organ, its abnormal position, or incomplete formation as well as subtle abnormalities such as congenital heart defects. The internal organs are dissected free, each organ is weighed, and the weight is compared

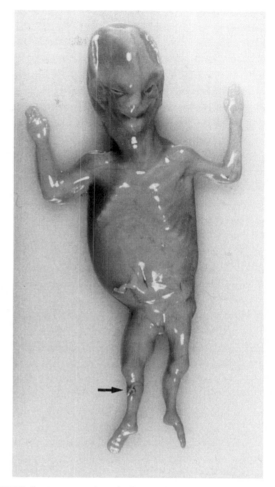

FIGURE 9 Macerated male fetus of 15 developmental weeks showing the collapse of head, distortion of facial features, and peeling of skin (arrow).

with normal values for the estimated gestational age. In addition to the external appearance, the cut surface of each organ is examined and compared to normal.

The examination of the brain is feasible only in nonmacerated fetuses, as it is usually liquefied in macerated fetuses.

D. Radiologic Examination

Morphologic assessment of the aborted fetus routinely includes radiographic examination for the determination of bone developmental age and documentation of skeleton abnormalities.

E. Photographic Documentation

Photographs and close-ups of all detected malformations enable identification of the range of abnormalities within a particular syndrome. Photographic documentation is also essential for comparisons, for reevaluations of the case, and for allowing consultation with other syndromologists.

F. Histological Examination

Sections of each organ are taken for histologic examination. Histologic examination is necessary for evaluating deviations from normal organ development and for identifying microscopic changes such as aspiration of infected amniotic fluid into the fetal lungs.

G. Examination of the Placenta

The morphology of the placenta of conceptions 9–18 developmental weeks is very similar to that of the placenta at term delivery. Essential steps in placental examination are establishing whether the placenta is a singleton or a multiple placentation such as in twins, its weight and shape, the site of the umbilical cord insertion, the length of the umbilical cord, the number of cord vessels, evaluation of the fetal and maternal surfaces and the parenchyma, and looking for any abnormalities in the appearance of the placental membranes. The umbilical cord, membranes, fetal, maternal surfaces, and full thickness sections of the placental parenchyma are examined histologically.

H. Cytogenetic Studies

Cytogenetic studies may be subdivided into *conventional* which involves tissue culture and *molecular* which can be performed on interphase, nondividing nuclei.

1. Conventional

Cytogenetic examination of abortion specimens consists of sampling the gestational sac, placental, or fetal tissue and processing them for cytogenetic studies. The tissue selected for culture must be viable and nonmacerated. Processing the tissue for cytogenetic studies involves tissue culturing and then manipulating the cell cycle of division so that the chromosomes are arrested in metaphase. The cells are put into hypotonic solution to cause the cells to swell and the cell membranes to rupture, releasing the chromosomes. Thus, a chromosome spread is generated to allow karyotype analysis. Among early spontaneous abortions, the incidence of chromosomal defects is high. Chromosomal defects are either numerical or structural. As many as 80% of specimens show a numerical chromosomal defect, e.g., trisomy (one extra chromo-

some), sex chromosome monosomy (absence of an X or Y chromosome), triploidy (three haploid sets of chromosomes, instead of two), tetraploidy (four haploid sets of chromosomes), or chromosomal mosaicism (the presence of two or more cell lines differing in chromosome number or structure). Chromosomal trisomies (one extra chromosome) are found in one-half of chromosomally abnormal specimens. Among all spontaneous abortions, chromosomal defects are estimated to occur in 50% of specimens; however, among late abortions, chromosomal defects are much less common, seen in <10% of specimens.

2. Molecular or Fluorescent *in Situ* Hybridization

Fluorescent *in situ* hybridization (FISH) is a relatively new cytogenetic technique which involves hybridization of fluorescent-labeled probes to specific gene sequences. DNA probes are a specific nucleic acid sequence which recognize complementary sequences on a specific chromosome. FISH can be performed on metaphase chromosomes as well as on cells in interphase. In metaphase FISH, a probe is bound to the homologous segment on a metaphase chromosome in a fixed preparation on a glass slide and visualized as a signal under a fluorescent microscope. FISH also allows nondividing cells to be analyzed in interphase, thus avoiding the need to culture the cells and manipulating cell division to obtain a metaphase spread. The ability to analyze nondividing cells is a marked advantage as it eliminates the problems associated with and the time required for tissue culture. FISH can be performed on samples from any viable placental or fetal tissue.

Disorders accessible to FISH diagnosis range from simple numerical chromosomal changes (e.g., Down's syndrome, also known as trisomy 21 because of an extra chromosome 21) to tiny microdeletions (such as in cases with a family history of di George syndrome in which members of that family have congenital heart defects, immunological problems, absence of the parathyroid glands, and abnormal facial features). In detecting numerical chromosome abnormalities, there can be either a gain or a loss of chromosomes. If, for example, trisomy 21 is suspected, a fluorescent-labeled probe to a sequence on chromosome 21 is applied to the metaphase or interphase preparation. Using a fluorescent microscope, the presence of two signals is contrary to the diagnosis of trisomy 21; however, if three signals are detected, this indicates trisomy 21. The same principle can be used for any other chromosome. In fact, several probes for each chromosome in a different color can be simultaneously demonstrated on the preparation for chromosomes 22, 21, 18, 16, 13, X, and Y. This is useful for detecting trisomies (Color Plates 10, 11, and 12), monosomy X, triploidies, and tetraploidies.

Thus, FISH is the most direct method for visualizing a specific sequence on the metaphase chromosome or an interphase cell. It allows both enumeration and localization of a DNA sequence underlying a cytogenetic abnormality.

I. Biochemical Studies

Biochemical studies are done on selected specimens usually to confirm the prenatal diagnosis in a specimen of pregnancy termination. They involve DNA and enzyme studies.

J. Microbiology Studies

Microbiology studies are done selectively to confirm or rule out bacterial, viral, or parasitic infections.

IV. BIOLOGICAL AND CLINICAL SIGNIFICANCE OF EMBRYOFETOPATHOLOGY INVESTIGATIONS

A miscarriage is an upsetting event for a family because the pregnancy is usually well planned and frequently the mother is >30 years of age, which means she is close to the end of her reproductive period. Following abortion, the parents want to know:

1. if they did anything wrong during the pregnancy,
2. if there is any hereditary problem in the family predisposing them to defective offspring,
3. if it is going to happen again,
4. what did the results of the studies on the aborted tissue show, and
5. if there are prenatal diagnostic tests available to detect the "defect" in future pregnancies.

The embryofetopathologist, through identification of developmental defects and with good documentation of the development of both gestational sac–placenta and embryo–fetus, can address these questions on a sound basis to direct genetic counseling. A meaningful report to the family physician and geneticist is the result of interpreting the findings on examination of the conceptus and ancillary studies in conjunction with information from a comprehensive maternal reproductive and family history. [*See* Genetic Counseling.]

To illustrate the clinical benefit of detailed examination of products of conception, consider four examples:

1. In early spontaneous abortion specimens, the identification of both morphologically and cytogenetically normal embryos with no evidence of infection clearly points to a defect in implantation due to some uterine abnormality or insufficient maternal hormonal support of the pregnancy. This finding will help the obstetrician in the management of future pregnancies.

2. The diagnosis of a growth disorganized embryo with a numerical chromosomal abnormality indicates a lethal defect due to abnormal gametes, abnormal fertilization, or cleavage. As all of these usually represent an accidental error, the parents can be reassured and encouraged to initiate a new pregnancy.

3. A different situation arises when chromosomal rearrangement is detected and is found to be inherited from one of the parents. Cytogenetic prenatal diagnosis is then offered in all future pregnancies.

4. The most significant finding from a genetic counseling point of view is the localized embryonic or fetal defect such as abnormal closure of the spine or the head (Fig. 10), cleft lip or cleft palate (Fig. 11), or congenital heart defect. These are congenital anomalies with variable pathogenesis and which can be carried to term and result in a newborn with a developmental defect. The detection of such defects in a spontaneously aborted embryo–fetus and the establishment of its etiology are used to identify the family at risk and to allow follow-up of any future pregnancies.

The detailed routine examination of human-aborted embryos and fetuses has advanced our understanding of intrauterine development and intrauterine diseases. It has established that pregnancy loss is a common event and that in the majority of cases it represents nature's mechanism of removing a defective conception. Combined morphological and cytogenetic studies have revealed that >80% of early spontaneous abortions show numerical chromosomal defects arising during fertilization (polyploidy) or due to abnormal chromosomes in either the maternal or the paternal gamete. The studies of aborted embryos have confirmed that most of the developmental defects

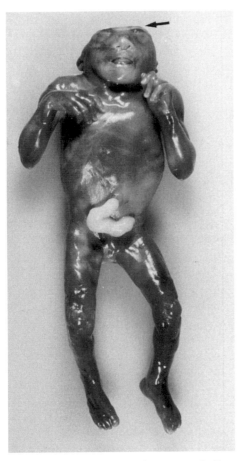

FIGURE 10 A 16-week fetus with missing skull and brain (arrow). This condition is known as anencephaly.

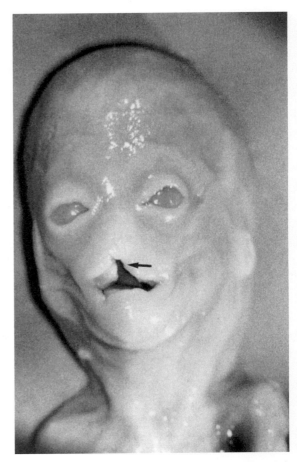

FIGURE 11 A macerated 11-week fetus with midline cleft of both the upper lip and palate (arrow).

are established during the time of organ formation and tissue differentiation.

Abortion studies have also indicated that the placenta plays an important role in maintenance of the pregnancy; when the placenta is normal, despite severe abnormality of the fetus, the pregnancy can survive until term and an abnormal infant may be born. This realization has led to the development of sophisticated prenatal techniques for the detection of a defective embryo–fetus. These include amniocentesis (a procedure by which amniotic fluid is removed at 12–16 weeks of gestation and examined cytogenetically, chemically, and for DNA studies), chorionic villus sampling (an alternative procedure to amniocentesis involving removal of a small piece of the placenta for DNA, biochemical, or cytogenetic studies at 10–11 weeks of gestation), ultrasound (a procedure in which detailed images of various structures in the embryo and fetus are obtained, e.g., the embryonic heart beat can be seen during the first trimester, growth parameters of the fetus can be assessed, the

presence of twins can be diagnosed, and the amount of amniotic fluid can be determined), fetoscopy (a fiber optic scope is inserted into the amniotic cavity for direct observation of specific fetal structures), and fetal blood sampling.

Following prenatal detection of a fetal defect and pregnancy termination, a detailed examination of the products of conception by a specialized pathologist is indicated. The pathologist is expected to confirm the presence of the defect identified by prenatal diagnosis, initiate ancilliary studies, and provide useful information to the obstetrician and genetic counselor for genetic counseling and management of future pregnancies.

BIBLIOGRAPHY

Kalousek, D. K., Fitch, N., and Paradice, B. A. (1990). "Pathology of the Human Embryo and Previable Fetus." Springer-Verlag, New York.

Mark, H. F. L. (1994). Fluorescent *in situ* hybridization as an adjunct to conventional cytogenetics. *In* "Annals of Clinical and Laboratory Science," Vol. 24, pp. 153–163. Philadelphia Institute of Clinical Science, Philadelphia.

Moore, K. L., and Persaud, T. V. N. (1993). "The Developing Human," 5th Ed. Saunders, Philadelphia.

Moore, K. L., and Persaud, T. V. N. (1994). "Color Atlas of Clinical Embryology." Saunders, Philadelphia.

Warburten, D., Byrne, J., and Cantii, N. (1991). "Chromosome and Prenatal Development: An Atlas." Oxford University Press, London.

Embryology and Placentation of Twins

FERNAND LEROY

Free University of Brussels and Saint Pierre Hospital

GLOSSARY

Blastocyst Preimplantation developmental stage consisting of a spherical and hollow structure formed by an outer layer of trophectoderm and an inner cell mass from which the embryo proper will be derived

Blastomeres Cells resulting from the first divisions of the fertilized egg

Gastrulation Developmental mechanism involving invagination of the upper embryonic layer (i.e., the ectoderm) and resulting in axial stratification of fundamental tissue components

Implantation Nestling of the blastocyst in the endometrium during the second week of development

Pituitary gonadotropin Peptide hormone controlling ovarian follicle growth and secretion, oocyte maturation, ovulation, and corpus luteum function

Oocyte Female germ cell undergoing maturation through two successive divisions, the first of which reduces the number of chromosomes by one-half (i.e., meiosis); both divisions are asymmetrical, producing the oocyte and a small "polar body"

Sex ratio Proportion of male infants among all births

Trophectoderm Outer cellular layer of the blastocyst, destined to differentiate into the trophoblast (i.e., future placenta and chorion)

Zona pellucida Acellular envelope of the preimplantation conceptus composed of glycoproteins

Zygosity Embryological origin of twins (i.e., one egg–one sperm or two eggs–two sperm)

BEING UNUSUAL, THE BIRTH OF TWINS HAS FASCInated mankind since the earliest historic times. From the viewpoint of biological evolution, twinning is an obvious exception to the general rule, by which only one fetus at a time is carried by the human female. This is also true in most monkeys and anthropoid apes and seems to have derived from natural selection, favoring an arboreal mode of life; that is, it would indeed have been more difficult for our simian ancestors to produce and raise a large litter up in a tree.

Some twins resemble each other very closely, and the confusion this can create has been a long-standing theme in fictional drama. Other twins, such as Jacob and Esau in the Bible, can be easily distinguished. The explanation is, of course, that some twins are derived from the division of a single fertilized ovum (i.e., a zygote) and others from the independent release and fertilization of two ova. Twins of the first type are genetically identical and hence of the same sex, while those of the second type are no more alike than ordinary brothers and sisters. The two types are therefore often called identical and fraternal, but in scientific usage they are termed monozygotic (MZ) and dizygotic (DZ).

Causal factors of DZ as well as MZ human twinning are still poorly understood. Both twin types, however, display a number of peculiarities, the study of which significantly broadens fundamental knowledge in genetics and in reproductive and developmental biology.

I. DIFFERENT TYPES OF TWINS

The simplest scientific argument in favor of the existence of two types of twins results from examining the combination of sexes within pairs. Since the general sex ratio is close to 0.5, it might be assumed that oocytes have nearly equal chances of being fertilized by an X or a Y sperm. Therefore, there is an equal probability for twins resulting from independent fertilization by two sperms to be sexually different or identical. Throughout the world an excess of like-sexed twins versus male–female pairs has been observed, and this difference is best explained by postulating that it corresponds to a subpopulation of MZ twins. Frequencies of both types of twins can hence be calculated by the well-known Weinberg formula

$$MZ = \frac{L - U}{N} \text{ and } DZ = \frac{2U}{N},$$

where MZ and DZ are mono- and dizygotic frequencies, respectively; L and U are like- and unlike-sexed pairs, respectively; and N is the total population.

There has been some debate about the validity of this method, namely on the basis that sexes among DZ pairs might not be determined independently. An excess of identically sexed pairs might, therefore, occur in this group. Conversely, the hypothesis of a preferential intrauterine loss of male twin fetuses, leading to the underestimation of like-sexed pairs, has also been raised against differential calculation. Be that as it may, it remains that versions of Weinberg's rule adapted to the prediction of zygosity distribution among triplets and quadruplets have given results that fitted remarkably well with frequencies observed. More recently, individual zygosity determination was carried out on about 2500 Belgian twin pairs, yielding figures in more than good agreement with those obtained by applying Weinberg's formula to the same material. Therefore, results of the numerous demographic and other studies using this method can be considered valid.

It has been hypothesized that a third type of twins can occur through fertilization by different sperm, of two cells arising from the same female germ cell (i.e., oocyte). Cases of dispermic mosaicism have been described in which the study of parental genotypes indicates that the two cellular types of which these individuals are composed differ not only in their paternal inheritance, but also partially in their maternal genes. In agreement with genetic exchange occurring between chromosomes at meiotic division (i.e., crossing over), it would appear that such anomalies are due to simultaneous fertilization of an egg and its second polar body by two different sperm. In view of the role of the zona pellucida, the occurrence of twins by this mechanism is difficult to visualize, but cannot altogether be excluded.

Obviously, although not identical, twins belonging to such pairs would be genetically closer than those originating from two unrelated oocytes. Also, if sufficiently frequent, they would form a sizable part of the group of physically (and even sexually) different pairs. Therefore, among all phenotypically dissimilar twins, the observed rates of identity within pairs for genetic markers would be higher than their theoretical percentages of concordance if these latter are calculated on the basis of an all-DZ hypothesis (i.e., if all physically different twins arise from different oocytes). Such studies have been carried out on a series of blood group types, showing that the observed and theoretical rates of genetic identity for these markers are always close. Therefore, it can be concluded that in humans the third type of twins, if existing at all, is rare.

As shown by the examination of genetic markers such as blood groups and human leukocyte antigen (HLA), some twins can undoubtedly arise from different fathers who have had intercourse with the same woman within a short period of time. It is interesting that in many archaic societies it was believed that all twins were conceived by this mechanism, called superfecundation. Superfetation, which implies differential fertilization of two oocytes released at a several-week interval, has not yet been conclusively demonstrated in humans.

Several pairs have been described in which, despite an MZ origin proven through genetic analysis, twins were phenotypically different, sometimes even to the point of being of opposite sexes. Such pairs have been termed heterokaryotes. They differ in their chromosomes, one chromosome being inherited in excess by one cell while lacking in the other, at an early stage of development. Most heterokaryotes are composed of a normal male or female, while the other partner shows the physical traits and the genotype of Turner's (absence of one X or Y chromosome) or Down's syndrome (three copies of chromosome 21). In some of these cases, both twins are, in fact, double mosaics, in each of which a different cell type predominates. As in other instances of monosomy or trisomy (i.e., lack or excess, respectively, of a chromosome), these anomalies are explained by an event of chromatid nondisjunction at the time of cell division. In the case of heterokaryotic twins, it is believed that abnormal chromosome segregation would have occurred at early embryonic cleavage rather than during gametogenesis, but in any case before splitting of the conceptus into the two twins. [See Chromosome Anomalies; Down's Syndrome, Molecular Genetics.]

Individual zygosity determination has become of paramount importance in modern twin studies. Besides sex combination, morphology of the placenta and the fetal membranes might give a clue (see Section III). Many genetic markers (e.g., blood groups, placental isoenzymes, and HLA antigens) have been used successfully for defining twin zygosity. However, precise methods using recombinant DNA technology are bound to become master tools for this purpose.

II. CAUSALITY OF DOUBLE OVULATION

Twinning rates show wide geographic variations. The highest values are found among black African populations, and the lowest are found in the Far East. Western European and North American countries, together with the Indian peninsula, exhibit intermediate frequencies. After breaking down into DZ and MZ figures, it appears that the MZ rate remains fairly constant everywhere (3.5–5.5 per 1000 births), whereas fluctuations are chiefly attributable to a strongly varying DZ frequency (from 2.5 in Japan to 45 in southwestern Nigeria, per 1000 births).

Factors that strongly influence the DZ rate are maternal age and pregnancy rank. Many studies have indicated that the chances of bearing fraternal twins rise steeply until about 40 years of age, after which ovaries seem to rapidly lose their capacity for double ovulation. It is clear, moreover, that at any given age the higher the number of previous pregnancies, the higher the DZ rate. In contrast, these factors have almost no influence on MZ frequency.

Nutrition also bears on DZ twinning. During World War II, twinning rates decreased markedly in several countries submitted to food shortage. It has also been shown that tall heavy women are more prone to conceive DZ twins.

Other socioeconomic factors play an obvious, though ill-understood, role in the frequency of DZ twinning. Scandinavian unmarried mothers exhibit higher DZ twinning rates, even allowing for maternal age, and there is no reason to believe that they enjoy better nutrition than other women. DZ rates are also significantly elevated among pregnancies occurring within the first few months of wedlock or after protracted separation of the partners. On the whole there are several arguments in favor of the view that proneness to DZ twinning reflects high overall fecundability and fertility.

Ripening and ovulation of ovarian follicles are un-

der the direct control of gonadotropins (i.e., sexual pituitary hormones), and it has been found that in mothers of twins the pituitary gland secretes higher amounts of these hormones, thereby favoring double ovulation. But why do these women show this disposition? The mere fact that there is a racial difference in DZ twinning rates suggests that genetic factors are involved in addition to the previously mentioned environmental influences. It is also popular knowledge that twinning tends to run in certain families. Also, mothers who have already delivered twins are prone to repeat this feat. From a host of investigations on the repeat frequency and familial incidence of twinning, it appears that proneness to double ovulation is genetically inherited. On the basis of these data, it has been proposed that the DZ twinning tendency is determined by a single pair of genes. Mothers carrying two recessive variants of this gene would be responsible for all spontaneous DZ pregnancies. These homozygous women represent 25% of the female population in Western countries. [See Follicle Growth and Luteinization; Pituitary.]

The use of gonadotropic hormones and clomiphene citrate (an ovulatory drug acting at hypothalamic level) in the treatment of anovulation is responsible for a number of DZ twins and higher-rank multiple pregnancies. Also, treatment by *in vitro* fertilization followed by multiple embryo placement in the uterus is responsible for a 20–25% rate of twins and multiple pregnancies. Therefore, infertility treatment has induced an increase of overall DZ twinning rates as recently observed in a series of developed countries. It has been suggested that *in vitro* fertilization as well as the use of ovulatory drugs might increase MZ frequency also, but factual arguments supporting these views remain scanty.

III. EMBRYOLOGY OF SINGLE-OVUM TWINS

Of the two membranes surrounding the fetus, the chorion develops toward the end of the first week of pregnancy, while the amnion, which lies inside the chorion and immediately surrounds the fetus, is not differentiated until the second week. Theoretically, there are, therefore, three stages at which division of the conceptus might give rise to MZ twins (Fig. 1). If it occurs before the end of the first week, the twins develop separated amnia and choria, as is the case in all DZ pairs. If only the embryonic (i.e., inner cell) mass divides in two before formation of the amnion,

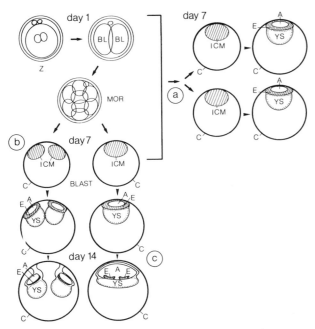

FIGURE 1 Early embryology of MZ twins. Modes of MZ twinning: (a) dichorial–diamniotic; (b) monochorial–diamniotic; (c) monochorial–monoamniotic. Z, zygote (i.e., a fertilized egg); BL, blastomeres; MOR, morula; ICM, inner cell mass; C, chorion; BLAST, blastocyst; A, amnion; E, embryo; YS, yolk sac.

centas belong to MZ pairs, that all DZ twins have dichorial placentas, but that not all dichorial twins are DZ.

A. Dichorial MZ Twins

In most textbooks and monographs on human placentation, it appears as an inescapable dogma that dichorial MZ twins must arise through separation of the two cells deriving from the first division of the fertilized egg (i.e., the blastomeres), because *in vitro* they can each give rise to a complete embryo. Close scrutiny indicates that such a mechanism is unlikely to be involved in the natural production of MZ twins, in view of the presence of the zona pellucida, a thick membrane that surrounds the embryo. No one has described two distinctly separated morulae or blastocysts within the same zona, as should be present if the first two blastomeres produced two separate embryos. Their independent development is unlikely because, when kept or put together, blastomeres always stick to one another and organize between themselves the formation of a single embryo. This is true even when blastomeres of two different species (e.g., a rat and a mouse) are experimentally combined *in vitro* to produce a single chimeric blastocyst.

the result will be a common chorion, but separate amnia. Finally, if the division takes place after differentiation of the amnion, twins not only share a common chorion, but are also contained in a single amniotic sac.

Three types of human MZ twins have indeed been described according to the morphology of fetal membranes and placenta, which might be dichorial–diamniotic (32%), monochorial–diamniotic (66%), or monochorial–monoamniotic (1–2%) (Fig. 2).

In practice, the chief difficulty lies in distinguishing between a secondarily fused dichorial placenta and a monochorial–diamniotic placenta. The components are, however, usually apparent from a naked-eye examination of membranes partitioning the gestational sacs. In a monochorial placenta the two layers of amnion appear translucent and peel away from each other, leaving nothing in between. In a dichorial placenta the septum is more opaque, and stripping the amnia leaves either a single fused layer or two separate layers of chorion attached to the fetal surface of the placenta. Histology of the dividing membranes is more time-consuming, but gives an almost infallible result.

It should thus be stressed that all monochorial pla-

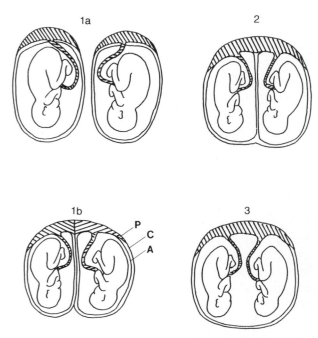

FIGURE 2 Placentation of twins: (1) diamniotic–dichorial, separated (a) or secondarily fused (b) (MZ or DZ); (2) diamniotic–monochorial (MZ); (3) monoamniotic (MZ, rare). P, placenta; C, chorion; A, amnion.

It is then possible that while traveling along the fallopian tube some eggs might escape from their zona, undergo separation of the early blastomeres, and give rise to MZ twins? Probably not, for it has been shown that naked mouse eggs at stages from one to eight cells, as well as single blastomeres isolated from such eggs, do not survive when transferred to the oviduct. Similar observations were reported in the rabbit, in which no blastocysts could be obtained from *in vivo* transfer of naked blastomeres originating from two- and four-cell eggs. Thus, culture *in vitro* appears to be the only possible way of achieving development of naked single blastomeres into blastocysts.

In addition to the possibility of a hostile effect of the tubal environment on unprotected eggs, the latter might also have some difficulty in achieving compaction (i.e., establishment of cohesive junctions between blastomeres when the embryo is composed of eight to 16 cells). It is known that eggs obtained from separated blastomeres compact easily when maintained under static *in vitro* conditions. However, it is possible that, *in vivo*, tubal wall motility prevents the establishment of tridimensional relationships, which are normally achieved through containment within the zona pellucida and are considered necessary for compaction.

A more likely explanation for the occurrence of dichorial MZ twins has been recently derived from the observation of *in vitro*-cultured cow eggs. Sometimes, when the embryo hatches out of the zona pellucida, instead of coming out through a large slit, the cow blastocyst progressively herniates through a small opening (Fig. 3). If both the inner cell mass and the trophectoderm participate in this process, an hourglass-shaped egg is formed, each part containing both cellular components. It is probable that the narrow connecting bridge will easily rupture, giving rise to separate twin blastocysts which can develop independently. In agreement with this interpretation, several births of dichorionic twin calves following transfer of a single day 7 blastocyst have been reported.

Experimental twinning has likewise been obtained in mice by simply nicking the zona pellucida at the blastocyst stage. Half hatching of human blastocysts through a small opening of the zona has also been observed *in vitro*.

B. Monochorial Twins

The logical explanation for the formation of monochorial twins is that they arise at the blastocyst stage through division or duplication of the inner cell mass

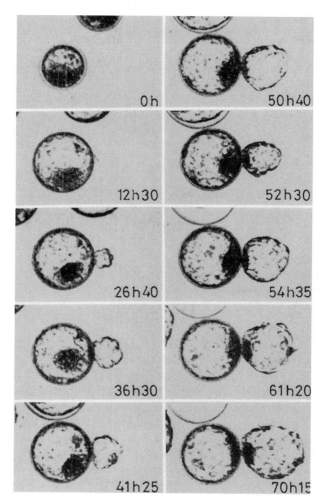

FIGURE 3 Time-lapse cinematography of a cow blastocyst undergoing atypical hatching, resulting in the separation of two complete twin blastocysts. [From A. Massip, P. Van der Zwalmen, J. Mulnard, and W. Zwijsen. *Vet. Rec.* **112**, 301 (1983), by permission of the authors.]

within a single trophectoderm. Such blastocysts have occasionally been found in species such as sheep or pigs, and the same type of duplication has also been obtained in mice by grafting a foreign inner cell mass into a host blastocyst.

In humans the earliest available specimen of monochorial twins belongs to the Carnegie Collection (No. 9009). It consists of twin didermic embryos (i.e., composed of two layers: ectoderm and endoderm), each of which has its own yolk sac and amnion and is appended by a separate connecting stalk to the inner wall of a single chorionic vesicle (Fig. 4). Although both are about 17 days old, one is noticeably smaller than the other. In somewhat older specimens a common yolk sac has sometimes been found.

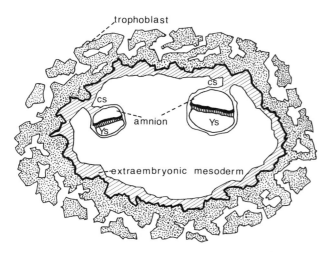

FIGURE 4 The earliest known monochorial twin embryos. cs, connecting stalk; Ys, yolk sac. [After G. W. Corner, *Am. J. Obst. Gynecol.* **70,** 933 (1955).]

An interesting problem related to monochondrial twins concerns egg orientation at the time of implantation. It is admitted that in normal single pregnancies the attachment of the umbilical cord in the center of the placenta reflects the primary site of implantation. Since the human conceptus penetrates the endometrium by the region including the inner cell mass (i.e., the future embryo), further development of the connecting stalk through which the umbilical vessels will run is centered in the middle of the area where the placenta will differentiate.

Therefore, it is believed that marginal insertion of the cord occurs because of abnormal orientation of the inner cell mass at the time of implantation. In the case of monochorial twins, at least one of the two inner cell masses located at different poles of the same blastocyst will, of necessity, become orientated away from the endometrial surface. It follows that a higher frequency of marginal insertions of the cord is to be expected in the group of monochorial twins. Data compiled from the literature confirm that monochorial twin placentas show abnormal cord insertion much more frequently than do those of dichorial pairs. [*See* Implantation, Embryology.]

Placentation of most monochorial twins entails a remarkable anatomical feature: formation of vascular connections (i.e., anastomoses) between the two fetal circulations. The most frequent types are the isolated arterioarterial communication and the combination of arterioarterial with arteriovenous anastomoses. The latter are almost always located at the capillary level in a so-called common villous district. Such areas

are centered on an umbilical artery of one fetus and a vein arising from its partner, which ramify and communicate into the depth of the placenta (Fig. 5). In contrast, arterioarterial and venovenous channels involve larger-caliber vessels running on the fetal surface of the placenta. Although the exchange of blood between the two fetuses raises a series of yet unsolved problems, it might be given a simplified interpretation based on correlations between anastomotic types and clinical observations. Broadly, several situations can occur. These are discussed below.

I. Hemodynamic Equilibrium

In placentas in which only arterioarterial and/or venovenous communications are present, blood can only circulate in these channels if pressures on each side are different. If such a difference occurs, blood pressure is readily equilibrated in both fetuses since these vessels are of large caliber. Any hemodynamic disturbance occurring in one twin is immediately transmitted to his partner and, other things equal, their intrauterine growths will be parallel.

In most cases, however, large-caliber anastomoses are associated with arteriovenous communications. It follows that the differences in pressure between artery and vein are continuously compensated for by a reverse blood flux occurring in arterioarterial or venovenous channels (Fig. 6). This situation is also hemodynamically balanced, but a third circulation permanently takes place. It is likely that this blood shunt is often responsible for fetal growth impairment since common villous areas might represent as much as one-fifth of the placental volume. This situation, however, is not the most detrimental.

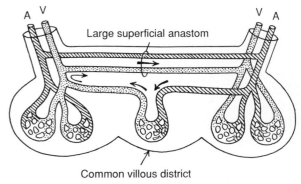

FIGURE 5 Vascular anastomoses in monochorial twin placenta. A, umbilical artery; V, umbilical vein. Dotted and hatched areas represent venous and arterial vessels, respectively.

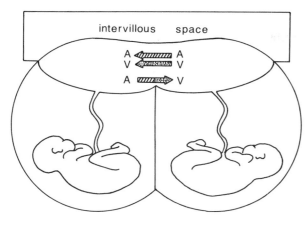

compensated A-V

anastomoses

FIGURE 6 Third circulation of monochorial twins. Hemodynamic equilibrium. A, arterial; V, venous. Dotted and hatched areas represent venous and arterial vessels, respectively.

2. Strong Hemodynamic Imbalance

In contrast, dramatic consequences arise from arteriovenous channels which are not compensated for (Fig. 7). Under such conditions one twin (i.e., the perfusor) will pump blood into the other (i.e., the perfused) until the perfusor's arterial pressure becomes equal to the perfused's venous pressure. Therefore, the transfused twin perishes from cardiac failure due to excessive plasmatic load, whereas the transfusor dies

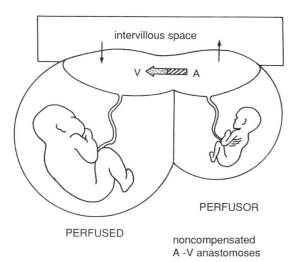

FIGURE 7 Third circulation of monochorial twins. Hemodynamic asymmetry. V, venous; A, arterial. Dotted and hatched areas represent venous and arterial vessels, respectively.

from hypoxia caused by blood depletion. For the same reasons excessive accumulation of amniotic fluid (hydramnios) develops on one side, while little of it forms on the other (oligoamnios). This is why, sometimes, the obstetrician comes across a case of very premature stillborn twins, one of which is hydropic, while the other appears underdeveloped.

3. Moderate Hemodynamic Imbalance

In most cases of the intertwin transfusion syndrome, the hemodynamic asymmetry is discrete, due to imperfect compensation of arteriovenous communications. The transfusion from one fetus to the other is then very slow, and pregnancy is able to proceed closer to term and the mother can give birth to living babies. The perfusing twin will, nevertheless, be pale and small. Besides marked anemia, plasmatic proteins are depleted and the blood contains many immature red blood cells as a sign of bone marrow reaction to anemia. For reasons still unknown, the corresponding placental area is swollen because of villous edema. On the other hand, the transfused fetus will often be heavier than his or her partner. He or she will also show a deep purple complexion because of an excess of red blood cells, which entails high levels of blood bilirubin derived from their destruction. This increased destruction of red blood cells in turn causes liver and spleen enlargement, which is often found in such infants.

4. Fetus Papyraceus

Sometimes after undergoing intrauterine death, a twin becomes completely dried out and flattened by his normal partner's growth. This so-called fetus papyraceus is often thought to have arisen from intertwin transfusion. This is unlikely, however, because in one-half of such cases the placenta is dichorial, and it is known that in such placentas vascular anastomoses are rare. Therefore, it is believed that a fetus papyraceus can result from a diversity of causes entailing intrauterine death around midpregnancy, among which intertwin transfusion is only a special case.

5. Acardiac Fetus

Another feature relating to twins' third circulation is the rare anomaly known as acardiac fetus. Such fetuses have no heart and often lack the upper part of the body, as if cut at waist level. The growth of this hugely malformed fetus is possible only because of the existence of two large vascular anastomoses through which the malformed twin can be perfused by the heart of his normal partner. One of the linking channels is arteriorarterial whereas the other is venovenous. It is clear that circulation in the acardiac fetus is running opposite the normal direction. He receives poorly oxygenated blood through an umbilical artery from his cotwin, blood returning to the normal partner through the venous anastomosis. This malformed fetus is thus permanently lacking oxygen.

According to one theory, these abnormal hemodynamic conditions would be responsible for the absence of heart and cephalic development. However, it is difficult to visualize how vascular anastomoses could have been present and played a teratogenic role at early stages of development. Therefore, the view that this anomaly results from another yet unknown cause appears more plausible. Accordingly, the association with a normal twin would be fortuitous, but nevertheless mandatory, to allow the acardiac fetus to survive and grow.

So far, no vascular communications between DZ twins have been directly demonstrated. There is evidence, however, that in rare instances they must have existed somehow. The presence of a mixture of blood cells with characteristics of both twins (i.e., chimerism) has been described in a series of otherwise normal twins, often of opposite sexes. In most of these cases, each member of the pair was endowed with two red blood cell types carrying different group antigens. Since the mixture was the same in each partner, its existence is best explained by the exchange of erythropoietic cells during fetal life. Immunological tolerance between such twins has been demonstrated by showing that skin grafts from one to the other were not rejected even when the sexes were different.

C. Monoamniotic Twins

Twins contained in a single pouch are rare (1–2% of MZ pairs). This is fortunate since not only do monoamniotic twins show a high rate of malformations, but often they perish through intrauterine entanglement of their umbilical cords.

Monoamniotic twinning is supposed to occur relatively late in development since it results from duplication of the embryonic rudiment of the germ disk (i.e., the ectodermal plate). Normally, the signal for the formation of the amnion, which appears during the second week of development, is related to the presence of the embryo.

Therefore, to give rise to monoamniotic twins, splitting of the embryo has to occur after the seventh day following fertilization. At that time, however, the cells

of the germ disk might not yet have differentiated into a single axial arrangement, which starts at about day 13 or 14, together with the appearance of the primitive streak. It is reasonable to assume that splitting occurring before axial differentiation is achieved will be incomplete, giving rise to conjoined twins. In technical terms of general embryology, the formation of monoamniotic twins corresponds to a process of double gastrulation (Fig. 8).

In some armadillos, nature offers a unique example of monoamniotic multiple embryogenesis. In these animals the embryonic disk splits regularly into several parts. Each of these, while developing into an embryo, moves away from the others and draws its part of the amniotic chamber. The amnion finally becomes a branched structure, and each embryo lies in an individual secondary sac connected to the original site of the amnion through a narrow duct. Two

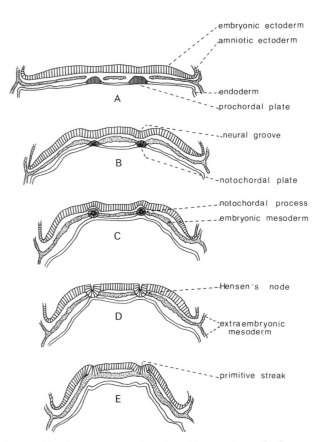

FIGURE 8 Putative mechanism of double gastrulation, leading to the formation of monoamniotic twins. (A–E) Successive transverse sections in the craniocaudal direction. (After Heuser's day 18 presomitic embryo. See W. J. Hamilton, J. D. Boyd, and H. W. Mossman. "Human Embryology," p. 55. Heffer, Cambridge, England, 1962.)

species of armadillo, *Dasypus novemcinctus* and *Dasypus hybrida*, are known to produce four and six to nine fetuses, respectively, through this mechanism of polyembryony. It was suggested that because of delayed implantation in armadillos, this condition might be involved in the production of MZ multiples. However, many species undergo delayed implantation without showing any polyembryony.

Conjoined twins occur no more than about once in 500 MZ twin pregnancies (i.e., once or twice in 10^5 births). For still unknown reasons, about two-thirds of these cases are girls. Seventy percent of so-called Siamese twins are joined by some part of their anterior thorax, but almost any imaginable type and degree of fusion along the longitudinal body axis have been described. This confirms that such anomalies and also monoamniotic twins must be generated by either partial or total parallel duplication of the axial blueprint defined at gastrulation.

D. Congenital Anomalies

In addition to the acardiac fetus and conjoined twins, which are clearly specific of MZ twinning, other malformations encountered in the general population are more frequent among twins. Since this increase is only found in the group of identically sexed pairs, it has been concluded that it pertains to MZ twins.

Galton's original hypothesis, that comparison between MZ and DZ twins might give a measure of environmental versus hereditary influences, has been applied to the analysis of birth defects. Such studies, however, are fraught with difficulties. A first obstacle relates to differential intrauterine conditions, to which most MZ twins are submitted. It is indeed difficult to evaluate which role such features as vascular anastomoses or abnormal implantation and cord insertion might have played in the occurrence of malformations and to differentiate their possible teratogenic effect on each twin. A low concordance rate for a given anomaly among MZ pairs does not, therefore, prove that genetic factors are unimportant determinants of the defect under study.

Another important limitation results from the rarity of both twinning and congenital malformations, which necessitates the analysis of large series of birth records. It has been estimated, for instance, that 40,000 births have to be monitored to detect just one twin with a cleft lip and/or palate. The lack of accurate individual zygosity determination has been another pitfall hampering the study of birth defects among twins.

It seems clear, nevertheless, that a series of malformative defects are more frequent among MZ twins than in corresponding populations of DZ pairs or singletons. These include neural defects (e.g., anencephaly and hydrocephalus), heart disease (namely, patent ductus arteriosus), cleft lip and/or palate, gut atresia, kidney malformation, syndactyly, and polydactyly. Except polydactyly, which shows a 100% concordance rate, all of these malformations appear to be discordant in more than one-half of the like-sexed pairs, with at least one twin carrying the defect. [*See* Birth Defects.]

It has been suggested that anomalies occurring more frequently in MZ twins resemble those caused by mutations of homeotic genes in animals (i.e., genes controlling spatial and temporal patterns of embryonic development). Proteins specified by these genes regulate the genetic control of early development. They are synthesized in embryonic cells, which might become unequally distributed when fission of the conceptus takes place, causing a congenital defect in one twin, but not in the other. This hypothesis, however, remains to be demonstrated.

The excess of congenital malformations linked to monozygosity suggests that there might be a common factor in the causation of both. The fact that MZ twinning is itself a deviation from normal embryonic development reinforces this view. Unfortunately, the precise origin of most birth defects remains ill explained, and their study, therefore, has not provided a clue as to the causes of embryonic splitting.

IV. POSSIBLE ORIGINS OF MONOZYGOTIC TWINS

The etiological factors giving rise to spontaneous MZ twinning in mammals remain unknown. Contrary to the case of DZ twinning, no clear genetic influence seems to be involved in the production of identical human twins.

In some invertebrates, as well as fish and birds, splitting of the embryo can be induced through cooling or through oxygen deprivation. No such results are available in mammals. It has been claimed that in humans the ratio of monochorial to dichorial placentation is particularly high among extrauterine twin pregnancies. Therefore, it was suggested that defective nutritional conditions might play a role in the formation of human MZ twins. However, this interpretation appears rather doubtful because the mechanism of monochorial twinning is probably set in motion well before blastocyst attachment.

Another etiological hypothesis relates to preovulatory overripeness or retarded fertilization of the oocyte. Partial or complete axial duplication of embryos has been obtained through delayed insemination in some amphibians and even in mammals. As regarding humans, it has been reported that mothers of MZ twins exhibit significantly more prolonged and irregular menstrual cycles than do those of DZ twins. Such abnormal ovulatory conditions might correspond to oocyte overmaturation.

Some authors have proposed that MZ twins are caused by a twinning impetus acting at random during early development. This would explain why the frequency distribution of the different types of placentation among MZ twins remains constant and in proportion to the duration of early developmental steps at which they are supposed to occur. However, as indicated, there is a fundamental difference in the embryological mechanism involved, namely, between the formation of mono- and dichorial MZ types. Therefore, a unitarian view on the causality of MZ twinning might not be realistic.

V. INTRAUTERINE GROWTH OF TWINS

Ultrasound screening in early pregnancy has evidenced the phenomenon known as the vanishing twin. Not infrequently, one of the gestational sacs of a twin or multiple gestation progressively shrinks and disappears. Only rarely does examination of fetal adnexa of the remaining twin show some morphological trace of this event. Since authors vary extensively in their evaluation of the vanishing twin frequency, it is difficult to estimate to what extent it might affect MZ and DZ indices among the general population.

Low birth weight is a major problem befalling twins and other multiples. An important cause of this defect is premature birth. The delivery of twins occurs, on average, at the 37th week of pregnancy, its duration being reduced to 35 and 34 weeks for triplets and quadruplets, respectively. Although the mechanism of labor onset in humans remains unclear, it might be assumed that uterine overstretching plays a role in the premature termination of multiple pregnancy. It is likewise possible that some humoral factor produced in greater amounts because of the existence of two fetoplacental units might reach a threshold level much sooner than in single pregnancies and induce early stimulation of uterine muscle activity.

However, the mean birth weight of multiple-pregnancy babies remains lower than that of singletons at equivalent gestational age since their intrauterine growth is significantly slowed during the last trimester of pregnancy. Twin babies are, therefore, not only premature, but also small for date. This has been confirmed by showing that their average biparietal diameter, as measured by ultrasound examination, lies somewhat below that of singletons at a similar intrauterine age.

The first explanation that comes to mind is that overcrowding of the uterus in multiple pregnancy prevents normal placental development. In twins, at all ages of gestation, placental weight and volume are lower than the double value of figures recorded in single pregnancies. However, at equivalent relative placental weight, twins are still lighter than singletons, even when allowing for gestational age. In other words, the percentage of placental weight reduction in twins (i.e., 12%) is less than their relative body weight defect (20%). Therefore, twins' growth retardation cannot be totally accounted for by impaired placental development. Evidence points to a role of insufficient blood supply to the uterus. It has been found that uteroplacental circulation in multiple pregnancies is much slower than normal and that twins have increased hemoglobin and red blood cell levels at birth. These data might be interpreted as a consequence of chronic intrauterine oxygen deprivation.

Curves of evolution using ultrasound assessment of biparietal diameters throughout pregnancy indicate that MZ twins on average are more growth retarded than are DZ twins. This difference is due to the existence of vascular anastomoses in monochorial placentas, which are present in about two-thirds of MZ pairs. This feature also explains why largest body weight differences within a pair are found among MZ twins. Hemodynamic disturbances related to vascular anastomoses in monochorial twins can be estimated to some extent by pulsed-wave Doppler application which allows one to study fetal blood flow velocity wave forms.

BIBLIOGRAPHY

Bryan, E. M. (1983). "The Nature and Nurture of Twins." Baillière Tindall, London.
Bulmer, M. G. (1970). "The Biology of Twinning in Man." Oxford Univ. Press (Clarendon), Oxford, England.

Doyle, P. E., Beral, V., Botting, B., and Wale, C. J. (1990). Congenital malformations in twins in England and Wales. *J. Epidem.* **45**, 43.

Divon, M. Y., and Hsu, H. W. (1992). Maternal and fetal blood flow velocity in intrauterine growth retardation. *Clin. Obst. Gynecol.* **35**, 156.

Hill, A. V. S. (1985). Use of minisatellite DNA probes for determination of twin zygosity at birth. *Lancet* **2**, 1394.

Källen, B. (1986). Congenital malformations in twins: A population study. *Acta Genet. Med. Gemellol.* **35**, 167.

Keith, L. G., Papiernik, E., Keith, D. M., and Luke, B. (1995). "Multiple pregnancy." Parthenon Pub. Grp., New York and London.

Leroy, F. (1985). Early embryology and placentation of human twins. *In* "Implantation of the Human Embryo" (R. G. Edwards, J. M. Purdy, and P. C. Steptoe, eds.), pp. 393–405. Academic Press, Orlando, FL.

MacGillivray, I., Nylander, P. P. S., and Corney, G. (1975). "Human Multiple Reproduction." Saunders, London.

MacGillivray, I., Campbell, D. I., and Thompson, B. (1988). "Twinning and Twins." Wiley, Chichester, England.

Vlietinck, R., Derom, C., Derom, R., Van den Berghe, H., and Thiery, M. (1988). The validity of Weinberg's rule in the East Flanders prospective twin survey. *Acta Genet. Med. Gemellol.* **37**, 137.

Emerging and Reemerging Infectious Diseases

FREDERICK A. MURPHY
University of California, Davis

I. The Emergence of New Diseases is Inevitable Yet Unpredictable
II. Variation and Evolution of Microbes and Viruses
III. Epidemiologic Considerations in the Emergence of Infectious Diseases
IV. Nature of Ecological and Zoonotic Factors in Regard to New, Emerging, and Reemerging Infectious Diseases
V. Nature of Behavioral and Societal Factors in Regard to New, Emerging, and Reemerging Diseases
VI. Prevention and Control of New, Emerging, and Reemerging Infectious Diseases

GLOSSARY

Mutation Heritable change in the nucleotide sequence of the genome of an organism

Epidemic Disease occurring in an unusually high number of individuals in a population at the same time

Acute infection In cells or the whole organism, infection with rapid onset and rapid resolution

Persistent infection In cells or the whole organism, infection that persists for a prolonged period after the primary infection

Risk assessment Scientific, data-based evaluation of data pertaining to factors contributing to the introduction and spread of disease

Disease surveillance Collection, collation, and analysis of data on disease incidence and disease dissemination

THE EMERGENCE OF A NEW INFECTIOUS DISEASE can be attributed to many elements, including microbial/viral determinants (such as mutation, recombination, natural selection, evolution), natural influences (such as ecologic and environmental influences), and factors pertaining to human activity (such as behavioral, societal, transport, commercial, and medical care factors). In general, there is no way to predict when or where the next important new microbial or viral pathogen will emerge; neither is there any way to reliably predict the ultimate importance of a microbe or virus as it first emerges. Given this reality, initial investigation at the first sign of the emergence of a new infectious disease must focus on characteristics such as mortality, severity of disease, transmissibility, and remote spread, all of which are important predictors of epidemic potential and societal threat. Clinical observations, pathologic examinations, and preliminary identification and characterization of the agent often provide early clues, as new, emerging microbes and viruses often resemble their closest genetic relatives in regard to their epidemiologic and pathogenetic characteristics. The elements of disease control and prevention programs include surveillance, diagnostics, and communications, linked to a professional response network. This network must be flexible; capable, for example, in one instance of emphasizing local professional infrastructure and in the next of emphasizing global epidemic aid. Given the nature and magnitude of the threat represented by new and emerging infectious diseases, the importance of such a global network seems clear.

I. THE EMERGENCE OF NEW DISEASES IS INEVITABLE YET UNPREDICTABLE

New or previously unrecognized microbes and viruses are constantly being identified. In most cases, there is

ENCYCLOPEDIA OF HUMAN BIOLOGY, Second Edition, VOLUME 3. Copyright © 1997 by Academic Press. All rights of reproduction in any form reserved.

no way to predict when or where the next important new infectious agent will emerge. Neither is there any way to predict its ultimate importance as it first emerges—it might emerge as the cause of a geographically limited curiosity, as the cause of intermittent disease outbreaks, or it might first be seen as the cause of a new epidemic. No one would have predicted the emergence of a new rodent-borne Hantavirus as the cause of a severe acute respiratory distress syndrome in the southwestern region of the United States in 1993 and certainly no one would have predicted the epidemic emergence of human immunodeficiency virus (HIV) as the cause of AIDS before the initial description of the disease in the early 1980s. [*See* Virology, Medical.]

New infectious diseases appear to be emerging with increasing frequency, as suggested by published reports of cases, outbreaks, and epidemics, and by the rate of identification of new pathogenic microbes and viruses. The list of newly emergent microbes and viruses of humans and animals is impressive, indeed, and is seemingly prophetic of more to come in the future (Tables I–III). There are several reasons why the emergence of new infectious diseases seems to be accelerating. The global human population has continued to grow inexorably, bringing increasingly

TABLE I
Some of the Most Important New, Emerging, and Reemerging Human Virus Pathogens

- *Crimean–Congo hemorrhagic fever virus*[a] (tick-borne; severe human disease with 10% mortality; widespread across Africa, Middle East, and Asia)
- *Dengue viruses*[a] (mosquito-borne; the cause of millions of cases of febrile disease in the tropics, dengue hemorrhagic fever, a life-threatening disease, especially in children)
- *Ebola,*[a] *Marburg,*[a] *and Reston viruses* (natural reservoirs unknown; Ebola and Marburg viruses are the causes of the most lethal hemorrhagic fevers known)
- *Group B and C rotaviruses* (rotavirus enteric disease is the second leading cause of death among infants in the world)
- *Guanarito virus*[a] (rodent-borne; the newly discovered cause of Venezuelan hemorrhagic fever)
- *Hantaviruses*[a] (rodent-borne; the cause of important rodent-borne hemorrhagic fever in Asia and Europe; now a new member virus is emerging as the cause of the severe, often fatal acute respiratory distress syndrome in the southwestern region of the United States)
- *Hepatitis C virus* (newly identified; the cause of much severe, chronic liver disease in the United States)
- *Hepatitis delta virus* (an unusual "helper" virus that makes hepatitis B more lethal)
- *Hepatitis E virus* (newly identified; the cause of epidemic hepatitis, especially in Asia; recently recognized as widespread along the U.S./Mexico border; the infection has a high mortality rate in pregnant women)
- *Human herpesviruses 6 and 7* (newly identified; the cause of a substantial proportion of febrile disease in children)
- *Human immunodeficiency viruses, HIV1 and HIV2* (the causes of AIDS, still emerging in many parts of the world)
- *Human papillomaviruses* (over 70 viruses; some associated with cervical, esophageal, and rectal cancers)
- *Human parvovirus B19* (the cause of roseola in children; a possible cause of fetal damage when pregnant women become infected)
- *Human T-lymphotropic viruses (HTLV1 and HTLV2)* (the cause of an adult leukemia and neurologic disease, especially in the tropics)
- *Influenza viruses* (the cause of thousands of deaths every winter in the elderly; the cause of the single most deadly epidemic ever recorded—the worldwide epidemic of 1918, in which over 20 million people died)
- *Japanese encephalitis virus* (mosquito-borne; very severe, lethal encephalitis; now spreading across Southeast Asia; great epidemic potential)
- *Junin virus*[a] (rodent-borne; the cause of Argentine hemorrhagic fever)
- *Lassa virus*[a] (rodent-borne; a very important, severe disease in West Africa; imported into a Chicago hospital in 1990)
- *Machupo virus*[a] (rodent-borne; the cause of Bolivian hemorrhagic fever)
- *Measles virus* (reemerging in urban centers and in college-aged adults in the United States)
- *Norwalk and related viruses* (major causes of outbreaks of severe diarrhea)
- *Polioviruses* (the cause of polio; still an important problem in developing countries of Africa and Asia; targeted by WHO for worldwide eradication by the year 2000)
- *Rabies virus* (transmitted by the bite of rabid animals; raccoon epidemic spreading across the northeastern United States)
- *Rift Valley fever virus*[a] (mosquito-borne; the cause of one of the most explosive epidemics ever seen in Africa)
- *Ross River virus* (mosquito-borne; cause of epidemic arthritis; has moved across the Pacific region several times)
- *Sabiá virus* (rodent-borne; virus from Brazil; newly discovered cause of hemorrhagic fever, including two laboratory-acquired cases)
- *Venezuelan encephalitis virus* (mosquito-borne; great potential for movement from its niche in Central and South America)
- *Yellow fever virus*[a] (mosquito-borne; one of the most deadly diseases in history)

[a] Viruses that cause hemorrhagic fevers in humans.

TABLE II
Some of the Most Important New, Emerging, and Reemerging Virus Pathogens of Animals

- *African horsesickness viruses* (mosquito-borne; a historic problem in southern Africa; now becoming entrenched in the Iberian peninsula; a major threat to horses worldwide)
- *African swine fever virus* (tick-borne and also spread by contact; an extremely pathogenic virus; present in Europe and South America; a major threat to swine in North America)
- *Avian influenza viruses* (spread by wild birds; a major threat to the poultry industry of the United States)
- *Bluetongue viruses (Culicoides-borne;* the isolation of several strains in Australia has become an important nontariff trade barrier issue)
- *Bovine spongiform encephalopathy agent* (recently recognized; the cause of a major epidemic in cattle in the United Kingdom, resulting in major economic loss and trade embargo)
- *Canine parvovirus* (a new virus, having mutated from feline panleukopenia virus; the virus has rapidly swept around the world, causing a pandemic of severe disease in dogs)
- *Chronic wasting disease of deer and elk* (a spongiform encephalopathy agent of unknown source, discovered in captive breeding herds in the United States)
- *Dolphin, porpoise, and phocine (seal) morbilliviruses* (epidemic disease first identified in 1988 in European seals was first thought to be derived from a land animal morbillivirus, such as canine distemper or rinderpest, but now it is realized that there are several important, emerging pathogens endangering these species)
- *Feline immunodeficiency virus and simian immunodeficiency viruses* (important new viruses, the one affecting cats in nature and the other serving as an important model in AIDS research)
- *Foot-and-mouth disease viruses* (still considered the most dangerous exotic viruses of animals in the world because of their capacity for rapid transmission and great economic loss; still entrenched in Africa and Asia)
- *Lelystad virus (mystery swine disease)* (a new virus, causing an important disease in swine in Europe and the United States)
- *Malignant catarrhal fever virus* (an exotic, lethal herpesvirus of cattle; an important nontariff trade barrier issue throughout the world)
- *Myxoma virus* [used to control rabbits in Australia, but with diminishing success; now a proposal has been advanced that genetically engineered myxoma virus carrying a gene for a sperm antigen be distributed to sterilize infected surviving rabbits; additionally, rabbit hemorrhagic disease virus (a calicivirus) is being proposed as a new way to control rabbits in Australia]
- *Rinderpest virus* (still considered very dangerous with potential for causing great economic loss; still entrenched in Africa and Asia)

larger numbers of people into close contact. There have been successive revolutions in transportation, making it possible to circumnavigate the globe in less than the incubation period of most infectious diseases.

Ecological changes brought about by human activity are occurring at a rapidly accelerating rate.

When a new infectious disease is suspected, it must be characterized (by the work of clinicians and pathol-

TABLE III
The Special Case of New, Emerging, and Reemerging Virus Pathogens of Endangered Species

- *Callitrichid arenavirus in tamarins (LCM/marmoset hepatitis)* (lethal disease; a major problem in the golden lion-tamarin, an endangered species)
- *Canine distemper virus in the black-footed ferret* (lethal disease; its control is crucial to the survival of this endangered North American species)
- *Eastern equine encephalitis virus in the whooping crane* (lethal disease; its control is crucial to the survival of this endangered North American species)
- *Feline infectious peritonitis virus in the cheetah* (lethal disease; its control is crucial to the survival of this endangered African species)
- *Phocid morbillivirus (seal distemper virus) and related viruses* (newly recognized; a complex of lethal diseases in several species of seals and sea lions, caused by several related morbilliviruses; control is crucial to the survival of several endangered species, worldwide)
- *Rabies in endangered free-living wild canid species* (lethal disease; its control is crucial to the survival of several endangered wild canid species)
- *Rinderpest virus in endangered free-living ruminant species* (several wild ruminant species of Africa serve as reservoir hosts of this virus, the source of virus causing epidemics in cattle; control is crucial to the acceptance of cohabitation of domestic and wild animals in large areas of Africa)
- *Simian hemorrhagic fever virus in captive macaques* (silent infection in several simian species serves as the reservoir of virus which causes lethal disease in macaques; very transmissible; emerging as a major problem in macaque breeding programs)

ogists and clinical support specialists) and assessed in regard to its impact on populations at risk (by the work of medical epidemiologists and support specialists). When a new microbe or virus is suspected, it must be isolated, identified, characterized, and methods must be developed for diagnosis and field investigation. All these activities must be integrated to construct a comprehensive view of the immediate problem at hand. Today, this kind of integrated investigative, problem-solving activity is usually thought to be the sole responsibility of governmental agencies, but, in fact, many professionals from throughout the health sector have played central roles in recent episodes. As it turns out, assessing risk and guiding intervention involve quite diverse activities; for example, in some cases complex field studies of the incidence of infection in the general population or in a selected subpopulation are necessary to determine such things as risk factors for infection, mode of transmission, and targets for intervention, whereas in other cases, studies of the pathogenetic mechanisms underpinning the clinical presentation in the individual patient might hold the key. Today, in nearly every instance, important clues lie in characterizing the molecular structure and replication strategy of the microbe or virus and the cellular pathobiology of the infection. Of course, the immediate goal of such an enterprise is to invent and guide prevention and control measures as these become necessary. All the while, such an enterprise is being observed by the public—the public must be rereminded of the impact that infectious diseases, or the infectious disease at hand, can have on societal well-being.

The sense of opportunity to advance the public's knowledge about public health and infectious diseases and to thereby facilitate prevention and control programs is, however, a two-edged sword: the public may be enchanted by quick, precise identification of problems and quick design of specific interventions. However, the same public also quickly experiences frustration and despair, as in the case of AIDS, where public health professionals are perceived as having largely failed in their attempts since the early 1980s to stem the tide of the global epidemic. For example, in 1992, AIDS became the leading cause of death in the United States among men aged 25–44 years and the fourth leading cause of death among women in the same age group. Here is a problem marked by continuing scientific disappointment and a realization that everything about HIV, the virus, and AIDS, the disease, is fundamentally difficult. At the same time, the public has come to understand that this is a problem that is marked by societal questions of increasing resource burden and declining political will. The worldwide AIDS epidemic, as the ultimate new, emergent infectious disease, the true "Andromeda Strain," offers lessons of many kinds that must be heeded. [*See* Acquired Immune Deficiency Syndrome, Epidemic.]

II. VARIATION AND EVOLUTION OF MICROBES AND VIRUSES

Variation in the genome of microbes and viruses can alter their character substantially, and certain variations can increase their disease potential incrementally. For this reason, it is important to recognize the capacity of microbes and viruses to continuously evolve. Several mechanisms drive evolution; first, there is a high frequency of mutation during replication. Second, during microbial or viral replication, there is often opportunity for recombination or reassortment of genes. This is most common in viruses with segmented genomes. Finally, certain viruses, such as the retroviruses, can recombine with genetic elements of their host cells, thereby acquiring new genes. Classical examples of variation include the definition many years ago of smallpox variants, variola major (Indian subcontinent and Europe, mortality up to 30%) and variola minor (South America, mortality about 1%) and naturally occurring attenuated poliovirus variants, some of which inspired the development of Sabin live-attenuated polio vaccines. Several other kinds of variations can contribute to the emergence of new infectious diseases. A change in host range can permit a microbe or virus to spread into a new species with devastating consequences. An increase in virulence can convert a nonpathogenic infectious agent into an important pathogen. A change in antigenic signature can permit a microbe or virus to infect a population already immune to parental strains of the same infectious agent.

III. EPIDEMIOLOGIC CONSIDERATIONS IN THE EMERGENCE OF INFECTIOUS DISEASES

Epidemics are classically divided according to their means of spread into two major categories: propagated and common source. Propagated epidemics depend on spread from host to host; they continue to expand as long as each infection gives rise to more than one new infection. Common source epidemics occur when an infectious agent is disseminated from a single focus; they usually result from contamination of air, food, water, drugs, medical devices, or the like.

A. Propagated Infections

In some instances, a new infectious disease is the consequence of the appearance of a microbe or a virus that is truly new. Recent dramatic instances of this phenomenon are the global epidemics caused by canine parvovirus and HIV. Canine parvovirus represents a true instance of recent virus mutation. A preexisting feline virus underwent a few key mutations that made it pathogenic and very transmissible in the canine population. Similarly, it seems that HIV-1 and HIV-2 evolved in subhuman primates from viral ancestors resembling simian immunodeficiency virus (SIV). In these instances, the new host populations had no background immunity or any resistance derived from long-term genetic selection. Without such natural barriers to dissemination, in each case international spread and global pandemics followed.

Continuing evolution of an infectious agent that is already present in a population can cause reemergence of epidemic disease. A familiar example is influenza, where virus variants are introduced regularly into the human population. Such evolution of influenza viruses may be accompanied by an increase in virulence and increased mortality. The greatest example of this phenomenon is the global influenza pandemic of 1918, in which more than 40 million people died.

On occasion, an infectious agent may emerge in epidemic form simply because of an alteration in its ecology. An example of historic importance is the emergence of epidemic poliomyelitis in the mid-19th century. This was not due to any change in the properties of the virus, as epidemiological studies strongly suggest that the same virus strains could be simultaneously associated with both pathogenic and nonpathogenic infection. Rather, this was due to an improvement in sanitation and personal hygiene, which led to a delay in the acquisition of enteric virus infections from infancy to childhood. Because infants were protected by maternal antibody, they usually underwent silent immunizing infections, whereas older children more often suffered paralytic poliomyelitis.

B. Common Source Outbreaks

If an infectious agent is introduced from a common source into a large population, disease can emerge on an epidemic scale. An example of this is the ongoing epidemic of bovine spongiform encephalopathy now underway in the United Kingdom. In this instance it appears that the scrapie agent of sheep contaminated protein supplements which were fed to cattle. This resulted in the apparent adaptation of the agent to cattle and in the emergence of a devastating new disease.

C. Persistent Infections, Long Incubation Periods, and Chronic Diseases

The dynamics of emergence of a new infectious disease can vary greatly, depending on the incubation period, whether the infection is acute or persistent, and whether the resulting disease (and shedding pattern) is acute or chronic. When associated with a short incubation period and acute disease, emergence can be a dramatic event. The most recent example is Hantavirus-associated respiratory distress syndrome in the United States, which first came to attention in 1993. However, when the emerging agent is associated with a long incubation period, the resulting epidemic may rise over a period of years before it reaches a peak. Furthermore, the long interval between infection and disease occurrence may obscure the identification of the causal agent or its mode of transmission. Finally, if the disease is chronic, the impact on the medical care system may spread over decades rather than weeks with grave consequences beyond those ordinarily associated with an acute epidemic.

A current example of long incubation period disease associated with emergence is AIDS. First, the length of the incubation period, which appears to average about 8–10 years, led to marked underestimates of the potential impact of the epidemic. Associated with this was uncertainty regarding the proportion of infections which would lead to death. The long incubation period led to the initial assumption that only a small proportion of infections, perhaps no more than 10%, would be fatal; projections of total AIDS incidence were correspondingly low. Another important factor was a failure to appreciate the impact of the lifelong persistence of infection and virus shedding. It proved difficult to model the number of potentially infectious individuals in the population at any given time and their impact on transmission dynamics. Finally, the chronic nature of AIDS, with an average survival of many years, produces a vast new burden on a health care system that was already failing to cope with patients whose needs exceed their means to pay.

The epidemic potential of newly emerging infectious diseases also varies depending on the mode of transmission, the immunological and genetic susceptibility of the host population, and the size of the potential population at risk. Epidemic potential is greatest for an agent that is readily transmitted from host to host, particularly via the respiratory route. Conversely, zoonotic agents and arthropod-borne agents are usually limited in their geographic range, although the latter are certainly not limited in their capacity for causing very large epidemics.

D. New Recognition of Previously Unrecognized Diseases

In some instances, a long existing infectious disease that has not been recognized as a specific entity may emerge as a newly recognized disease. Recognition may be triggered through a variety of circumstances, such as the occurrence of a specific outbreak which leads to intensive investigation or to the isolation of the agent and the subsequent development of laboratory methods permitting a specific etiologic diagnosis. New diseases that are clinically unique, such as AIDS, are more likely to be recognized early than diseases that closely resemble well-established clinical entities, such as hepatitis, diarrhea, or encephalitis. One example of "delayed emergence" due to late recognition is California encephalitis, caused by the La Crosse virus. This virus was first isolated in 1964 from a fatal case of encephalitis in a child who was hospitalized and died in 1960 in La Crosse, Wisconsin. Using this isolate as a source of antigen, retrospective serological surveys showed that the virus was an important cause of disease previously listed under the heading "viral meningitis of undetermined etiology." Since that time, La Crosse encephalitis has been reported each year from the midwestern United States, at an average annual incidence of about 75 cases, with no evidence of increasing or decreasing occurrence since the 1970s. This is consistent with an endemic infection that had been occurring regularly for many decades, i.e., long before the mid-1960s when the virus came to light.

IV. NATURE OF ECOLOGICAL AND ZOONOTIC FACTORS IN REGARD TO NEW, EMERGING, AND REEMERGING INFECTIOUS DISEASES

One set of factors contributing to the emergence of new microbes and viruses relates to their capacity to adapt to extremely diverse and changing econiches. One of the most complex sets of adaptive characteristics concerns the transmission of infectious agents by arthropods. The arthropod-borne viruses are examples *par excellence* where emergence and reemergence follow upon environmental manipulation or natural environmental change. When ecosystems are altered, infectious disease problems of humans and animals follow—in fact, much evidence suggests that the rate of ecologic and environmental change is accelerating and that the risk of emergence of new arthropod-borne diseases is increasing.

1. Population movements and the intrusion of humans and domestic animals into new arthropod habitats have resulted in many new emergent disease episodes, some of which are the stuff of fiction. The classic example of this was the emergence of yellow fever when susceptible humans entered the Central American jungle to build the Panama Canal—there are many contemporary examples suggesting that similar events will continue to happen, in most cases in unanticipated circumstances.

2. Ecologic factors pertaining to unique environments and geographic factors have contributed to many new, emergent disease episodes. Remote econiches, such as islands, free of particular species of reservoir hosts and vectors, are often particularly vulnerable to an introduced infectious agent. For example, the initial Pacific "island-hopping" of Ross River virus in the 1980s from its original niche in Australia caused "virgin-soil" epidemics of arthritis–myalgia syndrome in Fiji and Samoa—this virus will surely reemerge in a similar way again.

3. Deforestation has been the key to the exposure of farmers and domestic animals to new arthropods. The occurrence of Mayaro virus disease in Brazilian wood-cutters as they cleared the Amazonian forest in recent years is a case in point.

4. Increased long-distance travel facilitates the carriage of exotic arthropod vectors around the world. The carriage of the Asian mosquito, *Aedes albopictus*, a potential vector for dengue and California encephalitis viruses, into the United States in the water contained in imported used tires represents an unsolved problem of this kind.

5. Increased long-distance livestock transportation facilitates the carriage of infectious agents and arthropods (especially ticks) around the world. The introduction of the tick-borne agent, African swine fever virus, from Africa into Portugal (1957), Spain (1960), and South America (1960s and 1970s) is thought to

have occurred in this way—it is just a matter of time until this virus makes further international forays.

6. New routings of long-distance bird migrations, brough about by new man-made water impoundments, represent an important yet still untested new risk of introduction of arboviruses into new areas. This may be one key to the movement of Japanese encephalitis virus into new areas of Asia.

7. Ecologic factors pertaining to environmental pollution and uncontrolled urbanization are contributing to many new, emergent disease episodes. Arthropod vectors breeding in accumulations of water (tin cans, old tires, etc.) and sewage-laden water are a worldwide problem. Environmental chemical toxicants (herbicides, pesticides, residues) can also affect vector–virus relationships directly or indirectly. For example, mosquito resistance to all licensed insecticides in parts of California is a known direct effect of unsound mosquito abatement programs; this resistance may also have been augmented indirectly by uncontrolled pesticide usage against crop pests.

8. Ecologic factors pertaining to water usage, i.e., increasing irrigation and the expanding reuse of water, are becoming important factors in infectious disease emergence. The problem with primitive water and irrigation systems, which are developed without attention to arthropod control, is exemplified in the emergence of Japanese encephalitis in new areas of Southeast Asia.

9. Global warming, affecting sea level, estuarine wetlands, swamps, and human habitation patterns, may be affecting arthropod vector relationships throughout the tropics; however, data are scarce and many programs studying the effect of global warming have not included the participation of infectious disease experts.

Perhaps the best way to illustrate how ecological and zoonotic factors influence new, emerging, and reemerging infectious diseases is by examples.

A. Dengue

Dengue is one of the most rapidly emerging diseases in the tropical parts of the world, with millions of cases occurring each year. For example, Puerto Rico had five dengue epidemics in the first 75 years of this century, but has had six epidemics since 1983, at an estimated cost of over $150 million. At the same time, there has been a record number of cases elsewhere in the Americas; Brazil, Bolivia, Paraguay, Ecuador,

Nicaragua, and Cuba have experienced their first major dengue epidemics since the mid-1940s. These epidemics have involved multiple virus types; of the four dengue virus types, three are now circulating in the Caribbean region. These circumstances lead to dengue hemorrhagic fever—the lethal end of the dengue disease spectrum. Dengue hemorrhagic fever first occurred in the Americas in 1981; since then, 11 countries have reported cases, and since 1990 over 3000 cases have been reported each year.

Why are diseases like dengue emerging or reemerging, especially in the Americas? The answer is simple: urban mosquito habitats are expanding (the vector mosquito, *Aedes aegypti*, is extremely adapted to human proximity), mosquito density is increasing, and mosquito control is failing. This is occurring not just in the least developed countries, but in many developed countries, including the southern region of the United States. In all countries, financial resources for public health are severely limited and must be prioritized. Priority lists are political in nature and tend to emphasize day-in-day-out problems, not episodic problems—too often, mosquito control, which is very expensive, falls off the bottom of the priority list. Meanwhile, mosquito control is becoming more expensive as older, cheaper chemicals lose effectiveness or are banned as damaging to the environment and must be replaced by more expensive chemicals. As mosquito control fails, dengue follows quickly.

B. Yellow Fever

An even more frightening scenario associated with failing mosquito control is that yellow fever virus, which is transmitted by the same mosquito vector as dengue, *A. aegypti,* might also reemerge. Where dengue occurs the conditions are, *de facto,* appropriate for yellow fever—in this instance, the recipe for emergence is "just add virus" (by importation via an infected person or an infected mosquito). It is one of the mysteries of tropical medicine that yellow fever does not occur more often in such circumstances where vector density and a susceptible human population coexist. In fact, no one knows where, when, or even if yellow fever virus will reemerge in the kind of epidemics that were the scourge of tropical and subtropical cities of the Western hemisphere and Africa throughout the 17th, 18th, and 19th centuries; however, because this virus is so dangerous, the possibility is constantly on the mind of national and international health officials.

C. Hantavirus Diseases (Hemorrhagic Fever with Renal Syndrome and Acute Respiratory Distress Syndrome)

The first well-characterized disease of this kind was Korean hemorrhagic fever, which emerged during the Korean war of 1950–1952. Thousands of United Nations troops developed a mysterious disease marked by fever, headache, hemorrhage, and acute renal failure; the mortality rate was 5–10%. Despite much research, the agent of this disease remained unknown for 26 years; then, in 1976, a new virus, named Hantaan virus, was isolated in Korea from field mice. The discovery of this virus was, however, just "the tip of an iceberg." In subsequent years, related viruses have been found in many parts of the world in association with different rodents and as the cause of human diseases with more than 150 different local names. From an ecologic perspective, it has been found that there are seven or eight different subgroups of viruses and three different transmission patterns: rural, urban, and laboratory acquired. From a clinical perspective, two disease patterns are described: one marked by severe disease (hemorrhagic fever with renal syndrome, with significant mortality) and the other by mild disease (febrile disease without mortality). The rural, severe disease is widespread in the Far East (e.g., Korean hemorrhagic fever in Korea; "epidemic hemorrhagic fever" in China, causing more than 100,000 cases per year). A similar rural, severe disease is emerging in the Balkans (mortality rate of about 20%).

In May 1993, a new Hantavirus disease was recognized in the southwestern region of the United States. The disease appeared as an acute respiratory distress syndrome. Clinical signs include fever, headache, and cough, followed by acute pulmonary congestion and edema leading to hypoxia, shock, and, in many cases, death. Within a short time, cases were found in 23 states. As of June 1995, more than 110 cases had been confirmed and there had been more than 56 deaths. At the beginning of the investigation in 1993, even though the causative virus had not been isolated, serologic tests (using surrogate antigens from related Hantaviruses) and molecular biologic tests were developed and used to prove that the etiologic agent was a previously unknown Hantavirus. Viral RNA was amplified from patient specimens by the polymerase chain reaction (PCR), followed by sequencing. Comparing sequences obtained from specimens from different areas indicated that several different variant viruses (seemingly at least five variants), all new and previously unknown, were active in the United States. PCR sequences were extended until much of the genome of the new viruses had been sequenced; these sequences were then used in expression systems to provide homologous antigens for further studies and diagnostic services. The same serologic and molecular biologic methods were applied to large numbers of rodents collected in the areas where patients lived; this proved that at least eight species of rodents were involved, the primary reservoir host in the southwest being *Peromyscus maniculatus,* the deer mouse (about 30% of this species were found to harbor the viruses in several areas of the southwestern United States). These viruses, like other Hantaviruses, do not cause disease in their reservoir rodent hosts, but virus is shed in the saliva, urine, and feces of these animals probably for their entire lives. Human infection occurs by the inhalation of aerosols or dust containing infected dried rodent saliva or excreta. These viruses likely have always been present in the large area of the western region of the United States inhabited by *Peromyscus* species; they were recognized in 1993 only because of the number and clustering of human cases, which in turn were probably caused by a great increase in rodent numbers consequent to an increase in piñon seeds and other rodent food. As a result of this kind of rapid field and laboratory investigation, the public is being advised about reducing the risk of infection, mostly by reducing rodent habitats and food supplies in and near homes and taking precautions when cleaning rodent-contaminated areas. There is a lesson in this—even today, important diseases can remain hidden from scientists and public health officials. In this instance, a combination of classical clinical, pathological, and epidemiological methods, together with modern techniques of molecular virology, served to identify the agent rapidly, indicated the mode of human infection, and led to recommendations for control. This was a striking accomplishment, carried out in less than 6 months, and provides an example of the practical need for a comprehensive public health/biomedical research infrastructure.

D. Rabies

Rabies can serve as an example to illustrate ecologic factors that pertain to the adaptation and emergence of viruses in new econiches. These ecologic factors need not be envisioned as too mysterious. In some instances, the necessary ecologic elements are in place and the recipe for emergence simply involves the introduction of a virus. The most dramatic illustration of this in recent years has been the appearance of epidemic raccoon rabies in northeastern United States. This epizootic has been traced to the importation of raccoons from Florida to West Virginia in 1977. This epidemic demonstrates dramatically how human perturbation of the environment, in this instance involving the transport of wild animals, can lead to the emergence of a disease in a previously unaffected area. One key to our understanding of this episode was the discovery that rabies virus is not one virus; rather, rabies virus is a set of different genotypes, each transmitted within a separate reservoir host niche. In North America, there are about six terrestrial animal genotypes: the skunk in the northcentral states, the skunk in the southcentral states, the Arctic fox and red fox in Alaska and Canada, the gray fox in Arizona, the coyote and feral dog in southern Texas and northern Mexico, and the raccoon in southeastern, mid-Atlantic and now northeastern states. "Raccoons-bite-raccoons-bite-raccoons," and after some time their virus becomes a distinct genotype, highly adapted to the host cycle and inefficient if introduced into another host cycle. When this discovery was made (using monoclonal antibody and molecular biologic methods), many mysteries of rabies ecology were clarified.

V. NATURE OF BEHAVIORAL AND SOCIETAL FACTORS IN REGARD TO NEW, EMERGING, AND REEMERGING DISEASES

Diverse behavioral, commercial, societal, and medical care factors lead to the emergence of new diseases or new disease patterns. Many diseases that depend on such factors are emergent or reemergent. We have overarching societal problems concerning (1) sexually transmitted diseases; (2) diseases associated with daycare; (3) childhood diseases transmitted in the community; (4) diseases associated with the global movement of animals and animal products; (5) diseases associated with changes in agricultural production practices (preharvest food-safety issues); (6) diseases associated with food processing and retailing; and (7) diseases associated with medical care, such as hospi-

tal-acquired diseases, diseases associated with immunosuppressive therapy, diseases associated with organ transplantation, and diseases associated with blood banking. Again, the following examples illustrate how some of these factors influence new, emerging, and reemerging infectious diseases.

A. Sexually Transmitted Diseases

Sexually transmitted diseases had been declining over many years in developed countries, but now they are increasing at epidemic rates in certain populations, such as urban populations in the United States and many developing countries. This is especially disturbing because many of the sexually transmitted diseases enhance the risk of transmission of each other (via new bacteria and viruses entering through established lesions). Thus, risk of transmission is increased beyond expected rates, and "curable" diseases, such as gonorrhea and syphilis, support the spread of "incurable" diseases, such as genital herpes and HIV infection/AIDS. The major societal failing in regard to the reemergence of several of the sexually transmitted diseases stems from declining support for public health programs, but in some instances emergence reflects advances in diagnostics and the detection of new agents. One example is the emergence of human papillomavirus infection, genital papillomatosis, and cervical cancer. Multiple risk factors contribute to cervical cancer, including behavioral (sexual behavior), dietary, hormonal, and viral risk factors. There are several theories, but no real proofs as to how these risk factors, in concert, lead to cervical neoplasia. In any case, there are more than 13,000 new cases of cervical cancer reported in the United States annually, resulting in about 4500 deaths. Cervical cancer is the eighth most common cancer in women and 5-year survival is about 57–67%. Much of this mortality is preventable—early screening and early treatment of premalignant disease are known to be effective—this is the basis for the extensive cytological screening programs in place in many countries. Now that the viral risk factor for cervical cancer is known to be infection with certain types of human papillomavirus, DNA probes for these viruses are being added to cytological screening programs. Further research is urgently needed to answer the question of how these viruses relate to cervical cancer; further research must also be directed to developing a papillomavirus vaccine, which would be received well by women who are so familiar with the significance of the Pap smear. [*See* Sexually Transmitted Diseases.]

B. Day-Care and the Emergence of Virus Diarrhea

Shifts in the structure of the family in all developed countries have resulted in a dramatic increase in the proportion of children in day-care; currently, 11 million children in the United States spend a large part of their time in day-care. This trend is likely to continue; by the year 2000 it is estimated that more than 75% of mothers with children under 6 years of age will be working outside the home. The economic impact of day-care-associated illness is incredible: working mothers are forced to miss 1 to 4 weeks per year to care for their sick children—over 60% of employee absenteeism is due to child-care needs, particularly those of sick children. Infectious diseases represent the most important problems in day-care, with respiratory and diarrheal illnesses being most common. Additionally, children attending day-care facilities become silent reservoir hosts for some agents of disease, such as hepatitis A virus, several enteroviruses, and some *Salmonella* species. Depending on the disease, children attending day-care have a 2- to 18-fold increased risk of becoming ill compared with children at home. The most common diarrheal pathogens acquired in day-care centers are *Giardia lamblia*, *Shigella* species, rotaviruses, and other viruses.

Rotaviruses infect every child in its first 3 to 4 years of life; this leads to an estimated three million cases of diarrhea, 500,000 doctor visits, 70,000 hospitalizations for 300,000 inpatient days, 75–125 deaths, and costs of hospital care of $200–400 million per year in the United States. Worldwide, nearly one million children die each year of rotavirus diarrhea. It seems incredible that as late as 1973 the human rotaviruses were unknown—this was the case even though the viruses are now relatively easy to detect. This is a case of emergence being defined not so much by the leading edge of laboratory technology, but by the emerging application of conventional technology. Indeed, there probably are several more viruses and other microorganisms causing diarrhea that have yet to be discovered.

C. Food-Borne Diseases

Modern agricultural and food industry practices favor the emergence of new infectious disease problems.

Animal husbandry has changed in ways that increase stress and promote microbial and viral transmission and endemic infection cycles. Large numbers of animals are being confined in limited space and at very high density. Large numbers of animals are cared for by few, inadequately trained workers. Elaborate housing systems are used, but with inadequate evaluation of systems for ventilation, feeding, waste disposal, and cleaning. Additionally, some diseases are favored by the global expansion of agricultural markets, involving the global transport of animals, animal products, and animal semen and embryos. Yet other diseases can emerge as a consequence of changes in processing and distribution systems. [*See* Food Microbiology and Hygiene.]

We usually think of the importance of livestock diseases in terms of financial losses, as the capacity of the commercial livestock food industries of the developed countries are such that surpluses are more a problem than shortages. However, in developing countries this is not the case; livestock diseases, especially new, emerging, or reemerging diseases, cause immediate human suffering by substantially compromising scarce human food resources, especially high-quality protein.

D. Bovine Spongiform Encephalopathy

Bovine spongiform encephalopathy is a chronic fatal disease of cattle that was first diagnosed in the United Kingdom in 1986; as of 1995, more than 160,000 head of cattle in 30,000 herds (45% of all dairy herds) had become infected, and government expenditures had reached more than £300 million. Until recently, additional cases had been occurring at a rate of 1000 per week, mostly in dairy cattle between 3 and 5 years of age. The disease has also been diagnosed in Northern Ireland, Ireland, Oman, the Falkland Islands, Switzerland, Denmark, France, and Canada. Epidemiologic investigation has suggested that the epidemic was started by feeding cattle protein supplements derived from sheep meat and bone meal that was contaminated by the scrapie agent. Evidence shows that a change in the rendering process in the early 1980s resulted in a large increase in the exposure of cattle to the sheep agent. As the disease became established, it may have been amplified by inclusion in rendered products of meat and bone meal from infected cattle. In 1988, feeding ruminant-derived protein supplements and bone meal was banned, but the epidemic has left a number of important questions yet unanswered. Will the epidemic be controlled by prohibiting the inclusion of ruminant offal in meat and bone meal? Will the disease be maintained by vertical or horizontal transmission within the cattle population? When and how can British livestock be safely exported? Can biological materials derived from cattle be safely used in food, livestock, pharmaceutical, or health product industries? What is the risk of human consumption of beef or milk? What should the public be told about any potential risk? What lessons have we learned that will help to avoid epidemics like this in the future?

VI. PREVENTION AND CONTROL OF NEW, EMERGING, AND REEMERGING INFECTIOUS DISEASES

The world faces an immediate and urgent problem in regard to new, emerging, and reemerging infectious diseases. Our health systems are being challenged by these diseases, each of which seems to have qualities that defy established control methods. Nevertheless, new diseases are likely amenable to the development of prevention strategies based on innovative biotechnologic and epidemiologic research approaches. Certainly, these diseases could be dealt with more efficiently if we could better integrate the activities of all appropriate allied health disciplines (e.g., medicine, veterinary medicine, public health) into a single enterprise.

It seems clear that the world needs some kind of integrated infectious diseases surveillance enterprise, but how is this to be organized? All planning exercises conducted in the past few years, whether by the World Health Organization, the Centers for Disease Control, or the Institute of Medicine of the U.S. National Academy of Sciences, call for the establishment of a global surveillance/diagnostics/communications network, an "early warning system." The network would include the following activities:

1. Global disease surveillance.
2. Global diagnostics system.
3. Global integral research base.
4. Global communications system.
5. Technology transfer system.
6. Global prevention/intervention infrastructure.
7. Global emergency response system.
8. Global training program (teaching the teachers, teaching the leaders of the future).
9. Stable funding base.

At the grass-roots level, everywhere in the world, these surveillance/diagnostics/communications centers would influence the development of a generation of clinicians, who, when finding an unusual case or cluster of cases, would have a heightened "index of suspicion" for new or unusual diseases, diseases with epidemic potential. The centers would also set into place an improved worldwide public health epidemiologic and laboratory infrastructure, allowing better follow-up of the suspicions of these clinicians. Because we cannot predict future emergences of viruses and microorganisms, our strategy must be to detect suspicious events early and to respond quickly, intelligently, and comprehensively. The centers would be closely tied to leading clinical centers around the world so as to have immediate knowledge of new or unusual disease episodes, and they would be tied to a global communications network under the auspices of the World Health Organization. This concept has equal merit in regard to the early detection and communication of new, emerging, or reemerging veterinary microbial or viral problems—in fact, human disease centers and veterinary disease centers might be integrated into a single network. Surely, this proposal is worthy of national commitment and international development.

BIBLIOGRAPHY

Anonymous (1992). "Proceedings of the International Conference on Child Day-Care Health: Science, Prevention, and Practice." Centers for Disease Control, Atlanta, GA.

Donovan, P. (1993). A prescription for sexually transmitted diseases: Issues in Science and Technology, Summer 1993 Issue. U.S. National Academy of Sciences Press, Washington, DC.

Fenner, F., Gibbs, E. P. J., Murphy, F. A., Rott, R., Studdert, M. J., and White, D. O. (1993). "Veterinary Virology," 2nd Ed. Academic Press, Orlando.

Fenner, F., Henderson, D. A., Arita, I., Jezek, Z., and Ladnyi, I. D. (1988). "Smallpox and Its Eradication." World Health Organization, Geneva.

Garrett, L. (1994). "The Coming Plague." Newly Emerging Diseases in a World Out of Balance." Farrar, Straus and Giroux, New York.

Henderson, D. A. (1993). Surveillance systems and intergovernmental cooperation. In "Emerging Viruses" (S. S. Morse, ed.). Oxford University Press, New York.

Henig, R. M. (1993). "A Dancing Matrix: Voyages Along the Viral Frontier." Knopf, New York.

Lederberg, J., Shope, R. E., and Oaks, S. (eds.) (1992). "Emerging Infections, Microbial Threats to Health in the United States." U.S. National Academy of Sciences Press, Washington, DC.

Emotional Motor System

GERT HOLSTEGE
University of Groningen

GLOSSARY

Hemiplegia Paralysis of one side of the body

Lordosis Female receptive behavior

Nucleus raphe magnus Group of midline cells in the ventral tegmentum of the most rostral medulla oblongata

Nucleus raphe obscurus Group of midline cells in the dorsal tegmentum, dorsal and caudal to the nucleus raphe pallidus

Nucleus raphe pallidus Group of midline cells in the ventral tegmentum just caudal to the nucleus raphe magnus

Palsy Paralysis

Periaqueductal gray Gray matter around the aquaduct of Sylvius in the mesencephalon

Rhythm generators Group of cells that fire in a rhythmical way causing rhythmical movements

Vocalization Nonverbal production of sound

THE LIMBIC SYSTEM IS CLOSELY INVOLVED IN THE elaboration of emotional experience and expression and is associated with a wide variety of autonomic, visceral, and endocrine functions. The limbic system consists of several cortical and subcortical structures, although there is no agreement on exactly which structures belong to it. Some authors argue that the use of the term limbic system should be abandoned; nevertheless, many scientists still use this term. Subcortical regions usually included in the limbic system are the hypothalamus and the preoptic region, the amygdala, the bed nucleus of the stria terminalis, the septal nuclei, and the anterior and mediodorsal thalamic nuclei. The limbic system has extremely strong reciprocal connections with mesencephalic structures such as the periaqueductal gray (PAG) and the laterally and ventrally adjoining tegmentum (Nauta's limbic system–midbrain circuit). Many studies have shown that certain parts of the limbic system give rise to a descending motor system that is completely separate from the motor system, originating in structures such as the vestibular nuclei, red nucleus, or motor cortex. The last motor system, and especially the corticobulbospinal pathway, is well known by neurologists, and is called the somatic motor system. Holstege used the term "emotional motor system" for the other motor system, originating in parts of the limbic system or in structures strongly related to it.

I. INTRODUCTION

Hemiplegic patients with interruption of corticobulbospinal fibers show a complete central paresis of the lower face on one side, indicating that their somatic motor system is damaged. However, the emotional motor system is still intact because they are still able to smile spontaneously (e.g., when they enjoy a joke). Reversibly, in cases with postencephalitic parkinsonism, patients are able to show their teeth, whistle, and frown, indicating that their somatic motor system is intact. Their emotions, however, are not reflected in their countenance and they have a stiff, mask-like facial expression (poker face). Patients with irritative pontine lesions sometimes suffer from nonemotional laughter and crying, and patients with pseudobulbar palsy (e.g., with lesions in the mesencephalon) often suffer from uncontrollable fits of crying or laughter. Such fits are usually devoid of grief, joy, or amusement; they may even be accompanied by entirely incompatible emotions. Fits of crying and laughter may occur in the same patients; other patients may show only one of them. Apparently, all these patients suffer from damage of the emotional motor system.

Crying and laughter belong to an expressive behavior, which in animals is called vocalization. In many different species, vocalization can be elicited by stimulation in the caudal part of the PAG. Vocalization is based on a specific final common pathway, originating from a distinct group of neurons in the PAG that project to the nucleus retroambiguus (NRA), which in turn has direct access to all vocalization motonerons. In humans, this projection might form the final common pathway for laughing and crying. The vocalization neurons in the PAG receive their afferents from structures belonging to the limbic system and not from the voluntary system. All this clinical and experimental evidence shows that a complete dissociation seems to exist between the somatic or voluntary and the emotional motor system.

The somatic motor system, according to the concept of Kuypers, consists of a medial and a lateral component (Fig. 1). The medial component in mammals is involved in the maintenance of erect posture (anti-gravity movements), integration of body and limbs, synergy of the whole limb, and orientation of body and head. It not only controls motoneurons and premotor interneurons of axial and proximal muscles of the trunk, but also those of neck and extrinsic eye muscles. The lateral component controls the distal muscle premotor inter- and motoneurons. The descending systems involved are the rubro- and the corticospinal tracts. The most elaborated projections are the direct corticospinal projections to motoneurons. They are only present in highly developed mammals, such as monkey, ape, and humans.

The emotional motor system also has two components (Fig. 1). The medial component consists of the pathways from medial hypothalamus and mesencephalon to caudal brain stem, which, in turn, projects diffusely to the spinal cord. The lateral component represents several distinct pathways underlying specific emotional behaviors. According to the concept of Holstege, the premotor interneurons, which belong to the so-called basic motor system (Fig. 1), as well as the motoneurons themselves should be considered as tools of both the somatic and the emotional motor systems.

This article discusses the various pathways of the emotional motor system, beginning with the diffuse caudal brain stem–spinal projections.

II. DIFFUSE PATHWAYS TO THE SPINAL CORD

A. Projections from Nuclei Raphe Magnus, Pallidus, and Obscurus and Ventral Part of the Caudal Pontine and Medullary Medial Reticular Formation

Retrograde and anterograde tracing studies indicate that the nuclei raphe magnus (NRM), pallidus (NRP), and obscurus (NRO), with their adjoining reticular formation, send fibers throughout the length of the spinal cord, giving off collaterals to all spinal levels. These descending systems are extremely diffuse and are not topographically organized. Furthermore, a heterogeneity exists in these projections in which the NRM and adjoining reticular formation project to the dorsal horn (Fig. 2, left), whereas the NRP, NRO, and adjoining reticular formation project to the intermediate zone and the ventral horn, including the motoneuronal cell groups (Fig. 2, right).

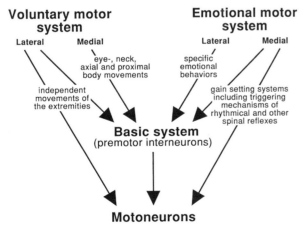

FIGURE I   Schematic overview of the three subdivisions of the motor system.

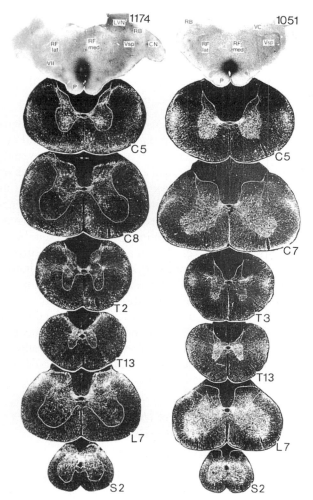

FIGURE 2 Bright-field photomicrographs of autoradiographs showing tritiated leucine injection sites in the raphe nuclei, and dark-field photomicrographs showing the distributions of the labeled fibers in the spinal cord. (Left) An injection is shown in the caudal NRM and adjoining reticular formation. Note that labeled fibers are distributed mainly to the dorsal horn (laminae I, the upper part of II and V), the intermediate zone, and the autonomic motoneuronal cell groups. (Right) The injection is placed in the NRP and the immediately adjoining tegmentum. Note that the labeled fibers are not distributed to the dorsal horn, but very strongly to the ventral horn (intermediate zone and autonomic and somatic motoneuronal cell groups. The following abbreviations pertain to these used throughout the figures: AA, anterior amygdaloid nucleus; AC, anterior commissure; ACN, nucleus of the anterior commissure; AD, anterodorsal nucleus of the thalamus; AH, anterior hypothalamic area; AL, lateral amygdaloid nucleus; AM, anteromedial nucleus of the thalamus; AP, area postrema; AV, anteroventral nucleus of the thalamus; BC, brachium conjunctivum; BIC, brachium of the inferior colliculus; BL, basolateral amygdaloid nucleus; BM, basomedial amygdaloid nucleus; BNSTL, lateral part of the bed nucleus of stria terminalis; BNSTM, medial part of the bed nucleus of the stria terminalis; BP, brachium pontis; CA, central amygdaloid nucleus; CC, corpus callosum; Cd, caudate nucleus; CGL, lateral geniculate body; CGM, medial geniculate body; CI, capsula interna; CI, inferior colliculus (Fig. 13); CL, claustrum;

Physiological studies are consistent with the anatomy of the descending pathways outlined earlier. Electrical stimulation in the NRM and the adjacent ventral part of the upper medullary medial reticular formation inhibits neurons in the caudal spinal trigeminal nucleus and spinal dorsal horn, producing inhibition of nociception. The diffuse organization of the raphe- and ventromedial medullary projections to the moto-

CM, centromedian thalamic nucleus; CN, cochlear nuclei; CO, cortical amygdaloid nucleus; CR, corpus restiforme; CS, superior colliculus; CSN, nucleus raphe centralis superior; CU, nucleus cuneatus; CUN, cuneiform nucleus; DBV, nucleus of the diagonal band of Broca; DH, dorsal hypothalamic area; DMH, dorsomedial hypothalamic nucleus; EC, external cuneate nucleus; ECU, external cuneate nucleus; En, entopeduncular nucleus; F, fornix; G, nucleus gracilis; GP, globus pallidus; HC, hippocampus; IC, inferior colliculus; IN, interpeduncular nucleus; IO, inferior olive; IVN, inferior vestibular nucleus; KF, nucleus Kölliker-Fuse; LL, lateral lemniscus; LOTR, lateral olfactory tract; LP, lateral posterior nucleus of the thalamus; LRN, lateral reticular nucleus; LTF, lateral tegmental field; LVN, lateral vestibular nucleus; MB, mammillary body; MC, nucleus medialis centralis of the thalamus; MD, nucleus medialis dorsalis of the thalamus; MesV, mesencephalic trigeminal tract; ML, medial lemniscus; MLF, medial longitudinal fasciculus; motV, motor trigeminal nucleus; MTF, medial tegmental field; MVN, medial vestibular nucleus; NCL, nucleus centralis lateralis; NLL, nucleus of the lateral lemniscus; NRM, nucleus raphe magnus; NRP, nucleus raphe pallidus; NRTP, nucleus reticularis tegmenti pontis; NTB, nucleus of the trapezoid body; NTS, nucleus tractus solitarius; nVII, facial nerve; OC, optic chiasm; OT, optic tract; P, pyramidal tract; PAG, periaqueductal gray; PBL, lateral parabrachial nucleus; PBM, medial parabrachial nucleus; PC, pedunculus cerebri; PEA, anterior part of periventricular hypothalamic nucleus; PON, pontine nuclei; PP, posterior pretectal nucleus; Pt, parataenial nucleus of the thalamus; Pu, putamen; PV, posterior paraventricular nucleus of the thalamus; PVA, paraventricular nucleus of the thalamus (anterior part); PVN, paraventricular hypothalamic nucleus; R, reticular nucleus of the thalamus; RE, nucleus reuniens of the thalamus; RF, reticular formation; RFmed, medial reticular formation; RFlat, lateral reticular formation; RM, nucleus raphe magnus; RN, red nucleus; RP, nucleus raphe pallidus; RST, rubrospinal tract; S, solitary complex; SC, suprachiasmatic nucleus (Fig. 10C); SC, nucleus subcoeruleus (Fig. 4F); SC, superior colliculus (Fig. 4A); SI, substantia innominata; SM, stria medullaris; SN, substantia nigra; SO, superior olivary complex; SCN, supraoptic nucleus; ST, subthalamic nucleus; STT, stria terminalis; TMT, mammillothalamic tract; VA, ventroanterior nucleus of the thalamus; VB, ventrobasal complex of the thalamus; VC, vestibular complex; VL, ventrolateral nucleus of the thalamus; VM, ventromedial nucleus of the thalamus; VMH, ventromedial nucleus of the hypothalamus; VTN, ventral tegmental nucleus; ZI, zona incerta; III, oculomotor nucleus; IV, trochlear nucleus; Vm, motor trigeminal nucleus; Vprinc., principal trigeminal nucleus; Vsp., spinal trigeminal complex; Vsp.cd., spinal trigeminal complex pars caudalis; VI, abducens nucleus; VII, facial nucleus; Xd, dorsal vagal nucleus; XII, hypoglossal nucleus.

neuronal cell groups suggests that they do not steer specific motor activities such as movements of distal (arm, hand, or leg) or axial parts of the body, but have a global effect on the level of activity of the motoneurons.

Many different neurotransmitter substances exist in the caudal raphe nuclei and ventromedial medullary reticular formation, of which serotonin (5HT) is the best known. Serotonin plays a role in the facilitation of motoneurons, probably directly by acting on the CA^{2+} conductance or indirectly by reduction of K^+ conductance of the motoneuron membrane. Thus, serotonin might change the excitability of the motoneurons for inputs from other sources such as the motor cortex. Several other peptides are also present in the spinally projecting neurons in the ventromedial medulla and NRP and NRO. Many neurons contain substance P, thyrotropin-releasing hormone, somatostatin, methionine, and leucine-enkephalin, whereas relatively few neurons contain vasoactive intestinal peptide and cholecystokinin. Most of these peptides coexist to a variable extent with serotonin in the same neuron. It must be emphasized that a major portion of the diffuse descending pathways to the dorsal horn and the motoneuronal cell groups is not derived from serotonergic neurons. Possible neurotransmitter candidates for these pathways are acetylcholine and somatostatin, but γ-aminobutyric acid (GABA) and glycine may also play an important role. Colocalization of serotonin and GABA in spinal cord-projecting neurons in the ventral medulla in the rat has been demonstrated. This means that some terminals taking part in this diffuse descending system may have inhibitory as well as facilitatory effects on the postsynaptic element (i.e., the motoneuron), although the majority is probably either facilitatory or inhibitory. Spinal motoneurons display a bistable behavior, i.e., they can switch back and forth to a higher excitable level. Bistable behavior disappears after spinal transection, but reappears after subsequent intravenous injection of the serotonin precursor 5-hydroxytryptophan. Thus, intact descending pathways seem to be essential for this bistable behavior of motoneurons, and serotonin is one of the neurotransmitters involved in switching to a higher level of excitation. GABA may be involved in switching to a lower level of excitation.

In summary, the diffuse descending pathways originating in the ventromedial medulla, including the NRP and NRO, have very general and diffuse facilitatory or inhibitory effects on motoneurons as well as on the interneurons in the intermediate zone. Although most of the terminals have either a facilitatory or an inhibitory function, some evidence suggests that terminals also exist with both facilitatory and inhibitory functions.

B. Projections from the Dorsolateral Pontine Tegmental Field

Many neurons in the locus coeruleus and/or nucleus subcoeruleus and ventral part of the parabrachial nuclei project diffusely to all parts of the spinal gray matter. Many of these neurons contain noradrenaline or acetylcholine, but not both. Electrical stimulation in the area of the locus coeruleus and subcoeruleus produces a decrease in input resistance and a concurrent nonselective enhancement in motoneuron excitability indicative of an overall facilitation of motoneurons. Furthermore, in the rat there is some evidence that noradrenergic fibers, derived from the locus coeruleus and descending via the ventrolateral funiculus, have an inhibitory effect on nociception. Thus, neurons in the area of the locus coeruleus and subcoeruleus have an effect on the spinal cord, which is strikingly similar to that obtained after stimulation in NRM and NRP and their adjacent reticular formation.

C. Projections from the Rostral Mesencephalon and Caudal Hypothalamus

Dopamine-containing neurons located in the border region of the rostral mesencephalon and dorsal and posterior hypothalamus, extending dorsally along the paraventricular gray of the caudal thalamus, project throughout the whole gray matter at any level of the spinal cord. The distribution of the dopaminergic fibers (using the neurotransmitter dopamine) in the spinal gray strongly resembles that of noradrenergic fibers in the spinal cord. Functionally, there is also a resemblance between noradrenergic and dopaminergic fiber projections to the spinal cord. Infusion of dopamine in the spinal cord increases (sympathetic) motoneuron activity and has an inhibitory effect on noxious input to the spinal cord.

III. PROJECTIONS FROM THE MESENCEPHALON TO CAUDAL BRAIN STEM AND SPINAL CORD

The PAG and adjoining tegmental field project to a large number of structures in the caudal brain stem and spinal cord. A short description of some of these projections will be given, which are summarized in Fig. 3.

A. Descending PAG Projections to the NRM and NRP and Adjoining Ventral Part of the Caudal Pontine and Medullary Medial Tegmentum

Anterograde and retrograde tracing studies indicate that an enormous number of horseradish peroxidase (HRP)-labeled neurons in the PAG and laterally and ventrolaterally adjoining areas project in the same basic pattern to NRM, NRP, and the ventral part of the caudal pontine and medullary medial tegmentum (Fig. 4). On their way to the medulla they give off fibers to the area of the locus coeruleus and nucleus subcoeruleus. Neurons in the ventrolateral portion of the caudal PAG and the ventrally adjoining mesencephalic tegmentum send fibers to the NRP. A mediolateral organization exists within the descending mesencephalic pathways. The main projection of the medially located neurons, i.e., neurons in the medial part of the dorsal PAG, is to the medially located NRM and the immediately adjacent tegmentum. On the other hand, neurons in the lateral PAG, the laterally adjacent tegmentum, and the intermediate and deep layers of the superior colliculus project mainly laterally to the ventral part of the caudal pontine and upper medullary medial tegmentum with virtually no projections to the NRM. Figure 5 schematically shows this mediolateral organization in the ipsilaterally descending pathways from the dorsal mesencephalon.

I. Involvement of the Descending Mesencephalic Projections in Control of Nociception

In animals as well as in humans the PAG is well known for its involvement in the supraspinal control of nociception. The strong impact on nociception is partly mediated via its projections to the NRM and adjacent reticular formation, because in cases with reversible blocks of the NRM and adjacent tegmentum, PAG stimulation results in reduced analgesic effects. However, the analgesic effects do not completely disappear after blocking the NRM and adjacent tegmentum, which suggests that other brain stem regions also play a role.

2. Involvement of the Descending Mesencephalic Projections in Locomotion

Just lateral to the brachium coniunctivum, just ventral to the cuneiform nucleus, and just rostral to the parabrachial nuclei is located the so-called pedunculopontine nucleus, which contains cholinergic neurons. Stimulation in the pedunculopontine nucleus induces locomotion in cats, which is the reason for its name "mesencephalic locomotor region" (MLR). The MLR not only comprises the pedunculopontine nucleus, but extends into the cuneiform nucleus, which is located just dorsal to the pedunculopontine nucleus.

The descending projections from this area are organized similar to those from the PAG and adjacent tegmentum. The mainly ipsilateral fiber stream first descends laterally in the mesencephalon and upper pons and then gradually shifts medially to terminate bilaterally, but mainly ipsilaterally in the ventral part

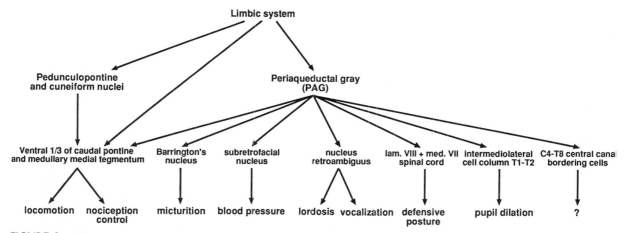

FIGURE 3 Schematic overview of the descending projections from the PAG and pedunculopontine and cuneiform nuclei to different regions of the caudal brain stem and spinal cord. The functions in which each of the projections might be involved are also indicated. It should be emphasized that these functional interpretations are only tentative.

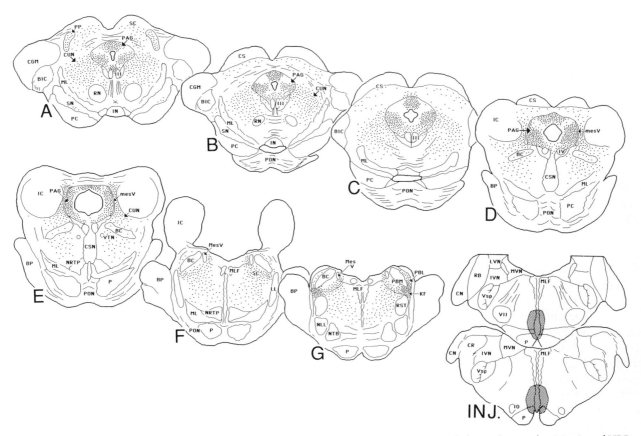

FIGURE 4   Schematic drawings of horseradish peroxidase (HRP)-labeled neurons in mesencephalon and pons after injection of HRP in the NRM/NRP region. Note the dense distribution of labeled neurons in the PAG (except its dorsolateral part) and the tegmentum ventrolateral to it. Note also the distribution of labeled neurons in the area of the ventral parabrachial nuclei and the nucleus Kölliker-Fuse (from Holstege, 1988a).

of the caudal pontine and medullary medial tegmental field. Electrical stimulation or injection of cholinergic agonists in this same caudal brain stem area also results in locomotion, and can control or override the stepping frequency induced by the mesencephalic locomotor region. Thus, locomotion, elicited in the MLR, seems to be based on the MLR projections to the medial part of the ventral medullary medial tegmentum, which, by way of their diffuse spinal projections, might control the rhythm generators in the spinal cord.

The afferent connections of the MLR are derived from lateral parts of the limbic system, such as the bed nucleus of the stria terminalis, central nucleus of the amygdala, and lateral hypothalamus. Strong afferent projections also originate in the entopeduncular nucleus, subthalamic nucleus, and the substantia nigra pars reticulata. In contrast, motor cortex projections to the MLR are very scarce. Thus, the MLR is influenced by extrapyramidal and lateral limbic structures and virtually not by somatic motor structures. Apparently, stepping should be considered as part of the emotional motor system and not as part of the somatic motor system.

B. PAG Projections to the Pontine Micturition Center; Involvement in Micturition Control

Neurons in the pontine micturition area in the dorsolateral pontine tegmentum in the cat (also called M-region or Barringtons area) coordinate bladder and bladder–sphincter contractions during micturition. The PAG projects to the pontine micturition area, and stimulation in the PAG has been shown to elicit micturition. The PAG seems to play an essential role in the supraspinal micturition reflex because, in contrast to the pontine micturition area itself, it receives direct specific afferents from the sacral cord. Appar-

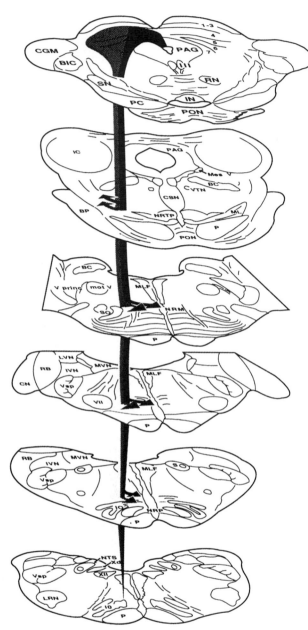

FIGURE 5 Schematic representation of the ipsilateral descending pathway, originating from the intermediate and deep layers of the superior colliculus and dorsal PAG. The mediolateral organization of this descending system is illustrated. The lateral (gray) component projects to the lateral aspects of the ventral part of the medial tegmentum of caudal pons and medulla oblongata. The medial (black) component projects to the medial aspects of the medial tegmentum, including the NRM. A similar mediolateral organization exists for the descending pathways originating in the more ventral part of the mesencephalic tegmentum (Holstege, 1991).

ently, the PAG serves as the basic integrator for micturition by receiving information concerning bladder filling and sending fibers to the pontine micturition area. [*See* Micturition Control.]

C. PAG Projections to the Ventrolateral Medulla; Involvement in Blood Pressure Control

Neurons in the subretrofacial nucleus (SRF), an area in the rostral part of the ventrolateral tegmental field of the medulla, are essential for the maintenance of the vasomotor tone and reflex regulation of the systemic arterial blood pressure. Cells in the rostral part of the SRF project specifically to the sympathetic preganglionic motoneurons in the intermediolateral cell column (IML), innervating the kidney and adrenal medulla, while neurons in the caudal part of the SRF innervate more caudal parts of the IML, which contain sympathetic preganglionic motoneurons innervating the hindlimb.

According to Bandler and co-workers, neurons in the dorsal portions of the caudal half of the PAG have an excitatory effect on the neurons in the SRF (increase of blood pressure), whereas neurons in the ventral part of the caudal PAG have an inhibitory effect (decrease of blood pressure). Furthermore, neurons in the subtentorial portion of the PAG project to the rostral part of the subretrofacial nucleus, which, in turn, sends fibers to the IML preganglionic motoneurons that innervate the kidney and adrenal medulla. On the other hand, neurons located slightly more rostral in the PAG, i.e., in the caudal part of the pretentorial PAG, project to the caudal subretrofacial nucleus, which in turn project to IML preganglionic motoneurons innervating the hindlimb. In conclusion, a precise organization exists in the mesencephalic control of blood pressure in different parts of the body.

D. PAG Projections to the Nucleus Retroambiguus; Involvement in Vocalization and Lordosis

I. Vocalization

Vocalization is the nonverbal production of sound and an example of emotional behavior. It should not be confused with speech. For speech direct cortical projections to mouth, pharynx, and larynx muscle motoneurons are needed and an enormous number of premotor cortical neurons to memorize how to pronunciate the different words. Only the human cortex is large enough to contain that number of neurons necessary for speech to take place. These neurons are located in Broca's area (Fig. 6). This is the reason that only the human species can speak. Patients with damage of the cortical areas involved in the produc-

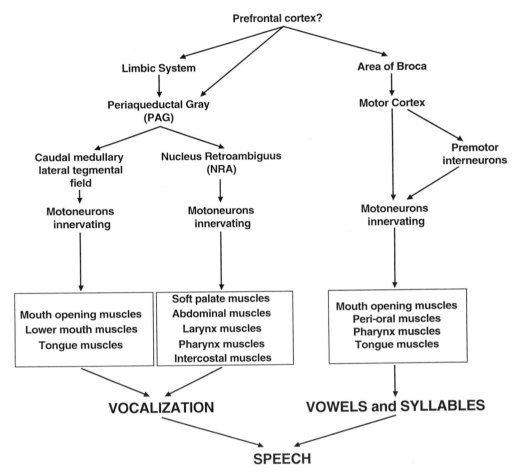

FIGURE 6　Schematic overview of the pathways possibly involved in the production of speech (from Holstege and Ehling, 1996).

tion or the perception of speech suffer from motor or sensory aphasia. These patients, however, can still vocalize, i.e., produce sounds, similar to animals. Apparently, their vocalization-producing apparatus is still intact. Possibly, speech might be considered as the (pre-)motor cortex modulation of vocalization (Fig. 6).

In many different species, from leopard frog to chimpanzee, stimulation in the caudal PAG results in vocalization. In humans, laughing and crying might be considered as examples of vocalization. Vocalization requires coordination between the activity of abdominal and expiratory intercostal muscles and muscles of the larynx, pharynx, and soft palate. The NRA is also involved in vocalization: NRA interneurons project to pharynx, larynx, and abdominal muscle motoneurons, and stimulation in the NRA produces noninte-

grated partial vocalization. A major portion of the NRA neurons involved in vocalization are located in the rostral part of the NRA because transections at −1 mm caudal to the obex abolished vocalization elicited in the PAG, whereas vocalization was not affected after transections at −4 mm caudal to the obex.

Holstege and co-workers have demonstrated in rat and cat that a specific group of neurons in the lateral and, to a limited extent, in the dorsal part of the caudal PAG send fibers to the NRA in the caudal medulla. The cell group in the PAG differs from the smaller cells projecting to the raphe nuclei and adjacent tegmentum or the larger cells projecting to the spinal cord. Direct PAG projections to the pharynx, larynx, soft palate, intercostal, and abdominal muscle motoneurons do not exist.

Although vocalization can also be elicited by stimulation in other parts of the lateral emotional motor system, including the lateral bed nucleus of the stria terminalis, central nucleus of the amygdala, and the lateral hypothalamus, the projection from the PAG to the NRA forms the final common pathway for vocalization (Fig. 7) because bilateral PAG lesions abolish vocalization completely.

2. Lordosis

Stimulation in the PAG also facilitates the lordosis reflex. Lordosis is species-specific, female receptive behavior. In the cat the full receptive posture consists of crouching (forelegs collapsed), lowering of the head, perineal elevation, tail deviation, and treading, often in combination with calling and vulval excretion. The PAG is essential for the control of lordosis. Stimulation of the lateral and dorsal PAG facilitates lordosis, whereas lesions suppress it in estrogen-primed rats. The question is how the PAG exerts its influence on lordosis behavior. In this respect its projection to the NRA might be of great importance, especially since the NRA has been shown to project to distinct motoneuronal cell groups in the lumbosacral cord (Fig. 8). Examples of motoneuronal cell groups receiving NRA afferents are axial muscles (e.g., medial longissimus), iliopsoas, hamstrings (biceps femoris, semimembranosus, and semitendinosus), adductors, and pelvic floor muscles. This combination of muscles seems to be activated specifically during lordosis behavior. Axial muscles extend the lower back, iliopsoas muscles fixate the lower spine to the pelvis, hamstring muscles cause extension of the hip and flexion of the knee, and adductor muscles stabilize the posture. Moreover, the medial longissimus muscle, when unilaterally activated, produces the typical deviation of the tail, and pelvic floor muscles contract rhythmically. Quadriceps (extensor of the knee) and triceps surae muscles (plantar flexors of the ankle) are not involved in this behavior. In the rat, mechanical stimulation of the cervix activates the iliopsoas muscles, and in the cat the m. semitendinosus is rhythmically active during mechanical stimulation of the cervix, whereas triceps surae muscles show sustained EMG activity during and after cervical stimulation.

One of the most important features of lordosis is that it is dependent on estrogen. The PAG cells projecting to the NRA might be estrogen-concentrating cells and they might receive a strong projection from estrogen-containing cells in the ventrolateral part of the ventromedial hypothalamus (VMH). Lordosis can also be facilitated and suppressed by stimulation and lesions in the VMH, respectively. Lesions in the VMH do not abolish PAG-facilitated lordosis in estrogen-primed rats, but when lesions are made in the PAG, VMH stimulation is not efficient for evoking lordosis. Apparently, the VMH is important for estrogen priming, but, in the same way as for vocalization, the PAG controls the final motor output of lordosis, and its projection to the NRA probably represents the final common pathway for lordosis behavior (Fig. 7).

The PAG also receives direct and specific ascending projections from the sacral spinal cord [See Micturition Control] and it is known to respond to lordosis-relevant somatosensory stimulation. Lordosis can still be elicited by somatosensory and vaginocervical stimulation in ovariectomized rats after precollicular decerebration. In freely moving animals, lordosis behavior is initiated easily by applying tactile stimuli to the skin of the flanks, posterior rump, tail base, and perineum (Fig. 9).

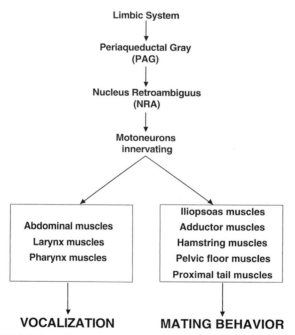

FIGURE 7 Schematic representation of the pathways for vocalization and lordosis from the limbic system to the vocalization and lordosis muscles.

E. PAG Projections to the Spinal Cord

Direct PAG projections to the spinal cord also exist. Some large neurons in the lateral PAG and laterally

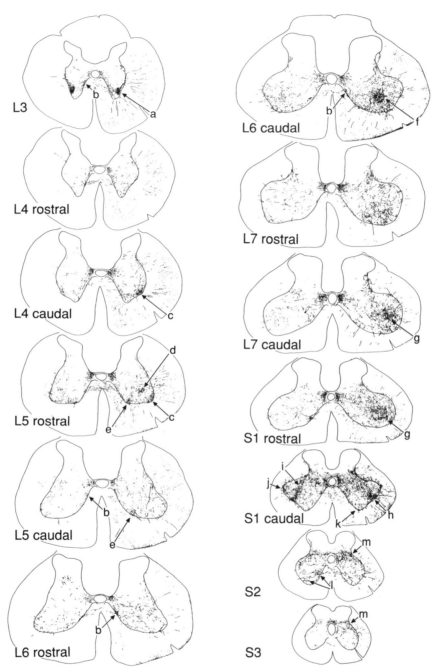

FIGURE 8 Schematic drawings of labeled fibers after injection of WGA-HRP in the NRA, combined with an ipsilateral hemisection. Each drawing represents the labeling of six consecutive collected (1:4) sections. Note the accumulation of labeled fibers in distinct portions of the ventral horn motoneuronal cell groups. The following motoneuronal cell groups receive NRA afferents: abdominal wall muscles (a), multifidus muscle (b), iliopsoas (c), adductor muscles (d), an axial muscle not yet clarified (e), semimembranosus (f), semitendinosus and biceps femoris (g), nucleus of Onuf, innervating the pelvic floor muscles (h), dendrites of Onuf motoneurons (i), intrinsic hindpaw muscles (j), caudal part of the medial longissimus (k), tail muscle (l), and the intermediolateral cell group containing autonomic (parasympathetic) motoneurons innervating bladder and sexual organs (m) (from VanderHorst and Holstege, 1995).

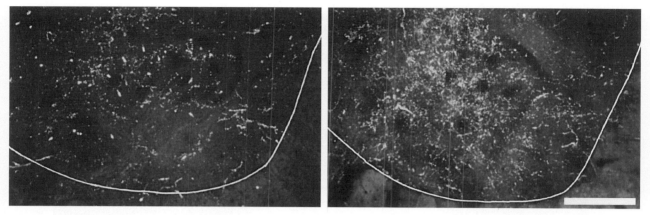

FIGURE 9   Dark-field polarized light photomicrographs of the wheat germ-agglutinin, labeled nucleus retroambiguus fibers in the motoneuronal cell groups of the semimembranosus horseradish peroxidase muscle in a nonestrous cat (left) and an estrous cat (right). Note the large difference in the density of the NRA fibers.

adjacent tegmentum send fibers via the ipsilateral ventral funiculus to terminate in laminae VIII and the adjoining part of VII throughout the length of the spinal cord. These projections are much denser to the cervical than to the lower parts of the spinal cord. The PAG–spinal neurons may play a role in the control of the axial muscles during threat display. A very few other fibers descend ipsilaterally in the lateral funiculus to terminate in the T1–T2 IML. This projection might be involved in the production of pupil dilation. Yet another pathway originates from a few hundred small neurons on the border between PAG and dorsally adjoining superior colliculus. Remarkably, these neurons terminate on cells bordering the spinal central canal at lower cervical and upper thoracic levels (Fig. 10). The function of this peculiar projection is completely unknown.

IV. PROJECTIONS FROM THE HYPOTHALAMUS TO CAUDAL BRAIN STEM AND SPINAL CORD

The descending hypothalamic projection systems differ greatly, depending on which part of the hypothalamus is considered. In this section the hypothalamus will be subdivided into anterior, paraventricular, posterior, and lateral parts.

A. Projections from the Anterior Hypothalamus/Preoptic Area

Neurons in the anterior hypothalamus/preoptic area project strongly to the caudal brain stem, but not to the spinal cord (Fig. 11). Neurons in the medial part of the anterior hypothalamus/preoptic area project (via a medial fiber stream) to central portions of the PAG and to the dorsal and superior central raphe nuclei in the pontine tegmentum. Further caudally, this pathway distributes many fibers to the ventromedial tegmentum of caudal pons and medulla, including the NRM and NRP. This medial preoptic/anterior hypothalamic projection clearly belongs to the medial component of the emotional motor system (Fig. 1) and, apparently, exerts a strong influence on the level setting system originating in the NRM/NRP and adjoining ventromedial medullary tegmentum. In functional terms, one could think that the medial preoptic area, which is known to play an important role in sexual behavior, many also set the level of general excitability. For example, during estrus the general somatic and autonomic motor activity of female rats or cats is much higher than when they are not in estrus. These differences in general excitability might be brought about by this medially descending pathway.

The anterior hypothalamus/preoptic area may also be involved in other survival mechanisms because application of cholinergic drugs in the anterior hypothalamus results in an emotional aversive response, which includes defense posture and autonomic (e.g., cardiovascular) manifestations. Possibly, these behaviors are also induced by the medially descending projections to PAG and ventromedial medulla.

B. Projections from the Paraventricular Hypothalamic Nucleus (PVN)

The PVN is best known for its projections to the hypophysis. Other neurons in the PVN project to the

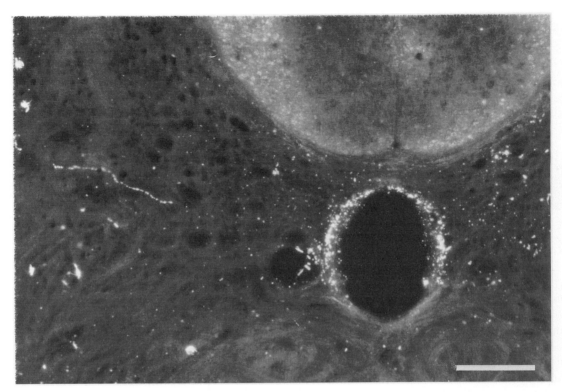

FIGURE 10   A dark-field photomicrograph combined with a polarized light photomicrograph showing the projection of neurons at the dorsal border area between PAG and superior colliculus terminating on cells bordering the central canal at level T2. Bar represents 1 mm.

caudal brain stem and spinal cord. These fibers go via the medial forebrain bundle and a well-defined pathway through the ventrolateral brain stem into the lateral and dorsolateral funiculus of the spinal cord throughout its total length. Via this pathway the PVN sends fibers to specific parts of the medullary lateral tegmental field, such as the noradrenergic brain stem nuclei A1, A2, and A5, the parasympathetic motoneurons in the salivatory nuclei, the rostral half of the solitary nucleus, in all parts of the dorsal vagal nucleus, and in the area postrema (Fig. 12). In the spinal cord (Fig. 13) the PVN projects bilaterally, but mainly ipsilaterally, to the marginal layer of the caudal spinal trigeminal nucleus and to laminae I and X throughout the length of the spinal cord, to the thoracolumbar (T1-L4) intermediolateral (sympathetic) motoneuronal cell group, and to the sacral intermediomedial and intermediolateral (parasympathetic) motoneuronal cell groups. Finally, the PVN projects to the nucleus of Onuf, which contains motoneurons innervating the pelvic floor muscles, including the urethral and anal sphincters. The function of this pathway is unknown, but one could speculate that in the light of the dif-

fuseness of the projection (the PVN projects to all autonomic neurons in brain stem and spinal cord), it is a general function, similar to, for example, the hormone ACTH. The PVN contains a large number of transmitter substances, and evidence shows that the PVN pathway to the caudal brain stem and spinal cord contains vasopressin and oxytocin as neurotransmitters. It is almost certain that other not yet identified transmitter substances also play a role in this pathway. The PVN is believed to play an important role in cardiovascular regulation and in feeding mechanisms.

C. Projections from the Medial Part of the Posterior Hypothalamic Area

The posterior hypothalamic area sends fibers via a medial pathway to the caudal raphe nuclei and adjoining reticular formation and to the spinal cord. The caudal hypothalamic projections to the NRM are weaker whereas those to the NRP are much stronger than more rostral hypothalamic projections to this area. In the spinal cord, caudal hypothalamic fibers

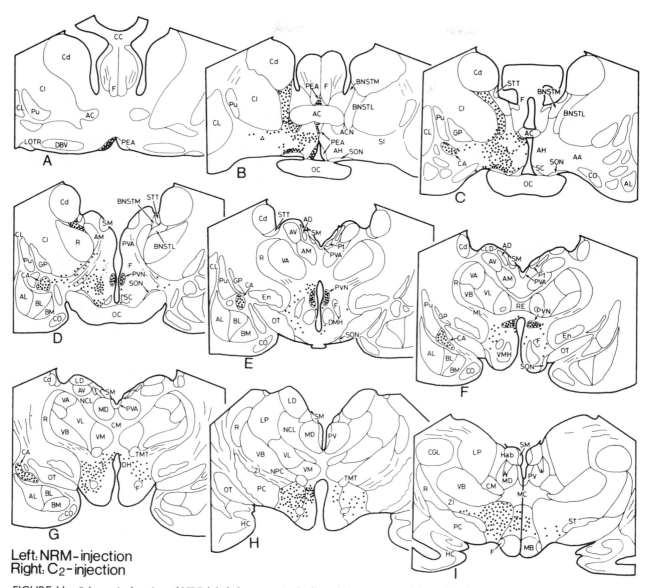

Left: NRM-injection
Right: C₂-injection

FIGURE 11 Schematic drawing of HRP-labeled neurons in the hypothalamus, amygdala, and bed nucleus of the stria terminalis. (Left) The pattern of distribution of labeled neurons after a large injection of HRP in the NRM, rostral NRP, and adjoining tegmentum is indicated. (Right) The pattern of distribution of HRP-labeled neurons after hemi-infiltration of HRP in the C2 spinal segment is shown (from Holstege, 1987).

terminate in the upper thoracic intermediolateral cell column and in lamina X.

The dorsomedial region of the caudal hypothalamus plays an important role in temperature regulation and contains the primary motor center for the production of shivering. Shivering is an involuntary response of skeletal muscles that are usually under voluntary control and all skeletal muscle groups can participate. Possibly, the strong caudal hypothalamic projections

to the NRP and adjacent tegmentum form part of this "shivering pathway."

D. Projections from the Lateral Hypothalamic Area

Functional and anatomic studies on the lateral hypothalamus have always been difficult because the fibers of the medial forebrain bundle pass through it. This

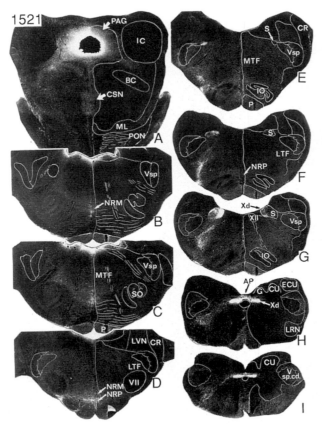

FIGURE 12 Dark-field photomicrographs of nine brain stem sections of a cat with an injection of [³H]-leucine in the area of the PVN of the hypothalamus. Note the distinct descending pathway in the area next to the pyramidal tract and its fiber distribution to the dorsal vagal nucleus, area postrema, and rostral solitary complex (from Holstege, 1987).

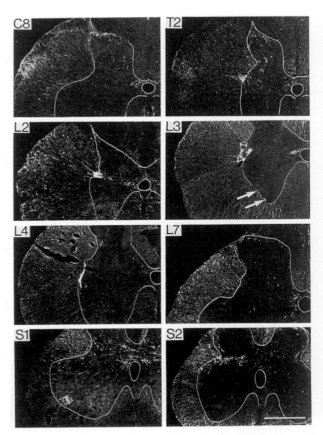

FIGURE 13 Dark-field photomicrographs of the spinal cord of the same cat as illustrated in Fig. 11, with a [³H]-leucine injection in the area of the PVN of the hypothalamus. Note the projection to lamina I (C8, T2, and L7), the sympathetic intermediolateral cell group (T2, L2, L3, and L4), the nucleus of Onuf (S1), and the parasympathetic intermediomedial and intermediolateral cell group (S2). The arrows in L3 probably indicate projections to distal dendrites of the motoneurons located in the sympathetic intermediolateral cell group (from Holstege, 1987).

important fiber bundle not only contains fibers originating in the lateral hypothalamus, but also in many other areas, and stimulation or lesions in this area not only affect lateral hypothalamic neurons, but also fibers derived from many other limbic structures. The lateral hypothalamus sends fibers to the PAG, the cuneiform nucleus, the parabrachial nuclei and nucleus Kölliker-Fuse, the nucleus subcoeruleus, the caudal pontine and medullary lateral tegmental field, and to the ventral part of the caudal pontine and medullary medial reticular formation. Only very few fibers terminate in the area of the NRM and none in the NRP. Some fibers terminate in the periphery of the dorsal vagal nucleus and in the rostral half of the solitary nucleus.

A distinct group of neurons in the lateral part of the rostral hypothalamus/preoptic area projects strongly to the pontine micturition area, and stimula-

tion of this part of the hypothalamus produces micturition-like bladder contractions. Only the caudal portion of the lateral hypothalamus sends fibers throughout the length of the spinal cord via the lateral and dorsolateral funiculi to the intermediate zone, lamina X, and the thoracolumbar sympathetic motoneurons. In summary, the lateral hypothalamus has direct access to autonomic motoneurons in brain stem and spinal cord, and indirect access, via premotor interneurons in the caudal pontine and medullary lateral tegmentum and spinal intermediate zone, to somatic motoneurons in brain stem and spinal cord. The lateral hypothalamus is involved in salivation and in feeding and drinking behavior, part of which might be brought about by its projections to the caudal brain stem lateral tegmental field, which contains premotor interneurons involved in activities such as swallowing, chewing, and licking. The lateral hypothalamus is probably also involved in cardiovascular control and defense behavior.

V. PROJECTIONS FROM AMYGDALA AND BED NUCLEUS OF THE STRIA TERMINALIS TO CAUDAL BRAIN STEM AND SPINAL CORD

It has been proposed that the central (CA) and medial amygdaloid nuclei and the lateral (BNSTL) and medial parts of the bed nucleus of the stria terminalis form a single anatomic entity, and many others have accepted this concept. The projections from CA and BNSTL to the caudal brain stem are virtually identical. Both structures send many fibers to the lateral hypothalamic area and, via the medial forebrain bundle, to the lateral part of mesencephalon, pons, and medulla oblongata (Fig. 14). At mesencephalic levels, fibers terminate in the PAG (except its dorsolateral part), the ventrolaterally adjoining nucleus cuneiformis and pedunculopontine nucleus, and the mesencephalic lateral tegmental field. In the pons, fibers, terminate laterally in the tegmentum, i.e., in the medial and lateral parabrachial nuclei, the nucleus Kölliker-Fuse, the nucleus subcoeruleus, and possibly the locus coeruleus. At caudal pontine and medullary levels the main termination area is the lateral tegmental field, including the rostral and caudal parts of the solitary nucleus and the peripheral parts of the dorsal vagal nucleus. An exception are the fibers that branch off from the laterally descending fiber bundle to terminate in the ventral part of the caudal pontine and upper medullary medial tegmentum. A few fibers terminate in the NRM, but none in the NRP (Fig. 13). Both CA and BNSTL send a few fibers to the intermediate zone of the C1 spinal cord, but not beyond that level.

Neurons in CA and BNSTL receive many afferent fibers from other (basolateral and basomedial) amygdaloid nuclei, but these connections are not reciprocal. Apparently, both CA and BNSTL serve as "output nuclei" for other parts of the anygdala/bed nucleus of the stria terminalis complex to reach the caudal brain stem. A great similarity exists between the caudal brain stem projections originating in the CA and BNSTL on the one hand and the lateral hypothalamic

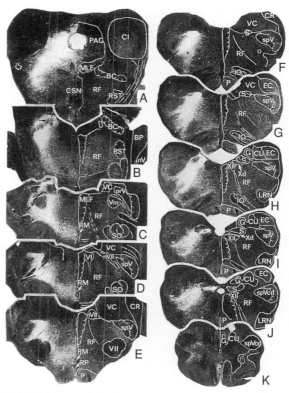

FIGURE 14 Dark-field photomicrographs of 11 brain stem sections of a cat with an injection of [³H]-leucine in the bed nucleus of the stria terminalis, which is almost identical to the projections from the central nucleus of the amygdala and the lateral hypothalamus. Note the strong projection to the PAG, with the exception of its dorsolateral part. Note also the strong projection to the bulbar lateral tegmental field and the projection to the ventral part of the caudal pontine and upper medullary medial tegmentum (from Holstege, 1991).

area on the other. In addition, all three areas have very strong mutual connections.

The direct projections from CA, BNSTL, and the lateral hypothalamus to the caudal brain stem lateral tegmental field may form the anatomic framework of the final output of the defense response of the animal. Electrical stimulation in the amygdala, bed nucleus of the stria terminalis, lateral hypothalamus, and PAG elicits defensive behavior. In fact, a column of electrical stimulation sites exists from CA, BNST, lateral hypothalamus, and PAG through the lateral mesencephalic tegmentum into the lateral tegmentum of the caudal brain stem, which elicits defensive behavior. The initial phase of such a response is arrest of all spontaneous ongoing activities, and the whole attitude of the animal changes into one of attention. The arousal is followed by orienting or searching move-

Descending Pathways from Limbic System to Caudal Brain Stem and Spinal Cord

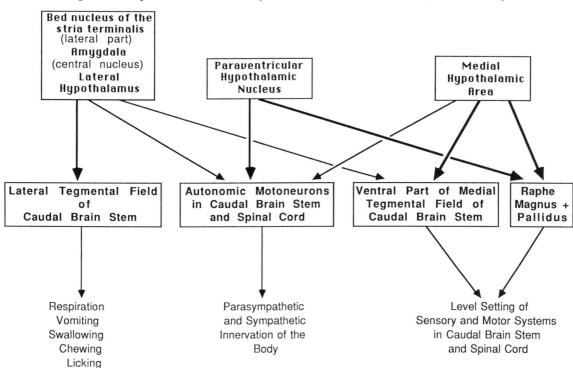

FIGURE 15 Schematic overview of the mediolateral organization of the limbic system pathways to brain stem and spinal cord and its possible functional implications. The strongest projections are indicated by thick arrows (from Holstege, 1988).

ments toward the contralateral side, frequently accompanied by sniffing, swallowing, chewing, and twitching of the ipsilateral facial musculature. Later in the defense reaction the cat retracts its head and crouches with the ears flattened to a posterior position. The cat growls or hisses, the pupils are dilated, and there is piloerection, elevation of blood pressure with bradycardia, an increased rate of breathing, and an alteration of gastric motility and secretion. On stronger stimulation an "affective" attack may take place, in which the cat strikes with its paw with claws unsheathed, in a series of swift, accurate blows. If the stimulus continues, the cat will bite savagely. Many of the activities in the beginning of the defense response are coordinated in the PAG and at a more primitive level in the caudal brain stem lateral tegmental field. The observation that part of this behavior appears to be ipsilateral corresponds with the predominantly ipsilateral projection of CA, BNSTL, and lateral hypothalamus to the PAG and caudal brain stem lateral tegmentum.

Figure 15 gives a schematic overview of the descending projections to the caudal brain stem from the hypothalamus, amygdala, and BNST. Similar to the descending projections from the mesencephalon, there is a mediolateral organization within this descending system in which the medial hypothalamus forms the medial component whereas the lateral hypothalamus, amygdala, and BNST form the lateral component. The PVN, with its direct projections to al preganglionic (sympathetic and parasympathetic) motoneurons in brain stem and spinal cord, occupies a separate position within this framework.

VI. PROJECTIONS FROM THE PREFRONTAL CORTEX TO CAUDAL BRAIN STEM AND SPINAL CORD

The prefrontal cortex projects directly to the caudal brain stem. In the rat, the medial frontal cortex sends fibers to the solitary nuclei, the dorsal parts of the parabrachial nuclei, the PAG, and the superior colliculus, while the insular cortex also projects to the solitary nuclei and PAG. Electrical stimulation of the rat's insular cortex leads to the elevation of arterial pressure and cardioacceleration. In the cat, the orbital gyrus, anterior insular cortex, and infralimbic cortex project to the solitary nuclei. In the monkey, the dorsolateral and dorsomedial prefrontal cortex projects to the locus coeruleus and nucleus raphe centralis superior.

VII. CONCLUSIONS

The emotional motor system represents the descending pathways from the limbic system-related structures to caudal brain stem and spinal cord. According to this concept (Fig. 1) the somatic motor system consists of the pathways that originate in the motor cortex, red nucleus, and brain stem nuclei such as vestibular nuclei, the dorsal part of the medial tegmentum, and the interstitial nucleus of Cajal. The basic motor system in this concept is formed by the premotor interneurons in brain stem and spinal cord, projecting to the motoneurons. The pathways belonging to the emotional motor system, which corresponds to some extent with the core and medial paracore of Nieuwenhuys, do not overlap with the pathways of the somatic motor system. An exception of this rule are the monoaminergic projections originating in the raphe nuclei and locus coeruleus/subcoeruleus complex, whose fibers terminate in many differnet structures in the central nervous system, including some belonging to the somatic motor system. Furthermore, with the exception of acetylcholine, glutamate, and aspertate, most of the many other neurotransmitters or neuromodulators within the central nervous system are found in the emotional motor system and not in the somatic motor system.

A mediolateral organization is not only present in the somatic motor system, but also in the emotional motor system, the medial component of which arises in the medial portions of hypothalamus and in major portions of the mesencephalon. Its main target is the ventral part of the caudal pontine and medullary medial tegmental field, including the caudal raphe nuclei.

The lateral component originates laterally in the limbic system, i.e., in the lateral hypothalamus, central nucleus of the amygdala, and bed nucleus of the stria terminalis. These structures project to the premotor interneurons in the lateral tegmental field of caudal pons and medulla, but not to the somatic motoneurons in this area. Also, the pathways originating in certain groups of PAG neurons, which project to laterally located structures in the caudal brain stem, but

not to motoneurons, belong to the lateral component of the emotional motor system. How far the prefrontal cortex plays a role within these systems remains to be determined.

A. Function of the Medial Component of the Emotional Motor System

The medial system, via its projections to the locus coeruleus/nucleus subcoeruleus and NRM and NRP/NRO and the diffuse coeruleo- and raphe-spinal pathways, has a global effect on the level of activity of the somatosensory and motoneurons in general by changing their membrane excitability. In other words, the emotional brain has a great impact on the sensory as well as on the motor system. In both systems it sets the gain or level of functioning of the neurons. The emotional state of the individual determines this level. For example, it is well known that many forms of stress, such as aggression, fear, and sexual arousal, induce analgesia, while at the same time the motor system is set at a "high" level and motoneurons can easily be excited by the somatic motor system or the lateral component of the emotional motor system. One might consider the brain stem structures, which project diffusely to all parts of the spinal cord, as tool for the limbic system controlling spinal cord activity. The diffuse descending system may also be used to trigger rhythmical (locomotion, shivering) or other in essence spinal reflexes.

B. Function of the Lateral Component of the Emotional Motor System

Some groups of neurons in the PAG are related to specific functions as vocalization, lordosis, micturition, and blood pressure control. These final common pathways belong to the lateral component of the emotional motor system. The same is true for projections from more rostrally located lateral limbic structures (central nucleus of the amygdala, lateral part of the bed nucleus of the stria terminalis, and lateral hypothalamus). They send fibers to the caudal brain stem lateral tegmental field, with basic system premotor interneurons involved in functions as respiration, vomiting, swallowing, chewing, and licking. These activities are displayed in the beginning of the flight or defense response and can be easily elicited by stimulation in the PAG or in the lateral parts of the limbic system. Thus, the structures belonging to the lateral component of the emotional motor system are responsible for the specific motor activities, related to emotional behavior.

BIBLIOGRAPHY

Carrive, P., and Bandler, R. (1991). Control of extracranial and hindlimb blood flow by the midbrain periaqueductal grey of the cat. *Exp. Brain Res.* **84,** 599–606.

Davis, P. J., Zhang, S. P., and Bandler, R. (1995). Midbrain and medullary regulation of respiration and vocalization. *Progr. Brain Res.* **107,** 315–325.

Heimer, L., de Olmos, J., Alheid, G. F., and Zaborszky, L. (1991). "Perestroika" in the basal forebrain: Opening the border between neurology and psychiatry. *In* "Role of the Forebrain in Sensation and Behavior" (G. Holstege, ed.). *Progr. Brain Res.* **87,** 109–165.

Holstege, G. (1987). Some anatomical observations on the projections from the hypothalamus to brainstem and spinal cord: An HRP and autoradiographic tracing study in the cat. *J. Comp. Neurol.* **260,** 98–126.

Holstege, G. (1988a). Direct and indirect pathways to lamina I in the medulla oblongata and spinal cord of the cat. *In* "Pain Modulation" (H. L. Fields and J. M. Besson, ed.). *Progr. Brain Res.* **77,** 47–94.

Holstege, G. (1988b). Brainstem-spinal cord projections in the cat, related to control of head and axial movements. *In* "Neuroanatomy of the Oculomotor System" (J. Büttner-Ennever, ed.), pp. 431–469. Elsevier, Amsterdam.

Holstege, G. (1991). Descending motor pathways and the spinal motor system: Limbic and non-limbic components. *In* "Role of the Forebrain in Sensation and Behavior" (G. Holstege, ed.). *Progr. Brain Res.* **87,** 307–412.

Holstege, G. (1996). The somatic motor system. *In* "The Emotional Motor System" (G. Holstege *et al.*, eds.), Vol. 107, pp. 9–26.

Holstege, G., and Ehling, T. (1996). Two motor systems involved in the production of speech. *In* "Vocal Fold Physiology, Controlling Complexity and Chaos" (P. J. Davis and N. H. Fletcher, eds.), pp. 153–169. Singular Publishing Group, San Diego.

Kaada, B. (1972). Stimulation and regional ablation of the amygdaloid complex with reference to functional representation. *In* "The Neurobiology of the Amygdala" (B. E. Eleftheriou, ed.), pp. 145–204. Plenum Press, New York.

Kuypers, H. G. J.M. (1981). Anatomy of the descending pathways. *In* "Handbook of Physiology" (R. E. Burke, ed.), Vol. II, pp. 597–666. Washington American Physiological Society.

MacLean, P. D. (1952). Some psychiatric implications of physiological studies on frontotemporal portion of limbic system. *EEG Clin. Neurophysiol.* **4,** 407–418.

Mouton, L. J., and Holstege, G. (1994). The periaqueductal gray in the cat projects to lamina VIII and the medial part of VII throughout the length of the spinal cord. *Exp. Brain Res.* **101,** 253–264.

Mouton, L. J., Kerstens, L., VanderWant, J., and Holstege, G. (1996). Dorsal border periaqueductal gray neurons project to the area directly adjacent to the central canal ependyma of the C4-T8 spinal cord in the cat. *Exp. Brain Res.* **112,** 11–23.

Nauta, W. J. H. (1958). Hippocampal projections and related neural pathways to the mid-brain in the cat. *Brain* **80,** 319–341.

Nieuwenhuys, R., Voogd, J., and Van Huijzen, C. (1988). "The Human Central Nervous System," 3rd Ed. Springer-Verlag, Berlin/Heidelberg.

Pfaff, D. W., Schwartz-Giblin, S., McCarthy, M. M., and Kow, L.-M. (1994). Cellular and molecular mechanisms of female reproductive behaviors. *In* "The Physiology of Reproduction," Chap. 36, pp. 107–220. Raven Press, New York.

Price, J. L., Russchen, F. T., and Amaral, D. G. (1987). The limbic region. II. The amygdaloid complex. *In* "Handbook of Chemical Neuroanatomy" (A. Björklund, T. Hökfelt, and L. W. Swanson, eds.), Vol. 5. pp. 279–388. Elsevier Science Publishers, Amsterdam.

Swanson, L. W., and Sawchenko, P. E. (1983). Hypothalamic integration: Organization of the paraventricular and supraoptic nuclei. *Annu. Rev. Neurosci.* **6,** 269–324.

VanderHorst, V. G. J. M., and Holstege, G. (1995). Caudal medullary pathways to lumbosacral motoneuronal cell groups in the cat; evidence for direct projections possibly representing the final common pathway for lordosis. *J. Comp. Neurol.* **359,** 457–475.

VanderHorst, V. G. J. M., and Holstege, G. (1997). Estrogen induces axonal outgrowth in the nucleus retroambiguus–lumbosacral motoneuronal pathway in the adult female cat. *J. Neurosci.* **17,** 1122–1136.

VanderHorst, V. G. J. M., and Holstege, G. (1997). Nucleus retroambiguus projections to lumbosacral motoneuronal cell groups in the male cat. *J. Comp. Neural.*, in press.

Endocrine System

HOWARD RASMUSSEN
Yale University School of Medicine

FRANKLYN F. BOLANDER, JR.
University of South Carolina

GLOSSARY

Acinar Pertaining to a cell with a small sac-like dilatation; can be found in various glands

Amino acids Any organic compound containing an amino (-NH$_2$) and a carboxyl (-COOH); building blocks for protein synthesis

Arachidonic acid Unsaturated fatty acid essential for human nutrition; precursor in the biosynthesis of leukotrienes, prostaglandins, and thromboxanes

Autonomic nervous system Portion of the nervous system concerned with the regulation of the activity of the cardiac muscle, smooth muscle, and glands

Chromosomes Structure in the nucleus containing a linear thread of DNA, which transmits genetic information and is associated with RNA and histones

Cytokine Hormone affecting the proliferation or differentiation of blood or immune cells

Cytosol Liquid medium of the cytoplasm (the protoplasm of the cell)

Ectopic Located or produced away from its normal position

Enzymes Protein molecule that catalyzes chemical reactions of other substances without itself being destroyed or altered upon completion of the reaction

Fatty acids Any straight-chain single carboxyl group acid, especially those naturally occurring in fats, generally classified as saturated when they have no double bonds, monounsaturated when they have one double bond, and polyunsaturated when they have multiple double bonds

Gene Segment of a DNA molecule that contains all the information required for synthesis of a product; biological unit of heredity

Glycerol Trihydric sugar alcohol (CH$_2$OHCHOHCH$_2$OH); alcoholic component of fats

Glycogen Chief carbohydrate storage material in animals, a long-chain polymer of glucose

Heterologous Different in either structure, position, or origin

Homologous Corresponding in structure, position, and origin

Ion Atom or radical having a charge (positive or negative) owing to the loss or gain of one or more electrons

Ketone Any of a large class of organic compounds containing the carbonyl group (C=O), whose carbon atom is joined to two other carbon atoms; ketone bodies include β-hydroxybutyric acid, acetoacetic acid, and acetone

Kinase Subclass of the transferase enzymes, which catalyze the transfer of a high-energy phosphate group from a donor compound (adenosine triphosphate or guanosine triphosphate) to an acceptor compound

Lactic acid End product of glycolysis involved in many biochemical processes

Mitochondria Small separate organelle found in the cytoplasm of cells, which is the principal site for the generation of energy (adenosine triphosphate)

Molar Measure of the concentration of solute, expressed as the number of moles of solute per liter

Multimeric Consisting of multiple subunits

Neurotransmitter Any of a group of substances that are released on excitation from the axon terminal of a presynaptic neuron of the central or peripheral nervous system and travel across the synaptic cleft to either excite or inhibit the target cell

Peroxidase Sub-subclass of enzymes of the oxidoreductase class that catalyze the oxidation of organic substrates by hydrogen peroxide, which is reduced to water

Phosphatases Any of a subclass of enzymes of the hydrolase class that catalyze the release of inorganic phosphate from phosphoric esters

Phosphodiesterases Any of a subclass of enzymes of the hydrolase class that catalyze the hydrolysis of one of the two ester linkages in a phosphodiester compound

Polymer Chemical compound consisting of repeating structural units

Proteolysis Splitting of proteins by hydrolysis of the peptide bonds with formation of smaller polypeptides

Pseudopodia Temporary cytoplasmic extrusion by means of which an ameba or other ameboid organism or cell moves about or engulfs food

Somatic Pertaining to or characteristic of the body

Transducer Translation of one form of signal to another

Very low-density lipoprotein Lipid–protein complex in which lipids are transported in the blood

TWO MAJOR SYSTEMS INTEGRATE CELLULAR AND tissue responses in multicellular organisms: the nervous and endocrine systems. Neural responses are characterized by their rapidity and their restricted nature; in contrast, endocrine responses are characterized by a wide range of temporal responses from relatively rapid to prolonged and by their diffuse nature. Even so, these systems are ultimately linked at the level of the hypothalamus, properly called a neuroendocrine organ. Also, the two systems possess important similarities. First, transmission of information from one cell to the other is via chemical messenger and, second, the target cells possess specific receptors, or recognition molecules, for these chemical messengers. The chemical messengers in the nervous system are known as neurotransmitters, and those of the endocrine system as hormones. Initially, neurotransmitter receptors were thought to be linked to ion channels, and hormone receptors to membrane-associated enzymes or to chromosomes; however, evidence now indicates that these sharp distinctions are not as complete as once believed. A particular entity can serve as a neurotransmitter in one setting and as a hormone in another. Furthermore, some hormones act on receptors linked to ion channels, and some neurotransmitters act on membrane-associated enzymes. In addition, some neurotransmitters are released into the bloodstream and have a wide distribution in the body. In fact, the same chemical entity can serve to convey information from one cell to another by a variety of pathways. [See Neuroendocrinology.]

I. HORMONES AND ENDOCRINE TISSUES

A hormone was originally defined as an agent released into the bloodstream from cells in one organ and then transported to another organ, where it interacted with target cells to alter their function. During the first 50–60 years of this century, the term hormone came to mean an agent released into the bloodstream from specific endocrine organs. These classic endocrine organs were recognized as specific tissues whose sole function was to receive chemical signals from one or more bodily organs, process this information, and then release a specialized chemical mediator to alter the function of other organs. Hence, the notion of endocrine mediator and hormone were synonymous. However, in the past 20–25 years, it has become clear that the same chemical entity can act either as a neurotransmitter or as an autocrine, paracrine, neurocrine, or endocrine mediator (Fig. 1). A neurotransmitter is an agent released from a nerve ending (synapse) that acts on a second neuron. An autocrine agent is released into the cellular environment and acts on the same cell from which it was released; a paracrine agent acts on adjacent cells in the same tissue or organ (without entering the bloodstream); a neurocrine agent is released from nerve synapses and acts on one or more postsynaptic cells that are not neurons; and an endocrine agent acts on distant target cells. Thus, a more general definition of a hormone is an agent released from one cell that acts to alter the function of that cell, an adjacent cell, or a distant cell.

More recently, the definition of what constitutes an endocrine organ has also expanded. The original definition encompassed 10 highly specialized organs, each of whose function appeared to be solely that of secreting a particular hormone(s). It is now clear that practically every tissue in the body secretes chemical messengers, which regulate either their own function or those of distant cells, tissues, or organs.

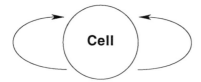

Cell

Endocrine

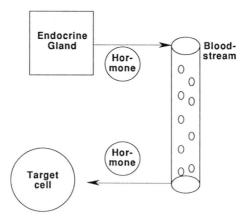

FIGURE 1 The various forms of hormone secretion: autocrine, in which the secreted product acts on its cell of origin; paracrine, in which the secreted product acts on an adjacent cell in the same tissue; neurocrine, in which a neuron secretes the product that acts on distant cells; and endocrine, in which the secreted product enters the bloodstream and is carried to a distant target cell.

Paracrine

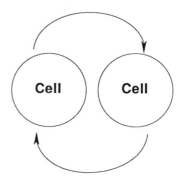

Cell   **Cell**

Neurocrine

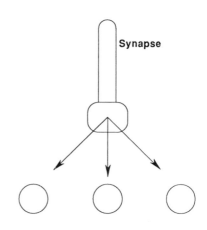

Synapse

A. Classic Endocrine Tissues and Hormones

The classic endocrine organs (Fig. 2) are (1) hypothalamus, (2) anterior pituitary, (3) posterior pituitary, (4) thyroid, (5) parathyroid, (6) islets of Langerhans in the pancreas, (7) adrenal cortex, (8) adrenal medulla, (9) ovary, and (10) testes.

Three types of chemical entities have been identified as the products of these tissues (Tables I–III). The steroid hormones are listed in Table I. In this and the succeeding tables, the name, source, site of action, and principal actions are listed for each hormone. In the case of the steroid hormones, most have one or two principal actions; however, cortisol has a wide spectrum of effects, and the sex hormones have effects not only on the uterus and testes but also on all the secondary sex organs such as the breast and vagina

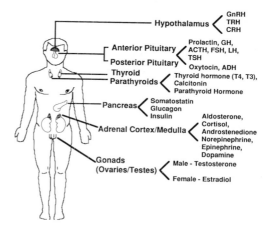

FIGURE 2 Classic endocrine glands and their hormones. ACTH, adrenocorticotrophic hormone; ADH, antidiuretic hormone; CRH, corticotropin-releasing hormone; FSH, follicle-stimulating hormone; GH, growth hormone; GnRH, gonadotropin-releasing hormone; LH, luteinizing hormone; TRH, thyrotropin-releasing hormone; TSH, thyroid-stimulating hormone.

TABLE I
Steroidal Hormones

Hormone	Source	Site of action	Action
Estradiol	Ovary	Uterus	Growth of endometrium
		Breast	Growth of breast and other secondary sexual organs
		Bone	Growth of skeleton
Progesterone	Ovary	Uterus	Maturation of endometrium
Testosterone	Testes	Spermatogonia	Spermatogenesis
		Male sex organs	Growth and development
		Muscle and bone	Growth
Dihydrotestosterone	Several tissues convert testosterone to dihydrotestosterone	Male sex organs	Growth and development
Cortisol	Zona fasciculata of adrenal cortex	Many tissues—liver, connective tissue, bone, fat cells, lymphoid tissue	Multiple effects that are part of the response to stress
Aldosterone	Zona glomerulosa of adrenal cortex	Kidney Intestine	Increased retention of sodium and decreased retention of potassium
Androstenedione	Zona reticularis of adrenal cortex	Secondary sex organs	Weak effects similar to testosterone
1,25-Dihydroxyvitamin D_3	Kidney	Intestine	Increased absorption of calcium and phosphate
		Bone	Increased resorption of calcium and phosphate
24,25-Dihydroxyvitamin D_3	Kidney and other tissues	Bone	Increased bone formation and mineralization

in the female, and the penis, skeletal muscle, and hair growth in the male. The steroid hormones include the female sex hormones estradiol and progesterone from the ovary, the male hormones testosterone and dihydrotestosterone from the testes and androstenedione from the adrenal cortex, and the adrenal cortical hormones aldosterone and cortisol. [See Steroids.]

The amine hormones (Table II) include epinephrine and norepinephrine from the adrenal medulla, serotonin and dopamine from the hypothalamus, histamine from mast cells, and thyroxine and triiodothyronine from the thyroid acinar cells. Again, only a few of the major effects of epinephrine, norepinephrine, and serotonin are listed. These substances are released

TABLE II
Amine Hormones

Hormone	Source	Site of action	Action
Epinephrine	Adrenal medulla Nerve endings	Liver, heart, blood vessels Fat cells, muscle, blood cells	Increases glucose production, breakdown of fat, heart rate, and many other actions
Norepinephrine	Adrenal medulla Nerve endings	Blood vessels Brain	Contraction Neurotransmitter
Dopamine	Nerve endings	Anterior pituitary Brain	Inhibits prolactin secretion Neurotransmitter
Serotonin	Gut endocrine cells	Microvessels Bronchial muscle	Initiates inflammatory response Contraction
Thyroxine (T_4)	Thyroid gland	Many different cells	Regulates basal metabolic rate
Triiodothyronine (T_3)	Thyroid and liver ($T_4 \rightarrow T_3$)	Many different cells	Regulates basal metabolic rate

TABLE III
Classic Protein and Peptide Hormones

Hormone	Source	Site of action	Action
Insulin	Beta cells of endocrine pancreas	Liver, muscle, fat cells	Increases storage of glucose, fat, and protein
Glucagon	Alpha cells of endocrine pancreas	Liver	Increases glucose production
Parathyroid hormone	Parathyroid gland	Kidney	Increases calcium retention, increases synthesis of 1,25-dihydroxyvitamin D_3, decreases phosphate retention
		Bone	Increases bone resorption
Calcitonin	C cells of thyroid gland	Bone	Inhibits bone resorption
Somatostatin	Hypothalamus, endocrine pancreas, gut endocrine cells	Anterior pituitary	Inhibits growth hormone secretion, inhibits insulin secretion, inhibits actions of several gut hormones
Angiotensin II	Bloodstream from a precursor	Blood vessels	Contraction
		Zona glomerulosa	Stimulates aldosterone secretion
		Brain	Neurotransmitter
		Kidney	Decreased sodium excretion
Atrial natriuretic peptide	Atria of heart	Kidney	Increased sodium and H_2O excretion
		Zona glomerulosa	Inhibits aldosterone secretion
		Blood vessels	Dilatation
Vasopressin	Hypothalamus Posterior pituitary	Kidney	Increased H_2O retention
Oxytocin	Hypothalamus Posterior pituitary	Breast	Milk secretion
		Uterus	Contraction
Thyroid-stimulating hormone (TSH)	Anterior pituitary	Thyroid gland	Stimulates synthesis and secretion of thyroid hormone
Adrenocorticotrophic hormone (ACTH)	Anterior pituitary	Adrenal cortex	Stimulates synthesis and secretion of cortisol, androstenedione, and aldosterone
Follicle-stimulating hormone (FSH)	Anterior pituitary	Ovary	Increased estradiol synthesis and secretion
Luteinizing hormone (LH)	Anterior pituitary	Ovary	Increased synthesis and secretion of progesterone
		Testes	Increased synthesis and secretion of testosterone
Prolactin	Anterior pituitary	Breast	Increased milk products
Growth hormone	Anterior pituitary	Liver	Synthesis and release of insulin-like growth factor
Melanocyte-stimulating hormone	Intermediate lobe of the pituitary	Melanocytes	Skin darkening
Thyrotropin-releasing hormone	Hypothalamus	Anterior pituitary	Release of TSH
Corticotrophin-releasing hormone	Hypothalamus	Anterior pituitary	Release of ACTH
Gonadotropin-releasing hormone	Hypothalamus	Anterior pituitary	Release of FSH and LH
Growth hormone-releasing factor	Hypothalamus	Anterior pituitary	Release of GH

(continues)

TABLE III (*Continued*)

Hormone	Source	Site of action	Action
Insulin-like growth factor	Liver	Muscle	Growth and proliferation
		Bone and cartilage cells	Growth and proliferation
		Other cells	Growth and proliferation
Nerve growth factor	Several	Nerve cells	Growth and proliferation
Epidermal growth factor	Several	Epidermal cells and many others	Growth and proliferation
Large and increasing number of tissue-derived growth factors such as platelet-derived growth factors	Many	Large number of cells—several growth factors often work in combination	Growth and proliferation of a variety of different cell types
Cholecystokinin	Gut endocrine cells	Exocrine pancreas	Stimulates secretion of digestive enzymes
		Endocrine pancreas	Stimulates secretion of insulin
		Gallbladder	Stimulates contraction
Gastric inhibitory peptide	Gut endocrine cells	Stomach	Inhibits motility and acid secretion
		Endocrine pancreas	Stimulates insulin secretion
Secretin	Gut endocrine cells	Exocrine pancreas	Stimulates waste and salt excretion
Vasointestinal peptide	Gut endocrine cells	Intestine	Stimulates fluid and electrolyte secretion
		Smooth muscle	Stimulates contraction
Erythropoietin	Kidney	Bone marrow	Stimulates red cell production
Bombesin	Gut endocrine cells (others)	Many	Stimulates acid secretion
Gastrin	G cells in stomach	Parietal cells in stomach	Stimulates insulin secretion
Glucagon-like peptide-1	Gut endocrine cells	Endocrine pancreas	Stimulates insulin secretion
Endothelin	Endothelial cells	Smooth muscle	Stimulates contraction

from nerve terminals in the central nervous system in the autonomic nervous system, and in the enteric (intestinal) nervous system where they exert a variety of effects on secretory processes and on the state of contraction of smooth muscle cells in blood vessels, bronchi, and the intestinal wall.

The peptide and protein hormones are the largest and most diverse group (Table III), and each year additional members of this group continue to be identified. The list in Table III is incomplete in the sense that many of these peptides serve different functions in different locations. For example, thyrotropin-releasing hormone (TRH) acts on the pituitary to stimulate thyroid-stimulating hormone (TSH) secretion, but in the central nervous system TRH acts as a neurotransmitter. Likewise, a complete listing of the numerous peptide growth factors has not been given. Many of these stimulate, either alone or in combination with other growth factors, the growth and proliferation of cultured mammalian

cells, but often the specific cell types they act on in an intact animal are unknown. Furthermore, amines, such a serotonin, or peptides, such as angiotensin, in addition to their classic actions, can also serve as one of the growth factors that stimulate the proliferation of certain classes of cells. It is particularly noteworthy that many intestinal peptides and amines that act as endocrine or neuroendocrine agents are also found in the central nervous system, where they are thought to act on neurotransmitters. The classic peptide hormones include insulin and glucagon from the islets of Langerhans; calcitonin from the thyroid C cells; parathyroid hormone (PTH) from the parathyroid gland; TSH, follicle-stimulating hormone (FSH), luteinizing hormone (LH), prolactin, adrenocorticotropin (ACTH), and growth hormone from specific cell types in the anterior pituitary; vasopressin and oxytocin from the posterior pituitary; and TRH, corticotropin-releasing hormone, growth hormone-releasing factor, gonadotrophin-releasing hormone, and somatostatin from specific loci in the hypothalamus. [*See* Peptide Hormones of the Gut.]

B. Nonclassic Endocrine Tissues and Hormones

Many tissues and organs that are not specific endocrine glands nonetheless manufacture and secrete hormones that are just as important to bodily function as are the classic hormones (Fig. 3). These nonclassic hormones fall into at least three classes: sterols, proteins and peptides, and eicosanoids (derivatives of the fatty acid arachidonic acid).

The only nonclassic steroidal hormones are derivatives of vitamin D_3 and include 24,25-dihydroxyvitamin D_3 and 1,25-dihydroxyvitamin D_3. The latter is manufactured in the kidney tubular cells.

By far the largest group of nonclassic hormones are peptides and proteins (Table III). Some of these are (1) atrial natriuretic peptide synthesized and secreted by modified heart muscle cells in the atria of the heart; (2) erythropoietin secreted by the kidney; (3) insulin-like growth factors, which are secreted by parenchymal cells of the liver; (4) a variety of peptide hormones secreted by cells in the intestinal tract, including cholecystokinin, gastrin, gastric inhibitory peptide, glucagon-like peptide, vasoactive intestinal peptide, secretin, and bombesin; (5) a large and increasing number of tissue growth factors, including nerve growth factor, epidermal growth factor, fibroblast growth factor, platelet-derived growth factor, and others; and (6) endothelin secreted by the endothelial cells lining the inner surface of blood vessels.

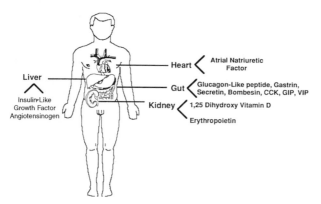

FIGURE 3   Some of the nonclassic endocrine organs and their hormonal products. CCK, cholecystokinin; GIP, gastric inhibitory peptide; VIP, vasoactive intestinal peptide.

Several other major peptide hormones are actually secreted as the inactive precursors hormogens into the bloodstream, where they undergo partial proteolysis to smaller active peptide hormones. Both bradykinin and angiotensin II are generated in this fashion.

Another large class of chemical messengers are the products of the further metabolism of arachidonic acid. The major classes of these compounds are prostaglandins, prostacyclin, thromboxanes, and leukotrienes. These are not thought to act as classic hormones but rather as paracrine and autocrine agents in the regulation of tissue function. Nonetheless, they interact with specific cell-surface receptors to initiate similar cellular responses, as do certain peptide hormones.

II. ORGANIZATION OF ENDOCRINE SYSTEMS

There are two general features of endocrine systems: (1) they involve multiple hormones acting on a number of cell types, thus achieving an integrated response, and (2) they display the property of feedback control.

If one oversimplifies somewhat, there is a direct correlation between the importance of the particular behavior to survival and the complexity of the endocrine inputs that regulate it. Thus, for example, it is essential to regulate plasma glucose concentration within certain limits. If the glucose concentration is too low, brain cell functions are impaired and coma occurs. If the plasma glucose is too high, glucose is lost in the urine, and high glucose also has toxic effects on the cells lining the walls of blood vessels.

When one eats a normal meal, the beta cell in the endocrine pancreas (the source of insulin) is informed by neural signals from both the head and intestine. These nerves release acetylcholine and cholecystokinin, respectively, and sensitize the beta cells so that when the glucose is absorbed from the intestine and causes an increase in blood glucose concentration, the beta cells respond with an increase in the insulin secretion. This effect of glucose is further enhanced by hormones such as cholecystokinin, gastric inhibitory peptide, and glucagon-like peptide-1, all of which act on the beta cell. Hence, a small increase in blood glucose concentration causes a large change in insulin secretory rate. Insulin acts on the three major organs to enhance the disposal of glucose. Insulin can be thought of as a storage hormone, which causes the cells in the liver, skeletal muscle, and adipose tissue to store the glucose in forms that can be utilized for energy production in times of fasting. In the liver, insulin causes some of the glucose to be converted into glycogen (a glucose polymer) and the rest into a form of fat (triglycerides), which is secreted into the bloodstream as very low-density lipoproteins (VLDLs). Insulin acts on the skeletal muscle to cause additional glucose to be stored in the muscle cell as glycogen. Insulin acts on the fat cell to convert glucose to α-glycerol phosphate and to break down the triglyceride (in VLDL) to glycerol and free fatty acid. These acids interact with α-glycerol to reform triglycerides in the fat cell. [*See* Insulin and Glucagon.]

During fasting, the blood glucose falls; however, it is essential that it not fall too low because the brain is nearly totally dependent on glucose for its energy metabolism, whereas most other tissues in the body can utilize free fatty acids or ketone bodies (partial breakdown productions of fatty acids) as their energy source. To maintain a minimal level of plasma glucose so that the brain continues to function, a variety of hormonal signals act in a coordinate way to (1) liberate free fatty acids from the triglyceride in fatty cells; (2) instruct the liver to metabolize some of the free fatty acids to ketone bodies; (3) instruct the skeletal and heart muscle cells to preferentially utilize free fatty acids and ketone bodies for their energy needs; (4) cause the breakdown of liver glycogen and release the glucose (resulting from this breakdown) into the bloodstream to maintain a supply of glucose for the brain; (5) cause the breakdown of skeletal muscle glycogen to lactic acid, which is released into the bloodstream; (6) cause the breakdown of muscle proteins to amino acids, which are also released into the bloodstream; and (7) instruct the liver to synthesize glucose from lactic acid, glycerol, and amino acids (gluconeogenesis).

The signals that bring about a net release of fatty acids from adipose tissue are (1) a decrease in plasma insulin concentration and (2) an increase in plasma epinephrine and growth hormone. The major signal that instructs the liver to metabolize fatty acids to ketone bodies is a decrease in plasma insulin concentration. The preferential utilization of fatty acids and ketone bodies by skeletal muscle cells occurs because of lack of insulin and as a direct consequence of the increase in ketone body and free fatty acid concentrations in the blood. The breakdown of liver glycogen results from the direct actions of glucagon and epinephrine on the liver cell, and the breakdown of skeletal muscle glycogen results from the action of epineph-

rine on muscle cells. An increase in protein breakdown in the muscle is related to a rise in plasma cortisol concentration. The increased synthesis of glucose from lactate, glycerol, and amino acids results from the combined actions of a decreased plasma insulin concentration and an increased concentration of glucagon, epinephrine, and cortisol (the so-called counterregulatory hormones). In addition, the increased supply of lactate and amino acids from muscle, and free fatty acids from adipose tissue, are necessary to sustain adequate rates of glucose production.

The relationship between the plasma Ca^{2+} concentration and the secretion of PTH can be utilized to illustrate a common property of endocrine systems: negative feedback control. When, for any reason, the concentration of Ca^{2+} in the blood plasma decreases slightly, this change is perceived by the cells of the parathyroid gland. These cells respond by increasing the secretion of PTH. As a result, the plasma PTH concentration rises. This rise is detected by PTH receptors on cells in kidney and bone. These cells respond in such a way that the plasma Ca^{2+} concentration rises. This rise is sensed by the parathyroid cells, which now reduce PTH secretion. [*See* Parathyroid Gland and Hormone.]

The control of endocrine response is usually much more complicated. A large number are hierarchical. They involve specific cells in the hypothalamus secreting a releasing hormone, which acts, in turn, on a particular class of cells in the anterior pituitary, inducing them to release a tropic hormone that acts, in turn, on a peripheral endocrine gland. These systems also display feedback control, which characteristically involves the hormone product of the peripheral endocrine glands as the feedback modulator of the cells in both the hypothalamus and the pituitary (Fig. 4). For example, TRH secreted by cells in the median eminence of the hypothalamus acts on thyrotrophs in the anterior pituitary to stimulate the secretion of TSH. The TSH acts, in turn, on the acinar cells of the thyroid gland to stimulate the release of the thyroid hormone thyroxine. The resulting increase in plasma thyroxine concentration acts as a negative feedback signal on TRH-secreting cells in the hypothalamus and, more importantly, on the thyrotrophs in the anterior pituitary to inhibit TSH secretion. [*See* Hypothalamus: Pituitary; Thyroid Gland and Its Hormones.]

Similar but more complex feedback relationships exist in the system involved in the regulation of cortisol secretion from the adrenal zona fasciculata. Corticotropin-releasing hormone from the hypothalamus acts on corticotrophs in the pituitary to release

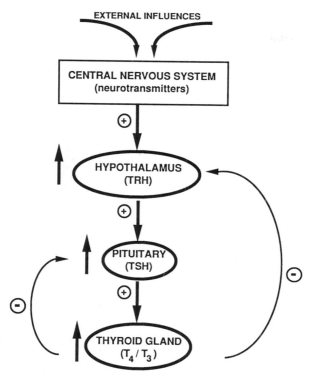

FIGURE 4 The principle of negative feedback control in an endocrine system as illustrated in the case of the interrelationship between thyrotropin-releasing hormone (TRH) from the hypothalamus, thyroid-stimulating hormone (TSH) from the pituitary, and thyroid hormones (T_4/T_3) from the thyroid. TRH stimulates TSH secretion, which in turn stimulates T_4 and T_3 secretion from the thyroid. T_4 and T_3 act as negative feedback controls at the level of the pituitary and the hypothalamus. In addition, neural inputs from the central nervous system affect TRH secretion in a positive way.

ACTH, which in turn stimulates cortisol production from the adrenal cortex. Cortisol acts as a negative feedback regulator of ACTH production by interacting with cells in both the hypothalamus and the pituitary. However, superimposed on this feedback relationship is an additional signal from the central nervous system that dictates a diurnal variation in ACTH and cortisol secretion.

Even more complex patterns of control are evident in the control of other pituitary hormones. In particular, many of these hormones are secreted in a pulsatile fashion rather than in a continuous manner.

III. CHARACTERISTICS OF HORMONES

Despite the diversity in their chemical structures, hormones share a number of common attributes. They are effective at very low concentrations (10^{-12}–10^{-8} M). Their actions depend on their combining with specific recognition proteins—receptors, which exist either on the surface or in the interior of the target cells. These receptors possess a high degree of specificity and affinity for a given hormone. The tissue and cellular distribution of these receptors determines which cells are targets for the action of the particular hormone. In some cases, the distribution of receptors for a specific hormone is quite limited (e.g., ACTH receptors are confined to adrenal cortical cells and some neurons in the central nervous system); in other cases, they are widely distributed (e.g., insulin receptors are present on a wide variety of cell types). The association of a hormone with its receptor leads to the activation of one or more transducing systems in the cells. As a general rule, specific cellular responses are regulated by multiple hormonal inputs, which are both stimulatory and inhibitory. Finally, the cellular response can be modulated by changing either the concentration of the particular hormone or the number, affinity, or cellular distribution of receptors.

Another common feature of hormones is their inactivation. If they are to serve as messengers, then both decreases as well as increases in their concentrations must be tightly regulated. In all cases, the enzymes responsible for their inactivation are different from the receptors for the given hormone. Also, the rate of change in plasma concentration is a reflection of the typical time course of action of the particular hormones. Many eicosanoids, amines, and peptide hormones have actions that are rapid in onset and in termination. Other peptides, thyroxine, and steroid and steroidal hormones act much more slowly and have prolonged effects. This difference is reflected in their biological half-lives in blood or bodily fluids. Eicosanoids and most amines have half-lives of seconds to minutes, peptide and protein hormones of minutes, and thyroxine and steroid hormones of hours to days.

Another class distinction is also apparent in terms of receptor location. Receptors for peptides and protein hormones, most amines, and eicosanoids are located on the cell surface, but receptors for thyroxine and steroidal hormones are located within the cell. [See Cell Receptors.]

IV. FUNCTIONS OF HORMONES

The endocrine system can be viewed as a messenger system that employs hormones to convey information from one cell, tissue, or organ to another. In the broadest sense, hormones coordinate and integrate the activities of the different organs and tissues of the body so that each serves the needs of the organism in a particular circumstance at a particular moment in time. From a physiological point of view, hormones regulate four major processes: reproduction, growth and development, maintenance of the internal environment, and regulation of nutrient utilization, energy storage, and energy metabolism.

A. Reproduction

In the case of reproduction, hormones not only regulate the functions of ovary and testes so that sperm and ova are produced at appropriate times and in appropriate numbers, but they act on a variety of other tissues to ensure the appropriate preparation of the uterus, for example, for implantation and successful survival of the fertilized ovum. Furthermore, in embryonic life the appropriate hormones appearing in the appropriate sequence are required for the development of the sexual organs.

B. Growth and Development

Endocrine control of somatic growth involves not only the production of growth hormone by the pituitary gland, but, as a consequence of growth hormone action, production of insulin-like growth factors in the liver. These growth factors act to stimulate the growth of bone, muscle, and other tissues. Appropriate concentrations of thyroid hormone, insulin, adrenal cortical hormones, and the sex hormones are also necessary for ordered skeletal growth and for the cessation of growth at an appropriate time. [See Tissue Factor.]

In addition to this aspect of growth regulation, hormones are involved in the regulation of cell proliferation in many organs in the adult. Multiple, distinct tissue growth factors that regulate the proliferation of fibroblasts, epithelial cells, liver cells, and a variety of other cells have been identified. Commonly, for any one type of cell, several growth factors, acting on different intracellular signaling systems in the proper sequence, are necessary to induce a proliferative response. Abnormalities in these growth and proliferative responses can lead to the appearances of either benign or malignant tumors or, as in the case of blood vessels, to the altered proliferation of smooth muscle

cells, which plays an important role in the pathogenesis of atherosclerosis. [See Atherosclerosis.]

C. Maintenance of the Internal Environment

A general function of the endocrine system is that of maintaining the constancy of the internal environment. Our cells, whether in liver, muscle, kidney, nervous system, or bone, live in a highly stable but dynamic environment in which key constituents such as sodium, hydrogen, potassium, calcium, chloride, bicarbonate, phosphate, and magnesium ions must be maintained within very precise limits for these cells to function properly. These ionic constituents are constantly being lost in urine and/or skin and are being replenished from the diet. The organism must be able to coordinately regulate the function of intestine, bone, kidney, and other organs so as to efficiently store these dietary constituents at feeding time and to conserve them during fasting. The endocrine system orchestrates this coordinate behavior.

D. Nutrient Utilization and Energy Production

No function of hormones is of a more immediate and constant importance than that of regulating the efficient storage of ingested carbohydrates, fats, and amino acids into glycogen, fats, and proteins and of mobilizing these stored substances during periods of fasting or starvation. These substances are metabolized to tissue proteins, to other tissue constituents needed for cell growth and repair, or to CO_2 and H_2O with the generation of adenosine triphosphate (ATP), the common currency of cellular energy transactions. This ATP is employed to drive the work functions of the cell such as contraction, secretion, the synthesis of proteins and other complex molecules, and ion transport. [See Adenosine Triphosphate (ATP).]

These different phases of energy metabolism are regulated by different hormones. Insulin, for example, can be considered the major energy storage hormone of the body. When its concentration increases, there is a more efficient uptake of glucose, amino acids, and fats into tissues such as liver, muscle, and adipose tissue. On the other hand, glucagon, epinephrine, and cortisol function to mobilize glucose directly from liver glycogen and indirectly from skeletal muscle glycogen, fatty acids from adipose tissue, and amino acids from muscle to supply adequate amounts of glucose and fatty acids for the use of other organs and tissues during periods of fasting or starvation.

V. BIOSYNTHESIS AND SECRETION OF HORMONES

Each chemically distinct class of hormones is synthesized by a different cellular route.

A. Peptides

The peptide and protein hormones are synthesized from amino acids via the classic DNA → RNA → protein pathway, and the finished product is packaged in secretory vesicles within the cell until an appropriate stimulus triggers its release (Fig. 5).

Release involves the fusion of the membranes of the vesicle with the surface (plasma) membrane of the cell, a process known as exocytosis (Fig. 5). A common feature of peptide hormone biosynthesis is that the initial protein product resulting from the transla-

tion of the specific messenger RNA is a larger molecule, a preprohormone, which then undergoes proteolytic modification to a prohormone and finally to the active hormone. For example, preproinsulin is the initial product synthesized in the beta cell of the islets of Langerhans. After its synthesis, it is converted to proinsulin, which enters the immature secretory vesicle, wherein it is converted by the action of specific proteases to insulin. In this active form, insulin is secreted into the bloodstream. In unusual cases, a secreted peptide is not biologically active until it is further processed in the bloodstream.

B. Steroids

In the case of the steroid hormones, the starting material is cholesterol. In the ovaries, testes, and adrenal, cholesterol is first converted to pregnenolone and then by successive modifications to the appropriate steroid hormone, estradiol or progesterone (ovary), testosterone (testes), and cortisol or aldosterone (adrenal). None of these is stored in secretory vesicles; rather, synthesis and secretion are concurrent (Fig. 6). [*See* Cholesterol.]

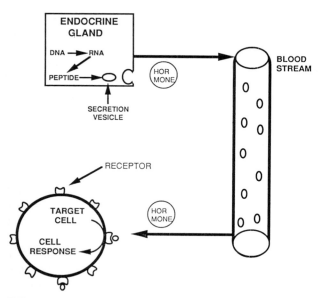

FIGURE 5 Secretion, transport, and cellular site of action of peptide hormones. The hormone is synthesized in the cells of the endocrine gland from amino acids. The product is stored in secretory vesicles, which on an appropriate stimulus fuse with the plasma membrane and thereby release their contents into the bloodstream. The hormone is transported in the blood, in the unbound state, to its target cell, where it binds to specific receptors on the cell surface to initiate a cellular response.

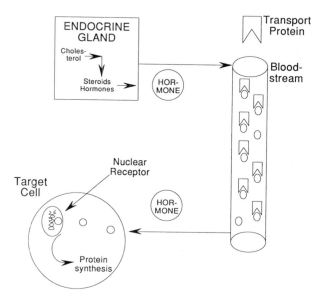

FIGURE 6 Secretion, transport, and site of action of steroid hormones. The hormone is synthesized from cholesterol and immediately secreted into the bloodstream, where it combines to a specific transport protein. In this form, it is carried to the target cell, where the hormone dissociates from the transport protein, enters the cell, and binds to an intracellular receptor, which in turn associates with sites on DNA within the nucleus.

The sterol hormone 1,25-dihydroxyvitamin D_3 is unique in that successive steps in its biosynthesis occur in different organs: cholesterol to 7-dehydrocortesterol to vitamin D_3 in the skin, vitamin D_3 to 25-hydroxyvitamin D_3 in the liver, and 25-hydroxyvitamin D_3 to 1,25-dihydroxyvitamin D_3 in the kidney.

C. Amines

The synthesis of the amine hormones such as epinephrine involves the conversion of the amino acid tyrosine into the particular amines by a sequence of several steps. The finished product is stored in secretory vesicles; that is, it is packaged and secreted like the peptide hormones.

The most unique pathway of biosynthesis and secretion is that found in the thyroid gland. This process takes place within acinar cells of the thyroid. These cells form into small spherical clusters within which an internal, extracellular space, the acinus or follicle, exists. The initial sequence of events involves the biosynthesis of a specific protein, thyroglobulin in the acinar or follicular cell. This is packaged in secretory vesicles and secreted into the acinar lumen. At the time of its secretion, the thyroglobulin is iodinated on tyrosine residues by a specific peroxidase. Closely aligned iodinated tyrosine residues in the protein molecule are then coupled to form peptide-bound tri- and tetraiodothyronines; that is, the hormone precursor is a large protein molecule in whose structure the small hormone (each composed of two modified tyrosine residues) is contained. Upon stimulation by TSH, the acinar cells send out pseudopodia from their luminal surfaces, which engulf and internalize the iodinated thyroglobulin. This intracellular thyroglobulin undergoes breakdown to release the free thyroid hormones thyroxine and triiodothyronine from the other (basal) side of the cell into the bloodstream.

VI. TRANSPORT TO TARGET CELLS

The peptide and amine hormones are readily soluble in bodily fluids and, in general, are transported in the blood plasma in unbound forms (Fig. 5). An exception to this generalization are the insulin-like growth factors, which are transported, in large part, bound to specific carrier proteins.

The steroidal hormones and the thyronines are not readily soluble in water and, hence, are transported from site of synthesis to target cells by specific carrier proteins in the bloodstream (Fig. 6). These carrier proteins are of two general types: albumin and prealbumins, which have the ability to bind a large variety of small molecules with moderate affinity and low specificity, and specific carrier proteins [e.g., thyronine-binding globulin (TBG) or cortisol-binding globulin (CBG)], which have high affinity and specificity for the particular hormone. These proteins bind their respective hormones tightly so that when a mixture of hormone and binding protein exists, as is the case in the blood plasma, a small amount of unbound (free) hormone and a large amount of bound hormone exist in equilibrium with each other.

In general, the binding capacity (a measure of the plasma content) of the carrier protein is considerably greater than the circulating concentration of the steroid hormone or thyronine. Hence only a very small amount of the hormone exists in the unbound state. It is this free (or unbound) form of the hormone to which the target cell responds. Furthermore, the content of a given carrier protein in the plasma can change independent of the content of the hormone that is transported.

VII. MEASUREMENT OF HORMONE CONCENTRATIONS

Measurement of the concentrations of the various hormones in blood plasma or serum and/or in the urine is an extremely important means by which the endocrine physiologist studies the behavior of endocrine systems and by which the endocrine physician (endocrinologist) studies disorders of endocrine gland function.

In the case of catecholamines and steroid hormones, chemical methods for measuring their contents or those of their breakdown products in urine have been developed. However, such chemical methods have not been of great utility in the measurement of hormone concentrations in small samples of blood plasma or serum. This is particularly true of peptide hormones, because their low concentrations combined with their common chemical features have made a direct chemical approach nearly impossible. However, by using a different method known as radioimmunoassay (Fig. 7), it has been possible to measure, with a high degree of accuracy, the plasma concentrations of all the peptide hormones, the steroid and sterol hormones, the thyronines, and many other substances, including a variety of commonly used drugs.

The method of radioimmunoassay depends on three reagents: (1) a specific, high-affinity antibody against

I

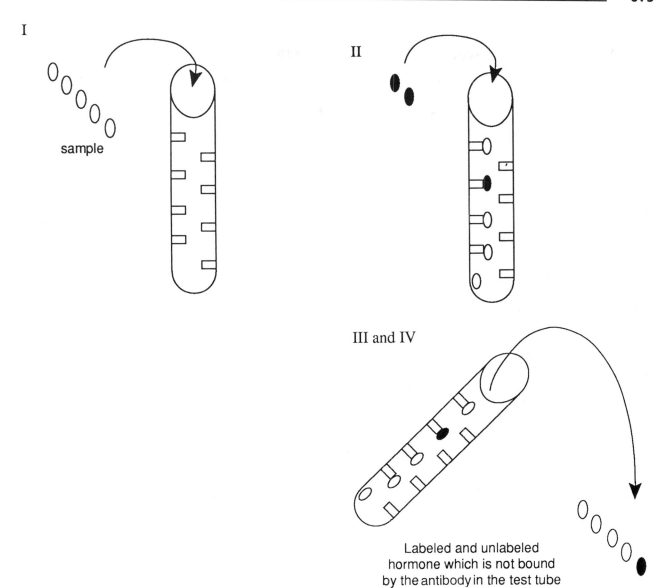

sample

II

III and IV

Labeled and unlabeled
hormone which is not bound
by the antibody in the test tube
is removed

FIGURE 7 Steps involved in performing a solid-phase radioimmunoassay.

I. Add blood sample with unknown amount of hormone to antibody-coated tube.

II. Add small amount of radioactively labeled hormone to test tube containing sample.

III. Test tube decanted and amount of radioactivity left in the test tube is determined.

IV. Amount of radioactively labeled hormone left in the test tube is proportional to the amount of unlabeled hormone contained in the unknown sample. This is determined by comparing it to the displacement of known amounts of unlabeled hormone (standard curve).

the specific hormone or drug; (2) the production of a radioactively labeled hormone (in the case of peptide hormones, this is commonly achieved by iodinating a specific amino acid residue, e.g., tyrosine); and (3) after allowing antibody and labeled and unlabeled hormone to interact, a method for separating antibody-bound from free labeled hromone. The basic principle of the assay is quite simple: Standard amounts of labeled hormone and antibody are added to a series of tubes containing either standard amounts of unlabeled hormone or the hormone contained in a particular plasma or serum sample. The more unlabeled hormone (standard or unknown) that is present, the less labeled hormone becomes bound by the antibody, because the two compete for the limited number of binding sites on the antibody. By developing a standard curve from the analysis of standard amounts of unlabeled hormone, it is possible to define where on this curve the unknown sample falls and, therefore, the concentration of hormone in that sample. [*See* Radioimmunoassays.]

VIII. CELLULAR EFFECT OF HORMONES

At the cellular level, hormones, after combining with specific receptors, regulate the secretion of other hormones, control the transport of nutrients and ions across cell membranes, control the transcellular transport of nutrients and ions across epithelial cells, regulate growth and proliferation, control metabolic rate, regulate the metabolism of sugar, fatty acids, and amino acids, and regulate the contraction and relaxation of smooth muscles.

In the case of peptide and amine hormones, hormone–receptor interactions occur at the cell surface. The interaction leads not only to an activation of some membrane transducer but also to the internalization of the hormone–receptor complex. Hence, the receptor population on the cell surface is constantly changing: they are lost by internalization and added by new synthesis or recycling. Chronic high extracellular concentrations of hormone lead to a steady-state decrease in the number of receptors, a process known as down-regulation. Conversely, chronic low concentrations of extracellular hormone lead to an increase in the number of surface receptors for the particular hormone, a phenomenon known as up-regulation.

In addition to the effects of the homologous hormone on surface receptor number, there are many circumstances where exposure of cells to heterologous hormones, even of a different class, can alter the number of surface receptors for a particular hormone. Thus, an important way that hormones exert their cellular effects is to change the expression of receptors for other hormones on their surface, thereby making them either more or less sensitive to the effects of these other hormones.

In any specific cell type, a given hormone usually exerts multiple effects, all of which are components of the integrated response of the cell. For example, when ACTH stimulates the synthesis of cortisol from cholesterol, the hormone stimulates cholesterol synthesis and the hydrolysis of stored cholesterol esters to ensure a sufficient supply of the precursor cholesterol; it also enhances the rate of production of ATP and nicotinamide adenine dinucleotide phosphate,

components needed in the synthetic process, and it stimulates the transport of cholesterol from the cytosol into the mitochondria—the site where initial transformation of cholesterol occurs.

IX. MECHANISMS OF ACTION

In spite of the diversity of effects of hormones on target cells, there are a few common mechanisms by which hormones act. The first distinction is between steroidal hormones and thyronines on the one hand and peptide and amine hormones on the other.

A. Steroids and Thyronines

Steroid hormones, 1,25-dihydroxyvitamin D_3, and thyronines exert their actions by combining with specific intracellular receptors in the cell (Fig. 8). These

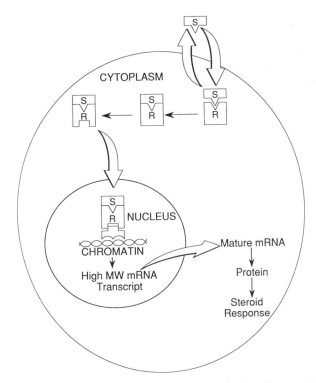

FIGURE 8   Mechanism of steroid hormone action. The steroid hormone (S) enters the cell by diffusion and binds with a cytoplasmic receptor (R). As a consequence, the receptor undergoes a conformational change, is taken up by the nucleus, and binds to specific DNA sequences, regulatory elements of specific genes. This results in transcription of the gene, and a high-molecular-weight messenger RNA (MW mRNA) is produced, which is processed within the nucleus to a mature, lower-molecular-weight mRNA. This is transported to the cytoplasm, where it is translated by ribosomes or polyribosomes into a potein. The increase in the content of the protein(s) leads to altered cell behaivor.

may be located in either the cytoplasm or the nucleus. When the hormone reaches the cell, it diffuses across the plasma membrane and binds to the receptor. The binding of the steroid (or thyronine) to the receptor leads to a change in its conformation so that it now binds to specific, high-affinity sites on the DNA in the nucleus. Within the nucleus, the steroid–receptor complex binds to specific regulatory elements with unique DNA sequences. Binding of the complex to the regulatory element initiates the synthesis of specific messenger RNAs, which in turn stimulate the synthesis of specific proteins. These proteins, in turn, bring about changes in cell function and/or cell growth.

B. Peptides and Amines

Peptide, protein, and amine hormones interact with receptors located on the cell surface. Interaction of hormone with receptors initiates events within the plasma membrane that lead to the activation of specific membrane transducers; that is, they transduce the original extracellular message into an intracellular message or messages.

Five types of intracellular messengers have been recognized: cyclic nucleotides, Ca^{2+}, inositol phosphates, certain lipids, and phosphoproteins.

I. The Cyclic AMP Messenger System

Two major cyclic nucleotides have been identified: cyclic 3′,5′-adenosine monophosphate (cAMP) and cyclic 3′,5′-guanosine monophosphate (cGMP). Each is synthesized by a specific enzyme, adenylate and guanylate cyclase, respectively; each is degraded by one or more phosphodiesterases to their respective products, 5′AMP and 5′GMP; and each appears to act mainly by controlling the activity of ion channels and a particular class of enzymes: cAMP-dependent and cGMP-dependent protein kinases. Both cAMP and cGMP have actions in multiple tissues, and their respective cyclases are activated by different hormonal receptors in different tissues. In addition, much more is known about the role of the cAMP in regulating cell function; its properties will be discussed in terms of the second messenger model of hormone action (Fig. 9).

Adenylate cyclase is associated with the plasma membrane. Interaction of the hormone with its receptor on the plasma membrane leads to an activation of the adenylate cyclase via a GTP-binding protein; the cyclase then catalyzes the formation of cAMP from ATP. The cAMP is released into the intracellular fluids, where it serves as a second or intracellular messenger. The cAMP binds to a specific cAMP receptor protein, which is the regulatory subunit (R) of an enzyme, cAMP-dependent protein kinase, formed by this subunit bound to the catalytic (C) subunit. When cAMP associates with R, R and C dissociate. The released free C is active in catalyzing the combination of phosphates to a number of protein substrates on either serine or threonine residues, using ATP energy. The phosphoproteins thus formed are different in conformation and, hence, in the catalytic efficiency from the original proteins. Classic examples of known proteins that become phosphorylated by cAMP-dependent protein kinase are phosphorylase kinase and glycogen synthase, key enzymes in the degradation and synthesis of glycogen, respectively. Phosphorylation of phosphorylase kinase increases its activity, whereas that of glycogen synthase decreases its activity. [*See* Protein Phosphorylation.]

Many peptide and amine hormones act by either stimulating or inhibiting adenylate cyclase. Examples of stimulating hormones are epinephrine, glucagon, serotonin, PTH, calcitonin, TSH, ACTH, FSH, gastric inhibitory peptide, secretin, vasoactive intestinal peptide, and some growth factors. Examples of inhibitory hormones are epinephrine (acting via a different class of receptors), adenosine, angiotensin II, somatostatin, and dopamine.

The hormones that regulate adenylate cyclase act rapidly, and once the hormonal signal is terminated, the response usually ceases rapidly because a class of enzymes known as phosphodiesterases rapidly breaks down cAMP to 5′AMP, and another class, known as phosphoprotein phosphatases, dephosphorylates phosphoproteins. The reversible phosphorylation and dephosphorylation of cellular proteins is a nearly universal way in which peptide and amine hormones regulate cellular function.

2. Ca^{2+} Messenger System

Along with cAMP, Ca^{2+} ion is a nearly universal second messenger in hormone action. The generation of the Ca^{2+} messenger can occur by several different mechanisms, and its intracellular actions are more diverse. The concentration of Ca^{2+} in the extracellular fluid is 10^{-3} M and that within the cell cytosol is 10^{-7} M. In addition, Ca^{2+} is stored within the cell in a specialized compartment within the endoplasmic reticulum of nonmuscle cells and the sarcoplasmic reticulum in skeletal and cardiac muscle. An increase in Ca^{2+} concentration within the cell can occur as a result of either an increase in Ca^{2+} entry from the extracellular fluids across the plasma membrane via specific

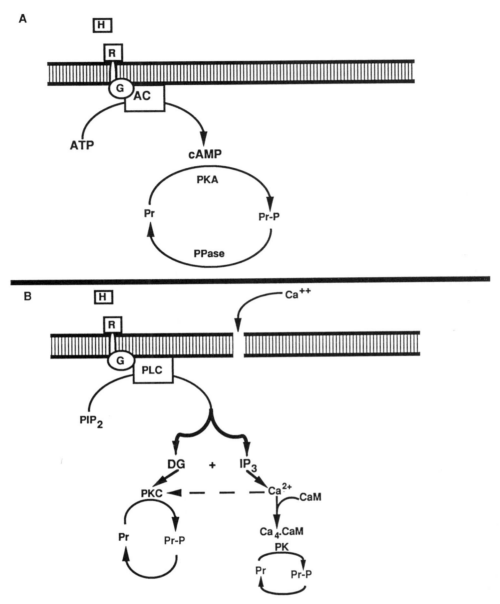

FIGURE 9 The cAMP and Ca^{2+} messenger systems. (A) Interaction of hormone (H) with receptor (R) on the plasma membrane activates the enzyme adenylate cyclase (AC) via a specific type of membrane protein, the guanine nucleotide-binding protein (G). As a result, cAMP is produced from ATP. cAMP then binds to and activates a specific class of enzymes, the cAMP-dependent protein kinases (PKA), thereby stimulating the phosphorylation of a group of cellular proteins (Pr $\Rightarrow$ Pr-P). Additionally, cAMP acts to inhibit another class of enzymes, the phosphoprotein phosphatases (PPases), which dephosphorylate these proteins. (B) Interaction of hormone (H) with receptor (R) acting via a "G" protein stimulates the activity of phosphoinositide-specific phospholipase C (PLC), causing the hydrolysis of phosphatidylinositol-4,5-bisphosphate (PIP_2) to give rise to diacylglycerol (DG) and inositol-1,4,5-trisphosphate (IP_3). The IP_3 causes the release of Ca^{2+} from an intracellular pool. In addition, hormone–receptor interaction leads to an increase in Ca^{2+} influx across the plasma membrane (hatched bars). These Ca^{2+} signals stimulate the activity of calmodulin (CaM)-dependent protein kinases and along with DG the activity of protein kinase C (PKC) to catalyze the phosphorylation of cellular proteins. The extent of phosphorylation of these proteins is also determined by the activities of phosphoprotein phosphatases (PPases).

Ca^{2+} channels in this membrane or by the release of Ca^{2+} from internal stores. In some cases, a hormone interacting with its surface receptor will cause only an increase in Ca^{2+} influx, but more commonly it causes both an increase in influx and a release of internal Ca^{2+}. In these latter cases (Fig. 9), hormone–receptor interaction almost always leads to the activation of the plasma membrane-associated enzyme phospholipase C, which catalyzes the hydrolysis of a specific membrane lipid, phosphatidylinositol-4,5-bisphosphate, resulting in the production of two messengers: water-soluble inositol-1,4,5-trisphosphate and membrane-soluble diacylglycerol. The inositol-1,4,5-trisphosphate acts as a signal to cause the release of internal Ca^{2+}. The diacylglycerol causes the soluble, Ca^{2+}-insensitive form of a specific protein kinase, protein kinase C, to associate with the plasma membrane and in this way become Ca^{2+}-sensitive.

A change in intracellular Ca^{2+} concentration serves as a messenger by binding to specific Ca^{2+} receptor proteins, or Ca^{2+}-activated enzymes. Binding of Ca^{2+} to a receptor protein leads to a change in the conformation of the protein, which then interacts with other proteins (enzymes), thereby altering their activity. In many cases, these enzymes are protein kinases. Hence, just as with cAMP, a common consequence of messenger Ca^{2+} is the activation of protein kinases.

Common Ca^{2+} receptor proteins are calmodulin and troponin C. Protein kinases activated by Ca^{2+}-calmodulin include myosin light-chain kinase, phosphorylase kinase, multifunctional protein kinase, and protein kinase C.

The specific receptors of many peptide and amine hormones, tissue growth factors, and eicosanoids are linked to the phosphoinositide–Ca^{2+} messenger system. The hormones include epinephrine, angiotensin II, bombesin, cholecystokinin, serotonin, certain prostaglandins, and thromboxanes.

3. Interactions between Ca^{2+} and cAMP

Although the two major messenger systems are presented as separate pathways for cell activation, they nearly always function together. They often regulate the same cellular response or components of a particular response. Hence, cAMP and Ca^{2+} serve as nearly universal, synarchic messengers in the activation of processes such as glycogen synthesis and breakdown, smooth muscle contraction and relaxation, aldosterone secretion, insulin secretion, the secretion of a wide variety of other peptides, amine, and steroid hormones, and the growth and/or differentiation of many cell types; however, their

relationship is not stereotyped. They can function in a coordinate, hierarchical, redundant, antagonistic, or a sequential relationship to modulate a specific cellular response.

4. Tyrosine Kinase-Linked Systems

A common consequence of peptide hormone–receptor interaction, for another class of peptide hormones, is the activation of a specific type of protein kinase, which adds a phosphate residue to tyrosine in proteins. In contrast, the cAMP- and Ca^{2+}-calmodulin-dependent protein kinases as well as protein kinase C catalyze the phosphorylation of either serine and/or threonine residues on their substrate proteins. The insulin receptor, and those of many other growth factors, possesses as an intrinsic part of its structure a protein kinase activity that catalyzes the phosphorylation of tyrosine residues in the protein substrate. The cytokine receptors do not have intrinsic tyrosine kinase activity but noncovalently associate with a soluble, cytoplasmic tyrosine kinase. In both groups of receptors, one of the substrates is the receptor itself; for example, the interaction of insulin with its receptor leads to the autophosphorylation of tyrosine residues on the receptor protein. This creates docking sites for other substrates that possess phosphotyrosine binding domains. One such substrate is a protein that brings together Ras, a small GTP-binding protein, and its activator. Activated Ras then initiates a protein kinase cascade that eventually stimulates the mitogen-activated protein kinase, or MAP kinase, which can mediate many of the effects of these hormones.

X. DISORDERS OF ENDOCRINE FUNCTION

Disorders of endocrine function result from abnormalities in hormone secretion or activity or from alterations in the responsiveness of target tissues to normal hormone levels.

A. Abnormal Secretion or Activity

The first mechanism for hormone deficiency would be the absence of the secreting gland. Congenitally absent glands have been shown to be secondary to defects in morphogenic molecules or transcription factors active during embryogenesis; postnatal destruction of glands is usually the result of an autoimmune reaction against some antigen in the gland, for example, hypothyroidism from thyroiditis caused by antibodies to thyroid peroxidase. Infections, tumors, and strokes have also been shown to be causes of glandular destruction. If the gland is intact, hormone deficiency can arise from deletions or mutations in the genes for the hormone itself (peptide hormones) or for synthetic enzymes (e.g., steroids). In some cases of hypopituitarism, the transcription factor responsible for transcribing the pituitary glycohormone genes is defective. Finally, sensors for hormone secretagogues can be genetically altered: insulin is released from the pancreatic islets in response to glucose, which is detected by glucokinase. Some forms of diabetes mellitus are caused by mutant enzymes that cannot signal the cell that glucose levels are elevated and insulin is not secreted. In another example, parathormone is released when calcium levels fall. In one form of hypoparathyroidism, the calcium sensor is mutated in such a way that it is constitutively active; the cell is erroneously informed that calcium levels are high and secretes insufficient parathormone.

The molecular basis of hormone overproduction has also been discovered for several endocrinopathies. Secretion of many hormones is stimulated by tropic hormones that utilize the cAMP pathway. In McCune–Albright disease, a mutant GTP-binding protein constitutively activates the adenylate cyclase; the resulting high levels of cAMP persistently stimulate the secretion of several hormones, including ACTH, melanocyte-stimulating hormone (MSH), TSH, and the gonadotropins. In some forms of hyperparathyroidism, the calcium sensor discussed earlier is inactivated; now the parathyroid gland cannot detect any calcium and continuously releases parathormone to "restore" levels of this cation. Activating mutations in the regulatory regions of hormone genes have also been reported and probably represent the major mechanism for the secretion of hormones by tumors, a phenomenon called ectopic production. Finally, there is a single report of a mutation that lies within the hormone itself and that increases activity by increasing the half-life of the circulating hormone.

In the foregoing examples, the defect resides within the secreting gland itself; this is primary hypersecretion (Fig. 10). Secondary hypersecretion results from overstimulation by a normal secretagogue. For example, vitamin D deficiency results in the impairment of calcium absorption from the intestine. The resulting hypocalcemia leads to the persistent secretion of parathormone, that is, secondary hyperparathyroidism.

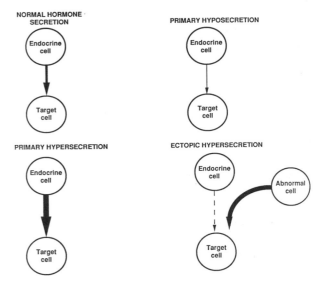

FIGURE 10   Disordered patterns of hormone secretion.

B. Altered Responsiveness of Target Tissues

Some endocrine diseases are characterized by normal serum hormone levels, in which case the problem lies within the target tissues. First of all, the hormone may not have access to its receptor; in some autoimmune diseases, blocking antibodies prevent hormones from binding to their receptors. In other "hormone deficiencies" the gene for the receptor itself is deleted or mutated; for example, mutations have been described in the insulin receptor in certain forms of diabetes mellitus and in the androgen receptor in some forms of male pseudohermaphroditism. End-organ resistance can also be a result of defects downstream of the hormone receptor. Inactivating mutants of the GTP-binding protein prevent both the stimulation of adenylate cyclase and the elevation of cAMP. Interestingly, although many hormones utilize cAMP, the effects of this disease primarily reflect an impairment in parathormone actions. In another example, vasopressin activates renal water channels that resorb water and concentrate urine. In some forms of nephrogenic diabetes insipidus, vasopressin and its receptor are normal, but the water channel is defective. Finally, end-organ resistance can be physiological; the numbers of receptors for many hormones are reduced in the presence of sustained elevations of their cognate ligands. This feedback regulation is called down-regulation and it protects the cell from overstimulation. In maturity onset diabetes mellitus, the peripheral tissues are relatively resistant to insulin because insulin levels are very high and insulin receptors are consequently low.

In contrast to end-organ resistance, target tissues can also act as if they were being overstimulated, even though serum hormone levels are normal or low. In Graves' disease, antibodies to the TSH receptor bind to the receptor; but instead of blocking the receptor, they activate it and cause the thyroid gland to overproduce thyronines. Activating mutations of receptors are also known; apparently the mutations change the receptor conformation to one that resembles the hormone-bound state. In differentiated tissues, this activation can result in premature development, for example, constitutively active LH receptors cause male precocious puberty. In undifferentiated tissues and stem cells, constitutively active receptors, especially for the growth factors, can lead to hyperplasia and even frank cancer. Finally, activating mutations of the GTP-binding protein have been documented in McCune–Albright disease. The persistant elevation of cAMP can result in the hyperplasia and oversecretion of several endocrine glands.

BIBLIOGRAPHY

Bolander, F. F. (1995). "Molecular Endocrinology," 2nd Ed. Academic Press, San Diego.

Hadley, M. E. (1996). "Endocrinology," 4th Ed. Prentice–Hall, Upper Saddle River, NJ.

Wilson, J. D., and Foster, D. W. (eds.) (1992). "Williams Textbook of Endocrinology," 8th Ed. Saunders, Philadelphia.

Endocrinology, Developmental

PIERRE C. SIZONENKO
University of Geneva Medical School

GLOSSARY

Perinatal period First 28 days of life
Prepuberty From the second to the tenth year of life
Puberty Period during which the functions of reproduction are developing

ENDOCRINOLOGY IN THE PEDIATRIC AGE GROUP differs from that in the adult, as it deals with a growing and maturing organism. The concept of growth should be associated with the maturation of the organs and, in particular, the endocrine glands. Each gland or organ has its own developmental patterns, scheduled to occur sharply at a preset time. Biological events are to be interpreted in view of embryology (morphogenesis and differentiation), chronology of events affecting growth, and maturation of organs and functions (progressive gene expression, synthesis of proteins, activity of enzymes). Growth and maturation depend on numerous hormones acting through endocrine mechanisms and growth factors, whose action is paracrine and/or autocrine.

Four developmental periods are described in the growing individual before adulthood is reached: fetal, perinatal, prepubertal, and pubertal.

I. PRENATAL ENDOCRINOLOGY

The understanding of prenatal endocrinology in humans remains in a primitive stage compared with that of several animal models such as the rat, rabbit, sheep, and monkey.

Growth and maturation of the fetus depend on maternal, placental, and fetal hormonal factors. Frequently, the origin of some of these factors is impossible to trace precisely, because many hormones or factors can pass through the placenta or even undergo transformation within the placenta. In addition, the placenta secretes several hormones and factors. Therefore, the role of the placenta has rendered the studies of prenatal endocrinology particularly difficult in the human.

A. Hypothalamus and Posterior Pituitary

The differentiation of the diencephalon of the embryo occurs around 34 days postconception and the posterior pituitary primordium appears toward the 49th day. The capillaries of the hypothalamo–pituitary portal system develop between 60 and 100 days. At 21 weeks gestation, the hypothalamo–pituitary portal system is complete and active. The transmission of neurosecretory material into the hypophyseal portal system with the concurrent development of the hypothalamus occurs between 14 and 18 weeks gestation. At this time, the hypothalamic nuclei and fibers of the supraoptic tract appear with further differentiation of the pars tuberalis and the median eminence. Three types of neurohormones or factors have been identified very early in the fetal hypothalamus: (1) releasing or inhibiting factors or hormones, which are peptides; (2) aminergic neurotransmitters such as dopamine, norepinephrine, and serotonin; and (3) oxytocin and vasopressin nonapepides, which are synthesized in the paraventricular and supraoptic nuclei of the hypothalamus and transported by neurons to the posterior pituitary. [See Hypothalamus; Peptides; Pituitary.]

I. Neurotransmitters

Neurotransmitters as detected by monoamine fluorescence were observed in the hypothalamus as early as the 10th week of fetal life and in the median eminence at the 13th week. Nothing is presently known of the metabolism of monoamines in the human fetal hypothalamus.

2. Oxytocin and Vasopressin

Grains of secretion of oxytocin have been detected in the hypothalamus of 19-week-old fetuses and in the posterior pituitary at 23 weeks. At time of labor and delivery, the human fetus secretes significant amounts of oxytocin. Oxytocin does not cross the placenta. Before birth, the high levels of oxytocin observed in the umbilical artery could play a role in the onset of labor. This is also suggested by the observation that anencephalic fetuses without a hypothalamus usually undergo prolonged pregnancy.

Vasopressin or antidiuretic hormone is found in the hypothalamus at 10–15 weeks gestation and after 19–23 weeks in the posterior pituitary.

Genes for many of the hypothalamic factors have been clearly identified. Knowledge about their structures, roles, and the regulation of their expression is growing at a tremendous rate, such as the growth hormone-releasing hormone and the gonadotropin-releasing hormone genes. This is also true for the processing of the hormones and their receptors.

B. The Anterior Pituitary and Target Organs

The capacity of the fetal anterior pituitary gland to synthesize and to store protein hormones is present during the first trimester of pregnancy.

I. Growth Hormone-Releasing Hormone, Growth Hormone-Release-Inhibiting Factor, Growth Hormone, and Growth Factors

Growth hormone secretion by the pituitary is under the control of two hypothalamic factors, the growth hormone-releasing hormone or factor and the growth hormone-release-inhibiting factor or hormone or somatostatin.

a. Growth Hormone-Releasing Hormone

Growth hormone-releasing hormone neurons have been detected in the arcuate nucleus of the human fetus from the 28th week of development. Concentration of growth hormone-releasing hormone is observed in the median eminence neurons from the 31st week. These neurons retain an immature morphology until birth, suggesting that the control of the growth hormone secretion by growth hormone-releasing hormone is effective only at a late stage of gestation or during the perinatal period in humans.

b. Growth Hormone-Release-Inhibiting Hormone

Growth hormone-release-inhibiting hormone has been found widely distributed throughout the central nervous system and in other organs such as the pancreas and stomach. Growth hormone-release-inhibiting hormone is observed in the hypothalamus at 20 weeks of pregnancy. There is a positive correlation between growth hormone-release-inhibiting hormone concentration in the hypothalamus and gestational age. Injection of growth hormone-release-inhibiting hormone in sheep fetuses in utero induces a decrease of plasma growth hormone at any time of gestation.

c. Growth Hormone and Growth Factors

i. Growth Hormone　Pituitary acidophilic cells containing growth hormone have been detected by the 9th week of fetal life. Its content in growth hormone increases throughout pregnancy. In fetal serum, concentrations of growth hormone increase until the 20th–24th week. Serum growth hormone decreases thereafter until the end of gestation, but the levels observed are higher than those of the normal child and adult. Maternal or fetal growth hormone does not cross the placenta. Fetal growth hormone secretion can be stimulated in vivo by stress such as anoxia or acidosis at midgestation or at delivery time. The role of growth hormone in fetal growth has been challenged by experiments of nature such as anencephalic or apituitary fetuses and by children with idiopathic hypopituitarism, who usually have normal birth length. Available evidence suggests that neither maternal nor fetal growth hormone is essential. Recently, a placental growth hormone has been isolated; its role remains unknown. The syncytiotrophoblastic cells of the placenta also secrete the human chorionic somatotrophic hormone, which has a considerable homology with pituitary growth hormone. This placental hormone, which circulates mainly in the maternal compartment, may play a major metabolic role in sparing energy in the mother, hence providing glucose and proteins necessary to the growth of the fetus.

ii. *Growth Factors* Growth factors in the fetus are essentially *insulin-like growth factor I* (IGF I), or *somatomedin C,* and *insulin-like growth factor II* (IGF II). Possible fetal variants of IGF I and IGF II have also been described. Both factors stimulate multiplication of fetal cells and probably depend on the placental lactogenic hormone secreted by the trophoblastic cells, insulin, and nutritional factors such as glucose. Growth hormone probably plays a minor role. IGF II is mainly secreted in the human fetus and is thought to be involved in the growth of the brain. Insulin, which is structurally similar to IGF I and IGF II, has also been postulated as a growth factor in the fetus. The exact role of factors such as insulin or growth factors on fetal skeletal growth still remains uncertain. In addition, several other growth factors, such as *epidermal growth factor, fibroblast growth factor, and nerve growth factor,* have specific actions on fetal tissues either alone, in synergism (between them), or in association with thyroid hormones, androgens, and IGF I and IGF II. [*See* Insulin-like Growth Factors and Fetal Growth; Growth Factors and Tissue Repair.]

2. Prolactin

Prolactin secretion is controlled by a hypothalamic inhibiting factor, the nature of which is still uncertain (the neurotransmitter dopamine or a gonadotropin-releasing hormone-associated protein named GAP). Presence of prolactin has been observed in the pituitary gland at 68 days gestation. Both prolactin pituitary content and serum concentration increase throughout gestation. The patterns of secretion of prolactin during pregnancy show a similar increase in the fetus and the mother, with a maximum occurring during late gestation and at term, in direct correlation with the increase of circulating estrogens during pregnancy. No placental transfer of prolactin has been reported. A role for fetal prolactin in water transport across the amniotic membranes to maintain a normal volume and composition of the amniotic fluid has been postulated.

3. The Hypothalamo–Pituitary–Thyroid Axis

The thyroid gland is controlled by thyrotropin, secreted by the pituitary. Thyrotropin itself mainly depends on a hypothalamic factor named thyrotropin-releasing hormone (TRH).

a. TRH

TRH has been detected in human fetal brain 4–5 weeks postconception. Its concentration increases progressively during gestation. Of the TRH present in the central nervous system, 20–30% is concentrated in the hypothalamus. Of the total brain TRH, 5% is found in the pituitary gland. Placenta and amniotic fluid contain important amounts of TRH. The role of placental TRH on the fetal pituitary–thyroid axis is yet unknown.

b. Thyrotropin

Thyrotropin or thyrostimulating hormone has been observed in the pituitary and in the plasma of human fetuses as early as the 11th–12th weeks, coincidental with the onset of iodine uptake by the fetal thyroid gland and the synthesis of iodothyronines. Serum thyrotropin concentration increases at midgestation, probably in relation to an increased secretion or an effect of thyrotropin. Thyrotropin does not cross the placenta.

c. The Thyroid Gland

i. *Ontogenesis of the Thyroid Gland* Intracellular colloid formation with synthesis of thyroid hormones appears between 73 and 80 days gestation. Follicular structures are present from 80 days. Growth of the thyroid gland does not seem dependent on fetal thyrotropin. The synthesis of thyroglobulin occurs by the 29th day of gestation.

ii. *Thyroid Hormone Secretion and Metabolism* Thyroxine (T_4) was observed at 78 days in the human fetus plasma. Serum total and free thyroxine as well as total and free triiodothyronine (T_3) increase rapidly during gestation; however, the levels of total and free triiodothyronine are lower in late gestation than in normal children. There is a relative triiodothyronine deficiency in the human fetus with high concentrations of reverse triiodothyronine. Thyroxine, triiodothyronine, and reverse triiodothyronine are present in the amniotic fluid. There is no correlation between fetal serum thyroid hormones and amniotic fluid levels. Consequently, this observation does not allow for prenatal diagnosis of congenital hypothyroidism by amniotic fluid sampling.

By the 12th week of gestation, the presence of specific thyroxine-binding globulin and prealbumin is observed. Liver synthesis of thyroxine-binding globulin depends on maternal estrogens. Thyroxine binds preferentially to thyroxine-binding globulin rather than to prealbumin.

iii. *Placental Transfer of Thyroid Hormones* In humans, placental transfer of thyroid hormones is limited. Less than 1% of a large dose of T_4 given to mothers in labor is transferred to the fetal circulation. Large doses of T_3 administered to women near term slightly decrease fetal serum T_4 concentration. No correlation exists between maternal and fetal serum concentrations of total and free T_3, T_4, or thyrotropin. There is a materno-to-fetal gradient of T_3 and free T_3 during pregnancy. A materno-to-fetal gradient of total and free T_4 is observed before 20 weeks gestation. Near term, this gradient reverses and tends to favor the fetal compartment. Thus, because of the placental barrier, the fetal thyroid system develops and functions autonomously of the maternal system. The placenta is able to concentrate iodine in the fetus. Administration of iodine at high doses or of radioactive iodine to a pregnant woman is a danger for the very active fetal thyroid gland. Similarly, antithyroid drugs, which cross the placenta, can produce a goiter in the fetus. In maternal autoimmune thyrotoxicosis, the presence of thyrostimulating immunoglobulins, which can pass the placental barrier, may cause neonatal hyperthyroidism.

iv. *Role of Thyroid Hormones on Brain Development* In spite of the numerous actions of thyroid hormones on metabolism and enzyme activities, they do not play a specific role in fetal somatic growth. However, they have a very important role in the developing nervous system of the fetus. Prenatal thyroidectomy in the monkey induces a decrease in the content of cerebral RNA and proteins. There is a decrease in the multiplication of neurons and glial cells both in the cerebrum and in the cerebellum. Thyroxine has a direct stimulatory effect on the maturation and the assembly of microtubules of the neuronal cells and on the development of synaptogenesis. Thyroid hormones increase the concentration of nerve growth factor in brain tissue.

4. Sexual Differentiation of the Fetus

Fetal sexual differentiation is an asymmetrical process, which consists of a series of events actively programmed at appropriate critical periods of fetal life and which leads to the sexual dimorphism observed at birth (Table I). Genetic factors and hormonal factors will alternate in this chain of programmed transformations of the gonads, the internal sex organs, and the external genitalia. *De facto,* the male genetic factors and the male hormones will orientate the fetus to maleness, femaleness resulting from the absence of

TABLE I

Timing of Sexual Differentiation in the Human Fetus

Fetal age[a] (weeks)	Crown–rump length (mm)	Sex-differentiating events
	Blastocyst	Inactivation of one X chromosome
4	2–3	Development of Wolffian ducts
5	7	Migration of primordial germ cells in the undifferentiated gonad
6	10–15	Development of Müllerian ducts
7	13–20	Differentiation of seminiferous tubes
8	30	Regression of Müllerian ducts in the male fetus
8	32–35	Appearance of Leydig cells; first synthesis of testosterone
9	43	Total regression of Müllerian ducts: loss of sensitivity of Müllerian ducts in the female fetus
9	43	First meiotic prophase in ovogonia
10	43–45	Beginning of masculinization of external genitalia
10	50	Beginning of regression of Wolffian ducts in the female fetus
12	70	Fetal testis is in the internal inguinal ring
12–14	70–90	Male penile urethra is completed
14	90	Appearance of first spermatogonia
16	100	Appearance of first ovarian follicles
17	120	Numerous Leydig cells; peak of testosterone secretion
20	150	Regression of Leydig cells; diminished testosterone secretion
24	200	First multilayered ovarian follicles; canalization of the vagina
28	230	Cessation of ovogonia multiplication
28	230	Descent of testis

[a]Fetal age in weeks after the last menstrual period.

any masculinizing genetic factor or hormone acting during the critical period of differentiation. Psychological sexual identity is acquired during postnatal life and is the result of psychosocial and hormonal imprinting.

a. Genetic Factors and Sex Determination

The critical role of the Y chromosome and of male hormones in male orientation is well documented, with the development of the female sexual differentiation occurring in the absence of male genetic determinants. Genetic sex is established at the time of fecundation by the nature of the chromosomal composition of the spermatozoon, whether it contains a Y chromosome, which has a dominant effect (23,Y constitution), or an X chromosome (23,X constitution). The development of such a gonad into a testis depends on the presence of the Y chromosome, whereas the absence of the Y chromosome will result in female development, irrespective of the number of X chromosomes. This effect was thought to be due to the presence of a unique gene located on the short arm of the Y chromosome. Several years ago, such an effect was supposed to be linked to a male-specific histocompatibility gene named the H-Y antigen, which was thought to be the primary testis inducer. However, it was soon found that such an antigen did not explain all the sex-reversal cases observed in nature in both human and mouse. Since then several genes have been proposed as candidates for the testis-determining factor. An interval of 140 kilobases (kb) located between 140 and 280 kb of the proximal border of the pseudoautosomal pairing region was first isolated as the region with the testis-determining factor (TDF in the human and Tdf in the mouse). Successive studies using recombinant DNA methods have tried to localize the TDF locus. From a 140-kb region, a highly conserved gene was located in the 1A2 region of the Y chromosome and named zinc finger protein-Y (ZFY), coding for zinc-finger-containing protein that could well function as a DNA-binding transcriptor regulator and be a good candidate for the testis-determining gene. However, a homologous sequence called zinc finger X (ZFX) was then found on the X chromosome, which questioned this hypothesis. From further deletion studies, it was found that the 1A1 region was the one most likely to contain the TDF gene. From this 60-kb region, a 35-kb region was deduced, in which a single-copy gene was found that is highly conserved and shows homologies both with the sexual mating-type protein Mc required for mating in *Schizosaccharides pombo* yeast and with the nonhistone nuclear HMG (high-mobility group) proteins expressed during embryogenesis; it was also thought to function as a DNA-binding transcription factor. This 14-kb gene has been named sex-determining region of the Y (SRY) in the human (Fig. 1). The possibility remains that the ZFY and the SRY genes are two separate but neighboring genes. However, testis development must be possible only through the interaction of the SRY gene with other genes, located on autosomal chromosomes, some of which are involved in the regulation of SRY expression, others possibly being downstream targets of SRY. Several of these genes have been found when studying sex-reversed patients presenting with malformations and ambiguous genitalia such as camptomelic dysplasia (SOX 9, named from SRY—related HMG—box gene 9) or Wilms' tumor (WT1) (Fig. 2). Duplication of one of these genes (DSS, dosage-sensitive sex) could also affect

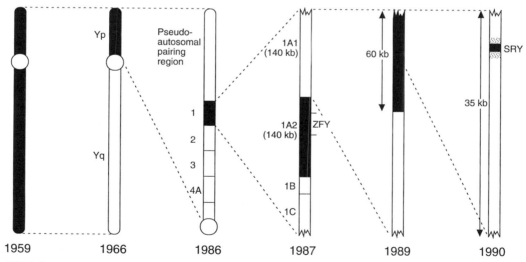

FIGURE 1 Schematic history of the 31-year hunt for the testis-determining factor. The chromosomal region thought to include the elusive factor is shaded. The search has narrowed from 30–40 million bases (1959) to fewer than 250 bases encoding for the conserved 80-amino-acid motif of SRY (1990). See text for further explanation. [Reproduced from A. McLaren (1990). *Nature* 346, 216–217.]

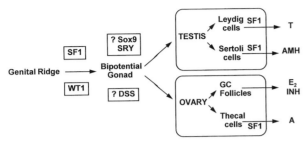

FIGURE 2   Differentiation of the gonad from the genital ridge into a bipotential gonad through several autosomal genes (SF1 located on chromosome 10, WT1 located on chromosome 11). In the presence of SRY (the sex-determining region of the Y), the bipotential gonad differentiates into the testis. Autosomal genes (such as SOX9 on chromosome 17, DSS on chromosome X) are probably necessary for the full differentiation of the testis. SF1 also plays an important role in the full functional development of the testis and the ovary during fetal life. A, A4-androstenedione; AMH, anti-Müllerian hormone; E2, estradiol; INH, inhibin; T, testosterone. See text for further explanation.

the differentiation of the testis. Another gene, SF1 (steroidogenic factor 1, an orphan nuclear receptor), would affect not only the testis but also the adrenal gland.

b. Gonadal Differentiation

The gonadal primordium, which is common in both sexes, develops on the ventral surface of the mesonephros, where primordial germ cells migrate. Testicular tissue, in particular seminiferous tubes, is observed in the embryo at 43–49 days. Sertoli cells secrete a glycoprotein, the anti-Müllerian hormone, from the 7th week of fetal life. Leydig cells are found at 8 weeks. Testosterone secretion by the Leydig cells starts at the same age and is maximal between 14 and 16 weeks. This peak is concomitant with the peak of the placental secretion of human chorionic gonadotropin, suggesting that the secretion of testosterone is mainly under its control. Male anencephalic newborns who do not synthesize pituitary gonadotropins have normal male genitalia. The fetal testis migrates during gestation from the upper part of the abdomen to the inguinal canal. At birth, the testes are usually present in the scrotum.

Ovarian differentiation is a passive procedure but needs the presence of the two X chromosomes. Ovarian organogenesis occurs at the 13th week and primordial follicles are observed from the 16th week. At 5 months gestation, the ovary contains 7 million germinal cells. At birth, the number falls to 2 million. The fetal ovary is capable of secreting steroids from the 8th week.

c. Differentiation of the Genital Ducts

Internal genital ducts are derived from the differentiation of the two pairs of ducts: the Wolffian ducts and the Müllerian ducts (Fig. 3). Experiments performed in gonadectomized male rabbit fetuses before the age of differentiation showed that Wolffian ducts degenerate and Müllerian ducts develop into tubes, the uterus, and the upper part of the vagina. In opposite experiments, locally implanted fetal testis induces in female fetuses regression of the Müllerian ducts and development of the Wolffian ducts. Local implants of testosterone induce development of the Wolffian ducts and no regression of the Müllerian ducts. These experiments led to the concept of the anti-Müllerian hormone, which was later identified. Müllerian ducts are sensitive to anti-Müllerian hormone during a short period of fetal development, up to the 8th week. The Wolffian duct structures are also stabilized during a critical fetal period and apparently only by high local concentrations of testosterone.

d. Differentiation of the Urogenital Sinus and the External Genitalia

In both sexes, the urogenital sinus and the external genitalia are similar up to the 9th week. Masculinization begins by a lengthening of the anogenital distance, followed by a fusion of the labioscrotal swellings in the midline, forming the scrotum, and of the

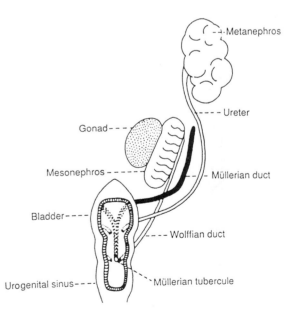

FIGURE 3   Undifferentiated stage of sex differentiation, with presence of fetal structures: Müllerian and Wolffian ducts and the urogenital sinus with the genital tubercle.

rims of the urethral folds leading to the formation of the primordium of the male urethra (Fig. 4). Penile organogenesis is completed by 12–14 weeks. The male differentiation of the external genitalia depends entirely on the secretion of testosterone by the fetal testis. In the female fetus exposed to androgens, different stages of the fusion of the labioscrotal swellings can be observed. Two main types of enzymatic abnormalities of testosterone secretion are known in the human: (1) excessive androgen secretion by the adrenal gland, which will cause virilization of female fetuses, as in congenital virilizing adrenal hyperplasia with female pseudohermaphroditism; and (2) enzymatic biosynthetic defects affecting either testosterone formation, or the end-organ sensitivity to testosterone, which will cause a defect in the development of the male ducts and the external genitalia leading to male pseudohermaphroditism. Penile growth contin-

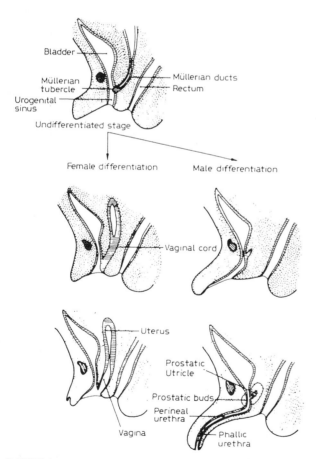

FIGURE 4 Differentiation of the urogenital sinus and the external genitalia. [Reproduced with permission from N. Josso (ed.) (1981). The intersex child. *In* "Pediatric and Adolescent Endocrinology," Vol. 8, p. 8. S. Karger AG, Basel.]

ues during pregnancy and normally depends on fetal testosterone. Testosterone acts directly on the differentiation of the epididymis, the vas deferens, and the seminal vesicle. Reduction of testosterone to dihydrotestosterone by 5α-reductase is necessary to obtain differentiation of the prostate, the prostatic utricule, the scrotum, and the penis (Fig. 5).

5. Hypothalamo–Pituitary–Gonadal Axis

As just described, secretions of the testis are necessary for the male sex differentiation and the hypothalamo–pituitary–gonadal axis matures during fetal life.

a. Gonadotropin-Releasing Hormone

Gonadotropin-releasing hormone (GnRH) was observed in the brain of a 4.5-week-old fetus and is mainly located in the hypothalamus. The concentration of GnRH in the hypothalamus varies with age of gestation and sex. Temporal series of observations have shown that the GnRH neurons move through a well-defined pathway from the olfactory placode to the basal forebrain and the hypothalamus through the basal septum as early as 19 weeks of gestation. In a human fetus affected with hypogonadotrophic hypogonadism (so-called Kallmann's syndrome), caused by a deletion on chromosome X (Xp 22.3), the GnRH neurons are packed at the top of the olfactory apparatus and never enter the forebrain, that is, the preoptic area and the hypothalamus.

b. Gonadotropins, Testosterone, and Estrogens

Both follicle-stimulating hormone and luteinizing hormone are synthesized by the fetal pituitary under the stimulation of gonadotropin-releasing hormone. Both gonadotropins have been detected in the pituitary cells as early as 10 weeks gestation. Only α-subunits of gonadotropins are found until the 10th week in the pituitary gland. Fetal pituitary follicle-stimulating hormone concentration increases between 150 and 210 days. In both sexes, fetal serum concentration of follicle-stimulating hormone peaks between 100 and 150 days, followed by a decline until term. Pituitary luteinizing hormone increases between 100 and 150 days gestation. In fetal serum, luteinizing hormone is also present but has been difficult to separate from human chorionic gonadotropin. Fetal human chorionic gonadotropin, which originates from the placenta, exhibits a peak by 90–120 days and then decreases. Interestingly, the pattern of change of human chorionic gonadotropin is related to that reported for serum testosterone in the male fetus. Tes-

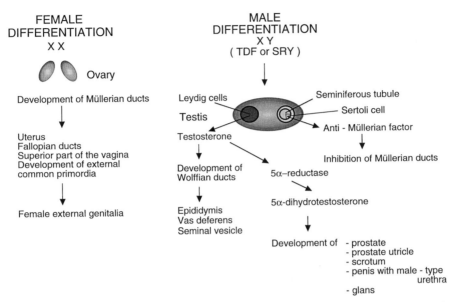

FIGURE 5   Genetic and hormonal factors acting on the differentiation of the gonads, internal sex organs, and external genitalia in female and male fetuses. Development of male sex structures depends on the presence of the anti-Müllerian factor, the secretion of testosterone, and its conversion to 5α-dihydrotestosterone. SRY, sex-determining region of the Y; TDF, testis-determining factor.

tosterone peaks between 11 and 18 weeks in the male fetus, corresponding to the time of the sex differentiation. A further decrease of testosterone is observed between 17 and 24 weeks gestation.

In the female fetus, testosterone secretion is very low. Fetal circulating estrogens are mostly part of the placental production. The exact proportion of the estrogens originating from the fetal ovary is unknown.

6. The Hypothalamo–Pituitary–Adrenal Axis, the Fetoplacental Unit

a. Corticotropin-Releasing Hormone

Corticotropin-releasing hormone is present in the median eminence as early as the 16th week of pregnancy.

b. Adrenocorticotropin Hormone and Related Peptides

A common precursor for adrenocorticotropin hormone (ACTH) and endorphins, named pro-opiomelanocortin, has been isolated. [*See* Pituitary Gland Hormones.] These ACTH-related peptides may play an important role during fetal life. A change occurs in the pituitary content of ACTH-related peptides. At early gestation (12–18 weeks), substantial amounts of ACTH, β-lipotropin (β-LPH), and β-endorphins

are found in contrast to relatively low contents of α-melanocyte-stimulating hormone (α-MSH) and corticotropin-like intermediate lobe peptide (CLIP). From midgestation, α-MSH and CLIP become the predominant forms together with ACTH. Just before birth, the amount of the two cleavage products α-MSH and CLIP sharply increases.

ACTH is present in pituitaries of 14-week-old fetuses. Transplacental passage of ACTH is very unlikely. High levels of serum ACTH are observed in 12-week-old fetuses. The concentrations in serum decreased by 35–40 weeks. Studies of human anencephalic fetuses with hypoplastic adrenals have given evidence for the critical role of ACTH in the normal growth of the adrenal cortex. A placental human ACTH may stimulate the development of the fetal adrenal cortex. ACTH and β-endorphins are present in amniotic fluid, and the levels are increased in case of fetal distress.

c. Fetal Adrenal Cortex

The fetal adrenal cortex plays a very important role in the fetus for its direct action and also for the supply of metabolites to the placenta. The maternal and fetal adrenal glands, the fetal liver, and the placenta constitute the *fetoplacental unit*.

Originating from the mesoderm, adrenocortical cells separate from the celomic epithelium and form two masses on either side of the aorta. Adjacent to the cortical cells are the medullary crest cells, which migrate from the neural crest; they will form the adrenal medulla glands. By 6–7 weeks gestation, the fetal adrenal cortex is constituted of a thin outer layer and a large inner layer named the *fetal zone,* which represents 80% of the total fetal gland.

i. *Inner Zone or Fetal Zone*   The principal characteristic of the fetal zone is the absence of the Δ^5-3β-hydroxysteroid dehydrogenase activity. Thus, the fetal zone is unable to synthesize Δ^4-3-ketosteroids from Δ^5-3β-hydrosteroids (Fig. 6). It has been suggested that the growth of the fetal zone during the first trimester of pregnancy depends on the human chorionic gonadotropin produced by the placenta. After the fourth or the fifth month, the maintenance of the adrenal cortex would depend on other trophic factors such as fetal ACTH and related peptides. In anencephalic fetuses, the adrenal glands develop normally during the first trimester of pregnancy and undergo involution later on, after the fifth month.

ii. *Outer Zone*   Cortisol secretion has been detected in the outer zone of the fetal adrenal cortex as early as 8–10 weeks gestation. Aldosterone synthesis is present from the 15th week of gestation. A rise of fetal cortisol has been implicated in the onset of labor. Clinical experience has suggested that fetal adrenal insufficiency, as observed in anencephalic fetuses, may be the cause of delayed parturition. Shortened pregnancy has been described in cases of fetal adrenal hyperplasia secreting large amounts of cortisol.

iii. *Placental Transfer of Cortisol*   Placental transfer of cortisol from the mother to the fetus has been demonstrated in humans. In cases of excessive maternal production of cortisol, adrenal insufficiency and low-plasma ACTH have been observed in the fetus. Fetal cortisol plays an important role in the development of surfactant factor and the maturation of pulmonary alveolar lining. Synthetic glucocorticoids acting like

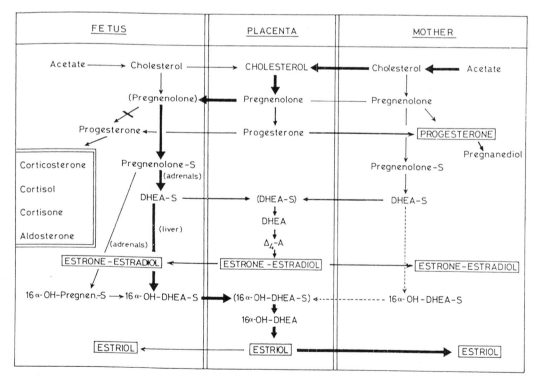

FIGURE 6   The fetoplacental unit: steroid biosynthetic pathways and interrelations of steroid metabolism among the mother, the placenta, and the fetus. [Reproduced with permission from P. C. Sizonenko and M. L. Aubert (1986). Pre- and perinatal endocrinology. *In* "Human Growth" (F. Falkner and J. M. Tanner, eds.), Vol. 1. Plenum, New York.]

cortisol, such as dexamethasone, betamethasone, or methylprednisone, cross the placenta and are used as inducers of fetal surfactant factor. Thyroid hormones, estradiol, prolactin, thyrostimulating hormone-releasing factor, and β-adrenergic compounds would also play a role in the production of lung surfactant.

d. Fetoplacental Unit

Because both the placenta and the fetus lack certain enzyme activities for complete steroidogenesis, the concept of a fetoplacental unit has arisen. Enzymes absent in the placenta apparently are present in the fetus and vice versa, and the integration of maternal, placental, and fetal functions can explain the production of the steroids made during the pregnancy.

The main enzymatic pathways are represented in Fig. 6. Cholesterol produced by the mother from acetate is transformed into Δ^5-pregnenolone in the placenta. Δ^5-pregnolone is then transferred into the fetal adrenal glands to be converted into Δ^5-pregnenolone-sulfate and into dehydroepiandrosterone-sulfate (DHEA-S). DHEA-S is then hydroxylated into 16α-OH-DHEA-S in the fetal liver, which in turn is transferred to the placenta, cleaved by sulfatases, and aromatized into estriol, which passes into the maternal compartment and is mainly excreted in the maternal urine. Hence, estriol synthesis is a complex process. Therefore, the assessment of fetal viability by urinary estriol will depend on functions of both fetal and maternal adrenal glands, fetal and maternal liver, sulfate cleavage, and aromatization by the placenta.

C. Calcium, Parathyroid Glands, Calcitonin, and Vitamin D

1. Fetal Calcium

The fetus has important needs for calcium, two-thirds of which has been acquired during the last trimester of gestation (26 mg/kg/day). In total, during pregnancy, the mother transfers 30 g of calcium to the fetus. Similarly, transplacental transfer of phosphorus during the last trimester is important (80 mg/kg/day). The plasma levels of 1,25-dihydroxycholecalciferol, the active metabolite of vitamin D, rise in the mother, permitting an increased intestinal absorption of calcium. Plasma total and ionized calcium levels in the fetus are higher than in the mother, suggesting an active materno-fetal gradient. Similar active transports of phosphorus and magnesium from the mother to the fetus have been suggested. Plasma levels of phosphorus and magnesium are higher in the fetus than in the mother.

2. Parathyroid Glands

The parathyroid glands derive from the endoderm of the 4th branchial pouch and the 3rd branchial cleft. Parathormone has been detected in parathyroid glands of fetuses at 14 weeks gestation. Parathormone does not cross the placenta, and its role during fetal life is unknown. Newborns with congenital hypoparathyroidism have a normal skeleton and do not present hypocalcemic manifestations before 2 days of life. Parathormone secretion seems to depend on the maternal levels of plasma calcium as maternal chronic hypocalcemia stimulates hyperplasia of the fetal parathyroid glands, and maternal chronic hypercalcemia induces hypoplasia of the fetal glands. [See Parathyroid Gland and Hormone.]

3. Calcitonin

Calcitonin-secreting cells, of neuroectodermic origin, named C-cells, have been found very early in the fetal thyroid gland. Synthesis of calcitonin is present at 14 weeks gestation. The role of calcitonin during fetal life is unknown.

4. Vitamin D and Its Metabolites

Vitamin D and its metabolites (25-hydroxycholecalciferol and 1,25-dihydroxycholecalciferol) are lower in the fetus than in the mother. Fetal plasma 25-hydroxycholecalciferol is highly correlated with the maternal levels and represents 60–70% of the maternal level. However, this gradient is low when the maternal level is low and is elevated when the maternal concentration is high, suggesting a protective action of the placenta against vitamin D deficiency or vitamin D intoxication. 1,25-Dihydroxycholecalciferol is present in fetal blood at 27 weeks gestation. This synthesis is taking place in both the fetal kidney and the placenta. [See Vitamin D.]

D. Pancreas

The fetal pancreas comes from an outgrowth of the duodenal endoderm. Differentiation of A-cells and B-cells occurs at 10 weeks gestation. D-cells secreting somatostatin or pancreatic polypeptide are present at 17 weeks.

1. Insulin

Insulin and C-peptide content of the human fetal pancreas is in direct correlation with the number of islet cells, and both peptides are present as early as the 12th week. Glucose is a poor stimulating agent of insulin release. On the contrary, amino acids such as arginine are a very potent factor for insulin secretion. These observations suggest different mechanisms for the release of insulin from the fetal B-cell. Only a very minimal fraction of plasma insulin passes the placenta. Insulin and C-peptide have been detected in the amniotic fluid. Amniotic fluid concentrations of both peptides originating mainly from the pancreas of the fetus are higher in diabetic pregnancies, supporting the observation that the pancreas of such a fetus is hyperplastic. [See Insulin and Glucagon.]

2. Glucagon

Glucagon has been detected as early as 10–12 weeks gestation. Alanine stimulates glucagon secretion in the fetus. Glucagon does not cross the placenta in the human.

E. Adrenal Medulla

Adrenal medullary gland derives from the neuroectodermal tissue. The chromaffin cells have been observed very early, at 8 weeks gestation. Some of them will differentiate as pheochromoblasts, invade the adrenal cortex, and give birth to the adrenal medulla. The fetal adrenal medulla develops mainly during the second half of gestation. However, the bulk of the chromaffin cells remain extra-adrenal and form the adrenergic neurons and ganglia along the aorta. This extra-adrenal tissue will involute later after birth. Catecholamines have been detected in the adrenal medulla at 15 weeks gestation. The maternal catecholamines do not cross the placenta. They are present in the amniotic fluid and have been found to be higher in case of intrauterine growth retardation, particularly in the case of maternal smoking.

II. PERINATAL ENDOCRINOLOGY

Birth is a *stressful event* in the sense that the newborn is separated from the mother and the placenta and must develop metabolic and hormonal mechanisms, which will permit the adaptation to extrauterine life (e.g., thermic control, food intake, day–night rhythm). The newborn has a complete potential endocrine system, which will become fully operational only during the first weeks of life. [See Endocrine System.]

A. Hypothalamus and Posterior Pituitary

1. Oxytocin

At birth, umbilical arterial concentrations of oxytocin are higher than venous ones, suggesting that the fetus secretes oxytocin. Concentrations decrease after 30 min but remain higher than in adults. Oxytocin role during the perinatal period remains unknown.

2. Vasopressin

At birth, the plasma levels of vasopressin are higher in infants born by vaginal delivery than those born by cesarean section. Stress such as fetal hypoxia or diminished placental blood circulation stimulates vasopressin secretion. However, no correlation exists between plasma vasopressin and the usual criteria of fetal distress. A positive correlation between vasopressin and stages of cervical dilatation has been observed. The levels of vasopressin decrease rapidly after birth. The neonate is able to respond adequately to a water load or to a hypertonic infusion, providing evidence that the posterior pituitary and the osmoreceptor systems are functioning appropriately. Immature renal function more likely explains the decreased ability of the newborn infant to concentrate urine.

B. The Anterior Pituitary and Target Organs

1. Growth Hormone and Growth Factors

In the newborn, plasma concentrations of growth hormone are high compared with those of adults. They rise at 48 hr of life and then decrease progressively during the next 4 weeks of life. In premature infants, the levels are higher than in normal babies. The newborn infant demonstrates a paradoxical response to glucose or to stress, but a normal response to hypoglycemia, amino acids, and growth hormone-releasing hormone. These observations suggest that the maturation of the secreting mechanisms for growth hormone secretion is not achieved. Sleep-associated secretion of growth hormone appears at 3 months of life.

IGF I levels in cord blood are correlated to birth weight and length of newborns between 24 and 42 weeks gestation. In infants with intrauterine growth retardation, IGF I levels are lower than those in normal infants. In some newborns with intrauterine growth retardation, a catch-up growth can be observed. Little is known of the mechanisms by which this postnatal catch-up occurs.

2. Prolactin

At birth, plasma prolactin is very high and decreases very quickly during the first 5 days of life but remains above the levels measured during childhood until 6 weeks of life. In anencephalic newborns, plasma concentrations of prolactin are similar to the normal newborn levels, suggesting that the secretion of prolactin does not depend on a hypothalamic hormone stimulating the pituitary synthesis and the secretion of prolactin.

3. The Hypothalamo–Pituitary–Thyroid Axis

a. Thyrotropin-Releasing Hormone

High levels of thyrotropin-releasing hormone are found in cord blood compared with maternal concentrations during the first 20–40 min after birth. These high concentrations are probably related to the rapid activation of the hypothalamo–pituitary–thyroid axis observed immediately after birth. In addition, the plasma of newborns contains a low amount of thyrotropin-releasing hormone degrading enzyme. This enzymatic activity appears in the newborn serum at 3 days of life.

b. Thyrotropin

In the newborn, thyrotropin is higher than in maternal blood, rises further 10–30 min after birth, plateaus during the following 3–4 hr, and decreases at 48 hr of life. Fall in the body temperature of the newborn as well as the section of the cord are probably the triggering mechanisms. However, the prevention of the drop of the body temperature of the newborn does not suppress the thyrotropin peak.

c. Thyroid Hormones

Free and total thyroxine concentrations in the cord blood are similar to the maternal levels; conversely, total and free triiodothyronine plasma levels are much lower than the maternal ones. The concentrations of reverse triiodothyronine are higher than those in maternal plasma. Following the rise of thyrotropin, thyroxine and triiodothyronine concentrations increase and remain elevated during the next 24–72 hr of life. The newborn switches from a biochemical triiodothyronine-deficient state to a condition of biochemical thyrotoxicosis with high triiodothyronine levels, which last 3–4 weeks. The exact mechanism of this biological hyperthyroid state is not known. In premature and small-for-gestational-age babies, total and free thyroxine plasma concentrations are lower than in normal infants. Although there is a similar neonatal rise in thyrotropin as in normal neonates, this paradoxical state of transient hypothyroidism is probably due to a delayed maturation of the hypothalamo–pituitary–thyroid axis. [See Thyroid Gland and Its Hormones.]

Because the hypothalamo–pituitary–thyroid axis is active at birth, screening programs for early diagnosis of congenital hypothyroidism based on the measurement of heel-prick blood thyrotropin at Day 5 of life have been established. Some programs are based on measurement of thyroxine. Such screening permits early therapy of the congenital hypothyroidism, leading to the normal psychomotor development of the affected newborns.

4. The Hypothalamo–Pituitary–Gonadal Axis

a. Gonadotropins

In the newborn cord blood, plasma concentrations of human chorionic gonadotropin and α-subunits of gonadotropins are high and similar in the two sexes. Intact follicle-stimulating hormone and luteinizing hormone are low. In anencephalic newborns, the gonadotropin content of the pituitary is low, suggesting that gonadotropin-releasing hormone is necessary for the synthesis of intact molecules of gonadotropins.

After birth, follicle-stimulating hormone is higher in female infants than in males, and it remains so during the first 2 years of life (Figs. 7 and 8). Levels of luteinizing hormone are low during the first week and increase later. During this period, gonadotropin

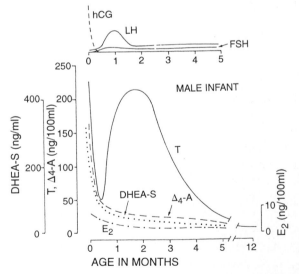

FIGURE 7 Serum concentrations of human chorionic gonadotropin (hCG), luteinizing hormone (LH), follicle-stimulating hormone (FSH), testosterone (T), Δ^4-androstenedione (Δ^4-A), dehydroepiandrosterone-sulfate (DHEA-S), and estradiol (E_2) in male infants during the first months of life. [Reproduced with permission from P. C. Sizonenko and M. L. Aubert (1986). Pre- and perinatal endocrinology. In "Human Growth" (F. Falkner and J. M. Tanner, eds.), Vol. 1. Plenum, New York.]

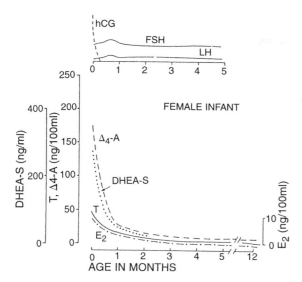

FIGURE 8 Serum concentrations of human chorionic gonadotropin (hCG), luteinizing hormone (LH), follicle-stimulating hormone (FSH), testosterone (T), Δ^4-androstenedione (Δ^4-A), dehydroepiandrosterone-sulfate (DHEA-S), and estradiol (E_2) in female infants during the first months of life. [Reproduced with permission from P. C. Sizonenko and M. L. Aubert (1986). Pre- and perinatal endocrinology. *In* "Human Growth" (F. Falkner and J. M. Tanner, eds.), Vol. 1. Plenum, New York.]

concentrations in plasma are in the range seen during puberty, suggesting that the sensitivity of the negative feedback control regulating the hypothalamo–pituitary–gonadal axis has not reached the degree of the childhood level. Gonadotropins are secreted during this period in a pulsatile fashion, as in the adult.

b. Gonadal Steroids

In cord blood, testosterone concentration is slightly higher in the male than in the female infant. In peripheral blood, the difference of testosterone concentration is more marked. In male infants, plasma testosterone decreases on Day 5, increases on Day 10, with a peak value at 2 months (Fig. 7). This testosterone rise is secondary to the elevation of luteinizing hormone. This rise is also observed in premature babies.

In the female newborn, the ovary contains numerous active follicles. In cord blood, high levels of estradiol and estrone are present without sex difference. They decrease rapidly during the first 72 hr of life (Fig. 8).

5. The Hypothalamo–Pituitary–Adrenal Axis
a. Perinatal Secretion of Cortisol

Secretion of the steroids depending on the fetal zone, such as dehydroepiandrosterone, DHEA-S,

16α-OH-DHEA-S, and estriol, decreases progressively during 8–10 weeks postnatally (Figs. 7 and 8). Plasma cortisol levels are higher in newborns born by vaginal delivery than in infants born by cesarean section, suggesting that the hypothalamo–pituitary–adrenal axis is operative at birth. Peripheral plasma levels are high during the first weeks postnatally. Cortisol response to stress is present in the young infant. Secretion of cortisol is higher in infants and reaches adult value at the end of the first year. Nycthemereal rhythms of cortisol appear during the second month of life.

b. 17α-OH-Progesterone

17α-OH-Progesterone and progesterone decrease rapidly after birth, without difference between sexes. Measurement of 17α-OH-progesterone on Day 5 of life has been proposed as a means to screen for congenital-virilizing adrenal hyperplasia due to 21-hydroxylase deficiency. In premature babies, levels of 17α-OH-progesterone are higher than in full-term newborns, rendering screening results often difficult to interpret.

c. Renin–Angiotensin–Aldosterone System

Secretion of aldosterone in 1 to 8-day-old newborns is low as compared with that of older infants. The aldosterone concentrations rose in newborns whose mothers were put on a low-sodium diet and remained high during the first 3 days of life on a low-sodium diet. These results suggest that the renin–angiotensin–aldosterone system is fully active in the newborn. Plasma renin activity in newborns is inversely proportional to the sodium balance and the urinary excretion of sodium.

C. Calcium Metabolism

At birth in normal newborns, plasma total and ionized calcium levels, which are higher than those in the mother, decrease to a nadir on Day 3 of postnatal life. Concomitantly, serum parathormone levels, which are low at birth, begin to rise, inducing a rise in plasma 1,25-dihydroxycholecalciferol. Plasma phosphorus also rises progressively, mainly depending on the milk formula. Breast milk maintains lower phosphorus levels than some milk formula rich in phosphorus. It has been suggested that transient neonatal hypocalcemia or tetany that can be observed during the first days of life, particularly in premature infants, can be caused by a functional hypoparathyroidism, sometimes exaggerated by a diet rich in phosphorus. It is possible that neonatal tetany can also be induced by a low body calcium mass, as observed in premature babies, and/or low maternal levels of vitamin D. Supplementation of pregnant women with vitamin D has been suggested. Calcitonin levels are high at birth and decrease progressively until the first month of life. This hypersecretion of calcitonin may play a role in the regulation of plasma calcium during the neonatal period. In addition, the kidney tubule responsiveness to parathormone seems to be impaired in newborn infants. A tubular maturation with an increase of phosphorus clearance is observed during the first weeks of life.

D. Pancreatic Hormones

1. Insulin

In cord blood as well as in peripheral blood at birth, insulin levels are positively correlated with birth weight. Administration of glucose, arginine, or glucagon to newborn infants induces a sluggish response of insulin as compared with older infants, suggesting a progressive maturation of the mechanisms of secretion of pancreatic insulin: B-cells become increasingly sensitive to usual stimuli.

2. Glucagon

Blood glucose decreases 1 hr after birth. This fall, reaching a nadir within hours of birth, is associated with a significant increase in plasma glucagon concentration. The rise of glucagon, despite low levels of blood glucose, occurs only 24 hr after and is subsequently followed by the expected rise of glucose to normal values. Glucagon secretion is normally stimulated by arginine and alanine. Glucose, at this period of life, is a poor suppressor of glucagon.

E. Adrenal Medulla

After birth, most of the extra-adrenal chromaffin tissues undergo atrophy. In contrast, the adrenal medulla develops rapidly. Norepinephrine is the main catecholamine secreted during the neonatal period. During the first 3 years after birth, epinephrine becomes the principal catecholamine secreted by the adrenal medulla, and urinary catecholamine excretion increases with age and body weight. The newborn is able to increase the secretion of norepinephrine and epinephrine in response to stress, hypoxia, or hypoglycemia.

III. ENDOCRINOLOGY OF PREPUBERTY

Prepuberty, from 2 to 10 years, is mainly characterized by (1) a continuation of growth on a slow rate compared with the growth during the first 2 years of life and the pubertal growth; (2) a period of quiescence of the hypothalamo–pituitary–gonadal axis, after the perinatal activity described earlier and before the activation of the hypothalamo–pituitary–gonadal axis, named the *gonadarche,* leading to the pubertal development; and (3) the occurrence at 7 years of the maturation of the adrenal axis, named the *adrenarche,* before sexual maturation.

A. Growth Velocity and Bone Age

Growth has been divided into four periods characterized by the growth velocity. The first one, from birth to 2–3 years, exhibits a fast growth velocity at birth, which declines progressively (Fig. 9). The second phase, from 2–3 to 10 years, is characterized by a slow decline of growth velocity. The third phase is the pubertal growth linked to sex hormones, which stimulate the bone growth and maturation. The fourth period consists of the rapid decline of the growth velocity, which arrests completely 3–4 years after the pubertal growth velocity peak. Bone growth is responsible for the height of the child. Growth of the individual is linked to the development of the epiphyseal cartilage, which is progressively calcified with age and incorporated to the metaphysis of the bone. Radiography of the bones, particularly of the hand and the wrist, permits the determination of the *bone age.* Based on the shape of the bones determined at each age, this concept of bone age has arisen. Usually, the bone age corresponds to the chronological age of the majority of individuals of the same sex. Prediction of the adult height can be calculated from the bone age. Several tables for such prediction are used, particularly in case of short or tall stature. [*See* Growth, Anatomical.]

B. Hormonal Control of Growth

Growth hormone plays the major role in the postnatal growth of bones and the different organs. It acts either directly on the bone cartilage or indirectly through *growth factors* (Fig. 10). These growth factors are synthesized locally either in the target organs or in the liver. They are transported in the blood either in a free form or linked to binding proteins. Synthesis

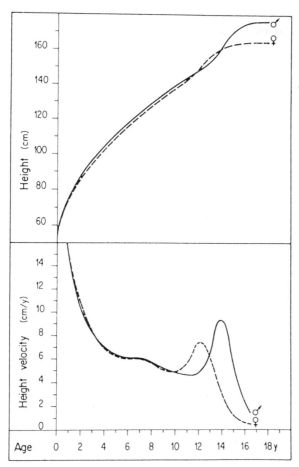

FIGURE 9 Normal growth: median values of height and height velocity for girls and boys. [Reproduced with permission from A. Prader (1990). Hormonal regulation of growth and the adolescent growth spurt. *In* "The Control of Onset of Puberty," p. 535. Williams & Wilkins, Baltimore.]

and secretion of growth factors are stimulated by growth hormone, insulin, and nutrition and are inhibited by glucocorticoids. *Insulin* has an important role in growth, although it is difficult to separate the direct effects of insulin (cell multiplication, energy metabolism, glucose and amino acid transport, lipogenesis) from its indirect effects through the growth factors. *Thyroid hormones* are necessary for the bone growth and the bone maturation as well as the brain development (see earlier). They act synergistically with growth hormone. Hypothyroidism in children induces growth retardation and delay in bone maturation. In hyperthyroidism, thyroid hormones stimulate excessive growth and advance bone age. *Sex hormones* (i.e., testosterone in males and estrogens, mainly estradiol, in females) stimulate growth. Exces-

sive production of sex hormones as observed in sexual precocity or in congenital adrenal hyperplasia accelerates growth and, moreover, bone age. This advanced bone maturation leads to adult short stature. Delay in pubertal development is associated with short stature and bone age retardation. Sex hormones act directly on the cartilage but also increase growth hormone secretion, which has a synergistic action on growth. *Glucocorticoids* in excess have an inhibitory effect on growth and on bone maturation. They act directly on the cartilage by inhibiting the mineralization of the epiphysis, indirectly by decreasing the production of growth factors, by inhibiting the action of the growth factors on the cartilage, and finally by possibly decreasing the secretion of pituitary growth hormone. *Genetic factors* remain the main regulators of growth and cell division. Hormone actions depend on these genetic factors acting through gene regulation mechanisms. Absence of tissue receptors to growth hormone and deficiency of enzymes necessary for the normal synthesis of cartilage are examples of such abnormal regulation. Genetic factors are the most frequent causes of short or tall stature. *Nutrition* is also the main factor regulating growth of the organism. Both caloric (marasmus) and protein (kwashiorkor) deprivation induces severe growth retardation (Fig. 11). *Psychosocial disorders,* such as maternal deprivation, called psychosocial dwarfism when severe, can induce growth retardation.

C. Endocrine Disorders

Many of the disorders observed in children are also found during adulthood. Therefore, only disorders specifically related to pediatrics will be briefly described.

I. Growth Hormone Disorders

Growth hormone deficiency can be either isolated or associated with other deficiencies of the anterior and/or posterior pituitary. It can be familial by recessive or dominant autosomal transmission (by deletion of the gene of growth hormone) or, more commonly, sporadic. Etiology includes malformations or tumors of the hypothalamo–pituitary region. The main symptom is growth retardation with reduced growth velocity. Therapy with genetically synthesized recombinant human growth hormone induces catch-up growth with general normal or near-normal adult height. Hypersecretion of growth hormone is rare during childhood.

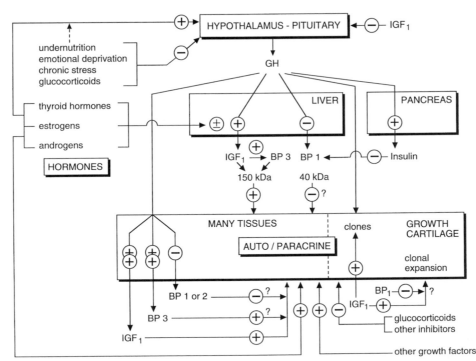

FIGURE 10 Hormonal regulation of growth. Pituitary growth hormone (GH) secretion is regulated by hypothalamic growth hormone-releasing hormone and growth hormone-release-inhibiting hormone. GH mainly has an indirect effect on the cartilage through the growth factor called insulin-like growth factor I (IGF I). Recently a direct action of GH on the cartilage has been postulated. GH and IGF I inhibit secretion of GH. GH also has a direct action on tissues like muscle and fat tissue and on carbohydrate metabolism. IGF I production is enhanced by nutrition and insulin. IGF I production is diminished by malnutrition, chronic diseases, and glucocorticoids. Thyroid hormones, estrogens, and androgens, at physiological levels, stimulate secretion of GH and IGF I. The latter is bound to several binding proteins, for example, BP1, BP2, and BP3. IGF-BP1 production is decreased by insulin, whereas IGF-BP3 production is stimulated by GH and IGF I itself. Glucocorticoids and estrogens at high doses inhibit IGF I production. Inhibitors of IGF I have been observed in malnutrition and renal insufficiency. [Reproduced with permission from J. L. van den Brande (1993). Postnatal growth and its endocrine regulation. *In* "Pediatric Endocrinology" (J. Bertrand, R. Rappaport and P. C. Sizonenko, eds.), p. 160. Williams & Wilkins, Baltimore.]

2. Thyroid Hormone Disorders

The main thyroid hormone disorder is congenital hypothyroidism due to either the absence of the thyroid gland (agenesis) or abnormal descent of the thyroid gland (ectopic gland). If not diagnosed during the neonatal period by screening (see earlier), symptoms appear slowly and constitute the classic aspect of myxoedema, with short stature and mental retardation. At this stage of the disease, therapy with thyroid hormones restores normal growth but does not prevent mental retardation. Hypothyroidism due to enzymatic defects of thyroid hormones synthesis is rare; the autoimmune disease of the thyroid gland, named thyroiditis (Hashimoto's disease), with hypothyroidism or hyperthyroidism, is more frequent. Hyperthyroidism

(Graves–Basedow's disease), goiters, and cancer of the thyroid gland are also observed during this period of life.

3. Adrenal Gland Disorders

a. Adrenal Cortex Disorders

Deficiencies of the adrenal cortex hormones (Addison's disease), hyperfunctions of the adrenal cortex (Cushing's syndrome, or disease), or tumors of the adrenal cortex are rare during childhood. Congenital adrenal hyperplasia is an autosomal recessive disorder affecting one of the enzymes of the metabolic pathway of cortisol. The most frequent disorder is 21-hydroxylase deficiency. Two forms have been observed: (1) one incomplete form with only excessive production

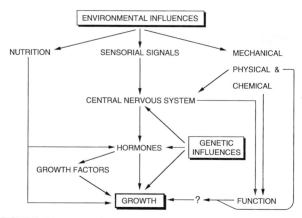

FIGURE 11 Main factors regulating growth. [Reprinted with permission from J. L. van den Brande (1993). Postnatal growth and its endocrine regulation. *In* "Pediatric Endocrinology" (J. Bertrand, R. Rappaport, and P. C. Sizonenko, eds.), p. 160. Williams & Wilkins, Baltimore.]

of androgens, inducing in females sexual ambiguity and in both sexes accelerated growth and sexual precocity, and (2) one complete form with, in addition to the cortisol deficiency and excessive secretion of androgens, a deficiency in the synthesis of aldosterone, provoking loss of sodium in the urine with possible severe dehydration of the infant. Hydrocortisone is administered as a substitute for cortisol. Mineralocorticoids should be added in the complete form. Cortisone suppresses the excessive ACTH secretion due to the absence of cortisol, reduces the abnormal production of the adrenal androgens, and, therefore, reduces the virilization process. Several other types of enzymatic deficiencies have been described. They can induce sexual ambiguity either by virilization of the female fetus or by nonvirilization of the male.

b. Adrenal Medulla Disorders

Tumors of the adrenal medulla are infrequently observed during childhood. They constitute benign tumors of the chromaffin tissue named pheochromocytoma with arterial hypertension, excessive sweating, high temperature, and abnormal behavior. High levels of catecholamines and their metabolites are found in blood and urine.

4. Disorders of Calcium Metabolism

Calcium metabolism disorders are similar to these diseases observed during adulthood. In the case of parathormone deficiency, tetany is the main symptom and constitutes the disease named hypoparathyroidism. It can be linked with growth retardation and

transmitted as a sex-linked recessive genetic defect or associated to aplasia of the thymus or to hypoplasia of the adrenal glands and/or moniliasis. Tetany can also be observed in the presence of high levels of plasma parathormone, suggesting a resistance to parathormone and constituting the pseudohypoparathyroidism.

5. Disorders of the Endocrine Pancreas
a. Diabetes Mellitus

Diabetes mellitus in the child is due to insulin deficiency in relation to an autoimmune destruction of the B-cells; therefore, it is insulin dependent. It is a lean diabetes in contrast to the diabetes observed in the obese patient (5% of all diabetic children).

b. Hypoglycemia Due to Hyperinsulin

Many abnormalities of carbohydrate metabolism cause hypoglycemia in children. Among them, excessive secretion of insulin due to B-cell tumor or hyperplasia has been reported, particularly in young infants, presenting with ectopic B-cell islets within the pancreas, called nesidioblastosis.

IV. ENDOCRINOLOGY OF PUBERTY

Puberty is defined as the period of life during which sexual maturation occurs (i.e., the growth of the gonads, the development of the external genitalia and the secondary sexual characteristics, and the establishment of the normal functions of reproduction). These changes are associated with increased growth, bone maturation with fusion of the epiphysis leading to the cessation of growth, and psychological, social, and behavioral modifications that are characteristic of adolescence. [*See* Puberty.]

A. Normal Pubertal Development

1. Age at Puberty

In females, normal puberty begins between the ages of 8.5 and 13.3 years, with a median age of 10.9 years. Total pubertal growth is between 9.5 and 14.5 years, with the mean pubertal growth spurt occurring around the median age of 12.2 years at 9 cm/year, from 6 to 11 cm; mean age of first menstrual bleeding, called menarche, is 12.9 years, between 11.7 and 15.3 years. In males, normal puberty starts between 9.2 and 14.2 years, with a median age of 11.2 years. Testicular volumes increase >4 ml; total pubertal growth lasts between 10.5 and 17.5 years, with a mean growth spurt observed at a median age of 13.9 years at 10.5 cm/year, from 7 to 15.5 cm. In both sexes, adolescents frequently follow an irregular pattern with some secondary sexual characteristics (e.g., axillary or pubic hair) appearing prematurely or lagging behind.

2. Sexual Characteristics and Hormonal Changes

In females, the first change in the sexual characteristics is breast budding, which is followed by the appearance of axillary and pubic hair (Table II). Vulvae will subsequently mature with the development of the labia minora and majora, and the vaginal mucosa becomes pink. Menarche appears 2–3 years after the first breast budding. Increase of the volume of the ovaries and mainly of the uterus can be followed by ultrasonography of the pelvis. Plasma gonadotropins, prolactin, and estradiol increase progressively. Maturation of the ovary leads to menstrual bleedings and the appearance of ovulatory menstrual cycles, several months after the menarche. Plasma progesterone increases during the luteal phase of the ovulatory cycle. The first cycles after menarche are usually anovulatory.

In males, volume of the testes increases with appearance of axillary and pubic hair, growth of the penis, and development of the scrotum (Table III). Moustache, voice deepening due to the enlargement of the larynx, and acne will appear around 14 years of age. Plasma gonadotropins and testosterone increase progressively. Spermatozoids can be found in the urine of boys at 14–15 years. First ejaculations appear at the same age.

In addition to the changes of circulating gonadotropins and sex steroids, sex steroid-binding globulins decrease. IGF I levels peak during the pubertal growth spurt, particularly in boys. Inhibin, which is secreted by the Sertoli cells of the testes and the granulosa cells of the follicle, increases in the serum as puberty develops.

B. Maturation of the Adrenal Cortex

The androgenic zone of the adrenal cortex (the zona reticularis) begins to mature as early as 7 years in girls and 8 years in boys. This is expressed by the rising levels of plasma DHEA and DHEA-S, followed by those of androstenedione. This early secretion of adrenal androgens represents a maturation of the adrenal cortex, possibly mediated by a yet poorly identified pituitary adrenal androgen-stimulating factor or by an intra-adrenal regulation of the secretion of the androgens by the adrenal cortex. Whether or not the

TABLE II
Pubertal Stages and Growth Spurt in Girls[a]

		Chronological age[b]	Bone age
Breast development			
B1	Prepubertal		
B2	Budding of breast areola enlarged	8.5–**10.9**–13.3	8.5–**10.5**–13.2
B3	Enlargement of the breast with palpable glandular tissue	9.8–**12.2**–14.6	10.2–**12.0**–14.0
B4	Additional enlarged areola above tissular breast tissue	11.4–**13.2**–15.0	11.5–**13.5**–15.0
B5	Adult breast	11.6–**14.0**–16.4	12.5–**15.0**–16.0
Pubic hair			
PH1	Absent pubic hair		
PH2	Few, scanty hairs	8.0–**10.4**–12.8	8.5–**11.5**–13.0
PH3	Thick, wiry hair	9.8–**12.2**–14.6	10.5–**12.2**–14.5
PH4	Triangle-shaped hair	10.8–**13.0**–15.2	11.2–**13.2**–15.2
PH5	Adult female	11.6–**14.0**–16.6	
Growth spurt		10.2–**12.2**–14.2	10.0–**12.5**–14.5

[a]Breast and pubic hair development and growth spurt in relation to chronological and bone ages.

[b]Chronological and bone ages are given with the 95% confidence limits. Bold indicates average age. [Reproduced with permission from P. C. Sizonenko (1987). Normal sexual maturation. *Paediatrician* **14**, 191–207. S. Karger AG, Basel.]

TABLE III
Pubertal Stages and Growth Spurt in Boys[a]

		Chronological age[b]	Bone age
Genitalia stages			
G1	Prepubertal testes (TVI[c] < 4)		
G2	Pubertal testes (TVI between 4 and 5)	9.2–**11.2**–14.2	9.0–**11.5**–13.5
G3	TVI between 7 and 11	10.5–**12.9**–15.4	10.5–**13.2**–15.0
G4	TVI between 9 and 17	11.6–**13.8**–16.0	12.5–**14.5**–16.0
G5	Adult genitalia	12.5–**14.7**–16.9	
Pubic hair			
PH1	Absent pubic hair		
PH2	Few, scanty hair	9.2–**12.2**–15.2	11.5–**13.5**–14.5
PH3	Thick, wiry hair	11.1–**13.5**–15.9	11.5–**14.2**–15.5
PH4	Triangle-shaped hair	12.0–**14.2**–16.4	12.5–**14.2**–16.5
PH5	Adult male	12.9–**14.9**–16.9	
Growth spurt		12.3–**13.9**–15.5	12.5–**14.5**–16.0

[a]Genitalia and pubic hair stages and growth spurt in relation to chronological and bone ages.
[b]Chronological and bone ages are given with the 95% confidence limits. Bold indicates average age.
[c]TVI, testicular volume index. [Reproduced with permission from P. C. Sizonenko (1987). Normal sexual maturation. *Paediatrician* **14**, 191–207. S. Karger AG, Basel.]

adrenal androgens have any effect on the subsequent maturation of the hypothalamo–pituitary–gonadal axis remains purely speculative, as many examples of nature show complete dissociation between the two maturational processes. Adrenal androgens are believed to be responsible for the development of axillary and pubic hair in adolescent girls.

C. Mechanisms of Puberty Onset

Numerous factors influence the age of puberty and its development: some are genetic, like familial delay in puberty, or endogenous; others are exogenous, like nutrition or sport. Nutritional factors play an important role in the onset of puberty. Many studies have shown that menarche is related to a critical body composition, particularly fat content and a critical weight of 47.8 ± 0.5 kg for American girls. In anorexia nervosa, puberty is delayed. The theory of a critical body weight has been challenged by many experiments of nature: tall girls have menarche at a greater weight than the critical weight but, in general, obese girls have menarche earlier. This critical weight theory probably reflects a temporal relationship rather than a real cause–effect relationship of the cerebral "appetite" center and an "onset of puberty" center. Very recently leptin secreted by the adipose tissue may play such a role.

Sexual maturation of the adolescent is under the endocrine control of the hypothalamus, which,

through its gonadotropin-releasing hormone, regulates the secretion of the pituitary gonadotropins. In turn, the gonadotropins control the growth of the gonads and their functions (Figs. 12 and 13). Pituitary growth hormone and thyroid hormones are necessary for complete pubertal development. Night and day

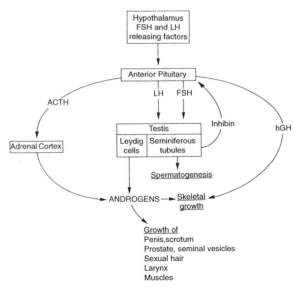

FIGURE 12 Endocrine control of sexual maturation in the male adolescent. [Reproduced with permission from P. C. Sizonenko (1987). Normal sexual maturation. *Paediatrician* **14**, 191–201. S. Karger AG, Basel.]

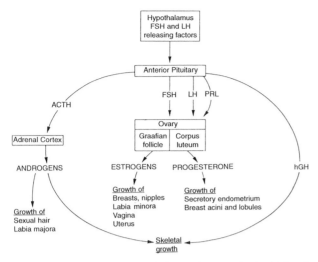

FIGURE 13   Endocrine control of sexual maturation in the female adolescent. [Reproduced with permission from P. C. Sizonenko (1987). Normal sexual maturation. *Paediatrician* **14**, 191–201. S. Karger AG, Basel.]

pulsatile secretion of growth hormone increases during this period. Growth factors such as IGF I increase considerably during sexual maturation. Adrenal glands through the adrenarche are also involved.

The central nervous system plays a key role in this endocrine mechanism. The nature of the impulses of the central nervous system, which controls the activity of the arcuate nucleus of the mediobasal hypothalamus, which in turn secretes gonadotropin-releasing hormone, is not yet understood. In animals, the pineal gland has been shown to exert an inhibitory action on the hypothalamus and the pubertal development by its secretion of melatonin. At the present time, whether or not this is also true in the human being is unknown. The hypothalamo–pituitary–gonadal axis is already operative before puberty, and its maturation gradually achieves its adult functional level during puberty. Therefore, puberty likely represents the result of a slow integrated maturational process rather than the sudden awakening of an organ function that directs sexual maturation. Puberty begins spontaneously when a certain bone maturation has been achieved, irrespective of chronological age.

The hypothalamic–pituitary maturation consists of the rising production of gonadotropins, which are secreted in a pulsatile fashion. Amplitude of the secretory peaks of gonadotropins increases, and the frequency of the secretory pulse changes. Several factors influencing the pulsatility have been described recently (Figs. 14 and 15). It is now clear that the prepubertal restraint on the gonadotropin-releasing hormone pulse generator cannot be accounted for by an ele-

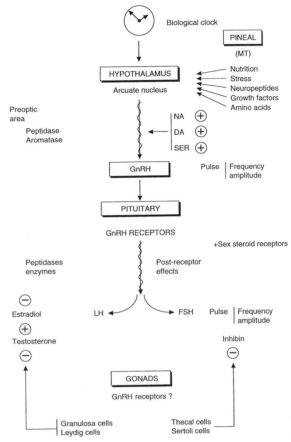

FIGURE 14   Neuroendocrine axis involved in pubertal development. Under the influence of a biological clock and the stimulating or inhibiting effect of cerebral factors such as the biogenic amines [norepinephrine (NA), dopamine (DA), and serotonin (SER)], environmental factors such as stress, nutrition, and the pineal gland [secreting, in particular, melatonin (MT)], the hypothalamus secretes gonadotropin-releasing hormone (GnRH) by pulses of higher amplitude and/or frequency. GnRH induces synthesis and stimulates secretion of pituitary gonadotropins [luteinizing hormone (LH) and follicle-stimulating hormone (FSH)]. These hormones are also secreted in a pulsatile fashion. Acting on the ovary or the testis, they induce the ripening of the follicle or spermatogenesis and the secretion of estradiol or testosterone. In turn, estradiol and testosterone act on the hypothalamic–pituitary axis through a negative (also, in the girl, a positive) feedback mechanism. Inhibin, secreted by the testicular Sertoli cells and the follicle, also plays a role in the feedback mechanism. GnRH, LH, and FSH act on target cells through specific receptors. [Reproduced with permission from P. C. Sizonenko (1987). Normal sexual maturation. *Paediatrician* **14**, 191–201. S. Karger AG, Basel.]

vated endogenous opioid (endorphin) tone. It is now suggested that several factors, such as an increase in excitatory tone at the glutamate (*N*-methyl-D-aspartate) receptor, a decrease in γ-aminobutyric acid inhibitory tone, or an increase in the transforming growth factor-α gene expression, would induce the onset of puberty. However, there is presently no inte-

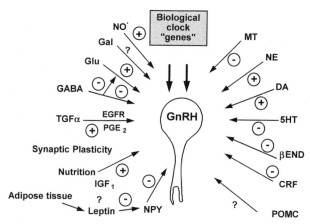

FIGURE 15 Stimulatory (+) or inhibitory (−) factors influencing the pulsatile release of gonadotropin-releasing hormone and the onset of puberty, in addition to biological clock "genes." CRF, corticotropin-releasing hormone; DA, dopamine; βEND, β-endorphins; EGFR, epidermal growth factor receptor; GABA, γ-aminobutyric acid; Gal, galanin; Glu, glutamate; 5HT, 5-hydroxytryptamine (serotonine); IGF I, insulin-like growth factor I; MT, melatonin; NE, norepinephrine; NO•, nitric oxide; NPY, neuropeptide Y; PGE₂, prostaglandin E₂; POMC, pro-opiomelanocortin; TGFα, transforming growth factor α.

grative model for the prepubertal restraint of pulsatile gonadotropin-releasing hormone release and the ontogeny of the gonadotropin-releasing hormone pulse generator that will also include other factors such as stress, season (i.e., the pineal gland), physical exercise and nutrition through leptin, and central mediators (neuropeptide Y, etc.). The results are the increased production of the gonadal steroids with a new resetting of the negative hypothalamo–pituitary negative feedback mechanism to the adult level (Fig. 16). The "gonadostat" theory, which suggests a low threshold of the hypothalamo–pituitary system before puberty and a decrease in the sensitivity at the onset of puberty as the primary mechanism for the onset of puberty, is no longer valid. This change in the sensitivity of the negative feedback mechanism is still present but is secondary to the main effect of the maturing hypothalamic–pituitary axis. In girls in late puberty, a positive feedback mechanism develops whereby increasing levels of estradiol trigger the pituitary secretion of luteinizing hormone, which induces rupture of the mature ovarian follicle and, subsequently, ovulation. This maturation leads to the cyclic pattern of gonadotropin secretion with monthly ovulation.

The pubertal development of the human is one step of the general evolution of the endocrine reproductive

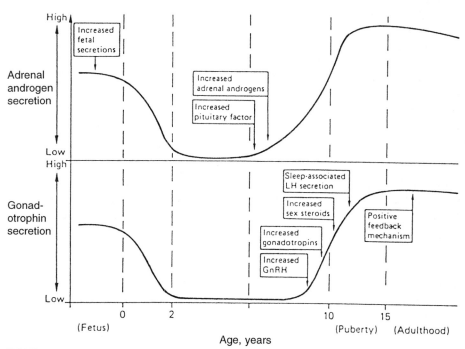

FIGURE 16 Current concepts of the maturation of the hypothalamic–pituitary–gonadal axis (gonadarche) and the secretion of the adrenal androgens (adrenarche) during prenatal life, infancy, childhood, and puberty. [Reproduced with permission from P. C. Sizonenko (1987). Normal sexual maturation. *Paediatrician* **14,** 191–201. S. Karger AG, Basel.]

axis: during fetal life, the axis is very active at the time of sexual differentiation of the fetus, as well as during the perinatal period with the development of a postnatal genital "activation," the exact role of which is not known. These two steps are followed by a "quiescent" period, before the pubertal "awakening" of the reproductive axis (Fig. 16). The triggering mechanism for the awakening of the pubertal development is not yet known. However, it seems that the gonadotropin-releasing hormone pulse generator activity is never completely suppressed during childhood. At puberty the reaugmentation in gonadotropin-releasing hormone drive is mediated, in part, by an acceleration of frequency.

D. Disorders of Pubertal Development

1. Pubertal Physiological Discrepancies

Discordant manifestations, which may be present before or during normal puberty, may cause concerns to children, adolescents, and their parents. *Premature thelarche* consists of the isolated development of breasts in girls usually between 1 and 3 years of age. This condition disappears spontaneously. *Premature adrenarche* is defined by the appearance of pubic and/or axillary hair before the age of 8 years in girls and 9 years in boys. This condition, which does not require any therapy, is followed by normal puberty at the normal age. Pubertal *gynecomastia* (i.e., development of some breast tissue) affects 40–60% of boys during pubertal development. Only a small percentage of the boys require ablation of the tissue. Usually, the breast tissue disappears spontaneously.

2. Precocious Puberty

Precocious puberty is defined as the development of sexual characteristics before the age of 8 years in girls and 10 years in boys. Two types of precocious puberty are described: (1) isosexual sexual precocity, or true precocious puberty, due to the premature activation of the hypothalamo–pituitary–gonadal axis (i.e., in the same direction as the genetic sex of the child), and (2) pseudoprecocious puberty, due to abnormal secretions of the adrenal gland or the gonad, particularly tumors of these two glands. In that case, the precocious puberty can be in the same direction as the sex of the child and the pseudoprecocious puberty is called isosexual pseudoprecocious puberty (estrogen-secreting tumors of the ovary in the girl, testosterone-secreting tumors of the testes in the boy). When in the opposite direction, it represents heterosexual pseudoprecocious puberty, such as androgen-secreting tumors of the adrenal gland or of the ovary, or congenital virilizing adrenal hyperplasia inducing masculinization in girls, or estrogen-secreting tumors of the adrenal gland or of the testis inducing feminization in boys. Each type of precocious puberty has its own therapy. Central precocious puberty is presently treated by administration of agonists of gonadotropin-releasing hormone, which induced a desensitization of the pituitary receptors to gonadotropin-releasing hormone and a complete reduction of the secretion of the gonadotropins.

3. Delayed Puberty

Delayed puberty is defined as the absence of pubertal development (absence of breast development in girls, or increase of volume of the testes in boys) after the age of 13 years in girls and 14 years in boys. Delayed puberty can be a simple delay in adolescence (often familial) but could represent the manifestation of abnormalities called hypogonadism. The disease can affect either the hypothalamo–pituitary axis (this condition, with low levels of gonadotropins, is called hypogonadotropic hypogonadism) or the gonad itself (called hypergonadotropic hypogonadism with high levels of gonadotropins due to the absence of sex steroids acting on the negative feedback mechanism).

V. CONCLUSIONS

This article on pediatric endocrinology discusses the complexity of the mechanisms involved in the overall process of growth in children, from the fetal life to adulthood. It schematically presents the many hormonal factors that disturb or enhance growth. It also includes some of the factors involved in tissue generation, differentiation, and maturation (such as sexual differentiation), which are essential components of growth.

BIBLIOGRAPHY

Bertrand, J., Rappaport, R., and Sizonenko, P. C. (eds.) (1993). "Pediatric Endocrinology," 2nd Ed. Williams & Wilkins, Baltimore.

Delange, F., Fisher, D. A., and Malvaux, P. (eds.) (1985). Pediatric thyroidology. *In* "Pediatric and Adolescent Endocrinology," Vol. 14. Karger, New York/Basel.

Falkner, F., and Tanner, J. M. (eds.) (1986). "Human Growth," Vol. 1, "Developmental Biology and Prenatal Growth," 2nd Ed. Plenum, New York/London.

Falkner, F., and Tanner, J. M. (eds.) (1986). "Human Growth," Vol. 3, "Methodology, Ecological, Genetic, and Nutritional Effects on Growth," 2nd Ed. Plenum, New York/London.

Grumbach, M. M., Sizonenko, P. C., and Aubert, M. L. (eds.) (1990). "Control of Onset of Puberty." Williams & Wilkins, Baltimore.

MacGillivray, M. H. (1987). Disorders of growth and development. *In* "Endocrinology and Metabolism," (P. Felig, J. D. Baxter, A. E. Broadus, and L. A. Frohman, eds.). McGraw–Hill, New York.

Plant, T. M., and Lee, P. A. (eds.) (1995). "The Neurobiology of Puberty." Society for Endocrinology, Journal of Endocrinology Limited, Bristol, England.

Sizonenko, P. C. (1993). Human sexual differentiation. *In* "Reproductive Health," Vol. 2, "Frontiers in Endocrinology." Ares-Serono Symposia Publications, Rome.

Sizonenko, P. C., and Aubert, M. L. (1986). Pre- and perinatal endocrinology. *In* "Human Growth," Vol. 2, "Postnatal Growth and Neurobiology" (F. Falkner and J. M. Tanner, eds.), 2nd Ed. Plenum, New York/London.

Endometriosis

KENNETH H. H. WONG
ELI Y. ADASHI
University of Utah School of Medicine

GLOSSARY

Assisted reproductive techniques Techniques including *in vitro* fertilization (IVF), gamete intrafallopian transfer, zygote intrafallopian transfer, and controlled ovarian hyperstimulation combined with intrauterine insemination

Endometrioma A solitary, nonneoplastic mass containing endometrial tissue

Gamete intrafallopian transfer Placement of human ova and sperm into the oviduct

Infertility Inability to conceive after 1 year of unprotected sexual intercourse

In vitro **fertilization** Fertilization of ova by sperm in a laboratory environment

Laparoscopic surgery Surgery of the interior of the abdomen by means of a narrow telescope to directly view the peritoneal cavity

Zygote intrafallopian transfer IVF with transfer of the fertilized ovum to the oviducts via transabdominal cannulation

ENDOMETRIOSIS IS DEFINED AS THE PRESENCE OF functioning endometrial glands and stroma outside the usual location of the uterine cavity. It is one of the most common disorders encountered by gynecologists today and has been identified in virtually all tissues and organs of the female body. A wide spectrum of clinical problems may occur with endometriosis which can be a vexing problem for both patient and physician. Considerable progress has been made involving the pathogenesis, diagnosis, and treatment of endometriosis.

I. PATHOGENESIS

A. Background

The first description of aberrant endometrial glands and stroma was published in the 1800s, but it was not until the 1920s that the term "endometriosis" was introduced by John Sampson. Although Sampson's originally proposed transplantation theory through retrograde menstruation remains the most popular theory to explain the pathogenesis of endometriosis, several other theories have been proposed.

B. Transplantation Theory

In 1927, Sampson proposed that retrograde flow of endometrial tissue through the fallopian tubes, which seed the abdominal cavity, is the probable cause of endometriosis. Subsequent experimental and clinical data have supported this hypothesis, including the presence and viability of endometrial cells in peritoneal fluid, indicating retrograde menstruation and documentation of retrograde menstruation as occurring in 70–90% of women. Also, a higher incidence of endometriosis is associated with obstruction of the outflow of uterine menstrual fluid in both baboons and women.

However, the transplantation theory does not explain why most women with retrograde menstruation do not develop endometriosis. Furthermore, there are

ENCYCLOPEDIA OF HUMAN BIOLOGY, Second Edition, VOLUME 3.

instances of extrapelvic endometriosis involving the lung and brain that cannot be attributed solely to retrograde menstruation.

C. Coelomic Metaplasia Theory

In contrast to the transplantation theory, the coelomic hypothesis states that the peritoneal mesothelium, which the mullerian duct also derives from, may undergo metaplastic transformation into endometrial tissue. This theory may explain the rare occurrence of endometriosis in men treated with high doses of estrogens. However, this theory to date has not been supported with strong clinical or experimental data.

D. Induction Theory

The induction theory is an extension of the coelomic theory and proposes that unknown biochemical substances, possibly released from shed endometrium, may induce undifferentiated peritoneal cells to form endometriotic tissue. This theory has been demonstrated experimentally in rabbits, but has not been substantiated in primates or women.

E. Immunologic Theory

Recent investigations have suggested an altered immune response in the pathogenesis of endometriosis. An inhibited clearance of viable endometrial cells due to an impaired immune system may lead to endometriosis development. Alterations in cell-mediated immunity, including natural killer cells, macrophages, T cells, and B cells, have been demonstrated in women with endometriosis. In addition, changes in humoral immunity, including abnormal autoantibodies, are associated with endometriosis. Whether endometriosis is related to systemic or local alterations in the immune system has yet to be established. [*See* Immune Response.]

F. Genetic Predisposition

Several investigators have demonstrated a familial disposition to endometriosis with groupings of mothers and their daughters being afflicted with the disease. The risk of endometriosis is seven times greater if a first-degree relative has been affected by endometriosis. Speculation by researchers is that the predisposition is inherited via multifactorial inheritance as no specific Mendelian inheritance has been established.

In summary, most authorities postulate that several of the previously mentioned factors are probably involved in the pathogenesis of endometriosis and that the degree of contribution for each mechanism is variable from patient to patient.

II. EPIDEMIOLOGY

A. Prevalence

The actual prevalence of endometriosis in the general population is difficult to determine as the diagnosis of endometriosis requires surgical confirmation. Published reports have varying figures of 1 to 50%, but most experts would estimate the prevalence at 3–10% of women in the reproductive age group. In women with infertility, the prevalence has been estimated at 25–35%.

B. Age Distribution

Endometriosis is found almost exclusively in women of reproductive age with the mean age at diagnosis reported between 25 and 29 years. However, endometriosis has been reported in adolescents and in premenarcheal girls as well as in postmenopausal women. The exact age-specific incidence or prevalence is not known.

C. Ethnicity and Socioeconomic Status

Early studies have suggested a higher incidence of endometriosis among white women and women with higher socioeconomic status. However, these studies did not control for such confounding variables as access to health care and contraception, cultural differences concerning childbearing patterns, and the incidence of sexually transmitted diseases. When these variables are considered, the incidence rate of endometriosis is similar among women of different races and socioeconomic backgrounds.

III. CLINICAL PRESENTATION

A. Pain

Pelvic pain is a frequent complaint among women with endometriosis. Pain may be manifested as dysmenorrhea, pain during menstruation, which typically begins before the onset of menstrual bleeding and continues throughout the menstrual period. In addition, dyspareunia or pain with intercourse may also be present. Finally, patients with endometriosis can develop chronic pelvic pain defined as pelvic pain of greater than 6 months of duration. Interestingly, a common observation is that the intensity of pelvic pain does not correlate with the severity of endometriosis and that some women with endometriosis remain asymptomatic.

B. Infertility

A common symptom associated with endometriosis is infertility. When endometriosis is severe enough to cause adhesions interfering with ovum pickup and fallopian tube motility, there is no question of its role in infertility. However, the association between peritoneal endometriosis of minimal or mild stages and infertility remains controversial. In women with mild endometriosis, some investigators have reported a lower spontaneous monthly fecundity rate, with the chance of conceiving in 1 month. However, long-term cumulative pregnancy rates approach 90% in untreated women with minimal or mild endometriosis. Furthermore, the role of endometriosis-associated infertility is questioned by the failure of demonstrating a benefit from treatment of infertility in women with minimal or mild endometriosis. The absence of benefit from therapy might in fact reflect problems with treatments rather than a lack of association between endometriosis and infertility.

C. Menstrual Irregularities

Abnormal bleeding has been noted in 15–20% of women with endometriosis; the most frequent complaint is premenstrual spotting. Although all types of menstrual disorders have been linked with endometriosis, there is no convincing data to associate endometriosis with menstrual dysfunction. [*See* Menstrual Cycle.]

IV. DIAGNOSIS

A. Examination

The findings on physical examination of a patient with endometriosis are dependent on the severity and location of the disease. Localized tenderness of the posterior cul-de-sac, the peritoneal area between the uterus and the rectum, and nodularity of the uterosacral ligaments can be found in one-third of patients with endometriosis. With advanced disease, there is

obliteration of the cul-de-sac and fixed retroversion of the uterus. Ovarian involvement may constitute tenderness and enlargement, a sign of an endometrioma or "chocolate cyst." However, in many women with endometriosis, no abnormality is ever detected with clinical physical examination. Due to the lack of definitive clinical signs associated with endometriosis, surgical investigation is mandatory in confirming the diagnosis of endometriosis.

B. Laparoscopy

The procedure of choice for the surgical confirmation of endometriosis is laparoscopy. Laparoscopy is performed on an outpatient basis and enables the surgeon to accurately inspect the peritoneal cavity. The most common sites of involvement include the ovary, posterior cul-de-sac, broad ligament, uterosacral ligament, rectosigmoid colon, and bladder.

Typically, endometriosis tissue appears as bluish–blackish "powder burn" lesions on the serosal surfaces of the peritoneum. In addition, endometriosis may vary in appearance as red, white, or yellow lesions as well as nodular, vesicular, or fibrotic lesions. Biopsy of the endometriotic lesions during laparoscopy may be helpful if the lesions are atypical in appearance. Microscopically, these implants consist of endometrial glands and stroma.

C. Classification

Accurate staging of the extent of endometriosis is critical during the diagnostic procedure. Staging facilitates the comparison of results from various treatments, consistent exchange of information among physicians, and possible serial assessment of the response to therapy. The most commonly used system is the revised American Fertility Society classification introduced in 1985 (Fig. 1). Although this system reflects the extent of endometriotic disease, it does not correlate with the severity of symptoms or infertility.

D. Nonsurgical Diagnosis

Currently, there is no specific blood test available for the diagnosis of endometriosis. However, several studies have investigated the role of CA-125, an ovarian epithelial tumor antigen found on derivations of the coelomic epithelium, as a possible clinical marker of endometriosis. Although an increased level of CA-125 has been associated with the severity of disease,

Patient's Name _____ Date _____

Stage I (Minimal)   • 1-5   Laparoscopy _____ Laparotomy _____ Photography _____
Stage II (Mild)   • 6-15   Recommended Treatment _____
Stage III (Moderate) • 16-40   _____
Stage IV (Severe)   • >40   Prognosis _____
Total _____

PERITONEUM	ENDOMETRIOSIS		<1cm	1-3cm	>3cm
	Superficial		1	2	4
	Deep		2	4	6
OVARY	R	Superficial	1	2	4
		Deep	4	16	20
	L	Superficial	1	2	4
		Deep	4	16	20

	POSTERIOR CULDESAC OBLITERATION	Partial		Complete
		4		40

	ADHESIONS		<1/3 Enclosure	1/3-2/3 Enclosure	>2/3 Enclosure
OVARY	R	Filmy	1	2	4
		Dense	4	8	16
	L	Filmy	1	2	4
		Dense	4	8	16
TUBE	R	Filmy	1	2	4
		Dense	4*	8*	16
	L	Filmy	1	2	4
		Dense	4*	8*	16

*If the fimbriated end of the fallopian tube is completely enclosed, change the point assignment to 16.

Additional Endometriosis _____   Associated Pathology _____
_____   _____
_____   _____
_____   _____

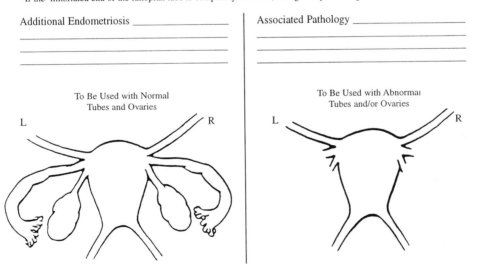

FIGURE I   The revised American Fertility Society classification of endometriosis introduced in 1985. Reproduced with permission of the publisher, the American Society for Reproductive Medicine (The American Fertility Society).

the majority of women with mild forms of endometriosis do not have elevated levels. For now, CA-125 assays appear to be useful in the follow-up of the effects of therapy.

Imaging modalities also appear to be of limited utility in the diagnosis of endometriosis. Ultrasonography, computed tomography, and magnetic resonance imaging have all been evaluated for the diagnosis of endometriosis. Because of the limited sensitivities of all these studies, imaging tests may only be helpful in identifying suspicious masses or tumors such as endometriomas.

V. TREATMENT

The appropriate management for endometriosis varies widely because of the spectrum of clinical symptoms and differences in severity of disease from one patient to the next. Thus, treatment plans must be individualized, being dictated by multiple variables including the patient's age, severity of symptoms, stage of disease, and future reproductive plans of the patient. The main goals of therapy for endometriosis include pain relief, resolution of endometriotic implants, and restoration of fertility. Currently, treatment options for endometriosis can be divided into four categories: expectant management, surgical therapy, medical therapy, and assisted reproductive techniques (ART).

A. Expectant Management

Traditionally, the initial treatment of minimal or mild endometriosis-associated infertility involved expectant or observational management. Studies report long-term pregnancy rates from 55 to 90% in patients with endometriosis who were treated expectantly. Furthermore, these results did not differ from results of patients treated with medical or conservative surgical therapies.

In patients with infertility or severe pain and extensive endometriosis, expectant management seems less justified. Because endometriosis appears to be a progressive disease, there appears to be no value in delaying the treatment of symptomatic endometriosis regardless of the stage of disease. In addition, surgery improves pregnancy rates in comparison with expectant management in patients with moderate or severe endometriosis.

B. Surgical Management

1. Conservative Surgery

Preservation of reproductive function is desirable in most women with endometriosis. Conservative surgical treatment is aimed at maintaining the fertility potential by destroying or removing endometriotic implants while preserving the organs needed for procreation. Laparoscopy is most commonly used today whereas laparotomy or surgery through an abdominal incision is reserved for patients with severe disease. At the time of laparoscopy, resection or destruction of endometriotic tissue can be concurrently performed using sharp dissection with laparoscopic scissors, electrocautery, and various lasers, including carbon dioxide (CO_2), argon, potassium–titanyl–phosphate (KTP), or neodymium yetrium aluminum garnet (Nd : Yag) types. The goal of surgery is to destroy or excise all visible endometriotic implants and to restore anatomy to normal.

If disease is more extensive, various adjunctive procedures may be performed, including endometrioma removal, adhesion (scar tissue) lysis, and presacral neurectomy or uterosacral ablation (transection of uterine nerves for pain relief). The value of these procedures in improving fertility or decreasing symptoms remains undetermined; however, previous studies have demonstrated implants recurring from 20 to 40% of patients treated. Furthermore, it has been estimated that one in four women will have a second operation for endometriosis recurrence.

Numerous studies have evaluated the efficacy of alleviating either pelvic pain or enhancing infertility, although most of these studies have not established an adequate control group for comparison. Pain relief has been reported in 60–80% of patients treated with laser laparoscopy and has been demonstrated to be superior to expectant management.

In the relief of infertility, patients with moderate endometriosis treated with conservative surgery can expect a 60% success rate whereas patients with severe disease have a success rate of 35%. As previously stated, there still is no evidence of fertility enhancement by conservative surgical treatment in patients with minimal or mild disease.

2. Definitive Surgery

Definitive surgery may be considered for patients with severe disease refractory to medical treatment, conservative surgery, or both and for whom future fertility is not a consideration. Total abdominal hysterectomy (removal of the uterus), bilateral salpingo-oopherectomy (removal of both fallopian tubes and ovaries), and removal of all visible endometriosis is considered the definitive treatment of endometriosis-related pain. Such surgery is associated with a 90% rate in alleviating pain, although when one or both ovaries are preserved the rate of recurrence of disease may approach 7%.

Estrogen replacement therapy (ERT) should be considered for all women in whom a definitive surgery was performed. After ERT, recurrence rates range from 0 to 5% in women treated with definitive surgery.

C. Medical Management

Clinical observations that endometriosis regresses after menopause and that symptoms are alleviated during pregnancy have been the basis of medical management of endometriosis. Hypoestrogenism has been noted to result in endometrial atrophy, whereas pharmacological doses of progestins or androgens decidualize endometrial tissue, resulting in atrophy. Therefore, current regimens employ these actions to alter the menstrual cycle by attempting states of pseudopregnancy, pseudomenopause, or chronic anovulation. The most commonly used medications in the treatment of endometriosis are oral contraceptives, danazol, progestins, and gonadotropin-releasing hormone agonists (GnRHa). Although these medical regimens have demonstrated efficacy in alleviating symptoms of endometriosis, current data only support a suppressive role rather than a curative one.

1. Oral Contraceptives

"Pseudopregnancy" by the use of a combination preparation of estrogen and progestin oral contraceptives was one of the first effective treatments of endometriosis. A state of pseudopregnancy is achieved by administering oral contraceptives continuously for 6 to 12 months. The objective of treatment is to induce amenorrhea (the absence of menses). The symptomatic relief of pain has been reported in 60–90% of patients. A 5–10% annual recurrence rate can be expected after an initial first-year recurrence rate of 17–18%. Contraindications and side effects are similar to those associated with cyclic oral contraceptive therapy.

2. Danazol

Danazol is an isoxazole derivative of the synthetic steroid 17α-ethinyltestosterone (Fig. 2) and has been used for the treatment of endometriosis since 1971. Initially prescribed for its "pseudomenopausal effect," danazol is an attenuated androgen (agent that possesses masculinizing activities) that is active when administered orally. The hormonal milieu created by danazol is responsible for the therapeutic effects against endometriosis. Danazol binds to androgen and progesterone receptors and decreases follicle-stimulating hormone (FSH) and luteinizing hormone (LH) levels, thereby decreasing estrogen and progesterone secretion by the ovary. Therefore, danazol produces a hypoestrogenic effect on end organs. Danazol also displaces testosterone from sex hormone-binding globulin, increasing free testosterone levels which promote its androgenic effects.

Treatment with danazol in patients with endometriosis results in up to 90% of patients achieving pain relief. However, no effect has been demonstrated for adhesions associated with endometriosis. In addition, no therapeutic effect of danazol has been reported in the treatment of endometriosis-associated infertility. Up to 80% of patients taking danazol will develop some side effects, including weight gain, fluid retention, acne, hot flushes, depression, muscle cramps,

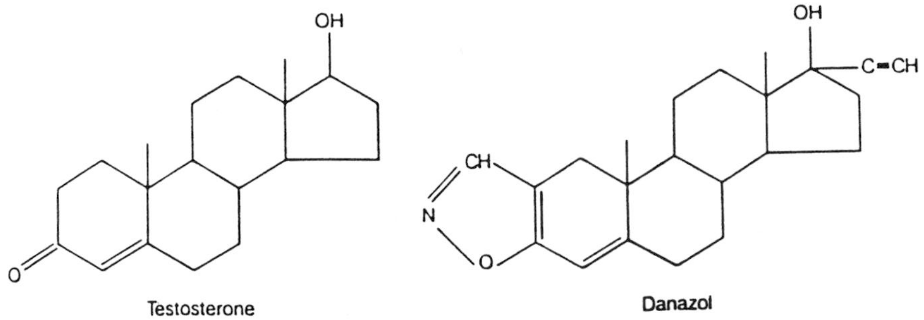

FIGURE 2 The chemical structure of danazol. The chemical structure of testosterone is shown for comparison.

decreased breast size, and hirsutism (abnormal hairiness in women.)

3. Progestins

Progestins are synthetic or natural pharmaceutical agents having similar effects to those of progesterone. The most commonly used progestin to treat endometriosis is orally administered medroxyprogesterone acetate. An intramuscular preparation is also available. Side effects include breakthrough bleeding, fluid retention, weight gain, depression, and breast tenderness. Progestin therapy has been reported to be as effective as danazol for pain relief in patients with endometriosis but is not effective for endometriosis-associated infertility. In light of the decreased side effects of progestins in comparison with danazol, progestins are often the first choice for medical treatment of endometriosis.

4. Gonadotropin-Releasing Hormone Agonists

The discovery in 1971 of the structure of gonadotropin releasing hormone (GnRH) has led to the development of GnRHa, the newest medical approach in the treatment of endometriosis. These decapeptides differ from the native GnRH molecules by specific amino acid substitutions resulting in longer half-lives of the compounds and greater receptor-binding affinities. The net result is a downregulation of pituitary GnRH receptors, low FSH and LH levels, and a reversible "medical oophorectomy." Approved GnRHa is administered as a nasal spray or in a monthly intramuscular injection.

The side effects associated with GnRHa therapy are primarily due to hypoestrogenism, similar to menopause. The most common side effects include hot flushes, vaginal dryness, insomnia, irritability, headache, and depression. However, more significant concerns are the effects of GnRHa on bone loss and serum lipoprotein levels. Various types of so-called add-back therapy, a combination of GnRHa with steroid hormone replacement, have been employed to combat the negative effects of GnRHa.

GnRHa therapy results in the amelioration of pain symptoms in 75 to 90% of patients with endometriosis, results similar to danazol in comparison trials. In contrast, the definitive enhancement of fertility in patients with endometriosis using GnRHa has not been demonstrated. Finally, although recent studies have reported comparable efficacy of add-back regimens, the role of add-back therapy for endometriosis remains to be determined.

D. Assisted Reproductive Techniques

Assisted reproductive techniques have been used with increasing frequency to treat endometriosis-associated infertility. The techniques include controlled ovarian hyperstimulation (COH), *in vitro* fertilization (IVF), gamete intrafallopian transfer (GIFT), and zygote intrafallopian transfer (ZIFT).

COH with clomiphene citrate, a synthetic compound used to induce ovulation, or injectable human menopausal gonadotropin has been used successfully to treat patients with endometriosis-associated infertility. Studies have demonstrated COH alone or in combination with intrauterine inseminations to enhance monthly fecundity rates. However, cumulative pregnancy rates (the observed pregnancy rate during a defined time interval of follow-up) were not increased with COH treatment in comparison to expectant management.

IVF, GIFT, and ZIFT have received considerable attention in the treatment of infertility in women with endometriosis. Several investigators have reported lower pregnancy rates with moderate or severe endometriosis patients treated with these techniques whereas others have not demonstrated any relationship. In contrast, comparable results with IVF in women diagnosed with endometriosis compared to those in women with only tubal infertility have been demonstrated. For now, the true impact of ART in the treatment of endometriosis-associated infertility remains to be established.

BIBLIOGRAPHY

Berek, J. S. (1996). "Novak's Gynecology," 12th Ed. Williams and Wilkins. Baltimore.

Herbst, A. L., Mishell, D. R., Jr., Stenchever, M. A., and Droegemueller, W. (1992). "Comprehensive Gynecology," 2nd Ed. Mosby, St. Louis.

Lu, P. Y., and Ory, S. J. (1995). Endometriosis: Current management. *Mayo Clin. Proc.* 70, 453.

Nezhat, C. R., Berger, G. S., Nezhat, F. R., Buttram, V. C., Jr., and Nezhat, C. H. (1995). "Endometriosis: Advanced Management and Surgical Techniques." Springer-Verlag, New York.

Olive, D. L., and Schwartz, L. B. (1993). Medical progress: Endometriosis. *N. Engl. J. Med.* 328, 1759.

Speroff, L., Glass, R. H., and Kase, N. G. (1994). "Clinical Gynecological Endocrinology and Infertility," 5th Ed. Williams and Wilkins, Baltimore.

Energy Metabolism

BRITTON CHANCE
University of Pennsylvania

GLOSSARY

Glycolytic activity Breakdown of sugars into simpler compounds such as pyruvate or lactate with the formation of small amounts of ATP

Metabolic controllers Inorganic phosphate (P_i); adenosine triphosphate (ATP), an "energy currency molecule"; adenosine diphosphate (ADP); phosphocreatine (PCr), an "energy currency molecule"

Metabolism Physical and chemical process common to all living organisms where energy is produced and transformed for survival purposes

Michaelis–Menten kinetics K_m, amount of a regulatory chemical to give half-maximal effect; V, velocity of adenosine triphosphate (ATP) synthesis necessary to equal that of ATPase; V_m, maximum velocity is defined as that of the maximal capability of the mitochondrial ATP synthesis as activated by the relevant control substrate [e.g., adenosine diphosphate (ADP)]; ATPase, total breakdown of ATP in metabolic function

Mitochondrial respiratory chain Mitochondrion is an organelle that is the principal site of the generation of metabolic energy in the formation of ATP

Nicotine adenenine dinucleotide, reduced Formed from pyruvate and regulated by the activity of glycolysis

CONTROL OF ENERGY METABOLISM REFERS TO the regulation of energy conservation to meet the needs of energy utilization.

I. INTRODUCTION

Years of work have gone into understanding the control of energy metabolism; it has been a literal "golden fleece" of most investigations on the biochemistry of human body organs. Energy metabolism is a key to the function of the replicative and metabolic systems and has been indicated a key to developmental potentiality of Darwinian characteristics. Enzymes, substrates, oxygen delivery, and energy conservation and degradation create a homeostatic state by which life can continue and multiply. The importance of energy in program cell death (apoptosis) suggests a role of energy metabolism in orderly cell death. Claude Bernard himself recognized *le millieu fixe* in 1878, and Walter B. Cannon recognized the physiological and some of the biochemical implications in 1939. A better understanding of feedback control in enzyme systems came with the quantitation of the exquisitely sensitive control of respiration by the adenosine diphosphate (ADP) level, with a K_m (the amount of a regulatory chemical to give half-maximal effect) of 10–20 μM, establishing a "tight" negative feedback loop in which adenosine triphosphate (ATP) production can be precisely matched to ATP utilization and further, Jenerson and colleagues have proposed higher order control by ADP in muscular contraction. As indicated in Fig. 1, the control properties of mitochondria in which equal rates of output of ATP from the mitochondrial respiratory chain and the breakdown of ATP into ADP and inorganic phosphate (P_i) by functional activity identifies the steady state. ADP and P_i are fed back directly into the respiratory chain to resynthesize ATP and to ensure that no deficit occurs. Oxygen is reduced to water, and substrates are oxidized to carbon dioxide by flow in the respiratory chain. [*See* ATP Synthesis in Mitochondria; Adenosine Triphosphate (ATP).]

The important parameter of control of the mitochondrial respiratory chain is the characteristics for

713

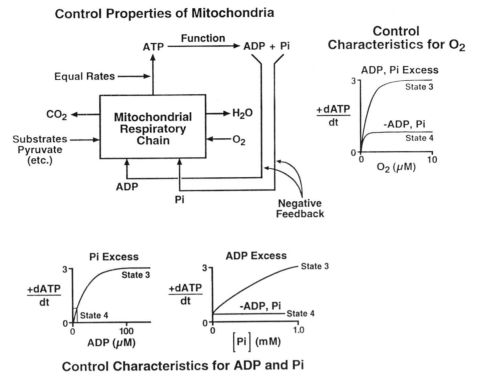

Control Properties of Mitochondria

Control Characteristics for ADP and Pi

FIGURE I   Feedback diagram for ADP (ATP?) control.

control by ADP and P_i (shown in the two bottom diagrams in Fig. 1); these diagrams are sometimes called "transfer characteristics." The rate of ATP synthesis is $+dATP/dt$. For excess P_i, this quantity is related to the ADP concentration in a first or higher order to finally give state 3, or maximal rate of ATP syntheses. Figure 1 expresses this by transfer function diagrams. Note that the scale for ADP control is 0–100 μM. The control characteristic for P_i control is on a millimolar scale, 0–1 mM, and for excess ADP has a K_m of >0.5 mM. However, if ADP is limiting, phosphate will have very little effect, as indicated by the state 4 transfer characteristic. Finally, we can have the control of respiration by the oxygen concentration. If ADP and P_i are in excess, the state 3 rate can be obtained under oxygen control; if ADP and P_i are at low concentrations, then the state 4 rate will be controlled with somewhat lower oxygen concentrations.

The controls of great interest are those where ADP is in control but oxygen may become limiting, under which condition ADP will rise to restore the equality of ATP synthesis and ATP breakdown, providing biochemical homeostasis over a wider range of conditions than would be possible if ADP were not under control and at a high gain feedback loop. One of the basic theorems of metabolic control is that the controller concentrations should be small compared with the affinity constants for optimal regulation (see below).

A striking and further recognition of the existence of feedback control in cells and tissues is provided by the experimental demonstration of biochemical oscillations; only systems that have feedback control operating beyond the bounds of stability can exhibit these oscillations.

Determination of the ADP level in tissues has been elusive because of the nearly inevitable breakdown of ATP in biochemical assays and the presence of ADP in different compartments within a single cell. The use of phosphorus magnetic resonance spectroscopy (MRS) as a basic research tool and as used in the clinical setting for the study of metabolizing systems in which the creatine kinase activity was sufficient to maintain creatine, ATP, phosphocreatine (PCr), and ADP near equilibrium, together with the MRS observation of PCr and P_i from the formation of P_i from PCr, revolutionized the study of metabolic control. In some organs of the body, tight feedback control is essential for organ function, such as heart and brain, and for the variation in demands for ATP synthesis, as in the rest–work transition for skeletal muscle.

II. ROLE OF CREATINE KINASE

Figure 1 illustrates the metabolic control loop for liver in which creatine kinase is of negligible activity. In brain, skeletal muscle, heart, etc., the coupling between the export of ATP and the breakdown of ATP passes through an energy store, phosphocreatine, and a very active enzyme system, creatine kinase, whose function is to combine ATP with creatine to form phosphocreatine and ADP. This provides a ready energy store for functional activity, minimizing changes of ATP. Certain groups have favored the idea that creatine kinase, localized on the mitochondrial membrane, handles all the traffic of ADP and P_i to mitochondria, and ATP from the mitochondria through phosphocreatine and creatine. They shuttle between the site of synthesis (the mitochondria) and the site of utilization (muscle, nerve, etc.). Transvection of different isoenzymes of creatine kinase and "knockout" of the mitochondrial creatine kinase clearly show that the creatine kinase bound to the mitochondrial membrane is not part of an essential creatine phosphocreatine shuttle; free diffusion of all partners seems adequate to support active metabolism.

III. CELL COMPARTMENTATION

The two main compartments of the eukaryotic cell, the highly organized cells of all higher organisms, are the cytoplasm and its mitochondrial spaces insofar as energy metabolism is concerned. Nuclei, endoplasmic reticulum, and vacuolar spaces have their own roles and their particular contribution to cell function. But we need principally to concern ourselves with the cytoplasmic/mitochondrial relations, recognizing that the cytoplasm can contain the functional organelles, the myofibrils, etc. In the neuron, the synaptic junction, dendrites, axons, and transmembrane ion gradients constitute the corresponding "compartment" where functional activities demand ATP breakdown and its resynthesis. The mitochondrion is an autonomous organelle that has its own matrix space concentrations, transport mechanisms, and proton and electron gradients. For the purposes of this discussion, however, we will be able to explain metabolic control as a two-compartment system: in the cytoplasm, where ATP breakdown is employed to drive functional activity (i.e., in muscle contraction and ion transport), and in the mitochondrial space, where ATP is generated from oxygen reduction. In the cytoplasmic space, the glycolytic system under normoxic

conditions is tightly regulated so that the ADP and phosphate levels are only sufficient to provide adequate substrate for the mitochondria. Thus, the reduced nicotinamide adenine dinucleotide (NADH) is available in adequate supplies so that ADP is the main regulator. Similarly, oxygen delivery is adequate under normoxic conditions, again resulting in ADP as the main regulator of oxidative metabolism. [*See* Mitochondrial Respiratory Chain.]

IV. ROLE OF CHEMICAL CONTROLLERS

Historically, phosphate was the main contender for regulation because, very obviously, it could be readily determined by the analytical biochemist and "made sense" as a regulator of metabolism. The idea that the affinity of ADP and phosphate for mitochondrial ATP synthesis would be different was pointed out, and, on the basis of simplistic kinetic arguments, namely the highest-affinity component would provide the most sensitive control, or ADP control. Researchers studied ascites tumor cells, yeast cells, skeletal muscle, and heart muscle, and sound evidence was produced in favor of ADP control. In the case of P_i, the remarkable discrepancy of the analytical chemical value and the actual changes in metabolic control was discovered in 1963.

The concept of perturbing the metabolic feedback loop by varying cell work and thus measuring the change of the metabolic control parameters, ADP and phosphate were realized only in the past decade with the use of phosphorus MRS (^{31}P MRS) of skeletal tissue where the work, ATP, P_i, and PCr could be for the first time directly measured. It was immediately observed that P_i varied over very wide ranges. Initially it was equal to or greater than the *in vitro* K_m of mitochondria, and during work rose to 10–20 times the rest values. Thus, P_i is not regulating oxidative metabolism in the aerobic muscle. The K_{ADP} values calculated on the basis of the *in vitro* values for the creatine kinase equilibrium coincided almost exactly with the *in vitro* K_{ADP}. Thus, the ADP values required to maintain the appropriate ATP synthesis rate are correlated with that of the ATP breakdown rate due to muscle contraction. On this basis, the following simple Michaelis–Menten-type relationship was proposed for first-order control in a new context:

$$V/V_m = \frac{1}{1 + (K_m/\text{ADP})}, \qquad (1)$$

where V is defined as that of ATP synthesis, V_m is defined as that of the maximal capability of the mitochondrial ATP synthesis, and the relevant control substrate is ADP. Equation (1) can be applied to the living system under conditions where the velocity of ATP breakdown is matched exactly by the velocity of ATP synthesis by adjusting the ADP level, phosphate being sufficiently high so that it is not a regulator of primary importance. For this relationship to hold, it is important to recognize that the measured quantity in skeletal muscle function is (external) mechanical work. However, the actual ATP breakdown is due to internal plus external work, and this work is provided by a complex system of muscle fibers of different types in different locations. Thus, the use of external work to measure internal ATPase can be precise only under the following conditions:

1. Where the work load is such that the same fibers are used throughout the exercise; thus, recruitment of fibers and executing the work with movement of other parts of the body must be completely avoided.

2. The pH should be no more than 6.9.

3. The essential criterion of the steady state must be strictly observed, and is characteristic of slow twitch oxidative fibers capable of endurance performance. Fast twitch fibers are not capable of a steady-state function, the lack of adequate respiratory activity results in a system in which PCr and, indeed, ATP are continuously declining during exercise. Thus, slow twitch oxidative fibers must be recruited for the particular exercise. This means, in detail, that finger flexor muscles, which are often incapable of steady-state operation, can give results that would not be expected to meet the requirements of the steady state. Under certain conditions, the wrist flexor muscles have met the criterion of the steady state, whereas the leg muscles of the quadriceps group of the gastrocnemius muscle are most appropriate in humans, and correspondingly large, highly oxidative muscles in endurance performance animals are appropriate for study.

In other organs such as the liver, significant success in metabolic load perturbation has been achieved by stresses such as urogenesis, whereas in the brain epilepsy has been studied.

In the heart, the steady-state characteristics are more pronounced, and the requirements for steady-state function are more essential than in any other organ. Nature has proliferated control mechanisms so that not only is ADP maintained low, and thus available as a significant controller, but supplementary controls apparently involving NADH and possibly even oxygen delivery are superposed on ADP control, so that under most circumstances ADP itself is maintained in a homeostatic condition by regulations that are postulated to involve NADH delivery and oxygen delivery, as mentioned earlier. Thus, nature has followed the basic tenants of feedback control, namely (1) to employ a high gain system [i.e., a small change of concentration results in a large change of ATP synthesis (this favors ADP control over phosphate control)] and (2) to afford a multiplicity of high gain regulatory systems if one is to assure homeostatic conditions over the widest range of stresses (e.g., the heart may have three superposed feedback loops). In this case, higher-order control may be obtained as illustrated in the diagram for ADP control (Fig. 1). In such cases, the analysis of the loop is difficult, except by "inhibitor" dissection, or more fruitfully by observing the development of metabolic control in neonate animals, e.g., the fetal sheep heart shows first-order metabolic controls, whereas the adult shows higher-order metabolic controls as illustrated in Fig. 1.

V. APPLICATION TO DISEASES

The significant step forward from the concept of homeostasis as a physiological phenomenon to one of biochemical nature affords a much better coupling of feedback theory to health and disease. One of the striking examples has been the study of neonate brain, in which nearly 1000 observations made in Philadelphia and in London have identified ADP values for which stability is obtained (measured as PCr/P_i of approximately 1 for which smaller values lead to brain atrophy and brain death). Thus, the margin between stability and life, and instability and death, is a close one for the neonate. The adult brain apparently operates at a greater margin of safety: PCr/P_i of 2–3 is usually observed.

Genetic and metabolic diseases, which interrupt the delivery of substrate or delete an enzyme directly involved in ATP synthesis, cause loss of control and severe lactic acidosis in patients with deficiencies of the respiratory chain.

The most important concept emerging from these studies is that of stability and instability and criticality of metabolic control. The management of illnesses that involve oxygen or substrate delivery to the mitochondria are characterized by instability phenomena, which can be directly quantified by ^{31}P MRS of the particular organ and based on a single measurement; namely the PCr/P_i value of brain, heart, and skeletal tissues can indicate whether ADP is low enough to be in a stable region or has risen to values near the plateau of the rectangular hyperbola for metabolic control, thus providing the system with marginal stability. [*See* Metabolic Regulation.]

VI. STABILITY/INSTABILITY

Coupled to the concept of feedback stability and instability is the heterogeneity of metabolic demand (e.g., in brain, where some neurons may be extremely active as evoked by a particular physiological function and adjacent neurons may be highly inactive). Under these conditions, hypoxic stress could have a more dramatic effect on the neurons in which V [see Eq. (1)] is high due to the necessary work level than on those in which V is low due to the lack of stimulation. Thus, hypoxic stress can inflict more damage on the highly functional neurons than on those that are substantially at rest.

Numerous other applications of these basic principles of metabolic control will emerge as greater localization of MRS is achieved and as noninvasive methods to measure oxygen delivery, particularly to heart and brain, become available for clinical study.

VII. FUNCTIONAL ACTIVITY

Mentioned earlier is the fact that metabolic activity, particularly of the brain, but also skeletal muscle, can be heterogeneous and that groups of neurons or portions of muscle can be activated well beyond the level of neighboring tissues. Specific localized activities are studied by modern imaging technology, particularly functional magnetic resonance imaging (fMRI). More recently, optical imaging has proved efficacious in identifying the demand for oxygen delivery by the capillary bed by functionally activated regions of the brain. Generally, MRI data have been interpreted in terms of increases of blood flow in functional regions of the brain that provide a luxury perfusion, i.e., oxygen delivery in excess of that required for mitochondrial activity. More recent studies with optical methods that discriminate delivery of increased blood concentration in the blood vessels from the oxygenation of hemoglobin. In most cases, the luxury perfusion idea is found and ADP control, as indicated in the two oxygen "excess" diagrams at the bottom of Fig. 1, is identified. However, at high levels of neuronal activation, oxygen delivery may be insufficient to account for the enhanced oxygen demand, and it is possible that the control characteristics for oxygen may evolve, in some physiological cases, as in the case of stroke, where the occlusion of portions of the vascular bed occurs. In summary, the possibilities of metabolic control exemplified by Fig. 1 can be shown to exist in studies of functional activity of brain and muscle.

BIBLIOGRAPHY

Beauvoit B., Kitai, T., and Chance, B. (1994). Contribution of the mitochondrial compartment to the optical properties of the rat liver: A theoretical and practical approach. *Biophys. J.* **67,** 2501–2510.

Blackstone, N. W. (1995). Units of evolution perspective on the emdosymbiat origin of mitochondrion. *Evolution* **49,** 785–796.

Chance, B., Leigh, J. S., Jr., Clark, B. J., Maris, J., Kent, J., and Smith, D. (1985). Control of oxidative metabolism and oxygen delivery in human skeletal muscle: A steady-state analysis of the work/energy cost transfer function. *Proc. Natl. Acad. Sci. USA* **82,** 8384–8388.

Chance, B., Leigh, J. S., Kent, J., McCully, K., Nioka, S., Clark, B. J., Maris, J. M., and Graham, T. (1986). Multiple controls of oxidative metabolism of living tissues as studied by ^{31}P MRS. *Proc. Natl. Acad. Sci. USA* **83,** 9458–9462.

Hamaoka, T., McCully, K. K., Iwane, H., and Chance, B. (1994). Non-invasive measures of muscle metabolism. *In* "Exercise and Oxygen Toxicity" (C. K. Sen, L. Packer, and O. Hanninen, eds.), pp. 481–510. Elsevier, Amsterdam.

Lehninger, A. L. (1965). "Bioenergetics." Benjamin Cummings, Menlo Park, CA.

Stryer, L. (1975). "Biochemistry." Freeman, San Francisco.

Wright, B. (ed.) (1963). "Control Mechanisms in Respiration and Fermentation." The Roland Press Co., New York.

Environmental and Occupational Toxicology

BERNARD DAVID GOLDSTEIN

Environmental and Occupational Health Sciences Institute

GLOSSARY

Carcinogen Chemical or physical agent capable of causing cancer: sometimes discussed as direct or genotoxic agents, which have the ability, by themselves, to cause cancer and indirect or promoting agents, which cause the promotion or progression of the cancer process initiated by some other mechanism

Regulatory standard In environmental and occupational toxicology, usually a numerical statement of the amount of a chemical or physical agent that is permitted to be present in air, water, food, or soil, or is allowed to be emitted from a specific pollutant source

Risk assessment Characterization of the risk of a given chemical or physical agent, or relevant mixtures, based on identification of its hazard, assessment of the extent of exposure, and understanding of the dose–response relation

Safety assessment Standardized testing of chemical or physical agents for toxicity performed in laboratory animals or *in vitro*

Teratogen Chemical or physical agent capable of producing a change resulting in a fetal abnormality

ENVIRONMENTAL AND OCCUPATIONAL TOXICOLogy is concerned with the adverse effects of chemical agents in the workplace and the general environment. The science of toxicology is at the interface between chemistry and biology, dealing with the interaction between chemical or physical agents and biological systems. Growing evidence of the adverse effects to humans and to ecosystems of chemicals in the workplace and general environment has led to increased emphasis on the science of toxicology, including an understanding of the processes by which toxic agents produce their effects, and a focus on the ability to predict and prevent toxicity.

The traditional background of toxicology is the study of poisons. As a science, toxicology is usually traced back to Paracelsus, a 16th century physician and alchemist who taught that the dose makes the poison. Simply stated, all chemicals can be conceived of as poisonous, there is only a question of the amount. Dose–response remains a central feature of modern toxicology.

I. INTRODUCTION

In the late 17th century, Ramazzini, the father of occupational medicine, published a treatise on diseases of workers. He clearly recognized that different diseases were associated with different work conditions, including problems caused by chemicals in workers such as chemists themselves. In the 19th century, with the growth of a chemical industry initially in Europe, the effects of chemicals on workers became even more apparent. Observations included aplastic anemia caused by benzene, methemoglobinemia, and bladder cancer in aniline dye factory workers, and the first deaths caused by arsine. Toward the end of the 18th century, Sir Percival Pott noted an increased incidence of scrotal cancers among chimney sweeps caused by the presence of cancer-causing components

ENCYCLOPEDIA OF HUMAN BIOLOGY, Second Edition, VOLUME 3.

of soot. In the next decades, there was a gradual recognition that good hygienic practices could prevent these tumors in chimney sweeps.

Today there are about 5 million known chemicals, approximately 75,000 of which are in commerce and therefore have the potential for human exposure at the workplace or in the general environment. A major focus of occupational health is the total prevention of exposure to all chemical agents. However, what may be called the first rule of public health is that accidents will happen and exposures will occur. Various national and international organizations have developed standards for occupational and environmental exposures to chemicals, creating guidelines for allowable exposures. These have greatly evolved through the years, with an increasing number of chemicals subject to more stringent limitations. There have been two driving forces for this phenomenon: (1) observations of the adverse impact of chemicals on workers and on the general population, and (2) a great increase in the extent of knowledge concerning the basic toxicology of chemicals.

Standards intended to protect the general public from chemical and physical agents have also been developed, with the breadth of chemicals covered and the stringency of the standards being greatly accelerated in the past three decades. With a rare exception, environmental standards are more stringent than occupational standards. This reflects the greater time of potential exposure (lifetime compared with 8 hr daily, 5 days weekly for 45 years) and the likelihood of increased susceptibility in the general population (children, the aged, and the infirm compared with individuals healthy enough to work).

Toxicology can be divided into two types of approaches for the purposes of this article: Safety assessment is the relatively formal approach in which chemicals are carefully tested in laboratory animals or *in vitro* for the likelihood that they will produce toxicity; toxicological science consists of approaches to understanding the mechanism by which chemicals produce adverse biological impacts. Safety assessment began with the simple counting of dead animals after exposure at different concentrations of the chemical being tested, a procedure known as the LD_{50}, the dose that kills 50% of the animals. As toxicological science has evolved, the insights obtained have been applied to safety assessment, leading to much more sophisticated animal tests of greater validity to the protection of humans. For noncarcinogens, protective factors are applied to the no observed effect level (NOEL). For carcinogens, a risk assessment (extrapolation) process

is used. The understanding of basic mechanisms of chemical action also has led to the development of highly useful *in vitro* tests applicable to safety assessment. However, such test tube approaches are currently only valid as supplements and, despite the wishes of everyone involved, cannot totally supplant the testing of chemicals in live animals.

II. BASIC CONCEPTS OF TOXICOLOGY

There are three fundamental concepts of toxicology: each chemical or physical agent produces a specific pattern of toxic effects; the response to the agent is dependent on dose; and the response in laboratory animals is useful in the prediction of response in humans. Factors affecting the toxicity of chemicals can be considered in terms of physicochemical variables (e.g., solubility and vapor pressure) or in terms of physiological variables. Particularly prominent among the latter are factors affecting absorption, distribution, metabolism, and excretion. Of note is that metabolism does not necessarily result in a less toxic agent. There are many instances in which it is the metabolic product or process that is responsible for toxic effects, including cancer causation.

Toxicological processes can be described in many ways. Temporally, exposures are often divided into acute, subchronic, and chronic. Effects may be described as immediate or delayed and reversible or irreversible. Immediate effects from acute exposure are usually recognizable and readily attributed to the chemical insult. It is much more difficult to detect the insidious development of subtle irreversible effects leading over long periods of time to chronic disease resulting from chronic exposure.

A limited number of final common pathways lead to changes recognized as disease. Chemical and physical agents are capable of initiating each of these pathways, including alteration of energy use; disruption of the synthesis of macromolecules such as protein and DNA; membrane damage leading to cell death; somatic mutations leading to cancer; a variety of different alterations producing immune dysfunction; and fibrosis and other changes in the intercellular matrix leading to scarring and organ dysfunction.

In addition to these general pathways, a variety of toxicological processes is specific to individual organs. For example, the iodine concentrating powers of the thyroid make thyroid cancer the endpoint to be feared after exposure to radioactive iodine, and the metabolic capabilities of the liver render it particularly susceptible to compounds with reactive metabolite intermediates.

To the public, cancer is undoubtedly the most feared endpoint of exposure to chemicals in the workplace or general environment. Approximately 85% of human cancer is due to environmental factors, but that is based on environment in the broadest sense, which includes anything that is not fully dependent on genetic factors. A much smaller percentage of human cancer is due to exposure to chemicals in the workplace or general environment. The mechanism of chemical carcinogenesis is generally thought of in terms of a mutation of genetic material in an individual cell followed by progression to a multicellular cancer. The initial cancer cell and its progeny are not responsive to usual stimuli leading to cell maturation and death. Exogenous chemicals and physical agents may play a role not only in causing the DNA change that initiates cancer, but also may exert their cancer-causing effects through genetic or nongenetic mechanisms that promote the biological progression and growth of the tumor.

A major concern in environmental and occupational toxicology is the determination of the causes of variations in human susceptibility. Advances in toxicological science have shown that there are four major reasons for enhanced susceptibility to a chemical. They can be thought of in terms of four mechanisms:

1. For a given ambient level, there can be an enhanced uptake of a chemical into the body. Such variations often occur because of life-style factors such as exercise leading to increased inhalation or outdoor activities leading to greater exposure to sunlight.

2. For a given uptake into the body, there can be an increased delivery of a chemical or its metabolites to a target organ. This occurs most frequently with compounds for which there is a variation in the extent to which the metabolism can occur, e.g., carcinogenic polycyclic aromatic hydrocarbons.

3. For the same target organ dose of the chemical or its metabolites, there can be an increase in effects in the target organ. Enhanced responsiveness can be seen with certain inherited disorders but may also reflect the simultaneous occurrence of exposure to another chemical or the presence of a disease.

4. For the same degree of organ damage, there can be an enhanced effect on the overall organism. Greater susceptibility of the individual to a given degree of organ dysfunction generally reflects impaired reserve.

For example, a certain extent of loss of lung function might impact on the elderly and those with chronic obstructive pulmonary disease without being noticeable in a healthy teenager.

III. ROUTES OF EXPOSURE

An important consideration in toxicology is the route of exposure. Compounds may have markedly different effects dependent on whether they are inhaled, ingested, or absorbed through the skin. For example, rapidly reactive compounds such as formaldehyde will not get past the initial barrier of the upper airways if inhaled, the gut mucosa if ingested, or the outer layer of the skin. However, a lipophilic compound such as benzene, for which toxicity is dependent on metabolism, will readily penetrate lung or gut and to some extent will also be absorbed through the skin. Thus, for benzene the route of exposure is not as important as it is for relatively reactive compounds such as formaldehyde or ozone or for compounds for which physical properties limit distribution (e.g., asbestos inhalation).

There is a need to be concerned about appropriately duplicating actual human exposure routes when extrapolating from laboratory data. A number of "surprises" have occurred (e.g., plasticizer components being found insoluble when hot tea is added to a plastic cup, but readily extractable when lemon is present; or fire-resistant chemicals in children's pajamas not being absorbed when tested directly on skin but readily extractable and absorbed when urine is present). Similarly, the bioavailability of a component of food can be greatly altered by such factors as pH dependence and the presence of binding materials in the food. Uptake into the body can also be altered by the physiological status of the ingesting organ. Thus, the relatively water-soluble gas sulfur dioxide on quiet breathing does not penetrate the human respiratory tract beyond the nasal pharynx, but on exercise with deep oral breathing can reach the lung; the pH of stomach contents will vary; and abraded or cut skin can be far more permeable than normal skin.

IV. CLASSES OF CHEMICAL AND PHYSICAL AGENTS

Toxic agents are classified in many different ways, including chemical class (e.g., metals, hydrocarbons);

physicochemical properties (e.g., gases, reductants); sources (air pollutants, food contaminants); functions (e.g., solvents, plasticizers); and degree of persistence in the human body or general environment. To facilitate discussion of the toxicity of chemical and physical compounds, an overview of certain classes of compounds is provided. This is by no means an exhaustive or comprehensive review of toxic agents. [See Carcinogenic Chemicals.]

A. Solvents

Human exposure to solvents at the workplace, the home, and the outdoor environment is common. A major class of solvents consists of those used to dissolve nonaqueous substances. In general, these are relatively hydrophobic agents that are required to have low vapor pressures for many uses (e.g., paints and glues). Many hydrocarbon solvents have anesthetic-like properties. In general, the higher the vapor pressure the more rapidly a compound will induce such central nervous system effects as dizziness, lack of coordination, euphoria, and giddiness, leading to narcosis and anesthesia. Chemical structure–activity relations for anesthetic-like endpoints can be readily discerned among solvent groups. Thus, among benzene and the alkyl benzenes, the lower molecular weight compounds more rapidly induce narcosis, whereas the higher weight compounds, once narcosis has been achieved, are more difficult to get out of the body. This represents the physicochemical characteristics of these compounds being translated into their ability to cross the blood–brain barrier and be dissolved in brain lipid. The limitations of using chemical structure–activity relation as the sole means to predict toxicity are also illustrated by this group of solvents. It is only benzene that produces severe bone marrow toxicity and leukemia, all the others are without hematological effects. Other solvents also have specific toxicity related to unique structural elements or metabolism. Thus, both hexane and methyl-*n*-butylketone are potent producers of human peripheral neuropathies, whereas compounds that have slightly different structures (e.g., heptane and methyl ethylketone) are not. This is due to the formation of a neurotoxic metabolite hexanedione.

B. Metals and Trace Elements

The toxicology of metals illustrates the diversity of human exposure to chemical and physical agents. All routes of entry into the body are possible: for example,

lead can be taken up after ingestion in food, water, or soil; can be inhaled as a small particulate from automobile exhaust or smelter output; or can be absorbed up through the skin when in the tetraethyl form. Some metals are essential nutrients but may be toxic in inappropriate concentrations or locations. Toxicity may differ depending on the chemical form of the element, particularly if the metal is complexed to a carbon. Further, although the toxicology of organic arsenic differs somewhat from inorganic arsenic, both are markedly different from arsine, the hydrogenated gas that causes a dramatic destruction of circulating red blood cells, an effect that does not occur with the other forms. Valence also can be important: for example, hexavalent chromium is carcinogenic, whereas trivalent chromium is an essential nutrient. Solubility also plays a major role in toxicity, as does charge, which determines whether the metal crosses biological membranes.

Lead is a thoroughly studied metal. It affects many organs. Allowable occupational and environmental exposures have repetitively been lowered as new research has indicated yet additional effects of lead. As an environmental poison, the major concern has been central nervous system effects in children. The ability of lead to cross the blood–brain barrier appears to be greater in young children. At high levels of chronic or subchronic exposure to inorganic lead, encephalopathic effects occur, which are characterized by a wide variety of symptoms including restlessness, irritability, headaches, poor performance in school, and eventual progression to convulsions, coma, and death. An impressive series of studies has produced compelling evidence that at much lower body lead levels, perhaps equivalent to those currently present in the general population or even lower, lead produces a decrease in cognitive abilities of the developing brain. Less conclusive evidence has led to a suggestion that low level body lead also causes an increase in blood pressure. At concentrations frequently present in the workplace, lead may produce peripheral neuropathies, renal and hematopoietic effects. Long-term renal toxicity and gouty nephropathy have been ascribed to lead. The hematopoietic effects, although less debilitating, are often used as biological markers for lead body burden and effect. At relatively low levels, lead interferes at different steps in heme synthesis leading to a buildup of δ-aminolevulinic acid and free erythrocyte protoporphyrins (FEP), the latter being commonly used in screening tests. At higher levels, lead causes anemia both through a decrease in hemoglobin synthesis and by shortening red blood cell survival. Despite a decrease in blood lead levels because of the substantial removal of lead from gasoline in the United States, which is only slowly being followed in Europe and elsewhere, many environmental and workplace sources of lead remain.

Mercury is among the many metals that have an affinity for the kidney. In the inorganic Hg^{2+} form, mercury tends to accumulate in the kidney, leading to chronic renal disease. The affinity of organic mercurals for the kidney is primarily a function of the rapidity with which the alkyl mercury bond is dissociated *in vivo*. Longer chain organic mercurals tend to dissociate rapidly and have a renal affinity similar to that of inorganic mercury. However, methyl mercury, although predominately ending up in the kidney, has a relatively high concentration in the brain, and the major manifestation is central nervous system toxicity. Toxicity to the brain is also foremost when considering elemental mercury, which is able to penetrate to the brain because it is uncharged. Once there, it is oxidized to Hg^{2+} and becomes trapped in the brain. "Mad as a hatter" is a phrase derived from the neuropsychiatric manifestations of mercury vapor poisoning among hatters. The central nervous system effects of methyl mercury differ somewhat from elemental mercury, being focused more on the sensory system. Both forms produce tremor. Renal effects, predominantly of mercury salts, range from death caused by acute tubular damage to chronic renal disease often characterized by proteinuria, nephrotic syndrome, and anemia.

C. Gases and Aerosols

The toxicology of inhaled substances is dependent on the intrinsic chemical and physical properties of the agent and the physiology of the lungs and upper airways. The depth of penetration of inhaled gases is to a large extent dependent on their solubility in the highly moist respiratory tract and their inherent reactivity. For example, because ozone is less soluble in water than sulfur dioxide, ozone is able to penetrate to the level of the terminal bronchiole rather than being removed in the upper airway. But ozone is so reactive that only perhaps 0.1% or less reaches the deep lung. The penetration of particles is to a large extent size dependent, with larger particles being removed in the upper airway. Of those particles that penetrate deeply into the lung, the responsiveness of the airway will depend on a number of factors including the pH, solubility, and chemical composition of the particulate.

The short-term pulmonary response of concern is bronchoconstriction, the narrowing of airways. This is a nonspecific and to some extent protective response against the inhalation of respiratory irritants but is also the basis for difficulty in breathing among asthmatics and individuals with chronic obstructive pulmonary disease. Bronchoconstrictive responses were presumed to be the major cause of the approximately 4000 deaths that occurred during the 1952 London smog episode. Occupational asthma covers a wide variety of acute bronchoconstrictive insults, mostly representing an allergic response to a workplace compound. Individual variation in susceptibility is common, but for some agents (e.g., toluene diisocyanate) a relatively large proportion of the population can be at risk. Typically, the acute response is greatest on return to work (Monday morning asthma) and ameliorates during weekends or vacations.

Two types of mechanisms of bronchoconstriction have been observed. One is a pharmacologic-type response occurring through direct chemical or vagus nerve mediation of airway smooth muscle tone such as occurs with allergies. This appears to be readily reversible and, in a situation such as sulfur dioxide exposure and certain allergens, modulates its effect on continued exposure. The second general mechanism of bronchoconstriction is associated with an overall inflammatory response (e.g., the response to phosgene and ozone), which worsens on further exposure and leaves more concern for eventual chronic effects.

Other acute responses to air pollutants include irritation of the throat and, particularly for photochemical air pollutants, the eyes. Of potentially greater health importance, a number of pollutants are able to potentiate bacterial respiratory tract infections, at least in laboratory animals. In the case of ozone and nitrogen dioxide, this appears to occur through interfering with the ability of alveolar macrophages to produce bactericidal free radicals after phagocytosis. [See Toxicology, Pulmonary.]

More difficult to determine, but of greatest consequences, are the potential long-term effects of inhaled compounds. Experience at the workplace has clearly demonstrated the relation of various dusts to chronic lung disease and to cancer. Most notable is asbestos exposure, which can lead to three diseases: asbestosis, a chronic fibrosing disease of the lung that can cause death and debilitation because of replacement of gas exchange tissue with scar; mesothelioma, a usually fatal tumor of the outer lining of the lungs or abdomen; and bronchogenic lung cancer, a tumor usually associated with cigarette smoking. Of note is that there is a synergistic multiplicative interaction between cigarette smoking and asbestos exposure. A similar interaction is observed with exposure to radon in miners, which has led to further emphasis on the cessation of smoking for those who live in homes in high natural radon areas. Other occupational groups with a high risk of chronic lung disease include coal miners and textile workers. The causation or exacerbation of chronic lung disease by outdoor air pollutants has occurred in the past because of the gas–aerosol complex caused by the uncontrolled burning of fossil fuels and may now be occurring because of ozone and other photochemical oxidants stemming primarily from the use of automobiles and from acid particulates. [See Asbestos; Tobacco Smoking, Impact on Health.]

Air pollutants are not restricted to the outdoor environment. For many common pollutants, particularly solvents, exposure levels are likely to be higher indoors than out. Such problems as radon, offgassing of pollutants from plastics and construction materials, the presence of asbestos, and household solvents are of potential health importance.

V. RISK ASSESSMENT

Risk assessment as a formal process has become an important part of decision making regarding the control of environmental and occupational chemicals. In theory, the assessment of the extent to which a chemical presents a risk should be able to proceed independently of decisions concerning the management of the chemical; management decisions often involve other factors such as technical feasibility, economic cost, and a host of political considerations. Part of the reason for the recent emphasis on risk assessment has been a perception among the public that the political process has interfered with the unbiased assessment of the risk of chemical and physical agents. In the United States, a major recent impetus to risk assessment in the federal regulatory process was a report from the National Academy of Sciences in 1983 that described and defined the process.

Risk assessment consists of four components: hazard identification, dose–response estimation, exposure assessment, and risk characterization.

A. Hazard Identification

Toxicity is an intrinsic property of a chemical or physical agent. Although all compounds are poisonous at a sufficient dose, the poison does not attack the same target organ in each case. For example, asbestos exposure leads to tumors in the lung but not in the brain. Identification of a human hazard usually occurs through high-dose human exposure in the workplace or through accidental or intentional poisoning. In most cases, hazard identification is reasonably straightforward, but in others, the lack of certainty greatly impinges on regulatory activities. For example, there are only approximately two dozen known human carcinogens defined as such on the basis of evidence in human beings. However, there are almost 200 compounds that have been identified to be carcinogenic in laboratory animals and thus have some degree of likelihood of also being carcinogenic in humans. Further, there are many other compounds for which there is concern about the potential for carcinogenicity based on such findings as similar chemical structure to known carcinogens, positive short-term tests for mutagenicity, etc. In essence, there is a continuum of evidence ranging from great likelihood but no proof of human carcinogenicity, down to the most minuscule of evidence. The regulator is forced to draw a line through that continuum in deciding which compounds deserve to be regulated to protect human health. As reasonable scientists will differ as to the likelihood that the compound will turn out to be a human carcinogen, and as a straight line through such a continuum means that some compounds will be just above or below the line that is used for regulation, the hazard identification step for carcinogens will always lead to controversy.

B. Dose–Response Estimation

The identification of hazard is a qualitative step. Quantitative risk assessment requires dose–response data that can be obtained from animal studies or epidemiological studies of humans. In either case, extrapolation is necessary to determine the pertinence of the observation to the dose range and the population of concern. The problem of extrapolating from high dose to low dose has been central to questions concerning the appropriateness of risk assessment as a tool. Models have been used to perform such extrapolations. Of first importance is to determine the expected shape of the dose–response curve, which will differ depending on the endpoint and should be based on an understanding of the basic toxicology. Understanding the mechanism by which a given chemical produces cancer or other diseases can have a major impact on extrapolation of the data base to lower level human exposure. For example, much attention has been directed to the possibility that a different model is needed for genotoxic carcinogens, which are known to react directly with DNA, and nongenotoxic agents for which the effects on genetic material are not obvious. Among the latter, there may be a specific need for different models such as a receptor-based model for 2,3,7,8-tetrachlorodibenzodioxin, which seems to represent a new class of carcinogens acting through specific receptor interactions, in this case with a receptor related to those for estrogenic and other growth factors. Advances in understanding dose have occurred through the use of physiologically based pharmacokinetics. This approach provides better information on the actual dose to target tissue, thus permitting extrapolation from animals to humans in a more realistic fashion.

C. Exposure Assessment

The assessment of exposure is critical to risk assessment. No matter how hazardous a material, without exposure there is no risk. Although much attention has been focused on the dose–response aspects of risk assessment, in many cases the scientific uncertainty associated with exposure is even greater and, as a rule, could be resolved much more readily. The methods for exposure assessment can be divided into three basic approaches. Using information about the sources and the location of the population at risk, we can use a variety of simple or sophisticated modeling techniques to determine the extent to which an individual might be exposed to the pollutant. We can also determine analytically the ambient concentration of the compound of interest in air, water, soil, or food. Advances in personal monitoring techniques have led to improvement in the ability to typify the exposure of individuals. Third, biological monitoring involves taking biological samples from the population at risk. Thus the body burden of arsenic can be estimated through knowledge of its urinary excretion, or of lead from determination of blood lead level. New techniques capable of determining exposure through biological markers are being rapidly developed based on advances in toxicology and molecular biology. The ability to measure tiny amounts of organic chemicals adducted to macromolecules such as hemoglobin protein or blood cell DNA, and advances in analytical technique to measure urinary metabolites, has put many additional tools in the hands of those interested in exposure assessment.

D. Risk Characterization

After completing the various analytical steps, the risk must be characterized. It is in this risk assessment step that attempts to separate completely risk assessment from risk management often fail. One current issue concerns the extent to which the risk assessment process is appropriate for characterizing risk to the maximally exposed individual rather than simply the risk to a population. Another area of controversy in risk characterization concerns whether it is appropriate to give a single number, often as a plausible upper boundary of the risk estimate, or whether the risk should be quantified as a maximum likelihood estimate. In either case, there are those who advocate that the error boundary be placed around risk assessment figures in the hopes of improving communication of the uncertainties. Yet although the effort is to communicate to nonscientists, an error boundary can also mislead by ignoring the central tendency of the data.

The risk assessment process has a number of uncertainties associated with it. First and foremost, risk assessment must be understood as a process leading to information useful for regulation. Of great importance is that risk assessments be performed in an open, explicit, and logical way that has the confidence of the scientific and regulatory community. To enhance the process, various national and international organizations have developed guidelines for the performance of risk assessment that describe the assumptions and processes. Deviation from the guidelines is possible but only if done openly and with a justification that will withstand rigorous scientific review.

The communication of risk to the public involves more than simply characterizing the risk. Individual perception of risk is conditioned by many factors, including the dread of the endpoint, the extent to which the individual feels a victim of external forces, the extent to which the individual participates or gains from the risk-taking activity, and, particularly in environmental issues, the sense of outrage reflecting the imposition of risk to individuals without their consent or involvement. These qualitative factors are subject to intense evaluation as the environmental regulatory community recognizes the need to obtain the informed consent of the public.

VI. RECENT TRENDS IN ENVIRONMENTAL AND OCCUPATIONAL TOXICOLOGY

A. Biological Markers

A biological marker is an indicator in a biological system of some event occurring elsewhere in that system. The use of biological markers is at least as old as Hippocrates, who diagnosed disease by inspecting the color of the urine. There has been renewed interest in the potential role for biological markers in environmental and occupational toxicology. This is due to recognition that advances in toxicology and in analytical sciences have led to the opportunity to use ethically obtained body samples from humans (e.g., blood, urine, saliva, and hair) as test organs to detect the presence or effect of relatively low levels of environmental agents.

Biological markers are divided into three types: markers of exposure, of effect, and of susceptibility. A continuum of events exists between exposure and disease so that certain markers can reflect both exposure and effect. It is ideal to be able to obtain information from one biological marker about both the extent of exposure and the degree to which an adverse effect is occurring. A reasonably good example is that of carboxyhemoglobin. Carbon monoxide combines with the oxygen-combining site of hemoglobin, thereby preventing the delivery of oxygen to the tissues. Measurement of carboxyhemoglobin provides an integrated level of exposure to carbon monoxide in the past 8–12 hr as well as being an endpoint related to the basic mechanism of effect and thereby predictive of adverse consequences. An increased understanding of the mechanism of action of chemicals will lead to new biological markers capable of indicating both exposure and effect. In the case of genotoxic carcinogens, by being able to know the intermediate(s) responsible for the adverse effects of this agent and to detect its presence bound to the nucleic acid of blood cells, it will be possible to predict effect.

Biological markers of exposure and effect should be particularly powerful tools for risk assessment. The importance of biological markers to exposure assessment has already been described. In addition, certain parts of the dose–response curve are not now approachable by standard epidemiological or animal toxicology techniques. For example, at levels of benzene that are present in the general atmosphere, or in the usual instance of water contamination, it is necessary to extrapolate the risk from studies performed at much higher levels. It is inconceivable that a sufficient number of laboratory animals could be tested at one time to measure a cancer endpoint at realistic levels of benzene exposure, and it is beyond any reasonable possibility that standard epidemiological techniques will be able to assess such risk. However, in the future it might be possible to relate the presence of an adduct to DNA of a benzene metabolite

with the extent to which benzene exposure had occurred. Determination of whether this occurs through a dose–response pathway consistent with a one hit model (i.e., every molecule has a finite and equal chance of causing cancer) or through some other model would then be possible. Until that time, the prudence that characterizes public health approaches will preclude any change from the conservative assumption that any one molecule of a carcinogen has a finite risk of causing cancer.

Epidemiological studies will undoubtedly be greatly improved through the use of biological markers of exposure. Much of the epidemiology relating chemicals to disease processes has relied on qualitative information (e.g., the presence of a worker in a specific location in the factory). The use of biological markers as indicators of the extent to which an individual has actually been exposed to a potential chemical hazard will greatly improve the strength of any observed epidemiological association and the power of the epidemiological study.

Biological markers of susceptibility may be of particular value in determining which individuals are at risk from exposure to chemicals. However, such markers will raise many ethical issues that should be addressed.

B. Multichemical Exposures

The focus in environmental and occupational toxicology has generally been on a single chemical. This is in keeping with the regulatory thrust, which tends to be one chemical in one medium. Unfortunately, we tend to be exposed to mixtures of chemicals from the same source (e.g., in gasoline), from different sources in the same medium (e.g., ozone and sulfuric acid in air), or from different sources in different media (e.g., the apparent interaction between inhaled benzene and dietary ethanol). Study of the toxicology of such mixtures has been recognized as a major challenge that has been relatively untouched. Chemical interactions can be described under a number of different headings. When the effects of two chemicals are equivalent to the sum of the effects seen with either chemical acting alone, then the effects are said to be additive. If greater than the sum of the individual effects, it is known as synergism. Antagonism refers to a situation in which the effects are less than additive, and potentiation refers to a situation in which a chemical seemingly without any effect enhances the activity of another chemical. As the number of potential pollutant interactions is almost infinite, the only hope we have to deal intelligently with this problem is to understand the basic toxicological mechanisms by which these effects occur.

VII. ECOTOXICOLOGY

Ectoxicology is a rapidly advancing field that is just beginning to get the attention it deserves. Although the chemicals that might affect human health are for the most part similar to the ones that affect ecosystems, this is not always the case. Furthermore, of central importance to understanding ecotoxicology is knowledge of the basic biology of the ecosystem.

There are four basic reasons to study the effects of environmental chemicals in ecological systems:

1. The canary principle—study the animal because it serves as a sentinel for humans.
2. The ecosystem as a source of nutrition—study it because we eat it or we compete with it for food.
3. The need for propitiation—study it because of human needs other than nutrition (e.g., hunting, fishing, and the care of pets or lawns).
4. For its own sake—we recognize the importance, on this planet of limited size, of the protection of all living things.

Much attention has focused on the first three of these reasons, but it is likely to be the fourth that is most important in the long run.

BIBLIOGRAPHY

Gallo, M., Gochfeld, M., and Goldstein, B. D. (1987). Biomedical aspects of environmental toxicology. *In* "Toxic Chemicals, Health and the Environment" (L. B. Lave and A. C. Upton, eds.), pp. 170–204. The John Hopkins University Press, Baltimore, MD.

Goldstein, B. D. (1988). Risk assessment/risk management is a three-step process: In defense of EPA's risk assessment guidelines. *J. Am. Coll. Toxicol.* 7, 543–549.

Klaassen, C., and Eaton, D. (1991). Principles of toxicity. *In* "Casarrett and Doull's Toxicology: The Basic Science of Poisons" (C. D. Klaassen, M. O. Amdur, and J. Doull, eds.), 4th Ed., pp. 12–49. Pergamon Press, New York.

National Research Council (1983). "Risk Assessment in the Federal Government: Managing the Process." National Academy Press, Washington, DC.

National Research Council (1989). "Biological Markers in Reproductive Toxicology." National Academy Press, Washington, DC.

Enzyme Inhibitors

MERTON SANDLER
Royal Postgraduate Medical School, University of London

H. JOHN SMITH
University of Wales College of Cardiff

GLOSSARY

Active site Area on the enzyme surface where the substrate is bound and catalysis occurs

Coenzyme Nonprotein component essential for the functioning of an enzyme

Enzyme Protein that, either alone or in combination with nonprotein components, catalyzes the reactions of its substrates

Substrate Substance that is modified by an enzyme

INHIBITORS ARE CHEMICAL AGENTS THAT DEcrease the ability of an enzyme to catalyze the reaction of its substrates. Enzyme inhibitors have been used extensively as tools to elucidate biochemical pathways in the body or, more importantly, as drugs to produce a useful clinical response by removal of the action of a particular enzyme (see Table I and examples in text). In the past, many drugs introduced into therapy as a result of microbiological or pharmacological screening tests have subsequently been shown to exert their action by inhibition of a specific enzyme in the body. On occasion, a drug introduced for one purpose has exhibited side effects due to its undesired inhibition of an enzyme in an unrelated biochemical pathway. Such findings have provided a "lead" compound as a starting point for the development of specific, potent inhibitors for the newly observed target enzyme. More

recently, the impetus resulting from increased knowledge of the biochemical processes involved in the disease state has led to the rational design and synthesis of specific inhibitors toward identified target enzymes in such pathways.

I. BASIC CONCEPTS

A useful clinical response may be achieved by inhibiting a suitably selected enzyme, the blockade of which subsequently leads to a decrease in concentration of the product of the enzyme-catalyzed reaction or an increase in concentration of the substrate. Which effect is clinically important depends on the biochemical system concerned, as illustrated in Eq. 1. In a biochemical chain (Eq. 1), consisting of its associated enzymes and substrates, where the final product (metabolite) has an action judged to be clinically undesirable, inhibition of the first enzyme in the chain, E_1, will reduce the metabolite concentration and alleviate the condition or disease (e.g., inhibition of xanthine oxidase to decrease conversion of purines to uric acid, which causes gout). Alternatively, where a metabolite is required for normal body functioning and is removed by a degradative enzyme, which is not part of the metabolites' synthetic chain (Eq. 2), inhibition of the degradative enzyme will lead to a desired clinical response as the concentration of the metabolite increases (e.g., inhibition of acetylcholinesterase to allow build up of acetylcholine at nerve endings in the treatment of glaucoma). This approach may also be used to preserve the action of a drug toward its target enzyme, where the drug is degraded rapidly either by the body or by a bacterial enzyme before it can exert its action (e.g., β-lactamase inhibitors).

729

TABLE I
Some Other Target Enzymes for Drugs

Enzyme inhibited	Drug	Clinical use
Xanthine oxidase	Allopurinol	Gout
Carbonic anhydrase II	Acetazolamide, methazolamide, dichlorphenamide, ethoxzolamide	Glaucoma, anticonvulsant
Carbonic anhydrase	Sulthiame	Anticonvulsant (epilepsy)
Prostaglandin synthetase	Indomethacin, ibuprofen, naproxen, aspirin	Anti-inflammatory
Na⁺,K⁺-ATPase	Cardiac glycosides	Cardiac disorders
Riboxyl amidotransferase	6-Mercaptopurine, azathioprine	Anticancer therapy
L-Dihydroorotate dehydrogenase	Biphenquinate	Anticancer agent
Aromatase	Aminoglutethimide, 4-hydroxyandrostenedione	Oestrogen-mediated breast cancer
H⁺,K⁺-ATPase	Omeprazole	Anti-ulcer agent
Trypsin and related enzymes	Gabexate mesylate, camostat mesylate	Pancreatitis and hyperproteolytic states
Plasmin	ε-Aminocaproic acid (EACA), p-aminomethylbenzoic acid (pAMBA), tranexamic acid (AMCA)	Antifibrinolytic agent
Cholesterol synthesis enzyme	Meglutol	Hypolipidaemic
Aldehyde dehydrogenase	Nitrefazole	Alcoholism
Sterol 14α-demethylase of fungi	Miconazole, clotrimazole, ketoconazole, triconazole	Antimycotic
Aldose reductase	Sorbinil	Diabetes mellitus complications
Reverse transcriptase	Zidovudine	Acquired immunodeficiency syndrome
Succinic semialdehyde dehydrogenase	Sodium valproate	Epilepsy
Thymidine kinase and thymidylate kinase	Idoxuridine	Antiviral agent
DNA, RNA polymerases	Cytarabine (Ara-C)	Antiviral and anticancer agent
Aspartate transcarbamylase	Sparfosic acid (PALA)	Anticancer agent
Dihydrofolate reductase	Trimethoprim, methotrexate, pyrimethamine	Antibacterial, anticancer, and antiprotozoal agent
Transpeptidase	Penicillins, cephalosporins, cephamycins, carbapenems, monobactams	Antibiotics
Pyruvate dehydrogenase	Organo-arsenicals	Antiprotozoal agents
Alanine racemase	D-Cycloserine	Antibiotic
Dihydropteroate synthetase	Sulfonamides	Antibacterial
Monoamine oxidase (MAO)	Iproniazid, phenelzine, isocarboxazid, tranylcypromine	Antidepressant
MAO A	Toloxatone, moclobemide, clorgyline	Antidepressant
MAO B	Selegiline [(−)–deprenyl]	Codrug with L-dopa for Parkinson's disease
Thymidylate synthetase	5-Fluorouracil	Anticancer agent
Aldehyde dehydrogenase	Disulfiram	Alcoholism
Bacterial urease	Acetohydroxamic acid	Chronic urinary infections
Formylglycinamide ribonucleotide aminotransferase	Azaserine	Anticancer
Peptidyl transferase	Chloramphenicol	Antibiotic
Ornithine decarboxylase	Eflornithine	Anticancer and antiprotozoal agent
Aromatic amino acid decarboxylase	Benserazide, carbidopa	Codrug with L-dopa for Parkinson's disease
Acetylcholinesterase	Neostigmine, physostigmine isofluorophate, ecothiophate	Glaucoma, myasthenia gravis

Here the enzyme inhibitor has the role of adjuvant drug.

$$A \xrightarrow{E_1} B \xrightarrow{E_2} C \xrightarrow{E_3} D(\text{metabolite}) \quad (1)$$

$$\underset{(\text{metabolite})}{D} \xrightarrow{E} \text{inert product} \quad (2)$$

Occasionally, in bacterial infections, two drugs that act at different points in an essential biochemical chain of the microorganism (Eq. 3) may be administered. This approach is necessary when resistance to a single drug has built up due to prolonged use and resistant bacterial strains have emerged.

$$A \xrightarrow{E_1} B \xrightarrow{E_2} C \xrightarrow{E_3} D \qquad (3)$$

Overall metabolite production by a biochemical chain may also be decreased by inhibiting an enzyme associated with production of a coenzyme essential for the functioning of an enzyme in the chain (Eq. 4).

$$A \xrightarrow{E_1} B \xrightarrow{E_2} C \xrightarrow{E_3} D$$
$$\text{coenzyme} \quad \text{modified coenzyme} \qquad (4)$$
$$E$$

II. TYPES OF INHIBITORS

A. Reversible Inhibitors

Reversible inhibitors bind to the enzyme by forming either an enzyme inhibitor (EI) or an enzyme-inhibitor substrate (EIS) complex through attractive forces, which do not involve covalent bond formation.

Competitive inhibitors compete with the substrate for the active site of the enzyme and, by forming an inactive EI complex, decrease the amount of enzyme available to combine with the substrate, as shown in Eq. 5:

$$E + S \underset{K_s}{\rightleftharpoons} ES \xrightarrow{k_2} E + P,$$
$$I \, \Vert \, K_i$$
$$EI \qquad (5)$$

where E is the enzyme, S is the substrate, I is the inhibitor, P is the product, K_s is the dissociation constant for the ES complex, k_2 is the rate constant for the breakdown of ES, and K_i is the dissociation constant for the EI complex. The rate (v) of the enzyme-catalyzed reaction in the presence of a competitive inhibitor is modified in accordance with Eq. 6:

$$v = \frac{V_{max}}{1 + \dfrac{K_m}{[S]}\left(1 + \dfrac{[I]}{K_i}\right)}, \qquad (6)$$

where concentration terms are shown in square brackets and K_m is the Michaelis–Menten constant.

The extent that the rate of the reaction is slowed is determined by the inhibitor concentration and the value of K_i. A low value (10^{-6}–10^{-8} M) for K_i is characteristic for a potent inhibitor. The inhibition may be removed by increasing the substrate concentration (see Eq. 6), as would be expected for a competitive relationship between inhibitor and substrate.

The potency of an inhibitor is reflected in the K_i value, and this may be conveniently determined by the Lineweaver–Burk method. The rate of reaction (v) is determined for increasing substrate concentration in the absence and presence of a fixed concentration of inhibitor. Rearrangements of Eq. 6 gives Eq. 7. A plot of $1/v$ versus $1/[S]$ for the normal and inhibited reactions gives two lines that intersect on the $1/v$ axis at $1/V_{max}$ and on the $1/[S]$ axis at $-1/K_m$ (inhibitor absent) and $-1/K_m (1 + I/K_i)$, from which the value of K_i can be calculated:

$$\frac{1}{v} = \frac{1}{V_{max}} + \frac{K_m(1 + [I]/K_i)}{V_{max}[S]}. \qquad (7)$$

Noncompetitive inhibitors do not compete with the substrate for the active site but bind to the ES complex and prevent its breakdown to products (Eq. 8). The rate of the enzyme-catalyzed reaction (Eq. 9) is modified through the V_{max} parameter and, for a fixed inhibitor concentration, cannot be reversed by increased substrate concentration.

$$E + S \rightleftharpoons ES \longrightarrow E + P$$
$$I \, \Vert \, K_i \qquad I \, \Vert \, K_i \qquad (8)$$
$$EI + S \rightleftharpoons EIS$$

$$v = \frac{V_{max}/(1 + I/K_i)}{1 + \dfrac{K_m}{S}} \qquad (9)$$

A plot of $1/v$ versus $1/[S]$ in this instance gives converging lines for the normal and inhibited reactions, which intersect on the $1/[S]$ axis at $-1/K_m$ and at the $1/v$ axis at $1/V_{max}$ and $(1 + I/K_i)/V_{max}$, respectively. Most reversible enzyme inhibitors used as drugs are competitive so that noncompetitive inhibitors as well as other classes of reversible inhibitor (e.g., uncompetitive and mixed types) are not well known.

Examples

A key component of the renin–angiotensin system responsible for maintaining blood pressure is angiotensin-converting enzyme (ACE), which converts the inactive angiotensin I to the vascoconstrictor angiotensin II (Eq. 10).

$$\text{Asp-Arg-Val-Tyr-Ile-His-Pro-Phe} \updownarrow \text{His-Leu} \xrightarrow{\text{ACE}}$$

angiotensin I

Asp-Arg-Val-Tyr-Ile-His-Pro-Phe

angiotensin II $\qquad (10)$

Captopril (I) is a potent, reversible, tight-binding inhibitor of ACE. It is an orally active drug that lowers

(I)

(II)

(III)

blood pressure in patients with hypertension. Further additions to this class of drugs are enalapril (II), which resembles captopril, and cilazapril (III), which was designed by computer-assisted graphic modeling to bind in a similar manner to captopril at the active site of ACE. Angiotensin I is considered to bind to the active site (Eq. 11) through an electrostatic bond formed between its terminal carboxylate ion and a protonated arginine residue on the protein, a hydrogen bond to the carbonyl of the amide and coordination with zinc of the carbonyl-oxygen atom of the amide bond to be broken. As illustrated in Eq. 11, ACE inhibitors with similar stereochemistry have the same spatial arrangement of groups for binding to these points on the enzyme surface.

A transition state analog is a special type of competitive reversible inhibitor that binds more tightly to the enzyme than the substrate, as reflected in the K_i value (10^{-9}–10^{-11} M).

$$E + S \underset{K_N\ddagger}{\rightleftharpoons} E' + S\ddagger' \longrightarrow E + P$$

$$K_s \updownarrow \qquad \qquad \updownarrow K_T$$

$$ES \underset{K_E\ddagger}{\rightleftharpoons} ES\ddagger \longrightarrow EP \qquad (12)$$

In an enzyme-catalyzed reaction, a high-energy activated complex, known as the transition state, is formed. The energy required for formation of the transition state is the activation energy and is the barrier to the reaction occurring naturally. A transition state analog is a *stable* compound, which resembles the substrate portion of the unstable transition state.

The transition state for an enzyme-catalyzed reac-

Substrate

Captopril $\qquad (11)$

$$(13)$$

tion can be shown to be more strongly bound to the enzyme than the substrate(s), as follows. Equation 12 shows a substrate-enzyme reaction and the corresponding noncatalyzed reaction with the same bond making and breaking mechanism, where ES‡ and S‡′ represent the transition states, respectively, K_E‡ and K_N‡ the respective equilibrium constants, and K_s and K_T the respective association constants for formation of ES and the hypothetical binding of S‡′ to E. At equilibrium, $K_T K_N$‡ $= K_S K_E$‡, which is modified to $K_T k_N = K_S k_E$, where k_E and k_N are the first-order rate constants for breakdown of the ES complex and nonenzymatic reaction, respectively. Rearrangement to $K_T = K_S k_E / k_N$ shows that because the ratio k_E / k_N is of the order 10^{10}, or more than the transition state, S‡′ must be bound more tightly to the enzyme than the substrate. Consequently, a transition state analog, which bears some resemblance to the substrate part of the transition state for the enzyme-catalyzed reaction, will be a potent inhibitor of the enzyme.

Examples

Emphysema is a disease of the lung associated with loss of elastic recoil due to destruction of elastin by elastase. Human elastase is released by neutrophils in the lungs and, whereas in normal subjects it is removed by the plasma protein inhibitor, α_1-antitrypsin, an acquired (due to excessive smoking) or inherited deficiency of the inhibitor leads to the presence of excess elastase in the lung and lung degradation.

Elastase is a serine protease in which a serine hydroxyl group at the active site is activated by adjacent Asp-His residues to react with substrate in the ES complex to form (through a tetrahedral intermediate) an acyl enzyme with cleavage of a susceptible alanine or valine peptide bond (Eq. 13). The acyl enzyme is hydrolyzed through the Asp-His catalytic system,

which activates water, on this occasion, to the carboxylic acid, with regeneration of the enzyme. The tetrahedral intermediate is stabilized by hydrogen bonding between its negatively charged oxygen and peptide NH groups (oxyanion hole).

The enzyme has an extended substrate-binding region of at least four or five subsites on either side of the bond to be hydrolyzed.

Certain aldehydes, such as elastatinal isolated from *Actinomycetes,* which contains an alaninal residue, and synthetic peptide aldehydes, which contain a valinal residue (e.g., AdSO$_2$-Lys-Pro-Val-H (AdSO$_2$, adamantanesulphonyl), are good inhibitors of the enzyme. The aldehyde inhibitors are considered to be transition-state analogs of the enzyme. They bind several orders of magnitude more strongly than the corresponding amide substrates by forming a hemiacetal linkage between the aldehyde carbonyl group and the active site serine, so forming a tetrahedral intermediate reminiscent of that formed in the normal enzyme-catalyzed reaction (Eq. 14).

$$(14)$$

MeO-Suc-Ala-Pro-Boro-Val, which has the terminal structure -Pro-NH-CH(CH$_2$(CH$_3$)$_2$)-B(OH)$_2$, is a trigonal boron compound, which is a transition-state analog inhibitor of the enzyme. Trigonal boron compounds readily form compounds with OH^-, and these compounds form a tetrahedral boron compound with the serine alkoxide $(Ser-O^-)$ group in a similar manner to that described for aldehydes, for example:

$$E-Ser-O-\overset{\overset{\displaystyle OH}{|}}{\underset{\underset{\displaystyle R}{|}}{B}}-O^-.$$

B. Irreversible Inhibition

1. Active Site-Directed Inhibitors

These resemble the substrate sufficiently closely to form an EI complex, resembling the ES complex, within which a reactive function on the inhibitor reacts with a functional group on an enzyme residue in or close to the active site, with formation of a covalent bond (Eq. 15). The inhibitor residue is firmly held at or near the active site of the enzyme and prevents access of substrate and subsequent catalysis.

$$E + I \underset{}{\overset{K_i}{\rightleftharpoons}} \underset{\text{complex}}{EI} \overset{k_{+2}}{\longrightarrow} \underset{\text{inhibited enzyme}}{E - I} \quad (15)$$

A comparison of the relative potency of inhibitors toward an enzyme is achieved by comparison of their K_i and k_{+2} values, where k_{+2} is the first-order rate constant for breakdown of the complex to inhibited enzyme (Eq. 15). A more revealing comparison is given by the ratio k_{+2}/K_i. These parameters may be determined as follows.

Integration of Eq. 15 gives

$$k_1 t = \ln E - \ln (E - x), \quad (16)$$

where k_1 is the observed first-order rate constant, E is the initial enzyme content, and x is the inhibited enzyme content, and

$$k_1 = \frac{k_{+2}}{1 + \dfrac{K_i}{[I]}}, \quad (17)$$

which, in the reciprocal form, is

$$\frac{1}{k_1} = \frac{K_i}{k_{+2}[I]} + \frac{1}{k_{+2}}. \quad (18)$$

A plot of $\log(E - x)$ versus t (i.e., enzyme activity remaining with time), using a known concentration of inhibitor, gives a regression line with slope $-k_1/2.303$. A secondary plot of $1/k_1$ versus $1/[I]$ for different inhibitor concentrations gives a line where the intercept on the $1/k_1$ axis is $1/k_{+2}$ and that on the $1/[I]$ axis corresponds to $-1/K_i$.

Examples

Thrombin is a serine protease responsible for the clotting of blood by conversion of fibrinogen to fibrin. It is not present in the blood until released from prothrombin by a physiological stimulus. Effective chloromethyl ketone peptide or peptide aldehyde inhibitors of the enzyme possess a combination of the Pro-Arg present at the cleavage sites of factor XIII and prothrombin, together with D-Phe, rather than the Gly-Val-Arg of the natural substrate, fibrinogen. The potent irreversible inhibitor D-Phe-Pro-Arg-CH_2Cl is selective toward thrombin and, at effective inhibitor concentrations, has little effect on other trypsin-like proteases such as factor Xa, plasmin, urokinase, and kallikrein, where an inhibitory action would upset normal body functions. Serine proteases are typically alkylated by peptide chloromethyl ketones on the nitrogen in the imidazole ring of the histidine residue, constituting the catalytic triad, Asp-His-Ser.

2. Mechanism-Based Enzyme Inactivators

These are also known as suicide substrates or k_{cat} inhibitors. These inhibitors act as substrates of the enzyme but are modified in such a way that *usually* a reactive function is formed, which covalently bonds with a reactive group present at the active site, so irreversibly inhibiting the enzyme. The reactive species (EI*) formed may degrade to innocuous products so that every turnover of the substrate may not lead to inhibition of the enzyme (Eq. 19). Partition between the inhibition event and product formation is measured by the partition ratio (k_{+4}/k_{+3}). Mechanism-based inactivators are not reactive species until armed by their specific target enzyme. Consequently, they show few toxic side effects due to this built-in specificity for their target.

$$E + I \underset{k_{-1}}{\overset{k_{+1}}{\rightleftharpoons}} \underset{\text{complex}}{EI} \overset{k_{+2}}{\longrightarrow} EI^* \overset{k_{+3}}{\longrightarrow} E - I \quad (19)$$
$$\downarrow k_{+4}$$
$$E + P$$

Kinetic analysis of Eq. 19 follows that for Eqs. 16 and 17, where active site-directed inhibitors were considered, except that the kinetic parameters for the reaction are more complex. In the simplest case, where partitioning does not occur ($k_4 = 0$), then

$$k_1 = \frac{k_{+2}k_{+3}/(k_{+2} + k_{+3})}{1 + K_{m^i}/[I]} \quad (20)$$

and

$$K_{m^i} = \frac{k_{-1}k_{+3} + k_{+2}k_{+3}}{k_{+1}(k_{+2} + k_{+3})}, \quad (21)$$

where k_1 is the observed first-order rate constant and K_{m^i} has replaced K_i in this more rigorous treatment.

Examples

Enzymes using pyridoxal phosphate as a coenzyme are concerned with amino acid modification and have provided a number of targets for the design of mechanism-based inactivators as potential clinical agents.

γ-Aminobutyric acid (GABA) is the main inhibitory neurotransmitter in the brain, and its levels are regulated by GABA transaminase (GABA-T) (Eq. 22). Inhibitors of the enzyme allow a build-up of GABA, an effect that is useful in the treatment of epilepsy.

(22)

Vigabatrin (IV; γ-vinyl GABA) is a time-dependent specific inhibitor of the enzyme and is used as an antiepileptic drug. One suggested mechanism for irreversible inhibition of the enzyme is shown in Eq. 23.

Vigabatrin forms a Schiff base with pyridoxal phosphate present at the active site of the enzyme, followed by an enzyme-catalyzed loss of a proton to form the vinylimine. Reaction between the terminal carbon of the vinylimine, which carries a positive charge, and a nucleophile (center of negative charge or excess electrons [e.g., $-NH_2$]) on the enzyme surface leads to covalent bond formation and irreversible inhibition of the enzyme.

Certain bacteria produce β-lactamase enzymes and are resistant to the antimicrobial action of β-lactam antibiotics [e.g., penicilllins (formula V) and cephalosporins]. The enzyme hydrolyzes the β-lactam ring of the antibiotic to give ring-opened products that are inert as antibiotics.

(v) Benzyl penicillin

The resistance of these bacteria to β-lactam antibiotics can be circumvented by the use of β-lactamase inhibitors as adjuvants in antibiotic therapy.

β-Lactamases are serine proteases and catalyze β-lactam ring hydrolysis, in the general manner previously described (Eq. 13), by initially forming an acyl enzyme (cf. Eq. 24).

Clavulanic acid, a mechanism-based inactivator of

(23)

(24)

β-lactamase, is a poor substrate of the enzyme. It forms an acyl enzyme, which is sufficiently stable to partition between the normal hydrolysis product and rearranged forms that are not hydrolyzed, leading to irreversible inhibition of the enzyme.

The rearrangements probably involve opening of the five-membered ring and removal of this species from equilibrium with the acyl enzyme by trapping of the oxygen anion as the more stable ketone form of the enolate anion. Clavulanic acid progressively inhibits the enzyme, and inhibition is complete after 150 turnovers. Here, chemical changes in the acyl enzyme, due to rearrangements, lead to irreversible inhibition, which is not the usual mechanism of action of this type of inactivator.

III. SELECTION OF A SUITABLE TARGET–ENZYME INHIBITOR COMBINATION IN DRUG DESIGN STRATEGY

The biochemical environment of a target enzyme may affect the selection of the type of inhibitor required satisfactorily to inhibit the enzyme.

In a biochemical chain of reactions in a steady state (Eq. 25), the concentration of the pool of the initial substrate (A) is under the influence of other chains and is reasonably constant. Reversible inhibition of E_1, which catalyzes the *first* step in the chain, could be expected successfully to decrease production of the undesired metabolite (D). Generally, inhibition of an enzyme elsewhere in the chain would be expected to be less successful due to a buildup in concentration of the substrate concerned, which would overcome the block (see Eq. 6) because the throughput of (A) is reasonably constant. Rapid fluctuations in the level of (A), which could theoretically affect inhibition of E_1, have not been noted in practice.

$$A \xrightarrow{E_1} B \xrightarrow{E_2} C \xrightarrow{E_3} D \,(\text{metabolite}) \qquad (25)$$

Irreversible inhibition progressively decreases the level of the target enzyme, wherever its position in the chain, and the biochemical environment of the enzyme is unimportant. However, the rate of synthesis of fresh enzyme by the body or pathogen to replace the inhibited enzyme must be slower than the rate of inhibition of the enzyme so as to maintain the desired effect. Consequently, the potency (k_2/K_i) of active site-directed inhibitors and the partition ratio for mechanism-based inactivators are important factors in their usefulness as potential drugs.

Specificity toward the target enzyme is an important consideration for potential clinical agents. Serious side effects could be apparent for all types of inhibitors where closely related enzymes with different biological roles (e.g., trypsin-like enzymes) are concerned. Active site-directed irreversible inhibitors, which could alkylate or acylate a range of essential tissue constituents containing amino or thiol groups (e.g., glutathione, proteins), could possess a more general toxicity.

The current emphasis on the design of mechanism-based inactivators as potential drugs revolves around their inherent lack of toxicity, specificity toward the target enzyme, and ability irreversibly to block a metabolic chain independent of the location of the target enzyme within the chain.

BIBLIOGRAPHY

Barrett, A. J., and Salvesen, G. (eds.) (1986). "Proteinase Inhibitors." Elsevier, Amsterdam.

Sandler, M. (ed.) (1980). "Enzyme Inhibitors as Drugs." Macmillan, London.

Sandler, M., and Smith, H. J. (eds.) (1989). "Design of Enzyme Inhibitors as Drugs." Oxford University Press, Oxford.

Sandler, M., and Smith, H. J. (eds.) (1994). "Design of Enzyme Inhibitors as Drugs," Vol. 2. Oxford University Press, Oxford.

Smith, H. J. (ed.) (1988). "Introduction to the Principles of Drug Design." 2nd Ed. Wright, London.

Enzyme Isolation

ROBERT K. SCOPES

La Trobe University

I. Enzymes of the Human Body
II. Purposes for Isolating Enzymes
III. Isolation of Enzymes from Mammalian Tissues
IV. Recombinant Techniques for Expression and Isolation of Human Enzymes

GLOSSARY

Adsorbent Solid material that attracts otherwise soluble molecules to its surface

Denaturation Loss of structural integrity of a protein molecule, resulting in loss of its biological activity

Enzyme assay A test to determine the amount of enzyme activity present in a sample

Gel Filamentous molecules making a three-dimensional net, with large water-filled pores between the strands, through which large molecules, such as enzymes, can penetrate

Homogenize To break up a tissue in the presence of an extracting buffer so as to release cellular components into the soluble fraction

Hydrophobic Having the character of low solubility in water and high solubility in organic solvents

Ligand A molecule, generally small (i.e., a molecular weight of less than 1000), that interacts in a specific way with a macromolecule such as an enzyme; generally, a "biospecificity," representing a natural physiological process

Recombinant technique Use of DNA manipulation to enable genes to be expressed (i.e., to function) in a foreign "host" organism, giving rise to the synthesis of the protein encoded by the gene

ENZYME ISOLATION IS A PHRASE THAT ENCOMpasses many methodologies in protein chemistry. Enzymes are catalytic protein molecules responsible for all of the chemical reactions that occur in life. The isolation of enzymes is a necessary first stage in any detailed study; "isolation" here refers to a purification that may not completely isolate the specific enzyme molecules from all other material, but should go a long way toward obtaining a homogenous preparation of enzyme molecules suitable for studying the enzyme's structural, catalytic, and physiological roles. Techniques for protein isolation are varied, and the behavior of an enzyme studied for the first time is unpredictable. It is only because the properties of different protein molecules vary so much that we are able to isolate individual enzymes from the other proteins of the cell. An isolation scheme usually involves several steps and will most likely be unique for each enzyme.

I. ENZYMES OF THE HUMAN BODY

A. Background

The whole of the life process is controlled by enzymes, molecules that catalyze the chemical reactions essential for all of the processes occurring in living organisms. The human body is no exception to this: Every move we make, every nutrient we digest, even every thought we have is ultimately controlled by enzymes and enzyme-like interactions.

Every cell in our body contains thousands of different enzymes, each responsible for a particular function. The differing amounts and intracellular distribution of the enzymes cause one cell to differentiate from another, so that, for example, a liver cell does not behave the same way as a kidney cell. Each of these cells has a different complement of enzymes (though many not involved in the specific processes of the cell are common to both). A liver cell, arguably the most complex multifunctional type of cell in the

ENCYCLOPEDIA OF HUMAN BIOLOGY, Second Edition, VOLUME 3.

body, has enzymes responsible for specific liver functions, such as regulating the glucose supply, detoxifying foreign substances, metabolizing fatty acids, and synthesizing serum proteins. The liver cell also has all of the enzymes necessary for the functioning of every cell in the body, e.g., enzymes involved in protein and nucleic acid syntheses (and breakdown), cell growth and division, and in energy (ATP) generation.

One of the principal activities of biochemists in the 20th century has been the study of enzymes. Departments of biochemistry evolved in universities both from physiologists and from organic (i.e., natural product) chemists, and these origins are reflected in the dual nature of biochemists—those who are interested in the biological processes and those who wish to understand the chemistry behind them. The study of enzymes is the link between biology and chemical activity; in 19th-century terms, between the "life force" and basic chemistry. To study an enzyme, it is necessary to isolate it from the other constituents of the cell or tissue where it is found; little detailed work can be performed on unfractionated living material. Thus, over the years, methods have evolved for separating pure enzymes from the complex mixtures in which they exist naturally.

B. Enzymes from Other Animals

The study of an enzyme from humans may raise both ethical and practical problems. Whereas there is no problem in obtaining samples of plasma and, in most places, placental tissue, the source of other tissues must be cadavers or amputation, neither of which is satisfactory with a view toward investigating living, fresh, undiseased tissue. In addition, the possibility that technicians may contract diseases such as hepatitis or acquired immunodeficiency syndrome is a further obstacle. Consequently, most of what we know of human biochemistry has come from studying that of other mammals from baboons to rabbits, from horses to mice, on the assumption that the correspondence is close.

At the enzyme level, fortunately, the biochemistries are close; it is relatively rare for an enzyme isolated from another mammal to differ significantly from its human counterpart. At higher levels of cellular function, however, differences become more marked, for, after all, if all of the cells and organs of a pig behaved exactly as those of a human, we should not be able to distinguish a pig from a human! The distribution and amounts of enzymes, rather than their fundamental properties, cause differentiation in cellular behav-

ior. Thus, we can be reasonably confident that studies of mammalian enzymes have direct relevance to human biology.

II. PURPOSES FOR ISOLATING ENZYMES

A. Structural Studies

The isolation and study of an enzyme from a mammalian source are usually motivated by reasons beyond mere scientific curiosity. It is abundantly clear that any such study can impinge on medicine, as enzymes are responsible for most of our body's behavior, both normal and pathological. Those in the field of medicine aim to restore abnormal bodily behavior to normal and, wittingly or otherwise, frequently use drugs that attack abnormalities in the distribution (and sometimes behavior) of human enzymes. The detailed study of a misbehaving enzyme's structure and catalytic activity can lead to the design of a drug to counteract its misbehavior. Generally, such drugs will be inhibitors, slowing down or suppressing the activity of an enzyme that, for pathological reasons, is overactive in a given tissue. To design such a drug, a full study of the enzyme at the atomic level (e.g., X-ray crystallography) is necessary; thus, the pure enzyme must be isolated in reasonable quantity from a relevant comparable source.

B. Therapeutic Uses

In a few cases, and perhaps increasingly in the future, the lack of a normal enzyme function can be treated by infusion or injection of the enzyme itself. However, this can only presently be done for extracellular enzymes; ideally, the source should be human to avoid immunological problems. Thus, factor VIII and other blood clotting factors are being used increasingly to treat hemophiliacs, and tissue plasminogen activator is being used to dissolve blood clots.

The problem of producing the proteins used in therapy is solved by DNA manipulation. It is possible to extract the human DNA (gene) responsible for the production of a particular protein or enzyme and introduce it to foreign cells in tissue culture or to bacteria. These "host" cells then produce the gene product in safe conditions and in quantities that would be impossible using human tissues.

III. ISOLATION OF ENZYMES FROM MAMMALIAN TISSUES

A. Introduction

All enzymes are proteins.[1] Thus, the techniques for isolating enzymes can be the same as for isolating other proteins. However, because of their subtle structural design, many enzymes are more sensitive to inactivation (i.e., loss of enzymic activity) than other proteins with simpler functions. Consequently, techniques that have recently been widely used to isolate many proteins of interest (e.g., peptide hormones, viral antigens, and growth factors, which are mostly extracellular and are relatively sturdy molecules) may not be applicable to enzymes that are mostly intracellular and are being protected from the environment (i.e., less sturdy). The techniques used also depend on the reason for isolating the enzyme, in particular, the amount of end product required and whether it needs to be in the active state.

Useful work, directed toward gene isolation, can be done on a few micrograms of inactive enzyme, provided that it is homogeneously pure. On the other hand, the study of an enzyme's structure requires milligrams, if not hundreds of milligrams, of pure active product, so the scale of operations must be vastly greater. Generally, enzyme isolation involves one or more "classical" techniques for protein purification, perhaps followed by a more specifically designed step to select the chosen enzyme. These methods are described below. As every enzyme is different, there can be no generally applicable methodology, and the most successful processes frequently involve a step that may be unique to that enzyme or a small class of proteins.

B. Enzyme Extraction

The majority of enzymes exist in a soluble form, i.e., in an aqueous environment: the cellular cytoplasm, the interior of organelles such as mitochondria, or extracellular fluids such as plasma or digestive juices. Other enzymes may be fixed in location, embedded in or attached to membranes or other solid material. To purify an enzyme, it is necessary to obtain an extract, an aqueous solution containing the dissolved enzyme. In most cases, all that is required (for intracel-

lular enzymes) is to homogenize the tissues so that the cells are broken and the soluble enzymes are released into the suspending liquid. For membrane-bound enzymes, however, treatment with detergent (usually after isolating the organelles concerned) is generally needed to separate the protein from the fatty component of membranes.

The first step is to identify the tissue that is the best, most convenient, source of the enzyme. It should contain the enzyme required in large amounts in a form easily released into the extraction buffer. Next, this tissue is broken down in a blender, using 2–5 volumes of an appropriate buffer to 1 volume of tissue, yielding a "homogenate." The extract is prepared by centrifuging the homogenate to obtain a (fairly) clear aqueous solution containing all of the soluble components of the tissues. The enzyme could be less than one part in 1000 of all the proteins present.

Enzymes are protein molecules that are particularly vulnerable to a variety of stresses that cause them to lose their natural structure—they become "denatured." Such stresses must be avoided during the extraction and subsequent processing. Many things can cause enzyme denaturation and inactivation, including extremes of pH (acid or alkaline), high temperatures, organic solvents, many types of salts, and oxidizing agents. Generally these conditions are avoided unless the particular enzyme is particularly resistant to such stresses.

C. Enzyme Activity Measurement

It is axiomatic that an enzyme isolation cannot proceed without a method for measuring the enzyme's presence with reasonable precision. In virtually all cases, this is a measure of the enzyme's catalytic activity. Each enzyme requires a separate assay; hence, we can only generalize the methods used for detecting and quantitating activity. The assay method should be quick and easy, even if this means a sacrifice of accuracy. The amount of total protein should be known, but this is less important than knowing how much enzyme is present. During purification, as much enzyme activity is retained as possible, whereas the total protein decreases to a small amount.

Enzyme assay methods can be divided in many ways. One way is to group them into "continuous" methods, in which the progress of reaction can be monitored continuously as it occurs, and "stopped" methods, in which the result of the enzyme's activity can only be determined after stopping the reaction

---

[1]Exceptions are recently discovered pieces of RNA (ribozymes) with catalytic function that can be described as enzymes. This is not surprising because RNAs can acquire three-dimensional folding similar to that found in proteins.

and measuring how much product has been formed. These methods are not considered further here.

D. Fractionation Procedures

The most used methods for fractionating protein mixtures fall into one of three categories: precipitation (differential solubility), adsorption (batch or column chromatography), and methods in which the proteins remain soluble at all times.

I. Precipitation

Precipitation methods rely on the fact that in any mixture of proteins the solubility properties of individual proteins differ substantially so that change in the properties of the solvent causes some proteins to aggregate, forming a precipitate, while others remain soluble.

Some of the ways in which solvent properties are changed include pH alterations (increased or decreased acidity), ionic strength (salt concentration), dielectric constant (organic solvent addition, which changes the strength of attraction between molecules), and water activity (addition of hydrophilic polymers to make less water available to the proteins).

A typical procedure is to take the extract containing all solubilized proteins and add a precipitant to a known concentration. For ionic strength increases (i.e., "salting out"), ammonium sulfate is normally used; up to 750 g of ammonium sulfate can be dissolved into 1 liter of extract. For changing dielectric constant, alcohol or acetone is added, up to 60% vol/vol, keeping the temperature close to 0° to minimize loss through denaturation.

After equilibration at a suitable concentration of precipitant, the aggregated proteins are removed by centrifugation and more precipitant is added to the supernatant. This may be repeated several times, until such a time when the enzyme of interest is found in the precipitate after it is redissolved in a suitable buffer. Such fractionation procedures typically result in only a modest amount of purification (i.e., a 3- to 10-fold increase in specific activity, which is the activity per unit weight of total protein), with about 80% recovery of activity. They are most useful, however, as a first step, reducing a large volume of crude extract to a more easily handled volume of mainly proteinaceous material.

The aim of all fractionation schemes should be to maximize the recovery of active material, while min-

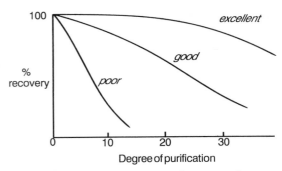

FIGURE I   Relationship between overall recovery of an enzyme's activity and the purification factor (see text) achieved in the isolation step. Precipitation methods are typical of "poor" steps; nevertheless, these play an important part in an overall process.

imizing the recovery of total protein (Fig. 1). The specific activity (i.e., the activity per milligram of protein) should increase at each step (the purification factor). The efficiency of each step is measured by the increase in specific activity, with due regard to maintaining a high percentage of recovery.

Precipitation steps in a fractionation procedure may be useful at any stage; they may be used at any step in purification simply to concentrate the protein mixture even if it does not contribute any significant purification.

2. Adsorption

The most valuable technique in enzyme purification is column chromatography. Selective adsorption of proteins to a suitable adsorbent (see below) due to some of the physicochemical properties of the proteins, followed by controlled elution procedures, can achieve purification factors of 10–100, with good recovery of activity. The basic equipment consists of a column (i.e., a glass tube filled with adsorbent), a magnetic stirrer to mix two salt solutions, thus creating a concentration gradient (see Fig. 2), and a means of collecting regular fractions emerging from the column A fraction collector, with continuous monitoring of column eluates for protein absorbance at 280 nm; in addition, a peristaltic pump to maintain a steady flow rate is also normally regarded as "basic" equipment (Fig. 2). More sophisticated automated equipment is available and is widely used, especially for enzyme purifications routinely repeated.

Methods under the category of high-performance liquid chromatography (HPLC) are frequently used, in which the quality of the equipment and materials is paramount. However, the most common form of HPLC, called "reverse phase," is often not appro-

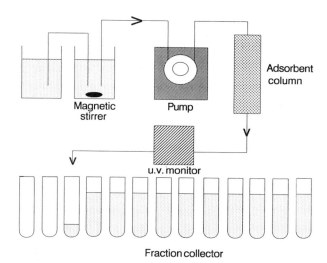

FIGURE 2 A column chromatographic step. A gradient of solvent is created with the aid of a magnetic stirrer by adding graded amounts of high-concentration solute to a vessel in which the concentration is initially low and is pumped through the column. Eluted proteins are monitored by ultraviolet absorption, typically at 280 nm, and pass into tubes in the fraction collector.

priate for enzymes because it requires the use of high concentrations of organic solvents and sometimes acid pH, both of which can destroy enzymes.

The heart of the process is the column packing material and the mode in which it operates to adsorb proteins selectively. This may be by ion exchange (cation or anion), electrostatic interactions; hydrophobic adsorbents, interactions with hydrophobic amino acids on the protein surface; affinity adsorbent, biospecific interaction with enzyme at the site that binds the substrate, the ligand-binding site; inorganic adsorbents such as hydroxyapatite, which interact by opposite charges and other modes not completely understood; dye adsorbents which are "multifunctional," interacting with protein surfaces in a variety of modes; metal chelates, forming coordinate metal–protein bonds between the column containing metal ions and the enzyme; and other specialized materials of unclear operational modes.

The column materials are mainly commercial products and consist of solid bead particles. The beads are sufficiently porous to allow protein molecules to diffuse through them. The functional adsorbent structure is chemically linked to the bead particle; as the main adsorption surface is internal, the effective surface area is large.

A protein mixture, typically obtained after a precipitation step, is applied to the column preequilibrated in an appropriate buffer. The enzyme required might not bind, in which case the column must adsorb most other proteins if the step is to be efficient. Normally, though, the desired enzyme binds, and, after a washing step, an elution procedure which separates the adsorbed proteins from each other is performed. Generally, a gradient of eluant (e.g., salt) is applied, i.e., the column is perfused with liquid of progressively increasing salt concentrations; the more weakly bound proteins are eluted at low salt concentrations, and the strongly bound proteins are eluted only toward the end of the gradient, at higher salt concentrations (Fig. 3).

Adsorption chromatography is mainly a trial-and-error technique when beginning an enzyme purification that has not previously been attempted. The adsorption properties of the enzyme protein cannot be predicted from its known catalytic activity. However, "designer adsorbents," which are created specifically for the enzyme's catalytic properties, have been successfully used in many cases. These are the true affinity adsorbents, which interact with enzymes through the catalytic site. For this purpose, a substrate, an inhibitor of enzyme activity, or another effector known to bind to the enzyme is used as an immobilized ligand on the column; ideally, the only proteins that bind to this column are the enzymes that recognize and bind to the ligand.

After washing the column, the specifically bound enzymes are eluted, either by using a buffer change (pH, ionic strength), which weakens the binding, or by displacing the enzyme with the buffer containing free ligand, which, by combining the enzyme, frees it from the immobilized ligand (affinity elution; Fig. 4). Affinity chromatography is theoretically an ideal method for isolating an enzyme. In practice, complications such as difficult chemistry of immobilization and

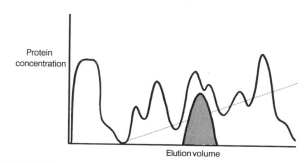

FIGURE 3 Idealized elution of proteins from an ion-exchange column. Solid line, ultraviolet adsorption at 280 nm; dashed line, salt concentration gradient; shaded area, enzyme activity.

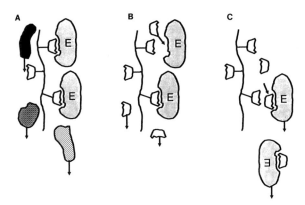

FIGURE 4   Principle of affinity elution from an affinity adsorbent. (A) The mixture of proteins is applied to the adsorbent, and the enzyme (E) binds to the immobilized ligand. (B) The column is then washed with buffer containing soluble ligand. (C) The enzyme exchanges from immobilized to soluble ligand, and so is specifically eluted from the column. In theory, the only enzymes eluted from the column are those that interact with the ligand being used.

nonspecific adsorption mean that it is not suitable for all purposes.

A final example of adsorption chromatography is the ultimate specific column, an immunoadsorbent. Antibodies raised against a previously purified enzyme will generally bind to that enzyme and no other protein. The antibodies, either monoclonal or polyclonal, can be covalently attached to a matrix such as agarose and can pull out the desired enzyme from a complex mixture. Purification factors of 1000 and more are possible for enzymes existing in only trace amounts. The major problem with immunoadsorbents is how to displace the enzyme from the column because the conditions required may be so harsh that the enzyme is inactivated. Relatively weakly binding monoclonal antibodies may be selected to overcome this problem. [See Monoclonal Antibody Technology.]

3. Techniques in Which the Proteins Remain in Solution

Both precipitation and adsorption techniques can alter the structure of unstable enzymes; the least damaging methods are those that allow all the proteins to remain in aqueous solution at a neutral pH. Chief among these are electrophoretic methods and gel filtration.

Electrophoresis is the movement of a charged particle in an electric field. For nearly a century, attempts have been made to exploit electrophoresis for separating proteins according to their charges. Because of numerous technical problems, there is still no widely

accepted electrophoretic technique, although there are many systems that have proven successful for individual cases. Nevertheless, electrophoresis, especially in laminar gels, has become the most widely used technique for analytical separations of protein. In an analytical system, only a few micrograms of sample is used, and there is no attempt to recover the separated proteins. For preparative work, several milligrams must be separated and the individual fractions must then be collected.

As a refinement of the electrophoresis system, a newer technique known as isoelectric focusing has been used successfully. Simple electrophoresis makes use of the fact that, at a fixed pH, a given protein has a certain charge that results in a certain speed of movement in an electric field. However, if the pH changes, the charge alters, so the protein's speed of migration also changes. At a particular pH, called the protein's isoelectric point, the charge on the protein is zero—all positives and negatives cancel out. As the pH approaches this value, the speed of migration in the electric field decreases, to zero when it reaches the isoelectric point. Isoelectric focusing is a system that creates a gradient of pH such that each protein component moves until it reaches its particular isoelectric pH within the gradient. At this point, having no net charge, the protein molecule stops moving. The protein molecules "focus" to their respective isoelectric points. After the electric field is switched off, the protein can be eluted from the isoelectric strips (Fig. 5).

A much more extensively used technique is gel filtration. This is perhaps the most "gentle" of all methods; each molecule experiences only a diffusion through gel pores during the separation process. Porous beads are used, which allow small molecules, including the smallest proteins, to diffuse freely through them. The sizes of the pores in the beads

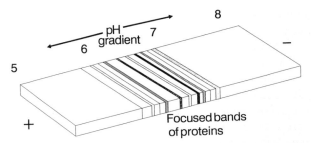

FIGURE 5   Separation of protein bands in a pH gradient generated by electrofocusing. If the separation medium is a porous gel material, it can be sliced and the separated bands can be eluted from the medium.

vary, however, and some are too narrow to allow middle-sized proteins to pass. In some cases, all pores are too narrow to admit the largest protein molecules. Thus, large proteins pass down a column containing the beads, flowing mainly outside of the beads, and emerge quickly. The smallest molecules pass in and out of the beads freely, and so are retained in the column for longer times relative to the large proteins. With the optimum material, different sizes of protein molecules mixed together in the applied sample emerge from the column at different times and so are separated. In gel filtration, there is no adsorption, and the protein molecules remain in solution at all times.

IV. RECOMBINANT TECHNIQUES FOR EXPRESSION AND ISOLATION OF HUMAN ENZYMES

The developments in molecular biology in the last decade have revolutionized our ability to produce large amounts of specific proteins and enzymes, including human products. Only a brief outline of the methods can be given here. In general, the procedure for isolating an enzyme making use of recombinant techniques involves the following:

1. A stretch of the base sequence of the gene encoding the enzyme must be known. This may be obtainable from general genome sequencing (e.g., The Human Genome Project will provide extensive information on enzyme genes) or by other genetic processes. If not, then the base sequence must be deduced, using the genetic code, from a portion of the amino acid sequence. An oligonucleotide (synthetic single-stranded DNA) is synthesized that contains all possible codon combinations for the amino acid sequence. To obtain a portion of the amino acid sequence that is unique to this enzyme, it is necessary to first purify the enzyme from its natural source by conventional techniques. It may be possible to purify it on a very small scale by less conventional methods, e.g., gel electrophoresis, because only a few micrograms are needed for an amino acid sequence and the enzyme need not be active.

2. A "complementary DNA library" is prepared from the mRNA of a tissue that has a lot of the required enzyme in it, and this is screened with the synthetic DNA to isolate a clone containing the desired gene.

3. The gene is transferred to a host organism or tissue culture system that enables active expression of the enzyme from the recombinant gene.

4. The enzyme is purified from the culture, separating it from all other proteins present in the host system.

Taking these four steps in turn, the first is fairly routine, the requirements being a purified sample of the enzyme (not necessarily from a human source) and facilities for sequencing the protein and synthesizing the DNA (both can be done commercially).

The second step may be troublesome in that fresh human tissue is needed at this point; complementary DNA libraries are of variable quality, although techniques for creating them have greatly improved. [See Chromosome-Specific Human Gene Libraries.]

The third step, expression in the host organism or tissue culture, is generally the most difficult. Human enzymes are often expressed in bacterial hosts as inactive protein, improperly folded. In eukaryotic cells such as yeast or animal tissue culture, the expression levels may be low. Often, there is also a problem of glycosylation; many proteins, mainly extracellular, have a carbohydrate addition that is not made in the host cell. Thus, the protein may be perfect as polypeptide chains, but not active in the correct way, due to incorrect glycosylation.

Finally, purification can be troublesome, mainly because, if these enzymes are to be used clinically, they must be ultrapure and demonstrably free of any detectable proteins from the host organism. To achieve this result, it is possible to modify the gene so that the exprssed protein has an additional portion (a tag) that enables its easy isolation, but the tag must subsequently be clipped off.

Molecular biology techniques have allowed the production of many proteins, especially hormones, in previously unimaginable amounts. Modern molecular biology also makes it possible to modify proteins and enzymes so that they differ from the natural protein, a procedure that can be used to avoid many problems encountered in the use of enzymes for human clinical therapy.

BIBLIOGRAPHY

Burgess, R. R. (ed.) (1987). "Proceedings of the UCLA Symposium on Protein Purification, Micro to Macro." Liss, New York.

Deutscher, M. P. (eds.) (1990). Guide to protein purification. In "Methods in Enzymology," Vol. 182. Academic Press, San Diego.

Harris, E. L. V., and Angal, S. (eds.) (1990). "Protein Purifications: Applications. A Practical Approach." Oxford Univ. Press, Oxford.

Pharmacia Fine Chemicals (PFC). "Affinity Chromatography: Principles and Methods." PFC, Uppsala, Sweden.

Pharmacia Fine Chemicals (PFC). "Gel Filtration: Theory and Practice." PFC, Uppsala, Sweden.

Pharmacia Fine Chemicals (PFC). "Ion Exchange Chromatography: Principles and Methods." PFC, Uppsala, Sweden.

Scopes, R. K. (1993). "Protein Purification, Principles and Practice," 3rd Ed. Springer-Verlag, New York.

Turkova, J., Chaiken, I. M., and Hearn, M. T. W., eds. (1986). Proceedings of the 6th International Symposium on Bioaffinity Chromatography and Related Techniques. *J. Chromatogr.* **376.**

Enzymes, Coenzymes, and the Control of Cellular Chemical Reactions

STEPHEN P. J. BROOKS

Food Directorate, Health Canada, Ottawa

I. How Enzymes Function
II. Enzyme Function–Structure Relationships
III. Coenzymes and Their Function
IV. Cellular Control of Enzyme Activity

GLOSSARY

E_a Activation energy for a reaction

ΔG Gibbs free energy for a system

$\Delta G^{o\prime}$ Standard free energy at pH 7.0

h Planck's constant (6.626×10^{-34} J/sec)

K_a Binding constant for an activator

K_{eq} Equilibrium constant for a reaction

K_i Binding constant for an inhibitor

K_m Michaelis constant for substrate binding to an enzyme active site; defined as the concentration of substrate that produces half-maximal enzyme activity

N Avogadro's number (6.023×10^{23} molecules/mol)

P Product of a reaction

R Gas constant for an ideal gas (8.3144 J/deg/mol)

S Substrate for a reaction

ΔS Change in entropy associated with a reaction

T Temperature (°K)

V_{max} Maximal velocity of an enzyme measured at saturating substrate concentrations

ENZYMES ARE EITHER PROTEIN OR NUCLEIC ACID molecules that act as catalysts to speed up reactions. Enzymes are so effective that, with respect to the multitude of chemical reactions that could take place in the cell, only enzyme-catalyzed reactions occur. This is be-cause, in comparison to enzyme-catalyzed reactions, the noncatalyzed reactions occur at too slow a rate. As a rule, each enzyme catalyzes only a single reaction. This means that enzymes effectively determine the types of reactions that occur in the cell; enzymes direct cellular metabolism. Because enzymes are so powerful they are also strictly regulated by other cellular compounds and, sometimes, by the reactants themselves. It is this combination of reaction specificity and regulation of activity that helps to maintain cellular homeostasis under widely different metabolic conditions and permits cells to perform various tasks. In animal liver, for example, cellular and blood glucose concentrations are regulated by controlling the enzymes responsible for storing glucose (in the form of glycogen) or releasing glucose (break down glycogen) depending on whether blood glucose levels are high (during eating) or low (in between meals). The liver regulates these processes through control of the enzymes that act specifically on glycogen.

In this article, the reader shall see how enzymes are specifically controlled (Section I) and how their physical structure relates to their function (Section II). In Section III, a review of the cofactors required for enzyme function is presented, and a short discussion of cellular mechanisms of enzyme control is presented in Section IV. Several references are provided at the end of this review and should be referred to for background information and for a more complete treatment of the subjects presented.

I. HOW ENZYMES FUNCTION

All biological processes, from the utilization of glucose, to the building of new proteins, to the reactions driving muscle contraction, are chemical reactions, and as such must obey the fundamental rules governing chemical processes. These rules determine (1) how fast a reaction happens and (2) how much starting material reacts to make product. A fundamental understanding of what governs chemical processes is essential to the understanding of how an enzyme works since chemical processes and enzyme action are intimately linked. A short discussion of factors that are enzyme independent (i.e., those that depend only on the chemical properties of the reactants) is important in understanding enzyme action.

A. Thermodynamic Aspects of Kinetic Function

1. Processes Not under Enzyme Control

In any chemical reaction, the degree to which a particular reaction proceeds (how much product is formed) is determined only by the nature of the starting material and the products. Quantification of the degree to which a reaction proceeds is measured by the equilibrium constant, defined as

$$K_{eq} = [Product]/[Starting\ material], \qquad (1)$$

where the square brackets denote concentration. The K_{eq} value is measured when product and starting material concentrations are constant (equilibrium condition) and is independent of the path of molecular transformation and the time required to reach equilibrium. It is possible to analyze the energetics of the reaction in terms of the energy change of all the components of the reaction as well as those of the surrounding system. This is because the extent to which a reaction occurs is a function of the starting materials, the products, and the properties of the system. In 1878, Willard Gibbs derived a function relating the change in free energy of a system (ΔG) to the change in enthalpy (ΔH, related to the internal energy of the system) and entropy (ΔS, related to the degree of randomness of the system):

$$\Delta G = \Delta H - T\Delta S. \qquad (2)$$

The ΔG value is an important criterion for determining reaction characteristics: a reaction can occur spontaneously (requires no energy input) only if ΔG is negative and will not proceed without an input of energy when $\Delta G \geq 0$. Equation (2) shows that a reaction can be driven by a decrease in the enthalpy

of the system (a negative ΔH results when the products have a lower energy than the starting material) and/or by an increase in the randomness of the system (ΔS). An example of the first type of process is the burning of wood, which releases a large amount of energy because the heat of combustion of the products is much lower than that of the starting materials. An example of the second type of process is the dilution of salt in water because the Na^+ and Cl^- ions are now no longer in a highly ordered crystal.

The K_{eq} constant [Eq. (1)] is related to the change in the total energy of the system (ΔG) and the energy change measured under standard conditions ($\Delta G^{\circ\prime}$):

$$\Delta G = \Delta G^{\circ\prime} + RT \times \ln(K_{eq}), \qquad (3)$$

where R is the gas constant and T is the temperature in $°K$. At equilibrium, $\Delta G = 0$ (by definition), and Eq. (3) gives

$$\Delta G^{\circ\prime} = -RT \times \ln(K_{eq}). \qquad (4)$$

Equation (4) shows that $\Delta G^{\circ\prime}$ and the equilibrium constant for a reaction are related by a simple equation that is independent of any kinetic parameters. This shows that it is possible to determine $\Delta G^{\circ\prime}$ values for any reaction simply by measuring the concentrations of the reactants at equilibrium. Tables of $\Delta G^{\circ\prime}$ values for several reactions (under standard conditions) have already been compiled and can be found in many standard biochemistry and chemistry texts. These values are useful for predicting equilibrium concentrations of reactants. Once equilibrium concentrations are known for the reaction, the predicted concentrations can be compared with actual concentrations measured in the cell. Therefore, it is possible to measure how far a reaction is from its equilibrium point by comparing predicted and experimental values.

2. Transition State Theory and Reaction Rates

The foregoing thermodynamic analysis shows that a reaction is spontaneous (i.e., can occur without the input of energy to the system) if ΔG is negative; however, it says nothing of how fast that reaction may occur. The factors determining the rate of the reaction are best illustrated by considering the energy diagram

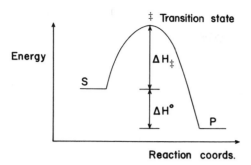

FIGURE I Idealized representation of a single reaction scheme. The energy level of the reactants is plotted versus hypothetical reaction coordinates (the reaction proceeds from left to right). *S* represents starting material and *P* represents products. The symbol ‡ represents the activated state, and $\Delta H^{\ddagger}$ is the activation enthalpy for this particular reaction.

for an idealized reaction scheme shown in Fig. 1. For starting material (*S*) to react to form product (*P*), additional energy is required to break the chemical bonds of the starting material and/or produce the required reactant configurations. Wood, for example, does not spontaneously combust without first lighting it, even though the burning of wood has a large negative ΔG.

The way in which the activation energy influences the rate of a reaction is best described by transition state theory. The transition state is defined as a configuration of maximum potential energy through which the reactants must pass before continuing to form products. It is commonly denoted by a "‡" symbol (Fig. 1). The transition state is clearly distinguished from an intermediate, which is a meta-stable minimum on the reaction profile. In thermodynamic terms, the equilibrium constant for the formation of the transition state is given by

$$K^{\ddagger} = [A^{\ddagger}]/[A], \qquad (5)$$

where $A^{\ddagger}$ represents the transition state. Using the thermodynamic principles developed earlier and quantum mechanics, it is possible to transform Eq. (5) into Eq. (6), an equation relating the enthalpy ($\Delta H^{\ddagger}$) and entropy ($\Delta S^{\ddagger}$) associated with transition state formation to the rate of the reaction:

$$k = (RT/Nh) \exp(\Delta S^{\ddagger}/R - \Delta H^{\ddagger}/RT), \qquad (6)$$

where k is the reaction rate, R is the gas constant, N is Avogadro's number, and h is Planck's constant. Equation (6) explains much about reaction properties.

First, it shows that reactions are temperature dependent, with rates increasing in direct proportion to the increase in temperature. Second, it shows that the reaction rate is directly influenced by the activation energy (E_a) of the reaction [in Eq. (6), $E_a = \Delta H^{\ddagger} + RT$], with a larger E_a value resulting in a smaller rate constant. Finally, it shows that the reaction rate is dependent only on the formation of the transition state (the transition state is regarded as the highest energy point on the reaction profile). Once the transition state is formed, the reaction quickly proceeds to form products.

3. How Enzymes Influence Reaction Rates

The transition state theory outlined in Section I,A,2 also applies to enzymes since enzymes are simply catalysts that act to speed up reaction rates. Equation (6) shows that there are two different ways by which an enzyme may increase the rate of a reaction. First, enzymes may lower the E_a by stabilizing a high-energy intermediate, possibly by covalent bonding. Since a more stable transition state is easier to form [$K^{\ddagger}$ in Eq. (5) is larger], the $\Delta H^{\ddagger}$ in Eq. (6) will be smaller (see Fig. 1) and a greater number of molecules can react to form product. This is illustrated in Eq. (6), where the magnitude of $\Delta H^{\ddagger}$ directly influences the rate constant; decreasing $\Delta H^{\ddagger}$ increases k. Coenzyme A, pyridoxyl phosphate, and thiamine phosphate are good examples of coenzymes that, when bound to enzymes, stabilize high-energy intermediates (see Section III). Second, enzymes may increase the reaction rate by providing a favorable reaction entropy (decrease $\Delta S^{\ddagger}$). This can occur when reactive groups on both substrates and enzymes are oriented to optimize catalysis. Equation (6) shows that an increase in transition state entropy will directly increase the reaction rate. Enzymes that hydrolyze sugar polymers are good examples of this type of catalysis (see Section II).

B. Kinetic Mechanisms for Enzyme Function

I. One-Substrate Reactions

As noted earlier, all enzymes are catalysts that increase reaction rates either by stabilizing high-energy intermediates or by orienting reactive groups so that they are in close proximity. The exact relationship between catalysis and the overall mechanism of enzyme-catalyzed reactions is the subject of this section. Consider the general reaction of Fig. 1 with *S* giving rise to *P*. The introduction of an enzyme catalyst to this process changes the details of the reaction mechanism because enzymes are physical entities that participate directly in the reaction mechanism. A general one-substrate reaction is

$$S + E \underset{k_{-1}}{\overset{k_1}{\rightleftharpoons}} ES \underset{k_{-2}}{\overset{k_2}{\rightleftharpoons}} EP \underset{k_{-3}}{\overset{k_3}{\rightleftharpoons}} E + P. \qquad (7)$$

Equation (7) illustrates all the steps that are necessary to describe any one-substrate reaction, including the binding of substrate to enzyme (the starting material in an enzyme reaction is usually referred to as substrate), the conversion of enzyme-bound substrate to enzyme-bound product, and the release of product from the enzyme. Note that Eq. (7) does not formally include the formation of a transition state as a discrete step in the reaction mechanism. Its existence is, however, encompassed in the conversion of enzyme-bound substrate into product (step 2). Equation (7) also demonstrates the temporal order for an enzyme-catalyzed reaction: (1) substrate binds to the enzyme to form an enzyme–substrate complex (*ES*), (2) *ES* reacts to give the enzyme–product complex (*EP*), and (3) *EP* dissociates to give free enzyme and product. This last step regenerates the catalyst so that free enzyme can now bind new *S* to continue the reaction.

Two different enzyme-catalyzed reaction profiles are shown in Fig. 2 to illustrate the wide variety of mechanisms that can account for a simple reaction such as that of Eq. (7). In the reaction of Fig. 2A, formation of the transition state is the slowest step as shown by the higher energy associated with its formation. In Fig. 2B, the formation of the *ES* complex is the slow step. It is important to identify the slow step because the overall reaction velocity will be determined only by the rate of this step; in an enzyme-catalyzed reaction, the slow step (and consequently the overall reaction rate) is not always related to the formation of the transition state.

In both reactions of Fig. 2, the formation of *ES* and *EP* requires the input of energy, as shown by the higher energy of these species. This is normally true for any enzyme reaction because of the solvation effects. Any solute (such as a substrate) that is dissolved in water is surrounded by a layer of tightly bound water molecules. Before substrate can bind to the enzyme, this water layer must be stripped away to expose the reactive groups on the substrate (the same is true for the enzyme as we shall see in Section II). The amount of energy required for this process depends on the substrate and enzyme species and may vary

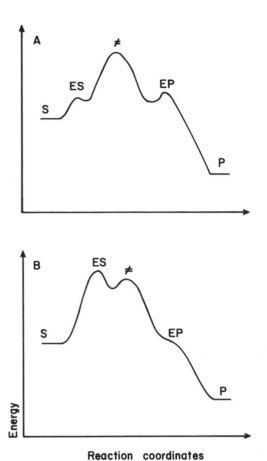

FIGURE 2 Energy diagram for idealized enzyme-catalyzed reactions. The energy level of the reactants and reacting species is plotted versus reaction coordinates. (A) Reaction sequence in which the transition state is the most energetically unfavorable species. (B) Formation of *ES* is the most energetically unfavorable step. *S* and *ES* represent the free and enzyme-bound substrates. *P* and *EP* represent free and enzyme-bound products, respectively.

considerably from reaction to reaction as shown graphically in Figs. 2A and 2B.

The reaction of Eq. (7) was drawn with arrows going in both directions to illustrate that all reactions are reversible; a finite flow of *S* to *P* and *P* to *S* always exists for any given reaction, even though at equilibrium the net flow is zero. Note also that each step of Eq. (7) has a rate constant associated with it, which can be calculated from the individual concentrations and Eq. (6), or more commonly is determined experimentally from enzyme studies. It is important at this point to distinguish between an overall reaction rate constant and the rate constants associated with the individual steps of Eq. (7). The overall rate of any reaction is measured by following the formation of product over time and it reflects the rate of the slowest step in the reaction mechanism. The microscopic rate constants (i.e., the individual rate constants associated with each step in the reaction mechanism) are often difficult to measure experimentally but correspond only to the step indicated in the reaction. These steps may or may not limit the rate of the overall reaction depending on their magnitude relative to that of the other steps of the reaction sequence. Both overall and microscopic reactions have a transition state equivalent (point of maximum energy), and so may be adequately described by Eq. (6). The microscopic rate constants are related to the K_{eq} values for formation of each species by

$$K_1 = k_1/k_{-1}, \ K_2 = k_2/k_{-2}, \ K_3 = k_3/k_{-3}. \quad (8)$$

It is possible to reduce Eq. (7) to a general scheme that describes enzyme behavior by assuming that the concentration of EP is zero:

$$S + E \underset{k_{-1}}{\overset{k_1}{\rightleftharpoons}} ES \overset{k_2}{\longrightarrow} E + P. \quad (9)$$

Experimentally it is relatively easy to generate this condition because most enzyme reactions are performed *in vitro* with an excess of substrate. When measuring the initial enzyme rate (measured within the first few minutes before product has a chance to accumulate) the concentration of EP is negligible and $[P] \approx 0$. This condition is referred to as the initial velocity assumption and greatly simplifies Eq. (7) and the resulting initial velocity equations.

To reduce the scheme of Eq. (9) into a usable equation describing an overall enzyme reaction, one must make an additional assumption about the specifics of the reaction mechanism. Depending on where the rate-limiting step lies, it is possible to make one of two different assumptions (see Fig. 2). Both assumptions produce an equation of the general form

$$v = V_{max} \times [S]/(K_m + [S]), \quad (10)$$

where v is the observed rate of reaction (v = change in product over time) and V_{max} is the maximal rate of the reaction measured at infinite substrate concentra-

tions. Equation (10) is shown graphically in Fig. 3. It is easy to see why an enzyme-catalyzed reaction has a maximum rate, even though a noncatalyzed reaction does not. Since S binds E to form an ES complex with an equilibrium constant K_1 ($=k_1/k_{-1}$),

$$S + E \underset{k_{-1}}{\overset{k_1}{\rightleftharpoons}} ES, \quad (11)$$

increasing the concentration of S while keeping E constant will eventually drive all the free E into the ES complex. At this point ES is the only enzyme species ($ES \approx$ [Total enzyme]): there is no more free E for S to bind. The rate of the reaction is now independent of S and is given by the concentration of ES (=[Total enzyme]) multiplied by the rate of its decomposition, k_2:

$$V_{max} = k_2 \times \text{[Total enzyme]}. \quad (12)$$

The K_m value (shown in Fig. 3) is defined as the concentration of substrate at which the velocity is equal to $V_{max}/2$. In practical terms the K_m value is a measure of the affinity of enzyme for substrate, but the exact relationship between the K_m value and the substrate affinity depends on what assumptions are made about the reaction mechanism. Michaelis and Menten assumed that the formation of the transition step was much slower than the rate of ES formation (Fig. 2A). In this case the K_m value (the m stands for Michaelis) is equal to the dissociation constant for ES formation:

$$K_m = 1/K_1 = k_{-1}/k_1. \quad (13)$$

A more general assumption, first made by Briggs and Haldane, is that a steady state is reached in which the concentration of intermediate is constant: $d[ES]/dt = 0$. This leads to a kinetic definition for the K_m value:

$$K_m = (k_2 + k_{-1})/k_1. \quad (14)$$

Equation (13) applies when the binding reaction is the slow step (Fig. 2B), or when the rate of ES formation is not much different from the rate of transition state formation. A priori choice between these two definitions is impossible since a detailed kinetic study must be completed before one or the other can be ruled out. Fortunately, studies of overall enzyme rates are

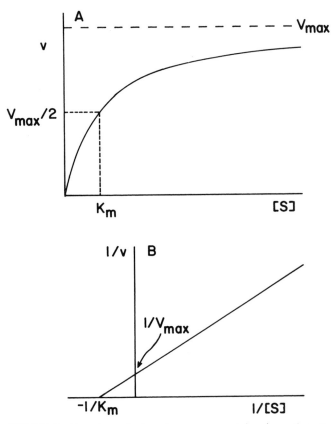

FIGURE 3 Reaction velocity of an enzyme-catalyzed reaction as a function of the substrate concentration. (A) Normal plot of the velocity versus substrate. (B) Lineweaver–Burk plot of $1/v$ versus $1/[S]$. The maximal velocity (V_{max}) is indicated as well as the concentration of substrate that produces half-maximal saturation (K_m). Enzyme reaction kinetics are most commonly graphed using the Lineweaver–Burk plot, even though some authors dislike this plot because it magnifies errors at low substrate concentrations. It is used here because the references at the end of this article use this graph exclusively.

not dependent on how the K_m is defined since Eq. (10) applies equally well to both definitions.

Figure 3B demonstrates how K_m and V_{max} values are graphically determined. The method is that of Lineweaver and Burk, who rearranged Eq. (10) into

$$1/v = K_m/V_{max} \times 1/[S] + 1/V_{max}. \qquad (15)$$

The V_{max} and K_m parameters are obtained by extrapolating the line through the $1/[S]$ and 1/velocity data pairs to its intersection with the x and y axes. Thus, at infinite substrate concentrations, $1/[S] = 0$ and $1/v = 1/V_{max}$. The V_{max} value is obtained from the reciprocal of the y-axis intercept. Equation (15) also shows that when $v = V_{max}/2$, $1/K_m = 1/[S]$, and the

K_m value is obtained from the reciprocal of the x-axis intercept. The Lineweaver–Burk plot will be used to illustrate the effects of inhibitors later in this article.

2. Multisubstrate and Cooperative Reactions

The preceding analysis gave a simple equation relating substrate concentrations to observed reaction velocity at a fixed enzyme concentration. The result of this analysis, Eq. (10), applies directly to systems where one substrate reacts to give one product under initial velocity conditions. Numerous such equations (of varying complexity) exist that describe enzyme reactions involving two substrates as well as those involving cooperative substrate binding. These equations are necessary because the majority of enzymes catalyze

TABLE I

Kinetic Equations Describing Common Multisubstrate and Cooperative Enzyme Models[a]

Pattern	Variable [S]	V_{max}^{app}	K_m^{app}
Random Ordered	$[S_1]$	$V_{max}[S_2]/(K_2 + [S_2])$	$(K_{12} + K_1[S_2])/(K_2 + [S_2])$
Substituted	$[S_1]$	$V_{max}[S_2]/(K_2 + [S_2])$	$K_1[S_2]/(K_2 + [S_2])$
Cooperative	$[S^h]$	V_{max}	K_m^h

[a] K_1 is the K_m value for S_1, K_2 is the K_m value for S_2, and $K_{12} = K_1 \times K_2$ (see Fig. 4). The values for patterns 1, 2, and 3 were obtained by assuming that the binding reactions are at equilibrium (Michaelis–Menten assumption). The equation describing pattern 4 is an empirical relationship derived by Hill to describe oxygen binding to hemoglobin. In cooperative enzyme kinetics, h represents the Hill coefficient, an arbitrary measure of the cooperativity of the system that has no physical significance.

reactions involving more than one substrate–product pair. Many enzymes also require coenzymes to complete their catalytic cycles (see Section III,A). When these coenzymes are soluble (i.e., they can reversibly bind to enzyme active sites), they are treated as substrates. Thus, reactions between substrates and coenzymes are equivalent to two-substrate reactions. Four of the most common mechanisms are presented in Table I for reference. The first three kinetic patterns are distinguished from the fourth, the cooperative model, by the linearity of the Lineweaver–Burk plots when $1/v$ is plotted against the reciprocal of the first substrate concentration. Cooperative enzymes have nonlinear Lineweaver–Burk plots. Figure 4 shows mechanisms for the three most common enzyme pathways with two substrates. The three reaction pathways are distinguished by the order that the substrates bind to the enzyme. In the random-equilibrium model (Fig. 4A), either substrate can bind to free enzyme. The ordered-equilibrium model (Fig. 4B) is observed when only the first substrate can bind to free enzyme; the second substrate can bind only the ES_1 complex. The substituted-enzyme model (or Ping-Pong mechanism, Fig. 4C) describes a situation where one substrate binds free enzyme and reacts to give a covalently modified enzyme (E'). The covalently modified enzyme now reacts with the second substrate to give free enzyme and product. This mechanism is often observed with phosphate transfer mechanisms.

Cooperative enzyme kinetics patterns often arise when an enzyme is made up of more than one subunit; enzymes are often composed of more than one polypeptide chain (see Section II,A) and each of the individual chains (called subunits) of a multisubunit enzyme can possess an active site. In simple cooperative models that can be described by the Hill equation, at least two of the enzyme subunits must possess binding

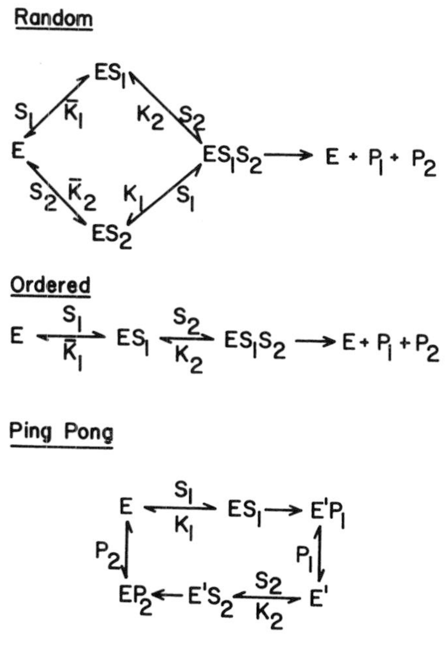

FIGURE 4 Common initial velocity two-substrate mechanisms for the random-equilibrium model, the ordered-equilibrium model, and the substituted enzyme model (Ping-Pong model). S_1, first substrate; S_2, second substrate; E, free enzyme. The $\overline{K}$ values represent true dissociation constants and the K values represent the dissociation constants in the presence of infinite amounts of the opposite substrate. The mechanisms assume that the reactions are at equilibrium (Michaelis–Menten assumption).

sites for the same substrate. As is the case for all cooperative enzyme reactions, the binding of substrate to one subunit affects the binding of substrate to other subunits. This means that the binding of one substrate molecule changes the affinity of the enzyme for the next substrate molecule. In the case of positive cooperativity, binding of one substrate increases the enzyme affinity for the second substrate. Negative cooperativity describes exactly the opposite effect.

Although the complete equations for the mechanisms shown in Fig. 4 are quite complex, they can be simplified to Eq. (10) with V_{max}^{app} (apparent V_{max}) replacing V_{max} and K_m^{app} (apparent K_m) replacing K_m. This means that the individual kinetic constants for these reactions can be obtained by manipulating substrate concentrations and plotting the resulting velocities in graphs similar to those of Fig. 3B. Table I shows the values of V_{max}^{app} and K_m^{app} for four common kinetic patterns. In the first three equations, V_{max}^{app} and K_m^{app} values are obtained from the intercepts when velocity (v) is plotted against the corresonding substrate value shown in Table I. The exact kinetic equation describing the fourth mechanism (cooperative) is complex, but may be adequately described by the Hill equation shown in Table I. Note that the Hill coefficient is simply an arbitrary measure of the increase in affinity that results from the binding of the first substrate and has no precise physical meaning.

C. Enzyme Inhibition and Activation

In general, the cellular concentrations of enzyme substrates and products do not change significantly even when the metabolic rate of the cell changes dramatically. If enzyme rates were controlled strictly by changes in substrate concentrations this would mean that cellular enzyme activities were constant; the enzyme velocity would always be equal to the value corresponding to the "unchanging" cellular substrate concentration (see Fig. 3). This means that other mechanisms must exist for controlling enzyme rates. It is especially important to control enzymes found at the beginning of metabolic pathways as well as those found at metabolic branch points (where several different pathways interact). By regulating the activity of these enzymes the cell can determine the activity of each pathway since the flux through the pathway is, in large measure, controlled by the amount of substrate provided by the initial enzyme. Enzyme inhibition and activation are important mechansms for cellular control of individual reactions. In general, inhibitors (defined as substances that decrease enzyme activity) and activators (substances that increase enzyme activity) are compounds that monitor the energy state of the cell (such as adenosine triphosphate or adenosine monophosphate), or are products or substrates of the enzymes themselves. These compounds serve to directly link enzyme activity to the energy state of the cell and to the demand for substrate or product. In this article we will consider only reversible inhibitors and activators.

Three general classes of inhibitors exist. They are classified according to which enzyme form they bind [either E and/or ES in Eq. (9)], and what kinetic effect they have on the velocity versus substrate plots. Competitive inhibitors compete directly with substrate for free enzyme and so affect the substrate's apparent affinity for enzyme (the K_m value increases with increasing inhibitor concentration). Uncompetitive inhibitors bind to the ES complex and reduce the rate of product formation ($k_2' < k_2$ giving a decreased V_{max} value). Mixed inhibitors bind both free and substrate-bound enzyme to give increased K_m values and decreased V_{max} values. The effects of inhibitors are best illustrated by considering a Botts–Moralis scheme, which is simply Eq. (9) expanded to include two new equilibria:

$$
\begin{array}{ccc}
E & \xrightleftharpoons[K_m]{S} ES \xrightarrow{k_2} & E + P \\
I \updownarrow K_j & \quad I \updownarrow K_j' & \\
EI & \xrightleftharpoons[K_m']{S} ESI \xrightarrow{k_2'} & EI + P
\end{array}
\tag{16}
$$

In Eq. (16), K_i corresponds to the inhibitor dissociation constant for competitive inhibitors, and K_i' to the inhibitor dissociation constant for uncompetitive inhibitors. The scheme of Eq. (16) can be reduced to a generalized equation by assuming that the E, EI, ES, and ESI species are all in equilibrium. This is equivalent to the Michaelis and Menten assumption described earlier. As expected, solution of the equilibria in Eq. (16) gives the generalized Eq. (10) with V_{max}^{app} replacing V_{max} and K_m^{app} replacing K_m. Equation (17) shows how these new values relate to the individual binding constants of Eq. (16):

$$
V_{max}^{app} = [\text{Total enzyme}]
$$
$$
\times (k_2 + k_2'[I]/K_i')/(1 + [I]/K_i') \tag{17a}
$$
$$
K_m^{app} = K_m \times (1 + [I]/K_i)/(1 + [I]/K_i'). \tag{17b}
$$

The V_{max}^{app} and K_m^{app} values are the apparent V_{max} and K_m values measured at finite concentrations of inhibitor. They are obtained in the same way as V_{max} and K_m values by extrapolating $[S]$ to infinite concentra-

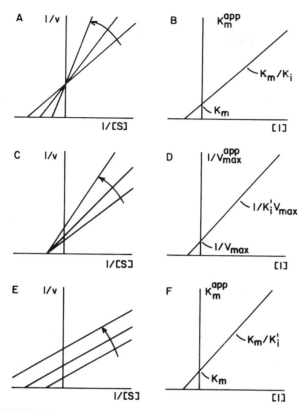

FIGURE 5 Graphical patterns for competitive, noncompetitive, and uncompetitive inhibition. (A, C, E) Primary plots of $1/v$ versus $1/[S]$ at increasing inhibitor concentrations (shown by arrow). (B, D, F) Secondary plots of indicated parameter as a function of increasing inhibitor concentration. Parameters are obtained as shown on the plots.

tions (see Fig. 5). Table II lists some of the more commonly occurring inhibition patterns and illustrates how each inhibition pattern arises. Figure 5 shows how the Lineweaver–Burk plots behave for some of these inhibitors. For example, competitive inhibition occurs when an inhibitor competes with substrate for free E only (the inhibitor does not bind the ES complex). This is expressed mathematically as $K_i' = K_m' = \infty$. Figure 5A shows the effect of competitive inhibition on the Lineweaver–Burk plots, and Fig. 5B shows how to obtain the kinetic constants from such an analysis. Other inhibition patterns are also presented in Table II and Fig. 5.

It is possible to use the Botts–Moralis scheme for analyzing reversible enzyme activation in the same way as enzyme inhibition, by replacing the inhibitor term (I) in Eq. (16) with A (for activator) and the inhibitor dissociation constant (K_i) with an activator dissociation constant (K_a). Depending on the value of the individual constants, kinetic patterns similar to those of Fig. 5 may be observed. However, activation is rarely as simple as inhibition and nonlinear secondary plots often result. It is, perhaps, for this reason that well-defined common activation patterns (analogous to competitive, uncompetitive, and mixed inhibition patterns) have not been characterized.

II. ENZYME FUNCTION–STRUCTURE RELATIONSHIPS

Enzyme kinetic studies provide a lot of information about the overall enzyme mechanism but they provide

TABLE II
Common Inhibition Patterns[a]

Pattern	Conditions	V_{max}^{app}	K_m^{app}
Competitive	$K_i' = K_m' = \infty$	V_{max}	$K_m \times (1 + [I]/K_i)$
Partially competitive	$K_m' > K_m$ $K_i' > K_i$ $k_2 = k_2'$	V_{max}	$K_m \times (1 + [I]/K_i)/(1 + [I]/K_i')$
Mixed (noncompetitive)	$K_m = K_m'$ $K_i = K_i'$ $k_2' = 0$	$V_{max}/(1 + [I]/K_i')$	K_m
Uncompetitive	$K_i = K_m' = \infty$ $k_2' = 0$	$V_{max}/(1 + [I]/K_i')$	$K_m/(1 + [I]/K_i')$
Partially noncompetitive	$K_i \neq K_i'$ $K_m \neq K_m'$ $k_2 \neq k_2'$	See Eq. (17)	

[a]Figure 4 lists the graphical patterns associated with these inhibition types.

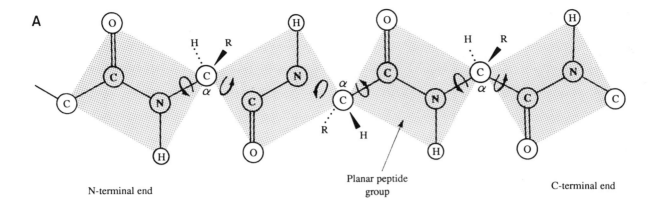

A

N-terminal end

Planar peptide group

C-terminal end

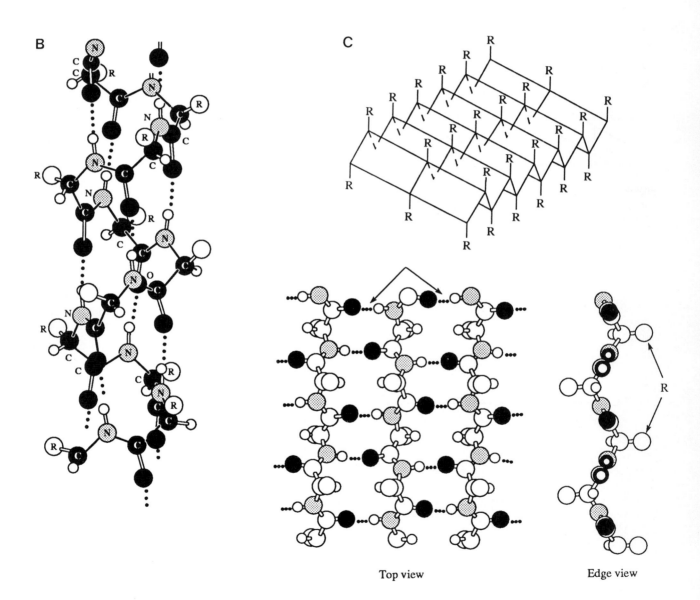

B

C

Top view

Edge view

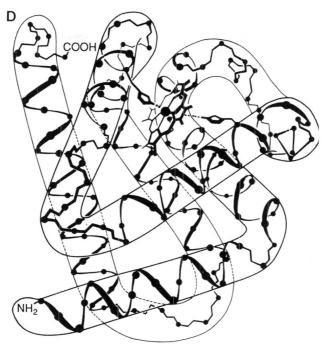

FIGURE 6 (*Continued*)

no useful information about how an enzyme specifically catalyzes any given reaction. In this section we examine the structure of an enzyme in some detail with the aim of observing how enzymes perform their function.

A. Enzyme Structure

The old maxim "All enzymes are proteins but not all proteins are enzymes" was long regarded as a universal truth, but recently, with the discovery of autocatalytic messenger RNA molecules (capable of splicing themselves), this old "truth" is no longer valid. Enzymologists have classically divided enzyme structure into four separate categories: primary, secondary, ter-

tiary, and quaternary. The ordering of these four categories reflects the order of influence of one type of structure on the next. For example, the primary structure influences the degree of secondary structure present, but not the reverse.

Protein-based enzymes are complex molecules composed of a chain of amino acids linked together through peptide bonds (Fig. 6A). Each enzyme (often called a polypeptide) can be identified by its unique sequence of amino acids and, for this reason, the amino acid sequence is called an enzyme's *primary structure*. The primary structure is determined by the nucleotide bases on the DNA molecule that code for the protein. The individual properties of the amino acids, and the relative amount of each amino acid in

FIGURE 6 Structural elements of enzyme conformation. (A) Protein primary structure. Amino acid components are commonly numbered starting with the N-terminal end. The peptide bonds (between two amino acids) are planar as indicated in the figure. [Reproduced, with permission, from A. L. Lehninger (1975). "Biochemistry," 2nd Ed. Worth, New York.] (B) Ball-and-stick model of α-helix. The α-helix shape is similar to a spring with the peptide bonds and α-carbon atoms forming the backbone of the structure. R groups stick out away from the α-helix. [Reproduced, with permission, from G. H. Haggis, D. Michie, A. R. Muir, K. B. Roberts, and P. H. B. Walker (1964). "Introduction to Molecular Biology." John Wiley & Sons, New York.] (C) Top: Schematic representation of three parallel polypeptides in a β-sheet structure. [Reproduced, with permission, from T. P. Bennett (1968). "Graphic Biochemistry." Macmillan, New York.] Lower Left: Ball-and-stick model of β-pleated sheet showing the interchain hydrogen bonding between the carboxyl oxygens (solid circles) and the amide hydrogen (open circles). Lower right: Ball-and-stick model of β-sheet showing R groups projecting out from the plane of the hydrogen bonds. [Redrawn, with permission, from H. D. Springall (1954). "The Structural Chemistry of Proteins." Academic Press, New York.] (D) Schematic diagram of the myoglobin backbone showing the high percentage of α-helical structure and heme prosthetic group. [Reproduced, with permission, from H. Neurath (ed.) (1964). "The Proteins." Academic Press, New York.]

the enzyme, play an important role in determining the secondary, tertiary, and quaternary structures of an enzyme. Amino acids can be divided into four different classes depending on their functional groups (called R groups): hydrophobic, polar, negatively charged, and positively charged (Table III). The properties of these R groups determine whether they will be highly solvated (polar, negatively charged, and positively charged R groups tend to be on the outside of a protein, in direct contact with water) or will be hidden in the interior of the protein away from water (hydrophobic).

The secondary structure refers to highly ordered substructures found within an enzyme. Two types of secondary structure have been identified: α-helices (Fig. 6B) and β-pleated sheets (Fig. 6C). The formation of these structures depends directly on the enzyme's amino acid sequence; relatively small, uncharged polar R groups organize themselves into α-helices and relatively small nonpolar R groups form β-pleated sheets. Both α-helices and β-sheets may

form a significant proportion of an enzyme's three-dimensional structure (see the picture of myoglobin, Fig. 6D). Proline is a special amino acid because of its unique structure (Table III). Introduction of proline into an amino acid sequence introduces a permanent bend at this position. For this reason, the presence of proline in an α-helix or β-sheet will disrupt the secondary structure at this point.

Tertiary enzyme structure refers to the final structure obtained when an amino acid chain is placed in water. Tertiary structure is largely determined by the interaction of R groups on the surface of the protein with the water and with other R groups on the protein surface. Polar R groups will move to the outside of the protein and nonpolar R groups will move toward the inside of the protein. The movement of certain parts of the polypeptide chain away from water will give a protein a characteristic three-dimensional shape. The final structure represents a minimum energy point and is a compromise between all the forces acting on it. Note that the reference point, in this discussion, is a water solution. It is possible to alter the tertiary structure of an enzyme by radically changing the solvating medium. For example, changing the solvating medium to an organic solvent or altering the pH, ionic strength, or temperature can radically change protein structure, often to the point of irreversible denaturation. These changes cause proteins to lose structure either by solvating nonpolar R groups (dissolving the protein in an organic solvent) or by causing R groups on the surface of the protein to become nonsoluble so that proteins stick together (pH changes).

The fourth level of enzyme structural organization, quaternary structure, exists only in multisubunit enzymes. In fact, most enzymes are not single polypeptide chains but are either dimers, tetramers, or polymers of several identical subunits (each polypeptide chain is termed a subunit in a multiple subunit enzyme). In the case of multiple subunit enzymes, the R groups on each polypeptide chain may interact with each other, as well as with the solvating medium, to give a final enzyme structure that is different from that for the isolated subunit.

B. How Enzyme-Catalyzed Reactions Occur: Lysozyme as a Model

When an enzyme has folded into its final structure, a specialized area or cleft on the protein surface exists termed the *active site*. It is here that substrate binds, and it is here that the R groups responsible for binding

TABLE III
Types of Amio Acids

Class	Name	R group[a]
Nonpolar	Alanine	$-CH_3$
	Valine	$-CH(CH_3)_2$
	Leucine	$-CH_2-CH(CH_3)_2$
	Isoleucine	$-CH(CH_3)-CH_2-CH_3$
	Proline	$^-OOC \diagdown \diagdown NH$
	Phenylalanine	$-CH_2-\phi$
	Tryptophan	$-CH_2$-indole
	Methionine	$-CH_2-CH_2-S-CH_3$
Polar, uncharged	Glycine	$-H$
	Serine	$-CH_2-OH$
	Threonine	$-CH(OH)-CH_3$
	Cysteine	$-CH_2-SH$
	Tyrosine	$-CH_2-\phi-OH$
	Asparagine	$-CH_2-CO-NH_2$
	Glutamine	$-CH_2-CH_2-CO-NH_2$
Polar, negative	Aspartic acid	$-CH_2-COOH$
	Glutamic acid	$-CH_2-CH_2-COOH$
Polar, positive	Lysine	$-CH_2-CH_2-CH_2-CH_2-NH_3^+$
	Arginine	$-CH_2-CH_2-CH_2-NH-C(NH_3^+)-NH_2$
	Histidine	$-CH_2$-imidazole$^+$

[a] ϕ represents benzene. The basic amino acid structure is R-CH (NH_3^+)-COO$^-$. As indicated, amino acids are usually zwitterionic at pH 7. Charges on amino acid R are measured at pH 7.1. The complete structure of proline is shown.

substrate and for catalyzing the reaction are located. The active site of an enzyme is a highly conserved region of the protein and is specifically configured to bind only one substrate (or one class of substrates). How well a particular substrate binds the active site depends on (1) which amino acids are present, (2) the positioning of the amino acids within the active site, and (3) other environmental factors such as the ionic strength, pH, and ion composition of the solvating medium. Although it has been convenient to think of the binding of substrate to active site as a "lock and key" fit, this is by no means a complete description of the interaction. Enzymes are not rigid structures but are flexible so that the binding of substrate induces a change in enzyme conformation (remember that this is the reason for cooperative enzyme kinetic mechanisms). This dynamic binding mechanism is called *induced fit* and can be illustrated by measurements of enzyme structure in the presence and absence of substrate. X-ray crystallography shows that enzyme structure is different when substrate is bound to the active site, a fact that is consistent with the proposed mechanism of enzyme action. The induced fit model does not imply that the enzyme and substrate in the *ES* complex form a perfectly complementary structure. Enzymes could not operate efficiently if substrate binding was optimized since a minimum in the energy graph, caused by a stable *ES* complex, would decrease the rate of catalysis; a large, negative $\Delta G^{\circ\prime}$ for *ES* formation would make the $\Delta G^{\circ\prime}$ for the catalysis step more positive. It is the combination of active site geometry and enzyme conformational change that gives the enzyme–substrate complex its specificity. Even small changes in substrate chemical composition will dramatically alter its affinity.

The binding interaction between enzyme (or more specifically amino acid R groups at the enzyme active site) and substrate can also contribute to the catalytic efficiency of an enzyme. A good example of this effect is seen in the action of lysozyme, an enzyme found in mammalian tears and mucus that lyses the sugars of bacterial cell walls. These walls are composed of alternating residues of *N*-acetylglycosamine (NAG) and *N*-acetylmuramic acid (NAM). Lysozyme binds six sugar residues at a time, three NAG and three NAM units. Catalytic action cleaves a single NAM–NAG linkage, NAM occupying the fourth spot in the active site. Cleavage is aided by physical strain on the NAM residue, forcing it to adopt an energetically unfavorable half-chair conformation (Fig. 7). The strain comes from the interaction between substrate and enzyme after binding has induced a conforma-

tional change. This destabilization means that the *ES* complex is at a higher energy than the free *S*, effectively lowering the activation energy and leading to a rate enhancement.

Once substrate is bound to the active site it is oriented so that amino acid R groups on the protein can directly participate in the catalytic mechanism. This is best illustrated in Fig. 7, which shows how aspartate 52 and glutamic acid 35 of lysozyme act to catalyze the NAG–NAM cleavage. Both groups are important to catalysis; glutamic acid provides the proton for acid catalysis and aspartate stabilizes the resulting glycosyl-oxocarbonium ion. Lysozyme action thus shows that enzymes may increase the rate of catalysis by stabilizing the transition state (aspartate stabilizes the positively charged transition state ion). This mode of enzyme action is a common catalytic mechanism of other enzymes as well. For example, tight complexes between enzymes and synthetically manufactured transition state analogs show that enzyme structures are optimized to bind and stabilize transition states (and not substrates). Lysozyme action also demonstrates that enzymes catalyze reactions using simple chemical species (glutamic acid and aspartic acid in this case). Thus, the active site is not an isolated environment where specialized chemistry occurs. Enzymes are, however, more efficient catalysts than simple chemical agents because (1) they orient the substrate so that optimal efficiency between catalyst (e.g., protons for lysozyme) and susceptible bond is achieved, and (2) the bind substrates specifically so that other molecules are not acted upon. This is in contrast to chemical catalysts, which are frequently not substrate-specific nor stereo-specific and, consequently, produce a significant proportion of undesirable products. Enzyme-catalyzed reactions are highly specific and so enzymes produce essentially 100% of a single product.

III. COENZYMES AND THEIR FUNCTION

Lysozyme is an example of an enzyme-catalyzed group transfer in which a glycosyl residue is transferred to water. Enzyme reactions, however, are not limited to simple acid-catalyzed water hydrolysis reactions. Several different types of enzyme reactions exist that can be loosely divided into four different classes. A short list of these is presented in Table IV along with examples of reactions and enzymes. Many of the reactions shown in Table IV can be catalyzed using

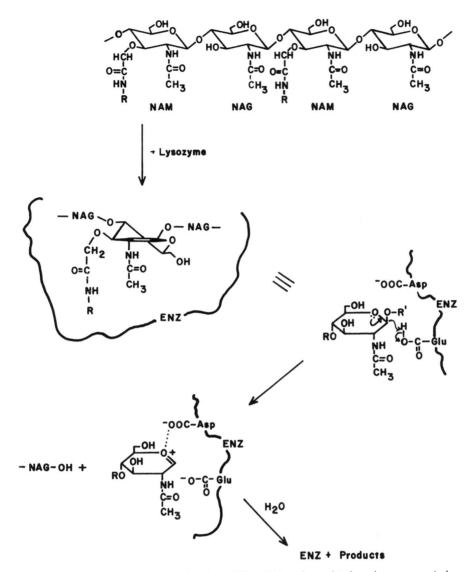

FIGURE 7   Lysozyme catalytic mechanism. NAM–NAG polymer binds to lysozyme to induce the half-chair conformation at position 4. Aspartate (Asp) and glutamate (Glu) participate directly in the reaction sequence. Glutamate donates a proton, and aspartate stabilizes the glycosyl-oxocarbonium ion (shown by dots). [Reproduced, with permission, from C. Walsh (1979). "Enzymatic Reaction Mechanisms." Freeman, San Francisco.]

simple acid–base catalysis and so can be carried out with positively or negatively charged amino acid R groups. The hydrolysis of NAM–NAG sugar residues is an example of such a reaction that is catalyzed by aspartic acid 35 of lysozyme. However, several of the reactions in Table IV cannot be carried out with simple acid–base catalysts and require additional reactive groups not available on amino acid R groups. When this is true, coenzymes act as additional carriers that can accept and donate chemical groups, hydrogen atoms, or electrons (Table V). Several of the reactions listed in Table IV indicate the requirement for coenzymes, demonstrating their importance in catalyzing cellular reactions. Coenzymes can be broadly classified into three groups: (1) compounds that transfer high-energy phosphate, (2) compounds that accept and donate electrons, and (3) compounds that activate substrates. [See Coenzymes, Biochemistry.]

TABLE IV
Classification of Enzyme Reactions[a]

Class	Types of reactions
Group transfers	Acyl transfer to water (proteases)
	γ-Glutamyl and amino transfer (glutamine synthetase)
	Phosphoryl transfer I (phosphatases, ATPases, phosphodiesterases)
	Phosphoryl transfer II (kinases)
	Nucleotidyl and pyrophosphoryl transfer
	Glycosyl transfer (glycogen synthetase, lysozyme)
Oxidations/ reductions	NAD(P)-requiring enzymes (alcohol dehydrogenase, lactate dehydrogenase, glucose 6-phosphate dehydrogenase)
	FAD-requiring enzymes (succinate dehydrogenase)
	Monooxygenases I (FAD, FMN, and pterin-dependent reactions)
	Monooxygenases II (copper-requiring enzymes)
	Metallo-flavoprotein oxidases
	Heme-containing oxidases (catalase, cytochrome c oxidase)
	Dioxygenases
Eliminations/ isomerizations/ rearrangements	Addition/elimination of water (aconitase, fumarase, enolase)
	Addition/elimination of ammonia (phenylalanine ammonia lyase)
	Isomerizations (phosphoglucose isomerase)
	Rearrangements (B_{12}-dependent enzymes)
C–C bond breaking	Decarboxylations (β-keto and β-hydroxy acids, α-amino acids)
	Carboxylations (biotin-dependent enzymes)
	Aldol reactions (aldolase)
	Pyridoxyl phosphate-requiring enzymes
	Tetrahydrofolate and S-adenosylmethionine-requiring enzymes

[a] Reactions are listed by common enzyme mechanism and are grouped according to Walsh. [Reproduced, with permission, from C. Walsh (1979). "Enzymatic Reaction Mechanisms." Freeman, San Francisco.]

A. Coenzymes That Transfer High-Energy Phosphate

Compounds that transfer high-energy phosphate are highly phosphorylated derivatives of the nucleotide bases (e.g., adenosine and guanosine), which exist in either mono-, di-, or triphosphate forms. Usually, the diphosphate acts as a phosphate acceptor and the triphosphate acts as a phosphate donor. The monophosphate is made during biosynthetic reactions re-quiring a pyrophosphate intermediate: ATP → AMP + PP_i. All of these compounds are extremely important to metabolic and biosynthetic reactions since they act as the energy currency of the cell. Adenosine triphosphate (ATP) is by far the most common cellular high-energy phosphate compound, occurring at concentrations approximately 100-fold higher than other high-energy phosphate intermediates. These compounds are considered as "high-energy phosphates" because hydrolysis of γ-phosphorous of ATP yields $\Delta G^{\circ\prime} = -30.6$ kJ/mole of energy corresponding to a K_{eq} value of approximately 10^8 in favor of adenosine diphosphate (ADP) and P_i formation [see Eq. (4)]. This means that many reactions can be forced to completion by coupling them to ATP hydrolysis; a reaction with a positive $\Delta G^{\circ\prime}$ value can be made negative if the reaction is coupled to ATP hydrolysis (the overall $\Delta G^{\circ\prime}$ for a reaction is the sum of the individual $\Delta G^{\circ\prime}$ values). In accordance with its function, ATP does not remain enzyme-bound in between catalytic cycles but diffuses from enzyme to enzyme to participate in other high-energy phosphate transfer reactions. For this reason ATP is usually treated as an enzyme substrate because, like any other substrate, it must bind

TABLE V
Coenzymes[a]

Classification	Coenzyme name (vitamin)	Abbreviation
High-energy phosphate transfer	Adenosine triphosphate	ATP
	Guanosine triphosphate	GTP
Electron acceptors	Nicotinamide adenine dinucleotide (niacin)	NAD
	Nicotinamide adenine dinucleotide phosphate (niacin)	NADP
	Flavin adenine dinucleotide (B_2)	FAD
	Riboflavin 5′-phosphate (B_2)	FMN
	Lipoamide	—
	Coenzyme Q (ubiquinone)	CoQ
Substrate activators	Coenzyme A (pantothenic acid)	CoA
	Acyl carrier protein (phosphopantetheine)	ACP
	Biotin	—
	Thiamine pyrophosphate (B_1)	TPP
	Pyridoxyl phosphate (B_6)	PLP
	Coenzyme B_{12}	B_{12}
	Tetrahydrofolate (folic acid)	THF
	S-Adenosyl methionine	SAM

[a] The biochemical names of the coenzymes are presented along with the vitamin they are derived from. Abbreviations are common and may not be universal.

enzyme to form an *ES* complex. In this way the ATP generated by glycolysis can be used in any number of energy-requiring processes such as biosynthetic pathways, membrane ion pumps, and protein synthesis. Because of its importance in cellular biochemistry, enzymes that use ATP are highly regulated since they control the cellular concentration of ATP through their action. These enzymes are usually found at the start of metabolic pathways and are often inhibited or activated by ATP, ADP, and AMP, as well as by the end products of the metabolic pathway. [*See* Adenosine Triphosphate (ATP).]

B. Coenzymes That Accept and Donate Electrons

Coenzymes that both accept and donate electrons participate in oxidation and reduction reactions, serving as two-electron acceptors or two-electron donors depending on the reaction. In accordance with this function, several of these coenzymes have delocalized π-ring systems that stabilize positive and negative charges; extended π systems help stabilize positive charges present in the oxidized form of the coenzyme. During the normal oxidation–reduction cycle, these coenzymes can either reversibly bind enzyme active sites [NAD, NADP, and coenzyme Q (CoQ) diffuse from enzyme to enzyme] or remain closely associated with one enzyme (e.g., FAD, FMN). In the former case, NAD, NADP, and CoQ act as substrates to transfer either reducing or oxidizing equivalents from reaction to reaction. NAD and NADP participate in water-based reactions and CoQ participates in lipid (membrane) reactions. In the case of FMN and FAD coenzymes, the reduced riboflavin coenzyme transfers electrons to a soluble coenzyme such as NAD(P) or CoQ to complete its oxidation–reduction cycle. NAD and NADP can be considered as high-energy electron carriers, much like ATP is a high-energy phosphate carrier; these coenzymes are similar to ATP in their mode of action. Nicotinamide coenzymes thus transfer electrons from substrates to either (1) reduce compounds during biosynthetic reactions (e.g., during fatty acid synthesis) or (2) provide the electrochemical energy necessary for ATP synthesis in the mitochondrion. Like ATP-utilizing enzymes, nicotinamide coenzyme-utilizing enzymes are often highly regulated. However, unlike ATP-utilizing enzymes, they are not regulated by NAD or NADP concentrations but are sensitive to substrate and product concentrations. [*See* ATP Synthesis in Mitochondria.]

C. Coenzymes That Activate Substrates

Although the class of coenzymes that activate substrates includes several different types of reactions, all of these coenzymes react with substrate at the enzyme active site to produce a new, more reactive product. For example, coenzyme A (CoA) and acyl carrier protein (ACP) attach to acetyl residues (CH_3-CO-) to form acetyl-CoA and acetyl-ACP compounds. Both CoA and ACP are good chemical leaving groups so that the reactivity of these acetyl derivatives is much higher than that of simple acetic acid ($\Delta G°'$ for hydrolysis to acetic acid and CoA or ACP is large and negative). Formation of these compounds effectively creates a high energy acetate derivative.

Biotin, a highly reactive double-ring system, is covalently attached to an enzyme by a long, flexible arm. Biotin serves as a carboxyl group carrier, adding -COOH groups to compounds. The reaction is energetically unfavorable and is coupled to the hydrolysis of ATP to drive the reaction to completion.

Thiamine pyrophosphate and pyridoxal phosphate catalyze several different types of reactions, including α-condensation and α-cleavage reactions (thiamine pyrophosphate), as well as transamination, decarboxylation, elimination, and aldol cleavage (pyridoxal phosphate). Both coenzymes are the reactive species in the enzyme active site although neither is covalently attached to the enzyme. The thiamine pyrophosphate mechanism proceeds via a thiazolium dipolar ion that can attack ketone or imide compounds. The newly formed adduct reacts readily to give products (carbon dioxide is often a product). Pyridoxal phosphate also forms an adduct with substrate (called a Schiff's base) that decomposes to give product. These coenzymes increase the rate of catalysis by stabilizing the transition state to lower the overall activation energy. In both cases, the substrate–coenzyme adduct is stabilized by the conjugated ring systems of the coenzyme that act as either an electron sink or an electron source during the reaction.

Coenzyme B_{12} is another example of a compound that catalyzes a reaction by stabilizing the transition state. B_{12} is an unusual coenzyme because it contains a copper atom in the center of a corrin ring system coordinated to adenosine below and ribose above the plane of the corrin ring. The copper is the reactive center of the coenzyme and can easily accept and donate a single electron; coenzyme B_{12} catalyzes reactions using radical ion chemistry. It is this property that enables B_{12} to catalyze a wide range of reactions, including group migrations, dehydrations, and oxidation–reduction reactions (often in conjunction with the oxidation–reduction coenzymes listed earlier).

Both *S*-adenosyl methionine (SAM) and tetrahydrofolate (THF) serve as methyl group carriers in much the same way that CoA and ACP serve as acetyl carriers. Since the chemistry of carbon–carbon bond breaking is complex, reactions involving THF and SAM often involve other coenzymes required for reduction or oxidation, as well as B_{12}. SAM and THF reactions are not common, but these coenzymes participate in two extremely important reactions: methylation of DNA to turn off gene transcription and the methylation of homocysteine to form methionine, 1 of the 21 amino acids that make up proteins (see Table III).

IV. CELLULAR CONTROL OF ENZYME ACTIVITY

The preceding sections dealt with control mechanisms that operated on enzyme reactions isolated within the test tube. However, researchers study individual enzyme control mechanisms as an indirect method for learning how metabolic pathways are regulated. Metabolic pathways contain several steps, each catalyzed by a single enzyme; metabolic pathways are really chains of enzymes that catalyze successive reactions:

$$S \underset{}{\overset{E_1}{\rightleftharpoons}} I_1 \underset{}{\overset{E_2}{\rightleftharpoons}} I_2 \underset{}{\overset{E_3}{\rightleftharpoons}} I_3 \underset{}{\overset{E_4}{\rightleftharpoons}} P. \qquad (18)$$

In Eq. (18), each enzyme in the sequence (E_1, E_2, E_3, and E_4) catalyzes a reversible reaction with the product of the each enzyme reaction (the pathway intermediates, I_j) serving as the substrate for the next enzyme. Equation (18) is represented as a linear sequence of reactions with a single substrate being converted exclusively into product. In reality, metabolic pathways often contain branch points so that an intermediate (such as I_2) could give rise to several different intermediates that will eventually be transformed into other products. The regulation of flux down the different metabolic pathways that arise from a single branch point is complex but consists of all the elements described earlier as well as those of the present section. For the purposes of simplicity, however, we confine ourselves to a linear sequence of enzymes. How might the rate of product formation be controlled in Eq. (18)?

A. Control of Metabolic Pathways *in Vivo*

In the classic view of pathway regulation, the important enzymes of the pathway are seen as those regulating the entry of S into the pathway (E_1), those producing the final product (E_4), and those that have exceptionally low activity. To determine how pathway flux was regulated, therefore, E_1 would be tested for sensitivity to inhibition by its product (I_1), as well as by any intermediate along the metabolic pathway. E_4 would be tested for activation or inhibition by any pathway intermediate as well as for inhibition by the final product. If one of the intermediary enzymes had a particularly low activity, it might also be tested for sensitivity to inhibitors or activators. Inhibition by a pathway product could provide a means of regulating the entry of substrate into the pathway to control the amount of product formed (feedback inhibition). Activation by a pathway intermediate could provide a means of clearing the pathway of high substrate or intermediate concentrations (feed-forward activation).

The enzymes could also be investigated for possible changes in covalent modification. Enzymes are commonly regulated by reversible covalent modification—the most common form being phosphorylation of a serine or histidine residue. These modifications can alter the protein secondary, tertiary, or quaternary structure to influence the binding of substrates or allosteric modifiers. Phosphorylation may also directly affect the enzyme active site. This enzyme control mechanism has several advantages: (1) it is readily reversible, (2) it can be quickly reversed, (3) it is a stable change (does not depend on cellular metabolite concentrations), and (4) it serves to provide a mechanism for linking enzyme activity to external (hormonal) signals. For these reasons, enzyme phosphorylation is one of the most common forms of regulating enzyme activity.

A final test would involve using the substrates, intermediates, and products of competing metabolic pathways as inhibitors and activators to look for possible effects on the regulatory enzymes. As you can see, an enormous number of experiments must be completed before a pathway can be adequately defined. Despite all this information, the experimenter still had no exact method for determining how changes in enzyme rates observed *in vitro* (where all these experiments have been performed) translate into control of cellular metabolic pathways *in vivo* (the area of interest).

Recently, theoretical mathematical frameworks

have been developed to help answer these questions. Three models have been developed, each with their own merits. It is beyond the scope of the present article to give a detailed treatment of these theories. However, they all share the same fundamental concept: that control of a metabolic pathway is shared among all the enzymes of the pathway. This can be demonstrated by measuring the change in the flux through the pathway (J) as a function of the change in the activity of the individual enzymes (v_j). Metabolic Control theory assigns the term *flux control coefficient* ($C_{E_j}^J$) to this ratio:

$$C_{E_j}^J = \left(\frac{\delta J}{J} \times \frac{v_j}{\delta v_j} \right)_{ss} = \left(\frac{\delta \ln J}{\delta \ln v_j} \right)_{ss}. \qquad (19)$$

Application of theories such as the Metabolic Control theory to metabolic pathways has shown that, in general, single enzymes do not control the flux through a metabolic pathway. The interaction between enzyme control and pathway flux is complicated. For example, reducing the rate of one enzyme will increase its flux control coefficient but will reduce control by other enzymes since the sum of the flux control coefficients for any single pathway is unity:

$$\sum_{j=1}^{n} C_{E_j}^J = 1.0. \qquad (20)$$

The magnitude of the changes in the flux control of the other enzymes will depend on the flux through the system and on the activities of the enzymes. Although at present it is technically difficult to design experiments that gauge the flux control of enzymes within pathways *in vivo*, Eqs. (19) and (20) provide the researcher with the basis for examining cellular control of metabolic pathways. As such they point the way toward the future of enzyme kinetics.

B. Formation of Multienzyme Complexes

Enzymologists used to think of the cell as a collection of soluble enzymes randomly distributed throughout the cytoplasm. As cellular organelles and structures were discovered (the mitochondrion, plasma membrane, microsomal membrane, nucleus, lysosome, peroxysome, Golgi apparatus), it became clear that some enzymes were localized to specific areas of the cell. For example, many enzymes of the citric acid cycle are found exclusively within the mitochondrion.

Later investigations revealed that, in isolated cases, enzymes could be bound together to form *multienzyme complexes*. This was discovered by the copurification of several different enzyme activities in a single complex. The enzymes in the complex complement one another so that they often represent sequential enzymes of a metabolic sequence as depicted in Eq. (18). The enzymes of fatty acid synthase, the pyruvate dehydrogenase complex, and cytochrome c oxidase are good examples of multienzyme complexes.

Although a few enzymes could be easily identified as belonging to a complex, the majority of enzymes were still thought of as freely soluble within the cytosol. More recently, enzymologists have begun to probe the physical organization of soluble enzymes within the cytosol. Many cellular structures exist that readily bind enzymes when tested *in vitro* (e.g., actin and myosin filaments, mitochondrial membranes, plasma membranes). One can also demonstrate that direct enzyme–enzyme binding occurs *in vitro*. The discovery that enzymes could bind to subcellular structures and to each other led some researchers to speculate on the functionality of these interactions. In particular, they proposed that the enzymes of a single metabolic pathway could be bound together to form a multienzyme complex. The potential benefits of a multienzyme complex would include: (1) enzymes in the complex would work at faster rates because their substrates would be produced in close proximity to the active site of the next enzyme in the sequence and their products would be quickly removed, (2) enzymes in the complex would respond more quickly to inhibitors and activators (metabolic intermediates would be in close proximity to inhibitor and activator sites), and (3) formation and destruction of the complex could serve as a mechanism to control overall pathway flux. Unfortunately, the experimental evidence gathered to test the function and existence of multienzyme complexes is not conclusive because of the problems encountered in reproducing the cellular conditions *in vitro*. For example, *in vitro* measurement of enzyme binding is carried out with purified components under conditions of low protein concentration and low ionic strength because it is extremely difficult to maintain cellular conditions when performing binding experiments. Both conditions are opposite to what is found in the cell and serve, largely, to promote interactions where none may occur. Detailed kinetic analysis and mathematical modeling have disproved the majority of these complexes. Nevertheless, they do appear to exist in isolated cases.

The most convincing evidence for a multienzyme

complex comes from experiments on the glycolytic enzymes (glucose-utilizing pathway) of smooth muscle. In smooth muscle cells, the glycolytic pathway appears to be divided between two cellular locations: one that is used to fuel muscle contraction and one that maintains plasma membrane enzyme activities. This was demonstrated using isolated cells and radioactively labeled glucose to follow product formation. Since both pathways contain glycolytic enzymes, they both use glucose as substrate. However, each glycolytic "complex" obtains its glucose from a different location (different pool). With isolated cells it was possible to label only one of these glucose pools and show that very little of the labeled glucose was utilized by the other "complex." To date, this is the only firmly established example of a multienzyme complex of glycolytic enzymes. The failure of most experiments to prove the existence of kinetically functional multienzyme complexes does not mean that cytosolic enzymes are freely soluble within the cytoplasm. However, it does mean that multienzyme complexes are not a universal method for controlling enzyme velocities and pathway flux.

BIBLIOGRAPHY

Cornish-Bowden, A. (1979). "Fundamentals of Enzyme Kinetics." Butterworths, Toronto.

Lehninger, A. H. (1975). "Biochemistry." Worth, New York.

Metzler, D. E. (1977). "Biochemistry. The Chemical Reactions of Living Cells." Academic Press, New York.

Segal, G. (1975). "Enzyme Kinetics." John Wiley & Sons, New York.

Sucking, K. E., and Sucking, C. J. (1980). "Biological Chemistry. The Molecular Approach to Biological Systems." Cambridge Univ. Press, London.

Suelter, C. H. (1985). "A Practical Guide to Enzymology, Vol. 3, Biochemistry." John Wiley & Sons, New York.

Walsh, C. (1979). "Enzymatic Reaction Mechanisms." Freeman, San Francisco.

Voet, D., and Voet, J. G. (1995). "Biochemistry," 2nd Ed. Wiley & Sons, New York.

Enzyme Thermostability

TIM J. AHERN

Genetics Institute, Inc.

I. Enzyme Conformation, Activity, and Stability
II. Thermal Inactivation
III. Reversible Thermoinactivation
IV. Irreversible Thermoinactivation
V. Enzyme Thermoinactivation *in Vivo*
VI. Enzyme Thermostabilization

GLOSSARY

Deamidation Hydrolysis of the amide ($CONH_2$) side chain of the amino acid asparagine and glutamine, resulting in the liberation of free ammonia

Enzyme conformation Three-dimensional structure of the polypeptide chain, amino acid side chains, and nonprotein adducts and cofactors of an enzyme

Enzyme thermoinactivation Loss of the biological activity of an enzyme that is accelerated by increasing temperature

THE STABILITY OF AN ENZYME AT HIGH TEMPERA-tures depends on many cooperative, intramolecular interactions maintaining its native structure. Enzymes can inactivate because of reversible processes that cause the protein to unfold or by irreversible processes that include aggregation, destruction of disulfide bonds, hydrolysis of peptide bonds, and deamidation of asparaginyl residues. These mechanisms occur both *in vitro* and *in vivo*. Enzymes exhibiting enhanced thermostability have been engineered as a result of our understanding how these processes inactivate enzymes.

I. ENZYME CONFORMATION, ACTIVITY, AND STABILITY

Within the structure of enzymes, the requirements for a certain degree of rigidity and integrity of composition are balanced against the need for flexibility and susceptibility to degradation. If enzymes had no rigidity, they would not be restricted to conformations suitable for binding specific molecules and catalyzing chemical reactions. Similarly, if enzymes were subject to rapid and random destruction of the covalent bonds comprising their primary structure, they quickly would be unable to maintain functional conformations. At the other extreme, if enzymes were *too* rigid, they also could not function because interaction with other molecules requires that portions of enzymes be able to make limited, concerted motions. Finally, if enzymes were too resistant to degradation, their persistence and accumulation would pose critical problems to living systems, which depend on the recycling of molecules for new synthesis to meet evolving needs. For these reasons, enzymes are held together by forces that are just strong enough to maintain them and their unique functions within the biological milieu for a limited time, after which they undergo inactivation and degradation. It is the inherent weakness of the chemical bonds and noncovalent interactions maintaining the structure of enzymes that determines the degree of enzyme thermostability.

The thermostability of enzymes relies on (1) the overall strength of noncovalent interactions maintaining the native conformation(s), (2) the integrity of the amino acid residues of which polypeptide chains are composed, (3) the amide bonds linking them, (4) the disulfide bonds (if present) that form

ENCYCLOPEDIA OF HUMAN BIOLOGY, Second Edition, VOLUME 3.

cross-links between cysteinyl residues, and (5) miscellaneous elements found in various proteins.

The noncovalent forces that maintain the higher levels of structural organization in proteins—hydrogen, ionic, and van der Waals (hydrophobic) interactions—are relatively weak. Furthermore, the stability of the native conformation apparently relies on the cooperative presence of all such interactions, because once unfolding begins, the process usually goes to completion.

With only a few exceptions, the common 20 amino acids comprising enzymes are so stable that they survive incubation in dilute hydrochloric acid at temperatures in excess of 100°C, conditions routinely used to hydrolyze proteins to form free amino acids in the course of composition analysis. However, the amide side chains of asparagine and glutamine, being similar to peptide bonds, also undergo hydrolysis not only under these but much less extreme conditions. The contribution of deamidation to enzyme thermoinactivation will be discussed below.

The peptide bond that unites amino acid residues of proteins in linear chains suits the requirements for limited stability well because it can be synthesized and broken at a relatively low expenditure of energy, therefore permitting rapid turnover of enzymes. This bond also provides considerable freedom of motion of the polymerized amino acid residues, thus allowing a degree of flexibility. Disulfide bonds are also susceptible to irreversible chemical degradation. Such observations further underscore the tenuous nature of protein stability.

II. THERMAL INACTIVATION

The behavior of enzymes at elevated temperatures illustrates the events leading to spontaneous enzyme destabilization, inactivation, and degradation. When an aqueous solution of an enzyme is heated, the following molecular events begin to take place. First, the enzyme molecules partly unfold because of a heat-induced disruption of noncovalent interactions maintaining the catalytically active conformation at room temperature. This process, which almost invariably leads to enzyme inactivation, is reversible for many enzymes; the native conformation and the enzymatic activity are completely recovered when the enzyme solution is promptly cooled. However, if heating persists, only a decreasing fraction of the enzymatic activity returns on cooling, signifying that other, *irreversible* processes take place.

III. REVERSIBLE THERMOINACTIVATION

Reversible thermal unfolding (denaturation) of enzymes has been a subject of intensive investigation for several decades. The phenomenon is amenable to straightforward and exact thermodynamic analysis and, as a consequence, is conceptually well understood. If the activity of an enzyme lost because of exposure to elevated temperature can be regained by a return to a lower temperature, then as the name implies, the inactivation is "reversible." Perhaps the most marked reversible effect of elevated temperature on an enzyme is the increase in motion of its constituent parts to such a degree that what is known as the ordered, "native" conformation can be said to be lost, replaced by largely disordered, "denatured" conformations. This can be represented as

$$N \rightleftharpoons U, \tag{1}$$

in which N is the native and U is the unfolded forms of an enzyme, respectively.

The native state is not a single, unique conformation because no enzyme is entirely rigid; even at subzero temperatures and in the crystalline state, the atoms comprising a protein undergo vibrations, rotations, and small translations of about 0.2–0.5 Å. At intermediate temperatures, the range at which most enzymes exhibit their optimal activity (0–60°C), reversible translations of whole segments of protein structure, called "breathing," are observed. These and other concerted motions required for substrate binding, catalysis, and product release define the displacements referred to collectively as native conformation. Nevertheless, despite the existence of some freedom of movement, the predominant conformations are restricted by a complex balance of intramolecular, noncovalent interactions.

At elevated temperatures (in most instances, above approximately 50°C), extensive cooperative intramolecular motions may take place, which effectively denature an enzyme: At a given temperature that varies for each enzyme under specific environmental conditions of pH, ionic strength, etc., the native structure can no longer prevail against the drive toward increased entropy of the unfolded state expressed as rapid, random motion, and the protein loses most of its ordered secondary and tertiary structure as it undergoes what is known as denaturation. The unfolding of an enzyme in solution changes many easily measured physical characteristics such as viscosity,

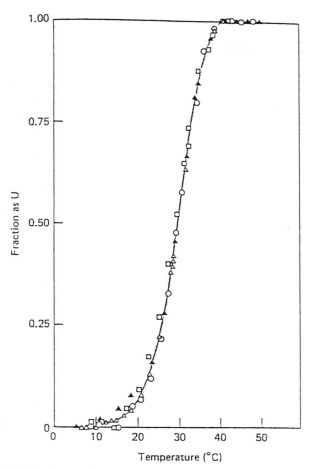

FIGURE I Equilibrium measurements of temperature-induced unfolding of bovine ribonuclease A in HCl–KCl at pH 2.1 and 0.019 ionic strength, measured by an increase in viscosity (□), a decrease in optical rotation at 365 nm (○), and a decrease of UV absorbance at 287 nm (△). Measurement of a second cooling from 41°C for 16 hr (▲). [From Ginsberg, A., and Carroll, W. R. (1965). Biochemistry **4**, 2159–2174, with permission. Copyright by the American Chemical Society.]

optical rotation, and ultraviolet absorbance (Fig. 1), which makes possible determination of the extent of unfolding.

In the course of denaturation, the amino acid residues comprising the active site of an enzyme are inevitably dispersed and, as a consequence, catalytic activity is lost. Loss of activity by unfolding can be regarded as the first step in nearly all enzyme thermoinactivation processes.

This mechanism has been designated as a reversible process in Eq. (1); an unfolded enzyme, once cooled below its characteristic transition temperature, should, according to the "thermodynamic hypothesis," refold to form the active, native state. This model

appears highly plausible because the native conformation is favored over all others during the *in vivo* synthesis of the disordered, nascent chain. It follows, then, that provided we choose the appropriate conditions for renaturation, even a randomly coiled protein should successfully refold to form the native state once again. This is generally true, at least for single-chained, monomeric enzymes, provided they have not undergone posttranslational modifications (e.g., selective proteolysis and excision of portions of the polypeptide chain, as in the case of the protein insulin). For example, when an aqueous solution of lysozyme is heated to 100°C, well above the transition temperature of the enzyme, the catalytic activity is immediately lost. The reversibility of the thermoinactivation is illustrated by the recovery of 100% of the activity when the enzyme is promptly cooled to 25°C, well below the transition temperature. The reversible equilibrium between native and disordered conformations has been demonstrated in this way for other enzymes as well.

IV. IRREVERSIBLE THERMOINACTIVATION

All enzymes undergo conformational transitions at elevated temperatures, but prolonged incubation results in a loss of activity that is not readily reversible once the solution is cooled. Irreversible thermoinactivation of enzymes is the result of destruction of covalent bonds within the enzyme, as well as some conformational processes that are for all practical purposes "irreversible." Equation (2) represents the framework, distinguishing the various processes leading to thermal inactivation of enzymes.

Once unfolded, many enzymes become insoluble and form large particulates. Aggregation is a concentration-dependent process because of its polymolecular nature, and it can be explained by the amphiphilic nature of the surface of a disordered enzyme. Portions of the polypeptide chain that are normally buried tend to be much more hydrophobic than those exposed to solvent in the native structure. Once an enzyme is denatured, the exposed hydrophobic surfaces have a tendency to avoid interaction with the aqueous solvent because water forms ordered clathrate structures around hydrophobic residues. The imposition of order on the solvent results in a decrease in entropy of the system as a whole. Thus, provided enzyme concentrations are high enough, such hydrophobic surfaces may form intermolecular interfaces via aggre-

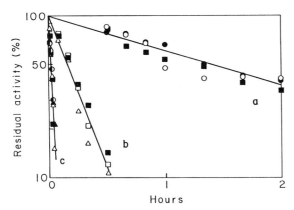

FIGURE 2 Time course of irreversible thermoinactivation of hen egg white lysozyme at 100°C as a function of pH and enzyme concentration. (a) pH 4 (0.1 M Na acetate); (b) pH 6 (0.01 M Na cacodylate); (c) pH 8 (0.1 M Na phosphate). The concentrations of lysozyme were 1000 (○), 100 (●), 50 (□), 10 (■), 5 (△), 1 (▲), and 0.5 μM (◐). After incubation for the time indicated, aliquots of enzyme solutions were cooled and assayed for residual catalytic activity. [From Ahern, T. J., and Klibanov, A. M. (1985). *Science* **228**, 1280–1284, with permission. Copyright by the AAAS.]

gation in an attempt to maximize the entropy of the solvent and thereby reduce the free energy of the system.

However, even the activity of dilute solutions of enzymes often cannot be recovered after prolonged heating followed by cooling. This irreversible thermal inactivation follows first-order kinetics, can be independent of the initial enzyme concentration (Fig. 2), and is not due to the formation of aggregates and can

therefore be attributed to monomolecular processes [see Eq. (2)].

It is important to determine whether monomolecular, apparently irreversible thermoinactivation is caused by covalent changes of the primary structure or by changes in higher orders of structure because the activity of conformationally altered enzymes that have undergone no irreversible deterioration has the potential to be regained. The existence of monomolecular, incorrect structure formation at high temperatures can be explained by the fact that there is more than one way to fold a protein: On denaturation, the tendency to bury hydrophobic groups, combined with the freedom of a protein to sample many conformational states, results in new, kinetically or thermodynamically stable structures that are catalytically inactive. Even after cooling, these incorrectly folded, "scrambled" structures may remain because a high kinetic barrier prevents spontaneous refolding to the native conformation. (Disulfide exchange, resulting in the mismatching of cysteinyl residues, can play a role in the formation of these scrambled structures. This mechanism is discussed separately below.)

Such processes can be distinguished from covalent mechanisms resulting in destruction of the polypeptide chain or chemical deterioration of the side chain residues. These latter processes are truly irreversible and define the basal rate of irreversible enzyme thermoinactivation.

Three techniques distinguish potentially reversible inactivation processes from covalent mechanisms affecting primary structure: (1) comparison of the rate of irreversible thermoinactivation in the presence and

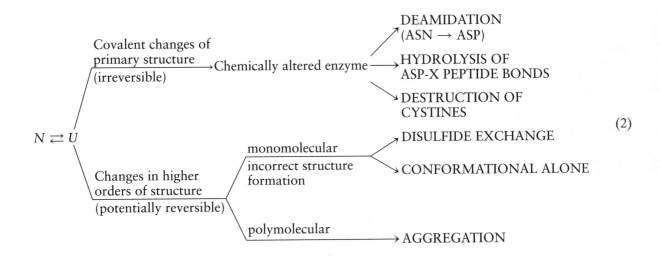

$$(2)$$

absence of strong denaturants such as guanidine hydrochloride or acetamide. Denaturants disrupt noncovalent interactions in protein and maintain the enzyme molecules in a highly unfolded form; therefore, the enzyme is unlikely to assume inactive conformations separated from the native state by high activation energies. On subsequent cooling and dilution, the random coil should refold to the native structure. Conversely, denaturing agents are not expected to affect the rates of most covalent reactions. The above reasoning constitutes the first criterion: If addition of a denaturant stabilizes an enzyme against irreversible thermal inactivation, conformational processes are involved; if there is no effect, the rate of inactivation is probably due to covalent processes affecting the primary structure of the enzyme.

Potentially reversible mechanisms of thermoinactivation may be distinguished from destructive, covalent processes in enzymes by a second approach: (2) determine whether unfolding and refolding of the previously heated and cooled enzymes result in at least partial recovery of the enzymatic activity lost. Whereas the first approach attempts to *prevent* incorrect structures from forming, the second approach attempts to *recover* activity lost by incorrect structure formation. (If S-S bonds are present, the inactivated, enzyme should be reduced and reoxidized during reactivation experiments.)

Finally (3), the magnitude of the rate of irreversible thermoinactivation of an enzyme can indicate whether the predominant mechanism is conformational or covalent. The half-lives of the covalent processes found to cause irreversible thermoinactivation are relatively large—of about 10 min to more than an hour at 100°C. Thus, if an enzyme inactivates irreversibly in less than 2 min at 100°C (e.g., lysozyme at pH 8) or inactivates rapidly at temperatures below 70°C at near-neutral pH, the inactivation is most likely predominantly due to conformational processes.

A. Potentially Reversible Mechanisms

In enzymes containing disulfide bridges, the formation of inactive monomolecular structures may be due to disulfide exchange, which results in incorrect pairing of cysteinyl residues. This reaction, which is known to occur in proteins at neutral and alkaline pH, requires the presence of catalytic amounts of thiols that promote the interchange by nucleophilic attack on the sulfur atoms of a disulfide. (How these thiols are produced in the course of heating an enzyme is de-

scribed in Section IV,B.) The contribution of disulfide exchange in the formation of incorrect structures can be prevented if the enzyme is heated in the presence of thiol scavengers (e.g., *p*-mercuribenzoate and *N*-ethylmaleimide) or copper ion, which catalyzes the spontaneous air oxidation of thiols. The rate of heat-induced destruction of S-S bonds is almost entirely independent of the nature of the protein.

Enzymes that contain no disulfide bridges also undergo potentially reversible thermoinactivation. For example, α-amylases from *Bacillus amyloliquefaciens* and *B. stearothermophilus* contain 0 and 1 cysteinyl residue each, respectively, yet both are stabilized at least threefold against irreversible thermoinactivation at 90°C, pH 6.5, by the presence of 8 *M* acetamide. Furthermore, activity lost when the amylases are incubated in the absence of denaturants can be partially regained if the enzyme is briefly treated afterward with hot, concentrated denaturant. Therefore, it appears that not all potentially reversible thermoinactivation is due to disulfide exchange. The detailed nature of incorrect folding in such cases is an intriguing topic for future research that may help explain how enzymes from organisms that live at elevated temperature are more resistant to inactivation processes compared with enzymes from organisms that live at moderate temperature.

As mentioned earlier, similar procedures of renaturation have been successfully applied to enzymes that are observed to aggregate on heating. The noncovalent interactions (primarily hydrogen bonds and hydrophobic interactions) believed to be responsible for maintaining aggregates are apparently disrupted by the presence of a denaturant; once redissolved, the enzyme can refold to the native conformation when the denaturant is removed.

B. Irreversible Mechanisms

Conformational processes alone cannot account for the irreversible loss of activity of enzymes. Despite all measures to prevent potentially reversible inactivation, enzymes are nevertheless observed to inactivate irreversibly at high temperatures in aqueous solutions throughout the entire range of pH. The covalent mechanisms responsible are treated separately below.

Conceptually, covalent processes can affect enzyme structure in the following ways: cleavage of the polypeptide chain by hydrolysis, destruction of individual amino acid residues, destruction of disulfide bonds, and reactions involving metal ions, cofactors, and adducts caused by glycosylation, etc.

1. Peptide Chain Integrity

Hydrolysis of the polypeptide chain adjacent to aspartyl residues can account for significant irreversible thermoinactivation of enzymes at elevated temperatures under mildly alkaline conditions (e.g., pH 4). The Asp-X bond (where X is the amino acid residue bound to the carboxyl group of Asp) is the most labile peptide bond under those conditions. If heating of an enzyme continues, the peptide bonds on both sides of aspartyl residues are hydrolyzed, and free aspartic acid is released into the solution. Several pathways result in the release of aspartic acid from proteins (Fig. 3a).

The time course of hydrolysis of peptide bonds comprising the polypeptide chain(s) of an enzyme can be monitored after the appearance of the resulting fragments. The mechanism is deduced from the identities of the amino acids at the new carboxyl and amino termini must be created by the intrapeptide chain hydrolysis.

2. Amino Acid Destruction

Deamidation of asparaginyl residues within enzymes occurs at elevated temperatures under acidic, neutral, and alkaline conditions. Release of ammonia as a result of deamidation is thought to occur during the formation of an unstable intramolecular cyclic imide intermediate (Fig. 3b). When the cyclic ring hydrolyzes, a carboxyl side chain is left in place of the original, uncharged asparaginyl residue. Except under very acidic conditions, carboxyl groups are negatively charged; as a result, deamidated enzymes can be physically separated from their normal, nondeamidated forms by means of isoelectric focusing and thus quantified. The release of ammonia can be determined as well.

Use of the methods described earlier makes possible monitoring the extent of deamidation during thermoinactivation. For example, the time course of the initial evolution of ammonia during the heating of lysozyme occurs at a significant rate relative to enzyme inactivation (Fig. 4). The ammonia is ascribed to deamidation of asparagine residues because studies of model peptides have shown that the 14 asparagine residues of lysozyme are much more labile than its 3 glutamine residues.

In some cases, the loss of activity caused by deamidation can be measured by assay for enzymatic activity of deamidated forms of the thermoinactivated en-

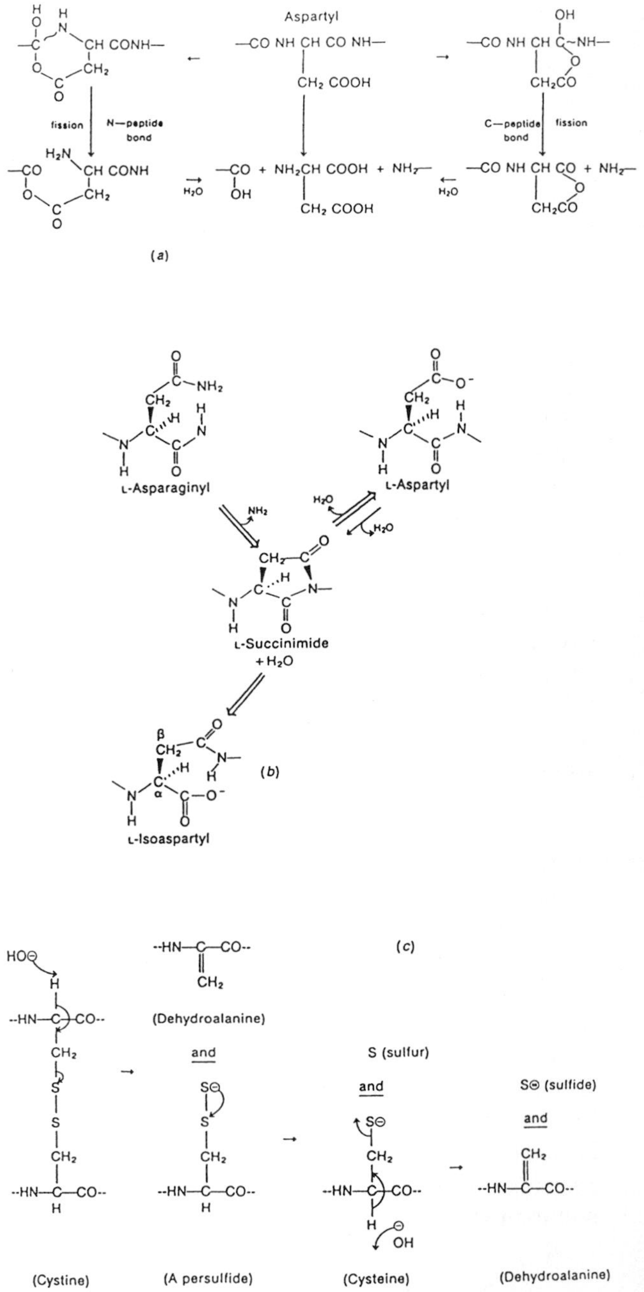

FIGURE 3   Proposed mechanisms of covalent reactions causing irreversible enzyme thermoinactivation. (a) Hydrolysis of peptide bonds adjacent to Asp residues. [From Inglis, A. S. (1983). *Methods Enzymol.* **91**, 324–332, with permission. Copyright by Academic Press.] (b) Deamidation of Asn residues, resulting in either L-aspartyl or L-isoaspartyl residues, depending on which amide linkage in the proposed succinimide intermediate is hydrolyzed. [From Clarke, S. (1985). *Annu. Rev. Biochem.* **54**, 479–506, with permission from Annual Reviews Inc.] (c) Destruction of disulfide linkages via base-catalyzed O-elimination. [From Whitaker, J. R. (1980). *In* "Chemical Deterioration of Proteins." ACS Symposium Series, Vol. 23, pp. 320–326, with permission. Copyright by the American Chemical Society.]

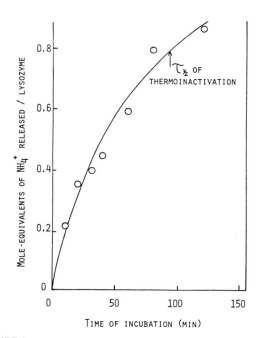

FIGURE 4 Release of ammonia during thermoinactivation of hen egg white lysozyme, 100°C, pH 4. Appearance of ammonia in solution was determined colorimetrically by a modified method of Forman, D. T. (1964). *Clin. Chem.* **10**, 497–508. Time at which only one-half of initial catalytic activity could be recovered after cooling to 23°C is indicated.

zyme. The relative specific activities of deamidated species of various enzymes are depicted in Table I.

Deamidation may cause not only the conversion of asparagine residues to aspartic acid, but also the formation of peptide bonds incorporating the remains of the side chain. This can be explained by the proposed mechanism of deamidation, illustrated in Fig. 3b; the nature of the resulting polypeptide linkage depends on which bond in the succinimide intermediate is hydrolyzed. Deamidation resulting in enzyme inactivation occurs during incubation in aqueous solutions at whatever pH chosen.

3. Destruction of Disulfide Bridges

In addition to the described disulfide exchange, cystines also undergo irreversible destruction known as β-elimination. The base-catalyzed abstraction of a proton from the α-carbon of a cysteine residue forming a disulfide bridge results in the cleavage of the cystine cross-link to form residues of dehydroalanine and thiocysteine (Fig. 3c). Dehydroalanine is a reactive species that can undergo an addition reaction with the ε-amino group of a lysine residue to form the novel intramolecular cross-link, lysinoalanine.

The degradative product of the complementary Cys residue, thiocysteine, undergoes further deterioration to yield the hydrosulfide ion (HS-) as one of many end products.

The formation of the degradative products described earlier accounts for the fate of approximately 90% of all the cystine lost during irreversible thermoinactivation of ribonuclease A. Furthermore, the contribution of β-elimination to loss of enzymatic activity was demonstrated by the finding that the degree of stabilization of ribonuclease was proportional to the degree of protection against β-elimination by reversible protection of the Cys residues by chemical modification.

4. Summary of General Mechanisms

Use of these techniques makes possible the calculation of the contribution of each inactivating process to the overall rate of irreversible thermoinactivation. Comparisons of the directly measured overall rates of enzyme thermoinactivation with the contributions of individual mechanisms are listed in Table II. Irreversible inactivation of lysozyme at 100°C is brought about (1) at pH 4 by a combination of deamidation and peptide hydrolysis; (2) at pH 6 by deamidation; and (3) at pH 8 by a combination of deamidation, destruction of disulfide bonds, and formation of in-

TABLE I

Effect of Deamidation on the Relative Specific Activity of Enzymes

Enzyme	Relative specific activity
Hen egg white lysozyme	
Native	1.00
Monodeamidated	0.53
Di- and trideamidated	0.21
Bovine pancreatic ribonuclease A	
Native	1.00
Monodeamidated	0.65
Dideamidated	0.38
Trideamidated	0.19
Cytochrome *c*	
Native	1.00
Monodeamidated	0.60
Dideamidated	0.20
Yeast triosephosphate isomerase	
Native	1.00
(Asn 78 → Asp 78)[a]	0.66

[a]The altered enzyme was produced by means of site-directed mutagenesis and expressed in *Escherichia coli*.

TABLE II

Rate Constants of Irreversible Thermoinactivation of Lysozyme and Ribonuclease: The Overall Process and Contributions of Individual Mechanisms to Thermoinactivation

	Rate constant (hr^{-1})		
Irreversible thermoinactivation	pH 4	pH 6	pH 8
Hen egg white lysozyme, 100°C			
Directly measured overall process	0.49	4.1	50
Deamidation of Asn residues	0.45	4.1	18
Hydrolysis of Asp-X peptide bonds	0.12		
Due to individual mechanisms			
Destruction of crystine residues			6
Formation of incorrect structures			32
Bovine pancreatic ribonuclease A, 90°			
Directly measured overall process	0.13	0.56	23.4
Deamidation of Asn residues	0.02	0.15	0.8
Hydrolysis of Asp-X peptide bonds	0.10		
Due to individual mechanisms			
Destruction of cystine residues		0.05	2.8
Formation of incorrect structures[a]		0.31	19.4

[a] Shown to be due to thiol-catalyzed disulfide interchange.

correct structures. Analogous findings were reported for the irreversible inactivation of ribonuclease at 90°C, with the additional finding that an incorrect structure formation, observed at both pH 6 and 8, was shown to be due to thiol-catalyzed exchange; also, β-elimination of cystine residues could account directly for approximately 10% of the loss of activity at pH 6.

Although these mechanisms adequately account for the irreversible loss of activity of the enzymes studied, it is likely that secondary mechanisms, resulting in only a small fraction of the overall rate of inactivation, are also at work. These may include (1) hydrolysis of the peptide chain adjacent to amino acid residues other than aspartate (peptide bonds involving glutamic acid, glycine, alanine, serine, and threonine have been reported to be cleaved during prolonged digestion in dilute acid); (2) deamidation of glutamine—as opposed to asparagine—residues; and (3) racemization of amino acid residues.

5. Other Mechanisms

In addition to the processes causing irreversible thermoinactivation outlined earlier, many additional degradative processes specific to enzymes containing unique constitutive elements may result in thermoinactivation as well. In addition to the 20 common amino acid residues, more than 100 unusual amino

acids also exist in proteins, and half of them are susceptible to chemical deterioration [e.g., hydrolytic scission of side chain groups bound to indole, phenoxy, thioether, amino, imidazole, and sulfhydryl residues and the derivatives of Ser and Thr (e.g., O-glycosyl and O-phosphoryl groups) and Gln and Asn (e.g., methylated and glycosylated)]. Of the nonamino acid moieties associated with enzymes, the carbon–nitrogen bonds in purines and pyridines, glycosidic bonds, and phosphodiester bonds undergo hydrolytic breakdown at rates comparable to the hydrolysis of peptide bonds. If present, reducing sugars and fatty acid degradative by-products can undergo the Maillard reaction with amino groups in enzymes to produce Schiff bases on removal of water. Metal ions can accelerate hydrolytic cleavage of peptide bonds.

In addition to these covalent, deteriorative reactions, simple dissociation of noncovalently bound prosthetic groups during thermally induced denaturation can be irreversible. For example, once molybdenum has been dissociated from the active center of sulfite oxidase during incubation at high temperature, no reactivation appears possible by cooling the enzyme solution and adding an excess of the metal ion. Nevertheless, it is possible that the activity of some enzymes that lose their cofactors during thermoinactivation may be regenerated. For example, enzymes requiring metal–sulfur compounds can be reactivated after loss of their cofactors by the addition of metal salts together with thiols or organic sulfides.

V. ENZYME THERMOINACTIVATION *IN VIVO*

Folding and degradation of enzymes *in vivo* is affected by features unique to the intracellular environment. Most pools of intracellular proteins are in a dynamic state of continual synthesis and degradation. The half-lives of proteins in the cell range from minutes to days and can even vary in response to external conditions (e.g., starvation *in vivo*) or withdrawal of serum during cultivation of cells.

The folding of proteins in eukaryotic cells occurs during or immediately after translocation of the peptide chains to the internal (luminal) side of the endoplasmic reticulum (ER). Covalent attachment of carbohydrates and fatty acids to the polypeptide chain may also contribute to the protein folding pathway. Several proteins in the ER and Golgi assist in the proper folding of nascent polypeptide chains. The formation of disulfide bonds in proteins that are eventually secreted is catalyzed by disulfide isomerase. Binding protein (BiP; also known as GRP78) has been implicated in the formation of oligomeric proteins and has been found to bind and retain improperly folded proteins within the ER. [See Proteins.]

Ubiquitin, a small protein found in all eukaryotic cells, covalently attaches to abnormal and some short-lived proteins via the ε-amino groups of lysinyl residues. Ubiquitination apparently marks the proteins for rapid degradation by intracellular proteases, the multicatalytic endopeptidase complex (the "proteasome"), and perhaps by ubiquitin itself. Specialized vesicles known as liposomes are the site of degradation of many membrane-bound proteins and long-lived cytosolic proteins. Other degradative pathways may be restricted to mitochondria, the ER, and Golgi, which degrade a portion of secretory and membrane proteins before they reach the cytosol.

The thermostability of enzymes may contribute to their susceptibility to intracellular degradative processes. Cellular proteins having disrupted structures, as a result of either posttranslational oxidative damage or substitution of amino acid residues by analogs, are more rapidly conjugated to ubiquitin than are native proteins. Research with temperature-sensitive mutants of proteins (e.g., the oligomeric VSV G glycoprotein) has shown that normal maturation and export from the Golgi occurs only below the nonpermissive temperature for the mutant; if the temperature is raised, oligomerization does not occur, and the monomer is retained within the Golgi.

Many proteins, especially structural proteins, are long-lived. Proteins of the eye lens can avoid total degradation and recycling until the death of the organism. Therefore, the initial effects of degradation that would trigger rapid protein turnover in most tissues may accumulate in lens proteins. For example, triosephosphate isomerase undergoes spontaneous deamidation of specific asparaginyl residues during the aging of erythrocytes and eye lens. Amino acid composition analysis has also shown that the aspartic acid content of tissues such as rat brain, liver, and heart also increases with age as the asparaginyl content concomitantly decreases. Deamidation of asparaginyl residues has also been observed in crystallized proteins by means of neutron-scattering studies. The rate of deamidation of folded proteins depends on the local conformation and bonding structure. If the asparaginyl residue is sterically prevented from forming the imide intermediate, the rate of deamidation is markedly lowered.

Once deamidation of asparagine occurs, hydrolysis of the cyclic intermediate can produce either an aspartyl residue in the place of the original asparagine or a novel, iso-aspartyl side chain from what had been the C' carbon of the original asparagine, and insertion of the remains of the asparaginyl side chain into the polypeptide chain (Fig. 3b). There may exist a mechanism in vivo for eliminating such β-carboxyl linkages in polypeptide chains. A mammalian enzyme, carboxyl methyltransferase, preferentially methylates such abnormal isoaspartyl groups; on demethylation, the cyclic intermediate is reformed; subsequent hydrolysis yields a mixture of both normal aspartyl and iso-aspartyl residues. By this means, the inclusion of abnormal β-carboxyl linkages into polypeptide chains can be reversed. Enzymatic pathways leading to the reversal of spontaneous damage to proteins are also capable of reducing methionyl residues in proteins that have undergone oxidation. The physiological importance of such reactions in the in vivo repair of proteins remains to be determined.

VI. ENZYME THERMOSTABILIZATION

A. Thermophilic Enzymes

When compared with the enzymes of organisms that normally live below 40°C, those from thermophilic bacteria are extremely resistant to thermodenaturation. Some remain active despite prolonged heating near 90°C. Although structural information concerning thermophilic enzymes is scanty, it appears increased thermostability is due to small stability-enhancing changes throughout the proteins. Because the overall stabilization free energy of the native states of proteins is low, relatively small increments in the strength of existing bonds can result in significant increases in resistance to thermal denaturation. Because hydrophobic interactions are stabilized by increases in temperature, it is not surprising that thermophilic proteins have increased internal—and decreased external—hydrophobicity. Similar increases in the stability of α-helices and β-sheet structures in thermophilic proteins have also been noted.

B. Protein Engineering

1. Enhancement of Thermostability by Design

It is reasonable to assume that the overall stability of enzymes can be increased by replacing those amino acid residues providing only weak contributions to the conformational stability of the molecule with others providing stronger interactions.

The superposition of hydrogen bond dipoles in α-helices, resulting in opposite charges at the ends of helical segments, was originally believed to be a relatively small electrostatic contribution to the sum of energetic interactions maintaining the native structures of proteins. Recently, however, researchers demonstrated that amino acid substitutions designed to increase stabilization via the helix dipoles in T4 lysozyme increased the melting transition temperature by as much as 4°C, reflecting a stabilization of the protein conformation of approximately 1.6 kcal/mol. The mutations introduced charged aspartyl residues that interacted electrostatically with the positively charged amino termini of helices in the protein. Earlier work showed that decreasing the charge from +2 to −1 on the amino-terminal residue of a helix in analogs of the S-peptide of ribonuclease S resulted in an increase of the melting temperature of the reconstituted enzyme by as much as 6°C compared with the protein containing the native S-peptide.

Secreted proteins are further stabilized by the presence of disulfide bonds. Engineering novel disulfide bonds into dihydrofolate reductase, T4 lysozyme, and subtilisin BPN′ have stabilized the active enzymes with respect to reversible unfolding. In one instance, the addition of three disulfide bonds to T4 lysozyme increased the melting transition temperature by as much as 23.4°C.

Another approach that enhances the stability of a protein is to replace flexible residues (e.g., glycine), which require greater free energy than other, sterically constrained residues to restrict its conformation. Provided the substitutions do not introduce undesirable steric interactions, the mutated protein will have a decreased entropy of unfolding, resulting in a higher temperature of denaturation. In accord with this theory, substitution of one glycine and one alanine residue in T4 lysozyme increased the melting transition temperature by as much as 2°C, reflecting an increase of approximately 1 kcal/mol to the free energy of folding. Similarly, substitutions of two glycines in λ repressor increased the melting temperature of the N-terminal domain by 3–6°C.

Discrepancies between the amino acid sequences of enzymes having high degrees of overall sequence similarity but widely varying thermostability can be "corrected" to increase the thermostability of the less stable enzymes within the family. This technique has yielded a mutant form of the neutral protease from *Bacillus stearothermophilus* of enhanced thermostability, although many mutants designed by this criterion exhibited decreased thermostability.

2. Enhancement of Thermostability by Selection

The examples cited previously illustrate rational approaches to stabilizing the conformation of proteins; by means of structural and functional data, the researchers conceived of specific changes that would theoretically improve the stability of a given enzyme. A second approach selects for improvement in the stability of an enzyme without prior knowledge of its structure. Mutants of kanamycin nucleotidyl transferase having increased thermostability were discovered by cloning and expressing the enzyme from a mesophilic organism in a thermophilic bacteria and selecting for colonies containing the activity of the enzyme at temperatures higher than the normal melting transition temperature of the enzyme. Such screening techniques also have yielded variants of T4 lysozyme and subtilisin BPN′ having enhanced thermostability.

Random mutations exist in staphylococcal nuclease that stabilize the enzyme against unfolding by guanidine hydrochloride. For example, whereas ~0.85 M guanidine hydrochloride results in the denaturation of the wild-type enzyme, as much as 1.3 M denaturant is required to unfold one of the mutants. Furthermore, three such stabilizing mutations were found to "correct" other mutations that had resulted in colonies having greatly reduced nuclease activity. One explanation is that the stabilizing substitutions may confer a global stability on the enzyme. In a similar fashion, the activities of nonfunctional and presumably unstable mutants of cytochrome c were partially restored by a second-site mutation (i.e., asparagine-57 replaced by isoleucine). Subsequent construction of the Asn57-Ile mutant by site-directed mutagenesis resulted in an extraordinary 17°C increase in the transition temperature, corresponding to a greater than twofold increase in the free energy change for thermal unfolding.

3. Preventing Inactivation of Proteins Caused by Covalent Mechanisms

Replacement of specific asparagine residues in triosephosphate isomerase resulted in a nearly two-fold decrease in the rate of irreversible thermoinactivation of the enzyme. Exposure to oxidative environments can result in the degradation of methionine. Subtilisin was stabilized against such oxidation by replacing a single methionine important for catalysis with serine, alanine, or leucine.

BIBLIOGRAPHY

Ahern, T. J., Casal, J. I., Petsko, G. A., and Klibanov, A. M. (1987). Control of oligomeric enzyme thermostability by protein engineering. *Proc. Natl. Acad. Sci. USA* **84,** 675–679.

Ahern, T. J., and Klibanov, A. M. (1985). The mechanism of irreversible enzyme inactivation at 100°C. *Science* **228,** 1280–1284.

Ahern, T. J., and Manning, M. C. (eds.) (1992). "Stability of Protein Pharmaceuticals." Plenum Press, New York.

Anfinsen, C. B. (1973). Principles that govern the folding of protein chains. *Science* **181,** 223–230.

Argos, P., Rossman, M. G., Grau, U. M., Zuber, H., Frank, K. G., and Tratschin, J. D. (1979). Thermal stability and protein structure. *Biochemistry* **18,** 5698–5703.

Clarke, S. (1985). Protein carboxyl methyltransferases: Two distinct classes of enzymes. *Annu. Rev. Biochem.* **54,** 479–506.

Cleland, J. L., Powell, M. F., and Shire, S. J. (1993). The development of stable protein formulations: A close look at aggregation, deamidation, and oxidation. *Crit. Rev. Therap. Drug Carrier Syst.* **10,** 1–100.

Das, G., Hickey, D. R., McLendon, D., McLendon, G., and Sherman, F. (1989). Dramatic thermostabilization of yeast iso-1-cytochrome c by an asparagine → isoleucine replacement at position 57. *Proc. Natl. Acad. Sci. USA* **86,** 496–499.

Karplus, M., and McCammon, J. A. (1981). The internal dynamics of globular proteins. *CRC Crit. Rev. Biochem.* **9,** 293–349.

Klibanov, A. M. (1983). Stabilization of enzymes against thermal inactivation. *Adv. Appl. Microbiol.* **29,** 1–29.

Lapanje, S. (1978). "Physicochemical Aspects of Protein Denaturation." Wiley, New York.

Matsumura, M., Signor, G., and Mathews, B. W. (1989). Substantial increase of protein stability by multiple disulphide bonds. *Nature* **342,** 291–293.

Pfeil, W. (1981). The problem of the stability of globular proteins. *Mol. Cell. Biochem.* **40,** 3–28.

Rothman, J. E. (1989). Polypeptide chain binding proteins: Catalysts of protein folding and related processes in cells. *Cell* **59,** 591–601.

Whitaker, J. R. (1980). *In* "Chemical Deterioration of Proteins" (J. R. Whitaker and M. Fujimaki, eds.), Vol. 23, pp. 145–164. American Chemical Society, Washington, DC.

White, R. H. (1984). Hydrolytic stability of biomolecules at high temperatures and its implications for life at 250°C. *Nature* **310,** 430–432.

Ependyma

J. E. BRUNI

University of Manitoba

GLOSSARY

Ependyma Single uninterrupted layer of ciliated squamous to columnar epithelial-like cells that line the cerebral ventricles and central canal of the spinal cord

Tanycytes Ependymal cell variants found only within circumscribed regions of the adult mammalian ventricular system. They are distinguished by the presence of a radially directed basal process that extends into the neuropil and terminates on fenestrated on blood vessels, neurons, glia, or the external glial limitans

THE EPENDYMA IS A SINGLE UNINTERRUPTED layer of mainly ciliated squamous to columnar epithelial-like cells that line the cerebral ventricles and cental canal of the spinal cord. However, its composition is heterogeneous, with some cells (tanycytes) having basal processes that extend into the subjacent neuropil and terminate on blood vessels, neurons, glia, or at the pial surface. Although it has been the subject of many studies over the years, much of our information about the ependyma is still incomplete and its functions remain largely speculative.

I. CELL TYPES

The lining of the cerebral ventricles and central canal of the spinal cord is similar in both form and construction to most epithelial membranes found in the body.

The first to characterize the epithelial nature of the ependymal cell lining was J. E. Purkinje (1836), and since then many investigators have recognized that it is of heterogeneous composition and in particular that some cells have basal processes that extend into the subjacent neuropil. In 1954, E. Horstmann first applied the descriptive term "tanycyte" (Greek for elongated or stretched-out) to such elongated ependymal cells seen in the nervous system of selachians using special gold sublimate staining methods. Particular interest in tanycytes originated with the observation that within a circumscribed region of the adult mammalian third ventricle they established an anatomical link between the cerebrospinal fluid (CSF) and the hypophysial-portal vasculature and that this peculiar structural arrangement possibly had significant functional implications. These nonneuronal cells together with the CSF were thought by Lötgren (1960), to be involved in a transport system that subserved some neuroendocrine function. Anand Kumar (1973), Knigge and Scott (1970), and others have proposed that hormones secreted into the CSF were transported to the hypophysial-portal vasculature via tanycytes of the infundibular recess of the third cerebral ventricle. Knigge and Scott (1970) have also suggested, however, that if such a transport exists, it was probably of limited physiological importance, whereas Pilgrim (1978) concluded that the transport hypothesis involving tanycytes and their putative neuroendocrine involvement was not supported by adequate evidence. Although the ependyma in general and tanycytes in particular have been studied for years, more information is still needed regarding their functions. [*See* Spinal Cord.]

A. Cuboidal-Columnar Ependymal Cells

According to the current literature, the lining of the cerebral ventricles and central canal of the spinal cord in a variety of adult mammalian species consists, under normal conditions, of a single uninterrupted layer of ciliated squamous to columnar ependymal cells or ependymocytes (Figs. 1–3).

The typical cuboidal-columnar ependymal cell has a conspicuous spherical or oval vesicular nucleus, often eccentrically placed and occupying a large part of the cell (Figs. 1 and 4). Where the ependyma is reduced to a thin flattened or squamous layer of cells, the nuclei are also elongated and widely separated. The cytoplasm of the cell has a fine granular appearance, with dispersed agranular endoplasmic reticulum and numerous spherical to elongated mitochondria (Figs. 4 and 5). Organelles and inclusions are principally confined to the apical and lateral parts of the cell. Ribosomes, although dispersed throughout the cytoplasm in clusters or rosettes, are only infrequently associated with the endoplasmic reticulum. The Golgi complex with its typical complement of vesicles and cisternae generally occupies a supranuclear position. Randomly dispersed filaments and microtubules are frequently observed within the cytoplasm and in some species dense bundles of filaments are found in a perinuclear array (Figs. 4 and 5). Membrane-bound multivesicular bodies and lysosomes are also common. Keep and Jones (1990) reported that ependyma like the choroid plexus shows a postnatal increase in the number of mitochondria, suggesting that ependymal cells increase their metabolic activity during the postnatal period for reasons that are still not clear. Ependyma also exhibit strong oxidative enzyme activity that correlates with the complement and distribution

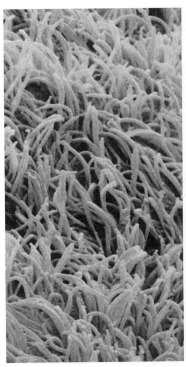

FIGURE 2   Scanning electron micrograph of the upper two-thirds of the rabbit third cerebral ventricle showing the densely ciliated nature of the luminal surface of the common mural ependymal cell surface. ×2800. [Reproduced with permission from J. E. Bruni (1974). *Can. J. Neurol. Sci.* **1**(1), 59–73.]

of the mitochondria. This has been interpreted by Nandy and Bourne (1965) to mean that these cells have a high oxidative metabolism and are more metabolically active than would be expected of inert cells, which simply form a mechanical lining. Nandy and Bourne (1965) also found high concentrations of acid phosphatase in the apical parts of ependymal cells with a distribution similar to that of the lysosomes within the cell. They considered this an indication that these cells might be involved in differential permeability and transport functions.

The luminal plasma membrane of ependymal cells is complexly organized into numerous microvillus-like projections and tufts of cilia (Figs. 1, 2, 4, and 5). The apical ependymal surface is frequently overlapped by folds of cytoplasm originating from the apical surface of contiguous cells. A variety of isolated profiles are also occasionally encountered on the ventricular surface of ependymal cells, some are supraependymal cells others exhibit the appearance of unmyelinated nerve fibers.

The lateral surfaces of adjacent ependymal cells

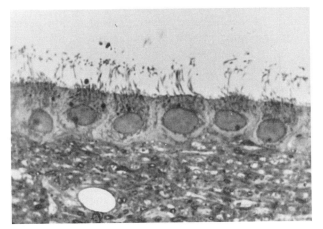

FIGURE 1   Cuboidal ciliated ependyma lining the lateral part of the floor of the fourth ventricle in a rat. Most common mural ependymal cells lack basal processes and abut directly upon the subjacent neuropil.

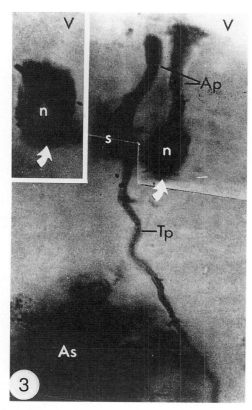

FIGURE 3 Common mural (n) and tanycyte (s) ependymal cells as seen in Golgi preparations. The cuboidal or columnar cell bodies of the former rest directly on the subjacent neuropil (arrows). In contrast, the tanycyte typically has a single process (Tp) that arises from the base of the soma (s) and projects for a variable distance into the neuropil. AP, apical process; As, fibrous astrocyte; V, third ventricle. ×1310; inset, ×1170. [Reproduced with permission from J. E. Bruni *et al.* (1983). *Anat. Anz.* **153**, 53–68.]

are usually without elaborate folds or interdigitations (Figs. 1, 3, 4, and 5) and cells are held together by junctional complexes of the zonulae adherents and occludentes variety (Fig. 5). Typical desmosomes do not occur between ependymal cells.

The basal surfaces of the common form of ependymal cell are without elaborate folds or processes. They characteristically abut directly upon the underlying network of neuronal and glial cells and fibers without the interposition of a basal lamina that typically underlies most other epithelial membranes (Figs. 1, 3, and 4). Infrequently, common mural ependymal cells are encountered either singly or in small clusters that do have a "tanycyte-like" basal process (see the following) that extends for only a short distance into the subjacent neuropil. The precise status of these particular cells is uncertain. For a more detailed description of the morphology of the ventricular lining,

readers are referred to articles by Bruni *et al.* (1985), Bruni (1974), and Peters *et al.* (1976).

B. Tanycyte Ependymal Cells

Interspersed among the common mural ependymal cells are populations of morphologically and presumed functionally distinct cells called "tanycytes," which are conceived to be specialized forms of ependymal cells (Figs. 3, 6, and 7a). Tanycytes are ubiquitous in the nervous system of nonmammalian vertebrates but are found primarily in the developing nervous system and localized regions of the mature mammalian nervous system. Although clusters of these cells are scattered throughout the mammalian spinal cord and ventricular system, they are particularly numerous along the ventrolateral walls and floor of the third ventricle. Their presence there has been demonstrated by Flament-Durand and Brion (1985), Fleischhauer (1972), and others using a variety of histological and histochemical procedures. The features that serve to distinguish tanycytes from common mural ependymal cells are somewhat species dependent and equivocal. Cell shape, cytoplasmic density, the absence of perinuclear bands of filaments, and the presence of long basal processes are the principal distinguishing features according to Millhouse (1972).

The cell bodies and nuclei of tanycytes are generally more elongated than those of the common mural cells (Figs. 3 and 7a), however, they do contain the normal complement of organelles and inclusions, some of which extend into their basal processes (Fig. 7). Mitochondria are particularly numerous in the basal cytoplasm and throughout the processes, as are microtubules.

The apical surfaces of tanycytes are furnished with very few cilia, rather they are characterized by an extensive elaboration of pleomorphic microvillus-like processes (Fig. 6). The lateral surfaces between adjacent cells may or may not be interdigitated. They generally contain intercellular junctional complexes similar to those described between common mural ependyma. Occluding junctions, however, most often occur between tanycytes, particularly in certain locations such as the median eminence of the third ventricle, which lacks a blood–brain barrier, and posesses instead a blood–csf barrier (Brightman *et al.*, 1975).

The most distinguishing morphological feature of tanycytes is their radially directed basal process, which extends for a variable distance (sometimes several hundred microns) into the subjacent neuropil (Figs. 3, 6, 7a, and 9a). It is frequently directed toward

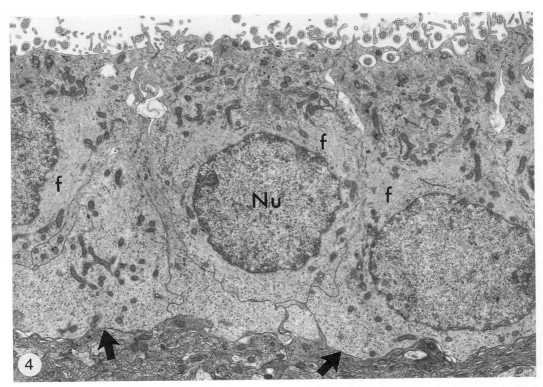

FIGURE 4 Transmission electron micrograph showing the cuboidal form of ependymal cell in the rat third ventricle, which sits directly on the subependymal neuropil (arrows). It is characterized by a large vesicular nucleus (Nu), which occupies a substantial part of the cell. The concentration of filaments (f) in a perinuclear location displaces the organelles and inclusions to the apical and lateral margins of the cell. The lateral surfaces of adjacent cells are without elaborate interdigitations. ×7900.

a blood vessel wall, where it may enwrap the vessel (Figs. 6, 8a and 8b) or equally typical may terminate on neurons (Fig. 7), glia (Figs. 3, 7b), or the external glial limitans (Fig. 8c). The process characteristically contains numerous axially oriented mitochondria, microtubules, and filaments (Fig. 7). Dilated cisternae of endoplasmic reticulum and rosettes of ribosomes are also present. Tanycyte processes are typically unbranched throughout their course, except at their termination, where they may divide into several slender branches (Figs. 6 and 8c).

Not only do mature common mural ependyma and tanycytes differ in structure, location, enzyme content, and function, but it has also been suggested that tanycytes may not themselves be a homogeneous population of cells. Akmayev and Fidelina (1976), Baun *et al.* (1983a), Flament-Durand and Brion (1985), and others have distinguished various forms of tanycytes on the basis of differences in structure, location, and enzyme composition. Despite the differences between common mural ependymal cells and tanycytes, Jarvis and Andrew (1988) found that both exhibit electro-

physiological properties such as high negative resting membrane potentials, responsiveness to extracellular K^+ levels, low input resistances, and extensive cell-to-cell interconnections via gap junctions that are characteristic of glial cells.

II. DEVELOPMENT AND DIFFERENTIATION

Neuroepithelial cells that develop into ependymal cells begin to proliferate soon after neural plate formation, before gestational day (GD) 10, in some parts of the rodent brain. Investigators have charted the temporal and regional histogenesis of ependyma using autoradiographic labeling techniques as well as immunocytochemical markers for various cytoskeletal (vimentin, cytokeratins, glial fibrillary acidic protein) and diffusible proteins (S-100).

During the developmental period, morphologically distinct ependymal cells are generated regionally from the undifferentiated neuroepithelium along a caudal

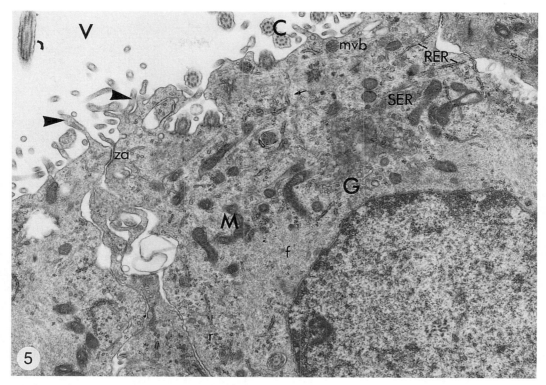

FIGURE 5 Supranuclear portion of cuboidal ependymal cells as in Fig. 4 illustrating the abundance of juxtraventricular organelles and inclusions. The apical membrane bordering the ventricle (V) is complexly organized into numerous microvillus-like projections (arrowheads) and cilia (C). At the lateral apices of contiguous cells, zonulae adherentes (za) comprise the junctional complexes. G, Golgi complex; M, mitochondria; r, ribosomes; RER, rough endoplasmic reticulum; SER, smooth endoplasmic reticulum; f, filaments; arrow, microtubules; mvb, multivesicular bodies. ×16,500.

to rostral gradient, reaching a peak on different days in different regions of the cerebral ventricles and then declining. In the rodent brain, the bulk of common ependymal cells are formed between GD 14 and 18, however, formation of the specialized tanycytes is not completed until the first or second postnatal week in some brain regions. Compatible with the late cytogenesis of tanycytes, studies by Bruni *et al.* (1983a, 1985) and others suggest that they may acquire adult form only postnatally and that even their function may be age dependent. Unlike common mural ependymal cells, in which glial fibrillary acidic protein (GFAP) immunostaining is absent, tanycytes were found by Edwards *et al.* (1990) and others to acquire GFAP antigenicity during postnatal development and, once acquired, it persisted throughout life. It should be noted, however, that there are discrepancies in the literature regarding the temporal and regional expression in ependyma of the various immunocytochemical markers, including GFAP. In the course of reptilian phylogenetic evolution, the astrocytic form of the neu-

roglial cell makes its appearance as the GFAP-positive tanycytic ependymal cells become scarce. Similarly, during ontogenetic development in higher vertebrates, Bodega *et al.* (1994) found that GFAP appears in astrocytes only as tanycytes that predominate during the earlier stages of development lose their GFAP expression and develop into GFAP-negative common mural ependymal cells.

In the human, Sarnat (1992b) observed that the earliest ependymal differentiation occurred in the floor plate of the neural tube at 4 weeks gestation but was not completed in either the third or lateral ventricles until 22 weeks gestation. He also noted that during the fetal period, most ependymal cells have basal processes that extend into the subjacent parenchyma (i.e., tanycytes or tanycyte-like) and are intensely GFAP positive. He was able to distinguish them from another GFAP-positive process-bearing cell type, known as radial glial cells, which are also ubiquitous in the nervous system during the developmental period. Whether tanycytes in the human nervous system arise from radial glia or

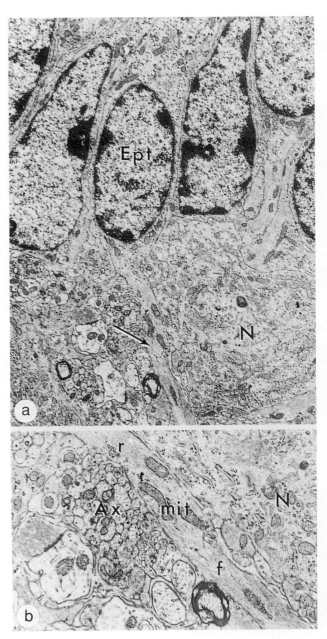

FIGURE 6 Scanning electron micrographs arranged as a montage illustrating the structure of a presumed ependymal tanycyte at the level of the rabbit infundibular recess. The luminal surface of the cell body (Ts) is nonciliated and its single tapering process (Tp) is unbranched throughout its course through the median eminence except at its termination, where multiple end-feet (Tfp) are given off. The two parts of the basal process (arrowheads) were joined before being severed by the electron beam. C, capillary (~9 μm diameter). ×2700. [Reproduced with permission from J. E. Bruni (1974). *Can. J. Neurol. Sci.* **1,** 59–73.]

FIGURE 7 (a) A basal process (arrow) arising from an ependymal tanycyte (Ept) bordering on the central canal can be seen projecting radially into the neuropil of the rat spinal cord. N, neuron ×7300. (b) An enlargement of a segment of this basal process. It contains mitochondria (mit), a few ribosomes (r), some profiles of smooth endoplasmic reticulum, a few microtubules, and a large number of filaments (f). A neuron (N), axons (Ax), and dendrites surround the tanycyte process and are partially separated from them by thin sheets of astrocytic cytoplasm. ×14,375. [Reproduced with permission from J. E. Bruni and W. A. Anderson (1987). *Acta Anat.* S. Karger AG, Basel **128,** 265–273.]

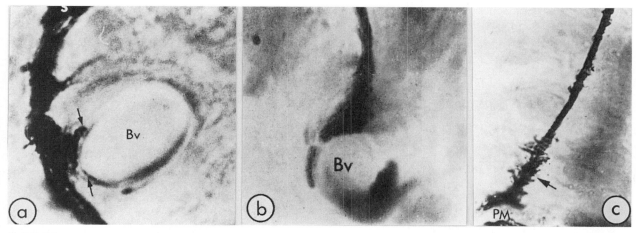

FIGURE 8 Tanycyte processes within the lateral wall of the rat third ventricle as seen in Golgi preparations. (a) Contact between the proximal irregular segment of a tanycyte process and a blood vessel (Bv) in the periventricular zone (arrows). ×2280. (b) Distally tanycytes also contact blood vessels by means of terminal expansions or foot-like processes. ×2350. (c) The basal processes of some tanycytes terminate on the pia mater (PM). A variation sometimes seen in the tanycytic form is the presence of spines along the terminal segments (arrow) ×1550. S, tanycyte soma. [Reproduced with permission from J. E. Bruni *et al.* (1983). *Anat. Anz.* 153, 53–68.]

from uncommitted ventricular zone cells has yet to be determined. By late gestation (≥15–19 weeks), tanycytes were observed by Gould *et al.* (1987, 1990) to gradually lose their immunoreactivity as they mature into common ependymal cells following acquisition of cilia and loss of basal processes. Tanycytes are believed by Gould *et al.* (1990) to represent a transitional cell during human fetal development and not a distinct population. Their processes retract with differentiation and they eventually give rise to common ependymal cells as they mature.

III. RESPONSE TO INJURY

Cell division is a conspicuous feature of ependyma during embryological and early postnatal development (Chauhan and Lewis, 1979). In normal adult animals, however, [³H]thymidine labeling of ependymal cells indicative of mitosis has only occasionally been observed in some regions of the ventricles and in the central canal of the spinal cord. Some investigators reported a postnatal decline in the mitotic activity of ependymal cells and the persistence of only a residual low level of ependymal cell replication in the adult brain. [*See* Mitosis.]

Following injury, most epithelial membranes undergo a process of repair that usually involves proliferation and migration of cells. The replacement of cells in the repair process usually occurs by mitosis of either preexisting cells or undifferentiated stem cells and

depends on the degree of specialization achieved by that epithelium. The proliferative capacity of ependyma during the postnatal period in response to pathological processes has been studied by Bruni *et al.* (1983b, 1985, 1987), Sarnat (1995), and many others. With few exceptions, lesions involving ventricular ependyma in adult animals including humans have been reported to be irreversible and observations suggest a lack of regenerative ability postnatally.

However, there are some reports that morphologically mature ependymal cells are capable of mitotic division and that mitosis occurs while well-formed junctions are maintained with adjoining cells (Figs. 9 and 10). Bruni *et al.*, (1983b) found that focal mechanical disruption of the ependymal lining of the floor of the rat fourth ventricle produced significant qualitative and quantitative changes and induced proliferation of ependyma, particularly at the wound margins, but ultimately incomplete reepithelization and wound closure. Bruni *et al.* (1983b, 1987) reported ependymal cell labeling indices seen at the wound margins, as determined by [³H]thymidine autoradiography, of from 0.08 to 1.73%.

In addition to mechanical-, chemical-, and radiation-induced lesions, there is also an extensive literature documenting the periventricular effects of ventriculomegaly induced by various means. For the most part, there is a consistent sequence and array of pathologic changes that correlates with the degree of ventricular dilatation. Sarnat (1995), Bruni *et al.* (1985) and others have described stretching, atrophy and

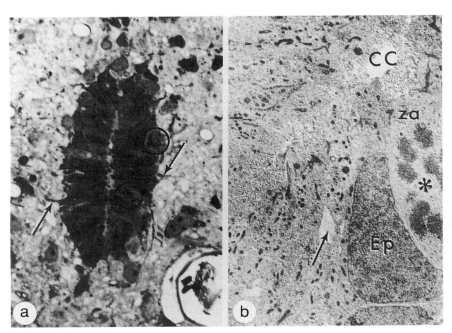

FIGURE 9 (a) Central canal of the thoracolumbar spinal cord in a rat 2 days after cord injury showing a collapsed lumen lined by an irregular array of radially elongated ependymal cells, several of which have their basal poles drawn out into processes (arrows). Note the presence of mitotic figures (circled) within the lining. ×670. (b) The apical cytoplasm of ependymal cells (Ep) lining the collapsed central canal (CC). Note the presence of a mitotic ependymal cell (∗) bordering directly on the central canal and attached to adjoining cells by zonulae adherentes (za) (arrow). ×3900 [Reproduced with permission from J. E. Bruni and W. A. Anderson (1987). *Acta Anat.* S. Karger AG, Basel **128**, 265–273.]

focal loss of ependymal cells within the lining, periventricular edema, and subependymal gliosis. Though some ependymal cell proliferation has been reported as part of the overall tissue response, Sarnat (1995) has determined that repair of the deficit appears to be undertaken entirely by subependymal cells with no evidence of their differentiation into ciliated ependyma cells.

In contrast to the cerebral ventricles, proliferation of ependymal cells has been more commonly reported in response to spinal cord injury, suggesting that inherent differences may exist in the proliferative capacity of ependyma in different regions of the neuraxis. The clinical literature also indicates that the incidence of ependymomas, often related to preceding trauma, is higher in the spinal cord than elsewhere. Proliferation of ependymal cells in the central canal of the spinal cord has been observed in response to a variety of insults such as neuronal chromatolysis following sciatic neurectomy, compression injury, and transection of the spinal cord. It is not clear, however, whether the tanycytic form of the ependymal cell or some other precursor cell type may be responsible for this proliferative capacity. Bruni and Anderson (1987) reported that, following spinal cord injury, ependymal mitotic activity (Fig. 9) reached a maximum of 3.34% 2 days after injury. [*See* Cerebral Ventricular System and Cerebrospinal Fluid.]

In contrast to mammalian vertebrates, more complete regeneration of the spinal cord has been reported by Norlander and Singer (1978), Simpson (1968), Bryant and Wozney (1974), and others in certain fishes, amphibians, and reptiles and the ependyma is believed to play a significant role in the initiation and maintenance of the regenerative process in these species. Ependyma undergoes hypertrophy and hyperplasia, becomes pseudostratified, and transforms into process-bearing tanycyte-like cells at regeneration sites. Channels form between adjacent ependymal cell processes in order to provide a substrate for guidance and direction of regenerating spinal cord axons. In addition to initiating and maintaining spinal cord regeneration, ependymal cells have also been reported by these and other investigators to play a role in limb and tail regeneration as well as cartilage differentiation in these species.

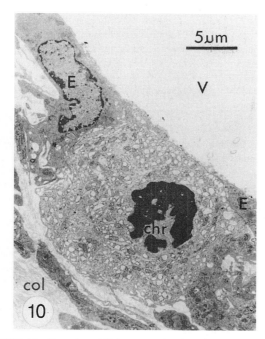

FIGURE 10 Ependymal lining near the site of an intraventricularly placed shunt tube in the rabbit lateral ventricle (V) showing that mitosis may occur while well-formed gap junctions are maintained with adjoining ependymal cells (E); col, collagen. [Reproduced with permission from M. R. Del Bigio and J. E. Bruni (1986). *J. Neurosurg,* **64**, 932–943.]

IV. FUNCTIONS OF EPENDYMAL CELLS

(1) Güldner and Wolff (1973), Scott and Paull (1979) have observed "Synapse-like" contacts between tanycytes and axon terminals of neurons within certain regions of the nervous system. These synaptic contacts, which can number in the hundreds per cell, contain clear and/or dense core vesicles but lack the typical postsynaptic densities seen in neuron-to-neuron synapses. This peculiar relationship has been interpreted to mean that some as yet unknown function(s) of these tanycytes may come under some form of nervous control. Curiously, however, Jarvis and Andrew (1988) were unable to detect electrophysiological correlates of these putative synapse-like contacts with neurons.

(2) The processes of tanycytes have been observed to ensheath preterminal or terminal portions of certain axons within the median eminence and to form "synapse-like" contacts with them. They also have along their surfaces receptors for various neurotransmitters (like dopamine) as well as growth factors. By virtue of these associations, Kozlowski and Coates (1985) and others have reported that tanycytes modulate hormone secretion into perivascular spaces of the hypophysial-portal vasculature by varying the extent of their investment of the axon terminals and hence their neurohemal contact, by altering their exposure to extracellular calcium, and by altering their membrane properties.

(3) According to Jarvis and Andrews (1988) ependyma-like neuroglial cells may take up and spatially buffer extracellular K^+ and tanycytes may shunt the K^+ to the CSF or into capillaries and, in so doing, influence the excitability of nearby neurons.

(4) Silverman *et al.* (1991) and others believe that tanycyte processes may act as a growth and guidance matrix in certain regions of the nervous system for migrating neurons during early development or they may provide a permissive substrate and/or trophic substances required to guide axons to their proper termination. In the mature nervous system, Bleier (1972) contends that tanycyte processes may delimit functional groups of neurons.

(5) Ependymal cells lining the ventricles may regulate the transport of water, ions, and molecules between the CSF and the brain extracellular fluid and accordingly function as a filtration barrier protecting the brain from potentially harmful substances. For details on the chemical and mechanical features of ependyma that would allow it to serve such a protective role, the reader is referred to the review by Del Bigio (1995).

(6) It has been suggested by Ma (1993) that tanycytes may react to the biochemical composition in the cerebral ventricle and plasma by altering nitric oxide synthesis and release, which in turn could influence neuronal activity and/or cerebral blood flow.

(7) Cuevas *et al.* (1991) and Tooyama *et al.* (1991) have localized fibroblast growth factor (FGF)-like immunoreactivity in both ependymal cells and tanycytes within the walls of the cerebral ventricles. Such presence has been related to their release into CSF after nervous system trauma, and it has been suggested that the ependyma may be involved in regulating FGF concentration in the CSF or some presently unknown function(s) of this growth factor on brain may be exerted through the CSF–ependymal pathway.

(8) Receptor sites for a plethora of ligands/factors (rat neural antigen-2, IGF-I, δ and μ opioid receptors, NGF receptors, glucose transporter protein, cell adhesion molecules, an anti-inflammatory protein lipocortin-1 (a determinant of lipid transport and metabolism, and endothelin receptors), as yet of unknown functional significance, have in recent years also been

localized within common ependymal cells and/or tanycytes.

(9) According to Sarnat (1992) and Gould *et al.* (1990) fetal ependymal cells, unlike their mature counterpart, exhibit features that denote a transport and/or secretory capacity. During the development of the nervous system, Sarnat (1992) reported fetal ependyma play an important role in the arrest of neurogenesis, axonal guidance, facilitation of motor neuron differentiation, maintenance and transformation of radial glia, and transport of nutrients before development of capillary networks. Gould *et al.* (1987, 1990) also suggested that ependyma may provide initial structural support for the germinal-layer blood vessels in the preterm infant and neuronal guidance.

BIBLIOGRAPHY

Akmayev, I. G., and Fidelina, O. V. (1976). Morphological aspects of the Hypothalamo-Hypophyseal System-VI. The Tanycytes: Their relation to sexual differation of the hypothalamus. *Cell. Tissue Res.* **173**, 407–416.

Anand Kumar, T. C. (1973). Cellular and humoral pathways in the neuroendocrine regulation of reproductive function. *J. Reprod. Fertil.* **20**, 11–25.

Bleier, R. (1972). Structural relationship of ependymal cells and their processes within the hypothalamus. *In* "Brain–Endocrine Interaction, Median Eminence: Structure and Function," pp. 306–318. Karger, Basel.

Bodega, G., Suarez, I., Rubio, M., and Fernandez, B. (1994). Ependyma: Phylogenetic evolution of glial fibrillary acidic protein (GFAP) and vimentin expression in vertebrate spinal cord. *Histochemistry* **102**, 113–122.

Brightman, M. W., Prescott, L., and Reese, T. S. (1975). Intercellular junctions of special ependyma: Brain–Endocrine Interaction II: The Ventricular System in Neuroendocrine Mechanisms. 2nd International Symposium, pp. 146–165. Karger, Basel.

Bruni, J. E. (1974). Scanning and transmission electron microscopy of the ependymal lining of the third ventricle. *Can. J. Neurol. Sci.* **1**, 59–73.

Bruni, J. E., and Anderson, W. A. (1987). Ependyma of the rat fourth ventricle and central canal: Response to injury. *Acta Anat.* **128**, 265–273.

Bruni, J. E., Clattenburg, R. E., and Millar, E. (1983a). Tanycyte ependymal cells in the third ventricle of young and adult rats. *Anat. Anz.* **153**, 53–68.

Bruni, J. E., Clattenburg, R. E., and Paterson, J. A. (1983b). Ependymal cells of the rat fourth ventricle: Response to injury. *Scanning Electron Microsc.* **2**, 649–661.

Bruni, J. E., Del Bigio, M. R., and Clattenburg, R. E. (1985). Ependyma: Normal and pathological. A review of the literature. *Brain Res. Rev.* **9**, 1–19.

Bryant, S. V., and Wozny, K. J. (1974). Stimulation of limb regeneration in the lizard *Xantusia vigilis* by means of ependymal implants. *J. Exp. Zool.* **189**, 339–352.

Chauhan, A. N., and Lewis, P. D. (1979). A quantitative study of cell proliferation in ependyma and choroid plexus in the postnatal rat brain. *Neuropathol. Appl. Neurobiol.* **5**, 303–309.

Cuevas, P., Gimenez-Gallego, G., Martinez-Murillo, R., and Carceller, F. (1991). Immunohistochemical localization of basic fibroblast growth factor in ependymal cells of the rat lateral and third ventricles. *Acta Anat.* **141**, 307–310.

Del Bigio, M. R. (1995). The ependyma: A protective barrier between brain and cerebrospinal fluid. *Glia.* **14**, 1–13.

Edwards, M. A., Yamamoto, M., and Caviness, V. S. (1990). Organization of radial glia and related cells in the developing murine CNS. *Neuroscience* **36**(1), 121–144.

Flament-Durand, J., and Brion, J. P. (1985). Tanycytes: Morphology and functions: A review. *Int. Rev. Cytol.* **96**, 121–155.

Fleischhauer, K. (1972). Ependyma and subependymal layer. *In* "The Structure and Function of Nervous Tissue" (G. H. Bourne, ed.), Vol. 6, pp. 1–46. Academic Press, New York.

Gould, J., and Howard, S. (1987). An immunohistochemical study of the germinal layer in the late gestation human fetal brain. *Neuropathol. Appl. Neurobiol.* **13**, 421–437.

Gould, S. J., Howard, S., and Papadaki, L. (1990). The development of ependyma in the human fetal brain: An immunohistological and electron microscopic study. *Dev. Brain Res.* **55**(2), 255–267.

Güldner, F-H., and Wolff, J. R. (1973). Neurono-glial synapse-like contacts in the median eminence of the rat: Ultrastructure, staining properties and distribution on tanycytes. *Brain Res.* **61**, 217–234.

Horstmann, E. (1954). Die faserglia des Selachiergehirns. *Z. Zellforsch. Mikros. Anat.* **39**, 588–617.

Jarvis, C. R., and Andrew, R. D. (1988). Correlated electrophysiology and morphology of the ependyma in rat hypothalamus. *J. Neurosci.* **8**(10), 3691–3702.

Keep, R. F., and Jones, H. C. (1990). A morphometric study on the development of the lateral ventricle choroid plexus, choroid plexus capillaries and ventricular ependyma in the rat. *Dev. Brain Res.* **56**(1), 47–53.

Knigge, K. M., and Scott, D. E. (1970). Structure and function of the median eminence. *Am. J. Anat.* **129**, 223–244.

Kozlowski, G. P., and Coates, P. W. (1985). Ependymoneuronal specializations between LHRH fibers and cells of the cerebroventricular system. *Cell Tissue Res.* **242**, 310–311.

Leonhardt, H. (1980). Ependym und Circumventriculäre Organe. *In* "Handbsuch der Microskopishen Anatomie des Menschen" (A. Oksche and L. Vollrath, eds.), Vol. 4, pp. 336–342. Springer-Verlag, Berlin.

Löfgren, F. (1960). The infundibular recess, a component in the hypothalamo-adenohypophyseal system. *Acta Morph. Neerl. Scand.* **3**, 55–78.

Ma, P. M. (1993). Tanycytes in the sunfish brain: NADPH-diaphorase histochemistry and regional distribution. *J. Comp. Neurol.* **336**, 77–95.

Millhouse, D. E. (1972). Light and electron microscopic studies of the ventricular walls. *Z. Zellforsch.* **127**, 149–172.

Nandy, K., and Bourne, G. H. (1965). Histochemical studies on the ependyma lining the central canal of the spinal cord in the rat with a note on its functional significance. *Acta Anat.* **60**, 539–550.

Norlander, R. H., and Singer, M. (1978). The role of ependyma in regeneration of the spinal cord in the urodele amphibian tail. *J. Comp. Neurol.* **180**, 349–374.

Peters, A., Palay, S. L., and de Webester, H. (1976). The ependyma. *In* "The Fine Structure of the Nervous System: The Neurons and Supporting Cells" (A. Peters, ed.). Saunders, Philadelphia.

Pilgrim, C. (1978). Transport function of hypothalamic tanycyte ependyma: How good is the evidence? *Neuroscience* **3**, 277–283.

Purkinje, J. E. (1836). Ueber Flimmerbewegungen im gehirn. *Muller's Arch. Anat. Physiol.* **289**.

Sarnat, H. B. (1992). Role of human fetal ependyma. *Pediatr. Neurol.* **8**, 163–178.

Sarnat, H. B. (1992). Regional differentiation of the human fetal ependyma: Immunocytochemical markers. *J. Neuropathol. Exp. Neurol* **51**(1), 58–75.

Sarnat, H. B. (1995). Ependymal reactions to injury. A review. *J. Neuropathol. Exp. Neurol.* **54**(1), 1–15.

Scott, D. E., and Paull, W. K. (1979). The tanycyte of the rat medion eminence—I synaptoid contacts. *Cell. Tissue Res.* **200**, 329–334.

Silverman, R. C., Gibson, M. J., Charlton, H. M., and Silverman, A. J. (1991). Relationship of glia to GnRH axonal outgrowth from the third ventricular grafts in hpg hosts. *Exp. Neurol.* **114**(3), 259–274.

Simpson, S. B. (1968). Morphology of the regenerated spinal cord in the lizard, *Anolis carolinensis. J. Comp. Neurol.* **134**, 193–210.

Tooyama, I., Hara, Y., Yasuhara, O., Oomura, Y., Sasaki, K., Muto, T., Suzuki, K., Hanai, K., and Kimura, H. (1991). Production of antisera to acidic fibroblast growth factor and their application to immunohistochemical study in the rat brain. *Neuroscience* **40**(3), 769–779.

Epidermal Proliferation

JAMES T. ELDER
University of Michigan

REINER LAMMERS
Max-Planck-Institut fur Biochemie

GLOSSARY

Autocrine In reference to cytokine–growth factor responsiveness, the cell responds to growth factors produced by the same cell; this response may take place intracellularly or on the cell surface

Basal cell Keratinocyte attached to the basement membrane of the dermoepidermal junction

Cytokine Any polypeptide that participates in intercellular signaling via secretion into the extracellular space and binding to receptors on target cells

Growth factor Any cytokine that induces cellular proliferation

Paracrine Responder cell is in the vicinity of the producer cell and may be a more or less differentiated cell of the same lineage, or a cell of a different lineage; no circulatory system is required to transport the factor from the producer to the responder cell

Stratum corneum Uppermost layer of the epidermis, consisting of plate-like squamous cells

THE EPIDERMIS IS A SPECIALIZED EPITHELIUM THAT forms the external portion of the largest organ of the body—the skin. The keratinocyte, a specialized epithelial cell, is the majority cell type of the epidermis. The keratinocyte must maintain a finely tuned balance between cell proliferation and loss through terminal differentiation in normal epidermal homeostasis and yet be able to greatly increase its proliferative rate in response to injury. A precise understanding of this regulation requires definition of the intercellular signals that regulate keratinocyte growth and differentiation. For each signal identified, it is necessary to define the cell type(s) responsible for its production, the target cell(s) upon which it acts, and the intracellular signaling pathways that mediate both its production and its ultimate effects on keratinocyte growth and differentiation. This article reviews our available knowledge about the molecular signals that regulate epidermal proliferation. For this purpose, we consider psoriasis an inflammatory and hyperplastic skin disease, as an "experiment of nature" that may be helpful in identifying the molecular signals that regulate epidermal homeostasis in normal epidermis.

I. PSORIASIS, WOUND HEALING, AND EPIDERMAL HOMEOSTASIS

Psoriasis is a common skin disease characterized by markedly increased proliferation and altered differentiation of keratinocytes in the context of multiple biochemical, immunological, inflammatory, and vascular abnormalities. Despite its complexities, psoriasis has much to teach us about the balance between proliferation and differentiation in self-renewing tissues. Psoriasis has been likened to an unregulated process of cutaneous wound healing, because epidermal hyperplasia, inflammatory cell infiltrates, and vascular alterations characterize both processes. One reason for this similarity could be that both processes are mediated by the same set of intercellular signals, including growth factors, cytokines, and extracellular matrix components. Another reason is that each cell type present in the skin is constrained by its developmental history to display only a subset of potential responses to extracellular signals. In contrast, the clinical features of psoriasis and healing wounds differ markedly from those of various malignant neoplasms. These differences may offer important clues to the molecular eitology of psoriasis and the normal regulation of epidermal proliferation.

A. Psoriasis Is Not Cancer

Although cancer is similar to psoriasis and healing wounds in that cellular proliferation is increased, cancer differs from psoriasis and healing wounds in several important ways. In normal individuals, cutaneous wound healing and psoriatic lesions are highly self-limited processes, which are confined to the skin. Psoriasis is often characterized by exacerbations and remissions, and lesions are frequently multifocal and often display considerable symmetry. In contrast, cancer usually arises in a single site and spreads locally or metastatically without remission or detectable symmetry. A substantial literature supports the concept that cancers arise owing to somatic mutations affecting either the structure or the appropriate regulation of critical cellular genes (protooncogenes). The expression of several protooncogenes (including *c-myc, c-fos, c-jun, c-erbB, c-Ha-ras*) in psoriasis is not significantly different from those in normal skin. The clinical behavior of psoriasis and healing wounds is more consistent with altered regulation of normal intercellular signaling processes, because the appearance of the skin can return to normal after wound healing is complete, after antipsoriatic therapy, or after spontaneous improvement of psoriasis. [*See* Oncogene Amplification in Human Cancer.]

B. Local Regulation of Keratinocyte Proliferation

The behavior of wounded skin suggests that at least part of the proliferative signal inherent to the repair process must arise locally within the wound, because epidermal hyperplasia is not observed at distant sites.

Overproduction of such a factor or factors required for keratinocyte growth could explain many features of the psoriatic phenotype, including its similarity to wound healing. Moreover, this local proliferative signal could result from either overproduction or enhanced responsiveness to factors that promote growth or to decreased production or responsiveness to factors that inhibit growth. [*See* Keratinocyte Transformation, Human.]

II. POSITIVE REGULATION OF EPIDERMAL PROLIFERATION

A. Transforming Growth Factor-α: An Autocrine Keratinocyte Growth Factor

Transforming growth factor-α (TGF-α) is a potent growth factor for keratinocytes and other epithelia (see the following). Whereas expression of protooncogene transcripts is comparable in normal skin and psoriatic lesions, TGF-α messenger RNA (mRNA) and protein levels are markedly increased in psoriatic lesions. These results identify TGF-α as a strong candidate for a locally produced growth factor hypothesized earlier.

TGF-α is the evolutionary homologue of epidermal growth factor (EGF) and binds to the EGF receptor with high affinity. Though TGF-α was initially described as a growth factor for transformed cells and is found in a variety of tumors, TGF-α and related EGF receptor-binding peptides are also produced by a variety of normal embryonic and adult tissues. In contrast, EGF itself appears to be synthesized in only a limited subset of tissues. There is no detectable EGF mRNA in epidermis or cultured keratinocytes under conditions in which TGF-α mRNA is easily detectable. Thus, TGF-α appears to be the endogenous epidermal ligand for the EGF receptor. TGF-α is synthesized as a 160-amino-acid precursor, which undergoes a variety of posttranslational modifications, including phosphorylation, glycosylation, and proteolytic cleavage. A higher-molecular-weight form of TGF-α is found in the plasma membrane of a variety of cell types (pro-TGF-α), from which TGF-α is released by proteolytic cleavage. TGF-α is mitogenic for many epithelial and fibroblastic cells, which express the EGF receptor. The membrane-associated form of pro-TGF-α has also been shown to have mitogenic activity, suggesting that cell-surface expression of TGF-α may be sufficient to stimulate the growth of adjacent cells. Moreover, several extracellular matrix (ECM) molecules, including thrombospondin, laminin, and tenascin, contain multiple EGF-like domains, suggesting that ECM molecules might directly stimulate the EGF receptor.

TGF-α induces its own expression in human and murine keratinocytes (autoinduction) and can also be induced in various epithelial tissues by other injurious stimuli such as ultraviolet light and partial hepatectomy. These observations suggest an important role for TGF-α in mediating epithelial proliferation in wound healing. The tumor-promoting phorbol ester tetradecanoyl phorbol acetate (TPA) is also a potent stimulus for expression of TGF-α, suggesting its possible role in the process of tumor promotion. Whether these various stimuli use common or divergent intracellular signal transduction pathways to induce TGF-α gene expression is not yet clear (see Section VI). [See Transforming Growth Factor-α.]

B. EGF Receptor

TGF-α exerts its pleiotypic effects on target cells by binding to and activating the EGF receptor (Fig. 1). The EGF receptor is a 170-kDa phosphoglycoprotein containing an extracellular domain that binds EGF or TGF-α, a short, hydrophobic membrane-spanning domain, and a C-terminal domain, which displays ligand-dependent tyrosine kinase activity. In vitro mutagenesis experiments have demonstrated that the tyrosine kinase activity of the EGF receptor is essential for the transduction of pleiotypic responses to EGF, including cell growth. It is likely that TGF-α activates the receptor kinase in a very similar manner, although blocking experiments using monoclonal anti-EGF re-

ceptor antibodies have indicated that EGF and TGF-α may recognize the receptor in slightly different ways. Normal cells regulate the number of EGF receptors via a combination of endosomal degradation (down-regulation) and transcriptional control. Transfection experiments in fibroblasts have shown that concurrent overexpression of TGF-α and EGF receptor leads to malignant transformation, whereas overexpression of either molecule alone is insufficient for transformation. In psoriasis, EGF receptor expression is comparable in normal and psoriatic skin. The lack of marked EGF receptor overexpression in psoriasis in the face of markedly elevated levels of TGF-α therefore distinguishes psoriasis and EGF receptor-mediated neoplastic transformation, is consistent with the clinical differences between psoriasis and cancer, and could help to explain the observation that the incidence of skin cancer is not higher is psoriatics than in the general population.

C. Paracrine Keratinocyte Growth Factors

When cultured in serum-free medium, epidermal keratinocytes require insulin in the culture medium for growth. This insulin requirement can be replaced by 100-fold lower concentrations of insulin-like growth factor 1 (IGF-1), suggesting that keratinocytes utilize the IGF-1 receptor rather than the insulin receptor. IGF-1 receptor mRNA is expressed by cultured keratinocytes and intact epidermis. However, neither cultured keratinocytes nor intact epidermis express detectable mRNA for insulin or IGF-1. Because circulating levels of insulin in vivo are much lower than those required for keratinocyte growth, it is likely that keratinocytes utilize IGF-1 as the ligand for the IGF receptor in vivo.

Keratinocyte growth is also stimulated by basic fibroblast growth factor (bFGF). Although some controversy exists at present, actively growing keratinocytes appear not to synthesize detectable mRNA for bFGF, acidic FGF, or a recently described homologue, keratinocyte growth factor (KGF). Cultured fibroblasts, however, express easily detectable IGF-1, bFGF, and KGF, and the size of each of these factors is such that they would be predicted to permeate the dermal–epidermal junction and gain access to keratinocytes. Yet fibroblasts do not express detectable TGF-α mRNA. Therefore, whereas TGF-α appears to be an autocrine growth factor for keratinocytes, IGF-1, bFGF, and KGF appear to be paracrine keratinocyte growth factors (Fig. 1). This model is sup-

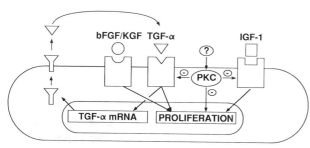

FIGURE 1 Signal transduction pathways mediating keratinocyte growth. The receptor depicted for TGF-α is the EGF receptor. The receptors for EGF/TGF-α, IFG-1, bFGF, and KGF possess intrinsic tyrosine kinase activity. Protein kinase C (PKC) negatively regulates the tyrosine kinase activities of the EGF and IGF-1 receptors. The extracellular mediator providing the physiological stimulus for PKC activation is unknown. Of the three growth factors depicted, only TGF-α is synthesized constitutively by keratinocytes.

ported by data from experiments with transgenic mice that overexpress a dominant negative KGF receptor in their basal keratinocytes. These animals are characterized by epidermal atrophy and delayed reepithelialization of a wound.

Studies involving transformed murine keratinocytes in serum-free medium have revealed that EGF acts synergistically with either bFGF or insulin to stimulate growth, whereas combinations of bFGF and insulin in the absence of EGF are inactive (Fig. 1). Since each of the receptors for these factors encodes a tyrosine kinase, these results suggest that phosphorylation of a certain combination of cellular substrates, which are differentially phosphorylated by the EGF, IGF-1, and bFGF receptors, is required for progression through the cell cycle. It has long been established in fibroblasts that EGF acts to render the cells competent to enter the cell cycle, whereas insulin acts as a progression factor, allowing the cells to traverse the G1 phase of the cycle and enter into the S phase. Thus, it would appear that a similar mechanism exists in keratinocytes, but that keratinocytes are an autocrine source of competence factors (TGF-α) and require a paracrine source of progression factors (IGF-1).

III. NEGATIVE REGULATION OF EPIDERMAL PROLIFERATION

A. Transforming Growth Factor-β

The requirement for a steady-state balance between proliferation and terminal differentiation in the epidermis and other epithelia suggests that some form of communication takes place between the proliferative and differentiated cell compartments. Several different polypeptides have been proposed to fill this role by acting as negative mediators of epidermal growth. Transforming growth factor-β (TGF-β) was discovered by virtue of its ability to promote the growth of transformed cells in semisolid medium. However, it was later appreciated to be a potent antiproliferative agent for a variety of nontransformed and transformed epithelial cells, including epidermal keratinocytes. This effect is associated with a dramatic reduction in expression of the protooncogene *c-myc* mRNA in human and murine keratinocytes and appears to be mediated by the product of the tumor suppressor gene, RB. Many different cell types and tissues express TGF-β, which is synthesized as a latent homodimeric precursor containing two interchain disulfide bridges. Mature, biologically active TGF-β is generated from

this complex by proteolytic cleavage and possibly by exposure to acidic environments *in vivo*. In many tissues, especially bone, TGF-β is an abundant component, but the bulk of the TGF-β appears to be in a biologically inactive state. It is now appreciated that there are multiple TGF-β species (TGF-β1 through TGF-β5 have been described), which are divergent members of a multigene family. In psoriasis, which is characterized by markedly increased epidermal proliferation, TGF-β mRNA levels are not significantly different in normal and lesional epidermis. Although the issue of how much biologically active TGF-β protein is present in psoriatic versus normal epidermis has not yet been adequately addressed, these data suggest that psoriatic epidermal hyperplasia is associated with overexpression of a growth stimulator (TGF-α) rather than loss of an antiproliferative stimulus (TGF-β). Moreover, whereas autoinduction of TGF-β has been observed in some cell types (e.g., fibroblasts), it has not been observed in keratinocytes. TGF-β causes a marked increase in keratinocyte spreading, and thus one of its normal biological roles in the epidermis may be to stimulate the flattening and spreading of keratinocytes, which occurs in the early phases of epidermal wound closure. Increased TGF-β immunoreactivity has also been observed in hair follicles, where it may play an important role in the regulation of the complex dermal–epidermal interactions involved in hair growth. [*See* Transforming Growth Factor-β.]

B. Interferon-γ

It is well documented that interferon-γ (IFN-γ), though not a product of keratinocytes but rather of T cells (see Section V,B), also displays growth inhibitory properties toward keratinocytes *in vitro*. Paradoxically, IFN-γ induces a 2.5-fold increase in TGF-α gene expression after 24–48 hr in cultured keratinocytes, even as the cells reduce their proliferative rate. This paradox may be explained by the fact that IFN-γ treatment concurrently decreases the expression of the EGF receptor gene, which would render the cells refractory to the increased TGF-α present. *In vivo*, it is possible that increased TGF-α produced by more-differentiated, EGF receptor-deficient cells stimulates the proliferation of less-differentiated, EGF receptor-rich basal and/or supra-basal cells. These results suggest that stimulation of autocrine growth factor expression in terminally differentiated cells may be an important mechanism by which the balance between proliferation and terminal differentiation is maintained in epidermal homeostasis. Although speculative, these results suggest a plausible model linking psoriatic epidermal hyperplasia and the increased numbers of T cells present in psoriatic lesions (see Section V,C). Because cyclosporin A (CsA) inhibits the production of IFN-γ by T cells, this model could also explain the potent antipsoriatic effect of CsA. [See Interferons.]

IV. EXTRACELLULAR MATRIX AND PROTEASES AS REGULATORS OF GROWTH FACTOR ACTION

It was mentioned earlier that certain extracellular matrix molecules have EGF-like domains, which could potentially stimulate the EGF receptor. In this regard, it is interesting that thrombospondin and fibronectin each promote migration of epidermal keratinocytes in vitro, because growth and migration of these cells appears to be closely linked. ECM molecules appear to have additional roles in modulating the effects of growth factors, principally by virtue of the ability of many growth factors and cytokines to bind to ECM molecules. Thus, both bFGF and granulocyte–macrophage colony-stimulating factor (GM-CSF) have been shown to bind to heparin-containing glycosaminoglycans, and heparin can potently potentiate or inhibit the effects of these growth factors in vitro depending on the experimental conditions employed. Finally, certain growth factors, including EGF and TGF-α and -β, can potently stimulate the synthesis and secretion of ECM molecules.

The structure and composition of the epidermal basement membrane and many other ECM-containing structures are dramatically altered under conditions of injury and certain disease states. These alterations are mediated in large part by the actions of extracellular proteases and their inhibitors, such as plasminogen activator and plasminogen activator inhibitor-1, and these molecules are components of the ECM itself. How this multifactorial network of ECM molecules, proteases, and growth factors operates to regulate the proliferation of the epidermis and other organs continues to be a challenging area for experimental analysis. [See Extracellular Matrix.]

V. THE CUTANEOUS CYTOKINE NETWORK

A rapidly expanding literature is documenting the many cytokines that can be expressed by keratinocytes and other cell types present in the skin, especially under in vitro conditions. In many cases, however, it is not currently clear whether or not active forms of these cytokines are actually present in normal skin and in disease states, and whether or not they are activating the same cellular responses in vivo as they are capable of in vitro. Though most cytokines have growth factor activity for certain immune and inflammatory cells, it is less clear in general that they have direct growth factor activity for epidermal keratinocytes.

Here we briefly review the cell types present in normal and psoriatic skin and consider the expression of selected cytokines likely to be produced by them in vivo as they pertain to epidermal proliferation (Table I). [See Cytokines in the Immune Response.]

A. Keratinocytes

Keratinocytes express a variety of cytokines in vitro, including interleukins-1α and -β, granulocyte, macrophage, and granulocyte–macrophage colony-stimulating factors (IL-1α, IL-1β, G-CSF, M-CSF, and GM-CSF, respectively), tumor necrosis factor-α (TNF-α), IL-3, and IL-8. Since keratinocytes are the majority cell type of the epidermis, more data are currently available to compare the in vivo and in vitro expression of these cytokines than for other cell types present in the skin.

There is general agreement that normal epidermis contains large amounts of IL-1α immunoreactivity and biological activity, as much as 60 μg of IL-1α per individual. Paradoxically, however, several laboratories have reported that IL-1α mRNA is undetectable in epidermal RNA by RNA blot hybridization. This apparent contradiction remains to be resolved. Moreover, epidermis contains large amounts of an inhibitor of IL-1 bioactivity, whose relevance to the regulation of epidermal IL-1 effects in vivo is currently unclear. Evidence from our laboratories indicates that IL-1α bioactivity and immunoreactivity are reduced, whereas IL-1β immunoreactivity and mRNA are increased in psoriatic lesions. Preliminary data from our laboratory indicates that mRNAs for GM-CSF and TNF-α are detectable in some samples from normal and psoriatic epidermis, whereas IL-8 mRNA is clearly increased in psoriatic lesions. In contrast to some recent reports, we find that IL-6 mRNA is undetectable in human epidermis. The roles of any of these cytokines in the stimulation of psoriatic epidermal hyperplasia remain speculative, although IL-3 and GM-CSF have been reported to stimulate keratinocyte

TABLE I

Regulators of Epidermal Growth

Factor	Produced by keratinocytes?	Effect on keratinocyte growth *in vitro*
Transforming growth factor-α	Y	+
Epidermal growth factor	N	+
Insulin-like growth factor-1	N	+
Insulin	N	+
Basic fibroblast growth factor	N(?)[a]	+
Keratinocyte growth factor	N	+
Transforming growth factor-β_1	Y	−
Transforming growth factor-β_2	Y	−
Interferon-γ	N	−
Tumor necrosis factor-α	Y	+/−
Epidermal mitosis-inhibitory pentapeptide	Y	−
Epidermal G1-chalone	Y	−
Interleukin-1α and -β	Y	0 (human)/+ (mouse)
Interleukin-3	Y	+(?)
Granulocyte–macrophage colony-stimulating factor	Y	+(?)
Interleukin-6	N(?)	0(?)
Interleukin-8	Y	?

[a](?) indicates conflicting results.

proliferation in defined medium. It is noteworthy that at least two of these cytokines (IL-1 and TNF-α) stimulate the expression of the adhesion molecule (ICAM-1 on keratinocytes. Expression of ICAM-1 allows binding of T lymphocytes to the epidermis, which may serve as a stimulus to epidermal proliferation (see the following section).

B. T Lymphocytes

As mentioned earlier, IFN-γ is expressed primarily by T cells and natural killer cells and exhibits growth inhibitory effects on keratinocytes. Since T cells are a prominent component of the infiltrating cells in psoriatic lesions, the effects of IFN-γ and other lymphokines on keratinocyte proliferation may be relevant to the pathophysiology of this disorder. It has been shown that keratinocytes from psoriatic patients are relatively refractory to the growth inhibitory properties of IFN-γ, and it was suggested that this could be responsible for enhanced keratinocyte proliferation in psoriasis. However, the refractoriness to IFN-γ-induced growth inhibition relative to normal keratinocytes is relatively small (about 1.5-fold) and seems

unlikely to be the sole cause of the marked (10-fold) increase in proliferation characteristic of psoriatic epidermis. At the present time, there is little convincing evidence for increased expression of IL-2, IL-3, IL-4, IFN-γ or other T cell-derived cytokines in psoriatic lesions. Moreover, with the exception of IL-3, currently no evidence indicates that any of these lymphokines directly stimulate keratinocyte growth *in vitro*. [See Lymphocytes.]

C. Antigen-Presenting Cells

Antigen-presenting cells include monocyte–macrophages, Langerhans cells (epidermal dendritic cells with antigen-presenting capacity), and dermal dendritic cells (dermal dendrocytes). These cells are thought to stimulate proliferation of antigen-specific T cells by a combination of signals: antigen bound to class II major histocompatibility complex (MHC) molecules on the cell surface of the APC (first signal) and elaboration of IL-1 (second signal). A dermal dendritic cell of bone marrow origin, distinct from Langerhans cells, was identified and shown to have antigen-presenting properties (see Section V,F). Increased numbers and functional alloantigen-presenting activity of APCs have been demonstrated in psoriatic lesions; however, the relevance of this observation to psoriatic epidermal hyperplasia has not been established. Cells of the monocyte–macrophage lineage produce a number of cytokines and growth factors upon activation by various stimuli; among these are IL-1, IL-3, IL-6, IL-8, TGF-α, and TGF-β. In evolving psoriatic lesions, it is possible that TGF-α elaborated by monocyte–macrophages or dendritic cells could trigger lesion development by stimulating keratinocyte expression of TGF-α. [See Macrophages.]

D. Other Blood Elements and Mast Cells

Polymorphonuclear leukocytes (PMNs) are abundant in psoriatic lesions, forming microabscesses just below the stratum corneum of the epidermis. PMNs express IL-1α and -β in surprising amounts, as well as a number of other inflammatory mediators such as eicosanoids, proteases, and peroxidases. The platelet α-granule and mast cell granules contain a diverse array of growth factors, including a TGF-α-like peptide, TGF-β, and, as the name implies, platelet-derived growth factor, as well as numerous protease activities. Psoriatic lesions are characterized by gross exudation of plasma into the epidermal compartment, and stimuli capable of causing platelet degranulation may release large quantities of these factors into the epidermis in a localized fashion. The role of these factors, as well as complement components such as C5a in the generation of psoriatic epidermal hyperplasia, is poorly understood.

E. Melanocytes

Although the primary role of melanocytes in the epidermis appears to be the synthesis of melanin and its transport to keratinocytes for purposes of photoprotection against ultraviolet light, its properties as a growth regulator for keratinocytes are largely unexplored. In part, this has been due to difficulties in establishing in vitro cultures of untransformed melanocytes. Recent advances in this area, however, have indicated that melanocytes display a proliferative response to bFGF and insulin/IGF-1. However, production of keratinocyte growth factors by melanocytes is only beginning to be explored in vitro.

F. Fibroblasts

The potential role of the fibroblast in the generation of paracrine keratinocyte growth factors was discussed previously. However, it is noteworthy that many of the spindle-shaped cells of the dermis previously called fibroblasts may actually be bone marrow-derived dermal dendrocytes of the monocyte–macrophage lineage; these dermal dendrocytes can be identified by surface staining for factor XIIIa and other bone marrow-derived cell markers and are found in increased abundance in psoriatic lesions. Fibroblastic cells derived from psoriatic lesions have been reported to stimulate the outgrowth of normal keratinocytes in collagen gels better than normal fibroblasts; however, these results have been difficult to duplicate under different culture conditions. Whether or not contamination of the fibroblastic cells with dermal dendrocytes is responsible for the conflicting results is, at present, unclear.

G. Endothelial Cells

Alterations in the cutaneous vasculature, specifically the superficial capillary plexus of the papillary dermis, occur very early in the development of psoriatic lesions and are the last morphological changes to normalize following antipsoriatic therapy or spontaneous improvement of psoriasis. These changes consist of elongation, increased tortuosity, and dilatation of the capillary loop as a result of increased endothelial cell proliferation. Therefore, it is of interest that the epidermis is a more potent source of angiogenic factors than is the dermis in various in vivo assay systems, and that TGF-α is a potent angiogenic factor in the hamster cheek pouch model. Since TGF-α is capable of inducing vascular endothelial growth factor (VEGF) in vitro it is probable that VEGF mediates these effects. However, a wide variety of cytokines possess angiogenic activity, and it would be premature to assign the psoriatic angiogenic activity solely to TGF-α at this time. Capillary permeability is increased in psoriasis, and this may allow increased egress of plasma growth factors into the epidermal milieu as well as provide a marked increase in total blood flow to the epidermal compartment. Recently, increased attention has also been given to the stimulation of binding to and diapedesis between endothelial cells by T cells and monocyte–macrophages in response to IL-1, TNF-α, IFN-γ, and other cytokines, which may be abnormally active in psoriatic lesions. Clearly, these factors may be very important in the regulation of epidermal growth in psoriasis as well as other inflammatory disorders, many of which are characterized by increased epidermal proliferation.

H. Neurons

Free nerve fibers reach the dermoepidermal junction and penetrate the epidermis, and these fibers contain tachykinins such as substance P, calcitonin generelated peptide, neuropeptide Y, and vasoactive intestinal peptide. These mediators have been shown to generate an efferent signal to the skin (the axon reflex) even though they are contained within sensory nerve fibers. It is the experience of every dermatologist that many dermatoses are exacerbated by psychic "stress." Elucidation of the molecular details of the phenomenon remains a major objective of dermatological research.

VI. SIGNAL TRANSDUCTION MECHANISMS AND GROWTH FACTOR RESPONSIVENESS

The major signal transduction pathways thought to be active in keratinocytes are (1) the receptor-associated tyrosine kinases; (2) G proteins, which couple extracellular receptors to intracellular effectors, including the adenylate cyclase–protein kinase A system and phospholipase C; (3) the phosphoinositide (PI) cycle, utilizing PI kinases and PI-specific phospholipase C to regulate release of calcium from intracellular stores; (4) calcium–phospholipid-dependent protein kinases [protein kinase C (PKC)]; and (5) calcium–calmodulin-dependent protein kinases.

One example of how signal transduction pathways may interact to regulate growth factor responsiveness has potential therapeutic relevance to psoriasis and involves the interaction of PKC with the EGF receptor tyrosine kinase (Fig. 1). PKC-mediated phosphorylation of the threonine-654 residue of the EGF receptor reduces the tyrosine kinase activity of the receptor without reducing its affinity for ligand. In psoriasis, PKC levels are markedly reduced relative to normal epidermis. Thus, it is possible that the EGF receptor tyrosine kinase is more active in psoriatic epidermis. Because all responses to exogenous EGF (or TGF-α) appear to involve the tyrosine kinase activity of the EGF receptor, it is likely that the autoinductive response to TGF-α does so as well. Therefore, loss of PKC activity in psoriasis may release a constraint that is normally placed on autoinduction, resulting in increased TGF-α production and responsiveness, as well as increased epidermal proliferation. The explanation for reduced PKC levels in psoriasis is currently unclear but could involve down-regulation of PKC due to an as yet unidentified, chronic stimulus. Such a stimulus could be provided by infiltrating immune and/or inflammatory cells or nerves in the form of histamine, bradykinin, platelet-activating factor, or tachykinins such as substance P. This model has the advantage of integrating several biochemical and cellular abnormalities known to be present in psoriasis. Future studies in our laboratories will test the many hypotheses raised by this model at the molecular level.

BIBLIOGRAPHY

Balkwill, F. R., and Burke, F. (1989). The cytokine network. *Immunol. Today* **10**(9), 299.

Bishop, J. M. (1987). The molecular genetics of cancer. *Science* **235**, 305.

Braverman, I. M., and Sibley, J. (1982). Role of the microcirculation in the treatment and pathogenesis of psoriasis. *J. Invest. Dermatol.* **78**, 12.

Burgess, W. H., and Maciag, T. (1989). The heparin-binding (fibroblast) growth factor family of proteins. *Annu. Rev. Biochem.* **58**, 575.

Chen, W. S., Lazar, C. S., Poenie, M., Tsien, R. Y., Gill, G. N., and Rosenfeld, M. G. (1987). Requirement for intrinsic protein tyrosine kinase in the immediate and late actions of the EGF receptor. *Nature* **328**, 820.

Derynck, R. (1988). Transforming growth factor-α. *Cell* **54**, 593.

Elder, J. T., Fisher, G. J., Lindquist, P. B., Bennett, G. L., Pittelkow, M. R., Coffey, R. J., Ellingsworth, L., Derynck, R., and Voorhees, J. J. (1989). Overexpression of transforming growth factor α in psoriatic epidermis. *Science* **243**, 811.

Fantl, W. J., Johnson, D. E., and Williams, L. T. (1993). Signalling by receptor tyrosine kinases. *Annu. Rev. Biochem.* **62**, 453.

Finch, P. W., Rubin, J. S., Miki, T., Ron, D., and Aaronson, S. A. (1989). Human KGF is FGF-related with properties of a paracrine effector of epithelial cell growth. *Science* **245**, 752.

Kupper, T. S. (1989). The role of epidermal cytokines. *In* "The Immunophysiology of Cells and Cytokines" (J. Oppenheim and E. Shevach, eds.). Oxford Univ. Press, New York.

Massague, J., and Polyak, K. (1995). Mammalian antiproliferative signals and their targets. *Curr. Opin. Genet. Dev.* **5**, 91.

Pardee, A. B. (1987). Molecules involved in proliferation of normal and cancer cells: Presidential address. *Cancer Res.* **47**, 1488.

Sporn, M. B., and Roberts, A. B. (1988). Peptide growth factors are multifunctional. *Nature* **332**, 217.

Werner, S., Smola, H., Liao, X., Longaker, M. T., Krieg, T., Hofschneider, P. H., and Williams, L. T. (1994). The function of KGF in morphogenesis of epithelium and reepithelialization of wounds. *Science* **266**, 819.

Epilepsies

DOUGLAS S. F. LING
LARRY S. BENARDO
State University of New York Health Science Center at Brooklyn

GLOSSARY

Antagonist Chemical substance which binds to a receptor and prevents the activity of the natural transmitter substance at a synapse

Electroencephalogram Recording of electrical activity of the brain as seen by electrodes fastened to the scalp

Receptor Site on a neuron that binds a specific chemical substance (neurotransmitter) released from another neuron; activation of a receptor initiates a cascade of events that causes either excitation or inhibition of the neuron

Seizure Clinical presentation of episodic stereotypic motor, sensory, autonomic, psychic, or cognitive events that result from a paroxysmal abnormal discharge of central nervous system neurons

Synapse Specialized connection between neurons for interneuronal communication commonly mediated through the release of a chemical neurotransmitter

THE EPILEPSIES REPRESENT A FAMILY OF NEURO-logical disorders characterized by recurrent, transient, self-sustained seizures. This group of disorders exhibits myriad symptoms, occurs in a complex array of patterns, and has many causes. Some epilepsies can progress from subtle disturbances to become generalized seizures, whereas others remain stable. Although arising from a wide variety of etiological factors, all conditions that induce seizures cause cerebral neurons to become excessively excited. The excessive discharge among groups of brain cells may be limited to a particular focal area in the brain or may involve wide regions of the central nervous system. The symptoms of epilepsy are diverse, and seizure manifestations take on many forms. The most common presentations include muscular spasms or jerks, episodes of partial or complete loss of consciousness, and impairment of awareness during seemingly conscious activity. Most epileptic seizures can be classified on the basis of their clinical pattern and the accompanying electroencephalographic (EEG) pattern.

I. CLASSIFICATION OF EPILEPTIC SEIZURES

The two most widely accepted classification systems for seizures were developed by commissions organized by the International League against Epilepsy to categorize epileptic seizures: the International Classification of Epileptic Seizures (ICES, Table I) and the more recently devised International Classification of Epilepsies and Epileptic Syndromes (ICE, Table II). These systems consolidate the wide diversity of epileptic presentations into concise schemes to allow for a more precise description of seizures beyond the traditional terminology of *"grand mal," "petit mal," "psychomotor,"* and *"focal."* The ICES classifies epileptic seizures by seizure type, emphasizing symptoms and location of seizure origin and spread. The ICE classifies epilepsies and epileptic syndromes by taking into account additional determinants such as etiology, age of onset, precipitating factors, and severity. Both schemes maintain a basic distinction between partial (i.e., focal, local) and generalized seizures. Partial seizures arise from the localized spread of activity from

TABLE I

The International Classification of Epileptic Seizures[a]

I. Partial (focal, local) seizures
  A. Simple partial seizures (consciousness not impaired)
    1. With motor symptoms
    2. With somatosensory or special sensory systems
    3. With autonomic symptoms
    4. With psychic symptoms
  B. Complex partial seizures (with impairment of consciousness)
    1. Beginning as simple partial seizures and progressing to impairment of consciousness
      a. With no other features
      b. With features as in I.A. 1-I.A.4
      c. With automatisms
  C. Partial seizures evolving to secondarily generalized seizures
    1. Simple partial seizures evolving to generalized seizures
    2. Complex partial seizures evolving to generalized seizures
    3. Simple partial seizures evolving to complex partial seizures to generalized seizures
II. Generalized seizures (convulsive or nonconvulsive)
  A. Absence seizures
    1. Absence seizures
    2. Atypical absence seizures
  B. Myoclonic seizures
  C. Clonic seizures
  D. Tonic seizures
  E. Tonic–clonic seizures
  F. Atonic seizures
III. Unclassified epileptic seizures
    Includes all seizures that cannot be classified because of inadequate or incomplete data and some that defy classification in hitherto described categories. This includes some neonatal seizures, e.g., rhythmic eye movements, chewing, and swimming movements.

[a]Source: Commission on Classification and Terminology of the International League Against Epilepsy, (1981). Proposal for revised clinical and electroencephalographic classification of epileptic seizures. *Epilepsia* **22**, 489–501. (© 1981 The International League against Epilepsy, used with permission.)

specific loci in the cerebral cortex. The symptoms of partial seizure range from impairment of sensation or thought to convulsive movements of a specific part of the body. Some seizures may begin as local epileptiform activity that quickly spreads throughout the brain (i.e., secondarily generalize). Primary generalized seizures involve both hemispheres of the brain simultaneously from the outset. As a rule, generalized seizures involve loss of consciousness. The symptoms may be limited to loss of consciousness (e.g., absence seizures) or may entail tonic–clonic (grand mal) activity.

A. Partial Seizures

Partial seizures are the most common seizure type overall and comprise the vast majority of adult onset epilepsy. These focal seizures initially affect only a limited area within the cerebrum, most commonly in the frontal or temporal lobes. The clinical characteristics of the seizure reflect the specific part of the brain affected, so a myriad of symptoms may occur. Partial seizures are divided into three main categories: (1) simple partial seizures, in which consciousness is not impaired; (2) complex partial seizures, in which there is some impairment of consciousness; and (3) partial seizures evolving to secondarily generalized seizures. Simple partial seizures may evolve into complex partial seizures, i.e., consciousness may be lost as the seizure progresses. Both simple and complex seizures may become "secondarily generalized," developing into a generalized seizure (usually a tonic–clonic, generalized tonic, or generalized clonic seizure). In these cases, the partial seizure may be experienced as an aura or warning immediately before the generalized seizure.

Simple partial seizures are focal events that may involve motor systems, sensory systems, autonomic systems, or psychic phenomena. The clinical manifestations are stereotypic and might include jerking (clonus) of any muscle group (motor); tingling or numbness (sensory); blood pressure or heart rate changes (autonomic); and déjà vu, jamais vu, or hallucinations (psychic). Simple partial seizures may occur at any age and signal a localized cerebral lesion. Any cortical region is vulnerable, with the temporal cortex being the most common site.

Complex partial seizures may begin as simple partial seizures which progress and are followed by an impairment of consciousness. In some cases, consciousness may be impaired from the outset. During the impairment of consciousness the patient is inaccessible, appearing vacant or glazed, and there may be confusion or motor symptoms, the latter often involving automatisms. An automatism is involuntary motor activity ranging from simple actions such as chewing or swallowing to more complex, semi-purposeful behavior as singing, gesturing, or walking. The duration of complex partial seizures is highly variable, lasting from a few seconds to several minutes. Amnesia for the event is common and usually encompasses the entire duration of the seizure. The patient may

TABLE II

International Classification of Epilepsies and Epileptic Syndromes[a]

1. Localization-related (focal, local, partial) epilepsies and syndromes
   1.1 Idiopathic (with age-related onset). At present, two syndromes are established:
       • benign childhood epilepsy with centrotemporal spike
       • childhood epilepsy with occipital paroxysms
   1.2 Symptomatic: This category comprises syndromes of great individual variability
2. Complex partial seizures (with impairment of consciousness)
   2.1 Idiopathic (with age-related onset, in order of age appearance)
       • Benign neonatal familial convulsions
       • Benign neonatal convulsions
       • Benign myoclonic epilepsy in infancy
       • Childhood absence epilepsy (pyknolepsy, petit mal)
       • Juvenile absence epilepsy
       • Juvenile myoclonic epilepsy (impulsive petit mal)
       • Epilepsy with grand mal seizures (GTCS) on awakening
   2.2 Idiopathic and/or symptomatic (in order of age appearance)
       • West's syndrome (infantile spasms, Blitz–Nick–Salaam Krämpfe)
       • Lennox–Gastaut syndrome
       • Epilepsy with myoclonic–astatic seizures
       • Epilepsy with myoclonic absences
   2.3 Symptomatic
       2.3.1 Nonspecific etiology
           • Early myoclonic encephalopathy
       2.3.2 Specific syndromes
           • Epileptic seizures may complicate many disease states. Included under this heading are those diseases in which seizures are a presenting or predominant feature.
3. Epilepsies and syndromes undetermined as to whether they are focal or generalized
   3.1 With both generalized and focal seizures
       • Neonatal seizures
       • Severe myoclonic epilepsy in infancy
       • Epilepsy with continuous spikes and waves during slow-wave sleep
       • Acquired epileptic aphasia (Landau–Kleffner syndrome)
   3.2 Without unequivocal generalized or focal features
4. Special syndromes
   4.1 Situation-related seizures (Gelegenheitsanfälle)
       • Febrile convulsions
       • Seizures related to other identifiable situations, such as stress, hormones, drugs, alcohol, or sleep deprivation
   4.2 Isolated, apparently unprovoked epileptic events
   4.3 Epilepsies characterized by specific modes of seizures precipitated
   4.4 Chronic progressive epilepsia partialis continua of childhood

[a]Source: Commission on Classification and Terminology of the International League Against Epilepsy. (1985). Proposal for classification of epilepsies and epileptic syndromes. *Epilepsia, 26,* 268–278. (© 1985 The International League against Epilepsy, used with permission.)

have no recollection of the seizure and must be informed by a witness that one has occurred.

B. Primary Generalized Seizures

This type of epilepsy has its onset mainly during childhood and adolescence. In generalized seizures, the clinical presentation and EEG recording show that both hemispheres of the brain are simultaneously involved in the seizure from the onset of the attack, with no apparent anatomical or functional focus. Consciousness is invariably impaired from the onset due to the extensive cortical involvement, and motor changes are bilateral and may be accompanied by autonomic discharge. The EEG patterns are bilateral and highly synchronous over both hemispheres, affecting most of the central gray matter. Generalized seizures are divided into six clinical patterns, but any combination of types may occur.

Absence seizures were the so-called *"petit mal"* attacks. They commonly develop during childhood or adolescence, and are characterized by a sudden, brief interruption of consciousness (the absence), often accompanied by a cessation of motor activity. The patient will be in a state of "suspended animation," unaware and inaccessible. Some absence seizures are accompanied by motor symptoms, such as automatisms. Typical absence seizures demonstrate a characteristic EEG pattern of an intermittent regular, symmetrical, synchronous paroxysmal activity. The attack usually occurs without warning and ends as abruptly as the onset, with previous activity resumed as if nothing happened. Often, the patient is unaware that an attack has occurred. Absence seizures may be precipitated by several factors such as drowsiness, photic stimulation, and hyperventilation.

In atypical absence seizures, tone changes are more severe than in typical absence seizures and the onset is not as sudden. Patient awareness and accessibility are not always completely impaired. The EEG pattern is distinct from that of typical absence seizures, showing slower paroxysmal activity. Atypical absences are often associated with mental retardation or other severe neurological abnormalities.

Myoclonic seizures consist of sudden, brief contractions of the limbs and/or trunk. They may occur singly or as a rapid succession of events and may vary in severity. Recovery is usually immediate.

Clonic seizures involve repetitive contractions of body musculature. The jerking is often asymmetrical and irregular, but can be bilaterally synchronous. Clonic seizures are usually found in infants and young adults.

Tonic seizures are characterized by a sustained contraction of body musculature. The onset is often accompanied by a loud cry as air is forced through the larynx (and vocal cords) from the involuntary contraction of the laryngeal and respiratory muscles. The patient falls due to the tonic contraction of axial and limb muscles. Tonic seizures are usually brief, lasting from a few seconds to a minute.

Tonic–clonic seizures represent the classical *"grandmal"* attack. They begin with a tonic phase and progress to a clonic stage. The entire seizure may last 3 to 5 min. During the tonic phase, the patient's body stiffens and respiration stops; cyanosis may occur. The tonic phase usually lasts less than a minute. As the seizure progresses to the clonic phase, convulsive movements of all four limbs, jaw, and facial muscles usually occur. The tongue is sometimes bitten due to the involuntary jaw contractions. Increased salivation may cause frothing at the mouth. Urinary incontinence may occur. The clonic stage lasts for 30 to 60 sec, after which the patient is often confused and sleepy for several minutes. There is often amnesia for the seizure.

Atonic seizures are characterized by a sudden loss of postural tone, sometimes causing the patient to fall abruptly to the ground. It usually begins in childhood and persists into adulthood. The episodes are very short in duration and frequently cause injury. Consciousness may be lost briefly and is usually regained within moments of the loss of tone. These seizures are usually observed in mentally retarded children with other evidence of brain abnormality.

C. Epileptic Syndromes

Febrile convulsions are generalized seizures associated with high fever in children between 6 months and 5 years of age. They occur in about 5% of the population and are generally benign. The seizures are of generalized tonic or tonic–clonic type. In general, uncomplicated febrile convulsions carry a moderate risk of recurrence (30–40%) and a low risk of developing epilepsy (about 1%).

Infantile spasms or West's syndrome is a disorder seen in infants characterized by seizures exhibiting abrupt flexions of the neck, trunk, and limbs, and a highly characteristic EEG pattern known as hypsarrhythmia. It occurs during the first 2 years of age. They may appear many times a day, frequently in clusters. The hypsarrhythmic EEG is a mixture of high-amplitude spikes, polyspikes, and high-amplitude slow waves occurring in a disorganized pattern. Mental retardation is present in more than 80% of infants exhibiting West's syndrome. Causes of the syndrome are varied.

Lennox–Gastaut syndrome is a complex condition characterized by multiple patterns of generalized seizures, <3 Hz slow spike-and-wave EEG patterns, and mental retardation. Its prognosis is uniformly poor. The age of onset is usually before 8 or 9 years. This syndrome is often difficult to treat.

Benign rolandic epilepsy is a relatively common disorder which usually begins between the ages of 3 and 13 years. Children are neurologically and mentally normal. The seizures occur infrequently and generally respond well to antiepileptic drugs. The seizures usually cease by age 16. The seizure begins as a simple partial seizure, with sensory and motor involvement of the face, mouth, tongue, and larynx; cessation of speech occurs. Loss of consciousness and generalized convulsions may ensue with nocturnal attacks. Diurnal episodes are usually partial simple seizures, whereas nocturnal episodes are often secondarily generalized seizures. The EEG is characterized by recurrent high-voltage spikes in the central midtemporal area of the brain.

Juvenile myoclonic epilepsy is a disorder that usually begins between ages 12 and 19 years of age. Symptoms include recurrent bilateral myoclonic jerks occurring after waking, sometimes followed by generalized (tonic–clonic) seizures. The EEG exhibits characteristic bursts of generalized 3- to 6-Hz spike-and-wave complexes. These seizures often respond well to drug therapy. However, attacks may recur after medication is withdrawn.

Reflex epilepsy describes seizures (either focal or generalized) precipitated by various sensory stimuli, such as touch, sound, music, reading, speaking, or lights. Reflex epilepsy is extremely rare.

Generalized tonic–clonic seizures may occur with awakening. It has been shown that a relationship exists between the circadian sleep–wake cycle. Most patients with awakening seizures exhibit an onset of the disorder between the ages of 10 and 25 years. The generalized tonic–clonic seizures are often associated with myoclonic seizures and, in some rare instances, absence seizures.

II. EPIDEMIOLOGY

Epilepsies occur in about 1–2% of the general population, and it is estimated that new cases appear at an annual rate of 40/100,000, suggesting that over 100,000 new cases are diagnosed annually in the United States alone. The rate is highest in children less than 5 years old, is relatively stable between the ages of 20 and 50 years, and then steadily rises again in elderly people. The incidence of partial (focal) seizures is generally higher than that of generalized seizures.

III. ETIOLOGY

The clincial causes of seizures are highly diverse and sometimes obscure. Epileptic seizures may be the result of a single idiopathic factor or a convergence of many factors which all contribute to seizure generation. Cerebral infections, head trauma, anoxia, cerebral vascular disease, progressive cerebral degeneration, brain tumors, stroke, metabolic and systemic diseases, and certain medications may all cause epileptic attacks of some type. However, in many patients no underlying cause can be determined. Moreover, the etiology varies considerably with the age of onset.

A. Childhood Seizures

Some newborn children (about 1%) exhibit symptomatic convulsions, usually partial seizures. Causes of

neonatal seizures include birth complications such as hypoxia or hemorrhage, cerebral infections such as meningitis, or metabolic problems such as hypoglycemia. Congenital birth defects may also play a role in childhood seizures.

B. Head Injury

Epileptic seizures are observed in a significant number of patients who sustain head injuries. Some of these partial seizures are seen in the first week ("early" posttraumatic seizures). The prognosis is favorable, with attacks subsiding in most patients. Convulsive episodes exhibited after the first week ("late" epilepsies) occur in a smaller proportion of cases. In most cases, these seizures do not subside. In general, the chance of epilepsy continuing in a patient who has sustained a traumatic head injury are increased by complicating factors such as depressed cranial fracture, open head injury, and intracranial hemorrhage.

C. Toxic States

Certain toxic states can lead to seizures, but not epilepsy, as reversal of the toxicity will stop the seizure. For example, the ingestion of legal (amphetamines, antidepressants, alcohol) or illegal (cocaine, PCP) drugs can lead to seizures by lowering seizure threshold. Likewise, certain metabolic states, such as hypo- or hyperglycemia, hypocalcemia, hypomagnesemia, or depressed hepatic or renal function, can also predispose individuals to seizures. [*See* Alcohol Toxicology; Cocaine and Stimulant Addiction.]

D. Genetic Factors

Seizures can accompany inherited disorders, congenital diseases, malformations, and chromosomal abnormalities. However, the precise role of inheritance in epileptic seizures remains unclear, as epilepsy is a consequence of the cerebral disorders which may accompany hereditary or genetic diseases. Furthermore, it can be difficult to distinguish the respective roles of inheritance and environmental factors on the development of conditions that may predispose an individual to epilepsy. However, increasing evidence shows that genetics is an important determinant in some epileptic syndromes, especially in cases of pediatric seizures. Family members of patients with generalized epilepsy have an increased probability of developing seizures. They are also more likely to develop seizures following head injury or cerebral infection. Better elucidation of the genetic factors of epilepsy will undoubtedly have a significant impact on the treatment of this disorder. [*See* Genes; Human Genetics.]

IV. PATHOPHYSIOLOGY

Most epileptic seizures occur in the cerebral cortex. Many of the characteristics of neurons and neuronal circuits of the cerebral cortex may contribute to apparent vulnerability of this region to seizures. The principal cells in the cortex are pyramidal cells. The dense population and organization of pyramidal cells confer a high degree of cell-to-cell connectivity within many cortical regions. Such a high-fidelity system may predispose cortical circuits to synchronous firing. The cellular arrangement appears to be conducive to the generation of large electrical field potentials which promote the spread of electrical activity. [*See* Cortex.]

Neurons in specific regions of the cortex have distinctive, inherent properties which allow them to fire high-frequency bursts of electrical activity. They are capable of repetitive, high-frequency firing when subject to prolonged activation, such as that induced by synaptic excitatory drive.

Another characteristic feature of the cortex is the high level of feedback or recurrent excitatory connections. In this arrangement, exciting one pathway may lead to a "chain reaction" consisting of regenerative excitation of an entire population of cells. The high degree of positive (i.e., excitatory) feedback circuits contributes to the amplification of excitation and ultimately to the spread of seizure activity.

A. Working Theories of Epileptogenesis

Many theories and models exist to describe the pathophysiological events that give rise to seizure generation. A common concept that underlies all working theories of epileptogenesis is that seizures arise, in part, from domination of activity by excitation, most likely caused by an imbalance between excitatory and inhibitory electrical activity in the central nervous system. Normally, inhibitory inputs to neurons moderate excitatory activity. When excitation overwhelms inhibition, a synchronous discharge of neurons may develop and be allowed to spread unchecked. This activity may be localized to a particular region (focal seizures) or can extend throughout the brain (generalized seizure). The basic physiological mechanisms underlying seizures have been studied at molecular, cel-

lular, and neuronal network levels. It is likely that clinical epileptic seizures arise from a combination of mechanisms. Much of the evidence for the basic mechanisms underlying seizure generation comes from studies in both animal models and human epileptic patients. [*See* Autonomic Nervous System.]

B. Enhancement of Glutamate-Mediated Excitation

Glutamate is the primary excitatory amino acid neurotransmitter in the central nervous system and consists mainly of three receptor types: N-methyl-D-aspartate (NMDA), kainate, and α-amino-3-hydroxy-5-methylioxyzole-4-propionic acid receptors. Enhancement of the glutamate receptor-mediated activity would contribute to hyperexcitability. Repeated activation of NMDA receptors can lead to lasting potentiation of excitatory drive, implicated in long-term plastic changes in neuronal circuits such as those involved in memory and learning. Experiments in animal preparations have shown that maneuvers which excessively increase NMDA-mediated neurotransmission give rise to repetitive, synchronous (epileptiform) firing. Thus, the high-frequency activity exhibited by cortical neurons can lead to both short-term and long-term enhancements of synaptic excitatory drive. This could predispose cortical networks to the generation and spread of regenerative, synchronous electrical activity.

C. Reductions in GABA-Mediated Inhibition

γ-Aminobutyric acid (GABA) is the primary inhibitory neurotransmitter in the brain. Activation of neuronal GABA receptors causes a depression of neuron firing. GABA-mediated inhibition exerts a moderating influence on the electrical activity of neurons and neuronal circuits, acting to counter excitatory drive to cells and, thus, check the spread of excitation. Studies have shown that reductions in GABA-mediated inhibition via the antagonism of GABA receptors, blockade of GABA-gated ionic currents, or elimination of GABAergic inhibitory interneurons can lead to hyperexcitability in the cerebrum. High-frequency electrical activity in turn causes a further decrease in the strength of inhibitory pathways. Thus, disinhibition of cortical circuits may also lead to hyperexcitability and may allow hypersynchronous excitation to spread unchecked throughout cortical regions.

D. Alterations in Extracellular Environment

Electrical activity in individual neurons is governed by the flow of ionic currents through ion channels. When activated, these channels allow the passage of specific ions to flow down an electrochemical gradient across the cellular membrane, causing the neuron to become depolarized and fire an electrical action potential. The electrical signal generated propagates down the length of the neuron and is transmitted to the other neurons via synaptic connections, usually by the release of chemical neurotransmitters. This causes the sequential activation of other neurons. Studies have shown that during the high-frequency activation of cortial neurons, changes in the extracellular environment further increase excitability. This may be one way in which subclinical (i.e., no seizure is manifested) epileptic discharges can initiate frank epileptic attacks in patients predisposed to seizures.

V. TREATMENT

A. Drug Therapy

Most seizures interfere with daily living; they often impair consciousness and render their victim vulnerable. Moreover, evidence shows that repeated, unchecked seizure activity itself is injurious and can be fatal. Luckily, most seizures can be successfully managed by anticonvulsant drug therapy. The basic approach is to select the appropriate antiepileptic agent for the specific type of disorder. In general, use of a single agent is preferable to multidrug therapy, but in some instances a second drug may be required to achieve control of seizures.

Several antiepileptic drugs are available. The choice of drug used is influenced primarily by the type of seizures manifested and the side effect profiles of the drugs considered. The drugs used fall into two general classes with respect to their activity. The first is composed of those medications which limit excessive neuronal firing and thereby restrict the spread of epileptic activity, such as phenytoin (Dilantin), carbamazepine

(Tegretol), and lamotrigine (Lamictal). The other group of drugs acts to enhance the actions of the inhibitory transmitter GABA [sodium valproate (Depakote), phenobarbital, primidone (Mysoline), and clonazepam (Klonopin)]. The new drug gabapentin (Neurontin) has a unique but unknown mechanism of action.

B. Surgical Management

Surgery is the treatment of choice in cases where the seizure activity is known to be caused by a demonstrable and resectable anatomic lesion (e.g., a tumor). In addition, in cases of refractory partial seizures, surgical removal of the focus of electrically abnormal tissue may be a viable option in treating recurrent attacks. Despite the efficacy of the antiepileptic agents, seizures may remain inadequately controlled in an estimated 30% of epileptic patients. The principle is to identify and remove the focal region of the brain which represents the site of seizure generation (usually a unilateral anterior temporal lobe). Surgical ablation of the seizure locus can, in carefully chosen cases, successfully eliminate seizures.

BIBLIOGRAPHY

Asbury, A. K., McKhann, G. M., and McDonald, W. I. (eds.) (1992). "Diseases of the Nervous System: Clinical Neurobiology," 2nd Ed. Saunders, Philadelphia.

Commission on Classification and Terminology of the International League against Epilepsy. (1981). Proposal for revised clinical and electroencephalographic classification of epileptic seizures. *Epilepsia* 22, 489–501.

Commission on Classification and Terminology of the International League against Epilepsy. (1985). Proposal for classification of epilepsies and epileptic syndromes. *Epilepsia* 26, 268–278.

Dichter, M. A., and Ayala, G. F. (1987). Cellular mechanisms of epilepsy: A status report. *Science* 237, 157–164.

Gutnick, M. J., and Mody, I. (eds.) (1995). "The Cortical Neuron." Oxford University Press, Oxford.

Penry, J. K. (ed.) (1986). "Epilepsy: Diagnosis, Management, Quality of Life," Raven Press, New York.

Rowland, L. P. (ed.) (1995). "Merritt's Textbook of Neurology," 9th Ed. Williams & Wilkins, Philadelphia.

Schwartzkroin, P. A. (ed.) (1993). "Epilepsy: Models, Mechanisms, and Concepts." Cambridge University Press, Cambridge.

Epstein–Barr Virus

JOAKIM DILLNER
Karolinska Institute

GLOSSARY

Episomal persistence The presence of the viral genome as covalently closed circular DNA (episomes), not directly associated with the cellular DNA

Immortalization The ability to confer an indefinite lifespan to infected cells; it should not be confused with transformation, which in addition implies that infected cells are tumorigenic and have lost contact inhibition

Latency A mostly silent state of viral infection characterized by low expression of only a few viral genes and replication of viral DNA only concomitantly with the cellular DNA; there is no production of infectious virus and minimal or no cytopathic effects

Viral lytic cycle The series of events that starts with activation of a latent infection, proceeds with expression of viral early genes, replication of viral DNA, expression of late genes, and finally lysis of the infected cell and release of the infectious virus

EPSTEIN–BARR VIRUS (EBV) IS A WIDESPREAD human herpesvirus that is the causative agent of infectious mononucleosis (IM) and is associated with a series of human malignancies, in particular Burkitt's lymphoma (BL), nasopharyngeal carcinoma (NPC), Hodgkin's disease (HD), and the EBV-carrying lymphoproliferative disorder in immunodeficient hosts. EBV replicates in the epithelial cells of the nasopharynx and the parotid gland. The virus is shed in the saliva and is frequently transmitted by kissing. From its replicative site in the oropharynx, the virus infects and immortalizes B lymphocytes, whereafter viral latency is established in the B lymphocytes for the remainder of the infected person's life. Because more than 90% of humans carry EBV in a latent form, it is assumed that interaction between the virus and the host's immune system will prevent disease for most infected people. In contrast to many other viruses, the virus does not become incorporated (integrate) into the cellular genome, but persists in EBV-infected cells and in EBV-carrying tumors in a free, circular form (episomes). The association of EBV to several major human malignancies, as well as its potent ability to immortalize B lymphocytes, has made EBV one of the most important examples for understanding the role of viruses in human cancer.

I. EPSTEIN–BARR VIRUS (EBV) INFECTION

A. Viral Tropism

EBV is spread by oral contact, usually by the ingestion of saliva from an EBV-infected individual. The virus replicates in epithelial cells of the upper respiratory tract, particularly in epithelial cells of the parotid gland. The virus persists in the infected individual throughout life, and infectious virus is intermittently released in the saliva of almost all healthy EBV-infected individuals. From the epithelial cells in the upper respiratory tract, the virus infects infiltrating B lymphocytes and establishes latency. The infectability by EBV is mainly restricted to B lymphocytes due to the fact that the EBV receptor, through which the virus gains entry into cells, is the B-cell-specific CR_2 receptor for the third component of complement (C3d). It is unclear whether the same receptor is also expressed on epithelial cells of the nasopharynx. An alternative mechanism for viral entry into epithelial cells has been proposed: the presence of IgA antibodies in the viral capsid has been found to increase the viral uptake of epithelial cells by endocytosis of virus–antibody complexes mediated by the IgA receptor. Certain lymphomas of T-cell origin also carry EBV, but the mechanism for viral entry into T cells is unclear.

B. *In Vitro* EBV Infection: The Lymphoblastoid Cell Line

In vitro EBV-infected B lymphocytes are immortalized into lymphoblastoid cell lines (LCLs). In contrast to the normal B lymphocyte, the LCL has acquired the ability to produce its own B-cell growth factor (BCGF) and possesses the ability to grow indefinitely in culture. Whereas the LCLs are dependent on BCGF for their growth, the Burkitt's lymphoma (BL)-derived cell lines may produce and respond to BCGF, but are in no way dependent on BCGF. Compared to the BL-derived cell line, the LCL is regarded as low malignant since it does not form tumors in nude mice and shows a low cloning frequency in soft agar. If EBV is inoculated into certain species of New World monkeys, such as cottontop marmosets, owl monkeys, and tamarins, polyclonal malignant lymphomas with phenotypic characteristics similar to LCL arise. Similarly, B-lymphoblastoid tumors resembling LCL may occur in EBV-infected patients with severe immunodeficiency syndromes.

II. EBV-ASSOCIATED DISEASES

A. Acute Primary EBV Infection: Infectious Mononucleosis

Infectious mononucleosis is a self-limiting disease in which EBV-infected B cells are stimulated to proliferate, which is followed by the appearance of characteristic, atypical cytotoxic T cells. The disease has a peak incidence between 17 and 25 years of age, but can also occur in children and older adults. In lower socioeconomic groups, infection with EBV occurs during the early years of life and is usually not accompanied by clinical illness, but results in the appearance of virus-specific antibodies (seroconversion) which induce immunity to EBV for life. Clinical IM symptoms develop in only about 50% of previously uninfected adolescents that become infected with EBV. IM is frequently transmitted by kissing, presumably as a result of ingestion of viral particles shed in the saliva. IM patients are particularly contagious since they shed much larger amounts of virus than do healthy EBV carriers.

Prodromal symptoms of IM are headache, chills, and lassitude. The typical IM syndrome lasts about 3 weeks and consists of fever, lymphadenopathy, skin rashes, pharyngitis, spleen enlargement, and some liver dysfunction. It is not the EBV-infected B cells that cause the symptoms, but rather the T cells that have become hyperactivated in response to the infection. After 3 to 4 weeks the symptoms resolve without treatment and complete recovery almost always ensues. Recurrencies have not been documented.

During the early stages of IM the patients develop the characteristic heterophile antibodies, which agglutinate sheep erythrocytes and are used in tests for the diagnosis of IM (Paul–Bunnel–Davidsson test). A heterogeneous population of antibodies is induced by the EBV infection of B cells (polyclonal stimulation) and a similar phenomenon is seen during the early stages of IM. At this stage up to 18% of circulating B cells may express EBNA, but this is diminished to some 0.1% already in the second week of illness. IgM antibodies to the viral capsid antigen (VCA) and antibodies to the early antigens (EA), mainly the diffuse subcomponent (EA-D), arise early in the course of IM. The IgM-VCA antibodies give way to IgG-VCA after the first weeks, whereas EA-D antibodies are usually not seen in healthy seropositive individuals. Antibodies to the EBNA antigenic complex mainly appear during convalescence, some 30–50 days after the onset of disease. Antibodies to the EBNA antigens 2–6 are indeed induced early in the course of IM, but it is the delayed appearance of antibodies to the major EBNA protein (EBNA-1) that is diagnostic for IM.

Chronic mononucleosis is a syndrome characterized by crippling lassitude and recurrent opportunistic infections. There is widespread confusion concerning chronic mononucleosis, as the syndrome has been grossly overdiagnosed in recent years. The criteria for diagnosis should include strongly elevated antibody titers to VCA, high antibody titers to EA, and the syndrome should also start as a normal IM, which does not disappear in the usual self-limiting course.

Typically, antibodies to EBNA are also low or absent due to a decrease in the EBNA-1 antibody titers, whereas antibodies to EBNA-2 and other EBNAs are normal or elevated. Patients presenting only with fatigue and marginally abnormal EBV titers should be classified as having a chronic fatigue syndrome. In a few extreme cases of chronic IM, the antiviral drug acyclovir has brought about almost complete recovery, whereas it has no effect on the chronic fatigue syndrome.

B. Oral Hairy Leukoplakia (OHL)

Hairy leukoplakia of the oral cavity is a disease of the immunosuppressed, particularly the AIDS patient. In OHL, there is extensive lytic EBV replication in the outer layers of the infected epithelium. That EBV is the cause of OHL can be inferred from the fact that OHL can be efficiently treated by inhibiting EBV replication with acyclovir treatment. Since EBV replication intermittently also occurs in the oropharynx of healthy EBV carriers, OHL is presumed to be the clinical result of an extraordinarily vigorous EBV replication.

C. EBV Genome Carrying Lymphoproliferative Diseases in Immunodeficient Hosts

1. Lymphoproliferative Disorder of the Immunosuppressed

EBV DNA-positive lymphoproliferative disorders of B-cell origin have been found in a number of other conditions associated with immune deficiencies, either genetically determined (e.g., ataxia telangiectasia) or medically induced (renal and cardiac transplant recipients). The lymphocyte proliferation is polyclonal at the onset. Monoclonal EBV-carrying lymphomas frequently arise in transplant recipients, but do not carry specific karyotypic alterations. The EBV-carrying lymphoproliferative disorders of transplant recipients, whether polyclonal or monoclonal, may show complete regression once the immunosuppressive therapy is removed. [See Immunobiology of Transplantation.]

2. Fatal Infectious Mononucleosis

Although IM is characterized by B-cell proliferation and occasionally presents with severe clinical symptoms, lethal cases of IM are very rare. Typical features of fatal IM are depletion of T cells, bone marrow plasmacytosis, infiltration of brain and viscera by lymphocytes, the presence of heterophile antibodies, and a widely disseminated proliferation of EBNA-positive B cells. The syndrome frequently has a familial setting and may be linked to the X-linked lymphoproliferative syndrome (XLP). This syndrome is characterized by the defective proliferation of B lymphocytes and a deficient immune response to EBV. Approximately 40% of XLP patients develop fatal IM, 40% develop malignant B-cell lymphoma, and 20% develop dysgammaglobulinemia, often associated with chronic EBV infection.

D. Burkitts Lymphoma

BL is a non-Hodgkin B-lymphocytic lymphoma of low differentiation. The disease accounts for nearly one-half of all cancer in children in the tropical regions of Africa. BL occurs with high frequency across tropical Africa and in Papua New Guinea, but only at altitudes <1500 m (endemic form). BL also occurs outside the endemic areas, but very rarely (sporadic form). The two forms differ in several phenotypical aspects. Also, the location of the endemic form of BL is usually in the jaws, whereas the sporadic form is frequently found in other locations, such as the long bones, kidneys, adrenals, thyroids, ovaries, or testes. The peak incidence of endemic Burkitt's lymphoma is 6–10 years of age, whereas the sporadic form usually occurs >15 years of age. Some 98% of endemic BLs carry the EBV genome, whereas only 20% of the sporadic form does.

In an EBV-carrying BL, all the cells express the EBV-determined nuclear antigen (EBNA). Other EBV-determined antigens first found to be expressed in these tumors are the VCA, EA, and the membrane antigen (MA). BL patients consistently have elevated antibody titers against VCA. The geometric mean anti-VCA titer is eightfold higher than in a control group with other malignancies. An African child with an elevated VCA titer has an elevated risk to develop BL of >30 times higher than the normal population. Moreover, BL patients exhibit antibody titers to the restricted type of the early antigen (EA-R) and to the MA, which are not normally seen among healthy EBV-seropositive donors. The EA-R and MA titers are correlated with the clinical activity of BL; high titers of anti-MA are observed in long-term survivors and dramatic rises in anti-EA-R levels correlate with disease onset or relapse.

The rare occurrence of BL outside the endemic area shows that some other factor, specific to the endemic area, must play a role in the etiology of BL as well. BL has been hypothesized to develop in three stages in the endemic regions. Phase I is the primary infection with EBV. Whereas uninfected B lymphocytes differentiate toward the plasma cell end stage when stimulated, EBV-infected B lymphocytes are arrested in an early differentiation stage and will continue to divide. These EBV-infected cells may risk chromosomal damage in direct proportion to the number of cell divisions. Phase II involves an environmental cofactor, most likely endemic malaria. Infection with *Plasmodium falciparum* may lead to an impairment of the ability of cytotoxic T cells to control the proliferation of the EBV-infected B cells, thereby further increasing the "pool" of EBV-infected B cells. Phase III involves the translocation of the distal part of chromosome 8 to chromosome 14 (or 2 or 22), which is a regular finding in BL. Since chromosome 8 contains the c-*myc* oncogene and the other chromosomes contain the loci of the strong enhancers for the immunoglobulin genes of the heavy chain or the λ and κ chains, respectively, the translocation leads to the constitutive activation of the c-*myc* oncogene and subsequent monoclonal B-lymphocyte proliferation resulting in BL.

E. Nasopharyngeal Carcinoma

Nasopharyngeal carcinoma is the most common tumor in the densely populated areas of southern China. The tumor is also common in east and north Africa and in Arctic Eskimo populations. Irrespective of its geographic origin, 100% of anaplastic NPC tumors carry multiple copies of EBV. The rare squamous cell carcinomas of the nasopharynx are, however, only occasionally EBV positive. In the tumor, both EBV DNA and EBNA can be detected in epithelial cancer cells. As in BL, antibody titers to VCA are elevated (approximately 10-fold). Antibodies against early antigen are regularly present in NPC patient sera; however, unlike in BL, they are mainly directed against the diffuse subspecificity (EA-D). Characteristically, these patients develop IgA antibodies to VCA, even before a detectable tumor mass has developed. Since an IgA response to VCA is not seen in other conditions, the IgA-VCA test can be used routinely for the early detection of NPC with high sensitivity and specificity. The IgA-VCA test has been employed on 185,000 inhabitants in the provinces Zang-Wu and Laucheng in Southern China, resulting in the early diagnosis of 132 NPC cases.

The 100% correlation between EBV and NPC indicates that EBV has an etiological role in this tumor, although the restricted geographical occurrence of the tumor indicates that genetic and/or environmental cofactors also contribute to its etiology. First-generation immigrants of southern Chinese origin maintain a high frequency of NPC. Offspring of mixed marriages between southern Chinese and non-Chinese groups show an intermediate frequency. Several reports also speak of familial aggregation of NPC. A high consumption of salted fish, containing nitrosamines that are able to reactivate EBV *in vitro*, has also been implicated as a strong risk factor for NPC, at least in southern China.

F. Hodgkin's Disease

Hodgkin's disease is a disorder primarily involving the lymphoid tissues. From a clinical point of view the disease behaves like a cancer, but it is not clear whether the disease is truly neoplastic or inflammatory in origin. The diagnosis of HD requires the demonstration of the so-called Reed–Sternberg (R-S) cells, which are atypical giant cells with abundant cytoplasm and a multilobate nucleus or multiple nuclei. The R-S cell is considered to be the primary pathogenic cell in HD. In the Western world, HD accounts for about 1% of cancers and is one of the most common cancers among young adults.

The epidemiology of HD has several characteristics, suggesting that it may be an uncommon consequence of a common infection. Notably, the incidence of the disease varies between different geographical areas, between social classes, between seasons, and is related to the social environment during childhood. An increased risk for HD is also associated with certain HLA haplotypes.

Most importantly, however, several large cohort studies have demonstrated that the risk for HD is elevated up to five times among subjects with a history of infectious mononucleosis. Also, both elevated VCA and EA antibody titers are associated with HD, indicating an association with the activation of EBV. The EBV activation is not a consequence of the disease since elevated EBV titers can be found more than a decade before the onset of HD.

The EBV genome can be demonstrated in 50–75% of HD cases and the EBV DNA resides within the R-S cells themselves. Occasionally, cells of the proliferative lymphocytes comprising the HD tissue may also be EBV positive. The EBV genome in the R-S cells is monoclonal, again arguing that EBV infection did not occur after disease onset. The extent of EBV detectability varies with the HD subtype, with the cases of mixed cellularity HD being EBV positive more regularly than the nodular sclerosis or lymphocyte predominance subtypes. The EBV-positive cells in HD express the EBNA-1 and LMP proteins, but none of the other EBNA antigens. Although it seems clear that EBV is somehow involved in the etiology of HD, the precise mechanism is unknown. Possible scenarios have focused on the fact that the EBV LMP protein, which has an *in vitro*-transforming ability, is expressed in the R-S cells or have hypothesized that an immune dysfunction involving an inability to keep the EBV infection latent will lead to a chronic inflammatory response.

G. Other EBV-Associated Cancers

I. Non-Hodgkin's Lymphomas (NHL)

A minority of NHL cases are EBV positive. The EBV positivity is a regular feature of certain forms of NHL, e.g., EBV is frequently found in B-cell lymphomas of the HIV-infected patient, particularly immunoblastic lymphomas and primary central nervous system lymphomas. EBV is also found in some extranodal T-cell lymphomas, most regularly among lethal midline granulomas. NHL is also associated with abnormal EBV antibody titers, which can also be demonstrated prior to the onset of disease.

2. Gastric Adenocarcinomas

A series of reports have described the presence of EBV in a minority (5–15%) of gastric adenocarcinomas. EBV is primarily found among undifferentiated cancers with lymphoid infiltrates.

3. Undifferentiated Carcinomas of the Nasopharyngeal Type

This tumor may in rare circumstances be found at other anatomical sites such as stomach, salivary glands, thymus, and lungs. These tumors are usually EBV positive.

4. Leiomyosarcomas in HIV-Infected Children

These cancers are regularly EBV positive, according to recent reports. Leiomyosarcomas from HIV-negative subjects are EBV negative. Since EBV was not previously known to be able to infect smooth muscle cells, these findings indicate that the cellular tropism of EBV may be wider than expected.

H. Other Diseases

A number of diseases of autoimmune character are associated with signs of EBV reactivation, notably rheumatoid arthritis, the Guillain–Barré syndrome, the Chediak–Higashi syndrome, and the Sjögren's syndrome. Whether the EBV reactivation is a consequence of these diseases or is important for the pathogenesis is debated. [*See* Autoimmune Disease.]

III. THE EBV GENOME

A. General Structure of EBV DNA

The EBV genome of the virus particle is a linear double-stranded DNA molecule of 173,000 bp (Fig. 1). At each end of the molecule there are 4 to 12 copies of a 500-bp terminal repeat (TR). A recognized role of the TRs is to facilitate the circularization of EBV DNA following infection. Inside the infected cell, the viral DNA becomes covalently linked and the genome persists as multiple covalently closed circular episomes. Multiple tandem repeats of a 3071-bp internal repeat (IR) sequence separate the genome into two domains: a short unique region (U_S) of 15,000 bp and a long unique region (U_L) of 150,000 bp (Fig. 1). Several other smaller repeat units have also been found. The EBV DNA has been cloned as a set of overlapping fragments and the complete nucleotide sequence of EBV has been determined. Possible protein-encoding regions, the open reading frames (ORFs), have been designated according to their position in the genome defined by fragments produced by the restriction enzyme *Bam*HI. The system is best explained by an example: "BKRF 1" means the *Bam*HI K fragment rightward open reading frame number 1. The locations of the *Bam*HI fragments in the EBV genome are depicted in Fig. 1. [*See* DNA and Gene Transcription.]

B. Transcription of the EBV Genome in Immortalized Cells

In the virus nonproducer immortalized cells, at least five distinct regions are transcribed. The most abundantly transcribed region is in the *Eco*RI J fragment and encodes two small nonpolyadenylated RNAs transcribed by RNA polymerase III. These RNAs do not code for protein, but exist in the cell complexed with a host cell protein termed La. They have been designated as EBERs, for EBV-encoded RNAs, and

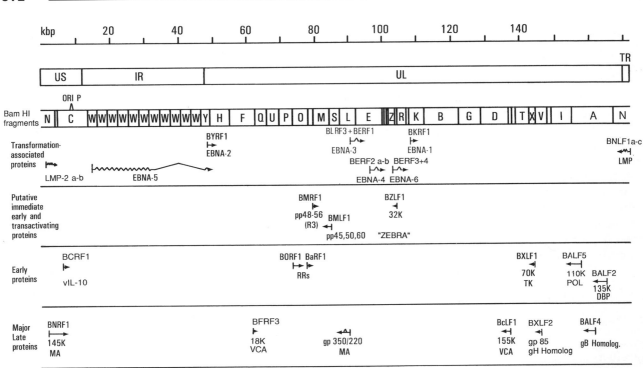

FIGURE 1   Genetic map of EBV. (Top) Relative size of the EBV genome in kilobase pairs (kbp). Upper bar: Organization of the genome in short unique sequence (US), internal repeat (IR), long unique sequence (UL), and terminal repeat (TR). Lower bar: Position of the *Bam*HI restriction enzyme fragments. (Gene map) The size, position, and direction of transcription of some major mRNA species are shown with arrows. The horizontal lines in arrows denote coding sequences (exons). The names of the open reading frames are given above each transcript, (e.g., BYRF1 is *Bam*HI fragment Y rightward open reading frame number 1). DBP, major DNA-binding protein; EBNA, EBV-determined nuclear antigen; gB homolog, a protein with homology to the HSV glycoprotein B; K, kilodalton; LMP, latent membrane protein; MA, membrane antigen; POL, EBV DNA polymerase; pp, phosphoprotein; R3, a monoclonal antibody used to define these proteins; RRs, ribonucleotide reductases; TK, thymidine kinase; VCA, viral capsid antigens; ZEBRA, *Bam*HI Z-encoded EBV virus reactivator protein; vIL-10, a viral protein functionally and structurally homologous to interleukin 10.

show structural similarities to the VA RNAs of adenovirus. It is highly likely that the EBERs, such as adenovirus VA RNAs, facilitate the translation of viral mRNAs. The *Bam*HI WYH region corresponds to the internal repeat and the adjacent 3′ region. The principal transcript from this region is a 3.0-kb mRNA that has a 1.6-kb exon toward its 3′ end, encoding a nuclear protein termed EBNA-2. Several mRNAs from this region also contain a long coding region resulting from several small open reading frames in the *Bam* W and Y fragments spliced together. This ORF encodes another nuclear protein, called EBNA-5. Because the EBNA-1 mRNA, the EBNA-2 mRNA, and possibly also the EBNA-3, EBNA-4, and EBNA-6 mRNAs include EBNA-5 coding sequences in their 5′ parts, EBNA-5 has therefore also been assigned the alternative designation EBNALP (EBNA leader protein). There is also an mRNA species described that includes only the

EBNA-5 encoding ORF. During primary EBV infection of B lymphocytes, transcription of the EBNA-encoding mRNAs is initiated at a promoter in the *Bam*HI W fragment, which also contains a lymphocyte-specific enhancer. In many established cell lines, transcription of the EBNA mRNAs is instead initiated at a promotor further upstream, in the *Bam*HI C fragment.

The *Bam*HI WYH region is associated with EBV transformation since viral strains with a deletion in this region are incapable of transformation and the transforming ability is restored if the *Bam* WYH region is reintroduced. Both the *Bam* WYH-encoded proteins, EBNA-2 and EBNA-5, contribute to the EBV-transforming ability.

The *Bam*HI E fragment contains three long open reading frames, which encode part of the nuclear antigens EBNA-3, EBNA-4, and EBNA-6. The *Bam*HI K fragment hybridizes to a 3.7-kb mRNA that has a

2.0-kb exon in this region which contains the BKRF1 ORF. This ORF contains the complete coding sequence for the nuclear antigen EBNA-1.

The *Bam*HI N$_{het}$ fragment gives rise to a 2.9-kb abundant, leftward mRNA which encodes a membrane protein expressed during latent infection, termed LMP. The *Bam*HI C fragment encodes two very low abundance spliced mRNAs which encode two membrane proteins, termed LMP-2a and LMP-2b.

IV. EBV-ENCODED PROTEINS IN EBV-IMMORTALIZED CELLS: EBNA 1–6 AND LMPI–2

A. EBNA

All latently EBV-infected cells contain the nuclear antigen EBNA, defined by anti-complement immunofluorescence (ACIF) with EBV immune sera. EBNA is an antigenic complex composed of at least six different proteins, designated EBNA 1–6. The EBNA ACIF test primarily measures EBNA-1.

I. EBNA-I

The EBNA-1 protein has a molecular mass between 65 and 92 kDa, depending on viral isolates, which vary in the length of the third internal repeat array (IR3) included in the BKRF1 ORF. The IR3 is a simple repeat array of only three triplet codons, GGG, GCA, and GGA, repeated in an irregular fashion. Due to the IR3, EBNA-1 contains a 20- to 40-kDa copolymer containing only glycine and alanine. The major antigenic determinant of EBNA-1 is present within the glycine–alanine region and can be synthesized as a 20 amino acid peptide (referred to as p107). Affinity-purified human antibodies to the p107 peptide are used in immunofluorescent staining as a routine method for the specific detection of EBNA-1. p107 is also used clinically for the detection of human anti-EBNA antibodies by the ELISA test: an IgM antibody response is diagnostic of recent EBV infection (infectious mononucleosis) whereas an IgG response signals past EBV infection (healthy carrier state) and IgA antibodies signal NPC.

The human chromosomes contain multiple sequences homologous to the IR3 repeat. The anti-p107 IgM antibodies seen during acute infection also react with cellular proteins, whereas the anti-p107 IgG antibodies that develop during convalescence are specific for EBNA-1. The close relationship between this major EBNA epitope and cellular proteins may explain

why the IgG antibodies to EBNA do not appear until 1–2 months after infection. It has also been speculated that this relationship may be a cause of EBV-associated autoimmunity.

The carboxy-terminal part of EBNA-1 is responsible for its DNA-binding activity. EBNA-1 has three specific binding sites on the EBV genome. Two of them are localized in the *Bam* C fragment, in a region termed oriP which is necessary for EBV plasmid maintenance. The oriP region also contains a transcriptional enhancer, dependent on the binding of the EBNA-1 protein. The functions of EBNA-1 thus include the regulation of plasmid maintenance as well as the regulation of transcription of EBNA-encoding mRNAs.

A unique feature of the EBNA-1 protein is that it binds diffusely to all the human chromosomes, as can be seen on ACIF staining on mitotic EBV-carrying cells. This is not the case for EBNA-2, EBNA-3, or for any other tumor antigens. A possible function of the EBNA-1 chromosome binding during metaphase may be to secure the equal distribution of episomes to the daughter cells. The EBV episomes are known to associate with the human chromosomes, in a random pattern. This could conceivably be mediated by the EBNA-1 protein, e.g., by simultaneous binding to the EBV genome and to a structural chromosomal protein.

Finally, an intriguing property of the EBNA-1 protein is that it appears to be nonimmunogenic for cytotoxic T cells (CTLs), thereby enabling EBNA-1 expression and episomal maintenance without attack from the cytotoxic T cells of the host. This property is dependent on the huge glycine-alanine repeat. Possibly, the repeat region could inhibit breakdown and/or processing of the EBNA-1 protein to peptides that are displayed to CTLs in the context of the MHC class I.

2. EBNA-2

EBNA-2 is a nuclear phosphoprotein, which to a very large part (26%) is composed of a proline polymer. The protein binds to both single- and double-stranded DNA and coprecipitates with a 32-kDa cellular protein. Antibodies to EBNA-2 appear early during the course of IM, in contrast to the EBNA-1 antibodies, which do not appear until several months after the onset of symptoms. An antibody titer to EBNA-2 exceeding the titer to EBNA-1 is indicative of primary EBV infection (IM) or of EBV reactivation (e.g., following immunosuppression).

All lymphoblastoid cell lines express EBNA-2, whereas biopsies of BL tumors and newly established

BL-derived cell lines do not. EBNA-2 may be required only for the proliferation of low malignant LCLs that have an activated B-cell phenotype, but is not necessary for the growth of the highly malignant BL-derived lines that have a B-cell phenotype resembling that of resting, "memory" B cells. Furthermore, the EBNA-2-encoding *Bam* WYH region is deleted in four different EBV-carrying BL-derived cell lines, indicating that this region may even impose a selective disadvantage to the cells of the BL phenotype.

When transfected into mouse fibroblasts, the EBNA-2 gene confers the ability to grow in a low serum concentration. When transfected into human EBV-negative BL-derived cells, the EBNA-2 gene induces expression of the surface antigen CD23, a B-cell activation marker. Since CD23 is related to the receptor for the B-cell growth factor (BCGF), this induction may be an important event in EBV-induced immortalization.

3. EBNA-3, EBNA-4, and EBNA-6

These high molecular mass nuclear antigens are encoded by the *Bam*HI E region. EBNA-3 is a 143- to 157-kDa protein encoded by a short ORF in the *Bam* L fragment (BLRF3) and by a long ORF in the *Bam* E fragment (BERF1). EBNA-4 has a molecular mass of about 180 kDa whereas EBNA-6 is a 160-kDa polypeptide. EBNA-4 is encoded by one short (BERF2a) and one long (BERF2b) ORF situated in the *Bam* E fragment. EBNA-6 is encoded by a short ORF (BERF3) and a long ORF (BERF4), both situated in the *Bam*HI E fragment immediately 3′ of the EBNA-4 gene. These three NAs have some amino acid homology to each other. They are primarily expressed in LCLs and in posttransplant lymphomas, most regularly EBNA-3.

4. EBNA-5

EBNA-5 is an extraordinary complex nuclear antigen: As many as 28 different molecular mass forms, ranging in size from 20 to 130 kDa, have been described. Although each EBV-infected cell line usually has only one or a few major EBNA-5 molecular mass species, as many as 11 species have been described in the same cell. Between each molecular mass species there is a regular spacing of 6 kDa to the adjacent species, although sometimes smaller molecular mass differences of 2 and 4 kDa are seen. The genetic basis for the generation of these polypeptides is complex. EBNA-5 is encoded by the internal repeat region (the multiple *Bam*HI W fragments) and by the *Bam*HI Y fragment. The EBNA-5-encoding messages contain multiple exons, where ORFs with a different coding capacity are made by different splicings which include various numbers of repeats. None of the exons contains an initiation codon, which is instead provided by a rare splicing event.

EBNA-5 is invariably expressed in LCLs, whereas many BL-derived cell lines do not express this protein. A phenotypic drift in the long-term-established BL-derived cell lines toward a more LCL-like phenotype is also accompanied by the appearance of EBNA-5 expression. EBNA-5 is a DNA-binding protein that binds more strongly to the nucleus than any other EBNA protein.

B. LMP

1. LMP-1

LMP-1 is an unglycosylated 62-kDa molecular mass protein containing a short hydrophilic N terminus, six strongly hydrophobic regions interspersed with short hydrophilic regions, and a long hydrophilic carboxy terminus. It spans the plasma membrane six times and has both a N and a C termini at the cytoplasmic side. The LMP-encoding mRNA from the *Bam*HI N fragment is the most abundant EBV-specific mRNA in the immortalized cell. In some cell types, transcription of the LMP gene is dependent on an EBNA-2-dependent enhancer.

By immunofluorescence, LMP is localized in "patches" at the cell membrane, in a pattern similar to what is found for the cytoskeletal protein vimentin. The two proteins may be directly associated. Transfection of rat fibroblasts or human keratinocytes induces transformation, including tumorigenicity in nude mice. In B cells, LMP may block apoptosis by induction of the Bcl2 oncoprotein. In epithelial cells, the transforming mechanism may involve the LMP-induced upregulation of the expression of the CD40 surface antigen.

2. LMP-2a and b

These proteins are encoded by the *Bam* C fragment. They share the same amino acid sequence, except for one small exon. LMP-2 expression is usually found concomitantly with LMP-1 expression. Antibodies to LMP-2 are found in NPC patients.

V. EBV TYPES A AND B: GENETIC VARIATION OF EBNA-2 and EBNA-3

DNA sequencing of the *Bam*HI WYH region from several different viral isolates has shown that the EBNA-2-coding region can be present in two different types with only about 25% homology in the DNA sequence and 50% in the proteins. The B-type EBNA-2 lacks most of the polyproline repeat region. Based on the type of EBNA-2 gene, EBV strains are subtyped into type A and type B strains. The two EBV strains also differ in their EBNA-3 genes. The type A virus is the most common type of EBV in isolates from Europe and the United States. The B-type virus has frequently been detected in isolates from BL-endemic areas. EBV isolates from immunosuppressed patients are frequently of the B type and are also found in isolates from the Western world. *In vitro*, lymphoblastoid cell lines immortalized by type B viruses show a slower growth and grow more in large clumps than LCLs immortalized by the type A virus.

VI. SIMIAN EBV-LIKE VIRUSES

EBV-like viruses have been isolated from Old World primate species, notably *Herpesvirus* (*H*). *gorilla* from gorillas (*Gorilla gorilla*), *H. pan* from chimpanzees (*Pan troglodytes*), *H. pongo* from orangutans (*Pongo pygmaeus*), *H. papio* from baboons (*Papio hamadryas*), and the herpesvirus of macaques (*Macaca fascicularis*). They are all B lymphotropic and are able to transform B cells into lymphoblastoid cell lines. These EBV-like viruses all share 30–40% DNA sequence homology to EBV. They express VCA and EA that are highly cross-reactive between the different species. Like EBV, The simian EBV-like viruses also express multiple nuclear antigen (NAs). Serologic cross-reactions are found with several of the nuclear antigens, especially EBNA-2. The macaque herpesvirus is an important model for lymphomagenesis in AIDS since inoculation of these monkeys with the simian immunodeficiency virus will regularly induce lymphomas carrying the EBV-related herpesvirus of macaques.

VII. LATENCY TYPES I, II, AND III

The expression of the immortalization-associated proteins in the EBV-associated diseases shows distinctively different patterns for the different diseases. The *in vitro* infection, the LCL, can express all six EBNAs and the LMPs. EBNA-1, -2, -3, and -5 and LMP-1 are invariably expressed in all LCLs. The EBV-carrying lymphoproliferative disease in immunodeficient hosts, which closely resembles the LCL phenotypically, has an expression of immortalization-associated proteins similar to that of the LCL. In contrast, biopsies of BL and newly established BL-derived lines express only EBNA-1, no other EBNAs, and no LMP. However, BL-

derived cell lines can alter their phenotype upon cultivation *in vitro* and concomitantly start to express some of these proteins. Like the BL, the NPC biopsies invariably express EBNA-1. However, NPC can also express the LMP proteins in about 50% of the cases. EBNA-2 is never expressed in NPC and other EBNAs have also not been detected. In Hodgkin's disease, the expression pattern is similar to NPC. Primary EBV infection *in vitro* leads to expression of all the immortalization-associated proteins, with EBNA-2 and EBNA-5 appearing somewhat earlier than the other proteins.

Based on these expression patterns, it has been hypothesized that the EBV-induced polyclonal proliferation of B cells, as seen during IM and EBV-carrying lymphoproliferative disease, requires all of the immortalization-associated proteins. This stage has been termed latency type III. The attack of the sensitized cytotoxic T cells, which are primarily directed against epitopes in EBNA-3, -4, -5, and -6, but also to EBNA-2 and LMP, may then induce a downregulation of these proteins and the return of the B cells to their resting state and a truly latent, nonproliferative EBV infection. This is called latency type I. This resting B cell, expressing only EBNA-1, would then in rare cases be the origin of a BL following the typical chromosomal translocation event. EBNA-1 stands out from the other EBV antigens since EBNA-1 is always expressed in EBV-infected cells. The function of EBNA-1 is known: it regulates the oriP plasmid maintenance region and actively prevents integration, a function that is vital to EBV and could not possibly be downregulated. Since EBNA-1 is not detected by cytotoxic T cells, it can be expressed in the resting cells of type I latency without risk of elimination of the virus-infected cell by CTLs. Type II latency is defined by the protein expression pattern seen in HD and in most cases of NPC, i.e., expression of EBNA-1 and the LMP proteins, but none of the other EBNA proteins. The role of type II latency in the natural history of EBV infection is unknown. A role in virus replication is unlikely since the viral lytic cycle can be started directly from any one of the three types of latency.

VIII. VIRAL PROTEINS OF THE LYTIC CYCLE

A. Detection, Classification, and Nomenclature of the Viral Lytic Cycle Proteins

In several EBV-carrying BL-derived cell lines and LCLs, a small fraction of the cells, usually 1 to 5%, spontaneously express early and late viral antigens. The percentage of cells entering the viral lytic cycle can be greatly increased by treatment of infected cells with chemical inducers or by other procedures. The appearance of viral lytic cycle-associated proteins in such cells can then be detected by serological or biochemical methods. Some 35 viral proteins ranging in size from 350 to 18 kDa have been identified. These proteins are classified as either early or late depending on whether they are synthesized before or after the viral DNA synthesis.

B. The EBV Gene Map

More than 80 major ORF have been identified in the EBV DNA, 15 of which code for late proteins. The virus particle contains approximately 30 different proteins, suggesting that many late genes have not yet been identified. Up to 100 viral proteins have been mapped on the viral DNA. The gene map shown in Fig. 1 lists only a few of the major proteins that have been conclusively mapped. Note that early and late genes are interspersed throughout the genome. Most of the lytic cycle genes consist of a single ORF with the promoter region located closely upstream. In contrast, the immortalization-associated genes have, in general, very complex structures.

C. Early Genes and Proteins

1. Viral Genes and Proteins Related to Activation of the Viral Lytic Cycle

Immediate early (IE) genes are operationally defined as a subset of early viral genes expressed in the absence of ongoing protein synthesis and, therefore, do not require virus-coded activation for their expression. To study them, protein synthesis is blocked by reversible inhibitors, such as cycloheximide. After a period of time, cycloheximide is removed and the transcription-blocking drug actinomycin D is added. Under these conditions, IE gene products accumulate, whereas early and late genes are inactive.

As many as seven IE proteins, ranging in size from 120 to 48 kDa, have been described. The *Bam*HI M fragment, ORF BMRF1, encodes a family of proteins that are abundantly expressed in lytically infected cells, are highly antigenic, and constitute the main component of the EA-D antigen. BMLF1, another ORF in the *Bam* M fragment, also encodes putative IE proteins.

Another approach to the identification of IE genes employs the defective EBV DNA produced by P3HR-1 cells, a long-time established BL cell line. When used for superinfection of cells carrying latent EBV genomes, this DNA induces the lytic cycle. The reactivating ability resides in the *Bam*HI Z fragment. The reactivating protein has been identified as a 33-kDa protein and has been designated ZEBRA for the *Bam*HI Z-encoded EB virus reactivator. Antibodies to the ZEBRA protein are present in only a few percentage of healthy EBV carriers, but are regularly found among patients with IM, NPC, or EBV reactivation, such as in AIDS. In summary, the two *Bam*HI M genes and the *Bam*HI Z gene seem to encode transactivators which are likely to be functionally related to the activation of virus replication in one way or another.

2. Viral Enzymes

The EBV DNA polymerase (110 kDa) is encoded by a single open reading frame: BALF5. An associated 45-kDa protein stimulates the enzymatic activity of the EBV DNA polymerase. Some regions of the EBV polymerase gene are homologous to the DNA polymerase locus of poxvirus, adenovirus, and cytomegalovirus. The EBV DNA polymerase is readily distinguished from its cellular counterparts by three properties: (i) stimulation by ammonium sulfate, (ii) preferential utilization of synthetic templates, and (iii) higher sensitivity to certain drugs such as phosphonoacetic acid and to acyclovir triphosphate. With respect to these three properties the EBV and the HSV DNA polymerases are very similar.

The viral exonuclease is associated with a 70-kDa molecule encoded by the *Bam*HI B fragment. It may provide nucleotide precursors by degradation of cellular DNA or may be involved in genetic recombination. Antibodies to the EBV exonuclease are preferentially found in sera of NPC patients.

Ribonucleotide reductases (RR), which catalyze the conversion of ribonucleoside diphosphates to deoxynucleoside diphosphates, are responsible for the synthesis of nucleotide precursors of DNA. Two subunits, encoded by the ORFs BORF1 and BaRF1, may be responsible for catalysis and regulation, respectively, and are highly homologous to the two subunits of the HSV enzyme.

The viral thymidine kinase (TK) is encoded by the BXLF1 open reading frame. The EBV TK (70 kDa) has a good overall homology with the HSV-1 TK and also cross-reacts with monoclonal antibodies to the HSV TK. The replication of EBV is highly sensitive to certain drugs, e.g., acyclovir, that require a viral TK for their phosphorylation and activation.

3. Other Major Early Proteins

The major DNA-binding protein (135 kDa) is the most abundant of the early proteins and is encoded entirely by the BALF2 ORF. It has significant sequence homology with the corresponding HSV protein (ICP 8). The major DNA-binding protein is required for replication of the viral DNA, possibly by keeping the DNA in an extended configuration. The BALF2 ORF is deleted in the EBV genome carried by the BL cell line Raji. When these cells are induced to enter the EBV lytic cycle, early antigens are produced, but no viral DNA synthesis and no late protein synthesis follows, presumably for lack of the 135-kDa protein. This phenomenon is also exploited in EBV serology: induced Raji cells are the standard source of EA for immunofluorescence.

An abundantly expressed early protein, encoded by the BCRF1 ORF, has a striking >80% amino acid homology to the cellular interleukin 10 (IL-10) protein. Both the cellular IL-10 and its viral counterpart function as a macrophage deactivation factor and also stimulate the growth and differentiation of B cells. The role of the virally encoded IL-10 protein in the biology of EBV might be to delay the immunological attack against the highly antigenic EBV proteins expressed during the lytic cycle until the lytic cycle has reached completion.

Several of the early proteins are very abundant in virus-producing B cells. The 48- to 55-kDa *Bam*HI M-encoded family of early polypeptides, discussed earlier, constitutes the major component of the diffuse subspecificity of the early antigen (EA-D). An 85-kDa early protein is a cytoplasmic protein detected in immunofluorescence as filamentous structures and is the main target for EA-R antibodies. The 135-kDa major DNA-binding protein is the major early antigen detected by immunoprecipitation with human sera, but is not detected by immunofluorescence.

D. Proteins of the Virus Particle

The Epstein–Barr virus particle is eicosahedral and enveloped, with typical herpesvirus morphology. More than 30 polypeptides are associated with it. The major late antigen of virus-producing cells is a 155-kDa polypeptide. A 145-kDa virus envelope protein, an 18-kDa late protein encoded by the BFRF3 ORF, and a 155-kDa polypeptide constitute the VCA complex in virus-producing cells. The 155-kDa protein has been mapped to the BcLF1 ORF.

The most abundant virion envelope proteins are two immunologically cross-reactive glycoproteins of 350 and 220 kDa (gp 350/220), both encoded by the *Bam*HI L fragment. The translation of differentially spliced transcripts yields two precursor polypeptides with molecular masses of 135 and 100 kDa, respectively; the higher molecular masses of the proteins are due to glycosylation. These proteins are the main target for neutralizing antibodies and are also responsible for the binding of the virus to its cellular receptor. Therefore, they are the main target for attempts to develop an EBV vaccine.

An unrelated immunogenic glycoprotein of 85 kDa (gp85) is encoded by the BXLF2 ORF in the *Bam*HI X fragment. gp85 has some limited homology to the glycoprotein H (gH) of HSV.

Another late protein of 110 kDa is encoded by the BALF 4 ORF and is structurally homologous to the glycoprotein B (gB) of HSV, which is implicated in the internalization of HSV virions.

A 145-kDa unglycosylated protein is a major constituent of the virus particle. It appears to be encoded by the BNRF 1 ORF of the *Bam*HI N and C fragments.

BIBLIOGRAPHY

Baer, R., Bankier, A. T., Biggin, M. D., Deininger, P., Farrell, P. J., Gibson, T. J., Hatfull, G., Hudson, G. S., Satchwell, S. C., Seguin, C., Tuffnell, P. S., and Barrell, B. G. (1984). DNA sequence and expression of the B95-8 Epstein–Barr virus genome. *Nature*, **310**, 207–211.

Dillner, J., and Kallin, B. (1988). The Epstein–Barr virus proteins. *Adv. Cancer Res.* **50**, 95–158.

Ernberg, I., and Kallin, B. (1984). Epstein–Barr virus and its association with human malignant diseases. *Cancer Surv.* **3**, 51–89.

Henle, W., and Henle, G. (1979). Seroepidemiology of the virus. *In* "The Epstein–Barr Virus" (M. A. Epstein and B. G. Achong, eds.), pp. 61–78. Springer Verlag, Berlin.

Farrell, P. J. (1995). Epstein-Barr virus immortalizing genes. *Trends Microbiol.* **3**, 105–109.

Khanna, R., Burrows, S. R., and Moss, D. J. (1995). Immune regulation in Epstein–Barr virus-associated diseases. *Microbiol. Rev.* **59**, 387–405.

Kieff, E., Hennessy, K., Fennewald, S., Matsuo, T., Dambaugh, T., Heller, M., and Hummel, M. (1985). Biochemistry of latent EBV infection and associated growth transformation. *In* "Burkitt's Lymphoma: A Human Cancer Model" (G. M. Lenoir, G. O'Connor, and C. L. M. Olweny, eds.), pp. 323–329. Scientific publication No. 60, International Agency for Research on Cancer, Lyon, France.

Knutson, J. C., and Sugden, B. (1989). Immortalization of B-lymphocytes by Epstein–Barr virus. *In* "Advances in Viral Oncology" (G. Klein, ed.), Vol. 8, pp. 151–712. Raven Press, New York.

Niedobitek, G., and Young, L. S. (1994). Epstein–Barr virus persistence and virus-associated tumours. *Lancet*, **343**, 333–335.

Ergonomics

K. H. E. KROEMER

Virginia Polytechnic Institute and State University (VA TECH)

I. History and Development
II. Goals and Scope
III. Contributing Sciences
IV. The Current Data Base in Ergonomics
V. Education in Ergonomics
VI. Professional Organizations

GLOSSARY

Anthropometry Science of measuring and describing the physical dimensions of the human body

Biomechanics Method of describing the physical behavior of the human body in mechanical terms

Ergonomics Discipline to study human characteristics for the appropriate design of the living and work environment; in the United States, often called human factors, or human factors engineering

Hawthorne effect Often subtle, certainly unplanned and unexpected, effect produced in persons taking part in an experiment by uncontrolled variables, such as by the attention paid to them

Humanization of work Goal of ergonomics to achieve ease and efficiency of technological systems involving humans

Industrial engineering Engineering branch concerned with the interactions among people, machinery, and energy

Industrial hygiene Control of occupational health hazards that arise as a result of doing work

Work physiology (work psychology) Application of theoretical knowledge about the human body (and mind) to the evaluation of occupational workloads to derive recommendations for improving the work situation

ALL MAN-MADE TOOLS, DEVICES, EQUIPMENT, machines, and environments should serve, directly or indirectly, to further the safety, well-being, and per-

formance of humans. Ergonomics (also called human factors in the United States) is the discipline to "study human characteristics for the appropriate design of the living and work environment."

Thus, ergonomics has two distinct aspects: (1) the study, research, and experiments used to determine the specific human traits and characteristics that one needs for engineering design and (2) the application and engineering in which tools, machines, shelter, environment, and work tasks and job procedures are designed to fit and accommodate the human. Of course, the actual performance of "human and equipment in the environment" is also a topic of study and research to assess the suitability of the design and to determine possible improvements.

I. HISTORY AND DEVELOPMENT

Fitting things to the human is as old as humankind. Primitive tools and weapons made of stone, bone, and wood were selected for their fit to the human hand and their suitability for the intended purpose. Purposeful shaping of these tools was the next step; creating and manufacturing followed. Fitting clothes and making shelters certainly were early and fundamental ergonomic activities.

With increasing complexity of the human society, organizational and management challenges developed. Purposeful training of workers and soldiers, for example, became necessary together with behavior formation and control. Roman soldiers underwent well-organized military conditioning until they could perform military exercises without sweat accumulating on their skin. "Drying the legions" relied, consciously or by experience, on the principle of training and adapting the physiological capabilities of the recruits to the physical requirements. For major proj-

ENCYCLOPEDIA OF HUMAN BIOLOGY, Second Edition, VOLUME 3.

ects, such as building the pyramids in old Egypt, assembling armies for warfare, and sheltering and supplying the inhabitants of ancient cities with food and water, sophisticated knowledge of human needs and desires was required, and careful planning and complex logistics needed to be mastered.

Artists, military officers, employers, and sports enthusiasts were interested in body build and physical performance. Medicine men treated illnesses and injuries. Anatomical and anthropological disciplines began to develop. About 400 B.C., Hippocrates described a scheme of four body types, which were supposedly determined by their fluids: the moist type was dominated by black gall; the dry type was influenced by yellow gall; the cold type was characterized by slime, and the warm type was governed by blood.

In the fifteenth, sixteenth, and seventeenth centuries, persons such as Leonardo da Vinci and Alfonso Borrelli mastered the complete knowledge of anatomy, physiology, and equipment design; they were artist, scientist, and engineer in one. In the eighteenth century, the sciences of anatomy and physiology diversified and accumulated specific and detailed knowledge; psychology began to develop as a separate science.

Well into the nineteenth century, the scientific disciplines tended to be theoretically oriented, devoted solely to understanding the complex human being. The stereotype is the scientist in a white coat leading a research-devoted life in the laboratory. But with increasing interest in industrial work and the old interest in military employment of the human, "applied" aspects of the pure sciences began to develop in the early 1800s. In France, Lamar, Lavoisier, Amar, and Duchenne researched energy capabilities and strains of the working human body; Marey developed methods to describe human motions at work and to determine work payment systems. In England, the Industrial Fatigue Research Board considered theoretical and practical aspects of the human at work. In Italy, Mosso developed dynamometers and ergometers to research fatigue. In Scandinavia, Johannsson and Tigerstedt developed the scientific discipline of work physiology with a so-named institute in Germany founded by Rubner in 1913. In the United States, the Harvard Fatigue Laboratory was established in the 1920s.

In the first half of the twentieth century, work (or industrial) physiology and psychology were well advanced and widely recognized, both in their theoretical work "to study human characteristics" and in the application of this knowledge "for the appropriate design of the living and work environment." Two distinct approaches to study human characteristics had developed: one was particularly concerned with physiological and physical traits of the human, and the other was mainly interested in psychological and social proprieties. Although these approaches greatly overlapped, the anatomical–physical and physiological aspects were studied mainly in Europe, and the psychological and social aspects in North America.

Based on a broad fundament of anatomical, anthropological, and physiological research, applied or work physiology had become of great concern, particularly during the hunger years associated with World War I in Europe. With marginal living conditions, the question of minimal or suitable nutrition to perform certain activities (e.g., the energy consumption while performing agricultural, industrial, military, and household tasks; the relationships between energy consumption and heart rate; the use of muscular capabilities; suitable body postures at work; design of equipment and workplaces to fit the human body) was researched for application purposes. Testing persons for their ability to perform physical and mental work, as well as vigilance and attention, mental workload, driver behavior and road sign legibility, and related topics, was known as psychotechnology in the 1920s in Europe.

In the United States, "most psychologists at this time [around 1900], were strictly scientific and deliberately avoided studying problems that strayed outside the boundaries of pure research" (Muchinsky, 1987, p. 13). However, activities such as sending and receiving Morse code, perception and attention at work, using psychology in advertising, promoting industrial efficiency, and other basically psychological aspects were applied to problems in business and industry. A particularly important step was the development of intelligence testing to screen military recruits during World War I and later industrial workers, for assigning them to jobs suitable for their mental capabilities. Thus, the terms intelligence testing and industrial psychology appeared. (Muchinsky explains that the term "industrial psychology" first appeared as a typographical error, actually meant to read "individual" psychology.) Gould (1981) provides a partly amusing, partly disturbing account of the early years of intelligence testing.

Among the best-known results in industrial psychology are those of the experiments at the Hawthorne Works in the mid-1920s. The study was designed to assess relationships between lighting and efficiency in work rooms where electrical equipment was produced. The bizarre finding was that the workers' productivity increased or remained at a satisfactory level whether or not the illumination was changed, obviously in response to the attention paid to the workers by the researchers. This became known as the Hawthorne effect.

Industrial psychology divided into several recognized branches, including personnel psychology, organizational behavior, industrial relations, and engineering psychology. During World War II, the "human factor" as part of a "man–machine system" became of major concern because, in many cases, the technological development had led to machines and operational systems that required more attention, more strength, and more endurance from individuals and teams than many humans could muster. For example, radar screens were to be observed over many hours with the intent of detecting some "blips" among others. High-performance aircraft made the pilot unable to operate hand controls properly under high *g*-loads, such as in sharp turns, or even made the pilot black out. Cockpits in tanks and aircraft were kept small and low, requiring that suitably small crew members be selected. Combat morale and performance were difficult to maintain under stressful conditions. Thus, military and related efforts at the home front generated the need to consider human physique and psychology, as an individual and in teams, purposefully and knowingly in the design of task, equipment, and environment.

Both in Europe and in North America, various names were used to describe the activities of psychologists, sociologists, physiologists, anthropologists, statisticians, and engineers who studied the human and used this information in design, selection, and training. On January 13 and 14, 1950, British researchers met in Cambridge, United Kingdom, and discussed among other topics, the name of a society to be founded that would represent their activities. Among other terms, "ergonomics" (Greek for *ergon-,* work or effort; *-nomos,* law or surroundings) was proposed, a word invented by K. F. H. Murrell in late 1949. The new word ergonomics was neutral, not implying priority to physiology, psychology, functional anatomy, or engineering. It was easily remembered and recognized and could be used in any language. Ergonomics was formally accepted as the name of the new society at its council meeting on February 16, 1950 (Edholm and Murrell, 1974).

Several aspects are worthy of a brief note. The original proposal for the name vote included two alternative name suggestions. One was the Human Research Society, the other the Ergonomic Society. Note that there is no "s" at the end of "Ergonomic"; apparently, the "s" somehow slipped onto the voting ballot. The attachment of the "s" has made the derivation of an adjective or adverb relating to ergonomics rather difficult. (Incidentally, there are several claims that similar words had been coined and used earlier, e.g., in Poland.) The alternately proposed term, Human Research Society, bears some similarity to the term human engineering.

In the United States, similar discussions went on about a proper name to describe the activities of the new discipline. In 1956, a group of persons convened to establish a formal society. The name ergonomics was rejected in 1957, and instead the term human factors was selected to describe the professional activities. Often, the additional word engineering is used to indicate applications, such as in human (factors) engineering.

There has been, and still is, some discussion on whether or not human factors is different from ergonomics, which of the two relies more on psychology or physiology, or which is more theoretical or practical. Today, the two terms are usually considered synonymous, as exemplified by the Canadian society: it uses human factors in its English name, and ergonomie in its French version. In 1992, the U.S. society changed its name to Human Factors and Ergonomics Society.

II. GOALS AND SCOPE

[Ergonomics] can be defined as the application of scientific principles, methods, and data drawn from a variety of disciplines to the development of engineering systems in which people play a significant role. Successful application is measured by improved productivity, efficiency, safety, and acceptance of the resultant system design. The disciplines that may be applied to a particular problem include psychology, cognitive science, physiology, biomechanics, applied physical anthropology, and industrial systems engineering. The systems range from the use of a simple tool by a consumer to a multiperson sociotechnical system. They typically include both technological and human components.

[Ergonomic] specialists from these and other disciplines are united by a singular perspective on the system design process: that design begins with an understanding of the user's role in overall system performance and that systems exist to serve their users, whether they are consumers, system operators, production workers, or maintenance crews. This user-oriented design philosophy acknowledges human variability as a design parameter. The resultant designs incorporate features that take advantage of unique human capabilities as well as build in safeguards to avoid or reduce the impact of unpredictable human error. (National Research Council, 1983, pp. 2–3)

Ergonomics adapt the man-made world to the people involved. Ergonomics focus on the human as the most important component of our technological systems, and span the range from basic research to engineering and managerial applications. Thus, the utmost goal of ergonomics is humanization of work. This goal may be symbolized by the "EE" of ease and efficiency for which all technological systems and their elements should be designed. This requires knowledge of the characteristics of the people involved, particularly of their dimensions, their capabilities, and their limitations.

There is a hierarchy of goals in ergonomics. The basic intent is to generate tolerable working conditions that do not pose known dangers to human life or health. This basic requirement assured, the next goal is to generate acceptable conditions upon which the people involved can voluntarily agree according to current scientific knowledge and under given sociological, technological, and organizational circumstances. Of course, the final goal is to generate optimal conditions, which are so well adapted to human characteristics, capabilities, and desires that physical, mental, and social well-being is achieved.

Ergonomics is neutral: it takes no sides, neither of employers nor of employees, neither for nor against technological progress. It is not a philosophy, but a scientific discipline.

III. CONTRIBUTING SCIENCES

Ergonomics is a growing and changing science. That development stems from increasing and improving knowledge about the human and is driven by new applications and new technological developments.

As discussed earlier, a number of classic sciences provide the fundamental knowledge about the human. The anthropological basis consists of anatomy, describing the build of the human body; orthopedics, concerned with the skeletal system; physiology, dealing with the functions and activities of the living body, including the physical and chemical processes involved; medicine, concerned with illnesses and their prevention and healing; psychology, the science of mind and behavior; and sociology, concerned with the development, structure, interaction, and behavior of individuals or groups. Of course, physics, chemistry, mathematics, and statistics also supply knowledge, approaches, and techniques.

From these basic sciences, a group of more applied

disciplines developed into the core of ergonomics. These include primarily anthropometry, the measuring and description of the physical dimensions of the human body; biomechanics, describing the physical behavior of the body in mechanical terms; work physiology, applying physiological knowledge and measuring techniques to the body at work; industrial hygiene, concerned with the control of occupational health hazards that arise as a result of doing work; and management, dealing with and coordinating the intentions of the employer and the employees. Of course, many other disciplinary areas have developed that also are part of ergonomics, contribute to it, or partly overlap it such as labor relations or safety engineering.

Several distinct application areas use ergonomics as basic components of their knowledge base or of their work procedures. Among these are industrial engineering, by definition concerned with the interactions among people, machinery, and energies; bioengineering, working to replace worn or damaged body parts; systems engineering in which the human is an important component of the overall work unit; safety engineering, which focuses on the well-being of the human; and military engineering, which relies on the human as soldier or operator. Naturally, other application areas are in urgent need of ergonomic information and data, such as computer-aided design in which information about the human must be provided in computerized form. Oceanographic, aeronautical, and astronautical engineering also rely intensively on ergonomic knowledge.

The development from the basic sciences to the applied ones in ergonomics, and the use of ergonomic knowledge in specific disciplines, is depicted schematically in Fig. 1. As more knowledge about the human becomes available, as new opportunities develop to make use of human capabilities in modern systems, and as needs for protecting the person from outside events arise, ergonomics change and develop.

IV. THE CURRENT DATA BASE IN ERGONOMICS

The existing basis of ergonomic knowledge is deep, wide, and incomplete. It is deep because much of the wealth of information gathered in the "classic mother sciences" over decades and centuries contributes to ergonomic information. For example, anatomical and

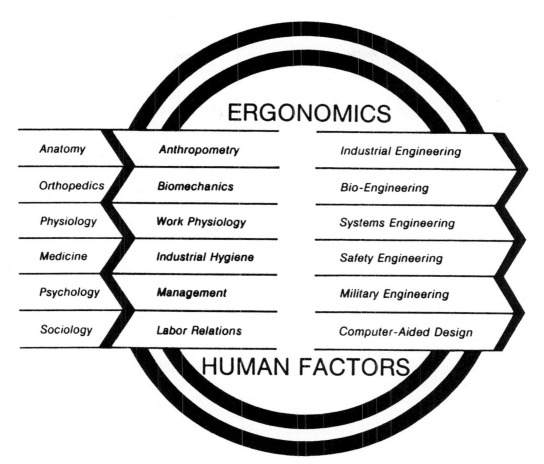

FIGURE I   Ergonomics: contributing sciences, specialty areas, and primary users.

physiological knowledge is far-reaching and fairly complete.

The knowledge base of ergonomics is also wide, because many classic and new sciences contribute to ergonomic knowledge. For example, the knowledge about the functions of the body is paralleled with information about mental and social traits and needs.

Nevertheless, the ergonomic knowledge base is incomplete, particularly in rapidly developing areas such as sociology and biomechanics and even on "old" topics such as the elderly. Furthermore, it is incomplete because of new demands on the human in interaction with a new technical system; consider the exciting problems and challenges associated with long-duration space exploration by humans. Also, much of the information gathered in the traditional mother sciences was not developed with an ergonomic application in mind, nor did the research address new

conditions such as in underwater habitat, modern warfare teams, or space exploration. However, one does not have to consider only "exotic" conditions of human–technology interactions. Even for rather mundane and everyday applications, information is not always available in appropriate quantity or quality. A 1985 evaluation of our knowledge base in anthropometry and biomechanics, for example, has shown that the information about the human body build and size is insufficient to develop a complete or correct computerized model of the human body. This lack of suitable knowledge is even more pronounced in biomechanics, where not enough is known, for instance, about active muscular exertions (as opposed to reactive responses) of the human body to model its behavior in an automobile accident or in sports activities (Kroemer *et al.*, 1988).

In attempting an overview about the currently available ergonomic data base, one finds that a large body of written information exists. A typical listing of printed sources of information, available in the United States in 1989, includes about two dozen books, a variety of standards, and about half a dozen journals (see lists following the Bibliography).

V. EDUCATION IN ERGONOMICS

Academic education in ergonomics/human factors in the United States is at the graduate level, that is, after several years of studying disciplines such as engineering, psychology, physiology, physics, and chemistry, usually to the bachelor's level.

In the United States and Canada, the formal curriculum in human factors or ergonomics is often a specific option in colleges of engineering or psychology, with about half of the programs in either one. The trend seems to be toward academic education in engineering, almost without exception in industrial engineering departments. Here, one or two courses in ergonomics/human engineering are usually offered to undergraduates, while M.S. or Ph.D. formal specialization and intensive instruction are provided on the graduate level. The graduate degree is, generally, in industrial engineering or psychology, at the masters level. Very few schools in North America, or in the United Kingdom, convey a degree specifically in ergonomics.

Many academic programs are housed in departments where some of the professors provide the core of ergonomic information, while additional instructors are called in from other departments to provide a wider perspective. Only a few universities have large enough departments in industrial engineering or psychology to rely exclusively or predominaely on the faculty drawn from that department. The Human Factors and Ergonomics Society provides detailed information on programs and courses in human factors of U.S. and Canadian universities. Similarly, the Ergonomics Society provides an overview of the programs in the United Kingdom and in many other countries.

The requirements for entering a graduate program in the ergonomics/human factors option vary from institution to institution, but typically a bachelor's degree is required. If the applicant to the program does not have an undergraduate degree in the specific field (either psychology or engineering), the student may have to take additional courses to make up existing deficiencies. The time from entering the program with a bachelor's degree to completing it with a master's degree is at least 2 years, but 4 years or longer to obtain a doctoral degree.

Employment of specialists in human factors in the United States used to be mostly in the military services, occupational safety and health, industrial hygiene, or government agencies. However, with increasing importance placed on human health and performance in all occupations and professions, the demand for ergonomists/human factors engineers is now widespread and still growing. This growth is expected to continue because even traditional disciplines such as mechanical or architectural engineering realize that the proper "interfacing" of equipment with the human is an important aspect to assure performance, safety, and human well-being.

VI. PROFESSIONAL ORGANIZATIONS

The largest single (national) professional organization in ergonomics/human factors is the Human Factors and Ergonomics Society (HFES) (P.O. Box 1369, Santa Monica, California 90406). The HFES has over 5300 members in 1994, with various technical interest groups. The HFES publishes its own journals, *Human Factors* and *Ergonomics in Design*, and organizes an annual professional congress.

The Ergonomics Society is located in the Department of Human Sciences, University of Technology, Loughborough LE11 3TU, United Kingdom. Founded in 1950, it is the oldest professional organization for ergonomics. The Ergonomics Society had about one thousand members in 1994, mostly in the United Kingdom. The Society supports two journals, *Ergonomics* and *Applied Ergonomics*, and organizes an annual professional congress.

More than two dozen national societies exist (Table I), most carrying the term ergonomics in their title. Most are members of the International Ergonomics Association (founded in 1959), whose Secretary General is presently Prof. ir. Pieter Rookmaaker (Netherlands Railways, SE ARBO/Ergonomics, P.O. Box 2025, 3500 HA Utrecht, The Netherlands). In all there are about 12,000 ergonomists worldwide, mostly in North America, Europe, Japan, and Australia/New Zealand, with fairly large contingents in Southeast Asia and South America. Ergonomists, often called "psychologists" with some modifying term, also work in Eastern Europe, the former Soviet Union, and China.

TABLE I
Ergonomic Societies and Recent Memberships[a]

Country/region	Founding date	Membership		Population (million)	Members (per million)
		Date	Number		
Australia	1964	1994	537	17.4	31
Austria	1976	1990	42	7.5	6
Belgium	1986	1993	140	9.9	14
Brazil	1983	1988	244	143.3	2
Canada	1968	1994	518	28.3	18
Czech Republic and Slovakia				15.7	
China	1989	1989	300	1151	0.3
Croatia	1974	1993	70	4.6	15
Denmark		1994	846	5.1	166
Finland	1985	1994	90	5	18
France	1963	1987	531	55.2	10
Germany	1958	1991	700	76.6	9
Greece	1988	1993	33	10.2	3
Hungary	1987	1988	90	4.6	20
India	1987	1993	52	866	0.1
Indonesia	1988	1988	120	176.8	0.7
Israel	1982	1992	120	4.3	28
Italy	1961	1994	320	57.2	6
Japan	1964	1993	1864	124.0	15
Korea (South)	1982	1988	250	43.9	6
New Zealand	1986	1994	114	3.3	35
Netherlands	1962	1994	631	15	42
Norway		1994	155	4.2	37
Poland	1977	1994	386	37.5	10
Russia	1989	1994	150	150.0	1
Singapore	1988	1990	36	2.6	14
South Africa	1984	1994	40	39.4	1
Southeast Asia	1984	1990	86	250	3
Spain	1988	1992	187	39.6	5
Sweden		1994	298	8.5	35
Taiwan	1993	1994	240	20.6	12
Ukraine	1992			52.0	
United Kingdom	1949	1992	1030	56.5	18
United States	1957	1994	5354	248.8	21
Former Yugoslavia	1973	1989	50		
Affiliated societies					
European Society for Dental Ergonomics		1994	37		
Human Ergology Society	1970	1994	230		
Sustaining member					
Hungarian Council of Industrial Design and Ergonomics					
Total			15,891		

[a]Adapted from *Ergonomics* 37(8), 1578 (1994).

BIBLIOGRAPHY

Edholm, O. G., and Murrell, K. H. F. (1974). "The Ergonomics Research Society. A History 1949 to 1970." The Council of the Ergonomics Research Society.

Gould, S. J. (1981). "The Mismeasure of Man." Norton, New York.

Kroemer, K. H. E., Snook, S. H., Meadows, S. K., and Deutsch, S. (eds.) (1988). "Ergonomic Models of Anthropometry, Human Biomechanics, and Operator–Equipment Interfaces." National Academy Press, Washington, D.C.

Muchinsky, P. M. (1987). "Psychology Applied to Work," 2nd Ed. Dorsey, Chicago.

National Research Council (ed.) (1983). "Research Needs for Human Factors." National Academy Press, Washington, D.C.

BOOKS

Astrand, P. O., and Rodahl, K. (1986). "Textbook of Work Physiology," 3rd Ed. McGraw–Hill, New York.

Bailey, R. W. (1982). "Human Performance Engineering: A Guide for Systems Designers." Prentice–Hall, Englewood Cliffs, NJ.

Boff, K. R., Kaufman, L., and Thomas, J. P. (eds.) (1986). "Handbook of Perception and Human Performance." John Wiley & Sons, New York.

Chaffin, D. B., and Andersson, G. B. J. (1991). "Occupational Biomechanics." John Wiley & Sons, New York.

Chapanis, A. (ed.) (1975). "Ethnic Variables in Human Factors Engineering." Johns Hopkins Univ. Press, Baltimore.

Eastman-Kodak Company (1983, 1986). "Ergonomic Design for People at Work," Vols. 1 and 2. Van Nostrand Reinhold, New York.

Grandjean, E. (1988). "Fitting the Task to the Man," 5th Ed. International Publications Service, New York/Taylor & Francis, London.

Helander, M. (ed.) (1981). "Human Factors/Ergonomics for Building and Construction." John Wiley & Sons, New York.

Helander, M. (ed.) (1988). "Handbook of Human–Computer Interaction." Elsevier, Amsterdam/New York.

Huchingson, R. D. (1981). "New Horizons for Human Factors in Design." McGraw–Hill, New York.

Konz, S. (1990). "Work Design: Industrial Ergonomics," 3rd Ed. Grid, Columbus, OH.

Kroemer, K. H. E., Kroemer, H. B., and Kroemer-Elbert, K. E. (1994). "Ergonomics: How to Design for Ease and Efficiency." Prentice–Hall, Englewood Cliffs, NJ.

NASA/Webb (1978). "Anthropometric Source Book," 3 vols.

NASA Reference Publication 1024. NASA, Houston. (NTIS, Springfield, VA 22161, order #79 11 734.)

Oborne, D. J., and Gruneberg, M. M. (eds.) (1983). "The Physical Environment at Work." John Wiley & Sons, New York.

Plog, B. A. (ed.) (1988). "Fundamentals of Industrial Hygiene." National Safety Council, Chicago.

Proctor, R. W., and Van Zandt, T. (1994). "Human Factors in Simple and Complex Systems." Allgn & Bacon, Boston.

Salvendy, G. (ed.) (1987). "Handbook of Human Factors." John Wiley & Sons, New York.

Sanders, M. S., and McCormick, E. J. (1993). "Human Factors in Engineering and Design," 7th Ed. McGraw–Hill, New York.

STANDARDS

ANSI/HFS 100 (1988). American National Standard for Human Factors Engineering of Visual Display Terminal Workstations. Human Factors Society, Santa Monica, CA (in revision).

ANSI/ASHRAE 55 (1981). Thermal Environment Conditions for Human Occupancy. American Society of Heating, Refrigerating, and Air-Conditioning Engineers, Atlanta, GA.

Department of Defense, Military Standard: Human Factors Engineering Design for Army Material, MII-HDBK-759A. U.S. Government Printing Office, Washington, D.C. (Available from Naval Publications and Forms Center, 5801 Tabor Avenue, Philadelphia, PA 19120.)

Department of Defense, Military Standard: Human Engineering Design Criteria for Military Systems, Equipment and Facilities, MIL-STD-1472C. U.S. Government Printing Office, Washington, D.C.

Department of Defense, Military Handbook: Anthropometry of U.S. Military Personnel (metric), DOD-HDBK-743. U.S. Government Printing Office, Washington, D.C. (Available from Naval Publications and Forms Center, 5801 Tabor Avenue, Philadelphia, PA 19120.)

JOURNALS

Applied Ergonomics, Elsevier Science Ltd, Oxford, UK.

Biomechanics, John Wiley & Sons, New York.

Ergonomics, Taylor & Francis, London.

Ergonomics in Design, Human Factors and Ergonomics Society, Santa Monica, CA.

Human Factors, Human Factors and Ergonomics Society, Santa Monica, CA.

Journal of the American Industrial Hygiene Association, AIHA, Akron, OH.

Spine, Lippincott, Philadelphia.

Estrogen and Osteoporosis

HARALD VERHAAR
SIJMEN DUURSMA
Utrecht University Hospital

GLOSSARY

Estrogen Normal female hormone produced by the ovary during the whole period of the menstrual cycle

Progestogen Hormone produced during the second half of the menstrual cycle

Osteoblasts Cells that produce new bone

Osteoclasts Cells that resorb bone

Osteoporosis Porotic bone, resulting from a negative balance between the activity of osteoblasts and osteoclasts; most common in postmenopausal women but also common in men with increasing age and in pathological conditions; it increases risk for fractures

F. A. ALBRIGHT (1940) HYPOTHESIZED A CAUSAL relationship between the postmenopausal state of women and the occurrence of vertebral and femoral fractures, which are observed in an increasing number with increasing age. The skeleton consists of long bones and irregular-shaped ones. The long bones, for example, the femur, have more compact (cortical) bone, which forms the solid outer shell of bones, giving them great strength. The irregular-shaped bones, for example, the vertebrae, have more trabecular (or so-called spongy or cancellous) bone, which is the interior meshwork of bones containing bone marrow. In healthy adults about 5% of bone is replaced yearly. This process starts with bone resorption by osteoclasts, cells that originate from stem cells that also produce the red and white blood cells. The bone defect is refilled with new bone, produced by osteoblasts, which originate from stem cells that also produce cells for the formation of cartilage, muscles, and fat tissue. Osteoblasts produce the organic matrix of bone tissue with collagen fibers, which are long protein chains. The local situation in the matrix permits calcium and phosphate ions to form crystals of hydroxyapatite, the major component of bone mineral. Osteoblasts also produce the enzyme alkaline phosphatase, which plays a role in the mineralization process. The serum activity of this enzyme gives information about the osteoblastic bone-forming activity. Serum osteocalcin, a bone matrix protein, is selectively secreted by osteoblasts and reflects osteoblastic activity. A new biochemical marker for bone formation is serum procollagen I C-terminal propeptide (PICP), which is released during the extracellular processing of procollagen to collagen. Recent studies show that serum collagen I C-terminal telopeptide (ICTP), liberated during degradation of type I collagen, correlates with osteoclastic activity. The amino acid hydroxyproline is a degradation product of collagen and is produced by the osteoclastic bone resorption. The urinary excretion of this amino acid provides information about the process of bone resorption. The urinary excretion of pyridinolines, cross-links of collagen, has also been validated as a marker of bone resorption. To get information about the amount of bone in a person, techniques have been developed to measure bone mineral content in parts of the skeleton, for example, the forearm or the lumbar spine, or in the total skeleton.

In normal persons, nearly all bone matrix has been mineralized and bone mineral content is measured by double-energy X-ray absorptiometry (DEXA), which measures the amount of bone. However, in mineralization defects, bone mineral content does not reflect the real bone mass. With these techniques the effect of therapeutical agents on bone cells and bone mass can be measured.

I. ESTROGEN AND BONE

The development and activity of bone-resorbing and -producing cells are regulated by a complex interaction of systemic factors, supplied by the circulation, and locally produced factors. In addition, osteoclasts and osteoblasts produce locally active factors that adjust the activity of the two types of cells with opposite functions. The ratio between the number of bone cells and the amount of bone tissue is much higher in trabecular than in cortical bone. Changes in the systemic factors, for example, estrogen and progestogen, that regulate bone cell metabolism, have different effects on trabecular and cortical bone. A negative balance between the activity of osteoblasts and osteoclasts results in bone loss. Postmenopausal women yearly lose about 2–4% of their trabecular bone mass during the early postmenopausal years. About 10 years after menopause, this bone loss results in vertebral fractures in about 25% of these women. Because of the lower number of bone cells per unit bone volume, cortical bone loss needs more time. The incidence of hip fractures increases sharply after the age of 65–70 years.

Menopause is the moment of the last menstruation, a consequence of the decreasing estrogen production by the ovaries. Most common is the spontaneous loss of estrogen production in women of about 50 years of age, but surgical oophorectomy, anorexia nervosa, and intensive physical training also lead to decrease of estrogen. A medical protocol to address the relationship between the postmenopausal state and bone loss is estrogen replacement therapy. In postmenopausal women treated with estrogen for 1 to 20 years, the progress of osteoporosis was arrested, as judged by measurements of total height. In another study, oophorectomized women were treated with an estrogen preparation. They showed no decrease in height and bone mineral content, measured on the hand and forearm. It was concluded that estrogen prevents postmenopausal bone loss in both the axial and the peripheral skeleton. In women who receive estrogen treatment since the time of menopause, bone density at 80 years of age may decrease by about 10% from bone density at menopause, as compared with a decrease of about 30% in women who have never been treated. This improvement in bone density is likely to reduce the risk of fracture by about two-thirds. The effect of estrogen is dose dependent. In a 25-year study of women with severe postmenopausal osteoporosis, those receiving 0.625 mg conjugated estrogens had a reduction in vertebral fracture rate from 60 to 25 per 1000 patient-years, but those who received 1.25 mg conjugated estrogens had a fracture rate of only 3 per 1000 patient-years. A does of 0.3 mg of conjugated estrogens was not sufficient to prevent postmenopausal bone loss, but with the addition of 1000 mg calcium, no decrease of bone mineral content was observed.

The most common prescribed oral estrogen preparations are conjugated estrogens, a mixture of some steroids extracted from the urine of mares in gestation; estradiolvalerate, a synthetic estrogen with a structure nearly equal to that of human estrogen; and ethinylestradiol, another synthetic preparation. Oral micronized 17β-estradiol is also used, as is transdermal 17β-estradiol. The lowest effective dose for prevention of postmenopausal bone loss is 0.625 mg for conjugated estrogens, 2 mg for estradiolvalerate, 25 μg for ethinylestradiol, and 1 mg of micronized 17β-estradiol. The individual dose needed depends on the body weight. Estrogen is also produced by peripheral conversion of androstenedione in fat tissue; the more fat tissue a women has, the more estrogen she produces. Measuring the serum concentration of estradiol can be a guideline to adjust the individual therapeutic dose. In a study with the use of estrogen administration via the skin with a special plaster and oral conjugated estrogens, a relationship was observed between serum estradiol concentrations <120 pg/ml and the urinary calcium/creatinine and the hydroxyproline/creatinine ratios. These ratios provide information about bone resorption. Serum concentrations >120 pg/ml provided no additional bone-sparing effect.

The administration of estrogen causes growth of the endometrium and if continued over a long time, it enhances the risk for endometrial carcinoma. The cyclic combination with a progestogen preparation prevents the development of endometrial carcinoma by periodic bleeding. Besides preventing estrogen-induced endometrial cancer, progestogens may have other positive effects in combination with estrogen in hormone replacement therapy. The production of progesterone by postmenopausal women is greatly reduced and the prevention of postmenopausal osteoporosis may not be unique to estrogens, since therapy with progestogens has also been found to protect bone loss. Evidence for an important role of progesterone in bone metabolism was shown in two previous studies; a decrease in the endogenous production of progesterone in premenopausal women, arising during menstrual cycles with insufficient and short luteal phases and anovulatory cycles, was associated with an excess of bone loss, despite normal production of estradiol. Other data relating progesterone to mineral metabolism showed higher progesterone levels in postmenopausal "slow losers" compared to "rapid losers" of bone mass from the same population. Thus, the addition of a progestogen to estrogen therapy may be important in the prevention and treatment of osteoporosis. Whereas most studies indicate that estrogen therapy inhibits the resorption of calcium from bone, at least three studies have shown that the addition of a progestogen to estrogen therapy may actually increase bone mass by promoting new bone formation.

L. E. Nachtigall *et al.* (1979) observed that estrogen/progestogen therapy, which began less than 3 years after the last menstruation, increased bone density. Although there was some loss of bone mass in the estrogen/progestogen-treated patients when therapy was started later than 3 years after menopause, the loss was significantly less than that in the placebo group. This study emphasized the importance of beginning estrogen/progestogen therapy early in the climacteric, but also indicated that these hormones are beneficial to osteoporotic women regardless of age. Most postmenopausal women are not charmed by a preventive regime with continuation of premenopausal cyclic bleedings and menstrual discomfort. In a study of 55 women who were 70 years old and receiving estrogen/progestogen replacement therapy, 44% left the 1-year study before its conclusion, mostly because of menstrual problems. However, in early postmenopausal women, the daily combination of estradiolvalerate and cyproteronacetate prevented bone loss and caused only short periods with spot bleeding during the first 6 months; later on no further bleeding occurred over the remaining 2-year study period. Continuous combined therapeutic regimes are considered in women with a contradiction for bleedings and are sometimes recommendable in elderly women.

II. HOW DOES ESTROGEN PREVENT POSTMENOPAUSAL BONE LOSS?

Because until recently it was not possible to demonstrate estrogen receptors in bone cells, an indirect effect of estrogen, mediated by a second factor, was assumed. It has been suggested that, after menopause, bone cells are more sensitive to parathyroid hormone, but no evidence supports this idea. Another suggestion was an increase in the synthesis of 1,25-dihydroxycholecalciferol after estrogen replacement therapy, which is the active metabolite of vitamin D and produced by the enzyme 1α-hydroxylase in the kidneys. It was found that a single injection of estradiol increased the 1α-hydroxylase activity in bird kidney homogenates. A rise in serum parathyroid hormone concentration, with secondary stimulation of the enzyme 1α-hydroxylase and the production of 1,25-dihydroxycholecalciferol, was also reported after estrogen treatment in patients with postmenopausal osteoporosis. In other studies, these observations could not be confirmed. It has generally been accepted that serum concentrations of calcium and phosphate, the activity of alkaline phosphatase, and the urinary excretion of calcium and hydroxyproline are significantly higher in postmenopausal women compared with premenopausal women. These changes cannot be explained by the supposed changes in the plasma concentrations of parathyroid hormone or 1,25-dihydroxycholecalciferol after menopause, and seem to be secondary to the modified bone metabolism rather than their cause.

III. ESTROGEN AND CALCITONIN

Calcitonin is a polypeptide hormone produced by the C cells in the thyroid gland. The production and secretion is stimulated by an increase in serum calcium concentration. Calcitonin caused an increase in the number of osteoblasts and in bone length of mouse radius rudiments in a culture system. It also stimulated the multiplication of cultured fetal chicken osteoblasts. Whether this stimulation of mitogenesis in osteoblasts precursors is relevant to human physiology is uncertain. Bone resorption, induced by parathyroid hormone in neonatal mouse vertebral bones, was inhibited by calcitonin. Calcitonin is the only hormone with a direct effect on osteoclasts, which possess calcitonin receptors. The hormone inhibited isolated osteoclasts from eroding the surface of bone slices in culture.

Estradiol stimulates the secretion of calcitonin in cultured rat thyroid C cells, but at concentrations higher than the serum estrogen level in postmenopausal women, using the normally advised dose to prevent postmenopausal bone loss. Administration of estrogen in postmenopausal women resulted in a significant rise in plasma calcitonin concentration at noon, but not at 9:00 A.M. In other clinical studies, no changes in plasma calcitonin concentration could be observed during estrogen replacement therapy. Physiological levels of estradiol in postmenopausal women showed no relationship to serum calcitonin levels. However, basal levels of serum calcitonin were lower in postmenopausal women when compared with premenopausal women, and were strongly correlated with circulating estrone levels. From this study it is concluded that the calcitonin secretion capacity appears to be modulated by circulating estrone levels. As with high doses of estradiol, low doses of progesterone increased the secretion of calcitonin in rat thyroid C cells. In estrogen/progestogen replacement regimes, usually with low doses of estrogen, the rise in serum calcitonin concentration may be at least partly the result of the progestogen preparation. An intermediate function for calcitonin between estrogen and bone metabolism remains possible, but it might be more plausible for progesterone. Further investigations about the interaction between calcitonin, estrogen, and progesterone are needed to clarify their role in bone metabolism.

IV. ESTROGEN AND PROSTAGLANDINS

Nearly all mammalian cells have the capacity to produce prostaglandins by cleaving arachidonic acid from membrane phospholipid. The prostaglandins are rapidly metabolized by enzymes found in nearly all tissues. Their importance lies in their function as locally active substances. It has been suggested that estrogen might affect bone metabolism indirectly by inhibiting endogenous prostaglandin E_2 synthesis. Prostaglandin E_2 causes a stimulation of osteoblastic collagen synthesis, but is also a potent stimulator of bone resorption and calcium release from bone. Prostaglandin E_2 is reported to increase the number and activity of osteoclasts in bone cultures. Osteoblasts produce significant amounts of prostaglandin E_2,

which may have a function in the coupling between the activity of osteoblasts and osteoclasts. Using advanced techniques, however, no direct effect of prostaglandin E_2 on bone resorption has been observed.

Prostaglandin E_2 can prevent bone loss due to estrogen deficiency in ovariectomized rats and add extra cancellous bone to these rats by stimulating bone formation that exceeds bone resorption. The effect of estrogen on prostaglandin E_2 synthesis was also studied in young rats. Oophorectomy prior to their sacrifice resulted in a twofold increase in prostaglandin E_2 release from parietal bones, and *in vivo* replacement therapy with estradiol inhibited the increased prostaglandin E_2 release. These results support the possibility that estrogen influences prostaglandin E_2 production, which can explain the development of osteoporosis and the preventive effect of estrogen replacement therapy in postmenopausal women.

V. DIRECT AND INDIRECT EFFECTS OF ESTROGEN/PROGESTOGEN ON BONE

A. Indirect Action via Growth Factors

Estrogen inhibits the production of insulin-like growth factor-I (IGF-I) in the liver, which results in an increase in plasma growth hormone concentration because of the negative feedback between these hormones. In a study with estrogen substitution over a 3-week period, these effects on IGF-I and growth hormone could be confirmed in postmenopausal women. Further investigations on IGF-I, growth hormone, and parameters of bone metabolism supported the idea that IGF-I and growth hormone have an essential function in the regulation of bone metabolism and in the interaction between estrogen and bone cells, but others could not confirm these observations.

For many years, it has been assumed that growth hormone needs the production of the intermediate substance IGF-I for its effects on tissues; however, during the last few years, growth hormone has been demonstrated also to have a direct effect on some types of cells, including osteoblasts in cultures of fetal chicken calvaria. The effect of growth hormone might be caused by the local production of IGF-I, because, in physiological concentrations, IGF-I stimulates the synthesis of DNA and collagen in fetal rat calvaria. However, other studies demonstrated a direct effect of growth hormone on osteoblasts, without the intermediate production of IGF-I. Very fast effects of growth hormone on the process of osteoblast prolif-eration make a direct effect plausible, because the IGF-I transcription needs much more time.

The decrease in IGF-I production and the rise in the serum growth hormone concentration in post-menopausal women using estrogen replacement ther-apy for 3 weeks were confirmed in other trials. Con-tinuous estrogen/progestogen replacement therapy during a 2-year period caused a stabile decrease in serum IGF-I by about 25% and a stabile increase in growth hormone concentration by about 200%. The concentrations of IGF-II did not change. This was expected because it is supposed that IGF-II plays a role in cell metabolism during fetal life, whereas IGF-I acts in adults. The effect of estrogen replacement ther-apy on the serum concentrations of IGF-I and growth hormone in postmenopausal women is firmly estab-lished. Such therapy in postmenopausal women causes a decrease in both bone formation and resorp-tion. In cultured osteoblasts, the effect of estrogen is mediated by an increased production of IGF-I, but in postmenopausal women a decrease in serum IGF-I concentration was observed after estrogen substitu-tion. About 80% of the circulating IGF-I is produced by the liver, the remaining 20% by peripheral cells like osteoblasts. Estrogen inhibits the IGF-I production in the liver and stimulates the production of this growth factor in osteoblasts. The inhibition in the liver cells is of more importance than the stimulating effect on bone cells, resulting in a fall in IGF-I available for osteoblasts and osteoclasts. The feedback between IGF-I and growth hormone causes a rise in plasma growth hormone concentration, which stimulates bone formation by the osteoblasts. In this model (Fig. 1), osteoporosis has been reduced to a disturbance between bone formation and resorption, as a result of a change in the balance between growth factors that regulate proliferation and differentiation of os-teoblasts and osteoclasts.

Besides IGF-I, osteoblastic cells also produce a num-ber of other growth factors that can be stored in the bone matrix, including IGF-II, transforming growth factor-β (TGF-β), platelet-derived growth factor (PDGF), and basic fibroblast growth factor (bFGF). Of these growth factors, IGF-II and TGF-β are the two most abundant polypeptides produced by human bone cells. In the rat osteoblastic cell line UMR-106, estrogen has been shown to increase the secretion of IGF-I/-II and TGF-β by these cells. Estrogen is capable of inducing binding proteins for IGF-I/-II in bone cell culture, which are able to influence the biological actions of these polypeptides in bone. IGF-I/-II and

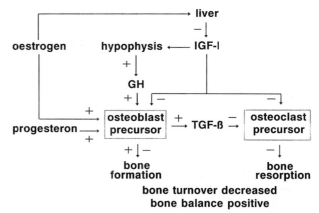

FIGURE 1 Schematic representation of the changes in insulin-like growth factor-I (IGF-I) and growth hormone (GH) after estrogen replacement therapy in postmenopausal women. Estrogen substitu-tion reduces the IGF-I production in the liver and causes a decrease in plasma IGF-I concentration. IGF-I stimulates osteoblast and osteoclast precursors. A fall in plasma IGF-I concentration leads to inhibition of bone formation and resorption. The decrease in plasma IGF-I concentration causes an increase in plasma GH level. GH and estrogen induce local production of IGF-I in osteoblasts. A direct effect of GH on osteoblasts, independent of IGF-I, has also been demonstrated. After estrogen substitution, the decrease in the systemic IGF-I concentration seems to be of more importance than the increase in the locally produced IGF-I by the osteoblasts. The net result is a decrease in bone formation and resorption with a change in their balance in a positive direction. However, proliferation and differentiation of bone cells also depend on other local and possibly systemic factors, as well as factors that regulate the coupling between osteoblasts and osteoclasts, like TGF-β. Interaction between these factors regulates the complex process of bone formation, bone resorption, and the maintenance of bone mass during adult life.

TGF-β may play a role in bone formation since these factors are released during bone resorption and are mitogenic to normal osteoblasts. They also affect bone cells by activation of membrane receptors on osteo-blastic cells. TGF-β has been shown to be capable of decreasing bone resorption by inhibiting formation of osteoclasts. These findings suggest once more that sex steroids may mediate their effects on bone by altering the production of IGF-I/-II (or their binding proteins) and other growth factors like TGF-β. [*See* Transforming Growth Factor-β.]

B. Direct Action on Bone Cells

In addition to the possible intermediate role of IGF-I/-II, TGF-β, and other growth factors, other data indicate that the steriods estrogen and progesterone may directly affect human bone cells. Since the demonstration of active estrogen receptors in osteoblastic cells, it was thought that this steroid possibly was able to act directly on bone to modulate remodeling. However, P. E. Keeting and colleagues could not identify direct effects of estrogen on the biology of normal adult human osteoblast-like (HOB) cellls. Neither HOB cell proliferation nor differentiation was altered by estrogen administration. Gray and colleagues found that estrogen treatment transiently decreased cell proliferation, but enhanced alkaline phosphatase activity in the transformed rat osteoblastic cell line UMR-106. In the human osteosarcoma cell line HBT-96, which had been transfected with the estrogen receptor gene, treatment with estrogen inhibited cell proliferation, but did not influence alkaline phosphatase activity. In osteoblast-like cells from new-born rat calvariae, M. Ernst and colleagues found that estrogen administration resulted in increased cell proliferation and procollagen I mRNA levels. Thus estrogen has been reported both to stimulate and to inhibit the proliferation and differentiation of osteoblast-like cells. These conflicting results probably relate to differences in responses of the various model systems used in the cited studies.

Data of direct progestogen effects on bone cells are very scarce. The first evidence for a direct action of progesterone on bone-forming cells comes from studies in which progesterone caused significant displacement of synthetic glucocorticoids from osteoblastic cells and competed with these steroids for glucocorticoid receptors. More recently, inducible receptors for progesterone were demonstrated to be present in human osteoblast-like cells. Estradiol was able to induce an increase in mRNA encoding for the progesterone receptor in these bone-derived cells. Histomorphometric investigations revealed that therapy with the progestogen preparation medroxyprogesterone resulted in the significantly increased bone formation of both cortical and trabecular bone in beagle dogs, with a 50% increase in the number of osteoblasts. Thus, progestogens seem to stimulate bone formation by osteoblastic cells.

VI. SUMMARY

Osteoporosis is a common disturbance in bone metabolism resulting in a decrease in the amount of bone and an increase in fracture incidence. Estrogen replacement therapy in postmenopausal women, mostly in combination with the prescription of progestogen, prevents postmenopausal bone loss. The mechanism by which estrogen and progestogens influence bone metabolism is still under discussion. A direct effect of estrogen and progestogen on bone cells is possible, but changes in mediator substances can also explain the effects. Calcitonin and prostaglandin E_2 are possible candidates for the role of mediator. Changes in growth factors, especially IGF-I in relationship to growth hormone, can also explain the alterations in bone metabolism after the menopause and after estrogen substitution. A combined effect of the mentioned and other factors is plausible. Proliferation and differentiation of bone cells is a complex process, with the interaction of systemic growth factors and locally produced factors. The details of this process are still unclear.

BIBLIOGRAPHY

Duursma, S. A., Raymakers, J. A., Van Beresteyn, E. C. H., and Schaafsma, G. (1987). Clinical aspects of osteoporosis. *World Rev. Nutr. Diet.* **50**, 92.

Duursma, S. A., Slootweg, M. C., and Bijlsma, J. W. J. (1988). How do estrogens prevent postmenopausal osteoporosis? *In* "Crossroads in Aging" (M. Bergener, M. Ermini, and H. B. Stähelin, eds.). Academic Press, London.

Eriksen, E. F., Colvard, D. S., Berg, N. J., Graham, M. L., Mann, K. G., Spelsberg, T. C., and Riggs, B. L. (1988). Evidence of estrogen receptors in normal human osteoblast-like cells. *Science* **241**, 84.

Heersche, J. N. M. (1989). Bone cells and bone turnover—The basis for pathogenesis. *In* "Metabolic Bone Disease: Cellular and Tissue Mechanisms" (C. S. Tam, J. N. M. Heersche, and T. M. Murray, eds.), pp. 1–19. CRC Press, Boca Raton, FL.

Jee, W. S. S. (1988). The skeletal tissues. *In* "Histology: Cell and Tissue Biology" (L. Weiss, ed.). Elsevier Biomedical, New York/Amsterdam.

Lindsey, R., Hart, D. M., Purdey, D., Ferguson, M. M., Clark, A. S., and Kraszewski, A. (1978). Comparative effects of oestrogen and a progestogen on bone loss in postmenopausal women. *Clin. Sci.* **54**, 193.

Lobo, R. A., Whitehead, M. I., and Wadler, G. I. (eds.) (1989). Consensus development conference on progestogens. *Int. Proceed. J.* **1**, 1.

Martin, T. J., and Raisz, L. G. (eds.) (1987). "Clinical Endocrinology of Calcium Metabolism." Dekker, New York.

Mundy, G. R., and Roodman, G. D. (1987). Osteoclast ontogeny and function. *Bone Min. Res.* **5**, 209.

Nachtigall, L. E., Nachtigall, R. H., Nachtigall, R. D., and Beckman, E. M. (1979). Estrogen replacement therapy. A 10-year prospective study in the relationship to osteoporosis. *Obstet. Gynecol.* **53**, 277.

Parsons, J. A. (ed.) (1982). "Endocrinology of Calcium Metabolism." Raven, New York.

Prior, J. C. (1990). Progesterone as a bone-trophic hormone. *Endocrine Rev.* **11**, 386.

Raisz, L. G., and Martin, T. J. (1983). Prostaglandins in bone and mineral metabolism. *Bone Min. Res.* **2**, 286.

Riggs, B. L., and Melton III, L. J. (1988). "Osteoporosis, Etiology, Diagnosis and Management." Raven, New York.

Sakamoto, S., and Sakamoto, M. (1986). Bone collagenase, osteoblasts and cell-mediated bone resorption. *Bone Min. Res.* **4,** 49.

Verhaar, H. J. J., Damen, C. A., Duursma, S. A., and Scheven, B. A. A. (1994). A comparison of the action of progestins and estrogen on the growth and differentiation of normal adult human osteoblast-like cells *in vitro. Bone* **15,** 307.

Evolving Hominid Strategies

PAUL GRAVES

Southampton University

GLOSSARY

Encephalization Size of brain relative to body size

Home base Fixed domicile at which offspring are cared for and to which subsistence materials are brought back from foraging

Hominid From the family name Hominidae, denotes those fossil species with whom we share common ancestry, exclusive of other primates

Social intellect Intelligence specifically adapted to the problems of social life

Taphonomy Study of the geological deposition and transformation of plant and animal remains

IN THE PAST 6 MILLION YEARS, THE BEHAVIOR and ecology of hominids have radically changed. Our earliest ancestors were very different in their anatomy and behavior from either living primates or modern humans. Traditional models of human evolution, which stressed the role of technology and the hunting of large mammals, are still found in current literature. But the rapidly expanding body of evidence from paleontology, archaeology, and primatology is fundamentally changing our understanding of the forces that shaped the biological and behavioral development of our species.

I. HOMINIDS AS BIPEDAL PRIMATES

A. Descent from Apes

Since the publication of Darwin's *Origin*, the descent of humans from apes has been the subject of controversy and debate. Over the last century, a number of alternative models for human ancestry have been proposed. At one time it was believed that humans were directly and separately descended from early prosimians similar to the modern Tarsir. More recently, paleontological evidence led many to decide that the hominid lineage shared a last common ancestor with the orangutan rather than the African great apes. However, it is now virtually certain that the hominid lineage is most closely related to the chimpanzees (genus *Pan*). Immunological comparison and comparison of DNA sequences by hybridization have shown a very close genetic link between *Homo* and *Pan,* and the use of estimated rates of genetic divergence as a molecular clock has now enabled researchers to put a time scale on the whole hominoid (ape) radiation. The ancestors of African apes and orangutan diverged around 15–17 million years (Myr) ago. Subsequently, the *Pan*–hominid and gorilla lineages diverged at 10–8 Myr ago. A number of separate studies corroborate a divergence of *Pan* and the early hominids at 6–7 Myr ago. [*See* Human Evolution.]

B. The Earliest Hominids

In recent years, our knowledge of the earliest hominids has been greatly enhanced by a number of fossil discoveries in the East African Rift Valley region. These hominids, of genus *Australopithecus*, exhibit an anatomy that shows strong affinity with other hominoid lineages. *Australopithecus* forelimbs exhibit a "retention" of features associated with a climbing, arboreal adaptation, whereas recent reanalysis of hominid body size suggests a considerable degree of sexual dimorphism (difference in body size between sexes). The earliest *Australopithecus,* species *afarensis,* has often been thought to be extremely small, at ca. 30 kg and ca. 4 ft (1.23 m) tall, but it now seems likely that some males weighed as much as 80 kg. This has considerable implications both for models of behavioral patterns and for estimates of relative brain size. The earliest hominids seem to have been less encephalized (had a lower brain–body ratio) than living apes such as the chimpanzee.

C. Bipedalism

The main distinguishing feature of all hominids is bipedalism. Despite debates concerning the efficiency of *afarensis* locomotion, archaeological evidence from fossil footprints and anatomical studies indicates that the earliest known hominids were effective bipeds. A number of explanatory models for the origin of this distinctive feature have been proposed.

1. Bipedalism may have freed the hand to use tools. The cognitive requirements of tool-making may, in turn, have set up a selective pressure for greater encephalization.
2. Bipeds would be able to carry food over some distance and, thus, efficiently exploit the carcasses of large animals. This would also allow hominids to occupy fixed home bases, because females would be provisioned by males.
3. Similarly, bipeds could carry their offspring over considerable distances. This would facilitate longer periods of infant dependency and, thus, greater encephalization.
4. An upright posture would have minimized the effects of mid-day solar radiation on body temperature, because an upright posture reduces the body surface area directly exposed to the sun.

However, the growing body of paleontological and archaeological evidence serves to cast doubt on several of these models.

D. Early Hominids as Bipedal Apes

The tendency to see hominid evolution as progressive has led many students of human evolution to expect their ancestors to run before they can walk! Tool use, although a feature common to hominids and apes, is not evident in the archaeological record until ca. 2.3 Myr ago (the earliest Oldowan pebble tools derive

from Koobi Fora in Kenya). This would be some 3–4 Myr after the appearance of the first hominids. Moreover, we know that other primates use tools without being obligate bipeds. Some forms of tool use might be facilitated by bipedalism, such as the use of weapons, but there is no obvious causal connection between tool use and an upright posture. (For the relationship between tool-making and intelligence, see below.)

For the earliest hominids, similar problems exist in relating the origins of bipedalism to the carrying of food. No evidence indicates that the earliest hominids carried objects over any distance. It might be argued that *Australopithecenes* were carrying perishable objects such as plant materials or digging sticks; however, in the absence of archaeological remains, no positive inference can be made. Nevertheless, we know that, as in the case of tool use, other primates are capable of transporting food or other objects without bipedal locomotion. The claim by M. D. Leakey for the existence of fossil "living floors" at Olduvai Gorge (Tanzania) is controversial and relates to a later date (ca. 1–8 to 1.5 Myr ago). It seems unlikely that any hominids used fixed home bases until the Middle and Upper Paleolithic [100–20 thousand years (Kyr) ago]. Even modern hunter–gatherers tend to use rather ephemeral camps as part of their settlement system, which might only be used for a matter of days or weeks. Such camps would be unlikely to be preserved in the archaeological record.

The carrying of infants presents similar problems. Other primates carry their infants in a variety of ways and often for considerable periods. Young chimpanzees may be carried by their mothers for up to 4 years, with the main period of dependency in the first 2 years of life. Infant-carrying hypotheses depend on the assumptions that (1) bipedalism led directly to greater encephalization and (2) the earliest hominids did not have the same body hair cover as living primates. It is now clear that hominids were bipedal for at least 2–3 Myr before they developed significantly larger brains. Moreover, because we cannot be sure when human body hair evolved to its present form, it is possible that *Australopithicene* or even early *Homo* infants were carried clinging to their mothers' fur, as is observed in other apes and monkeys.

Of all the progressively adaptive models proposed, only temperature regulation is at all convincing. Experimental studies suggest that temperature regulation would have been a significant factor for hominids living in the open savannah, if they were active in the middle of the day when the tropical sun is at its zenith.

Furthermore, it is clear that several features of human skin, such as the reduction of body hair and the development of sweat glands, may be associated with the requirements of keeping the body cool. Experiments have shown that while humans are capable of long periods of continued physical exertion such as running, other tropical mammals such as the large carnivores and herbivores can run only limited distances without succumbing to heat exhaustion.

However, the earliest hominids are likely to have inhabited more closed "mosaic" environments, consisting of open woodland and riverline gallery forest. Here they would not have been exposed to the same intensity of solar radiation, nor, indeed, would they have needed a capacity for sustained physical exertion. It seems more likely, then, that temperature regulation only became a significant selective factor when hominids colonized more open habitats (at ca. 2.5–2.3 Myr ago).

Thus, the combined evidence of anatomy, ecology, and archaeology does not support any particular progressive adaptive traits among the earliest (genus *Australopithecus*) hominids; rather, they should simply be considered as a bipedal form of ape. Bipedalism may be seen as an adaptation to the varied locomotor requirements of a mixed arboreal and terrestrial environment, under the constraints of an anatomy originally evolved for an arboreal existence. Hominids, like all apes, have a postcranial skeleton which reflects the requirements of brachiation, i.e., locomotion by swinging from the arms, as observed in contemporary gibbons. This has precluded a return to quadrupedal locomotion in all terrestrial apes, including the knuckle-walking gorilla and chimpanzee and the bipedal human. As bipedal apes, the earliest hominids exhibit an anatomy fairly typical of the great apes in terms of encephalization, sexual dimorphism, and forelimb specialization. Bipedalism might be seen as a preadaptation for later developments, but, given our current understanding of the chronology of hominid evolution, it was a preadaptation that took considerable time to become significant. This suggests that the action of other parameters was necessary for radical changes in intelligence and subsistence to occur.

II. TECHNOLOGY

A. Technological Determinism

Technology has generally been seen as a central causal factor in human evolution, mediating changes in sub-sistence, leading to increased intelligence and enhancing human physical capabilities. As such, Bergson's argument that man should be renamed *Homo faber* has continued to dominate analyses of hominid behavioral strategies. However, as in the bipedalism debate, the growth of the paleontological and archaeological record (and a more refined chronological framework) has cast some doubt on the causal efficacy of technological determination in human evolution.

B. Development of Technology

The earliest, Lower Paleolithic, tools date from ca. 2.3 Myr ago and are generally termed Oldowan (after the site of Olduvai Gorge). These tools are generally crude chopping or cleaving implements, made by striking four or five flakes from a cobble or pebble. Oldowan types persist until ca. 1.2 Myr ago, although similar crude implements are found in contexts as late as the Neolithic (ca. 3000 BC). By about 1.6 Myr ago, the first Acheulian assemblages appear, characterized by a tool type known as the hand ax, a large ovate or wedge-shaped tool formed by striking numerous parallel flakes along the axis of a cobble or core. Acheulian assemblages continue to dominate the archaeological record until ca. 3–200 Kyr ago. The hand ax itself is found in assemblages as late as 40 Kyr ago, albeit in a more refined form. However, the function of such objects is unknown; some are so large as to be practically useless for most subsistence tasks.

At around 200 Kyr ago, these "core tools" (formed on a large core of rock) were supplemented by new forms of flake-based implements. The earliest of these were produced using the Levallois technique, in which a circular, triangular, or oval flake was struck from a prepared core. Many of these flakes would have been retouched (shaped by further flaking) for use in cutting and scraping, although some appear to have been intended as the points of crude spears (although they would have been too heavy for throwing spears). Levallois or Mousterian Middle Paleolithic techniques persist until ca. 40 Kyr ago, but already by this time many regions such as southwest Asia and Africa yield evidence of a more refined, blade-based technique normally associated with the Upper Paleolithic (generally considered to be the period between 30 and 10 Kyr ago).

Blades are essentially long, parallel-sided flakes struck vertically around the circumference of a prepared cylindrical or conical core of flint, or some similar microcrystalline rock. A blade would generally be used as a "blank" for the preparation of more

A

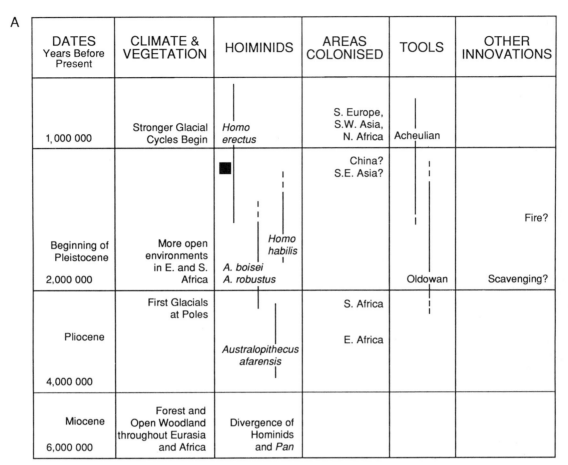

DATES Years Before Present	CLIMATE & VEGETATION	HOIMINIDS	AREAS COLONISED	TOOLS	OTHER INNOVATIONS
1,000 000	Stronger Glacial Cycles Begin	Homo erectus	S. Europe, S.W. Asia, N. Africa	Acheulian	
Beginning of Pleistocene 2,000 000	More open environments in E. and S. Africa	Homo habilis A. boisei A. robustus	China? S.E. Asia?	Oldowan	Fire? Scavenging?
Pliocene 4,000 000	First Glacials at Poles	Australopithecus afarensis	S. Africa E. Africa		
Miocene 6,000 000	Forest and Open Woodland throughout Eurasia and Africa	Divergence of Hominids and Pan			

FIGURE 1   Chronological table of human evolution: (A) 6 Myr to 500 Kyr and (B) 500 Kyr to the Holocene (present interglacial).

refined leaf-shaped spearheads or arrowheads or be broken into pieces to form smaller "bladelets," which would be mounted in a wooden or bone haft. Later Paleolithic stone technologies are also accompanied by the use of bones and ivory and by the appearance of art, both in cave paintings and in carvings of bone and ivory. Postglacial lithic technology is largely a refinement of late glacial upper Paleolithic techniques.

C. Technology and Chronology

The chronological development of technology is indicative of its part in human evolution. The first stone tools appear in the archaeological record more or less simultaneously with the first members of the genus *Homo*. No *Australopithicene* has ever been found in exclusive and incontrovertible association with tools of any kind, although the lineage persists until ca. 1

Myr ago. This suggests the idea that higher intelligence is associated with the development of tools and associated subsistence strategies; however, the subsequent development of technology contradicts this.

The persistence of the Acheulian assemblage for around 1 Myr, with a few minor changes, does not correlate with other patterns of change and stasis. During this same million year period, the hominid brain and body developed to roughly their modern size, and the intermediate *Homo erectus* and early *Homo sapiens* chronospecies colonized the subtropical parts of the Old World and entered the temperate zone. Although the occurrence of Levallois and Mousterian technologies represents something of a departure, the major changes in tool form and function do not occur until the Upper Paleolithic. Only here do we have evidence of effective projectile points, the use of the spear thrower and the bow, and the construction of substantial dwellings.

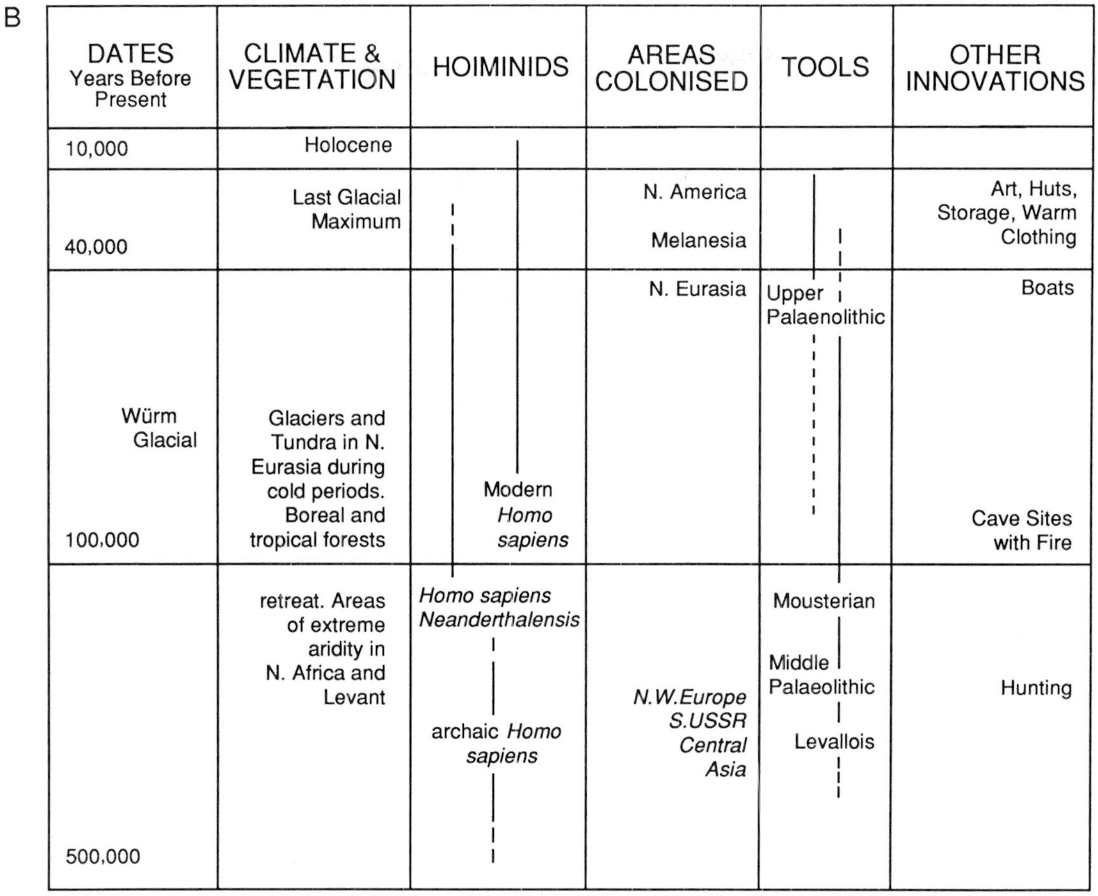

B

DATES Years Before Present	CLIMATE & VEGETATION	HOIMINIDS	AREAS COLONISED	TOOLS	OTHER INNOVATIONS
10,000	Holocene				
40,000	Last Glacial Maximum		N. America / Melanesia		Art, Huts, Storage, Warm Clothing
Würm Glacial / 100,000	Glaciers and Tundra in N. Eurasia during cold periods. Boreal and tropical forests	Modern *Homo sapiens*	N. Eurasia	Upper Palaenolithic	Boats / Cave Sites with Fire
500,000	retreat. Areas of extreme aridity in N. Africa and Levant	*Homo sapiens Neanderthalensis* / archaic *Homo sapiens*	N.W.Europe S.USSR Central Asia	Mousterian / Middle Palaeolithic / Levallois	Hunting

FIGURE I (*Continued*)

D. Function of Technology

Clearly, then, chronology suggests that, for most of the past 6 Myr, technological development has not been a central causal agent in hominid evolution (Fig. 1). Ideas about the association of tools with hunting have been questioned (see below), and it now seems that technological development was in no way focal to the initial stages of the colonization of the globe. Indeed, in many areas of the Far East, there are no stone tools before ca. 50 Kyr ago, whereas hominids had been present in these areas for more than 1 Myr. If anything, we may say that technology has had a significant impact on human biology for the last 50–100 Kyr. Evidence for this is clear in the appearance of more gracile (less heavily built) "modern" *Homo sapiens sapiens* some time between 100 and 40 Kyr ago. It is believed that these hominids were able to survive the conditions of glacial northern Eurasia because of superior technology, and that the use of such items as throwing spears, bows, and hafted tools of all kinds relieved the selective pressure that had produced their more robust ancestors such as *H. erectus*.

III. EARLY HOMINID SUBSISTENCE

A. Man the Hunter

Much as traditional accounts of human evolution have dwelled on the role of technology, they have also focused on the role of hunting as a uniquely human adaptation among the primates. However, as with other theories, recent evidence contradicts this view, and new accounts of hominid subsistence are emerging. Over the last 30 years, the concept of "man the hunter" has been in gradual retreat, as hominid capacity for hunting and uniqueness as a hunter have come to be doubted.

B. Hunters or Hunted?

Early ethnographic work led anthropologists to assume that hunting was, in some sense, the primitive state of humanity. Hence, when the first fossils of early *H. sapiens,* and later of *Australopithecus,* were discovered, it was assumed that these creatures were hunters. Man, in Ardrey's words, was seen as the "killer ape"; his fossil remains surrounded by the stone tools he had used to kill his prey and the bones of those prey. The first doubts about these hypotheses arose from analysis of the faunal remains at early hominid sites and, in particular, their taphonomy (the conditions of deposition).

Work in South Africa and at East African sites has revealed that much of the fossil deposition at these sites was due to fluvial or carnivore activity. Not only were the bones of other animals the remains of carnivore kills, early hominids were also preyed upon by either the ancestors of the modern leopard or some form of the saber-toothed cat. Moreover, only limited evidence from "cut marks" on bone indicates that hominids were themselves butchering large animals. In many cases, the coincidence of large herbivore carcasses with lithic material is the work of fluvial activity in the kinds of river margin environments preferred by early hominids.

It is, in fact, hardly surprising that early hominids were not big game hunters; they were simply too small to tackle large game and were not equipped, as their descendants were, with effective weapons. Thus, most authorities would now argue that early hominids were scavengers; foraging for carnivore kills or preying upon large animals that had become trapped or disabled. This scavenged material might be supplemented by the hunting of small game, as is observed among other primates such as the chimpanzee and baboon. Indeed, it could be said that all primates have a basic hunting adaptation, given that they have traits such as binocular vision, which probably evolved to facilitate insectivory. [*See* Carnivory.]

C. Seed Eaters

Most living primates are largely herbivorous, eating leaves or fruit, and, given that a hunting adaptation now seems less likely, the role of plant matter in the hominid diet has an increasing significance for the study of hominid strategies. Work conducted in the 1960s and 1970s has revealed that even modern hunters and gatherers are not as reliant on large game as was once thought. Hunting is a highly unpredict-

able business, and many societies rely on gathered material or the products of horticulture to sustain themselves when no meat is available. Although no reliable figures are available, it seems that meat contributes only 25–30% of the diet of most hunter–gatherer societies (although exceptions exist), compared with estimates of perhaps 6% among the East African chimpanzee.

Unfortunately, plant remains are rarely preserved in the Paleolithic archaeological record, and much remains speculative. Analysis of early hominid dentition reveals that at least some early hominids such as *Australopithecus robustus* and *Australopithecus boisei* may have specialized in eating dry savannah plant foods such as seeds and corms. Moreover, it seems likely that all early hominids would have gathered a varied diet of plant materials in the mosaic environments of savannah river and lake margins. We can be fairly certain that meat eating was a part of hominid subsistence from about 2 Myr ago onward, but the relative importance of plant and animal matter continues to be in doubt.

D. Woman the Gatherer

Many, if not most, researchers in the domain of hominid gender roles have assumed, as Lovejoy does, that early hominids were monogamous and pair-bonded. Indeed, this pair bond has been seen as the fundamental basis for more effective care of infants, hence creating the conditions for longer periods of parental dependence and, thus, larger brains. However, the biological and sociological evidence for the role of monogamy as a fundamental human trait is equivocal. It has been suggested that various primary and secondary sexual characteristics of humans, such as relatively small testicle size or large breasts, are to be associated with monogamous-pair bonding, but in fact, these traits cannot be causally linked to monogamy. Humans are virtually continually sexually active throughout adult life, whereas sexual activity among monogamous species tends to be infrequent. Moreover, it can be shown ethnographically that attitudes toward sexual partners and sexual practices reflect a tendency toward polygyny that is only suppressed, for social reasons, in modern societies.

Given this evidence, the fact that early hominids were not big game hunters has considerable implications for their socioecology. Traditionally, it has been assumed that male hunters or indeed scavengers would have provisioned their female partners, who would reside, for safety from predators, at a central home base; however, ethnography reveals that much if not most of the gathered material in hunter–gatherer diets is provided by women. In fact, it seems likely that the economic dependence of females on males is a fairly recent and context-specific phenomenon. One might suggest that some division of labor between sexes, or other groups, occurred quite early in the evolution of *Homo*. For each gender, we might expect differing life history tactics, reflecting differing reproductive strategies. Thus, for males, highly nutritious, but unpredictable, sources of meat from hunting and scavenging would be preferred, enhancing an individual's capacity to compete for social status and for mates. Females, however, would concentrate on lower value but more predictable sources of nutrition such as plant material. This would reflect the need for a more sustainable, long-term reproductive history, as the female reproductive capacity is limited by the time required for gestation. Social sharing of these materials, either with kin groups or potential mates, would constitute the basis for an ecology of social division of labor.

IV. ICE AGE HUNTERS

A. Middle Paleolithic Adaptations

As noted earlier, stone tools from before the Middle Paleolithic are unlikely to have been used for hunting. These probably were used for the butchery of scavenged carcasses and the preparation of plant materials. It has been argued that even Middle Paleolithic populations were not hunters. In particular, faunal remains from sites such as Klasies River Mouth (in South Africa, dated ca. 150–90 Kyr ago) do not seem consistent with hunting. However, most authorities would agree that some form of large game hunting appeared during this period. The heavy Levallois Mousterian points were probably used in hunting large game by what is termed a confrontational method. Middle Paleolithic hominids such as the Neanderthals were extremely robust, while the projectile points suggest that they were used to form heavy thrusting spears. It is therefore suggested that large fauna were killed either by being mobbed by a large number of hunters or by being driven to their deaths (as at La Cotte in Jersey) over a convenient precipice. Such techniques would be highly dangerous, and, notably, Middle Paleolithic skeletons show a very high incidence of traumatic injuries, a surprising number of which were not actually fatal.

Most early evidence of hominid hunting derives from the last glacial cycle (the Wurm, or Wisconsin), which was the first period in which hominids continued to occupy temperate latitudes, despite deteriorating climatic conditions. In previous cycles, hominids probably only entered these latitudes in intermediate periods, between the cold glacial maxima and the warm interglacials when heavy boreal forest cover made most of northern Eurasia impassable. In this context, hunting may be seen as an adaptation to cold and, more generally, to arid climatic conditions or regions. Thus, in tropical regions we find that hunting populations occupied the more arid areas of Africa, the Middle East, and Asia, while in northern latitudes hominid populations took advantage of the open glacial habitat to develop increasingly effective hunting strategies.

Traditionally, paleoanthropology has regarded glacial habitats as inhospitable, but in fact the reduction of tree cover in these periods led to a massive increase in populations of large fauna. In the early Wurm (125–30 Kyr ago), climatic conditions would have been mild enough to sustain the confrontational Middle Paleolithic hunters. Anatomical studies have shown that European *H. sapiens neanderthalensis* had several specialized adaptations to cold conditions. Their relatively short limbs minimized body surface area and, thus, heat loss, whereas particularly large nasal cavities were probably adapted to the warming of cold air and the retention of moisture.

However, the more extreme conditions of the later Wurm (30–10 Kyr ago) probably required more sophisticated hunting strategies and social organization. In northern temperate latitudes, these more complex strategies are generally thought to be associated with the appearance of Upper Paleolithic technologies and the distinctive, anatomically "modern" *H. sapiens sapiens*. This apparent replacement of indigeneous populations raises a number of difficult questions concerning the relationship among anatomical differences, behavior, and intelligence.

B. Modern Humans and the Upper Paleolithic

The so-called Middle–Upper Paleolithic transition is perhaps the most controversial area in hominid research. For at least half a century, paleoanthropology has debated the relationship between European and Near Eastern Neanderthal populations and the apparently immigrant populations of modern *H. sapiens sapiens*. The controversy has centered on whether indigenous populations evolved into "moderns" or whether they were replaced by modern humans who had evolved separately in Africa. Recent research in physical anthropology and microbiology is significantly resolving this debate, although a number of ecological and social questions remain open.

Research on human mitochondrial DNA has, through analysis of mutations and estimation of mutation rates, suggested that all living humans are descended from one ancestral population or even individual, originating in Africa at about 200 Kyr ago. This scenario is consistent with an origin of modern humans in Africa, as represented by fossils of early *H. sapiens sapiens* such as the Omo 1 and 2 skulls. It has been suggested that these modern *H. sapiens sapiens* evolved separately from the Eurasian *H. sapiens neanderthalensis,* diverging from a common root represented by the Petralona, Steinheim, and Swanscombe skulls of early *H. sapiens.* The modern types first entered southwest Asia around 100 Kyr ago (much earlier than had previously been thought) and colonized northern Eurasia and South Asia between 50 and 30 Kyr ago.

However, a number of questions remain open. Although, traditionally, modern types have been associated with the more advanced technology and subsistence techniques of the Upper Paleolithic, many of the earliest modern-type fossils are associated with Middle Paleolithic tools. Moreover, the blade-based lithics identified with the Upper Paleolithic are found in very early contexts in both North and South Africa. Therefore, it is difficult to decide on the nature of the link between anatomical and behavioral differences. Were the anatomical differences between moderns and Neanderthals actually a causal factor in behavioral differences? As noted earlier, most of the anatomical change in human evolution has taken place without a concomitant change in technology. Given the fact that mitochondrial DNA is only transmitted through the female line, we may never know if living humans have inherited Neanderthal nuclear DNAs through the male line.

C. Hominid Mobility and Global Colonization

Certainly it would seem that the behavioral changes associated with the European Upper Paleolithic were directed toward a strategy for survival in cold arid climates. Growing evidence indicates very high levels of mobility among hominid populations in the last glacial, not least from the fact that all of northern Eurasia was colonized at this time, and, in the late glacial period (15–11 Kyr ago), hominids first crossed Beringia (now the Bering Strait) into Alaska. Moreover, it is clear that between 50 and 30 Kyr ago hominids were already successfully crossing open seas to colonize Australasia and Melanesia.

This high level of mobility, essential to survival in the severe climates of glacial northern Europe and Siberia, was associated with complex networks of social interaction. The distribution of raw materials, tool types, and art objects across the western Soviet Union and eastern and western Europe demonstrates that artifacts were either carried or traded over great distances. These trading links may be associated with networks for the exchange of mates and information. Moreover, the existence of Upper Paleolithic sites at the very fringes of the northern ice caps suggests that annual or seasonal migrations may have involved journeys over many hundreds of kilometers, e.g. from southwest France to Britain and Belgium and The Netherlands, or from the Amur River in southeast Siberia to the northeast Chukchi peninsula, although such journeys are hard to substantiate.

D. Art and Language

The coincidence of the appearance of art and of modern humans in the Upper Paleolithic is often thought to reflect the cognitive differences between *H. sapiens sapiens* and archaic *H. sapiens* such as the Neanderthals. The appearance of art perhaps implies the emergence of language; however, this may be doubted on archaeological, biological, and psychological grounds. Although the array of parietal (= cave) and mobilary (= carved) works of art in the Upper Paleolithic is impressive, many later archaeological periods are notable for the virtual absence of representational art. To claim that art is a necessary indicator of linguistic competence is, thus, problematic with respect to these later periods. No doubt the Upper Paleolithic origin of art does reflect considerable social change, but this need be no more significant than the appearance of cubism and nonrepresentational art in the 20th century.

Biologically it has been argued that the laryngeal tract of modern humans is different from that of early *H. sapiens.* Lieberman has modeled the throat and tongue of various hominids through inference from cranial and mandibular posture. He argues that Neanderthals would not have been able to form all of the vowels and consonants of modern human speech. However, most modern humans do not use >50%

of possible phonetic sounds, suggesting that Neanderthals would have been perfectly capable of developing a language. Moreover, other physical differences, such as in the structure of the hyoid bone, have now been cast into doubt. Studies of endocranial casts suggest that lateral asymmetry of the brain (generally associated with language in humans) begins early in the hominid line, if not before.

It is likely that the evolution of the hominid lineage would have seen progressive development of linguistic communication, rather than a unitary "origin" of language at a particular temporal locus. The development of language should reflect such factors as division of labor, mobility, and social organization, where explicit language would be needed to structure activity in time and space. It is likely, then, that the evolution of language would track the evolution of society. [*See* Language, Evolution.]

V. SOCIAL EVOLUTION

A. Why Are Humans Intelligent?

Traditionally, explanations of human intelligence have concentrated on the cognitive requirements of tool-making and subsistence strategies involving hunting or scavenging from home bases, but as we have seen, these hypotheses do not totally accord with the evidence as it now exists. The development of technology does not keep pace with the development of the human brain to the extent that the brain of early *H. sapiens* had already reached its modern size well before the appearance of Upper Paleolithic complexities of technology. The alternative to a technodeterminist model for the origin of intelligence would be to consider the requirements of subsistence. In the early 1970s, Isaac suggested that social organization of parental care, or the sharing of resources, might explain the need for greater intelligence. These notions hint at the developments now taking place in psychology and primatology, which suggest that the requirements of the social axis of life may be more important than has previously been considered.

B. Social Intellect Hypothesis

Studies of primate behavior have revealed that monkeys and apes have limited capacities to deal with practical tasks; however, recent research has demonstrated that primates have much more subtle capacities when confronted with social tasks. Primates have very developed capacities for recognizing social status and, in the case of chimpanzees, recognizing individual identities. In the wild, many species of monkeys are capable of complex communication concerning the nature and whereabouts of threats from predators. Moreover, they are also observed to use subtle techniques of deception to achieve social or sexual ends.

These observations and social psychological theories concerning the development of human intelligence have led to the formulation of what is now called the social intellect hypothesis. This suggests that the primary function of primate and human intellect is to deal with the complexities of social life. Most practical problems (except perhaps those encountered by modern humans) are relatively simple compared with the difficulties of achieving and maintaining social status or controlling the effects of agonistic (threatening or violent) encounters, which place stress on group cohesion. Practical problems are often confronted through the medium of social life, as in the case of avoiding predators or in finding subsistence material.

The social intellect hypothesis, therefore, suggests that the development of the human brain over the last 2 Myr may have had more to do with social causes than practicalities such as tool-making. This is hardly surprising when one considers the complex mechanisms required by social living. Humans have a highly developed ability to recognize and remember the identity of, and their relationship to, other members of their own species. This ability is absent in all other primates, except in a less-developed form in the chimpanzee and perhaps the gorilla. Moreover, of course, the large part of the human brain given over to the understanding and production of language is quintessentially a social function. [*See* Brain Evolution; Reasoning and Natural Selection.]

C. Social Intellect and Subsistence

The sharing of food may be a fundamental part of hominid adaptation. The unpredictability of diverse foraging strategies may be compensated by the sharing of resources, whereas large items such as carcasses may only be effectively exploited by groups. In fact, other primates do exhibit limited sharing behaviors, particularly with offspring or as "sexual favors" to potential or actual mates. However, the regular and habitual sharing of resources within a group requires recognition of complex individual relationships, if inequity and agonistic disputes are to be avoided. Concomitant analyses of social status and relationship are exactly what is predicted by the social intellect hypothesis.

Similarly, division of labor is accomplished by social means. The organization of society in time and space permits the efficient exploitation of resources within the home range. Where subsistence activities require more than individual foraging, task groups may be formed for specific purposes (e.g., to seek lithic raw material), and the products of such expeditions are shared within the group. The collective appropriation of resources contrasts with most primate foraging, where individuals forage for themselves within a protective group, which must move around its home range en masse.

D. Social Organization and Colonization

The development of social organization may be inferred from the archaeological record. From the first appearance of genus *Homo* onward, the appropriation of raw materials and subsistence resources develops in geographical scale and temporal displacement. Even *H. erectus* is likely to have obtained lithic raw material from as far as 50 km from its home range. The exploitation of more arid, more inaccessible, and more seasonal environments would have required an increasingly logistic social organization of time and space. Similarly, the eventual appearance of large game hunting would have required considerable social organization in the production of effective weapons, the prosecution of the hunt, distribution of products, and the control of resources.

In the Middle and Upper Paleolithic, we find clear and considerable development of social organization. Raw materials are procured over increasingly long distances, and eventually distinctive regional cultures and practices appear. The burial of the dead and care for the old or infirm are already apparent in some Neanderthal groups, suggesting that social relationships were permanent and not simply forgotten as soon as an individual no longer was productive or had died.

The Upper Paleolithic provides further ample evidence of the existence of social networks and of social structures extending over greater ranges of time and space. Colonization of land masses separated by sea probably required the maintenance of social ties of great distances if colonists were not to be isolated and thus driven to extinction by the unpredictable nature of novel environments. Moreover, some of the caves of southwest Europe, such as Altamira, or the Upper Paleolithic villages of the Ukraine may represent centers of aggregation for populations, which would disperse in the spring and summer to exploit resources in more inhospitable northern latitudes. These aggregation centers were also foci of artistic activity, suggesting that the appearance of art is associated with the large-scale complexity of Upper Paleolithic social organization.

All phases of hominid colonization, therefore, may be seen as the product of developing forms of social organization, where time and space are manipulated at the social level to maximize the exploitation of resources. The development of social complexity would have had wide ranging consequences, as in the development of language and the appearance of art. The increasing mobility of groups and individuals would lead to changes in the genetic structure of populations, with less likelihood of local isolation. It is probable that the encounter between modern *H. sapiens* and Neanderthals may also be understood in social terms. Differences in social organization and individual recognition might explain how, and indeed whether or not, the two populations interacted and intermingled.

BIBLIOGRAPHY

Almquist, A. J., and Boaz, N. T. (1996). "Biological Anthropology: A Synthetic Approach to Human Evolution." Prentice–Hall, Englewood Cliffs, NJ.

Binford, L. R. (1983). "In Pursuit of the Past." Thames and Hudson, London.

Byrne, R., and Whiten, A. (1988). "Machiavellian Intelligence." Oxford University Press, London.

Foley, R. (1987). "Another Unique Species: Patterns in Human Evolutionary Ecology." Longman Scientific and Technical, London.

Gamble, C. S. (1987). "The Paleolithic Settlement of Europe." Cambridge University Press, Cambridge.

Noble, W., and Davidson, I. (1996). "Human Evolution, Language and Mind: A Psychological and Archaeological Inquiry." Cambridge University Press, Cambridge.

Richards, G. (1987). "Human Evolution: An Introduction for the Behavioural Sciences." Routledge and Kegan Paul, London.

Excitatory Neurotransmitters and Their Involvement in Neurodegeneration

AKHLAQ A. FAROOQUI
LLOYD A. HORROCKS
Ohio State University, Columbus

I. Classification and Properties of Excitatory Amino Acid Receptors
II. Possible Mechanism of Cell Injury by Excitatory Amino Acids
III. Excitatory Amino Acid Receptors and Neurological Disorders
IV. Summary

GLOSSARY

Alzheimer disease A neurodegenerative disease characterized by memory loss, deposition of β-amyloid, and a large number of plaques and tangles in the cerebral cortex and hippocampus

Excitotoxicity A process by which an excess of glutamate and its analogs excites neurons and brings about their death

Ischemia An insult to brain that interrupts its blood supply

G protein A family of homologous guanine nucleotide-binding proteins involved in signal transduction

Neurodegeneration Degeneration or destruction of neurons in trauma or in the disease process

EXCITATORY AMINO ACIDS, SUCH AS GLUTAMATE and aspartate, are major neurotransmitters in the mammalian central nervous system. Although these amino acids are responsible for normal excitatory transmission, they also represent a potential source of neurotoxicity. For this reason, the concentration of glutamate is regulated by several mechanisms. Abnormally low levels of glutamate can compromise normal levels of excitation, whereas excessive levels can produce toxic effects. Treatment of neurons with glutamate, aspartate, and their related analogs causes cell swelling, vacuolization, and eventual cell death. This observation prompted Olney and his associates to propose an "excitotoxin" concept of neuronal cell death. Excitotoxicity refers to a paradoxical phenomenon whereby the neuroexcitatory action of glutamate and related compounds becomes transformed into a neuropathological proces that can rapidly cause neuronal cell death.

I. CLASSIFICATION AND PROPERTIES OF EXCITATORY AMINO ACID RECEPTORS

Glutamate exerts its effect by interacting with certain receptors called glutamate receptors or excitatory amino acid receptors (EAA). Based on biochemical, electrophysiological, and molecular studies, EAA are classified into two broad classes: ionotropic receptors are made up of oligomeric proteins which form ligand-gated ion channels (ion channel linked) and metabotropic receptors which are coupled through G-proteins to second messenger systems (G-protein linked).

Ionotropic receptors are further classified into (1) N-methyl-D-aspartate (NMDA) receptors, (2) α-amino-3-hydroxy-5-methyl-4-isoxazole propionate (AMPA) receptors, (3) kainate receptors, and (4) the AP_4 receptor. AMPA and kainate receptors mediate fast exitatory synaptic transmission, are rapidly stimulated and desensitized by glutamate, and are permeable to Na^+ and K^+. NMDA receptors, however, are

ENCYCLOPEDIA OF HUMAN BIOLOGY, Second Edition, VOLUME 3. Copyright © 1997 by Academic Press. All rights of reproduction in any form reserved.

stimulated and desensitized more slowly by glutamate and are permeable to Ca^{2+}. NMDA receptors are selectively antagonized by α-amino-ω-phosphonocarboxylic acids, including 2-amino-5-phosphonovalerate. AMPA and kainate receptors are markedly antagonized by 6-cyano-7-nitroquinoxaline-2,3-dione (CNQX), 6,7-nitroquinoxaline-2,3-dione (DNQX), and 6-nitro-7-sulfamoylbenzoquinoxaline-2,3-dione (NBQX). Selective antagonists of kainate receptors [(2-amino-3-[3-(carboxymethoxy)-5-methylisoxazol-4-yl] propionic acid and 2-amino-3-[2-(3-hydroxy-5-methylisoxazol-4-yl)methyl-5-methyl-3-oxisozoline-4-yl]propionic acid] have also been reported (Table I).

A. Ionotropic Receptors

1. NMDA Receptor

The NMDA receptor consists of four domains: (1) the transmitter recognition site with which NMDA and L-glutamate interact; (2) a cation-binding site located inside the channel where Mg^{2+} can bind and block transmembrane ion fluxes; (3) a phencyclidine (PCP)-binding site that requires agonist binding to the transmitter recognition site, interacts with the cation-binding site, and at which a number of dissociative anesthetics (PCP and ketamine), Σ opiate N-allylnormetazocine (SKF-10047), and MK-801 bind to function as open channel blockers; and (4) a glycine-binding site that appears to allosterically modulate the interaction between the transmitter recognition site and the PCP-binding site. It also contains a polyamine-binding site. In addition, evidence shows that Zn^{2+}, acting at a separate site near the mouth of the ion channel, acts as an inhibitory modulator of channel function. The NMDA receptor complex also contains an arachidonic acid-binding site. The amino acid sequences of this binding site resemble fatty acid-binding proteins. Other modulators of NMDA receptors include sulfhydryl redox reagents and H^+ ions. Normal functioning of the NMDA receptor complex depends on a dynamic equilibrium among various do-

TABLE I
Pharmacological Properties of Excitatory Amino Acid Receptors

Receptor	Agonist	Antagonist	Ion permeability	Second messenger
Ionotropic				
NMDA receptor	NMDA, L-Glu, L-Asp Gly, D-Ser, polamines, Ibo, (1R,3R)-ACPD	D-AP5, CPP, CGP39653, LY233053, CGS19755, PCP, TCP, (RS)-CPP, (R)-CPP, L689, 560, MK810, SKF10047, ketamine	Ca^{2+}, K^+, Na^+	Ca^{2+}
Kainate receptor	KA, Dom, QA, L-Glu	CNQX, DNQX, AMOA, AMNH	Na^+, K^+	Ca^{2+}
AMPA receptor	AMPA, QA, L-Glu, (S)-Willardiine	CNQX, DNQX	Na^+, K^+	Ca^{2+}
L-AP4 receptor	L-AP4 Lsop, L-Glu	Not known	Not known	Not known
Metabotropic				
trans-ACPD receptor	trans-ACPD, L-Glu, QA, (RS)-3,5-dihydroxy phenylglycine, (2S,3S,4S) isomer (L-CCG-I), L-BMAA	(S)-4-carboxyphenyl, glycine, MCCG, MTPG	none	(1,4,5)IP$_3$/DAG Ca^{2+}

Abbreviations: ACPD, *trans*-1-amino-cyclopentyl-1,3-dicarboxylate; AMPA, α-2-amino-3-hydroxy-5-methylisoxazole-4-propionate; AP4, 2-amino-4-phosphonobutyrate; AP5, 2-amino-5-phosphonovalerate; ASP, aspartate; CNQX, 6-cyano-7-nitroquinoxaline-2,3-dione; CPP, 3-(2-carboxypiperazin-4-yl)propyl-1-phosphate; DNQX, 6,7-dinitroquinoxaline-2,3-dione; Glu, glutamate; Gly, glycine; IBO, ibotenate; IP$_3$, inositol 1,4,5-trisphosphate; DAG, diacylglycerol; KA, kainate; MK-801, dibenzocyclohepteneimine; NMDA, *N*-methyl-D-aspartate; PCP, phencyclidine; QA, quisqualate; SER, serine; TCP 1-[1-(2-thienylcylclohexyl]piperidine; LY-233053, *cis*($\pm$)-4[2*H*-tetraazol-5-yl)-methyl]piperidine-2-carboxylic acid; CGS19755; *cis*-4-phosphonomethyl-2-piperidinecarboxylic acid; CGP39653, (D,L-(*E*)-2-amino-4-propyl-5-phosphono-3-pentenoic acid; AMNH, 2-amino-3-[2-(3-hydroxy-5-methylisoxazol-4-yl)methyl-5-methyl-3-oxoisoxazolin-4-yl]propionic acid; AMOA, 2-amino-3-[3-(carboxymethoxy)-5-methylisoxazol-4-yl]propionic acid; DOM, domoate; (2S,3S,4S) (L-CCG-I), (2S,3S,4S) α-(carboxycyclopropyl)glycine; L-BMAA,β-*N*-methylamino-L-alanine; SKF-10047, Σ-opiates, *N*-allylnormetazocine; (1R,3R) ACPD, (1R,3R)-1-aminocyclopentane-1,3-dicarboxylic acid; (RS) CPP, 3-(RS)-2-carboxypiperazin-4-yl)propyl-1-phosphonic acid; (R)-CPP, 3[(R)-2-carboxypiperazin-4-yl]propyl-1-phosphonic acid; L-689,560, *trans*-2-carboxyl-5,7-dichloro-4-phenyl-aminocarbonylamino-1,2,3,4-tetrahydroquinoline; MCCG, (2S,3S,4S)-2-methyl-2-(carboxycyclopropyl) glycine; MTPG, (RS)-α-methyl-4-tetrazolylphenylglycine.

main components. Loss of equilibrium during membrane perturbation may cause the entire system to malfunction and result in an expression of excitotoxicity.

There is growing evidence for the existence of multiple subtypes of NMDA receptors. Molecular cloning studies have identified at least four complementary DNA species in rat brain, encoding NMDA receptor subtypes NMDAR1 (NR1), NMDAR2A (NR2A), NMDAR2B (NR2B), and NMDAR2C (NR2C). NR2A, NR2B, and NR2C channels resembles each other 50 to 70% in amino acid sequences. NR2A and NR2C channels differ in gating behavior, magnesium sensitivity, and regional distribution in rat brain. All NMDA receptor subtypes contain phosphorylation sites for Ca^{2+}–calmodulin-dependent kinase and protein kinase C at the cytoplasmic domains. These kinases play a crucial role in the induction and maintenance of long-term potentiation.

2. AMPA Receptor

Using radioligand [³H]AMPA, the AMPA receptor was solubilized from chick brain. The rank order of potency for competitive ligands in displacing [³H]AMPA binding is reported to be AMPA = quisqualate > 6-cyano-7-nitroquinoxaline-2,3-dione > L-glutamate > kainate. Purified AMPA receptor from rat brain has a molecular mass of 105 kDa. Treatment of rat brain membranes with phospholipase A_2 significantly increases the binding of [³H]AMPA to the AMPA receptor. Kinetic analysis has indicated that phospholipase A_2 treatment increases the affinity of the AMPA receptor without changing the maximum number of sites. Quinoxalinediones (CNQX, DNQX, and NBQX) are selective and potent competitive antagonists, and the 2,3-benzodiazepine muscle relaxant GYK1 52466 (1-(4-aminophenyl)-4-methyl-7,8-methylenedioxy-5H-2,3-benzodiazepine HCl) is a highly selective noncompetitive antagonist. Collective evidence from binding studies indicates that AMPA receptors mediate those neurotoxic events that involve postsynaptic neuronal membranes.

3. Kainate Receptor

Ligand-binding studies with [³H]kainate have demonstrated specific saturable and high-affinity binding to brain membranes. Displacement studies with analogs of kainate have indicated a good correlation between excitatory and neurotoxic potencies of kainate.

Kainate-binding sites have been solubilized and purified from rat and frog brain. Gel filtration on Sepharose 6B indicates a molecular mass of 550 kDa. Poly-

acrylamide gel separations give a molecular mass of 48 kDa. The large differences in molecular mass determined by the previously mentioned procedures may result from the breaking of the kainate receptor complex into subunits.

Xenopus brain is an exceptionally rich source of both kainate- and AMPA-binding sites, and the relationship between these sites has been studied in detail. In this tissue the functional interaction between kainate and AMPA can be unequivocally correlated with a physical colocalization of the two types of sites in a single protein complex. The two sites coexist in a 1 : 1 ratio and cannot be separated by various physical and biochemical procedures. The purified protein shows high-affinity binding for both AMPA and kainate and they are mutually and fully competitive, with K_i values identical to the K_d values for the radioligand (34 nM AMPA and 15 nM kainate). Molecular cloning studies have identified at least six kainate/AMPA receptor subtypes (GluR1, GluR2, GluR3, GluR4, GluR5, and GluR6). The AMPA and kainate receptor subunits are large, with a subunit molecular mass of about 100 kDa, and show little sequence homology to subunits of other receptors. All of these subtypes exhibit an individual distribution in the nervous system and also show alteration in expression during development. It is interesting to note that GluR6 is markedly stimulated by kainate, but not by AMPA, and is therefore, strictly speaking, a kainate rather than a kainate/AMPA receptor.

4. L-AP4 Receptor

In contrast to NMDA, AMPA, and kainate receptors, the L-AP4 receptor is characterized by the agonistic action of L-AP4 on certain glutamate-using synapses. Based on various pharmacological studies, it has been suggested that L-AP4 may act at both pre- and postsynaptic sites.

Although the occurrence of the L-AP4 receptor has been clearly established in various preparations by physiological studies, its biochemical characterization has been elusive. Thus, very little is known about the molecular properties of this receptor.

B. Metabotropic Receptors

This receptor acts through a GTP-binding, protein-dependent mechanism to elicit polyphosphoinositide hydrolysis. Its activation leads to increased cellular levels of inositol 1,4,5-trisphosphate and diacylglycerols with mobilization of Ca^{2+} from intracellular stores. Because it responds to quisqualate and not to AMPA, it is AMPA insensitive. It was called a QP receptor.

Potent stimulation of polyphosphoinositide turnover in mammalian brain slices also occurs with the rigid glutamate analog (±)*trans*-1-aminocyclopentyl-1,3-dicarboxylate (ACPD). The Qp receptor has been renamed the *trans*-ACPD receptor. The *trans*-ACPD receptor is also stimulated by (2S,3S,4S) α-(carboxy-cyclopropyl) glycine (L-CCG-1) in hippocampal synaptoneurosomes.

The most effective inhibitors of metabotropic receptors are 2-amino-3-phosphonopropionate (AP3) and 2-amino-4-phosphonobutyrate (AP4), both of which block responses in a noncompetitive fashion.

cDNA for the metabotropic glutamate receptor has been cloned from the rat cerebellum cDNA library. These studies have demonstrated that the cloned receptor (mGluR1) stimulates phosphatidyl-inositol turnover through a pertussis toxin-sensitive G-protein. In transfected Chinese hamster ovary cells, the rank of phosphatidylinositol hydrolysis potency is quisqualate > L-glutamate > ibotenate > L-homocysteine sulfinate ≥ (±)*trans*-ACPD. This receptor also evokes the stimulation of cAMP formation and arachidonic acid release with comparable agonist potencies. The mRNA for this receptor is predominantly expressed in hippocampal and cerebellar neuronal cells. This receptor shares no sequence similarity with other members of G-protein-coupled receptors and possesses a unique structure (M_r 113 kDa) with a large hydrophilic sequence preceding the seven putative transmembrane domains. The occurrence of four additional subtypes of mGluR (mGluR2, mGluR3, mGluR4, and mGluR5) has also been reported. The four newly identified receptors share a high degree of sequence similarity with mGluR1 and possess a large extracellular domain preceding the seven putative transmembrane segments. mGluR2, mGluR3, and mGluR4, unlike mGluR1, are coupled to an inhibitory cAMP cascade. Although the precise signaling pathway of mGluR3 and mGluR4 remains to be elucidated, no linkage has been found between these receptors and polyphosphoinositide turnover. mGluR5 has 60% sequence homology with mGluR1 and is coupled to the phosphoinositide response. Thus, among the five metabotropic receptors reported, mGluR1 and mGluR5 are the only ones capable of stimulating polyphosphoinositide turnover.

The function of the *trans*-ACPD receptor remains unknown. However, there is growing speculation that it plays a key role in synaptic growth and synaptic stabilization. The metabotropic excitatory amino acid receptors are also implicated in long-term potentiation (LTP), a long-lasting increase in synaptic efficacy induced by high-frequency stimulation of specific pathways in the hippocampus. It is now well established that protein kinase C is necessary for the maintenance of LTP. The observation that metabotropic excitatory amino acid receptors are able to activate protein kinase C suggests that they may be involved in the initiation of the late component of LTP. [*See* Neurotransmitters and Neuropeptide Receptors in the Brain.]

II. POSSIBLE MECHANISM OF CELL INJURY CAUSED BY EXCITATORY AMINO ACIDS

The molecular mechanisms of cell injury caused by excitatory amino acids are becoming evident. *In vitro* studies indicate that excitotoxin-induced neuronal injury may involve two distinct events. First, the exposure of neurons to an excitotoxin (30 min) may cause an acute neuronal swelling (acute neurotoxicity) resulting from the depolarization-mediated influx of Na^+, Cl^-, and water. This process is reversible if the excitotoxin is removed from the system. The degree to which this event contributes to neuronal injury is unclear, but it has been suggested that water entry causes osmotic lysis. The second event is characterized by an excessive calcium influx primarily via NMDA receptor channel activation. This is brought about by exposing neurons to glutamate for briefer periods (5 min) and observing neuronal degeneration over a 24-hr period (delayed neurotoxicity).

A rise in intracellular Ca^{2+} may affect the activities of a number of enzymes, including the activation of lipolytic (lipases and phospholipases) and proteolytic (Calpain I and other calcium-dependent proteases) enzymes. The activation of lipolytic enzymes releases arachidonic acid from neuronal membrane phospholipids and sets "the arachidonic acid cascade" in motion. The latter includes the synthesis of prostaglandins, leukotrienes, and thromboxanes.

The arachidonic acid cascade also potentiates the formation of free radicals and lipid hydroperoxides. The latter are known to inhibit reacylation of phospholipids in neuronal membranes. This inhibition may constitute an important mechanism whereby peroxidative processes contribute to irreversible neuronal injury and death.

The activation of proteases by calcium may cause a breakdown of the cytoskeleton, leading to severe cellular damage. [See Calcium, Biochemistry.] Thus, exposure of hippocampal neurons to an excitotoxin also causes the degradation of spectrin and MAP2 which correlates well with subsequent neuronal injury. The cytosolic protease calpain induces the conversion of xanthine dehydrogenase to xanthine oxidase and may help in the production of free radicals. Another target of calcium-induced injury may be protein kinase C. This enzyme catalyzes the phosphorylation of many synaptic membrane proteins and is regulated by Ca^{2+}, diacylglycerol, free fatty acids, and phosphatidylserine. Protein kinase C plays an important role in delayed neurodegeneration; its activators are toxic to cultured human cortical neurons and its inhibitors diminish neuronal loss. Calcium may also stimulate calcineurin, a calcium/calmodulin-dependent phosphatase that is involved in the dephosphorylation of DARPP-32 (a dopamine- and cAMP-regulated 32-kDa phosphoprotein) and MAP2 (a microtubule-associated protein).

A recent advancement in understanding the role of Ca^{2+} in delayed neurotoxicity is the observation that inhibitors of nitric oxide synthase block the development of delayed NMDA-induced neuronal death in cortical neurons and hippocampal slices. Nitric oxide is a short-lived, intra- and intercellular second messenger whose release is dependent on increases in intracellular Ca^{2+}. For nitric oxide to play an important role in delayed neurodegeneration would require that this second messenger activate other longer-lived processes or be released in an ongoing fashion for a critical period of time. Nitric oxide stimulates guanylate cyclases, leading to increases in intracellular cGMP. Thus, cGMP could provide a means for longer-lasting changes in intracellular function initiated by nitric oxide. Nitric oxide can also induce neuronal degeneration by virtue of its free radical property.

Thus, two major processes may be involved in neuronal injury caused by the overstimulation of excitatory amino acid receptors. One is the large Ca^{2+} influx (neuronal injury occurring when a certain threshold intracellular Ca^{2+} concentration is attained for a certain duration) and the other is the accumulation of free radicals and lipid peroxides as a result of neural membrane phospholipid degradation. Another mechanism that may be involved in glutamate neurotoxicity is the inhibition of cystine uptake by glutamate and its analogs. The reduced availability of cystine reduces the availability of glutathione, an important component of cellular defense against free-radical injury, and therefore increases vulnerability to oxidative stress. Free radicals can disrupt membrane integrity by reacting with proteins and unsaturated lipids in the plasma membrane. These reactions lead to a chemical cross-linking of membrane proteins and lipids and to a reduction in membrane-unsaturated lipid content. This depletion of unsaturated lipids may be associated with alterations in membrane fluidity and permeability and changes in activities of membrane-bound enzymes and receptors. It has been suggested that calcium and free radicals may act in concert to induce neuronal injury.

Another important process that may be involved in neuronal degeneration in neurodegenerative diseases is the impairment of energy metabolism. Experimental evidence for impaired energy metabolism in excitotoxicity has been obtained. Inhibitors of oxidative phosphorylation or Na^+, K^+-ATPase allow glutamate or NMDA to become neurotoxic at concentrations that ordinarily produce no toxicity. How delayed onset and slow progression of neurodegenerative disease is related to a defect in energy metabolism and excitotoxicity remains to be seen.

III. EXCITATORY AMINO ACID RECEPTORS AND NEUROLOGICAL DISORDERS

Excitatory amino acid receptors may be involved in a number of neurogenerative states (Table II). The levels of glycerophospholipids (plasmalogens and polyphosphoinositides) and the number of excitatory amino acid receptors are altered in Alzheimer disease, ischemia, spinal cord trauma, status epilepticus, acute stress, and HIV-related dementia. This decrease in glycerophospholipids is accompanied by marked elevations in phospholipid degradation metabolites such as glycerophosphocholine, phosphocholine, and phosphoethanolamine. Furthermore, marked increases have been reported to occur in levels of prostaglandins and lipid peroxides in ischemia, epilepsy, and various neurodegenerative diseases. Marked changes observed in phospholipids and their catabolic metabolites may be coupled to the elevated activities of lipolytic enzymes in Alzheimer disease, ischemia, spinal cord trauma, and animal models of status epilepticus. The stimulation of lipases and phospholipase A_2 is reported in ischemia and Alzheimer disease. These enzymes are coupled to excitatory amino acid receptors and are involved in the turnover of neural membrane phospholipids.

TABLE II
Involvement of Excitatory Amino Acid Receptors in Neurodegenerative States

Condition pathologic	Excitotoxin	EAA receptor involved
Ischemia	Glutamate	NMDA and *trans*-ACPD
Alzheimer's disease	Glutamate	NMDA, AMPA
Spinal cord injury	Glutamate	NMDA
Head injury	Glutamate	NMDA
Epilepsy	Glutamate	NMDA, AMPA
Huntington disease	Quinolinate	NMDA
Guam-type amyotrophic lateral sclerosis/Parkinsonism dementia	β-N-Methylamino L-Alanine	NMDA
AIDS dementia	Quinolinate	NMDA
Olivopontocerebeller atrophy	Quinolinate	*trans*-ACPD
Acute stress	Quinolinate	AMPA
Schizophrenia	Glutamate	NMDA

Free radical formation and excitatory amino acid release are mutually related and cooperate in a series of molecular events that link ischemic injury to neuronal cell death. Furthermore, lazaroids (21-amino steroids) are powerful antioxidants and can attenuate the neuronal injury induced by the exogenous application of glutamate or its structural analogs. Free radicals are attractive candidates for mediating the expression of excitotoxicity. Thus excitotoxic injury and free radical-mediated injury may involve molecular events that overlap substantially but not completely. These changes in neural membrane phospholipids and their metabolites suggest that the breakdown of neuronal membrane phospholipids induced through excitatory amino acid receptors may play a major role in the pathophysiology of neural trauma and neurodegenerative diseases. [*See* Free Radicals and Disease.]

The cytotoxicity of glutamate in cultured cerebral cortical neurons as well as in cerebellar slices is apparently related to increased intracellular Ca^{2+} which stimulates and regulates phospholipases, proteases, protein kinases, and endonucleases. During intermittent transsynaptic stimulation with excitatory amino acid receptors, the resulting increase in the intracellular free calcium concentration is rapidly reversed by the intervention of several mechanisms that regulate intracellular Ca^{2+} homeostatis. However, under pathological situations, the resulting increase in the extracellular glutamate concentration elicits a constant stimulation of excitatory amino acid receptors.

This results in a persistent increase in intracellular-free calcium which may be responsible for the abnormal metabolism of neural membrane phospholipids and the generation of high levels of free fatty acids, diacylglycerols, eicosanoids, and lipid peroxides. These lipid metabolites, along with abnormal ion homeostasis and lack of energy generation, may be responsible for cell death in a variety of neurological disorders such as Alzheimer disease, epilepsy, spinal cord trauma, and amyotropic lateral sclerosis/Parkinson complex. It is interesting to note that all these neurological disorders show an increase in intracellular-free Ca^{2+}, abnormal phospholipid metabolism, and accumulation of free fatty acid, diacylglycerols, prostaglandins, and related compounds. This suggests that excitotoxin-induced membrane phospholipid metabolism may be a common mechanism involved in the cellular response in the previously described neurological disorders. It is not known at this time whether the involvement of membrane phospholipids in neural trauma is a primary process or is induced by secondary factors.

The emphasis on excitatory amino acids and their receptors in cell trauma does not rule out the participation of other mechanisms involved in cell injury and degeneration. However, it is timely and appropriate to apply the concept of excitotoxicity to the degradation of membrane phospholipids. Excitotoxin-stimulated membrane phospholipid degradation results in the generation of high levels of free fatty acids, diacylglycerol, and eicosanoids. These lipid metabolites at low levels act as second messengers, but at high concentrations produce cytotoxicity which may be involved in cell injury and death. In acute trauma (such as ischemia and spinal cord injury) where blood flow and O_2 are cut off and ATP levels decrease rather rapidly, the excitotoxin-induced brain damage may be rapid (days), while in other disorders such as Alzheimer disease where oxygen and nutrients are available to the nerve cell and ATP levels are maintained almost to normal levels, it may take a longer time period (years). The evidence just described shows that the involvement of excitatory amino acids and their receptors in neural membrane phospholipid metabolism is, for the most part, indirect and circumstantial. Further studies are required on the molecular mechanisms of exitotoxin-induced membrane abnormalities.

IV. SUMMARY

Excitatory amino acids and their receptors play an important role in membrane phospholipid metabolism. Persistent stimulation of excitatory amino acid receptors by glutamate may be involved in neurodegenerative diseases and in brain and spinal cord trauma. The molecular mechanism of neurodegeneration induced by excitatory amino acids is, however, not known. Excitotoxin-induced calcium entry causes the stimulation of phospholipases and lipases, proteases, and endonucleases. Some of these enzymes act on neural membrane phospholipids and their stimulation results in the accumulation of free fatty acids, diacylglycerols, eicosanoids, and lipid peroxides in neurodegenerative diseases and in brain and spinal cord trauma. Other enzymes, such as protein kinase C and calcium-dependent proteases, may also contribute to the neuronal injury. The excitotoxin-induced alteration in membrane phospholipid metabolism in neurodegenerative diseases and neural trauma can be studied in animal and cell culture models. These models can be used to study the molecular mechanisms of the neurodegenerative processes and to screen the efficacy of therapeutic drugs.

ACKNOWLEDGMENTS

This work was supported in part by NIH Research Grants NS-10165 and NS-29441.

BIBLIOGRAPHY

Choi, D. W. (1988). Glutamate neurotoxicity and diseases of the nervous system. *Neuron* **1**, 628–634.

Dawson, V. L., Dawson, T. M., Bartley D. A., Uhl, G. R., and Snyder, S. M. (1993). Mechanism of nitric oxide-mediated neurotoxicity in primary cultures. *J. Neurosci.* **13**, 2651–2661.

Dawson, R., Beal, M. F., Bondy, S. C., DeMonte, D. A., and Isom, G. E. (1995). Excitotoxins, aging, environmental neurotoxins: Implications for understanding human neurodegenerative diseases. *Toxicol. Appl. Pharmacol.* **134**, 1–17.

Farooqui, A. A., and Horrocks, L. A. (1991). Excitatory amino acid receptors, neural membrane phospholipid metabolism and neurological disorders. *Brain Res. Rev.* **16**, 171–191.

Farooqui, A. A., Haun, S. E., and Horrocks, L. A. (1994). Ischemia and hypoxia. In "Basic Neurochemistry" (G. Siegel, R. W. Albers, B. W. Agranoff, and D. Molinoff, eds.), 5th Ed., pp. 867–883. Raven Press, New York.

Gasic, G. P., and Hollman, M. (1992). Molecular neurobiology of glutamate receptors. *Annu. Rev. Physiol.* **54**, 507–536.

Henley, J. M. (1994). Kainate-binding proteins: Phylogeny, structures and possible functions. *Trends Pharmacol. Sci.* **15**, 182–190.

Lipton, S. A., and Rosenberg, P. A. (1994). Mechanism of disease: Excitatory amino acids as a final common pathway for neurological disorders. *N. Engl. J. Med.* **330**, 613–622.

Mattson, M. P. (1990). Second messengers in neuronal growth and degeneration. *In* "Current Aspects of Neurosciences" (N. N. Osborne, ed.), Vol. 2, pp. 1–48. Macmillan, New York.

Nicoletti, F., Bruno, V., Copani, A., Casabona, G., and Knopfel, T. (1996). Metabotropic glutamate receptors: A new target for the therapy of neurodegenerative disorders. *Trends Neurosci.* **19**, 267–271.

Nishizuka, Y. (1989). Studies and prospectives of the protein kinase C family for cellular regulation. *Cancer* **63**, 1892–1903.

Olney, J. W., and Farber, N. B. (1995). Glutamate receptor dysfunction and schizophrenia. *Arch. Gen. Psychol.* **52**, 998–1007.

Regan, R. F., Panter, S. S., Witz, A., Tilly, J. L., and Giffard, R. G. (1995). Ultrastructure of excitotoxic neuronal death in marine cortical cultures. *Brain Res.* **705**, 188–198.

Schoepp, D. D., and Conn, P. J. (1993). Metabotropic glutamate receptors in brain function and pathology. *Trends Pharmacol. Sci.* **14**, 13–20.

Seeburg, P. H. (1993). The molecular biology of mammalian glutamate receptor channel. *Trends Pharmacol. Sci.* **14**, 297–303.

Exercise

FRANK W. BOOTH

BRIAN S. TSENG

University of Texas Medical School at Houston

GLOSSARY

Adaptation Change in the chemical architecture of cells that minimizes disruption in the cellular or tissue biochemistry during a given intensity of physical exercise

Fitness Ability to undertake physical exercise or daily activities without undue fatigue. There are multiple types of fitness (e.g., aerobic fitness, strength fitness, coordination fitness, and flexibility fitness)

Health adaptations to physical training Sufficient daily exercise to decrease the risk of acquiring many chronic diseases and to enable many elderly people to remain self-sufficient, thus avoiding the need for nursing home care

Maximal oxygen consumption The most oxygen that the body can use in aerobic exercise; synonymous with "maximal aerobic fitness"

Physical training Exercise bouts repeated for numerous days, leading to adaptations

PHYSICAL EXERCISE IS THE MOVEMENT OF BONES caused by muscular contraction. The type of exercise and its resulting adaptation are dependent on the frequency, intensity, and duration of the physical work. A single bout of exercise can be identified as a daily workout. The frequency within the exercise bout would be the number of muscle contractions per minute. For example, the frequency of leg movements, and thus contractions of leg muscles, is greater in rapid running than in slow walking. The intensity of the exercise bout is the amount of work done per minute. For example, contracting arm muscles the same distance against 100 kg is of greater intensity than contracting them against 25 kg. Another example is that running is more intense than a slow walk because a greater distance, and thus more physical work, is done per unit of time when running. Duration is the length of the exercise bout.

An acute, or single, bout of exercise is distinguished from physical training, which is the summation of repeated daily bouts of exercise. A minimum frequency of two resistance training and three aerobic training bouts of exercise per week, repeated for many weeks, is necessary to produce training adaptations. Resistance training consists of contracting skeletal muscle against a near-maximal load (high intensity) for 30 contractions (low frequency) per day. As an adaptation the skeletal muscle enlarges after a few months. In contrast, aerobic exercises are low-intensity high-frequency activities that involve large masses of skeletal muscle (e.g., running, swimming, cycling, and rowing). Whereas resistance training requires near-30 maximal contractions per day for a training adaptation, submaximal aerobic exercise requires a minimum of 20–30 min/day to produce adaptations such as bradycardia, increased maximum cardiac output, and increased mitochondria density in skeletal muscle.

I. AEROBIC EXERCISE

A. Quantification of Aerobic Exercise Intensity

In aerobic exercise oxygen is used by the muscles as they convert energy in glucose (sugar) and fatty acids (fats) to an energy form called ATP, which provides the direct source of energy for muscle fibers to shorten, thus causing limbs to move and exercise to occur. Measurement of the quantity of oxygen consumed during aerobic exercise indicates how many calories of sugar and fats were used and indirectly quantifies the caloric cost of the exercise. Greater oxygen consumption during aerobic exercise results in more calories being used and more physical work being done by the rhythmic activity (i.e., movement) of the limbs. For example, oxygen consumption increases each time the number of steps taken per minute increases. Normal walking uses less oxygen than a slow jog, which in turn requires less oxygen than a fast run. [See Adenosine Triphosphate (ATP).]

B. Quantification of Fitness to Perform Aerobic Exercise

Knowledge of oxygen consumption during maximal aerobic exercise can also be used to classify the work fitness of the heart. The maximal number of steps that can be taken by the legs when running during a defined period depends on the fitness of the heart to do work. For example, an Olympic marathon runner will have a higher maximal oxygen consumption per minute of aerobic exercise than a sedentary person, who in turn will have a higher oxygen consumption than a bedridden individual. The marathon runner will have the highest heart fitness of the three, while the bedridden person will have the lowest fitness of the heart to work. The more heart-fit a person is, the greater number of calories he or she can use per minute before fatigue. ("Aerobic fitness" or "aerobic work capacity" is defined as the maximal amount of oxygen that can be used to make energy for work during physical exercise.)

C. What Limits Maximal Aerobic Fitness (Maximal Oxygen Consumption)?

Three aspects of maximal oxygen consumption (maximal caloric expenditure per minute) must be considered. They are: What limits it? What is its significance to an exercise task? How can it be altered? The factor limiting the maximal amount of oxygen that can be used per minute in converting sugar and fats to ATP during aerobic exercise in a normal young person at sea level is the maximal capacity of the heart to pump blood. Aerobic exercise requires the transfer of oxy-

gen from air to muscle. Thus, oxygen in the air must be breathed into the lungs, where oxygen moves from alveoli (small air sacs in the lungs) to the pulmonary blood (blood leaves the right side of the heart and flows through the lungs, known as pulmonary blood flow). Following exit from the lungs, the blood is "arterialized" blood (i.e., it is rich in oxygen) and flows into the left heart, from which it is ejected by the left ventricle to flow to the rest of body (including the working muscles).

Oxygen is transferred to muscle cells through small blood vessels (i.e., capillaries), and oxygen then diffuses through the capillary membrane and the sarcolemma to mitochondria (the power plants of cells, where oxygen is used to convert foods to ATP). Thus, oxygen movement from the air to muscle mitochondria could be limited at numerous sites during its transfer. In healthy young humans at sea level, the limiting site for oxygen transfer in maximal aerobic exercise is the maximal ability of the heart to eject blood per minute.

D. What Is the Significance of Maximal Aerobic Fitness?

The second important aspect of the concept of maximal oxygen consumption in aerobic exercise is that its value determines how long a person can work before becoming fatigued (i.e., the person is unable to continue to exercise at that work magnitude). For example, if persons A and B both weighed 160 lb and walked at 3 mph, did light industrial work, or did housework, they would expend 5 cal of energy per minute and use 1 liter of oxygen per minute. If the maximal oxygen uptake for person A was 2 liters of oxygen per minute and for person B was 4 liters of oxygen per minute, then person A would be working at 50% of his maximal aerobic work capacity (1 liter of oxygen used per 2 liters of oxygen consumption capacity) and person B would be working at 25% of his maximal aerobic potential (1 liter of oxygen used per 4 liters of oxygen consumption capacity) when they both walk at 3 mph.

The intensity of aerobic exercise for a given person is determined by the percentage of maximal aerobic work done, not by the absolute number of calories or oxygen used. Person A would fatigue sooner than person B if they were walking together at 3 mph, because person A would be working at a higher percentage of his capacity to use oxygen or calories. The reason for fatiguing sooner when the percentage of

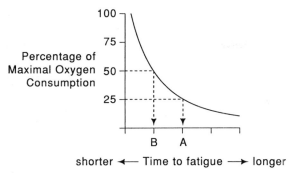

FIGURE I The time of work until the work cannot be continued any longer (due to fatigue) is inversely related to the percentage of maximal oxygen consumption invoked by the work being done. When person A has one-half the absolute value of maximal oxygen consumption of person B, and when persons A and B work at the same intensity or oxygen consumption, then person B will work a shorter time before fatigue, because person B is working at a higher percentage of his maximal oxygen consumption. [Reprinted with permission from R. H. Strauss (1984). "Sports Medicine," p. 44, Saunders, Philadelphia.]

aerobic effort is higher is that the time one can exercise aerobically at a given oxygen consumption (in this example, using 1 liter of oxygen per minute) is inversely related to the percentage of maximal oxygen that can be consumed per minute (Fig. 1).

Thus, person B can continue exercising at 25% of his maximal ability to use oxygen or calories when person A has fatigued, walking at the same speed. Thus, the maximal oxygen or calories that can be used per minute sets the comfort zone for work. A person with a high maximal oxygen consumption can either do the same workload for a longer period or undertake higher quantities of physical aerobic work for the same time than a person with a lower maximal oxygen consumption.

E. Can Aerobic Fitness for a Given Person Change?

Since maximal oxygen consumption per minute in a given person can be increased (by physical training) or decreased (by detraining, by sickness or disability, or by aging), a person's ability to work aerobically is determined by his state of physical training, health, and age. A healthy 20-year-old who experimentally undergoes a forced continuous period in bed for 20 days will experience physical deconditioning or detraining and a decrease in his potential to use oxygen or calories during aerobic exercise.

Likewise with aging from the age of 20 years, maxi-

mal oxygen consumption declines about 10% per decade. Thus, housekeeping chores will require a greater percentage of the maximal aerobic capacity of an 80-year-old than of a 20-year-old. The reason is the difference in their maximal oxygen consumptions. Housework requires the same aerobic cost of calories or oxygen at both ages, but the 80-year-old has a lower aerobic fitness, works at a higher percentage of his maximal oxygen consumption, and thus fatigues sooner than does a 20-year-old doing the same work.

F. Caloric Costs of Work

Work is defined as force times distance. In running, force is the weight of the body being lifted against gravity as a step is taken. Thus, work during running is the weight of the body times the distance moved. If small allowances for differences in body weight and speed of covering a distance are not made, approximately 100 kcal are used for each mile completed. The conversion of calories used to the amount of fat needed for this caloric expenditure follows. If a person were to add 1 mile more of distance to his usual daily exercise for 1 year, the caloric equivalent of 10 lb of body fat (36,500 kcal) would be used. Since only about 0.03 lb of stored fat is used for each additional mile on a single day, it is obvious that reduction of body fat by aerobic exercise is a long-term process (i.e., months to years) and is related to the additional distance covered, rather than to the speed at which the exercise is performed. The effect of the speed is small: completing 1 mile by running uses only about 10 kcal more than walking.

Table I gives the caloric expenditures for certain sports. In brief, more calories are burned when larger numbers or masses of skeletal muscle are contracting (e.g., cross-country skiing uses more kilocalories than throwing a dart), when the exercise causes the body to be lifted more against gravity (e.g., walking upstairs uses almost three times as many kilocalories as walking downstairs), and when the body is moved over a longer distance (e.g., walking 5 miles uses almost five times as many kilocalories as running 1 mile).

G. Appetite and Aerobic Exercise

As discussed earlier, increased amounts of aerobic exercise use more calories. As a general rule, the increase in caloric intake because of increased appetite does not compensate for the increase in caloric expenditure, and some loss in body weight always occurs.

TABLE I

Caloric Ladder for a 154-lb Person[a]

kcal/hr	Activity
1100	Running at 9 mph, walking upstairs, cross-country skiing at 8 mph
1000	—
900	Wrestling, rowing at 6 mph, running at 7 mph
800	Vigorous rope-skipping, boxing, soccer, swimming at 50 yd/min
700	Rapid ice-skating, bicycling at 13 mph, playing competitive squash
600	Running at 5 mph, playing paddleball, vigorous downhill skiing, cross-country skiing at 4 mph, playing basketball
500	Weight training; playing football, volleyball, or singles tennis; skating leisurely, bicycling at 11 mph
400	Swimming at 25 yd/min, walking downstairs, doing light calisthenics, playing doubles tennis
300	Walking at 3 mph, playing softball, bicycling at 8 mph, sailing, leisurely rope-skipping

[a]Caloric expenditure increases 10% for each additional 15 pounds in body weight.

Weight loss, however, is greater if caloric intake is maintained at the original level. [*See* Appetite.]

II. HEALTH BENEFITS OF EXERCISE

A. Aerobic Exercise

The regimen of aerobic exercise (e.g., cycling, swimming, or walking) in daily activities decreases the risk of acquiring numerous chronic diseases. Sedentary people who do not engage in regular aerobic exercises have an increased susceptibility to the following significant and chronic diseases; breast and reproductive tract cancers, claudication, colon cancer, congestive heart failure, coronary artery disease, depression, hypertension, adult-onset diabetes, obesity, osteoporosis, sleep apnea, joint disease, and stroke. Since regular aerobic exercise decreases these risks and can be therapeutic in treating many of these diseases, exercise is an important and inexpensive component to preventative, pediatric, primary care, internal, rehabilitative, and geriatric medicine. The exercise durations recommended to obtain the aerobic health benefits of decreasing the risk of the aforementioned diseases are

given in Table II. For minimally active people, one should gradually work toward expending the target of 1000 cal (1 cal = 1000 kcal) each week in exercise for health benefits. These exercise durations can be done in segments throughout a given day. For example, three separate stair climbs done at different times could be combined with two walks at other times to expend calories each day.

In general, the person who weights twice as much burns two times as many calories with exercise because they carry twice the weight during exercise, but diet and body composition must also be considered. For example, a high-fat-diet or obese person would ideally restrict fat calories and exercise for durations progressively exceeding these recommended times for maximal health benefit. Body weight itself is not an index of body composition because it does not distinguish lean body mass from body fat. With increasing percentage of body fat, particularly abdominal fat, there is even a higher risk to acquire chronic cardiovascular diseases.

TABLE II

Daily Exercise That Is Required to Decrease Risk of Chronic Diseases

Exercise	Body weight	
	110 lb	220 lb
Minimal health benefits obtained by exercise expenditure of 143 cal/day (1000 cal/week)		
Cycling, 5.5 mph	36 min	18 min
Gardening, raking	53 min	27 min
Running		
9 min/mile	15 min	8 min
11.5 min/mile	21 min	11 min
Stair climbing/hill climbing		
Up	12 min	7 min
Down	30 min	17 min
Swimming, slow crawl	23 min	11 min
Walking, normal pace	36 min	18 min
Additional health benefits, including longevity, obtained by exercise expenditure of 286 cal/day (2000 cal/week)		
Cycling, 5.5 mph	72 min	36 min
Gardening, raking	106 min	53 min
Running		
9 min/mile	29 min	15 min
11.5 min/mile	42 min	21 min
Stair climbing/hill climbing		
Up	23 min	13 min
Down	60 min	34 min
Swimming, slow crawl	45 min	22 min
Walking, normal pace	72 min	36 min

Important factors in developing life-long exercise habits are the following: (a) begin exercise training with short durations and low intensities; (b) progressively add duration and intensity in small increments (do not push to the discomfort point); (c) avoid injury (injury is a sure way to discontinue the training regimen); (d) make exercise a part of your normal routine (try to set regular times in your daily schedule); (e) select an exercise, or variety of exercises, that you enjoy and maintain it for life); and (f) if over 35 years of age, or with existing health problems, obtain physician clearance for exercise.

In addition to minimizing the risk of many diseases, an important outcome of expending 2000 cal per week (Table II) is that these exercises will prevent some of the loss in cardiovascular fitness that typically occurs after age 25 years. The total cumulative effect of annual losses of cardiovascular fitness can be extremely significant after decades, contributing to a severe loss of stamina. In the elderly, this effect contributes to an inability to care for oneself, and hence loss of independent living. Thus, proper aerobic exercise improves the health and wellness of people through adaptations that both decrease the risk of acquiring significant diseases and allow the elderly to perform activities of daily living with less difficulty. The health benefits that are derived from the daily exercise durations listed in Table II are, in part, a result of decreasing body fat. The caloric equivalent of at least 29 lbs of fat is used each year when an exercise program expends 2000 cal per week for one year (Table II). To achieve and maintain this health benefit, exercise must be a part of the daily routine. Clearly, nonvigorous exercise is preferable to inactivity. Recent studies have even shown that vigorous exercise (a minimum of 7 cal/min for a total of 1500 cal/week) is associated with an additional attribute of enhanced longevity, unlike nonvigorous exercise and sedentary life-styles.

B. Resistance Exercise

From the age of 25 to 50 years, 10% of skeletal muscle mass disappears. Between 50 and 80 years of age, another 30% of muscle mass is lost in sedentary and exclusively aerobic people. Resistance training can significantly attenuate the loss of both skeletal muscle mass and strength that occurs with aging. With the exception of osteoporosis, it is not known whether strength training provides any protection against developing chronic diseases. However, strength, mobility, balance, and thus the quality of life are enhanced by resistance training, particularly in individuals older than 50 years.

C. Diet

The Committee on Diet and Health of the National Research Council recommends that a balance be made between food intake and physical activity to maintain appropriate body weight. They further recommend that total fat intake be reduced to 30% or less of total calories and saturated fatty acid intake to less than 10% of total calories, eating five or more servings of vegetables and fruits every day, and consuming between 1.2 and 1.6 g of protein per day/kg body weight. Daily total caloric intake depends on age, gender, body size, health status, and goals. For example, a sedentary, 110-lb, adult male and female require 1880 and 1627 cal, respectively, per day. If they weighed 220 lbs, the male and female need 2625 and 2155 cal, respectively, each day. Long-term body weight is a function of calories eaten minus calories burned. Healthful diet and increased activity are essential to obtain and maintain ideal body weight over a life span.

III. CARDIOVASCULAR SYSTEM

A. Effect of Aerobic Exercise on the Heart

Aerobic exercise continued daily for weeks improves the ability of the normal healthy heart to pump blood. For humans the minimum amount of aerobic exercise needed to change the pumping capacity of the heart is 20–30 min/day, 3–4 days/week, at a heart rate corresponding to about 70% of the maximal rate. An approximate and easy method for estimating the maximal heart rate is to subtract a person's age from 220. Thus, a 20-year-old has a maximal heart rate near 200 beats per minute and an 80-year-old has a maximal heart of about 140 beats per minute. However, 33% of people have a true maximal heart rate 10 beats either higher or lower than these estimates. A better way to obtain the maximal heart rate is to measure it during maximal exercise, but this is more difficult to do outside of a laboratory. The heart rate corresponding to 70% of the maximum is calculated by the formula

70% of maximal heart rate
= 0.7 × (maximal heart rate − resting heart rate)
　　　　　　　　　　　　+ resting heart rate.

TABLE III

Magnitude of Increase in Heart Functions for Various
Workloads after Aerobic Exercise Training

Heart function	At rest	Running at 5 mph	Running at maximal speed
Cardiac output (liters of blood per minute)	Same	Same[a]	Increase
Heart rate (beats per minute)	Less	Less[b]	Same
Stroke volume (milliliters of blood per beat)	More	More	More

[a]The increase in cardiac output when going from rest to running at 5 mph is the same before and after training.

[b]The increase in heart rate when going from rest to running at 5 mph is less after aerobic training than prior to it.

For example, for a 20-year-old it would be

$0.7 \times (200 - 70) + 70$
$= 161$ beats per minute at 70% of maximal exercise.

If the person in this example ran 3–4 days/week for 20–30 min/day at a heart rate of 161 beats per minute for 1 month (aerobic exercise training), the person's maximal heart rate would not change, but his heart rate at rest and at the same running intensity would be lower (the lower heart rate is termed bradycardia). Heart rate is one of two components determining the amount of blood ejected from the heart per minute (cardiac output). The formula for cardiac output is

Cardiac output = heart rate $\times$ stroke volume.

The effects of aerobic exercise training on these three components are given in Table III.

After aerobic exercise training, stroke volume (i.e., the amount of blood ejected by a single heart beat) increases at rest, at 70% of maximal aerobic work, and at maximal heart rate. Other heart functions are also increased after 4–8 weeks of aerobic exercise training, as shown in Table IV.

Maximal cardiac output (i.e., the heart's reserve to pump blood) was enhanced by 6 liters/min after aerobic exercise training. Such a reserve could assist survival to an absolute challenge (e.g., if 10% of myocardial tissue stopped contracting from an illness); moreover, it permits the transfer of more oxygen from the lungs to the tissues at the maximal heart rate. The

TABLE IV

Heart Function Values at Maximal Exercise before and after
Aerobic Exercise Training

Heart function	Before training	After training
Cardiac output (liters of blood per minute)	20	26
Heart rate (beats per minute)	200	200
Stroke volume (liters of blood per beat)	0.10	0.13

consequences of increasing maximal cardiac output are to make the caloric cost of housework a smaller percentage of the maximum number of calories that can be used per minute.

Information is becoming available to explain the mechanism by which aerobic training improves the fitness of the heart to work. One of the factors is a change in the activity of the sympathetic nervous innervation of the sinoatrial node, the heart pacemaker. The sympathetic system increases heart rate. During aerobic exercise its activity increases, causing an increase in heart rate. However, after aerobic training the sympathetic stimulation of the sinoatrial node is reduced, and the heart rate increases less when running at the same speed (Table V).

It is not known how the body "senses" the percent-

TABLE V

Degree of Lessening of the Sympathetic Discharge Related to
the Percentage of Improvement in Maximal
Oxygen Consumption[a]

Heart function	Before aerobic training	After aerobic training
Maximal oxygen consumption (liters of oxygen per minute)	3.0	4.0
Maximal heart rate	200	200
When running at 5 mph		
Oxygen consumption (liters of oxygen per minute)	2.1	2.1
Heart rate (beats per minute)	160	140
Percentage of maximal oxygen consumption	69%	54%
Percentage of maximal heart rate	70%	53%

[a]See Table VI.

age of maximal aerobic work effort and then modifies the activity of the sympathetic nervous system to adjust the heart rate. Aerobic training might improve the contractility of the heart (i.e., the amount of muscle tension at a given heart length) by increasing calcium influx during sarcolemmal depolarization. Higher intracellular free calcium would increase muscle tension, which would cause the improved stroke volume. [See Exercise and Cardiovascular Function.]

B. Exercises Improving the Fitness of the Heart to Work

By definition, all exercises use more calories than the basal metabolic rate. As indicated in Section I,F, some types of exercise use more calories than others. Not all types of exercise produce aerobic or heart fitness (i.e., bradycardia and increased stroke volume). For example, if a healthy 20-year-old with a maximal oxygen consumption of 4 liters of oxygen per minute walks, he will use extra calories but will not develop heart fitness, because he will not be working at 70% of his maximal aerobic work capacity. On the other hand, if an 80-year-old with a maximal oxygen consumption of 2 liters of oxygen per minute walks, he would be working above 70% of his maximal aerobic capacity and will develop heart fitness. Thus, walking uses calories and produces heart fitness for an 80-year-old, whereas for a normal 20-year-old, walking only uses calories but does not improve the aerobic work capacity.

Another exercise, weight-lifting, produces increased strength of those skeletal muscles that are trained, but does not improve the capacity of the heart to work. The reason is that weight-lifting and power training do not maintain an increased heart rate that is 70% of maximal heart rate for 30 consecutive minutes. Only circuit training, in which numerous strength exercises are performed sequentially without rest periods in between, produces some heart fitness. Weight training uses extra calories. On the other hand, aerobic exercises usually produce little or no increases in the strength of skeletal muscles.

These examples indicate the specificity of the type of exercise on the adaptation of training. Other examples are given in Table VI.

C. Partitioning of Cardiac Output to Tissues during Aerobic Exercise

The tissue that most needs the increased blood flow during exercise is the contracting skeletal muscle.

Blood flow supplies the oxygen and metabolic fuels for the synthesis of ATP, the energy source required for muscle contraction. The normal blood flow to muscle is insufficient to supply the energy needs for vigorous contraction; hence, an increase in cardiac output is required in aerobic exercise. If all tissues were to share similar percentage increases in cardiac output during aerobic exercise, skeletal muscles would still not have sufficient arterial blood flow to perform maximal work. During exercise blood flow is shifted to the exercising skeletal muscle from other organs, by the following mechanisms. Sympathetic stimulation of arterioles of all tissues but the brain is increased, causing the smooth muscle around the arterioles to contract, constricting the radius of the arteriolar lumen. This decrease in radius increases resistance to blood flow, so that less blood reaches the tissue. For example, in maximal aerobic exercise the amount of blood flowing to the liver, kidney, stomach, and intestines is reduced to 20% of nonexercise blood flow. [See Skeletal Muscle.]

A similar process occurs in inactive skeletal muscle, whereas blood flow increases in heart and contracting skeletal muscle. In working muscle the metabolic byproducts of contraction (e.g., adenosine, carbon dioxide, and hydrogen and potassium ions) counteract the symaphetic system-induced constriction, resulting in dilation of the arterioles. The consequence of these events is to redistribute the flow resulting from the increase in cardiac output to the working skeletal muscle. When the requirement for blood flow to exercising skeletal muscle begins to exceed maximal cardiac output, a reflex from muscle activates the sympathetic nervous system so that the increase in muscle blood flow is lessened and arterial blood pressure is maintained in intense exercise. This reflex is attributed to protons (i.e., hydrogen ions) arising in part from an increased glycolytic rate in the working skeletal muscle.

D. Aerobic Exercise Is an Example of the "Fight or Flight" Syndrome

The increased sympathetic nervous system stimulation of arteriolar smooth muscle is a common survival event in the body. For example, it occurs in blood loss (i.e., hemorrhage), during which blood volume and arteriolar blood pressure decrease. If blood pressure falls too much, the force driving blood flow through tissues becomes insufficient, and tissues become hypoxic and die. A reflex survival mechanism

TABLE VI
Fitness Elements

Category of exercise training	Example of exercise	Uses calories?	Is heart fitness improved?	Is strength of skeletal muscle improved?	Is coordination improved?
Aerobic training	Jogging	Yes	Yes	No	Yes
Walking 20-year-old	Walking	Yes	No	No	Sometimes
Strength training	Weightlifting	Yes	No	Yes	Yes
Anaerobic training	Sprinting	Yes	No	No	Minimal
Coordination training	Somersaulting	Yes	No	No	Yes

released by the drop in blood pressure during hemorrhage signals the sympathetic nervous system to become more active, so that arterioles constrict and blood pressure increases.

During aerobic exercise the engagement of the sympathetic nervous system prevents a fall in arterial blood pressure. As opposed to hemorrhage, the signal to increase arterial blood pressure during exercise comes from the working muscle and some areas of the brain. However, as in hemorrhage, activation of the sympathetic nervous system by the survival mechanism of fight or flight prevents fainting due to hypotension during aerobic exercise (the flight).

IV. METABOLIC RESPONSES

A. Communicative Roles of the Sympathetic Nervous System during Aerobic Exercise

Another example of how the sympathetic nervous system integrates all tissues to cooperate to maximize the ability to continue aerobic exercise is the regulation of sugar utilization. During aerobic exercise, sympathetic nervous system discharges to the β cells of the pancreas cause a decrease in insulin secretion into the blood, with a consequent decrease in blood insulin levels.

As in juvenile diabetes, if insulin is low in the blood, sugar (glucose) will not enter skeletal muscle. However, contraction of the muscle fiber causes the recruitment of glucose transporters from an internal storage site to the cell membrane. Increased numbers and activity of glucose transporters in the cell membrane override the effect of low blood insulin. Glucose uptake increases in the working, but not the resting, skeletal muscle. In this way blood sugar is redirected to muscles needing it. [*See* Insulin and Glucagon.]

This situation is analogous to the redistribution of blood flow toward exercising skeletal muscle. Low blood insulin also allows an increased breakdown and release of fatty acids from fat cells into the blood in a normal person during exercise (Fig. 2). When this occurs, a larger amount of fatty acids is used as fuel for muscle work, sparing the limited stores of glucose in the form of glycogen in the body until later during exercise. The net result of this regulatory role of the sympathetic nervous system is that the duration of aerobic exercise is longer before fatigue. Aerobic training accentuates this effect. When running at the same speed after training, less muscle glycogen is used as compared to before training. [*See* Fatty Acids.]

B. Heat as a By-product of Muscle Contraction

The formation of ATP from glucose and oxygen releases heat as well as other by-products into the muscle, and this heat is carried from the muscle to the rest of the body by the blood flow, thus causing body temperature to increase during exercise. Some of this is lost to the environment, preventing body temperature from rising above 41°C, and consequently preventing heat stroke.

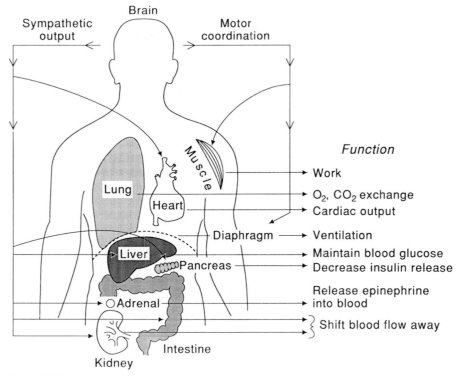

FIGURE 2 Integrative control of multiple organs during aerobic exercise. Both the nervous and hormonal systems communicate to all organs and tissues in the body during exercise and cause each organ system to undertake a specific function, so that the summation of all organ systems allows sufficient oxygen and caloric fuel for the contraction of large masses of skeletal muscle during exercise and allows for the elimination of metabolic by-products, such as carbon dioxide (from the lungs), heat (from the skin), and hydrogen ions (from the kidneys).

Both overheating and dehydration (i.e., loss of body water via sweat) can limit exercise. Thus, fluid replenishment during exercise and limitation of exercise at temperatures above 27°C are important. In addition, maximal aerobic capacity is reduced during exercise at high environmental temperature. For example, maximal oxygen consumption can decrease from 4 to 3 liters of oxygen per minute in the heat, while maximal cardiac output remains unchanged. The mechanism is the following. Exercising in the heat causes vasodilation of the skin, where heat from blood flowing from the core of the body can be lost to the external environment by convection (wind), by radiation (heat waves), by transfer from the skin to cooler external objects (conduction), or by the evaporation of sweat. The vasodilation of skin arterioles and small veins during exercise in the heat diverts some of the blood flow needed by the exercising skeletal muscles to the skin. Thus, the amount of blood supplied to skeletal muscle is reduced, and muscle work is limited by the lowered supply of oxygen.

V. ADAPTATIONS

A. Plasticity or Adaptability of the Body to Aerobic Exercise Training

If aerobic exercise is performed 3–4 days/week for many weeks, two things are perceived by the person who is exercising. First, the performance of a certain amount of aerobic work (e.g., running at 5 mph) does not seem as stressful or as hard as before training. Second, performance of an absolute amount of aerobic work after training can be done for a longer duration before fatigue forces a stop to the exercise. Part of the reason for these perceptions is given in Section I,D and Table V. Aerobic training increases maximal oxygen consumption with the consequent effect that an absolute workload requires a smaller percentage of the maximal oxygen that the body can use. Essentially, the body adapts to repeated daily bouts of exercise, so that there is less disruption of the internal environment surrounding and within cells (homeostasis). Biochemical and structural changes occur after training (adaptations), so that a given amount of exercise produces less stress on cells and on the organism. [See pH Homeostasis.]

Because the composition of cells and tissues is altered by the training, cells and tissues are said to exhibit plasticity. Two examples can be given. First, maximal cardiac output increases with aerobic training, which has been mentioned in Table IV. Since cardiac output for a given submaximal aerobic exercise is unchanged by aerobic training, the heart works at a smaller percentage of its reserve after training. A second example is weight-lifting training. Skeletal muscle enlarges so that a 100-lb load is distributed over a larger cross-sectional area of the muscle after training. Thus, the amount of the 100 lb supported by a 1-square-inch area is less after weight training of this muscle. In both of these examples, homeostasis for an absolute workload is less disrupted after cellular and organ adaptations to training.

B. Biochemical Basis of Adaptations to Aerobic Exercise

The tissue for which most is known about plasticity or adaptability to aerobic exercise is skeletal muscle. In response to aerobic training, the density of mitochondria and capillaries in skeletal muscle increases, but the muscular mass remains relatively unchanged. Adaptations to training are specific to the skeletal muscle that is trained. If a single leg is trained to bicycle, while the opposite leg does no aerobic training (on a specially constructed bike), mitochondria increase only in the muscle of the leg that trained. The opposite nonexercised leg has no change in mitochondria.

One of the major functions of the increased mitochondria is to spare the utilization of the body's stores of glucose during aerobic exercise. The mechanism is the following. The process of muscle shortening requires an energy source, ATP, to power the sliding of the muscle's contractile elements. Thus, during muscle contraction, ATP falls and ADP rises. After aerobic training, if the density of mitochondria in the trained skeletal muscle doubles, then the oxidative phosphorylation sites per unit volume of muscle doubled. ADP rises only one-half as much as its increase in nontrained muscle to obtain the same number of produced ATP molecules. The smaller increase in ADP in the aerobically trained skeletal muscle spares the utilization of glucose for ATP production. Glycolytic flux from glucose to pyruvate during exercise is increased less by the smaller ADP increase in skeletal muscle with higher mitochondrial density. Therefore, the trained skeletal muscle uses less glucose. [See ATP Synthesis in Mitochondria.]

The trained muscle uses fat as an energy source to make ATP. Because the amount of glucose that is stored as glycogen is limited, the body usually begins to run low on glucose for fuel after running about 18 miles. When blood glucose concentrations begin to fall as glycogen stores are used up, the body is unable to continue exercising at the same speed, probably because nerves and red blood cells can only use glucose to make ATP. (Muscle can oxidize both glucose and fatty acids to make ATP.) Thus, the training adaptation of glycogen sparing permits running longer at the same speed before fatiguing due to glycogen depletion. It is well known that eating sugar during running to exhaustion delays the onset of fatigue.

Present information suggests that exercise increases mitochondria by increasing the synthesis of mitochondrial proteins, owing to an increase in the concentration of mRNAs for mitochondrial proteins.

VI. SUMMARY

Muscle contraction permits animals and humans to move, and thus to survive. "Physical exercise" is a term employed to describe this activity. Physical exercise stresses the body and requires the integration of multiple organ systems to supply oxygen and fuel to the working skeletal muscles. Injury-free daily exercise results in lifelong adaptations that diminish the stress of physical activity and enable more work prior to fatigue. We recommend a combination of both aerobic and resistance training. Aerobic training produces adaptations that decrease the risk of acquiring many chronic diseases. Resistance and aerobic training both improve the quality of life and offset frailty. Properly done exercises on a daily regimen increase workforce productivity, reduce health care costs, and ultimately decrease human suffering, morbidity, and mortality.

BIBLIOGRAPHY

American Academy of Orthopaedic Surgeons (1991). "Athletic Training and Sports Medicine." American Academy of Orthopaedic Surgeons, Park Ridge, IL.

Åstrand, P. O. (1986). "Textbook of Work Physiology." McGraw–Hill, New York.

Booth, F. W., and Thomason, D. B. (1991). Molecular and cellular adaptation of muscle in response to exercise: Perspectives of various animal models. *Physiol. Rev.* **71,** 541–585.

Booth, F. W., and Tseng, B. S. (1995). America needs exercise for health. *Med. Sci. Sports Exerc.* **26,** 462–465.

Brooks, G. A., Fahey, T. D., and White, T. (1996). "Exercise Physiology: Human Bioenergetics and its Application," 2nd Edition, Mayfield, Mountain View.

Coyle, E. F. (1995). Integration of the physiological factors determining endurance performance. *Exerc. Sport Sci. Rev.* **23,** 25–63.

Johnson, B. D., Badr, M. S., and Dempsey, J. A. (1994). Impact of the aging pulmonary system on the response to exercise. *Clin. Chest Med.* **15,** 229–246.

Lamb, D. R., and Murray, R. (1988). "Perspectives in Exercise Science and Sports Medicine. Volume I: Prolonged Exercise." Benchmark, Indianapolis, IN.

Lee, I.-M., Hseih, C.-C., and Paffenbarger, R. S. (1995). Exercise intensity and longevity in men. *JAMA* **273,** 1179–1184.

Lloyd, E. L. (1994). ABC of sports medicine. Temperature and performance—II. Heat *Br. Med. J.* **309,** 587–589.

Pate, R. R. *et al.* (1995). Physical activity and public health. A recommendation from the Centers for Disease Control and Prevention and the American College of Sports Medicine. *JAMA* **27,** 402–407.

Rogers, M. A., and Evans, W. J. (1993). Changes in skeletal muscle with aging: Effects of exercise. *Exerc. Sport Sci. Rev.* **23,** 65–102.

Rowell, L. B. (1993). "Human Cardiovascular Control." Oxford Univ. Press, New York.

Saltin, B., and Gollnick, P. D. (1983). Skeletal muscle adaptability: Significance for metabolism and performance. *In* "Handbook of Physiology, Section 10: Skeletal Muscle" (L. D. Peachey, ed.). Williams & Wilkins, Baltimore.

Vollestad, N. K., and Sejersted, O. M. (1986). Exercise in human physiology. *Acta Physiol. Scand.* **128** (Suppl. 556), 7–166.

Exercise and Cardiovascular Function

GEORGE E. TAFFET
ROBERT W. HOLTZ
CHARLOTTE A. TATE
Baylor College of Medicine and The University of Houston

GLOSSARY

Duration Length of time that the exercise intensity can be maintained per unit of time or distance

Dynamic exercise Isotonic exercise; movement or physical activity in which work is performed; force is generated for external work

Exercise Force or tension developed by muscle; movement; physical activity above the basal, or resting, state

Intensity Amount of exercise that can be exerted per unit of effort; it implies a quantity of effort per unit of effort

Static exercise Isometric exercise; force is generated in muscle, but no external work is performed; resistance exercise

IN THE BASAL, OR RESTING, METABOLIC STATE THE cardiovascular system of a healthy adult pumps blood at only a small percentage of its potential capacity. The blood pumped from the heart (i.e., cardiac output) is primarily directed to the visceral organs, the brain, and the heart itself, with modest flow to the peripheral muscles. With the abrupt onset of physical activity (i.e., exercise), the increased energy demands of the working muscle require an augmentation of blood flow to the musculature for sustained physical activity. During exercise, therefore, a greater propor-tion of the cardiac output must be directed toward the working muscle for the delivery of oxygen and nutrients and for the removal of metabolic waste products and heat. Thus, the function of the cardio-vascular system must be precisely regulated to meet the increased demands for blood flow by the working muscles. The adjustments and regulation of the car-diovascular system with exercise are considered here, and we describe the cardiovascular adjustments to dynamic exercise in which a large muscle mass is used. A brief description of the response to static exercise is given. The mechanisms underlying the physiological adjustments with exercise cannot be deeply analyzed, however, and the reader is directed to works in the Bibliography for an in-depth treatment.

I. CARDIOVASCULAR ADJUSTMENTS TO DYNAMIC EXERCISE

A. Introduction

Whole-body oxygen consumption per unit of time ($\dot{V}O_2$) increases in the transition from rest to physical activity. The amount of oxygen consumed depends on the imposed absolute workload and can go from 2–3 ml/kg/min at rest to 60 ml/kg/min at maximal capacity in an active young adult male. The oxygen consumed for gardening, for example, is lower than that consumed for the lifting of heavy boxes. The intensity of work therefore determines in part the absolute oxygen consumption, which reflects the aero-bic (i.e., oxygen-consuming) energy expenditure nec-essary to perform the exercise. The ability to sustain the increased workload with the new higher oxygen consumption is a measure of endurance, which depends in part on the relative workload (i.e., the percentage of one's maximal capacity to consume oxygen). The maximal capacity to consume oxygen (i.e., maximal oxygen consumption) is determined by progressively increasing the external workload until the amount of oxygen consumed no longer increases, despite an increase in the workload. The additional energy demands for the workloads higher than that required to elicit maximal oxygen consumption are met through anaerobic processes.

Maximal oxygen consumption is the best measure of cardiovascular fitness and provides a quantitative index of the maximal capacity of the cardiovascular system to deliver blood, with its oxygen and nutrients, to the working muscles and to remove metabolic waste products and heat. Maximal oxygen consumption is higher in an endurance-trained individual, although submaximal oxygen consumption (i.e., the oxygen consumed at a submaximal absolute workload) is equal in trained and untrained people of comparable age, body weight, and health status. The oxygen cost of doing submaximal tasks does not vary much among individuals of comparable body weight. This means that the same absolute submaximal workload taxes a greater percentage of the untrained person's maximal oxygen consumption. The importance of this concept lies in the quality of life for the endurance-trained person. The tasks of everyday living tax the maximal capacity of the trained person to a lesser extent, so that the task is relatively easier after an endurance training program.

B. The Fick Equation

Oxygen consumption results from numerous processes, which are best represented by the Fick equation,

$$\dot{V}O_2 = \dot{Q} \times (a - \bar{v})O_2 \text{ difference},$$

where $\dot{V}O_2$ is oxygen consumption, $\dot{Q}$ is cardiac output, and $(a - \bar{v})O_2$ difference is the ability of the working muscles to extract and utilize oxygen from the perfusing blood. An increase in cardiac output and/or the $(a - \bar{v})O_2$ difference augments oxygen consumption. Most exercise physiologists agree that maximal oxygen consumption is limited by cardiac output (i.e., the primary determinant), whereas the ability to perform prolonged work at a submaximal workload might be limited by the $(a - \bar{v})O_2$ difference. The maximal capacity of the individual to perform

aerobic work is limited by processes related to the cardiac output, rather than peripherally at the skeletal muscle. The components of the Fick equation are discussed in the following sections.

I. Cardiac Output

Cardiac output is the product of the heart rate times the stroke volume (i.e., the amount of blood pumped by the heart per beat). The systolic function of the heart is the actual pumping of the blood filling the ventricles via the contraction of the cardiac muscle cells. The diastolic function of the heart is the relaxation of the cardiac muscle cells following contraction, which allows the ventricles to fill with the incoming blood. Both systole and diastole, or contraction and relaxation, are important determinants of cardiac output. They are regulated by a number of factors that increase the chronotropic state of the heart (i.e., heart rate) and the positive inotrophy (i.e., the increased rate of contraction and relaxation). Since cardiac output is so central to the adjustments of the cardiovascular system to dynamic exercise, much more emphasis is placed on the central regulating factors than on the peripheral adjustments. [*See* Cardiovascular System, Physiology and Biochemistry.]

Cardiac output linearly increases as the work intensity increases. In a young adult male, cardiac output increases from 5 liters/min at rest to 30 liters/min at maximal oxygen consumption. Females of comparable age, health, and degree of physical fitness typically have lower cardiac outputs at all intensities of work, primarily because of their typically smaller body size.

a. Heart Rate

Although both heart rate and stroke volume increase with exercise to increase cardiac output, the augmentation of heart rate is the most important variable in most individuals. With the onset of dynamic exercise, there is a rapid increase in heart rate in proportion to the increase in $\dot{V}O_2$ and the external work. Multiple control mechanisms produce the increase in heart rate, including the parasympathetic and sympathetic nervous systems. Initially, the withdrawal of parasympathetic stimulation produces an increased rate, which is further augmented by sympathetic stimulation as exercise continues. Even at mild levels of dynamic exercise (i.e., 20–30% $\dot{V}O_2$ max), the sympathetic nervous system increases heart rate in proportion to demand. This effect is mediated by the β_1-adrenergic receptor, which responds to the neurotransmitter norepinephrine released at nerve terminals, because it can be blocked with β_1-selective antagonists such as atenolol. The circulating neurotransmitter, in contrast, does not appear to have much effect on heart rate. In fact, the heart rate of denervated or transplanted hearts, which have no sympathetic nervous input, cannot increase with exercise despite elevated circulating epinephrine and norepinephrine via endogenous overflow from the adrenal glands or from exogenous epinephrine infusion.

The other control of heart rate is intrinsic to the myocardium. The maximum heart rate appears to be approximately 220 beats per minute in young adult males, which is perhaps 10 beats per minute higher in females of comparable age. This limit is characteristic of the heart's internal pacemaker system, the sinoatrial node, and the conduction pathways. Electrical stimulation of the heart that circumvents the pacemaker system, either internal (e.g., arrhythmias) or external (e.g., pacemakers), can easily exceed the maximum heart rate, indicating that the intrinsic control of the maximum heart rate, indicating that the intrinsic control of the maximum heart rate is at the sinoatrial node, which generates the signals, rather than the atrioventricular conduction pathway, which distributes the signal throughout the heart.

Exercise training does not alter the maximum heart rate; however, the submaximal heart rate at any given absolute workload and at rest is lower following an endurance training program. The resting bradycardia (i.e., low heart rate, 35–40 beats per minute in some athletes) and lower submaximal heart rate in a trained individual result primarily from a relative decrease in exercise-induced sympathetic tone, although an augmentation of parasympathetic tone has been implicated in some studies. This means that in an endurance-trained athlete, in whom the oxygen cost of performing a submaximal task is the same as in an untrained person, carciac output can be maintained only by increasing the stroke volume. Furthermore, the increased maximal oxygen consumption in an endurance-trained person must be gained primarily by an increased cardiac output via an augmentation of stroke volume. Thus, stroke volume is critical to the oxygen consumption of an endurance-trained person.

b. Stroke Volume

The stroke volume is the amount of blood ejected from the heart into the aorta with each contraction. The increase in stroke volume that occurs with dynamic exercise is quite small when compared to a two- or three-fold increase in heart rate. However, in a trained person whose submaximal heart rate is lower with an unchanged maximal heart rate, changes in stroke volume are the major way that cardiac output can increase with exercise, as stated earlier. Stroke volume during exercise is regulated by three primary factors: the Frank–Starling mechanism (preload), afterload or peripheral resistance, and contractility with (to a lesser extent) cardiac hypertrophy in highly trained athletes.

i. *Frank–Starling Mechanism* The force of contraction of the heart's ventricles varies as a function of the end-diastolic size, which is the absolute volume of the ventricle at the end of the filling period. This means that the larger the end-diastolic size, the greater is the force of contraction. This phenomenon is called the Frank–Starling mechanism, or Starling's law of the heart, and is the major way in which stroke volume is increased during exercise in which stroke volume is an important determinant of cardiac output, such as in an endurance-trained athlete. This mechanism is based on the relationship between length and active tension of the sarcomere, which is the basic contractile unit at the molecular level. Basically, this mechanism means that as the volume of the ventricle expands, the muscle is stretched passively, and this passive stretch increases the force of contraction. However, there is a point at which an increasing passive stretch actually decreases the force of contraction. This ordinarily does not happen in a normal heart.

For our purposes here the Frank–Starling curve is the integration of all of the responses of the heart to changes in its preload. Preload is the reflection of the filling of the heart, which is obviously dependent on venous return of blood to the heart. The venous return is altered by the position of the body. For example, there is less filling (per unit time) in the upright position than in the supine position. When dynamic leg exercise is performed in the upright position, the blood pooled in the legs is forced back to the heart by the contraction of the leg muscles. Furthermore, the effort of ventilating the lungs produces negative pressure within the chest and pulls blood into the heart. This increased venous return results in an increased right-sided preload, and thus left-sided preload. The increase in left ventricular filling is greater for supine exercise than for upright exercise. The left ventricular end-diastolic volume might be greater for supine exercise than for upright exercise, but similar peak left ventricular end-diastolic volumes occur in both postures at close to maximum levels of exercise intensity.

In an endurance-trained person, the total blood volume can increase as much as 10%. This expansion of total blood volume increases stroke volume via the Frank–Starling mechanism. The expansion of blood volume can also be achieved by the infusion of a physiologically compatible liquid (e.g., a dextran solution), which in turn augments stroke volume.

ii. *Afterload*   Afterload is the tension or stress placed on the ventricular wall while the ventricle is ejecting blood. In part, it is determined by the systolic pressure in the ventricle; therefore, it is influenced by the systolic arterial blood pressure and the peripheral vascular resistance. One can envision afterload as a "wall of pressure" against which the heart must pump blood. When the wall is high, the heart must pump harder to overcome this resistance by pressure. By inference, the amount of blood pumped with each beat is lower when the wall is higher. Thus, mechanisms must be in place to ensure that the pressures developed at the periphery are not excessive and can be reasonably overcome during dynamic exercise.

With the onset of dynamic exercise, systolic blood pressure increases from 120 to 140–160 mm Hg, depending on the intensity of the exercise. This increase in systolic blood pressure results primarily from the increased blood flow within the first few minutes of physical activity, and it correlates positively with both the increased cardiac output and the increased relative intensity of exercise (i.e., the percentage of maximal oxygen consumption). Diastolic blood pressure, on the other hand, does not increase much during dynamic exercise. Therefore, there is a minor change in mean arterial blood pressure with dynamic exercise, at least when the intensity is less than 75% of one's maximal oxygen consumption. At the same time, the peripheral resistance to blood flow, or vasomotor tone, is substantially reduced in the initial stages of all intensities of dynamic exercise. Overall, there is an exercise-induced reduction of peripheral resistance with a minor rise in arterial blood pressure, so that stroke volume is increased rapidly in the early stages of dynamic exercise. (The mechanisms underlying the decreased peripheral resistance with exercise are discussed in Section IV.)

iii. *Contractility*   Contractility, the intrinsic state of the heart, is difficult to measure *in vivo*, primarily because there are numerous confounding variables. A true measure of contractility is possible only with an isolated heart in which these variables are controlled or with isolated heart muscle strips (e.g., papillary muscles). Experimental evidence supports the idea that contractility is increased to some degree in exercising animals, which is produced in part by β_1-adrenergic sympathetic stimulation. In experimental animals, in which preload and afterload can be manipulated independently, the maximum rate of pressure development ($+dP/dt$) is increased during exercise in exercise-trained hearts, indicating that exercise training results in an augmentation of intrinsic contractility. Also, when the heart of an exercising dog has the same preload and afterload as that of a resting animal, it ejects more blood, again suggesting an augmentation of contractility with increased physiological demand. The effect of exercise training on contractility in the unstressed state (i.e., at rest) is ill defined.

The speed of contraction ($+dP/dt$) has a subcellular basis at the level of the basic contractile apparatus (i.e., the sarcomere) in the cardiac muscle cell. The sarcomere is composed of two primary contractile proteins, actin and myosin, and two regulatory proteins, troponin and tropomyosin. Numerous studies have shown that the speed of contraction is related directly to the different forms of myosin (i.e., slow to fast forms), so that the faster myosin form is observed in the hearts of animals that have a faster $+dP/dt$. Relaxation (or $-dP/dt$) of the cardiac muscle cell after contraction is an important component of the contractile properties of cardiac muscle cells, especially for the optimal filling of the heart during diastole. However, relatively few studies have indicated any change in the relaxation of the cardiac muscle cells after exercise training, except under certain circumstances in which relaxation is impaired (e.g., old age).

iv. *"Physiological" Hypertrophy*   Although the evidence is not concrete, another mechanism that can increase the stroke volume is physiological cardiac hypertrophy. ("Physiological" is used in the sense that the underlying mechanism is not caused by disease.) The left ventricular mass of highly trained athletes can be 50% higher than that of age-matched controls. Unlike the hypertrophy found in hypertension, an athlete's heart is enlarged primarily in end-diastolic cavity dimension, with lesser changes in end-systolic dimension and posterior wall and septal wall thickness. The hypertrophied heart has a 20–60% increase in stroke volume at rest and during exercise. With 3 months of vigorous training by either swimming or running, this hypertrophy can occur and then revert back to baseline values with as little as 3 weeks of detraining. This hypertrophy is an absolute increase in cardiac mass that does not result from an increase in the number of cardiac muscle cells (i.e., hyperplasia), but rather from the size of the individual cardiac muscle cells (i.e., hypertrophy).

2. Arteriovenous Oxygen Difference

Along with the increased blood flow to the working muscles, the ability of the muscle to extract oxygen and nutrients increases. This occurs through a variety of mechanisms, one of the most important physiological mechanisms being the increased surface area of the capillaries (via their openings) for the facilitation of this exchange. Since the $(a - \bar{v})O_2$ difference arises from both the arterial oxygen saturation and the mixed venous oxygen content, it is important to realize that the arterial oxygen saturation is nearly complete at 95% and does not change much with exercise. Therefore, the ability of the working muscle to extract and utilize the oxygen and nutrients coming to it from the heart is the primary determinant of the increased $(a - \bar{v})O_2$ difference observed during exercise.

II. CONTROL OF PERIPHERAL BLOOD FLOW DURING DYNAMIC EXERCISE

The regulation of blood flow during dynamic exercise is precisely matched to the energetic needs of the working muscles. At rest, approximately 15% of the cardiac output is directed toward skeletal muscle. With the onset of exercise, the amount and percentage of the cardiac output going to the active musculature will increase as exercise intensity increases, such that at $\dot{V}O_2$ max approximately 85% of the blood flow is going to the active skeletal muscle. This, coupled with the fact that during intense exercise cardiac output can increase to over five times its resting value, indicates that the blood flow in the active skeletal muscle can be 25-fold higher during exercise than during rest. Similarly, the amount of blood that the heart delivers to itself will change with exercise intensity. Although the actual percentage of the cardiac output that is delivered to the heart remains more or less the same (i.e., about 5%), since the absolute value of the cardiac output increases with exercise, the absolute volume of blood delivered to the heart also increases (about 250 ml/min at rest versus 1 liter/min at $\dot{V}O_2$ max). In addition, when exercise proceeds longer than 5 minutes, blood flow increases to the skin for the necessary dissipation of the heat produced during the muscular effort. In contrast, with the onset of exercise a large portion of the blood flow to the viscera (and inactive musculature) is diminished, which allows for the increased delivery to the active muscles. Blood flow to the brain, however, is not compromised during exercise.

The amount of blood delivered to a tissue can basically be modulated by two factors. First, the heart can be stimulated to increase cardiac output, as described earlier, which increases the amount of blood available for perfusing the various tissues. However, the tissues that can make use of the increased cardiac output are determined by the second factor, the circulatory system, namely, the capillaries. For example, at rest most of the capillaries in skeletal muscle lie dormant. The opening of these capillaries with exercise increases the number of channels through which the blood can perfuse the tissue, thus increasing the volume of blood flow without necessarily increasing the velocity of blood flow (which can decrease the efficiency of nutrient and oxygen exchange between the blood and the tissue). In this way, perfusion of the working muscles can increase by 20-fold or more. However, since cardiac output can increase by only a finite amount, and since this amount must be shared by the entire body, maximal perfusion can only occur when the mass of the working muscle is small.

The opening of dormant capillaries to increase perfusion of the working muscles is thought to occur primarily through competition between these two mechanisms: local control factors, which are vasodilators, and neural factors, which can vasoconstrict (e.g., adrenergic) or vasodilate (e.g., cholinergic). The main local factor involved in dilatation of the coronary vasculature is adenosine. Since the heart relies heavily on aerobic metabolism, when the myocardial oxygen supply decreases (e.g., during exercise, when its activity increases), adenosine triphosphate (ATP) cannot be regenerated and adenosine diphosphate (ADP) will begin to accumulate. Accumulated ADP is subsequently broken down to adenosine, which acts as a signal to dilate the coronary vasculature and increase the oxygen delivery, via increased blood flow, to the heart. Interestingly, adenosine will not act to vasodilate the coronary vasculature when oxygen levels are normal. Since diastolic blood pressure is not increased much during exercise, the perfusion pressure in the coronary arteries cannot, by itself, improve supply. Thus, the local autoregulation that the heart uses to dilate resistance vessels is the major mechanism for increasing myocardial perfusion. Adenosine involvement has also been implicated in controlling vasodilation in skeletal muscle as well, although it may not be as important as it is in the coronary vasculature.

Other factors that act locally to open dormant capillaries include changes in local temperature and pH, and the accumulation of metabolites (e.g., Krebs cycle intermediates and lactate) and ions (e.g., potassium and phosphate), which will start to accumulate when the muscle begins to work. There is some question as to whether the factors listed here stimulate vasodilation for the entire duration of an exercise bout or merely act to initiate vasodilation, with some unknown factor acting to maintain it. The changes in pH and the accumulation of metabolites and ions are usually maximal after a few minutes of exercise, but then gradually return to resting levels with continued exercise, yet the degree of vasodilation will remain more or less constant over the duration of the exercise bout. This vasodilation may be maintained by nitric oxide, also known as endothelial-derived relaxing factor or EDRF, which has been demonstrated to be released in response to sheer stress to the walls of the vasculature, as would occur in areas where blood flow has been increased. Prostaglandins have also been implicated as local vasodilators in skeletal muscle during exercise.

Competing to a degree with the local factors are the neural control factors, which arise from the sympathetic stimulation that occurs with the onset of exercise. As alluded to earlier, the sympathetic stimulation of blood flow to the working muscles comes from two subsystems of the sympathetic nervous system: the adrenergic constrictor system and the cholinergic dilator system. The adrenergic constrictor system works through the release of the neurotransmitter norepinephrine and results in vasoconstriction (thus, vasomotor tone), whereas the cholinergic dilator system is effected by the release of the neurotransmitter acetylcholine with subsequent vasodilatation. In the heart and the skeletal muscle, the cholinergic nervous system acts to vasodilate the arterioles. Working in concert, these two systems act to vasodilate the working muscle and to vasoconstrict the viscera and the nonworking muscles. However, the local metabolic control can override any neural influence in the working muscle that might act to vasoconstrict.

With exercise training, capillarization of both skeletal muscle and heart increases in many cases. When this is observed, the increase is in capillary density, rather than in the diameter of the vessels. This augmentation of capillarization allows for the greater perfusion of the trained muscle, thereby facilitating exchange between the blood and the muscle fibers.

III. CARDIOVASCULAR ADJUSTMENTS TO STATIC EXERCISE

Unlike dynamic exercise, in which the energy expenditure can be determined by the amount of oxygen consumed during work, the intensity of static exercise is determined by the amount of force or tension the muscle produces to perform the exercise. In practice this is determined by having a person resist a weight (e.g., lifting a barbell and holding it). (The lift, of course, is dynamic exercise.) Such exercise is also called resistance exercise. Another method in the laboratory setting is to use a hand dynamometer, which allows a person to squeeze to his or her maximum force (i.e., maximal voluntary contraction).

The most remarkable response to an acute resistance effort is seen in the response of blood pressure (pressor response), particularly the diastolic blood pressure, which does not change much during dynamic exercise. Both systolic and diastolic blood pressures increase in relation to the intensity of the voluntary contraction, as a percentage of the maximal voluntary contraction. Thus, mean arterial pressure increases. The heart rate increases as it does during dynamic exercise, but not to the same relative extent, unless the voluntary contraction exceeds 40–50% of the maximal voluntary contraction. For example, when a heavy weight is lifted, both heart rate and blood pressure increase, although not linearly, with an increase in active muscle mass. During a double leg-press in which a highly trained lifter is allowed to hold his breath, the heart rate can reach as high as 170 beats per minute (i.e., close to the maximum), with blood pressures of 320 mm Hg systolic and 250 mm Hg diastolic. The goal of the increased pressures is to overcome the high intramuscular pressures to maintain perfusion.

Components contributing to the elevated blood pressure include an increase in cardiac output, elevated intrathoracic and intra-abdominal pressures (especially while holding one's breath), and a mild increase in peripheral resistance. During a static contraction, the heart rate and blood pressure increase as the contraction is held. Depending on the muscle, contractions that attain only 40% of the maximal voluntary contraction can reduce muscle blood flow through occlusion of the blood vessels by the static contraction of the force-generating muscles. Receptors in the muscle respond to the products of metabolism and stimulate the increases in systolic pressure through feedback mechanisms via unmyelinated (group IV) afferent nerves. In contrast, the receptors that stimulate the increase in heart rate appear to be tension receptors.

Functionally, the hearts of resistance-trained athletes are not different from those of sedentary controls at rest. Although the stroke volume for athletes is larger than for the controls in most studies, perhaps through cardiac hypertrophy, the difference is attributed to an increase in body size (i.e., lean body weight or body surface area) in the trained group. Systolic function might be slightly enhanced by resistance training at rest; however, the effect of training on systolic function during exercise is unknown, perhaps because of the difficulty in experimentally deriving functionally meaningful data. On the other hand, an impairment in diastolic function is characteristic of most types of pressure overload-induced cardiac hypertrophy. This has not been shown to be the case in noninvasive studies of resistance-trained athletes. Peak filling rates and other measures of diastolic function are greater than expected in the trained group; however, when normalized for the larger body size, there is no loss of diastolic function. This conclusion is made with some hesitancy. Only small numbers of subjects have been studied, and the tragic presence of anabolic steroids in these male subjects might be confounding the interpretation.

IV. REGULATION OF THE CARDIOVASCULAR SYSTEM DURING EXERCISE: CENTRAL VERSUS REFLEX COMMAND

The regulation of the cardiovascular system during exercise is extremely complex and under precise control, so that under ordinary circumstances the cardiovascular response is in concert with the intensity of the exercise bout. Although the intensity of dynamic exercise is matched by an augmentation of cardiac output, and the intensity of static exercise is matched by increased arterial blood pressure, both types of exercise invoke an increased activity of the sympathetic nervous system and decreased activity of the parasympathetic nervous system. The underlying neural mechanisms are unknown with any precision, despite considerable effort since the mid-1800s. The difficulty in deriving data that can be interpreted without ambiguity is that the response to exercise is a multifac-

torial process, involving numerous systems besides the cardiovascular system. The homeostatic mechanisms necessary for survival of the organism necessitate this complexity. However, two general theories have evolved in the past 100 years.

The first theory holds that the response of the cardiovascular system to exercise arises from the direct action of a central command descending from the motor centers of the cardiovascular control centers in the brain (i.e., the central neural command theory). The other theory hypothesizes that the response comes from reflexes emanating from afferent neural activity of receptors in the muscles and joints that are activated during exercise (i.e., the reflex neural command theory). Both mechanisms might have a final common pathway from the brain to the cardiovascular system. Both mechanisms can work separately or together, and it is probable that both mechanisms are operant during exercise to make the cardiovascular adjustments appropriate to the intensity of the exercise.

BIBLIOGRAPHY

Armstrong, R. B. (1988). Distribution of blood flow in the muscles of conscious animals during exercise. *Am. J. Cardiol.* **62**, 9E–14E.

Astrand, P. O., and Rodahl, K. (1977). "Textbook of Work Physiology. Physiological Bases of Exercise." McGraw-Hill, New York.

Berne, R. M., and Levy, M. N. (1992). "Cardiovascular Physiology." Mosby Yearbook, St. Louis.

Braunwald, E., Ross, J. R., Jr., and Sonnenblick, E. H. (1976). "Mechanisms of Contraction of the Normal and Failing Heart." Little, Brown, Boston.

Colan, S. D. (1992). Mechanics of left ventricular systolic and diastolic function in physiological hypertrophy of the athlete heart. *Cardio. Clin.* **10**, 227–240.

Hanson, P., and Nagle, F. (1987). Isometric exercise: Cardiovascular responses in normal and cardiac populations. *Cardiol. Clin.* **5**, 157–170.

McArdle, W. D., Katch, F. I., and Katch, V. L. (1986). "Exercise Physiology, Energy, Nutrition, and Human Performance." Lea & Febiger, Philadelphia.

Mitchell, J. H. (1985). Cardiovascular control during exercise: Central and reflex neural mechanisms. *Am. J. Cardiol.* **55**, 34D–41D.

Mitchell, J. H. (1990). J. B. Wolffe memorial lecture. Neural control of the circulation during exercise. *Med. Sci. Sports Exerc.* **22**, 141–154.

Rowell, L. R. (1986). "Human Circulation Regulation during Physical Stress." Oxford Univ. Press, New York.

Rowell, L. R. (1993). "Human Cardiovascular Control." Oxford Univ. Press, New York.

Saltin, B. and Rowell, L. B. (1980). Functional adaptations to physical activity and inactivity. *Fed. Proc. Fed. Am. Soc. Exp. Biol.* **39**, 1506–1513.

Extracellular Matrix

KENNETH M. YAMADA

National Institutes of Health

GLOSSARY

Amino acid abbreviations used Ala, alanine; Arg, arginine; Asn, asparagine; Asp, aspartic acid; Cys, cysteine; Gln, glutamine; Glu, glutamic acid; Gly, glycine; His, histidine; Ile, isoleucine; Leu, leucine; Lys, lysine; Met, methionine; Phe, phenylalanine; Pro, proline; Ser, serine; Thr, threonine; Trp, tryptophan; Tyr, tyrosine; Val, valine

Basement membrane Thin, noncellular sheet of extracellular matrix underlying epithelia; the region closest to cells is termed the basal lamina, and the more loosely structured peripheral zone is termed the reticular layer

Cytoskeleton Cytoplasmic structural elements serving as an intracellular framework or "skeleton." For this article, the most relevant elements are actin microfilament bundles, composed of F-actin polymers and associated proteins

Neural crest cells Migratory cells derived from the dorsolateral edge of the closing neural tube that form sensory and sympathetic ganglia and other structures

EXTRACELLULAR MATRIX IS THE NONCELLULAR material that provides the supporting, structural, externally multifunctional, regulatory environment of cells throughout the body. Its remarkably diverse constituents range from the structural collagens to adhesive glycoproteins, and from clear, jelly-like gels rich in hyaluronate to dense, rigid bone; it can vary in organization from locally homogeneous to elaborately structured. Major functions of the extracellular matrix include (1) providing structural support, tensile strength, or cushioning, (2) providing substrates and pathways for cell migration, and (3) regulating cellular differentiation and metabolic functions. Individual extracellular matrix molecules are often multifunctional, and their functions are now understood in terms of specific functional binding or recognition sites. Specific cellular receptors provide the mechanisms for cell-surface interactions with these diverse and critically important structural and regulatory molecules.

I. INTRODUCTION

The extracellular matrix consists of a wide variety of specialized proteins, complex carbohydrates, and other molecules that surround, support, and organize cells and tissues. Without connective tissue, bone, and the unique local extracellular matrix, the human body would be little more than an amorphous mass of unorganized, poorly differentiated cells. Human growth and development, as well as tissue maintenance and wound repair, all require a host of essential extracellular molecules and their specific cellular receptors, which are regulated by complex patterns of gene expression. Table I presents a list of constituents of extracellular matrices and some binding functions of each.

II. FUNCTIONS OF THE EXTRACELLULAR MATRIX

A. Functions in Structural Organization

As structural elements, extracellular matrix molecules provide support for cells and dictate the external boundaries of most tissues. The epithelia of the body

TABLE I

Extracellular Matrix Molecules (partial listing)

| Molecule | Molecular weight[a] | | Binds to[b] |
	Complex	Subunits	
Collagens	(Variety of types[c])		Cells, PGs, Fn, Ln, Vn, Nd, SPARC
Fibronectin	~500,000[d]	~250,000[d]	Cells, Col, PGs, Fbn, amyloid P component
Laminins	~900,000[c]	400,000[c], 210,000, 200,000	Cells, PGs, Col type IV, En, sulfatides, Fln
Proteoglycans	(Wide range of types[c])		Cells, PGs, Col, Fn, Ln
Elastin	Polymer	63,000[d]	Cells, microfibrils
Vitronectin (S-protein)		75,000	Cells, heparin, Col (contains somatomedin B)
Thrombospondins	~450,000	~140,000[c]	Cells, PGs, Col, Ln, Fbn, sulfatides, calcium
Tenascins	Hexamer	~230,000[c,d]	Cells, Fn
Entactin (nidogen)		150,000	Cells, Ln, Col type IV, Fln
Osteonectin (SPARC)		33,000	Calcium, other?
Fibulins[c]	~400,000	105,000 monomer or 195,000	Fn, En, Ln
Agrin[c]		95,000	
Netrins[c]		75,000–78,000	
Chondronectin	176,000	56,000	Cells, Col type II
Link protein		39,000	Cartilage PG, hyaluronate
Osteocalcin		5,800	Hydroxylapatite
Bone sialoprotein		34,000	Cells, hydroxylapatite
Osteopontin		33,000	Hydroxylapatite
Epinectin		70,000	Epithelial cells, heparin
Hyaluronectin		59,000	Hyaluronate, neurons
Amyloid P component	250,000	25,000	Fn, ? elastase

[a]If the molecules exist as regular complexes, the total size of the complex is indicated at the left; the right column indicates the size of each subunit or monomer. Molecular weights are generally derived from sequence data (lacking weight of carbohydrates) or from sodium dodecyl sulfate polyacrylamide gel electrophoresis.

[b]Many of these molecules are bound by cells, as well as binding to the indicated ligands. PG, proteoglycan; Fn, fibronectin; Ln, laminin; Vn, vitronectin; En, entactin; Col, collagen(s); Fbn, fibrin/fibrinogen; Fln, fibulin. Many matrix molecules also bind various cytokines and growth factors.

[c]Molecular weights vary substantially depending on the type.

[d]Molecular weights vary owing to alternative splicing of precursor messenger RNA, which produces a family of closely related variants.

that cover skin, mucous membranes, and the interior of glands all rest on extracellular structures termed *basement membranes*. These thin sheets consist of an organized matrix of type IV collagen, laminin, and other molecules; one hypothesis about how basement membrane components are organized is shown in Fig. 1A. Basement membranes provide surfaces on which epithelial cells adhere and assume a polarized orientation. They also constitute a barrier separating epithelial cells from underlying connective tissue, and tumor cells that cross this boundary are generally considered malignant.

Other types of extracellular matrices are organized differently. Loose connective tissue produced by fi-broblasts is composed of extracellular matrix molecules such as type I collagen and proteoglycans. The sheets of collagenous material forming fascia, the cords of collagen forming ligaments and tendons, cartilage, bone, and even the clear vitreous humor of the eye all consist of different types of extracellular matrix secreted by specific cells as structural materials. The diversity of such matrices is particularly obvious when comparing the molecular organization of basement membrane with that of cartilage (Fig. 1). The specific molecules involved in such structural roles and the molecular interactions that organize them are considered in this article. [*See* Collagen, Structure and Function; Elastin; Proteoglycans.]

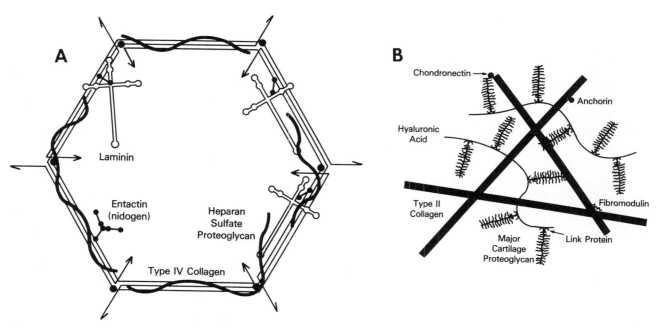

FIGURE 1 Comparison of the organization of two distinctive types of extracellular matrix. Schematic models of the molecular organization of (A) basement membrane (upper basal lamina portion) and (B) cartilage matrix are shown as examples of complexity and diversity of extracellular matrices. Basement membrane consists of a scaffolding of type IV collagen, hypothesized by some to exist as sheets of hexagonal arrays (a single component hexagon is shown, which is connected to its neighbors by single-sided arrows); particularly functionally important three-dimensional interconnections are formed by binding of tips of collagen molecules to one another (*regular arrows*). Major additional constituents are laminin, heparan sulfate proteoglycan (perlecan), and entactin (nidogen). In contrast, cartilage (B) is more loosely organized, consisting of type II and other collagens, a major chondroitin sulfate-containing proteoglycan termed aggregan plus others, hyaluronic acid, and other, more minor protein constituents.

B. Functions in Cell Migration

Extracellular matrix molecules such as fibronectin and laminin can provide pathways for cell migration. For example, studies of gastrulation (an early stage of embryonic development) in animals reveal that fibronectin provides a substrate for migration of cells during establishment of the basic body plan. Fibronectin and probably several other matrix proteins are also important for the subsequent migration of embryonic neural crest cells throughout the body to form many of the structures of the face, the sympathetic and parasympathetic nervous systems, and skin pigment. To migrate, cells use specific receptors that bind to key sites in these proteins (see Section VI). Inhibiting the receptor-mediated interaction of cells with such extracellular substrates halts migration.

During wound healing, a meshwork of fibronectin linked to fibrin provides an excellent migratory substrate for epithelial cells as they close the wound, as well as a general scaffolding that migrating fibroblasts penetrate; these fibroblasts then secrete collagen and other molecules to form replacement connective tissue. The initial migratory/scaffolding proteins generally disappear after the wound is healed and are replaced by stable basement membrane and mature connective tissue. [*See* Connective Tissue.]

C. Functions in Cell Growth and Differentiation

Specific extracellular matrix molecules such as collagen, fibronectin, laminin, and heparan sulfate proteoglycan can each stimulate the growth or differentiation of different cell types. For example, collagen gels and artificially reconstituted basement membranes promote epithelial cell differentiation. In contrast, excess fibronectin blocks chondrocyte differentiation to cartilage while stimulating sympathetic neuron differentiation from neural crest cells. As yet, little is known about the actual mechanisms of action of these molecules nor how they are regulated and then coordinate with one another to help regulate cellular behavior.

III. COLLAGENS AND ELASTIN

Collagens are the most abundant class of protein in the human body. At least 19 different types of collagen have been identified to date, which fulfill a variety of different functions including providing crucial supportive and overall structural stability to tissues. Elastin is another major structural extracellular protein; collagens and elastin will be considered only in the context of other matrix proteins and their receptors. A notable characteristic of these proteins is their ability to self-assemble into lengthy fibrils after secretion; this property simplifies the assembly of extracellular matrices. [See Collagen, Structure and Function; Elastin.]

IV. FIBRONECTIN

In recent years, a number of noncollagenous glycoproteins have been identified in the extracellular matrix that can begin to account for its wide range of biological activities. Fibronectin is presently the best studied of these glycoproteins, which can mediate or modulate cell adhesion, morphology, and migration. Although produced from only a single gene, fibronectin molecules can differ by the insertion or absence of three regions of protein termed ED-A (EIIIA), ED-B (EIIIB), or IIICS, which occurs by differences in processing (alternative splicing) of fibronectin precursor messenger RNA (Fig. 2). This diversity of fibronectins derived from a single gene provides a simple mechanism for generating a potential diversity of functions.

Fibronectin can mediate a host of adhesive and binding functions; for example, fibronectins can produce adhesion to surfaces of most cell types except erythrocytes. As shown in Fig. 2, each fibronectin molecule contains a linear array of modular binding units that bind to cells as well as to other key extracellular molecules, including collagen, fibrin, and heparin (heparan or dermatan sulfate proteoglycans). Various pairings of these binding domains can account for the ability of fibronectin to form cross-links between different extracellular matrix proteins, as well as its ability to mediate cell adhesion to collagen or fibrin. It can even bind to a heparan sulfate proteoglycan in the plasma membrane to mediate binding to cell surfaces.

Most interactions of fibronectin with cells, however, appear to be mediated by specific receptors that can bind to certain key sequences in fibronectin as indicated in Fig. 2 and Table II. A key recognition unit

in fibronectin and several other extracellular adhesion proteins needed for receptor binding is the amino acid sequence Arg-Gly-Asp (e.g., as seen within the fibronectin Gly-Arg-Gly-Asp-Ser sequence). Full binding specificity and strength of binding to the major fibronectin receptor, however, requires another region containing the sequence Pro-His-Ser-Arg-Asn, which functions in synergy with this basic adhesion sequence.

Cell-type specificity of recognition of fibronectin is mediated by yet another two sites in the IIICS region, present in only certain fibronectin molecules; they contain distinct amino acid recognition sequences (Table II). For example, although most cells adhere to the former Arg-Gly-Asp and synergy sequences, adhesive recognition of the latter two IIICS sites appears much more restricted (e.g., to cells derived from the neural crest, such as peripheral nervous system cells, and to lymphocytes). Besides these four adhesion regions, a fifth general region is involved in binding to heparin and to cells (Fig. 2 and Table II).

Besides displaying striking adhesive activity, fibronectin can also promote the migration of many types of cells. It has been identified as a migratory substrate for embryonic gastrulating cells, neural crest cells and their derivatives, and primordial germ cells, plus mature corneal and epidermal cells and a variety of tumor cells.

The identification of crucial binding sequences in fibronectin has permitted the development of synthetic peptide inhibitors that competitively inhibit function. These nontoxic inhibitors can be used to probe the functions of such recognition sequences in intact organisms. Such studies have implicated the Arg-Gly-Asp sequence in embryonic cell migration as well as in tumor cell invasion and metastasis. For example, certain peptides from fibronectin can inhibit metastasis in animal model systems. [See Metastasis.]

Finally, fibronectin can modify the morphology, growth, and differentiation of certain cells. Many of its morphological effects can be attributed to its adhesive activity, including cell flattening, decreased cell-surface microvilli, cell–cell alignment, and organization of intracellular actin microfilament bundles. Moreover, fibronectin can stimulate certain cells to proliferate, as well as inhibiting differentiation (e.g., blocking the synthesis of proteoglycans by cartilage cells). How fibronectin mediates its effects on growth and differentiation remains to be determined. Another important function of fibronectin and other extracellular molecules is to serve as carriers of growth and differentiation factors such as transforming growth

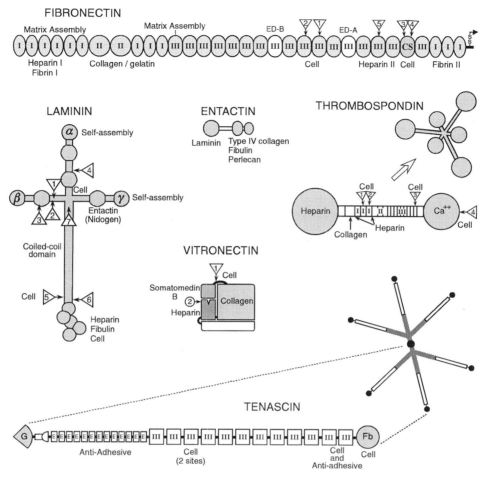

FIGURE 2 A variety of individual extracellular matrix molecules and their functional sites. These molecules often have multiple functional domains as indicated by labels (e.g., collagen- or heparin-binding domains), as well as recently identified specific amino acid sequences recognized by cellular receptors (*numbered triangles*). See Table I for characteristics of these molecules, and Table II for a listing of the specific recognition sequences identified to date. Fibronectin is composed of three types of repeating unit (i.e., types I, II, and III); it also contains a series of functional binding domains (e.g., for binding to heparin or to collagen). Two type III units termed ED-A and ED-B can be present or absent because of alternative splicing. The IIICS region can undergo more complex alternative splicing at three sites. Fibronectin forms dimers or multimers by disulfide bonding at the carboxyl terminus (*right end*). Laminins often contain three noncovalently associated chains, termed α, β, and γ. The long arm consists of all three subunits entwined in a rigid coiled-coil structure terminated by a lobulated terminal domain (G). Thrombospondin assembles into a trimer with the larger globular domains outward. Vitronectin is globular and contains somatomedin B, which can be released by proteolysis. A novel cryptic heparin-binding domain (2) is exposed and active in only a small subpopulation of vitronectin molecules but is revealed by experimental denaturation (e.g., with urea). Entactin (nidogen) is a nodular molecule that can sometimes form clusters by self-aggregating. Tenascin is composed of a variety of domains including EGF-like repeats (E), fibronectin type III units, and a fibrinogen-like globular domain (Fb). Tenascin monomers often assemble into hexamers (hexabrachions).

factor-β (TGF-β). Local accumulations of fibronectin, or its focal destruction by proteases, may lead to local release of such regulatory factors from such extracellular matrix reservoirs. [*See* Transforming Growth Factor-β.]

V. LAMININ

At least seven structurally distinct forms of the adhesive protein laminin have been identified from different cell sources (e.g., classic laminin, merosin, S-lami-

TABLE II
Specific Cell Adhesive Recognition Sequences in Matrix Proteins

Number[a] and name	Amino acid sequence recognized[b]	Cells binding the sequence
Fibronectin		
1. Cell-binding domain	Gly-*Arg-Gly-Asp*-Ser	Most cells tested
2. Synergistic sequence	Pro-His-Ser-*Arg*-Asn	Cells using the major fibronectin receptors $\alpha_5\beta_1$ or $\alpha_{IIb}\beta_3$
3. IIICS: CS1 site	Asp-Glu-Leu-Pro-Gln-Leu-Val-Thr-Leu-Pro-His-Pro-Asn-Leu-His-Gly-Pro-Glu-Ile-*Leu-Asp-Val*-Pro-Ser	Most neural crest derivatives, lymphocytes
4. IIICS: REDV[c] site	Arg-Glu-Asp-Val	Melanoma cells, others?
5. Peptide I	Tyr-Glu-Lys-Pro-Gly-Ser-Pro-Pro-Arg-Glu-Val-Val-Pro-Arg-Pro-Arg-Pro-Gly-Val	Melanoma cells
Peptide II	Lys-Asn-Asn-Gln-Lys-Ser-Glu-Pro-Leu-Ile-Gly-Arg-Lys-Lys-Thr	Melanoma cells
Laminin		
1. YIGSR[c]	Tyr-Ile-Gly-Ser-Arg-	Many cell types
2. PDSGR[c]	Pro-Asp-Ser-Gly-Arg	Fibrosarcoma, melanoma, others
3. F9	Arg-Tyr-Val-Val-Leu-Pro-Arg-Pro-Val-Cys-Phe-Glu-Lys-Gly-Met-Asn-Tyr-Val-Arg	Fibrosarcoma, melanoma, glioma, endothelial cells
4. RGD[c] site	*Arg-Gly-Asp*-Asn	Endothelial, others?
5. p20	Arg-Asn-Ile-Ala-Glu-Ile-Ile-Lys-Asp-Ile	Neurons
6. IKVAV[c]	Ile-Lys-Val-Ala-Val	Many cell types
7. LMA-5	(Cys)-Tyr-Phe-Gln-Arg-Tyr-Leu-Ile	Neurons
8. PA22-2	Ser-Arg-Ala-Arg-Lys-Gln-Ala-Ala-Ser-Ile-Lys-Val-Ala-Val-Ser-Ala-Asp-Arg	Neurons
9. LRE[c] site	Leu-Arg-Glu	Ciliary ganglion neurons
Entactin (nidogen)		
1. RDG[c] site	Ser-Ile-Gly-Phe-*Arg-Gly-Asp*-Gly-Gln-Thr-Cys	Mammary tumor cells
Vitronectin		
1. RGD[c] site	*Arg-Gly-Asp*-Val	Many cells

[a]Numbers correspond to numbering of open triangles in Fig. 2.
[b]See glossary for full name of amino acids; italics show putative minimal sequences, although in some cases the minimal sequence required for activity has not yet been determined.
[c]These sites are named according to the single-letter abbreviations of their crucial amino acids.

nin, and kalinin). The apparently most abundant form depicted in Fig. 2 is a prominent constituent of basement membranes, where nearly all laminins are found. They play important roles in mediating adhesion of epithelial cells to flat surfaces. Laminins also contribute to the migration of certain embryonic cells near basement membranes (e.g., of certain neural crest cells and cardiac mesenchymal cells). Laminins can also appear transiently elsewhere for use in other functions (e.g., in directing neuronal outgrowth). Like fibronectin, laminin can mediate the adhesion and migration of cells, although it appears to be more specialized for epithelial and neuronal cells. Laminin also contains a series of amino acid sequences recognized by distinct cell-surface receptors (Fig. 2 and Table II). Besides a functional sequence that resembles the key fibronectin

Arg-Gly-Asp sequence, others contain the sequence Tyr-Ile-Gly-Ser-Arg, Ile-Lys-Val-Ala-Val, and more as indicated in Fig. 2. Many of the numerous reported peptide adhesive sites appear to be cryptic, and probably require proteolysis or denaturation for exposure or release. At least two such synthetic peptides have been reported to inhibit tumor cell metastasis in animal studies modeled on those with fibronectin peptides. Unlike fibronectin, laminin also has the ability to adhere to cells by novel mechanisms involving binding to sulfated glycolipids or cell-surface glycosyl transferases in the plasma membrane. [*See* Laminin in Neuronal Development.]

Like fibronectin, laminin can bind to heparin and heparan sulfate proteoglycan; it also binds to type IV collagen and entactin (nidogen). These latter two binding activities are probably quite important in the assembly of basement membranes from these molecules (Fig. 1).

VI. RECEPTORS FOR EXTRACELLULAR MATRIX PROTEINS

A. Overview

It has become increasingly apparent that extracellular matrix molecules interact with cells via specialized receptors in the plasma membrane (Fig. 3). These receptors can differ in type and specificity depending on the type of cell examined, and various combinations of receptors can produce distinctive patterns of adhesion. The largest group of receptors belongs to the integrin family (Table III), although other receptor mechanisms are also important. The existence of these receptor systems can account for cell specificity of interactions with different matrix proteins, and they provide straightforward mechanisms for mediating and regulating cell interactions

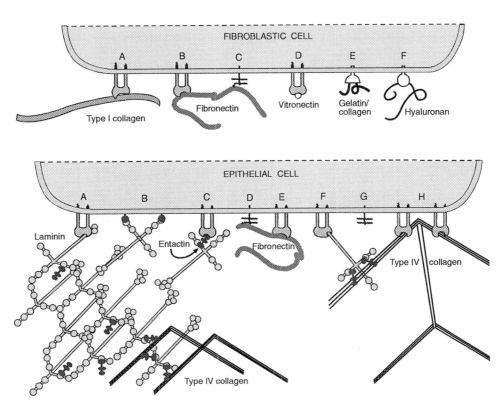

FIGURE 3 Cellular receptor systems for extracellular matrix molecules. Major receptors used by fibroblastic cells are compared with those used by epithelial cells. On fibroblastic and mesenchymal cells (top), at least six types of receptor can be found: (A) integrins binding to type I and other collagens (e.g., the $\alpha_2\beta_1$ and $\alpha_3\beta_1$ integrins); (B) integrins that bind fibronectin (e.g., $\alpha_5\beta_1$); (C) heparan sulfate proteoglycan(s) such as syndecan in mesenchymal cells that bind to fibronectin's heparin-binding site or collagen; (D) vitronectin receptors (e.g., $\alpha_V\beta_3$); (E) other collagen- and gelatin-binding receptor(s); and (F) a hyaluronic acid receptor, for example, CD44 or RHAMM. On various epithelial cells (bottom), major adhesive interactions can occur with laminin and type IV collagen: (A) integrins that bind laminin at a site just above the large globular domain (e.g., $\alpha_3\beta_1$); (B) sulfated glycolipids (sulfatides) bind to the short arms of laminin; (C) the Arg-Gly-Asp site of entactin (nidogen) is bound by an unidentified integrin; the entactin then binds to laminin; (D, E) in migrating epithelial cells separated from the basement membrane (e.g., during wound closure), integrins (E) and/or syndecan (D) can bind and use fibronectin as a migration substrate; (F) integrins bind laminin, which binds entactin (nidogen), which in turn mediates binding to type IV collagen; (G) syndecan is also present on mature epithelia; (H) integrins can also bind to type IV collagen directly. Not all cells of each type possess each type of receptor, and specialized cells may also display other additional receptors not shown.

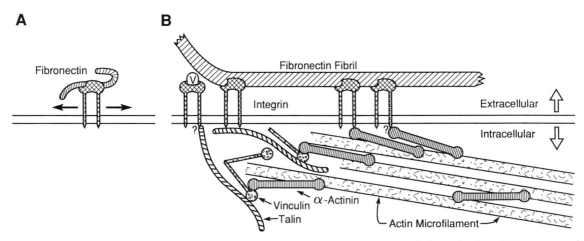

FIGURE 4 Transmembrane interactions of extracellular matrix receptors with intracellular molecules. (A) In a region of plasma membrane on a rapidly migrating cell, a single fibronectin molecule is bound by an integrin that is free to diffuse laterally and is not encumbered by complexes with intracellular molecules. (B) On the membrane of a more stationary cell, integrin receptors bind to vitronectin (V) or to fibronectin fibrils on the extracellular face of the plasma membrane, as well as to intracellular cytoskeletal molecules such as talin, α-actinin, or tensin as part of large complexes; the receptors are immobile when participating in these large transmembrane complexes. Question marks indicate interactions that have been suggested but that may require further proof.

TABLE III
Integrin Family of Cell Adhesion Receptors

Name	Alternative names	Molecular weight of α (reduced)[a,b]	Ligands bound[c]	Putative adhesion function
β_1 integrins (VLAs)[d]				
$\alpha_1\beta_1$	VLA-1	210,000	Col, Ln	Cell–matrix
$\alpha_2\beta_1$	VLA-2, GP$_{Ia-IIa}$	165,000	Col, (Ln)	Cell–matrix (and –cell?)
$\alpha_3\beta_1$	VLA-3	130,000 + 25,000	Fn, Ln, Col, Epi	Cell–matrix (and –cell?)
$\alpha_4\beta_1$	VLA-4	150,000	Fn, V-CAM	Cell–cell and –matrix
$\alpha_5\beta_1$	VLA-5, fibronectin receptor, GP$_{Ic-IIa}$	135,000 + 25,000	Fn	Cell–matrix
$\alpha_6\beta_1$	VLA-6	120,000 + 30,000	Ln	Cell–matrix
β_2 integrins (leukocyte adhesion receptors)				
$\alpha_L\beta_2$	LFA-1	180,000	I-CAMs	Leukocyte cell–cell
$\alpha_M\beta_2$	Mac-1, CR-3	170,000	iC3b, Fb	Leukocyte cell–cell
$\alpha_X\beta_2$	p150,90	150,000	iC3b	Leukocyte cell–cell
β_3 integrins (cytoadhesions)				
$\alpha_{IIb}\beta_3$	GP$_{IIb-IIIa}$	120,000 + 25,000	Fb, Fn, vWF, Vn, Tsp	Platelet cell–cell and cell–matrix
$\alpha_V\beta_3$	Vitronectin receptor	125,000 + 24,000	Vn, Fb, vWF, Tsp, Fn	Cell–matrix
Other integrins[e]				
$\alpha_6\beta_4$		120,000 + 30,000	Laminins	Cell–matrix
$\alpha_V\beta_5$	Alternative vitronectin receptor	125,000 + 24,000	Vn, Fn	Cell–matrix

[a] After chemical reduction of disulfide bonds.
[b] Molecular weights of β subunits are: β_1 = 110,000; β_2 = 90,000; β_3 = 90,000; β_4 = 220,000; β_5 = 110,000.
[c] Col, collagen; Fn, fibronectin; Ln, laminin; Fb, fibrinogen; vWF, von Willebrand factor; Vn, vitronectin; Tsp, thrombospondin; Epi, epiligrin (kalinin).
[d] Additional members of the family now include the following α subunits: α_7, α_8, α_9, and α_V.
[e] Additional β subunits identified now include β_6, β_7, and β_8.

with extracellular molecules. [*See* Receptors, Biochemistry.]

B. Integrin Receptor Family

There are at least 20 receptors in the human integrin family, with more members likely to be discovered (Table III). Each receptor has a distinctive pattern of expression during development and in different adult tissues, and each has its own specialized functions. In general, these receptors function in cell adhesion and migration. For example, many cells use the $\alpha_2\beta_1$ collagen receptor for adherence and migration in collagen matrices, while using the $\alpha_5\beta_1$ or other fibronectin receptors on fibronectin substrates. Epithelial cells can adhere to collagen directly, which suggests that they can attach directly to the structural meshwork of type IV collagen in basement membranes by a mechanism distinct from adhesion mediated via laminin (Fig. 3).

Adhesion to fibronectin can occur through at least eight different receptors, which may produce differing effects on cell behavior. For example, the $\alpha_5\beta_1$ receptor is required for migration of certain tumor cells, although it functions instead to form strong adhesions and an organized fibronectin matrix around certain normal fibroblasts. At least five other integrin receptors can also mediate cell interactions with laminin.

Many integrin receptors can mediate transmembrane effects on the organization of molecules in the cytoplasm of cells (Fig. 4). Specifically, several integrins, including $\alpha_5\beta_1$, $\alpha_v\beta_3$, and $\alpha_{IIb}\beta_3$ (GPIIb-IIIa), are associated with the intracellular accumulation of organized arrays of actin-containing microfilaments; they may do so by forming receptor aggregates that bind linking proteins such as tensin, talin, and α-actinin (Fig. 4). The functional consequence is that organized cytoskeletons form inside cells, affecting cell shape and function (e.g., promoting tenacious adhesion and halting cell migration).

C. Other Receptors and Binding Proteins

Other possible receptors besides integrins have been identified for collagens, laminin, elastin, hyaluronan, and proteoglycans. A 67,000-dalton peripheral membrane protein binds both laminin and elastin and is variously reported to function by binding to the peptide sequences Tyr-Ile-Gly-Ser-Arg or Leu-Gly-Thr-Ile-Pro-Gly or galactoside sugars. This protein is increased on many tumor cells and is thought to contribute to the capacity of these cells to metastasize.

Hyaluronan can be bound to cells by at least two distinct receptors, termed CD44 and RHAMM. Surprisingly, certain intracellular proteins of the annexin family can also bind to the cell surface after release from cells, and serve as receptors for collagen type II (annexin V, also termed anchorin CII) and tenascin (annexin II).

The wide variety of integrin and nonintegrin cell-surface receptors can begin to account for the complexities of cell interactions with various extracellular matrices (Fig. 3). The regulation of these receptors and of the many known extracellular matrix molecules is under the control of a host of growth factors and cytokines such as TGF-β, as well as developmental regulatory systems. Such complexity of regulation might be expected, given the direct roles of extracellular molecules in a wide variety of cellular functions.

BIBLIOGRAPHY

Hay, E. D. (ed.) (1991). "Cell Biology of Extracellular Matrix," 2nd Ed. Plenum, New York.

Horwitz, A. F., and Thiery, J. P. (eds.) (1994). Cell-to-cell contact and extracellular matrix. *Curr. Opin. Cell Biol.* 6, 645–779.

Mayne, R., and Burgeson, R. E. (1987). "Structure and Function of Collagen Types." Academic Press, Orlando, FL.

McDonald, J. A., and Mecham, R. P. (1991). "Receptors for Extracellular Matrix." Academic Press, San Diego.

Mecham, R. P., and Wright, T. N. (1987). "Biology of Proteoglycans." Academic Press, Orlando, FL.

Piez, K. A. (ed.) (1984). "Extracellular Matrix Biochemistry." Elsevier, New York.

Sandell, L. J., and Boyd, C. D. (1990). "Extracellular Matrix Genes." Academic Press, San Diego.

Yamada, K. M., and Gumbiner, B. M. (eds.) (1995). Cell-to-cell contact and extracellular matrix. *Curr. Opin. Cell Biol.* 7, 615–767.

Eye, Anatomy

BRENDA J. TRIPATHI
RAMESH C. TRIPATHI
K. V. CHALAM
University of South Carolina School of Medicine

GLOSSARY

Diopter Measure of the power of a lens, equal to the reciprocal of its focal length in meters; a converging lens is taken as positive, a diverging lens as negative

Dioptrics Science of the refraction of light by transmission

Emmetropia State of proper correlation between the refractive system of the eye and the axial length of the eyeball, with rays of light entering the eye parallel to the optic axis being brought to a focus exactly on the retina

Fornix conjunctivae Conjunctival cul-de-sac; retrotarsal fold; concave recess formed by the junction of the bulbar and palpebral portions of the conjunctiva, that of the upper lid being the f. conjunctivae superior, and that of the lower lid the f. conjunctivae inferior

Fovea centralis retinae Central fovea of retina, foveal centralis, fovea; conical central depression in the macula retinae where the retina is very thin so that rays of light have free passage to the layer of photoreceptors, mostly cones; this is the area of most distinct vision to which the visual axis is directed

Glaucoma Disease of the eye characterized by excavation and degeneration of the optic disk, and nerve fiber bundle damage producing characteristic defects in the field of vision; increased intraocular pressure due to restricted outflow of the aqueous through the trabecular meshwork and Schlemm's canal system is a major risk factor

Melanocyte Cell bearing or capable of forming melanin; located in the skin and many structures of the eye

Myopia Shortsightedness; nearsightedness; near or short sight; a condition in which, in consequence of an error in refraction or of elongation of the globe of the eye, parallel rays are focused in front of the retina

Tonofilament Structural cytoplasmic protein, bundles of which together form a tonofibril; the protein of epidermal tonofilaments is keratin; the filaments are 8–10 nm in thickness

Zonular fiber (fibrae zonulares) Fine, elastic filaments that arise from the surface of the epithelium of the ciliary body as far back as the ora serrata and especially from the corona ciliaris and extend to the equatorial anterior and posterior surfaces of the lens of the eye; through their attachments, the ciliary muscle produces changes in the curvature of the lens during accommodation; collectively they form the zonula ciliaris or zonules of Zinn

VISUAL PERCEPTION OF OUR ENVIRONMENT involves two discrete processes: transduction of radiant energy into electrical impulses and the subsequent interpretation of these impulses by the brain. Our vision, defined as the translation of afferent neuronal impulses into meaningful symbols and patterns, can be shaped by our experience; however, we are born with sight, defined as the neuronal input of the retina to the brain.

The eye represents a higher degree of specialization than is found in any tissue in the body. It is the sensory apparatus for radiant energy of wavelengths between 400 and 750 nm, and it is responsible for some 40% of the total sensory input to the human brain. The main function of the eyeball is to provide an environment that is optimal for the function of its photoreceptive layer, the retina. All of the other complex structures of the eye, including the appendages, diop-

ENCYCLOPEDIA OF HUMAN BIOLOGY, Second Edition, VOLUME 3.

tric system, and protective and motor mechanisms, as well as the vascular and nerve supplies, facilitate the accurate and efficient functioning of the retina.

The anatomic structures mediating human photoreception can be classified under four main headings: the eyeball, its protective apparatus, its motor and supporting apparatus, and the visual pathway.

I. EYEBALL

The human eye, connected to the brain by the optic nerve, is recessed in a bony orbit. Although the eyeball is generally referred to as a globe, it is only approximately spherical, consisting of segments of two spheres placed one in front of the other (Fig. 1). The anterior one-sixth of the surface of the globe comprises the cornea; the posterior five-sixths is occupied by the sclera, which is less curved in the anterior region than it is behind the equator of the eyeball. The change in the curvature of the two spheres at the corneoscleral junction produces a shallow groove, the external scleral sulcus, which, *in vivo*, is filled largely by the conjunctiva and bulbar fascia. Originating from the sides of the pyramidal bony orbit and inserting on the sclera are the four rectus and two oblique extraocular muscles.

The geometric axis of the eye joins the central point of the cornea anteriorly and that of the scleral curvature posteriorly. However, the visual axis joins the

fixation point and the fovea centralis of the retina (Fig. 1). The dimensions of the adult emmetropic human eye are fairly constant; the average diameter is 24 mm sagitally (anterior–posterior), 23 mm vertically, and 23.5 mm horizontally. The sagittal diameter may vary the most, ranging from 21 to 26 mm in the normal eye, depending on its refractive state and, with a high degree of axial myopia, this diameter may be as great as 29 mm. Overall, the male eye is about 0.5 mm larger than that of the female, and the eyes of Blacks are somewhat larger than those of Caucasians. The eye of the newborn is more nearly spherical than is the adult eye, with a sagittal diameter of 16–17 mm, but this increases rapidly to about 23 mm by 3 years of age, and the adult size is attained by puberty. The average eyeball weighs about 7.5 g and its volume is about 6.5 ml.

A. Surface Anatomy

Anteriorly, the white opaque sclera is demarcated from the transparent cornea by the bluish gray junctional zone of the limbus (2 mm wide vertically and 1.5 mm wide horizontally). In this zone, numerous small vessels emerge from the sclera (Fig. 2a). An imaginary line, the spiral of Tillaux, connects the insertions of the four rectus muscles. In front of the insertions of the rectus muscles into the sclera, anterior ciliary arteries and veins pierce the sclera; two arteries accompany each rectus muscle, except the lateral rectus, which has only one associated artery (Fig. 2).

Posteriorly, the optic nerve emerges from the eyeball. The center of the optic nerve is displaced from the center of the geometric axis 3 mm nasally and 1 mm inferior to the posterior pole; this is responsible for the asymmetry of the three tunics (see below) of the eye. The optic nerve is covered by meningeal sheaths and is surrounded by 20 short posterior ciliary arteries and 10 short posterior ciliary nerves, all of which pierce the sclera in a ring shape (Fig. 2b). Two long ciliary nerves enter the sclera in the horizontal plane, one on the nasal and one on the temporal side, 3.6 and 3.9 mm, respectively, from the optic nerve. The four or more vortex veins, which drain the venous system of the uveal tract, emerge 3–6 mm behind the equator. In general, two superior and two inferior vortex veins pierce the sclera obliquely, above and below the optic nerve on each side, in the vertical plane. Each pair is separated from the other by the width of the superior or inferior rectus muscle, respectively. A posterior view of the globe also reveals the

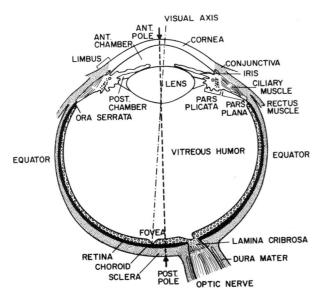

FIGURE 1 Diagrammatic representation of topography of a left human eye as seen in horizontal (meridional) section.

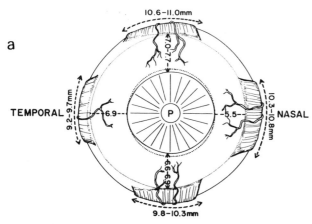

FIGURE 2a General topography of the eyeball seen from the front. The cornea is elliptical in shape, with the limbal zone (dotted line) being narrower horizontally than vertically. The spiral of Tillaux (dotted circle) forms an imaginary line by connecting the insertions of the four rectus muscles, which corresponds approximately to the ora serrata. The width of the rectus muscle tendons (in mm), the distance of their insertions with respect to the corneal margin (in mm), and the vascular supply are also shown. P, pupil.

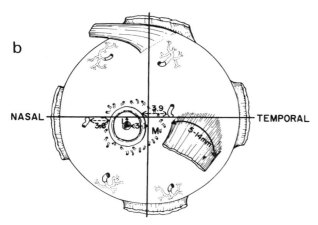

FIGURE 2b Diagrammatic representation showing general topography of the excised eyeball seen from the posterior aspect. The center of the optic nerve is located approximately 3 mm nasal and 1 mm inferior to the posterior pole. Some 20 short posterior ciliary arteries and 10 short posterior ciliary nerves pierce the sclera around the nerve. The two long ciliary arteries are located along the horizontal plane, 3.6 mm nasal and 3.9 mm temporal to the optic nerve. The macula (M) is located between the optic nerve and the curved insertion of the inferior oblique muscle tendon (IO), which may vary from 5 to 14 mm in width at its termination. The oblique scleral insertion (7–18 mm wide) of the superior oblique muscle tendon is temporal to the vertical meridian, mostly behind the equator. The approximate location of the two superior and the two inferior vortex veins is shown in relation to the vertical plane.

insertions of the oblique muscles. The muscular insertion of the inferior oblique is close to the fovea, whereas the tendinous insertion of the superior oblique is about 13 mm posterior to the limbus (Fig. 2b).

B. Tunics of the Eyeball

Bisection of the eyeball along a meridian discloses the fundamental architecture of the globe (Fig. 1). The three tunics of the eyeball, from the outside to the inside, are the fibrous coat of sclera and cornea, the uveal tract, and the retina.

I. Fibrous Coat

a. Sclera

The sclera is the opaque, dull white, collagenous, viscoelastic outer coat of the eye. In addition to maintaining the intraocular pressure, the sclera keeps the ocular dimensions stable for optimum visual function. Enlargement of the eye, as occurs in congenital glaucoma and myopia, is caused by a gradual distention of the collagen bundles in the sclera beyond their elastic limit. Externally, the sclera is covered by the episclera and Tenon's capsule (fascia bulbi), which are connected by fine trabeculae. The anterior portion of the sclera is visible through the bulbar conjunctiva as the "white" of the eye. Broad collagen lamellae of variable diameter and of irregular distribution interweave with elastic fibers in an intricate fashion in the scleral stroma, thus increasing the tensile strength of the globe. The opacity of the sclera can be attributed to its high water content, the haphazard arrangement and variable diameter of its collagen fibrils, and a low concentration of ground substance. The sclera contains sparse fibroblasts, and the orientation of the collagen bundles varies in different regions. The innermost layer of the sclera, which separates it from the choroid and ciliary body, is called the lamina fusca, a layer rich in elastic tissue and pigmented cells. Its fine collagen bundles merge with the fibrous lamellae of the suprachoroidea and supraciliaris.

The sclera has two large openings, the anterior and posterior scleral foramina, and numerous smaller openings through which ocular nerves and blood vessels pass. The anterior foramen is not a true opening because there is structural continuity between the cornea and the sclera at the limbus. The posterior scleral foramen is formed by a sieve-like apparatus, the lamina cribrosa, through which axons of the optic nerve pass (Fig. 1).

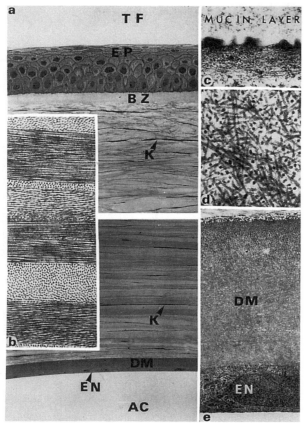

FIGURE 3   (a) Light micrograph of cornea showing its layered architecture. EP, multilayered squamous epithelium; BZ, Bowman's zone of stroma; K, keratocytes in anterior and posterior stroma; DM, Descemet's membrane; EN, endothelium. The cornea is bathed anteriorly by tear fluid (TF) and posteriorly by aqueous humor of the anterior chamber (AC) ×190. (b) Corneal stroma showing regular and parallel arrangement of collagen fibrils in individual lamellae that cross each other at right angles. ×12,440. (c) A superficial cell of corneal epithelium showing microvillous surface projections covered by mucinous layer of the tear film. ×15,890. (d) Bowman's zone of stroma constituted by a feltwork arrangement of collagen fibrils. ×19,140. (e) Descemet's membrane (DM) is the thickened basement membrane of the endothelium (EN). ×3470. Figures 3b–e are electron micrographs.

b. Cornea

The cornea consists of a clear, transparent, avascular, viscoelastic tissue with a smooth, convex external surface and a concave internal surface (Fig. 3). The main function of the cornea is optical; it forms the principal refracting surface of the dioptric system of the eye, accounting for 70% of total refractive power (45 diopters). When the eyelids are open, the cornea is separated from the air only by the precorneal tear film, a physiologic secretion that covers the external surface of the corneal and conjunctival epithelium.

By filling minor surface irregularities, the tear film provides a smooth air–cornea interface for refraction. The precorneal tear layer is the main vehicle for the supply of nourishment to the cornea and for the maintenance of a nonkeratinized state of the corneal epithelium.

Structurally, the cornea consists of four layers: an epithelial layer with its basement membrane, the stroma with its anterior-modified Bowman's zone, Descemet's membrane (the thick basement membrane of the endothelium), and the corneal endothelium (Fig. 3).

The corneal epithelium is the most sensitive and probably the most highly organized stratified squamous epithelium in the body. It is composed of five to six cell layers. The superficial layers consist of flattened, polygonal cells, which disintegrate or are wiped away with the movement of the lids and by the precorneal tear film. Approximately 14% of these surface cells are exfoliated each day. The middle zone of the epithelium contains polygonal or "wing" cells; this zone is two or three cell layers thick and is derived from the mitotic activity of the basal cells. The basal zone, the deepest layer of the corneal epithelium, is responsible for its weekly turnover. The intercellular spaces of the corneal epithelium are filled with ground substance, and adjacent cells are attached by numerous desmosomal junctions. The cytoplasm of the cells is rich in tonofilaments, which are oriented along the long axis of the cells and are condensed at the desmosomes. The tall columnar (20 × 10 μm) cells of the basal (germinal) layer are attached to their basement membrane by strong hemidesmosomal junctions. At the limbus, the corneal epithelium is continuous with that of the conjunctiva.

The corneal stroma is predominantly collagenous and constitutes approximately 85% of the corneal thickness (Fig. 3). Anteriorly, it is composed of a homogenous, acellular, collagenous latticework called Bowman's layer, or zone (8–14 μm thick), that is highly resilient and resists deformation and trauma as well as the passage of bacterial organisms and foreign bodies. Bowman's zone is perforated in many places by unmyelinated nerves in transit to the corneal epithelium. The corneal stroma deeper to Bowman's zone is formed by parallel lamallae (numbering 200–250) of collagen fibrils. The overall arrangement of the collagen fibrils mimics a three-dimensional diffraction grating, with individual fibrils separated by less than half the wavelength of light, thus providing the structural basis for the transparency of the cornea. If this orderly arrangement of collagen fibrils is destroyed by

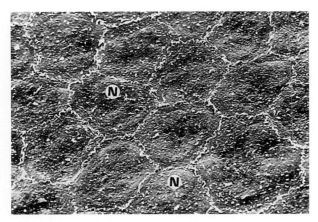

FIGURE 4 Scanning electron micrograph of normal human corneal endothelium. Individual cells are arranged regularly and are approximately hexagonal in shape and contain an oval nucleus (N). Microvillous projections are present on the surface and especially along the cell borders. ×1455.

Because the endothelial cells normally do not undergo mitosis after birth, cell loss is generally compensated for by the spreading and thinning of adjacent cells; this loss occurs with age and after trauma, inflammation, and degenerative insults. The structural and functional integrity of the corneal endothelium is vital to the maintenance of corneal transparency.

c. Corneoscleral Limbus

At the limbus, the corneal epithelium loses its regular structure and continues with the conjunctival epithelium. The compact transitional zone of corneoscleral tissue that forms the midlimbus is traversed by intrascleral vascular plexuses and nerves (Fig. 5). The limbal vessels provide nourishment for the peripheral, avascular cornea. The deep or inner limbus contains the canal of Schlemm and the trabecular meshwork, which pro-

pathologic processes, the architecture of the stromal fibrils is not renewed, and replacement occurs by the formation of scar tissue. The stromal ground substance consists primarily of two sulfated glycosaminoglycans, keratan sulfate and chondroitin-4-sulfate, which are secreted by native stromal keratocytes. The sparsely distributed keratocytes or stromal fibroblasts that constitute 3–5% of the stromal volume are flattened, thin, irregularly shaped cells.

Descemet's membrane, one of the thickest basement membranes in the body, forms the posterior boundary of the corneal stroma (Fig. 3). This layer, composed primarily of collagen, polysaccharides, and glycoproteins, is secreted by the corneal endothelium. The anterior (embryonic) zone is banded, but the posterior region is amorphous. The thickness of Descemet's membrane increases with age; it measures 3 μm at birth and 9 μm during adulthood.

The monolayer of corneal endothelium that lines the posterior surface of Descemet's membrane demarcates the anterior boundary of the aqueous cavity of the anterior chamber (Fig. 3). Approximately 500,000 closely fitted, hexagonal, flattened cells, which are 18–20 μm wide and 5–6 μm thick, comprise the corneal endothelium (Fig. 4). The cells contain a centrally placed, oval nucleus, as well as abundant mitochondria and other organelles, signifying the active metabolic state of the cells. Junctional complexes, consisting of zonulae occludentes and gap junctions, connect adjacent endothelial cells and preclude the influx of aqueous humor into the corneal stroma, thus maintaining corneal turgescence and transparency.

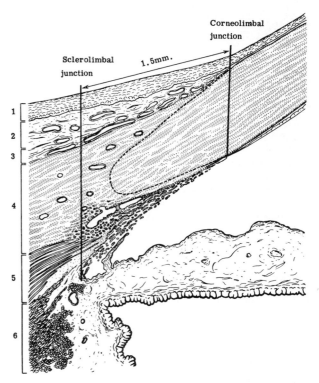

FIGURE 5 Diagrammatic representation of the limbus showing various structures from superficial to deep regions. (1) Conjunctiva, (2) conjunctival stroma, (3) Tenon's capsule and episclera, (4) limbal or corneoscleral stroma containing intrascleral plexus of veins and collector channels from Schlemm's canal, (5) meridional portion of ciliary muscle, and (6) radial and circular portions of ciliary muscle. The so-called pathologist's limbus is about a 1.5-mm-wide zone extending posteriorly from a line joining the peripheral termination of Bowman's zone and Descemet's membrane of the cornea. The histologist's limbus (dotted line) is represented by a cone-shaped termination of the corneal lamellae into the sclera.

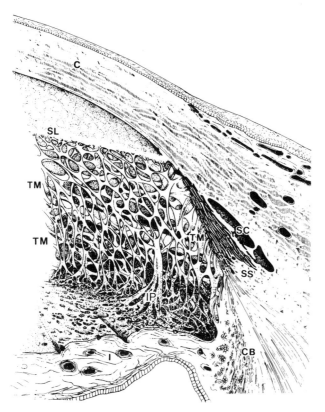

FIGURE 6   Semidiagrammatic representation of the angle of the anterior chamber of the human eye, as seen in a composite sectional view and three-dimensional gonioscopic view. The trabecular meshwork (TM) is made up of superimposed, circumferentially oriented fibrocellular sheets. Perforations in individual sheets (intratrabecular spaces) and spaces between adjacent sheets (intertrabecular spaces) provide the pathway for bulk flow of aqueous humor from the angle of the anterior chamber. The aqueous humor is drained finally into the canal of Schlemm (SC) and collector channels (CC). C, cornea; SL, line of Schwalbe; SS, scleral spur; CB, ciliary body; IP, iris process; I, iris.

vide the major pathway for the drainage of aqueous humor. Dynamic and structural alterations in this region are responsible for decreased aqueous outflow which leads to elevated intraocular pressure.

The trabecular meshwork is composed of a lattice of superimposed, perforated sheets or beams that extend from the scleral spur and the anterior face of the ciliary muscle to the line of Schwalbe and deep corneal lamellae (Fig. 6). The core of each beam is formed by fibrous tissue that is covered by flattened trabecular cells. Changes in the tone of the ciliary muscle can alter the porosity of the meshwork. In aging and glaucomatous eyes, degenerative changes in the trabecular beams and increased accumulation of extracellular materials contribute to increased resistance to aqueous outflow.

The canal of Schlemm, located in the inner part of the limbus, is a toroidal structure that is supported on its inner aspect by the trabecular meshwork and on its outer aspect by the compact corneoscleral tissue. It is lined by a continuous single layer of endothelial cells, which contain giant vacuoles (Fig. 7). This unique structural configuration forms transcellular channels and is implicated in the bulk flow of aqueous humor into the canal. After draining into a deep scleral plexus, aqueous humor enters the venous circulation via the intra- and episcleral veins (Fig. 8). Several collector channels (aqueous veins of Ascher) bypass the deep scleral plexus and terminate directly in the subconjunctival veins after coursing through the sclera (Fig. 8).

2. Uveal Tract

The uveal tract forms the pigmented vascularized tunic of the eye. It is divided into three discrete regions: the choroid, ciliary body, and iris (Fig. 1).

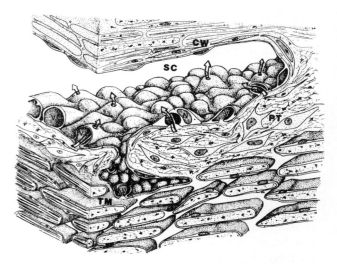

FIGURE 7   A composite three-dimensional schematic rendering of the walls of Schlemm's canal (SC) and adjacent trabecular meshwork (TM). The spindle-shaped endothelial cells lining the trabecular wall of Schlemm's canal are characterized by luminal bulges corresponding to unique macrovacuolar configurations (V) and nuclei (N). The macrovacuolar configurations are formed by surface invaginations on the basal aspect of individual cells which gradually enlarge to open eventually on the apical aspect of the cell surface, thus forming transcellular channels (arrows) for the bulk flow of aqueous humor down a pressure gradient. The endothelial lining of the trabecular wall is supported by a variable zone of cell-rich pericanalicular tissue (PT) below which lie the organized superimposed trabecular sheets having inter- and intratrabecular spaces that allow the flow of aqueous humor from the anterior chamber to the canal of Schlemm. The compact corneoscleral wall (CW) of Schlemm's canal is formed by the lamellar arrangement of collagen and elastic tissue.

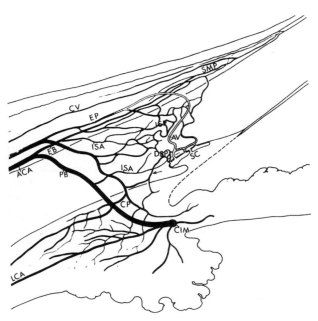

FIGURE 8 The limbal and ciliary vasculature shown diagrammatically. The deep scleral plexus (DSP) drains aqueous humor from the canal of Schlemm (SC) and is connected anteriorly to the intra- and episcleral plexuses of veins (ISP and EP). The aqueous veins (AV) arise directly from the canal of Schlemm or occasionally from the deep scleral plexus and join the episcleral veins. ISA, intrascleral arterial branches; PB, major perforating and EB, episcleral branches of anterior ciliary artery (ACA); CIM, circulus iridis major; SMP, superficial marginal plexus; CP, ciliary plexus; LCA, long posterior ciliary artery; CV, conjunctival vessel.

a. Choroid

The choroid has four main layers: the suprachoroidea, the stroma, the choriocapillaris, and Bruch's membrane. The suprachoroidea is a superficial layer of the choroid made up of collagen lamellae that blend into the sclera. The potential space between the sclera and choroid (suprachoroidal space) is traversed by the long and short ciliary arteries that supply the uveal tract. Before leaving the sclera, the exiting veins converge to form ampullae that drain into the vortex veins. The choroidal stroma consists of an extensive network of vessels as well as loose collagenous and elastic tissue, fibroblasts, melanocytes, lymphocytes, and smooth muscle "stars." The choriocapillaris, the capillary layer of the choroid, is anterior to the choroidal stroma. It consists of the largest vessels, which are flattened, closely packed, and highly permeable (fenestrated). These vessels supply nutrients and oxygen to the retinal pigment epithelium and the outer sensory retina and, exclusively, to the vessel-free area of the fovea. Bruch's membrane, the innermost com-

ponent of the choroid, has an anterior cuticular layer, which is derived from the retinal pigment epithelium, and a posterior layer, which functions as the basement membrane of the choriocapillaris. The core is formed by a network of collagen and elastic tissue.

b. Ciliary Body

The ciliary body is an anterior continuation of the choroid and retina and, as such, is divisible into uveal and neuroepithelial portions. The uveal portion includes the supraciliaris, the ciliary muscle, and the vessel layer.

The supraciliaris is a continuation of the suprachoroidea and is analogous to the suprachoroidal space in that it is only potentially present, coming into existence only in certain pathologic conditions. The supraciliaris has fewer collagen lamellae and less fibroelastic tissue than does the suprachoroidea. This space, along with the uveovortex veins, provides auxiliary drainage for aqueous humor that passes through the extracellular spaces of the anterior face of the ciliary body. The ciliary muscle is a ring of smooth muscle tissue that consists of longitudinal (meridional), radial, and circumferential fibers. This muscle has an integral role in slackening the suspensory ligaments of the lens, which permits an increase in the convexity of the lens (accommodation). The vessel layer of the ciliary body supplies blood to the ciliary muscle and consists almost exclusively of fenestrated capillaries.

The neuroepithelial portion consists of the double-layered ciliary epithelium: A pigmented layer, which is continuous with the retinal pigment epithelium, and the cuboidal, or low columnar, nonpigmented epithelium, which is a direct continuation of the sensory retina. The apical surfaces of the two epithelial layers are juxtaposed. Aqueous humor is secreted by the epithelial cells of the ciliary processes, which are supplied by fenestrated capillaries. The ciliary epithelium is smooth posteriorly (this portion is called the pars plana), but becomes markedly convoluted anteriorly (pars plicata) and forms the finger-like projections of ciliary processes.

c. Iris

The iris is located in front of the lens and is a continuation of the ciliary body. It forms a delicate diaphragm between the anterior and the posterior chambers of the eye (Fig. 9). The color of the iris varies among individuals, depending on the amount of pigment in the stromal cells. Stromal pigmentation increases rapidly during the first year of life, and hence

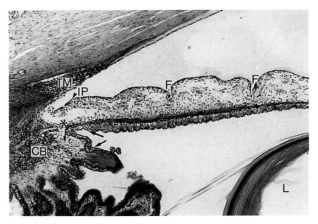

FIGURE 9 Transverse section through the mid and peripheral part of the iris. The dilator muscle ends (asterisk) near the iris root which is the thinnest part of the iris. The posterior layer of the iris epithelium begins to lose its pigment and becomes continuous with the nonpigmented posterior epithelial layer of the pars plicata; the transitional zone is marked by arrows. Due to partial dilation of the pupil, concentric furrows (F) are present anteriorly. The anterior border layer is partly continuous with the root of the iris process (IP). TM, trabecular meshwork; CB, ciliary body; L, lens. Partially bleached section. Photomicrography ×39.

many infants who were born "blue-eyed" gradually acquire darker pigmentation. The central aperture of the iris (the pupil) controls the amount of light entering the eye and provides an opening for the free flow of aqueous humor from the posterior chamber to the anterior chamber. The medial portion of the iris rests directly on the anterior surface of the lens.

The iris has two main components: the uveal portion (stroma) and the neuroepithelial portion. The stroma consists of a narrow anterior avascular border layer, with crypts of variable size, and a wider vascularized posterior stromal layer. The anterior border layer is formed by interlacing fibroblasts and pigmented melanocytes in an extracellular ground substance. In addition to these randomly distributed cells, clump cells, mast cells, macrophages, and lymphocytes are also present in the stroma. The collagen fibrils are arranged in a lattice pattern, which provides a scaffold during pupillary dilatation and constriction.

The posterior surface of the iris (neuroepithelial portion) is formed by a layer of pigmented cells, which are a continuation of the nonpigmented layer of the ciliary epithelium (Fig. 9). The myoepithelial layer of the iris, which constitutes the dilator pupillae muscle, is a continuation of the pigmented layer of the ciliary epithelium. The smooth muscle of the sphincter pupillae is located in the iris stroma around the pupil and controls the entry of light to the retina.

3. Retina

The retina is derived from neuroectoderm and represents an extension of the brain in the eye. This thin, delicate layer of stratified nervous tissue is composed of the retinal pigment epithelium and the sensory retina (Fig. 10). [*See* Retina.]

a. Retinal Pigment Epithelium

The retinal pigment epithelium is a single layer of hexagonal, cuboidal epithelial cells, 14–60 μm wide and 10–14 μm long, depending on the region of the retina. The cells are narrow, tall, and uniform in the foveal region, but they are wide and irregular toward the ora serrata. The retinal pigment epithelium extends from the margin of the optic nerve posteriorly to the ciliary epithelium anteriorly. The apices of the cells are joined by zonulae occludentes and, together with the cell borders, form the external blood–retinal barrier. The apical region of the cell surface is thrown into microvillous projections, which enclasp the outer segments of the photoreceptors. Granules of melanin, which are responsible for absorbing excess light, are concentrated mainly in the apical region of the cells. The primary functions of the cells include transport of nutrients from the choroidal vasculature to the photoreceptor layer, removal of metabolites, and active phagocytosis and recycling of photoreceptor disks that are shed daily from the outer segments of the photoreceptors. The cells secrete a basal lamina (the internal layer of Bruch's membrane) to which they are attached firmly. Because the sensory retina and the pigment epithelium are attached to each other only through the extracellular matrix, a separation of the two layers occurs in clinical conditions of retinal detachment.

b. Sensory Retina

The sensory retina consists of a layer of photoreceptor cells, the axons of which synapse with intermediate cells for modulation of the photic response. Impulses modified by these modulator cells are relayed to an inner layer of ganglion cells, which transmit spike discharges (action potentials) through the optic nerve to the brain.

The general architecture of the retina is represented by 10 layers, including the retinal pigment epithelium (Figs. 10 and 11). The neuronal layers are not strictly distinct and should be considered merely as a theoretical paradigm of retinal organization:

1. The retinal pigment epithelium (described earlier).

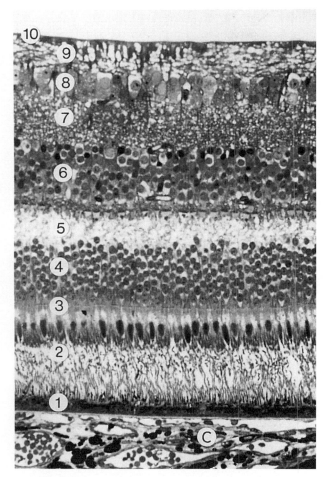

FIGURE 10 Transverse section of retina showing pigment epithelium (1) attached to choroid (C) through Bruch's membrane and sensory retina consisting of (2) photoreceptor layer, (3) external limiting membrane, (4) outer nuclear layer, (5) outer plexiform layer, (6) inner nuclear layer, (7) inner plexiform layer, (8) ganglion cell layer, (9) nerve fiber layer, and (10) internal limiting membrane. ×280.

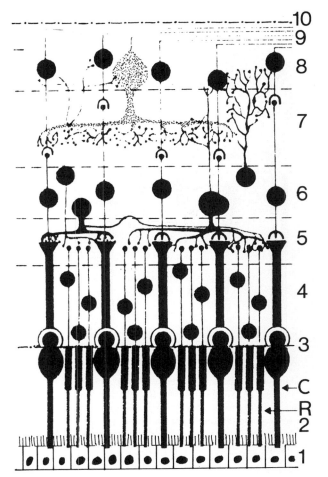

FIGURE 11 Diagrammatic representation of elements comprising the retina. (1) Pigment epithelium, (2) photoreceptor layer consisting of rods (R) and cones (C), (3) external limiting membrane, (4) outer nuclear layer, (5) outer plexiform layer, (6) inner nuclear layer, (7) inner plexiform layer, (8) ganglion cell layer, (9) nerve fiber layer, and (10) internal-limiting membrane.

2. The photoreceptor layer of the retina. These neuronal cells are distinguished functionally and structurally as rods and cones. Each of these cells have three components: the end organ, consisting of a stack of membranous disks containing the visual pigment (the outer segments); the cell body with the nuclei; and the inner terminal fibers (Fig. 12). The 110–125 million rod photoreceptors are concentrated near the retinal periphery, whereas the cones (6.3–6.8 million), which are involved in photopic vision and color perception, have their maximum density at the fovea in the center of the macula (Fig. 13).

3. The external limiting membrane is formed by the attachment of photoreceptors to supporting elements called Muller cells.

4. The outer nuclear layer consists of densely packed cell bodies of photoreceptors with their nuclei and cytoplasm.

5. The outer plexiform layer is defined by the synapses of first- and second-order retinal neurons. In this region, dendrites of the bipolar, horizontal cells and associated cells, which modulate and transmit stimuli from photoreceptors, form synapses with photoreceptors.

6. The inner nuclear layer consists of nuclei belonging to bipolar and associated cells; it also contains the deepest capillaries of the central retinal artery.

7. The inner plexiform layer is defined by synapses

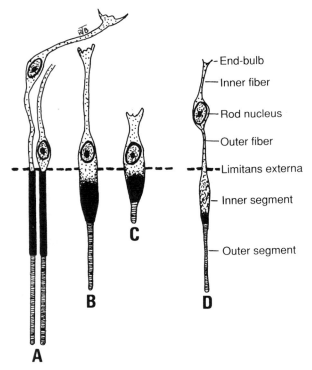

FIGURE 12 Schematic representation of photoreceptors of human retina. (A) Cones from the foveola, (B) cone from midway between the ora serrata and the optic disc, (C) cone from near the ora serrata, and (D) rod. Note the variations in the shape and size of these photoreceptor elements depending on their location in the retina.

between the second (intermediate) and the third (cerebral) retinal neurons located in this layer. The axons of bipolar and amacrine cells synapse with dendrites from ganglion cells.

8. The ganglion cell layer contains cell bodies (designated as M or parasol, P1 or midget, and P2 or small bistratified) of cerebral neurons and their supporting neuroglial cells.

9. The nerve fiber layer is composed of axons of the ganglion cells, which run parallel to the retinal surface and converge at the optic disk.

10. The internal limiting membrane is a basement membrane of both retinal and vitreal elements. Muller cells extend from the internal to the external limiting membranes and, together with astrocytes and accessory glia, provide skeletal support for the retina.

The structural organization of the retina shows variations that depend on the topography. The retina is thickest (about 0.5 mm) at the margin of the optic disc, and thinnest at the fovea (between 0.2 and 0.3

mm) and at the ora serrata (0.1 mm). The central retina, a specialized region approximately 6 mm in diameter (this includes 15° of the visual field), has a central region, the fovea (1.5 mm in diameter, 5° of the visual field). Because of its yellow tinge, this central area is called the macula lutea. The cell layers present at the fovea are the pigment epithelium, the photoreceptors (exclusively cones, which are tall and slender), the outer nuclear layer, and the outer plexiform layer, which is designated as the layer of Henle (Fig. 13). The nerve fibers arising from the fovea are directed to the temporal region of the optic disc and constitute the papillomacular bundle (Fig. 14). At the edge of the fovea, the retina has full thickness (abun-

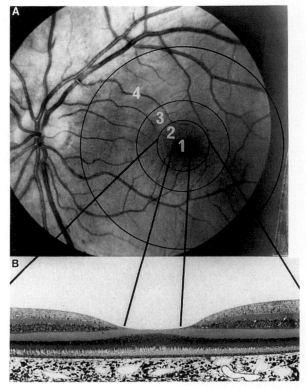

FIGURE 13 (A) Fundus photograph of left human eye showing topographic demarcation of the area centralis that measures 5.5–6 mm in diameter (outermost circle) and its subdivisions (inner circles). (From an anatomic standpoint, the zones demarcated are in fact horizontally elliptical rather than circular as shown diagrammatically here.) The central area of the macular region is represented by the fovea centralis (2) approximately 1.85 mm in diameter, which has a central pit, the foveola (1), 0.35 mm in diameter. The anatomically distinguishable retinal belts surrounding the fovea centralis are the parafovea (3), 0.5 mm wide, and the perifovea (4), 1.5 mm wide. (B) Transverse section of foveal retina matched topographically to the fundus photograph shown in A. Photomicrograph ×43.

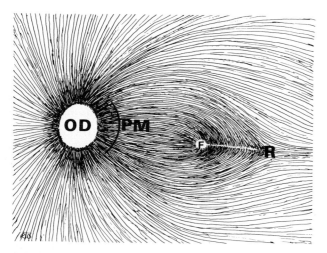

FIGURE 14   Schematic representation of the course and arrangement of the nerve fibers to the optic nerve. OD, optic disk; PM, papillo-macular bundle; F, foveola; R, raphe.

dant ganglion cells are multilayered, Henle's layer is thick, and the cones are less slender and are intermingled with rods).

Except in the foveola (0.4 mm diameter), the retina is permeated by blood vessels to the level of the innermost part of the outer nuclear layer (Fig. 10). The structural and functional characteristics of these vessels are such that a competent blood–retinal barrier is normally maintained.

C. Chambers of the Eye

The eyeball contains three distinct chambers: the anterior chamber, posterior chamber, and vitreous cavity (Fig. 1).

1. Anterior Chamber

The anterior chamber is an ellipsoidal cavity bounded anteriorly by the corneal endothelium and peripherally by the trabecular meshwork (Figs. 1, 5, and 6). Posteriorly, the anterior chamber is demarcated by the anterior surface of the iris and the pupillary portion of the lens. The apex of the chamber lies at the anterior face of the ciliary body. The average volume of aqueous humor in the anterior chamber is about 0.2 ml in the human eye.

2. Posterior Chamber

The posterior chamber is the space bounded anteriorly by the posterior surface of the iris, laterally by the ciliary processes, posteriorly by the anterior face of the vitreous, and medially by the lens equator (Fig. 15). It contains the freshly secreted aqueous humor (0.5 μl) formed by the ciliary processes.

3. Vitreous Cavity

The vitreous cavity of the eye occupies almost four-fifths of the total volume of the globe (5.2 ml). It is filled with a transparent, avascular, gel-like substance of embryonic microcollagenous structure and is rich in hyaluronic acid. The vitreous is undoubtedly the most fragile connective tissue structure in the body. The boundaries of the vitreous cavity are demarcated by the internal surface of the retina and optic nerve posteriorly, by the pars plana laterally, and by the posterior surface of the lens anteriorly. The main sites of attachment for the vitreous fibrils are the ora serrata, the optic nerve, and the macular region (Fig. 16). Age-related degeneration, especially in myopic eyes, causes "floaters" to form in the vitreous.

D. Lens

The human lens is a transparent, biconvex, semisolid, avascular structure (9 mm in diameter and 4 mm in central thickness) located behind the iris in a shallow depression of the anterior vitreous (Fig. 1). A system of zonular fibers that extend from the ciliary epithelium to the equatorial capsule holds the lens in place.

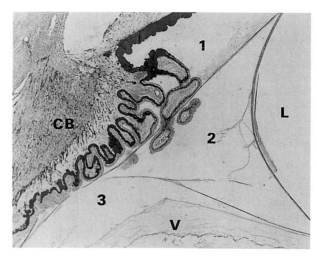

FIGURE 15   The three subdivisions of the posterior chamber in the human eye as seen in the meridional section of the globe. (1) Posterior chamber proper or prezonular space, (2) zonular circumlental space or canal of Hanover, and (3) retrozonular space or canal of Petit. Note the widening of the latter toward the ciliary body (CB) and narrowing toward the lens (L). V, condensed anterior vitreous. Photomicrograph ×30.

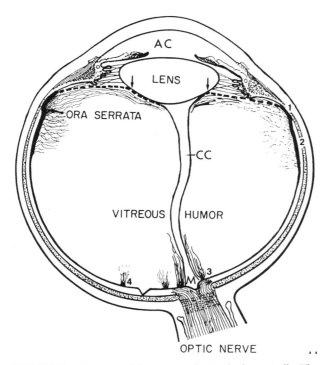

which continue as far as the equator. The tapering and elongated lens cells or fibers (7–10 μm long, 8–10 μm wide, and 2–5 μm thick) comprise the lens substance; mature, anucleated lens fibers evolve from young, nucleated epithelial cells at the equator. The nuclei of the cells migrate to a location somewhat anterior to the equator to form the nuclear bow, and they eventually disintegrate. The superficial cortical cells are rich in microfilaments, especially actin, which have a major role in the accommodative process because of their inherent elasticity. With the superimposition of new cells, the older fibers progressively become displaced toward the center of the lens. Because the lens cells or fibers grow from the entire equator toward the anterior and posterior poles of the lens, they meet along radiating lines or sutures. Adjacent fibers interdigitate by forming complex ball-and-socket junctions.

Several zones or "nuclei" may be distinguished in the adult lens. The embryonic nucleus is formed by the posterior primordial lens cells, which elongate rapidly to fill the lens vesicle and lose their cell nuclei;

FIGURE 16 Anatomy of the vitreous depicted schematically. The vitreous body conforms to the shape of the cavity in which it is contained. Centrally, it is traversed by Cloquet's canal (CC) containing hyaloidean vitreous that extends in an S-shaped fashion from a tortuous funnel-shaped expansion on the posterior aspect of the lens to the optic nerve head with a smaller funnel-shaped expansion, the so-called area of Martegiani (M). Anteriorly, the vitreous face is in contact with the posterior lens surface separated only by the capillary space of Berger which extends peripherally into the retrozonular space of the posterior chamber, the so-called canal of Petit. In young eyes, the vitreous face is adherent to the posterior lens capsule in the region of the hyaloideocapsular ligament (arrows). The vitreous body is firmly attached (solid line) to about a 2-mm zone of the pars plana (1) and continues posteriorly to about a 4-mm span of the peripheral retina (2). The dotted line anteriorly represents the anterior vitreous face which, in places, is in contact with the posterior zonular fibers. Posteriorly, the vitreous face is attached to the edge of the optic disk (3), although not as firmly as at the vitreous base. A similar annular attachment, 3–4 mm in diameter, is seen in young eyes at the macular region (4). AC, anterior chamber.

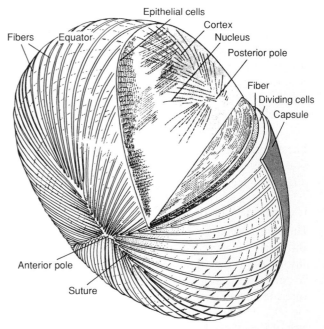

FIGURE 17 Schematic representation of the human lens. Growth of the lens occurs from the nuclear bow region near the equator. The new cells elongate and form lens fibers that wrap around the periphery of the lens and meet at the sutures where complex interdigitations occur. The outer fibers are nucleated and hexagonal in shape; as they are displaced inward by newer cortical fibers, they gradually lose their nuclei and other intracellular organelles. [From R. Van Heyningen (1975). *Sci. Am.* **233**, 70–81. Copyright Scientific American, Inc. All rights reserved.]

The lens is composed of three parts: the lens capsule, which exhibits regional variations in its thickness; the lens epithelium; and the lens substance, which consists of cortical and nuclear fibers (Fig. 17). The lens capsule (one of the thickest basal laminae in the body) completely encloses the lens substance. It is a thick, transparent, smooth, reflective, elastic, periodic acid Schiff-positive, acellular, collagenous basement membrane. Anteriorly beneath the capsule there is a single layer of cuboidal cells (the lens epithelium), hence, there is no epithelial cell layer beneath the posterior capsule as there is present anteriorly. The fetal nucleus has two distinctive Y-shaped sutures (erect anteriorly and inverted posteriorly) to which the lens fibers are joined. The sutures formed after birth become increasingly complex as the eye ages. The adult nucleus is formed by older lens fibers, which have an increased density. Histologically, the outlines of individual cells in the lens nucleus are indistinguishable. On slit-lamp biomicroscopy, a zone of external fibers shows a refractive index less than that of the nuclear zone and is distinguished as the lens cortex.

The zonular fibers (suspensory ligaments) form a circumferential suspensory apparatus, which maintains the lens in position (Fig. 15). Contraction of the ciliary muscle loosens the tension on the zonular fibers, which induces a greater curvature in the lens; this increases the accommodative power of the lens (by up to 17 diopters in the young eye).

II. PROTECTIVE APPARATUS

A. Orbit

The eyeball is housed and is freely mobile in the bony cavity, which is filled with a semispecialized fibrofatty tissue (Fig. 18). The orbit is formed by six bones: the

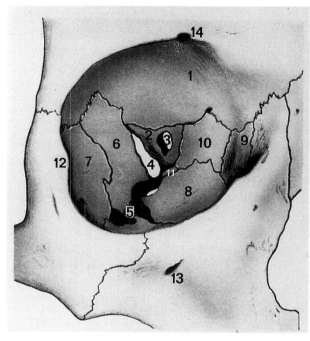

FIGURE 19   The right orbit viewed along its axis. (1) Frontal bone, (2) lesser wing of sphenoid, (3) optic canal, (4) superior orbital fissure, (5) inferior orbital fissure, (6) greater wing of sphenoid, (7) zygomatic arch, (8) orbital plate of maxillary bone, (9) lacrimal bone and fossa, (10) ethmoid bone, (11) orbital process of palatine bone, (12) lateral orbital tubercle, (13) inferior orbital foramen, and (14) supraorbital notch. [From P. Henkind (1982). The Eye: Anatomy and Measurements, Lippincott.]

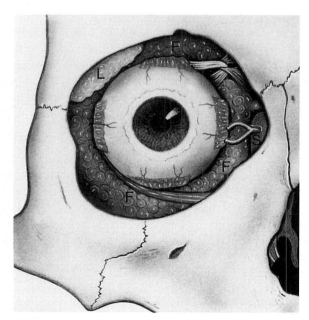

FIGURE 18   Globe in the orbit supported by orbital fat (F) and extraocular muscles. The lacrimal system, consisting of gland (L) and sac (S), is also shown. [From P. Henkind (1982). The Eye: Anatomy and Measurements, Lippincott.]

maxilla and the palatine, frontal, sphenoid, zygomatic, and ethmoid bones (Fig. 19). The orbital cavity has a quadrilateral pyramid shape, with a triangular pyramid near its apex. The two principal posterior openings of the bony orbit, the optic foramen and the superior orbital fissure, allow passage of the optic nerve and blood vessels.

B. Eyelids

The eyelids consist of skin, muscles, and fibrous tissue (Fig. 20). They are lined internally by the mucous membrane of the conjunctiva and externally by thin skin without any subcutaneous fat. The muscles of the eyelids are the striated orbicularis oculi and the levator palpebrae superioris, as well as the small palpebral smooth muscle of Muller. Internally, a compact, fibrous tissue (the tarsal plate) contains modified sebaceous glands (the Meibomian glands), which open at the lid margin. The glands associated with the eyelashes are the glands of Moll (sudiferous) and of Zeis (sebaceous).

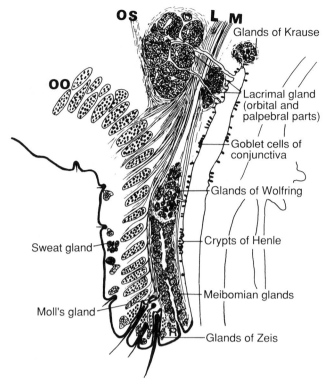

FIGURE 20 Diagrammatic representation of upper eyelid in vertical section. L, levator muscle; M, Muller muscle; OO, orbicularis oculi; R, muscle of Riolan; OS, orbital septum.

C. Conjunctiva

The conjunctiva is a thin, translucent mucous membrane that lines the posterior surface of the eyelids (palpebral conjunctiva), is reflected on itself to form the fornices, and lines the anterior surface of the sclera (Figs. 20 and 21). At the lid margin, the conjunctiva joins the skin (the mucocutaneous junction), and at

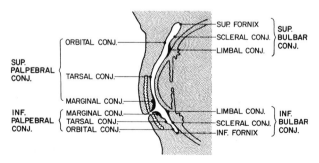

FIGURE 21 Diagrammatic representation of the conjunctival sac and its regional topography as seen in a vertical section of a closed eye.

the corneal margin, it is continuous with the corneal epithelium (Fig. 20). At the lacrimal punctum, the mucous membrane is continuous with the membrane lining the lacrimal canaliculi. On the nasal side of the globe and at the edge of the caruncle, the conjunctiva forms a crescent (plica semilunaris), which represents the third eyelid or nictating membrane of many animals. The bulbar conjunctiva is connected loosely to the underlying fascia bulbi, which, in turn, is attached loosely to the underlying episclera and sclera.

The epithelial lining of the conjunctiva and the cornea form the external protective covering of the open eye. Both are kept moist constantly by the tears, which prevent keratinization. The conjunctiva, like other mucous membranes in the body, consists of epithelium and substantia propria. The thickness of the epithelium (two to five cell layers) varies from region to region, being thinnest in the bulbar region and thickest at the mucocutaneous junction of the lid. Melanin granules are present in the cuboidal basal cells and occasionally in the middle layer of the epithelium. Mucus-secreting goblet cells are distributed irregularly in the conjunctiva but are most abundant in the fornices and sparse in the bulbar conjunctiva. The connective tissue of the stroma contains blood vessels, nerves, and glands; its compactness varies with the topography. The superficial layer is a fine, fibrous reticular network, which, after the age of 3 months, contains nodules of lymphoid tissue. The vascularized deeper layers are made up of irregularly arranged bundles of collagen and some elastic tissue, as well as randomly distributed fibroblasts, melanocytes, mast cells, lymphocytes, and plasma cells. The fibrous tissue is continuous with the attached margin of the upper and lower tarsal plates and contains the smooth palpebral muscle of Muller.

D. Lacrimal Apparatus

The main lacrimal gland (of the tubuloalveolar type) is located in the anterior lateral portion of the roof of the orbit (Fig. 18) and, together with its palpebral portions, opens into the superior fornix through several excretory ducts. The accessory lacrimal glands of Krause and Wolfring are distributed irregularly in the conjunctival stroma (Fig. 20). The blinking action of the lids distributes the tear layer over the keratoconjunctival surface and sweeps away foreign particles. The drainage system for the tears consists of the puncta, canaliculi, lacrimal sac, and the nasolacrimal ducts that open into the inferior nasal meatus (Fig. 22).

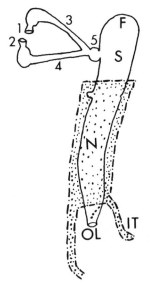

FIGURE 22   The lacrimal drainage system. (1) Superior punctum, (2) inferior punctum, (3) superior canaliculus, (4) inferior canaliculus, and (5) common canaliculus (sinus of Maier). F, fundus (5 mm); S, sac (8 mm); N, nasolacrimal duct (intraosseus part = 12–15 mm; intrameatal part = 5–6 mm); OL, osteum lacrimale opening in the inferior meatus; the constriction is formed by the plica lacrimalis or so-called valve of Hasner. IT, inferior turbinate bone.

III. MOTOR AND SUPPORTING APPARATUS

A. Extraocular Muscles

The motor apparatus of the eye is responsible for the accurate and rapid mobility of the globe. Six striated extraocular (extrinsic) muscles comprise the most important components of the motor apparatus (Figs. 2, 3, and 23). These muscles differ from other striated muscles in their unique two-fiber system: the fast-acting white fibers are involved in rapid ocular movements, and the slow-acting red muscles maintain a steady gaze. The extraocular muscles are also unusual in their high ratio of nerve to muscle fiber; this facilitates the precise coordination of muscle activity that is essential for binocular vision.

Four of the extrinsic muscles (medial, lateral, inferior, and superior rectus muscles; Figs. 2b and 23) are concerned primarily with inward, outward, downward, and upward eye movements, respectively. All of these muscles have a common origin in a ring-shaped fibrous structure, the annulus of Zinn, that surrounds the optic nerve. The two oblique extrinsic muscles, the superior and inferior, are involved in intermediate movements of the globe. The superior oblique passes from the upper side of the posterior orbit through a fibro-cartilagenous pulley before inserting into the sclera. The inferior oblique muscle inserts directly at about the posterior pole of the eyeball. [*See* Eye Movements.]

B. Vasculature of the Eye and Its Adenxa

The ophthalmic division of the internal carotid artery is the principal source of nutrients for the eye and the contents of the orbit. The eyelids and conjunctiva are well supplied with anastomotic branches of both the ophthalmic artery and divisions of the external carotid artery.

1. Ophthalmic Artery

The ophthalmic artery is the first intracranial branch of the internal carotid artery, after it passes through the cavernous sinus. Branches of the ophthalmic artery include

1. The central retinal artery, which supplies the optic nerve and, after emerging from the optic disc, branches to nourish the retina as deep as the inner nuclear layer.
2. The posterior ciliary arteries, which consist of 6–20 branches. The majority (short ciliary arteries, Fig. 2b) enter the choroid to supply the choriocapillaris. Two branches (long posterior ciliary arteries) anastomose with branches of the anterior ciliary arteries to supply the ciliary body.
3. The recurrent meningeal artery, which anastomoses with the middle meningeal artery.
4. The terminal divisions of the lacrimal artery, which supply the lacrimal gland, eyelid, and conjunctiva.
5. The anterior ciliary arteries, which are derived from arteries that supply the extraocular muscles.
6. The supraorbital artery, which supplies the upper eyelid, levator muscle, and periorbita.
7. The medial palpebral arteries.
8. The anterior and posterior ethmoid arteries.
9. The dorsalis nasal artery, which supplies the lacrimal sac.
10. The supratrochlear artery.
11. The episcleral and conjunctival arteries.

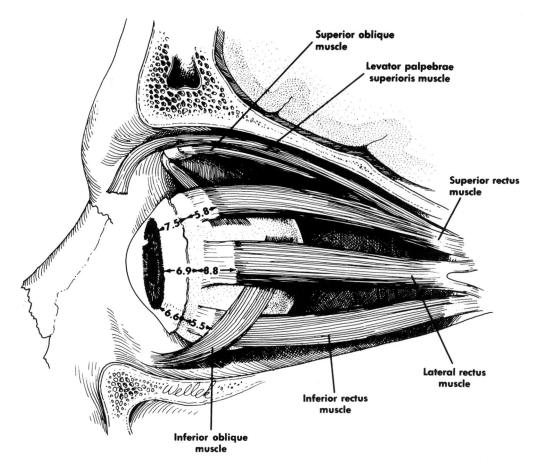

FIGURE 23 Extrinsic muscles of the eye. Both oblique muscles insert behind the equator of the globe. The inferior oblique muscle passes over the body of the inferior rectus muscle but beneath the lateral rectus muscle. The numbers indicate the distance of the insertion from the corneoscleral limbus and the length of the muscle tendon. [From F. W. Newell (1982). "Ophthalmology: Principles and Concepts," 4th Ed. C. V. Mosby, St. Louis.]

2. External Carotid Divisions to the Orbit and Adnexa

The blood supply to the eye and eyelids comes from the external facial artery, the superficial temporal artery, and the internal maxillary artery.

3. Venous Drainage of the Eye

The superior and inferior orbital veins are the principal pathways for the return of venous blood from the eye and its adnexa. The veins twist tortuously through the orbit before emptying into the intracranial cavernous sinus, an endothelium-lined venous cavity between the meningeal and the periosteal layers of the dura mater. The internal carotid artery, together with the abducens, oculomotor, trochlear, and trigeminal nerves, passes through or adjacent to the cavernous sinus.

C. Innervation of the Eye

1. Motor Innervation of Intrinsic and Extraocular Muscles

a. Cranial Nerves III, IV, and V

Eye movements are under the control of three major cranial nerves: the oculomotor (III), trochlear (IV), and abducens (VI) nerves (Fig. 24). Nuclei for the oculomotor and trochlear nerves are in the midbrain and that for the abducens in the pons. All three nerve nuclei have paramedial locations.

The efferent fibers of the oculomotor nerve pass ventrally in the midbrain through the red nucleus and exit in the subarachnoid space. The nerve then passes under the origin of the posterior cerebral artery and lateral to the posterior communicating artery. This is an important relationship because a third nerve palsy is frequently associated with an aneurysm in these

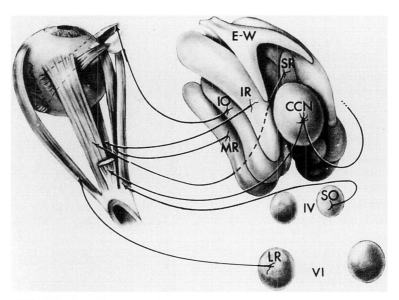

FIGURE 24   Diagrammatic representation of the organization of the oculomotor nuclear complex as viewed from above, left posterior aspect. The superior rectus (SR) and the superior oblique (SO) motor pools are crossed. CCN, caudal nucleus to both lid levator muscles; LR, abducens nucleus for lateral rectus muscle; E-W, Edinger–Westphal parasympathetic subnucleus; IR, inferior rectus nucleus; IO, inferior oblique nucleus; MR, medial rectus nucleus. [From R. Warwick (1953). *J. Comp. Neurol.* **98,** 449.]

arteries. The third nerve runs in the cavernous sinus. It lies inferior and medial to the trochlear and abducens nerves in the anterior part of the sinus. In certain conditions, the pupillomotor fibers of the third nerve in the sinus may be spared preferentially so that although eye movements may be restricted, the pupil may be normal.

As the third nerve enters the orbit, it divides into two branches: (1) the superior branch which innervates the superior rectus muscle and the levator palpebrae superioris, and (2) the inferior branch that gives off twigs to the inferior and medial rectus muscles and a long twig to the inferior oblique muscle. A division from the twig to the inferior oblique passes through the ciliary ganglion and supplies the sphincter of the pupil and the ciliary body by way of the short ciliary nerves.

The most common conditions that affect the third nerve are aneurysm, tumor, inflammation, and vascular disease.

The trochlear nerve is the only cranial nerve that exits the brain stem dorsally and crosses to supply the contralateral superior oblique muscle. After passing through the cavernous sinus with the other oculomotor nerves, the trochlear enters the orbit through the superior orbital fissure.

The most common condition that affects the fourth nerve is trauma. Contrecoup contusions occur because the nerve exits dorsally and passes around the midbrain. Palsies of the trochlear nerve are significantly less common than of the abducens or oculomotor nerves.

The nucleus of the abducens nerve is located caudally in the paramedial pontine tegmentum. Efferent fibers from this nucleus course toward the lateral rectus muscles and pass it ventrally. The nucleus of this sixth nerve is involved frequently in vascular disease occurring in the brain stem. After passing through the superior orbital fissure (annular segment), the sixth nerve supplies the lateral rectus muscle.

The most common conditions affecting the sixth nerve are tumor, vascular disease, multiple sclerosis, infection, diabetes, and trauma.

2. Sensory Innervation

The sensory innervation of all ocular tissues is provided by components of the ophthalmic division of the trigeminal nerve.

3. Autonomic Innervation

The sympathetic innervation of the eye and orbit is derived from preganglionic fibers of the superior cervical ganglion, which pass through the ciliary ganglion and then along branches of the ophthalmic artery (the external carotid artery) and along the nasociliary nerve (a branch of the ophthalmic division of the trigeminal nerve). Sympathetic fibers innervate sweat glands, muscles of the face and eyelids, uveal blood vessels, and Muller's muscle of the eyelid. The nerves for the dilator pupillae muscle of the iris do not pass through the ciliary ganglion.

The parasympathetic nerves, or efferent branches of the oculomotor nerve, synapse in the ciliary ganglion. Postganglionic efferent fibers innervate the ciliary muscle for the control of accommodation and of the sphincter pupillae muscle.

IV. VISUAL PATHWAY

A. Optic Nerve

The optic nerve is formed by an aggregation of approximately 1.3 million axons that arise from the retinal ganglion cells and extend to the lateral geniculate body in the brain, where they synapse. The optic nerve may be divided into three portions. The intraocular part, known as the optic disc or papilla, is about 1.5 mm in diameter and is formed by unmyelinated axons. This axonal aggregation is more compact on the temporal side (the papillomacular bundle) than it is on the nasal side (Fig. 14). Because there are no photoreceptor elements in the optic disc, this area represents the blind spot in the visual field. The central region of the disc has a physiologic pit through which the central retinal artery and vein emerge. In the intrascleral portion of the optic nerve, the axonal bundles acquire a myelin sheath and, together with the astroglia and oligodendroglia, pass through perforations in the sclera (the lamina cribrosa, Fig. 25). Within the orbit, the optic nerve (25–30 mm in length and about 3 mm in diameter) forms an S-shaped loop, which permits its extension during movement of the eyeball.

The optic nerve is covered by a pia mater, an arachnoid mater, and a dura mater similar to that of the brain (Fig. 25). The dura mater merges with the sclera, but the pia and arachnoid maters fuse at the lamina cribrosa, thus obliterating the flow of cerebrospinal fluid around the nerve. The orbital portion of the

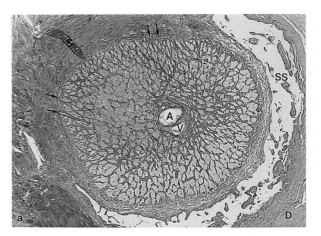

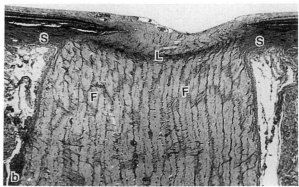

FIGURE 25 (a) Flat section of sclera passing through the lamina cribrosa. Note that the connective tissue strands of the lamina cribrosa are continuous with the sclera (S). Due to a slightly tilted plane of section, the posterior continuation of the sclera with the dura mater (D) is seen separated from the optic nerve by arachnoid matter and subarachnoid space (SS). There is a dense connective tissue matrix around the central artery (A) and vein (V), which share a common adventitia. Wilder silver stain × 20. (b) Transverse section showing continuity of septal system of lamina cribrosa (L) with that of the sclera (S). Note the myelination of the nerve bundles (F) that occurs as the axons proceed posteriorly from the lamina cribrosa. Photomicrograph × 30.

optic nerve continues posteriorly into the cranium through the optic foramen in the bony orbit.

B. Optic Radiations and Visual Cortex

In the brain, the optic nerve passes in a dorsal and medial direction to enter the formation of the optic chiasma (Fig. 26). The length of the cranial portion of the nerve varies (6–23 mm). The optic chiasma is a transversely oval structure that forms a bridge-like junction between the terminal portion of the two optic nerves. The nerves from the two eyes enter at the antero-lateral angles of the chiasma. The optic chi-

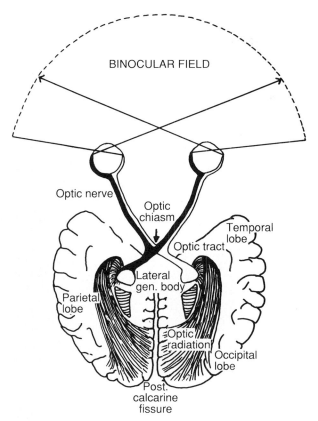

FIGURE 26 Diagrammatic representation of uniocular (145°) and binocular (120°) visual fields and the visual pathways in humans.

molopgous with the entire geniculate body of lower vertebrates. In humans, it is poorly developed and has no direct implications for vision. The dorsal nucleus is an oval structure with an elevation on the postero-lateral aspect. It serves as a relay station in the projection of the visual fibers to the occipital cortex. The geniculate body has a laminated structure, composed of alternate layers of white (medullated nerve fibers) and gray (synaptic junctions) matter. This lamination provides a sharp separation for the termination of the crossed and uncrossed fibers of the optic nerves so that layers 1, 4, and 6 receive fibers from the contralateral eye and layers 2, 3, and 5 receive fibers from the ipsilateral eye. From the lateral geniculate body, the efferent nerve fibers pass to the occipital cortex as the geniculo-calcerine pathway or optic radiations (Fig. 26). The fibers terminate in the striate cortex and synapse primarily with the granular cells, which are situated in layer IV. These cells serve as primary receptor elements, as well as interneurons through which visual impulses are transmitted to other layers, especially layers V and VI. Integration of visual stimuli occurs in higher-order visual associative areas, which are linked directly with the striate cortex. [*See* Visual System.]

The remainder of the nerve fibers in the optic tract enter the hypothalamus (these fibers are implicated in circadean rhythm) and the superior collicullus, which is involved in coordinating reflex ocular movements.

BIBLIOGRAPHY

Bron, A. J., Tripathi, R. C., and Tripathi, B. J. (1997). "Wolff's Anatomy of the Eye and Orbit," 8th Ed. Chapman and Hall, London.

Fine, B. S., and Yanoff, M. (1979). "Ocular Histology: A Textbook and Atlas," 2nd Ed. Harper and Row, New York.

Hogan, M. J., Alvarado, J., and Weddell, J. E. (1971). "Histology of the Human Eye: An Atlas and Textbook." Saunders, Philadelphia.

Jacobiec, F. A. (1982). "Ocular Anatomy, Embryology, and Teratology." Harper and Row, Philadelphia.

Tripathi, R. C., and Tripathi, B. J. (1984). Anatomy of the human eye, orbit, and adnexa. *In* "The Eye" (H. Davson, ed.), 2nd Ed., Vol. 1. Academic Press, London.

asma measures about 12 mm transversely, 8 mm sagitally, and 3–5 mm dorso-ventrally. Partial cross-over of the axons from the two optic nerves occurs in the optic chiasma, so that axons from the nasal portion of the retina on the contralateral (opposite) side and from the temporal retina on the ipsilateral (same) side combine to form the optic tracts, which emerge from the postero-lateral angles of the chiasma. Initially, the optic tracts are round, but they continue laterally and posteriorly as flattened bands. The nerve fibers from the retina end in two masses of gray matter: the lateral geniculate body and the superior colliculus.

The lateral geniculate body is subdivided into the dorsal and ventral nuclei. The ventral nucleus is ho-

Eye Movements

BARBARA A. BROOKS

University of Texas Health Science Center at San Antonio

GLOSSARY

Brain stem Area at the base of the brain containing most of the cranial nerve nuclei, including the oculomotor nuclei

Conjugate movement Movement in which the lines of gaze in the two eyes move synchronously in the same direction, as during a saccade to the right or left

Oculomotor neurons Neurons with cell bodies in the brain stem oculomotor nuclei, whose axons constitute the oculomotor nerves and which synapse directly onto the eye muscle fibers; the "final common pathway" between the nervous and muscular systems

Premotor neurons Neurons in the brain stem that receive and process information from higher brain centers and feed directly onto the oculomotor neurons

Saccade Fast, conjugate eye movement that directs the line of gaze to objects of interest

Smooth pursuit Slow tracking eye movement that enables continuous clear vision of a moving object

Vestibulo-ocular reflexes Compensatory eye movements that stabilize retinal images during head and body movements

Vergence movement Slow movement occurring when the lines of gaze in the two eyes move in opposite directions, as in changing focus from a far object to one nearby (convergence), or vice versa (divergence)

Visual angle The arc described by the eyeball as it swivels around its orbital center (specified in degrees of amplitude)

THE IMPORTANCE OF EYE MOVEMENTS IN THE maintenance of clear vision has long been appreciated, but only within the last century was the existence of several different types of human eye movement first recognized. By 1902, a differentiation was established between five major functional classes of movement, all of which work to achieve a stable visual image at the central retina (or fovea) of each eye. Two types of movements stabilize the visual image and the eyeball during head and body movement (the vestibulo-ocular reflex and the optokinetic reflex), and three others serve to place or maintain a visual target at the center of gaze (saccades, smooth pursuit, vergences). During the first four movements the two eyes move in the same direction by the same amount (conjugate or versional movements); the fifth class requires that the eyes move in different directions, sometimes by different amounts (vergence or disconjugate movements). Finally, there are the tiny micromovements that occur during visual fixation, which are largely unconscious but important in maintaining a clear visual image. The five basic classes of eye movement have different but sometimes overlapping neural control networks in the brain. They all share a final common pathway between the nervous system that "commands" the eye movements and the eye muscles that actually move the eyeball: the brain stem oculomotor neurons, which travel to the muscles in cranial nerves III, IV, and VI (Table I).

I. THE GLOBE AND ITS EXTRAOCULAR MUSCLES

Eye movements are produced by six muscles attached to the outside of each eyeball or globe (Table I). The muscles act to produce three basic kinds of globe rotations or eye movements: horizontal, vertical, and torsional. The muscle fibers are driven very precisely by the oculomotor neurons, which have cell bodies residing in the oculomotor, abducens, or trochlear

ENCYCLOPEDIA OF HUMAN BIOLOGY, Second Edition, VOLUME 3.

TABLE I

Main action of each eye muscle on eye movement, when starting from direction of gaze straight ahead. Brain stem cranial (or oculomotor) nerves activate the eye muscles. Main action is shown for the *right* eye (next to the arrows).

Eye Muscle	Cranial Nerve # and Nucleus	Main Action on Eyeball	Illustration
Medial Rectus	III (Oculomotor)	Horizontal Adduction	
Lateral Rectus	VI (Abducens)	Horizontal Abduction	
Superior Oblique	IV (Trochlear)	Depression; Down and Outward	
Superior Rectus	III (Oculomotor)	Elevation; Up and Inward	
Inferior Oblique	III (Oculomotor)	Elevation; Up and Outward	
Inferior Rectus	III (Oculomotor)	Depression; Down and Inward	

nuclei, located bilaterally in the brain stem and collectively known as the oculomotor nuclei (Table I); the neurotransmitter acetylcholine is active at the neuromuscular junction. Individual muscle fibers (and the oculomotor neurons) show bursts of action potentials during fast muscular contractions, which produce fast eye movements, slowly changing rates of discharge during slow movements, and an exact relation between ongoing firing rate and position of the eye during steady gaze. The oculomotor neurons signal impartially for all types of eye movement; for example, there are no special oculomotor neurons (or muscle fibers) devoted only to smooth pursuit or to saccades, or to vergence movements. The oculomotor nuclei are under the influence of a variety of brain centers, most directly by several classes of nearby brain stem premotor neurons that process information from higher command centers and connect monosynaptically to the oculomotor neurons. [*See* Eye, Anatomy.]

The oculomotor system can be contrasted with the skeletal motor system in several important respects. The extraocular muscles are not required to pull against gravity; they operate against a fixed load determined by the viscoelastic properties of the globe and its contents. There is no stretch reflex in the eye muscles, although there are many centrally projecting muscle afferents, whose function is not yet fully understood. During movement, the muscles act in complementary pairs that rotate the eye around a fixed "joint," which is the geographical center of the globe. The specific contribution of each muscle to an individual eye movement is a complicated function of the initial position of the eye in the orbit. Muscle tone is determined by a high rate of centrally controlled ongoing discharge in the oculomotor neurons. Contraction of a specific force in an eye muscle depends on an exact input from its oculomotor neurons and produces a predictable angular displacement of the globe, associated with reciprocal inhibition of the antagonist muscle(s) of the same eye. Synchronous, conjugate movements of the two eyes require that appropriate inhibitory and excitatory neuronal signals cross the brain stem midline to coordinate the oculomotor nuclei for synergistic muscular action (Fig. 1). Because of their relative simplicity, the peripheral neural control of eye movement and the movements themselves

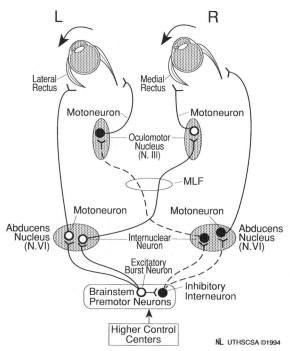

FIGURE 1 Neuronal activity required to produce a conjugate saccadic movement to the left. Pathways with unbroken lines and white neurons are excited, and pathways with broken lines and black neurons are inhibitory or silent during the movement. For this leftward movement, the lateral rectus muscle of the left eye (L) and the medial rectus of the right eye (R) must contract simultaneously. Brain stem premotor "burst" neurons discharge the excitatory chain, on command from a higher center. At the same time, the left medial rectus and the right lateral rectus must relax, so that the movement is unopposed. To accomplish this, an inhibitory interneuron in the brain stem is also activated by the burst neuron, which silences activity in the appropriate pathways for the duration of the movement. Excitatory and inhibitory information that must cross over the midline does so in a structure called the medial longitudinal fasciculus (MLF).

are easier to measure and quantify than in the skeletal motor system, with its complicated trajectories and variable loads.

II. TYPES OF EYE MOVEMENTS

The basic eye movement types occur continuously in alert, active primates possessing binocular vision, such as humans, monkeys, and apes. Clear, three-dimensional vision (stereopsis) requires that the image of a single visual object affect corresponding areas of best detail and color vision in each retina (the foveas); the directions of gaze in both eyes must be coordinated properly to avoid seeing double (diplopia). Animals

lacking foveas, such as rabbits, with laterally placed eyes and panoramic vision, have almost no stereopsis and a very limited range of eye movements, which usually are a reflexive accompaniment of body or head orientation. The achievement of highly detailed, stereoptic vision in the primates is of evolutionary advantage in all aspects of visual perception.

A. Saccades

A saccade is a fast eye movement that rapidly shifts the line of sight. Conjugate saccadic movements ensure that the fovea in each eye is stimulated by the same visual target during visual inspection, or fixation. Saccades may be involuntary or voluntary and range from the primitive reflexes that stabilize images of the visual world during body and head motion (e.g., the fast "reset" phases of vestibular and optokinetic nystagmus; Fig. 2b) to the consciously initiated movements of active scanning with the head stationary (Fig. 2a). Scores of saccades, separated by fixation intervals, are sequentially performed every minute during scanning of a busy visual environment and during close activities such as reading. The eyes can

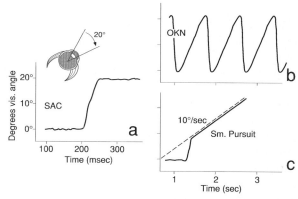

FIGURE 2 Electrically recorded waveforms of different eye movements (the electrooculogram or EOG). (a) A saccadic movement (SAC) moves the eye through 20° of visual angle. Entire movement requires 50 msec. (b) Optokinetic nystagmus (OKN) generates a "saw-tooth" waveform as the gaze follows vertical stripes moving smoothly in one horizontal direction. During the slow pursuit phase, a stripe is fixated until the eye deviates to its farthest extent; a saccadic fast phase then "resets" the globe back in the opposite direction, another stripe is fixated, and the pattern is repeated. A similar waveform is seen during vestibular nystagmus (see text). Note the time base. (c) The dashed line represents a target moving at 10°/sec across the visual field. The eye (solid line) makes a small "catch-up" saccade to fixate the target. Thereafter, the velocity of smooth pursuit eye movement matches the target to keep its image on the fovea.

perform up to four saccades each second separated by an irreducible minimum of about 200 msec. Saccades are the fastest of all ocular movements, attaining maximum velocities of 900°/sec and amplitudes of 90° of visual angle. Most of our everyday saccades are less than 20° of arc and 60 msec of duration. The trajectory of a saccade is stereotyped and smooth, with a rapid initial acceleration, a peak velocity halfway through the movement, and a gradual deceleration into a new orbital position (see Fig. 2a). Saccades have been called "ballistic"; following initial acceleration, their trajectories, like that of a bullet or cannonball, seem to be passively determined. Though it is a fact that saccades are usually too fast for modification by visual feedback, they are not truly ballistic because they are under continuous control by the oculomotor nervous system. Velocity is not usually under voluntary control, but can be affected by disease, inattention, fatigue, and drugs. As a general rule, larger saccades require more time (in msec) and are faster in velocity than small ones. The response time or latency of a saccade to a visual target flashed on the retinal periphery is usually around 200–500 msec; during this time the target position is registered by the visual system and assessed by several coordinated brain centers, which then issue a command to the brain stem to move the eyes by an amplitude appropriate to place the target image at the foveas. A special class of saccades, elicited in special circumstances, may have latencies as low as 110 msec ("express saccades").

B. Smooth Pursuit

The object of smooth pursuit eye movements is to maintain clear vision of a moving visual target (Fig. 2c). Smooth pursuit can be caused by a moving image that affects only the fovea and its immediate surrounding retina; it usually is impossible to voluntarily produce smooth pursuit in the dark without a visual target. Smooth pursuit may occur in conjunction with head movement (e.g., as we watch players in a basketball game) or with the head stationary. Eye velocity closely matches retinal image velocity up to a maximum of around 100°/sec of visual angle. Smooth pursuit is best developed in animals such as the higher primates, who have binocular vision and well-developed foveas. Some bird and lizard species have excellent foveal vision but no pursuit system; afoveate animals, such as rabbits, cats, and goldfish, are essentially incapable of smooth pursuit movement. Smooth pursuit is affected by a variety of drugs and by level of alertness, and there is a decline in pursuit performance in old age. Because of the importance of visual input,

the visual pathways from retina to visual cortex must be intact for it to occur. Motor signals must be generated to drive the muscles proportionately to the velocity of the target and to adjust eye position when the target "slips" off the foveal area (retinal error). The neural control pathways for smooth pursuit are being worked out; early on it became clear that the brain pathways for smooth pursuit were different from those controlling saccades, because damage to the saccadic system did not seem to affect smooth pursuit, and vice versa.

C. Vestibulo-ocular Reflex

Smooth, conjugate eye movements can be induced by input from the vestibular sensory system. During active or passive rotational head movements, which stimulate the semicircular canals and otoliths of the vestibular system, the eyes will deviate in the opposite direction to stabilize the eyeball in visual space. During prolonged bodily rotation, slow compensatory eye movements are interrupted by rapid reset movements as the eye reaches maximum deviation in the socket, in a pattern called vestibular nystagmus (see the "sawtoothed" pattern in Fig. 2b). One of the most important functions of the vestibulo-ocular reflex (VOR) is the stabilization of retinal images during the many head and body motions of ordinary locomotion, such as walking or running; patients with vestibular damage report that during walking, the visual world appears to jiggle up and down in a very confusing way. In the dark, vestibular input to the oculomotor neurons is relatively direct through a three-neuron pathway from the semicircular canals to the vestibular nuclei, from vestibular nuclei to the oculomotor nuclei, and from the oculomotor nuclei to the eye muscles. In lighted environments, visual information relayed through the cerebellum may modulate the direct vestibular circuit so that the vestibular reflex may be modified or suppressed, as happens when a whirling dancer fixates the horizon. Research indicates that visual input is continuously modifying the vestibulo-oculomotor circuits in awake, active animals. Certain brain stem premotor neurons that connect to oculomotor neurons may integrate information from the visual and vestibular systems, and may be the site of a simple kind of "motor learning." For example, activity in premotor neurons can be measured in alert monkeys while they make head and eye movements for reward. The animals can be fitted with special lenses that either slow down or speed up the retinal image of the visual world during head turning. The normal VOR is unable to compensate for this exaggerated image motion. However, after wearing the lenses for a few days, the brain stem premotor neurons modify their output to the oculomotor neurons and the VOR adapts to the new visual input. When the lenses are removed, the VOR slowly reverts to its former normal state as does the output of the premotor neurons. This type of experiment shows a remarkable adaptability of function, even for this relatively reflexive process. The cerebellum is a crucial link in the pathway from visual system to premotor neurons, since damage to it destroys the adaptive response.

The slow phase of vestibular nystagmus survives the anatomical destruction of brain stem areas that regulate saccadic movement; the fast phase does not, indicating that nystagmic fast phases at least partially depend on the same brain stem control centers as voluntary saccades.

D. Optokinetic Reflex

Optokinetic nystagmus can be elicited by purely visual stimulation with the head stationary, such as occurs when viewing the passing scene from a moving train (railway nystagmus; see Fig. 2b). The reflex is elicited by wholesale movement of the entire visual environment and comprises, like vestibular nystagmus, a slow phase followed by a fast reset movement after the eye has reached maximum possible angular deviation. After visual simulation stops, there is a period of optokinetic "after-nystagmus." This pattern obviously depends critically on the visual system; the cerebellum also plays an important role in these visually guided compensatory movements. Fast reset phases, like the fast phases of vestibular nystagmus, depend on the same brain stem mechanisms as saccades.

E. Vergence (or Disconjugate) Eye Movements

Vergence eye movements are visually mediated, are mainly horizontal, and occur when we change focus from a near object to a distant one (divergence), or from a far object to one close by (convergence). The net result is that the gaze axes of the two eyes move in opposite directions, as compared with conjugate (or versional) eye movements. Viewing an index finger as it moves toward and away from the nose will illustrate convergent and divergent movement. Horizontal convergence requires simultaneous activation of the medial rectus of both eyes and inhibition of the lateral recti, divergence the opposite pattern. The stimulation for vergence is either *disparity* of images affecting the two retinas (the same object affecting noncorrespond-

ing retinal points and producing diplopia, as during voluntary crossing of the eyes) or *blur* (lack of proper focus). In ordinary conditions, both blur and disparity interact to generate vergence movements; fusional (disparity) vergence and accommodative (blur-induced) vergence are tightly linked anatomically. Most vergence movements are a combination of slow and fast components; a small saccade in combination with the vergence command points the eye in the general direction of the visual target, before the slower vergence movement places the target in good focus on each fovea. Vergences utilize the same motoneurons as the conjugate saccadic and smooth pursuit systems but have three sets of special premotor command neurons located in the midbrain near the oculomotor nuclei. These premotor cells are exclusively devoted to either convergence or divergence and encode vergence velocity and vergence eye position. Obviously, higher command centers for the vergences require an intact visual system.

F. Miniature Eye Movements of Fixation

The eyes are never motionless. Even during steady fixation of an object, continuous, small, unconscious movements occur, rarely exceeding 15–20 minutes of visual angle. Slow ocular movements are called drifts and move the fovea over, and sometimes slightly away from, the target of interest. Tiny corrective saccades that return the fovea to the target are called flicks. Experiments have shown that drifts are essential for continuous clear vision and may be consciously controlled, though with considerable effort. Images that are totally motionless on the retina, such as those produced by the vasculature of the retina, or while wearing special contact lenses with visual patterns that move with the eye, become invisible after a few seconds. Neural activity in the visual system that signals motionless images shows a similar decline. Frequent motion of the retinal image, whether produced by eye movements or externally moving images, is essential to clarity and continuity of visual experience.

III. BRAIN STEM CONTROL OF EYE MOVEMENT

A. The Role of the Oculomotor Neurons

The two basic variants of conjugate volitional eye motion, slow pursuit and fast saccadic, require different patterns of discharge in the oculomotor neuron. Smooth pursuit and vestibulo-optokinetic slow phases require a *velocity*-coded signal from the oculomotoneurons that is proportional to target velocity and retinal error during smooth pursuit, and for complementing or canceling head movement during eye–head coordination. The velocity signal must regulate the eye muscles very exactly during movements that may variably accelerate and decelerate. The motoneurons are controlled by tonically firing premotor cells in the brain stem reticular formation that code eye velocity for both smooth pursuit and vestibular slow phases.

Saccadic eye movements need (1) a powerful muscular contraction at movement onset to overcome the viscous drag of orbital tissues and (2) a steady muscular contraction to maintain the new eye position in the orbit against elastic restoring forces. The burst of motoneuron firing needed to initiate muscle contraction during saccades has been termed the pulse of innervation, wheras the tonic signal associated with new eye position following saccades is the step of innervation; therefore, a saccadic movement is controlled by a "pulse–step" of innervation. During steady fixation with the eyes looking straight forward, oculomotor neurons discharge at a relatively high but regular rate that varies from neuron to neuron. For a given motoneuron, this rate increases in a strong burst about 5–8 msec before a saccade, if the neuron innervates a muscle that contracts during the movement (the "on," or pulling, direction of the muscle). The velocity of the saccade is related to the burst frequency, whereas the duration of the burst determines its amplitude or size. The firing rate following the saccade is precisely matched to the new eye position and the amount of contraction required to hold the eye steady. (The transform of eye *velocity* signal to eye *position* signal occurs at the brain stem premotor level in a circuit that performs a mathematical integration for all conjugate eye movements, called the neural integrator; see Fig. 3.) During saccades in the opposite, or "off," direction, the motoneuron is inhibited and then assumes a new lower tonic discharge corresponding to the new eye position (sufficient deviation in the off direction may silence the motoneuron). The slope of the function relating eye position and discharge rate varies from neuron to neuron, as does the eye position at which the neurons begin to fire.

B. Brain Stem Premotor Mechanisms

Premotor control of saccadic eye movements is relatively well known, and exemplifies the type of circuit

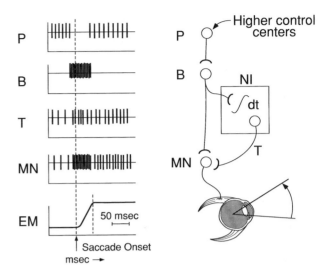

FIGURE 3 Simplified diagram relating activity in brain stem premotor neurons to the "pulse–step" discharge of motoneurons (MN) during saccadic eye movement. Higher control centers inhibit the pause cell (P) just before movement, permitting the burst cells (B) to generate the pulse. The pulse is integrated by the neural integrator (NI-*dt*) to provide the step change which is observed in tonic neurons (T). The burst and tonic neurons converge on the motoneuron, resulting in saccadic eye movement (EM) followed by a new steady eye position.

required for many voluntary versional eye movements. The pons and mesencephalon of the brain stem contain premotor neurons that provide input to the oculomotor nuclei. Evidence from single-cell recordings in alert monkeys trained to saccadic eye tracking shows essentially three types of cells: pause cells, burst cells, and tonic cells (Fig. 3). Burst cells begin discharging at high frequency just prior to saccades; pause cells are tonically active during eye fixation and cease firing just before saccades; tonic cells alter their steady discharge immediately after a saccade, assuming a rate proportional to the new eye position. The silence of burst cells during fixation is thought to result from tonic inhibition by pause cells; just before a saccade, a higher center inhibits the pause cells, thus disinhibiting the burst neuron, which begins to discharge 7–14 msec prior to onset of the saccade and continues for its duration. Burst cells provide the pulse in the pulse–step pattern delivered to the motoneuron. The discharge of the burst cells is integrated mathematically by the neural integrator to change the rate of firing in tonic cells, thereby providing the step of the pulse–step command to the motoneurons (Fig. 3); the integrator is located in the nucleus prepositus hypoglossi. During fixation, the pause neurons theoretically prevent burst neurons from extraneous firing with the resultant unintended ocular oscillations (opsoclonus and flutter).

Brain stem premotor neurons of all kinds—pause, burst, and tonic—have been described in anatomical locations that are clinically correlated with disturbances in saccadic eye movement. Horizontal saccades are controlled by the paramedian pontine reticular formation (PPRF), whereas vertical saccades depend on an intact rostral mesencephalon [rostral interstitial nucleus of the medial longitudinal fasciculus (riMLF)]. Unilateral lesions of the PPRF eliminate saccades to the ipsilateral side; bilateral PPRF damage removes all fast horizontal eye movements. Bilateral lesions of the riMLF produce a loss of all vertical saccades; bilateral lesions of the medial and caudal PPRF disrupt fast movements in all directions, including nystagmic reset phases. Most of these lesions do not drastically affect slow movements.

Brain stem regulation of smooth pursuit movements is not as thoroughly understood as it is for saccades. Structures containing neurons that encode signals important for smooth pursuit (e.g., eye velocity) include several pontine nuclei, the vestibular complex, the nucleus prepositus hypoglossi, parts of the mesencephalic reticular formation, and PPRF. The cerebellum has a large role in coordinating various types of slow eye movements.

Studies of brain stem neurotransmitters suggest that several excitatory and inhibitory amino acids (glutamate, aspartate, glycine) play a role in the premotor pathways.

IV. HIGHER CONTROL CENTERS

Anatomical and neurophysiological studies have revealed a large, often reciprocally connected, cortical and subcortical network of neural centers and pathways that participate in eye movement control. Since most voluntary eye movements are related to visual stimulation, one job of the higher control centers is to coordinate visual input with motor output ("sensorimotor integration"); cells in many oculomotor areas respond to visual stimulation as well as fire during eye movement. The brain also contains internal representations or "maps" of the external world based mainly on learning, so that another job of the higher centers is to relate such internally maintained sensory and motor representations to current, ongoing sensorimotor activity. Not surprisingly, direct electrical

stimulation of neurons in many of these brain areas *causes* eye movement.

The control of ocular motility requires circuits ranging from relatively simple and subcortical (for more reflexive movements such as the VOR) to the highly complex cortical pathways activated during voluntary, willed movements without a sensory stimulus (e.g., saccades to a remembered or imaginary visual location). The full range of ocular motility depends on cortical areas in occipital, temporal, parietal, and frontal lobes (Fig. 4), connecting with subcortical centers in the thalamus, basal ganglia, vestibular nuclei, superior colliculus, and the cerebellum.

Circuits for the control of slow eye movement are illustrated by the neural pathway for smooth pursuit. Starting with excitation of the visual pathways by a moving target, activity spreads from the occipital (visual) cortex to secondary visual areas for analysis of moving objects in the temporal cortex. (Damage to the temporal cortex affects smooth pursuit and optokinetic nystagmus but not simple vestibular reflexes.) Subcortical centers subsequently activated include the dorsolateral pons, the cerebellar flocculus, vestibular nuclei, separate premotor areas in the brain stem reticular formation, and finally the oculomotor neurons. These pathways are continuously activated during the constant adjustment of eye velocity to changing target velocity. The eye fields of the frontal cortex also contribute to smooth pursuit, especially during predictable target motions.

Subcortical neurons with discharge patterns time-locked to fast eye movement (e.g., saccades) are found in the cerebellum, the vestibular nuclei, the superior colliculus, the basal ganglia, and the intralaminar nuclei and pulvinar of the thalamus. Cortical areas involved in the production of voluntary saccades include the visual cortex, the parietal eye fields, the frontal eye fields (area 8), and the dorsolateral prefrontal and supplementary motor areas of the frontal cortex (see Fig. 4). Many of these areas project *directly* to the superior colliculus, an important center for sensorimotor integration, which in turn triggers the brain stem premotor neurons that feed onto the motoneurons. Such parallel, descending pathways suggest that many eye movements involve neural circuits that may substitute for each other if necessary. An illustration is seen in the roles of the frontal eye fields, parietal eye fields, and superior colliculus in triggering saccades; all of these areas receive visual information and project independently to the conjugate gaze centers of the brain stem. Destruction of any area alone produces only temporary deficits in saccadic movement.

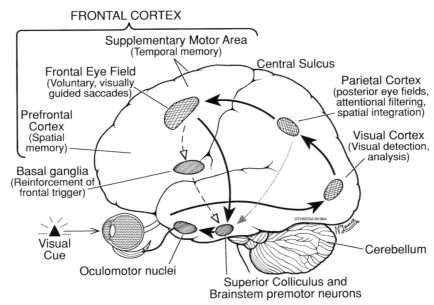

FIGURE 4 Probable route of neural activation for a saccade to a cue in the visual field. The visual cue activates the retina of the eye and the visual pathway up to the visual cortex, which projects to the parietal eye fields. If the cue is interesting or meaningful, parietal connections to the frontal eye fields are excited. The frontal eye fields then trigger the superior colliculus and brain stem premotor neurons, which instruct neurons of the oculomotor nuclei to activate the eye muscles, causing a saccadic movement to the cue. Main excitatory route is shown by heavy black lines and arrows. Alternate triggering route from parietal eye fields is shown by the gray line and arrow. Auxiliary route through basal ganglia for inhibitory and excitatory reinforcement of frontal eye field activity is indicated by the dashed line and white arrows. Other areas mentioned in text.

Lesions of both frontal eye fields *and* superior colliculus produce permanent, severe deficiencies, as do lesions of both the frontal and the parietal eye fields. The frontal eye fields may be able to trigger saccades through its independent connections to the brain stem saccadic generator, in the absence of both the parietal fields and the superior colliculus.

Higher control centers are being studied in ingenious experiments using alert, trained animals (usually primates because their visual and oculomotor systems most strongly resemble our own). In experiments studying saccades, monkeys are trained to make eye movements of specific amplitudes and directions for food reward, while neural activity is simultaneously monitored. The test situations are designed to elicit saccades in different contexts such as visually guided tracking saccades, saccades to a remembered visual target, reflexive saccades to novel objects, or random, "spontaneous" saccades. It has been found that visual and oculomotor activity in the higher control centers is frequently conditionally dependent on behavioral context and significance for the animal. For instance,

the parietal cortex contains neurons that answer *only* to visual stimuli that signify that a reward will be given for a subsequent saccade to the location of the stimulus; meaningless stimuli in the same cell's visual receptive field elicit no activity at all. Some of these cells will also fire during or before saccades that are performed in order to obtain eventual reward; the same saccades made coincidentally or "spontaneously" without reward are not signaled by the cell. The parietal cortex also contains neurons that fire before an *intended* saccade, whether or not the saccade is actually made.

Such experimental evidence suggests that (1) the frontal eye fields (area 8) trigger intentional, self-generated saccades, visually guided saccades, and saccades to remembered targets in a spatial reference system; (2) the parietal eye fields regulate visually guided saccades, provide attentional filtering of visual information, and function in the internal representation of the external world; (3) the eye fields of the supplementary motor area are important in planning temporal sequences of saccades (short-term temporal memory); (4) the prefrontal eye fields participate in "spatial" short-term memory used for programming memory-guided saccades to specific locations in the visual field, and also in inhibition of unwanted saccades during fixation; and (5) the basal ganglia work with frontal eye fields by providing auxiliary phasic excitation and tonic inhibition of the superior colliculus. All of these areas have additional saccade-related functions. (See circuit illustration in Fig. 4.)

Experiments using the methods of single-neuron recording in alert trained animals, systems analysis, the new anatomical tracing and biochemical techniques, as well as clinical studies in humans, are providing data for elegant models of the oculomotor control systems. The results are often pertinent to general motor function. Yet many questions remain to be answered, including the exact relations among brain stem and subcortical and cortical components.

BIBLIOGRAPHY

Anderson, T. J., Jenkins, I. H., Brooks, D. J., Hawken, M. B., Frackowiak, R. S. J., and Kennard, C. (1994). Cortical control of saccades and fixation in man: A PET study. *Brain* 117, 1073–1084.

Becker, W. (1989). Metrics. *In* "The Neurobiology of Saccadic Eye Movements: Reviews of Oculomotor Research (R. H. Wurtz and M. E. Goldberg, eds.), Vol. III, pp. 13–66. Amsterdam; Elsevier.

Bruce, C. J. (1990). Integration of sensory and motor signals in primate frontal eye fields. *In* "Signal and Sense: Local and Global Order in Perceptual Maps" (G. M. Edelman, W. E. Gall, and W. M. Cowan, eds.), pp. 261–314. Wiley–Liss, New York.

Carpenter, R. H. S. (1988). "Movements of the Eyes," 2nd Ed. Pion Limited, London.

Goldberg, M. E., Eggers, H. M., and Gouras, P. (1991). The ocular motor system. *In* "Principles of Neural Science" (E. R. Kandel, J. H. Schwartz, and T. M. Jessel, eds.), 3rd Ed., pp. 660–676. Elsevier, Amsterdam.

Gross, C. G., and Graziano, M. S. A. (1995). Multiple representations of space in the brain. *Neuroscientist* 1, 43–50.

Keller, E. L., and Heinen, S. J. (1991). Generation of smooth-pursuit eye movements: Neuronal mechanisms and pathways. *Neurosci. Res.* 11, 79–107.

Leigh, R. J., and Zee, D. S. (1991). "The Neurology of Eye Movements." 2nd Ed. Davis, Philadelphia.

Lisberger, S. G. (1988). The neural basis for learning of simple motor skills. *Science* 242, 728–735.

Moschovakis, A. K., and Highstein, S. M. (1994). The anatomy and physiology of primate neurons that control rapid eye movements. *Annu. Rev. Neurosci.*, 17, 465–488.

Pierrot-Deseilligny, C. (1994). Saccade and smooth-pursuit impairment after cerebral hemispheric lesions. *Eur. J. Neurol.* 34, 121–134.

Robinson, D. A. (1981). The use of control systems analysis in the neurophysiology of eye movements. *Annu. Rev. Neurosci.* 4, 463–503.

Sparks, D. L., and Barton, E. J. (1993). Neural control of saccadic eye movements. *Curr. Opin. Neurobiol.* 3(6), 966–972.

Wurtz, R. H., Komatzu, H., Düsteler, M. R., and Yamasaki, D. S. (1990). Motion and movement: Cerebral cortical visual processing for pursuit eye movements. *In* "Signal and sense: Local and Global Order in Perceptual Maps" (G. M. Edelman, W. E. Gall, and W. M. Cowan, eds.), pp. 233–260. Wiley–Liss, New York.

Famine: Causes, Prevention, and Relief

JOHN W. MELLOR
John Mellor Associates, Inc.

GLOSSARY

Chronic hunger Continuing absence of adequate food supplies to support an active, healthy life

Famine Widespread mortality caused by sharply reduced food consumption

Famine early warning system System to collect, coordinate, analyze, and interpret data on a wide range of indicators—especially meteorological, economic, agricultural, demographic, and nutritional—to predict famine or conditions predisposing to famine

Ketone bodies Products of incomplete oxidation of fatty acids whose production by the liver increases when fatty acid breakdown becomes rapid, as in prolonged fasting; their acid forms may serve as fuel for basic body functions in the absence of the ingestion of carbohydrates

FAMINE IS A CONDITION OF WIDESPREAD MORtality caused by sharply reduced food consumption on the part of a substantial population. A sharp decline in real purchasing power for a large number of poor people is the immediate cause of famine. But that decline in purchasing power is usually caused by decline in food production in successive years brought about by poor weather, war, or both. The consequent complex interactions between prices, employment, and assets impoverish victims and lead to sharply increased mortality. Government policy is a key determinant as to whether or not these conditions mature into widespread famine. Famine was long ago abolished from most of the Western world and more recently so in Asia. Unfortunately, famine remains common in much of Africa because of civil unrest and a paucity of resources, including trained people, institutions, and infrastructure. The same general economic development and political consensus that has eliminated famine in the rest of the world is needed to abolish famine from Africa. In this context, judiciously provided foreign aid can be of immense help.

Famine should be distinguished from chronic hunger. Although both are symptoms of poverty and a lack of food, famine lies at the extreme end on the hunger continuum. It is distinguished by large-scale loss of life, social disruption, and an economic chaos that destroys production potential. Symptoms include migration, distress sales of land, livestock, and other productive assets, the division and impoverishment of society's poorest families, crime, and the disintegration of customary moral codes. Breaking this downward spiral requires special efforts to eliminate the causes of famine and mitigate its effect.

Famine has been ubiquitous throughout history and across all regions of the globe. With economic development, the global incidence of famine declined sharply, except in Africa, where such disasters continue to threaten with increasing regularity. Both preparedness for these remaining recurrences of famine and their eventual elimination require thoughtful actions after and between crises. Tackling famine also requires political commitment by rich nations that can

ENCYCLOPEDIA OF HUMAN BIOLOGY, Second Edition, VOLUME 3.   Copyright © 1997 by Academic Press.   All rights of reproduction in any form reserved.

contribute know-how and resources and by poorer nations that must respond to the needs of their citizenry.

I. A BRIEF HISTORY OF FAMINES

One of the earliest accounts of famine was recorded on stone more than 6000 years ago, but the misery and deprivation of famine certainly predate written history. Since AD 10, the United Kingdom suffered more than 180 famines, and between 106 BC and 1929, China had 1828 famines, an average of 90 famines in each 100 years. In famines associated with the bubonic plague (1345–1348), as many as 43 million Europeans died, including as much as two-thirds of the Italian population. Five hundred years later, because of a blight of the potato crop, approximately 1 million (roughly 12%) Irish perished and an equal number migrated. As related in the Bible, when Joseph predicted that seven fat years would be followed by seven lean years, the Pharaoh ordered the granaries stocked, "that the land perish not through famine."

The modern era certainly has not been free of famine. Because famine typically strikes impoverished areas with a history of high mortality rates and poor record-keeping, death counts are mostly guesswork. More than 3 million people may have perished in the great Bengal famine of 1943; in 1974 another 1.5 million from that region starved in newly created Bangladesh. China suffered a horrific famine from 1959 to 1961 that was kept secret from the outside world. Estimates of mortality range from 16 to 64 million, depending on the definitions and methods of calculation, but it was clearly the worst famine of the twentieth century.

In Africa, famines continue to occur. But famine is not a newcomer to Africa, which has suffered from localized famines for centuries. Recent famines have caught world attention as the scale of death and media coverage have grown. In the early 1970s, famine struck many corners of the continent simultaneously; death estimates were 100,000 in the sparsely populated Sahelian countries of Mauritania, Senegal, Burkina Faso, Chad, Niger, and Mali, and 200,000 in Ethiopia. Estimates vary from 75,000 to 1.5 million civilians in the 1968–1969 war-related famine in Biafra. Accurate information on the 1983–1985 famine in Africa is scant, and disaster appears to be continuing as drought, locusts, and civil war plague certain countries. Such problems suggest that small famines frequently may go unnoticed in isolated areas.

Indeed, progress against famine is evident today.

In the face of recurring famines in Africa, the immense progress in famine reduction should not be ignored. That progress and the lessons from it hold hope for the still famine-prone areas. Mortality estimates indicate that, in much of the world, famines are lessening in their frequency and intensity. More areas of the world are famine-free, and even the severest droughts in developed countries do not lead to famines. Furthermore, India, Bangladesh, and a few African countries have demonstrated that with careful planning and management, very poor developing countries can block the chain of events that traditionally leads from crop failure to widespread death.

As our knowledge of causes, prevention, and preparedness expands, we must take action to abolish future famines. Today's global food surpluses, advancements in international transportation, and economic development generally offer opportunity for guaranteeing short-run emergency supplies and for supporting the many tasks involved in developing long-run food security.

II. BIOMEDICAL CONSEQUENCES OF FAMINE IN INDIVIDUALS[1]

The death associated with severe and prolonged food deprivation is an agonizing one. In the absence of intake of adequate dietary energy, proteins essential for brain activity and other vital functions may be broken down for fuel. In prolonged food deprivation, the body calls upon metabolic and hormonal mechanisms to protect against this breakdown.

In brief and simple terms, the steps in the starvation process from the time of the last meal may be characterized as follows: (1) food is digested and absorbed; (2) adipose (fatty) tissues stores in the body are used to fuel liver functions leading to production of glucose, the body's main sugar fuel source, and to fuel muscle activity; (3) as adipose stores are reduced (over 2 or 3 days), the liver calls upon amino acids (the constituents of protein) from muscle proteins to produce glucose to fuel brain function; the liver also begins *ketogenesis,* a process that increases the efficiency of use of remaining fatty acids from adipose tissue through incomplete fatty acid breakdown, leading to the production of *ketone bodies,* which in turn help to fuel brain function and spare the muscles to some degree; (4) after about a week of fasting, the acid

---

[1]This section is drawn from the article "Famine" by Joanne Csete in the 1991 edition of "The Encyclopedia of Human Biology."

forms of ketones, or *ketoacids,* increase in the blood and help provide fuel to the brain; and (5) as the second week of fasting nears, the brain is fueled preferentially by ketoacids rather than glucose, preventing the breakdown of remaining muscle protein, at least for a limited period. At this point and in the absence of other complications, the duration of survival depends on the quantity of remaining adipose stores.

Over an extended period, the brain appears not to function at its best when fueled by ketoacids rather than by glucose. Emotional changes, especially depression, are usually pronounced in persons suffering from starvation. As the body is focused on the protection of basic brain functions, so emotions and instincts become focused on simple survival, and social and kinship interactions decrease in value. Personal appearance also changes as muscle tone and body weight are rapidly lost. Edema or swelling of some body parts may occur. Many other signs of physiological and psychological deterioration have been documented in a range of famine situations.

In many famines, however, death can be attributed to infectious disease rather than to starvation itself. (It is often difficult to pinpoint the more important causal factor since undernutrition and infection exacerbate each other.) The calamitous events that may trigger famine are also likely to cause the breakdown of sanitation and basic health-care systems, increasing the likelihood of epidemics among persons in the famine region. In addition, undernutrition, even of a less severe degree than that experienced in famine, undermines the body's immune system that defends against infections. Diarrheal diseases gain an added foothold with deterioration of the gastrointestinal tract in the starved individual. Moreover, in response to famine, affected populations may gather in overcrowded, undersupplied makeshift camps or feeding sites that are likely settings for transmission of infectious disease. Thus, death from dysentery, cholera, and other diarrheal disease, diphtheria, louse-born typhus, other vector-borne parasitic disease, and respiratory infections are commonly reported in famine situations. That famine often occurs in areas already suffering from high prevalence of infectious disease and low levels of preventive service delivery only increases the likelihood of high mortality from infectious disease. [*See* Infectious Diseases.]

III. CAUSES OF FAMINE

The underlying cause of famine is crop failure, which undermines incomes of the already very poor. The most obvious causes of crop failure are bad weather, civil disruption, or both. When a crop fails, the subsistence farmer is robbed of his or her sole source of income, the food grown largely for consumption and for meager sales. For other poor groups, such as landless laborers, wage earners, and pastoralists, the crop failure reduces real income indirectly through sharply higher prices of food, reduced employment, or a combination of the two. These underlying forces can be exacerbated by market disruptions due to hoarding, inflation, or war, and by further changes in the income distribution. Government errors of omission and commission may permit the natural disaster to mature into famine. In his landmark publications, Amartya Sen has emphasized how lack of income to purchase food can exacerbate famine, but attention should not be drawn away from the critical role of agricultural production in determining income of the poor in developing countries and from monitoring a vital link in the chain of causality of famine.

A. Succession of Poor Crop Years

Seldom does famine arise from a single bad growing season. Rural people can usually absorb a single-year shortfall by depleting stocks or borrowing food and money from wealthier relatives and neighbors. Of course, large farmers are usually better insulated than small ones. History has illustrated that the second and subsequent years of crop failure portend disaster. For example, in 1966 and 1967, droughts and floods struck India, causing two successive years of crop failure. For India as a nation, foodgrain production fell by roughly 14% in each year from its predrought high. In the state of Bihar, farmers responded to the dryness by switching from rice to the more drought-resistant maize and millet crops. In the first year of the disaster, total foodgrain production fell only 5%. In 1966 and 1967, even hardier crops could not compensate for the rice crop failure, and production fell by 45% from the 1964–1965 level. The extremely good crop year of 1964–1965 helped protect consumption in the following year, but 1966–1967 was characterized by prefamine conditions of migration, high prices, hoarding, profiteering, and hunger.

The disaster that struck the Sahelian states in 1974 was the culmination of 8 years of poor weather and crops. In 1983 and 1984, drought, floods, and civil war once again crippled Ethiopia, Sudan, and the Sahel.

The production drop may be localized, and aggregate production may not fall in a famine year. For example, in both 1974 and 1984, floods in Bangladesh destroyed a portion of the rice crop; however, the same rains boosted the harvest in upland regions. Edward Clay cautions that "the paradox of hunger without severe loss of production depends on the level of aggregation." If a region is defined broadly enough, surpluses will outweigh deficits, although any given locality may go hungry. As long as markets are fragmented by poor transport and government inhibitions to trade, and income maintenance efforts fail, the production loss remains threatening.

B. Prices

Shocks to agricultural production often send agricultural prices soaring. The poor are squeezed from both sides as prices rise for the goods that they buy and fall for the goods that they sell. Because food expenditures represent a major share of a poor family's budget (71%), a leap in food prices puts a survival diet out of reach. A 10% decline in food supply leads to declines that differ by class in income, expenditures, and ultimately consumption; the poorest reduce total food consumption by 35%, whereas the wealthiest tighten their belts by only 8%. In absolute terms, the poor decrease their consumption of foodgrains more than 10 times as much as the wealthy in response to the same decline in supply. For those already living on the edge, this is famine.

To preserve their foodgrain consumption, the rich adjust by decreasing their demand for luxury foods and nonfood goods. The price of such goods in comparison with grain prices tends to fall during periods of famine. The demand for livestock is especially vulnerable. Consumer demand for the emaciated beasts collapses, while at the same time herders and farmers sell off livestock to purchase cheaper grain calories, further saturating the shrunken market. Predictably, nomadic herders have been especially hard hit by African droughts; in some regions of Ethiopia in 1974, the exchange rate between animals and grain deteriorated as much as 73%.

Likewise the large decline in nonfood expenditures by the prosperous precipitates dramatic price shifts for discretionary purchases vis-à-vis grain prices. In the Bengal famine, the terms of trade for common goods such as cloth, bamboo umbrellas, milk, fish, and haircuts, all produced by the poor, deteriorated as much as 70 to 80% between June 1942 and July 1943.

C. Employment

Employment crumbles as producers and consumers adjust to the decline in agricultural production, fur-

ther impoverishing marginal groups. As an example, in the drought years 1982–1984 compared to the subsequent normal year, in North Arcot in the state of Tamil Nadu, employment averaged 35% lower. Losses were especially severe for the landless; employment fell 40% for agricultural laborers and 45% for other laborers.

Several forces contribute to these declines. With less farming activity, landowners do not hire as many farm laborers. In North Arcot, India, in a major drought, agricultural employment declined for all but the largest farmers. Given the futility of tending crops, small farmers, agricultural laborers, and other laborers shifted to low-productivity tasks, such as additional tending of livestock. Concurrently, the fall in demand of well-to-do consumers for nonfood purchases has effects on employment as well as on prices. In the North Arcot case, the nonagricultural sector lost proportionately more jobs than the agricultural sector. In a situation of economic distress, large farmers cornered the few jobs in both the agricultural and nonagricultural sectors.

Changes in the labor intensity of agriculture following price disturbances represent a third potential interaction between crop failure and employment. As shortages forced up the price of rice in Bangladesh in 1974, that crop became more profitable in comparison with jute. In response, farmers shifted into rice from the far more labor-intensive jute production, which further contributed to unemployment for landless laborers. Thus, even while the food supply increased, the effects of the drought resulted in less purchasing power and famine for the poor, laboring class.

As these examples make clear, famine conditions redistribute incomes away from the poor, dealing them a smaller portion of a shrinking pie. The more employment falls, the less prices have to rise (and vice versa) in order for the economy to reach a new equilibrium following the agricultural disruption; either way, the poor bear the burden of adjustment. Furthermore, this process has a lasting effect because the poorest, with minimal stocks, resources, and access to credit, are forced to sell productive assets such as livestock, tools, and land. Household surveys under drought conditions in India show that such assets are usually sold on a falling market and reacquired after the famine at one-half to three times more than the sales price. Because they are able to protect their food consumption, employment, and productive assets, the more prosperous members of the community do not starve and rebound more quickly after famine.

A change in the distribution of income not only results from famine conditions but can exacerbate them. Added income in the hands of one group in society would, in the face of fixed food supplies, cause them to bid up food prices. The poor, with no protection against inflated food prices, would lose their access to food. Income gains by a subset of the poor are more likely to disrupt the food balance than income gains to the well-to-do; poor people use the additional income to satisfy their demand for basic food staples, whereas the well-to-do buy goods and services with a high employment content. Certainly changes in the income distribution contributed to the Bengal famine of 1943 as incomes generated by the wartime boom and government welfare policies protected some but not all groups of the poor from rapidly escalating food prices.

Nonetheless, demand is but one side of the food equation. Policies that focus solely on the distributional issues implied by a lack of purchasing power can worsen the claim on existing supplies (as occurred in Bengal). Furthermore, they distract attention not only from the long-term problem of abysmally low agricultural production, but also from the conditions of war, poor roads, and inadequate institutions that impede famine relief.

IV. GOVERNMENT POLICY TO PREVENT FAMINE

Government policy can ensure that natural disaster does not evolve into famine. Conversely, poor policies and armed conflicts heighten a nation's vulnerability to natural disaster. In some cases, policies and wars spark the crisis.

During China's Great Leap Forward, inflated 1958 grain statistics led the government to reallocate acreage coercively to nongrain production in 1959, while at the same time increasing purchases from the rural areas to be transferred to the cities. In addition, technical mismanagement on the part of overzealous and poorly qualified cadres and officials brought serious damage to farmlands by deep plowing, close planting, and water conservation errors. When 2 years of terrible weather hit the country in 1960 and 1961, the famine took on unprecedented proportions.

Although war cannot be blamed solely on the national government, the government's response to war greatly influences the evolution of a food crisis. War can contribute to food crises in several ways, including drawing labor from food production, dis-

rupting the marketing of agricultural inputs and crops, destroying fields, creating refugees, and hindering relief efforts.

In certain cases, war has induced famine in the absence of adverse weather conditions. The food crisis in Kampuchea peaked in 1979 after a decade of war-induced upheaval. Following the U.S. bombings and the genocidal and agricultural policies of the Pol Pot regime, the Vietnamese invasion of January 1979 dealt a crowning blow. The fleeing Khmer Rouge pursued a "scorched earth policy," both destroying and confiscating seeds, crops, and draught animals. Chaos hindered planting and the 1979–1980 rice harvest was 60% below normal. That then led to the price and employment effects noted earlier and the onset of famine.

An oft-cited example of war-related famine in a developed country is the Netherlands during World War II. Approximately 10,000 people of all social classes starved during the winter of 1944–1945 when the occupying Nazis blocked the import of food supplies.

Armed conflicts coincided with drought conditions in Chad, Angola, Ethiopia, Sudan, and Mozambique during the famines of 1983 through 1985. The worst hit regions of Ethiopia, Eritrea and Tigray, had been at war with the government for 22 and 9 years, respectively. Although it is unclear whether the Mengistu government intentionally tried to starve those regions, its refusal to permit safe passage for relief teams through the war zones and its controversial resettlement program impeded relief efforts.

At times it is the combination of war, mismanagement, and poor weather that leads to famine. In Bengal, the policy of removing all boats carrying more than ten passengers from the coast of Bengal and the limitation of cereal and rice trade between provinces hampered the flow of food to needy areas in 1942. In Ethiopia, some food was exported out of famine regions in 1973. In 1983, the Ethiopian government levied taxes in some drought-afflicted areas as part of the funding drive for the 10-year anniversary of the revolutionary government.

It was man-made disasters, not natural ones, that absorbed more than three-quarters of the disaster assistance channeled through the World Food Programme in the 1980s. Although famines may have been more controllable in recent years, some investigators suggest that the incidence of war-induced famines is on the upswing and that the worst famines in recent times are those associated with wars.

Although governments make errors, famine relief cannot be left to the vagaries of the marketplace.

As has already been shown, market forces, working through price and employment, place practically the whole weight of adjustment to a decline in food supply on the poor. The poor reduce food consumption by four times as much as the rich, and it is the poor who are already the least well fed. The effect is a major redistribution of income away from the poor. Positive government action must break the links that lead from natural disaster to death.

V. PREPAREDNESS AND RELIEF

There are positive steps that governments can take to ward off famine conditions by insulating the population from psychological or mental damage, migration, impoverishment, loss of assets, and death. To avoid the haphazardness of last-minute efforts to relieve such widespread suffering, governments must put in place measures to cope with the early stages of an incipient famine.

Because famines seldom arise overnight, policy makers must be alert to conditions that threaten the food system or public health. The best indicator of all is a democratic political system with its network of politicians sensing conditions at all levels of society. Their observations of social distress can be supplemented with statistical analyses of prices, hoarding, smuggling, transportation bottlenecks, food imports, refugees, or malnutrition, and even satellite photography. Such indicators are specific to the particular location and require a thorough understanding of the dynamics of the local food system, which is why local politicians are especially well placed. But for local government to be heard, a central government must be listening. A free press has a major role in amplifying the messages of distress, carrying them to the capital for action by the bureaucracy, and harassing the bureaucracy for its inattention. Famines rarely afflict democratic countries with a free press. The more intransigent problem of chronic hunger does afflict such countries.

An early warning system sounds the famine alert. Once a food crisis has been detected and diagnosed, a complementary plan of programs and previously established emergency measures must be quickly implemented. As long ago as the 1880s, India designed its Famine Codes in great detail to respond quickly and efficiently to disaster conditions. Similar ideas are now embodied in the Indian government's Drought Management Plan. Following the food crises of the early 1970s, the Food and Agriculture Organization

of the United Nations set up the Global Information and Early Warning System and the International Emergency Food Reserve.

By formalizing such procedures, several low-income countries have controlled natural disasters from developing into widespread famines. For example, in Bihar, India, symptoms of distress mounted following successive crop failures. Late in 1966, local administration was reinforced, communication lines improved, radio famine bulletins issued, and voluntary and official relief bodies formed. Existing public works and relief schemes grew from an average employment of 8000 to a peak of 700,000 by the end of 1967. The system of fair price shops and rationing was expanded, supported in part by massive quantities of food aid. Inoculations were given against smallpox and cholera. Water goals were set at 2 gallons per person per day and supported by well-drilling and water transport projects. Agricultural support included loans, efforts to save crops, pumps, and road projects. Out of a population of 53 million, estimates of starvation deaths ranged only from the hundreds to several thousands: more severe and widespread famine was averted. Effective national policies and foreign aid each played an important role.

Another state of India, Maharashtra, suffered three consecutive years of especially hostile weather from 1970 to 1972. Foodgrain production was 18, 29, and 54% below normal from 1971 to 1973, respectively, and there were great scarcities of food and water. A massive relief effort was undertaken, again based on employment projects with a productive, development focus on soil conservation, afforestation, canal excavation, irrigation, and drinking water. Nearly 5 million people were employed in relief work during the height of the crisis in 1973 and 1974, including large numbers of women who were employed at the same wages as men. Human and animal lives were protected by food and fodder distribution schemes, in addition to medical relief to combat epidemics. Given the efficiency of the state government, one may take seriously official reports that of the 25 to 30 million people threatened in the rural areas, there were no famine deaths during the drought period. Of course, chronic hunger and its effects continued through the period.

Ten years after the 1974 calamity, the Bangladeshi government was far better prepared for natural disaster. When monsoon flooding in 1984 wiped out as much as 30% of the crops in some districts, famine conditions were minimized. Edward Clay points to better institutions, some early warning mechanisms, better agricultural data, and the government's greater commitment and ability to contain the famine. Since the previous disaster, storage capacity had been expanded and distributed throughout the country to hasten response time. Employment and supplemental feeding programs were already in place, supporting living standards for the poor, and were expanded when the crisis mounted. More plentiful agricultural and foreign exchange resources better protected farmers from floods and crop failures and enhanced the country's capacity to import from the world market. Improved donor relations, including the presence of multilateral World Food Programme and multiyear commitments of food aid, kept reliable supplies in the pipeline.

Botswana, within Africa's Kalahari Desert, suffers from an average of three severe drought years each decade. In 1975 the government took the initial steps toward designing a drought prevention plan and in 1978 a private group, the Botswana Society, convened a national symposium on drought. In the same year the government established the Interministerial Drought Committee; it meets monthly to review data on rain, agricultural conditions, national grain storage levels, donor commitments, and nutritional standards. The agricultural indicators signal the need for agricultural support measures such as credit, seed distribution, and cattle buy-outs, and the nutritional data trigger expansion of ongoing feeding programs and employment schemes. During the 1983 drought, Botswana was able to limit the increase in child malnutrition from 26 to 31%, and in this context reportedly withstood the drought better than its South African neighbors. Several factors contributed to Botswana's success: disaster measures in drought-prone Botswana are a political necessity; Botswana's proximity to South Africa benefits its trade and transportation network; diamond mine revenues have strengthened the government's budget and foreign exchange positions; and the country has a politically stable, liberal democracy. In addition to and as a result of these factors, Botswana receives a generous share of foreign aid to support its programs.

As these examples show, success in warding off famine requires tremendous coordination and efficiency. The international community should not despair, to the further loss of the poor, that aid is delivered inefficiently. Rather aid donors should work to ameliorate the nearly impossible conditions that accompany disaster situations. Famine prevention, preparedness, and relief complement each other. Data collected on crops, climate, prices, and so on provide

early warning indicators that can be applied to development planning. Research at the International Food Policy Research Institute has shown that unexpected changes in rural public works employment can indicate potential famine conditions months in advance, as people abandon their failing crops for stopgap employment. In addition to signaling an alert, such programs simultaneously provide jobs and purchasing power to likely victims before the situation deteriorates, while supporting the road, port, and bridge building projects that enhance both relief and prevention efforts.

VI. FAMINE PREVENTION

Famine preparedness and relief necessarily emphasize distribution. Prevention returns us to the underlying cause of famine—decline in food supply and the means of purchasing food. To prevent famine, both the supply side and the demand side of the food equation must balance at a higher level, one that provides a margin of safety in food supplies, employment, and income for the poor. This process requires economic development and political stability.

Thirty years of experience have demonstrated that the transformation from an agrarian to an industrial society must proceed through the development of the agricultural sector. Policies must focus on increasing the productivity of the rural workers who make up the majority of the country's laborers. As agricultural surpluses grow, producers sell rather than consume a greater portion of their output. Livestock numbers grow, absorbing surplus grain and providing animal protein. In time, the farm workforce shrinks, liberating resources for industrial expansion. As wealth grows, food becomes a smaller portion of the household budget. In the event that disaster strikes, food stocks are adequate, livestock are slaughtered, foreign trade and finances are adjusted, unemployment is limited, and price swings are tolerable. The conditions of scarcity that characterize famine cease to arise.

Initiating agricultural productivity growth is a complex and time-consuming process. To oversimplify, there are three priorities: agricultural research, the increased use of purchased inputs in farming, and the development of a comprehensive physical and human infrastructure.

Modern science offers new opportunities for eliminating famine. By radically increasing yields per acre, science can generate food margins that are adequate to withstand weather shocks. Plants can also be bred for drought resistance; development of short-cycle varieties in Senegal, for example, saved the groundnut and millet crops from total devastation by erratic rains in recent years. Science can also help generate the employment and hence income needed to buy this food. Evidence from India indicates that high-yielding wheat varieties require 60% more labor per hectare, but less labor per unit of output, than traditional varieties. However, technology does not transfer directly. National research institutions must be established in developing countries to hasten the adaptation of technologies from other agricultural zones.

Similarly, modern farm inputs, and the institutions to provide them, are needed to complement agricultural research, which relies heavily on fertilizers, herbicides, insecticides, and commercial seed. Mechanization may also be necessary in a few specific cases where seasonal labor shortages or low labor productivity prevail. However, in general, appropriate technology for low income countries is labor intensive. Of course, traditional agricultures continually generate improved technology, but typically provide only a 0.5% rate of growth, which is not sufficient to keep up with population growth. It is only from the efficiencies of modern, science-based research that rates sufficient to reduce famine can be achieved.

Transportation, communications, and administrative capacity are obviously essential in the long run for preventing famines and in the short run for relieving them. The image of the isolated, self-sufficient village may be romantic but, certainly is not practical, particularly as growing populations press against the limited land base. Economic development demands and provides an infrastructure capable of moving goods in and out of the farming sector. Agricultural commercialization generates the incomes and surpluses necessary to absorb occasional disruptions. Relief measures require roads, communication, and political ties or else pockets of famine remain inaccessible, as in Eritrea and Tigray, Ethiopia. Food aid used for relief measures can be extended into the post-famine period to build infrastructure and provide administrative capacity.

Of course, since rapidly growing populations exacerbate the need for increased food production, family planning can also contribute to long-term famine prevention. However, particularly in rural areas, family planning seems to accompany or follow the processes of agricultural growth and commercialization.

VII. TACKLING FAMINES IN AFRICA

Africa has become the quintessential famine arena. In 1972 and 1973, continuing drought in Ethiopia and the Sahel caused water holes to dry up, crops to fail, and humans and livestock to perish. Disaster again struck sub-Saharan Africa in 1983. According to the Food and Agriculture Organization, abnormal food shortages were reported in more than half of Africa's countries, placing at risk as many as 30 million people, or 16% of the total population. By 1985, 11 million people were displaced and herds were decimated. Death counts for the region are still uncertain: a United States Senate staff report estimated 1 million deaths in Ethiopia and the Sudan together.

Even in periods without famine, the continent is characterized by poverty, malnutrition, high infant mortality, and high birth rates. It is estimated from a sample of 12 African countries that almost 25% of all preschool children are chronically undernourished (below the 90% reference of height-for-age). Data show that Africans tend to gain and lose weight with the seasons, since the most strenuous energy demands occur just before harvest when grain bins are bare. Data from Zambia show that even when people eat more calories during the planting and weeding seasons, their body weight still declines. Because area planted decreases with the debilitation of the labor force after a poor crop year, food shortages become a self-perpetuating process, trapping Africa in a downward spiral. Although infant mortality rates have shown significant declines in recent years, sub-Saharan Africans lose on average about ten times as many infants per thousand population as North Americans or Europeans. Birth rates, however, are among the highest in the world, and the rate of population increase is the highest of all continental regions.

The African food situation is discouraging, and much of the continent remains vulnerable to famine. Food production growth has been very poor in Africa as compared with Asia, and the gap between the two is widening. In Asia (excluding China), annual food production growth rates per capita were 0.2% in the 1960s and 0.8% in the 1970s. The corresponding rates for sub-Saharan Africa were −0.4 and −1.5%. Increased food imports and dependency on food aid failed to compensate for this shortage, and per capita consumption of grain decreased at an annual rate of 0.4% during the 1970s while world consumption increased by 0.8%.

Reflecting this low food production, marketable surpluses for the peasant sector are lower in Africa than those in Asia. This large sector consumes nearly all of what it produces, bringing to market only the remaining 10 to 20%. In addition, labor productivity in the dominant peasant sector is less than half that in the comparable sector of India.

Reasons for the low marketing and productivity are complex. Weather certainly has played a role in recent years. It is unclear whether frequent drought of recent decades is part of a long-term trend or short-term cycle, but it has been drier than in the 1950s and early 1960s. Increased population and livestock pressure on marginal lands has accelerated desertification.

Earning a profit in farming is difficult when poor infrastructure hinders the movement of farm inputs and produce and when poor policies limit incentives. Africa has an especially poor infrastructure. A comparison of kilometers of road and percentage of paved road in several African nations, India, and the United States shows that road infrastructure in African nations is very bad—typically one-tenth as much road length per square kilometer as India, which is two-thirds that of the United States. And India has only half the proportion paved as the United States, with African countries typically one-fourth to one-half that of India.

Popular focus with respect to poor economic performance in Africa has emphasized the need for policy reform, including higher agricultural prices, substitution of rural for urban bias, elimination of state-controlled marketing systems, and state and other types of large-scale farms. Most governments in Africa are proceeding with such policy reforms, returning attention to the positive needs of research, reliable input supply, and infrastructure emphasized here.

Good policy is difficult in the face of political unrest as governments try to achieve national unity in the wake of a purposely divisive colonial history. Cooperative efforts to achieve regional food security, such as the Southern African Development Coordination Conference, have promise, but are limited by problems of consensus. Given these political problems, it is still difficult for many African governments to focus even on relief measures, let alone on long-term projects of agricultural development.

In addition to the need for peace and sound policies, Africa must develop its own combination of green revolution technologies akin to those that launched Asian development. This is no minor task, considering the well-known paucity of trained people, resources, and institutions in Africa.

VIII. FOREIGN ASSISTANCE

Foreign assistance has an important role to play in solving the problems of famine. Food and financial aid can be used by African nations to build infrastructure and support relief, employment, and nutrition projects, as well as to provide foreign expertise and training for citizens. Foreign assistance must be applied in the context of the national political environment, an especially difficult task in war-torn or once-colonized countries. Thoughtful application of foreign assistance can help build nations if care is taken to avoid the divisive tendencies of power politics. Though the application of foreign assistance may be inefficient, it is by no means dispensable.

Curing famine, especially in Africa, will take time. We must not look for miracle shortcuts to a problem that has plagued humankind for centuries, yet we must hesitate no longer. By adopting lessons learned elsewhere, we will gradually be able to treat disaster situations as they occur and finally achieve the development that will eradicate famine.

BIBLIOGRAPHY

Clay, E., and Everitt, E. (eds.) (1985). "Food Aid Emergencies: A Report on the Third IDS Food Aid Seminar," Discussion Paper No. 206. Institute of Development Studies, University of Sussex, Brighton, England.
Currey, B., and Hugo, G. (1984). "Famine as a Geographical Phenomenon." Reidel, Amsterdam.
Dando, W. (1980). "The Geography of Famine." Arnold, London.
Ezekiel, H. (1987). "A Rural Employment Guarantee as an Early Warning System." International Food Policy Research Institute, Washington, D.C.
Joachim, von Braun, and Kennedy, Eileen (1995). "Economic Development and Nutrition." Baltimore.
Mellor, J. W. (1978). "Food Price Policy and Income Distribution in Low Income Nations." *Econ. Dev. Cultural Change* 27(1), 1.
Mellor, J. W. (1986). Prediction and prevention of famine. *Fed. Proc.* 45(10), 2427.
Mellor, John W. (ed.) (1995). "Agriculture on the Road to Industrialization." Johns Hopkins University Press, Baltimore.
Mellor, J., and Johnston, B. (1984). "The World Food Equation: Interrelations among Development, Employment, and Food Consumption." *J. Econ. Lit.* 22, 531.
Robson, J. R. K. (ed.) (1981). "Famine: Its Causes, Effects, and Management." Gordon & Breach, New York.
Scrimshaw, N. S. (1987). The phenomenon of famine. *Annu. Rev. Nutr.* 7(1).
Sen, A. (1981). "Poverty and Famines: An Essay on Entitlement and Deprivation." Clarendon Press, Oxford, England.
Torry, W. I. (1984). Social science research on famine: A critical evaluation. *Hum. Ecol.* 12(3), 227.

Fatty Acids

STUART SMITH

Children's Hospital, Oakland Research Institute

FATTY ACIDS ARE COMPOSED OF A HYDROCARbon chain, from which the properties of lipid solubility derive, and a terminal carboxyl group that imparts acidic properties. The commonest fatty acids contain from 4 to 24 carbon atoms, although some with more than 30 carbon atoms have been described. When all of the carbon atoms in the hydrocarbon chain are bonded to the maximum number of hydrogen atoms, the fatty acid is said to be saturated and when adjacent carbon atoms are bonded to only one hydrogen atom, they are said to be unsaturated; all of the unsaturated fatty acids synthesized by humans have the double bonds in the *cis* orientation. In humans, fatty acids represent a major storage form of energy, they provide an essential structural component of membranes, they are used to modify and regulate the properties of many proteins through direct covalent linkage, and certain of the polyunsaturated fatty acids are the precursors of essential signaling molecules. All saturated and many of the common unsaturated fatty acids can be synthesized by humans, but certain essential polyunsaturated fatty acids must be supplied in the diet. The influence of dietary fatty acids on the risk of cardiovascular disease continues to be a vigorously debated and controversial topic. A number of genetic defects in the metabolism of fatty acids have been identified, most involving an enzyme deficiency that results in abnormal lipid storage in a tissue. [*See* Lipids.]

I. STRUCTURE AND NOMENCLATURE

When all of the carbon atoms in the hydrocarbon chain are each bonded to two hydrogen atoms, three in the case of the terminal carbon atom, the fatty acid is said to be a saturated or alkanoic acid. When two adjacent carbon atoms are each bonded to only one hydrogen atom, an ethylenic double bond exists between the carbons and the fatty acid is said to be an unsaturated or alkenoic acid. Fatty acids containing only one double bond are said to be monounsaturated, whereas those containing more than one double bond are polyunsaturated. Most naturally occurring unsaturated fatty acids have the *cis* configuration about the double bond, which introduces a kink into the molecule. The rarer *trans*-isomers lack the kink and physically resemble the saturated fatty acids (Fig. 1).

Many fatty acids are still referred to by trivial names adopted from the original source of the compound, e.g., palmitic acid (palm oil), lauric acid (laurel tree), oleic acid (Latin, *oleum*, oil), and linoleic acid (linseed oil). The scientific names, however, indicate the number of carbon atoms as well as the location of any double bonds present and their orientation, *cis* or *trans*. Shorthand forms in which the fatty acids are identified by the number of carbon atoms and the number of double bonds are commonly used. Thus palmitic (hexadecanoic) acid is 16:0, or $C_{16:0}$. Three different shorthand forms have found common usage in describing the structures of unsaturated fatty acids (Fig. 1). One based on the use of the Greek letter delta (Δ) indicates how far a double bond is from the carboxyl carbon. Thus linoleic acid, *cis*-9, *cis*-12-octadecadienoic acid, can be abbreviated $C18:2\Delta9c,12c$. Because the double bonds are all *cis* and are separated by a carbon with only single bonds (unconjugated, or methylene interrupted) in the com-

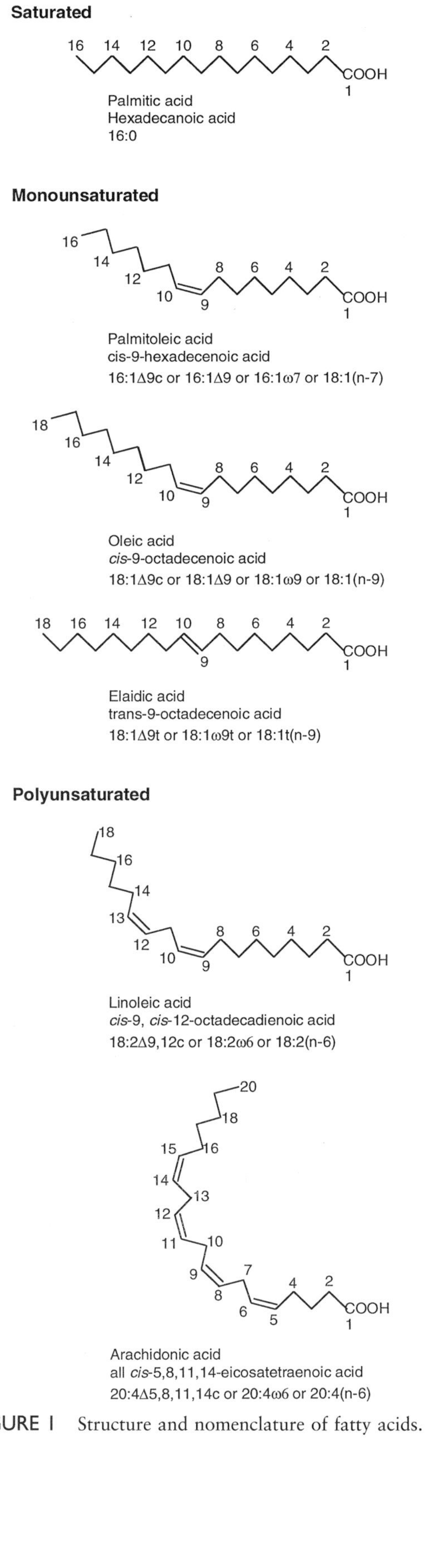

Saturated

Palmitic acid
Hexadecanoic acid
16:0

Monounsaturated

Palmitoleic acid
cis-9-hexadecenoic acid
16:1Δ9c or 16:1Δ9 or 16:1ω7 or 18:1(n-7)

Oleic acid
cis-9-octadecenoic acid
18:1Δ9c or 18:1Δ9 or 18:1ω9 or 18:1(n-9)

Elaidic acid
trans-9-octadecenoic acid
18:1Δ9t or 18:1ω9t or 18:1t(n-9)

Polyunsaturated

Linoleic acid
cis-9, *cis*-12-octadecadienoic acid
18:2Δ9,12c or 18:2ω6 or 18:2(n-6)

Arachidonic acid
all *cis*-5,8,11,14-eicosatetraenoic acid
20:4Δ5,8,11,14c or 20:4ω6 or 20:4(n-6)

FIGURE I Structure and nomenclature of fatty acids.

monest type of polyunsaturated fatty acid, frequently only the position of the first double bond is given. Two alternative shorthand systems indicate the position of the double bond closest to the methyl, or ω carbon atom. In the ω and 'n' systems, which are essentially interchangeable, linoleic acid becomes C18:2ω6 or 18:2(n-6).

II. BIOLOGICAL FUNCTION

A. Energy Storage

In humans, fatty acids represent a major storage form of energy. When caloric intake exceeds energy demand and glycogen stores are saturated, excess dietary carbohydrate is converted to fatty acid and is stored in adipose tissue as triglyceride, together with excess dietary fat. When energy demand cannot be met by the utilization of stored glycogen, fat is mobilized from adipose and is transported to tissues where it is taken up and oxidized for energy production. Muscle is particularly well adapted for using fatty acid oxidation as an energy source and the capacity of this tissue for fatty acid oxidation is enhanced by exercise.

B. Membrane Structure

Fatty acids provide an essential structural component of all biological membranes, including the plasma membrane and internal membranes of eucaryotic cells. These membranes consist of a lipid bilayer containing phospholipid, glycolipid, and cholesterol. Various protein receptors are embedded in the membrane that receive and transduce environmental signals as well as transport proteins that facilitate the flux of nutrients, ions, and waste materials. The fatty acid composition of membrane lipids play a critical role in determining the fluidity and functional properties of the membrane: fluidity is decreased by saturated fatty acids and increased by unsaturated fatty acids.

C. Signaling

The essential polyunsaturated fatty acids linoleic and linolenic acids are the precursors of hormone-like molecules prostaglandins and leukotrienes which regulate a variety of physiological process.

D. Covalent Modification of Proteins

Covalent attachment of long chain saturated fatty acids to many proteins serves to direct subcellular localization and/or influence function. The commonest modifications involve myristoylation (14:0) of an N-terminal glycine and palmitoylation (16:0) of internal cysteine residues. N-terminal myristoyl moieties do not appear to undergo independent turnover whereas palmitoylation is often a dynamic process. Many palmitoylated proteins are membrane associated, and the dynamic acylation of receptor and signaling proteins can affect both association with the membrane and interaction with other proteins. The transferases and thioesterases that catalyze the turnover of protein-bound palmitoyl residues have not been well characterized.

III. NUTRITIONAL ASPECTS

A. Dietary Sources of Fatty Acids

The average diet of adult Americans contains about 160 g of fat per day, amounting to about 40% of caloric needs. Most of the fat is ingested as triglyceride that typically is derived from butter, margarine, cheese, shortenings, cooking and salad oils, nuts, and fat associated with meat, poultry, and fish. The fatty acid compositions of some common dietary constituents are presented in Table I.

B. Polyunsaturated Fatty Acids

Two dietary polyunsaturated fatty acids, linoleate (18:2 Δ9,12) and linolenate (18:3 Δ9,12,15), members of the n-6 and n-3 series, respectively, are essential to humans because they cannot be synthesized by the body. Linoleate and linolenate are essential in part because they are precursors for prostaglandins and leukotrienes, hormone-like compounds that regulate a number of cellular activities. In addition, polyunsaturated fatty acids regulate the fluidity of cell membranes and influence their functional properties. Di-

TABLE I
Fatty Acid Composition of Some Dietary Fats

Shorthand	Common name	Systematic name	Human milk	Cow milk	Olive oil	Corn oil	Sunflower oil	Cod liver oil	Salmon fillet
4:0	Butyric	Butanoic		3.3					
6:0	Caproic	Hexanoic		1.6					
8:0	Caprylic	Octanoic		1.3					
10:0	Capric	Decanoic	1.3	3.0					
12:0	Lauric	Dodecanoic	3.1	3.1					
14:0	Myristic	Tetradecanoic	5.1	9.5		1		4.9	2.9
16:0	Palmitic	Hexadecanoic	20.2	26.3	12	14	6	12.4	10.7
18:0	Stearic	Octadecanoic	3.8	14.6	2	2	6	1.8	3.6
20:0	Arachidic	Eicosanoic							
22:0	Behenic	Docosanoic							
16:1(n-7)	Palmitoleic	*cis*-9-Hexadecenoic	5.7	2.3	1			11.6	5.0
18:1(n-9)	Oleic	*cis*-9-Octadecanoic	46.4	29.8	72	30	33	22.6	24.5
18:2(n-6)	Linoleic	*cis*-9, *cis*-12-Octadecadienoic	13.0	2.4	11	50	52	1.4	5.2
18:3(n-3)	α-Linolenic	all *cis*-9,12,15-Octadecatrienoic	1.4	0.8	1	2			5.3
18:4(n-3)		all-*cis*-6,9,12,15-Octadecatetraenoic						1.9	1.5
20:1(n-9)		*cis*-11-Eicosaenoic						7.6	1.0
20:4(n-6)	Arachidonic	all-*cis*-5,8,11,14-Eicosatetraenoic						0.5	5.3
20:4(n-3)		all-*cis*-8,11,14,17-Eicosatetraenoic							2.3
20:5(n-3)		all-*cis*-5,8,11,14,17-Eicosapentaenoic						12.6	4.5
22:1(n-11)		*cis*-11-Docosaenoic						5.2	
22:4(n-6)		all-*cis*-7,10,13,16-Docosatetraenoic							2.2
22:5(n-3)		all-*cis*-7,10,13,16,19-Docosapentaenoic							5.0
22:5(n-6)		all-*cis*-4,7,10,13,16-Docosapentaenoic							2.3
22:6(n-3)		all-*cis*-4,7,10,13,16,19-Docosahexenoic						10.6	19.0

etary n-3 and n-6 polyunsaturated fatty acids lower blood levels of cholesterol and have antithrombic properties. The n-3 fatty acids 20:5 and 22:6 are significant components of fish oils and some evidence indicates that the ingestion of fish oil leads to decreased mortality from coronary heart disease.

C. *trans* Fatty Acids

These unusual fatty acids enter the food supply in small amounts from dairy and ruminant fats but mainly through the industrial hydrogenation of vegetable and marine oils in the manufacturing of margarine and shortening. The most common *trans* fatty acids found in these products are positional isomers of *trans*-octadecenoic acid with the double bond in the $\Delta 9$-$\Delta 12$ position, the exact location depending on the conditions of hydrogenation. *Trans* fatty acids have been reported to increase serum concentrations of low density lipoprotein cholesterol, triglycerides, and lipoprotein(a) and to decrease high density lipoprotein cholesterol, all considered as risk factors for coronary heart disease. In this regard the consumption of dietary *trans* fatty acids appears to have consequences more akin to those associated with the consumption of saturated rather than *cis*-unsaturated fatty acids. These effects have been attributed to the structural similarity between *trans* and saturated fatty acids (Fig. 1). Nevertheless, many scientists regard data on the health effects of dietary *trans* fatty acids as insufficient to warrant an outright band on their introduction into the food supply.

D. Digestion and Absorption of Fat

Fatty acids are ingested predominantly in the form of triglyceride. Digestion begins in the mouth by the action of a lingual lipase, continues in the stomach through the action of a gastric lipase, and is completed in the upper small intestine by the action of pancreatic lipase. The products of digestion, mainly free fatty acids and monoglycerides, are solubilized by the action of bile salts and are absorbed into the epithelial cells lining the intestine where they are reesterified to form triglycerides. The small droplets of triglycerides are stabilized by the addition of a layer or protein, mainly apoA and apoB-48, phospholipid and cholesterol, forming "chylomicrons." The chylomicrons pass through the epithelial cells into the lymphatic system and enter the blood at the angle of the left jugular and subclavian veins. The chylomicrons give

blood a milky appearance, a condition called postprandial lipemia. Fatty acids of less than 10 carbon atoms are water soluble and can enter the hepatic portal vein directly from the intestinal epithelial cells.

IV. METABOLISM

Fatty acids are stored and transported mainly as triglyceride in which three fatty acids are esterified to the three hydroxyls of glycerol. In plasma the triglycerides form the hydrophobic core of the transport lipoproteins. In biological membranes the fatty acids are incorporated primarily into phospholipids. A third type of lipid contains fatty acid esterified to cholesterol. The cholesterol esters are synthesized mainly in the liver and are transported in the hydrophobic core of plasma lipoproteins. A small amount of fatty acid is present in the cell and in blood as the free acid. In blood these unesterified fatty acids are transported mostly bound noncovalently to albumin. Within the cells of the body the fatty acids are transported bound noncovalently to fatty acid-binding proteins or, as the CoA esters, are associated noncovalently with acyl-CoA-binding protein. Fatty acids undergo metabolic conversions primarily as their CoA thioesters, which are formed by fatty acyl-CoA synthetases associated with the endoplasmic reticulum, the outer mitochondrial membrane, and the cytosolic face of the peroxysomal membrane. [*See* Coenzymes, Biochemistry.]

V. BIOSYNTHESIS

A. The *de Novo* Pathway

The major sites of synthesis of saturated fatty acids are in the soluble cytoplasm of liver, adipose tissue and, during lactation, the mammary gland. Acetyl-CoA carboxylase catalyzes the initial conversion of acetyl-CoA to malonyl-CoA, which subsequently is utilized as the carbon source for the elongation of an acetyl primer moiety to a long chain fatty acid in a series of iterative condensation and reductive reactions catalyzed by a single protein, the fatty acid synthase (Fig. 2). At each condensation step the third carbon atom of the malonyl moiety is lost as carbon dioxide. The fatty acid synthase protein consists of two identical large multifunctional polypeptides, each

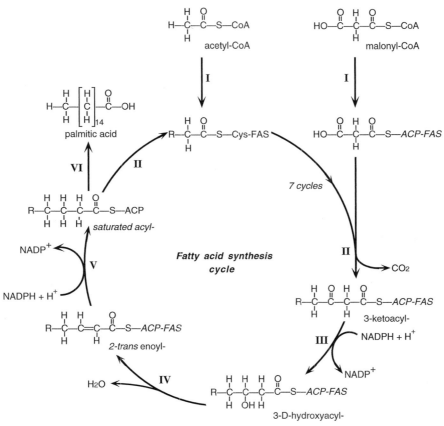

FIGURE 2 Pathway for the biosynthesis of fatty acids *de novo*. All reactions are catalyzed by a single multifunctional protein, the fatty acid synthase. The individual functional domains of the fatty acid synthase (FAS) are: ACP, acyl carrier protein; I, malonyl-CoA/acetyl-CoA : ACP acyl transferase; II, β-ketoacyl synthase; III, β-ketoacyl reductase; IV, dehydrase; V, enoyl reductase; VI, thioesterase. R may be H or $CH_3(CH_2)_n$, where n = 1,3,5,7,9, or 11.

containing six discrete enzymes and an "acyl carrier protein."

The kinetics and specificities of the component enzymes are well adapted to ensure that the iterative condensation of an acetyl moiety with successive malonyl moieties and complete reduction of the intermediates normally results in the formation of palmitic acid as the major product. Acetyl-CoA carboxylase is the pace-setting enzyme for the pathway and is subject to short-term regulation in response to changes in metabolic and nutritional status. This acute regulation is achieved through both the binding of allosteric ligands such as citrate and fatty acyl-CoAs and the phosphorylation of several serine residues that shift the equilibrium existing between the low activity dimeric form of the enzyme and the high activity polymeric form. Longer term regulation of the fatty acid biosynthetic pathway is mediated at the level of transcription of the two genes encoding acetyl-CoA carboxylase and fatty acid synthase.

B. Elongation and Desaturation

Fatty acids with more than 16 carbon atoms can be synthesized by either of two elongation systems present in liver and other tissues. One system associated with the endoplasmic reticulum uses malonyl-CoA as the elongating carbon source, whereas the other system associated with mitochondria uses acetyl-CoA. Unlike the *de novo* system for palmitic acid biosynthesis, the elongation systems do not utilize an acyl carrier protein, but appear to couple the intermediates directly to CoA. The individual enzymes of the elongation systems have not been well characterized.

The desaturation of acyl-CoAs involves the insertion of a *cis* double bond into the acyl chain catalyzed

(ω–6) family

18:2 Δ9,12
(linoleic)

↓ *Δ6 desaturase*

18:3 Δ6,9,12

↓ *elongase*

20:3 Δ8,11,14

↓ *Δ5 desaturase*

20:4 Δ5,8,11,14
(arachidonic acid)

(ω–3) family

18:3 Δ9,12,15
(α–linolenic)

↓ *Δ6 desaturase*

18:4 Δ6,9,12,15

↓ *elongase*

20:4 Δ8,11,14,17

↓ *Δ5 desaturase*

20:5 Δ5,8,11,14,17
all-*cis*-eicosapentaenoic

↓ *elongase*

22:5 Δ7,10,13,16,19
all-*cis*-docosapentaenoic

↓ *Δ4 desaturase*

22:6 Δ4,7,10,13,16,19
all-*cis*-docosahexaenoic

FIGURE 3 Pathways for the elongation and desaturation of the essential fatty acids.

by microsomal enzymes that require NADH and molecular oxygen. One of the commonest desaturases, the Δ9 desaturase, inserts a double bond between the 9th and 10th carbon atoms, e.g., in the conversion of palmitate (16:0) to palmitoleate (16:1Δ9) and stearate (18:0) to oleate (18:1Δ9).

Two important pathways involving the participation of both desaturates and elongation enzymes effect the conversion of linoleate and linolenate to important biologically active compounds (Fig. 3). In the ω-6 family of unsaturated fatty acids, linoleate is converted to arachidonate (20:4Δ5,8,11,14) and in the ω-3 family, linolenate is converted to eicosapentenoic (20:5 Δ5,8,11,14,17) and docosahexenoic acid (22:6 Δ4,7,10,13,16,19). These compounds are the immediate precursors of the prostaglandins and leukotrienes. The three major classes of unsaturated fatty acids, ω-9 (e.g., oleic), ω-6 (e.g., linoleic), and ω-3 (e.g., linolenic), are not metabolically interconvertible.

VI. OXIDATION

A. The Mitochondrial β-Oxidation Pathway

Fatty acyl-CoA thioesters formed by acyl-CoA synthetases located on the outer membrane of the mitochon-

dria cannot permeate the mitochondrial membrane directly. Thus long chain (<22 C-atoms) fatty acids are translocated across the inner mitochondrial membrane by a carnitine (L-3-hydroxy-4-trimethylammonium butyrate) carrier. Carnitine acyl-CoA transferase I, located on the outer side of the inner mitochondrial membrane, and transferase II, located on the inside of the membrane, catalyze the exchange of acyl moieties between CoA and carnitine. A translocase located in the membrane shuttles the acyl carnitine in and free carnitine out. Fatty acid oxidation involves the sequential action of four types of enzyme: a FAD-containing acyl dehydrogenase that produces a 2-*trans*-unsaturated derivative, an enoyl hydrase that hydrates the double bond to a L-3-hydroxy derivative, a hydroxyacyl dehydrogenase that converts the L-hydroxyl to a keto group, and a β-ketoacyl thiolase that splits off acetyl-CoA, yielding an acyl-CoA with two fewer carbon atoms (Fig. 4). Several of these enzymes exist in multiple forms having different substrate specificities. The short, medium, and long chain acyl-CoA dehydrogenases, short chain enoyl hydrase, short chain 3-hydroxyacyl-CoA dehydrogenase, and 3-ketoacyl-CoA thiolase are discrete, monofunctional proteins, whereas the long chain enoyl hydratase and long chain 3-hydroxyacyl-CoA dehydrogenase are integrated into a single polypeptide, the α-subunit. Four copies of the α-subunit together with four copies of the β-subunit, a long chain 3-ketoacyl-CoA thiolase, form a trifunctional multienzyme complex. This complex is associated with the inner mitochondrial membrane, whereas most of the other enzymes are located in the soluble matrix.

Oxidation of unsaturated fatty acids proceeds in a similar manner, except that additional enzymes are required: Fatty acids with double bonds at odd numbered carbon atoms ultimately yield *cis*-3 enoyl-CoA intermediates, e.g., by three cycles of β-oxidation of oleoyl-CoA (18:1Δ9c). These intermediates are then converted by a Δ3,Δ2 enoyl-CoA isomerase to *trans*-2 enoyl-CoA compounds that can be utilized by the enoyl-CoA hydratase. An alternative pathway proposed for the oxidation of fatty acids with double bonds at odd numbered carbon atoms involves reduction at the Δ5 stage. A 2,4-dienoyl-CoA reductase catalyzes the reduction of 2-*trans*, 4-*cis*-dienoyl-CoA intermediates to 3-*trans*-enoyl-CoA compounds which can undergo isomerization catalyzed by the Δ3,Δ2-enoyl-CoA isomerase.

Each turn of the fatty acid oxidation cycle generates five molecules of ATP; if the acetyl-CoA is oxidized through the citric acid cycle, 12 ATPs are generated

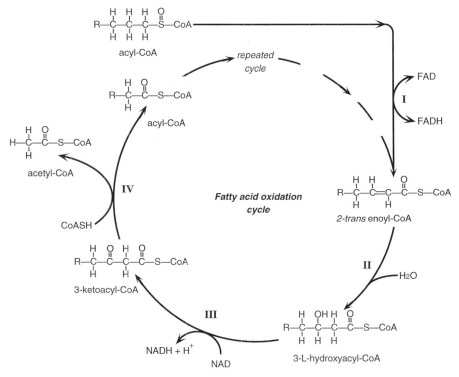

FIGURE 4 Mitochondrial pathway for the β-oxidation of fatty acids. I, FAD-dependent acyl-CoA dehydrogenase; II, enoyl-CoA hydrase; III, hydroxyacyl-CoA dehydrogenase; IV, β-ketoacyl-CoA thiolase.

per turn. The overall rate of fatty acid oxidation is controlled by the carnitine acyl-CoA transferase I, which regulates the rate of entry of fatty acids into the mitochondria. This enzyme is extremely sensitive to inhibition by malonyl-CoA, the concentration of which normally is low during fasting and elevated following feeding.

B. The α-Oxidation Pathway

A minor pathway for the metabolism of 3-methyl-branched and 2-hydroxy fatty acids involves the removal of only a single carbon at each step and, unlike the β-oxidation pathway, does not involve formation of a CoA thioester. In humans the primary site for α-oxidation appears to be the peroxysomes. Fatty acids containing a methyl substitution at carbon 3 are first hydroxylated at carbon 2 and then the carboxyl carbon is removed by a decarboxylase leaving the methyl group at carbon 2. The 2-methyl-branched fatty acid can then enter the β-oxidation pathway. 2-Hydroxy fatty acids also occur naturally mainly as components of brain glycolipids known as cerebrosides. These

fatty acids are decarboxylated directly by the α-oxidation pathway.

C. The ω-Oxidation Pathway

A minor, less well-characterized pathway for fatty acid oxidation, catalyzed by enzymes located in the endoplasmic reticulum and soluble cytoplasm, is used primarily for the oxidation of medium chain fatty acids. A cytochrome P450-mediated monooxygenase uses NADPH and molecular oxygen to hydroxylate the ω-carbon atom or, somewhat less efficiently, the ω-1-carbon atom. The hydroxyl group is then oxidated by, successively, NAD-dependent alcohol and aldehyde dehydrogenases in the endoplasmic reticulum and soluble cytoplasm. The resulting medium chain dicarboxylic acids can then enter the mitochondria and are degraded by β-oxidation.

D. The Peroxysomal β-Oxidation Pathway

Peroxysomes catalyze the β-oxidation of a variety of fatty acids and fatty acid derivatives essentially by the same mechanism as the enzymes of the mitochondrial system. Some substrates are oxidized exclusively in peroxysomes, including very long chain (>22 carbon atoms) fatty acids, long chain dicarboxylic acids, phytanic acid, prostaglandins, leukotrienes, and certain mono- and polyunsaturated fatty acids. Fatty acids are first activated to the CoA ester by a variety of acyl-CoA synthetases, having different chain-length specificities, that are located on the cytosolic side of the peroxysomal membrane. The acyl-CoAs are then transported across the membrane by a poorly understood mechanism and are subjected to oxidation by acyl-CoA oxidases. Like the mitochondrial acyl-CoA dehydrogenases, these enzymes are flavoproteins. However, unlike the mitochondrial system, the reoxidation of the reduced flavine occurs via direct interaction with molecular oxygen, yielding H_2O_2. Two mechanisms have been proposed to account for the ability of peroxysomes to oxidize polyunsaturated fatty acids. Because the Δ2-cis-enoyl-CoA intermediates formed during β-oxidation are hydrated to the D-3-hydroxyacyl-CoAs, configuration first must be inverted to the L-form before normal β-oxidation can resume. This epimerization occurs by a sequential dehydration/hydration mechanism involving two hydratases with opposite stereospecificities. The first step involves dehydration of the D-3-hydroxyacyl-CoA to a 2-*trans*-enoyl-CoA, catalyzed by a novel dehydratase known as D-3-hydroxyacyl-CoA dehydratase. In the second step the 2-*trans*-enoyl-CoA is rehydrated by enoyl-CoA hydratase, giving the L-3-hydroxyacyl-CoA, which can reenter the normal β-oxidation cycle. This epimerase-dependent pathway ocurs uniquely in the peroxysomes.

VII. DISORDERS OF FATTY ACID METABOLISM

A. Defects in Mitochondrial β-Oxidation

Genetic abnormalities in the functioning of the carnitine acyltransferases or carnitine deficiency can block the entry of fatty acids into the mitochondria and lead to various degrees of muscle pathology that are often precipitated by prolonged exercise or fasting. Patients are advised to adopt a regimen consisting of frequent meals of a low-fat, high-carbohydrate diet. Three types of inherited acyl-CoA dehydrogenase deficiency have been reported involving the long-chain, medium-chain, and short-chain-specific enzymes. The primary defect is the accumulation of acyl-CoA

intermediates in the mitochondria during fasting. The resulting reduced availability of CoA for other mitochondrial reactions has broad-ranging metabolic consequences, often producing hypoglycemia, vomiting, and cerebral edema. Patients must avoid fasting and consume frequent, low-fat, high-carbohydrate meals.

B. Defects in α-Oxidation

Refsum's disease is a rare inherited neurological disorder characterized by retinal pigment degeneration, chronic polyneuropathy, and ataxia that results from an inability to oxidize the methyl-branched fatty acid, phytanic acid (3,7,11,15-tetramethylhexadecanoic acid). This fatty acid accumulates in the triacylglycerol fraction of tissues and plasma lipoproteins. The normal pathway for the degradation of phytanic acid involves an initial α-oxidation step to yield first α-hydroxyphytanic acid (2-hydroxy-3,7,11,15-tetramethylhexadecanoic acid) then, after decarboxylation, pristanic acid (2,6,10,14-tetramethylpentadecanoic acid). The metabolic disorder lies in a failure to convert phytanic acid to α-hydroxyphytanic acid by phytanic oxidase, which is either missing or defective. A typical diet contains 50–100 mg phytanic acid per day, derived mainly from dairy products and ruminant fats. Adherence to a highly restricted diet can improve the clinical condition, particularly when initiated prior to the onset of severe neurological symptoms.

C. Defects in Peroxysomal Fatty Acid Oxidation

Zellweger syndrome is a rare but serious disorder of peroxysome biogenesis that is usually lethal within the first year of life. The severe clinical symptoms associated with this disease include hypotonia, psychomotor retardation, and deafness. Cells of Zellweger patients do not contain stable peroxysomal structures apparently because the peroxysomal proteins are rapidly degraded. Biochemically, the disorder is characterized by the accumulation of very long-chain fatty acids in plasma and tissues. Adrenoleukodystrophy is another peroxysomal disorder associated with a decreased ability to degrade very long-chain fatty acids, which can be of both dietary and endogenous origin. The biochemical defect has been located at the level of the very long-chain acyl-CoA synthetase enzyme which may be altered or missing. Although this disease is usually less severe than Zellweger syndrome, the clinical symptoms include brain demyelination and the accumulation of very long-chain fatty acids in plasma and tissues. Dietary therapy, which includes restriction of the very long-chain fatty acids in the diet and supplementation with a mixture of glyceryltrioleate and glyceryltrierucate (Lorenzo's oil), has been successful in minimizing neurological impairment only in patients with a mild form of the disease or in those treated at the onset of clinical symptoms. Current research is focusing on the identification of inhibitors of the fatty acid elongation system that may prevent accumulation of very long-chain fatty acids.

D. Essential Fatty Acid Deficiency

A deficiency of linoleic or linolenic acid leads to skin lesions, fragile red blood cells, hair and weight loss, and kidney damage. The intake of 1–2% of the total dietary energy requirement as essential fatty acids is sufficient to prevent the deficiency.

BIBLIOGRAPHY

Chow, C. K. (1992). "Fatty Acids in Foods and their Health Implications." Decker, New York.

Gunstone, F. D., Harwood, J. L., and Padley, F. B. (1986). "The Lipid Handbook." Chapman and Hall, London.

Gurr, M. I. (1992). "Role of Fats in Food and Nutrition." Elsevier, London.

Marinetti, G. V. (1990). "Disorders of Lipid Metabolism." Plenum, New York/London.

Smith, S. (1994). The animal fatty acid synthase: One gene, one polypeptide, seven enzymes. *FASEB J.* 8, 1248–1259.

Unsaturated Fatty Acids, Nutritional and Physiological Significance: The Report of the British Nutrition Foundation Task Force. (1992). Chapman and Hall, London.

Vance, D. E., and Vance, J. (1996). "Biochemistry of Lipids, Lipoproteins and Membranes." Elsevier, New York.

Feeding Behavior

ANDREW J. HILL
JOHN E. BLUNDELL
University of Leeds

I. Normal and Abnormal Eating
II. Determinants of Human Feeding
III. Approaches to Measurement

GLOSSARY

Behavior Those activities of an individual that can be observed and recorded by another individual, including verbal reports made about subjective conscious experiences

Bingeing Rapid consumption of a large amount of food, accompanied by feelings of loss of control, and normally followed by a purge (e.g., vomiting)

Dietary restraint Individual's motivation to limit intake (or diet) to maintain or lose weight

Energy intake Amount of food consumed, expressed in terms of its potential supply of energy to the body to maintain physiological processes

Hunger Urge to eat, often accompanied by characteristic bodily sensations, normally indicating that eating is imminent

Nutrient selection Intake of basic nutrients such as protein, carbohydrate, and fat (macronutrients) and vitamins and minerals (micronutrients) from the choice of foods available

Satiety Inhibition over food consumption resulting from the consequences of ingestion

FEEDING BEHAVIOR IS A FREQUENT AND FAMILIAR human activity. Eating reflects a variety of physiological processes and is expressed within a psychological, social, and cultural environment. Human feeding behavior embraces the total amount of energy consumed and the distribution of intake over time, or the pattern of eating episodes such as meals and snacks. Human feeding behavior also involves food choices and selection from the products available. A variety of eating practices and habits exists that could be regarded as normal within their own context. In Western society, abnormal eating (e.g., in bulimia nervosa or anorexia nervosa) has life-threatening medical consequences. One important area for study, therefore, is the distinction between normal and abnormal eating. In addition, the methods developed to assess the characteristics of human feeding behavior are an important part of this area. The detailed analysis of feeding behavior can be used as a tool to investigate abnormal conditions, as well as to disclose the operation of natural mechanisms. Because the measurement of feeding behavior includes monitoring those events that lead up to, and follow, the actual intake of food, a variety of techniques have been developed for the use of the researcher. These range from the observation of the specific behaviors engaged in while eating to the description of long-term food consumption patterns.

I. NORMAL AND ABNORMAL FEEDING

A. Overview

This article examines the scientific approach to the understanding of human feeding behavior as a motivated behavior, rather than describing the cultural patterns of cooking and eating that constitute the phenomenon we call cuisine. For humans, eating is much more than simply the provision of fuel and nutrients in adequate quantities for our bodily processes. Eating assumes a place of great importance in our lives. It is typically a social activity and can form the core of a major celebration or cultural ritual. We

ENCYCLOPEDIA OF HUMAN BIOLOGY, Second Edition, VOLUME 3.

usually spend 60 minutes or more per day engaged in eating or drinking, and a great deal of pleasure is derived from our contact with food. However, in considerations of human eating behavior, more attention is normally given to what is eaten than to the behavior itself, its antecedents, its consequences, and its expression.

In many areas of research, the assessment of behavior is reduced to a single number. For both theoretical and experimental convenience, researchers often settle for a coarse level of measurement and end up measuring the consequences of behavior rather than the behavior itself. In feeding research this usually involves measuring the amount of food consumed, by weight or energy value. These variables, however, are not themselves behavior—they are the results of behavior. Naturally, for a complete understanding the consequences of behavior and the behavior itself both must be analyzed and interpreted. Consequences cannot properly be understood if we are ignorant of the behavior itself. [*See* Behavior Measurement in Psychobiological Research.]

Behavior has a form and a structure. It is composed of elements or acts arranged in particular sequences. The analysis of this structure can take place at various levels, ranging from the molecular to the molar. It is important to recognize that the structure of behavior is not arbitrary, but functional. In other words, behavior is expressed in such a way as to achieve certain objectives. For all organisms (i.e., animals and humans) behavior is central to the adaptive capacity of the organism. Most frequently, the function of behavior is to knit together physiological and environmental demands. Accordingly, the structure of behavior reflects the operation of these physiological and environmental influences and their interactions.

The concept of behavioral structure implies the existence of changes with time (i.e., temporal changes). Thus, structure cannot be assessed by a single instantaneous measurement, but only by successive or continuous measurements over time. The temporal dimension reveals the complexity and power of behavioral analysis. The sequence of elements that make up a behavior can change, and the rates of expression of these elements can also change. It follows that the way behavior is shaped in bringing about a change in intake is as important as the change itself. The behavioral adjustment can result from a physiological disturbance, a nutritional challenge, or a change in environmental demands. In each case the analysis of behavior can be used as an aid in diagnosing underlying causes. This principle is particularly relevant in clinical research, in which

abnormalities in the physiological or psychological (i.e., environmental–social) domains are reflected in the altered structure of behavior as well as in the outcome of behavior.

The idea of using behavioral structure as a tool to disclose the operation of natural mechanisms, as well as to diagnose pathological conditions (or to assess the capacity of a pathological system), places demands on methodology. It is important that techniques and procedures can produce a valid representation of behavioral structure. It is for this reason that the approaches to the measurement of human eating behavior are of importance in their own right.

B. Expression

Human feeding behavior is a pattern of actions expressed to ensure our adequate supply of energy and specific nutrients for survival and adequate functioning. Feeding behavior can be characterized as a series of large (i.e., meals) and small (i.e., snacks) eating episodes, which leads to the ingestion of a certain amount of food (i.e., energy).

Feeding behavior also usually involves a selection of foods from an available range; this is often required in order to maintain an appropriate intake of essential macro- and micronutrients. This selection of foods can be broad or relatively restricted, depending on the adequacy of the supply and on individual food preferences. Given the familiar circumstance of a broad range of foods to choose from that are also affordable, what factors influence how people choose what to eat? A range of determinants arising from within the individual has been identified as potent forces guiding food selection. These include acquired preferences (acquired through learning mechanisms), conditioned aversions (i.e., powerful habitual rejections of particular foods), palatability (often assessed through an individual's preference for a food), and cravings (i.e., special preferences or yearnings for food). The mechanisms underlying these determinants of food choice are varied, and these factors are themselves worthy of more consideration than can be given here. [*See* Attitudes as Determinants of Food Consumption; Food Acceptance: Sensory, Somatic, and Social Influences.]

C. Normal Eating

What is meant by "normal" when we consider a behavior such as eating? Despite the universality of eating in the human behavioral repertoire, eating behavior is subject to major cultural variations in its expression. The most overwhelming cultural determinant is the adequacy of food supply. The developed and developing worlds differ in the quantity and diversity of foods available to their inhabitants. Symptoms of chronic malnutrition might reflect normal eating by many people, but only as the result of a desperately limited food supply. [See Malnutrition.]

Another, more subtle, cultural determinant of eating behavior is cuisine. The cuisine of a given culture not only refers to the style of food preparation, but places constraints on the acceptability of particular foods (e.g., pork, raw foods, insects, and spices) for its members and on its mode of presentation and consumption. The concept of normal eating might therefore be of questionable validity as it embraces such a diversity of attitudes, practices, and provisions.

The developed, or Western, world is marked by an abundant food supply (apart from the poorest members of society) and by the international nature of its cuisine. The more than adequate supply and variety available mean that weight gain might be the norm more often than weight maintenance or loss. However, the need to limit the amount of food eaten is felt by many people in technologically developed societies. Some of these people can be obviously overweight, or, in comparing themselves with society's advertised ideal body image, they might feel overweight or dissatisfied with certain parts of their body.

Dieting, or limiting the amount of food ingested, is therefore a major factor influencing the normal eating behavior of such people. For example, evidence suggests that the majority of women aged 18–25 at some time diet in an attempt to reduce or at least maintain (rather than increase) their body weight. These dieters do this by eliminating snacks or even meals (often breakfast) from their behavioral repertoire. In addition, they choose particular types of foods that are naturally low-energy foods or manufactured low-energy equivalents of normal foods and avoid other, often high-fat and highly desired, foods. Dieting, however, is not a naturally successful strategy even for weight maintenance. Episodes of food avoidance, which are seen as the most efficient way of losing weight and can last over days, rather than hours, are interspersed with periods of "overeating" to an extent that invalidates the entire process of abstinence. [See Food-Choice and Eating-Habit Strategies of Dieters.]

The wish to restrict food intake or to diet has been called dietary restraint. This restraint is regarded as a personal disposition or trait typical of a certain type of person. Research on restraint shows that being motivated to diet confers a particular pattern on behavior. Thus, the dieter is characterized by a susceptibility to certain specific events that upset the imposed inhibition over eating. This is called disinhibition (i.e., inhibition of inhibition). The variety of these disinhibitors provides an insight into the reasons for the general lack of success of self-motivated dieting. Such disinhibitors can include being "tricked" into eating something, seeing someone overeating, drinking alcohol, and feeling anxious or depressed, among others. Clearly, although this can represent a frequently encountered form of eating behavior of many individuals, it is not necessarily a desirable pattern of behavior.

D. Abnormal Eating

Aberrations of human feeding behavior include overconsumption (i.e., hyperphagia), underconsumption (i.e., aphagia or anorexia), the altered intake of nutrients, and inappropriate food choices. Attention is paid briefly here to circumstances in which human feeding behavior can be considered to be abnormal in clinical terms. Strictly defined, obesity is a disorder of weight. Whether or not obesity is a disorder of feeding behavior, however, is a debated issue. One important point of consensus is that excess weight derives from an imbalance between energy input and energy expenditure. Obesity can result, therefore, from a surplus of incoming energy or a deficit in energy expenditure, or both. It is clear that, if sustained over a long period, a fairly small imbalance can lead to substantial fat deposition. [See Eating Disorders; Obesity.]

Unfortunately, even if this is totally a product of excessive intake, it can be difficult to detect by monitoring food intake or eating behavior. Several investigations have followed the suggestion that the obese have defects in their perception of hunger or satiety, or that they display a characteristic obese eating style that lends itself to overconsumption. In general, the latter possibility has been hard to sustain. Although they do not display a unique eating style, the obese do display some deficiencies in their expression of hunger and satiety. In addition, the obese seem to be especially sensitive to the palatability of food. However, the extent to which other variables (e.g., dieting) influence the behavior of the obese is unknown, as those mildly or moderately obese people who might frequently diet are rarely distinguished from nondieters or severely overweight individuals.

Some obese people claim to be compulsive eaters, and the disturbance at a behavioral level in bulimia nervosa is in fact part of the clinical definition. Episodic food binges, in which eating is typically out of control, are followed by some form of purgation, most commonly by vomiting or laxative abuse. These binges, which are often interspersed with periods of near-starvation, are fundamental, but not exclusive, diagnostic criteria for bulimia. Most bulimics are women who might binge several times a day. A week's ration might be consumed in a single sitting, followed by purging of what has been eaten in order to prevent any gain in weight. Indeed, bulimics are usually of normal body weight. Bingeing could become the dominant feature of the life-style of a bulimic individual. Frequently, bingeing is a private activity, often going undetected by close family members.

Food and eating are also the dominant forces in the life of someone with anorexia nervosa, but they are manifest in a very different way from those of the bulimic. The restrictor form of anorexia nervosa is superficially the opposite of bulimia, as the anorexic imposes on him- or herself a near-starvation regimen. A lettuce leaf dressed with vinegar might constitute a meal, yet this does not imply a lack of concern with food and could be associated with considerable time spent cooking for other people. The sense of self-achievement and mastery derived from resisting hunger can reward this abstinence. Some of the physiological changes that take place in anorexia actually strengthen the resistance over eating, and this downward spiral can be difficult to break. Outwardly, the anorexic appears emaciated, a reflection of the much-reduced caloric intake. Also recognized now is the clinical picture of the anorexic who periodically binges. Here, the behavior incorporates both types of strategies, with long periods of starvation punctuated by episodes of bingeing and purging.

A substantially less life-threatening, but still abnormal, category of eating behavior is pica. Pica is the consumption of a substance not normally regarded as appropriate for human consumption. Thus, it is not the *behavior* that is abnormal so much as the *target* of the behavior. It has been suggested that the consumption of substances such as ice, mineral clay, bone, washing starch, or lead can actually have some adaptive value. When there is a high nutritional need (e.g., during pregnancy, childhood, or malnourishment), people might derive some nutritive benefit from these apparently nonnutritive sources. This notion is attractive, although somewhat speculative. [*See* Dietary Cravings and Aversions during Pregnancy.]

II. DETERMINANTS OF HUMAN FEEDING

A. Biological Determinants

Eating is most often viewed as the behavioral expression of a biological demand for nutrition. Eating is part of the homeostatic process regulating our internal bodily state. Hunger is simply the subjective manifestation of the body's need for food. Therefore, a simple sequence is achieved: a physiological demand is experienced through hunger and is met by eating food. The elements that make up this closed system are varied, operate at different levels of physiological functioning, and interact in a complex manner.

The basic physiological processes that operate when food is ingested and that serve to digest the material and make its constituent nutrients available to the body are well understood. Various physiological routines are involved in fuel utilization, and several routes are used to provide feedback about the state of the system. Physiological signals inform the brain so that behavior (i.e., eating or not eating) can be adjusted accordingly. The brain is the central integrator of information concerning the availability of glucose and other nutrients, stomach distension, and other signals. It is on the basis of this information that the subjective motivation to eat is believed to be activated. [*See* Brain.]

B. Biopsychological System

To assume that human feeding behavior is solely under the control of physiological processes ignores the power of psychological and environmental determinants of behavior. As already implied, environmental forces can shape and limit the expression of eating behavior. These ideas suggest that human feeding behavior is not the product of one particular mechanism, but is the output of a system—in this case a biopsychological system. The eating repertoire cannot be said to be the response of a single factor, but is the expression of a number of interacting factors. These factors exist in the biological or environmental domains. Moreover, the system has an integrity and acts in order to fulfill certain functions (e.g., the adequate supply of energy and nutrients or the preservation of the organism). A simplified systems view of feeding behavior is shown in Fig. 1. This conceptualization draws attention to the interrelationship between particular spheres of interest, including the external environment (cultural and physical), the behavioral act of

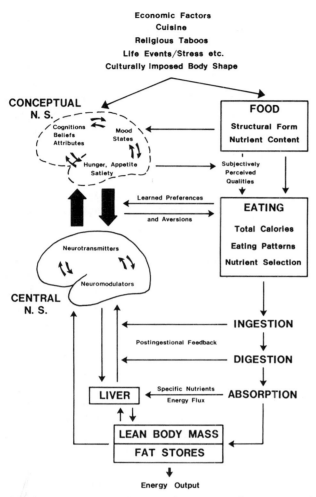

Economic Factors
Cuisine
Religious Taboos
Life Events/Stress etc.
Culturally Imposed Body Shape

CONCEPTUAL N. S.

Cognitions
Beliefs
Attributes

Mood States

Hunger, Appetite
Satiety

FOOD
Structural Form
Nutrient Content

Subjectively Perceived Qualities

Learned Preferences and Aversions

EATING
Total Calories
Eating Patterns
Nutrient Selection

Neurotransmitters

Neuromodulators

CENTRAL N. S.

INGESTION

Postingestional Feedback

DIGESTION

Specific Nutrients
Energy Flux

ABSORPTION

LIVER

LEAN BODY MASS
FAT STORES

Energy Output

FIGURE I A conceptualization of certain significant aspects of the biopsychological system underlying the control of human feeding behavior. N.S., nervous system.

eating (quantitative and qualitative components), the processes of ingestion and assimilation of foods, the storage and utilization of energy, and brain mechanisms implicated in the control system and mediating subjective states such as attributions and cognitions. Human feeding behavior therefore reflects an interaction between physiological events and environmental circumstances.

The strength of a systems approach lies in the capacity to demonstrate links among different levels of the system. Central to this view is the proposal that alterations in any one domain provoke changes elsewhere. One example of this principle is dieting, and the change that ensues is unwanted and counterproductive. The strategy of reducing food intake does not simply deny energy to the body, but also leads to an

adjustment in metabolism. Caloric restriction leads to a depression of metabolic rate, the change being severe when there is a marked and abrupt reduction in intake. This lowering of the metabolic rate is generally seen as an adaptive response to food restriction or a form of biological defense in the face of scarce commodities. The system is therefore integrative, pulling together biological events and psychological events, and demonstrates the intimate relationship in the interactions between these domains.

III. APPROACHES TO MEASUREMENT

A. Overview

As human feeding behavior has a definable structure, its description and assessment can be carried out at different levels of analysis. The scientific study of human feeding behavior demands that detailed investigation requires the analysis of microevents in the behavioral repertoire as well as the study of large-scale, long-term trends in consumption. Naturally, different types of procedures are required for these distinctive tasks.

The principle division is between direct and indirect techniques. The direct measurement of behavioral events is possible over short intervals, whereas long-term assessment usually depends on the indirect recording of food intake. Another distinction is between monitoring in a free natural environment, with attendant problems of control and precision, and measuring in artificial laboratory or clinic situations, in which accuracy is clearly easier to achieve. The particular limitations of data collected under these different circumstances present researchers with the dilemma of assigning importance to either the relatively inaccurate recording under natural circumstances or the very accurate recording in unnatural situations.

B. Dietary Studies of Individuals

Many of the methods for describing individual and group dietary practices are long established and have been extensively described and reviewed. Generally, individual dietary studies fall into two categories.

In the first category subjects are asked to record their present intake (e.g., by maintaining detailed records of all food eaten). These might be weighed records, in which all of the ingredients used in the preparation of food are precisely weighed, together with wastage at the end of the meal. Alternatively, they might be diaries of foods eaten, in which the quantities are described in terms of household measurements or estimation. The major drawback with these methods is a consequence of self-monitoring. It is possible, especially in certain groups, that constantly drawing attention to the process of selection and consumption of food leads some people to change their eating behavior.

The second category groups techniques used to aid the subject in remembering and describing their past intakes. The most common procedure elicits an inventory of the food eaten over the previous 24 hours. This is done either by asking subjects to note the foods on a checklist or by detailing the meals together with estimated amounts of food. This recall can be extended over a few days or longer.

Generally, recall techniques are quick and inexpensive and do not require specialist supervision. Their cost-effectiveness, however, might vary according to the required accuracy of the study. Many more 24-hour records are needed to accurately establish protein or fat intake than to describe energy intake.

Records of food intake, collected prospectively or retrospectively, are valuable sources of information concerning a person's eating behavior. Not only do they provide 24-hour summaries of nutrient intake, but they could provide a topography of eating during this period. Intake can be broken down into discrete episodes of eating, with average meal sizes and frequencies computed and distributions of snacks and meals plotted. However, it should also be recognized that the accuracy of data derived from food intake records is potentially compromised by the tables of food composition used to convert food records into quantities of nutrients, which represent the average composition of a particular food item.

C. Observational Studies

One apparently uncomplicated procedure for describing eating behavior is to observe and subsequently classify the entire sequence of behavior. In principle, because behavior is recorded in its totality, this strategy should be a powerful tool for describing human feeding.

Behavior can be monitored in naturalistic settings (e.g., cafes, restaurants, dining halls, or homes) or in the laboratory under controlled conditions. Analysis can be from live behavior or carried out from video recordings. Often, the observational method demands some form of sampling. Event sampling requires that

every occurrence of a specified event during the course of the observational period is recorded. Time sampling means that whatever event is occurring at specified (brief) intervals during the observational period is monitored. For eating behavior, which normally spans a relatively short period (usually a meal), a number of significant events (e.g., taking a bite, pausing, and swallowing) can be continuously recorded by an observer. It is necessary to check the accuracy of the ratings of this observer by comparing them with an independent observer rating the same sequence (i.e., the coefficient of concordance). This is essential since even events such as taking a bite of food, which can be defined fairly unambiguously, can be recorded differently by two independent observers. This is particularly important when the start and end of an event are recorded as well as the overall frequency of events. Used carefully, observational procedures can be sensitive research devices, despite the intensive nature of the data collection and analysis.

D. Specialized Apparatus

During the last 20 years a number of specialized devices have been developed or adapted to improve the sensitivity, accuracy, or reliability of measuring food consumption. Most provide continuous monitoring of intake. Some are designed for liquid rather than solid food, and others allow a degree of food choice. Some demand a somewhat unnatural eating response, whereas others attempt to allow unhindered eating to take place. No device is perfect; they all have strengths and weaknesses.

Several researchers have devised forms of liquid food reservoirs, in which liquid food is either sucked or pumped at a steady rate during the depression of a button. The technique has been used in a variety of circumstances, including examining the consequences of oral and intragastric feeding on subsequent intake, the eating behavior of normal and obese subjects, and the effects of drugs on voluntary intake. This approach provides an objective method for studying certain parameters of ingestion. However, the dependence on liquid food, with the consequent restriction on variety of taste and texture, obviously limits the extent to which results can be generalized to more natural eating circumstances and situations. In summary, the technique scores high for internal validity, but somewhat lower for external validity.

A very different device is based on the recognition that human eating is composed of a sequence of contacts between the mouth and the eating utensil, and that the number of contacts is proportional to the amount consumed. This system of monitoring eating via eating utensils operates through specially constructed spoons and forks with handles that contain miniaturized telemetering equipment. Each contact between utensil and mouth creates a signal that is transmitted to a recording device. Studies with a prototype have shown the utensils to provide good records of two parameters: the number of bites (or spoonfuls) and the interbite interval. At the moment the full potential of this system is unevaluated, but it is clearly an unusual and original approach to measurement.

An alternative technique shifts attention away from the structure of eating behavior to the actual food being eaten. Thus, monitoring eating via the plate demands the accurate measurement of alterations in the weight of the food being eaten. This is achieved by the continuous weighing of the individual's plate with a concealed electronic balance on which the plate rests. The device is called the universal eating monitor, and it can be used with either solid food on plates or liquid foods (e.g., soups) in dishes. Periodic readings of the weight of the plate (e.g., every 3 seconds) are made by an on-line computer during the meal. From these readings a cumulative intake curve (i.e., the weight of the food removed from the plate over time) can be plotted. This curve is the major parameter of eating provided by this technique. Several investigations using this device have demonstrated its usefulness and sensitivity. The accurate readout of adjustments in the weight of the food consumed is similar to that obtained with the reservoir method. The great advantage of this technique, however, is that the food can be eaten normally from a plate instead of being sucked or pumped through a pipe.

Many of the techniques now used to study human eating are derived from strategies used to monitor feeding patterns in animals. The human equivalent of pellet dispensers and eatometers are food-dispensing machines. The solid-food dispenser is basically a commercial food vending machine modified to provide small food units (e.g., quarter-sandwiches) with weight, nutrient content, and caloric value accurately controlled. The removal of each food unit from the dispenser can be monitored automatically, and a cumulative record can be obtained of the individual's behavior (i.e., feeding profile) and the weight of the food consumed (i.e., energy intake). This device has the advantage that it uses common solid foods that are likely to be regularly consumed and does not put

unusual demands on the eater. The resolving power of the device—its capacity to detect subtle or small adjustments in intake—is obviously restricted by the size of the individual food units. In its favor, however, is its capacity to monitor food selection. By stocking the machine with items varying in nutritional composition, for example, it becomes possible to measure individual preference for particular nutrients or tastes.

Returning once more to the behavior of the eater, one further approach has been to automate the recording of certain behavioral events which are available normally only through careful observation. Monitoring chewing and swallowing via a device called an edogram permits an objective insight into the microstructure of human eating. Swallowing is measured by changes in pressure in a small balloon resting on the Adam's apple and kept in place by an elastic collar. Chewing is simultaneously measured by a strain gauge that monitors jaw movements. Normally, a standardized test food is used with the apparatus (often a number of small open sandwiches). The output permits the calculation of microparameters such as the rate of chewing, duration of chewing between successive swallowing movements, and intrameal pauses without chewing. In addition, since the test food is composed of small consumable units of known weight, volume, and caloric value, the procedure allows continuous tracking of caloric intake. Consequently, the technique combines some of the best aspects of the utensil monitor and the plate monitor (i.e., the universal eating monitor).

E. Subjective Experience

The human capacity to communicate our internal experiences provides a further valuable source of information on eating behavior. Personal observations regarding the presence of certain bodily sensations or the intensity of hunger can be regarded as verbal behavior and, as such, are entirely available for quantification and measurement. As suggested earlier, subjective reports of hunger motivation are useful, as they might provide further insight into the processes guiding eating behavior. For this reason simultaneous recording of both subjective experience and overt eating behavior constitutes a powerful investigative strategy.

Subjective experiences concerning eating fall roughly into three categories: sensations, dispositions, and mood. Certain bodily sensations are well associated with times before or after eating. Commonplace experience associates gastric motility with hunger and gastric fullness with satiety. However, though sensations originating from the area of the stomach can be particularly prominent, they form only one component of a range of accompanying sensations. Discrete sensations such as light-headedness, a bad taste in the mouth, dry throat, or a rumbling stomach, together with an overall feeling of bodily weakness, can occur if the individual has not eaten for several hours. At the same time the person might experience characteristic changes in affective state or mood (e.g., irritability and restlessness). Eating brings about a radical change in both sensation and affect, bodily sensations generally being more pleasant and accompanied by feelings of contentment and well-being.

The measurement strategies used to quantify these experiences are of varying degrees of complexity. The two most common methods are fixed-point scales and visual analogue scales. Fixed-point scales are quick and simple to use, and the data they provide are easy to analyze. However, the scales can vary greatly in complexity. In considering the appropriate number of points to be included in this type of scale, the freedom to make a wide range of possible responses (i.e., multipoint scales) must be balanced against the precision and reliability of the device (i.e., scales with few points). However, research seems to indicate that scales with an insufficient number of fixed points can be insensitive to certain changes in subjective experience. In addition, the fixed points themselves are important determinants of the way people use the scales and distribute their ratings.

One way of overcoming some of the failings of fixed-point scales is to abolish the points completely. Thus, visual analogue scales are horizontal lines (often 10 cm long), unbroken and unmarked except for word anchors at each end. The user of the scale is instructed to mark the line at a point that most accurately represents the intensity of the subjective feeling at that time. The researcher measures the distance to that mark in millimeters from one end, thus yielding a score of 0–100. By doing away with all of the verbal labels except the end definitions, visual analogue scales retain the advantages of fixed-point scales, while avoiding many of the problems with response distributions.

It should be recognized that scales asking people to rate their present level of hunger are very general questions. It is possible that individual subjects might use the term "hunger" differently from the manner intended by the experimenter. It might therefore be appropriate to present more than one scale that taps into different components of hunger motivation, such as "satiety," "desire to eat," "amount of food desired," and even "thirst." A highly specific measure of hunger motivation is to measure an individual's willingness to eat, or preference for, particular foods. A quantification of hunger motivation might be derived from a preference rating for a particular food or food category from the range of food preferred at that moment, the number of portions desired, or the number selected from a finite list.

F. Postscript—Changing Styles of Feeding Behavior

It is clear that certain cultures impose patterns on eating activities and modulate the expression of physiological mechanisms. Changes in cultural attitudes (e.g., about body shape) and in the nature of the food supply might also be expected to adjust the style of eating (i.e., feeding repertoire). In technologically developed cultures (e.g., European, North American, and Asian), nutritional and catering developments are adjusting the pattern of overall consumption and the selection of nutrients. For example, instantly available (i.e., "fast") foods, often made palatable through taste and texture, can encourage an adjustment in the distribution of eating episodes. This could mean a shift from the meal–snack–meal pattern to a grazing style of ingestion, in which people ingest small amounts more or less continuously. If the attractive taste and texture are produced at the expense of macronutrient balance, then this grazing style can lead to an inappropriate intake of essential nutrients.

Human feeding behavior is not a rigid immutable form of human activity. It is adaptable and highly responsive to shifts in physiological, cultural, and nutritional conditions. It is therefore important to have reliable methods to monitor human feeding behavior and to be aware of alterations that could signal inappropriate physiological or cultural influences.

BIBLIOGRAPHY

Bellisle, F. (1979). Human feeding behaviour. *Neurosci. Biobehav. Rev.* **2**, 163.

de Castro, J. M. (1991). Social facilitation of the spontaneous meal size of humans occurs on both weekdays and weekends. *Physiol. Behav.* **49**, 1289.

Hill, A. J., Rogers, P. J., and Blundell, J. E. (1995). Techniques for the experimental measurement of human eating behaviour and food intake: A practical guide. *Int. J. Obesity* **19**, 361.

Meiselman, H. L. (1992). Methodology and theory in human eating research. *Appetite* **19,** 19.

Polivy, J., and Herman, C. P. (1987). Diagnosis and treatment of normal eating. *J. Consult. Clin. Psychol.* **55,** 635.

Sheppard, R. (1989). "Handbook of the Psychophysiology of Human Eating." John Wiley & Sons, Chichester, England.

Walsh, B. T. (ed.) (1988). "Eating Behaviour in Eating Disorders." American Psychiatric Press, Washington, D.C.

Fertilization

W RICHARD DUKELOW
KYLE B. DUKELOW
Michigan State University

GLOSSARY

Acrosome Lysosome-like structure located on the tip of the sperm head. It contains enzymes that are released during the sperm penetration process

Acrosome reaction Phenomenon occurring on the sperm head whereby the sperm penetration enzymes are released

Capacitation Period of time of sperm incubation in the female reproductive tract (or under culture conditions) when sperm receptors are exposed as the sperm "achieves the capacity" to penetrate the egg

Egg Ovum or female gamete released from the follicle on the ovary. The sperm penetrates the egg to bring about fertilization

Embryo Product of sperm : egg fusion that develops into a new being. In the very early stages the embryo is called the zygote

Epididymis Strap-like organ attached to the testes in the male. The sperm mature and are stored in this organ until ejaculation

Fallopian tube Oviduct; a tube-like structure that allows transport of the egg from the ovary to the uterus. It consists of two parts, the ampulla (where fertilization occurs) and the isthmus proximal to the uterus

Fertilization General term that includes preparation of the sperm in both the male and female reproductive tracts, penetration of the egg by the sperm, and approximation and disintegration of the pronuclear membranes within the egg to allow intermingling of the male and female chromosomal components

Oviduct See **Fallopian tube**

Ovum See **Egg**

Sperm plasma membrane Outermost membrane of the surface of the sperm. It plays important roles in capacitation, fusion, and release of the sperm-penetrating enzymes

Spermatozoa Sperm, the male gamete that contains the male chromosomal material and carries this material to the female gamete or egg

Zygote Fertilized egg or very early embryo

THE ROLE OF THE SPERM IN CREATING A PREGnancy was first recognized from unique dog experimentation in the 1700s. By the middle of the nineteenth century, the egg had been described and its role in the fertilization process recognized. In the late 1800s, using marine invertebrates that were easy to study, scientists first observed the actual penetration of the egg by a sperm to create a zygote or potential embryo. Since then the process of fertilization has been studied extensively in a wide variety of animals. Because of the importance of reproduction and genetics in the domestication of animals, much of the early work in the reproductive field was carried out in domesticated species. In the 1930s a great deal of interest was directed at the fertilization process of our close relative, the monkey (a nonhuman primate), and many pioneering efforts were made at that time. [*See* Reproduction.]

An obvious aid to investigations of fertilization was the ability to observe fertilization and early embryonic development under *in vitro* ("culture dish") conditions. Studies on *in vitro* fertilization began in the

1890s but were not consistently successful until after the discovery of the capacitation phenomenon in 1951. The classic research of Dr. M. C. Chang resulted in the first production of mammals by *in vitro* fertilization. In 1969 the first successful *in vitro* fertilization of human eggs was reported by Dr. R. G. Edwards, and in 1978 the first child was born by this same process. In the nearly two decades since that event, thousands of children conceived by *in vitro* fertilization have been born, and the technique has been applied to a large number of laboratory, domestic, and primate animals. These findings have allowed scientists to delve deeply into the biochemical and molecular aspects of fertilization so that today we have a sound understanding of the complicated process by which all animal life begins. [*See In Vitro* Fertilization.]

For the present description of the fertilization process, we will first examine the development of the sperm in the male reproductive tract and the characteristics that the sperm acquire as they pass through the male and female reproductive tracts to reach the site of fertilization (the oviduct or fallopian tube) of the female.

I. MATURATION OF THE SPERM IN THE MALE REPRODUCTIVE TRACT

Sperm that are produced in the testes do not have the ability to move or to fertilize eggs. They pass, by fluid movement, into a strap-like organ called the epididymis. The epididymis, composed of convoluted tubules, has three distinct areas. The area closer to the testes is called the head (caput) of the epididymis; distal to this is the body (corpus) and, finally, the tail (cauda) of the epididymis. It is within the epididymis that the sperm becomes vigorously motile and achieves the ability to fertilize.

The maturation process varies in different species. Generally, however, the sperm first start making circular movements in the body or proximal tail portions of the epididymis. Sperm are then stored in the tail portion of the epididymis after they normally have achieved full motility and fertilizing ability. During the passage through the epididymis, several biochemical events cause these changes. These changes are called "maturation" of the sperm; they involve mainly the sperm plasma membrane surrounding the acrosome and nucleus of the sperm, and also some changes occur in the nucleus or tail portion of the sperm, reflecting rearrangement of disulfide bonds. The epididymis is a very active organ, capable of absorbing and secreting fluids that alter the sperm. As the sperm passes through the epididymis it is sequentially exposed to a number of active macromolecules. As a result it changes the molecules that constitute the sperm membrane surface. The most evident changes are in the glycoproteins, that is, proteins to which complex sugars are attached, and are brought about by the enzymes galactosyl-transferase and sialyltransferase present in the epididymal fluid. The saccharide residues of the glycoproteins are altered and the net negative surface charge of the sperm plasma membrane is changed. [*See* Sperm.]

Some membrane lipids also undergo change at this same time. The body of the epididymis has a high cholesterol synthesis rate, and choleesterol is one of the lipid molecules integrated into the sperm plasma membrane at this time. [*See* Cholesterol.]

At ejaculation the sperm pass through the vas deferens, the ejaculatory duct, and then the urethra for deposition into the female reproductive tract. In some species (e.g., the Chinese hamster) the sperm are stored not only in the epididymis tail, but also in the vas deferens. During passage from the male reproductive tract the sperm are mixed with fluid secretions of the seminal vesicles (primarily fructose), the prostate gland, and the bulbourethral or Cowper's gland. These secretions serve to neutralize the acidity of the urethra and provide nutrients and protection to the sperm until it reaches the site of fertilization in the female. The combination of these fluids and the sperm represent the semen or ejaculate of the male.

II. CAPACITATION OF THE SPERM

In some species (e.g., rodents and the pig) the semen is ejaculated directly into the uterus of the female. In most other species (including the human), the semen is deposited in the anterior portion of the vagina and then must traverse through the cervix to reach the lumen of the uterus and thence to the oviduct or Fallopian tube. For many years it was assumed that any delay from the time of sperm deposition until fertilization was caused by the transport of the sperm to the oviduct. Some classic experiments on the rabbit by Walter Heape in England at the turn of the century indicated that there was a 6-hour delay between sperm deposition and egg penetration. This delay was ascribed to sperm transport and this theory persisted in the literature for nearly 60 years. Subsequent research on cattle and guinea pigs showed that the time of initial transport was not measured in hours but rather 2.5 to 4 minutes! A secondary phase of sperm migration also occurs and is aided by sperm motility. The discrepancy of these early results remained unexplained until the discovery of the phenomenon that we now call "capacitation" by M. C. Chang in Massachusetts and C. R. Austin in Australia. This means that sperm, as they are produced by the male, are not capable of penetrating the eggs, and to "achieve the capacity" (thence the name) they must first spend a period of residence within the female reproductive tract. It turns out that the time required for capacitation in the rabbit is 6 hours, in agreement with Walter Heape's finding.

The phenomenon of capacitation has been speculated to allow control of the speed with which sperm gain fertilizing ability and thereby promote the presence of freshly capacitated sperm to all ovulated eggs.

Capacitation can be divided into two separate events. The first of these (for which the term capacitation is now used) appears to be unique to mammals and occurs during the early passage of the sperm through the cervix, uterus, and oviduct. The second event is termed the "acrosome reaction" and will be treated later.

Capacitation is a requirement of all mammals that have been studied, whether fertilization is natural (*in vivo*) or *in vitro*. Because the phenomenon is of short duration (3–4 hours in nonhuman primates and estimated at 7–11 hours in humans) it is difficult to study. Under natural conditions capacitation is normally achieved in the female reproductive tract, probably through initial incubation in the uterus and subsequent incubation in the isthmal portion of the oviduct, just prior to fertilization. Under conditions of *in vitro* fertilization, capacitation also occurs *in vitro*. This requirement gives rise to a delay before fertilization can be achieved, after freshly ejaculated sperm are added to eggs. Capacitation involves the biochemical activation of the sperm, probably through the production of cyclic nucleotides, which alter the metabolic processes of the sperm and allow the acrosome reaction to proceed. Indeed, addition of cyclic AMP to the fertilization medium reduces the time required for capacitation in both mice and nonhuman primates. Original descriptions of capacitation indicated that the process might involve the removal of a "coating" from the sperm plasma membrane, and modern molecular biological techniques suggest that this "coating" could represent removal of inhibitors from receptor sites on the sperm membrane that would allow the acrosome reaction and attachment of the sperm

to the egg to occur. Electron microscopic studies have failed to show distinct changes in the membrane surface due to the capacitation phenomenon alone, whereas the acrosome reaction is clearly evident later. However, using freeze-fracture techniques, intramembraneous particles have been found to be altered during capacitation. Unlike surface-coating substances, these are intrinsic proteins of the sperm membrane.

A substance present in the seminal plasma of most species, termed "decapacitation factor" or "acrosome stabilizing factor," has the ability to reverse the capacitation phenomenon and render capacitated sperm as "decapacitated." These sperm are unable to fertilize eggs until they have undergone a second capacitation period. This shows that capacitation is reversible.

The time and endocrine requirements for capacitation vary with different species. Much of the early work was done with the rabbit, which requires a fairly long period of 4 to 6 hours for capacitation. Four technical requirements must be fulfilled for studying capacitation: (1) the ability to predict accurately the time of ovulation; (2) a knowledge of the fertilizable life of the ovum; (3) the ability to place sperm in the reproductive tract and recover it; and (4) the ability to recover the egg for verification of fertilization. Ideally both *in vitro* and *in vivo* fertilization systems should be used to confirm capacitation. In most species, one or more requirements cannot be fulfilled, explaining why the rabbit was used so extensively in the early days. Today it is possible to study other species as well. Among the studied species the required capacitation time appears to be: mouse, 1 hour; hamster, 3–4 hours; ferret, 3.5–11.5 hours; cat, 30 minutes to 1 hour; sheep, 1–5 hours; pig, less than 2 hours; primates, 3–4 hours; and humans, 7–11 hours (estimated).

Capacitation can occur in the vagina, the uterus, or the oviduct, with different time requirements, for instance, 4–6 hours in the rabbit uterus but 10 hours in the rabbit oviduct. *In vitro* capacitation occurs in high-ionic-strength media. There have been reports of the need for certain heat-stable, low-molecular-weight substances present in follicular fluid and blood serum and very rich in the adrenal gland, but in general research to identify specific capacitation factors has not been productive.

During natural fertilization *in vivo* the female must be under the influence of estrogen for capacitation to occur. In females in the follicular phase of the cycle, capacitation will occur, but not in females receiving progesterone, or pseudopregnant females. Progesterone has an anticapacitation effect but only in the uterus, not in the oviduct. Some of the action of the "minipill" contraceptive agents may be against capacitation. [See Uterus and Uterine Response to Estrogen.]

Many events occur in sperm during capacitation. As already indicated, adenyl cyclase and cAMP-dependent protein kinase play a role relative to the motility of capacitated sperm, which is markedly increased during capacitation. This effect is directly related to increased adenyl cyclase activity and increased cAMP availability. Activation of cAMP-dependent protein kinase appears to alter the sperm membranes through phosphorylation of membrane proteins. Sperm membrane phospholipids are also altered during capacitation. This may explain why capacitated sperm incorporate vital dyes (such as nigrosin-eosin live-dead stain) whereas freshly ejaculated sperm do not. Capacitated sperm also exhibit an increase in oxygen consumption and glycolytic activity after incubation in the female reproductive tract or in media containing tract fluids. [See Phosphorylation of Microtubule Protein.]

Changes also occur in intracellular ions during capacitation. Sperm normally have an ionic gradient across the sperm plasma membrane with the concentration of potassium much higher inside than outside, and the reverse is true for sodium. This ionic gradient is maintained by an ATPase-mediated sodium/potassium ion-exchange pump. In the guinea pig, there is a reduction in the intracellular potassium concentration after 2 hours of incubation when the sperm are ready for the acrosome reaction. There appears to be little change in the concentration of intracellular free calcium in rabbit sperm as a result of *in vitro* capacitation, in contrast to the acrosome reaction, which involves a massive influx of extracellular calcium. In the latter case this facilitates membrane fusion.

During capacitation a number of proteins are either attached to or released from the whole-sperm surface. They are identified by immunological techniques, but their precise nature and their involvement in the capacitation process are unknown. The activation of cAMP-dependent protein kinase and phosphorylation of sperm proteins are important for the initiation and maintenance of sperm motility. Cholesterol in the sperm membrane may play an important role in capacitation. This substance has profound effects on membranes and affects active and passive ion permeability by regulating orientation, fluidity, and thickness of the membrane lipids. It has been postulated that during *in vivo* capacitation there is a gradual removal of cholesterol by albumin or some other com-

ponent of the female genital tract, but firm evidence is lacking.

Many workers have attempted to capacitate sperm *in vitro* with various combinations of uterine, oviductal, or follicular fluid. At first, when *in vitro* fertilization of the human egg was announced (using freshly ejaculated sperm), many deduced that capacitation is not required for human sperm. In all cases, however, there is a delay from the time of sperm and egg mixing until sperm penetration occurs and this reflects the capacitation phenomenon. Capacitation could be attributed to the small amount of follicular fluid with high concentrations of steroids that accompanies the restricted eggs (with the associated cumulus oophorous cells), although many carefully defined media have now been used to culture sperm and eggs for *in vitro* fertilization. Nevertheless, a number of interesting enzyme activities are present in the female reproductive tract at the time of capacitation, which may play a role.

Several clinical applications of the capacitation process are important in humans, although human sperm maintain their fertility for at least 72 hours and perhaps longer, the fertilizable life span of the human egg being shorter, probably from 6 to 12 hours. Under perfect conditions, fertilization should therefore occur when sperm have just reached the capacitated state and when the egg has just ovulated. Such a timing should yield the highest conception rate with the lowest possibility of polyspermy, fertilization with aged gametes, or other factors that might yield congenital birth defects. The application of this principle to timed coitus for low-fertility couples or for intrauterine insemination suggests that sperm deposition should occur from 7 to 11 hours prior to the expected time of ovulation. Because this is suggestive evidence that estogrens have a beneficial effect on the fertilizable life of sperm and they enhance capacitation, such therapy could be useful.

The fact that once sperm have been capacitated their fertilizable life span in the oviduct is shortened is of interest from the standpoint of fertility enhancement. It would be of obvious value to provide a continuing supply of capacitated sperm at the site of fertilization in patients in whom the exact time of ovulation is unknown. Multiple insemination would thus be indicated. From the standpoint of contraception, a naturally occurring antifertility factor in the oviduct could be developed artificially to block fertilization.

Sperm can be capacitated *in vitro* by removal of sperm from seminal plasma and incubation in a culture medium. Placement of the *in vitro* capacitated sperm in the reproductive tract to achieve conception has application for the treatment of some cases of infertility. Such manipulation is utilized where there is immunologically mediated sperm inactivation or when the uterus is incapable of capacitating sperm. The uterus and oviduct can be bypassed to bring about fertilization. Of course there is still the problem of implantation in a hostile uterine environment. Two treatment programs are in current use. The first, intra-uterine insemination, is performed in women having hyperstimulated cycles closely monitored by ultra-sound to determine the state of the ovarian follicles, and to measure estradiol serum levels. Washed incu-bated sperm are inserted into the uterine cavity when the egg is in the oviduct. Multiple inseminations with washed sperm significantly increase the pregnancy rate. A second technique, gamete intrafallopian tube transfer, utilizes washed incubated sperm (capaci-tated) placed together with the egg in the oviduct. In experimental studies this has been termed "xenogen-ous fertilization" since it often uses a foreign species for the fertilization. Finally, as discussed elsewhere, *in vitro* fertilization uses washed sperm to successfully fertilize human oocytes.

The use of antiestrogens and progestation agents has strong implications for capacitation. The antiovu-latory properties of progesterone are well known. Lower levels (100–500 μg) of progestins prevent preg-nancy but in many cases ovulation does occur. Three mechanisms have been postulated to account for the antipregnancy effect: (1) an anticapacitation effect, (2) rendering of the cervical mucus spermicidal, and (3) alteration of the mechanism of gamete transport through the reproductive tract. Evidence has been pre-sented to substantiate all three effects and probably all act in concert to inhibit fertility.

III. THE ACROSOME REACTION AND ACTIVATION OF THE ACROSOMAL ENZYMES

Before discussing the acrosome reaction it is im-portant to identify the specific organelles involved. The head of the sperm is composed of nuclear material and contains the genetic material necessary for the successful propagation of the species. Located on the tip of the sperm is a sac-like structure similar to a lysosome or to the zymogen granule of the pancreatic cells. This structure has been termed a "bag of en-zymes" and has an inner acrosomal membrane next

to the sperm head and an outer acrosomal membrane approximating the sperm plasma membrane that sur-rounds the complete sperm head. After capacitation and attachment to the zona pellucida of the egg, and stimulated by molecules associated with the oocyte or its investments, the sperm undergoes the acrosome reaction. This is a vesiculation reaction between the outer acrosomal membrane and the plasma mem-brane of the sperm. This event results in the formation of pores on the surface of the sperm, which allows the release of the enzymatic contents of the acrosome.

A number of enzymes have been isolated from the acrosome but the two that have received the most attention are hyaluronidase and acrosin. Recently a PH-20 protein, found on the human and macaque monkey plasma membrane, has been shown to have hyaluronidase activity. About one-half of the hyal-uronidase is bound to the inner acrosomal membrane and the other half is free within the arosome itself. The acrosome reaction occurs in a short period of time (from 2 to 15 minutes) and consists of an initial loss of free hyaluronidase and then a gradual and sequential loss or activation of enzymes bound to the inner acrosomal membrane. These events occur during the time that the sperm penetrates the outer vestments of the egg. At the present time there is some confusion on whether the acrosomal reaction occurs after contact with the zona pellucida or during pas-sage through the cumulus oophorous and corona radi-ata layers surrounding the egg. Most likely the former is the case, but there may be species variations. There is evidence that both cAMP and diacyl glycerol (DAG) second messenger pathways are involved in the acro-some reaction.

At this time we must briefly review the morphology of the egg as it is released from the follicle on the ovary. The egg cytoplasm is surrounded by the egg plasma membrane and has a nucleus containing the genetic material. The egg plasma membrane is sur-rounded by three different structures or layers, the closest to the plasma membrane being the zona pellu-cida. This glycoprotein layer serves to protect the egg and, in most species, incorporates a "block to poly-spermy" in that once it is penetrated by a single sperm other sperm are prevented from gaining entry. This mechanism prevents the incorporation of more than a single set of chromosomes and therefore the produc-tion of polyploid (multiple sets of chromosomes) indi-viduals, which would not survive. Around the zona pellucida is a small layer of cuboidal cells called the corona radiata and around this a layer of larger cells, more loosely packed and held together by a hyaluronic acid matrix, called the cumulus oophorous. For a sperm to penetrate to the cytoplasm of the egg it must first pass through these layers. In some species, such as the sheep, cow, monotremes, and marsupials, the cumulus oophorous layer is shed shortly after ovulation, commonly by changes in the bicarbonate content of the oviductal fluid. When this occurs, the corona radiata and the zona pellucida represent the only egg vestments that the sperm must pass before reaching the egg plasma membrane. In other species, including the human, the cumulus cells and the less conspicuous corona radiata must be penetrated before the zona pellucida is reached. This is accomplished through release of free hyaluronidase and that bound to the inner acrosomal membrane of the sperm.

The cumulus cells from eggs incubated *in vitro* are readily dispersed by a hyaluronidase solution. Hyaluronidase is also released by dying sperm and this release must be differentiated from the "true" acrosome reaction. In some species, such as the guinea pig and the musk shrew, it is very easy to visualize the acrosome reaction. In other speices, such as the golden and Chinese hamster, the acrosomes are smaller and phase-contrast microscopy is required. The acrosomes of the rabbit, mouse, and human are thin and the acrosome reaction is more difficult to detect. The techniques used for studying the acrosome reaction in these species may result in death of the sperm; they include the acridine orange–UV method, the triple-stain technique, the p-lectin method, the chlortetracy-cline–UV method, and the naphthyl yellow/erythrosine B method.

There are species variations in the site from which the acrosome reaction initiates. In the rabbit it appears to be on the tip of the acrosome, whereas in the golden hamster, ram, and human, there is closer attachment at the equatorial segment of the acrosome cap region. In all cases, however, the release of enzymatic contents and the process of the acrosome reaction remain basically the same. Reference has already been made to the role of the zona pellucida and the cumulus oophorous in inducing the acrosome reaction. There is little doubt that the zona pellucida does have this ability in most species. The mouse has three zona pellucida proteins, one of which, ZP3, binds to the sperm plasma membrane over the acrosomal cap. The polypeptide chain in ZP3 molecules seems to serve as an initiator of the acrosome reaction. Whether the cumulus oophorous can induce the acrosome reaction is unknown.

The acrosome reaction may occur by two parallel pathways. One is independent of calcium ion concen-tration and results when the sperm plasma membrane interacts with the egg plasma membrane. There is then an increase in intracellular pH by sodium ion influx and hydrogen ion efflux. The second pathway is calcium dependent and represents membrane depolarization. Both of these methods open calcium channels, resulting in a large influx of extracellular calcium. The calcium goes through the plasma membrane and induces the fusion between the plasma membrane of the sperm and the outer acrosomal membrane, resulting in an exocytosis (release) of acrosomal contents. A large number of molecules have been implicated in the sperm acrosome reactions, including ions, sperm membrane proteins (including enzymes), cyclic nucleotides, calmodulin, and the acrosomal enzymes themselves.

Reference has already been made to the hyperactivation of sperm motility following capacitation. This hyperactivation plays an important role not only in the transport of the sperm the short distance from the distal region of the isthmus to the ampulla, but also in the penetration of the outer vestments of the egg. The biochemical constituents of the medium will influence the hyperactivation and the calcium uptake is essential for initiation of sperm hyperactivity in the guinea pig, hamster, and mouse. Bicarbonate seems to be necessary for hyperactivation of hamster sperm, and potassium, energy substrates, and albumins are all known to control hyperactivation in a number of other species. There is substantial evidence that the level of intracellular cAMP controls hyperactivation. The exact molecular mechanism is not known, but obviously there are molecular activities that allow sperm to change from the rigid stiffness of tail-beating characterized by precapacitation sperm to the more active state. Intake of calcium may be a reflection of methylation of membrane phospholipids in the sperm tail membrane, which enhances the entrance of calcium into the sperm and may stimulate membrane-bound adenylate cyclase, thus increasing the intracellular concentration of cAMP.

IV. SPERM PENETRATION OF THE EGG

Reference has already been made to the role of hyaluronidase from the acrosome or perhaps also from the surface membrane of the sperm in penetration of the cumulus oophorous layer of the egg. There appears to be varying need for the hyaluronidase in various species. Thus in cattle, where the cumulus cells are normally dispersed within hours of ovulation, the role of hyaluronidase can be less than in other species. Originally, it was believed that masses of sperm reached the oviduct and swam in heavy swarms about the egg. Although this may be true for *in vitro* fertilization, in natural fertilization very few sperm actually reach the site of fertilization. Thus the action of hyaluronidase probably is not a mass release to disperse the cumulus oophorus cells but rather acts on the head of an individual sperm to ease penetration. Little is known of penetration through the corona radiata and many workers assume that penetration is by hyaluronidase. There is evidence, however, that a distinct corona-penetrating enzyme (CPE) exists and is attached to the inner acrosomal membrane. This esterase-type enzyme facilitates penetration of the corona radiata.

The zona pellucida has a number of very important functions. One of these has already been mentioned, the "block the polyspermy" (sometimes called the "zona reaction") that occurs soon after penetration by a single sperm, caused by a release of material from the cortical granules near the surface of the egg plasma membrane. The release modifies the composition of the zona pellucida. In the mouse this reaction is caused by the partial hydrolysis of the two zona pellucida proteins, ZP2 and ZP3. A second function of the zone pellucida is to prevent fertilization of eggs with sperm from a different species. This rarely occurs unless the zona pellucida is removed artificially from the eggs, allowing sperm penetration. These findings suggest that through proteins on its head the sperm binds to species-specific receptor sites located in the zona pellucida of the unfertilized egg. Once fertilization has occurred, these receptors are probably inactivated to prevent further penetration.

The zona pellucida of different species varies from 2 to 25 μm in thickness and contains from 2 to 35 ng of protein. It is a loose network of fibrillogranular strands and microvilli; processes extend from the egg and the surrounding follicle cells into it. The zona pellucida is also permeable to large macromolecules, including immunoglobins and enzymes, as well as small viruses.

Several proteins have been isolated from the zona pellucida of eggs of different species. In the pig it has four glycoproteins, whereas in the mouse and hamster it has three. The most extensive work on receptors found in the zona pellucida has been carried out in the mouse. The ZP3 molecule, which is a glycoprotein with an apparent molecular mass of 83 kDa, appears to serve both as a sperm receptor and as an acrosome reaction inducer. ZP3 consists of a 44-kDa polypeptide chain with three or four oligosaccharides covalently bound to asparagine residues (N-linked) and additional oligosaccharides attached to serine and threonine residues (O-linked). The ZP3 is associated with ZP2, another glycoprotein, to form the long filaments that constitute the zona pellucida. The third glycoprotein, ZP1 (200 kDa), is a dimer composed of identical polypeptide chains that cross-link the filaments of the other two proteins. Recent evidence suggests that the ZP3 O-linked oligosaccharides are important in sperm receptor function. Since there are many possible oligosaccharide structures in terms of composition, sequence, branching patterns, and conformation, they probably determine the species specificity of the sperm receptors. It is assumed that these oligosaccharides are altered after fertilization so that other sperm are not recognized or fail to bind to the modified receptors present. This modification is brought about by enzymes released from the cortical granules of the egg. Because the O-linked oligosaccharides in ZP3 appear to play the primary role, the glycosidase of the cortical granules is the most likely enzyme.

The ZP3 O-linked oligosaccharides, however, do not induce the acrosome reaction in sperm *in vitro*. It has been suggested that after binding to sperm, the oligosaccharides must associate with a polypeptide chain for the acrosome reaction to occur. The ZP3 polypeptide chain itself may serve this function.

Several different sperm proteins may be involved in egg binding. These include lectins, glycosyltransferases, glycosidases, proteinases, a tyrosine kinase, D-mannosidase, and proacrosin/acrosin. The first three of these specifically interact with sugars. The protein must be associated with the plasma membrane surrounding the sperm head. As an example, galactosyltransferase, an enzyme that normally stimulates the transfer of galactose from uridine-5′-diphosphate-galactose to terminal N-acetylglucosamine residues to form N-acetylgalactosamine, is a potential mediator of sperm binding to mouse eggs. The sperm galactosyltransferase would bind to N-acetylglucosamine on the egg. Under this scenario, sperm–zona pellucida binding would be similar to an enzyme–substrate reaction. Similarly, on the sperm, sialyltransferase and the zona pellucida sialic acid may interact to participate in the sperm–zona pellucida binding reaction. The concept that the sperm bind to the zona pellucida receptors through lectin-like proteins with saccharide-binding activity is interesting. Some proteins isolated from sperm bind to zona pellucida molecules, but it is not clear to which groups, proteins, or oligosaccharides.

Regardless of the mechanism, all sperm must complete the acrosome reaction before passing through the zona pellucida. The sperm progress through the zona pellucida by vigorous movement of the sperm tail, which leaves a thin hole, sometimes called the "penetration slit." It is assumed that enzymes bound to the inner acrosomal membrane "soften" the zona pellucida and allow penetration. This function has been attributed to a number of membrane-bound enzymes, including acrosin, nonacrosin proteinase, hyaluronidase, arylsulfatase, glycosulfatase, and N-acetylhexosaminidase. Thus two mechanisms are proposed for zona pellucida penetration, one mechanical, by the power of the sperm tail, and the second enzymatic. Probably both mechanisms are involved. The passage of the sperm through the zona pellucida is comparatively rapid, taking from 3 to 20 minutes. Once past the zona pellucida the sperm crosses the perivitelline space and immediately attaches itself to the vitelline membrane, in which it is gradually incorporated. This attachment involves the sperm plasma membrane (not the inner acrosomal membrane) and occurs above the equatorial segment. The surface of the egg has many microvilli except in the area above the metaphase spindle of the second meiotic division. The sperm head attaches itself in an area rich in microvilli and is gradually taken in. After attachment, tail movement immediately ceases, probably because of a calcium ion influx or depolarization of the sperm tail membrane.

V. ACTIVATION OF THE EGG

After incorporation of the sperm head (and in most species the midpiece and tail) into the egg cytoplasm, the metabolically quiescent egg initiates a series of events referred to as "activation." This is probably triggered by the PH-20 protein.

The first step is the exocytosis of the cortical granules and the resumption of meiosis. The haploid nucleus of the egg then transforms into the female pronucleus. At the same time the sperm nucleus decondenses and becomes the male pronucleus. These two pronuclei come into close approximation near the center of the egg, their membranes disintegrate, and the chromosomes mix for the first mitotic division. This intermingling of chromosomes is considered the end of the fertilization process and the beginning of embryonic development.

The biochemistry of egg activation is not well understood in mammalian species. There is evidence of a strong release of intracellular calcium in the hamster egg 10–30 seconds after attachment of the sperm to the egg plasma membrane. This release continues for over an hour, but its exact significance is unknown. Also, after sperm–egg fusion, the conversion of ATP to ADP increases with concomitant stimulation of the respiration and metabolism of the egg.

Upon penetration of the sperm into the egg cytoplasm and release of the contents of the egg cortical granules, the zona pellucida, as already mentioned, becomes refractory to sperm penetration (zona reaction). A secondary block occurs at the level of the egg plasma membrane, known as the "vitelline block." The nature of this is poorly understood but it is very rapid. Both of the blocks to polyspermy at the zona pellucida and vitelline membrane levels effectively prevent penetration by more than one sperm. Human eggs seem to undergo a very strong zona reaction. Even when they are exposed to a large number of sperm, as during *in vitro* fertilization, many sperm can be bound to the surface of the zona pellucida, but few penetrate the inner regions of the zona. Nevertheless, in humans, polyspermic fertilization does occasionally occur, perhaps due to a delayed cortical granule exocytosis and a subsequent delay in the zona reaction. A number of factors could cause this anomaly, including egg immaturity at the time of sperm penetration, excessive aging of the eggs in culture, or zona pellucida defects.

VI. CONCLUSION

During the past four decades a great deal has been learned about the basic mechanisms of sperm maturation, capacitation, the acrosome reaction, and penetration of the egg. The application of molecular techniques to ascertain the exact physiology and biochemistry of these events has been very useful. Many unanswered questions still remain, but new techniques are becoming available that will answer these questions.

These studies will not only yield a better understanding of the overall process but will provide important clues to the regulation of fertilization in humans and other animals. This knowledge will have important consequences with regard to the curing of human infertility and contraceptive applications.

BIBLIOGRAPHY

Austin, C. R. (1985). Sperm maturation in the male and female genital tracts. *In* "Biology of Fertilization" (C. B. Metz and A. Monroy, eds.), Vol. 2, pp. 121–155. Academic Press, New York.

Dukelow, W. R., and Williams, W. L. (1988). Capacitation of sperm. *In* "Progress in Infertility" (S. J. Behrman, R. W. Kistner, and G. W. Patton, Jr., eds.), 3rd Ed., pp. 673–687. Little, Brown, Boston.

Oliphant, G., Reynold, A. B., and Thomas, T. S. (1985). Sperm surface components involved in the control of the acrosome reaction. *Am. J. Anat.* **174,** 269–283.

Overstreet, J. W., and VandeVoort, C. A. (1993). Sperm–zona pellucida interaction in macaques. *In* "*In Vitro* Fertilization and Embryo Transfer in Primates" (D. L. Wolf, R. M. Brenner, and R. L. Stouffer, eds.), p. 103–109. Springer-Verlag, New York.

Wassarman, P. M. (1988). Eggs, sperm and sugar: A recipe for fertilization. *News Physiol. Sci.* **3,** 120–124.

Wassarman, P. M. (1988). Fertilization in mammals. *Sci. Am.,* Dec., pp. 78–84.

Yanagimachi, R. (1994). Mammalian fertilization. *In* "The Physiology of Reproduction" (E. Knobil and J. Neill, eds.), 2nd Ed., Vol. 1, pp. 189–317. Raven, New York.

Fetus

FRANK D. ALLAN

The George Washington University Medical Center

I. Introduction
II. Maternal–Fetal Relationships
III. Fetal Growth and General Features
IV. Fetal Physiology

GLOSSARY

Amniotic fluid Fluid surrounding the embryo/fetus and within the amniotic cavity; produced by the lungs and the kidneys (in the form of urine)

Conceptus Product of conception (i.e., fertilization) and derivatives thereof, including the embryo and the extraembryonic membranes. The zygote, morula, blastocyst, and chorionic vesicle are early states of the developing conceptus

Differentiation Structural and functional elaboration or modification of cells, tissues, organs, or structures to perform a particular task

Embryo That portion of the conceptus that becomes the fetus or, at term, the newborn infant. Its development can be divided into stages designated the embryonic disk (bilaminar, then trilaminar) and the tubular, then definitive, embryo

Endometrium Lining (i.e., mucosa) of the uterus into which the conceptus implants. After implantation the endometrium becomes known as the decidua, a part of which forms the maternal component of the placenta

Extraembryonic membranes Nonembryonic tissues derived from the outer layer (i.e., the trophoblast) of cells of the blastocyst. Originally unilaminar, the trophoblast becomes the bilaminar (i.e., an outer syntrophoblast and an inner cytotrophoblast) and, finally, the trilaminar, chorion. A separate bilaminar delamination of trophoblast forms the amnion, which surrounds a cavity enclosing the embryo

Fetal circulation Pathway taken by blood within the fetus, the umbilical vessels, and the placenta

Fetal circulatory shunts Structures permitting fetal blood to pass to the placenta, then reenter the fetus (i.e., the umbilical arteries, vein, and ductus venosus); also, within the fetus, they permit blood to bypass the lungs (i.e., the interatrial foramen and the ductus arteriosus). Closure of each shunt is a normal consequence of birth

Fetus Developing organism, from the onset of the ninth postfertilization week to birth

Meconium Product of the bowel, including epithelial debris, mucus, and bile

Oxygen saturation Relative content of oxygen carried in the blood, expressed as a percentage

Placenta Disk-shaped specialized region of the chorion that interfaces with and includes the basal decidua of the endometrium; site of the transfer of substances between maternal and fetal bloodstreams

Primordium Earliest evidence or indication of an organ or structure constituting part of the embryo

Surfactant Lipid substance produced by cells lining the air spaces in the lung that reduces surface tension and makes inflation easier at the onset of respiration

DEVELOPMENT OF THE HUMAN ORGANISM AND its associated membranes from conception (i.e., fertilization) to birth can be divided into three principal periods: preembryonic, embryonic, and fetal. The term "conceptus" designated all of the products of conception (the organism proper and the associated membranes).

During the preembryonic period the conceptus grows from a single cell (i.e., the zygote or fertilized ovum) to a multicellular sphere containing an embryonic disk. The sphere invades the lining of the uterus (i.e., implants), where it can develop, using nutrients that this lining is designed to provide. The preembryonic period requires most of the first 2 weeks of prenatal life.

ENCYCLOPEDIA OF HUMAN BIOLOGY, Second Edition, VOLUME 3.

The embryonic period encompasses development of the third through eighth weeks, during which growth and definition of both the embryo and the extraembryonic membranes are accomplished, including the establishment of the definitive placenta from the latter.

The fetal period encompasses the 9th week through the 38th week, or birth.

I. INTRODUCTION

A. Overview and Definitions

During the first week the conceptus undergoes repeated cell divisions, which form a hollow sphere of cells called the blastocyst. Its outer layer (i.e., the trophoblast) surrounds a cavity containing an internal mass of cells (i.e., the embryoblast) attached on one side. At 6 days the conceptus burrows into the endometrium, or lining of the uterus, where it continues development. The trophoblast gives rise to fetal membranes, including a major portion of the placenta, which isolate the embryo, yet provide maternally derived nutrients, oxygen, and other necessary substances.

Implantation and elaboration of the trophoblast during the second week are paralleled by growth and differentiation of the embryoblast into a bilaminar plate of cells. The upper layer of cells is designated the epiblast, the lower hypoblast. During the third week the plate or disk becomes trilaminar and elongated. The intermediate layer of the plate is called the mesoblast (i.e., mesoderm) and becomes vascularized (i.e., penetrated by vessels) at the end of the week. A primitive heart derived from the mesoblast initiates circulation. The plate then becomes tubular, with an expanded cranial end. In Week 4 an infolding of surface cells (i.e., the epiblast) along the midline gives rise to the neural ectoderm, which forms primordia (i.e., precursors) of the brain and the spinal cord. The lower layer (i.e., hypoblast/endoderm) helps form a primitive gut within the tubular embryo as well as primordia of the internal organs (e.g., the lungs, liver, and pancreas). Bilateral swellings of the tubular embryo become the buds for upper and lower limbs. [*See* Implantation, Embryology.]

Thus, 1 lunar month after fertilization the conceptus is an implanted chorionic vesicle (derived from the trophoblast) containing a tubular embryo (derived from the embryoblast) with primordia of most body parts. Constituent cells are supplied with water, oxygen, minerals, and nutrients by an embryonic circula-

tory system linked to vessels within the chorion. The embryonic period continues to the end of the second lunar month, with elaboration of different tissues, organ primordia, systems, and other components. An embryo of 32 days is shown in Fig. 1. The embryo is clearly recognizable as human at the onset of the eighth week.

The embryo becomes a fetus at 57 days and measures about 32 mm (i.e., 1.25 inches) from crown to rump. Thereafter, the fetus grows about 1 cm/week throughout the remainder of prenatal life. At term, the end of 38 weeks, or 9.5 lunar months, the fetus, or neonate, measures about 32.5 cm (i.e., 13 inches) from crown to rump and 50 cm (i.e., 20 inches) from crown to heel.

B. Early Fetal Period (Third Month)

At the onset of the early fetal period, the head is relatively large (nearly one-half of the embryo's

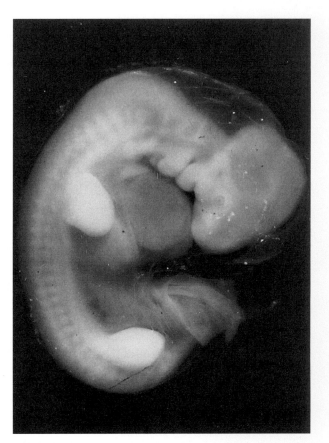

FIGURE I A 32-day embryo. Head at upper right overrides the heart and the liver (to the left and below), showing through the thin body wall. Upper and lower limb buds are paddle-like. The junction of the umbilical cord with the body wall is at lower center. © Frank D. Allan, Ph.D.

crown–rump length), the segments of upper and lower extremities are discernible, and digits are separated. Eyelids are just about to close, fusing temporarily over the eye. Sex of the embryo is not obvious from the external genitalia; however, by 10 weeks this is possible. All internal structures are present. Many of these are tiny replicas of the organ or structure they are to become. The fetal period permits each existing body part (e.g., organ, gland, muscle, and bone) to enlarge by adding additional cells, increasing the size of existing cells, elaborating intercellular materials that add shape or form, and establishing functions in these parts. [*See* Embryo, Body Pattern Formation.]

At this time the eye shows all its basic components: the bulb with outer cornea and sclera, internal lens, retina, iris with pigment, and primitive muscle fibers. Extrinsic eye muscles have even developed microscopic characteristics (i.e., striations) that mark voluntary muscles in postnatal life.

For more than 1 month, muscle cells in the wall of the heart have been contracting, pumping blood through vessels permeating the embryonic tissues and, via the umbilical cord, into the placenta. There, vital substances carried by maternal blood surrounding the chorionic villi lie just two thin membranes away from the blood of the fetus. Division of the heart into chambers is complete, and a definitive vascular system carries blood to and from all body parts. As these parts grow, vessels within match this growth, hence providing a constant supply of substances supporting cell function, replication, migration, and differentiation.

Primordia of musculoskeletal structures of the extremities, head, and body proper (i.e., trunk) are present, although most of the skeleton is still cartilage. Even so, there are areas where cartilage is being converted to bone by the deposition of calcium. [*See* Bone, Embryonic Development.]

All components of the brain and the spinal cord are formed, and nerves link the stem of the brain and the spinal cord to all tissues and organs of the body. Nerves in the head and the neck link the skin and the primordia therein with the brain stem. Skin of the body wall, or trunk, and muscles and tissues within the trunk and of the extremities are linked by nerves to the spinal cord. Hollow organs within the body's cavities have nerves that link them with the brain or the spinal cord, and in some instances both. [*See* Spinal Cord.]

The skin at this time is a delicate membrane, its epidermis two cells thick, supported by an embryonic connective tissue called mesenchyme, the precursor of the dermis. Invaginations from the epidermis into the dermis have established primitive hair follicles and glands. [*See* Skin.]

Oral and nasal cavities lead from a primitive mouth (i.e., stomodeum) to the primitive pharynx, from which tubular respiratory and digestive tracts arise. Within the wall of each tract, layers comparable to those of the adult are present, but thin. The larynx, tracheobronchial tract, and lungs are formed as a diverticulum of the primitive pharynx. Similarly, the liver, biliary tract, and pancreas arise as diverticula of the gut (continuous with the pharynx) below the segment destined to become stomach. The elongating stomach and intestine outgrow the available space inside the body cavity at this time and herniate, temporarily, into an extension of the cavity in the umbilical cord. By 11 weeks the herniated portion of the bowel has returned to the body cavity and the extension is obliterated. A fetus of 11 weeks is shown in Fig. 2. [*See* Digestive System, Anatomy; Larynx.]

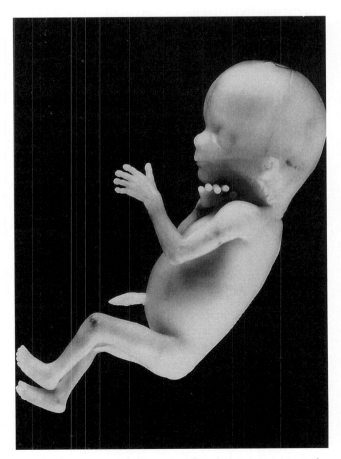

FIGURE 2 An 11-week fetus. Note the relative proportions and definiton of the body parts. Note, too, the translucency of the skin. © Frank D. Allan, Ph.D.

Urinary and reproductive systems develop in intimate association and share certain primordia. By the fetal period the definitive kidneys are present and functioning, as are the ureters and the urinary bladder. The final segment of this system, the urethra, completes changes related to a definition of the external genitalia by the 11th week. Gonads have differentiated as testes or ovaries, and the primordia of both tracts (male and female) have appeared. Whether the male or female genital tract is to dominate and persist is dependent on genetic sex and is effected by the action of hormones produced in the testes or ovaries. [*See* Reproductive System, Anatomy; Urinary System, Anatomy.]

At this time primordia of all endocrine organs have been established. The hypophysis, or pituitary gland at the base of the brain, thyroid and parathyroid glands in the neck, and the suprarenal glands (intimately associated with the kidneys and the gonads) are well defined as highly vascularized clusters of epithelioid cells. By the onset of fetal life, differentiation of the initial cellular inhabitants of the embryo's germ layers has given rise to distinctive cell lineages that restrict the ultimate fate and purpose of component cells. Increasing numbers of cells representing each of the basic tissues (e.g., epithelium, connective tissue, supporting tissues, blood, muscle, and nerve) have migrated to positions where they will best serve the organism.

II. MATERNAL–FETAL RELATIONSHIPS

Consideration of the physiology of the fetus requires an understanding of its anatomical and functional relationship to maternal tissues. As noted earlier, a portion of the conceptus, the trophoblast, gives rise to the so-called fetal membranes. After implantation the trophoblast forms the chorion, which serves as the fetal component of the interface. The endometrium (and blood from its intrinsic vessels) adjacent to the chorion serve as the material component.

A circumscribed portion of chorion and adjacent endometrium form the placenta, where maternal blood (from vessels in the endometrium opened during implantation) bathes finger-like extensions of the chorion (e.g., chorionic villi). Fetal blood circulates through villous capillaries, with access through the villus wall to fluid and materials in maternal blood filling the intervillous space (Fig. 3).

Within the chorionic vesicle the amnion, also derived from the trophoblast, forms a fluid-filled sac enveloping the embryo/fetus. The amnion is reflected from the sac onto the umbilical cord and is continuous at the umbilicus with the skin of the fetus. In its fluid bath the fetus is protected from fluctuations in temperature, effects of gravity (in large part), compression, or external mechanical stimuli.

The placenta is a remarkable organ and serves a multitude of functions: (1) a lung or a respiratory organ, which accomplishes the transfer of gases; (2) a digestive tract, providing for absorption of fluids, minerals, and nutrients; (3) a liver for the storage of glucose; (4) a kidney, functioning to eliminate nitrogenous waste substances; (5) an endocrine gland, producing several hormones affecting maternal and fetal tissues; and (6) a circulatory pump, since the contraction of smooth muscle in larger villi increases pressure within the intervillous space and expedites return of maternal blood to uterine veins. In addition, the so-called placental barrier provides an important, although limited, protection, preventing, in variable degree, the entry of noxious substances into the fetal bloodstream. This is particularly true during the first trimester of intrauterine life, but not of viruses.

III. FETAL GROWTH AND GENERAL FEATURES

Most fetal organs or structures at 4 months have established relationships and characteristics that will be carried into postnatal life (Figs. 4 and 5). Relative proportions, however, change appreciably. The head, with the brain and special organs, usurps a large proportion of the total organism during prenatal life. Initially, the upper limbs and trunk are markedly larger and better developed than the lower limbs and trunk. Proportions still favor the head at birth (one-quarter of the body length compared to one-seventh in the adult). Within the thorax, lungs occupy but a small portion of the available space compared to the heart, great vessels, and thymus gland. Organs in the abdominal cavity are relatively large, especially the liver and the spleen. Kidneys and adrenal glands are also relatively large structures and, with the stomach and the bowel, fill the cavity. In contrast, the pelvic cavity is small and the urinary bladder rises well above the pelvic brim. Genital organs are small and crowded into the limited space with a small terminal bowel and rectum.

Weeks 13–16 of prenatal life are marked by rapid

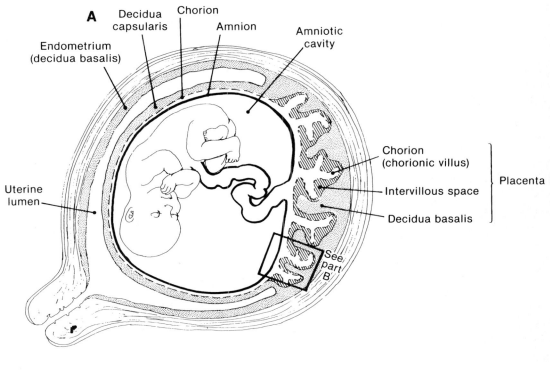

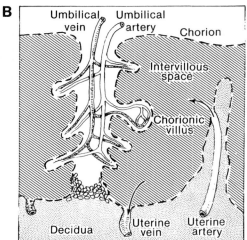

FIGURE 3 (A) Diagram of a uterus containing a fetus at the end of the third month, showing the extraembryonic tissues (i.e., chorion, amnion, and umbilical cord) within the uterine lining (i.e., decidua). Note the fetal and maternal components of the placenta. (B) Inset from lower right of (A) shows higher magnification of the chorionic villi and the intervillous space (cross-hatched space filled with maternal blood).

growth of the fetus. Ossification of skeletal structures initiated earlier proceeds rapidly. During the following weeks (17–20) growth slows, but fetal movement increases. Hair on the head, eyebrows, and body (i.e., lanugo) appears. By Week 20 the vernix caseosa, a cheesy substance that coats and protects a thin still-translucent skin, is formed (Fig. 6). Near the end

of the next 4-week period (i.e., Week 24), fingernails have appeared, and, within the lung, cells elaborate surfactant, the presence of which marks the point at which the lungs can inflate and the fetus could survive premature delivery. During Weeks 26–29 the respiratory apparatus matures and increases the likelihood of survival following premature delivery.

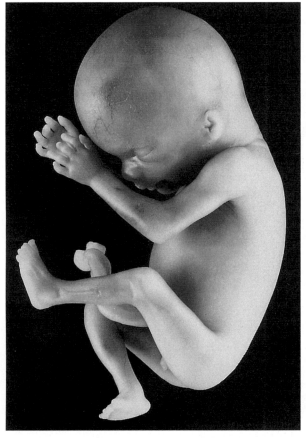

FIGURE 4 A 14-week fetus. Note flexion of the trunk and the lower extremities to fit the confines of the uterus. © Frank D. Allan, Ph.D.

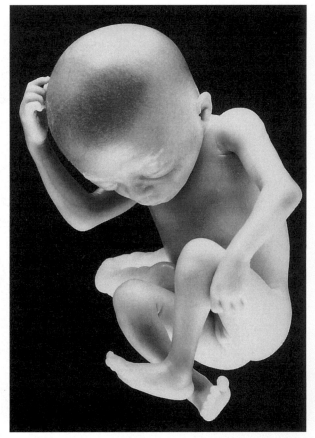

FIGURE 5 An 18-week female fetus. Note the beginning of eyebrows. © Frank D. Allan, Ph.D.

The eyelids reopen and fat begins to fill and thicken the trunk. Skin becomes less translucent and toenails appear. Hair on the head and the amount of lanugo increase. The final weeks of life before birth (i.e., Weeks 30–38) permit continued growth and an increased amount of fat in the skin. The fetus can grasp, and the breasts (both male and female) are well formed and often secrete colostrum at birth. Lanugo then decreases.

IV. FETAL PHYSIOLOGY

The basic function of cells and tissues of the developing organism is coexistent with growth and specialization of these cells and tissues before birth. Thus, all functions of cellular metabolism that affect growth,

cell division, respiration, motility, and specialization of cells as tissue components are ongoing throughout embryonic and fetal life. Individual cells composing the conceptus, embryo, or fetus must maintain their own well-being while performing specialized tasks benefiting the whole organism.

After cells specialize as one of the basic tissues and are incorporated into an organ, functions of the embryo/fetus can be considered at another level. It is not possible to deal with all such functions here. A few, however, are especially worthy of examination.

A. Early Fetal Function

In very early fetal life the function of cells, tissues, and the organs they constitute runs hand in hand with anatomical maturation. The fetus swallows amniotic fluid beginning with the third month. The

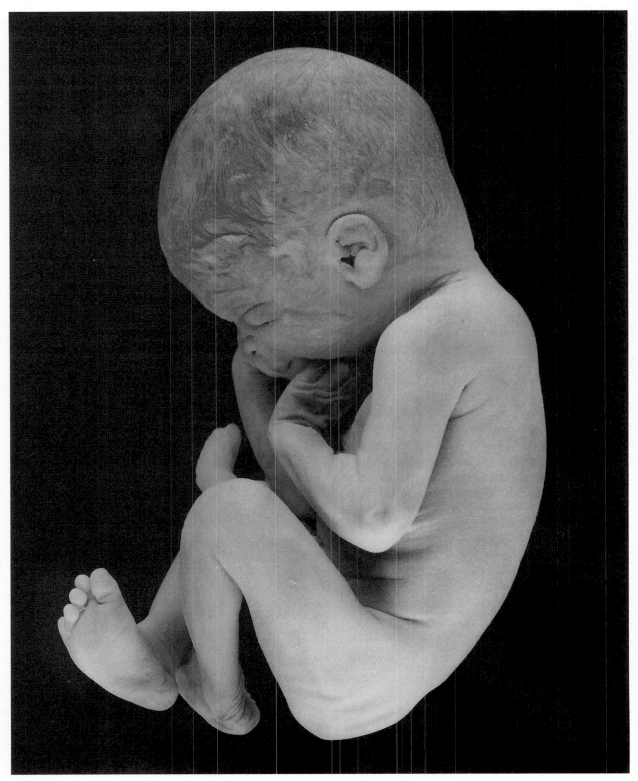

FIGURE 6 A fetus of 20+ weeks. Note the vernix caseosa smeared in the hair of the head. Also, note the opacity of the skin. © Frank D. Allan, Ph.D.

fluid and some of its components pass through the lining of the gut and walls of adjacent capillaries, thereby entering the bloodstream. The absorbed fluid reaches the kidneys, which produce urine even at this early stage. Urine is expelled into the amniotic fluid, where it mixes with fluid produced by the lungs and sloughed epithelial cells from the amnion, or surface ectoderm.

During the first month of the fetal period, muscles twitch spontaneously and in response to reflex impulses initiated by occasional stimuli. Protected from the usual surface stimuli of postnatal life, the fetus' hands are, nevertheless, often held close to the face. Should finger or thumb touch the area around the mouth, a pursing of lips or even suckling movement might result.

B. Fetal Circulation

By the onset of the fetal period, the heart has four chambers: right atrium, right ventricle, left atrium, and left ventricle (Fig. 7). Blood returning to the heart enters its atria; blood leaving is expelled from and by contraction of the ventricles. Blood leaving the left ventricle enters the aorta and is distributed to all fetal tissues (except the lungs) and to the placenta by way of impaired umbilical arteries. The lungs are supplied

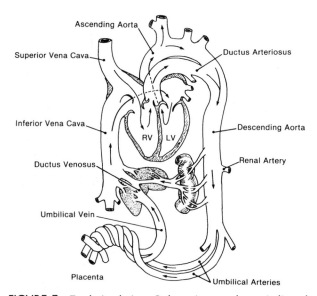

FIGURE 7   Fetal circulation. Only major vessels are indicated, with an attempt to represent smaller vessels that link with vascular plexuses in organs or tissues of the fetus and the associated placenta. RV, right ventricle; LV, left ventricle.

by outflow from the right ventricle via the pulmonary arteries.

Having perfused fetal tissues, blood flows back to the right atrium using tributaries of superior and inferior venae cavae (draining the upper and lower halves of the body, respectively). Aortic blood flowing through umbilical arteries to chorionic vessels in the placenta is drained by tributaries of a single umbilical vein, returning it to the fetus via the umbilical cord. This flow is shunted by the ductus venosus through the liver to enter the inferior vena cava. Another shunt, the ductus arteriosus, transmits blood from the pulmonary trunk to the descending aorta.

Blood entering the right atrium from the superior vena cava (relatively unoxygenated) passes directly into the right ventricle, whereas blood entering from the inferior vena cava (including the oxygenated umbilical flow) is shunted through an opening (i.e., the interatrial foramen) into the left atrium. Mixing of the two streams in the right atrium is minimal. Blood within the left atrium is joined by blood returning from the lungs and passes into the left ventricle. This blood is pumped into the aorta to be distributed by branches arising therefrom. Simultaneous contraction of the right ventricle pumps blood into the pulmonary trunk, where a small portion passes to the nonfunctioning lungs, whereas the major portion passes through the ductus arteriosus into the aorta. And so the circuit is completed.

Fetal blood aerated within the placenta achieves 80% oxygen saturation (the most highly oxygenated) and returns to the heart through the umbilical vein and the ductus venosus. When this blood joins the inferior vena cava, carrying oxygenated depleted blood from the lower part of the body, the oxygen saturation that enters the right atrium decreases to about 70% saturation. Blood entering the right atrium from the superior vena cava is only 30% saturated. Separation of the streams flowing through the right atrium from the two venae cavae is, therefore, significant. Blood passing through the foramen ovale, left atrium, and left ventricle reaches the aorta with an oxygen saturation of about 60%, whereas blood passing into the right ventricle and into the pulmonary trunk is about 50% saturated. Beyond the junction of the aorta and the ductus arteriosus the oxygen saturation is approximately 55%. Since most of the terminal aortic blood flow enters the umbilical arteries, blood to the placenta has the same oxygen saturation (i.e., 55%). [*See* Cardiovascular System, Anatomy; Cardiovascular System, Physiology and Biochemistry.]

C. Circulatory Changes at Birth

A remarkable series of changes in circulation takes place at birth, intimately related to the onset of respiration. Changes involve structure and function of the lungs and the circulatory pattern (Fig. 8).

During passage through the birth canal, umbilical blood flow is usually compromised to some degree. The fetus becomes hypoxic (i.e., receiving a diminished amount of oxygen), or, in severe cases, anoxic (i.e., deprived of oxygen). Obviously, when the umbilical cord is clamped or cut, the source of oxygen from maternal blood via the placenta is lost and carbon dioxide builds up in the fetal blood (i.e., hypercapnia). The fetal respiratory center in the brain is stimulated, and inspiratory efforts (i.e., contraction of the diaphragm and the intercostal muscles) result.

With inspiration, the air sacs (i.e., alveoli) within the lungs increase in size and number, and the capacity of the capillary network surrounding the sacs increases. Resistance in the pulmonary bed decreases; hence, blood flow through the bed increases. As a result the volume of blood returning to the left atrium from the lungs also increases.

With cessation of blood flow through the umbilical circuit, the umbilical vein and the ductus venosus collapse. Constriction of the umbilical arteries increases pressure within the aorta and perfusion of all body tissues and organs is thereby increased. Blood pressure within the pulmonary system increases due to back-pressure and flow from the aorta via the ductus arteriosus. The result is increased perfusion of the expanding lungs. Left atrial pressure also increases, and the valve of the interatrial foramen is functionally closed. Venous return to the right atrium increases as tributaries of the venae cavae receive the greater volume of blood perfusing the fetus. The right and left atrial pressures equalize, and the valve remains closed.

Continued respiration increases lung capacity, and oxygen saturation of the blood gradually increases. At an oxygen saturation of about 75% in aortic blood, the ductus arteriosus closes, and the entire output of the right heart passes to the lungs.

D. Pulmonary Apparatus

Remarkable as the changes in circulation are, they are no more remarkable than mechanisms that permit the lungs to change from a relatively compact cellular mass to one with increasing numbers of thin-walled "air" sacs, permitting an intimate relationship between blood-filled capillaries and the air sacs they embrace (the use of air sac is presumptuous for air is not present until birth).

Structural changes and cellular maturation of the fetal lungs at 6 months lead to functional competency of this organ. Thus, the occasional air sacs found at the ends of the bronchial tree at that time increase in numbers. The expanded capillary network intimately related to the air sacs enlarges as more and more alveolar spaces are formed. Bronchial and bronchiolar passageways are lined with simple columnar or cuboidal ciliated epithelium. Similar cells constitute the end pieces of the bronchopulmonary tree and surround small fluid-filled sacs. Like any moist juxtaposed membranes, the cells lining the spaces are held together by surface tension. By the 24th week of intrauterine life, certain of the cells forming these end pieces elaborate a surfactant that minimizes surface tension and permits inflation (expansion) of the potential spaces following repeated inspiratory efforts.

To survive delivery, the fetal lungs must be able to support respiration. At the end of the sixth lunar

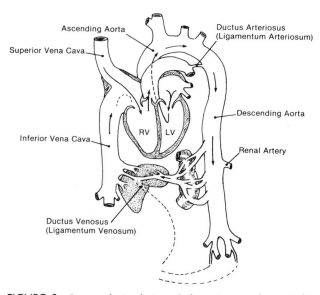

FIGURE 8 Postnatal circulation. Only major vessels are indicated. Note constriction of the fetal shunts (umbilical arteries, vein, and ductus venosus; ductus arteriosus) previously carrying blood to and from the placenta and those bypassing the lungs. Also, note the closure of the interatrial foramen (dashed lines in the right atrium). RV, right ventricle; LV, left ventricle.

month, this ability, although marginal, is sometimes sufficient. Before this time few, if any, fetuses survive removal from the uterus. Obviously, those born later have increasing chances of survival. Effective respiration is directly related to the maturation of the alveolar–capillary interface and production of sufficient surfactant. Neural regulatory mechanisms for respiration are present much earlier (i.e., at the end of the first trimester). Respiratory movements are common *in utero* and increase in frequency as term approaches. Nevertheless, it appears that such mechanisms are suppressed until birth, when the stress on the fetus ends this suppression. [*See* Respiratory System, Anatomy; Respiratory System, Physiology and Biochemistry.]

E. Urinary Apparatus

The kidneys, ureter, urinary bladder, and urethral passage are all functioning in simple fashion early in the fourth lunar month. Waste products of fetal metabolism are also being cleared from the blood within fetal capillaries of the placental chorionic villi. Nonetheless, functioning nephrons (i.e., functional units of the kidney) are present from this time on. Nephrons present in the fetus have small glomeruli (where arterial blood is filtered) and relatively short tubules (where the glomerular filtrate is processed). Research on animal fetuses suggests that the nephron of the definitive kidneys is functional by midgestation and awaits only the challenge of an increased load of metabolic waste following birth to fully mature.

F. Alimentary System

Cellular and tissue mechanisms for absorption and digestion are present by the end of the fourth month. Thus, microvilli of cells on intestinal villi (both villi and microvilli increase surface area of the lining of the intestine for absorption) and many intestinal digestive enzymes are present at this time. Gastric gland products appear in the fifth month. Swallowing of amniotic fluid occurs as early as the third month, and the frequency of this action increases as the fetus ages. The swallowed fluid is absorbed in the gastrointestinal tract and cleared by the placenta or the kidneys. Peristalsis of the tract is present from the fourth month, although the muscular layers of the gut wall are relatively thin and its contractions are weak. Attempts to suckle have been seen *in utero* and in aborted fetuses of 3 months. Premature infants suckle effectively shortly after delivery and are able to ingest and process a relatively large volume of milk within the first few days following birth.

G. Regulatory Systems

Obviously, some regulation of organs, tissues, and cells is required to integrate various activities and responses during fetal life, even though this integration might not be as essential for survival as it is after birth. Regulation and integration are achieved by the nervous and endocrine systems.

Initial nervous activity begins in the third fetal month. Neuromuscular reflexes are more obvious in the upper part of the fetus, but soon spread to the trunk and the lower extremities. Such reflexes are responses (i.e., muscle contractions) to stimulation of tactile and pressure receptors within or near these muscles. Reflexes that develop during the third month involve the mouth and the head (e.g., sucking, swallowing, gagging, and moving the mouth and, therefore, the head toward a perioral stimulus). Electrical activity of the nervous system is discernible at the same time, but periods of electrical silence can occur up to midterm. Movements of the fetus (i.e., quickening) are usually perceived by the mother at 4 months. By 7 months the electrical activity of the nervous system becomes continuous and rhythmical. Electric potentials appear following visual stimulation at the end of 6 months. Apparently, some visual function is present, and the response of the pupil to light can be elicited by this time.

Taste buds are functional at 6 months, and the modality for sweetness is well differentiated. Increased "drinking" of the amniotic fluid is effected when sweet substances are introduced.

Each of the endocrine glands is established early in the fetal period (i.e., the third month), and all are functioning, in variable degree, by midgestation. The pituitary gland, or hypophysis, has differentiated by the beginning of the second trimester and elaborates a number of hormones that have control over body tissues generally (growth) or specifically (trophic hormones) over other endocrine glands. The pars neuralis of the hypophysis produces its hormones (which cause the contraction of smooth muscle) at the same time. [*See* Endocrine System.]

The thyroid gland elaborates the thyroid hormones within the fourth month and has established a feedback system with the hypophysis by midgestation. Under hypophyseal control the thyroid hormone controls the general metabolism of cells. Parathyroid glands give evidence of functioning at the beginning of the fetal period and are active throughout pregnancy. Parathyroid hormone helps regulate the circulating level of calcium in the bloodstream. [*See* Parathyroid Gland and Hormone; Thyroid Gland and Its Hormones.]

The cortex of the adrenal gland is formed and actively producing steroid hormones at the end of the fourth month. A special zone of fetal cortex is lost at birth, and definition of the permanent structure occurs postnatally. The adrenal medulla is an extension of neural tissue (embedded in the cortex) that forms neurohumors (epinephrine and norepinephrine) identical to those produced by the autonomic nervous system.

Other endocrine cells exist as islets within the pancreas and are distinguishable (i.e., component cells can be differentiated) at 3 months. They are elaborating hormones (e.g., glucagon and insulin) by the fourth month.

BIBLIOGRAPHY

Barnes, A. C. (ed.) (1968). "Intrauterine Development." Lea & Febiger, Philadelphia.

England, M. A. (1991). "Color Atlas of Life before Birth." Mosby Year Book, Inc., Wolfe Publishing, St. Louis.

Moore, K. L., and Persaud, T. V. (eds.) (1993). "Developing Human: Clinically Oriented Embryology," 4th Ed. Saunders, Philadelphia.

Tanner, J. M., and Preece, M. A. (eds.) (1989). "The Physiology of Human Growth." Cambridge Univ. Press, Cambridge, England.

ISBN 0-12-226973-X